Advances in Industrial Ergonomics and Safety II

ADVANCES IN INDUSTRIAL ERGONOMICS AND SAFETY

General Editor

Anil MITAL

Ergonomics Research Laboratory
Department of Mechanical and Industrial Engineering
University of Cincinnati
Cincinnati, Ohio, U.S.A

ADVANCES IN INDUSTRIAL ERGONOMICS AND SAFETY II

Proceedings of the Annual International
Industrial Ergonomics and Safety Conference
held in Montreal, Quebec, CANADA, 10-13 June 1990

The Official Conference of the International Foundation
for Industrial Ergonomics and Safety Research

Edited by

Biman DAS
Department of Industrial Engineering
Technical University of Nova Scotia
Halifax, Nova Scotia, CANADA

Taylor & Francis
London · New York · Philadelphia

UK Taylor & Francis Ltd, 4 John St., London WC1N 2ET

USA Taylor & Francis Inc., 1900 Frost Road, Suite 101, Bristol, PA 19007

British Library Cataloguing in Publication Data

International Industrial Ergonomics and Safety Conference
(1990, Montreal, Quebec, Canada)
Advances in industrial ergonomics and safety II
1. Ergonomics
I. Title II. Das, Biman
620.8′2

ISBN 0-85066-748-8

Library of Congress Cataloging-in-Publication Data is available

Printed in Great Britain by Burgess Science Press, Basingstoke.

Preface

The *Annual International Industrial Ergonomics and Safety Conference,* sponsored by the *International Foundation for Industrial Ergonomics and Safety Research,* has become a major forum for the exchange of current knowledge and information in industrial ergonomics and safety among international researchers and industrial practitioners. The main goal has been to create productive, comfortable and safe workplaces. One hundred and ninety-six abstracts were submitted, out of which one hundred and forty-five abstracts were accepted by the International Program Advisory Committee for the presentation of papers at this Conference. It is always a major concern to maintain a proper balance between the quality of papers and the quantity of papers.

Advances in Industrial Ergonomics and Safety II contains one hundred and thirty-five papers, authored by two hundred and seventy-five researchers and practitioners. The papers are grouped under twenty section topics, plus conference addresses. The section topics are arranged in alphabetical order. It is believed that the wide range of research topics and variety of industrial applications included in this volume will prove to be a valuable resource of knowledge and information to all concerned with industrial ergonomics and safety. I wish to convey my sincere thanks to all the authors for their contribution.

Special thanks are extended to Mr. John Kozey and Ms. Dawn Lewis for their help in the preparation of this volume. The cooperation of Dr. Robin Mellors, Taylor and Francis Ltd. (Publishers), is duly acknowledged.

Last, but not least, I shall remain ever grateful to my wife Diane, daughter Anjali and son Bijon for their support and understanding during the preparation of this volume. To them, my apologies for being away many evenings and weekends, when this time really belonged to them. With genuine pleasure, I wish to dedicate this volume to them.

Department of Industrial Engineering
Technical University of Nova Scotia
Halifax, Nova Scotia, Canada
June 10, 1990

Biman Das, Ph.D., P.Eng.
Book/Proceedings Editor
and Conference Chairman

Dedication

This volume is dedicated to
my wife *Diane,* daughter *Anjali* and son *Bijon*

CONTENTS

XV. OCCUPATIONAL HEALTH AND SAFETY

XVI. ROBOTS AND AGRICULTURAL MACHINERY SAFETY

CONFERENCE ADDRESSES

Advances in Industrial Ergonomics and Safety II
Edited by Biman Das
Taylor & Francis 1990

GREAT PROBLEMS, GREAT PROSPECTS FOR SPACESHIP EARTH: AN ERGONOMICS AGENDA FOR THE 90'S

Harold P. Van Cott

National Academy of Sciences, National Research Council
2101 Constitution Avenue, N.W.
Washington, D.C. 20418

As part of a plan to prepare a human factors and ergonomics agenda for the National Academy of Sciences' Committee on Human Factors, leaders from our field in Western Europe were asked for their views on the issues and challenges that face our discipline in the decade ahead. Preliminary findings from the survey suggest that there is substantial agreement on many common problems and on the research and applications challenges they present to our field.

BACKGROUND

The first task undertaken by the Committee on Human Factors after it was formed in 1980 was to develop an agenda for the human factors and ergonomics research needed to meet the challenges of the 80's. Throughout that decade a report containing that agenda (NRC 1983) was used by the committee to plan its own projects and by a number of U.S. government agencies to establish directions and funding levels for federally supported human factors and ergonomics programs.

Today, in retrospect, we all realize that in the brief span of ten years astonishing changes have taken place in every aspect of human life: in the structure of societies and governments, in population demographics, geo-politics, economics, technology, the environment, education, and in the nature of human work and its organization. In just ten years our planet has shrunk from a vast globe in which disasters and catastrophes were isolated events to a tiny spaceship in which local events can have repercussions that affect the entire planet. We have also reluctantly learned to accept the fact that, while science and technology have had beneficial effects, many of the problems we face are also the consequences of science and technology.

In view of these massive changes, the Committee on Human Factors has decided that a new agenda will be needed for the 90's. As a part of the project to update that agenda, I surveyed opinion leaders in our field in Western Europe on the challenges that they perceive face our profession. I am pleased to report on the findings of this survey and hope that they will form a useful backdrop to the sessions that follow at this conference.

Why did I look to Europe to help inform an agenda for the United States? First, I wanted an independent check on our views. I wanted to guard against the parochialism to which nations - like people - can become unwitting victims. Second, I wanted to determine whether or not our perceptions in the United States of the problems and needs of our field are shared by other nations. Since Europe has tended to lead in industrial ergonomics and the U.S. on defense human factors, I wanted to identify areas to which the U.S. should turn its attention as military spending is reduced. Finally, I wanted evidence to support or refute the opinion held by some that the U.S. has fallen behind European nations in key areas of human factors and ergonomics.

THE SURVEY

Last Fall I wrote to 55 individuals in Belgium, Denmark, France, Holland, Norway, Sweden, Switzerland and West Germany with an invitation to answer five questions:

o What are the most compelling social and technological challenges that face the human factors and ergonomics European community in the 90's?

o What are the greatest and most critical gaps in the existing human factors knowledge base?

o Which nations lead in human factors and ergonomics research and in applications? What are these areas?

o It is said by some that the U.S. has fallen behind in human factors research and applications. Is this true? If so, in what areas? Why do you think that this has happened?

o Are there research methods or areas of application in Europe that we are unaware of in the United States or that we have ignored or rejected because of the rigidity of our paradigms or our provincialism?

As I write this today, I have before me replies from over a third of those whom I contacted. By the time we gather in Montreal I will have received additional replies and will have updated this initial summary.

The replies I have received so far have been thoughtful, sometimes detailed, and always interesting: a rich source of information, insights and opinions that varied from one respondent to another in detail and emphasis, but contained common themes and issues that concern those of us in North America as well as our colleagues from Western Europe.

The remainder of this paper is organized around the five sets of questions and the replies to them. I used these replies, not as a data set to be tabulated, but as a source of ideas for the Committee on Human Factors' planned agenda.

A SCENARIO FOR THE 90'S

Forecasting the future is a dicey business, but if one is to devise plans and to write agendas, then the context within which they will be implemented must be characterized. So, using the letters I received from Europe, and after many deep looks into my computer screens and crystal ball, I constructed the following scenario:

Economics

By the end of the 90's a single, more efficient economic market will have come into being: first in Western Europe and toward the end of the decade in Eastern Europe. During the same period the economies of Japan and the Pacific rim nations will continue to flourish. The U.S. economy, after a period of stagnation, will pick up momentum as it shifts from military spending to consumer products and to the export of high technology and services such as university training.

Demographics:
The 90's will see major changes in population demographics: aging populations in Europe and the U.S., major migrations of labor within the continent and from Third World nations to Europe and the U.S. The proportion of females in the workforce of the developed nations will increase and challenge existing job and workplace designs. As never before this population will be mobile: crisscrossing and jamming every available means of transportation.

Geopolitics
Of all of the changes that will shape the 90's, none are as likely to have more impact than glasnost and peristroika. While it is impossible to predict how these forces will shape the future, the arrow of freedom is not easily turned back. Decentralization of governments, a reduced threat of global war, a turning to common problems of justice, equity, and human rights: these are things that we have all dreamed about but had despaired of ever seeing. They may yet come to pass.

The Environment
Threats to the environment will come from combustion byproducts, pollution, acid rain, toxic and nuclear accidents, and changes in air composition and quality. Water and some material resources will continue to diminish. Global warming and the redistribution of temperature around the earth from deforestation and land development may be accompanied by major shifts in global weather and storm patterns. Increased attention will be given to measurement research and climate management throughout the world.

Technology
Continued growth will occur in telecommunications, automation robotics and telerobotics. More highly automated and flexible manufacturing will come into being. Expanded applications for satellite communications, telecommunications and teleconferencing, information systems, networking and data base accessability will occur. Increased use of computers by the work force and a trend to work on a decentralized basis are very likely. As never before technology will become a part of everyone's life, but the risk of technological accidents will continue to plague industrial nations.

Energy
A reduction in supplies of fossil fuels and a trend toward a greater reliance on new, smaller scale and safer systems for nuclear power generation and coal gassification will come about as a result of rising fuel imports and costs.

Jobs
There will be a reduction in the numbers of jobs and demands for high skill levels in the manufacturing and service sectors will increase. There will be pressure for greater humanization

of work and the workplace for all workers. Massive migrations will take place of skilled and skilled labor from one country to another.

Education and Training

Meeting the demand by high tech economies for a highly trained and flexible work force will become a major goal of all nations. The locus of this training will gradually shift from industry to public education as means are found to rejuvenate our outmoded school systems.

THE FINDINGS

Here are some of my preliminary findings:

The Social and Technical Challenges

The greatest social challenge of the 90's will be to ensure that in all aspects of creating a single European Market there is a parallel social dimension in which a key factor will be worker health and safety and the quality of life. During this period all member states will be challenged to raise standards and guidelines for the application of human factors; to help design for the female work force; to help support an increasing decentralization of work (down to one person working at home); and to place special priorities on human factors issues in high risk industries.

The greatest technical challenges of the 90's are to support the growth and productivity of the micro-electronics industry; to help manage the information explosion by developing cognitive aids for managing information access and use; and to plan the human factors redesign of an overburdened system of train, car and plane transportation.

Several respondents noted that technological progress and economic development will occur only with the harmonization of technology with human factors. Dr. Ahmet Cakir, at the Ergonomic Institut fur Arbeits- und Sozialeforschung in Berlin, stated that human factors is regarded by the EC-Parliament as one of the key issues for technological progress.

Not all of the old challenges have been resolved, and many respondents commented on the problems of noise, light, heat and vibration in the workplace that still need improvement.

Gaps in the Existing Human Factors Knowledge Base

Many respondents felt that the biggest gap in human factors was its understanding of cognitive processes: of how people develop and use models of remote systems to solve problems, react in crisis; take emergency action and maintain performance over a work period.

A major shift is needed in order to build a valid and comprehensive behavioral and cognitive framework from which to address problems of safety, cognitive load and stress induced behavior.

Of general concern to my correspondents is the lack of understanding of human error, its causation and correction or remediation. This concern is especially acute for those colleagues working with high risk industries such as nuclear plants.

Several replies underscored the wide gap that exists in their countries between human factors as a producer of knowledge, data and applications principles and the user of this information. J.G. Fox, of the Commission of the European Communities in Luxembourg, said: "I believe there is a technical gap to be bridged in determining the modus operandi of the successful transfer of our kind of technology if our profession is to develop."

Leading Nations in Human Factors

Most Europeans believe the U.S. generally leads other nations in human factors research and application. Reasons given are the support for military research and the numerical size of the profession. Credits were given, however, to European nations in a number of areas: in road and traffic system research to the Netherlands; in tracked vehicle research to Japan; in process control research to the U.K., Denmark and Sweden; in environmental design for homes to Sweden, Switzerland and France; and in civilian work place/industrial research to West Germany, the U.K. and Sweden. The U.S. was cited as being the leader in military human factors research.

Position of the USA in Human Factors

Most observers believe the U.S. is still a leader, citing as evidence the fact that a large number of articles in European English Language journals are by Americans. Other respondents felt that while this indicated that the U.S. may be a leader in research, it may be behind in applications. One respondent commented that aside from success in the application of human factors to military systems, in many industrial applications labor unions in Europe have been a positive force in supporting human factors legislation that does not yet exist in the States. Some observers noted that even in the area of U.S. military human factors, the armed forces had difficulty in justifying human factors to congress.

Professor Rohmert at the Institute fur Arbeitswissenshaft in Darmstadt, Germany, put it nicely when he said, "If there is a distinction to be made between human factors and ergonomics one can say that human factors is more U.S.-oriented and ergonomics more European oriented."

Areas of Research and Application that are Ignored or Rejected

Few respondents felt that the U.S. had failed to use research methods or application techniques that were in use in European countries. However some commentators noted that the U.S. should "become more aware of the importance of physiological and neurophysiological aspects of human work. Others commented on the fact that quasi-experimental field research and field trials seem to be more widely used in several European countries than in the U.S. These methods are judged to be particularly well suited to the study of manufacturing processes. Finally, in the opinion of some, there is a gap in the U.S. between accumulated knowledge and its application that is due to the more uncertain support accorded to human factors applications by many industries and the government.

SUMMARY

Perhaps the best summary of the findings from this informal survey of European leaders in our field comes from Donald Broadbent, at the University of Oxford, who said: "In my opinion the main challenges to our profession are common to all developed societies, rather than specifically European. A very few emphases may be created by geography: we have more road safety problems than you because our population density makes flying less useful as an alternative and bigger roads are impractical. There are indeed challenges of other kinds that are peculiar to this continent: in particular, we are in the middle of debates very similar to those in the first Continental Congress in your country. These concern the relative rights of states and of continental decision-makers, the interaction between different legal and professional systems, the proper organization of a continent-wide economy, and so on. These debates are given a greater urgency by technology; in Sweden, last October, I had a choice of Soviet, French, U.S. and two kinds of British T.V., so who is deciding what is too sexually explicit or racist? The debates can perhaps produce some human factors priorities at the margin that are different from those in North America... but frankly they do not seem to me to make much difference in the kind of problems our profession faces."

This then is a summary of what I have found to date. It by no means covers many topics discussed by my correspondents, but it does represent some of those topics accorded the greatest attention.

Advances in Industrial Ergonomics and Safety II
Edited by Biman Das
Taylor & Francis 1990

FUTURE OF ERGONOMICS RESEARCH AND APPLICATION

Walter Rohmert

Department of Ergonomics,
University of Technology Darmstadt
Petersenstraße 30, 6100 Darmstadt, West Germany

Structuring human work with the aid of ergonommics is becoming involved in more recent times in attempts of achieving a humanisation of work, an improvement in the quality of life. This initiative has far-reaching consequences as far as the four main tasks of ergonomics are concerned, namely the analysis, measurement, evaluation and design of human-work-systems. The involvement of ergonomics in these broad activities is easy to demand, but it brings with it greatly increased responsibilities, which we need to recognize and to accept.

INTRODUCTION

The concept of ergonomics was probably first introduced by Jastrzebowski (1857): "The significance of the deployment of our vital forces ... motivates us to approach the problem of work in a scientific manner and even to establish a separate discipline devoted to it ... so that we reap the best fruits in this life using the least possible effort with the maximum possible satisfaction both for our own and the common good and thus proceed in a just manner both towards our own consciences and towards others." (Translated by the author).

Efforts to improve the conditions of human labour are as old as human labour itself; in more recent times,however, attempts at a more systematic approach have been accelerated by two facts: (1) The Division of Labour and Responsibilities and increasing Mechanisation and Automatisation have resulted in work tasks which depart further and further from human's natural way of life. Thus it is necessary to possess knowledge about the relationship between humans and their work at an early stage of job planning; (2) In addition, Job Evaluation is in state of flux. The end of this changing state is at present the attempt to achieve a Humanisation of Work, an improvement in the quality of life. What does this mean for a science and technology like ergonomics?

ERGONOMICS PROBLEM SOLVING

It is usually possible to distinguish three phases in the evaluation and exploitation of a resource wherever and whenever people or mankind use it, be it a fruit of the soil or of the sea, be it material, energy, or human labour power: in the first phase, the joy at discovering the resouce means that it is usually used in an unaccounted and uncontrolled manner;

with reference to human labour power, one can speak here of the first stage in the exploitation of human labour power. Experience with the deployment of the resource shows how and to what extent its value depreciates; thus a second, restrictive phase is initiated in which effort is expended to prevent the negative consequences of exploitation caused by the arbitrary setting of job tasks. This restrictive phase can be enforced by means of national legislation (e.g. youth employment protection laws and laws for the protection of working women) or international conventions (e. g. those relating to the equality of sexes). The third phase, which could be termed the rational development of productive forces, is governed by reason. Here the strivings for maximum economy of the first phase and the striving for more humanity in the second are harmonised so that the negative results of exploitation can be eliminated, while at the same time by seeking optimum development, sub-optimisations can be avoided. The methods, theories and applications of ergonomics are directed towards achieving this harmony. To this end, ergonomics requires studies of human beings, their living and working conditions, and finally some understanding of the future requirements of human working conditions, based on a definite conception of the human condition.

Ergonomics involves the analysis, measurement, evaluation and design of human-work-systems (see Figure 1, Rohmert, 1983). By taking account of the theory of work-systems, the tasks of analysis and measurement can be made verifiable and thus be satisfactorily conducted. The stress and strain theory ensures that the ergonomic functions of measuring and evaluating human labour take sufficient account of the demands of humanity and productivity which are essential for any consideration of human personality. The four-stage hierarchy for assessing the degree of Humanisation of Work evaluates the criteria of ability, tolerability, acceptability and satisfaction with the aim of organising work in such a way that the human's sense of wellbeing within the work-system can be harmonised with the demands of the work-system itself.

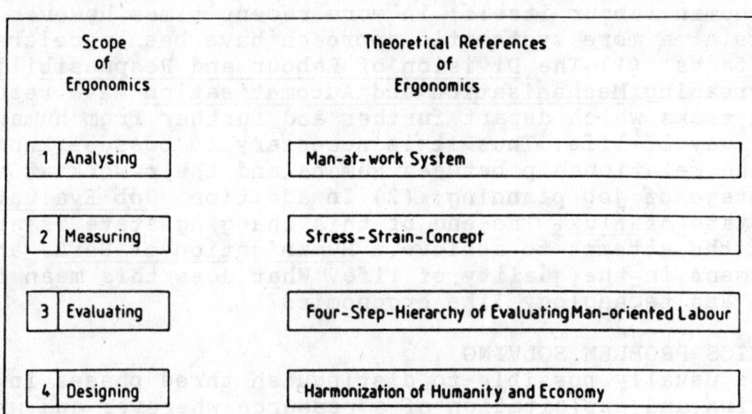

Figure 1: Four main tasks of Ergonomics and their Theoretical Fundamental Principles

MAKING A FRESH ATTEMPT AT STRUCTURING OF FORMS OF LABOUR

If human work is to be structured with the aid of ergonomics, the science starts by examining people at work and considering various boundary conditions of technology, science, the individual and society. Figure 2 shows various modes of examination and different modes of work.

The pysiological and engineering modes of examination tend towards approaches from the natural sciences, whereas the philosophical-historical and the philosophical-materialistic modes tend to use approaches from the social sciences and humanities. The physiological approach is primarily structural with reference to humans and only secondarily functional with reference to methods. The organic systems of human nerves and muscles are used as a systematic basis for looking at human work from the point of view of the demands placed on workers by their tasks. In an engineering approach, material flow, energy flow and information flow from humans are modelled as a contribution to the desired work effect in a human-work-system.

The humanities and social science based approaches of the philosophical-historical, and philosophical-materialistic

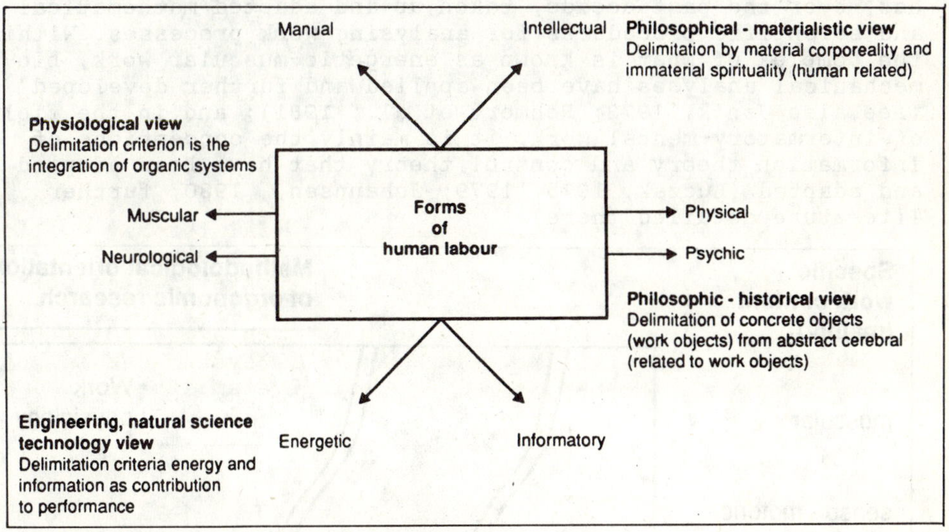

Figure 2: Connections Between Various Ways of Perceiving Human Labour and the Structuring of Forms of Labour

modes mainly consider human labour from the aspect of material output and human input, which are in contradiction to each other with regard to their social objectives. The two decisive factors for a physiological mode of examination are: (1) the concrete manifestation of human labour in its technical and organisational aspects; (2) the degree of matching between work and men, and between men and work. The physiological mode considers the effect of work on the human organism first and foremost; this effect is called strain. Thus ergonomics categorises human labour by forms of strain (Rohmert, 1983) and not, as business administration does, by organisational forms (Pfeiffer et al., 1977) or by legal, training, spatial or other factors (Hilf, 1957). Owing to the central significance of strain, the theory of the concept of stress and strain is particulary important for ergonomics.

MULTIDIMENSIONAL ERGONOMIC RESEARCH METHODS

Ergonomic analysis (see Figure 1) relates to the structuring and functioning of work systems. With reference to the structural aspect of analysis, Frieling at al. (1982) presented an inventory of methods of analysis from research projects in the field of work organisation. Mention could also be made here of the Arbeitswissenschaftliches Erhebungsverfahren zur Tätigkeitsanalyse - AET (Ergonomic Job Analysis Procedure for Activity Analysis, Rohmert and Landau, 1979, which also contains further bibliographical information) including subsequent developments in evaluation methods for comparing work systems and classifying work systems (Landau and Rohmert, 1981 a; Knorz et al., 1981) together with a large number of case studies in work analysis (Landau and Rohmert, 1981 b). With respect to the functional aspect, ergonomics has, over the past decade, taken up and adapted mathematical and scientific procedures for analysing work processes. Within the complex of what is known as energetic-muscular work, biomechanical analyses have been applied and further developed (see also Jenik, 1973; Rohmert et al., 1981); and in the field of informatory-mental work, it is mainly the conceptions of information theory and control theory that have been adopted and adapted (Luczak, 1975, 1979; Johannsen, 1980, further literature detailed there).

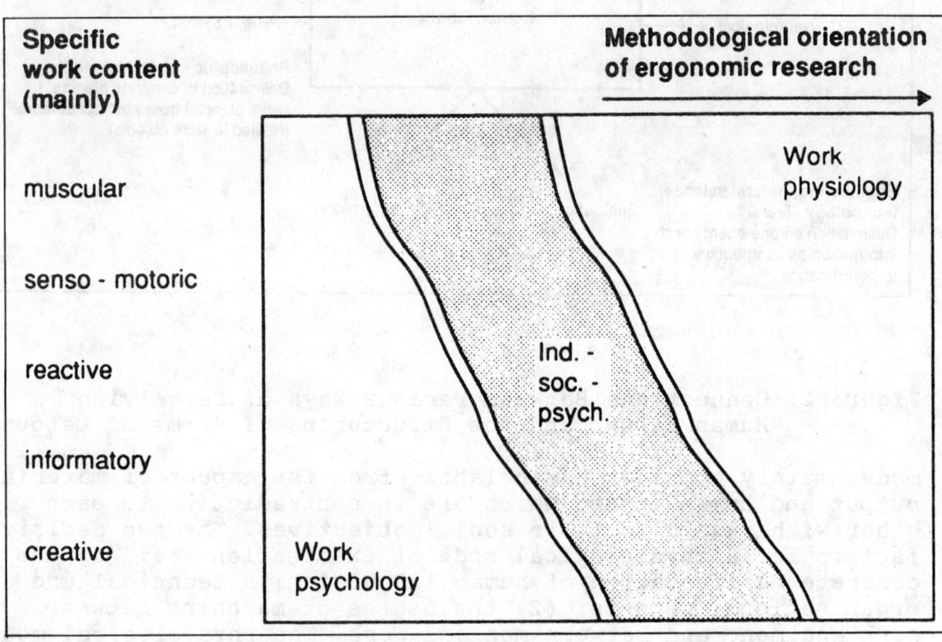

Figure 3: Classification of Human Work according to Methodological Demands

Ergonomic measurement (see Figure 1) is normally integrated into the the concept of stress and strain. The first basic concepts applied principally to forms of muscular work without significant freedom of action (Rohmert and Laurig, 1971). Further developments have taken account of the demands placed by the problems of information-oriented work, including the aspects of human behaviour. Individual characteristics and needs are extended by considering conditions of motivation. Today, the extensions also include action theory (Hacker, 1973) and activation theory (Lindsley, 1960). Thus too, the basic conditions have been fulfilled for approaching ergonomic questions from stress superpositions (Rohmert, 1982), i.e. for examining the effects of combined stresses, which previously were dealt with by the psychology of work or work physiology in isolation. As Figure 3 shows, every human work activity is normally only examined using multidimensional methods of measurement and evaluation and the results are then used to analyse and structure it. Figure 3 shows how the problems of ergonomic measurement challenge the ergonomic sub-disciplines of work physiology, work psychology, individual and social psychology to break out of their isolation. For a comprehensive analysis, evaluation and structuring of human activity, methods must be used which are able to appreciate the complex interaction of the five specific types of work content, the muscular, senso-motoric, reactive, informatory and creative, each of which imposes different strains (Rohmert, 1983). Thus for an adequate description of specific human work-contents, methodological superpositions must be applied (Rohmert, 1982). These have become a focus of ergonomic research in recent years.

If the concept of stress and strain is used as a basis for measurement in human work systems, ergonomic procedures start with a scientific assessment of the phenomena described, data registered or findings derived from these for examining the relations between men and work. The assessment becomes an ergonomic evaluation (see Figure 1) by having to be carried out in accordance with a hierarchy of levels referring to the person at work (see Figure 4). The use of the term "hierarchy" here means that the lower levels of evaluation must be realised before the next higher one can be approached. The evaluation must establish under which technical, ergonomic and organisational conditions work is executable, tolerable, acceptable and satisfying (Rohmert, 1983).

Using measurements in the sense understood by natural science, ergonomic evaluation can cover the short-term feasibility and boundary of long-term tolerability of human work, whereas the limits of acceptability and satisfaction as levels of evaluation are subject to large degrees of deviation which are connected with the relevant economic and social conditions (e. g. varying attitudes between employees and employers, or if the labour market situation is scarce or there is unemployment). The basic ideas behind the 4-level-theory are accepted by both employees and employers (e.g. cf. Birkwald, 1973; Sponder, 1973). Present discussions are concerned with aspects of the concept of the human being, which is at the basis of the evaluative hierarchy, and with the extension of the concept of work satisfaction. Here work satisfaction is sometimes equated with a maximum of well-being as expressed in

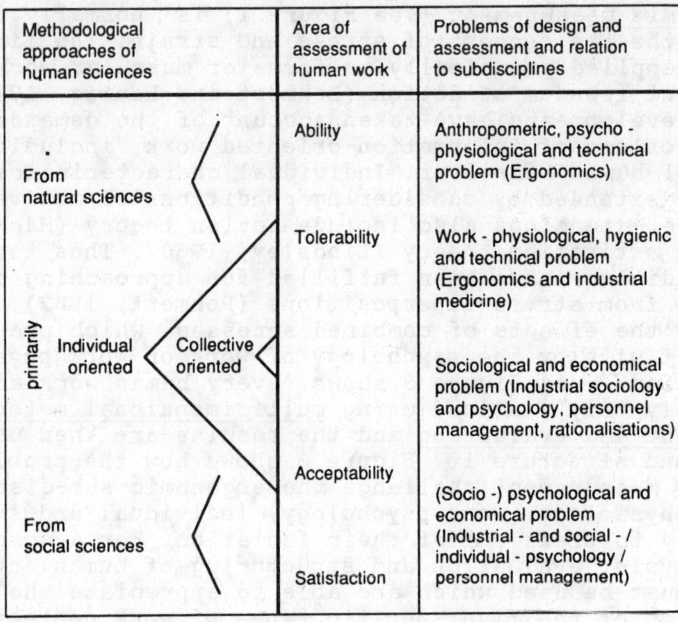

Figure 4: Areas of Assessment of Human Work

the WHO's concept of health (see Weil, 1981). Hacker and Macher (1978) substitute the concept of personality promotion for work satisfaction.

WORK DESIGN BY ADAPTATION OF MAN AND WORK

The fourth and most important task of ergonomics research is work design (see Figure 1). Work design is the only one answer to the question: "Why is ergonomics needed?" This question would not be asked if the subject of ergonomics were not mainly human. But because everybody is human, each person tends to think that he automatically knows most of what needs to be known about man when matching up a working situation to him. On the other hand, the amount of specialist knowledge already available about man is such that there is probably no one man, even a specialist, who can learn and know it all.

If a problem arises which is outside the limits of our general knowledge, we usually call upon the specialist for assistance. But if we have a complex problem to deal with involving humans, we still tend to use our own subjective opinion or we are going to consult the subjective opinions of other men instead of calling in a specialist perhaps because the existence of specialists in human factors is not yet widely known.

Low output, poor quality, errors and accidents are the usual symptoms of a problem which, among other things, may need ergonomics attention. The more the operator is stressed, the greater is the need to ensure a good match between man and his work to minimize the risk of error and maximize accuracy and output. This is the fundamental reason why the specialist knowledge from ergonomics is needed in modern industry, and particularly during the process of design.

Following the concept of ergonomics which was shown in the light of this paper work design means "work structuring using validated ergonomic findings" (see also Rohmert and Weg, 1976).

The updating of the German Betriebsverfassungsgesetz - BetrVG (Works Constitution Act) in 1972 (and the subsequent laws, regulations and guidelines) made it mandatory for "validated ergonomic findings" to be employed for "ensuring the humane structuring of work" (Humanisation of Work). These initiatives were intended to ensure that person-based criteria are used at the planning stage for Work Structuring and in the pre-stages of Occupational Safety where there is as yet no state regulation. If changes are made to work stations, work processes or the working environment in such a way that they clearly contradict validated ergonomic findings for ensuring the humane structuring of work and cause undue stress for employees, then measures can be demanded for removing, alleviating or compensating for this extra stress.

If particular stresses are to be diagnosed, (non-existent) phenomenological evidence is assumed, and this truly represents a new and important task for metrological ergonomics. Thus ergonomic laboratory studies are increasingly being complemented by field studies in order to gain analyses of work and to translate the research findings into practical Work Structuring. It is this legal obligation for Humanisation of Work which continually forms a topic both of academic discussion (Fürstenberg et al. 1983) and of disagreement between employees and employers.

REFERENCES

Birkwald, R., 1973, Menschengerechte Arbeitswelt. Arbeitsheft 013 der Industriegewerkschaft Metall für die Bundesrepublik Deutschland (Frankfurt a.M.: IG Metall).

Fürstenberg, F., Hanau, P., Kreikebaum, H. and Rohmert, W., 1983, Menschengerechte Gestaltung der Arbeit. In: Gesellschaft, Recht, Wirtschaft, Band 9, (Mannheim: Bibliographisches Institut).

Frieling, E., 1982, Bestandsaufnahme arbeitsanalytischer Methoden in Forschungsvorhaben aus dem Bereich "Arbeitsorganisation". (München: Engelhardt).

Hacker, W., 1973, Allgemeine Arbeits- und Ingenieurpsychologie. (Berlin: VEB Verlag der Wissenschaften).

Hacker, W. and Macher, F., 1978, Grundlagen und Möglichkeiten der projektionierenden und korrigierenden Gestaltung von Arbeitstätigkeiten. (Dresden: Universitätsaufträge).

Hilf, H.H., 1957, Arbeitswissenschaft, (München: Hanser).

Jastrzebowski, W., 1857, Rys Ergonomji czyli Nauki o Pracy. Przyroda o przemysl, 2, 277-281.

Jenik, P., 1973, Biomechanische Analyse ausgewählter Arbeitsbewegungen des Armes. (Berlin: Beuth).

Johannsen, G., 1980, Manuelle Regelung in Mensch-Maschine-Systemen. (Aachen: Habilitation).

Knorz, G., Landau, K. and Reus, J., 1981, VISA, ein Informationssystem zur Arbeitsanalyse. Part 2: Realisierung. Fortschrittliche Betriebsführung/Industrial Engineering, 30, 176-184.

Landau, K. and Rohmert, W., 1981a, VISA - ein Informationssystem zur Arbeitsanalyse. Part 1: Konzeption. Fortschrittliche Betriebsführung/Industrial Engineering, 30, 171-175.
Landau, K. and Rohmert, W. 1981b, Fallbeispiele zur Arbeitsanalyse. (Bern: Huber).
Lindsley, D.B., 1960, Attention, Conscriousness, Sleep and Wakefulness. In: Field, J., Magoure, H.W., Hall, V.E. Handbook of Physiology, 3 vols., section 1: Neurophysiology. (Washington: American Physiology Society, pp. 1533.
Luczak, H., 1975, Untersuchungen informatorischer Belastung und Beanspruchung des Menschen. (Düsseldorf: Verein Deutscher Ingenieure).
Luczak, H., 1979, Arbeitswissenschaftliche Untersuchungen von maximaler Arbeitsdauer und Erholzeiten bei informatorisch-mentaler Arbeit nach dem Kanal- und Regler-Mensch-Modell sowie superponierter Belastungen am Beispiel Hitzearbeit. (Düsseldorf: Verein Deutscher Ingenieure).
Pfeiffer, W., Dörrie, U. and Stoll, E., 1977, Menschliche Arbeit in der industriellen Produktion. (Göttingen: Vandenhoeck & Ruprecht).
Rohmert, W., 1982, Ergonomie der kombinierten Belastungen. (Köln: Schmidt).
Rohmert, W., 1983, Formen menschlicher Arbeit. In: Rohmert, W., Rutenfranz, J.: Praktische Arbeitsphysiologie. (Stuttgart, New York: Thieme, pp. 5-29.
Rohmert, W. and Laurig, W., 1971, Work Measurement. In: International Journal of Production Research, 9, 157-168.
Rohmert, W. and Weg, F.J., 1976, Organisation teilautonomer Gruppenarbeit. Betriebliche Projekte - Leitregeln zur Gestaltung. (München, Wien: Hanser).
Rohmert, W. and Landau, K., 1979, Das Arbeitswissenschaftliche Erhebungsverfahren zur Tätigkeitsanalyse - AET-Handbuch und Merkmalsheft. (Berlin: Huber).
Rohmert, W., Jenik, P., Mainzer, J., 1981, Biomechanik der menschlichen Arbeit. In: Bundesamt für Wehrtechnik und Beschaffung: Handbuch der Ergonomie. (München, Wien: Hanser), A 4.-1-12, 1-5, 1-10, 1-19, 1-4.
Sponder, J., 1973, Belastung und Beanspruchung bei Muskelarbeit als Grundlage der Arbeitsgestaltung und Erholzeitermittlung. In: Mitteilungen des Instituts für angewandte Arbeitswissenschaft e.V. (Köln), Sondernummer 39/1.
Weil, R., 1981, Arbeit und Wohlbefinden. (Köln: Institut für angewandte Arbeitswissenschaft e.V.).

I

AGING AND INDUSTRIAL PERFORMANCE

Advances in Industrial Ergonomics and Safety II
Edited by Biman Das
Taylor & Francis 1990

HEAT TOLERANCE AND AGING WITH APPLICATION TO INDUSTRIAL JOBS

Kent B. Pandolf

U.S. Army Research Institute of Environmental Medicine
Kansas Street, Natick, MA 01760-5007, USA

Exercise-heat tolerance has been shown to be reduced in older individuals. Some authors have suggested that aerobically fit older individuals have fewer performance decrements during exercise-heat stress than less fit individuals of the same age. However, none of these studies matched older and younger individuals for any pertinent physiological and/or morphological variables. Recently, experimental results were reported where men differed in age by 25.2 years, but were matched for body weight, surface area, surface area-to-weight ratio, percent body fat and maximal aerobic power. Evidence from this study indicated little impairment of the thermoregulatory system during exercise-heat stress at least through the fifth decade of life for these aerobically trained older men. These findings show the importance of aerobic training and other morphological considerations in selecting older individuals for jobs in hot industrial environments.

INTRODUCTION

In general, exercise-heat tolerance is thought to be reduced in older individuals (Drinkwater and Horvath, 1979; Wagner et al., 1972). Older men have been reported to have higher heart rates, mean skin and core temperatures, and lower sweat rates compared to younger men during exercise-heat stress both pre- and post-acclimation (Wagner et al., 1972). It has also been shown that older men start to sweat later and/or sweat less during exercise in the heat compared to younger men (Hellon and Lind, 1956). However, active older women had the same sweating capacity and core temperature at sweat onset as younger women while resting in the heat (Drinkwater et al., 1982).

Some authors have suggested that aerobically fit older men had fewer decrements in exercise-heat performance than less fit men of the same age (Robinson et al., 1965). However, this study actually compared the exercise-heat acclimation responses for fit men to the responses of these same individuals some 21 years later. These older men acclimated to exercise in the heat at the same rate and to about the same degree as when they were younger. However, none of the above-mentioned studies matched older and younger individuals for any pertinent physiological and/or morphological variables. It has been suggested by many that body weight, surface area, body fat, and/or maximal aerobic power are closely related to an individual's ability to thermoregulate during exercise in the heat. In addition, the above studies on aging do not answer whether the exercise-heat intolerance observed with aging is due to age per se or related to other factors such as decreased physical activity and lowered maximal aerobic power. Thus, these early studies dealing with aging and exercise-heat tolerance have only descriptively compared various age groups.

Recently, experimental findings were reported from a study where men differed in age by 25.2 years (mean age: 21.2 vs 46.4 yr), but were matched for body weight, surface area, surface area-to-weight ratio, percent body fat, and maximal aerobic power (Pandolf et al., 1988). The greater regular weekly aerobic activity for these older men was associated with their better initial performance during exercise in the heat; however, heat acclimation negated this advantage. Evidence from this study indicated little impairment of the thermoregulatory system at least through the fifth decade of life for these aerobically trained older men. These findings demonstrate the importance of aerobic training and other morphological considerations in selecting older individuals for jobs in hot industrial environments.

EXPERIMENTAL OBSERVATIONS

As depicted in Figure 1, Drinkwater and Horvath (1979) reported that the tolerance time to perform light exercise in dry heat (48°C) was reduced for both young girls and older women. These authors studied 38 non-acclimated females that ranged in age from 12 to 68 years. For the young girls, the reduced tolerance was related to the instability of an immature cardiovascular system. For the older women, exercise-heat intolerance was associated with an inadequate sweating response, cardiovascular instability and a marked decrement in maximal aerobic power.

As shown in Figure 2, Wagner et al. (1972) reported the physiological responses of ten younger (20-29 yr) and seven older (46-67 yr) men to dry heat (49°C) before and after exercise-heat acclimation. Both before and after exercise-heat acclimation, the younger men maintained lower mean skin and rectal temperatures, lower heart rates, and higher evaporation rates than the older men. The younger men also tolerated the exercise-heat stress longer than the older men. These authors suggest that the older men displayed a more limited sensitivity and secretory capacity in their sweating mechanism compared to the younger men which contributed to their inferior temperature regulation.

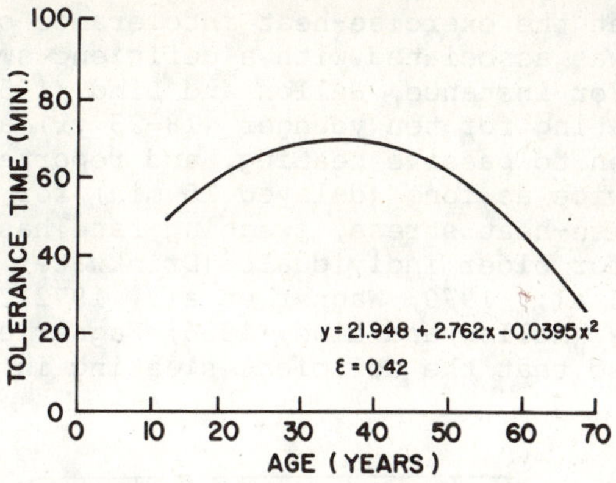

Figure 1. Tolerance time of females to light exercise at 48°C in relation to age (reproduced with permission from Drinkwater and Horvath, 1979).

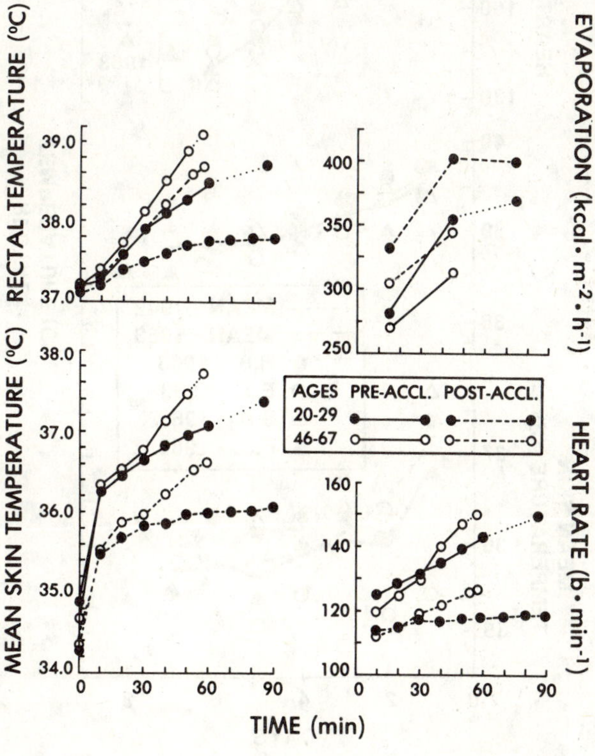

Figure 2. Effects of dry-heat acclimation on physiological responses contrasting young and old men (redrawn from Wagner et al., 1972).

Several authors (Drinkwater and Horvath, 1979; Hellon and Lind, 1956; Lind et al., 1970; Wagner et al., 1972) have suggested that the exercise-heat intolerance observed in older individuals was associated with a deficient sweating mechanism. For instance, Hellon and Lind (1956) compared the onset of sweating for ten younger (18-23 yr) and ten older (44-57 yr) men to passive heating, and reported that sweating onset took twice as long (delayed 15 min) for the older men. During exercise-heat stress, sweating rate has been reported to be lower for older individuals (Drinkwater and Horvath, 1979; Lind et al., 1970; Wagner et al., 1972). Several of these authors (Hellon and Lind, 1956; Wagner et al., 1972) have concluded that the deficient sweating in older

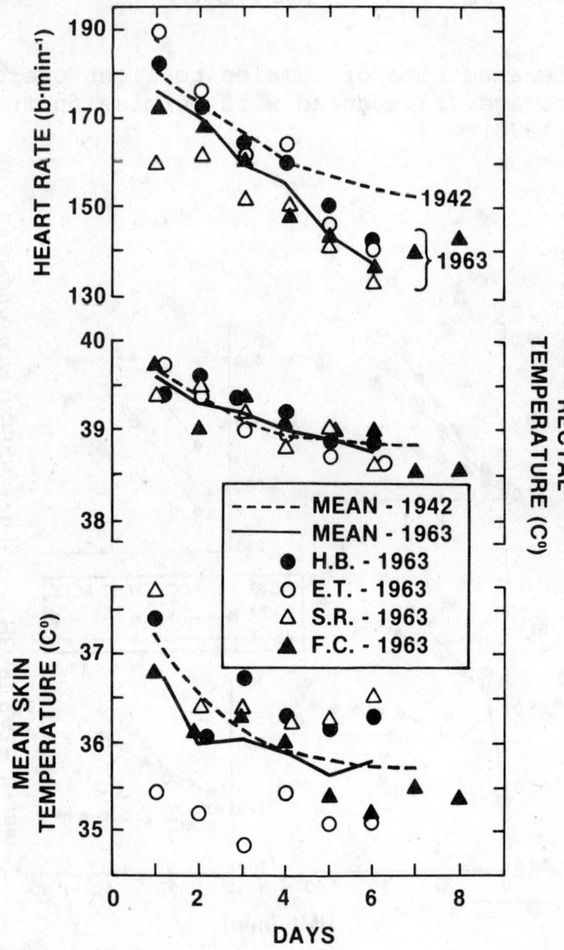

Figure 3. Physiological responses of four men evaluated over eight days of exercise-heat acclimation in 1942 and the responses of these same men evaluated in 1963 (redrawn from Robinson et al., 1965).

individuals was associated with neural degeneration and/or a decline in the functional ability of their sweat glands.

As illustrated in Figure 3, Robinson et al. (1965) compared the exercise-heat acclimation responses (40°C, db; 23.5°C, wb) of four men during 1942 (mean age 31 yr) to the responses of these same individuals some 21 years later. By most of the usual physiological criteria (mean skin and rectal temperatures, and heart rate), these older men showed similar patterns of exercise-heat acclimation and acclimated to about the same extent as when they were younger. The tolerance of these older men on the first day of acclimation was no less compared to their responses 21 years earlier. However, the initial cardiovascular strain observed on the first day of acclimation was greater for the older men in terms of relative cardiac cost given our knowledge concerning the reduction in maximum heart rate with increasing age. Overall, these older men displayed about the same degree of physiological strain during repeated exercise-heat stress as they did 21 years earlier and acclimated about as well. The habitually active lifestyle of these four individuals has been hypothesized to help explain why their thermoregulatory responses were not affected by age.

More recently, Pandolf et al. (1988) compared the thermoregulatory responses during exercise-heat acclimation (49°C, 20% rh) between nine younger (mean age 21.2 yr) and nine older men (mean age 46.4 yr) who were matched ($P>0.05$) for body weight, surface area, surface area-to-weight ratio, percent body fat and maximal aerobic power. Figure 4 compares the performance times between these middle-aged and young men during a comfortable (C) environment day (22°C, 50% rh) and during each of the ten exercise-heat acclimation days. During the first day of exercise-heat acclimation, performance time was 27 min longer for the middle-aged compared to the young

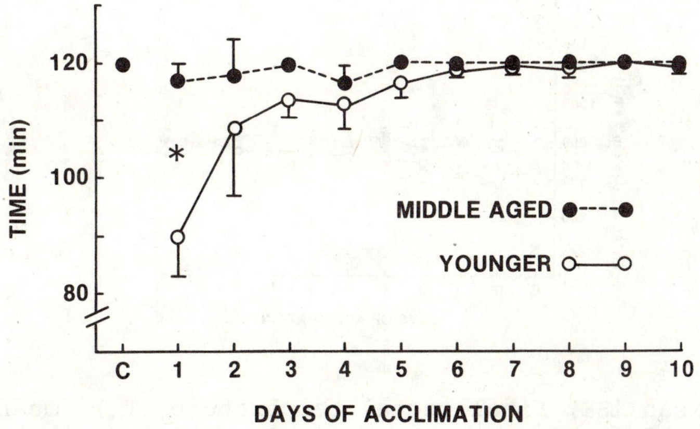

Figure 4. Mean(±SE) performance time of middle-aged and young men during control (C) day and each of 10 exercise-heat acclimation days. * Significant difference ($P<0.05$). All subjects completed 120-min on control day (from Pandolf et al., 1988).

men. However, performance time did not differ between these groups during the remaining days of exercise-heat acclimation.

Figure 5 shows that final rectal temperature, mean skin temperature and heart rate were lower while final total body sweat loss was higher for these middle-aged compared to the young men on the first exercise-heat acclimation day (Pandolf et al., 1988). These thermoregulatory advantages for the middle-aged men persisted for the first few days of exercise-heat acclimation. However, by the end of the acclimation

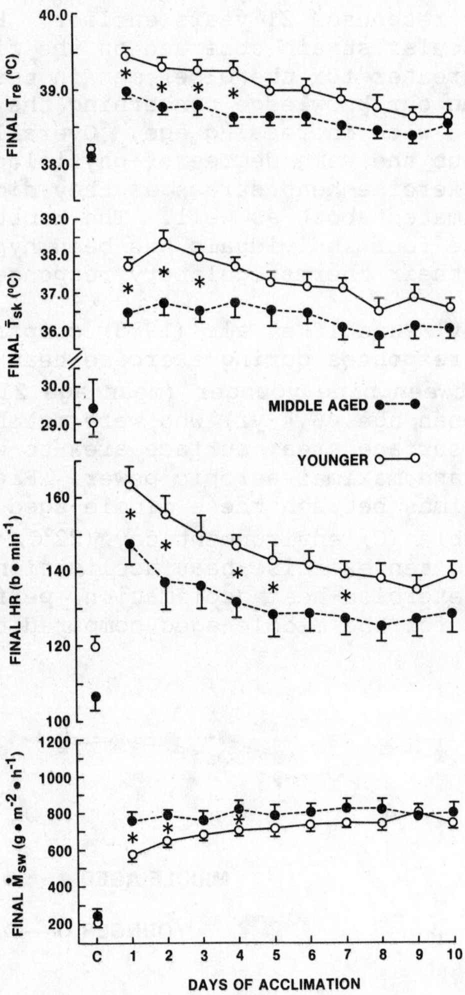

Figure 5. Mean(±SE) final rectal temperature (T_{re}), mean skin temperature ($\bar{T}_{sk}$), heart rate (HR), and total body sweat loss ($\dot{M}_{sw}$) of middle-aged and young men during the control (C) day and each of 10 exercise-heat acclimation days. Asterisk indicates significant (P<0.05) difference (from Pandolf et al., 1988).

period, no thermoregulatory differences were observed between these two age groups.

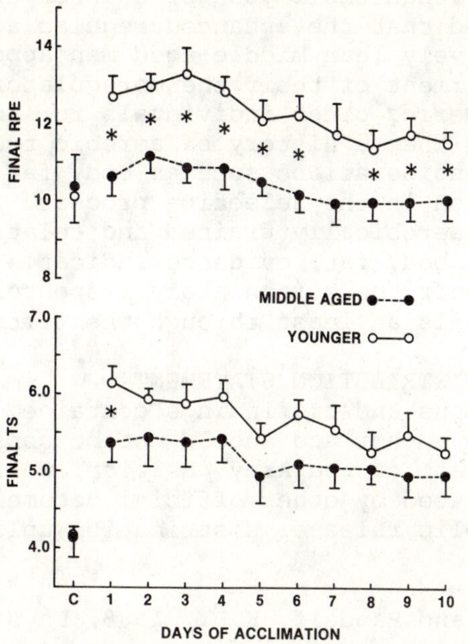

Figure 6. Mean(±SE) final rated perceived exertion (RPE) and thermal sensation (TS) of middle-aged and young men during the control (C) day and each of 10 exercise-heat acclimation days. Asterisk indicates significant (P<0.05) difference (from Pandolf et al., 1988).

Figure 6 illustrates that final rated perceived exertion (RPE) and final thermal sensation (TS) were generally lower for the middle-aged men throughout the ten days of acclimation (Pandolf et al., 1988). Final RPE was significantly lower for the older men on all acclimation days except for day 7 while final TS was significantly lower only on day 1. Pandolf et al. (1988) concluded from this study that there was little impairment of the thermoregulatory system at least through the fifth decade of life for aerobically trained middle-aged men.

CONCLUSIONS

For the general population, exercise-heat intolerance appears to be increased for older compared to younger individuals. However, the greater exercise-heat intolerance observed with aging does not appear to be related to age per se, but does seem to be associated with other factors such as decreased physical activity, lowered maximal aerobic power, obesity and certain disease states related to aging.

Researchers generally agree that high aerobic fitness achieved through physical training will reduce the physiological strain associated with exercise in the heat (Armstrong and Pandolf, 1988). The older men studied by Pandolf et al. (1988) averaged nearly 20 miles/week more aerobic activity than their younger counterparts. These authors concluded that the enhanced regular aerobic activity for their relatively lean middle-aged men appeared to have offset any impairment of their thermoregulatory systems.

When considering older individuals for hot industrial jobs, it appears that a history of aerobic training and other morphological considerations such as body fat should be given some consideration in the selection process. If older individuals are aerobically trained and relatively lean in terms of percent body fat, evidence indicates little impairment of their thermoregulatory responses compared to younger individuals at least through the fifth decade of life.

DISCLAIMER AND DISTRIBUTION STATEMENT

The views, opinions and/or findings contained in this report are those of the author and should not be construed as an official Department of the Army position, policy or decision, unless so designated by other official documentation. Approved for public release; distribution unlimited.

REFERENCES

Armstrong, L.E. and Pandolf, K.B., 1988, Physical training, cardiorespiratory physical fitness and exercise-heat tolerance. In Human Performance Physiology and Environmental Medicine at Terrestrial Extremes, edited by K.B. Pandolf, R.R. Gonzalez and M.N. Sawka (Indianapolis: Benchmark Press Inc.), pp. 199-226.

Drinkwater, B.L., Bedi, J.F., Loucks, A.B., Roche, S. and Horvath, S.M., 1982, Sweating sensitivity and capacity of women in relation to age. Journal of Applied Physiology, 53, 671-676.

Drinkwater, B.L. and Horvath, S.M., 1979, Heat tolerance and aging. Medicine and Science in Sports, 11, 49-55.

Hellon, R.F. and Lind A.R., 1956, Observations on the activity of sweat glands with special reference to the influence of ageing. Journal of Physiology (London), 133, 132-144.

Lind, A.R., Humphreys, P.W., Collins, K.J., Foster, K. and Sweetland, K.F., 1970, Influence of age and daily duration of exposure on responses of men to work in heat. Journal of Applied Physiology, 28, 50-56.

Pandolf, K.B., Cadarette, B.S., Sawka, M.N., Young, A.J., Francesconi, R.P. and Gonzalez, R.R., 1988, Thermoregulatory responses of middle-aged and young men during dry-heat acclimation. Journal of Applied Physiology, 65, 65-71.

Robinson, S., Belding, H.S., Consolazio, F.C., Horvath, S.M. and Turrell, E.S., 1965, Acclimatization of older men to work in heat. Journal of Applied Physiology, 20, 583-586.

Wagner, J.A., Robinson, S., Tzankoff, S.P. and Marino, R.P., 1972, Heat tolerance and acclimatization to work in the heat in relation to age. Journal of Applied Physiology, 33, 616-622.

Advances in Industrial Ergonomics and Safety II
Edited by Biman Das
Taylor & Francis 1990

AGING AND WORK: IMPLICATIONS FOR HUMAN FACTORS RESEARCH

Sara J. Czaja

Department of Industrial Engineering
University of Miami
Coral Gables, Florida

INTRODUCTION

There are several reasons why there is a sudden resurgence of interest in the "older worker" including: 1) an increase in number of older people in the population; 2) an increase in number of "older workers" (persons age 45+) in the labor force 3) legislation protecting individuals in this age group and 4) a decrease in number of younger workers especially in entry level positions.

Recent census estimates indicate that by the year 2000 persons aged 65+ will number at lest 32 million and represent 12 percent of the population; by the year 2030 this number will increase to 18 percent. Generally the labor force participation rates of persons over age 55 are declining. It is projected that by 1995 workers age 55 and older will only represent 11% of the labor force as compared to 14% in 1981. This may be problematic from several prospective including an increase in the economic dependency burden and reduced quality of life for older persons. Thus there is a need to understand why the labor force activity of older adults is declining and to identify strategies which maintain these people in the labor force.

While the work activity of persons over age 55 is declining the number of workers aged 45-54 is increasing. During the 1980's and 1990's the "baby boom" generation will move into middle age and by the year 2000 the number of workers over age 40 will increase by 19 percent (OTA, 1985). At the same time the number of younger workers is declining which implies a labor shortage especially for entry level positions. One way to address this problem is to encourage older people to remain in the work force or re-join the labor market after retirement. In essence all of these trends point to a need to consider theaging workers in job and work place design. Unfortunately there is a paucity of data relating age-related changes in function to work activity. This type of data is needed in order to insure that older workers will be successfully integrated into the work force. Successful integration implies worker efficiency, safety, comfort and satisfaction.

Aging and work performance

When considering how aging affects work current data is limited and inconclusive. General aspects of employee performance include: accident rates, absenteeism, turnover, grievances and productivity. The impact of aging on these measures depends on the particular measure in question, the type of data collected and the specific occupation. Regarding accidents, data indicate that older workers have lower accident rates then younger workers. However, older workers tend to remain off the job longer if they are injured. Absenteeism and turnover rates are also lower for older people, but this may be due, at least in part, to the fact that they would find it difficult to find another job (Stagner, 1985).

Considering productivity or work output it is hard to draw any conclusions on howaging affects work. Rhodes (1983) conducted a comprehensive review of the existing data on aging and productivity and concluded that there was equal evidence supporting that job performance increases with age, decreases with age or remains the same. She concluded that the result depends on the type of performance measures uses, the type of job and other factors such as experience. Similarly, Waldman and Avolia (1986) conducted a meta-analysis on over 40 samples that included data pertaining to the relationship between age and job performance. They classified the samples into three categories: a) supervisory ratings b) peer ratings and c) individual productivity. They found different results according to the type of performance measure. For objective indices performance was found to increase with higher ages whereas supervisor ratings showed a tendency to be lower for older employees. They found, however, that the relationship between ratings and age varied according to job type; ratings were higher for professionals than non-professionals. They concluded that age only accounts for a small amount of the variance in the range of individual differences in work performance and thus is not useful for estimating performance potential.

Examining the studies that have been done there appears to be little, if any bases for the widespread belief that performance declines with age. Instead, it may be that some tasks are unsuitable for older workers because the demands inherent in these tasks exceed their capabilities. The data that are available suggest that tasks which involve heavy physical demands or those which are externally paced maybe unsuitable for older workers (Davies and Sparrow, 1985). However, these conclusions are general and not specific to jobs or occupations. For example, although there appears to be age deficits in visual search performance it is unclear to what extent this affects performance on industrial inspections tasks. Czaja and Drury (1981) found age-related declines in inspection performance whereas Evans (1951) reported no age related effects on inspection performance. Clearly, research is needed in this area as inspection is prevalent and an important task in many industries including the microelectronics industry, garment industry and automobile industry.

With respect to production tasks available data suggests

that performance peeks in the 30's or 40's and declines gradually thereafter (e.g. Clay, 1956). The results are similar for white collar workers. However in many instances age only accounts for a small portion of the variance and factors such as training and experience often moderate age-related effects. For example, Sparrow and Davies (1988) examined the effects on performance of age, tenure, job complexity and training among a sample of service engineers. They found an inverted-U relationship between age, quality and speed of performance. However, three other important findings emerge from their study. One is that job complexity did not moderate the effect of age and performance and in fact for all age groups performance was higher for the more complex tasks. Secondly, age only accounted for a small portion of the variance. Finally, training is an important mediating variable between age and performance. Older engineers who had received recent task training performed better then those who had not received recent training. This suggests that training can mediate age declines in performance. This is a critical finding given the rapid influx of new technologies into the work place.

In summary, the relationship between age and job performance is complex and far from understood. One problem is that the data which is available is sometimes unreliable because of methodological shortcomings in data collection procedures. For example, studies which rely on supervisory ratings of performance may be unreliable if the rater has negative attitudes toward older workers. Also, many studies involve small samples, restricted age ranges or are cross-sectional which may confound age effects with factors such as experience, education or exposure to technology. Finally, the results vary according to type of task or type of measure. What's lacking is a systematic body of knowledge that links age related changes in skills and abilities to skill demands of jobs. Tasks analysis represents a powerful tool to accomplish this goal. Clearly, more research is needed in this area especially for information processing tasks such as computer-interactive tasks or supervisory control tasks which are becoming increasingly prevalent within today's work environment.

Relevant data

The above discussion is not meant to suggest that we are totally lacking in knowledge with respect to the process of aging. There is an abundance of data in the gerontological and psychological literatures which is relevant to work activities. However, there has been little done in the way of translating this data into design guidelines and principles. Also there is a general lack of communication between gerontologists and engineers. Thus we often find ourselves at a loss when attempting to accommodate the needs of older people in the work arena. As suggested by Smith (1989) more effort needs to be directed towards extrapolating findings from the lab to real life settings, generating data models and producing quantitative data which is not yet available.

Of the findings regarding aging and performance there are several which have relevance to work activities. These include declines in visual and auditory functioning and strength; a general slowing of behavior and changes in perception and cognition. It is important to point out that the elderly as a group are more heterogeneous then younger people with respect to health, social, economical and other characteristics. Thus age is an over-simplistic and often unmeaningful predictor of functional capacity. It is beyond the scope of the paper to provide a detailed discussion of the aging process. However specific examples of age effects are discussed in order to highlight potential implications for job activities and needed research in the area of aging and work.

For example age related declines in visual functioning are well documented. They include: a reduction in the range of accommodation; a loss of light transmission which produces significant changes in the amount of light impinging on the retina; a loss of contrast sensitivity and a loss of dynamic visual acuity. These changes in visual function have important implications for job and equipment design. Older people may require larger characters on screens or different color combinations between the characters and background to successfully interact with computer technologies. Also older workers may find it difficult to perform tasks which require discrimination of detail without some type of visual aids. Finally, older workers may require different types or levels of lighting to efficiently perform jobs.

While some guidelines exist regarding designing visual environments to accommodate age changes in vision our knowledge of the practical implications of vision losses with age is limited. There is a need for research directed towards understanding the relationship between age related changes in vision and job performance, and to identify and evaluate the suitability of intervention strategies to compensate for visual decline.

There are a number of age-related changes in cognitive abilities which also have important implications for work activities. These include a general slowing of behavior as age increases, declines in perceptual flexibility and speed of encoding information, difficulty processing complex information, shortening of the immediate memory span and difficulty retrieving information from long term memory (Welford, 1985).

These changes in cognitive abilities may have relevance to tasks such as computer-interactive tasks which are primarily characterized by their information processing requirements. For example it may be more difficult for older people to recall complex operational procedures involved with retrieval and manipulation of databases. Also these types of tasks may be more stressful for older people because of changes in cognitive skills. Currently few efforts have been directed towards understanding the effects of aging on more complex information processing requirements such as those inherent in computer-interactive tasks. In addition to this type of research it is important to identify what types of strategies older people use to compensate for declining abilities. This type of data would be useful in designing intervention strategies which aid the older worker.

Another area in which insufficient research has been

conducted is in the training of older workers. Documented declines in cognition have often been based on situations where older people had limited opportunity for learning. Several investigators have shown that if appropriate training strategies are employed, age differences in performance can be reduced or eliminated. To date there are only limited data available on task specific training strategies appropriate not only for older learners, but which benefit people of all age groups. It will also be useful to determine the extent to which providing training on component skills, such as visual search, benefits task performance on tasks such as inspection.

Similar issues can be raised with respect to the physiological demands of tasks. We know that the elderly as a group have reduced joint mobility, stamina and strength. This implies that many tasks, such as those performed in manufacturing settings, involve lifting and carrying and/or require stamina and strength are likely to be difficult for older workers. However, we only have limited data on problems encountered by older people in these occupations or to what extent performance is affected by training and motivation. Questions of this type need to be answered before appropriate job design strategies can be developed.

CONCLUSION

The above discussion highlights some of the age-related changes in abilities which may have relevance to work. The intent was to demonstrate the need for research in this area. In general, research attention to those aspects of work that become more difficult, less productive, or less satisfying with age could make a worthwhile contribution to the value of older workers who stay employed, and to the quality of their lives. The discipline of Human Factors Engineering can make valuable contributions in this area. Human Factor's research techniques such as task analysis can be applied to determine the skill requirements of jobs and the degree of fit between these requirements and the capabilities of older persons. This will allow us to identify which tasks are suitable for older people to perform, to target needed areas of intervention in job design, equipment and environmental design and also to identify areas where retraining will be required.

In addition Human Factors Engineers should study and collect normative data on the basic capabilities and limitations of older people as well as ecologically valid data on performance of the elderly on real world tasks. This type of data will allow us to eliminate gaps in existing knowledge and accommodate the needs of older adults in work environments.

REFERENCES

Clay, H.M., 1956, A study of performance in relation to age at two printing works. Journal of Gerontology, 11, 417-424.

Czaja, S.J. and Drury, C.G., 1981, Training programs for inspection. Human Factors, 23, 485-494.

Davies, D.R. and Sparrow, P.R., 1985, Age and work behavior. In N. Charness (Ed.). Aging and human performance, edited by N. Charness (Chichesler, England: Wiley), pp. 293-332.

Office of Technology Assessment, 1985, Technology and aging in America. (Washington, D.C.: Author).
Rhodes, S.R., 1983, Age-related differences in work attitudes and behavior. A review and conceptual analysis. Psychological Bulletin, 93, 328-367.
Smith, D.B., 1989, Human factors and aging: An overview of research needs and application opportunities. Unpublished manuscript.
Sparrow, P.R. and Davies, D.R., 1988, Effects of age, tenure, training and job complexity on technical performance. Psychology and Aging, 3, 307-314.
Waldman, D.A. and Avolio, B.J., 1986, A meta-analysis of age differences in job performance. Journal of Applied Psychology, 11, 33-38.
Welford, A.T., 1985, Changes of performance with age. In N. Charness (Ed.), Aging and human performance, edited by N. Charness (Chichesler, England: Wiley), pp 333-369.

Advances in Industrial Ergonomics and Safety II
Edited by Biman Das
Taylor & Francis 1990

UPDATING SKILLS, TRAINING, RETRAINING AND THE DECISION TO RETIRE OR WORK.

Harvey L. Sterns, Ph.D.
Department of Psychology
Institute for Life-Span
Development and Gerontology
THE UNIVERSITY OF AKRON
Office of Geriatric Medicine and Gerontology
NORTHEASTERN OHIO UNIVERSITIES COLLEGE OF MEDICINE
Akron, Ohio 44325-4307

INTRODUCTION

A primary purpose of this paper is to examine issues, theory and research relating to updating of skills, training, retraining and the decision to work or retire in the 1990's. The work environment is continuing to undergo dramatic change with plant closing, reduction in force and the introduction of new technology. The initial and continuing career choices made by workers is considered within the context of life-span development. Major opportunity for career growth and continuing skills enhancement comes from training and development. We focus here on job relevant training and older adult specific training. The development of effective training continues to be a major concern.

Older adults of the 1990's will need to continually update skills in a competitive working environment or accept employment in less demanding and low paying jobs in the service area. Capable older workers should be able to compete.

This paper draws on a number of recent book chapters by Sterns and Patchett (1984), Sterns, (1986), Sterns and Alexander (1987) and Sterns and Doverspike (1988, 1989). Readers may want to consult the full chapters for a more complete discussion.

Extension of work life or early retirement should come about by choice on the part of the older adult worker. The value of an older worker should be judged on his or her merit not by chronological age. Opportunities should be made available to older adult workers that are the same as any other worker of similar competence, vigor and ambition. Age neutral policies need to be part of current personnel policies and will assure that older persons who elect to continue working will have access to career opportunities consistent with his or her aspirations, abilities and potential (The Future of Older Workers In America: New Options for an Extended Working Life, 1980).

Technology innovation may affect the workplace in several ways, some of which 1) render skills and knowledge obsolete, 2) require the development of new knowledge and skills, 3) change employee attitudes and satisfaction regarding the work place positive and negative), 4) create new job openings and potentiality for career mobility, and 5) create unemployment Burack and McNichols, (1973).

The issues of employment and unemployment are important when discussing the older worker because older workers tend to have greater difficulty finding new employment and may not have the same retraining opportunities. For example, older adults may lack the educational background to compete with younger adults in certain occupations even though they may have a wealth of on-the-job knowledge and experience. Self-doubts and fear of failure may present the aging adult from taking advantage of retraining opportunities when their skills are not readily transferable to a new situation.

CAREER DEVELOPMENT--ONE RESPONSE TO TECHNOLOGICAL CHANGE

Although career development should be a lifelong process, the individual may consider plans to simply update current job knowledge, to engage in continuing education, or to undergo retraining.

Super (1980, 1984) presents career development with an emphasis on the individual and his or her environment. Occupational choice or assignment is not something that happens once in a life time. People and situations develop and career decisions become a series of mini decisions. These mini decisions add up to a series of occupational choices which appear to be one maxi-decision.

Discussions of adult development and work need to take into account the sequential nature of such decision making. Super (1984) states that readiness for adult career decision making might better be called career adaptability rather than career maturity. Re-exploration and reestablishment of careers have thus attracted a great deal of attention and the concepts of transition and crisis is now used to describe these processes.

A focus on individual developmental changes, as well as a more rapidly changing technology and economy highlight issues of transition for certain older adult workers. According to Super (1984), the timing of transition is a function of the individual's personality, attributes, and the situation and is not tied to age. He also feels that established models of career may continue to be valid in this period of increasing ethnic, social and sexual equality, changing technology and changing occupations. The frequency of different types of career patterns may change. Stable and conventional patterns of career may be the norm for a small number of highly educated capable individuals. Unstable, multiple-job and full time leisure (by choice, displacement or failure to be hired or rehired) may become the dominant career patterns for others.

Occupational choice processes can be reopened in a person's mature years as well as after retirement in a second or third career. The concept used here is optimization. People seek to find the best fit between changing interests and goals, their changing circumstances and the realities of the world of work. Occupational choice becomes a lifelong process of decision making for those who seek major satisfactions from their work Ginzberg, (1984).

It is clear that general statements regarding older adult workers that do not emphasize large individual differences in people and large differences in work demands continue to perpetuate the negative balance for older adult career development. Normal aging does not mean major changes in ability to work.

Sterns and Patchett (1984) and Sterns (1986) have attempted to develop a model of adult and older adult career development that is non-age specific. The model assumes that transitions in worklife may occur many times throughout a career (see Figure 1).

Looking at the model, the decision to change jobs or careers or to exit the system, is directly influenced by attitudes toward mobility and success or failure in previous career development activities. Numerous factors are hypothesized to affect mobility attitudes such as employment, tenure or stage in career, growth need strength, fear of stagnation, marketability perceptions, job market conditions, and chance encounters. The decision to change jobs or careers also may affect one's attitudes toward entering or re-entering the workforce. Any of these variables also could be mediated by the effects of various personality variables.

The model incorporates Hall's (1971) model of career growth which conceptualizes career planning from a goal setting perspective. Once a career goal decision such as the decision to engage in training or re-training is made, then goal attainment should lead to identity growth and enhanced self-esteem. Such enhanced self-esteem may then lead to greater commitment to future career development goals. Goal attainment enhances self-esteem which may increase perceptions of self-efficacy and future commitment to career development activities. Note that for an older adult with doubts about his/her abilities to perform well in training and/or retraining situations, successes early in the program may be crucial. Not only would successes enhance one's feelings of self-efficacy, but decisions to engage in future retraining may be enhanced as well. Failure could easily result in a decision to drop out of the retraining program as well as future programs, jeopardizing future career growth opportunities and decisions.

Age and/or life stage are important determinants of behavior. Many career models suggest that mobility rates of

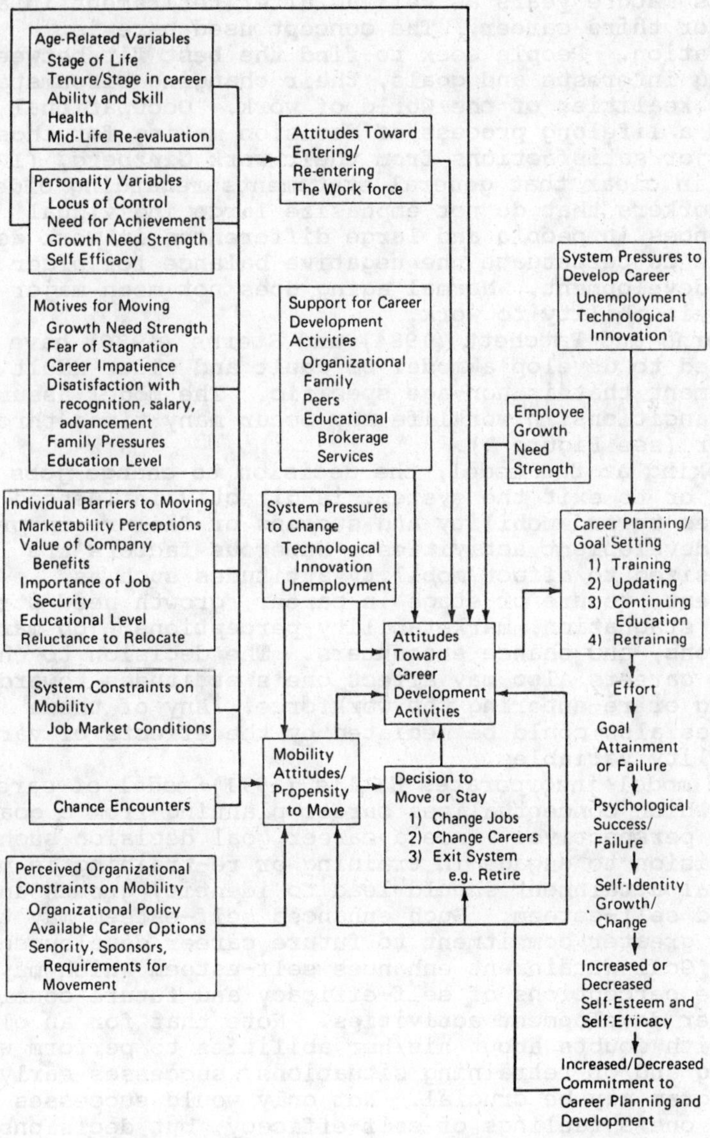

Figure I Career Progression in Middle and Later Adulthood (Sterns, 1986)

younger persons (up to age 30 or so) are much higher than those of older persons. Viega (1983) found that age correlated significantly with propensity to change jobs. Tenure (seniority) or stage in one's career has been a variable of interest in the mobility literature.

Individual barriers to moving are related to age and seniority. Viega (1983) states that perceptions of one's own marketability may strongly influence one's effort to explore alternative career opportunities. The longer a person remains with a company and the older the person becomes, the more likely it is that a person will be conservative about risking benefits accrued through the years in order to move to a new organization.

Viega (1983) identifies five motivations for moving which significantly influence propensity to leave a job, fear of stagnation, career impatience and dissatisfactions with one's salary, recognition and/or advance. A number of additional personality variables are hypothesized to affect mobility and career development attitudes, need for achievement and locus of control.

Adults make multiple career transitions throughout the life span. There are people 60, 70, or older who are still interested in working, changing jobs and further development of their careers. Past gerontological discussions have almost exclusively focused on the decision to retire. A great deal more attention needs to be given to multiple career transitions which may occur within a working life. This leads to a renewed emphasis on the important role that training and retraining can play in career development.

Atchley (1988) describes retirement as a change in roles from that of worker to retired person. Further, he traces the decision-making process surrounding retirement as including: 1) how individuals arrive at the point of considering retirement; 2) the factors that influence the consideration of the retirement choice; and 3) the factors that affect the actual timing of retirement.

Atchley (1988) stresses the importance of an individuals's felt need to continue employment as an influencing factor in the consideration of retirement. This need can be either internal or stem from external or system pressures. He suggests that two comparisons are vital when considering retirement. First, expected financial needs in retirement are compared with expected financial resources. Second, the individual compares his/her social situation in retirement with what it was while working. It is noted that access to information, availability of and accuracy of information, and personality variables will have an impact on these comparisons. The actual timing of retirement is proposed to be dependent upon the expected income comparison. According to this discussion, an individual whose perceived financial resources in retirement are less than his/her perceived economic needs, is expected to continue working if physically able to do so.

Another approach to the retirement issue centers around specific factors associated with the decision to retire. Both individual and institutional or work-related factors can impact upon the decision to retire.

The assumption often held by employers and workers is that retirement marks the end of paid employment. This is not always the case, however. Research on labor-force participation rates of retirees indicates that work during retirement is more common than had been previously assumed Beck, (1985). As a relatively new area of study, students and professionals interested in gerontology will find many avenues of research into the causes and consequences of post-retirement employment.

A compilation of existing evidence results in the profile that retirees who work tend to be well-educated, white collar rather than blue collar workers, self-employed, and have a positive attitude toward work. Most of the existing research reveals a strong influence of financial variables in post-retirement employment decisions.

TRAINING AND DEVELOPMENT OF THE OLDER ADULT WORKER

Education and training can offer new opportunities as well as optimization of individual development across the life span. As workers continue to learn and change, education and training is an important opportunity to facilitate career development at all periods in the work life. The need to integrate work and learning continues to be a major challenge

Many of today's older adult workers who are in need of training are casualties of a failure over many decades for industry to carry out appropriate responses to future training needs. A recent review of private sector training Lillard and Tan, (1986) reports that post school training is pervasive. Close to 40 percent of both men and women in the <u>Current Population Survey</u> reported having taken training to improve skills on the current job, a proportion that rises with time on the job. When Lillard and Tan averaged across job tenure, they found the training in company programs, on the job training and training in regular schools reported 10 to 15% of the time as a source of job-relevant training. Data from the National Longitudinal Surveys, for a two year period, indicate that the percentage of young men, career women and mature men (45-59) reporting some training were 30, 24, and 10 percent. For these groups, the employer was the single most important source of training.

The likelihood of receiving training was found to increase for men and women with level of school attainment with the exception of the most educated workers post graduate). Lack of formal education was found to limit access to post school investments in most kinds of job training. Likelihood of training was found to rise with time on the job. Inability to develop enduring job attachment increasingly reduces the likelihood of getting training, especially for older workers.

The major issue today is assuring equitable access to training opportunities for older adult workers. Policies and practices regarding training may be influenced by informal age restrictions, standards, and assumptions, which may exclude older employees. In-service training and re-training provides older adult workers with the opportunity to strengthen knowledge and skills Rosen and Jerdee, (1985).

The payoff period for training costs is often quite short. A worker's age may not be relevant. Rather the need for maintenance or improvement of skills should be the primary focus. The life of a training investment will likely be the same for most older workers as for younger workers. Very often the return is better for older workers since they are more often likely to stay with the organization (Doering, et al., 1983); Tucker, (1985).

Training of older adult workers may require changes in traditional training methods in order to accommodate special needs and limitations of some individuals. Training techniques directed to the needs of the older worker have existed for over two decades. Belbin and Belbin (1972) have used a number of training approaches--the discovery method, activity learning, and programmed instruction. Sterns (1986) provides a more detailed review of these techniques which are important precursors to designing present training approaches. Particularly important are a thorough task analysis and training needs analysis Goldstein, (1974).

The adult and older adult training and retraining literature has documented a number of dimensions for successful training programs. Seven areas emerge that should be considered when designing training programs. These include motivation, structure, familiarity, organization, time, active participation, and learning strategies.

CONCLUSION

The adult and older worker of today is faced with the continuing need to retrain and update skills. Education and training in organizations and informal education settings, should emphasize the optimization of individual development for continued career growth.

Managers and supervisors need to be sensitive to the individual needs, aspirations, and career goals of older workers. At the same time, older workers need to take advantage of training and retraining opportunities in order to continue to maintain a current knowledge base and to be competitive with workers of all ages. Displaced workers will need extremely supportive recruitment, training and placement programs for re-entry into the work role.

REFERENCES

Atchley, R. 1988. Social forces and aging: An introduction to social gerontology. (Belmont, CA: Wadsworth Publishing Co.)

Beck, S. 1985. Determinants of labor force activity among retired men. Research on Aging, 7, 251-280.

Belbin, E. and Belbin, R. I. 1972. Problems in adult retraining. (London, England: Heineman Publishing Co.)

Burack, E. H. and McNichols, T. J. 1973. Human resource planning: Technology policy, change. (Kent, Ohio: Center for business and economic research.)

Doering, M., Rhodes, S. R. and Schuster, M. 1983. Eds., The aging worker. (Beverly Hills, CA: Sage Publications)

The future of older workers in america: New options for an extended working life 1980. (New York: Work in America Institute, Inc.)

Ginzberg, E. 1984. Career development. In D. Brown, L. Brooks and Associates (Eds.), Career choice and devlopment. pp. 169-191. (San Francisco, CA: Jossey-Bass)

Goldstein, I. I. 1974. Training: Program development and evaluation. (Monterey, CA: Brooks/Cole Publishing Co.)

Hall, D. T. 1971. Potential for career growth. Personal Administration, 34, 18-30.

Lillard, L. A. and Tan, H. W. 1986. Private sector training: Who gets it and what are its effects? U. S. Department of Labor. (Santa Monica, CA: Rand Corp.)

Rosen, B. and Jerdee, T. H. 1985. Older employees: New roles for valued resources. (Homewood, IL: Dow-Jones-Irwin.)

Sterns, H. L. 1986. Training and retraining adult and older adult workers. In J. E. Birren, P. K. Robinson and J. E. Livingston Eds.. Age, health, and employment. (Englewood Cliffs, N. J.: Prentice-Hall), pp. 93-113.

Sterns, H. L. and Alexander, R. A. 1987. Industrial gerontology. The aging individual and work. In K. W. Schaie Ed., Annual review of gerontology and geriatrics, (New York: Springer Publishing Co.), pp. 243-264.

Sterns, H. L. and Doverspike, D. 1988. Training and developing the older worker: Implications for human resouce management. In H. Dennis (Ed.) 14 steps in managing an aging workforce. (Lexington, MA: Lexington Books.), pp. 97-110.

Sterns, H. L. and Patchett, M. 1984. Technology and the aging adult: Career development and training. In P. K. Robinson, J. E. Livingston, and J. E. Birren Eds.. Aging and technological advances, pp. 261-278, (New York: Plenum Press.)

Super, D. E. 1980. A life span approach to career development. Journal of Vocational Behavior, 16, 282-298.

Super, D. E. 1984. Career and life development. In D. Brown, L. Brooks and Associates Eds., Career choice and development, pp. 192-234. (San Francisco, CA: Jossey-Bass, Inc.)

Tucker, F. D. 1985. A study of the training needs of older workers: Implications for human resources development planning. Planning Personnel Management Journal, 14, 85-95.

Viega, J. F. 1983. Mobility influences during managerial career stages. Academy of Management Journal, 26, 1, 64-85.

Advances in Industrial Ergonomics and Safety II
Edited by Biman Das
Taylor & Francis 1990

EFFECTS OF AGING ON HUMAN THERMOREGULATION IN COLD ENVIRONMENTS

Andrew J. Young

US Army Research Institute of Environmental Medicine
Natick, Massachusetts, USA 01760-5007

The risk of hypothermia is considered greater for older than younger persons due to a reduced ability to maintain body temperature during cold exposure with aging. Epidemiological surveys of body temperature in older persons provide little support for this notion, and probably reflect inadequate technique and uncontrolled ambient conditions during measurements. Generally, older men appear less able than young to defend their core temperature during experimental cold exposures. Cold may elicit slightly lower metabolic heat production and the cutaneous vasoconstrictor response to cold appears slower in old than younger men. These aging effects may, however, be limited to men, since in a recent study, older women appeared to defend core temperature as well or better than younger women. The possibility that preventable changes in body composition and physical fitness rather than aging per se may account for impaired thermoregulatory responses to cold has not been adequately studied.

INTRODUCTION

It is generally accepted that ability to maintain body temperature during cold exposure is diminished with aging. However, the basis for this conclusion should be re-examined, since many individuals today are electing to remain in the work force until an older age than previously. Job requirements in many industries limit the protection from the cold that clothing and auxiliary heating can provide. Thus, whether or not there is an age-related decrement in physiological responses to cold is an important question. This paper will review the scientific evidence relevant to this issue.

HYPOTHERMIA IN THE ELDERLY

Much of the concern about the problem of hypothermia in aged individuals may be related to the observation that mortality rate increases during the winter (Collins, 1987; Keatinge, 1986), and the increase is more pronounced for populations over 60 years of age than for younger populations (Collins, 1986). Although hypothermia has been suggested as a factor for the increased winter mortality in the elderly (Horvath and Rochelle, 1977; Taylor, 1964), this may be unwarranted speculation. Most surveys of body temperature in elderly persons have reported oral temperatures which are easily influenced by environmental temperature and require the use of very precise technique to accurately reflect deep body temperature. Estimates of core temperature were obtained by Fox et al. (1973) who measured the urine temperature of over 1000 subjects aged 65 and older in their own homes at about the same time of day. Core temperatures were below 35 °C in less than 0.6% and below 35.5 °C in less than 10% of the subjects (Fox et al., 1973). In fact, hypothermia is only rarely observed as a primary cause of death (Collins, 1987; Keatinge, 1986).

The occurrence of acute myocardial infarction increases with sudden decreases in air temperature (Teng et al., 1955), and myocardial infarction and stroke account for most of the "excess" mortality in elderly persons during winter (Collins, 1987; Keatinge, 1986). Furthermore, hypothermia is often secondary to loss of consciousness in cold conditions due to poisoning or drug/alcohol intoxication (Keatinge, 1986; Murphy and Lipton, 1983), and the ethanol-induced impairment of thermoregulatory responses to cold may be more pronounced in older than younger persons (Murphy and Lipton, 1983). It is important to note that Fox et al. (1973) measured body temperatures of persons in their own homes. Since the ambient conditions were uncontrolled, this type of study is inadequate to determine if thermoregulatory responses to cold are influenced by aging.

Older persons often maintain their homes at ambient temperatures below the level considered optimal for thermal comfort (Fox et al., 1973). This probably reflects economic limitations (Fox et al., 1973), but it has also been suggested (Horvath et al., 1955; Krag et al., 1950) that the perception of cold is less pronounced in older than younger persons. If older persons are less sensitive to cold, then not only behavioral, but also physiological responses to cold could be blunted as compared to young persons. Results from controlled laboratory experiments comparing young and older persons acutely exposed to cold will be considered next.

PHYSIOLOGICAL RESPONSES TO ACUTE COLD EXPOSURE

Body temperature in humans is defended during cold exposure by two major physiological responses. Acute cold exposure elicits cutaneous vasoconstriction which increases tissue insulation and limits conductive heat loss from the body to the environment (Toner et al., 1988; Young, 1988; Young et al., 1989). In addition metabolic heat production increases by involuntary shivering of the skeletal muscles, a voluntary increase in physical activity or both (Toner et al., 1988; Young et al., 1989). In addition, body heat loss and metabolic heat production during cold exposure are influenced by factors related to the individuals' physical fitness and anthropomorphic characteristics. Lean individuals having a high body surface area relative to mass tend to lose body heat more rapidly and shiver more vigorously than those having a thick layer of subcutaneous fat and a low body surface area relative to mass (Toner et al., 1988; Young et al., 1989). Physically fit individuals may exhibit more pronounced shivering thermogenesis during cold exposure than less fit persons (Bittel et al., 1988).

The conclusions of several early studies of aging effects on thermoregulation in the cold do not all agree, possibly due to flawed design, methodology or data analysis. For example, Krag et al. (1950) reported that rectal temperature (T_{re}) fell more in older than younger subjects during cold exposure. However, exposure conditions varied as much as 10 °C between individuals in that study, and the younger subjects appeared to be fatter than the older subjects who were all hospital patients suffering from varying degrees of degenerative disease (Krag et al., 1950). Horvath et al. (1955) observed that during exposure to 10 °C air, 52-75 year old subjects experienced about a 0.3 °C fall in T_{re} as compared to unchanged temperatures in 22-27 year old subjects. The older subjects in this study were also hospital patients, and furthermore the young subjects were all males whereas the older subjects included both males and females. Therefore, the differences ascribed to aging may, in fact, reflect unconsidered gender and/or anthropomorphic differences. Lee et al. (1965) reported that older subjects achieved lower T_{re} than younger subjects during a three hour immersion in cool water, but subcutaneous skinfold thickness was two-fold greater in the younger than older subjects. Wagner et al. (1974) compared pre-pubertal (10-13 yr) and post-pubertal (14-16 yr) boys as well as young (19-29 yr) and older (46-67 yr) adult men during a 30 min exposure to 17 °C air, and concluded that older subjects experienced a greater loss of body heat stores and a more pronounced fall in T_{re}. In fact, there was no statistically significant difference in the fall in T_{re} of the older (-0.2 °C) and younger (-0.1 °C) adult men, although the very youngest boys experienced a slight increase (+0.2 °C) in T_{re}.

In contrast, there are reports that older persons maintain core temperature during cold exposure as well or better than young persons. Bernstein et al. (1956) observed no differences in T_{re} between older (60-72 yr) and younger (21-28 yr) females during 180 minutes of exposure to 17 °C air. The older women in that study, however, were fatter (43 vs. 35%) and had a smaller body surface area relative to mass than the young women, both of which would favor conservation of body heat, and may have obviated any age related difference in ability to defend core temperature in the cold. O'Hanlon and

Horvath (1970) reported a positive correlation between age and T_{re} after 60 minutes of exposure to 8 °C air, suggesting that older subjects maintained higher T_{re} than younger subjects. However, actual temperature data were not reported, and it can be estimated from the mean and standard deviation of the subjects' ages that only two subjects studied were over 44 years of age.

Two recent studies have attempted to separate the effects of aging on thermoregulatory responses to cold from the effects of disease, gender differences and variations in body composition. Wagner and Horvath (1985a) observed that healthy older (51-72 yr) and younger (20-30 yr) men experienced equivalent (-0.2 °C) declines in T_{re} during two hour exposures to 10 °C air, despite the fact that the older subjects had smaller body surface areas relative to their mass and were considerably fatter (24%) than the younger (11%) men. These investigators concluded that aging effects offset the protective effect of body fat for thermoregulation in the cold. This appears to be confirmed by the data of Collins et al. (1985) who studied the effects of a two hour exposure to 6 °C air on body temperature in healthy older (63-70 yr) and younger (18-24 yr) men having very similar body compositions (18.4 and 19.7% fat, respectively). Deep body temperature of the older men fell by 0.3 °C as compared to less than a 0.1 °C drop in the young men. Thus, older men may be less able than younger men to defend body temperature during exposure to cold. (As will be discussed later, there are some data to indicate that, in contrast to men, older women are able to defend their core temperature during cold exposure as well as young women.) The apparent decline in the ability of older men to defend their core temperature may be due to an impaired cutaneous vasoconstrictor response to cold, an impaired ability to generate body heat by shivering thermogenesis or both.

Upon acute exposure to cold, cutaneous vasoconstriction ensues thereby reducing blood flow and the transfer of heat from the body's core to shell. The resulting fall in skin temperature (T_{sk}) is often used to indicate the magnitude of the vasoconstrictor response. In studies employing less than 60-min of cold exposure, higher T_{sk} (Horvath et al., 1955; Wagner et al., 1974) and slightly (not statistically) higher finger blood flow rates have been observed in older than in younger subjects (Wagner et al., 1974). On the other hand, in studies which employ relatively long duration cold exposures (i.e. > 60 min), young and old subjects maintain similar T_{sk} (Bernstein et al., 1956; Mathew, 1986), unless the young subjects are extremely lean in which case they exhibit higher T_{sk} during cold exposure than the older subjects (Wagner and Horvath, 1985a). It has also been reported that young subjects achieve steady state T_{sk} more rapidly during longer duration cold exposures than the older subjects (Wagner and Horvath, 1985a). Lee et al. (1965) reported that during a 3-hr immersion in water cool enough to elicit maximal cutaneous vasoconstriction, older and younger subjects achieved similar values of maximal physical insulation, which is the resistance to body heat loss measured at the skin. Maximal insulation varied as a function of the thickness of subcutaneous fat and aging did not influence the relationship (Lee et al., 1965). These findings suggest that old and young persons can achieve similar levels of maximal vasoconstriction during cold exposure; however, older persons appear to require more time to achieve this degree of vasoconstriction.

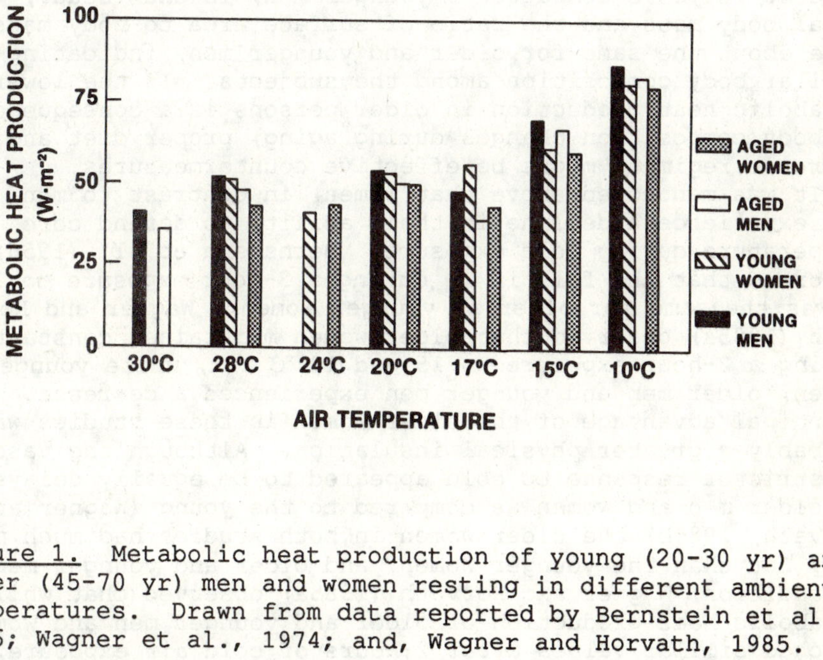

Figure 1. Metabolic heat production of young (20-30 yr) and older (45-70 yr) men and women resting in different ambient temperatures. Drawn from data reported by Bernstein et al., 1956; Wagner et al., 1974; and, Wagner and Horvath, 1985.

The influence of aging on the capacity for shivering thermogenesis is also controversial. Most reports indicate that metabolic heat production during cold exposure, which is probably almost completely due to shivering in humans (Toner and McArdle, 1988; Young, 1988), is less in older than younger persons (Bernstein et al., 1956; Horvath et al., 1955; Krag et al., 1950; Wagner et al., 1974), but data from other studies indicate little or no difference (O'Hanlon and Horvath, 1970; Wagner and Horvath, 1985a), or even greater metabolic heat production (Mathew et al., 1986) in older than younger persons during cold exposure. As discussed previously, differences among studies with regard to subject selection (age, health, body composition), exposure conditions and experimental procedures contribute to the disparate findings, and generalizations from inter-study comparisons may be tenuous. Figure 1 shows data from several different studies in which resting metabolic heat production of healthy subjects was measured at the end of exposure to a number of different air temperatures (or could be calculated from other data reported). It can be seen that there is a tendency for a lower metabolic heat production in resting older than younger individuals in both temperate as well as cold conditions. The effect is fairly small, and probably reflects a smaller muscle mass and greater mass of less metabolically active fat tissue relative to total body mass in the older persons studied. Mathew et al. (1986) observed substantially higher values of oxygen uptake during cold exposure in men aged 40-70 years compared to younger men; in that study, total body mass and the ratio of surface area to body mass were about the same for older and younger men, indicating a similar body composition among the subjects. If the lower metabolic heat production in older persons is a consequence of body composition changes during aging, proper diet and exercise regimens might be effective countermeasures.

It was mentioned above that women, in contrast to men, may not experience a decline in their ability to defend core temperature during cold exposure. Bernstein et al. (1956) reported that the fall in T_{re} during a 3-hour exposure to 17 °C was the same for older as younger women. Wagner and Horvath (1985a) observed that older women maintained constant T_{re} during a 2-hour exposure to 15 and 10 °C air, while younger women, older men and younger men experienced a decrease. The principal advantage of the older women in these studies was probably a greater physical insulation. Although the vasoconstrictor response to cold appeared to be equally delayed in older men and women as compared to the young (Wagner and Horvath, 1985b) the older women in both studies had much more body fat than the younger women, and older and younger men. In addition, Wagner and Horvath (1985a) observed that while metabolic heat production of older and younger men and women reached similar values after 2-hours of cold air exposure, the increase in metabolic heat production during the first 15-minutes of cold exposure was much more pronounced in the older women than in the other subjects.

SUMMARY

Although ability to thermoregulate in cold environments is generally considered diminished in older persons, a review of the scientific findings suggests that this belief may need to reconsidered. Controlled experiments do indicate that older men experience a greater fall in T_{re} during cold exposures lasting up to 3-hours than younger men. However, before concluding that older male workers are at risk, research is needed to determine whether or not even more pronounced body cooling results during exposures lasting 8-hours. Older males may defend their body temperature more poorly in the cold as a result of relatively slower vasoconstrictor response to cold and slightly diminished shivering thermogenesis. The later may reflect a loss of active muscle mass in the men studied, thus may be preventable with diet and exercise. The findings regarding older women suggest that they defend body temperature as well as younger women and possibly better than older men. However, this may reflect the much higher level of body fat content in the women studied as compared to the other group. Whether or not women who do not gain large amount of body fat as they age are able to adequately thermoregulate in the cold requires more study. It should be noted that all of the studies reviewed have compared the responses of old and young subjects while they rested quietly in the cold, and the thermoregulatory responses of older persons exercising in the cold have not been reported. Finally, the role of physical fitness and the effects of exercise training programs on thermoregulatory responses to cold have never been studied in older persons.

REFERENCES

Bernstein, L.M., Johnston, L.C., Ryan, R., Inouye, T. and Hick, F.K., 1956, Body composition as related to heat regulation in women. Journal of Applied Physiology, 9, 241-256.

Bittel, J.H.M., Nonotte-Varly, C., Livecchi-Gonnot, G.H., Savourey, G.L.M.J. and Hanniquet, A.M., 1988, Physical fitness and thermoregulatory reactions in a cold environment in men. Journal of Applied Physiology, 65, 1984-1989.

Collins, K.J., 1987, Effects of cold on old people. British Journal of Hospital Medicine, 38, 506-514.

Collins, K.J., Easton, J.C., Belfield-Smith, H., Exton-Smith, A.N. and Pluck, R.A., 1985, Effect of age on body temperature and blood pressure in cold environments. Clinical Science, 69, 465-470.

Fox, R.H., Woodward, P.M., Exton-Smith, A.N., Green, M.F., Donnison, D.V. and Wicks, W.H., 1973, Body temperatures in the elderly: A national study of physiological, social and environmental conditions. British Medical Journal, 1, 200-206.

Horvath, S.M. and Rochelle, R.D., 1977, Hypothermia in the aged. Environmental Health Perspectives, 20, 127-130.

Horvath, S.M., Radcliffe, C.E., Hutt, B.K., and Spurr, G.B., Metabolic responses of old people to a cold environment. Journal of Applied Physiology, 8, 145-148.

Keatinge, W., 1986, Medical problems of cold weather. Journal of the Royal College of Physicians (London), 20, 283-287.

Krag, C.L. and Kountz, W.B., 1950, Stability of body function in the aged. I. Effect of exposure of the body to cold. Journal of Gerontology, 5, 227-235.

Lee, D.Y., Hong, S.K. and Lee, P.H., 1965, Physical insulation of healthy men and women over 60 years. Journal of Applied Physiology, 20, 51-55.

Mathew, L., Purkayastha, S.S., Singh, R. and Sen Gupta, J., 1986, Influence of aging in the thermoregulatory efficiency of man. International Journal of Biometeorology, 30, 137-145.

Murphy, M.T. and Lipton, J.M., 1983, Effects of alcohol on thermoregulation in aged monkeys. Experimental Gerontology, 18, 19-27.

O'Hanlon, J.F. and Horvath, S.M., 1970, Changing physiological relationships in men under acute cold stress. Canadian Journal of Physiology and Pharmacology, 48, 1-10.

Taylor, G., 1964, The problem of hypothermia in the elderly. Practitioner, 193, 761-767.

Teng, H.C. and Heyer, H.D., 1955, The relationship between sudden changes in weather and the occurrence of acute myocardial infarction. American Heart Journal, 49, 9-20.

Toner, M.M. and McArdle, W.D., 1988, Physiological adjustments of man to the cold. In Human Performance Physiology and Environmental Medicine at Terrestial Extremes, edited by K.B. Pandolf, M.N. Sawka and R.R. Gonzalez (Indianapolis: Benchmark), pp. 361-399.

Wagner, J.A. and Horvath, S.M., 1985a, Influences of age and gender on human thermoregulatory responses to cold exposures. Journal of Applied Physiology, 58, 180-186.

Wagner, J.A. and Horvath, S.M., 1985b, Cardiovascular reactions to cold exposures differ with age and gender. Journal of Applied Physiology, 58, 187-192.

Wagner, J.A., Robinson, S. and Marino, R.P., 1974, Age and temperature regulation of humans in neutral and cold environments. Journal of Applied Physiology, 37, 562-565.

Young, A.J., 1988, Human adaptation to cold. In: Human Performance Physiology and Environmental Medicine at Terrestial Extremes, edited by K.B. Pandolf, M.N. Sawka and R.R. Gonzalez (Indianapolis: Benchmark), pp.401-434.

Young, A.J., Sawka, M.N., Neufer, P.D., Muza, S.R., Askew E.W. and Pandolf, K.B., 1989, Thermoregulation during cold water immersion is unimpaired by low muscle glycogen levels. Journal of Applied Physiology, 66, 1809-1816.

DISCLAIMER AND DISTRIBUTION STATEMENTS

The views, opinions and/or findings contained in this report are those of the author and should not be construed as an official Department of the Army position, policy, or decision unless so designated by other official documentation. Approved for public release; distribution is unlimited.

Advances in Industrial Ergonomics and Safety II
Edited by Biman Das
Taylor & Francis 1990

ON THE RELATIONSHIP BETWEEN AGE AND RESPONSIVENESS TO REHABILITATION

Elsayed Abdel-Moty, Ph.D.
Tarek Khalil, Ph.D., P.E.
Shihab Asfour, Ph.D.
Myron Goldberg, Ph.D.
Renee Rosomoff, C.R.N., M.B.A.
Hubert Rosomoff, M.D., D. Med. Sci.

University of Miami
Comprehensive Pain and Rehabilitation Center
at South Shore Hospital and Medical Center
600 Alton Road, Miami Beach, Florida 33139 USA

Seventy two patients with chronic low back pain (CLBP) participated in study of the relative responsiveness of people of different ages to a physical restoration and back pain rehabilitation program. Patients were grouped according to age as: 35 or less, 36-50, and above 50 years old. Results showed that CLBP impairs isometric muscle strength in all age groups. All patients improved significantly after 2 weeks of rehabilitation. Older female patients achieved most of their strength gains during the first 2 weeks of treatment whereas all others continued to show improvement throughout the 4-week program. Average strengths for patients in all groups were restored to strength levels of a population of healthy subjects.

INTRODUCTION

The relationship between aging and physical capacities in healthy subjects has been documented repeatedly. Similarly, the effects of physical conditioning and fitness programs on performance capabilities in old and young persons have been demonstrated by many investigators (e.g. Spirduso, 1981). However, a comparative study between older and younger patients with chronic low back pain (CLBP) upon their rehabilitation has not been documented.

The purpose of this study was to use a quantitative battery of tests in order to evaluate and monitor the strength capabilities of a population of CLBP patients at three stages of an aggressive physical medicine rehabilitation program. Furthermore, the evaluation tools were used to assess the relationship between age and responsiveness to the rehabilitation program.

MATERIALS AND METHODS

A pool 72 patients with CLBP, who were classified as Workers' Compensation, participated in this study. Subjects were grouped according to age in three categories: 1) Group #1 consisted of patients who were 35 years old or less; 2) Group #2 consisted of patients whose age was in the range between 36 and 50 years old; and 3) Group #3 consisted of patients who were 50 years of age or older. For the purpose of statistical analysis, groups were constructed of an equal number of 24 patients per group. Furthermore, each group comprised the same number of male and female patients (12 each).

All subjects were selected from the patients population of the University of Miami Comprehensive Pain and Rehabilitation Center (CPRC) at South Shore Hospital, Miami Beach, Florida. The CPRC utilizes a multidisciplinary comprehensive aggressive physical medicine approach to the management of chronic pain conditions. The components of this multidisciplinary inpatient/outpatient pain program are: neurosurgery, physiatry, physical therapy, occupational therapy, vocational counseling, nursing, psychology, psychiatry, biofeedback, muscle reeducation, and ergonomics. Treatment goals are: reduction/elimination of drug intake, physical restoration, vocational rehabilitation, behavioral modification, and return to work and productive life style (Rosomoff et al., 1981; Rosomoff and Rosomoff, 1987; Khalil et al., 1983, 1984, 1986).

The Ergonomics and Bioengineering Division, among its many contributions to pain management at the CPRC, evaluates patients performance and functional abilities upon admission to the Center and throughout rehabilitation in order to determine functional capacities, degree of functional loss, progress during rehabilitation, and treatment outcome. A battery of quantitative measures of human performance and functional capacities has being designed for this purpose (Khalil et al., 1984, 1986, 1987, 1988).

For the purpose of this study, the three groups described above were compared on the basis of trunk flexibility and isometric strength (of the arms, shoulders, composite, legs and back). Testing was performed at three intervals:

a) initial: upon patients admission to the CPRC;

b) midterm: after two weeks into the program (end of the inpatient phase); and

c) final - prior to discharge from the 4-week rehabilitation program.

For the evaluation of static muscle strength, the concept of Acceptable Maximum Effort (AME) was used (Khalil et al., 1987). Results are reported as average scores, percentage change from initial to final, and in terms of the Functional Restoration Index, FRI. The FRI is the ratio of patients' performance to the corresponding average values of healthy subjects. A FRI index in the proximity of unity indicates restoration of strength to the levels of healthy individuals.

The average strength levels for the healthy population were obtained from various sources (Battie et al., 1989; Chaffin et al., 1975; Asfour et al., 1984).

For data analysis, a factorial design with repeated measures was adopted (Kirk, 1968), with one grouping factor and two within factors with repeated observations on the within factor. There were three levels of the grouping factor AGE (3 age groups), two levels of the factor SEX (males and females), and three levels for the factor TIME (3 evaluation periods). Separate analysis of variance was conducted for each dependent variable (each strength testing maneuvers).

RESULTS

Descriptive information of the patients in each of the three groups stratified by sex is given in Table 1.

Table 1. Characteristics of the Various Groups Stratified by Gender.

Parameter	<35 Male	<35 Female	36-50 Male	36-50 Female	>50 Male	>50 Female
Sample Size	12	12	12	12	12	12
Average Age	29.0	24.6	41.2	41.6	58.9	55.6
Occupation (%ge):						
Light Work	28	60	21	44	50	91
Medium Work	16	40	29	52	21	9
Heavy Work	56	0	50	4	29	0
%age Injury due to:						
Slip&Fall	40	60	34	56	36	46
Lifting	28	20	26	32	14	9

When groups were compared on the basis of job-categories, it was observed that more than half the male patients in Group #1 were involved in heavy-type jobs at the time of injury. This number decreased significantly for patients in the older groups. Female patients consistently performed light to medium type jobs irrespective of the age group. In general, older patients performed jobs that are not heavy in nature (Fig. 1). These findings collectively reflect the pattern of employment in industry where younger males still dominate heavy type jobs and as employees grow older the majority become more involved in light-type jobs.

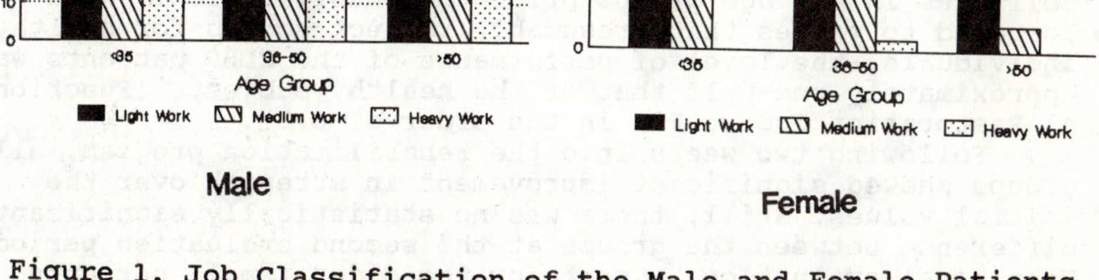

Figure 1. Job Classification of the Male and Female Patients in the Various Groups.

The causes of injury in this sample are presented in Table 1 and shown in Fig. 2. As was reported earlier in another study by our group (Khalil et al., 1984), a high percentage of the injuries in this sample was attributed to slipping and falling followed by lifting and carrying as a distant second. This confirms our previously reported statistics on a different sample of CLBP patients that manual materials handling, contrary to common beliefs, is not the leading risk factor in low back injuries. It is rather slips and falls that result in the injury in 34 to 60% of the time.

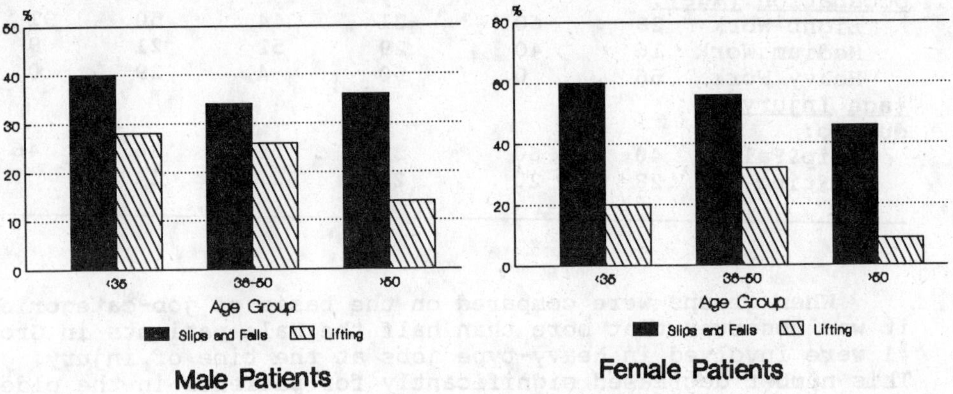

Figure 2. Comparison of the Two Major Causes of Low Back Injury in the Male and Female patients.

For all the dependent variables, ANOVA results showed a significant interaction effect. However, no discernible pattern of improvement could be detected. Analysis of the simple main effects and simple interaction effects was conducted in order to test for significance between groups, sexes, as well as over time. Duncan multiple range test was used to examine the difference between group means.

The average scores in each category of functional abilities (flexibility and strength) at the three evaluation periods are presented in Table 2 and Figure 3 (a-J) for all groups. Examination of the results showed that all patients demonstrated significant reduction in isometric strength abilities in all age groups prior to rehabilitation. When compared to values that presumably reflect scores for healthy individuals, the level of performance of the CLBP patients was approximately one-half that of the healthy subjects (Functional Restoration Index, FRI in the order of 0.5).

Following two weeks into the rehabilitation program, all groups showed significant improvement in strength over the initial values. Still, there was no statistically significant difference between the groups at the second evaluation period. Upon final evaluation, except for the older female patients who did not demonstrate significant change, all other groups continued to show improvement trends in strength. In general, the final scores were significantly higher than the initial ones, but not higher than strength levels at the midterm evaluation. For some groups, strength levels were higher than for other groups. However, This difference was not statistically significant. This can be attributed to the high standard deviation resulting from the small sample size used in this study. For all groups, the final FRI index approached, and in some cases exceeded, unity indicating restoration of function to average strength levels of healthy subjects.

Table 2. Measures of Performance of the Groups Stratified by Gender. Percentage changes (initial to final), "Norms", Initial FRI (FRI 1) and Final FRI (FRI 2) are provided.

Parameter	Age Group <35 Male	<35 Female	36-50 Male	36-50 Female	>50 Male	>50 Female
<u>Trunk Flexion</u>						
Initial	77.9	75.0	78.8	82.4	85.0	82.3
Final	122.0	115.0	117.6	118.2	106.4	116.8
%change	**56.6**	**53.3**	**49.2**	**43.5**	**25.2**	**41.9**
Norm	(105)	(105)	(105)	(105)	(105)	(105)
FRI 1	**0.74**	**0.71**	**0.75**	**0.70**	**0.81**	**0.78**
FRI 2	**1.16**	**1.10**	**1.12**	**1.13**	**1.01**	**1.11**
<u>Arm Strength</u>						
Initial	48.1	29.4	52.2	27.3	54.9	27.9
Final	86.1	48.8	83.8	40.9	79.8	50.2
%change	**79.0**	**66.0**	**60.5**	**49.8**	**45.4**	**79.9**
Norm	(91)	(49)	(83)	(50)	(78)	(46)
FRI 1	**0.53**	**0.60**	**0.63**	**0.55**	**0.70**	**0.61**
FRI 2	**0.95**	**1.00**	**1.01**	**0.82**	**1.02**	**1.09**
<u>Shoulder Strength</u>						
Initial	52.1	31.4	52.5	25.9	54.1	16.9
Final	89.8	54.4	91.2	37.2	81.6	42.4
%change	**72.4**	**73.3**	**73.7**	**43.6**	**50.8**	**150.9**
Norm	(110)	(53)	(99)	(48)	(88)	(42)
FRI 1	**0.47**	**0.59**	**0.53**	**0.54**	**0.62**	**0.40**
FRI 2	**0.82**	**1.03**	**0.92**	**0.78**	**0.93**	**1.01**
<u>Composite Strength</u>						
Initial	102.9	59.8	99.8	41.6	82.4	37.8
Final	212.6	154.2	227.7	112.8	206.4	104.8
%change	**106.6**	**157.9**	**128.2**	**171.2**	**150.5**	**177.3**
Norm	(235)	(118)	(209)	(115)	(190)	(104)
FRI 1	**0.44**	**0.51**	**0.48**	**0.36**	**0.43**	**0.36**
FRI 2	**0.91**	**1.31**	**1.09**	**0.98**	**1.09**	**1.01**
<u>Back Strength</u>						
Initial	53.4	44.2	64.0	37.6	68.6	24.3
Final	139.0	96.8	157.7	71.9	139.4	74.3
%change	**160.3**	**119.0**	**146.4**	**91.2**	**103.2**	**205.8**
Norm	(131)	(83)	(105)	(84)	(113)	(65)
FRI 1	**0.41**	**0.53**	**0.61**	**0.45**	**0.61**	**0.37**
FRI 2	**1.06**	**1.17**	**1.50**	**0.86**	**1.23**	**1.14**
<u>Leg Strength</u>						
Initial	115.2	82.2	126.7	55.0	123.1	43.4
Final	229.7	164.2	267.1	109.1	238.5	145.6
%change	**99.4**	**99.8**	**110.8**	**98.4**	**93.8**	**235.5**
Norm	(266)	(118)	(239)	(106)	(213)	(94)
FRI 1	**0.43**	**0.70**	**0.53**	**0.52**	**0.58**	**0.46**
FRI 2	**0.86**	**1.39**	**1.12**	**1.03**	**1.12**	**1.55**

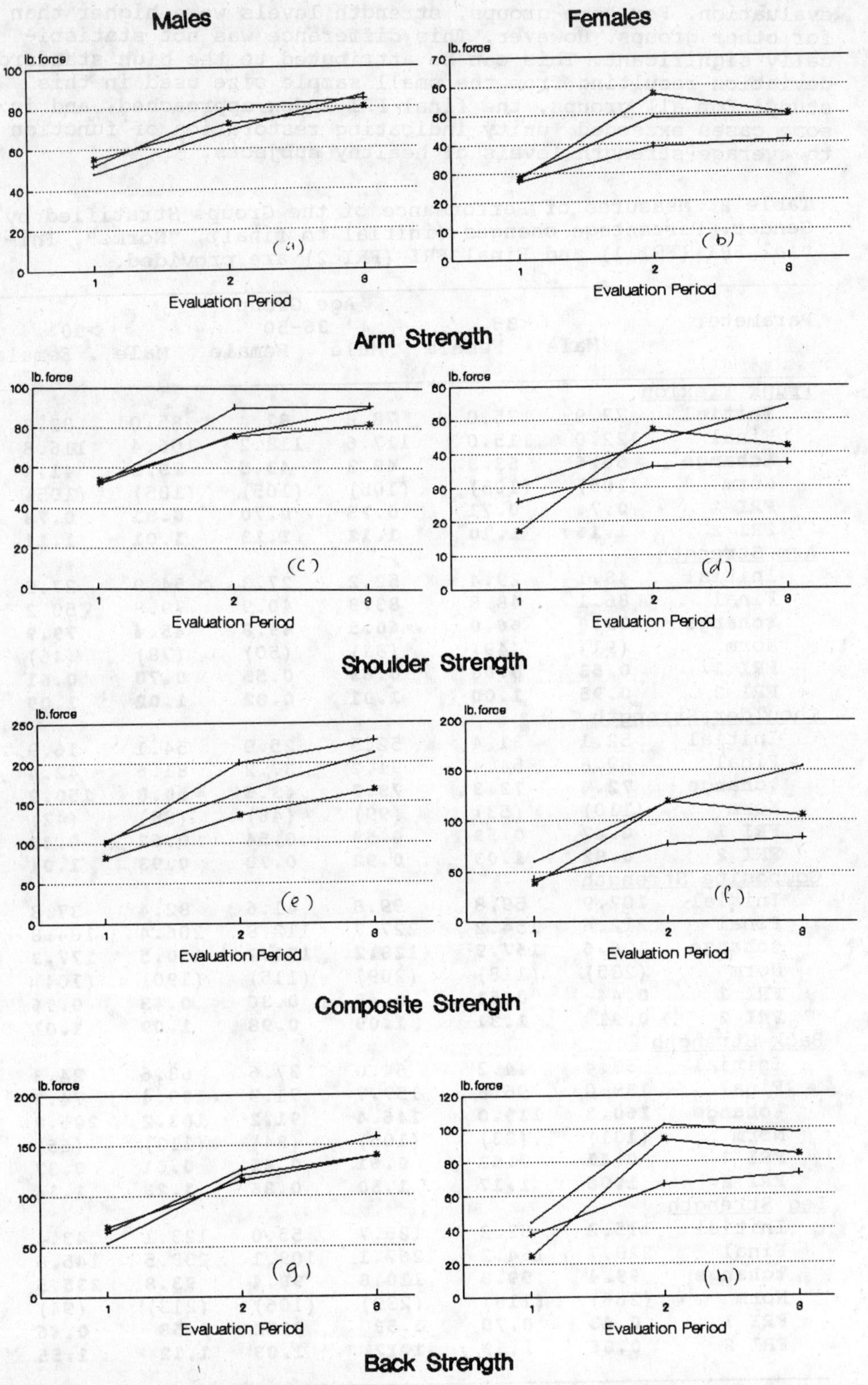
Males
Females
lb. force
Evaluation Period
(a)
(b)
Arm Strength
(c)
(d)
Shoulder Strength
(e)
(f)
Composite Strength
(g)
(h)
Back Strength

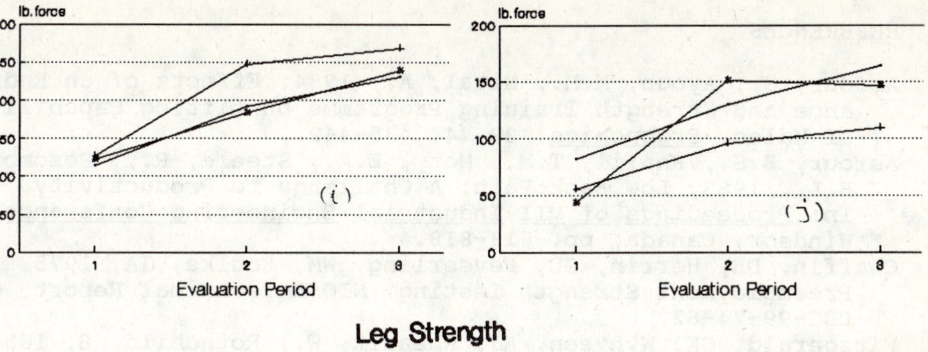

Figure 3. Changes in the Measures of Isometric Strength for Males and Females in the Various Groups.

(●) < 35 (+) 36-50 (*) > 50

DISCUSSION AND CONCLUSIONS

Chronic pain is associated with a decline in strength abilities of the injured individual. For people with CLBP, the deterioration of strength abilities can be reversed and function can be restored if the person undergoes a comprehensive back pain rehabilitation and physical restoration program. Exercise, treatment, and participation in activity contribute beneficially to muscular capacities of the CLBP individual.

The change in strength capacities upon rehabilitation was studied in this investigation for 72 individuals with CLBP in three age groups. Prior to treatment, patients in all groups exhibited significant reduction in isometric strength as compared to healthy individuals in the respective age groups. The physical restoration and rehabilitation at the CPRC enabled the patients to restore their strength capabilities following a 4-week program.

The findings indicate that, for this sample, age does not seem to be a critical factor in successful rehabilitation. Whether patients are old or young, they seem to follow similar trajectory on their way to strength recovery. In general, patients in all samples showed significant improvement during the first two weeks of rehabilitation. During the second two weeks, while all patient groups continued to improve, the older female patients did not demonstrate significant change in strength. Irrespective of age or sex, the average strength values of all patient groups reached the average strength values for the healthy population in their respective age group.

The direction for future research in this area will be towards studying a larger sample size.

REFERENCES

Asfour, S., Ayoub, M.M., Mital, A., 1984. Effects of an Endurance and Strength Training Programme on Lifting Capability of Males. Ergonomics, 27, 4, 435-442.

Asfour, S.S., Khalil, T.M., Moty, E.A., Steele, R., Rosomoff, H.L., 1983, Low Back Pain: A Challenge to Productivity. In: Proceedings of VII Industrial Engineering Conference, Windsor, Canada, pp. 813-818.

Chaffin, DB, Herrin, GD, Keyserling, WM, Foulke, JA, 1975. Preemployment Strength Testing. NIOSH Technical Report CDC-99-74-62.

Fitzgerald, GK, Wynveen, KJ, Rheault, W., Rothchild, B, 1983. Objective Assessment With Establishment of Normal Values for Lumbar Spinal ROM. Physical Therapy, 63:11:1776-1781.

Khalil, T.M., Asfour, S.S., Moty, E.A., 1985. New horizons for ergonomics research in low back pain. In Trends in ergonomics/human factors edited by R.E. Eberts, C.G. Eberts (North Holland: Elsevier Science Publishers B.V.), pp. 591-598.

Khalil, T.M., Asfour, S.S., Moty, E.A., 1985. The Contributions of Ergonomics to Pain Programs. In: Abstracts of the American Pain Society 5th General Meeting, p. 69.

Khalil, T.M., Asfour, S.S., Moty, E.A., Rosomoff, H.L., Steele, R., 1983. The management of low back pain: A comprehensive approach. In: Proceedings of the Annual Industrial Engineering Conference, Louisville, Ky.

Khalil, T.M., Asfour, S.S., Moty, E.A., 1984. Case studies in low back pain. In: Proceedings of the Human Factors Society 28th Annual Meeting, pp. 465-468.

Khalil, T.M., Goldberg, M.L., Asfour, S.S., Abdel-Moty, E., Steele, R., Rosomoff, H.L., 1987. Acceptable maximum effort (AME): A psychophysical measure of strength in back pain patients. Spine, 12, 4, 372-376.

Kirk, R.E., 1968. Experimental Design: Procedures for the behavioral sciences. Wadsworth Publishing Co., pp. 248.

National Institute for Occupational Safety and Health (NIOSH), 1981. A work practice guide for manual lifting, Technical report # 81-122, US Dept. of Health and Human Services, Cincinnati, Ohio.

Rosomoff, H.L., 1987. Comprehensive pain center approach to the treatment of low back pain. In: Low back pain, Report of a workshop, Rehab Research and Training Center, Dept Orthop Rehab, Univ of Virginia, pp. 78-85.

Rosomoff, H.L., Green, C., Silbert, M., and Steele, R., 1981. Pain and low back rehabilitation program at the University of Miami School of Medicine. In: New approaches to treatment of chronic pain : A review of multidisciplinary pain clinics and pain centers, edited by K.Y. Lorenzo (NIDA Research Monograph 36, Department of HHS).

Advances in Industrial Ergonomics and Safety II
Edited by Biman Das
Taylor & Francis 1990

APPLYING ERGONOMIC KNOWLEDGE ABOUT THE AGING WORKER TO THE FUTURE OF WORK AND TECHNOLOGY

Ulrika C. Wallersteiner

Ergo Systems Canada Inc.
Toronto (Mississauga)/Vancouver

As society grows older, employers will have to adapt the workplace to an aging workforce in order to remain competitive in world markets. This paper explores visual, auditory, physical and cognitive aging problems; actions employers can take to modify the impact of the aging process on productivity; and what factors need to be addressed in futuristic work activities if employers are to be successful in future economies.

INTRODUCTION

One thing we all have in common is that 'each of us is getting older by the minute'. Every day five thousand Americans reach age 65 and by 1990 more than 31 million Americans will be 65 or older. By 2030 more than 20% of all Americans will be 65 or older.

The aim of human factors/ergonomics is to create situations which maximize human capabilities, deemphazise their limitations, and ensures their safety. One area which hasn't been adequately integrated into ergonomic environments is the 'aging population'. Frequently environments are planned for a static user group.

If employers do not adapt their workplace to an aging workforce, it has been suggested that the aging of the North American population will pose a serious threat to our ability to compete in world markets. Not only is today's younger generation comparatively small, owing to the low birth rates of the last two decades, but an alarming proportion of North American youth lack the basic skills that employers require.

At the same time, the trends toward early retirement and increased life expectancy threaten employers with exploding pension and health care costs. We must now invest in the workplace to adapt to an aging workforce and to a declining younger workforce.

AGING PROCESS

Aging is not a disease that suddenly strikes us down but it is a natural biological process which needs to be accommodated rather than ignored. Koncelik (1982) pointed out that information about aging is often misconstrued as being synonymous with information about handicaps or persons with disabilities. Compared to traumatically acquired physical impairments, aging is gradual and frequently the chronological, physiological and psychological age do not coincide. Growing physical and psychological disabilities are compounded by poor design of surrounding environments and products leading to frustration and withdrawal of the individual.

Normal losses that occur with age appear as early as thirty. It has been well documented that individuals age at different rates from one another. Differences within the group aged 65-90 are frequently greater than the differences that exist between the middle-aged group, 40-64 years old, and the older group (Weg, 1983). They differ with regard to mobility, energy level, work activity, health and whether they are employed, active in the community or housebound. The following sections describe some aspects of aging that employers should understand in order to meet the needs of the older workers.

SENSORY AGING

In general, sensitivity to stimuli diminish with age. The response threshold to stimuli is increased; that is, the intensity and duration of a stimulus must be increased to evoke a sensory receptor response. Visual and hearing deficits in particular can lead to sensory isolation if left uncorrected.

Visual Aging

Of primary concern for the employer is the older worker's need for additional lighting. Over time the pupil of the eye contracts and the lens yellows resulting in the retina of a 60-year-old receiving about one-third as much light as that of a 20 year old. Work area illumination, especially for inspection type tasks, may need to be increased by as much as 60 percent. This can be accomplished by increasing the overall lighting levels or by providing task lighting at an individual's workstation. Because the lens of the eye hardens and is less elastic some people gradually lose their ability to focus upon objects. This can be easily corrected through use of eyeglasses, contact lenses, or by using a magnifying glass.

Auditory Aging

Hearing loss begins in the high frequencies, and can gradually move to the lower frequencies with age. Consonant sounds are affected more so than vowel sounds which explains why older people have more difficulty understanding speech, especially on the phone. A stiffening of the basilar membrane and a deterioration of cochlear hair cells are responsible for most hearing disorders. Try to reduce the background noise by dampening or isolating the noise source, or, increase the loudness of the signal to be heard. (However, in using the latter approach be careful not to deafen everyone else in the process!) Alternatively, it may be possible to switch the noise signal to a flashing warning light.

PHYSICAL AGING

Systemic aging shows a diminished ability to respond to stress (physical and/or emotional) and increased time is needed to return to prestress levels. Physical strength, endurance, speed and agility decline, especially if workers are not engaged in physically active jobs or pursue a sport.

Physiological & Physical Aging

Changes in muscle, bone and nervous tissue lead to decreases in work capacity and muscular strength, endurance, speed and flexibility. Furthermore, degenerative processes in cartilage and muscle lead to decreased mobility and increased risk of injury. Although some studies suggest that older workers have fewer accidents, once they are injured they are off the job for a longer period of time compared to their younger counterparts. Therefore it is critical that the workload be modified both in terms of lifting and of maintaining a proper working posture. Adjusting the workload can mean creating more smaller loads to lift, introducing a lifting assist, altering the work-rest cycle through job rotation or, in the case of a standing job, introducing a sit/stand stool whereby the worker can periodically rest on a stool while doing their task.

Age-related anthropometric changes occur on a sliding scale. Anthropometric graphs and tables do not reflect the dimension of age. If age is considered then most anthropometric information is either at the high end for younger adults or at the far low end for aged adults, especially for women (Koncelik,1982).

Stoudt (1981) reviews anthropometric data on aging. Data from a longitudinal study of 2000 males showed significant decreases in height start in the cohort aged 40 to 44 years (Friedlander et.al., 1977). Much of the shrinkage in stature takes place in the trunk, leading to reduced sitting and sitting eye heights. Anatomic arm and leg length remain relatively unchanged by the aging process. However, the measurement of functional reach may be diminished because of decreased mobility at the various joints of the body.

Shoulder width decreases while pelvic width and seat area show increases which are most likely related to increased fat deposition.

COGNITIVE AGING

Psychological changes do occur with aging in the way older people process information, respond emotionally, and behave in the context of a particular social and physical environment.

Due to changes in the neurological system (the slowing of electrical impulses) older people's reaction time increases. This can be considered to be an asset. If an employee is working more carefully then the likelihood of errors will decrease. Similarly, short-term memory capacity declines with age (Fozard, 1981); however, this is offset by the greater abilities in judgement and experience of the older worker. Welford (1977) suggests that as task complexity increases the amount of slowing of old adults relative to young ones increases more rapidly than the amount expected on the basis of greater task complexity in younger adults. However, the environment can be modified to optimize functioning by enhancing and amplifying stimulus qualities and by providing feedback mechanisms for the sensory- or cognitively impaired older person (Lawton, 1978).

When it comes to introducing new technology it is important that unnecessary task complexity is reduced, by ensuring that ergonomic principles have been applied during the development of the technology, that memory aids such as code lists are available, and that adequate feedback is provided. Furthermore, training programmes should be designed considering the previous experience of the aging worker.

CONCLUSION

It is critical that we replace our gerontophobia with a more positive and healthier image of aging. Ending age discrimination against older workers and discouraging the trend toward early retirement are important parts of today's challenges. In the United States, of all the retired workers covered by corporate pensions, half are not yet even 65 years old. In Canada, 20,000 people turn 65 years old every month! We simply cannot afford to go on wasting the talent and experience of men and women who still, have the potential to have many active, productive years ahead of them.

The changing nature of the occupational structure during the next 15 years needs to be examined more closely. The duration and organization of work schedules during the final years of occupational activity should be examined with the aim of providing a progressive form of retirement. Examples being: shorter daily or weekly work period, limited overtime, longer paid holidays in the final years preceding retirement, shared work formulas, and encouragement of part-time work in order to redistribute existing jobs and to ease the work load of certain job categories (Breton,1980, Naisbitt and Aburdene, 1990).

Age should not be used as a criterion to preventing workers from being retrained or from changing careers. Often employers are unsure as to whether older people will be productive members of the work force or whether they will be able to meet job demands, particularly since technical advances in many jobs require constant learning of new skills. While the learning rate slows down, the ability to learn does not decline among the mentally alert aged.

Technology can benefit also the older workers, especially those who have physical impairments. Assistive devices, computers, robots and other workplace technologies can compensate for waning physical strength, dexterity, eyesight or hearing and can improve. However, if the operation of the technologies is not designed with the older worker in mind, the results will be negative.

According to Maggie Kuhn, the Gray Panther leader, " People should have a lifelong opportunity to work. Businesses have found that, with mandatory retirement, they have thrown away many irreplaceable skills. But the answer is in the workplace. If you humanize the workplace, restructure it around flextime, around mentoring, people will be less eager to retire. Personally, I want to die in an airport, briefcase in hand, mission accomplished." (Dychtwald and Flower, 1989). (Those readers who know of my globe-trotting lifestyle and commitment to ergonomics will agree that Maggie's last sentence is my epitaph.)

Plan for your tomorrow by changing your attitudes towards the aging worker. How you treat the older person today will determine how the next generation will treat you; designing a safer workplace for the older worker will also better protect the younger worker. And by joining the Human Factors Association of Canada and the Human Factors Society (U.S.) Technical Interest Group on Aging to become more informed and by supporting research on work, aging and injuries you will be investing in your own future.

REFERENCES

Brenton, F., 1980, The Adaptation of Older Workers. In: Adaptability and Aging 2, IX International Conference of Social Gerentology. Paris: International Center of Social Gerentology.

Dychtwald, K, and Flower, J,1989, Age Wave, Jeremy P. Tarcher, Inc. L.A.

Fozard,J.L., 1981, Person-Environment Relationships in Adulthood: Implications for Human Factors Engineering. Human Factors, 23(1),7-27.

Friedlander,J.S., Costa,P.T., Bosse,R., Ellis,E., Rhoads,J.G., & Stoudt, H.W., 1977, Longitudinal physique changes among healthy white veterans at Boston. Human Biology, 49, 541-558.

Henderson, C., 1988, Old Glory: America Comes of Age. The Futurist, 22 (2):36-40.

Koncelik, J.A., 1982, Aging and the Product Environment. Stroudsburg: Hutchinson Ross Pub. Co.

Lawton,M.P., 1978, Sensory Deprivation and the Effect of the Environment on Management of the Senile Dementia Patient. In: Clinical Aspects of Alzheimer's Disease Conference,National Institute of Mental Health.

Naisbitt, J. and Aburdene, P., 1990, Megatrends 2000.William Morrow and Company, N.Y.

Office of Technology Assessment, 1984, Technology and Aging in America, Washington, D.C.

Statistics Canada,1977, Perspective Canada II, Catalogue 11-508E.

Statistics Canada, 1984, The Elderly in Canada, Catalogue 99-932

Stoudt, H.W., 1981, The Anthropometry of the Elderly. Human Factors, 23(1), 29-37.

Weg, R.B., 1983, Changing Physiology of Aging: Normal and Pathological. In: D.S.Woodruff & J.E.Birren (Eds.) Aging: Scientific Perspectives and Social Issues, Second Edition. Monterey: Brooks/Cole Pub. Co.

Welford, A.T., 1977, Motor Performance. In: J.E.Birren & K.W. Schaie (Eds.) Handbook of the psychology of Aging. New York: Van Nostrand Reinhold.

Advances in Industrial Ergonomics and Safety II
Edited by Biman Das
Taylor & Francis 1990

AN ERGONOMIC STUDY OF THE EFFECT OF TASK FREQUENCY ON WORK-REST SCHEDULES

A. M. Genaidy, S. Al-Rayes and A. Dickinson

Departments of Industrial Engineering and Psychology
Western Michigan University, Kalamazoo, MI 49008, USA

This study was conducted to determine work-rest schedules for individuals engaged in upper extremity, high frequency manual handling tasks using a new psychophysical procedure. The results suggest that, among other things, the total working time decreased with an increase in frequency. Moreover, resting periods should be applied more often for higher frequencies of handling.

INTRODUCTION

In a review of the ergonomics literature, Genaidy (1990) pointed out that (1) most studies have concentrated on the effects of short pauses of work and rest (less than 5 minutes) on the physiological responses of individuals performing on standard laboratory tests (e.g: bicycle ergometer and motor-driven treadmill), and (2) very little research has been done to examine the effects of work-rest schedules on individuals engaged in industrial tasks. Moreover, he argued that there is a need to develop a methodology for determining the length and frequency of working and resting periods in industrial settings and to test their use in the control of work-related injuries.

The objectives of this paper were (1) to test whether rest allowances could be ascertained using a new psychophysical procedure developed in this study and (2) to determine the effect of frequency of handling on work-rest schedules.

METHODS

Subjects

Five male college students, inexperienced in manual handling tasks, volunteered to participate in this study. Subjects were briefed on the purpose of the experiments and were medically examined prior to its start. The physical characteristics of the subjects were: (1) age: 24.6 $\pm$ 2.3 years, (2) body weight: 79.5 $\pm$ 7.9 kg, and (3) stature: 175.1 $\pm$ 4.4 cm.

Training

Subjects were trained for flexibility and muscular endurance during a total of sixteen sessions. During each session, the subject was trained for muscular endurance by simultaneoulsy handling a load in each hand for the task performed in this study until exhaustion due to local muscular fatigue and/or overall body fatigue. The details of the training program are described elsewhere (Genaidy et al., 1990).

Experimental procedures

Following flexibility and muscular endurance training, each subject was asked to handle a 4 kg load in each hand through a sequence of motions at three different height levels, namely, 76 cm, 127 cm and 178 cm above the floor level, respectively. These heights simulated the average knuckle, shoulder, and reach heights for the U.S. population. The height levels were set by mounting three wooden shelves on an industrial steel rack. Three racks were used: one on each side of the subject and one in front of him.

Each subject performed an upper extremity manual handling task requiring several steps in the following sequence:

1. Both hands grasped the loads which were placed on the sides of the subject on the 76 cm high shelves, then moved them through horizontal motions to the front 76 cm high shelf.
2. Both hands moved the loads up from the front 76 cm high shelf to the side 127 cm high shelves.
3. The loads were moved simultaneously through horizontal motions to the front 127 cm high shelf.
4. Both hands repeated the motions undertaken in steps 2 and 3 except that the shelves were located at 178 cm above the floor.
5. Both hands moved the loads down to the side 76 cm high shelves at their initial positions.

Each subject was asked to participate in two sessions. Only one session was conducted per day, and each session lasted four hours. During one session, the material handling task was performed at a frequency of 3 times/minute and during the other session the task was performed at a frequency of 6 times/minute. A metronome was used to pace each subject. As soon as the subject heard the sound of the metronome, he handled the load then waited for the next sound.

A new psychophysical methodology was developed to determine work-rest schedules. Within each experimental session, the subject was given control of the duration of the working and resting periods. He stopped working when he felt any discomfort in any part of his body. Then, he resumed work when he felt that the various parts of his body had recovered reasonably well from the previous working cycle. He was instructed not to work until complete exhaustion nor to rest until full recovery.

During heart rate measurements, three surface electrodes were affixed on the subjects' chests. The electrodes were then connected to an R-612 SensorMedics Dynograph physiological recorder. The ratings of perceived exertion were measured according to Borg's scale (1962).

Experimental design

A within-subject factor experimental design was used. Frequency of handling was the within-subject factor. Maximum acceptable working time (MAWT), maximum acceptable resting time (MART), number of cycles performed, ratings of perceived exertion (RPE) of the shoulder, arm, wrist and overall body, total working time (TWT), and total resting time (TRT) were the dependent variables.

The various dependent variables that were analyzed statistically were defined as follows:

1. MAWT was defined as the average of individual working times within a given experimental session.
2. MART was the average of individual resting times within a given experimental session.
3. The working heart rate was calculated as the average of the heart rate values that were recorded at the fifth and last minutes of the working cycles at the beginning, middle, and end of the 4-hour duration.

4. The recovery heart rate was the average of the heart rate values recorded at the fifth and last minutes of the resting cycles at the beginning, middle, and end of the 4-hour duration.

Despite the fact that the duration of each session was controlled to 4 hours, it varied slightly to allow a complete record of the final cycle. Therefore, TWT and TRT were computed as a percentage of the total duration of each experimental session.

RESULTS

The results of the analysis of variance showed that the effect of frequency on the number of cycles, total working time, total resting time, working heart rate, recovery heart rate, working and resting RPE of the arm, and working and resting RPE of the wrist was statistically significant at the 5% level.

Additional analyses were conducted to test whether the maximum acceptable working time decreased and maximum acceptable resting time increased during the course of the experimental session. A linear regression equation was developed for each treatment performed by each subject. While all regression coefficients were negative for the working periods and positive for the resting periods, they were not signicantly different from zero at the 5% level.

The means and standard deviations of the response varables are given in Table 1. The results indicate the following:

1. The total working time values for the frequencies of 3 and 6 times/min were 81% and 66% of the 4-hour session duration, respectively.
2. The average working heart rate values for the frequencies of 3 and 6 times/min were 102 and 116 beats/min, respectively.
3. The average recovery heart rate values for the frequencies of 3 and 6 times/min were 79 and 87 beats/min, respectively.
4. The number of work and rest cycles doubled from 7 for the frequency of 3 times/min to 14 for the frequency of 6 times/min.
5. The average working RPE (ratings of perceived exertion) values for the overall body was 10.6 for 3 times/min and 13.0 for 6 times/min.

It should be noted that the standard deviation of maximum acceptable working time at a frequency of 3 times/minute was quite high because one of the subjects performed the experimental task during the whole session without taking any rest.

DISCUSSION

Traditionally, rest allowances have been estimated in industry based on company practice or labor negotiations. These allowances have not taken into consideration human variables, task variables, and environmental variables. Thus, this study was conducted to examine the effect of the frequency of handling on work-rest schedules using a new psychophysical approach.

The results showed that the total working time for the frequency of 3 times/minute was, on the average, 81% of the total shift duration (approximately 4 hours in this study). For the 6 times/minute, the total working time was about 66%, a value that is 15% less than that for the 3 times/minute. The total resting time doubled from 3 to 6 times/minute.

The above mentioned findings were supported by the heart rate and ratings of perceived exertion data. The average working heart rate was 102 beats/minute for 3 times/minute and 116 beats/minute for 6 times/minute. Also, the working RPE for the overall body was 10.6 for 3 times/minute and 13.0 for 6 times/minute.

While the total working time for the frequency of 6 times/minute was significantly less than that of 3 times/minute, the amount of load

Table 1. Means and standard deviations (SD) of dependent variables.

Response Variable	Frequency (times/minute) 3 Mean	SD	6 Mean	SD
NC	7.2	4.5	13.8	6.1
MAWT	70.3	96.9	15.3	10.8
MART	4.9	3.3	6.3	1.7
TWT	81.1	15.8	65.7	20.1
TRT	19.9	15.8	34.3	11.2
RHR	69.0	6.0	70.0	7.8
WHR	101.6	7.3	116.4	11.2
RCHR	78.8	6.6	87.0	6.6
WSRPE	11.6	7.3	13.4	4.0
WARPE	9.8	6.2	13.0	3.7
WWRPE	8.6	5.6	11.4	4.0
WORPE	10.6	6.1	13.0	3.2
RSRPE	9.2	5.5	11.6	3.5
RARPE	8.4	4.9	11.2	3.4
RWRPE	6.8	4.3	10.0	3.8
RORPE	9.0	5.2	11.4	3.4

Note
NC: number of cycles, MAWT: maximum acceptable working time (minutes), MART: maximum acceptable resting time (minutes), TWT: total working time (%), TRT: total resting time (%), RHR: resting heart rate (beats/minute), WHR: working heart rate (beats/minute), RCHR: recovery heart rate (beats/minute), WSRPE: working RPE of the shoulder, WARPE: working RPE of the arm, WWRPE: working RPE of the wrist, WORPE: working RPE of the overall body, RSRPE: resting RPE of the shoulder, RARPE: resting RPE of the arm, RWRPE: resting RPE of the wrist, RORPE: resting RPE of the overall body.

handled by each subject at 6 times/minute was 1.8 times higher than that at 3 times/minute. Thus from a productivity standpoint, it is recommended that individuals perform at a frequency of 6 times/minute and take more rest allowances. It should be noted, however, that higher productivity may be achieved at higher frequencies. Genaidy (1990) reported that maximum physiological efficiency for arm lifting tasks is attained when handling a 20 kg load at a frequency of 9 times/minute for a 4-1 work-rest ratio.

CONCLUSIONS AND RECOMMENDATIONS

The conclusions that can be drawn from this study are:

1. The total working time decreased with an increase in the frequency of handling.
2. The total resting time increased with an increase in the frequency of handling.
3. The working heart rate and ratings of perceived exertion increased with an increase in the frequency of handling.
4. Resting periods should be applied more often for high frequencies of handling.
5. It is possible to determine rest allowances for manual handling tasks using the psychophysical approach.

It is recommended that the new psychophysical methodology developed in this study be used by ergonomists to establish optimum work-rest schedules for various human, task, and environmental conditions.

REFERENCES

Borg, G.A.V., 1962, Physical Performance and Perceived Exertion, Unpublished Thesis (Lund, Gleerup, Sweden).

Genaidy, A., 1990, The physiological effects of work-rest schedules on manual lifting tasks, In: Proceedings of the Ergonomics Society, London.

Genaidy, A.M., Agrawal, A., Houshyar, A., Dawson, M., Zabik, R., and Mousavinezhad, S.H., 1990, An ergonomic training program for industrial manual handling jobs. In: Proceedings of the Ergonomics Society, London.

Advances in Industrial Ergonomics and Safety II
Edited by Biman Das
Taylor & Francis 1990

SCHEDULING PHYSICAL WORK AND REST REGIMES UNDER TEMPERATE ENVIRONMENTS

M.A. EL-NAWAWI*, S.B. MOSHREF*, AND M.S. EL-MOHANDES**
* Professor, ** Lecturer

Department of Mechanical Engineering, Faculty of Engineering, Al-Azhar University, Cairo, Egypt.

This investigation was conducted to study the stressing effect of work-rest regimes when working under temperate environments. Subjects participated in a series of testing work-rest cycles scheduled within a 48 min session at 50% VO_{2max} obtained on a cycle ergometer followed by a 16-min recovery session. Four cycles of 6/6, 8/4 and 9/3 min/min work-rest and two cycles of 12/12 min/min were performed under three different levels of ambient temperature (20, 25 and 30 ° C WBGT). Heart rate, tympanic temperature and oral temperature were monitored during the work-rest and recovery sessions. The results obtained were extensively analysed and discussed.

INTRODUCTION

Many jobs in hot industries involve alternate periods of work in the heat and rest in nearly the same environment. In such fluctuating heat stress (metabolic or environmental), the prediction of the resulting strain has obvious relevance if permissible exposure limits have to be determined. Early investigations showed that the use of a time weighted average of the environmental and metabolic stresses could provide a simple and valid basis for that prediction of heat strain (Mairiaux et al., 1986). Guide-lines and criteria recommended by several investigators suggest an upper limit for work under environmental heat stress, but accept that in some industries, work may have to continue under conditions which exceed this limit. In this case, designing suitable work-rest regimes, depending on the conditions and the metabolic rate, is highly recommended. Optimizing work-rest regimes was reported to be one of the most important methods of reducing the overall heat strain for the individual worker (NIOSH, 1986).

The NIOSH (1980) working group has decided that there are not enough data on optimum work-rest ratios for different levels of heat stress and levels of metabolic work in terms of length of the rest cycle versus length of the work cycle. This study investigated the suitability of work-rest regimes while working under thermal stressing environment.

EXPERIMENTAL WORK AND DESIGN

The work load was obtained by pedalling on a cycle ergometer in a sitting position. This task was conducted through performing repeated cycles under one of four different work-rest regimes (Figure 1) for a total work-rest session of 48 minutes. The mechanical work load was maintained constant during work periods at 108 W (equivalent to approximately 50% VO_{2max} of subjects). The rate of pedalling was kept constant at 50 r.p.m. Three work/rest (W/R) levels (6/6, 8/4 and 9/3 min/min) were selected to represent the work-rest ratios of 1/1, 2/1 and 3/1 respectively. Two levels of W/R durations (6/6 and 12/12 min/min) were chosen to compare the work-rest duration effect. Four levels of cycle effect (C_1, C_2, C_3 and C_4) were considered for conducting the three W/R ratios. A recovery session of 16 min duration into four 4-minute periods (P_1, P_2, P_3 and P_4) was scheduled immediately after the 48 min work/rest.

The independent variables of the ambient temperature, work/rest ratio, work-rest duration, cycle and recovery period were subjected to statistical techniques of analysis of variance. Heart rate (HR) and tympanic temperature (T_{ty}) data were averaged for the last 2-minutes of either the work or recovery periods, while oral temperature (T_{or}) data were obtained after each work cessation or before the end of the recovery period.

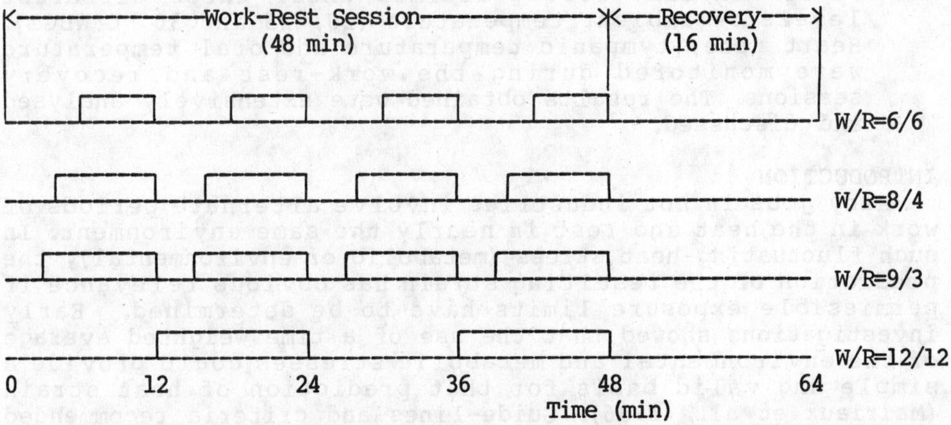

Figure (1) Schematic diagram of work-rest regimes

Four male students were selected as subjects from larger student population on the basis of their age and physical condition. They were in good health and aged between 20 and 22 years. None was athlete nor following regular physical training programmes. All subjects were dressed in similar clothing for each experimental run. The clothin included a T-shirt and trousers.

RESULTS AND DISCUSSIONS

The following discussion of results highlights the main finding obtained by the analyses of variances.

Effect of ambient temperature

The ANOVA results yielded highly significant effects of ambient temperature upon heart rate, tympanic and oral temperatures (0.01 level). Such significance was obtained during work and recovery sessions. Figures (2) and (3) indicate a progressive rise of heart rate with ambient temperature. At higher levels of temperature, such heart rate rise is more pronounced (e.g. the difference between the mean

HR values yielded under 30 and 25°C WBGT were twice those noticed between 25 and 20°C WBGT). Such findings are in agreement with those obtained by Brouha et al. (1960), and El-Nawawi et al. (1984). Kamon (1980), proposed one beat/minute (bpm) heart rate increment for every 1°C of environmental db temperature rise above the 25°C. The results shown in Figure (2) indicate that the mean HR increment imposed by the 25°C WBGT (T_{db}=31°C) ambient temperature above that of the 20°C WBGT (T_{db}=24.7°C) is equivalent to 0.83 bpm for every 1°C of ambient temperature rise. On the other hand, the mean HR increment imposed by the 30°C WBGT (T_{db}=38.5°C) yielded 1.2 bpm for every 1°C of ambient temperature rise.

As illustrated from the effect on heart rate, the tympanic temperature increments are not linearly proportional to the ambient temperature (Figure 4). The increments of tympanic temperature are more pronounced at higher levels of ambient temperature. The maximum mean T_{ty} at the end of work session (last two minutes of the fourth cycle) under the three levels of ambient temperature were 37.43, 37.69 and 38.67°C (corresponding to 20, 25 and 30°C WBGT respectively). These values were observed under the 9/3 W/R ratio level (the resting T_{ty}=36.58°C). The absolute maximum values for tympanic temperature may not be attained during these last two minutes. On the other hand, increments or stabilization of tympanic temperature were observed after work (Figure 5). However, in studies considering the effect on rectal temperature (Givoni and Goldman, 1972) a decline in rectal temperature occuring after certain time lag during sessions subsequent to work was reported. During this time lag, the body temperature continues to rise but at a lower rate compared to the last part of work.

The oral temperature showed a progressive rise with ambient temperature, where the mean values have risen from 36.81 to 36.90 and to 37.17°C under 20, 25 and 30°C WGBT respectively (Figure 6). Such values indicate a decrement of 0.26, 0.40 and 0.75°C in comparison with the corresponding tympanic temperature (Figure 4). The oral temperature values and differences from the tympanic are more pronounced under the higher ambient temperature levels. Testing between oral temperatures' treatments indicated significant differences between the results obtained for the 30 and each of the 20 and 25°C WBGT (0.01 level). No significance was observed between the 20 and the 25°C WBGT results. Such findings reassure the observation that the oral temperature response to high ambient temperature levels is more marked. The T_{or} recovery means obtained under the different levels of ambient temperature were slightly higher than those of work.

Effect of work/rest ratio

The Analyses of variances indicated significant effect of the work/rest ratio on heart rate (0.05 level); while indicated highly significant effect upon the tympanic and oral temperatures at the 0.01 level of significance. Figure (8) illustrates a plot of means and standard deviations of heart rate versus the three levels of work/rest ratio (6/6, 8/4 and 9/3). This figure shows a rise taking place under the 9/3 work/rest ratio level compared to the other two levels (6/6 and 8/4). This difference was tested and found significant, while no significance was detected between the 6/6 and 8/4

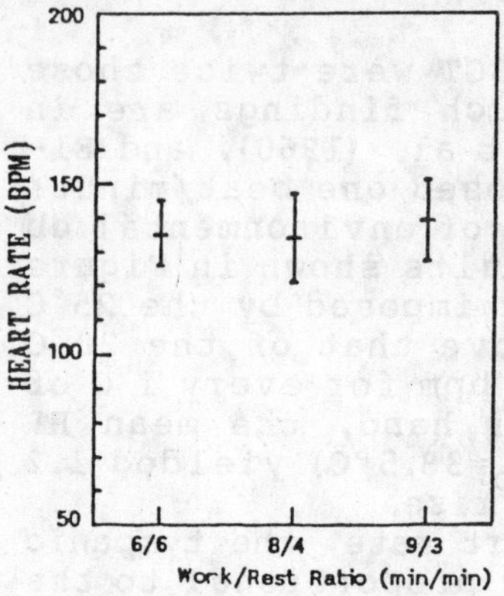

Fig. (2) Plot of Means and Standard Deviations of HR Versus Ambient Temperature.

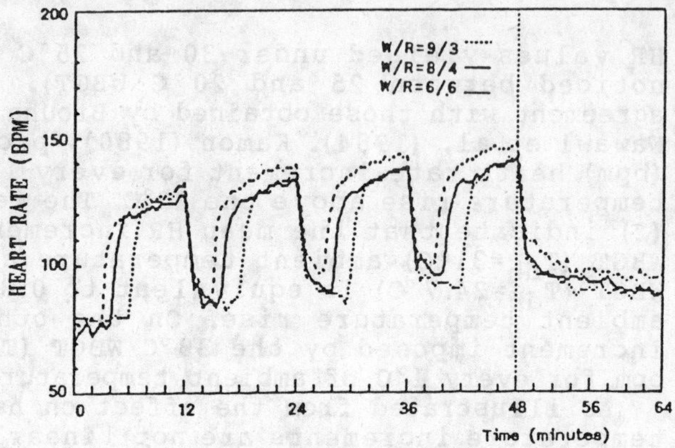

Fig. (3) Effect of Amient Temperature upon Heart Rate (HR). (Data for Working Under 9/3 Work/Rest level).

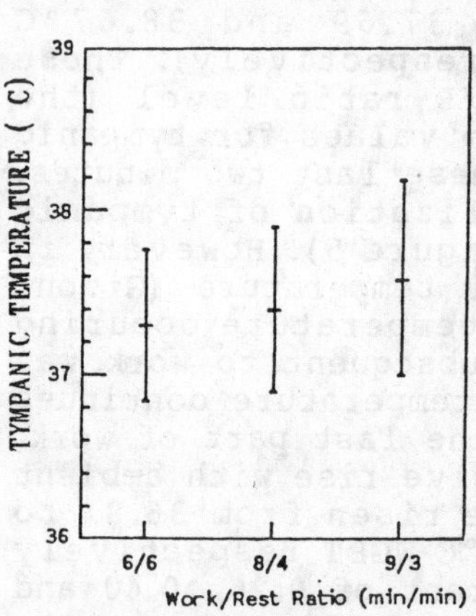

Fig. (4) Plot of Means and Standard Deviations of Tty Versus Ambient Temperature.

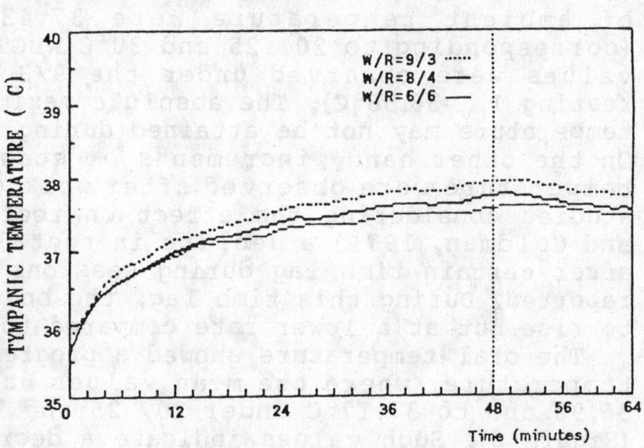

Fig. (5) Effect of Amient Temperature upon Tympanic Tempertatre (Tty).

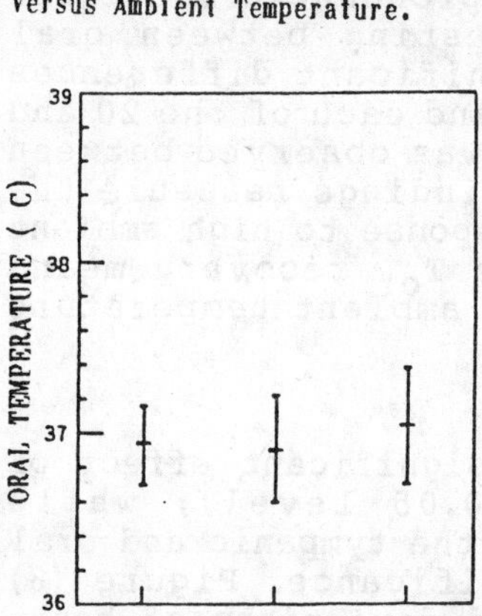

Fig. (6) Plot of Means and Standard Deviations of Tor Versus Ambient Temperature.

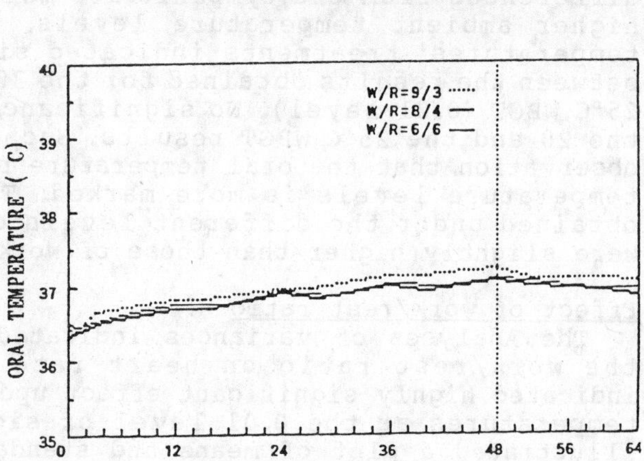

Fig. (7) Effect of Amient Temperature upon Oral Tempertatre (Tor).

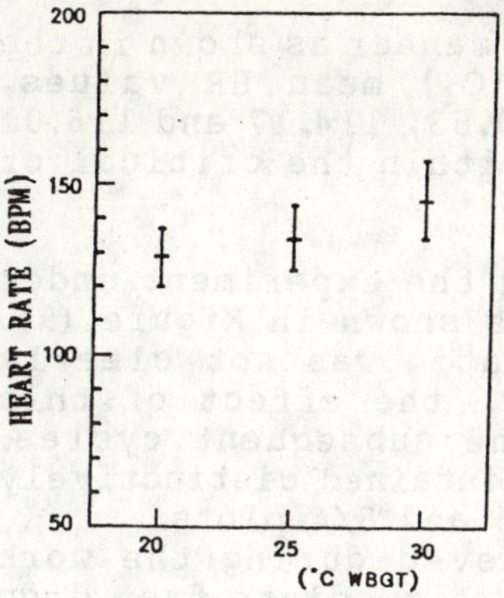

Fig. (8) Plot of Means and Standard Deviations of Heart Rate Versus W/R Ratio.

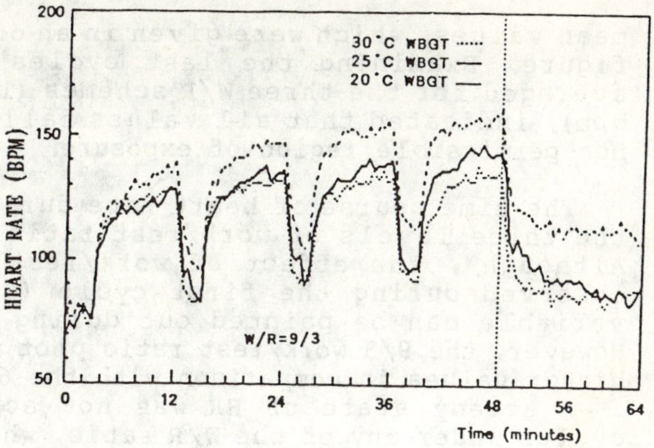

Fig. (9) Effect of Work/Rest Ratio upon Heart Rate (HR).

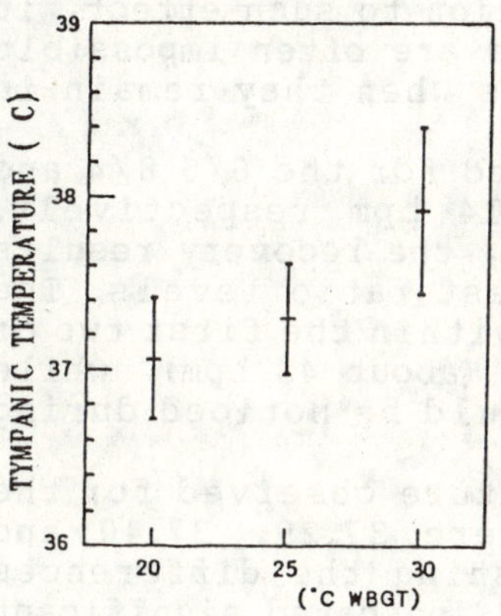

Fig. (10) Plot of Means and Standard Deviation of Tympanic Temperature Versus W/R Ratio.

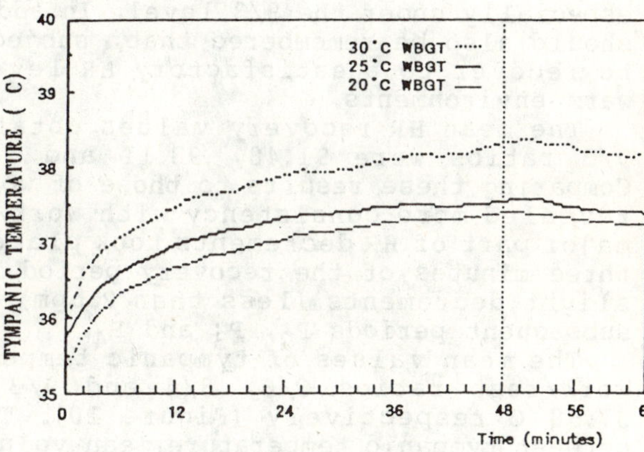

Fig. (11) Effect of Work/Rest Ratio upon Tympanic Temperature (Tty).

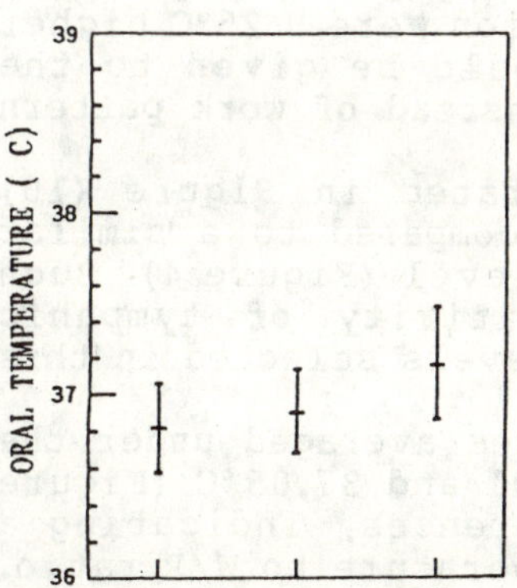

Fig. (12) Plot of Means and Standard Deviation of Tympanic Temperature Versus W/R Ratio.

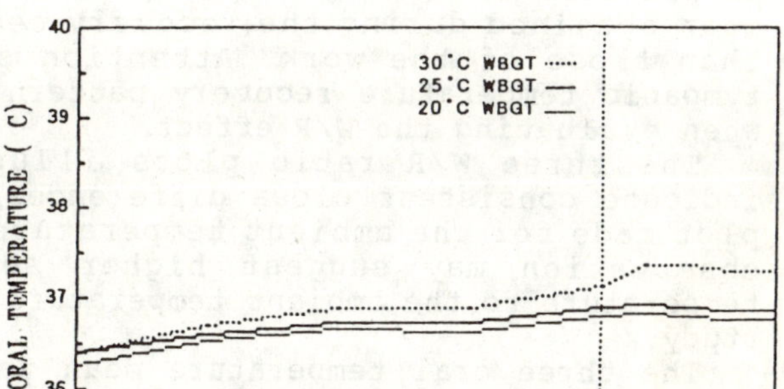

Fig. (13) Effect of Work/Rest Ratio upon Oral Temperature (Tor).

mean values, which were given in an odd manner as shown in the figure. Examining the last cycles' (C_4) mean HR values, averaged for the three W/R schemes (110.63, 114.17 and 116.01 bpm), indicated that all values fall within the critical or non-permissible region of exposure.

The time course of heart rate during the experiment under the three levels of work/rest ratio is shown in Figure (9). Although , the effect of work/rest ratio was not clearly observed during the first cycle (C_1), the effect of this variable can be pointed out during the subsequent cycles. However, the 9/3 work/rest ratio plot maintained distinctively higher values in comparison with the 6/6 and 8/4 plots.

A steady state of HR was not achieved during the work cycles under any of the W/R ratio, while immediate fast drop during the first minute of resting, then a much slower rate through the remainder of the rest period took place. The latter cycles illustrate the accumulative effect of the lack of achievement of the resting HR prior to resumption of work, especially under the 9/3 level. In addition to such effect, it should also be remembered that, subjects are often impossible to recover to a satisfactory HR levels when they remain in warm environments.

The mean HR recovery values obtained for the 6/6 8/4 and 9/3 ratios were 91.48, 93.11 and 98.14 bpm respectively. Comparing these results to those of work, the recovery results revealed more consistency with work/rest ratio levels. The major part of HR decrements took place within the first two or three minutes of the recovery period P_1 (about 43 bpm), while slight decrements (less than 2 bpm) could be noticed during subsequent periods P_2, P_3 and P_4.

The mean values of tympanic temperature observed for the work/rest ratios 6/6, 8/4 and 9/3 were 37.29, 37.40 and 37.58 C respectively (Figure 10). Testing the differences between tympanic temperature mean values indicated significant differences between the 9/3 and each of the 8/4 and 6/6 W/R ratio results only. Once again, the higher W/R ratio (9/3) produced the more pronounced effect on tympanic temperature as in previous results of effect on HR. The tympanic temperature mean obtained during the recovery session were 0.25°C higher than those of the work. Attention should be given to the tympanic temperature recovery pattern instead of work pattern when evaluating the W/R effect.

The three W/R ratio plots illustrated in Figure (10) indicate consistent close differences, compared to a similar plot made for the ambient temperature level (Figure 4). Such observation may suggest higher sensitivity of tympanic temperature to the ambient temperature levels selected in this study.

The three oral temperature mean values averaged under the three W/R ratio levels were 36.93, 36.91 and 37.05°C (Figure 12). These values showed closer differences, indicating a possible less sensitivity of oral temperature to W/R ratio. The oral temperature trend is similar to that obtained for the corresponding tympanic excluding the early stage of the first cycle, which is characterised by special factors affecting the collected measurements. High correlation was found (r=0.97) between the recovery of tympanic and oral temperature values, reflecting the reliability of using either oral or tympanic temperature for evaluating W/R effect.

Effect of cycle

The order of cycle indicated significant effect on HR, T_{ty} and T_{or}. A remarkable heart rate rise has taken place during each cycle reaching a mean value of almost twice the resting mean value for the subjects (63.5 beats/min). The heart rate during rest remained at a progressively higher level as the number of cycles increased. These results indicate not only the increased strain of performing the work activity (50% VO_{2max}) under the designed combination of W/R ratio and the environmental temperature, but also the rapid accummulation of strain induced with each successive cycle (Brouha and Maxfield, 1962). A progressive rise for the tympanic temperature with the cycle order was also obtained under the prescribed experimental conditions. Most of the tympanic temperature increments took place during the first cycle, while less increments were observed during the subsequent ones. Similarly, the first cycle results yielded a remarkable oral temperature rise (about 0.7 °C) above the resting value (36.44 °C). Less increments were observed for the subsequent cycles C_2, C_3 and C_4. A maximum mean value of 37.12 °C oral temperature was observed at C_4.

Effect of work-rest duration

The analyses of variances results indicated no significant effect of Work-Rest duration (6/6 and 12/12 min/min regimes) upon HR as well as oral temperature. A significant effect at 0.05 level was indicated upon the tympanic temperature. It seems that the HR is an accepted measure for the effect of W/R ratio, rather than for the work-rest durations. The tympanic temperature has thus shown to be more responsive to W/R durations than either of the two measures selected in this study.

CONCLUSIONS

1- Heart rate has proven to be the established predictor of the work/rest ratio effect under hot environments.

2- Tympanic temeprature showed higher sensitivity to thermal environment when examining the W/R ratio effect for the conditions used in this investigation.

3- Oral temperature values were lower than the tympanic, however, a linear correlation between tympanic and oral temperature was found, reflecting the reliability of using oral temperature as a predictor to heat content due to metabolic and thermal environment.

4- Thermal recovery was not attained under higher level of thermal environment (30°C WBGT). This finding was more emphasised under higher levels of W/R ratio.

5- A special consideration should be given resulting in the selection of comfortable W/R ratio under the higher levels of ambient temperature (30°C WBGT). Such consideration seems to be less critical under the lower levels of temperature (20 and 25°C WBGT).

6- The work-rest duration indicated minor effect upon HR, T_{ty} and T_{or}. However, longer work periods seemed to be more annoying to the subjects, especially under the higher level of ambient temperature.

REFERENCES

Brouha L., Smith P.E. and Delanne R., 1960, Physiological reactions of men and women during muscular activity and recovery in various environments, Journal of Applied Physiology, 16(1), pp. 133-140.

Brouha L. and Maxfield M.E., 1962, Practical evaluation of strain in muscular work and heat exposure by heart rate recovery curves, Development, Vol. 3, pp. 857-865.

El-Nawawi, M.A., Moshref, S.B. and El-Mohandes, M.S., 1984, Heart rate and oxygen consumption responses under combined physical and thermal loads, Proceedings of the 1st International Ain Shams University Conference On Prod. Eng. and Design for Development (Egypt), Vol. 3, pp. 857-865.

Givoni B. and Goldman R. F., 1972, Predicting rectal temperature response to work, environment and clothing, Journal of Applied Physiology, 32(6), pp. 812-822.

Kamon E., 1980, Design of work rest schedule for hot jobs, In the proceedings of NIOSH workshop on recommended heat stress standard, Cincinnati,U.S.; DHHS (NIOSH) publication no. 81-108, pp. 71-95.

Mairiaux P., Libert J.P., Candas V. and Vogt J.J., 1986, Prediction of strain for intermittent heat exposure. Ergonomics, 29(7), pp. 913-923.

NIOSH, 1980, Proceedings of NIOSH workshop on recommended heat stress standards. Dukes-Dobos F.N. and Henschel A., eds. - Cincinnati, U.S.; DHHS (NIOSH) publication No. 81-108, pp. 32-45.

NIOSH, 1986, Criteria for a recommended standard occupational exposure to hot environments, Revised criteria 1986, DHHS (NIOSH) Publication No. 86-113.

II

BACK INJURY AND REHABILITATION

Advances in Industrial Ergonomics and Safety II
Edited by Biman Das
Taylor & Francis 1990

THE DEVELOPMENT OF THE ERGONOMICS INFORMATION ANALYSIS SYSTEM FOR VISION IMPAIRED WORKERS

Jen-Gwo Chen and Luella Hwang[1]

Department of Industrial Engineering
University of Houston
4800 Calhoun Blvd.
Houston, Texas 77204-4812

[1] The Lighthouse of Houston
3530 West Dallas
P.O. Box 13435
Houston, TX 77019, USA

Outlined herein is a research report for the development of a system which encompasses a variety of functions which are essential to the successful placement of visually impaired workers. These functions include an Ergonomics Information Analysis System for Vision Impaired Workers (EIAS-VIW) to store, retrieve, and organize worker's performance data, and a Disability Index (DI) to measure vision impaired worker's residual capabilities in performing industrial assembly task. EIAS-VIW includes four data sections: case identification, physical characteristics and capacities, vocational test results, and industrial bench assembly task results. Forty-two vocational reports from the Lighthouse of Houston is analyzed to develop the basic database structure for EIAS-VIW.

INTRODUCTION

The vision impaired represent a relatively large population within our society whose contribution potential has not been realized, nor even fully evaluated. Through rehabilitation and training, the vast majority of these individuals can be taught the skills necessary to support and perform many meaningful jobs. Unfortunately, even after rehabilitation and training, significant numbers of vision impaired workers are either not matched to jobs, or are placed to positions which require skills incompatible with their abilities and training. Even worse, most companies evaluate vision impaired worker's performance based on the standard obtained from the normal individuals.

We feel that if the rehabilitation and training efforts are to be successful, they must be augmented by a job placement strategy that considers such factors as:

- the measurement of the vision impaired workers residual capabilities (e.g., previous job experience, physical and mental scope and limitations, mechanical skill, interest, temperament), and the consideration of these capabilities when considering job assignments
- the determination of existing job categories that can be performed by the vision impaired either through proper training of the individual, economically sound redesign of the job, or a combination of both
- a program which serves to demonstrate the worker's abilities to prospective employers, and which may be used to document the cost-effectiveness of employing the properly trained and properly matched vision impaired
- a program which serves to monitor and evaluate the worker's progress (and identify any problems, their causes, and methods for the alleviation of such problems) after employment is found

Consequently, a system entitled **VITAL**, for **V**ision **I**mpaired **T**ask and **A**ssignment **L**exicon is being developed. This system can be used in assisting the vision impaired in their return to the work-force; and in their pursuit of finding a job assignment that is rewarding, productive and meaningful. Currently, VITAL consists of two primary elements:

- A **D**isability **I**ndex (**DI**) tailored specifically for the vision impaired.
- A supporting database management system for use in the storage and manipulation of the vision impaired vocational evaluation test and industrial bench assembly task performance data.

DEVELOPMENT OF THE DATABASE MANAGEMENT SYSTEM

A database management system, which we term the **E**rgonomics **I**nformation **A**nalysis **S**ystem for **V**isually **I**mpaired **W**orkers (**EIAS - VIW**), is constructed for storage and manipulation of vision-impaired vocational evaluation test and industrial bench assembly task performance data. The primary input is the evaluation and training information obtained from the Vocational Evaluation Unit at the Lighthouse of Houston over the past four years.

Descriptions of EIAS-VIW System Components

The database module includes four data sections: case identification, physical characteristics and capacities, vocational test results, and industrial bench assembly task results. A brief outline of the section contents is given here. The case identification section deals with the identification of a case with regard to date and general personal information (such as age, gender, education, etc.). Referral information is recorded here for tracking cases and for analyzing the interaction between the case identifier (e.g., the vocational therapist) and problem analysts (e.g., the vocational rehabilitator). Summary information records from other sources may also be stored in this section. The physical characteristics and capacities section includes basic physical stamina (such as height, weight, lift strength, visual acuity), disability, and physical characteristics data for the worker (e.g., stoop, kneel capability), and additional data reflecting work environment parameters (e.g., noise, wet, poorly lighted). The vocational tests section contains information collected from vocational tasks testing — especially, the Wide Range Achievement Test (WRAT), Wide Range Interest Opinion Test (WRIOT), AAMD Becker Interest Inventory Test (AAMD-II), Career Occupational Preference System (COPS) Interest Inventory, MDC Behavior Identification Test (MDC-BI), and the VALPAR Component Work Samples Test (VALPAR). The industrial bench assembly tasks section includes results of tests designed to measure the worker's performance in performing a variety of industrial tasks.

Statistics of EIAS-VIW System Components

Forty-two vocational reports obtained from the Lighthouse of Houston were analyzed based on the EIAS-VIW system structure. Twenty-Six male and Sixteen female with age range from 21 to 59 are involved in this analysis. Most subjects are able and willing to perform sedentary or/and standing work. Most of them are able but not willing to work under bad work condition (e.g., around fumes or gases, around poorly ventilated areas, around noise). Five vocational tests (WRAT, WRIOT, COPS, MDC-BI, and VALPAR) are selected in the EIAS-VIW since those tests have been widely used among several Lighthouses in the nation. These tests also demonstrate their reliability in reflecting the worker's characteristics (e.g., aptitude, temperament).

System Configuration

EIAS-VIW is implemented in Foxbase+, a commercial relational database management system. Figure 1 shows the basic database structure which has been established in the EIAS-VIW.

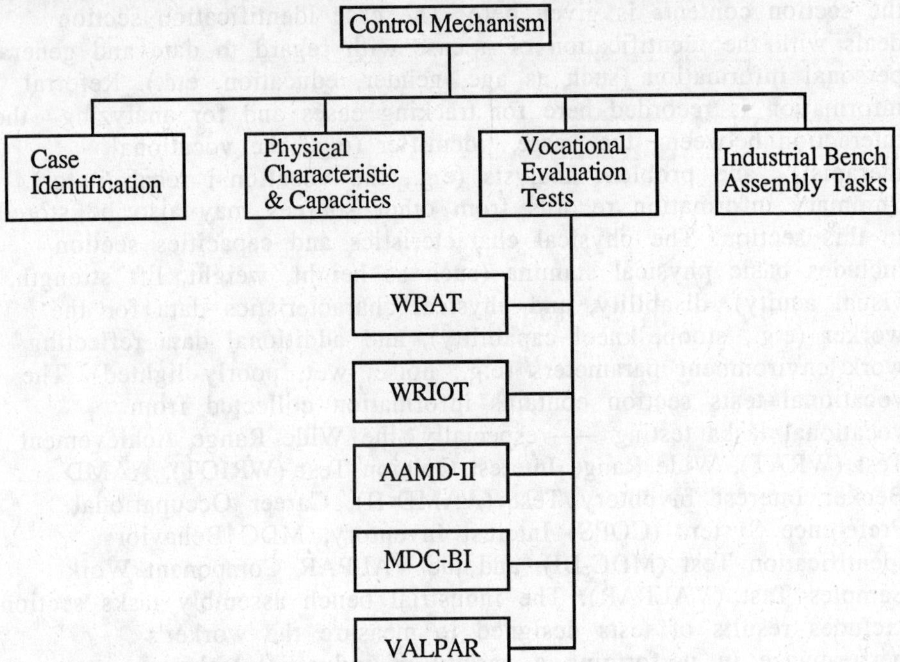

Figure 1. EIAS-VIW Database Structure.

The software structure for the EIAS-VIW is shown in Figure 2. The main menu includes eight functions: Append, Browse, Edit/View, Labels, Pack, Help, Report, and Quit. Eight subfunctions has been established under the function Edit/View.

DEVELOPMENT OF THE DISABILITY INDEX

The Disability Index Modeling

A Disability Index, used to measure the vision impaired worker's residual capabilities in performing industrial assembly task, is described in terms of the following components:

1. Achievement and Aptitude Index - It reflects the worker's achievement and aptitude characteristic.
2. Interest and Opinion Index - It reflects the worker's interest and opinion characteristics.

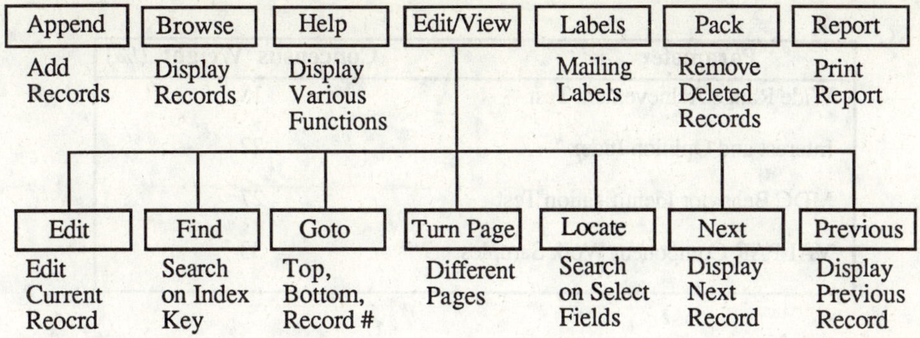

Figure 2. Basic Software Structure of EIAS_VIW

3. Temperament Index - It is used to determine the worker's behavior related to different job activities.
4. Work Sample Index - This index is compiled from five VALPAR COMPONENT WORK SAMPLE - B KITs. B-kit series is designed to measure the vision-impaired worker's ability to work at different work situations.

These components reflecting achievement and aptitude, temperament, interest, and worker's physical and mechanical skill levels related to a specific job are compiled for a sample of visually impaired persons and a mean computed for each basic job component. Each of these means will then be standardized by comparison with mean parameter values for the "normal" worker population. Finally, index weights obtained from expert concensus can be applied to arrive at an overall disability index descriptive of the particular job of interest. The index is scaled between 0 and 1.

The weighting factors of components were derived from expert concensus. The concensus result is shown in Table 1.

Table 1. Standardized Parameter Weight for Disability Index

Parameter	Concensus Weight (%)
Wide Range Achievement Test	18
Interest and Opinion Index	22
MDC Behavior Identification Test	27
VALPAR Component Work Samples	33

Disability Index Validation

The predictive capabilities of the disability index is evaluated through case study data — i.e., to determination of how well the computed DI value associated with a particular job agrees with actual visually impaired workers' performances in that job.

The degree of function performed by a worker is the ratio of the worker's actual job performance value and the job performance standard obtained from "normal" workers. For example, a client was assigned to the workshop division at the Lighthouse of Houston after receiving evaluation and training. His job primarily involved handling 2-ft. aluminum bars, selection of nuts and bolts, attaching bolts to bars, and tightening these with an air-impact wrench. The standard time for the "normal" worker is 1.06 min./unit. The impaired worker's actual average performance was 1.24 min./unit, calculated from weekly productivity reports. Thus, this worker can perform at the rate of 85.4% (1.06/1.24) that of the "normal" worker. Seventeen industrial bench assembly tasks with various worker characteristics, which have been conducted in the Lighthouse of Houston, are used in this validation process. Relationships between the disability index and the mean of these ratios for several such workers could then be established by Pearson coefficient analysis. The analysis result shows an r value of 0.67 for all tasks. This value demonstrates the predictability of the developed disability index.

DISCUSSION

The potential for implementation of the system/methodology described herein, as with any concept, depends on need, inherent value of the idea, and on the efforts devoted to disseminating information regarding its utility. In the introductory paragraphs of this document we addressed the need for a comprehensive job placement strategy for the vision impaired — that is, one which would incorporate traditional rehabilitation, training, and counseling efforts, but which would also lend emphasis to actual placement of individuals. We feel that this research work is a logical first step towards formulation of a practical system for assisting the rehabilitation community and the visually handicapped in their efforts to realize their placement goals.

ACKNOWLEDGEMENTS

This research was sponsored by the Research Initiation Grant (RIG) of the University of Houston. The authors gratefully acknowledge the Office of Sponsored Program at the University of Houston and the Lighthouse of the Houston for their supports.

Advances in Industrial Ergonomics and Safety II
Edited by Biman Das
Taylor & Francis 1990

THE WORK ENVIRONMENT VISUAL DEMANDS PROTOCOL: USING ERGONOMICS TO IMPROVE THE WORK PERFORMANCE OF SEVERELY VISUALLY IMPAIRED WORKERS

William H. Graves, Ed.D.
John H. Maxson, M.S.

Rehabilitation Research and Training Center
on Blindness and Low Vision
P. O. Drawer 6189
Mississippi State, MS 39762

INTRODUCTION

Environmental assessments are recommended by eye care professionals as a part of the low vision examination to improve the visual functioning of low vision persons in their work environments (Barraga, 1976; Carter, 1983; Faye, 1976; Jose et al., 1980; McCaa and Graves, 1986). The influence of environmental factors such as color, contrast, illumination, space, and time on visual functioning is well documented (Corn, 1983; Fry, 1962; Miles et al., 1984; Skoff and Pollack, 1969). Low vision aids that meet the visual functioning needs in the work environment can increase performance and comfort of persons with limited vision (Jose, 1983; McCaa and Graves, 1986, Miles et al., 1984, Marmion, 1986).

Determining the visual work needs of the visually impaired worker requires an assessment of vision and the visual characteristics of the work place. Other than the patient's report on his or her visual needs, direct information regarding visual elements of jobs often has not been available during the low vision examination process. Consequently, eye care professionals have experienced difficulty in identifying visual elements of the job. The *Work Environment Visual Demands Protocol* (WEVD) (Graves et al., 1987) is a procedure to identify visual demands of job tasks. The information obtained with the WEVD may be used in the low vision examination process by eye care professionals.

Research hypotheses

The purpose of the study was to evaluate the usefulness of the WEVD in the low vision examination process. The hypotheses were:

Low vision patients receiving low vision services from the low vision services team using data from the WEVD will report: a) fewer difficulties performing job tasks; and b) more use of the low vision device(s) on the job than those patients receiving low vision services where the low vision team did not have access to WEVD information.

METHODOLOGY

Fifty-eight individuals were identified and referred by a state vocational rehabilitation agency for blind persons to the low vision clinic of a teaching hospital. Thirty individuals (19 males, 11 females) participated in all phases of the project, completing the entire low vision process.

Procedure

Case records were reviewed to identify subjects eligible to participate in the study. Selection criteria included presence of no additional physical, emotional, or mental disabilities, and regular remunerative employment. Thirty subjects meeting selection criteria and volunteering to participate were randomly assigned to the experimental and control groups. All subjects were informed that if a low vision aid (LVA) would benefit them, especially as related to the performance of their jobs, appropriate LVAs would be provided at no cost to the subjects. Sixteen subjects were assigned to the control group and 14 were assigned to the experimental group.

After the subjects agreed to participate, a member of the WEVD research team scheduled an appointment with each subject and employer to collect data about the work environment of the subjects. The WEVD team member went to the subjects' places of employment and collected job task data using the WEVD. The information from the WEVD was analyzed, and a report was generated for the low vision treatment team.

The experimental group's low vision treatment team (ophthalmologist, low vision instructor, and optician) reviewed the WEVD protocol report prior to the low vision examination and the development of the low vision prescription(s), fitting, and instruction. The subjects in the control group received all services the experimental group received except that the low vision treatment team did not have the information from the WEVD protocol at the time of the low vision examination, the development of the prescription(s), fitting, and instruction.

Within 12 months after each subject had completed the low vision process, data were collected on: a) subjects' on-the-job comfort level; b) subjects' use of the low vision devices; c) low vision examination time; d) subjects' fatigue during the low vision examination; e) number and nature of follow-up visits; and f) number and types of adaptive devices.

Within the 12-month period after the subjects had received the LVA, the control group's WEVD data were presented to the low vision treatment team. The low vision treatment team was given the following options:

- reexamining the control group subjects;
- interviewing the control subjects to determine if modifications were needed in the low vision prescription; and
- not scheduling the control group subjects for either a follow-up examination or a follow-up interview.

During the 12-month follow-up period, the experimental and control groups were followed to determine comfort level on the job, use of the low vision aid on the job, and number and nature of follow-up visits.

RESULTS

Rate of low vision aid prescription

Of the 30 subjects referred to the low vision clinic (LVC), 87 percent (26) received an LVA. This percentage is higher than that reported by Verma (1975), Hoover and Kupfer (1959), and McCaa and Graves (1986). This higher rate may have been caused by the study's referral criteria, i.e., the subject is likely to benefit from LVAs.

Low vision aid use

The authors hypothesized that the experimental group would report more frequent use of the LVA on the job than would the control group. Half of the experimental group reported using their LVAs more than four hours per day, and 44 percent of the control group reported using their LVAs more than four hours per day. The experimental group showed more use (4.92 hours per day) than the control group (4.29 hours per day), but the difference was not statistically significant ($t(28)=.58$, $p>.05$).

The experimental group received fewer LVAs ($M=1.14$; $SD=.66$) than did the control group ($M=1.25$; $SD=.78$). The experimental group used more LVAs ($M=2.21$; $SD=.24$) to accomplish job tasks than did the control group ($M=2.00$; $SD=.73$).

Findings do not support the notion that environmental analysis improves the utility of the LVA for the low vision patient. However, these findings should be interpreted cautiously; the results could have been influenced by such factors as:

- subjects had long-standing visual disabilities and were likely to use nonvision-dependent techniques to accomplish job tasks; and
- subjects already used at least one LVA and the new LVAs may have been only supplemental.

Job task performance

The subjects in both experimental and control groups reported less difficulty performing near tasks after receiving the LVA(s) (Table 5). However, those in the experimental group reported a greater decrease in difficulty performing near tasks (from 79% to 14%) than the control group (from 62% to 19%). The 65 percent reduction in difficulty in performing near tasks of the experimental group compared with the 43 percent reduction of the control group suggests that the information provided by the WEVD contributes to reduction of difficulty of performing near tasks on the job.

The experimental group reported significantly fewer difficulties performing job tasks in the intermediate range ($t(12)=4.53$, $p<.01$) than did the control group ($t(14)=1.40$, $p>.05$). While these gains support use of information from the WEVD in the low vision examination process, it must be noted (Table 5) that the decrease in reported difficulty performing intermediate tasks was less dramatic than for those requiring near vision. This finding may be attributed to the job tasks of the subjects and/or the lack of responsiveness of LVAs to tasks performed in the intermediate range (Faye, 1986; McCaa and Graves, 1986).

All of the experimental group and 81 percent of the control group experienced difficulties performing distance tasks prior to the receipt of the LVA(s) provided in this study. After the LVA, 79 percent of the experimental group and 63 percent of the control group continued having difficulties. These improvements are not statistically significant for the experimental group ($t(12)=.61$, $p>.05$) or the control group ($t(14)=.30$, $p>.05$). It should be noted that distance tasks were not job requirement tasks for any member of the experimental group and were required for only one subject in the control group.

CONCLUSIONS

The purposes of this study were to develop and to validate a method to assess the visual characteristics of the work environment of the visually impaired individual. The evaluation of the WEVD and its software indicates that when systematically gathered information about the work environment of the person with a visual disability is used by the low vision treatment team, the individual who receives the LVA benefits.

Use of information from the WEVD appears related to a) reduction in near and intermediate task problems in the workplace, and b) the kind and function of low vision aids prescribed.

While all reported they frequently used the LVAs provided, the hypothesis that members of the experimental group would use their devices more often in a workday than those in the control group was not supported. This finding may be interpreted in at least two ways. The first is that the LVAs prescribed were task-specific. The small differences in time between the two groups while using the LVA may relate more to the time it took the LVA user to perform the task for which the aid was prescribed than to the question of how many hours in a day the prescribed LVA was used. It should also be recognized that the LVA prescribed was supplementary. That is, all subjects were using an LVA at referral. Therefore, it is likely that the question of hours of use per day was confounded by current or prior use of an LVA.

The subjects had long-standing visual disabilities. They had also learned to perform many tasks with nonsight-dependent techniques. Several were rehabilitation professionals who either taught newly blinded persons alternative techniques or who had received extensive training in the use of nonsight-dependent techniques. This history may have affected how often the LVAs were used.

A WEVD evaluation is a useful process for a number of situations. Information yielded from the workplace analysis may assist the eye care professional in identifying LVAs to assist the worker or student with limited vision. Current results and previous literature indicate this information would be most helpful when intermediate distance tasks are to be performed. The data indicate that the WEVD is helpful to low vision individuals in performing a variety of different job tasks, especially those in the intermediate range, which might be helped by an LVA.

Information generated by the WEVD process might be used by rehabilitation engineers in modifying job tasks for which appropriate and efficient LVAs are unavailable, and by rehabilitation counselors and placement specialists in identifying the visual demands of a job for employers and workers with limited vision. How the worker can be trained to meet the demands, and/or how the job can be modified to accommodate the worker, are valid concerns of the WEVD process.

REFERENCES

Barraga, N., 1976, Visual handicaps and learning: A developmental approach. (Belmont, CA: Wadsworth).

Carter, K., 1983, Comprehensive preliminary assessments of low vision. In Understanding low vision, edited by R. Jose. (New York: American Foundation for the Blind).

Corn, A., 1983, Visual function: A theoretical model for individuals with low vision. Journal of Visual Impairment & Blindness, 77, 373-377.

Faye, E. E., 1986, Characteristics of the target population. In Low vision and aging workshop, edited by D. L. Winfield and L. H. Trachtman. (Chapel Hill, NC: Research Triangle Institute).

Fry, G. A., 1962, Assessment of visual performance. Illuminating Engineering, 57, 426-437.

Graves, W. H., Maxson, J. H., and McCaa, C., 1987, Work environment visual demands protocol (final report). (Mississippi State, MS: Mississippi State University, Rehabilitation Research and Training Center on Blindness and Low Vision).

Hoover, R., & Kupfer, C., 1959, Low vision clinics: A report. American Journal of Ophthalmology, 48, 177-187.

Jose, R. T. (ed.), 1983, Understanding low vision. (New York: American Foundation for the Blind).

Jose, R., Smith, A., and Shane, K., 1980, Evaluating and stimulating vision in the multiply impaired. Journal of Visual Impairment & Blindness, 74, 2-8.

Marmion, S., 1986, Low vision visual performance as a function of environmental and task characteristics (technical report). (Mississippi State, MS: Mississippi State University, Rehabilitation Research and Training Center on Blindness and Low Vision).

McCaa, C., and Graves, W. H., 1986, The low vision clinic as an employment enhancement service (technical report). (Mississippi State, MS: Mississippi State University, Rehabilitation Research and Training Center on Blindness and Low Vision).

Miles, S. M., Cook, D., Huertas, V., and Lyon, S., 1984, Productivity and comfort of the visually impaired worker as a function of low vision aid usage and illumination/color contrast modifications (technical report). (Mississippi State, MS: Mississippi State University, Rehabilitation Research and Training Center on Blindness and Low Vision).
Verma, S. B., 1975, Analysis of 85 subnormal vision cases. Optometric Weekly, 1019-1026.

Advances in Industrial Ergonomics and Safety II
Edited by Biman Das
Taylor & Francis 1990

A SIMULATION PROGRAM FOR BACK INJURY REHABILITATION ANALYSIS

Jen-Gwo Chen and Susan Schellhase

Department of Industrial Engineering
University of Houston
Houston, Texas 77204-4812

A simulation program using SIMPAK has been developed to investigate the frequency of the utilization of each type of lumbar pain treatment. The output obtained from the simulation can help the physical therapy department schedule the number of modalities to keep in stock, the number of physical therapists required per day, and the number of patients that could be serviced per day.

INTRODUCTION

Back pain is the second most common reason for visiting a doctor (throat symptoms being the most common). Backaches are second only to upper respiratory infections as the cause of absenteeism from work (Cantu, 1987). Unfortunately, physical therapists are unable to predict how many treatments they can handle per day or per week. Thus, the treatment may be delayed because of lack of equipment or personnel.

Simulation is a technique for developing a representation of an actual system, in order to replicate or project the effects of certain changes in the system. This technique has been widely used in other areas with success. A simulation program is developed to provide the information related to the rehabilitation length for each type of lumbar pain treatment and the monthly or yearly treatment recovery rate. The output obtained from the simulation can help the physical therapy department schedule the number of modalities to keep in stock, the number of physical therapists required per day, and the number of patients that could be serviced per day.

REHABILITATION PROCESS

Injury due to accidents in sports or automobile related, requires that a preliminary diagnosis be made on-site prior to moving the injured subject. If it is determined that the injury could be a fracture-dislocation or dislocation of the vertebra, then the injured subject should be transported to a hospital on a fracture board (Cantu, 1987). Once the subject is stabilized a complete medical history is taken. When the subject visits the doctor in a non-emergency situations, the examination begins with complete medical history of the subject. The following questions need to be answered (Steinmann, 1982).

- How did the pain start?
- Were you injured?
- Where is the exact location of the pain?
- How often does it bother you?
- Have you had previous treatment?
- What are your daily routines?
- What is your state of mind?

Next, the medical personnel performs a physical examination on the subject. Observation of the subject's general posture, muscle tone, range-of-motion as follows: (1) as is, (2) against resistance, and (3) with support to negate gravity, and areas of local tenderness and muscle spasm is recorded. Muscular weakness and sensory changes in the lower extremities are also tested. General abdominal, rectal, and pelvic examinations may be required.

These results may indicate that several additional tests in some specific areas may be required (Reilly, 1981; O'Donoghue, 1970). These tests include:

- routine x-rays to reveal bone changes and to isolate the bone problems from the muscle and disc problems
- myelogram and discogram to check for disc problems
- electromyography (EMG) to investigate nerve and muscle problems
- Magnetic Resonance Imaging (MRI) to image the soft tissue, ligaments, tendons, spinal cord and nerve roots, and disc material
- Computerized Axial Tomography (CAT) to localize the portion being injured

In general, lumbar pain can be classified into strain and sprain problems. A strain problem refers to the injury of the muscle and tendon units, and can be classified into the following categories (Reilly, 1981):

- A first degree (mild) strain is an irritation to the muscle, tendon, or attachment that should be treated with protection to avoid re-injury.
- A second degree (moderate) strain is damage to the muscle, tendon, or attachment that should be treated with protection.

- A third degree (severe) strain is complete disruption of strength that should be treated with restoration of the unit.

A sprain refers to a ligament injury. The function of the ligament is to prevent abnormal motion of a joint while permitting normal functional motion. It can be chronic or acute with three categories:

- A first grade (minor) sprain is when some portion of the ligament is torn (with no function loss) with resultant hemorrhage into the ligament, and should be treated symptomatically (pain prevention)
- A second grade (severe) sprain is when some portion of the ligament is torn with some degree of functional loss, and should be treated with protection.
- A third grade (rupture) sprain is when there is complete loss of function, and should be treated with reapposition of the ligament ends.

Treatment starts with bed rest for the first 24 to 48 hours. During this time an ice pack or a cold spray may be used. A corset is worn to immobilize and protect the patient if necessary. Depending on the diagnosis a muscle relaxant could be used at this time to relieve the muscular tension. Some patients may be re-injured after receiving the pain relief medicine. Thus, pain relieving drugs are prescribed only for those with severe pain at the onset.

After two days, heat in the vicinity of the injury may be the prescribed treatment. A sprain should be stabilized with elastic strapping or a corset.

As the length of time from the onset of injury increases, the treatment would be varied depending on the subject's recovery progress.

A general outline of the non-surgery and post-surgery treatments are listed as follows:

ONSET (relief of pain and spasm)

- rest
- ice, cold packs, sprays
- analgesics
- muscle relaxants (different drugs for different people)
- anti-inflammatory agents
- heat (hot packs)
- muscle stimulation and/or ultrasound
- Transcutaneous Electrical Nerve Stimulation (TENS)
- short wave diathermy
- interferential current
- mobilization and manipulation
- physical therapy
- massage

- traction/countertraction

PAIN SUBSIDES (adopt appropriate exercise program)

- complete physical examination
- brace or lumbosacral corset
- exercise to restore range-of-motion and strength
- exercise to stretch out contracted muscles and ligaments (note: exercise to be started gradually)

EDUCATIONAL PROGRAM

- proper postural principles (standing, walking, sitting, driving, lifting, and sleeping)

It has become clear that nearly all of the methods used to treat the epidemic-sized ailment *do* work - at least for some people, some of the time. The question that has never been answered correctly with any consistency is " Which treatment works for whom and when ?"

SIMULATION MODEL

A simulation program is developed using SIMPAK to investigate the frequency of the utilization of each type of lumber pain treatment. Statistics for number of arrivals, length of recovery, and initial pain level were supplied by the outpatient physical therapy department of a local hospital.

Input Information

The program user selects the length of simulation (with a recommendation of 365 days). The patient arrival rate is uniformly distributed with a minimum of 1.8 days and a maximum of 3.2 days between arrivals. The initial pain level, in a scale of zero to ten, is normally distributed with a mean of 6 and a standard deviation of 1. A scale of zero represents pain free while the scale of ten means the worst pain the subject has ever felt. The recovery time has an exponential distribution with a mean of 3 months. The subject is checked every 2 weeks by the physician and possibly receives a new therapy program each time. The simulation program assumes that the subject is receiving therapy 7 days per week. The subject will continue with therapy as prescribed by the physician until he is pain free or when he enters a pain management program. Some patients will always be in the physical therapy program for lower back pain, unless the therapy department has just started receiving clients, Therefore, this simulation program has a lead time of 28 days.

Model Development

To develop the model, the rehabilitation activities (e.g., light massage, deep massage, exercise, etc.) required by the patients at different recovery stages should be identified. After the patients have been admitted into the program, different physical therapy recommendations will be prescribed as shown in Figure 1. Values of input parameters (arrival rate, recovery time, pain level, and length of the simulation) will be initiated based on the statistics described at the previous section. The system will accept the patients, diagnose their pain level, and simulate their rehabilitation process. Different treatment prescription will be assigned based on their recovery progress until the patients release from the hospital. A maintenance program is designed for monitoring the patients to prevent re-injury. The frequency of utilization of different types of modalities used in treatment will be computed during the simulation period. The flow of the rehabilitation is shown in Figure 2.

Output Information

The output includes two parts: input information generated by the system based on different distributions, and the simulation results. The system generated input information includes the length of simulation, the number of arrivals for lumbar pain, the min/avg/max initial pain level, and the min/avg/max recovery time. The simulation results show the number of recoveries and the utilization frequency of different types of modalities. It also gives statistics related to different treatment length per day.

The results concurred with expectations with a slight increase in the use of some of the types of treatment. This is due to the fact that each type of treatment may not be used in conjunction with the other possible prescribed treatments at that pain level. Therefore, the results give the expected average use of each type of treatment if the prescribed treatments were all used.

After correlating the average modality time usage per treatment with the average number of treatments per day, the output may be used to schedule: the number of modalities need to keep in stock per day, the number of physical therapists required per day, and the number of patients that can be serviced per day. Also the number and frequency of subjects entering a pain management program can be estimated.

DISCUSSION

Many variables exist when one is referring to human beings. In fact, we know that no two persons are alike. When a person has injured his/her body, the doctor can only rely on past experience for prescribing a treatment program that will help that person return to a pain free situation. The treatment may not always work.

The physical therapy department must also rely on past experience for determining the number and frequency of new patients that will require treatment. The simulation program will

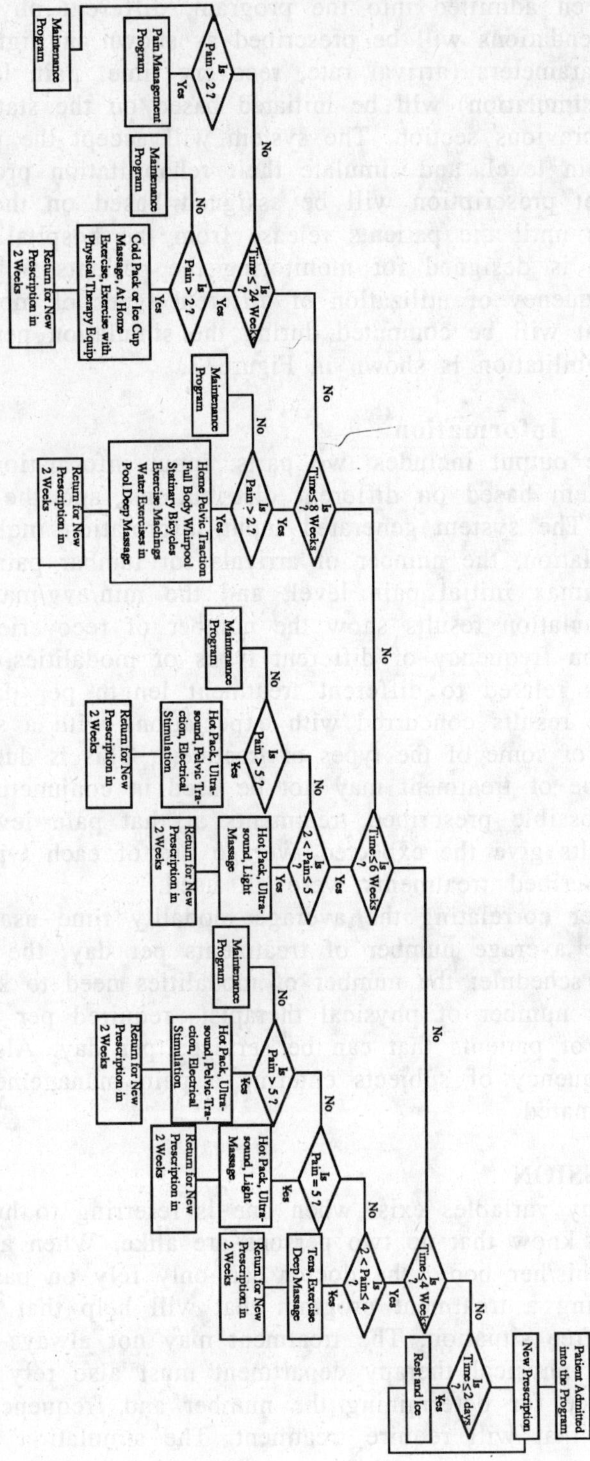

Figure 1. Physical Therapy Recommendations.

Stage 1	Values of Length of Simulation, Patient Arrival Rate, Recovery Time, Pain Level Initiation as the Input Information Generated by the System from Estimated Distributions
Stage 2	Prescribing Different Physical Therapy Recommendations Depends on the Pain Levels and Rehabilitation Progress
Stage 3	Assigning the Appropriate Types of Equipment Based on the Prescription and Availability
Stage 4	Compute the Length of Recovery, Types of Modalities ThatHave Been Used, Frequency of the Utilization, and Number of Recovered Patients
Stage 5	End of the Simulation, Present the Simulated Results on the Screen and/or to the Printer

Figure 2. The Simulation Process.

help the physical therapy department make a better estimation of the number of patients that will require treatment and the types of treatments those patients will probably be receive. This will allow them to schedule the patients so that there will be no waiting time for the different types of treatment (with adequate number of therapists available as well as sufficient equipment for the treatments).

REFERENCES

Cantu, R.W., 1987, The Exercising Adult, (New York: Macmillan Publishing Co.).

O'Donoghue, D.H., Treatment of Injuries to Athletes, 1970, (Philadelphia: W.B. Saunders).

Reilly, T., Sports Fitness and Sports Injuries, 1981, (London: Faber and Faber).

Steinmann, M., 1982, The American Medical Association Book or Back Care, (New York: Random House).

Advances in Industrial Ergonomics and Safety II
Edited by Biman Das
Taylor & Francis 1990

BACK PROBLEMS AFFECTING PRESCHOOL SPECIAL EDUCATION STAFF

Aghazadeh, F., Broussard, J.F., Waikar, A.M.

Industrial Engineering Department, Louisiana State University, Baton Rouge, LA 70803 U.S.A.

Bittner, A.C.

Battelle Human Affairs Research Center, Seattle, WA 98433, U.S.A

This study was conducted to determine the extent of back injuries among preschool special education personnel. Preschool special education staff have to frequently handle physically handicapped children which requires bending, lifting, and twisting. It was found that 84 percent of the staff had sustained back injuries during the course of their employment. Thirty-three percent of these injuries were reoccurrences.

INTRODUCTION

Back injuries are the most prevalent type of occupation related injuries. In 1988, according to the National Safety Council (Accident Facts, 1989) there were 400,000 back injury cases in the U.S.A., comprising on the average 22 percent of all injuries. Many occupations have twice this amount (Button and Pater, 1988). Lundgren (Grandjean, 1988) reports that 60 percent of the Swedish population suffers from low back pain during their working life. This figure is in agreement with the figures reported by Astrand (1977) and Hirsch (1966). The costs associated with these injuries are astronomical. Average costs per injury for workers' compensation cases closed in 1983 were estimated at $4,489 in wage compensation and $1581 in medical payments (Accident Facts, 1988). If we calculate only the costs of compensation for 400,000 injuries, the total amount is about 2.5 billion dollars. Although back injuries comprise 22 percent of all injuries, the associated cost is 32 percent of all injuries. In 1988, the cost of all work related injuries were 47.1 billion dollars; 32 percent of this figure is about 15 billion dollars. This number includes wage losses, medical expenses, insurance costs, and indirect costs and indicates the magnitude of the back injury problem.

Back injuries affect a wide range of professions. These injury incidents occur in diverse industries including agriculture, forestry, fisheries, mining, construction, manufacturing, transportation, public utilities, wholesale and retail sales, finance, services and public administration. An epidemiological study of semi-skilled construction workers showed that 65 percent suffered from low back pain (Damlund et al., 1986). The same study concluded that 53 percent of warehouse workers had low back pain (LBP). Low Back pain has become a serious occupational health problem for container truck drivers; a new occupation produced by containerization of ocean cargoes. Compared to the truck drivers, the container tractor drivers complained more frequently of low back pain (Konda et al., 1985). Back problems are also prevalent in the armed forces and health care systems. They are the most frequent type of injuries in the U.S. Navy (Hirstein, 1980). The data gathered for 1977-1978 show back injuries made up 22 percent of navy civilian and military injuries documented. Dehlin et al. (1976) investigated the incidence of back pain in 267 female nursing aids in a geriatric hospital; about 50 percent of them were afflicted with back problems. In a recent study Gagnon and Lortie (1988) reported that an analysis of occupational accidents affecting nursing aids in a geriatric hospital indicated that 62 percent of them suffered from back injuries. An analysis of the New York workers' compensation agency cases showed that nursing aides, orderlies, and attendants have the highest incidence ratio of all occupation categories. The occupations with the next highest incidence ratios were construction laborers, licensed practical nurses, garbage collectors, truck drivers, and registered nurses (Accident Facts, 1988). Evidently workers involved in manual lifting of objects, boxes or people, suffer more from back injuries than the workers who do not lift objects as part of their daily work. Examples of these workers are nursing aids who lift people and garbage collectors who handle objects, etc.

A search of literature indicates that lifting has been the main contributor to back injury and has been investigated a great deal. Occupations involving lifting, such as nursing and garbage collecting, have been studied in detail. One segment of the work force that appears to has been neglected is the educational systems. The literature does not contain much information in regard to the incidence of back injuries among teachers of preschool handicapped students who lift and handle handicapped students (Nagira et al. 1981). Management of orthopedically handicapped students is one of the least understood aspects of providing services for exceptional children (Parette and Hourcade, 1986). Back problems seem one of the inevitable consequence of working in the profession due to deep bending and high degree of hands-on manipulation required to manage students.

The objective of this pilot study was three folds: to determine the existence and extent of back injuries among preschool special education and handicapped teachers, to identify possible causal factors, and to suggest solutions to the problems.

PROCEDURE

In order to achieve the objectives of this study a questionnaire was developed and distributed among 91 special preschool personnel in the Baton Rouge area schools. The questionnaire was designed to determine the percentage of workers with back problems, possible causes of back problems, and information about workers, etc. In addition to the questionnaire, a field study was conducted to interview and observe the personnel and identify the problem areas.

RESULTS

Questionnaire

A total of 73 special preschool and handicapped teachers and aides responded to the questionnaire and participated in this study. Participants ranged from 24 to 55 years of age with education level of high school to a Masters degree. A surprisingly high large number of participants, 61, indicated that they had sustained back injuries as a result of their occupation; twelve had reoccurring injuries. The injuries ranged from muscle strains to ruptured disks. The age distribution indicated that 25 were in the 31-40 age group and 48 in the 41-60 age group. Response of injured workers indicated that they were injured while lifting the students. More specifically the injury resulted while

a. lifting a student to or from the floor,
b. twisting while moving a student between a wheelchair and a RIFTON chair (a special chair for handicapped children),
c. and lifting students to and from the school bus.

The participants were also asked to check the factors which they thought had contributed to their injuries. Table 1 depicts the responses.

Table 1. Tasks contributing to the back injury.

	Lifting an ambulatory student		Lifting an nonambulatory student
While twisting	8		15
From the floor	10		15
From a chair	2		4
From an elevation	3		7
Helping a student off the bus		21	
Helping a student out of the swimming pool		8	

Only 31 of the questionnaire respondents had received some form of formal training in a program conducted by the schools' physical therapist, while 25 had received training outside the school system. Most indicated that while the courses were helpful they were geared primarily to lifting sturdy, inanimate objects, not children.

Interestingly enough apparel also seemed to play a role in back injuries and back pain. Of the 12, who admitted to recurring back pain all wore high heel shoes and dresses to work. While the 12 workers who claimed no back pain or injury, all wore either flats or tennis shoes and pants.

Observations

Upon analyzing the response of participants to the questionnaire, it becomes apparent that back injury among special preschool teachers is prevalent. As the second phase of this study, a field investigation was conducted to observe these workers and determine the reasons for the high number of injuries. Subjects were observed in the classroom and at morning bus duty. Figures 1 and 2 show a worker transferring a student from the bus attendant to a waiting wheelchair. In Figure 1, she is in a twisted posture, while receiving the student, with one foot off the ground. After a carry of 3 feet, she bends over the wheelchair foot rest and positions the students in the wheelchair (Figure 2). This entire maneuver took less than a minute, but is repeated many times per day at this one center.

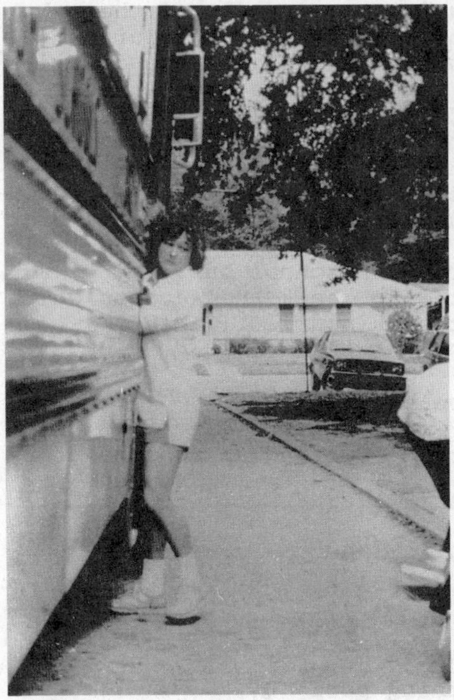

Figure 1

Figure 2

In contrast, Figure 3 shows a second bus duty teacher unloading a student from a Lift Gate Bus. These buses contain a hydraulic platform that extends out to raise or lower the student in the wheelchair safely. The teacher never has to bend or lift the student, and all controls for the lift gate are at shoulder height.

Figure 4 shows a worker performing gross motor testing and playing with a student. She is in an off balance position, while bending and holding a student (a 4 year old with Down's syndrome) to give him confidence while ascending and descending the stairs.

The most severe lost time back injury to date was to an aid on a bus. The injury occurred while lifting an eighty pound orthopedically handicapped child from the steps of the bus up to the floor of the bus. Bus aids are not allowed to leave the bus. Ordinarily, a child with this problem would be transported in a special lift gate bus which eliminates the need to perform this type of lift. But a lift gate bus was not mandated in this child's individual evaluation program.

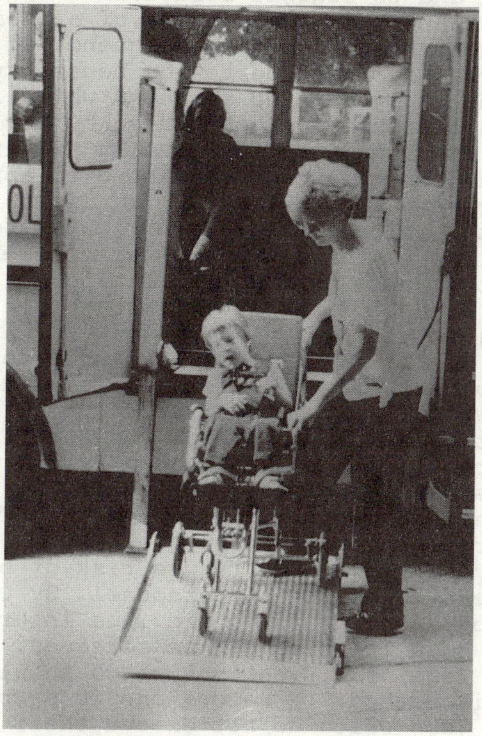

Figure 3

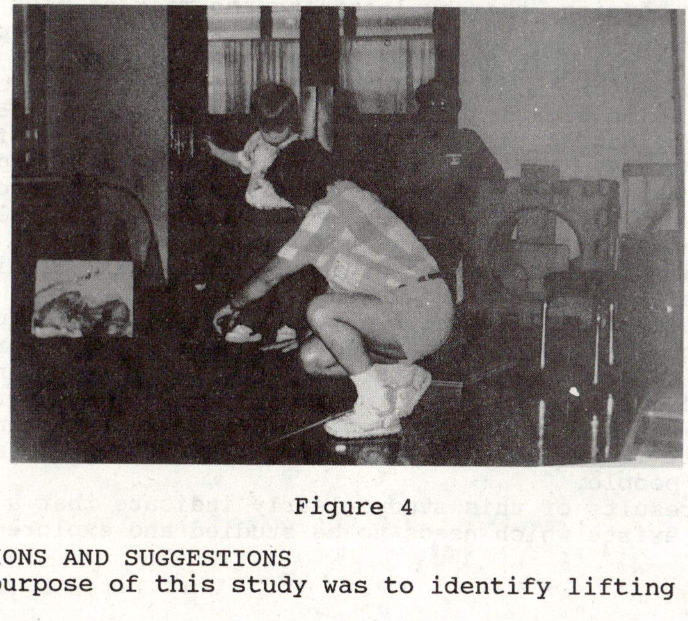

Figure 4

CONCLUSIONS AND SUGGESTIONS

The purpose of this study was to identify lifting problems

faced by teachers of preschool special education and handicapped children and to suggest solutions to these problems. Lifting problems and back pain/injury of a preschool special education teachers seem to be a consequence of their work. There are inherent hazards due to the necessity to closely manage developmentally delayed students which involve lifting and bending. This problem is compounded by having to handle a much less than ideal object from usually awkward positions. Also evident, is the of lack of training in appropriate lifting techniques.

Another factor which may have contributed to the high number of injuries is the way these workers perceive the difficulty of the work. A NASA TLX rating scale (Byers et al., 1988) was used to find out the subjective difficulty of the preschool jobs.

The results of the NASA TLX survey indicate that a majority of preschool special education teachers consider their workloads to be low. This may be desirable from a job satisfaction standpoint, but as can be seen from the data, lifting these students incorrectly can be hazardous. From a safety standpoint this may indicate a laxness on the part of the teachers to the need for proper lifting techniques. low-cost

Solutions are many and run the gamut from low-cost to costly. One way to reduce the future risk of back pain/injury is pre-employment screening of personnel. By using this well known methodology, workers with previous injuries or workers susceptible to back injuries can be screened. Of course there may be certain legal ramifications which should be considered in applying this method. A second one is proper placement of workers on the jobs which do not exceed their mental and physical capacities. The last approach is the modification of the job itself. Provision for the use of auxiliary equipment such as rolling stools, raised platforms, special short chairs and devices such as hula-hoops enables teachers to get physically closer to the students. This greatly reduces the need for bending, thereby lessening the risk of back injury. A more expensive method is the use of specially equipped buses. For years it has been the province of parents rather than professionals to bring about the change in the school system. While back injuries account for only about 25 percent of all accidents, they account for over 90 percent of the costs (Morris, 1985). It is the responsibility of legislative bodies to regulate state agencies. A helpful and necessary regulation is that lift-buses be provided for the multi-handicapped, seizure disordered, cerebral palsy or orthopedically handicapped students. This would eliminate hundreds of potentially damaging lifts per day.

Training is an effective way of reducing back injuries. Even though a number of injured workers trained before the injury, apparently the training was not sufficient and did not simulate a real life job situation. Participants were trained in lifting boxes, not children. Instead of practicing lifting boxes, it is suggested that dummies be used to simulate lifting people.

The results of this study clearly indicate that a definite problem exists which needs to be studied and explored further.

REFERENCES

Accident Facts, 1933-1989 editions. Chicago, Illinois: National Safety Council.

Astrand, P. and Rodahl, K., 1977, Text Book of Work Physiology: Physiological Bases of Exercise. New York: McGraw Hill, Inc.

Byers, J.C., Bittner, A., Christ, R., Hill, S., and Zaklad, A., 1988, Workload assessment of a remotely piloted vehicle (RPV) system. Proceedings of the Human Factors Society.

Button, R. and Pater, R., 1988, Pre-training is essential tool for fall and back injuries. Occupational Health and Safety, Vol. 56 (7).

Damlund. M., S. Goth, P. Hasle and K. Munk, 1986, Low back strain in Danish semi-skilled construction work. Applied Ergonomics, 17.1, pp. 31-39.

Dehlin, O.; Hedenrud, B.; and Horal, J., 1976, Back symptoms in nursing aids in a geriatric hospital. Scandinavian Journal of Rehabilitation Medicine, 8(2), pp. 47-53.

Gagnon, M., and Lortie, M., 1985, A biomechanical approach to low back problems in nursing aids. Trends in Ergonomics/ Human Factors IV. Elsevier Science Publishers.

Grandjean, E., 1988, Fitting The Task To The Man: An Ergonomic Approach. Taylor and Francis Ltd.

Hirsch, C., 1966, Low back pain etiology and pathogenesis. Applied Therapeutics, 8, pp. 857-862.

Hirstein, R.V., 1980, Back facts. Lifeline March/April, pp 15-18.

Konda, Y., Mito, H., Kadowaki, I., Yamashita, N., and Hosoka wa, M., 1985, Low back pain among container tractor drivers in harbor cargo transportation. Ergonomics International 1985,pp. 802-804.

Morris, A., 1985, Identifying workers at risk to back injury is not guesswork. Occupational Health and Safety, Vol. 55(12).

Nagira, T., Suzuki, J., Oze, Y., Ohara, H., and Aoyama, H., 1981, Cervicobrachial and low back disorders among school lunch workers and nursery school teachers in comparison with cash register operators. Journal of Human Ergology, Vol. 10 (2),pg. 117-124.

Parette, H., and Hourcade, J., 1986, Management strategies for orthopedically handicapped students. Teaching Exceptional Children, Vol. 18(4).

Advances in Industrial Ergonomics and Safety II
Edited by Biman Das
Taylor & Francis 1990

AN INVESTIGATION OF LOWER BACK INJURIES AND THE FORCES ACTING ON THE LOWER BACK DURING TASKS PERFORMED BY HOSPITAL WORKERS

John W. Kozey
Department of Industrial Engineering
Technical University of Nova Scotia
Halifax, N.S., B3J 2X4

Cheryl L. Hubley-Kozey
School of Physiotherapy
Dalhousie University
Halifax, N.S., B3H 3J5

John W. Grainger
School of Occupational Therapy
Dalhousie University
Halifax, N.S., B3H 3J5

INTRODUCTION

The literature is replete with research indicating the high incidence of low back pain and its associated costs both in terms of economics and human suffering (Andersson, 1979, 1981). Hospital workers like many other groups have been identified as having a high frequency of reported low back associated pain and injuries (Lortie, et.al. 1987, Hubley-Kozey et.al., 1985). An overview of the literature identifies over-exertion during lifting, pushing, pulling and carrying loads as the common tasks associated with many low back problems.

Although education programmes on correct lifting techniques, physical conditioning programmes and pre-employment screening have been utilized in various industries, neither the costs of, nor the numbers of low back related injuries has been significantly reduced. This has resulted in research being directed at addressing mechanisms of injury and a greater understanding of job tasks in an attempt to provide information which may be useful to reduce low back injuries (Gagnon and Lortie, 1987; McGill and Norman, 1985,1987).

In this regard the forces of the major muscle groups, in particular the back extensors and the forces acting on the vertebral discs are of primary importance to understanding the mechanisms of low back pain and injury. After extensive research acceptable and maximal permissable limits have been proposed for certain lifting tasks in a guide by NIOSH (1981). These limits at least in part, are based on the magnitude of the compressive forces which act on the lower back.

The purposes of this project were:

1. to identify the groups of individuals within a hospital who may be at risk of low back problems using retrospective injury data.
2. to identify by interviews with hospital workers, the tasks considered to be associated with lower back injuries.
3. to analyze the forces acting on the lower back during the performance of these tasks and compare these values to the criterion proposed by NIOSH.

The rationale for this approach lies in the specificity of the information gained to the actual tasks performed by hospital workers in their unique environment.

METHODS

The methods will be subdivided into sections dealing with the three stated purposes.

Injury data

The injury data were provided by the personnel department of the hospital for a 12 month time period. The main variables of interest were: total number of injuries, total number of low back injuries, job classification, age, sex and tasks associated with low back injuries. Descriptive statistics were calculated using a Hewlett Packard 9845 desktop computer and statistics software package.

Task identification

The injury data were then used as a guide for task identification and from these data 8 units within the hospital were identified which consisted of 7 nursing units and central escort. Two units with high injury frequencies were omitted from the study because of ethical and practical considerations. These units were both intensive care units. Interviews with the head nurses of each unit, the supervisor for central escort and tours of each of the units were used to identify the specific tasks the workers considered to be associated with the cause of low back pain or injury.

Task analysis and laboratory simulation

Volunteers from each of the units were identified and informed of the benefits, risks and procedures through a written consent which was approved by the hospital research committee and the Faculty of Health Professions Ethics Committee, Dalhousie University. The subject's age, sex, mass, height and job classification were recorded. A portable videotape system was used to film the subject performing various tasks identified in the previous section. Where possible 35 mm slides of the tasks were taken. Measures of various objects involved in the tasks were done using a standard scale and spring loaded cable tensiometer.

All 35 mm. films from the lab simulations and the appropriate hospital films were digitized using an HP9845 desktop computer and HP 9874 digitizer. The WATBAK software package was then used to calculate the major muscle moments of force and the shear and compressive forces acting on the L4/L5 disk using a static linked segment analysis. In all patient handling tasks the mass was kept conservatively low, the mass of the subjects being lifted only varied from 50 to 53 Kg.

The specific tasks were grouped into Patient handling, Transportation and Other tasks for ease of presentation. Descriptive statistics were calculated to provide an overview of the data.

RESULTS

The results of this study are presented in 3 sections corresponding to the purposes related to injury data, task identification and task analysis. The task analysis section includes both the field tasks and lab simulated tasks.

<u>Injury statistics</u>

Table 1 provides the actual and percentage information of low back injuries with respect to the total number of reported injuries and the number of reported injuries with respect to the number of employees in each job category. There were 505 nurses employed at the hospital during this study this represents 37.5% of all hospital employees. They incurred 125 (25%) injuries and 18 (72%) of the total number of low back injuries. The total number of injured employees with low back pain within the injured workers for each group varied with Central escort reporting 2 (33.3%) and nursing reporting 18 (14.4%).

Table 1. Employee injury statistics for a 12 month period.

Job Title	Number of employees N	Total # of injuries N	Total # of LBP injuries N	Percentage of employees injured	Percentage of employees with LBP
Aides	52	15	0	28.8	0.0
Cook	10	7	0	70.0	0.0
Dietician	8	-	-	-	-
Engineer	6	-	-	-	-
Escort	23	6	2	26.1	33.3
Instructor	3	-	-	-	-
Management	136	8	-	5.9	-
Nurse	505	125	18	24.8	14.4
Office	227	10	1	4.4	10.0
Pharmacy	17	-	-	-	-
Rehab Therapy	15	1	-	6.7	-
Technician	151	21	1	13.9	4.8
R&E Therapy	17	1	-	5.9	-
Security	1	-	-	-	-
Soc. & Pall.	5	-	-	-	-
Workers	171	87	3	50.9	3.4
Unknown	-	2	-	-	-
Total	1347	283	25	21.0	8.8

Table 2 provides the gender and age data for the reported

injuries and low back injuries. Percentages show that low back injuries made up 5.8 % of the total reported injuries for males and 9.5% for females with the combined figure being 8.8%. The data showed that the average age of individuals both male, female and totals were 2 to 3 years younger for low back injuries compared to the overall reported injuries.

Table 2. Injury statistics by age and gender.

Catagory	Total number of injuries N	Total Number of LBP injuries N	%	Average age all injuries X	sd	Average age only LBP X	sd
Males	52	3	5.8	31.4	9.2	29.1	6.2
Females	231	22	9.5	33.8	11.5	30.8	9.5
Combined	283	25	8.8	33.4	11.2	30.6	9.1

It was noted that patient handling was responsible for 68% of the total number of LBP injuries. Also, from the brief descriptions provided on the injury report forms, the majority of the LB injuries did not occur in a completely uncontrolled environment. Only 3 (12%) of the 25 low back injuries were reported as occurring in an uncontrolled manner. These injuries involved workers helping patients who were falling out of control and an individual slipping on a wet floor.

Twenty four percent of the injuries occurred in the intensive care units. However, because of practical and ethical considerations it was not deemed appropriate to collect information from either SICU or the MICU therefore the focus of the second part of this study was on the other units identified within the hospital.

Identification of tasks

Interviews with the head nurses, supervisor of central escort and a small sample of staff nurses from the 7 nursing ward identified and central escort provided the following task identifications (see Table 3).

Task analysis

Thirty-two slides from the field and laboratory simulations were analyzed. Table 4 presents the results for the patient handling, the transportation and the other tasks. The values presented are the range, mean and standard deviation data for the muscle moments of force and the compressive forces (Fc) acting on the lower back (L4/L5).

In general the patient handling tasks resulted in higher moments of force being produced by the erector spinae muscle group. The mean score for the patient handling tasks was 161 Nm. versus 86 Nm. for transportation and 116 Nm. for the other task category.

The compressive forces on the L4/L5 disk were also very high for the patient handling tasks. The mean compressive force was 3347 N representing 92% of the AL established by NIOSH. There was a large range of Fc values with the lowest value of 1568 N. occurring during the four person transfer from bed to stretcher and the highest value calculated for the fireman's lift of a patient from a wheelchair, 5187 N. The former value representing 46% of the AL and the latter 151 % of that limit.

Table 3. Tasks perceived by workers to be associated with LBP.

Department	Job Tasks
Opthomology	using surgery lifts putting drops in patient's eyes getting patient out of the bed lifting boxes of saline solution moving fainting patients a) in wheelchair b) geri chair to bed c) from floor to bed
Emergency	lifting patient from ambulance transfer patient to hospital stretcher pushing stretcher assisting patient out of car onto stretcher
Medicine	patient handling - pulling patient up in bed - transfer from bed to chair - transfer bed to stretcher
Operating room	lifting heavy sets (eg. book walter, ortho.) dumping buckets of water (aides) holding patients limbs while prepping oxygen tank transfer (aides)
Orthopaedics	patient transport bed making with water mattresses transfer patient from bed to chair bed making at various levels putting footwear on patients pushing geri chairs
Gynaecology & Neurology	no unique tasks identified
Central escort	transporting stretchers, wheelchairs, beds and mobilizers up various ramps

* It should be noted that some of the tasks identified in one unit may have been identified in another unit and thus were not repeated.

Table 4. Muscle moments of force and compressive forces acting during hospital tasks.

Activity	Muscle moments of force range (Nm)	mean	sd	Compressive forces range (N)	mean	sd
Patient Handling	108-246	161	36	1568-5187	3347	852
Transportation	24-154	86	57	762-3084	1948	1030
Other	60-223	109	44	1184-4814	2222	937

The results of the transportation tasks showed the lowest mean value for erector spinae muscle moments of force 86 Nm. (±57 Nm). The largest moment of force for the transportation tasks was recorded for the task of pulling the stretcher from the floor in Emergency. The lowest moment, 24 Nm. was calculated while pulling the stretcher up the nuclear medicine ramp. These results are minimum task values as they were calculated using the masses or measured force of the transportation device without a patient. Further discussion of the effect of adding a patient will be provided in the next section. The mean compressive force was 1948 N. (±1030 N) which represents 67% of the AL. The range was from 762 N. for the nuclear medicine ramp task to 3084 N. for the emergency stretcher lifting task. The latter represents a value of 90% of the AL.

The results for the other task categories had a mean muscle moment of 109 Nm (±44 Nm) and a mean compressive force of 2222 N (±937 N). The range of scores varied from 60 to 223 Nm for the muscle moments of force and the compressive forces ranged from a low of 1184 N to 4814 N. The highest value was measured while the subject incorrectly used the surgery lift and the lowest values was found while making a bed when the bed was positioned at the lowest height possible.

CONCLUSIONS

The purpose of this study was to identify occupational groups within a hospital who had higher reported frequencies of LBP, identify the tasks associated with these groups and lastly, to analyze the forces acting on the lower back during the performance of the job tasks. The injury data demonstrated that the occupational groups of nursing and central escort reported the greatest frequencies of LBP within the hospital. The two hospital units with the highest frequencies were not studied due to ethical and other problems. It was also found that people reporting LBP were on average, 2-3 years younger than the injured population of the hospital workers.

Subsequent interviews with employees and supervisors highlighted the fact that patient handling tasks were the primary activity at the time of the injury. This finding is consistent with the previously published literature of Hubley-Kozey, (1985) plus many other studies and was also supported by the hospital records. There was also a wide variety of tasks other than patient handling, which were proposed as being contributing factors in LBP which included making beds, lifting boxes and other operational tasks involving prolonged static postures. Surprisingly there were very few cases of sudden loading due to patients falling or slippery work surfaces. This means that most of the reported activities were routine to the workplace and considered as part of the normal tasks. It could be suggested that perhaps one method to reduce the frequency of LBP injuries would be through administrative controls on the work methods and the job tasks. For example, one of the areas of concern dealt with the handling of patients who had fainted and were either in a chair or on the floor. It was common for these patients to be lifted onto a bed. An alternative approach maybe to attend to the fainted patient without a change in location of the patient until such time that the patient is revived and can assist in the movement or until a suitable number of people are available to assist in the patient handling.

The estimation of the moments and forces acting on lower back were made with conservative loads. In the case of patient handling tasks the mass of the subjects varied from 50 to 53 Kg. The reason for using subjects with such a low mass was to present the minimum or a best case example to the hospital workers. Many of the patients lifted in a hospital would be much larger and it is conceivable that the mass could easily be doubled. Therefore when comparing the associated Fc values to a criterion value such as AL in the NIOSH guidelines these conservative estimates should strongly demonstrated the need to reduce the true loads in many hospital patient handling tasks. The same arguments could be made for all the transportation tasks analyzed in which only the force necessary to move the transportation device were used in the analysis.

In summary this study has presented data which is in agreement with previously published data regarding LB injuries among hospital workers. Interviews used to determine the nature of the hospital tasks most closely associated with injuries indicated that patient handling was the principle factor in the majority of cases. Other tasks analyzed had a large range of scores for the muscle moments and compressive forces acting on the lower back. In some cases these forces could be reduced through the use of sound biomechanical information on lifting and lifting mechanics. But equally important would be the implementation of administrative controls on the work methods and the work tasks.

REFERENCES

Andersson, G., 1979, Low back pain in industry: epidemiological aspects. Scandinavian Journal of Rehabilitation Medicine, 11, 163-168.

Andersson, G., 1981, Epidemiological aspects of low back pain in industry. Spine, 6, 53-60.

Gagnon, M. and M. Lortie, 1987, A biomechanical approach to low-back problems in nursing aides. In: Trends in Ergonomics and Human Factors IV-part B. edited by S.S. Asfour, (Oxford: North-Holland), 795-802.

Hubley-Kozey, C.L., Westers, B.M., Stanish, W.D. and Wall, J.C., 1985, An investigation into the incidence of low back pain in hospital workers. The Nova Scotia Medical Bulletin, February, 8-10.

McGill, S.M. and Norman, R.W., 1985, Dynamically and statically determined moments during lifting. Journal of Biomechanics, 18(12), 877-885.

McGill, S.M. and Norman, R.W., 1987, Effects of an anatomically detailed erector spinae model on L4/L5 disc compression and shear. Journal of Biomechanics, 20(6), 591-600.

Niosh (1981), Work practice guidelines for manual lifting. Department of Health and Human Services, National Institute for Occupational Safety and Health, Cincinnati, Ohio.

Advances in Industrial Ergonomics and Safety II
Edited by Biman Das
Taylor & Francis 1990

POSTURE AND PAIN: HEALTH EFFECTS AND ERGONOMICS INTERVENTIONS

Elsayed Abdel-Moty, Ph.D.
Tarek Khalil, Ph.D., P.E.
Myron Goldberg, Ph.D.
Renee Rosomoff, R.N., M.B.A.
Hubert Rosomoff, M.D., D. Med Sci.

University of Miami
Comprehensive Pain and Rehabilitation Center
South Shore Hospital and Medical Center
600 Alton Road, Miami Beach, Florida 33139 USA

This paper demonstrates the effects of faulty postures and poor work habits on health. A case study is used to represent a typical example of patients who sustain chronic pain due to cumulative stress. During patients's rehabilitation, Ergonomists intervened by conducting job analysis/simulation in order to identify the risk factors and implement appropriate corrective measures. The relationship between postural adjustments and health is discussed. Ergonomics evaluation and intervention methods are detailed. Results and their implications for prevention are also presented.

INTRODUCTION

The lumbar and cervical regions of the spine are common sources of pain and suffering. Low back pain (LBP) affects about 80% of people in industrialized countries, accounts for 25% of all disabling injuries, and consumes about 80% of total health care expenditures associated with musculoskeletal injuries. Frequently, pain involves, not only the back and neck, but also the arms, shoulders, legs, as well as the head resulting in a rather complex disability (a combination of chronic pain, work loss, sickness, hospitalization, compensation, and inability to lead normal life). CailIiet (1984) categorized the potential sources of LBP as follows: 1) soft tissue injury 2) dysfunction of spinal structures; 3) discogenic / neurological factors; and 4) disorders such as infection, metabolic and congenital. Of these, the Quebec study (SAAASD, 1987) cited injury to the soft tissue (tension, sprain, spasms, weakness, strain, contracture, inflammation); as the most common cause of pain. This type of injury can result from activities such as lifting heavy objects, slipping and falling, bending and twisting, driving, and accidents; with slipping and falling incidents accounting for about 40% of the cases (Khalil et al., 1984; Abdel-Moty et al., 1990).

Soft-tissue injury can also develop gradually as a result of the "wear and tear" that takes place when daily activities are carried out forcefully, awkwardly, or repetitively without considering human capabilities and limitation. This is being referred to as Cumulative Trauma Disorders (CTD). It is now being widely accepted that it is not only the level of physical exertion that causes LBP, rather it is the manner by which work activities are carried out. Our data have shown that those who perform light-type job tasks (sedentary) are equally likely to sustain back injury as those who perform heavy labor (Khalil et al., 1984; Abdel-Moty et al., 1990). Moreover, in about 20% of the time, patients with chronic LBP were not able to attribute their painful condition to a specific incident, and indicate that pain developed gradually over time due to unknown causes. Frequently, this group of patients have a common characteristic: their job tasks require them to perform such activities as excessive bending, twisting, reaching, and severe deviation from neutral postures repetitively.

Any approach to the management of this type of injury should emphasize the reduction of musculoskeletal stresses as the prerequisite to controlling re-injury and reducing pain and suffering. The Ergonomics body of knowledge provides numerous guidelines and considerations that deal with both sides of this subject: the design of the environmental, and the capabilities of its user. A poorly designed environment can result in poor work habits by not allowing the operator to assume good posture and use proper body mechanics. It may place undue stresses on the musculoskeletal system and sensitive body tissues due to static loading. Awkward repetitive movements can be hazardous to health, and consequently can affect productivity and the quality of life in general. On the other hand, a properly designed workplace does not assure stress free environment if the user does not practice good posture and proper body mechanics.

Engineering considerations deal with the design and layout of the physical workplace (equipment, tools, and workstations) with special attention to human characteristics. Poorly designed work stations are known to contribute to pain, trauma, and accidents by not allowing the operator to perform safely. Good posture is achieved through the alignment of the musculoskeletal system in its neutral, balanced position. Poor or awkward postures cause fatigue, strain, and eventually pain, and should be corrected. Poor posture can also result in loss of stability, slips, falls, and other related accidents. Faulty postures and poor alignment of the musculoskeletal system develop slowly and do not induce an immediate feeling of discomfort. Poor posture can develop due to obesity, weakened muscles, emotional tension, poor postural habits, and sometimes due to the improper design of work tools and job tasks. Body mechanics (also known as posture in motion) is the safest way to perform activities of daily living (sitting, standing, lifting, carrying, squatting, climbing stairs, etc.). Improper use of body mechanics taxes the muscular system, thus leading to poor skeletal alignment and faulty postures. Ideally, all three components: engineering of the workplace; postures of the operator; and body mechanics for performing job tasks; have to be integrated in ergonomic job design and analysis.

The case presented here demonstrates the value of implementing these principles in the design of workplaces and job tasks in order to reduce stresses on people who already have injury.

CASE PRESENTATION

This is a 42 year old female. She is a biology research scientist (biochemist) with a State University. She is right handed, stands 65 inches tall, weighs 128 pounds and uses eye glasses. In 1973 she began to develop (no accident involved) episodes of LBP with pain extending down the left leg. She was treated with bedrest for 2 weeks. In 1979 she developed right LBP which remained intermittent thereafter for several months. Pain was increased by activity. In 1983 she had a major exacerbation with pain bilaterally in the low back, again treated with bedrest and medications. In 1987 she received physiotherapy 2 times a week for six weeks. In 1988, she entered the CPRC for evaluation and treatment. Upon admission, she reported pain in the low back and left leg. She reflected:

"pain is almost constant in low back - both sides hurt but not together - when pain is on the right side then right knee and arch of right foot hurt - when pain is on left side, calf and heel of left leg hurt - activity involving standing increases the pain - back pain usually is not 'bad' at night - I start the day relatively fine but it (the pain) gets worse as the day goes on - I would like to be able to stand up and play tennis again!". She clearly recognized that she has been depressed, and angry.

Initial neurological examination showed normal gait, good equilibrium and coordination, intact motor and sensory system, equal and active reflexes bilaterally, good ranges of back motion, tenderness over the right ischial tuberosity, and tight hamstring and hip rotators. The impression gathered was that there was no evidence of active nerve root compression, no surgical indications, and she was given the diagnosis of lumbar myofascial syndromes, the type of condition which responds to aggressive physical medicine program to which she has not been exposed to prior to her admission to the CPRC. Upon admission to the rehabilitation program at the CPRC, her treatment program consisted of : physical therapy, occupational therapy, vocational rehabilitation, behavioral modification, detoxification, and ergonomics consultation for Functional Capacity Assessment (FCA) and Ergonomic Job Analysis (EJA) (Rosomoff et al., 1981; Khalil et al., 1985). The EJA is the scope to this paper.

ERGONOMIC METHODS OF EVALUATION AND INTERVENTION

Ergonomics Job Analysis (evaluation and intervention methodologies) consisted of 4 components: preliminary data collection, an initial evaluation, an intervention phase, and a final evaluation. These are presented below.

<u>1. Data Collection</u>

During the first week of patient's admission to the CPRC an initial evaluation was performed in order to obtain both individual and job-related information. The components of this evaluation were:

a. basic biographic information and physical characteristics of the individual: such as age, weight, height, occupation, cause and course of injury, and employment status. Summary of this information was given previously.

b. Qualitative description of the physical environment within which the job was performed: In this category the layout of the workplace and equipment parameters (locations, dimensions, priorities, frequency of use, etc.) were obtained through patient's description. She was asked to obtain photographs of the work environment.

c. Patient's self-description of a "typical" work day: From this description, job tasks were identified, the average time required to perform each task was estimated, and the sequence of tasks performance was determined. She described the day as follows: meeting with staff, paper work, telephone calls in one office, word processing and data analysis in another office, in addition to general laboratory activities. Tasks were classified as follows:

<u>Overall:</u> sitting - 60% standing - 30% walking - 10%
<u>While Sitting:</u>
writing - 30% typing - 10% meeting - 5%
talking on the phone 10% working with VDT 40%
Viewing cells using microscope - 5%
<u>While Standing and Walking</u>
reaching - 30% lifting - 10% carrying - 20%
Other lab activities - 40%

<u>2. Initial Evaluation</u>

The purpose of this evaluation was to analyze patient's performance, especially muscular work. At this stage, the job (the physical environment as well as the task) was simulated and assessment was performed as follows:

a. Task/job simulation: The information gathered from the initial evaluation was used, in conjunction with the patient's input, to construct a workplace that closely simulates the actual work station. It was recognized that exact reproduction of the environment is impossible since there are many factors (heat, noise, fumes, etc.) that are difficult to emulate. The goal here was to analyze patient's patterns of movement (body mechanics) within the environment. The bench, the office, and the computer work stations were simulated.

b. Analysis of Task Performance: In order to evaluate the manner by which the patient performed job tasks prior to admission to the CPRC, the patient performed tasks in the simulated environment which were similar to those encountered on the job. Emphasis was on activities that are done repeatedly or for long periods of time such as VDT tasks, viewing the microscope, talking on the phone, doing regular desk tasks, and performing bench-type activities (sampling, test tubes, mixing compounds, etc.). The goal was to identify the tasks which contribute more to stress production. For objective evaluation of performance, electrical activity of the lower back muscles were recorded (Fig. 1a). No feedback or suggestions for modifications or adjustments were given at this stage. Patient's self-report of discomfort/pain at the beginning and end of this 30 minutes evaluation session were obtained. Overall, she reported an increase in pain level from an initial 6/10 to about 9/10 at the end of this session. The overall average EMG level during this simulation was 4.2 uV*sec.

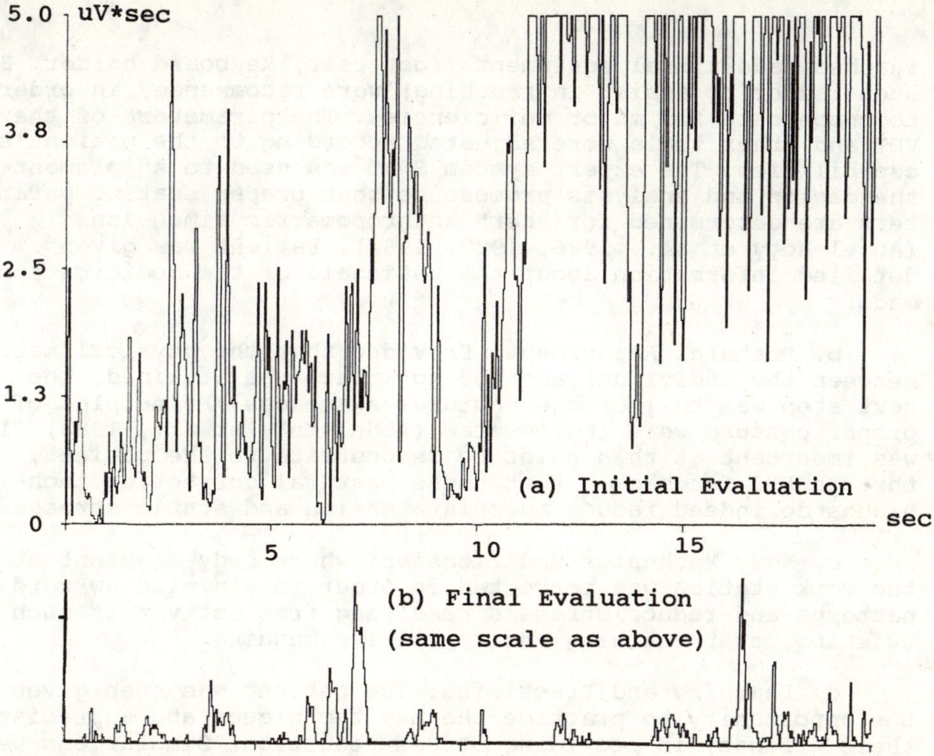

* Saturation at this level of amplification.

Fig. 1 EMG activity of right lumbar paraspinals during the initial (a) and final (b) evaluations.

c. Identification of the Risk Factors: While the simulated tasks were being performed, the evaluator observed patient's body mechanics and postural adjustment, identified critical motion patterns, and detected any mismatch between the patient's physical characteristics and the dimensions of the workplace. Also levels of EMG activity were correlated with the different movements. Inadequate elbow rest and chair parameters in general while sitting, work surface and keyboard heights relative to elbow height, lack of foot rest while sitting or standing, neck postures while using the microscope and while on the phone, over reaching to cabinet, posture while sitting on a high stool to work at the bench, and VDT screen height were all identified as potential problem areas.

3. Ergonomic Intervention

Following the identification of the risk factors, the intervention included the following strategies:

a. Reengineering of the Workplace: where sitting heights were adjusted based on anthropometric dimensions of the individual and work surface height, the layout of the work station was modified so that all tools were within functional reaches, additional equipment (foot rest, keyboard holder, 3-step ladder to assist in reaching) were recommended in order to compensate for major deficiencies. The parameters of the VDT and other tools were adjusted according to the patient's capabilities. The expert system SWAD was used to supplement the design and analysis process so that proper seating parameters are determined for "her" anthropometric dimensions (Abdel-Moty et al., 1986, 1987, 1988). Patient was given detailed information about the rationale of the modifications made.

b. Postural Adjustment: Provided that the physical match between the individual and the workplace was attained, the next step was to increase postural awareness. Principles of proper posture were implemented (Abdel-Moty et al., 1988). It was important at this point to demonstrate to the patient, through EMG feedback, that these postural correction techniques do indeed reduce muscular tension and static stresses.

c. Body Mechanics Modification: where body movement at the work station was corrected in order to minimize awkward patterns and reduce stresses resulting from activities such as twisting, over reaching and repetitive bending.

d. Learning and Practicing: The patient was then given the opportunity to practice the new techniques and appreciate their efficacy in providing stress reduction. Simulations were repeated and body mechanics were refined. Most often this type of activity was performed as a joint effort between ergonomics, occupational therapy, vocational rehabilitation, and biofeedback. The patient was taught to review work activities, execute her job tasks properly, receive suggestions for modifications or adjustments, and implement recommendations in the simulated environment. All these activities assured the patient that she is capable of carrying them out when she returns to the actual work site.

The impact of these adjustments, modifications, and education on the muscle activity of the lumbar paraspinals muscles is shown in Fig. 2 (a-d) and is summarized in Table 1. No doubt, there was dramatic reduction in muscle work upon the application of each intervention strategy.

4. Final Evaluation

A second evaluation was performed in order to determine, quantitatively, the changes in the outcome measures (EMG and self-report of pain) as a result of intervention. This evaluation session lasted 30 minutes. The overall average EMG value was 0.95 uV*sec (Fig. 1b and Table 1) reflecting a tremendous reduction in muscle activity as compared to the initial average level (4.2 uV*sec) while performing the same tasks. The combination of intervention and learning new techniques has allowed the patient to work at much lower physiological demand in the same environment. Furthermore, she reported no significant change in pain level at the end of the final evaluation session (initial evaluation increased pain by 50%).

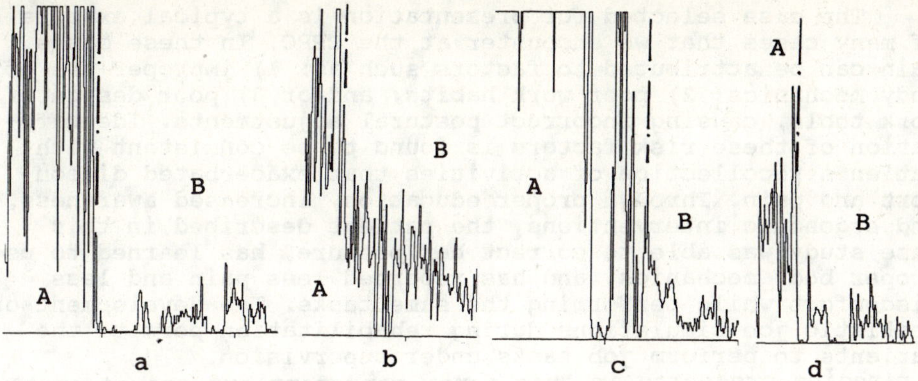

(a) using microscope (b) using keyboard
(c) reaching overhead (d) sitting on chair

A - No Intervention B - After Ergonomic Intervention

Fig. 2 Examples of EMG activity of the right lumbar paraspinals during before and after ergonomic intervention.

Table 1. Average Level of the Electromyographic Activity (in micro volt * sec) of the Right Lumbar Paraspinals Muscles during two type of Activities.

	Average EMG Level	
	Static*	Dynamic**
1. No Intervention	1.3	4.4
2. Re-Engineering of the Workplace	0.5	1.6
3. Correcting Posture	0.1	1.1
4. Employing Proper Body Mechanics	NA	0.9

* Static: Assuming selected work posture, no tasks performed;
** Dynamic: Performing job tasks.
NA Condition not applicable

SUMMARY AND CONCLUSIONS

The relationship between posture and musculoskeletal stresses is a rather complex one due to the large number of human and environmental factors involved. It is now recognized that poor and awkward postures can result in a multitude of health problems ranging from aches and pain to permanent deformation of the skeletal system.

Ergonomic contributions to the prevention and management of this type of injury have proven to be most valuable. Application of ergonomic guidelines for proper posture, task performance, and workplace design can lead to significant pain reduction. This can contribute to a reduction in medical and compensation costs. Ergonomic intervention methods in clinical settings are effective and valuable to the comprehensive rehabilitation of chronic pain patients. The intervention

The case selected for presentation is a typical example of many cases that we encounter at the CPRC. In these cases, pain can be attributed to factors such as: 1) improper use of body mechanics; 2) poor work habits; and/or 3) poor design of work tools, causing incorrect postural adjustments. Identification of these risk factors is found to be consistent with patients recollection of activities that exacerbated discomfort and pain. Through proper education, increased awareness, and ergonomic interventions, the patient described in this case study was able to correct her posture, has learned to use proper body mechanics, and has reported less pain and less discomfort while performing the same tasks. The development of realistic job simulations during rehabilitation permits the patients to perform job tasks under supervision. strategies presented in this paper represent but one of ergonomics contributions to the management of LBP. The inclusion of human engineering post treatment is needed to prevent aggravation of the problem.

REFERENCES

Abdel-Moty, E., Khalil, T.M., 1986. Computer aided design of the sitting workplace. Computers and Industrial Eng, 11,23-26.

Abdel-Moty, E., Khalil, T.M., 1988. A computerized expert system for work simplification and workplace design. In Expert Systems, edited by N.A. Botton and T. Raz (Industrial Engineering and Management Press), pp. 221-226.

Abdel-Moty, E., Khalil, T.M., 1987. Microcomputers in the design and analysis of the VDT sitting workplace. In Trends in ergonomics/human factors, edited by S.S. Asfour (North Holland: Elsevier Science Publishing B.V.), pp. 113-120.

Abdel-Moty, E., Khalil, T.M., Asfour, S.S., Rosomoff, R.S., Rosomoff, H, 1988. Ergonomics considerations for the reduction of physical task demands of low back pain patients. In Trends in ergonomics/human factors, Vol. V, edited by F. Aghazadeh (North Holland: Elsevier Science Publishing B.V.), pp. 959-967.

Khalil, T.M., Asfour, S.S., Abdel-Moty, E., 1985. New horizons for ergonomics research in low back pain. In Trends in ergonomics/human factors edited by R.E. Eberts and C.G. Eberts (North Holland: Elsevier Science Publishers B.V.), pp. 591-598.

Khalil, T.M., Asfour, S.S., Abdel-Moty, E., Steele, R., Rosomoff, H.L., 1984. The management of low back pain: A comprehensive approach. In: Proceedings of the Annual Industrial Engineering Conference, Louisville, pp. 199-204.

Khalil, T.M., Asfour, S.S., Abdel-Moty, E., Rosomoff, R.S., Rosomoff, H.L., 1988. Ergonomics Contribution to Low Back Pain Rehabilitation. Pain Management, Sep/Oct, 225-230.

Khalil, T.M., Goldberg, M.L., Asfour, S.S., Moty, E.A., Steele, R., Rosomoff, H.L., 1987. Acceptable maximum effort (AME): A psychophysical measure of strength in back pain patients. Spine, 12, 4, 372-376.

Rosomoff, H.L., Green, C., Silbert, M., Steele, R., 1981. Pain and low back rehabilitation program at the University of Miami School of Medicine. In New approaches to treatment of chronic pain : A review of multidisciplinary pain clinics and pain centers, edited by K.Y. Lorenzo (NIDA Research Monograph 36, Department of HHS).

Advances in Industrial Ergonomics and Safety II
Edited by Biman Das
Taylor & Francis 1990

A SUCCESSFUL APPLIED ERGONOMICS PROGRAM FOR PREVENTING OCCUPATIONAL BACK INJURIES

David T. Ridyard, CIH, CSP

Corporate Safety & Industrial Hygiene
Merck & Co., Inc.
West Point, PA 19486

Back injuries resulting from overexertion were the major cause of lost workdays at a chemical plant. An applied ergonomic methodology was used to recommend appropriate engineering or work practice controls for 59 manual materials handling tasks. Significant ergonomic improvements were implemented at the site, and job-specific materials handling training was provided to all manufacturing employees. A review of the normalized back injury data for an 8 year follow-up period shows that the incidence of overexertion back injuries has been reduced 62%. More importantly, there have been no lost-workday back injuries at the plant during the same follow-up period.

INTRODUCTION

The National Safety Council (NSC,1989) reports that back injuries account for approximately 22% of all occupational cases involving disability in the United States. Based on this percentage, approximately 400,000 disabling back injuries occur at work each year. The NSC estimates that the cost to industry ranges from $6 billion to $8 billion in direct compensation, legal expenses, employee training, and other administrative costs. Nelson (1987) cites figures reported by the 1984 International Conference on Injuries in the Workplace that work-related injuries and their associated costs amounted to an estimated $14 billion per year.

A review of the 1979-1981 Worker's Compensation records at a chemical company located in Western Pennsylvania indicated that back injuries associated with materials handling tasks were the major cause of lost workdays. For this reason, a comprehensive back injury prevention program was started at the site in 1982. The objectives of the program were:

1. to identify the manual materials handling tasks that may be associated with potential overexertion hazards;
2. to evaluate the relative degree of overexertion risk for each identified materials handling task; and,
3. to recommend appropriate engineering or work practice controls to either eliminate the manual handling task or reduce the degree of exertion to acceptable levels.

An applied ergonomic methodology was used to prevent back injuries by redesigning hazardous manual materials handling tasks to control the overexertion hazard. Snook (1978) discusses the effectiveness of the ergonomics approach to controlling back injuries. His analysis of 191 low back injuries indicated that, at best the ergonomics approach can reduce back injuries associated with heavy materials handling tasks by 67%. This can be accomplished by designing materials handling tasks so that more than 75% of the working population can perform the task without overexertion.

Background

This back injury prevention program was conducted at a 36 acre chemical plant located in Western Pennsylvania. The main products are liquid, polymer, or dry powder water treatment chemicals packaged as 30 and 55 gallon (200-600 pound) drums, or as 50 and 100 pound bags. Some products are shipped as bulk quantity liquids. Production (measured as pounds of product produced per year) has increased approximately 65% since 1982.

There were approximately 90 hourly employees in 1982 and employment increased to the current (December, 1989) level of 100. The average age of the hourly workforce was 31 in 1982, and 38 in 1989. The job classifications (production operators, mechanics, shippers, and general service workers) have remained constant since 1982, as have the average 8 hours per week (4 hours extra per shift, twice per week) of overtime work. Most employees are healthy males. Approximately 51% of the employees are production operators, 25% are mechanics, 18% are shippers, and 6% are general service workers.

The back injury prevention program was started when an injury analysis showed that approximately 22% of all injury claims; 47% of all lost-time injuries; and 64% of all lost-workdays during the 1979-1981 time period were the result of back injuries.

METHODS AND PROCEDURES

The program was designed to utilize the problem solving abilities of all employees at the site. Employees worked with the Corporate Safety & Industrial Hygiene Department to identify, evaluate and eliminate the overexertion hazards. In most cases, the control measures were suggested by hourly employees, supervisors, and managers at the site. The program consisted of the following eight general steps:

1. Analyzing Injury Data

Worker's Compensation accident data were reviewed for a three year time period (1979-1981) to determine the incidence, severity and distribution of materials handling back injuries at the site. The following information was recorded for each back injury:

- Worker's Compensation file number, and date filed;
- Employee name, job title, and department;
- Brief description of the accident;
- Accident type defined as lift, lower, push, pull, carry;
- Specific overexertion job stress factors such as "lifting 100# bags"; pulling 500# drums; or pushing frozen valve";
- Number of lost workdays if applicable.

There were 7.8 overexertion-related back injuries associated with materials handling activities per 100 manufacturing employees. Slip/trip/fall injuries not associated with materials handling activities (ie., back injuries from falls while walking to the parking lot, falls from ladders, etc.) were not included in the statistics. Note that slip/trip/fall injuries accounted for 19% of all back injuries at the site.

Seventy-one percent (71%) of the back injuries occurred while lifting bags, drums, pumps, and other materials weighing between 40 - 660 pounds. The remaining 29% occurred while rolling drums, pushing pallets of bags on floor conveyors, and pushing to tighten down a filter press.

Fifty-two percent (52%) of the back injuries occurred to chemical operators, 26% to shippers, 15% to mechanics, and the remaining 7% to general service workers. The relative risks of overexertion back injuries (percent injuries / percent population) for the four job classifications are shown in Table 1. Based on this analysis, all job classifications were included in the program.

Table 1. Relative risk of overexertion back injuries for each job classification, when compared to all four job classifications at the site.

Job Classification	Percent of Injuries	Percent of Population	Relative Injury Risk
Operators	52%	51%	1.02
Shippers	26%	18%	1.44
Mechanics	15%	25%	0.60
Service Worker	7%	6%	1.17

2. Interviewing Hourly and Salaried Employees

The list of task variables (see step 4, below) was reviewed with supervisors and employees to collect additional information regarding the injuries which occurred from 1979 to 1981. In addition, the list was reviewed with supervisors and employees to identify overexertion hazards associated with current materials handling hazards. The interviews were conducted in the employees' work areas, and care was taken to avoid placing blame for injuries.

3. Observing Materials Handling Activities

The next step of the hazard identification process was to observe the 59 materials handling tasks classified as potential overexertion hazards during the injury analysis and employee interviews.

4. Evaluating the Overexertion Risk for each Task

The following task variables, which are summarized by Ayoub et al. (1978) and the AIHA Ergonomics Guide to Manual Lifting (1970), were collected for each of the tasks:

- weight, size, stability, and configuration of the object;
- height, distance, rate, and rhythm of lift, lower, push, or pull;
- initial and sustained pushing and pulling forces;
- distance of carry;
- Frequency of task (daily, weekly, monthly or yearly);
- Postural constraints (reaching, bending and twisting).

The task variables were analyzed to determine the action limit (AL) weight and the maximum permissible limit (MPL) weight for all symmetric, two-handed lifts per the NIOSH Work Practices Guide for Manual Lifting (1982).

The task variables associated with all lowering, pushing, pulling, and carrying tasks were analyzed using the tabled psychophysical data reported by Snook (1978).

The use of the NIOSH (1982) and Snook (1978) methods proved to be an easy, effective mechanism for classifying the tasks as low, medium, or high exertion hazards. The following example demonstrates how the NIOSH method was used to analyze lifting task variables:

Example Task: An employee lifts twenty (100 #) bags from floor level to a 38" high conveyor. One bag is lifted approximately every 6 minutes. The bags are lifted with a smooth, symmetric (two-handed) lift.

The NIOSH Action Limit (AL) weight, which is defined as the weight that greater than 75% of women and 99% of men in industry can lift without overexertion, is calculated using the following equation:

$$AL(lb) = 90(6/H)\,(1-0.01\,|V-30|)\,(0.7+3/D)\,(1-F/F_{max}) \qquad (1)$$

where:

- H = horizontal location (inches) forward of the midpoint between the ankles at origin of lift;
- V = vertical location (inches) at origin of lift. Note the absolute value sign for the "V-30" value;
- D = vertical travel distance (inches) between origin and destination of lift;
- F = average frequency of lift (lifts/minute); and
- F_{max} = the maximum frequency which can be sustained per the data in Table 2.

Table 2. Maximum Frequency (F_{max}).

	Average Vertical Location (V > 30")	(V < 30")
Less than 1 hr. lifting period	18	15
More than 1 hr. lifting period	15	12

In the Example Task, H = 15" since the bags were held close to the body; V = 3" since the hand grip was located at the midpoint of the 6" high bags, and the bags were lifted from floor level; D = 38" since the bags were lifted onto a 38" high conveyor; F = 0.17 lifts/min.; and F_{max} = 12, per Table 2, since the lifting period exceeded 1 hour, and the average vertical location was less than 30". Solving for the action level:

$$AL\ (lb) = 90(6/15)(1-0.01|3-30|)(0.7+3/38)(1-0.17/12);$$
$$AL\ (lb) = 90(0.4)(0.73)(0.78)(0.96);$$
$$AL\ (lb) = 19.7$$

The NIOSH Maximum Permissible Limit (MPL) weight, which is defined as the weight that less than 1% of women and 25% of men in industry can lift without overexertion, is set at three times the Action Limit (AL) weight.

The MPL in the Example Task = 3(AL), = 3(19.7#), = 59.1#. (2)

The Example Task is classified as a high hazard task since the weight of the bag (100#) exceeds the NIOSH MPL weight of 59.1#. Note that this task would be classified as a medium hazard if 50# bags were lifted since the bag weight would fall between the NIOSH AL and MPL weights. This, in fact, may be the most approprite method for redesigning this high hazard task.

The tasks were classified as high, medium, or low overexertion hazards per the hazard classification system in Table 3.

Table 3. Hazard classification System for the 59 Manual Materials Handling Tasks.

Hazard Class	Analysis of Task Variables
High Hazard	• Weight lifted exceeds the NIOSH MPL. • Less than 25% of the working population can lower, push, pull or carry the load.
Medium Hazard	• Weight lifted is less than NIOSH MPL, but greater than the NIOSH AL. • More than 25% of the working population can lower, push, pull or carry the load.
Low Hazard	• Weight lifted is less than the NIOSH AL. • More than 75% of the working population can lower, push, pull or carry the load.

5. Reviewing Data with all Employees

The hazard classifications were reviewed at a meeting with hourly employees, supervisors and managers in production, maintenance, and engineering. This meeting was used to verify data, collect additional data, and solicit ideas regarding feasible task modifications to reduce the risk of overexertion.

6. Assigning a Numerical Hazard Ranking to Each Task

Each task was then assigned a numerical hazard priority ranking between 1 and 59. The hazard classifications, as well as more subjective information including the amount of twisting, reaching, bending, and the yearly frequency of the task, were used to rank the hazards.

7. Recommending Corrective Actions
The following corrective actions were recommended:

- The high hazard tasks should either be eliminated with an automated handling system, or redesigned to a lower hazard task by reducing the amount of physical exertion.
- The medium hazard tasks should be redesigned to a lower hazard task, if possible. Employees should receive job-specific materials-handling training.
- No actions were recommended for the low hazard tasks.

8. Implementing Corrective Actions
The final step of the program was the implementation of automated handling systems, task modifications, job-specific training and other controls. Management used the numerical hazard rankings to prioritize corrective actions.

RESULTS

The methodology used in this program resulted in the classification of 25 high, 25 medium, and 9 low hazard tasks.

Table 4 lists the types of work activities and task variables associated with each classification of hazards. The most common high hazard tasks involved lifting 100# bags, rolling and palletizing 400# to 650# drums, or pulling pallets of materials with hand trucks. The most common medium hazard tasks involved lifting 50# bags, or opening reactor lids. Low hazard tasks involved lifting and carrying light (<20#) loads.

Table 4. Summary of the 59 materials handling tasks.

Hazard Class	Types of Work Activities and Task Variables
High Hazard	• Lifting and lowering 100# bags. • Rolling and palletizing 450#-600# drums. • Pushing and pulling hand trucks with a fully-loaded (2400#) pallet of bags. • Lifting and lowering motors (40#-250#) in the maintenance shop motor room. • Lifting/positioning 92# dryer end-boards.
Medium Hazard	• Lifting and lowering 50# bags. • Opening and closing reactor lids.
Low Hazard	• Lifting and carrying 15# pails. • Lifting small 20# drums.

Corrective actions

Table 5 summarizes the ergonomic improvements implemented between 1982-1985 for the 25 high hazard tasks.

The task of lifting and lowering 100# bags at the product filling lines was eliminated by installing a hydraulic pallet loading turntable that rotates 90 degrees each time a bag is added to the pallet. The turntable also maintains the pallet at a constant height so the operator can slide the bag across a stainless steel table onto the pallet. The handling of 100# bags was totally eliminated in some areas by installing pneumatic transfer systems for powders, and by reducing the weight of certain 100# bag products to 50#.

The task of rolling and palletizing 450# to 600# drums at the filling areas was eliminated by installing 4-drum weigh scales at a new liquid blending area. The drums are now moved on pallets with a lift truck. In addition, drum handling equipment was purchased to lift drums of raw materials.

Electric lift trucks were installed to eliminate the task of pushing and pulling fully-loaded (2400#) pallet hand trucks.

A hydraulic lift was purchased for the maintenance shop to lift and lower motors, pumps, valves and other heavy pieces of equipment. The amount of bending and reaching required to retrieve heavy objects in the maintenance storage areas was reduced with improved housekeeping.

The task of lifting and positioning 92# end-boards on rotary dryers was not eliminated with engineering controls. Attempts to reduce the weight of the end-boards or provide mechanical assistance proved infeasible due to material specifications and inaccessibility of the work area. This task is performed approximately 6 times per year and, due to production reasons, is scheduled for elimination. The task is now performed by two workers using the good materials handling techniques outlined in a training session by their supervisor.

Table 5. Ergonomic improvements for the 25 high hazard tasks.

Type of Ergonomic Improvement	Total #
• Purchased hydraulic pallet turntables, drum dumpers, electric hand trucks, and other materials handling equipment.	7
• Eliminated production at areas with poor layout or inadequate handling equipment.	6
• Installed fully automated materials handling systems for powders and liquids.	4
• Completed miscellaneous engineering projects (improved maintenance & storage)	4
• Reduced weight of bags and drums from 100# to 50#.	3
• Implemented administrative controls (training, assigned additional worker)	1

Some of the medium hazard tasks have been eliminated by using the same materials handling equipment and systems which were installed for the high hazard tasks (ie., 50# and 100# bags are generally handled on the same product filling lines). Employees performing the medium hazard tasks were included in task-specific training programs to develop the skills and techniques required to perform the task without overexertion.

DISCUSSION

The methodology used in this program resulted in a systematic identification and analysis of the overexertion hazards associated with materials handling tasks. The program was effective in reducing both the incidence and severity of back injuries at this chemical plant.

Figure 1 shows the 62% average reduction in the incidence of total overexertion back injuries which occurred at the site since the program was started in 1982. In addition, severe back injuries resulting in lost workdays have been eliminated during the same follow-up period (1982-1989).

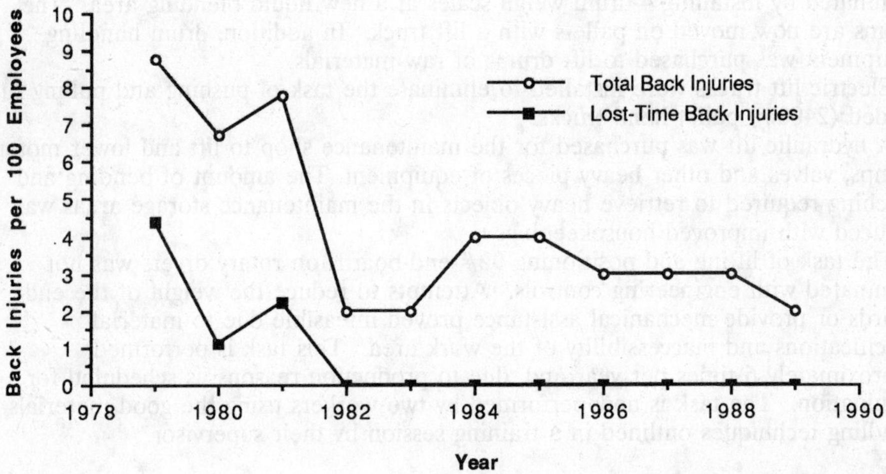

Figure 1. Annual incidence of overexertion back injuries for the time period 1979-1989.

The 62% reduction in the normalized back injury incidence rate was accomplished by eliminating all but one high hazard task and most of the medium hazard tasks. The magnitude of this reduction appears to be consistent with those estimated by Snook (1978). However, it is important to note that all tasks were not redesigned to the low hazard category that served as the basis for Snook's (1978) estimations.

REFERENCES

American Industrial Hygiene Association, 1970, "Ergonomic Guide to Manual Lifting." AIHA Ergonomics Guide Series.

Ayoub, M.M., Dryden, R.D., McDaniel, J.W., Knipfer, R.E., and Aghazadeh, F., 1978, Modeling of lifting capacity as a function of operator and task variables. In: Safety in Manual Materials Handling, edited by C.G. Drury. NIOSH Publication No. 78-185, 1978a. pp. 120-130.

National Institute for Occupational Safety and Health, 1981, Work Practices Guide for Manual Lifting. Technical Report No. 81-122, Cincinnati, Ohio.

National Safety Council, 1989, "Accident Facts".

Nelson, R.M., 1987, Prevention - a government perspective. Ergonomics, 30:2, 221.

Snook, S.H., 1978, The design of manual handling tasks. Ergonomics, 21, 963-985.

III

BIOINSTRUMENTATION AND ELECTROMYOGRAPHY

Advances in Industrial Ergonomics and Safety II
Edited by Biman Das
Taylor & Francis 1990

COMPARISON BETWEEN THE ONE AND TWO INCLINOMETER TECHNIQUES FOR MEASURING BODY RANGES OF MOTION USING THE ORTHORANGER II

A. Zaki, M. Goldberg, T. Khalil, J. Jarett, E. Rodriguez, A. Linial, R. Rosomoff, and H.L. Rosomoff

Comprehensive Pain and Rehabilitation Center
University of Miami

The objective of the present study was to develop and assess a one inclinometer technique that could substitute for the two inclinometer method recommended by the American Medical Association for measurement of spine segments range of motion. The reliability of the two techniques were compared utilizing the Orthoranger II electronic inclinometer. The results showed that both methods had the same level of variability. It was concluded that the one inclinometer technique can be substituted validly for the two inclinometer technique, without loss of reliability.

INTRODUCTION

Measuring the ranges of motion of different body segments is used in the determination of the degree of physical impairment, the development of treatment goals, and the evaluation of patient progress in treatment and rehabilitation programs. The Universal goniometer has been traditionally the most commonly used device to measure joint ranges of motion. The popularity of the Universal goniometer is due to its versatility, ease of use and low cost.

Universal goniometric measurements have been a topic of considerable concern, and several studies have been conducted to investigate its reliability. The results varied considerably. Some studies reported high reliability, and others reported low reliability. Hellebrandt et al (1949) indicated that the Universal goniometer is a dependable device in measuring joint motions. Rothstein et al (1983) examined the reliability of three Universal goniometers (large metal, large plastic, and small plastic) to measure the flexion and extension of the knee and elbow joints. They found that the three devices had high intratester and intertester reliabilities for all the measurements except for the knee extension. However, Wilmer et al (1947) indicated that up to 10 degrees error exists in the Universal goniometer measurements. Likewise, Hewitt (1928) found that a 5 degree variation existed in measuring wrist joint motion; similar findings were reported by Cobe (1928). Fish et al (1985) studied the sources of goniometric error at the elbow. They reported that improper alignment of the goniometer, misidentification of bony landmarks, and variation in manual force all contributed to the goniometric errors. Petherick et al (1988) conducted a study to compare the Universal goniometer and a fluid-based goniometer in measuring the elbow range of motion. They recommended the use of the fluid-based goniometer over the Universal goniometer.

The inconsistency in the reported reliability for the Universal goniometer can be attributed to the variation in the measurement procedures and the design of the device. We observed in a study conducted by our group that the data reported from using the Universal goniometer were reported by therapists in increments of 5 degrees, a circumstance which reduces the discrimination and accuracy of measurements. Moreover, it is difficult to place the Universal goniometer accurately and repetitively to measure the range of motion of individual spine segments.

Problems associated with the use of the Universal goniometer, along with technological advances, have led to the development of new devices, such as electronic inclinometers. The new devices are intended to correct the drawbacks and limitations experienced with the Universal goniometer and, thus, are expected to improve the accuracy, discrimination and reliability of goniometric measurements. The Orthoranger II* is one such example. The reliability and accuracy of this device were tested by Pisciota (1987). Two studies were performed, a benchmark test at three known angles (60, 90, 150), and a comparative study with a Universal goniometers on 15 normal subjects. It was reported from the benchmark test that the device is accurate to within one degree. The investigator concluded from the comparative study that the Orthoranger II was more accurate and repeatable than the Universal goniometer. A benchmark study conducted by our group confirmed that the device was measuring with an accuracy of one degree, thereby enhancing discrimination as well.

The most recognized guidelines for range of motion measurements are found in the "Guides to the Evaluation of Permanent Impairment" published by the American Medical Association (A.M.A.). In earlier publications, the measurement of spinal range of motion utilized the Universal type goniometer. The A.M.A. modified the guidelines for range of motion measurements in its recent publication (1988). The guidelines specify new techniques for spinal range of motion measurements using mechanical inclinometers and electronic inclinometers. As recommended in the guidelines, one of two techniques may be used to complete the assessment of spinal ranges of motion. The techniques include a single inclinometer technique and the two inclinometers technique. The reported single inclinometer technique requires the use of a costly electronic inclinometer with memory capabilities. The two inclinometers technique allows the use of simple, less costly mechanical inclinometers or electronic inclinometers without memory capabilities.

---

* Courtesy of M.I.Technologies, Daytona Beach, Florida.
The Orthoranger II is a hand-held electronic inclinometer without memory capability. It is composed of a rectangular box (5"X 2.75"X 0.75") with a digital display, a zero button, and a power switch. The device weighs 1 lb.

The aim of the present study was to investigate the potential of a third technique for performing spinal ranges of motion measurement based upon the A.M.A. guidelines. It was felt that a third technique was needed in light of two factors: (1) the relatively high costs associated with the type of device required for the single inclinometer technique and (2) possible methodological difficulties associated with the two inclinometers technique. The methodological difficulties surfaced during the implementation of the two inclinometers technique in our Center, the Comprehensive Pain and Rehabilitation Center (CPRC) of the University of Miami. Both physicians and therapists reported consistently that holding two devices simultaneously was cumbersome, leading to spurious movement of the devices during measurement. In response to these difficulties and the need to keep costs to a minimum, the Bioengineering and Ergonomics Division of the CPRC in conjunction with M.I.Technologies, Inc. designed a new measurement technique which permits the use of only one inclinometer or electronic inclinometer without memory to complete the measures of spinal ranges of motion as prescribed in the A.M.A. guidelines. The new technique actually represents a modification of the two inclinometers technique specified by the A.M.A. In the present study, the reliability of the proposed technique was determined and compared to that of the two inclinometer method. A complete description of the new technique is given in the Methods section.

METHOD

Eleven subjects volunteered to participate in this study (7 males and 4 females). The subjects were healthy with no reported history of back pain or musculoskeletal disorders. The subjects had an average age of 30.81 ± 7.8 years.

All measurements were performed by two registered physical therapists. For the purposes of this study, four spinal ranges of motions were tested: cervical flexion; cervical right lateral flexion; lumbar flexion; and lumbar right lateral flexion. Measurements were performed by use of the Orthoranger II. Each measurement was repeated 3 times.

To eliminate the effect of stretching and varying subject's effort during a measurement, subjects were not allowed to reach their maximum range of motion for each maneuver instead, movements were stopped at fixed angles. This was used to ensure that each subject moved through the same angle every time the measurement was repeated. For a particular measurement, the testing was performed in a randomized order, i.e. the order of rater, maneuver, and measurement method were random.

The methods followed for the two inclinometer technique were those recommended by the A.M.A. (1988). They were the following:

<u>Cervical Flexion</u> - <u>Two devices Method:</u> (Fig. 1)

The subjects were asked to sit on a stool with the back straight, arms beside the body and the head in the neutral position. The T1 spinous process and the occiput were located. The skin was marked at the T1 spinous process. The two Orthorangers were aligned with the sagittal plane and were placed over the two landmarks. Both devices were zeroed while the head was still in the neutral position. The subject was

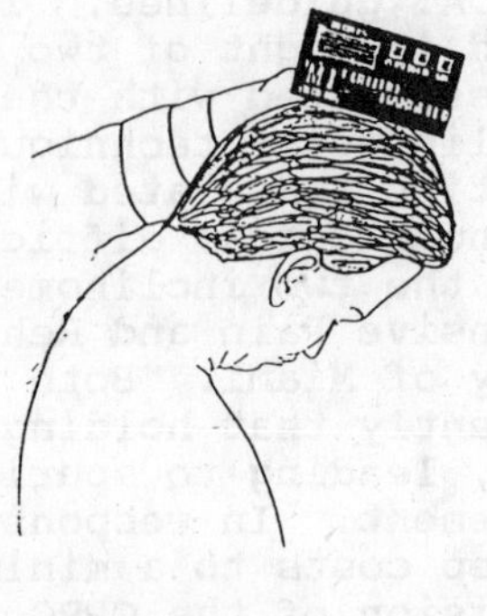

Fig.5: Cervical Flexion
One Device Method

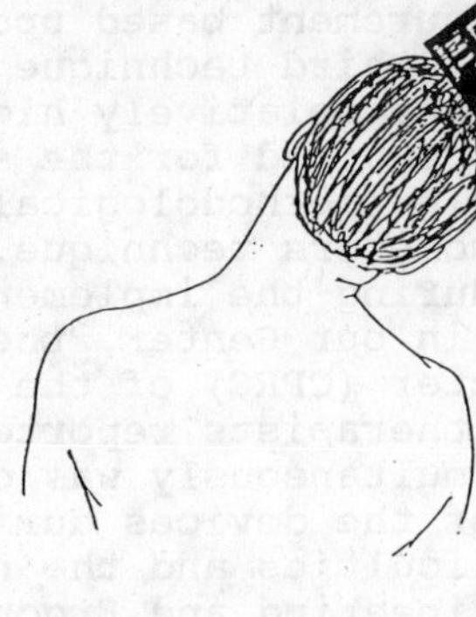

Fig.6: Cervical Lateral Flexion
One Device Method

Lumbar Lateral Flexion - One Device Method: (Fig. 8)

The subjects were positioned and the skin was marked as explained in #3 above. A ruler was placed over T12 aligned in the coronal plane perpendicular to the spine (the ruler should not be moved till the measurement was completed). The Orthoranger was placed on the upper edge of the ruler. The device was zeroed in this position and then was moved to the SMP landmark while still displaying a zero angle. The subject was asked to incline to the right and the angle at the SMP was recorded in this position [Angle 1]. The device was then moved to T12 and was realigned with the upper edge of the ruler (while the trunk was still inclined) and the angle was recorded [Angle 2]. The lumbar lateral flexion angle was calculated from this formula:
Angle of lumbar lateral flexion= Angle 2 - Angle 1

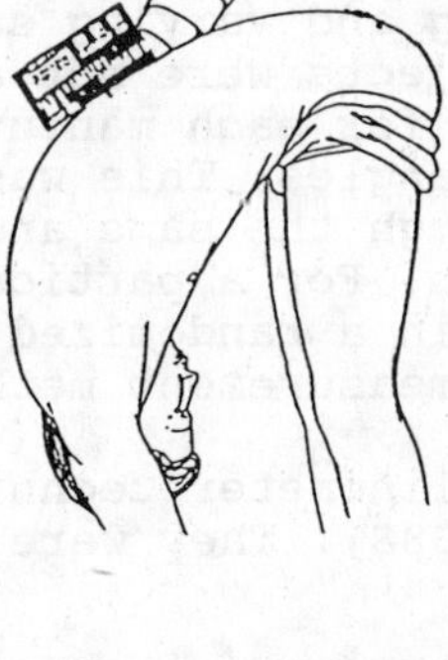

Fig.7: Lumbar Flexion
One Device Method

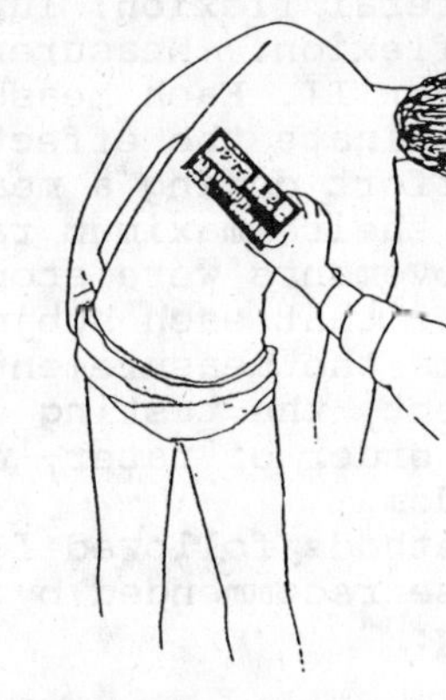

Fig.8: Lumbar Lateral Flexion
One Device Method

RESULTS

The data were analyzed to determine the comparability of the two methods. An analysis of variance (ANOVA) to determine if the main effects (Raters, Method, Repetitions) and their interactions differed from each other. Results from this

asked to flex the head forward to the predetermined fixed angle and the angles from the two devices were recorded. The Orthoranger reading at T1 was subtracted from the Orthoranger reading at the Occiput to obtain the angle of cervical flexion.

<u>Cervical Lateral Flexion</u> - <u>Two Devices Method:</u> (Fig. 2)

A procedure similar to the one described in #1 above was used. The difference between the two methods is that the two Orthorangers were aligned in the coronal plane instead of the sagittal plane and the subjects were asked to incline the head to the right instead of flexing forward.

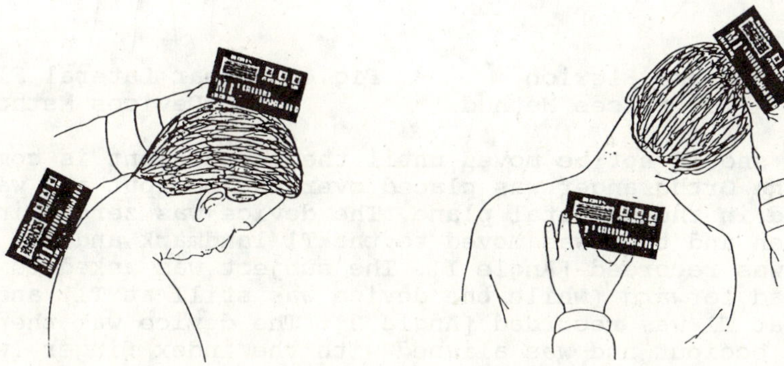

Fig.1: Cervical Flexion Two Devices Method

Fig.2: Cervical Lateral Flexion Two Devices Method

<u>Lumbar Flexion</u> - <u>Two Devices Method:</u> (Fig. 3)

The subjects were asked to stand straight, with the arms beside the body and the weight balanced on both feet (neutral position). Two landmarks were located and marked over the skin with a marker (the T12 spinous process and the Sacral midpoint-SMP). The two Orthorangers were then aligned in the sagittal plane and were placed over the two landmarks. Both devices were zeroed in the neutral position then the subject was asked to flex and the angles from the two devices were recorded. The Orthoranger reading at the sacral midpoint was subtracted from the Orthoranger reading at T12 to obtain the angle of Lumbar flexion.

<u>Lumbar Lateral Flexion</u> - <u>Two Devices Method:</u> (Fig. 4)

A procedure similar to the one described in #3 above was used. The difference between the two methods is that the two Orthorangers were aligned in the coronal plane instead of the sagittal plane and the subjects were asked to incline the trunk to the right instead of flexing forward.

The developed one inclinometer method for the same spinal ranges of motion are specified below:

<u>Cervical Flexion</u> - <u>One Device Method:</u> (Fig. 5)

The subjects were positioned as explained in the measurement of cervical flexion using the two devices method. The T1 spinous process was located and marked, the occiput was located and the index finger was placed on it (the index

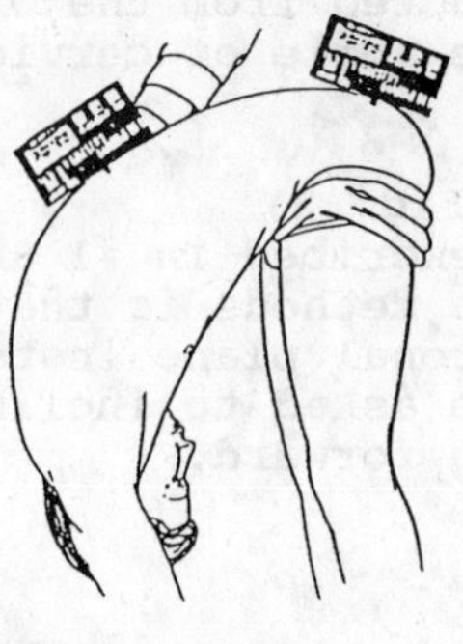

Fig.3: Lumbar Flexion
Two Devices Method

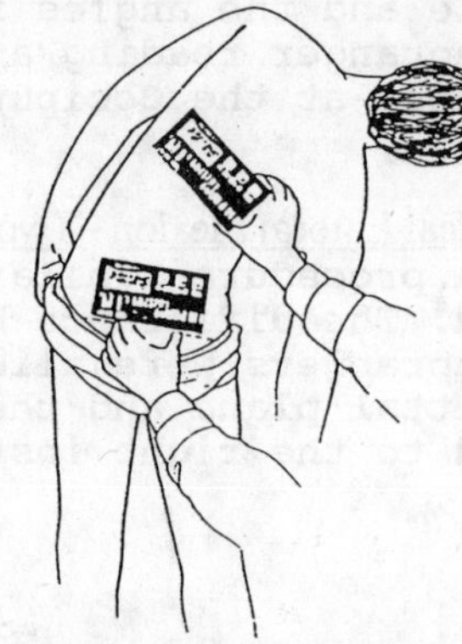

Fig.4: Lumbar Lateral Flexion
Two Devices Method

finger should not be moved until the measurement is completed). The Orthoranger was placed over the occiput and was aligned in the sagittal plane. The device was zeroed in this position and then was moved to the T1 landmark and the angle at T1 was recorded [Angle 1]. The subject was asked to flex the head forward (while the device was still at T1) and the angle at T1 was recorded [Angle 2]. The device was then moved to the occiput and was aligned with the index finger (while the head was still flexed) and the angle was recorded [Angle 3]. The cervical flexion angle was calculated from this formula: Angle of cervical flexion= Angle 3-(Angle 2-Angle 1)

Cervical Lateral Flexion - One Device Method: (Fig. 6)

The subject was positioned and the T1 process was marked as explained in #5 above. The occiput was then located and a ruler was placed on it aligned in the coronal plane perpendicular to the sagittal plane (the ruler should not be moved until the measurement is completed). The Orthoranger was placed on the upper edge of the ruler. The device was zeroed in this position and then was moved to the T1 landmark while still displaying a zero angle. The subject was asked to incline the head to the right and the angle at T1 was recorded in this position [Angle 1]. The device was then moved to the occiput and was placed on the upper edge of the ruler (while the head was still inclined) and the angle was recorded [Angle 2]. The cervical lateral flexion angle was calculated from this formula:

Angle of cervical lateral flexion= Angle 2-Angle 1

Lumbar Flexion - One Device Method: (Fig. 7)

The subject was positioned and the skin was marked as explained in #3 above. The Orthoranger was placed over T12 aligned in the sagittal plane. The device was zeroed in this position and then was moved to the SMP and the angle at SMP was recorded [Angle 1]. The subject was asked to flex (while the device was still at the SMP) and the angle at the SMP was recorded [Angle 2]. The device was then moved to T12 (while the subject was still flexed) and the angle was recorded [Angle 3]. The lumbar flexion angle was calculated from this formula: Angle of lumbar flexion = Angle 3-(Angle 2-Angle 1)

analysis showed that there were no significant differences for the main effects or their interactions. Intraclass correlation coefficients were calculated across methods and raters to determine the variability across the three repetitions . The correlation coefficients across the three repetitions were high for all four maneuvers. A test of homogeneity for correlation coefficients (between therapists) was conducted as recommended by Steel and Torrie (1980). This test showed that the correlation coefficients were homogeneous which allowed us to pool the correlation coefficients of the two raters together for each method and obtain a single correlation coefficient (Table 1). The correlation coefficient values were high (0.71-0.93).

Raters were questioned regarding the ease of use of and preference for the two measurement methods. Raters reported that the two devices method was cumbersome . They reported having difficulty in holding the two devices and zeroing them simultaneously. Raters preferred using the one device technique over the two devices technique. It was also observed that the overall time taken to perform a measurement using the one device technique is less than that for the two devices technique. This was attributed to the higher frequency of repeating measurements excessively when using the two devices technique, due to the difficulty in holding the two devices simultaneously in place which caused their movement with body motion, and the necessity for another measurement.

Table 1. Intraclass correlation coefficients with rater CF's pooled together.

Maneuver	One Device	Two Devices
Lumbar Flexion	0.87	0.86
Lumbar Lateral Flexion	0.75	0.72
Cervical Flexion	0.90	0.93
Cervical Lateral Flexion	0.71	0.82

DISCUSSION

Results of this study showed that when using the Orthoranger II to measure human spine range of motion, there was no significant difference in performing the measurements using the one device method proposed in this paper and the two devices method proposed by the A.M.A. Both methods had the same level of variability across repetitions and across raters. However, the one device method was more efficient in the time needed to perform a measurement. In addition, the raters preferred using the one device method over the two devices method due to the difficulty in holding the two devices simultaneously while performing a measurement.

When spinal ranges of motion were measured using both methods, none of the two methods was found to be superior to the other.

CONCLUSION

Both the two inclinometer and the one inclinometer methods can be used interchangeably to measure ranges of motion. However, the one device method may be preferred by many therapists since it is easier to implement, may save time, and requires only a single device which makes the procedure more cost effective. Additional studies using larger therapists' population are needed to investigate alternative methods of measurement as well as variation in instruments designs. Such studies should prove useful in the effort to obtain valid, reliable and cost effective methods for measuring body joints and spine ranges of motion.

REFERENCES

American Medical Association, 1977, Guides to the evaluation of permanent impairment.

American Medical Association, 1988, Guides to the evaluation of physical impairment (Chicago).

Beattie, P., Rothstein, J.M., and Lamb, R.L., 1987, Reliability of the attraction method for measuring lumbar spine backward bending. Physical Therapy, 67, 364-369.

Cobe, H.M., 1928, The range of active motions at the wrist of white adults. Journal of Bone and Joint Surgery, 26, 763-774.

Fish, D.R., and Wingate, L., 1985, Sources of goniometric error at the elbow. Physical Therapy, 65, 1666-1670.

Hellebrandt, F.A., Duvall, E.N., and Moore, M.L., 1949, The measurement of joint motion; Part III- Reliability of goniometry. The Physical Therapy Review, 29, 302-307.

Hewitt, D., 1928, The range of active motion at the wrist of women. Journal of Bone and Joint Surgery, 26, 775-787.

Petherick, M., Rheault, W., Kimble, S., Lechner, C., and Senear, V., 1988, Concurrent validity and intertester reliability of universal and fluid based goniometers for active elbow range of motion. Physical Therapy, 68, 966-969.

Pisciotta, J.C., 1987, Comparing the reliability of the Orthoranger II and the standard goniometer. Unpublished study, Ohio State University, Gait Analysis Laboratory.

Rothstein, J.M., Miller, P.J., and Roettger, R.F, 1983, Goniometric reliability in a clinical setting. Physical Therapy, 63, 1611-1615.

Wilmer, H.A., and Elkins, E.C., 1947, An optical goniometer for observing range of motion and joints; a preliminary report of a new instrument. Archives of Physical Medicine, 28, 695-705.

Advances in Industrial Ergonomics and Safety II
Edited by Biman Das
Taylor & Francis 1990

THE DEVELOPMENT AND TESTING OF AN ISOKINETIC STRENGTH TESTING DEVICE

G. Weisman, J. Baumhauer, R Seroussi, S. Reinecke, M. Pope
Vermont Rehabilitation Engineering Center
University of Vermont, Burlington, Vermont 05401

A device has been designed, fabricated and tested that can measure a person's whole body strength. The present study demonstrates the repeatability of results when testing vertical lifting strength isokinetically. Lifts were performed by 5 male subjects over the course of 3 days. None of the variables had an effect on measured strength, with the exception of height of the lift. Isometric strength values were generally 37% greater than the isokinetic values determined for the same height. Strength can be depicted graphically as iso-strength contours.

INTRODUCTION

Low back pain is the most common musculoskeletal disorder, the greatest source of compensation payments, and is the second most common cause of work loss (Kelsey et al. 1979). Epidemiologic studies (Hult, 1954; Horal, 1969; Kelsey, 1975) have shown that 40-80% of adults will experience low back pain at some time during their lives. The result, in terms of economic impact in the United States alone, is estimated to be in excess of 20 billion dollars per year.

In order to minimize fatigue or the risk of injury, the job should be fit to the worker. Work capacity evaluations play an important role in this process, as well as in the successful rehabilitation of injured workers. Success of any screening or work capacity program requires accurate information on actual job demands as well as precise measurements of worker capacities related to key job demands.

The literature abounds with techniques and methods for isometric, isotonic and isokinetic strength testing. The NIOSH Work Practices Guide for Manual Lifting (1981) recognizes isometric strength testing as a reasonable way to identify workers who might be at a higher risk of low back pain or overexertion injury. However, Mayer et al. (1985) point out that isometric muscle activity is usually associated with higher rates of injury than dynamic activity. Pytel et al. (1981) and Kroemer (1983) have criticized isometric strength testing because it does not reflect the fact that lifting is a dynamic activity. Marras et al. (1984) state that isometric strength testing can be misleading and endorse an isokinetic approach. They suggest that isokinetic evaluation is a more realistic procedure for evaluation of tasks related to manual material handling conditions. Kumar et al. (1988) point out the importance of determining strength capability throughout the range of motion. They point out that various postures affect strength. Isokinetic testing allows measurement of the actual force output capability at every point along the lifting arc.

The purpose of this study was to determine the repeatability of results, using a method of measuring isokinetic vertical lifting strength.

METHOD

Subjects: The test group consisted of five male subjects between the ages of twenty-three and thirty-five. No subject had a previous history of low back injury, pathology or surgery all and were recruited for participation based on these criteria. All subjects worked at a sedentary job or school and participated in various types of recreational activities. Prior to the initial testing, each subject read a laymen's summary of the experiment and test protocol and signed a consent form. Table 1 lists the age, height, weight and arm length for the five male subjects.

Table 1. Anthropometric variables of subjects

AGE (years)	28.8	+/-	5.3
HEIGHT (cm)	179.8	+/-	13.5
WEIGHT (kgs)	77.3	+/-	1.6
ARM LENGTH (cm)	66.0	+/-	3.0

Apparatus: A physical evaluation device was designed and fabricated that can test maximum isokinetic strength from any direction, without inhibiting range of motion. A dynamometer (Cybex II, Lumex Inc, New York) provided the main source of isokinetic resistance for the device. Through a number of fixed and adjustable pulleys, force can be applied to a cable attached to a handle. The speed for the device ranges from no movement, as in an isometric test, to 152.4 cm/sec of linear motion. Forces were measured at the handle by a strain-gauged load cell, in order to bypass any frictional forces inherent in the device. A potentiometer connected directly to the drive system measures the displacement of the cable. Both of these instruments output a voltage that is linearly proportional to the quantity being measured. The load cell and potentiometer are connected to a micro-computer (IBM PC) via an analog-to-digital converter (Data Translation Board).

Procedure: Each test session for a subject consisted of two sets of five vertical lifts in the mid-sagittal plane. The lifts were from floor to head height at each of two horizontal distances from the lifting handle, at a test speed of 30.5 cm/second. One-minute rests between lifts and five-minute rests between bouts were given to eliminate fatigue. The two horizontal distances were determined as a percentage of the subject's arm length as measured from the acromion process to the first metacarpal-phalangeal joint. At the completion of the isokinetic tests each subject rested five minutes, then lifted isometrically at the height of peak exertion as assessed by the isokinetic data. Three five-second isometric lifts were completed for each distance: 90% arm length first, then 30% arm length. Each subject completed three test sessions within a seven-day period, with only one session per day allowed. Each test session lasted approximately one hour. Subjects were questioned concerning muscle soreness and comfort after each session and prior to the following session. No subject verbalized discomfort during or after lifting. Subjects were instructed to lift vertically as hard and as fast as possible while maintaining their foot positions. Any lifting technique was acceptable and subjects were given a period of time before each session to become familiar with their technique and the machine. This warm-up period typically consisted of five to ten lifts. Once the lifting technique was chosen, subjects were requested to use that technique throughout the testing procedure. A video camera and recorder was used to tape each subject's lifting technique.

RESULTS

Statistical analyses of the data were made using a within-subject, completely crossed ANOVA program. The dependent variables were the force values at each of five heights (14.3%, 34.3%, 54.3%, 74.3% and 94.3% maximal height). The two horizontal distance groups (30% and 90% arm length) were analyzed independently. This repeated measures model tested for daily differences as well as differences between repetitions, sets of repetitions (bouts) or interactions between these measures for each individual. A summary of the ANOVA results is presented in Table 2. Using a t-test, a significant difference was found between the isokinetic force values for each individual at each of five heights for Bout 1 at two horizontal distances between Days 1, 2 and 3 (N=50). Correlation coefficients ranged from 0.833 to 0.912. A complete listing of correlation coefficients with Days can be found in Table 3.

The isometric force value was significantly larger than the isokinetic force measurement ($p<0.001$) at the same location. Interestingly, when the isokinetic force was examined as a percentage of the peak isometric force, the mean was 73.19% +/- 3.05% SE.

A computer program has been developed that uses the force values at four horizontal distances (30%, 50%, 70% and 90% arm length) to construct a two-dimensional strength field model for an individual. A sample contour plot is depicted in Figure 1.

Table 2. Analysis of Variance (ANOVA): Results for 30% and 90% Horizontal Arm Length

Source	Degrees of Freedom	Tail Probability 30%	Tail Probability 90%
D	2,8	0.1926	0.0998
B	1,4	0.0564	0.9062
DB	2,8	0.6177	0.2298
R	4,16	0.4702	0.6880
DR	8,32	0.5784	0.0808
BR	4,16	0.2354	0.2633
DBR	8,32	0.3295	0.3777
H	4,16	0.0322*	0.0001*
DH	8,32	0.5797	0.1011
BH	4,16	0.2556	0.1920
DBH	8,32	0.4297	0.3668
RH	16,64	0.9797	0.2541
DRH	32,128	0.9280	0.8265
BRH	16,64	0.4864	0.8280
DBRH	32,128	0.8847	0.5410

*significant values
(D=Day 1,2,3; B=Bout 1,2; R=Repetition; H=Height)

Table 3. Repeatability of Bout 1

Day 1 vs Day 2	r = 0.833
Day 2 vs Day 3	r = 0.912
Day 1 vs Day 3	r = 0.875

DISCUSSION

Although a number of studies have used devices that test trunk muscles as isolated functional groups to assess trunk strength isokinetically and isometrically, few have examined isokinetic and isometric strength through a functional lifting task. Kumar et al. (1988) have suggested the isolated testing of trunk muscles as is done with certain Cybex machines, may not provide information adequate to predict whole-body lifting capability. The present lift testing device addresses "whole body" lifting strength. Therefore, trunk, arm, shoulder, leg and chest musculature, all important in workplace lifting, are tested.

ANOVA results show that the only significant difference found among the force values at different heights. This is consistent with the results of Kumar et al. (1988) even though the present study represents a greater range of motion. Recognizing that strength varies throughout a range of motion, it is extremely important to identify where in space the lift will occur. In general, a person is strongest when lifting close to the body and around the knees. People with low back pain or other disabilities may demonstrate unique patterns of strength that can be assessed in light of the job requirements.

The present device allows for repeatable measures of lifting strength. No significant differences were found between Days or Repetitions for either horizonal distance. The correlation coefficients determined for the force values at each of five heights for Bout 1 between Days 1,2 and 3 (N=50) suggest high test-retest reliability. Therefore, using one bout in one day of testing will give reliable force measurements indicative of an individual's force production.

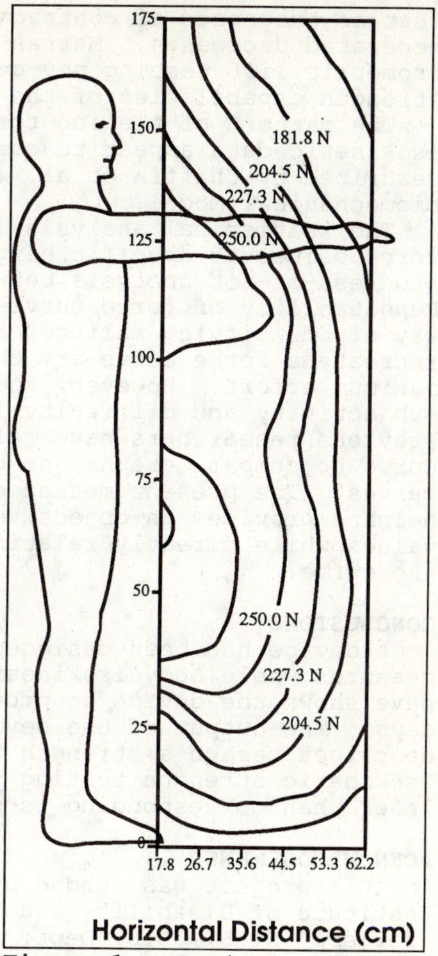

Figure 1. Typical contour plot of iso-strength lines.

It is interesting to note that, at the 30% arm length distance, the differences between Bouts approached significance ($p<0.0564$; df=1,4). Cell mean values were averaged over Day and Repetition calculation an average force value for Bout 1 as 42.0kg, and Bout 2 as 43.4kg. It appears that the higher force value for Bout 2 is due to a warming-up effect rather than a training effect due to lack of significance of the Bout vs Day interaction.

The significant difference between the isometric force measurements and the corresponding isokinetic values are consistent with the findings of other researchers (Kumar et al. 1988) This suggests that isometric values used to assess the maximal strength ability of a worker in the workplace may overestimate strength capabilities. Isokinetic values, which on the average are 73% of the corresponding isometric measurement, suggest that when moving at a speed of 60 degrees/sec (30.5cms/sec) an individual can not generate the force that can be generated during an isometric (0 degrees/sec) contraction. This would support the findings of Murray et al. (1980), Osternig et al. (1977) and Kumar et al. (1988) who reported that as the speed of contraction increases the force generated decreases. Marras et al. (1984) suggested that isometric lift testing may overestimate the functional strength capabilities of the worker.

The pattern of the iso-force bars constructed with the isokinetic data appear to correlate well with the pattern described by Chaffin et al. (1969) who used a theoretical biomechanical model.

The statistical analysis is based on comparisons among force values at specific heights, thus allowing for a goodness-of-fit analysis between force-generated curves. Repeatability of force curve generation is often used as a way of identifying malingerers. Those subjects not able to recreate a force curve are thought not to be giving a maximum effort. However, there is a great deal of subjectivity and difficulty in interpreting this data. Previous researchers have relied on the area under the curve to compare graphs, or basically "eyeballing" the curves. The present method of gathering data at specific heights provides an objective means of comparing force values while directly relating those values to points on the curve.

CONCLUSIONS

A device has been designed, constructed and tested that measures whole body isokinetic lifting strength. Tests have shown the device to provide repeatable results over 3 days. The output of the device is a contour plot that describes person's strength through a range of motion. Isokinetic strength testing yielded values significantly lower than corresponding isometric strength values.

ACKNOWLEDGEMENTS

This project was funded by a grant from the National Institute of Disability and Rehabilitation Research (formerly NIHR), U.S. Dept. of Education. (G008303001-03).

REFERENCES

Chaffin, D.B., 1969, A computerized biomechanical model-development of and use in studying gross body action. Journal of Biomechanics, 2, 429-441.

Horal, J., 1969, The clinical appearance of low back pain in the city of Gothenburg, Sweden. Acta Orthopaedica Scandinavica (Supplement), 118, 1-109.

Hult, L., 1954, The munkfors investigation. Acta Orthopaedica Scandinavica (Supplement), 16.

Kelsey, J.L., 1975, An epidemiological study of acute Herniated lumbar intervertebral discs. Rheumatology and Rehabilitation, 14, 144-159.

Kelsey, J.L., White, A.A. III, Pastides, H., & Bisbie, G.E., 1979, The impact of musculoskeletal disorders on the population of the United States. Journal of Bone & Joint Surgery, 61A, 959-964.

Kroemer, K.H.E., 1983, An isoinertial technique to assess individual lifting capability. Human Factors, 25, 493-506.

Kumar, S., Chaffin, D.B., Redfern, M., 1988, Isometric and isokinetic back and arm lifting strengths: Device and measurements. Journal of Biomechanics, 21, 135-144.

Marras, W.S., King, A.I., Joynt R.L., 1984, Measurements of loads on the lumbar spine under isometric and isokinetic conditions. Spine, Vol 9, 2nd edn., pp. 176-187.

Mayer, T.G., Gatchel, R.J., Kishino, N., Keeley, J., Capra, P., Mayer, H., Barnett, J., Mooney, V., 1985, Objective assessment of spine function following industrial injury. Spine, 10(6), 482-493.

Murray, M.P., Gardner, G.W., Mollinger, L.A., Septic, S.B., 1980, Strength of isometric and isokinetic contractions. Physical Therapy, 4, 412-419.

Osternig, L.R., Bates, B.T., James, S.L., 1977, Isokinetic and isometric torque force relationships. Archives of Physical Medicine and Rehabilitation, 58, 254-257.

Pytel, J.L., Kamon, E., 1981, Dynamic strength as a predictor for maximal acceptable lifting. Ergonomics, 24, 663=672.

Work Practices Guide for Manual Lifting, Technical Report, DHHS (NIOSH) Publication Number 81-122, March 1981.

Advances in Industrial Ergonomics and Safety II
Edited by Biman Das
Taylor & Francis 1990

SUBMAXIMAL PINCH FORCES MEASURED USING MINIATURE FORCE TRANSDUCERS ATTACHED TO THE DISTAL FINGER PADS

Robert G. Radwin, Todd R. Jensen, Seoungyeon Oh, and John G. Webster

University of Wisconsin-Madison
1513 University Avenue
Madison, Wisconsin 53706, U.S.A.

Small conductive polymer force transducers were attached to the distal phalangeal pads for measuring individual finger forces exerted during submaximal pulp pinch against two parallel plates attached to a strain gage dynamometer. Total resultant and individual finger forces were measured for 10 N, 15 N, and 20 N load weights and for 45 mm and 65 mm pinch spans. Total resultant pinch force increased from 13 N for the 10 N load to 29 N for the 20 N load weight. The index finger exerted the greatest force followed by the middle finger, ring finger, and small finger. Average contribution of individual fingers to the resultant total pinch force was 34% for the index finger, 24% for the middle finger, 21% for the ring finger, and 21% for the small finger. As load weight increased from 10 N to 20 N, the contribution of the index finger decreased as the contribution of the middle finger increased.

INTRODUCTION

Measurement and prediction of individual finger forces during submaximal grip exertions are important for developing functional biomechanical models and for designing work equipment and manual activities. Although individual finger forces have been studied in numerous investigations involving maximal grip exertion levels (Swanson, Matev, and DeGroot, 1970; Dickson, et al., 1972; Hazelton, et al., 1975; Ohtsuki, 1981; Ejeskär, et al., 1981; An, et al., 1985; Amis, 1987), little data is available concerning submaximal finger exertions. Westling and Johansson (1984) found the applied grip force between the index finger and thumb was critically balanced to optimize motor behavior so that slipping between the skin and the gripped object did not occur. Grip force control was heavily influenced by the weight of the objects held plus a safety margin factor specific to individual subjects.

Conventional methods for measuring pinch and grip force include electromyography and directly instrumenting handles with force transducers. Armstrong and Chaffin (1979) studied pinch forces exerted by workers performing manual work at submaximal levels using electromyograms. Surface electromyography, however, is limited to static exertions, is not specific to individual fingers, and is not practical for measuring individual finger forces during complex manual work activities.

Various instruments have been devised and used for measuring individual finger forces during maximal grip exertions. Dickson, et al. (1972) constructed a mechanical device for assessing postoperative individual finger strength by measuring displacement of a cantilever spring. Hazelton, et al. (1975) measured forces simultaneously at the middle and distal phalanges, for all four fingers. Their dynamometer consisted of flexible wire mesh straps attached to a strain gage ring for each finger. Ohtsuki (1981) measured individual finger forces by restraining the forearm and attaching rings to each finger, attached to individual strain gage load cells. Amis (1987) developed a dynamometer for simultaneous measurement of both normal forces and tangential shear forces for each of the three phalangeal segments of finger when gripping a cylinder. The instrument, however, didn't measure finger forces simultaneously.

A major difficulty in accurately predicting internal hand forces in biomechanical models is the determination of external finger forces for in different joint configurations and grasping functions. This is because adequate force transducers are needed for measuring individual finger forces applied during grasping activities. Due to a lack of force measurement instruments, biomechanical models of the fingers and hand have had to use assumptions concerning external finger forces. Armstrong (1982) assumed that individual finger forces are distributed in proportion to the relative strengths of the fingers. An, et al. (1985) assumed that the force of extrinsic finger flexor muscles were distributed in proportion to their physiological cross sectional areas. Another complicating factor is that individual finger exertions are not independent of one another. Ohtsuki (1981) concluded that activation of one finger may inhibit the activity of adjacent fingers.

METHODS

Individual finger forces were measured using thin force transducers taped to the distal phalangeal pads of the four fingers. The transducers were constructed using 0.3 mm thin Interlink conductive polymer sensors. A 1 mm high epoxy dome was placed over the sensing area to equally distribute applied forces over the pressure sensor, making it into a force transducer. Microfoam® tape was used for attaching the transducers directly to the fingers. These transducers had a useful range of 0 N to 30 N with a resolution of 1 N.

Figure 1 illustrates the apparatus and experimental organization. A strain gage dynamometer constructed from an aluminum beam was fixed at one end and loaded with forces acting perpendicular to its long axis. The dynamometer measured the transverse force acting within a beam causing a shear stress in the cross-section (Pronk and Niesing, 1981). This resulted in a force measurement instrument which was independent of point of application permitting the individual finger forces to sum linearly.

Static force sensor calibration on the finger was performed by placing each finger opposite the thumb against the suspended dynamometer. Maximum pinch force was measured for each finger and ten static calibration points were selected at equally spaced intervals up to 50 percent maximum voluntary exertion. Calibration levels were obtained for each sensor in random order. A second-order polynomial calibration curve was fit for each transducer using linear regression.

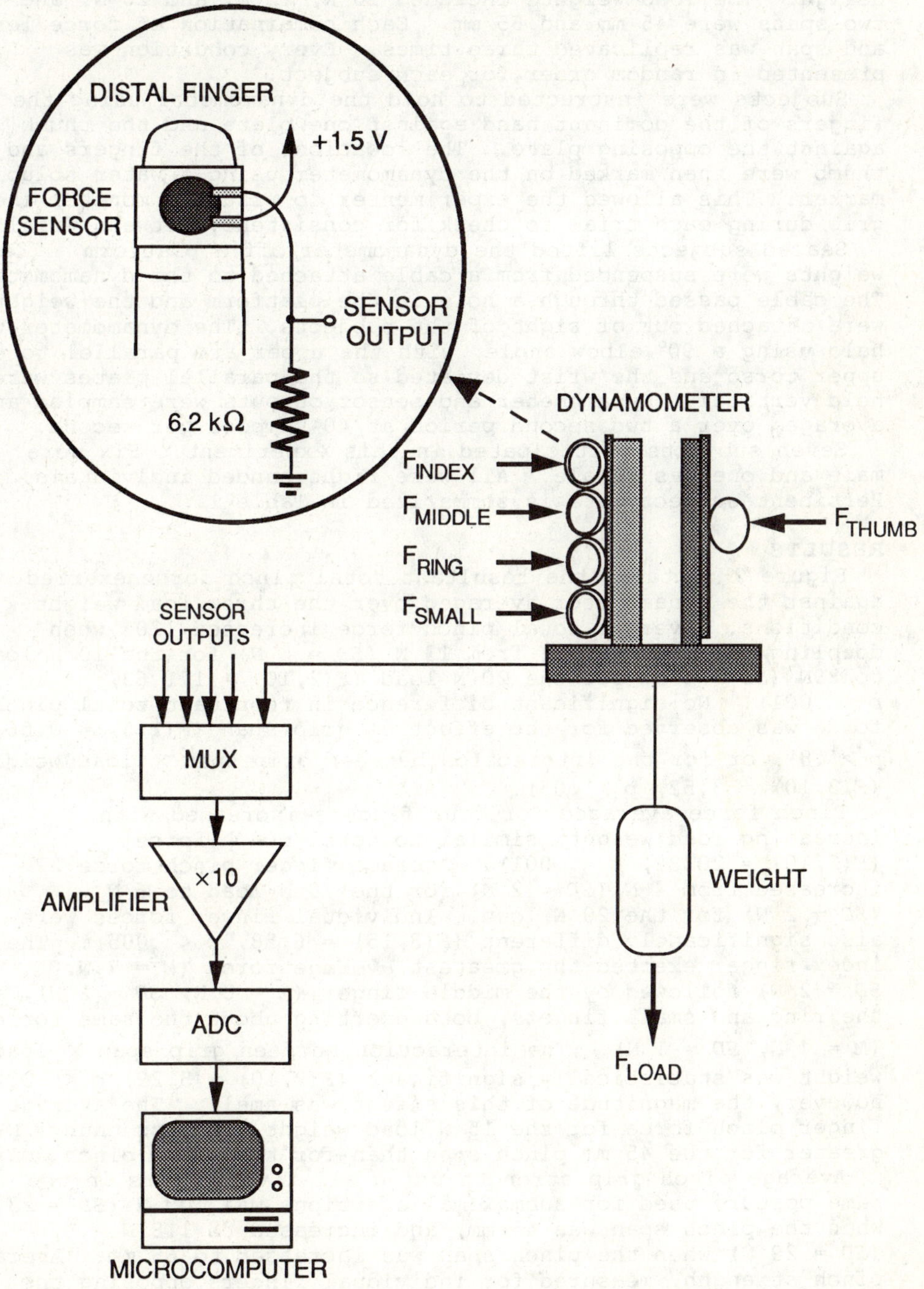

Figure 1: Block diagram of experimental apparatus and laboratory organization

Pinch compression force and individual finger forces were measured while pinching the dynamometer at three load weight levels and two pinch spans for a full-factorial experimental design. The load weights included 10 N, 15 N, and 20 N, and the two spans were 45 mm and 65 mm. Each combination of force level and span was replicated three times. Every condition was presented in random order for each subject.

Subjects were instructed to hold the dynamometer using the fingers of the dominant hand against one plate and the thumb against the opposing plate. The locations of the fingers and thumb were then marked on the dynamometer using a water soluble marker. This allowed the experimenter to visually monitor the grip during each trial to check for consistency between trials.

Seated subjects lifted the dynamometer off a platform. Load weights were suspended from a cable attached to the dynamometer. The cable passed through a hole in the platform and the weights were attached out of sight of the subjects. The dynamometer was held using a 90° elbow angle, with the upper arm parallel to the upper torso and the wrist deviated so the parallel plates were held vertical. Dynamometer and sensor outputs were sampled and averaged over a two second period at 40 samples per second.

Seven subjects participated in this experiment. Six were male and one was female. All were right-handed individuals. Pertinent subject data is summarized in Table 1.

RESULTS

Figure 2 contains the resultant total pinch force exerted against the dynamometer averaged over the three load weight conditions. Average total pinch force increased 120% when doubling the load weight from 13 N (SD = 3 N) for the 10 N load to 29N (SD = 4 N) for the 20 N load ($F(2,10) = 171.63$, $p < .001$). No significant difference in resultant total pinch force was observed for the effect of grip span ($F(1,5) = 0.06$, $p > .8$), or for the interaction between grip span × load weight ($F(2,10) = 3.52$, $p > .05$).

Pinch force averaged for four fingers increased with increasing load weight, similar to total pinch force ($F(2,10) = 20.89$, $p < .001$). Average finger pinch force increased from 4 N (SD = 2 N) for the 10 N load to 6 N (SD = 2 N) for the 20 N load. Individual finger forces were also significantly different ($F(3,15) = 6.58$, $p < .005$). The index finger exerted the greatest average force (M = 7 N, SD = 2 N) followed by the middle finger (M = 5 N, SD = 2 N), and the ring and small fingers, both exerting about the same force (M = 4 N, SD = 1 N). The interaction between grip span × load weight was statistically significant ($F(2,10) = 5.29$, $p < .05$), however, the magnitude of this effect was small. The average finger pinch force for the 15 N load weight was less than 1 N greater for the 45 mm pinch span than for the 65 mm pinch span.

Average pinch grip strength using all five fingers in the same posture used for submaximal exertions was 109 N (SD = 23 N) when the pinch span was 45 mm, and increased to 118 N (SD = 29 N) when the pinch span was increased to 65 mm. Average pinch strength, measured for individual fingers opposing the thumb, are summarized in Table 2.

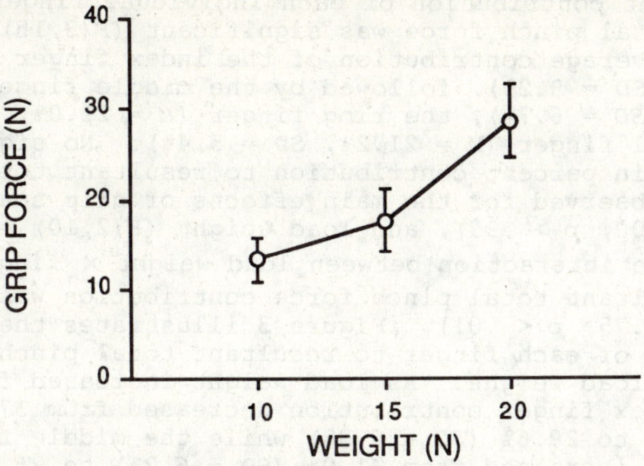

Figure 2: Average resultant pinch grip force plotted against load weight (6 subjects). Error bars represent ± one standard deviation.

Table 1

Summary of Subject Data

Subject	Gender	Age	Hand Length (cm)	Hand Width (cm)
1	M	24	19.9	8.3
2	M	28	18.3	8.3
3	M	22	18.6	9.1
4	M	18	20.4	8.9
5	M	20	20	8.4
6	F	26	16.7	6.9

Table 2

Individual Finger Strengths (6 Subjects)

Finger	Mean (N)	SD
Index	63	15
Middle	60	22
Ring	39	15
Small	32	10

The percent contribution of each individual finger to the resultant total pinch force was significant ($F(3,15) = 6.14$, $p < .01$). Average contribution of the index finger was greatest (M = 34.0%, SD = 9.2%), followed by the middle finger (*M* = 23.9%, *SD* = 6.7%), the ring finger (*M* = 21.0%, *SD* = 4.6%), and the small finger (M = 21.2%, SD = 5.4%). No significant differences in percent contribution to resultant total pinch force were observed for the main effects of grip span ($F(1,5) = 0.00$, $p > .99$), and load weight ($F(2,10) = 1.32$, $p > .3$). The interaction between load weight × finger for percent resultant total pinch force contribution was significant ($F(6,30) = 3.75$, $p < .01$). Figure 3 illustrates the percent contribution of each finger to resultant total pinch force as a function of load weight. As load weight increased from 10 N to 20 N the index finger contribution decreased from 37.3% (*SD* = 10.6%) to 29.6% (*SD* = 6.3%) while the middle finger contribution increased from 21.0% (SD = 6.2%) to 26.9% (SD = 5.8%). Average percent contribution of each individual finger to the resultant pinch grip strength at maximum voluntary exertion levels are given in Table 3.

DISCUSSION

Average finger and resultant total pinch force using all five digits increased with increasing load weight (See Figure 2). Westling and Johansson (1984) found similar results when gripping small objects between the index finger and thumb. The individual contribution of the fingers to total resultant pinch force, however, was not constant at submaximal exertion levels for increasing load (See Figure 3).

The two strongest fingers, the index and middle fingers (See Table 2), produced the greatest forces at submaximal levels. On the average, the index finger produced 34% of the grip force which corresponds well to the 32% contribution of the measured grip strength. At low exertion levels (load weight 10 N) the index finger contributed 37% and at higher exertion levels (load weight 30 N) the index finger contributed only 30%. Conversely, while the middle finger contributed only 21% of the total grip force for the low exertion level, the middle finger contributed 27% of the total pinch force at the higher exertion level. Hence the assumption that individual finger external forces are exerted in direct proportion the strength is not completely accurate.

Hazelton, et al. (1975) found the percentage distribution of strength of individual fingers indicated the middle finger produced approximately 33.5% of the total force, the index and ring fingers 25.0 % of the total force, and the little finger 16.5% of the total force. Ohtsuki (1981) also showed the relative share of individual fingers during maximal grip exertions was greatest for the middle finger (33%) and smallest in the little finger (15%). The ring finger strength (27--28%) was slightly greater than the index finger (24--25%). Furthermore, the individual finger strength decreased in proportion to the number of other fingers participating when performing simultaneous exertions. The results of this investigation indicated that as pinch force requirements increased, the middle finger contributed a greater percentage of the total pinch force while the index finger contribution declined (See Figure 3). As the the middle finger recruitment increased, the index finger contribution may have decreased due to inhibition, or at least interaction with increased middle finger activity. This interaction is apparently dependent on grip exertion level.

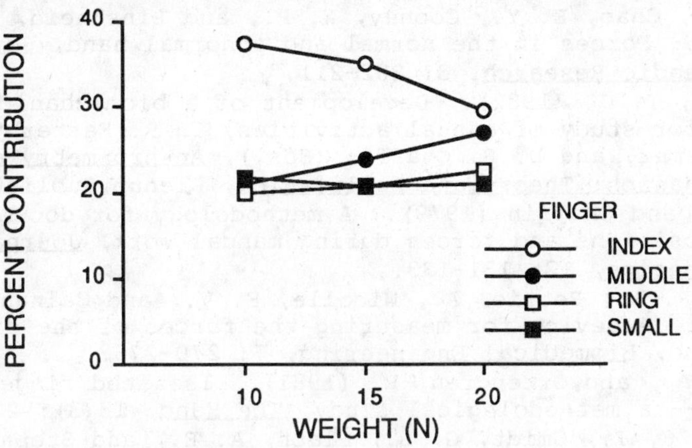

Figure 3: Percent contribution to resultant pinch force for each finger plotted against load weight.

Table 3

Individual Finger Force Expressed as Percent Contribution to Total Pinch Strength (6 Subjects)

Finger	Mean (%)	SD
Index	32.8	6.0
Middle	30.1	3.9
Ring	20.5	3.9
Small	16.5	2.0

This study measured distal phalangeal pinch force for five finger static grip using a relatively simple prehension pattern. This grip is often using occupationally for carrying large objects, such as plates of glass or for carrying tote pans that do not provide cut-outs for handles. Interphalangeal forces will be studied, as well as more complex prehension patterns, in future investigations. Transducers are now being developed in our laboratory for attachment to other locations on the fingers and hands. These transducers will be useful for studying static and dynamic forces involved with activities of daily living as well as manual work activities such as holding handles and tools.

ACKNOWLEDGEMENTS

This work was supported, in part by NIH Grant No. 1 R01 NS 26328-01.

REFERENCES

Amis, A. A. (1987). Variation of finger forces in maximal isometric grasp tests on a range of cylinder diameters, Journal of Biomedical Engineering, 9: 313-320.

An, K. N., Chao, E. Y., Cooney, W. P., and Linscheid, R. L. (1985). Forces in the normal and abnormal hand, Journal of Orthopaedic Research, 3: 202-211.

Armstrong, T. J. (1982). Development of a biomechanical hand model for study of manual activities, In R. Easterby, K. H. E. Kroemer, and D. B. Chaffin (Eds.), Anthropmetry and Biomechanics: Theory and Application, Plenum Publishing Corp.

Armstrong and Chaffin (1979). A methodology for documenting hand positions and forces during manual work, Journal of Biomechanics, 12: 131-133.

Dickson, R. A., Petrie, A., Nicolle, F. V., and Calnan, J. S. (1972). A device for measuring the forces of the digits of the hand, Biomedical Engineering, 7: 270-273.

Ejeskär, A., and Örtengren, R. (1981). Isolated finger flexion force -- a methodological study, The Hand, 13(3): 223-230.

Hazelton, F. T., Smidt, G. L., Flatt, A. E., and Stephens, R. I. (1975). The influence of the wrist position on the force produced by the finger flexors, Journal of Biomechanics, 8: 301-306.

Ohtsuki, T. (1981). Inhibition of individual fingers during grip strength exertion, Ergonomics, 24(1): 21-36.

Pronk, C. N. A. and Niesing, R. (1981). Measuring hand-grip force, using a application of strain gages, Medical and Biological Engineering and Computing, 19:127-128.

Swanson, Matev and DeGroot (1970). The strength of the hand, Bulletin of Prosthetics Research, Fall: 145-153.

Westling, G. and Johansson, R. S. (1984). Factors influencing the force control during precision grip, Experimental Brain Research, 53: 277-284.

Advances in Industrial Ergonomics and Safety II
Edited by Biman Das
Taylor & Francis 1990

THE EFFECT OF LOAD EXPECTATION ON MUSCLE RECRUITMENT

T.M. Khalil, S.M. Waly, and A.M. Zaki

Department of Industrial Engineering

University of Miami

This study investigated the effect of load expectation on EMG response recorded from the arm muscles when an individual lifts a load. The paper describes exploratory investigation to establish the existence of such a relationship, as well as a controlled follow-up study to verify initial findings. Subjects were asked to lift a closed container with an announced load which at times was different from its actual load. Peak to peak EMG was recorded at the moment when the lifting action was initiated. The results showed that the EMG activity was influenced significantly with the subjects' expectation of the load inside the container as opposed to its actual load.

INTRODUCTION

Musculoskeletal injuries are one of the most prevalent types of injuries in the work environment. It is ranked second in NIOSH's list of the ten leading work related injuries and diseases (NIOSH, 1986). It has been a topic of considerable concern due to both humanitarian and economic dimensions associated with it. Expenses of treatment and compensation due to low back injuries alone were estimated to be more than 14 billion dollars per year in the United States (Taber, 1982, Asfour, 1983, Nordby, 1981, NIOSH, 1986). This does not include indirect expenses associated with injuries which are usually estimated to be up to four times the direct expenses (Heinrich, 1959). The magnitude of the problem has inspired numerous researchers to investigate the causes of musculoskeletal injuries and methods of preventing them (NIOSH, 1981, Khalil et al, 1984, 1985, Asfour et al, 1983, Chaffin and Anderson , 1983). Many of the studies assumed that the load causing the injury was an important factor and that the body responds in proportion to the amount of load handled. Little is known about the body's response to unknown or unexpected loading. Unknown or unexpected loading occurs in situations involving slipping, struck by, suddenly shifting loads, etc.

Several studies suggested that sudden unexpected loading while performing a manual material handling task can be a major factor contributing to back pain and musculoskeletal injury in the workplace. Marras el al (1987) studied the effect of expected and unexpected loading on trunk muscles response. Their study showed that there is a difference in trunk muscles response between both types of loading. They indicated that for the peak component of muscle force (peak EMG within the period of muscle activity) , peak EMG for the unexpected loading was 1.7 times that of expected loading, and for the mean component of muscle force (average EMG within the period of muscle activity), the EMG for the unexpected loading was 2.4 times that of expected loading. They suggested that this unexpected loading can cause over-reaction of trunk muscles which loads the spine and causes back injury. Stobbe et al, 1988 studied the effect of unexpected loading on industrial back injuries. They interviewed workers suffering from low-back injuries in three different industries. This interview was meant to give them more information on the causes of injury other than what's available in the employer's injury reports. They concluded from their study that a large percentage of industrial back injuries are the result of sudden-movement/unexpected-loading. A study conducted by Manning et al (1981) confirmed the previous results. The study investigated the effect of slipping on low-back pain. The results showed that slipping accidents are an important cause of low-back pain. Magora (1973) investigated the relationship between low-back pain and certain physical activities required by preselected occupations. The effect of bending, rotation, reaching, and sudden maximal effort on low-back pain was studied. He concluded that sudden unexpected activities were related to a high incidence of low-back pain. Other activities showed a much lower incidence.

The objective of the present study was to investigate the EMG response of individuals due to load expectation. The hypothesis is that, in load handling, individuals anticipate the heaviness of a particular weight before lifting that weight, thus prepare their muscles to respond with a certain magnitude of force proportional to the expected load. The underlying assumption is that in the case of unexpected loading, an individual can misjudge the actual weight to be handled. Understanding the effect of load expectation is essential in answering some vital questions that are pertinent to the prevention effort of musculoskeletal injuries.

To test this hypothesis, an exploratory study was initially conducted to investigate if a relationship exists between load expectation and EMG activity. A follow-up study was also conducted to verify the findings of the initial study.

EXPLORATORY STUDY

Six subjects volunteered to participate in this study. A storage oscilloscope, a Polaroid camera, and an EMG amplifier were used to conduct this experiment. Surface EMG was recorded from the right biceps muscle of the subjects. Each subject was told that he/she is required to lift a container with two different loads of 20 lbs. and 30 lbs. respectively. The container was to be lifted from a table using a forearm flexion movement while keeping the upper arm tucked to the sides. Subjects were allowed to become accustomed to the feel of the 20 lbs. load by lifting the container three or four times after which a 3-4 minutes recovery time was allowed. After recovery, each subject was told that he/she will lift 20 lbs., then was asked to proceed with the first lift, the initial EMG response for the first lift was stored on the oscilloscope screen and a photograph was taken for that response. A recovery time of one minute was allowed during which the subject was told that the weight would be increased to 30 lbs. but actually remained unchanged. The subject was then asked to proceed with the second lift and the initial EMG response for the second lift was stored on the oscilloscope screen and a photograph was taken. Figures 1a and 1b show the corresponding response for the two lifts for one of the experimental subjects.

The actual load contents of the container in the second lift was achieved by removing 2 bags of weights from the container and adding 5 bags while subjects observed the exchange. The subjects were not told that the two bags removed were 5 lbs. each, bringing the total weight removed to 10 lbs., and that the five bags added were 2 lbs. each, bringing the "adjusted weight" of the container to its original weight. All the bags looked alike so the subjects would not differentiate between the 2 lbs. bags and the 5 lbs. bags. The photographs showed that the EMG activity was higher for the second lift than the first. Similar trends were observed for the six subjects, however the magnitude of the change differed from one subject to the next. This difference was attributed to the subject's anticipation of lifting a bigger load in the second lift. To further investigate the effect of the magnitude of the expected load on the EMG response, a seventh subject was tested. The same procedure was repeated but the subject was told that the load he will be lifting was 50 lbs. The actual load was only 20 lbs. The results are shown in Figure 2a &b.

The findings of the exploratory study were quite revealing that it motivated a second investigation which is described below.

EXPERIMENTAL PROCEDURE

Fifteen subjects volunteered to participate in this study. The subjects consisted of 11 males (ages 19 - 46) and 4 females ages (20-22). The subjects were healthy with no reported history of back pain or musculoskeletal disorders. The subjects were asked to sit on a chair with a straight back. The right arm was flexed at a 90 degree angle (Figure 3). The electrodes were connected to an EMG amplifier with a chart recorder. A five or twenty five lbs. load was placed inside a closed container with a top handle. The container did not allow the subjects to see its contents, thus did not allow the subjects to guess its weight. Subjects did not know the actual load contents of the container referred to as actual load. They were told that they will lift the container according to the following sequence of loads: 5, 25, 5, 25, 25. The first four lifts had the correct load. The fifth lift did not have the load the subject was expecting to lift. The actual load was 5 lbs. instead of an expected load of 25 lbs. (Table 1).

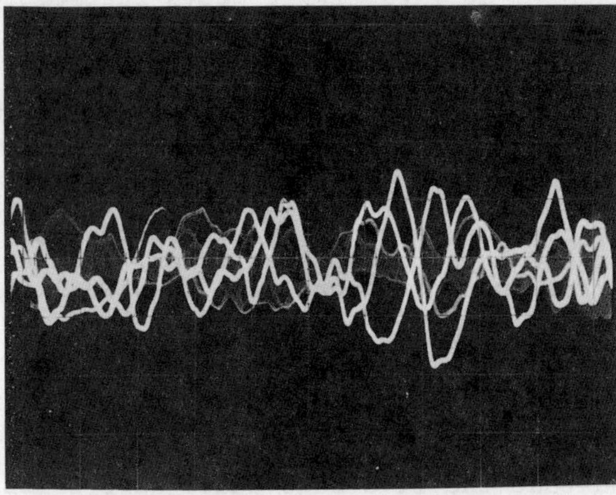

Figure 1a: Actual load=20 lbs., Expected load=20 lbs.

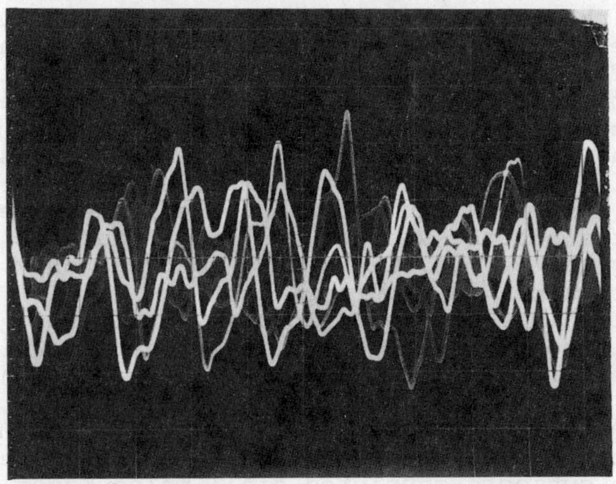

Figure 1b: Actual load=20 lbs., Expected load=30 lbs.

The weights were adjusted inside the container without the subjects knowledge of its actual weight contents. The experimenter transported the container to the subject on a cart in order not to allow the subject to perceive the load content of the container by looking at the experimenter's effort while carrying it. The subjects were requested to lift the container from the cart and the EMG at the start of the lift was recorded. A one minute rest period was allowed between lifts.

RESULTS

One way analysis of variance was used. The independent variable was the lift number (1,2,3,4, and 5) as shown in Table 1. The dependent variable was the EMG peak to peak amplitude values at the beginning of the lift. Tukey test was used to determine which lift number was significantly different from the others.

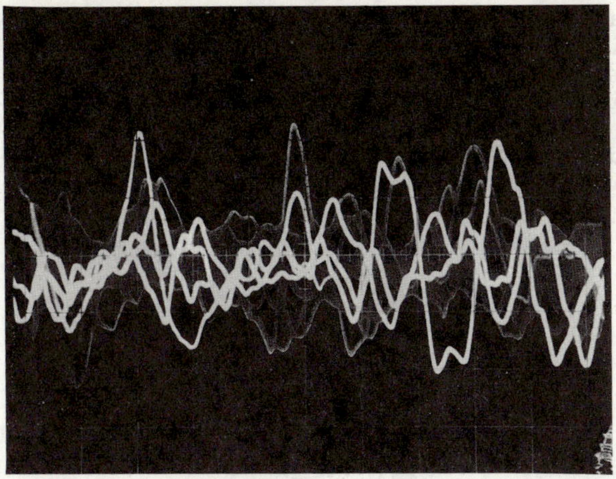

Figure 2a: Actual load=20 lbs, Expected load=20 lbs.

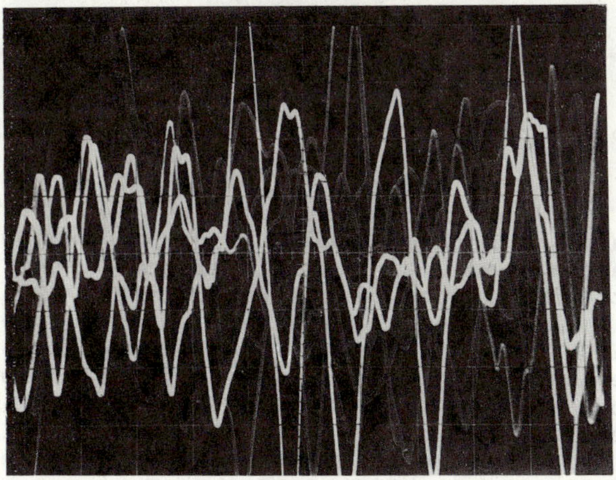

Figure 2b: Actual load=20 lbs, Expected load=50 lbs.

Table 1. Lift number and the corresponding Actual loads, expected loads, Mean and Standard deviation for EMG values

Lift No.	Actual load (lbs.)	Expected load (lbs.)	Mean EMG (mv)	Std. Dev. (mv)
1	5	5	7.11	1.63
2	25	25	15.16	4.58
3	5	5	8.61	3.87
4	25	25	14.47	3.90
5	5	25	14.00	4.65

The results of the analysis of variance showed that the lift number was significant at the 0.01 level. The Tukey test showed that lift numbers 1 and 3 were not significantly different from each other. Also, lift numbers 2, 4, and 5 were not significantly different from each other but were significantly different from the other two lifts. The means and standard deviations of the EMG values for the different lifts are shown in Table 1.

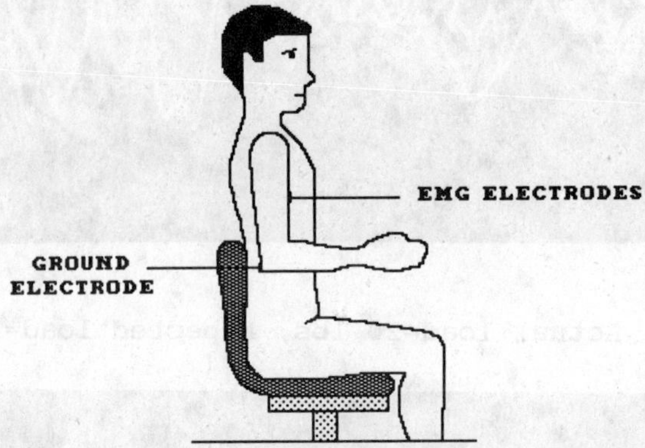

Figure 3: Subject positioning and electrode placement

DISCUSSION

The results of both studies showed that subjects' expectation of the load had a significant effect on the initial muscle output as indicated by the recorded EMG. In the experiment, lifts number 1 and 3 where the subjects expected to lift only 5 lbs. showed a drastically lower EMG value than those obtained when the subjects lifted the same load but were expecting a higher value of 25 lbs (Figure 4). The value of the EMG recorded during lift number 5 was not significantly different from those obtained when the subject expected to lift 25 lbs. and actually lifted 25 lbs. The mean EMG values recorded when the subjects were expecting a load of 25 lbs. were very close to each other whether the subjects actually lifted 25 lbs. or 5 lbs. loads. This indicates that the Central Nervous System (CNS) initiates neural signals to the muscle in proportion to the magnitude of load it expects. It is believed that it was not until a feedback was given to the CNS that the magnitude of the signals were adjusted appropriately to the magnitude of the load. This was shown by the immediate decline of the EMG signal following the initial peak on the chart. These results may have strong implications pertaining to musculoskeletal injuries and particularly back injuries. They indicate that there may exist a time lag between the motor (efferent) signals initiated by the CNS and the adjustment to those signals based on sensory (afferent) feedback from the muscles to the CNS. This apparent lag between the initial and adjusted signals can create conditions where the load demands and the corresponding muscle activities are either not fully matched in magnitude or

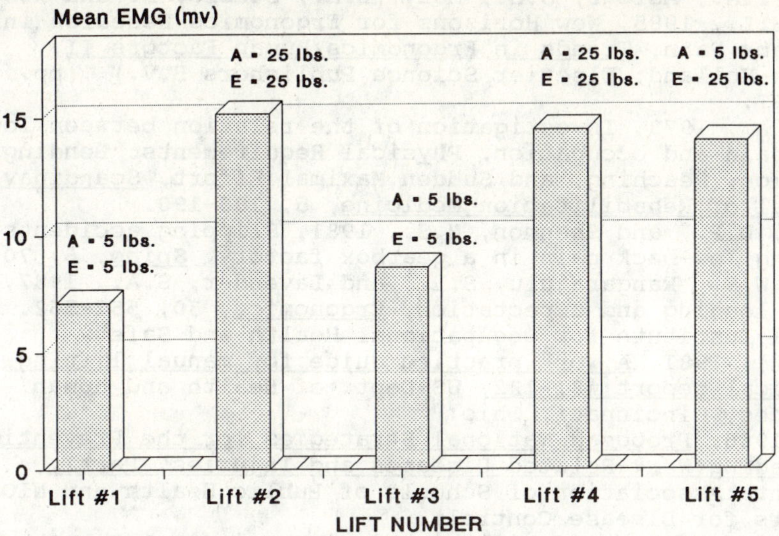

Figure 4: Lift number vs. EMG for actual and expected loads

synchronized in phase. Marras et al (1987) have shown that sudden external loading create higher EMG response values than those created by expected loading. The difference between findings in this study and those of Marras et al is that the neural signals here seems to have been initiated before loading as opposed to being initiated after loading in the Marras et al study. It is quite plausible that the initial lag in adjustment of the neuromuscular control loop can create an instantaneous imbalance in the system. This could be the instance where the body is most vulnerable to sustaining a musculoskeletal injury. Several studies have shown that a great majority of musculoskeletal injuries are due to slips and falls (Manning, 1981, Khalil et al, 1984). It is evident that slips create sudden unexpected movements where the muscle is unable to respond quickly to the magnitude of the unexpected external forces created. It seems that when the body is expecting the load, it can mobilize adequate motor signals in advance in order to prepare the human body for handling an external load and meanwhile protect the musculoskeletal structure from sudden (unexpected) loads. Additional research is needed in order to support this proposed theory of injury and to further explain its mechanism.

REFERENCES

Asfour, S.S., Khalil, T.M., Moty, E.A., Steele, R., and Rosomoff, H.L., 1983, Low back pain: a challenge to productivity, Proceedings of the 7th international conference on production research, Windsor, Canada.

Chaffin, D.B., and Anderson, G.B., 1984, Occupational Biomechanics, (New York: John Wiley & Sons).

Heinrich, H.W., 1959, Industrial accident prevention, 4th edition, (New York:McGraw-Hill Book Co.).

Khalil, T.M., Asfour, S.S., and Moty, E.A., 1984, Case Studies in Low Back Pain, Proceedings of the Human Factors Society- 28th Annual Meeting, pp. 465- 470.

Khalil, T.M., Asfour, S.S., Moty, E.A., Steele, R. and Rosomoff, H.L., 1985, New Horizons for Ergonomics Research in Low Back Pain, Trends in Ergonomics/Human Factors II, (North-Holland: Elsevier Science Publishers B.V.) , pp. 591-597.

Magora, A., 1973, Investigation of the relation between low back pain and occupation. Physical Requirements: Bending, Rotation, Reaching, and Sudden Maximal Effort, Scandinavian Journal of Rehabilitation Medicine, 5, 186-190.

Manning, D.P., and Shannon, H.S., 1981, Slipping accidents causing low-back pain in a gearbox factory, Spine, 6, 70-72

Marras, W.S., Rangarajulu, S.L., and Lavender, S.A., 1987, Trunk loading and expectation, Ergonomics, 30, 551-562.

National Institute for Occupational Health and Safety (NIOSH), 1981, A work practice guide for manual lifting, Technical report #81-122, US Dept. of Health and human Services, Cincinnati, Ohio.

NIOSH, 1986, Proposed National Strategies for the Prevention of Leading Work-Related Diseases and Injuries, Part1, (Atlanta:Association of Schools of Public Health and NIOSH, Centers for Disease Control).

Nordby, E., 1981, Epidemiology and diagnosis in low back injury. Occupational Health and Safety, Jan, 38-41.

Stobbe, T.J., and Plummer R.W.,1988, Sudden-Movement/Unexpected loading as a factor in back injuries, Trends in Ergonomics/Human Factors V, (North-Holland:Elsevier Science Publishers B.V.), pp. 713- 720.

Taber, M., 1982, Reconstructing the scene, back injuries, Occupational Health and Safety, 51, 16-22.

Advances in Industrial Ergonomics and Safety II
Edited by Biman Das
Taylor & Francis 1990

COMPUTERIZED ELECTROMYOGRAPHY FOR ERGONOMICS

SHRAWAN KUMAR AND DOUG GARAND
DEPARTMENT OF PHYSICAL THERAPY

UNIVERSITY OF ALBERTA
EDMONTON, ALBERTA, CANADA

The motion and force requirement of the industrial tasks are important indicators of musculoskeletal stress on the workers. Their qualification will allow a characterization of jobs. A comparative survey of tasks using electromyography will assist task stress reduction, task re-evaluation and task redesign. A PC based comprehensive EMG recording and analysis program has been developed and is described here.

INTRODUCTION

Musculoskeletal stress has been a leading cause of occupational injuries. Their reduction in workplace has not only significantly enhanced the perception of stress but has impacted positively on reduction injuries and absenteeism (Corlett, and Bishop 1978, Wesgaard et al 1986 and others). Musculoskeletal loads can be predicted through the medium of kinematics and kinetics using Newtonian mechanics. However, due to elaborate set-up and sophisticated analytical techniques it becomes a less practical method for such a use. Electromyography being a noninvasive and easy to use technique is an attractive alternative.

Use of electromyography is based on the premise of quantitative relationship between EMG and force applied. Studies reported by Lippold (1952) claimed linear relationship between IEMG and external force in isometric conditions under submaximal and maximal efforts from human gastrocnemius muscles. Bigland Lippold (1954) found that when force of contraction varied, with the rate of muscle shortening kept constant, a linear relationship between IEMG and the load was present. They found this relationship to be valid throughout the range of constant velocities tested by them. However, they reported that the gradient of this slope was different, not only for different people, but between trials within the subject. Milner-Brown and Stein (1975) reported that the EMG potentials and contractile responses may sum nonlinearly at moderate to high force levels, but in such a way that the rectified surface EMG still was linearly related to the force produced by muscle. However, Vredenbreaht and Rau (1973) reported quite nonlinear relationship between tension and EMG in the elbow flexors over a wide range of joint angles. Other studies have also indicated a nonlinear relationship between these two variables (Boussiet 1973, Kumar et al 1989).

While the controversy about the quantitative relationship between the force and EMG is far from settled, many authors have advocated the value of EMG in muscle stress reduction strategy (Westgaard et al 1986, Silverstein et al, 1986, Kumar, et al, 1987, Luttman et al 1989 and others). Kumar (1987) contends that in most modern industries or occupations excessive physical demands have been eliminated. This leaves the majority of work stations with low level static load which may be intermittent, prolonged or continuous. In these cases the load can be assessed generally quantitatively and semiquantitatively by electromyography. In carefully controlled studies well defined static components of the tasks can be quantitatively evaluated using electromyography. This technique however continues to be employed in different ways by different workers. The purpose of this computer software is to standardize the electromyographic data acquisition and its analysis in a systematic way. This software has been designed to determine the phasic and quantitative relationships, with relative magnitudes of muscle loads, slope of rise and fall of the myoelectric activity. Further, it will determine the time of onset and finish (if needed) of each channel separately giving a picture of a relative biomechanical hierarchy of muscle involved in well defined physical activities. A further comparison with a maximal voluntary contraction will give a normalized magnitude of the biomechanical load. The latter will provide a mechanism of quantitative comparison of job stresses for their reduction for ergonomic task designs.

THE SOFTWARE

The software has two modules. The first is responsible for data acquisition and the second for data analysis. The data acquisition module (figure 1) has three submodules. The first submodule takes the calibration signals and sets the gain. The second sets up subject directory and generates the randomized test sequence depending on testing protocol. The third allows to set recording parameters, acquire data and finally produce a graphic output and save the file.

The data saved from the data acquisition module acts as the input to the analysis module. The input data files contain calibrated values of torque, angular displacement, and up to 14 channels of EMG. The analysis program is an interactive module processing user selected files. The data for each trial is windowed according to the thresholds chosen. The torque, postural angle and EMG is analyzed and graphically displayed on a synchronized time scale. The sequence of events of the analysis program are shown in the flow charts (figures 2, 3, and 4). The analysis program allows to perform the following:

1. Activity windowing (onset and/or finish) based on the event criteria.
2. Activity parameters, such as: duration; torque; peak torque; time-torque relationship; range; angular angle; time and torque at onset and finish; time and angle at the peak torque; average torque;

3. EMG characteristics of every muscle sampled: running magnitude value in most peak EMG activity, time of peak, torque at peak, total peak EMG activity of all muscles combined, peak of individual muscles as angular posture at the peak, percent of force average EMG per second, area under the curve (EMG time integral), total area of EMG of all muscles sampled combined, EMG area of this muscle as percent of total EMG area of all muscles combined, total area of EMG of all agonist muscles, total area of EMG of all antagonist muscles, peak EMG as percent of reference activity, average EMG as percent of the reference activity, biomechanically corrected force output for each muscle sampled, and EMG-torque relationship for each muscle with gross and iomechanically corrected torques.
4. Activity profile of all variables sampled on time scale at preselected discrete intervals as percent of the activity cycle within the activity window.
5. EMG fatigue analysis for each muscle looking at the median frequency and its shift.

DATA ACQUISITION FLOW CHART

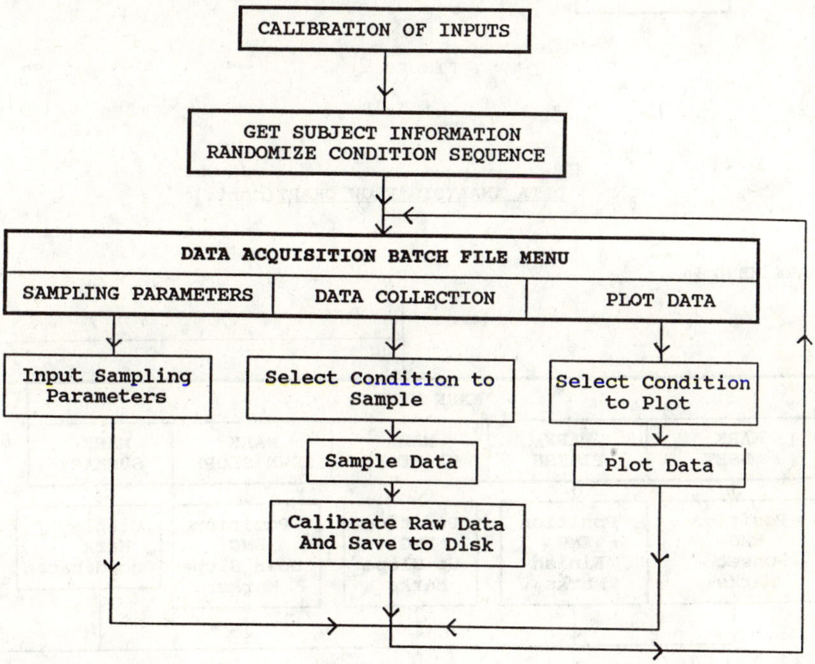

figure 1

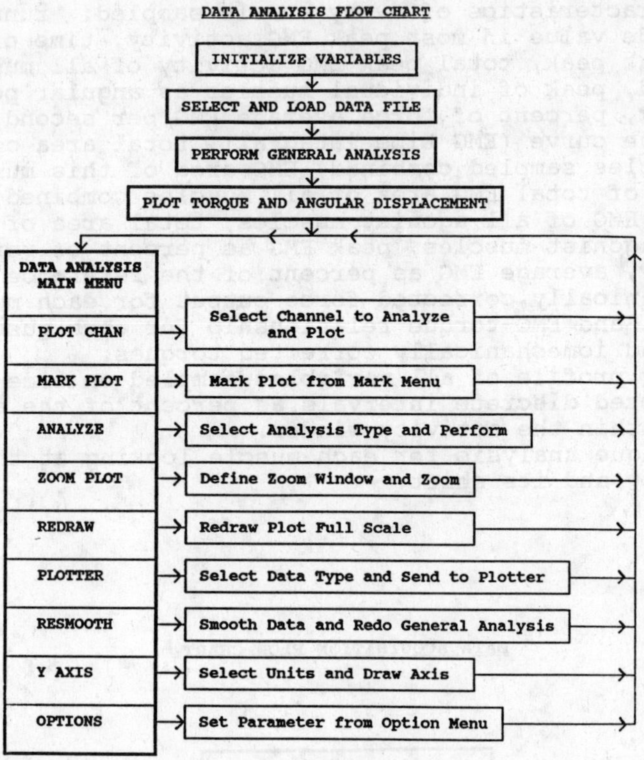

figure 2

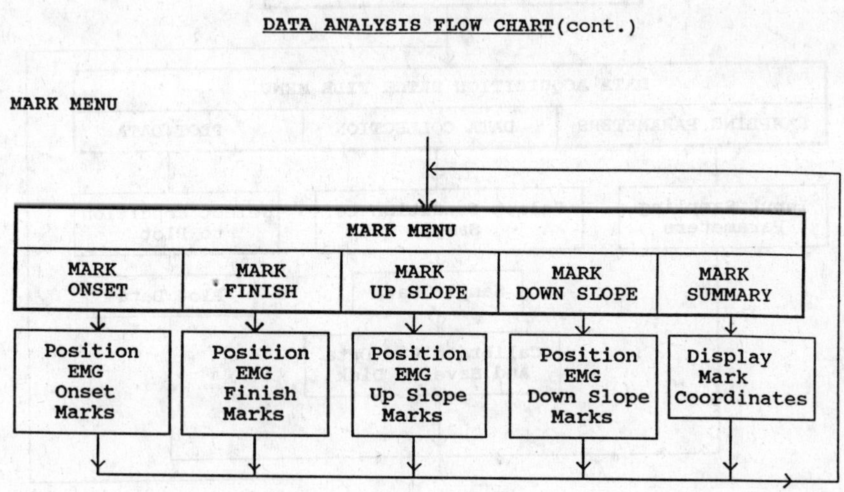

figure 3

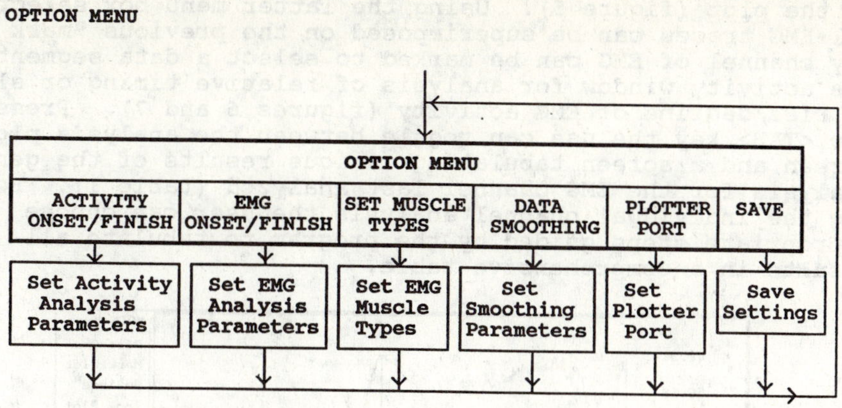

figure 4

The analysis program has six modules. The first module has help menu. The second and third submodules initialize the analysis program by accepting appropriate analysis criteria and setting up analysis. One of these sets up the analysis for kinesiological variables and the other for determination of fatigue depending on the option chosen. The fourth module performance analysis of EMG data along with the selected motion. The fifth module allows graphic superimposition of EMG on torque, angular displacement, smoothing of EMG data, calculation of slopes, zooming in, changing the axes and producing plots. The sixth and the final module allows to save the results and make next selection. To run this program one will require a minimum of 256 kB of memory and EGA graphics capability resident on IBM pc/xt/at. Graphic output of the analysis and the data can be obtained using an HP 7475A (or 100% compatible) plotter. The text analysis results are output to a file with formatting control codes embedded for an HP Laserjet+ printer.

PROGRAM EXECUTION

The program is activated by entering ANALYSIS (return). It will proceed in the following sequence. A header screen with an option of help or continuing with program will appear. The progression of the program is menu driven and is userfriendly. The user will be prompted for a data file directory. The current directory will be considered the default directory. If the entered directory is not located the user will be prompted to enter the path of the data files. If the directory is not found the directory selection prompt will repeat until a valid directory name is entered. From this directory appropriate file can be chosen by toggling the highlight bar and pressing the enter key. If the data needs smoothing it can be done by choosing the option. The program will proceed with general analysis displaying the results graphically. The first screen will contain a plot of torque and angular motion with a prompt/use input located along the bottom of the screen, and a menu box area along the right side of the plot (figure 5). Using the latter menu box selected or all EMG traces can be superimposed on the previous "mark plot" any channel of EMG can be marked to select a data segment as the activity window for analysis of relative timing or slope of rise/decline of EMG activity (figures 6 and 7). Pressing the <TAB> key the use can toggle between the analysis plots screen and a screen tabulating various results of the general analysis for the EMG channel last analyzed (table 1). Following the individual channel analysis the user can choose appropriate steps guided by the program to tabulate all results in a comprehensive table.

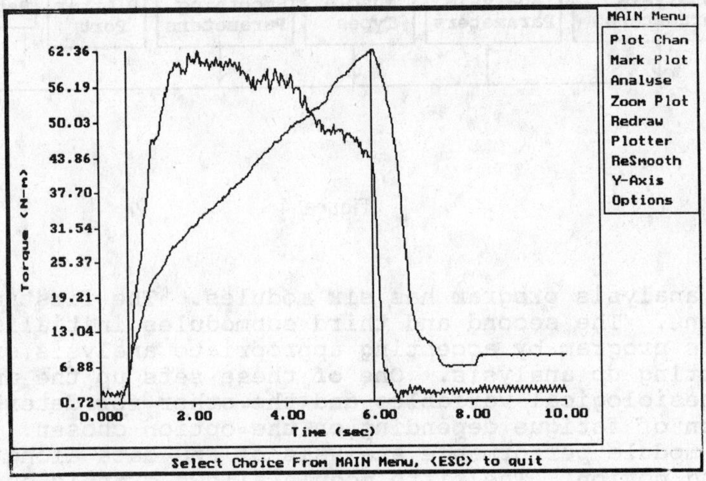

figure 5

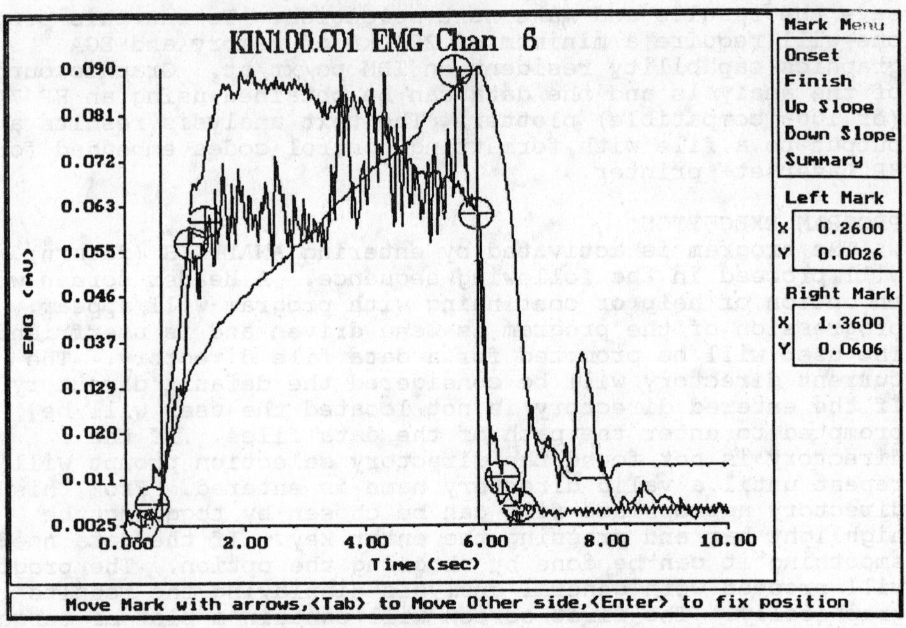

figure 6

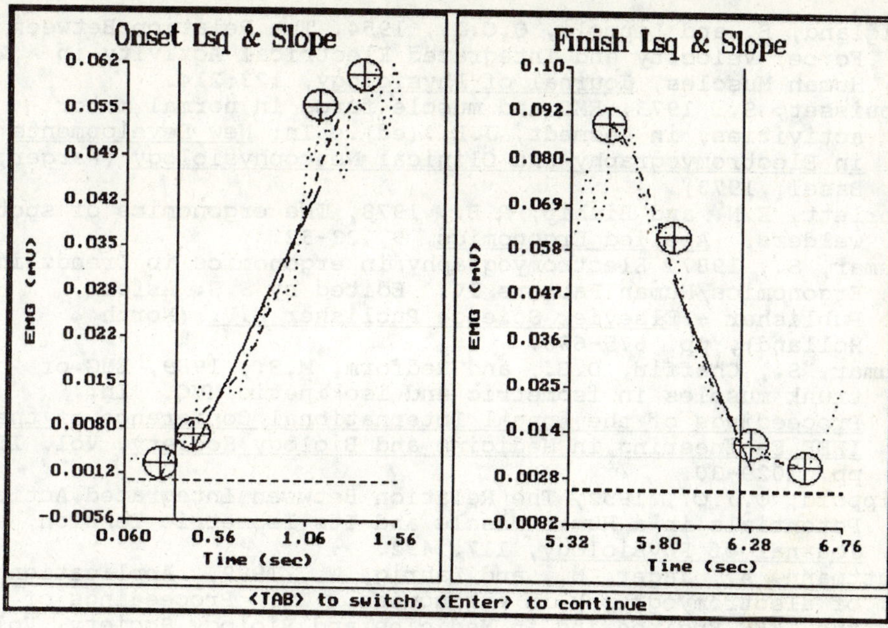

figure 7

General Analysis

Activity Onset :			
Time	=	0.6000	sec
Torque	=	5.38	N-m
Angle	=	-9.37	deg
Activity Finish :			
Time	=	5.8800	sec
Torque	=	43.72	N-m
Angle	=	39.29	deg
Duration	=	5.28	sec
Max Torque	=	62.0000	N-m
Avg Torque	=	51.60	N-m
Max Angle	=	39.29	deg

Analysis Results Channel 6

Activity Window :		(10 %)	
Start Time	=	-0.2600	sec
End Time	=	0.3400	sec
Slopes :			
Rise	=	0.0550	mV/sec
Decay	=	-0.1157	mV/sec
At EMG Maximum :			
Time	=	4.9400	sec
Torque	=	46.8300	Newtons
Angle	=	37.2300	degrees
Maximum	=	0.0889	mV
Average	=	0.0620	mV
Avg/sec	=	0.0117	mV/sec
Area	=	0.3278	mV-sec

table 1

REFERENCES

Bigland, B. and Lippold, O.C.J., 1954, The Relation Between Force, Velocity and Integrated Electrical Activity in Human Muscles, Journal of Physiology, 123:214.

Bouisset, S., 1973, EMG and muscle force in normal motor activities, in Desmedt, J.E. (ed). In: New Developments in Electromyography and Clinical Neurophysiology (Karger, Basel, 1973).

Corlett, E.N. and Biship, R.B., 1978, The ergonomics of spot welders. Applied Ergonomics, 9, 23-32.

Kumar, S., 1987, Electromyography in ergonomics in Trends in Ergonomics/Human Factors IV. Edited by S.S. Asfour, Publisher - Elsevier Science Publisher B.V. (North Holland), pp. 675-683.

Kumar, S., Chaffin, D.B., and Redform, M.S., 1989, EMG of trunk muscles in isometric and isokinetic MVC. In: Proceedings of the Annual International Conference of the IEEE Engineering in Medicine and Biology Society, Vol. II, pp. 1029-30.

Lippold, O.J.C., 1952, The Relation Between Integrated Action Potentials in a Human Muscle and Its Isometric Tension, Journal of Physiology, 117: 492.

Luttmann, A., Jager, M., and Laurig, W., 1989, Application of electromyography in ergonomics. In: Proceedings of the IEEE Engineering in Medicine and Biology Society, Vol. 11, pp. 1031-32.

Milner-Brown, H.S., and Stein, R.B., 1975, The relation between the surface electromyogram and muscular force. Journal of Physiology, 246, 549-569.

Silverstein, B., Fine, L., Armstrong, T., Joseph, B., Buchholz, B., and Robertson, M., 1986, Cumulative trauma disorders of the hand and wrist in industry. In: Corlett, E.N., Wilson, J., and Manenica, I., (eds) The Ergonomics of Working Posutres (Taylor & Francis, London 1986).

Vredenbregt, J., and Rau, G., 1973, Surface electromyography in relation to force, muscle length and endurance. In: New Developments in Electromyography and Clinical Neurophysiology. Ed. by J.E. Desmedt Publisher-Kerger, Basel, Vol. 1, pp. 607-622.

Westgaard, R.H., Waersted, M., Jansen, T., and Aaras, A., 1986, Muscle load and illness associated with constrained body postures. In: Corlett, E.N., Wilson, J., and Manenica, I., (eds) The Ergonomics of Working Postures (Taylor & Francis, London 1986).

Advances in Industrial Ergonomics and Safety II
Edited by Biman Das
Taylor & Francis 1990

Evaluation of muscular utilization percentage in trunk muscles for different conditions of load handling involving torsion

Denis Marchand[1], Martin C. Normand[1],
Yves Beauchamp [2], Daniel Imbeau [2]

[1] Département des sciences de l'activité physique
[2] Département d'ingénierie

Université du Québec à Trois-Rivières
P.O. Box 500, Trois-Rivières, QC, Canada, G9A 5H7

During the execution of an industrial handling task, mechanical stresses of low back muscles come from flexion, extension and torsion movements. This study was conducted to measure the level of activation of trunk muscles. The results suggest that the amplitude of trunk's torsional movements involves several changes in muscular utilization which support the view that the characteristics of the movement are as important as the handled load to explain muscular activation level.

INTRODUCTION

The mechanization and automatization of many industrial tasks are contributing to eliminate heavy load handling tasks. However a lot of specific and repetitive material handling tasks still remain. A study conducted by IRSST (Institut de Recherches sur la Santé et la Sécurité au Travail du Québec) revealed that repetitive body movements were responsible for 63 % of all reported back injuries. Indeed, muscular overuse which is attributable to high movement frequency and manipulation of heavy loads produce important stresses on the musculo-skeletal system. Most studies on manual load handling have considered movements in the sagittal plan. In industry, however, several handling tasks constrain the workers to use asymmetric movements of the trunk when moving loads. Schultz A., Crownwell R., Warwick D., Andersson G., (1987) studied muscular activation of the erector spinae and obliquus externus during isometric flexion, extension, lateral bending and rotation of the trunk. They found that erector spinae were more importantly solicited during extension movements than during lateral bending, whereas rotation movements induced a relatively low muscular activation. Furthermore, the obliquus externus abdominis were more importantly solicited during lateral bending, after flexion and rotation movements. Trunk muscles have different insertions and fiber orientations and therefore, are more or less active within different planes of motion. In order to know the implication of trunk muscles during handling tasks it will be preferable to use dynamic movements. This study compares different muscular loadings of the trunk muscles during material handling tasks involving dynamic torsion.

Most studies have shown their results by comparing the muscular activation levels of different loading conditions. Such an evaluation permits to follow the evolution of muscular demand according to the load increase or to the different movements, but does not tell about the percentage of maximal capacity the person is working. For instance, we do not know if the person is working at 30% or at 70 % of his maximal capacity. To obtain this percentage we need a reference value. This reference value is the activation (100 %) recorded during maximal voluntary contractions (Normand et al. 1986; Richards, 1985). The muscular utilization percentage (MUP) is obtain by dividing the EMG activity recorded during the execution of the task by the maximal value.

METHOD

Subjects

Twelve males subjects, ranging in age from 19-32 years, volunteered to participate in this study. They were all informed of the potential risks and hazards of their participation in this study and were required to sign an informed consent form. All subjects were examined to detect back pathology.

Tasks

The task consisted to manipulate three different loads (2, 10 and 20 kg) over three different trunk rotation amplitudes (0, 45 and 90 °). The movement speed was kept constant (40 °/s ± 5 °/s) for all movements. The tasks were as follows: (a) Sagittal lift: the subject takes a loads in front of him and bring back the load close to him (Figure 1-a); b) Torsion at 45 °: the subject lifts a load, placed on his left side at 45 ° from the sagittal axe (Figure 1-b), and carry the load over a motion range of 90° in order to place the load at 45 ° on his right side; c) Torsion at 90 °: the subject lifts a load, placed on his left side at 90 ° from the sagittal axe (Figure 1-c), and carry the load over a range of motion of 180° in order to place the load at 90 ° on his right side.

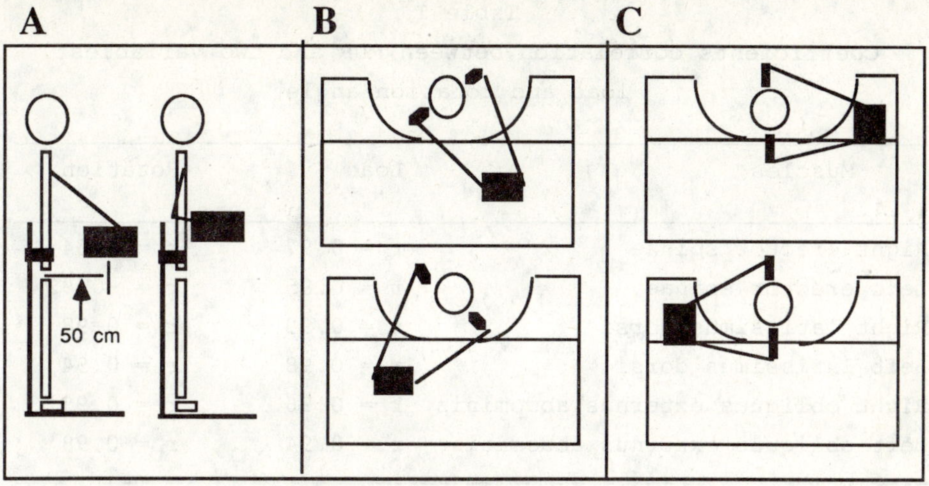

Figure 1: This figure illustrate the three movements. (A) the sagittal lift; (B) the torsion at 45° and (C) the torsion at 90 °.

Muscles

The muscle analyzed in this study were the right and the left erector spinae, latissimus dorsi and obliquus externus abdominis. For the erector spinae the electrodes were placed at three centimeters on each side of the spinous process at the L3 vertebra level. For the latissimus dorsi the electrodes were placed at three figerbreadths distal to and along the posterior axillary fold. For the obliquus externus abdominis, the electrodes were placed at five centimeters above the anterior superior iliac spine.

Apparatus

The myoelectric signals were recorded using Ag/AgCl disposable electrodes applied to the muscles. The signals were amplified and filtered (bandpass 30-600 hz) using a six channels Grass system (series #5), and were then full-wave rectified and low-pass filtered (6 Hz) to obtain a linear envelop of the EMGs. All signals were A/D converted (12 bits) using a multi-channel system (Watscope from Northen Digital inc.) connected to an IBM AT compatible computer.

Table I

Coefficients correlation between MUP and two variables: load and rotation angle

Muscles	Load	Rotation
Right erector spinae	r = 0.97	r = 0.64
Left erector spinae	r = 0.95	r = -0.92
Right latissimus dorsi	r = 0.99	r = 0.98
Left latissimus dorsi	r = 0.98	r = 0.94
Right obliquus externus abdominis	r = 0.96	r = 0.95
Left obliquus externus abdominis	r = 0.94	r = 0.98

The analysis of variance (Table II) reveals that the MUP of all right muscles increase significantly by increasing the load. For the left side muscles, only the erector spinae shows significant differences. The trunk tend to flex forward as the load increase. Since the erector spinae control the forward bending these muscles are more affected by an increase of the load.

Rotation effect

The results (Table I) indicate that the MUP of the latissimus dorsi and the externus obliquus abdominis muscles are highly correlated with the rotation angle. At the beginning of the lifting motion, the MUP of all contralateral muscles increase with the angle of rotation, except for the right erector spinae MUP which decrease between 45 ° and 90 ° of rotation. This decrease in the utilization of the right erector spinae is compensated by the two other right muscles (Figure 2).

For the muscles involved in the last portion of the movement (left side muscles) only the MUP of the left erector spinae decreased with increasing the angle of rotation (Figure 2). This decrease in the MUP is again compensated by an increase of MUP of the two other left muscles. The left externus obliquus abdominis is greatly utilized probably because he contributes to stop movement inertia and balance the load at the end of the movement. The ANOVA showed that the MUP of the right muscles are significantly increased by increasing the angle of rotation

Experimental procedure

First the subject was fixed in a special apparatus in order to avoid pelvic rotation. The loads were placed on a table in front of the subject. This table was adjustable in order to obtain zero degree of forward flexion of the trunk. The subject had to realize nine different experimental conditions (three loads x three angles of trunk rotation). The experimental conditions were completely randomized in order to avoid an effect of sequence. For each condition, EMGs of the six muscles were recorded. Three trials were preformed under each condition. The mean of these three trials was used for all subsequent statistical analyses.

Once the experimental conditions were completed the subject was required to perform maximal dynamic voluntary contractions (n=3) for each muscle groups involved in this study. The maximal contraction were recorded using an isotonic strength device (Dynatrac). The maximal activation was recorded during trunk extension in the sagittal plane for the erector spinae, during arm adduction in the frontal plane for the latissimus dorsi and during lateral flexion of the trunk in the frontal plane for the obliquus externus abdominis.

RESULTS AND DISCUSSION

Load effect

The results of this study reveal a linear relationship between load increase and MUP. As showed on table I, all muscles are highly correlated. These results are consistent with those reported by Petrofsky, J.S., 1979, Bigland-Ritchie & Woods, 1974, Bouisset & Goubel, 1973, Andersson et al., 1977 and Schultz et al., 1987

Since the movement was executed from the left to the right side, the contralateral muscles (right side) produced high MUP at the beginning of the lifting motion, while the left side muscles produced high values at the end of the movement. Figure 2 shows a greater load effect on the right erector spinae. The MUP is 36.9 % with a mass of 20 kg. However, for the muscles involved at the end of the lifting motion, the externus obliquus abdominis produce the greatest amount of work. This muscle plays an important role to stop the motion especially when the rotation amplitude is large. The Figure 2 indicates that the externus obliquus abdominis produced a MUP of 37.6 % with a mass of 20 kg.

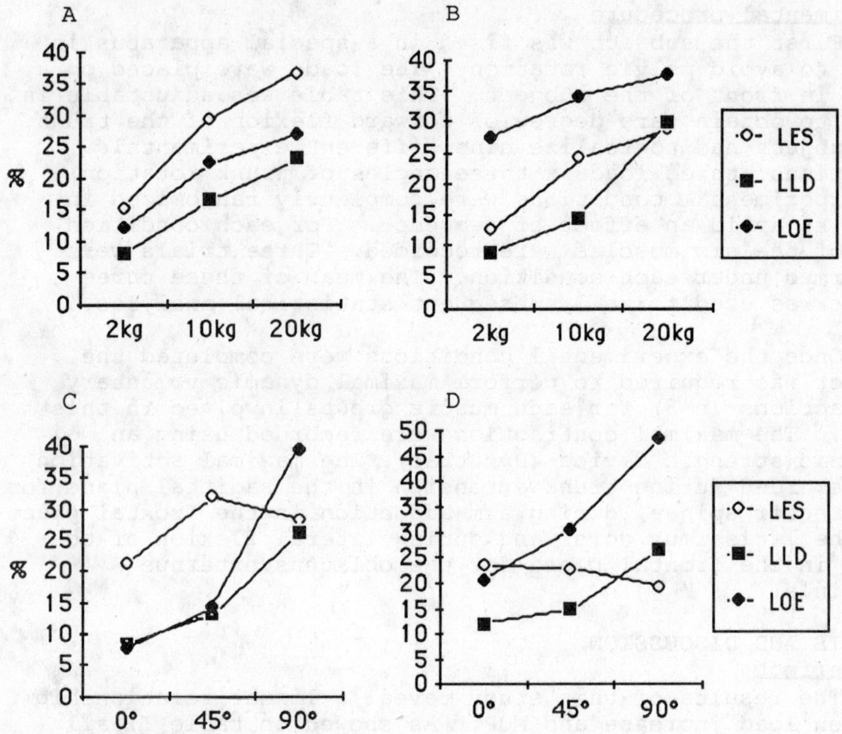

Figure 2: This figure compare the MUP for the right (A) and left (B) trunk muscles in function of the three loads manipulated and for the right (C) and left (D) trunk muscles for the three angles of rotation.

CONCLUSION

These results suggest that amplitude of trunk's torsional movement involves several changes in muscular utilization which supports the view that the movement's characteristics are as important as the handled load to explain muscular activation level. Two factors seem to have an influence on the activation of the erector spinae, mainly when the angle of rotation is 90 °. First, at this angle the torsion and the lateral bending of the trunk generate a great tension on lateral and posterior ligament which limit vertebra movements.

Table II

ANOVA results for the six muscles

	Load	Rotation
RES	F(2,10)=29.5 **	F(2,20)=24.9 **
LES	F(2,7)=36.9 **	F(2,14)=8.1 *
RLD	F(2,8)=25.1 **	F(2,16)=21.6 **
LLD	F(2,4)=6.1 NS	F(2,8)=5.6 NS
ROE	F(2,10)=29.3 **	F(2.20)=20,3 **
LOE	F(2,8)=3.2 NS	F(2.16)=5.9 NS

RES, right erector spinae; LES, left erector spinae; RLD, right latissimus dorsi; LLD, left latissimus dorsi; ROE, right obliquus externus abdominis; LOE, left obliquus externus abdominis

** p < 0.001 and
* p < 0.005;
NS not significant

This tension produced by the ligament help the mechanical work of the erector spinae by producing a decrease in the MUP (Gracovetsky, 1988). Second, the increase of MUP for the externus obliquus abdominis and the latissimus dorsi seem to demonstrate the implication of these muscles in the production of forces needed to move the load. Therefore, the fact that these muscles are more required by increasing the angle of rotation may be attributable to their anatomical positions. Indeed, considering that a muscle can produce strength only in their fibers alignment and that they have a good arm lever in order to balance the opposite force, it seems evident that loads took laterally implies more work for the externus obliquus abdominis and the latissimus dorsi.

REFERENCES

Andersson, G.B.J., Ortengren, R., Nachemson, A., 1977
Intradiskal pressure, intra-abdominal pressure and

myoelectrique back muscle activity related to posture and loading. Clinical. Orthopaedic. 129, 156-164

Bigland-Ritchie, B., Woods, J.J., 1974, Integrated EMG and oxygen uptake during dynamic contractions of human muscles. Journal of Applied Physiology, 36, 475-479.

Bouisset, S., Goubel, F., 1973 Integrated electomyographical activity and muscles work, Journal of Applied Physiology, 35, 695-702.

Gracovetsky., S., 1988, The spinal engine, New-York-Wien, Springer-Verlag.

Normand M.C., C.L. Richards, M. Filion, 1986, Muscle utilization in gait determined by a physiological calibration of EMG. Locomotion III, 263-264, 1986.

Petrofsky, J.S., 1979, Frequency and amplitude analysis of the EMG during exercise on bicycle ergometer. European. Journal of Applied Physiology., 41, 1-15.

Schultz, A., Cromwell, R., Warwick, D., Andersson, G.,1987, Lumbar trunk muscle use in standing isometric heavy exertions. Journal of Orthopaedic Research, 5, 320-329.

Advances in Industrial Ergonomics and Safety II
Edited by Biman Das
Taylor & Francis 1990

EVALUATION OF UPPER LIMB STRESSES USING MUSCULOSKELETAL LOADS DURING A ROTATING LIGHT ASSEMBLY TASK

R. Wells, A. Moore and J. Cholewicki

Department of Kinesiology, University of Waterloo
Waterloo, Ontario, Canada

This study investigated the utility of using musculoskeletal loads to assess the difference in chronic injury potential between tasks in rotating light assembly work. The musculoskeletal stresses were evaluated using predicted hand forces, as well as wrist and finger kinematics. These were combined to form estimates of soft tissue loads. Comparisons of musculoskeletal stresses, between jobs within the rotation sequence, showed one of the tasks to be much more stressful than the others. The method also showed the specific parts of this task which were stressful. This leads to suggestions for ergonomic interventions to reduce musculoskeletal stresses.

INTRODUCTION

This paper reports the upper limb musculoskeletal stresses experienced by rotating assembly workers on an light electronic assembly line, particularly the differences between jobs within the rotation sequence. Job rotation is often used to compensate for stressful work of short cycle time. Hagberg (1986) suggests that slightly different tasks may lead a decrease in the rate of fatigue. Jonsson, (1988) however, showed that for assembly of telephones, the load on the trapezius muscle showed only small differences between tasks whereas assembling and inspecting printed circuit boards did show differences in loading. The effectiveness of rotation therefore depends, in part, on the difference in all musculoskeletal loads between task. These previous investigators considered the shoulder musculature only but in these hand intensive tasks the musculoskeletal loads on the forearm and hand must also be considered. Wells et al. (1989) showed that using electromyography of the shoulder and forearm musculature was useful in ranking the stressfulness of jobs in a rotation sequence within a subject.

Armstrong (1986) reported a methodology for describing, quantitatively, upper limb posture and loads. In this paper we have extended this methodology to include the combined effect of hand loads, hand grip type and wrist posture in a manner which predicts measures relevant to soft tissue injuries.

METHODOLOGY

The tasks investigated here were line paced assembly operations (cycle time of 20-28 seconds), on small(weights of less than 500gm) electromechanical products. To relieve the physiological and psychological monotony, job rotation every two hours had been practiced routinely. The complete line was clustered into four groups on the basis of the dominant activities, Table 1.

Table 1. The four representative tasks. Right hand.

TASK#	ATTRIBUTE
17	Standing. Pick up part from belt using tweezers. Place in machine. Use tweezers to pick up second part from box and place onto first part. Press button on machine with fingers. Place complete part on conveyor.
18	Sitting. Take part from laterally placed conveyor and place in front. Take second part and place on the first. Place in machine and press palm buttons for 3 seconds. Replace on conveyor.
19	Sitting. Take part from conveyor onto workbench. Place three other parts into first. Place four screws into holes. Drive screws with inline, balanced air screwdriver. Replace part on conveyor.
20	Sitting. Take part from conveyor. Attach adhesive label onto part. Hold and inspect. Place in test fixture. Press button to initialise test and observe results. Replace part on conveyor.

Data on musculoskeletal stresses were collected using the instrumentation described by Wells et al., (1988). The system allowed the recording of quantitative estimates of the hand, wrist and shoulder loads during manipulative tasks. Four types of information were recorded: a standard video image of the activity, electromyographic activity of musculature (EMG) and the kinematics of the fingers and of the wrist. Kinematic data on hand motion were obtained via a lightweight instrumented glove (DataGlove) which transduced finger flexion/extension. The data were stored on a portable microcomputer. A non- restrictive goniometer (Penny and Giles) was used to transduce wrist flexion/extension and ulnar/radial deviation. The linear envelope EMG and wrist angle data (10 channels, bandwidth 0-10Hz) were collected on a battery powered signal processing unit worn as a waist pack. The EMG and wrist transducer data were then encoded and transmitted in digital form to one stereo sound channel of the video camcorder via an infra-red fibre optic link. The other sound channel contained timing pulses to synchronise the finger kinematics to the rest of the data. Upon playback of the video cassette all the above information could be retrieved, synchronized to the video image. Data on muscle activity(force), posture and repetition were recovered in machine readable form after analog to digital conversion at 30Hz.

Muscle activity of the workers was monitored bilaterally using surface electrodes over the trapezius(below the angle of the neck), the forearm flexors(midpoint of a line between medial epicondyle and radial styloid) and forearm extensors (over the bulk of the muscle just distal to the lateral epicondyle). The kinematics of the right wrist and fingers were also determined as previously described.

One female worker on the assembly line was selected as a subject in this pilot study. This volunteer had at least one years experience on the job and read and signed approved informed consent documents. Subject calibration involved eliciting maximal contractions from the forearm and shoulder musculature as well as calibrating the wrist and finger transducers. The relationship between EMG and grip force was established using sustained contractions against grip and pinch dynamometers.

After subject preparation and calibration the worker was monitored for at least ten minutes per workstation as she rotated through the four representative workstations. She worked at her regular places, at the line pace during a regular shift. The worker did not consider that the equipment hindered her in the performance of her task.

The linear envelope EMG data for the trapezius and forearm flexors were normalized to an MVC and analyzed over two minutes to produce the Amplitude Probability Distribution Function(APDF). These data and their interpretation for a cohort of six workers have been reported by Wells <u>et al</u>., (1989). The forearm electromyograms were further processed. Li <u>et al</u>., (1989) have shown that suitably calibrated myoelectric signals from surface electrodes can be used to predict hand loads. They also found that low pass filtering the full wave rectified EMG signal with a cutoff frequency of 1.5Hz gave the best fit between predicted and measured hand loads. In this study the calibration contractions were used to scale the subjects EMG data. The video tape was replayed in slow motion and the hand function defined as "out of contact with an object" or one of the grips defined below. The hand force was subsequently set to zero when the hand was not in contact with an object otherwise spuriously large loads can be predicted. Hand force was then estimated from a normalised model which represented hand force as quadratic function of EMG, wrist flexion/extension and wrist ulnar/radial deviation for different grip types. Six grip types were modelled; power grasp(30mm cylinder), "suitcase grip", one finger and two finger pinches, lateral "key pinch" and three finger pressure on a flat surface.

These force estimates were then used to compute a number of soft tissue loads and biomechanical parameters logically related to upper limb injury, (Wells <u>et al</u>., 1990). These included flexor tendon displacements, intra carpal tunnel loads and frictional work on the wrist flexor tendon sheaths. These, it will be noted, include the main risk factors of force, posture and repetition either singly or, in the case of frictional work, all together. These variables were computed for each of the four tasks in the rotation sequence.

RESULTS AND DISCUSSION

Table 2. summarizes the integrated effect of the calculated variables. Figure 1 depicts the time histories of the variables calculated. Complete cycles are also indicated on the figure.

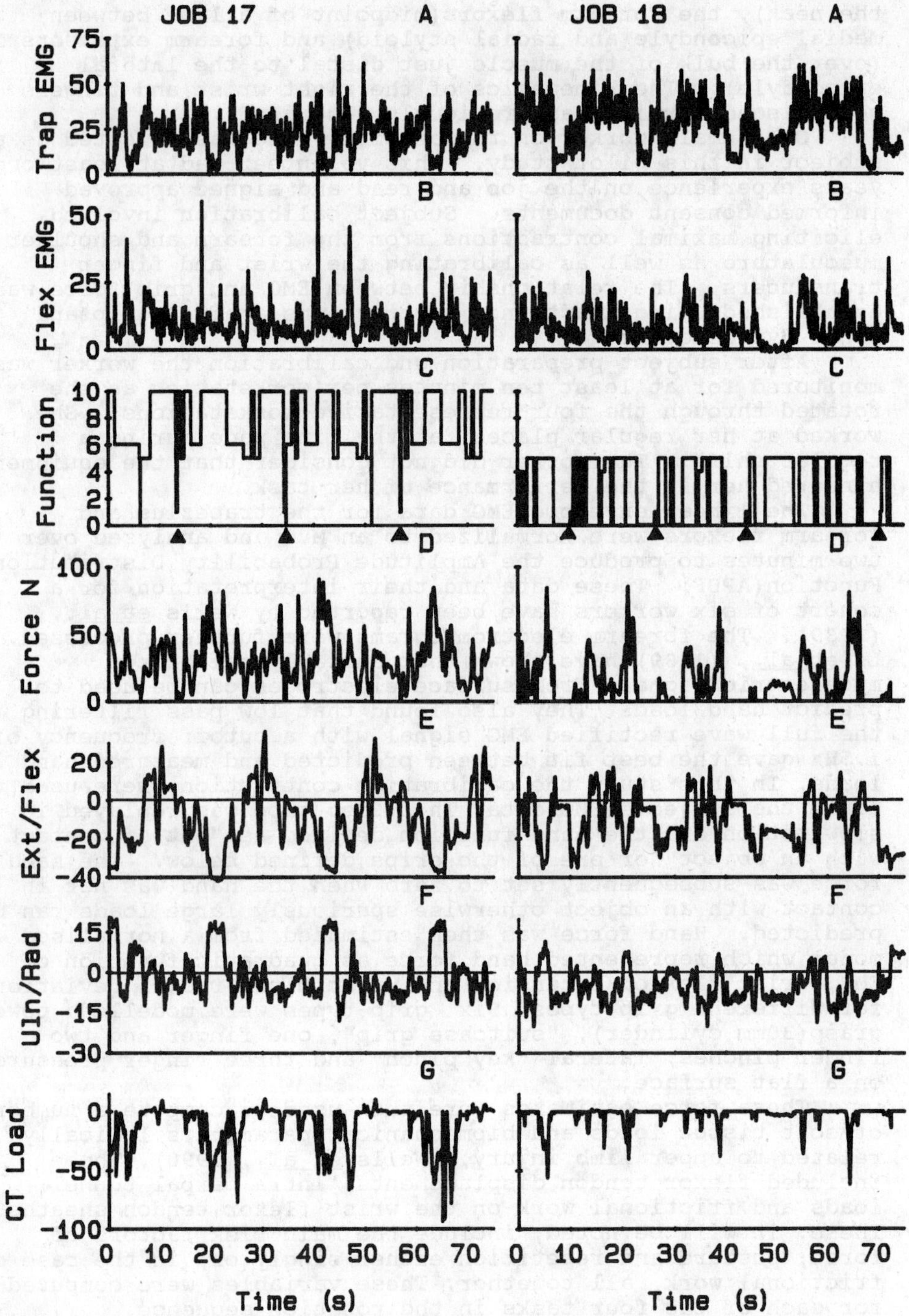

Figure 1. Time histories of the calculated variables for each of the four tasks. a)Trapezius and b)Forearm flexor EMG, low pass filtered at 6Hz, %MVC; c)hand function,0 -no contact, 1 - cycle marker, 2 - one finger tip pinch, 4 - one finger pulp pinch, 5 - two finger pulp pinch, 6 - lateral key pinch, 9and 10 - small diameter power grasps;

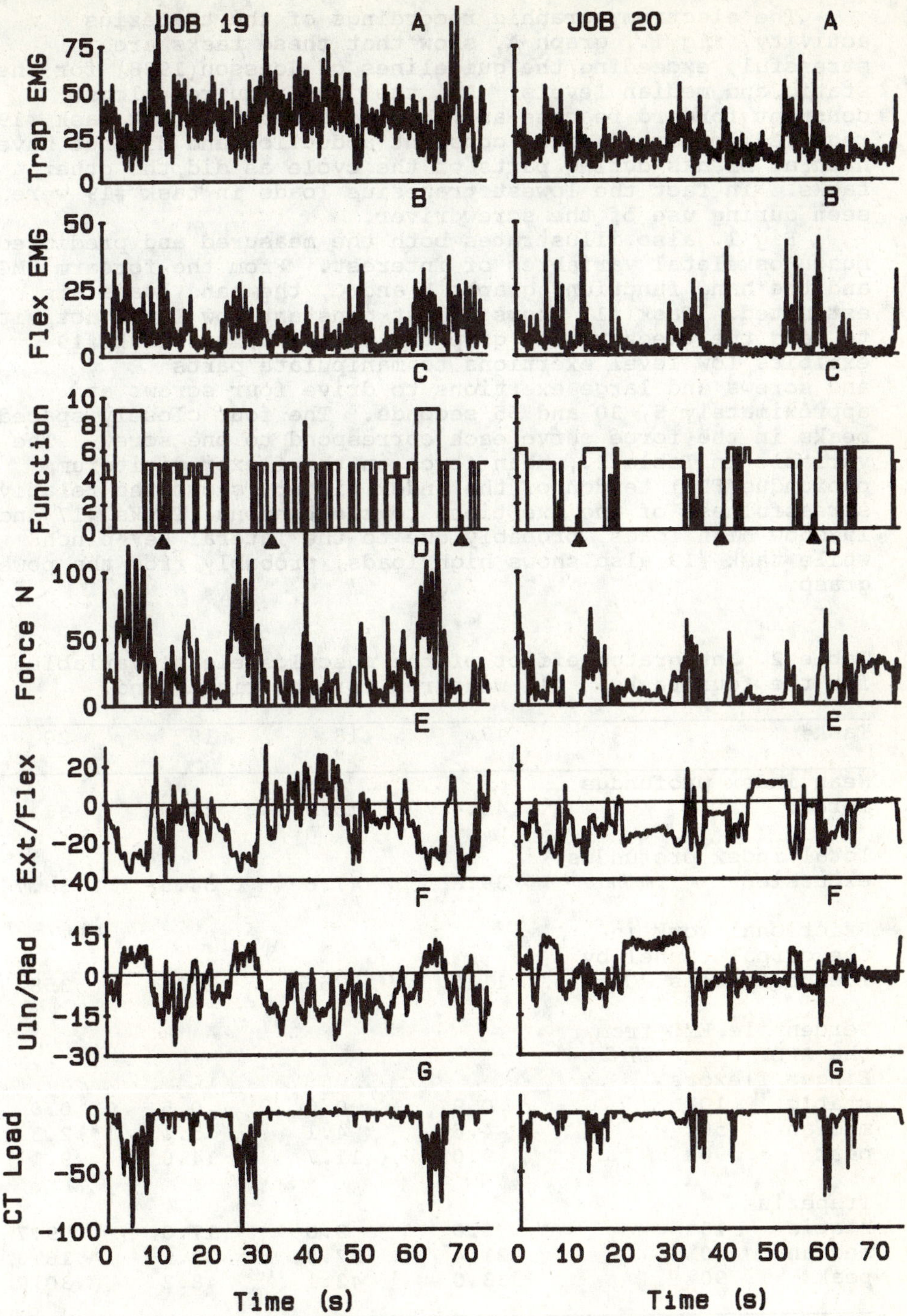

Figure 1.(Continued), d)predicted hand force, N; e) wrist angle, positive is flexion, degrees; f)wrist angle, positive is radial deviation, degrees; g)predicted intra carpal tunnel loads, negative is towards carpal bones, positive towards the flexor retinaculum, N/cm/cm.

The electromyographic recordings of the trapezius activity, Fig 1., graph A, show that these tasks are stressful, exceeding the guidelines of Jonsson(1988) for the static and median levels. All the jobs required almost constant forward reaches and shoulder abduction but task #19 required longer reaches, constant abduction and did not have natural breaks during parts of the cycle as did the other tasks. In fact the lowest trapezius loads in task #19 were seen during use of the screwdriver.

Fig 1. also illustrates both the measured and predicted musculoskeletal variables of interest. From the forearm EMG and the hand function, graphs B and C, the hand loads are estimated. Task #17 shows almost constant low level activity to hold the tweezers and grip the parts whereas task #19 exhibits low level exertions to manipulate parts and screws and large exertions to drive four screws at approximately 5, 30 and 65 seconds. The four closely spaced peaks in the force curve each correspond to one screw. The variable in Table 2., mean force on the flexor digitorum profundus(FDP) tendon of the index finger, shows the relative stressfulness of the exertions to the tendons. Tasks #17 and 19 show high loads, probably due to the lateral key pinch while task #19 also shows high loads, probably from the power grasp.

Table 2. Integrated effect of the musculoskeletal variables for the four tasks. All measures for the right hand.

Task#		17	18	19	20
Mean index profundus force.	*N*	45.4	15.9	34.0	31.3
Total index profundus excursion	*m/hr*	39.8	47.6	54.5	39.7
Frictional work in the carpal tunnel by flexor tendons	*J/hr*	1140	375	2824	358
Percentile EMG from the APDF	*%MVC*				
Finger flexors					
static	10%	0.9	0.0	0.5	0.0
median	50%	4.6	4.1	5.1	2.3
peak	90%	9.0	11.7	14.0	9.1
Trapezius					
static	10%	6.0	9.0	17.0	9.7
median	50%	21.4	27.8	33.1	16.3
peak	90%	33.0	42.1	48.2	30.9

The wrist angle in both flexion/extension and ulnar/ radial deviation are then represented as graphs E and F. The tasks show differences in range of motion used: task #20 demonstrates moderate wrist extension with wrist deviations close to neutral with many static periods. During this time the part is being held and inspected. In contrast task #19 shows two types of activity; manipulation and placing of parts and screws in slight flexion and in ulnar deviation and screw driving in about 25 degrees of extension and in radial deviation. Given that the functional position of the hand for these types of grip is in about 30 degrees of extension and slight ulnar deviation,(Wells et al., 1989) it can be seen that the tasks force the workers into undesirable positions. In task #19, for example, the position of the workpiece, flat on the bench, means that to install the screws a flexed wrist posture is required.

However, one cannot consider force and posture independently. A poor posture with minimal force exerts only small loads on the tendon sheaths as they pass through the carpal tunnel whereas a large load will be more stressful. Graph G plots the pressure exerted on the inside of the carpal tunnel, either towards the carpal bones (negative on the graph) or towards the flexor retinaculum (and median nerve). It reveals that for these jobs at least, little load was exerted on the flexor retinaculum. This worker was almost always in extension(task #19) or near neutral(task #18) at times when large forces were exerted.

Two additional variables are presented which help to characterise the stressfulness of the tasks, total tendon excursion and frictional work. These are presented in Table 2. Total tendon excursion through the carpal tunnel accounts for movements of the wrist and the finger joints. This therefore accounts for both repetitiveness and postural changes. Task #19 shows the highest excursions, most likely from the many movements of the wrist and fingers to place the parts and screws.

Frictional work includes contributions from force, posture and repetition: one can think of this measure as the product of the intra carpal tunnel loads(graph G) and the tendon excursion. For this measure task #19 was clearly the highest, probably because of the large intra carpal tunnel loads and the large tendon excursions.

In our previous laboratory simulations, (Wells et al., 1990) we found that two variables, frictional work and the median level EMG of the forearm flexors, followed reported injury risk most closely. On this basis we would rank task #19 as the most stressful on the rotation sequence followed by task #17. Considering the trapezius loads would support task #19's ranking. The relative musculoskeletal stresses determined here agree with the experience of the workers who also reported task #19 to be most stressful. Workers did not rotate from one "screwrunning" job to another similar task without a term on one of the other tasks.

Our recommendations for task #19 would be to reduce trapezius loads, perhaps by reorientation of the part, better matching of heights or arm rests and reducing wrist flexion by inclining the part towards the worker. The suggested contribution of the screw driving to the tasks stressfulness is more difficult to correct, given that most ergonomic features were already implemented on this tool. A reanalysis after changes would show whether the desired reduction in load on ALL components had taken place or whether it had simply moved the problem to another structure.

This study also demonstrated that, for this worker, the four tasks differed in the demands on the arms but that only tasks #17 and #20 allowed natural breaks in trapezius activity and even these tasks exceeded published guidelines for muscle load.

CONCLUSIONS

The method of evaluating the stressfulness of manipulative tasks described here has been shown to be sensitive to differences between manipulative tasks and to be helpful in pinpointing specific undesirable features of the tasks. This preliminary study also shows that it is feasible to perform these measures on workers during their regular work with minimal interference. At present we cannot interpret the values obtained for the musculoskeletal loads but larger scale studies, currently underway, will demonstrate the relationship between these measures and musculoskeletal disorders of workers.

ACKNOWLEDGEMENTS

This work was supported by grants from the Ontario Ministry of Labour, 224/R and NSERC A2785. The authors would like to acknowledge the technical help of Sharon Orr as well as the help, patience and good will of the subjects, union and management.

REFERENCES

Armstrong, T.J., 1986, Upper extremity posture: Definition, Measurement and control. In: The Ergonomics of Working Postures, edited by N. Corlett, J. Wilson and I. Manenica, (Taylor and Francis, London), pp.59-73.

Hagberg, M., 1986, Optimizing occupational muscular stress of the neck and shoulder. In: The Ergonomics of Working Postures, edited by N. Corlett, J. Wilson and I. Manenica, (Taylor and Francis, London), pp. 109-114,.

Jonsson,B., 1988, The static component in muscle work. European Journal of Applied Physiology, 57, 305-310.

Li, Y., Wells, R.P., Ranney, D. and Moore, A., 1989, Prediction of the pinch and grasp force exerted by the hand using surface and fine wire electromyography. In: Proceedings of the XII Congress of the International Society of Biomechanics, Los Angeles, June 26-30.

Wells, R.P., Moore A.E. and Ranney, D., 1990, Evaluation of hand intensive tasks using biomechanical load factors. In: Work Design in Practice, edited by C. Haslegrave, N. Corlett, J. Wilson and I. Manenica, (Taylor and Francis, London), (In Press).

Wells, R., Moore, A.E. and Ranney, D., 1989, Musculoskeletal stresses during light assembly. In: Proceedings of the 22nd Annual Conference of the Human Factors Association of Canada, Toronto, Canada, November 26-29.

Wells, R., Moore, A.E. and Ranney, D., 1988, Development of a system for recording occupational hand and wrist movement. In:Proceedings of the 21st Annual Conference of the Human Factors Association of Canada, Edmonton, Canada, Sept 13-16.

Advances in Industrial Ergonomics and Safety II
Edited by Biman Das
Taylor & Francis 1990

The effect of tilting the seat sideways on muscular activity of the back and on perceived comfort

Yves Beauchamp[1], Daniel Imbeau[1], Martin C. Normand[2], and Paul-André Courtois[1]

[1]Département d'ingénierie
[2]Département des sciences de l'activité physique

Université du Québec à Trois-Rivières
P.O. Box 500, Trois-Rivières
Québec, G9A 5H7, Canada

The sideways inclination of the seat on back muscular activity and on subjective perception of comfort were investigated. In the first part of the experiment, EMG activity of the left and right erector spinae muscles at the L3 level were measured in four subjects. In a second part, perceived comfort was rated on a nine-point rating scale by eight subjects. Results show a significant asymmetric muscular activity in the back as tilt angle increases. Perceived comfort is significant for relatively small angles of tilt. These results are presented and discussed.

INTRODUCTION

Although few, there are situations where an operator's seat may not be perfectly level in the frontal plane. For instance, an exploratory study on back pain in relation with work equipments was performed among nearly 600 police officers employed in the province of Québec. Among other results, this study revealed that car seat tilt angles of up to 4.67 degrees were present in fairly recent patrol cars (Côté, 1989).

The literature on seating generally assumes that a seat is level in the frontal plane (Chaffin and Andersson, 1984; Sanders and McCormick, 1987; Rodgers and coworkers, 1983; Scherrer and coworkers, 1981). Therefore, no research data appears to exist on the effect of tilting a seat sideways on the muscular activity of the back and on perceived comfort. It can be expected that such seating conditions are uncomfortable because they produce an imbalance in back muscular activity which stresses the worker's back, as the tilting angle increases. Such an imbalance in back muscular activity can lead to muscular pain in the short term and eventually to chronic problems (Scherrer and coworkers, 1981).

The question remains as to whether there is a minimum tilt angle above which muscular activity of the back muscles and perceived comfort, differ significantly from a level seat condition (i.e., zero degree seat tilt angle). An pilot study was thus conducted to investigate this question.

METHODS AND PROCEDURE

This experiment was conducted in two parts: back muscle electromyographic (EMG) measurements and perceived comfort. In the first part, EMG was measured on erector spinae muscles for seven values of seat tilt angles ranging from zero to three degrees. In the second part, perceived comfort was rated on a nine-point rating scale for seat tilt angles ranging from 0 to 4.5 degrees.

Subjects

Four male subjects (mean age of 27.7 years; S.D.= 1.7 years), volunteered to participate in the first part of the study. The mean stature of the subjects was 1.77 m (S.D.= 8.9 cm), and mean body weight was 77.6 Kg (S.D.= 6.4 Kg). Eight different male subjects (mean age of 26.6 years; S.D.= 3.7 years) participated in the second part of the experiment. The mean stature of these subjects was 1.78 m (S.D.= 8.6 cm), and mean body weight was 73.9 Kg (S.D.= 7.8 Kg).

To participate in the experiment, the subjects were required to not have experienced back problems in the last few years. They were college students and were paid for their participation on an hourly basis and were asked to sign an informed consent form.

Equipment

EMG was measured using surface pellet electrodes affixed to the skin over the left and right erector spinae muscles at the L3 level. The signals were pre-amplified, filtered, rectified, and integrated using a Nihon Kohden amplification and recording unit. Various support equipments (e.g., oscilloscope, chart recorder, tape recorder, etc.) were also used.

The tilt angle was generated using a variable slope electric treadmill on which a hard seat pan chair was fixed. The chair did not have a back support in order to minimize noise problems in the EMG measurements. The tilt angle of the chair was carefully controlled using a goniometer.

A shelf was installed in front of the subject as a home position for the hands to simulate use of a computer keyboard.

Procedures

In the first part of the study, the subject was shortly briefed about the purpose of the experiment and then asked to sign the consent form. The electrodes were then affixed to the skin of the subject. Next, he was asked to sit on the chair (seat tilt angle = 0 degree) and to rest his hands on the shelf installed in front of him. The shelf height was then adjusted so that the elbow angles were approximately at 90 degrees. Once the experimenter was satisfied with these adjustments, the experiment proceeded by presenting the subjects with the various seat tilt angles. A total of 7 seat tilt angles ranging from zero to three degrees (0.5 degree steps) were presented randomly to the subject. At each angle value, EMG measurements were taken during a 25 second period. The subject was notified of the forthcoming measurement period and hence was instructed not to move. To help the subject keep a static position, a mark to concentrate on was put on the wall in front of him.

Two subjects experienced the "uphill" side of the slope on their right hand side and the other two, on their left hand side. This was to ensure that the results would be independent of the slope orientation. To accomplish this, the chair was rotated 180 degrees on the treadmill so that an equal number of subjects experienced tilt angles on both sides. The "uphill" side of the slope is the side on which the muscles sampled were under compression while the "downhill" side was the side on which the muscles were under tension.

At each seat tilt angle, the subject was instructed to adopt a posture that was as comfortable as possible while keeping his shoulders level. Markers affixed on the skin at the shoulder height were used to ensure that these would remain level throughout the measurement periods.

Between each seat tilt angle condition, the subject was allowed to rest for a five minute period at a seat tilt angle of zero degrees to minimize fatigue effects. The duration of the first part of the experiment was about 1.5 hour.

In the second part of the experiment, the same equipment and procedure were used to generate the seat tilt angles. However, the EMG measurements were replaced by a measure of perceived comfort. After having experienced the tilt condition for about 1.5 minute, the subject was asked to report his level of perceived comfort by calling out the most appropriate corresponding number on the following nine-point rating scale:

1 - comfortable
2 -
3 - slightly uncomfortable
4 -
5 - uncomfortable
6 -
7 - very uncomfortable
8 -
9 - extremely uncomfortable

Subjective measures being much easier and quicker to obtain than EMG activity, ten tilt angles were used in the second part of the experiment instead of seven as in the first part (i.e., 0 to 4.5 degrees in 0.5 degree steps). Based on the results from the first part, it was decided that all subjects would experience the tilt angles on the same side, in the second part.

<u>Experimental design</u>

Each subject experienced seven trials in the first part of the experiment and ten trials in the second part. For each of these parts, a two-way ANOVA (analysis of variance) design was used with subjects and seat tilt angle serving as independent variables throughout the experiment. The dependent variables were slope of the integrated EMG signal and rated perceived comfort in the first and second part, respectively.

RESULTS AND DISCUSSION

<u>Effects of independent variables</u>

The side of the seat tilt angle (right/left) was not found significant at the $\alpha = 0.05$ level of confidence therefore, the data were pooled.

The summary of experimental results are shown in Tables 1-3. The ANOVAs revealed that there was a significant effect of seat tilt angle on all dependent measures: "uphill" side erector spinae EMG activity ($p < 0.0045$); "downhill" side erector spinae EMG activity ($p < 0.019$); and rated perceived comfort ($p < 0.0001$).

Table 1. Anova summary table for the "uphill" side erector spinae EMG activity.

Source	df	SS	MS	F	p
Subjects (S)	3	0.96	0.32		
Seat tilt angle (A)	6	1.28	0.21	4.76	.0045
S * A (Error)	18	0.81	0.05		

Table 2. Anova summary table for "downhill" side erector spinae EMG activity

Source	df	SS	MS	F	p
Subjects (S)	3	0.22	0.08		
Seat tilt angle (A)	6	0.31	0.05	3.46	.019
S * A (Error)	18	0.27	0.02		

Table 3. Anova summary table for perceived comfort.

Source	df	SS	MS	F	p
Subjects (S)	7	222.24	31.74		
Seat tilt angle (A)	9	842.51	93.61	47.17	<.0001
S * A (Error)	63	125.01	1.98		

Newman-Keuls post-hoc multiple range tests were also performed on the means of the three dependent variables. To obtain a better idea of the relative increase ("uphill" side) or decrease ("downhill" side) of back muscle EMG activity, the means were divided by their corresponding zero degree seat tilt position value. The results of these tests are shown in Figures 1 and 2. On the figures, different letters indicate means which are significantly different from each other at the $\alpha = 0.05$ level.

Figure 1 shows that the "uphill" side relative EMG activity began to differ significantly (at the $\alpha = 0.05$ level) from that of the level seat value at a two degree seat tilt angle. For the the "downhill" side, a corresponding significant difference is observed at a seat tilt angle of three degrees. Concerning the rated perceived comfort, the corresponding difference is significant at a two degree seat tilt angle. Interestingly, this is the same tilt angle at which a significantly higher relative EMG was observed on the "uphill" side.

<u>Prediction of relative EMG activity</u>

An attempt was made to develop simple regression equations in order to predict the relative EMG activity based on the seat tilt angle. These equations are shown in Figure 3.

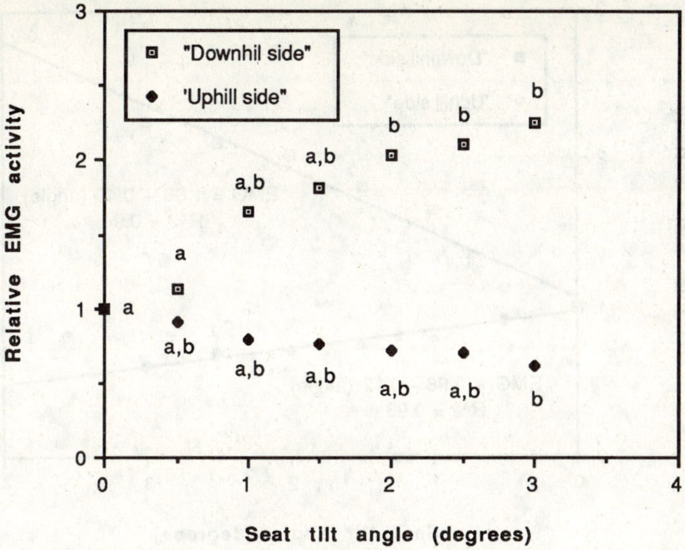

Figure 1. Relative EMG activity for the "uphill" and "downhill" erector spinae muscles as a function of seat tilt angle.

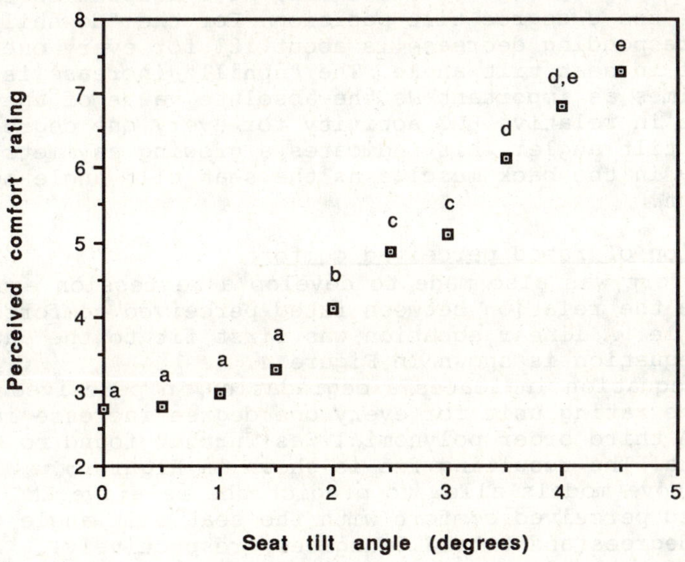

Figure 2. Perceived comfort rating as a function of seat tilt angle.

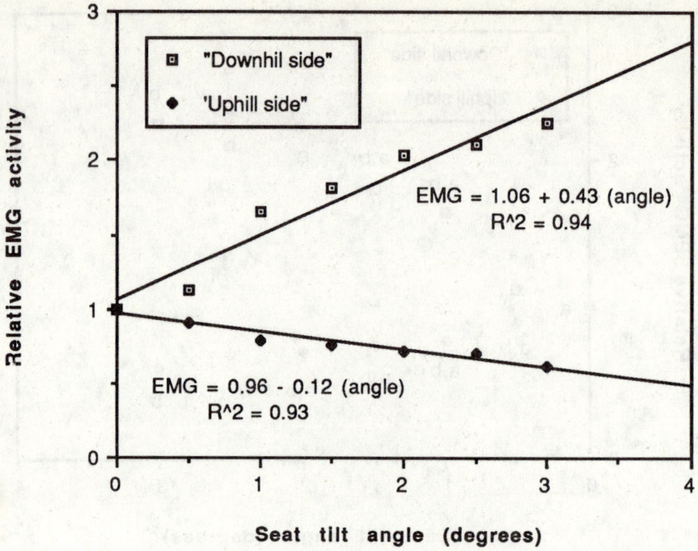

Figure 3. Regression equations for the relative EMG activity as a function of seat tilt angle.

The resulting equations indicate an increase of about 40% in EMG activity for every one degree increase in seat tilt angle for the "uphill" side of the chair, with respect to the activity level at the 0 degree tilt position. For the "downhill" side, the corresponding decrease is about 12% for every one degree increase in seat tilt angle. The "uphill" increase is thus over three times as important as the absolute value of the "downhill" decrease in relative EMG activity for every one degree increase in seat tilt angle. This indicates a growing asymmetrical activity in the back muscles as the seat tilt angle becomes more important.

Prediction of rated perceived comfort

An effort was also made to develop a regression equation to describe the relation between rated perceived comfort and seat tilt angle. A linear equation was first fit to the data. The simple equation is shown in Figure 4.

This equation indicates a degradation in perceived comfort of about one rating unit for every one degree increase in seat tilt angle. A third order polynomial was further found to best fit this data. The resulting fit is shown in Figure 5.

The above models allow to predict the relative EMG activities and rated perceived comfort with the seat tilt angle range from 0 to 3 degrees and 0 to 4.5 degrees, respectively.

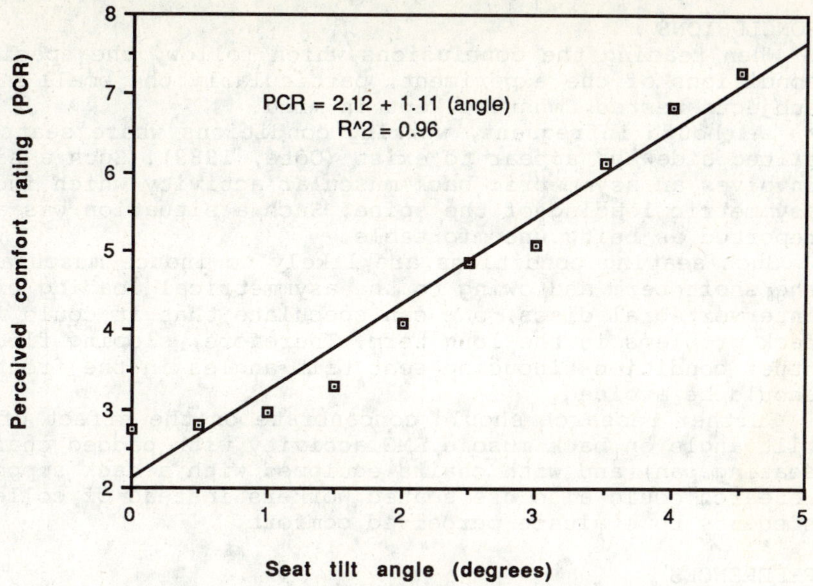

Figure 4. Linear regression equation for the rated perceived comfort as a function of seat tilt angle.

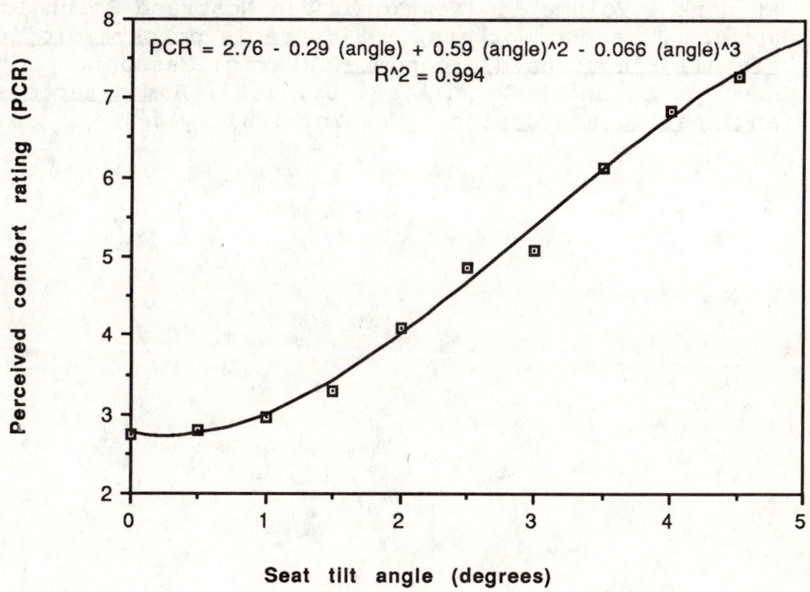

Figure 5. Third order polynomial regression equation for the rated perceived comfort as a function of seat tilt angle.

CONCLUSIONS

When reading the conclusions which follow, the specific conditions of the experiment, particularly the small number of subjects tested, must be kept in mind.

Although infrequent, working conditions where seats are tilted sideways appear to exist (Côté, 1989). Such a situation involves an asymmetric back muscular activity which indicates an asymmetric loading of the spine. Such a situation was also reported as being uncomfortable.

Such seating conditions are likely to induce muscular pain in the short term and owing to the asymmetrical loading of the intervertebral discs, one can speculate that it could lead to back problems in the long term. Therefore, sloping floors or other conditions inducing seat tilt angles in the frontal plane should be avoided.

Further research should concentrate on the effect of seat tilt angle on back muscle EMG activity with padded chairs (soft seating pan) and with chairs equipped with a back support. Such research could also use seated workers instead of college students to evaluate perceived comfort.

REFERENCES

Chaffin, D. B. and Andersson, G. 1984, Occupational biomechanics. (New York: Wiley).

Côté, M.-M., 1989, Auto-patrouille et maux de dos chez les policiers du Québec. (Montréal: Institut de recherche en santé et en sécurité du travail du Québec).

Rodgers, S. H. and coworkers, 1986, Ergonomic design for people at work - Volume 1. (New York: Van Nostrand Reinhold).

Scherrer, J. and coworkers, 1981, Précis de physiologie du travail - notions d'ergonomie. (Paris: Masson).

Sander, M. S. and McCormick, E. J., 1987, Human factors in engineering and design. (New York: Wiley).

IV

CUMULATIVE TRAUMA DISORDERS

Advances in Industrial Ergonomics and Safety II
Edited by Biman Das
Taylor & Francis 1990

CARPAL TUNNEL SYNDROME UPDATE

Fariborz Tayyari, Ph.D., P.E.

Department of Industrial Engineering
Bradley University
Peoria, Illinois 61625

and

Abdul-Karim Sohrabi, MD, MPH, FACS

Veteran Administration Medical Center
Danville, Illinois 61832

INTRODUCTION

Carpal tunnel

Carpal tunnel is a tunnel-like structure in the wrist which is enclosed by eight carpal bones (forming an arch) and the transverse carpal ligaments (roofing the arch). Nine tendons which control fingers movement and the median nerve (which enervates the thumb, index and middle fingers, and half the ring fingers pass through the tunnel, from the forearm into the hand. The carpal tunnel is illustrated in Figure 1 (Schenck, 1988; Dionne, 1984a; US Congress, 1985).

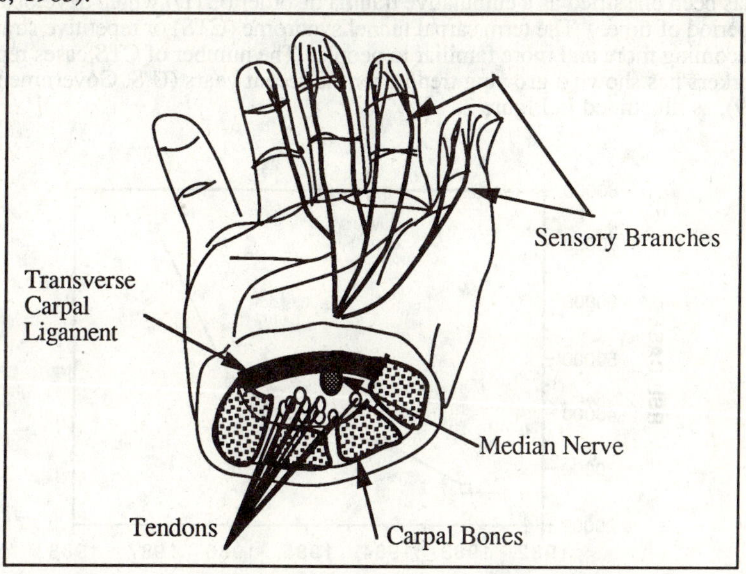

Figure 1: An illustration of the carpal tunnel, which is made up by carpal bones and transverse carpal ligament.

Carpal tunnel syndrome

When the wrist tendons become irritated and swollen, a resulting compression on the median nerve impairs the sensory and motor function of the hand. The nerve compression can be induced by wrist fractures, rheumatoid arthritis, tumors or diabetes mellitus (NIOSH, 1989). The increased pressure in the bony tunnel can trap and/or pinch the median nerve causing occasional numbness, pain and tingling in the thumb, index and middle fingers. This condition is called carpal tunnel syndrome (CTS).

Synonyms

The following synonyms have been used for CTS (Durham, 1960):

1. Tenovaginitis stenosan at the carpal tunnel;
2. Compression neuropathy of the median nerve in the carpal tunnel;
3. Partial thenar atrophy;
4. Median neuropathy;
5. Tardy median palsy.

Symptoms of carpal tunnel syndrome

A review of literature (e.g., Durham, 1960; Armstrong, 1983; Dionne, 1984a; U.S. Congress, 1985) as well as a recent pilot study conducted by the authors on hospital records of several patients indicated that any combination of the following symptoms may end up with the diagnose of the CTS:

1. Numbness, tingling and/or painful burning in the thumb, index and middle fingers in one or both hands. These symptoms occur most frequently at night and may be so severe that awaken the patient. They may also be present in the morning upon waking.
2. Maintaining the arms in certain positions will also cause the symptoms to occur.
3. Difficulty in moving the fingers.
3. Loss of grip strength.
4. Loss of power to squeeze bottles or open jars.
5. Loss of fingers sensation.
6. Subjective feeling of swollen fingers while little or no swelling is apparent.
7. In advance cases, the thenar muscle at the base of the thumb atrophies.

Reasons for increased number of occupational cases

CTS has been classified as a cumulative trauma disorder (CTD) which develops over an extended period of time. The term carpal tunnel syndrome (CTS) or repetitive strain injury (RSI) is becoming more and more familiar to people. The number of CTS cases reported among workers has shown a growing trend over the recent years (U.S. Government, 1984; Roel, 1989), as illustrated in Figure 2.

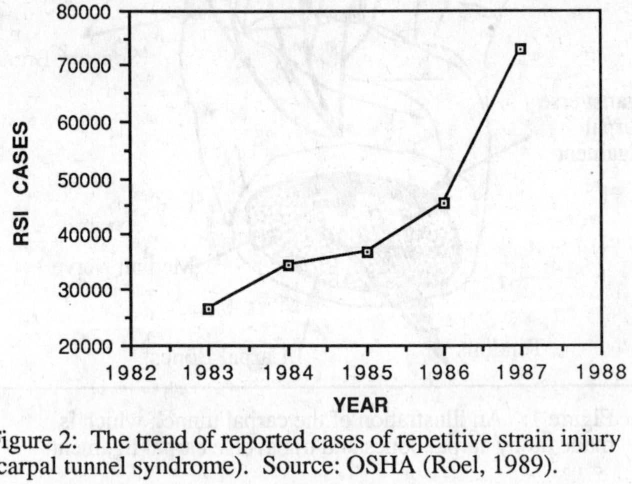

Figure 2: The trend of reported cases of repetitive strain injury (carpal tunnel syndrome). Source: OSHA (Roel, 1989).

NIOSH states that this increase may be explained by the following two reasons:

1. Automation and job specialization have limited the workers to performing a certain job task involving only a few manipulations, which are repeated an enormous number of times during a work-shift.
2. The increased awareness among workers regarding work tasks as a factor in the onset of carpal tunnel syndrome.

CTS RISK FACTORS

An exact causal relationship has not yet been developed for CTS. However, a review of literature revealed that the following factors should be given a significant consideration when studying CTS cases. The risk factors can be classified as occupational and non-occupational:

I. Occupational risk factors

The following occupational activities have been found to have connections with CTS (e.g., U.S. Government, 1984; Armstrong, 1986):

1. Repetitive movements: Performing tasks involving highly repetitive manual acts and repetitive use of small hand tools.
2. Force: Performing manual tasks requiring force.
3. Awkward hand motions: Performing tasks requiring the use of hand in an awkward position, excessive wrist deviations, or other stressful wrist postures in which the tendons will have to operate while bent over bones or tough transverse ligaments.
4. Pinch-hand position: Carrying thin items (e.g., books) held in a "pinch" position between the thumb and fingers which creates strain in the hand.
5. Low frequency vibration: Using vibrating tools (e.g., air- or motor-powered drills, chippers, sanders, saws, and drivers), especially under cold conditions.
6. Inadequate tool handle: Using tools that dig into the palm of the hand (e.g., short-handled screwdrivers) that irritate the tendons.
7. Gloves: Using gloves that do not fit properly.
8. Low temperature.

When the hand is used in an awkward position, the tendons are bunched together and will have to operate while bent over the carpal bones (bending wrist dorsally, Figure 3-a) or the transverse carpal ligament (bending the wrist palmarly, Figure 3-b). Repetitive and forceful wrist deviations (flexion or extension movements) gradually damage the median nerve and result in an increase of carpal tunnel pressure on the median nerve.

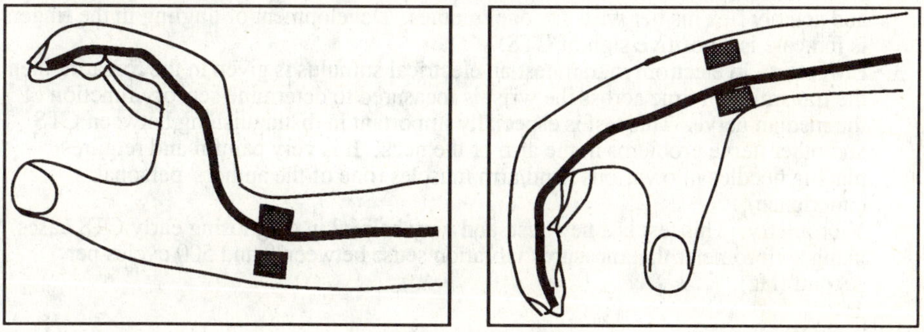

(a) (b)

Figure 3: When the wrist is bent dorsally (a) or palmarly (b), the tendons will have to operate while bent over the carpal bones or the transverse carpal ligaments, respectively. The friction may cause the tendons or their sheaths to swell.

Silverstein et al. (1987) found that CTS was strongly associated with high-force/high-repetitive jobs and to a lesser extent with low-force/high-repetitive jobs and high-force/low-repetitive jobs.

II. Non-occupational risk factors

The following non-occupational factors have been observed to be associated with many CTS cases (e.g., Durham, 1960; Massey, 1985; Daven, 1982; Yate et al., 1981; Phalen, 1970; Johnson et al., 1983; Bennett and Crouch, 1982; Sikka et al., 1983):

1. Pregnancy and breast feeding in women: However, these cases are usually transient and the symptoms should disappear after giving birth and stop of breast feeding.
2. Certain medications or therapies: Estrogen-progestin medication, or high dose progestin medication, danazol therapy or other steroidal medication for the treatment of endometriosis. Sikka et al. (1983) have stated that, in such conditions, fluid retention has been invoked as the mechanism by which the nerve is compressed.
3. Age, gender, wrist size and wrist structural variations: The risk of developing CTS is higher among individuals of age over 40 years. Risk is higher in women than men.
4. Certain diseases, such as gout, rheumatoid arthritis, multiple myeloma, myxedema, and primary amyloidosis disease.
5. Acute trauma.
6. Congenital defects.
7. Oral contraceptive use.
8. Gynecological surgery.

OTHER SIMILAR CONDITIONS

As Durham (1960) suggested, carpal tunnel syndrome must be differentiated from several other conditions that may affect the median nerve. Examples of these conditions are:

1. Progressive atrophy;
2. Syringomyelia;
3. Multiple Sclerosis;
4. Protrusion of cervical intervertebral disc;
5. Periarthritis of the cervical spine or shoulder joint; and
6. Scalenus anticus syndrome.

CLINICAL DIAGNOSIS OF CTS

CTS is clinically diagnosed (Schenck, 1988, 1989) by:

1. Tinel's test: In this test the median nerve at the wrist is tapped and sometimes elicits a tingling response in one or more fingers.
2. Phalen's test: In this the patient is asked to put the backs of his/her hands together and acutely flex his/her writs for one minutes. Development of tingling in the fingers is indicate is a positive sign of CTS.
3. EMG test: In electromyogram test an electrical stimulus is given in the forearm, then the transmission time across the wrist is measured to determine sensory function of the median nerve. This test is especially important in distinguishing between CTS and other nerve problems in the arm or the neck. It is very painful and requires placing needles into various hand/arm muscles (one of the authors' personal experience).
4. Vibrometry: This test is a new test, and very helpful in diagnosing early CTS cases using a vibrometer that measures vibration sense between 8 and 500 cycles per second (Hz).

PREVENTION OF CTS

Two solutions for preventing CTS and/or reducing its risk have been proposed, engineering and administrative.

Engineering solutions

These methods are based upon ergonomic principles used to analyze repetitive motion tasks, identify the stressful ones, and reduce such stressors. Examples of these methods are (e.g., Dionne, 1984b):

1. Ergonomically designed tools should be used to maintain a more natural position of the wrist and to ensure better distribution of grip forces during task performance. Examples of such designs are bent-handled tools that help keep the wrist straight.
2. Provide and orient fixtures and jigs to permit task performance within the normal range of arm movement.
3. Provide arm rests to support forearm, which enhances the hand manipulative ability.
4. Store parts in containers designed and situated within the worker's reach to minimize hand flexion and extension.

5. Provide downward force exertion through flanged support on tool handles.
6. Vibration dampening should be provided for vibrating hand tools.
7. Long-handled screwdrivers should be used because short handles dig into the palm of the hand, which irritate the tendons.
8. Hand tools with fluted handle surface should be avoided. They concentrate stress over a small surface in the hand.
9. Suspend power and heavy tools on balancers to reduce the static load on the hand-arm muscles during prolonged work periods.
10. Thin items (e.g., books) should be carried in cupped hand. In contrast, holding such items in a "pinch" position between the thumb and fingers causes strain in the hand.
11. The layout of work stations should be modified and fixtures should be used to mount work-pieces at angles which reduce the requirement of wrist deviations.
12. The work methods should be modified to provide for more frequent rest breaks and rotating of workers across jobs.

Administrative solutions

When engineering solutions fail to be economically feasible, administrative solutions should be tried. Examples of these methods are (e.g., Dionne, 1984b):

1. Provision of employee rotation within the work group exposed to stressful and repetitive tasks.
2. Train and supervise workers to use tools appropriately.
3. If symptoms of CTS prevail, a physician should be consulted to prevent any permanent damage to the nerve.

TREATMENT OF CARPAL TUNNEL SYNDROME

The treatment of CTS depends on its progressive stage, and can be divided into conservative and surgical:

Conservative treatment

The conservative or non-surgical treatment of CTS, also, depends on the stage of its progress (e.g., NSC, 1984; Ellis et al., 1976; Wolpow, 1984; Folkers, 1986; Schenck, 1988):

1. Wearing a night splint, reinforced with metal, to hold the wrist in slight hypertension. The splint allows time for a reduction in the synovial swelling that causes compression on the median nerve.
2. A day and night splint to prevent forceful contraction of the thumb and finger flexors.
3. Administering anti-inflammatory medication taken by mouth or injected at the wrist to reduce tendon swelling in the carpal tunnel.
4. There is a report of CTS treatment by administering pyridoxine (Vitamin B_6) for 9-11 weeks to patients with pyridoxine deficiency (Ellis et al., 1976). A similar claim has been made by Folkers (1986).
5. Physical therapy including controlled laser systems, interference currents, or bio-conductive therapy.
6. Switching job and rest, especially in the early stage of development, maybe helpful.

Schenck (1988) states that with the splint the wrist can rest. He also believes that a splint will solve the problem in many early detected cases of CTS.

Surgical Treatment

In cases where non-surgical treatments are not effective, surgery treatment should be carried out. The surgery procedure involves dividing the ligament across the palmar aspect of the wrist which releases pressure on the median nerve (Wolpow, 1984; Schenck, 1988). Schenck states that surgery is not necessarily a successful treatment and even, reoperation is only successful in small portion of patients. In pilot study of several CTS patients' medical records, it was found some cases in which a second surgery had been indicated because of reconnection of the splitted transverse carpal ligament.

REFERENCES

Armstrong, T.J., 1983, An Ergonomics Guide to Carpal Tunnel Syndrome, Ergonomics Guide Series, American Industrial Hygiene Association, Akron, Ohio.

Armstrong, T.J., 1986, Ergonomics and Cumulative Trauma Disorders. Hand Clinics, 2(3): 553-565.

Bennett, J.B. and Crouch, C.C., 1982, Compression Syndrome of the Recurrent Motor Branch of the Median Nerve. Journal of Hand Surgery, 7(4): 407-409.

Daven, A., 1982, Practical Considerations in the Treatment of Carpal Tunnel Syndrome. Journal of Medical Society of New Jersey, 79(4): 295-299.

Dionne. E.D., 1984a, Carpal Tunnel Syndrome, Part I: The Problem. National Safety News, 129(3): 42-45.

Dionne. E.D., 1984b, Carpal Tunnel Syndrome, Part II: Some Answers. National Safety News, 129(4): 53-57.

Durham, R.H., 1960, Encyclopedia of Medical Syndromes, Paul B. Hoeber, Inc., Medical Division of Harper & Brothers, New York.

Ellis, J.M., Kishi, T., Azuma, J., et al., 1976, Vitamin B_6 Deficiency in Patients with a Clinical Syndrome including the Carpal Tunnel Defect. Res. Commun. Chem. Pathol. Pharmacol., 76: 743-751. (Cited in Harrison et al., 1983).

Folkers, K., 1986, Contemporary Therapy with Vitamin B6, Vitamin B2, and Coenzyme Q10. Chemical & Engineering News, 64(16): 27-30 and 56-56.

Harrison, W.; Stewart, J.; Lovelace, R.; and Quitkin, F., 1983, Case Report of Carpal Tunnel Syndrome Associated with Tranylcypromine. American Journal of Psychiatry, 140(9): 1229-1230.

Massey, E.W. and Riley, T.L., 1985, Carpal Tunnel Syndrome in Pregnancy. Obstetrical and Gynecological Survey, 33(3): 145.

NIOSH, 1989, Carpal Tunnel Syndrome: Selected References. National Institute for Occupational Safety and Health (NIOSH), Centers for Diseases Control, Public Health Service, U.S. Department of Health and Human Services, Cincinnati, Ohio, March 1989.

NSC, 1984, Therapy Can Relieve CTS Symptoms. National Safety News, 129(4): 58.

Phalen, G.S., 1970, Reflections on 21 Years' Experience with the Carpal Tunnel Syndrome. JAMA, 212(8): 1365-1367.

Roel, R.E., 1989, Offices, Factories Hit by Repetitive Strain Injury. Newsday, (The Long Island Newspaper), Sunday, April 30, 1989, p. 5.

Schenck, R.R., 1989, Prevention and Treatment of Hand Injuries. Professional Safety, 34(5): 30-33.

Schenck, R.R., 1989, Keep in Touch with Pain. Safety & Health, 138(6): 39-42.

Sikka, A.; Kemmann, E.; Vrablik, R.M.; and Grossman, L., 1983, Carpal Tunnel Syndrome Associated Danazol. American Journal of Obstetrics and Gynecology, 147(1): 102-103.

Silverstein, B.A.; Fine, L.J.; and Armstrong, T.J., 1987, Occupational Factors and Carpal Tunnel Syndrome. American Journal of Industrial Medicine, 11: 343-358.

U.S. Congress, April 1985, Ergonomics and Prevention of Musculoskeletal Injuries, Chapter 7 in: Preventing Illness and Injury in the Workplace. U.S. Congress, Office of Technology Assessment, OTA-H-256.

U.S. Government, 1984, Effects of Carpal Tunnel Syndrome and Tendonitis on Postal Employees. Hearing Before the Subcommittee on Postal Personnel and Modernization of the Committee on Post Office and Civil Service House of Representatives, Ninety-Eighth Congress, Second Session, June 8, 1984, Serial No. 98-50, U.S. Government Printing Office, Washington, DC.

Wolpow, E., 1984, No Rest from the Wrist: The Carpal Tunnel Syndrome. Harvard Medical School Health Letter, 9: 3-5.

Advances in Industrial Ergonomics and Safety II
Edited by Biman Das
Taylor & Francis 1990

EVALUATION OF WORKPLACES FOR THE RISK OF CARPAL TUNNEL SYNDROME

A. Waikar, K. Lee, S. Sanyal, C. Parks and F. Aghazadeh

Industrial and Manufacturing Systems Engineering
Department, Louisiana State University
Baton Rouge, LA, USA

This paper presents a development of a prototype knowledge based system for evaluating a workplace to reduce the risk of carpal tunnel syndrome. The system inquires about the details of the workplace design and uses the information stored in its knowledge base to draw conclusions about the risk potential. It also provides limited suggestions for improving the design. MicroExpert shell was used for this system. The system was validated using the data obtained from two case studies. The predictions were found to be consistent with the actual incidents.

INTRODUCTION

Upper extremity trauma disorders such as carpal tunnel syndrome (CTS), tendonitis etc., have been causes of lost worker-hours and workers' compensation in labor intensive industries (ABC News, 1987). As much as 25% of the workforce in some jobs is afflicted with these disorders requiring serious medical attention (Armstrong and Chaffin, 1977; Armstrong et al., 1986). Of these, CTS is the most serious disorder (Chaffin, 1976).

CTS is a nervous ailment caused by repetitive compression of the median nerve located inside the carpal tunnel of the wrist. Common complaints include numbness, feelings of tingling sensation, weakness, paraesthesis, severe nocturnal pain, and swelling sensation in the thumb, index and middle fingers. The symptoms are progressive in nature and can lead to the total disability of the hand functions (Chaffin, 1976).

Research (Armstrong et al., 1986; Birbeck and Beer, 1975; Chaffin, 1976; Groneman, 1985; Tichauer, 1966) indicates that CTS may be regarded as an occupational disease. The disorder is noticed frequently among workers whose jobs involve repeated performance of the same motion patterns. Assembly line work, meat cutting, drilling in rocks and concrete, etc., are some of the occupations where the arms and hands are susceptible to CTS.

Presently in industry, predicting the potential risk for CTS involved in the tasks being performed is not straight forward (Armstrong, 1978). Workers performing the same task react differently. Also, a person's susceptibility to CTS depends on the individual's physical condition and other factors that can not be generalized (Bernard, 1979). Hence, predicting the chance of an individual being affected by CTS is open to subjective reasoning and is the task of an expert. However, services of experts are very expensive and difficult to obtain. Therefore, it would be very useful to industry if there is a mechanism which can evaluate the potential risk of CTS in a workplace without relying upon the expert.

The objective of this research was to develop a microcomputer based prototype consultant knowledge-based system. A knowledge base containing characteristics, symptoms, causes of CTS, etc., was constructed. It addressed various occupational and non-occupational factors which pose a risk of CTS. The expert system generates a series of questions and searches the knowledge base for an answer that best fits the situation. The system applies inference strategies on the knowledge base to arrive at the proper answer.

A work place designer who is not an expert in CTS can use this knowledge-based system to decide or assess the risk of CTS for a given work design. He can screen individuals as well as evaluate workplace and tools to estimate likelihood of an incidence of CTS. This would result in considerable savings through lower lost work-time and reduced medical and legal expenses.

PROTOTYPE SYSTEM

An expert system shell (MICROEXPERT) was used to develop the knowledge-based system described in this paper. It uses a goal driven inference strategy to drive the consultation session (MicroExpert User Guide, 1985). The knowledge-base contains about 150 rules. The number of rules permissible is limited only by the working memory of the machine.

The research findings and conclusions that have been used to form the knowledge base were obtained from extensive literature review (see bibliography). These indicate that CTS is caused by

- -repetitive compression of the median nerve
- -injury to the median nerve due to
  - -overexertion of the wrist muscles
  - -excessive extension or flexion of the wrist
- -reduction in the size of the carpal tunnel due to fluid accumulation (such as in pregnancy)
- -cuts, mal-aligned fractures
- -reduction in the supply of blood due to any of the above
- -certain diseases such as rheumatoid arthritis
- -changes in the hormone balance
- -use of certain long term medications

Injuries to the median nerve may be caused by the nature of the subject's occupation. However, it is important to take certain non-occupational factors into consideration before a judgment is made about the workplace or tool design. Certain domestic habits (mopping, washing dishes/washing clothes manually, knitting, sewing) and hobbies (gardening, fishing, playing certain musical instruments, playing tennis, golf, racquet ball, etc.) have been suspected to contribute to the incidence of CTS.

The knowledge base is divided into two segments; one relating to workplace design and other to tool design. Figure 1 shows that the overall analysis knowledge-base consists of two bases, each of which is complete by itself and has a goal which forms a sub-goal for the analysis knowledge base. For example, if the user wants to analyze a workplace design for stress concentration considerations only, then "Stress Conc" may be given as the final goal of the consultation. Figure 1 shows the architecture of the prototype knowledge-based system.

<u>Knowledge-base for the Evaluation of Workplace Design</u>

Bad workplace design has been known to be one of the occupational factors of CTS (Armstrong and Chaffin, 1977; Armstrong et al., 1986; Chaffin, 1976; Groneman, 1985). This knowledge base evaluates a workplace design from the responses provided by the user (Figure 2). It has been based upon the following considerations:

- Excessive wrist deviation required due to improper work place layout (especially repeated exertions while bending the wrist)
- Excessive stress concentration in the wrists resulting from lifting heavy loads and from use of badly designed handtools
- Gender effect. Females have been found to be more susceptible to CTS than males.
- Task rotation. Rotating tasks among workers and thus reducing the rate of repetition, has been found to be a good measure against CTS. In general, strenuous tasks should be interspaced with less strenuous tasks. The limitation to this is set by the type of skill the tasks demand.
- Vibrations. The workplace should not expose the worker's hands to vibrations. Vibrating parts should be separated by proper bushings and shock absorbers. Safe frequency of unbalanced vibrations has been found to lie in the range of 3 to 125 Hertz. The corresponding amplitudes should be less than 100 microns.
- Temperature. Low temperatures have been known to cause or aggravate CTS. Workplace temperature should be at least 20 degrees centigrade.
- Persons with rheumatoid arthritis, diabetes, or any ailment which may cause or aggravate CTS should not be assigned the use of tools requiring stressful exertion.

<u>Knowledge-base for the Evaluation of Tool Design</u>

Often, badly designed hand tools and improper use of hand tools are the primary causes of CTS (Armstrong et al., 1986; Tichauer, 1966; Tichauer, 1978). Following tool design considerations are important for reducing the risk of CTS.

- Designs which call for ulnar deviation of the wrist along with supination should be avoided. Attempt must be made to keep the wrist deviation at a minimum value at all times.
- Tools that are used while performing repetitive motions should be designed so as to require minimum force while using them. If possible, the motion should be mechanized.
- An attempt should be made to use the thumb for exerting forces more than any other finger. Thrust should always be

directed towards the base of the thumb instead of the palm.
- Heavy tools should have their center of gravity located as close as possible to the handle and to the body of the user. This reduces the tendency of the tool to rotate and slip.
- Excessive use against resistance and use while in abnormal position should be avoided.
- Repetitive motions that produce impact shocks should be avoided.
- The forces applied should be concentrated on the workpiece and very little reaction should be transmitted to the wrists or the palms.
- Contact surfaces between the tool and the palms should be sufficiently large to avoid stress concentration.
- Form fitting handles should be sparingly used.
- Transmission of excessive vibrations to the hands must be prevented. The frequency of unbalanced vibrations should be in the range of 3 to 125 Hertz and the corresponding amplitudes should be less than 100 microns.
- The tool should not blow cold air on the hands of the user.

The organization of this knowledge base is similar to the one for workplace evaluation. The primary goal of the consultation is "risk factor." Figure 3 shows the relationship of the various subgoals involved and their attributes. The main goal is represented by certain attributes which become the sub-goals in the latter part of the consultation.

Figure 2 shows the relationship of the various sub-goals with the goal "Risk Factor." The user has to go through a number of queries, the responses to which are used by the system to assign appropriate values to the goals.

The following non-occupational (personal) factors are also included in both knowledge-bases (workplace design and tool design).
- Medical conditions. The risk of CTS is found to be greater if the worker suffers from certain disorders like rheumatoid arthritis, fluid imbalance problems, diabetes, improperly healed carpal fractures, etc. (Armstrong, 1978; Bernard, 1979; Birbeck and Beer, 1975; Gray and Gottlieb, 1977; Groneman, 1985).
- Reoccurrence. Workers who have been affected by CTS in the past should not be assigned strenuous tasks as recurrence is common.
- Age. Persons above 50 years of age are more susceptible to CTS. If possible, tasks that are repetitive in nature and involve lifting should be assigned to younger persons.

This knowledge based system interactively inquires about the details of the proposed workplace and tool design and evaluates the designs for risk of CTS. Based on the responses of the user, the system is able to point out the risk factors in the design and suggest modifications to reduce their effects. Due to the nature of the problem, a rule based strategy was found to be the most convenient form for representing the knowledge. The prototype therefore, was developed using the MicroExpert shell, an expert system building tool. The knowledge base in the system is a separate entity and can be modified easily.

VALIDATION AND CONCLUSIONS

In an effort to make a workplace design cost effective, engineers often overlook the possibility of an occupational injury to the worker. Occurrence of CTS is common in hand intensive industries and its cause has been, in most cases, attributed to improperly designed hand tools and workplaces. Many designers are not aware of this disease and its causes. In this research, a knowledge based system was developed to evaluate workplace and tool design, and to provide limited suggestions for reducing the risk of carpal tunnel syndrome.

The system was validated with the data obtained from the case studies of Walter Walter and Cannon, 1981 and Hymovich Hymovisch and Lindholm,). The system's predictions were found to be quite consistent with actual incidents. The system presents its conclusion about the design as being "good" or "bad" and states whether the risk factor is "low", "moderate", "high", or "very high." However, it can not assign any probabilistic values to its conclusions. This is the limitation of the present version of this knowledge-based system. This limitation can be attributed to the absence of certain epidemiological data in its knowledge base and the inability of the (MicroExpert, 1985) shell to return non Boolean values to the data list created by the system. However, the prototype knowledge-based system developed in this research could form a valuable tool for workplace designers as well as the users.

BIBLIOGRAPHY

ABC News, 20/20, The Most Dangerous job in America, Transcripts of Show # 743, November 6, 1987.

Armstrong, T.J., and Chaffin, D.B. An Investigation of Chronic Wrist Injuries in Women, presented at American Industrial Hygiene Conference, New Orleans, LA. May 1977.

Armstrong, T.J., Radwin, R.G. and Hansen, D.J., Repetitive Trauma Disorders: Job Evaluation and design, Human Factors, 1986, 28 (3), pp 325-336.

Armstrong, T.J., 1978, Carpal Tunnel Syndrome - What is it ?, Technical Report, The University of Michigan, Ann Arbor.

Bernard M. L., Carpal Tunnel Syndrome: Identification and Control, Occupational Health Nursing, June 1979.

Birbeck, M.Q. and Beer, T.C., Occupation in Relation to the Carpal Tunnel Syndrome, Rheum. and Rehab., 1975, 14:218.

Chaffin, D.B., An Investigation of Occupational Wrist Injuries in Women, 1976.

Gray, R.G. and Gottlieb, N.L., Hand Flexor Tenosynovitis in Rheumatoid Arthritis, 1977 Arthritis and Rheumatism, pp 1003-1008.

Groneman, L., Carpal Tunnel Syndrome Can Be Lessened With Early Treatment, Occupational Health and Safety, Oct. 1985, pp 39-46.

Hymovisch, L. and Lindholm, M., Hand, Wrist and Forearm Injuries, The Result of Repetitive motions, Journal of Occupational Medicine, vol. 8, pp 573-577.

MicroExpert User Guide, 1985, McGraw Hill, New York.

Tichauer, E.R, Some Aspects of Stress on Forearm and Hand in Industry, Journal of Occupational Medicine, vol. 8, 1966, pp-63-71.

Tichauer, E.R., Ergonomic Principles Basic to Tool Design, American Industrial Hygiene Association, 1978.
Walter, S.D., Cannon, L.J. and Bernacki, E.J., Personal and Occupational Factors Associated with CTS, Journal of Occupational Medicine, April 1981, Vol. 23 No.4.

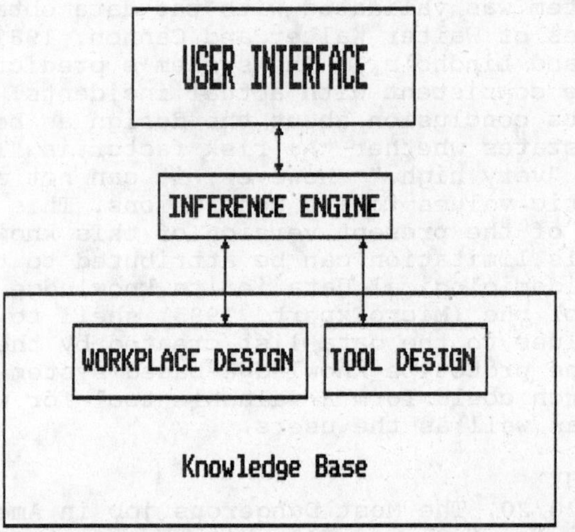

Figure 1. Architecture of the Expert System

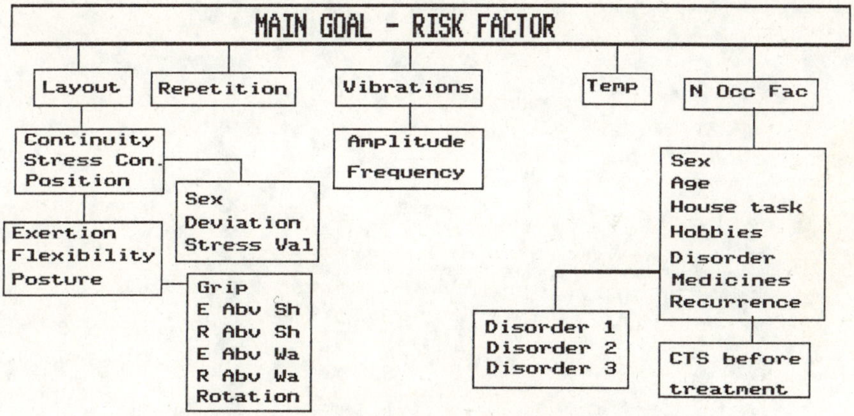

Figure 2. Workplace Design Knowledge Base

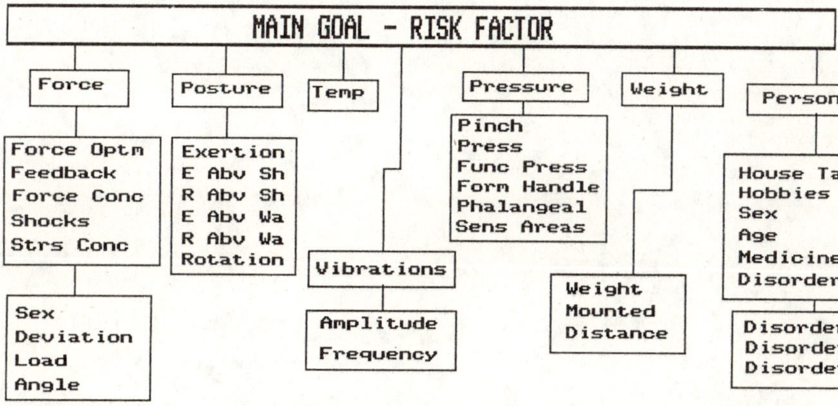

Figure 3. Tool Design Knowledge Base

Advances in Industrial Ergonomics and Safety II
Edited by Biman Das
Taylor & Francis 1990

A FATIGUE MECHANICS APPROACH TO CUMULATIVE TRAUMA DISORDERS

RAM R. BISHU, S. G. MANJUNATH, AND M. SUSAN HALLBECK.

DEPARTMENT OF INDUSTRIAL AND MANAGEMENT SYSTEMS ENGINEERING
UNIVERSITY OF NEBRASKA-LINCOLN, NEBRASKA

Cumulative trauma disorders encompass a wide range of musculoskeletal disorders and are the results of repeated microtrauma. High repetition, insufficient rest, awkward posture and high force are some of the risk factors for CTD. No metric or method exist to define the safe limits of repetition. This research addresses the question of "What is the safe limit for repetition?". Using the analogy between CTD and fatigue mechanics an exponential model linking force and cycles of repetition is postulated. A pilot investigation was performed to test the model. Five subjects participated in a factorial experiment comprising 2 hand condition (cotton glove and bare hand), 3 weights (3, 5, 11 lbs) and 2 replications. Results indicate the presence of an exponential relation between the load exerted by the hand and the cycles of repetitions. The results are discussed in light of providing design recommendations for allowable repetitions in tasks.

INTRODUCTION

Cumulative trauma disorders in the soft tissues of the upper extremities such as tendonitis, synovitis, bursitis, carpal tunnel syndrome, and others have been associated with jobs involving frequent hand exertions. Past studies have shown that such disorders are the results of repeated microtrauma, and can be caused, precipitated or aggravated by repeated exertions with certain hand and wrist postures--particularly in combination with high forces (Armstrong et al., 1982). Similar findings have been reported in a number of other studies as well (Armstrong and Chaffin, 1978; Silverstein et al., 1986). High repetition, insufficient rest, awkward posture and high force are some of the risk factors for CTD. Summaries of reported occupational factors for CTD have been provided by Armstrong et al. (1982), and Vern Putz Anderson, (1988). The most common factors reported by the investigators are unnatural postures, high force and repeated exertion.

Although repetition has been established as one of the causative factors for carpal tunnel syndrome, no metric or method exist to define the safe limits of repetition. This research attempts to address the questions: "What is the safe limit for repetition?", "What is the relationship between number of repetition and force?", and "How does one provide design recommendation?"

Metal fatigue is an important issue in mechanical design as it constitutes more than 80% of the reported failures. It is a complex phenomenon that has been well researched but not understood completely yet. In general, metal fatigue is the result of repeated application of stress, even though all stresses lie below the ultimate tensile strength. Fatigue strength is expressed as the maximum stress that can be tolerated for a given number of cycles and is derived from Stress-Number of cycle

(S-N) diagrams. Generally, fatigue strength is determined in the laboratory through S-N tests. The strength is defined by the expression,

$$NS^m = A \quad (1)$$

where

N = cycle to failure
S = stress range or amplitude
m = fatigue strength exponent
A = fatigue strength coefficient

For CTD there is sufficient epidemiology to suggest that combinations of force, repetition and posture are high risk factors. Among the various CTDs carpal tunnel syndrome (CTS) is the one that is most associated with repetition. In CTS the median nerve is compressed as it passes through the carpal tunnel at the base of palm and wrist, resulting in loss of sensation, pain, skin dryness and atrophy of muscles in the nerve distribution area. It can be argued that the factors leading to the compression of the median nerve to the point of initiation of the symptoms of CTS are perhaps similar to the phenomenon of failure due to "incremental collapse" in metal fatigue. Incremental collapse occurs when a static load is superimposed on some cyclic loading, which normally happens when forceful wrist exertions are made repeatedly.

According to the belt and pulley biomechanical model of the wrist (Armstrong and Chaffin, 1979) the tension developed in the tendon(F_T) is approximately three to four times the load on each finger(F_L), depending on hand posture, load orientation, and load application points. The supporting forces(F_R) of the tendon acting normal to the arc of the pulley depends on tendon tension and radius of curvature of the wrist. It has been suggested that when a person is squeezing an object with maximum effort, a typical finger load (F_L) would be about 150 N (Ohtsulu, 1981). Thus, the effector flexor tendon force would be between 420 and 645 N for the profundus and superficialis tendons combined from each finger (Chaffin and Anderson, 1984). Repetitive force of this magnitude is known to stretch the tendon due to residual strain (Goldstein, 1981). It is possible that this strain either weakens the tendon, or inflames it. Such an inflammatory reaction can cause the coefficient of friction to increase, thereby introducing some shear forces.

The total radial supporting force(F_R) depends on the tendon tension(F_T) and the wrist angle(θ), and represents the force acting on adjacent structures--for example, the ligaments, bones, and median nerve enclosed in the carpal tunnel. The magnitude of this force depends on the amount that the tendon wraps around the pulley. The resulting relationship is

$$F_R = 2F_T \sin \frac{\theta}{2} \quad (2)$$

The effect of the wrist angle θ in calculation of the radial force is, therefore, important. Thus, Armstrong and Chaffin (1978) relate the grip force to the radial supporting force through tendon tension and wrist angle. What is unclear is the relationship of the radial supporting force F_R to the incidence of trauma.

Therefore what we have in a belt and pulley model of the wrist is a constant static load (F_T), and an alternating load (F_R). Consider a simple wrist flexion, extension task. If we adopt the simple belt-pulley model, then there is a directional change at each alternation between flexion and extension, with the normal forces on the length of the synovium supporting the tendon alternating with each cycle. There would exist a mean stress due to the tension in the tendon (F_T) and an alternating stress (F_R) due to the radial force. It has been argued earlier that the incremental collapse (a fatigue failure mode) occurs when a static load is superimposed on some cyclic loading.

The actual nature of the cyclic stresses acting on the wrist would be far more complex than the simple case presented here. The cyclic reaction force can result in strain hardening or strain softening of the tendon. There is evidence for some amount of residual strain on the tendon (Abrahams, 1967). Such a residual strain is perhaps the sum effect of axial strain caused by F_T and cyclic strain caused by F_R. In mechanical parts, combinations of static and cyclic loading such as this leads to continuous energy absorbtion (strain hardening) culminating in nucleation of crack, which then propagates. For the human it is not unreasonable to state that at some point in the ratchetting cycle pain and discomfort will commence.

Therefore, using the analogy between CTD and fatigue mechanics the following model is postulated for repetitive work.

$$NL^m = G \qquad (3)$$

where N = number of cycles to voluntary limit at load L
L = gripping load
m and G are cumulative trauma parameters

This is somewhat similar to the conceptual model suggested by Tanaka and McGlothlin (1989). For purpose of simplicity the gripping load(L) is assumed in this model, instead of the normal reaction force(F_R), and the tendon tension (F_T). Such a model would provide a method for establishing safe limits for repetitions and would also give upper bounds for repetitions as a function of load and other task parameters. To test this model a pilot investigation was performed. The details of this experiment are presented next.

METHODS

The experimental design of this study was 5 subjects x 2 hand conditions (cotton glove and bare hand) x 3 weights (3, 5 and 11 lbs) x 2 replications. The order of presentation of the 30 treatment conditions was randomized. Five male subjects, all chosen from the student population at the University of Nebraska participated in this experiment. Initially the anthropometric dimensions were measured for each subject. The subjects were then asked to squeeze a hand dynamometer as hard as they could, once with their dominant bare hand, and once with a cotton glove on their dominant hand. The purpose was to determine their maximum voluntary contraction (MVC) for the bare hand and gloved condition.

For each trial the subjects had their forearm pronated, and in contact with a flat surface. Each subject held a 3, 5, or 11 pound dumbbell in a power grip and alternated between flexion and extension continuously. The pacing was achieved through a metronome set at 120 beats per minute. This required the subject to perform one full cycle (wrist flexion to wrist extension) every second. The subjects were instructed to keep up with the pace set by the metronome until they could no longer do so. After each trial the subjects were required to rest for at least two hours before the next trial could be administered. For each trial the number of cycles to voluntary cessation was recorded and treated as the dependent variable.

RESULTS

Initially an overall analysis of variance was performed on the frequency. The ANOVA summary is shown in Table 1. All the main effects, namely Hand, Weight, and Replication were significant. Figure 1 shows the effect of hand condition on the frequency, while Figure 2 shows the effect of weight condition on the frequency. The inverse relation of the 'maximum voluntary frequency' with weight is not surprising, and is consistent with published literature on force endurance relations (Rohmert, 1960). Similarly, the performance decrement with gloves as seen from Figure 2 is also consistent with earlier studies (Wang et. al.,1987).

The next stage of analysis was to develop an endurance (S/N) diagram. A regression analysis was performed with frequency to exhaustion as the dependent variable and proportional MVC as the independent variable. Two types of regression has been performed.

SOURCE	DF	SUM OF SQUARES	MEAN SQUARE	F-VALUE	PR>F
MODEL	4	141627.90	35406.975	40.05	0.0001
ERROR	55	48621.0833	884.01		
Total	59	190248.9833			

SOURCE	DF	ANOVA SS	F-VALUE	PR>F
HNCD	1	3792.1500	4.29	0.0430
WTCD	2	134881.7333	76.29	0.0001
REP	1	2954.0166	3.34	0.0730

Table 1. ANOVA summary

1. Polynomial Regression (cubic)

For bare handed condition:

$$\text{Frequency} = 373.43 - 52.06\ \text{MVC} + 2.655\ \text{MVC}^2 - 0.045\ \text{MVC}^3 \quad (4)$$

$(R^2 = 0.7456)$

For gloved condition:

$$\text{Frequency} = 290.76 - 36.46\ \text{MVC} + 1.70\ \text{MVC}^2 - 0.026\ \text{MVC}^3 \quad (5)$$

$(R^2 = 0.7942)$

Figure 3 shows the plot of this relation. It is seen from Figure 3 that the two curves are nearly parallel, indicating a uniform decrease in performance, and a lack of interaction. According to these equations the maximum frequency is at 373.43 for the bare handed condition, while it is at 290.76 for the gloved condition. More investigation is needed to determine if these numbers can be used as upper bounds for repetition in tasks involving frequent repetitions.

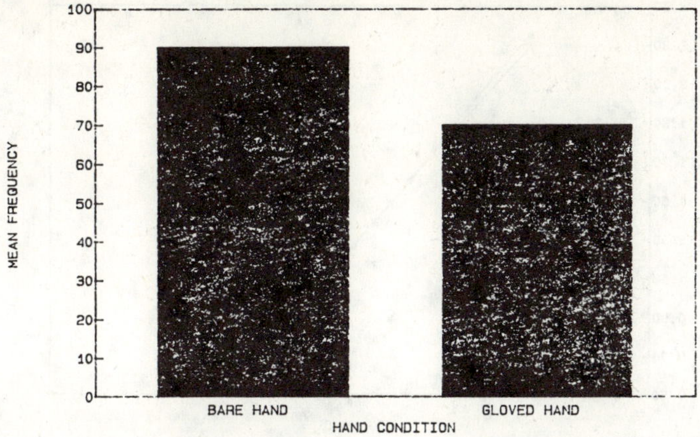

Figure 1. Effect of hand condition

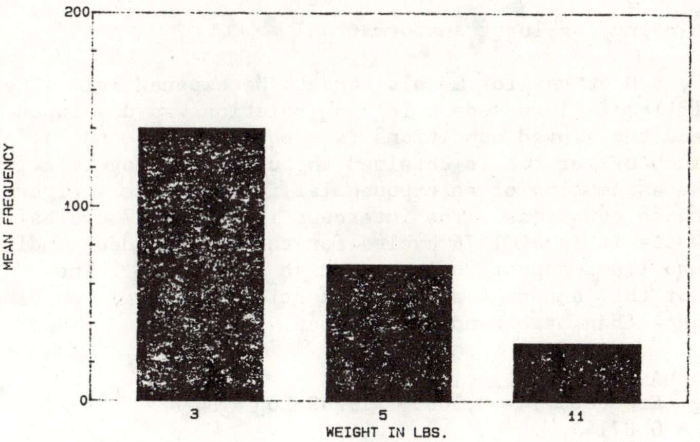

Figure 2. Effect of weight

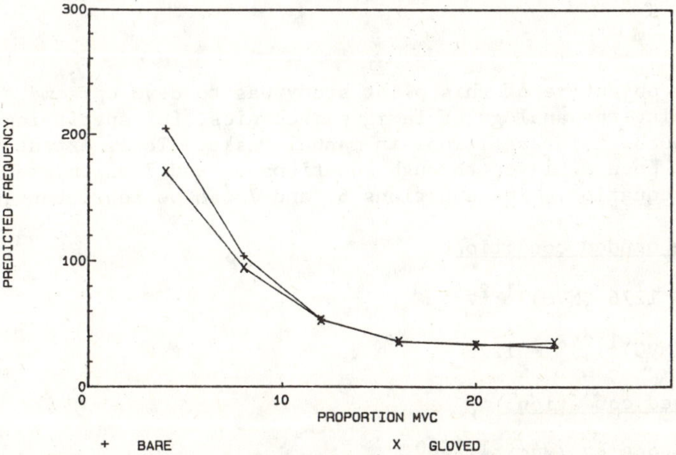

Figure 3. Polynomial function of frequency and MVC

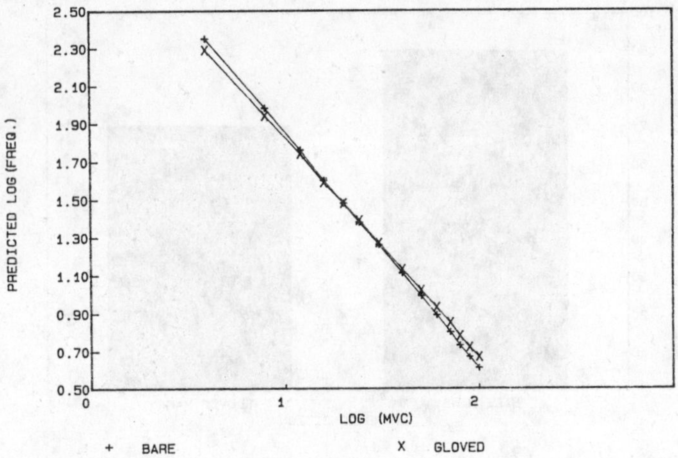

Figure 4. Log-log function of frequency and MVC

2. Regression using log-log transformation

Typically S-N curves for metals tend to be exponential. Therefore, assuming a similar failure mode a log-log relation was developed for the bare handed and the gloved conditions (see equations 6, and 7). It is seen that a much better fit is obtained through a log-log relation, validating the assumption of an exponential failure mode. Figure 4 shows the plot of these equations. The intercept is at 968.27 cycles for the gloved hand while it is at 1276 cycles for the bare handed condition. Further the two lines appear to intersect at .182 (MVC). The implications of this are that at lower exertion levels gloved hand has better endurance than bare handed condition.

For bare handed condition:

$$\text{Log (Frequency)} = 3.1036 - 1.248 \text{ Log (MVC)} \qquad (6)$$

$R^2 = 0.8714$

For gloved condition:

$$\text{Log (Frequency)} = 2.986 - 1.1580 \text{ Log (MVC)} \qquad (7)$$

$R^2 = 0.8910$

DISCUSSION

The main objective of this pilot study was to develop some rationale, using the analogy of fatigue mechanics, for specifying the permissible number of repetitions in manual tasks. To an extent the set objective has been achieved through equations 6, and 7 which are similar to a fatigue equation. The equations 6, and 7 can be rewritten as:

For bare handed condition:

$$N = 1276 \text{ (MVC)}^{-1.248} \quad \text{or}$$

$$N \text{ (MVC)}^{1.248} = 1276$$

For gloved condition:

$$N = 968.27 \text{ (MVC)}^{-1.1580} \quad \text{or}$$

$$N \text{ (MVC)}^{1.1580} = 968.27$$

The right hand side of these equations are the 'G' of equation 3. They denote the upper bounds for the repetitions under minimum load. The slopes are the 'm' of equation 3, and represent some measure of the relation between load and cycle. N, for any load, would denote the upper bound for permissible repetitions for that load.

A number of cautions are in order here. The behavior of biological material may be different from that of metals, and hence the analogy may not be valid. This model is very simplistic and is restricted to flexion and extension and does not include other forces. Further the recovery mechanisms have not been included in this model. The strain hardened tendons would recover with some rest. Real-life tasks are much more complex and include other forces and moments. Nevertheless the methodology proposed here could be a valid preventive tool for preventing CTDs. It can be used for specifying design guidelines for allowable number of repetitions in jobs involving repetition. For example equations could be derived for 'flexion-extension', and 'radial-ulnar deviation', and the lower of the two upper bounds could be used as a design parameter. In concept the methodology proposed here is similar to the one suggested by Tanaka and McGlothlin (1989). Subsequent objectives are to build on this model to include multiple stresses, and other parameters such as deviation angle, hand anthropometry and isometric strength.

REFERENCES

Abrahams, M., 1967 "Mechanical properties of tendons in-vitro". In Medicinal Biological Engineering, 5, 433-443.

Armstrong, T. J., and Chaffin, D. B., 1978, "An investigation of the relationship between displacements of the finger and wrist joints and the extrinsic finger flexor tendons." J. Biomech., 11, 119-128.

Armstrong, T. J., and Chaffin, D. B., 1979, "Some biomechanical aspects of the carpal tunnel, J. Biomech., 12, 567-570.

Armstrong, T. J., Foulke, J. A., Joseph, B. S., and Goldstein, S. A., 1982, "Investigation of cumulative trauma disorders in ma poultry processing plant", American Industrial Hygiene Association Journal, 43(2), 103-116.

Chaffin, D. B., and Andersson, G. B. J., 1984. Occupational Biomechanics, John Wiley and Sons, New York.

Goldstein, S., "Biomechanical aspects of cumulative trauma to tendons and tendon sheaths," 1981, unpublished Ph.D. thesis (bioengineering), University of Michigan, Ann Arbor.

Ohtsuki, T., 1981, "Inhibition of individual fingers during grip strength exertion," Ergonomics, 24(1), 21-36.

Rohmert, W., 1960. Ermittlung von Erholungspausen fur Statische Arbeit des Menschen. Int. Z. Angew. Physiol. Einsehl. Arbeitsphysiol., 18, pp. 123-164.

Silverstein, B. A., Fine, L. J., and Armstrong, T.J., 1986, "Hand wrist cumulative trauma disorders in the industry", British Journal of Industrial Medicine, 43, 779-784.

Tanaka, S., and McGlothlin, J. D., 1989, " A conceptual model to assess musculoskeletal stress of manual work for establishment of quantitative guidelines to prevent hand and wrist cumulative trauma disorders" In Advances in Industrial Ergonomics and Safety I, Anil Mital (Editor), Taylor and Francis, London, 419-426.

Vern Putz-Anderson, 1988, Cumulative trauma disorders. A manual for musculoskeletal diseases of the upper limbs, Taylor and Francis limited, London.

Wang, M. J., Bishu, R. R., and Rodgers, S. H.(1987). Grip strength changes when wearing three types of gloves. In the Proceedings of INTERFACE 87, Rochester, NY. pp. 349 - 354.

Advances in Industrial Ergonomics and Safety II
Edited by Biman Das
Taylor & Francis 1990

ASSESSING CUMULATIVE TRAUMA DISORDER RISK

James T. Burnette, Ph.D

Post Office Box 522
Oak Ridge, NC, USA 27310

CTD-123 is a computerized risk assessment model applicable to Cumulative Trauma Disorders (CTD). It integrates information from company health records, on-site and videotaped job analyses, and from employee questionnaires. It can serve as a unifying tool for team-oriented CTD investigations. During validation studies of CTD-123, it accurately predicted the relative CTD risk for several anatomical joints across eight different jobs. By including non-traditional risk factors it also accounted for different CTD rates in similar jobs at two different geographical sites.

INTRODUCTION

Cumulative Trauma Disorders continue to plague workers in spite of recent advances in our understanding of their causes and treatments. There is no model or technique sufficient in integrating and coordinating the diverse people, disciplines, and information necessary to reduce CTD risk at any given company. Such coordination requires a computer-based process of data collection and evaluation. The overall process can then include risk assessment, job/workplace design, employee selection, worker strengthening, employee training, and a review of administrative policies and practices, all designed to reduce the frequency and severity of CTD problems. For a more complete discussion of CTD's and the need for an integrating risk-assessment model see Burnette and Ayoub (1989a).

MODEL DEVELOPMENT

To be applicable across a broad range of jobs and industries, the model upon which the CTD prevention process is based must have specific characteristics and must meet certain requirements.

Characteristically, it must deal with microanalytic issues, such as task-related forces and frequencies, as well as macroergonomic issues, for instance, management policies and practices. It must be useful for both descriptive and predictive purposes. Because the model's information may come from several departments or disciplines (health records, job

interpersonal communications and had a longer history of health and safety problems. Probably as a result, they had earlier CTD complaints and the Medical Staff at Site #2 was better prepared to recognize and treat complainants.

With an accuracy rate of 92%, CTD-123 predicted which site would have the greater rate of CTD's. The latest version of CTD-123 (which employed the non-traditional CTD Risk Factors) had a greater correlation between calculated Risk Scores and actual CTD Prevalences than did the CTD-123 version using only the traditional CTD Risk Factors (r=0.55 and r=0.48, respectively).

SUMMARY

CTD-123 is a significant improvement over existing CTD risk assessment models. It accurately predicts the relative risk of developing CTD symptoms for the major anatomical joints over a variety of people, jobs and working situations. It is robust enough for use in real-life industrial applications. Further use will allow the development of absolute standards, based on CTD-123, useful in understanding and preventing CTD problems in the future.

REFERENCES

Burnette, J., 1989, CTD-123: A Cumulative Trauma Disorder Risk Assessment Model, Unpublished doctoral dissertation, North Carolina State University, Raleigh, NC.

Burnette, J. and Ayoub, M., 1989a, Cumulative Trauma Disorders. Part I: The problem. Pain Management, 2(4), 196-209.

Burnette, J. and Ayoub, M., 1989b, Cumulative Trauma Disorders. Part II: Assessment of risk. Pain Management, 2(5), 256-264.

Advances in Industrial Ergonomics and Safety II
Edited by Biman Das
Taylor & Francis 1990

ANALYSIS OF SELECTED HIGH RISK OPERATIONS IN A GARMENT INDUSTRY

Shrawan Kumar

Department of Physical Therapy
University of Alberta
Edmonton, Alberta, Canada T6G 2G4

Cumulative trauma disorders (CTD) were found to be prevelent among sewing machine operators of a jean manufacturing plant. Previously identified 11.8% of the operations designated as the high risk operations (HRO) were analysed for their cycle time, activity elements, repetition and duration of motion and nonneutral posture of shoulder, elbow, wrist joints and hands. The severity and duration of deviation of joints from neutral postures were found related to the frequency of occurrence of C.T.D.

INTRODUCTION

Incidences of cumulative trauma of the upper extremities have been frequently reported by many authors. Armstrong and Chaffin (1979) investigated selected personal attributes of female sewing machine operators employed in the same production sewing jobs with and without carpal tunnel syndrome. They concluded that use of forceful exertions, deviated wrists and pinch hand positions were associated with carpal tunnel syndrome. Wrist deviation was shown by Tichauer (1976) to be associated with increased incidence of tonosynovitis and other disorders in a prospective study of electronics assembly trainees. Keyboard operators with nonspecific muscle pain and cramping were reported by Duncan and Ferguson (1974) to be using wrist extension and ulnar deviation. Among poultry processing workers Armstrong et al (1982) found significantly elevated incidences of tendinous disorders of hands and wrists who required to flex and ulnarly deviate their wrists particularly in forceful exertions. Punnet et al (1985) in a cross sectional study found that the soft tissue disorders of the upper extremities among women garment workers were significantly higher than those of the hospital women workers. They also observed that the garment assembly jobs required stereotyped and repetitive hand motions contrary to those of the hospital workers. Vihma et al (1982) also found higher rates of soft tissue pain among garment assembly workers compared to stitchers whose tasks were less repetitive. Wick and Drury (1986) stated that cumulative trauma disorder (C.T.D.) continues to be a major health problem in garment industry. Accepting this as a fact, they designed to alter gross body posture of the operators and achieved 20% improvement in perceived postural stress. In addition, neck and shoulder problems among sewing machine workers in Sweden was reported to be 75% compared to 54% among the controls, (Blader et al 1987). These authors also stated the prevalance rate of the hand and wrist disorder, in an age and sex matched sample, was 51% among sewing machine workers compared to 28% of the control population. Punnet and Keyserling (1987) studied exposure to ergonomic stressors in 14 garment assembly jobs. In characterising the biomechanical features of these tasks, they reported frequent wrist extension, ulnar deviation and finger pinches.

Kumar (1989) studied the upper extremity morbidity in a garment industry employing 454 workers working under piece payment method. In this company in a six month period in 1987 a total of 138 days were lost to cumulative trauma disorders. In the same period an additional 243 days were lost due to C.T.D., reported in the year before. The total projected cost due to the C.T.D. in this plant for the year 1987 was $165,000. From the health records of the company he reported that 25.2% of production personnel were suffering from C.T.D. The highest frequency of occurrence was found in the age group of 40-49 years (34.6%). The employees under the age 30 years had 18.9% frequency rate of C.T.D. The prevalance of the injuries above the age of 40 years was higher than the percentage of employees in this category. Kumar (1989) reported a 34.3% of incidence rate among employees with less than one year experience. Beyond this experience period the incidence rate declined to one third rising significantly as the experience rates declined to one third rising significantly as the experience on the job increased. Of a total of 76 operations in the plant 11 jobs were responsible for 69% of all the C.T.D.'s. Most of the injuries (81.3%) reported in the plant were to wrists, 9.3% of which were confirmed carpal tunnel syndrome. Elbows and shoulders represented 13.08% and 5.6% respectively of all the injuries. Due to significantly higher incidence rate of the C.T.D.'s these eleven operations were designated as high risk operations and were analysed with a view to implement ergonomic modifications.

MATERIALS AND METHODS

All high risk operations were videotaped in the plant during the performance of the actual work for several cycles. This videotape was subsequently played repeatedly in the laboratory to clock the cycle time, identify activity elements and determine their duration. All time measurements were made on each of the three recorded cycles before averaging.

For recording upper extremity posture during every activity element of each high risk operation, the videotape was projected frame by frame, the relevant frames were frozen and printed on a videoprinter. From these hard copies the postures of shoulder, elbow, wrist and fingers were recorded about their respective axes of rotation according to Armstrong et al (1982). For the shoulder, three axes were used: sagittal - flexion/extension, coronal - abduction/adduction, and axial - lateral/medial (external/internal rotation). For elbows, the two axes used were sagittal - flexion/extension and axial - pronation/supination. Sagittal - flexion/extension, and corononal - radial and ulnar deviation were used for the wrist. For hand postures, six basic positions were used. These were pulp grasp, medial grasp, pulp pinch, lateral pinch, palm pinch and finger press. The postures of the left and right upper extremities were measured and coded separately. These postural data for all upper extremity joints for all activity elements of each selected high risk operation were plotted for standardised movements about all relevant axes of shoulder, elbow, and wrist as well as the hand on time scale.

These were selected on the basis of the operations having a similar efficiency rating in terms of numbers of units produced over an eight hour work period, preferably having a different injury rate. From these plots the postural information of the high risk operations was extracted and analysed for stressful postures assumed and the movements performed.

RESULTS

The process of high risk operation selection yielded a set of three jobs according to established criteria (Table 1). Their injury profile with the breakdown of the categories of injury is present in Table 2. The activity element and cycle time analysis revealed that the Fell inseam/outseam, back seam felling and Band-on operations had four, six and six activity elements respectively. They in turn required 13.5, 17.0 and 25.0 seconds respectively for completing the cycles (Table 3).

Table 1

Operations	Efficiency Rating Bundles/day (A)	Cycle time see (B)	Exposure Index (A x B)	CTD Percent injured	profile on job
Fell inseam/ outseam	11.2	13.5	151	21	
Back seam felling	12.0	17.0	204	36	
Band-on	11.1	25.0	275	46	

Table 2. Injury Profile of selected high risk operations

Operation	Number of Workers	Number of workers with CTD					CTD as % Workers
		CTS	WR	ELB	SH	T	
Fell inseam/ outseam	33	-	7	-	-	7	21
Back seam felling	11	-	3	1	-	4	36
Band on	26	1	10	-	1	12	46
Total	70	1	20	1	1	23	33

CTS - Carpal tunnel syndrome, WR - wrist, ELB - elbow, SH - shoulder, T - Total

Table 3. Activity elements, elemental times and the cycle time for selected high risk operations.

Fell inseam/outseam		Back seam felling		Band-on	
Activity elements	Elemental time sec.	Activity elements	Elemental time sec.	Activity elements	Elemental time sec.
Pick from right	1.0	Pick left /right	1.0	Pull	1.5
Fold & feed	5.0	Match, feed/ stitch	4.0	Fold	7.0
Guide & feed	6.5	Match, feed/ stitch	4.0	Feed	10.0
Stack	1.0	Turn	1.0	Cut	1.5
		Pick, match, feed/ stitch	6.0	Fold	3.0
		Stack	1.0	Stitch	2.0
Cycle time	13.5		17.0		25.0

A. Fell inseam/outseam:

In this operation, the worker picked the partially sewn jean trousers from her right and rolled one of the leg sleeves matching the edges fed to the sewing machine to sew the inner seam. This was then fed through the entire length from one leg to the other in a continuous motion.

The shoulders were maintained in a neutral posture for 85% of the cycle as far as flexion/extension and adduction/abduction were concerned. However, during 15% of the cycle, the shoulder deviated (Fig. 1). Similarly the right and left elbows were maintained in 45^0 extended for 92.5% and 85% of the cycle respectively. The left elbow started the cycle in 90^0 flexion for 7.5% of the cycle and in the final 7.5% of the cycle it was fully extended. The right elbow, though, was extended for the first 7.5% of the cycle stayed pronated throughout the cycle while the left elbow was supinated for the first 7.5% of the cycle (Fig. 2). The right wrist, in this task, deviated radially for 92% of the cycle, and the left wrist deviated on the ulnar side for 48% of the cycle. The right wrist remained fully extended for 37% of the cycle, and 45^0 extended for 55% of the cycle. It was only for 7.5% of the cycle that it was in neutral posture. The left

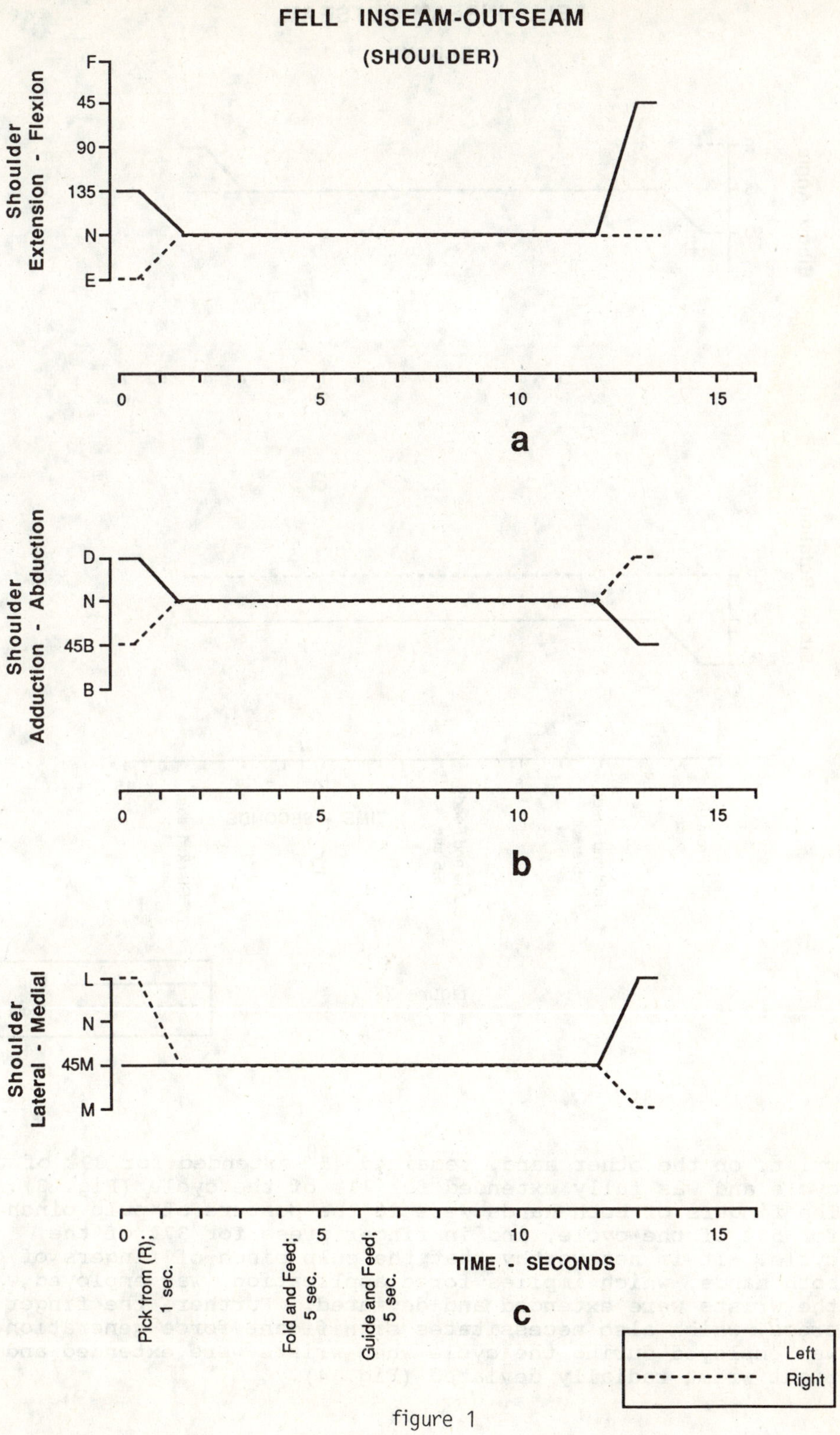
FELL INSEAM-OUTSEAM
(SHOULDER)
Shoulder
Extension - Flexion
F
45
90
135
N
E
0
5
10
15
a
Shoulder
Adduction - Abduction
D
N
45B
B
0
5
10
15
b
Shoulder
Lateral - Medial
L
N
45M
M
0
5
10
15
TIME - SECONDS
c
Pick from (R); 1 sec.
Fold and Feed; 5 sec.
Guide and Feed; 5 sec.
Left
Right

figure 1

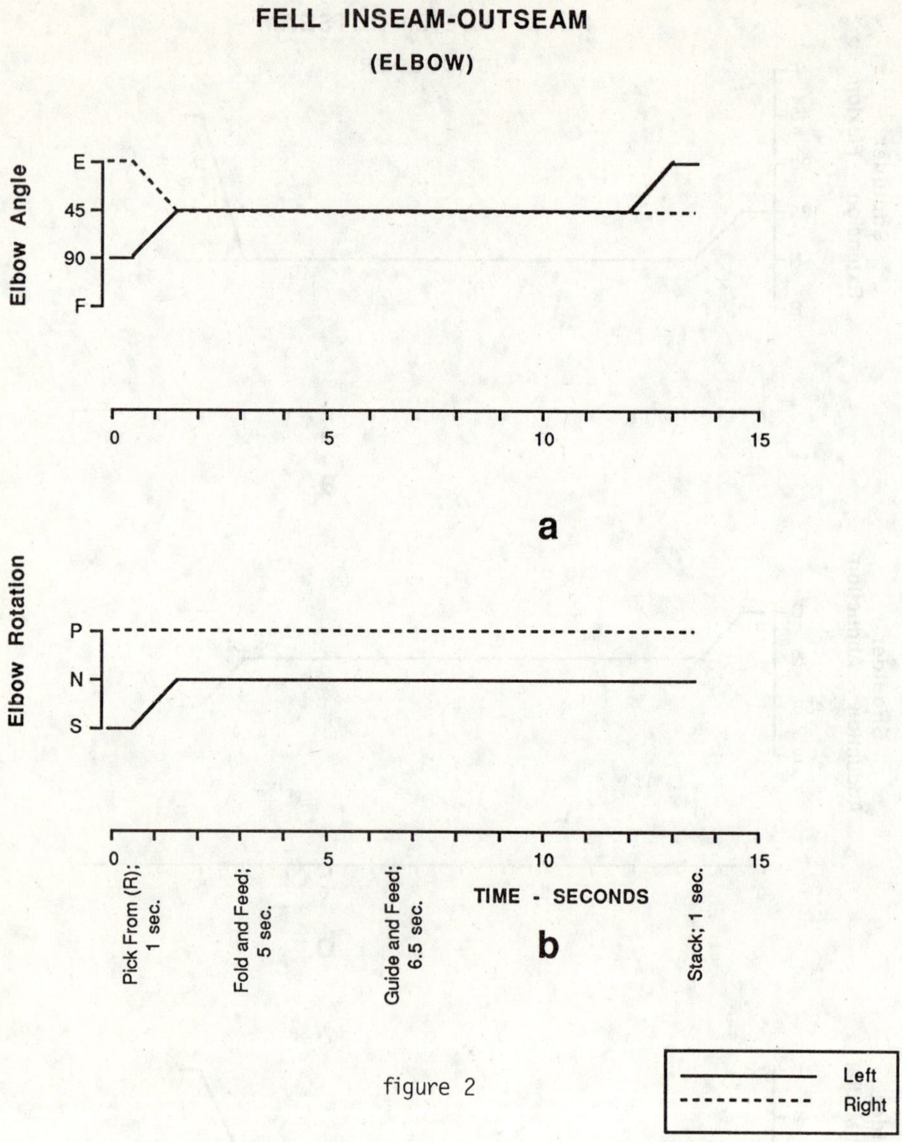

figure 2

wrist, on the other hand, remained 45^0 extended for 89% of the cycle and was fully extended for 11% of the cycle (Fig. 3). The fingers of both hands were in the posture of pulp pinch for 55% of the cycle, and in finger press for 37% of the cycle. It is noteworthy that the pulp pinch of fingers of both hands, which implies force application, was employed when the wrists were extended and deviated. Further, the finger press, which also necessitates significant force generation, was employed during the cycle when wrists were extended and right wrist radially deviated (Fig. 4).

of the cyc
deviated c
a posture
the cycle.
grasp posi

DISCUSSION
The ass
upper extr
Armstrong
However, t
upper extr
addressed.
the same r
pieces), t
Fell insea
21%, 36% a
operation
respective
repetitive
significan
from the s
similar re
may also b
maintenanc
connective
in additic
nerves and
due to sta
from the r
This appea
operation
injuries w
operation
deviated p
extenison
exposure w
the cycle.
degree of
less (45^0
operations
Due to the
controlled
observatio

CONCLUSION
The stu
associated
disorders.
duration o
exploratio

REFERENCES
Armstrong,
aspects
567-570

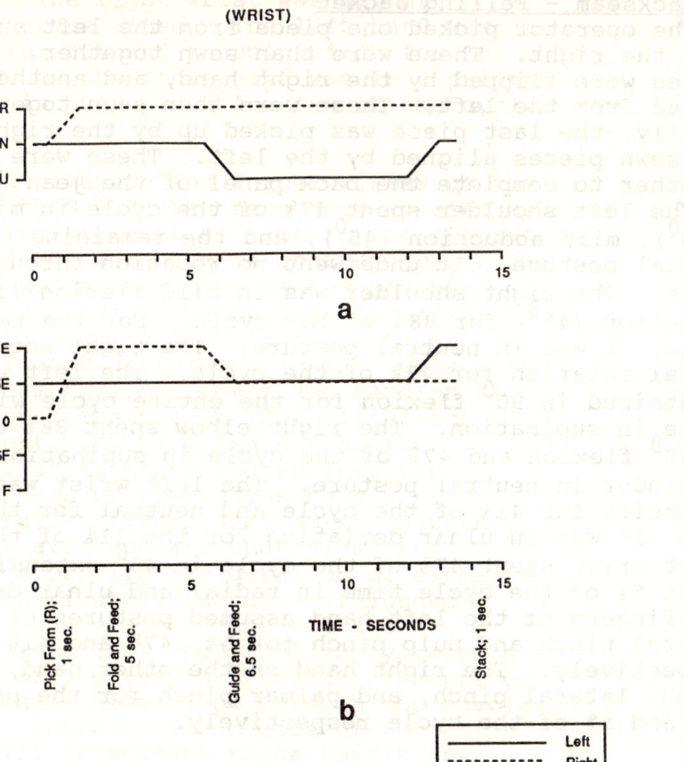

figure 3

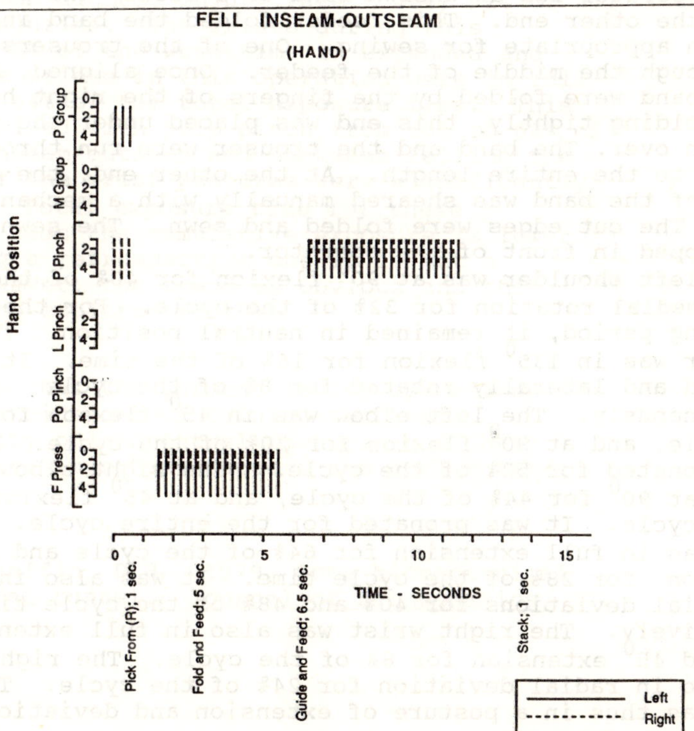

figure 4

Armstrong, T.J., Foulke, J.A., Joseph, B.S. and Goldstein, S.A., 1982, Investigation of cumulative trauma disorders in a poultry processing plant. American Industrial Hygiene Association Journal, 43, 103-116.

Blader, S., Holst, U.B., Danielsson, S. et al., 1987, Neck and shoulder complaints among sewing machine operators. Frequencies and diagnosis in comparison to a control population. In: Musculoskeletal Disorders at Work, edited by P. Buckle (London: Taylor & Francis), pp. 110-111.

Duncan, J. and Ferguson, D., 1974, Keyboard operating posture and symptoms in operating. Ergonomics, 17, 651-662.

Kumar, S., 1989, Upper extremity morbidity in a garment industry. In: Advances in Industrial Ergonomics and Safety I, edited by Anil Mital (London: Taylor & Francis), pp. 47-55.

Punnett, L. and Keyserling, W.M., 1987, Exposure to ergonomic stressors in the garment industry and critique of job-site work analysis methods. Ergonomics, 30, 1099-1116.

Punnett, L., Robins, J., Wegman, D.H. and Keyserling, W.M., 1985, Soft tissue disorders of the upper limb in women garment workers. Scandinavian Journal of Work Environment and Health, 11, 417-425.

Silverstein, B., Fine, L., Armstrong, T., Joseph, B., Robertson, M., 1986, Cumulative trauma disorders of the hand and wrist in industry. In: The Ergonomics of Working Postures, edited by N. Corlett, J. Wilson, I. Manenica, (London: Taylor and Francis) pp. 31-38.

Tichauer, E.R., 1976, Biomechanics sustains occupational safety and health. Ind Eng, Feb:46-56.

Vihma, T., Nurminen, M. and Mutanen, P., 1982, Sewing-machine operators' work and musculo-skeletal complaints. Ergonomics, 25, 295-298.

Wick, J. and Drury, C., 1986, Postural change due to applications of a sewing workstation. In: The Ergonomics of Working Postures, edited by N. Corlett, J. Wilson and I. Manenica (London: Taylor & Francis), pp. 375-379.

Advances in Industrial Ergonomics and Safety II
Edited by Biman Das
Taylor & Francis 1990

CUMULATIVE TRAUMA DISORDERS IN GLASS MANUFACTURING OPERATIONS --A CASE STUDY

Sheik N. Imrhan

Department of Industrial Engineering
University of Texas at Arlington
Arlington, Texas 76019

A study was made at a traditional glass manufacturing plant to determine work related factors which may have contributed to or caused musculoskeletal ailments among workers. Questionnaire responses and video-recordings of workers at work were analyzed. The results indicated several factors relating to workplace design, seating and tool design which needed design improvements.

INTRODUCTION

Glass manufacturing operations involve a great deal of manual work in traditional manufacturing plants. The impact of automated procedures has not been felt in many plants, mainly because of the perception that the benefits may be outweighed by the cost, at least in the forseeable future. The various jobs in a traditional low-tech plant are physically demanding. In one area -- the glass manufacturing proper -- workers are exposed to heat stress, compounded by noise and often heavy muscular work. In other areas heat and noise are not problematic, but the plague of modern repetitive tasks, albeit in an older plant, -- cumulative trauma disorders (CTD's) -- persist. This paper concerns the ergonomic aspects of these repetitive tasks in a traditional glass manufacturing plant.

Repetitive tasks were involved in the handling of the glasses from the time they left the molds and were cooled to the time they were packaged and stored for shipping. These tasks were almostly exclusively manual and performed mostly by females. The higher risk ones involved selecting and packing glasses into boxes, decoration, resorting and repacking, carton assembly, carton labeling (hot-end decoration), and machine shop operations (mainly mold cleaning). The tasks are characterized either by repetitive motions of the arms, constrained sitting postures, extended periods of standing, excessive forward and lateral bending at the hip, or the application of great hand forces. The jobs typically involve a combination of more than one of these. Such an environment is conducive to musculoskeletal stresses and to the development of Cumulative Trauma Disorders (CTD's) (Ayoub, 1973; Tichauer, 1978; and Putz-Anderson, 1988). Worker complaints have been very frequent.

A study was therefore made to identify work-related factors which could have caused or contributed to the incidences of the CTD's. The purpose of this paper is to describe that part of the study which dealt with CTD's from repetitive manual tasks. The specific objectives were:

1. To determine the distribution of the occurrence of worker self-assessed aliments from a questionnaire, according to
   a) the type of job performed (the department)
   b) the body part affected
   c) the duration of employment at that job
2. To draw inferences on the possible causes of the ailments, by comparing the questionnaire results with motion and posture analysis from vadeotaped material of the workers at work.

METHODS

Each of the high risk jobs, mentioned above, was videotaped for about 15 minutes, from at least 2 positions -- right side, left side, front or back -- depending on the clarity of the visual path. In all cases, at least one video-recording was done from the side which offered a clearer view of the angular deviations of the arm. A questionnaire was later administered in each of the high risk departments, to elicit responses concerning workers' assessment of their ailments. They were required to state whether they were experiencing muscular fatigue, tingling sensation, pain or numbness at various named parts of the body. The questionnaire items were explained by an industrial nurse who also recorded the responses, during a personal interview.

Subjects

Subjects were selected according to their availability. All were from the high risk jobs, listed above. The interview was conducted during the shift. Sixty-three worker completed the questionnaire (58 females and 5 males). The average age was 37.0 yr, and the average duration of employment on their present job was 7.2 yr. The breakdown of these two characteristics according to department is shown in Table 1.

Tasks Characteristics

A brief description of the characteristics of each of the tasks investigated is as follows:

Carton assembly: Workers stand and assemble cardboard boxes. The right hand retrieves the unassembled boxes and the left hand places the finished boxes onto a conveyor. Repetitive motions of the right hand involve extreme wrist flexion and ulnar deviation, and of the left hand extreme wrist extension and radial deviation. Both shoulders are also elevated repeatedly.

Select and Pack: Workers stand and remove glasses from a conveyor. The conveyors at different sites vary in height from 3 to 5 ft above floor level. The glasses are then fixed into storage boxes by applying a little force with the hand. Repetitive motions include shoulder elevation, and

Table 1. Subjects Characteristics According to Department simultaneously, with the fingers of each hand grasping 3 or 4.

		Department					
		Carton Assembly (n=5)	Select and Pack (n=21)	Deco. (n=8)	Hot End Deco. (n=7)	Resort and Repack (n=12)	Machine Shop (n=10)
Age	$\overline{X}$	43.5	34.5	34.8	35.1	40.6	37.9
	R	31-54	21-49	28-52	29-44	29-58	32-48
No. Yr of Work in Present Job		8.9	6.2	7.9	7.4	8.36	6.6
		8.0-9.5	1.0-12.0	2.0-11.5	2.5-12.0	1.0-14.0	1.5-13.0

wrist flexion and ulnar deviation. For some types of conveyors also, workers must repeatedly reach foward while bending forward and laterally at the hips.

<u>Resort and Repack</u>: Workers typically sit. The repetitive motions are very similar to the Select and Pack jobs.

Decoration: The worker sits in front of the decorating machine and manually feeds glasses into the machine with the right hand, then removes them with the left hand. The left hand then places them on a rotary conveyor which feeds into a linear one. The rate of feeding/removal averages 28 per minute. The two hands work simultaneously. Likewise the two feet work simultaneously at foot controls. Repetitive motions include right hand wrist flexion and ulnar deviation; left hand extension and radial deviation; raising both arms; and plantar flexion of the feet. Workers must frequently bend laterally at the hips. The eyes must maintain precise visual contact with the machine to fix the glasses in place.

<u>Hot end decoration</u>: The worker stands and manually feeds paper labels into a machine, then removes them immediately. Repetitive motions include wrist flexion in both arms.

<u>Machine shop operations</u>: These operations are of several types. The ones investigated involved workers sitting and cleaning glass molds positioned on a table. Most of the cleaning is done manually by rubbing abrasive materials (eg. sandpaper) fixed onto the ends of rubbing sticks on the molds. Part of the job is done by using small light weight rotary power tools. Some parts of the molds cannot be accessed properly by powered rotary tools. Poor sitting postures are compounded by poorly designed chairs and workplace. While using rubsticks, great muscular forces are generated -- for gripping with the fingers or whole hand; for alternate shoulder flexion and extension; and for pressing down with the hand. The wirst is usually in extreme extension and radial deviation, and the upper body bent forward or sideways at the hips.

RESULTS AND DISCUSSION

Since the sample was non-random and small within each department (Table 1) sophisticated statistical analyses were avoided. Instead the data were analysed by subjective evaluation of the descriptive statistics (Tables 2 and 3).

Table 2. Percentage of Persons in Each Department with Self-Assessed Musculoskeletal Ailments

	Department					
	Carton Assembly (n=5)	Select and pack (n=21)	(n=8)	Hot End (n=7)	Resort and Repack (n=12)	Machine Shop (n=10)
Finger	60	33	88	57	58	80
Palm	0	14	13	29	8	50
Wrist	60	29	88	71	58	60
Elbow	40	19	38	57	25	30
Shoulder	40	55	88	71	50	80
Neck	60	24	38	57	33	70
Upper Back	60	33	63	43	25	50
Lower Back	60	29	75	57	42	80
Knee	40	10	38	57	25	20
Ankle	80	33	25	29	33	40
Foot	80	10	38	57	42	10
Average	53	26	54	53	36	52

Table 3. Frequency Distribution of The Occurrence of Self-Assessed M sculoskeletal Disorders in Different Parts of the Body for 63 Workers.

	Type of Disorder			
	Muscular Fatigue	Tingling Sensation	Pain	Numbness
Body Part				
Finger	15	19	15	21
Palm	3	6	5	2
Wrist	12	9	28	4
Elbow	7	2	9	1
Shoulder	27	2	21	1
Neck	24	0	3	1
Upper Back	21	2	8	1
Lower Back	19	1	0	16
Knee	8	0	11	1
Ankle	7	1	5	0
Foot	19	3	6	1

Overall, all the departments had at least 50% of the respondents indicating at least one ailment, with the worst cases being Decoration (88%), Carton Assembly (80%), and Machine Shop (80%) (Table 2). For Decoration, Hot End Decoration (71%) and Machine Shop, the body parts with the greatest percentage complaints were the fingers, wrist, shoulders and lower back; for Carton Assembly, it was the feet and ankles.

More than 50% of the respondents complained of at least one ailment in the fingers (63% respondents complaining), wrist (61%), shoulders (64%), and the lower back (57%). The palm of the hand was the only listed body part which did not show a substantial average percentage complaints; though the figure for the machine shop was 50%. The other body regions had 32-47% complaints. Finger problems were worst in Decoration (88%) and Machine Shop (80%); wrist in the Decoration (88%) and Hot End Decoration (71%); shoulder in Decoration (88%), Hot End Decoration (71%) and Machine Shop (88%); and lower back in the Decoration (75%) and Machine Shop (80%).

Certain body parts which had smaller overall percentage complaints had substantial numbers of complaints in certain departments. For example, for the feet and ankles 80% (4/5) of the Carton Assembly workers complained of at least one ailment; the upper back had 60% (3/5) from Carton Assembly and 63% (5/8) from Decoration; the neck had 70% (7/10) from the Machine Shop, 60% (3/5) from Carton Assembly and 57% (4/7) from Hot End Decoration; and the elbow had 57% (4/7) for Hot End Decoration.

There were, as expected, different frequencies for different types of complaints for specific body parts, since the types of complaints were indicative of the severity of the ailment. It was also possible for a given worker to have more than one ailment to a given body part; for example, pain and tingling sensation of the wirst simultaneously. Table 3 shows this bivariate distribution. The data in that table indicate that muscular fatigue was most frequent in the region of the trunk, and the ends of the extremities, the fingers and feet. Tingling sensation was confined to the fingers; pain to the fingers, wrist and shoulders; and numbness to the fingers and lower back. The fingers seemed to be the worst affected part of the body.

RECOMMENDED SOLUTIONS

An analysis of the videotaped tasks indicated many conditions which could have caused or contributed to the the musculoskeletal problems summarized in Tables 2 and 3. Several solutions were offered which required workplace and tool redesign instead of more expensive automated process redesign.

In general, for sitting tasks, better chairs which conformed to workers' anthropometry and allowed adjustability were recommended because they were lacking. Management was also advised to ensure that workers knew how to change all adjustable features in chairs. For standing work, better padded floor mats were recommended. Recommendations were also made for shorter work cycles and worker rotation; this aspect will not be discussed further in this paper.

The section below briefly describes the most important

recommendations made for each department.

Carton Assembly

Better padded floor mats were suggested to help with problems in the ankles while standing. Repositioning the pallets on which the unassembled boxes were stacked, and the conveyor on which the assembled boxes were placed were suggested to reduce the flexion, ulnar deviation and extension of the wrist. This rearrangement was also expected to reduce forward and lateral bending at the waist and alleviate back problems.

Select and Pack

1. Providing a simple platform for short subjects to minimize shoulder elevation and abduction, and wrist flexion.
2. Reorienting the boxes in which the glasses are packed so that wrist flexion is minimized, and hand push force places less stress on the tendons in the carpal tunnel.

Decoration

1. Reversing the direction of the turntable or placing it a few inches forward of the operator to minimize wrist deviations of the left hand.
2. Reorienting the feeding conveyor to minimize wrist deviations of the right hand, and lateral bending of the trunk.
3. Using a chair with a larger backrest, and encouraging operators to adjust their chair at the beginning of the shift to assume good sitting posture, especially with respect to trunk, lower leg (angle at knee) and foot (angle at ankle) postures.
4. Redesigning decoration machine so that wrist deviations are minimized when inserting or extracting glasses from it. Relocating the foot controls a few inches forward.

Hot End Decoration

Using a mechanical device to press glued labels onto sides of unassembled boxes. The operator may then be able to sit and work.

Resort and Repack

The recommendations were basically the same as for Select and Pack operations. In addition, better design of the glass feeding mechanism on conveyors to bring glasses closer to the packers to minimize bending at the hips and reaching out with the arms.

Machine Shop

1. The use of powered hand-tools as much as possible to minimize muscular forces in pinching, power-gripping and rubbing.
2. Redesigned rubbing sticks aimed at keeping the wrist straight during rubbing actions, and at reducing muscular forces by bringing the hand closer to the rubbing tip.
3. Better chairs to conform to workers' anthropometry and to allow individual adjustment. A sit-stand chair

(Grandjean, 1980) was also recommended since it gives the upper body better mechanical advantage for hand-arm forces on the molds.

REFERENCES

Ayoub, M.M., 1973., Workplace design and posture. Human Factors,1973, 15 (3), 265-268.

Chaffin, D.B. and Andersson G.B.J., 1984, Occupational Biomechanics, New York, John Wiley and Sons.

Grandjean, E., 1980, Fitting the Task to the Man: An Ergonomic Approach. London, Taylor and Francis.

Putz-Anderson, V., 1988, Cumulative Trauma Disorders: A Manual for Musculoskeletal Diseases of the Upper Extremity. New York, Taylor and Francis.

Tichauer, E.R., 1978, The Biomechanical Basis of Ergonomics. Anatomy Applied to the Design of Work Situations, New York, Wiley Interscience.

CONCLUSIONS

The tasks involved in glass manufacturing operations in a traditional low-tech plant proved to be predominantly of the highly repetitive manual type. Among the work-related factors contributing to or causing Cumulative Trauma Disorders were poor postures, imposed by poor workplace and chair designs; extreme joint deviations, especially in the arm; poorly designed hand tools; and application of great arm forces with deviated joints. Since automation was not being considered, recommendations to combat musculoskeletal stresses were of the standard ergonomic types, especially in workplace and handtool redesign.

Acknowledgment

The author would like to thank Ms. Ann von der Heide, Department of Industrial Engineering, UTA, for her help in typing this paper.

Advances in Industrial Ergonomics and Safety II
Edited by Biman Das
Taylor & Francis 1990

AN ERGONOMIC STUDY FOR THE CONTROL OF UPPER EXTREMITY CUMULATIVE TRAUMA DISORDERS IN A MANUFACTURING INDUSTRY

A. M. Genaidy, K. M. Bafna, E. Delgado and A. Mhidze

Department of Industrial Engineering
Western Michigan University
Kalamazoo, MI 49008, U.S.A.

Ergonomic job analysis was used in a manufacturing industry to identify operations which posed employees to a high risk of CTDs. Moreover,the risk factors associated with these operations were recognized and quantified using ergonomic job design principles. Based on the analysis, engineering controls were recommended.

INTRODUCTION

The main objective of this study was to present the results of ergonomic job design principles used in (1) the analysis of high risk operations in a manufacturing industry and (2) the providing of design solutions for the control of upper extremity cumulative trauma disorders (CTDs) associated with these types of tasks.

The ergonomic job design approach adopted in this study consisted of several steps: (1) review of injury and illness records over a six-year duration (1984- 1989) for identifying high risk operations; (2) task analysis using video recordings and observations; (3) ergonomic risk analysis for identifying occupational factors which result in upper extremity cumulative trauma disorders, and quantifying the severity of each factor; and (4) developing recommendations for ergonomic job redesign. Each of the above steps is described below.

REVIEW OF INJURY AND ILLNESS RECORDS

An analysis of injury and illness records for a 6-year period revealed that the line inspector, line packer, inspector C, and punch press operations had the highest number of recordable medical cases, with the punch press operation being number one. Furthermore, these operations were also among the highest in terms of the number of lost work days. Hence, it was decided to examine these operations and redesign them using ergonomic job design principles. In this paper, the results of the line packer and punch press operations are given below.

TASK ANALYSIS

The operations were analyzed through video recordings and observations, and the job description of each operation was documented.

The various elements of the line packer operation are:

1. A labeled box is taken and positioned on a loading table.
2. Three clips are placed to hold down the flaps of the box.
3. Small cardboard trays are taken from a carton and positioned inside the packaging box.
4. When enough bearings to fill one tray have been accumulated on the conveyor, they are moved with both hands and positioned in one of the trays inside the box. When all the trays in one level have been filled, a cardboard insert is placed on top. The procedure of taking trays and filling them with bearings is then repeated.
5. While the bearings are accumulating on the conveyor, the worker prepares labels for use on the next box and applies them to two of its outer surfaces.
6. Once the box is filled, a cardboard divider is placed on top of the trays. The clips are then removed and placed in their storage place. The flaps on the box are closed.
7. The worker reaches for a stapler located above the shoulders and applies two staples to seal the box.
8. The box is carried by hand and placed on a pallet.

The various elements of the punch press operation are:

1. The operation starts with the right hand reaching and grasping a number of unprocessed bearings from the incoming container.
2. The right hand places the bearings on a table in front of the vertical press.
3. A bearing is then positioned vertically in the die with the right hand.
4. After one end is punched, the bearing is removed from the die with the left hand, rotated around and transferred to the right hand, and repositioned in the die with the right hand for punching the second end.
5. Each completely processed bearing is removed from the die with the left hand and placed on the working table. This is repeated for each group of bearings that were picked up by the right hand at the start of the cycle. The completed group is then placed in the outgoing container with the left hand.

RISK ANALYSIS

In an epidemiological study, Silverstein et al. (1986) reported that any job requiring a hand force of more than 4 kg is classified as a high force job and can lead to some forms of CTDs. In another study, Hammer (1934) found that repetitions greater than 1,500 to 2,000 times/hour produce tendon sheath irritation in the hands. Thus, the values of 4 kg and 1,500 times/hour were considered as the ergonomic tolerance limits for the force and repetitions, respectively.

Excessive forces, high repetitions, and awkward postures were identified as major risk factors for the five operations studied. The detailed analysis of each risk factor is given below.

Force

The forces exerted on the right and left forearm flexors were estimated from the analysis of the myoelectric signals obtained according to the procedures described by Armstong et al. (1979). A SensorMedics R-612 Dynograph physiological monitor was used to record the myoelectric signal.

Table 1 shows a summary of the forces exerted on the left and right forearm flexors for both operations. In general, forces increased with an increase in the number and size of bearings. Moreover, for both operations the forces were substantially higher than the 4 kg force limit suggested by Silverstein et al. (1986).

Repetition

Table 2 gives an overview of the analysis of the repetition risk factor for the two operations. This analysis was performed for different parts of the upper extremity (shoulder, elbow, wrist, hand, and finger) and different types of motions (abduction, adduction, flexion, extension, pronation, supination, ulnar deviation, radial deviation, pulp grasp, pulp pinch, palm pinch, finger press, and medial grasp).

Repetition was not found to be a risk factor for the line packer operation. For the punch press operation, very high shoulder flexion/extension repetitions, extremely high repetitions of forearm (around the elbow) pronation/supination and flexion/extension motions, very high repetitions of wrist ulnar deviation/radial deviation and flexion/extension motions, and very high repetitions of pulp grasp motions were present.

Posture

Posture was analyzed for different parts of the upper extremities and different types of motion. The results are shown in table 3. For the line packer, excessive shoulder abduction and flexion, forearm pronation, and wrist flexion and ulnar deviation were found. For the punch press operation, there was excessive shoulder abduction and flexion (only right), left forearm (around the elbow) supination, left wrist ulnar deviation and flexion, right wrist radial deviation, pulp grasp, and left medial grasp.

RECOMMENDATIONS

The following recommendations were proposed for the operations that were analyzed.

Line Packer

1. Eliminate hand stamping of labels and provide ready-to-use computer printed labels.
2. Modify the workplace to avoid twisting of the upper torso as well as the lifting of bearings with both hands to place in the box.

Place a special table at the end of the conveyor which has the capability of being lowered below the conveyor level to a depth equal to the height of the carton in which the bearings are packaged. When enough bearings are accumulated on the conveyor to fit in one tray, they are moved (by sliding) onto the table. A cardboard tray is placed on these bearings in an inverted position and the entire stack is moved to the

Table 1. Force (kg) exerted in the forearm flexors

Operation	Bearing Size	Number of Bearings	Force Left Hand	Force Right Hand
Line Packer	Small	15	12.90	7.67
		25	14.99	13.82
	Large	10	18.58	11.34
		15	25.89	14.98
		20	29.39	18.40
Punch Press	Small	10	26.78	12.21
	Large	10	26.50	19.03

Table 2. Repetitions (times/minute) for various operations

Operation	Number of Bearings Handled per hour	Body Part	Type of Motion	Repetitions LH	Repetitions RH
Line Packer	675	Shoulder	AB/AD	115	115
			F/E	125	125
		Elbow	P/S	180	225
			F/E	370	395
		Wrist	UD/RD	80	135
			F/E	170	190
Punch Press	1440	Shoulder	AB/AD	144	144
			F/E	5904	7344
		Elbow	P/S	17424	4896
			F/E	16560	29376
		Wrist	UD/RD	3312	11664
			F/E	10368	15552

Note: LH -- left hand; RH -- right hand; AB/AD -- abduction/adduction; F/E -- flexion/extension; P/S -- pronation/supination; UD/RD -- ulnar deviation/radial deviation.

Table 3. Posture (% work elements) analysis

Operation	Body Part	Type of Motion	Repetitions LH	Repetitions RH
Line Packer	Shoulder	AB	93	91
		AD	0	0
		F	96	90
		E	4	2
	Elbow	S	23	19
		P	72	71
	Wrist	UD	45	58
		RD	0	2
		F	48	54
		E	9	18
	Hand	PG	0	2
		PP	3	7
		PaP	7	15
		FP	28	23
		MG	25	17
Punch Press	Shoulder	AB	82	79
		AD	0	0
		F	25	55
		E	1	22
	Elbow	S	49	0
		P	23	18
	Wrist	UD	80	10
		RD	1	49
		F	55	40
		E	35	19
	Hand	PG	61	65
		PP	0	0
		PaP	0	0
		FP	0	0
		MG	89	1

Note: LH -- left hand; RH -- right hand; AB -- abduction; AD --adduction; F -- flexion; E -- extension; P -- pronation; S -- supination; UD -- ulnar deviation; RD -- radial deviation; PG -- pulp grasp; PP -- pulp pinch; PaP -- palm pinch; FP -- finger press; MG -- medial grasp.

edge of the table (in a direction perpendicular to the conveyor). This procedure is continued until there are enough rows of bearings for one layer. A cardboard divider is placed on top of this layer and the table is lowered so that the top of the divider is in line with the top of the conveyor. The above procedure is repeated to get the required number of rows of bearings in the new layer. This is continued until there are enough layers of bearings to fit in a carton.

The operator then places an empty carton upside down over the stack of bearings. The entire stack of bearings with the carton is moved (by sliding) to an adjacent fixture. This fixture has a jaw to clamp the bottom of the carton and rotate the entire carton (with contents) by 180 degrees. The flaps of the carton, which are now in an upright position, can now be sealed and the carton removed to the pallet by a vacu-hoist.

Punch Press

1. Redesign the die so that both edges of the bearing can be punched simultaneously. This will eliminate the need to flip the bearing with both hands and will cut the repetitions by almost 50%.
2. Eliminate the movement of the upper extremities to the sides by redesigning the workplace layout. The bearings can be fed automatically to and from the die by letting them slide on a tilted tray placed on either side of the die. Also keep the upper extremity motions in a way so that the elbow is maintained at a 90 degree angle.

DISCUSSION

In this study, two operations were identified as high risk operations based on a review of injury and illness records. Each one of these operations was analyzed and evaluated using the principles of ergonomic job design.

For the line packer operation, excessive forces and awkward postures were evident through the following elements: the operator stamps the labels and places them in the box, and the operator carries the bearings and places them inside the box. The punch press operation was extremely repetitive in nature and, coupled with awkward postures, resulted in high forces on the upper extremities. Examples of awkward postures were reaching to the right side of the body to obtain the bearings and placing the bearings in the box to the left side of the body.

Following the risk analysis, recommendations were provided on the basis of engineering controls which aim at the modification of the environment by altering work tools and workstations. These modifications resulted in job requirements that are within the capability of the workforce.

REFERENCES

Armstrong, T.J., Chaffin, D.B., and Foulke, J.A., 1979, A methodology for documenting hand positions and forces during manual work. Journal of Biomechanics, 12, 131-133.

Hammer, A.W., 1934, Tenosynovitis of the hand and the wrist. Medical Record, 140, 353-355.

Silverstein, B., Fine, L.J., Armstrong, T.J., 1986, Cumulative trauma disorders in industry. British Journal of Industrial Medicine, 43, 779-784.

Advances in Industrial Ergonomics and Safety II
Edited by Biman Das
Taylor & Francis 1990

WORKPLACE FACTORS CONTRIBUTING TO MUSCULOSKELETAL DISORDERS DURING ORTHOTIC FABRICATION

Noreen Hayashi, BSc(OT), OT(c), OCOT
Consultant, Prosthetics & Orthotics
Ministry of Health, Assistive Devices Branch
7 Overlea Blvd., 6th Floor
Toronto, Ontario
M4H 1A8 Canada

Susannah Gray, BSc(OT), OT(c)
Occupational Therapist
Toronto Grace Hospital
Toronto, Ontario
M4K 3C6 Canada

Pat McKee, MSc, BSc(OT), OT(c), OCOT
Asst. Professor, Div. of Occupational Therapy
University of Toronto
256 McCaul Street
Toronto, Ontario
M5T 1W5 Canada

A need was established for health care professionals involved in orthotic fabrication to apply sound ergonomic principles that minimize personal injury and maximize productivity. This study identifies various risk factors associated with orthotic fabrication; compares the incidence and nature of musculoskeletal complaints amonst occupational therapists and orthotists; and, makes practical recommendations for task redesign, tool considerations and handling techniques that minimize physical stressors. Since education and training were associated with decreased injuries amongst both study groups, the authors recommend that future orthotic continuing education courses and manuals emphasize safe biomechanics to help prevent costly cumulative trauma disorders.

INTRODUCTION

As rehabilitation professionals, occupational therapists (OTs) regularly educate their clients in self-management and prevention of physical and psychosocial disorders using human factors principles supported by the current research. This recent study conducted by OTs in the Metropolitan Toronto area strongly identified a need for health care professionals involved in orthotic fabrication to apply sound ergonomic principles that minimize personal injury and maximize productivity.

An orthosis is defined as any device applied directly and externally a person to support, correct, or compensate for an anatomical deformity or weakness. In addition, it may be assist, allow, or restrict body movement. Such devices include splints and braces which are either pre-fabricated or custom-made to support the limbs or spine during daily activities. A wide variety of materials, fabrication and fitting techniques as well as functional positions may be selected to suit the individuals' particular needs. Certified orthotists (COs) and occupational therapists (OTs) provide the majority of orthotic devices.

Occupational therapy is the assessment of function and adaptive behaviour and the treatment and prevention of disorders which affect function or adaptive behaviour to develop, maintain, rehabilitate or augment function in the areas of self-care, productivity and leisure (Ontario College of Occupational Therapists, 1989). OTs receive a four year academic education through university programmes accredited by the Canadian Association of Occupational Therapists. Subjects studied include anatomy, physiology, kinesiology, medicine and surgery, social sciences, psychiatry, research techniques, administration and therapeutic techniques. A minimum of 1200 hours of satisfactory performance in clinical fieldwork placements divided between physical medicine and psychiatry plus a passing grade on a written certification examination are necessary to practice occupational therapy in Canada.

Orthotists and prosthetists must have a previous related degree or diploma to receive specialized training in the fabrication, fitting and modification of specific orthotic and prosthetic devices via a two year technical programme at three Canadian colleges of applied arts and technology. Following the programme, a two year residency supervised by a Canadian Board Certified prosthetist or orthotist and satisfactory completion of a written and practical examination are required to practice orthotics or prosthetics in Canada.

Most current undergraduate programmes for OTs and certified orthotists (COs) appear to be remiss in addressing orthotic fabrication techniques which minimize joint and soft tissue stress. A literature review demonstrated that orthotic manuals and texts expound primarily on methods for the achievement of a well-fitting, esthetic orthotic device (Barr, 1975; Fess and Phillips, 1987; Hopkins and Smith, 1978; Hunter et al., 1984; Smith and Nephew, 1988; Ziegler, 1984). A separate study (Gwosdy, 1989) however, of Canadian occupational therapy programmes revealed that some universities are incorporating some personal safety techniques in the orthotic aspect of their physical medicine component.

Cumulative trauma disorders (CTDs), otherwise known as repetitive motion diseases, repetition strain injuries or overuse syndromes are physical disorders which may develop in tendons, muscles, nerves or the vascular systems in response to repeated over-exertions and excessive movements of the human body (Bear-Lehman and Blair, 1987). Disorders of this type can be extremely debilitating and may result in chronic or permanent disability creating loss of function or employment.

Bear-Lehman (1986) pointed out that in many instances, sufferers do not receive appropriate treatment and little is done to correct the underlying causes. Much of this neglect may be related to the difficulty in placing a definitive medical diagnosis upon often vague symptoms which demonstrate cycles of remission and exacerbation over many years. Traditional medical-model approaches tend to treat symptoms rather than address the root causes. An ergonomic or biopsychosocial intervention or model of treatment and prevention aims to improve performance and safety through the study and application of general principles which govern the interaction of humans with their working environments.

The literature does present a number of risk factors which may be associated with CTDs. Occupational factors such as repetitive exertions, high forces, mechanical stress, awkward postures, vibrations and low temperatures have repeatedly been suggested as contributors to repetitive motion disorder (Armstrong and Chaffin, 1979; Armstrong et al., 1988; Arndt, 1987; Cannon et al., 1981; Hoyt, 1984; Poole, 1988; and Tichauer, 1966).

Hand positions during activities are often cited as contributors to nerve compression disorders. Armstrong and Chaffin (1979) suggested that positions of extreme wrist flexion and extension reduce wrist tunnel volume significantly. These positions increase pressure on the median nerve transversing this tunnel, thereby resulting in numbness or tingling sometimes with reduced ability to pinch and grip (carpal tunnel syndrome, median nerve compression syndrome). As well, repeated hand activities which combine prolonged pinch and wrist flexion may predispose a person to carpal tunnel syndrome (Phalen, 1966; Smith et al., 1977).

All of the above studies emphasize that repetitive motion of the wrist and fingers engaged in certain positions may lead to repetition strain injuries. Professionals who engage in physical activities such as those associated with orthotic fabrication, therefore, are placed at risk of joint and soft tissue stress and injury if they fail to incorporate sound biomechanical principles and joint protection techniques into their daily practice.

For the above reasons plus the authors' personal observations and experiences, it was suspected that there may exist a high incidence of occupational-related hand and wrist problems associated with orthotic fabrication. The hypothesis was: <u>biomechanical stressors experienced during the fabrication of orthoses may lead to increased risk of personal injury</u>. A thorough literature search did not yield any previous studies which examined the relationship between orthotic fabrication and repetition strain injuries.

METHOD

A detailed self-administered subjective questionnaire was developed and refined for mailing to all Metro Toronto area physical medicine OT and orthotic services. Some of the questions which were incorporated examined the types and incidence of various cumulative trauma disorders and reported injuries associated with orthotic fabrication. Were the problems severe enough to cause people to take time from their work? Were some persons modifying their job demands and work techniques to accommodate their problems? Other questions extracted basic demographic information such as age, sex, years of work experience, degree of work time expended in orthotic practice, and level of formal instruction received concerning personal safety techniques.

Injured respondents were requested to detail the type and location of their particular injuries, duration, perceived sources of physical stress during the fabrication process, amount of down-time, whether disability benefits were awarded, and methods of alteration of work practises and their effectiveness.

Three questionnaires with cover letters and self-addressed stamped envelopes were mailed to each Metropolitan Toronto area physical medicine occupational therapy department manager. The directors were instructed to distribute these to the OTs who most frequently constructed orthoses.

Respondents had to fabricate/fit a minimum of one device per week. Unused questionnaires were marked "not suitable" and returned so that they could be tracked. The same procedure applied to orthotic departments accredited by the Canadian Board for Certification of Prosthetists and Orthotists. Cover letters assured complete anonymity.

Follow-up telephone reminders were completed at four weeks and five weeks post-mailing. All questionnaires received within six weeks of the mail-out were accepted into the study. Of the 92 questionnaires sent to OT departments, 82 were returned (89%), 31 were not suitable since they did not fit the inclusion criteria, leaving 51 questionnaires appropriate for study (55%). Of the 35 surveys sent to orthotic departments, 16 (46%) were returned, 2 were unuseable leaving 14 or 40% useful for the study. Thus, the overall inclusion rate was 65 out of 127 or 53%.

RESULTS

Of the 51 occupational therapists, 49 (96%) were female and 2(4%) were male. With the orthotists, 10 (71%) were male and 4 (29%) were female.

Of the 65 respondents, 48 (74%) reported some type of injury. Thirty-six per-cent of the 107 injuries reported by both groups combined were cumulative trauma type disorders such as thumb joint arthropathy, joint inflammation, ligament strains and sprains, tendonitis, tenosynovitis, nerve compression, and joint deformity. A surprisingly high incidence of single-trauma injuries due to lacerations and burns were reported by 43% of all respondents. Only 10 problems such as dermatitis, osteoarthritis, schaphoid fracture, lateral epicondylitis, carpal tunnel syndrome, and tendonitis were acquired prior to orthotic fabrication. It was reported, however, that all of these problems were aggravated by the fabrication process.

Of the orthotists, 78% (11/14) reported a total of 84 orthotic-related injuries. Of the injured, 10 (43.4%) reported skin lacerations, 1 (4.3%) suffered from knee and spinal pain, 7 (30.4%) had skin problems, and 5 (22%) had miscellaneous hand problems.

Of the OTs, 73% (37/51) reported a total of 84 orthotic-related injuries. Thirty-four (36%) of the injured group reported cumulative-trauma injuries, 38 (43%) were single-insult injuries, and 12 (11%) were muscle cramps.

OT results demonstrated what appeared to a be an increase in numbers of injuries with a corresponding increase in number of hours of work per week. Of the group that worked with orthoses up to 5 hours weekly, only 55% (11/20) reported hand injuries, whereas, in the 5 to 10 hours per week group, injuries were reported by 71% (10/14). Among the therapists who worked 10 to 15 hours per week in orthotics, 92% (13/14) experienced hand injuries and of those who worked 15 to 20 hours per week, 100% (3/3) reported injuries.

The amount of work experience also appeared to have an influence on the rate of injury of the OTs with 1 to less than 3 years of orthotic experience, 46% (6/13) reported hand injuries, whereas with the group with 5 years or more of work experience, 82% (19/23) were injured.

Perceived sources of stress for OTs included fabrication techniques such as scoring thick thermoplastic sheets; cutting out orthotic patterns from plastic sheets; trimming and finishing the orthoses; punching holes; bending wires; and use of improperly maintained or inappropriate tools and equipment.

For the orthotists, reported sources of stress were associated with postural strain from bending over benchwork, lifting heavy materials and plaster molds, transferring patients, and standing on hard floors for most of the day. Their physical complaints tended to centre on the neck, back, and knees in contrast to primarily hand joint and soft tissue problems with the OTs.

In comparison of stress-reduction factors, both groups recognized that properly maintained and appropriate tools and body postures helped to decrease physical stress. The orthotists further identified that the use of electrical equipment such as saws, grinders, and sanders as well as bench vises to stabilize materials proved to be effective methods of decreasing stress. No one received WCB benefits and only one person took one hour from work to treat a burn. Seventeen OTs who suffered from thumb joint pain, tendonitis, and carpal tunnel syndrome wore hand orthoses to restrict them from painful or aggravating movements. Ice, heat or local steroid injections provided short-term relief in some cases.

Respondents who had received training in appropriate tool and material use, proper body mechanics and general safety precautions reported less injury than those who had not received such instruction.

DISCUSSION AND RECOMMENDATIONS:

This pilot study being retrospective in nature lacked a control group for comparison of demographic groups. Neither self-report based on questionable personal recall or perceptions of injuries nor volunteer bias could be ruled out. Currently disabled OTs and COs who were unable to continue their orthotic practice were not examined. Some individuals, for example, may have switched to a less physically demanding aspect of their practice.

Questions intentionally focused on upper extremity complaints and therefore limited the scope of complaints for the orthotist population. The eligibility criteria may have been too leniant in asking that a minimum of only one orthotic device be fabricated by the respondent each week; is the manufacture of only one orthotic device per week likely to create a CTD? Avocational activities such as sports and household maintenance as well as other demanding job-related non-orthotic stressors may have contributed to the high injury rate of 74%. Inevitably, some of the questions were misinterpreted by the respondents.

Although demographic data had been gathered about hand size and hand dominance, the limited size of the orthotist group did not permit comparison to the OT group.

The results of this pilot study reveal some important implications for the professionals involved in the fabrication of orthoses. Both groups in the study exhibited dramatically high incidence of injuries, however, each group differed in the types of reported injuries such that orthotists had more single trauma type injuries and the occupational therapists experienced more cumulative trauma type problems.

The finding of the increased incidence of CTDs amongst OTs with increased number of hours worked supports that of Nathan (1988) who found that as the time on the job increased, the accumulating damage decreases the tissue tolerances, even though the tasks remain unchanged. OT problems such as thumb joint arthropathy, joint inflammation, ligament strains and sprains, tendonitis, tenosynovitis and nerve compressions increased in prevalance with number of years of work experience. Nathan (1988) found that increased work experience irrespective of occupation to be the best predictor of increased development of carpal tunnel syndrome.

Of interest, certified orthotists spend all of their work time fabricating orthoses and most occupational therapists spend only a portion of their job engaged in this activity, yet the OTs reported a much higher number of CTDs than the COs. What could account for this difference may be that OTs tend to adapt carpentry tools to handle the orthotic materials, while the COs use specialized heavy duty machinery designed specifically for handling orthotic materials.

An unanticipated finding was the high incidence of single trauma type injuries such as lacerations and burns reported by both groups. These injuries are due to lack of caution and poor safety techniques which are readily prevented by personal safety education. Training and instruction in safety techniques were seen to reduce the amount of injury incurred by the professionals in this study.

The study also revealed that as a result of their injuries, over half of both professional groups introduced modifications to their work methods. These job modifications included interspacing of heavy with lighter duties and slowing of the workpace. Although the COs reported total effectiveness with their job alterations, the OTs found that their adjustments were effective 89% of the time.

These findings speak to the need of professionals such as occupational therapists and certified orthotists to become more aware of the potential risk factors associated with orthotic fabrication. These groups as well as others should implement work techniques which minimize joint and soft tissue stress and in turn may improve productivity.

Using study results, the authors have compiled a set of practical recommendations for this purpose:

(1) For scoring of thick thermoplastic sheets hold the cutting tool in a dagger-like position to take advantage of the larger elbow muscles and reduce wrist deviation and flexion.
(2) Pre-heat thermoplastic materials to soften them and ease cutting effort. Pre-cut splint blanks may be used.
(3) Use leverage whenever possible. Use the open palm of the hand and body weight to crack orthotic sheets over a table.
(4) To improve grip on hand tools, use padded long handles to disperse the forces on the fingers and palmar tissues over a wider area (Chaffin and Andersson, 1984).

(5) Avoid exerting large forces over the centre of the palm to reduce damage to the underlying nerves, arteries, and tendons. Use smaller hand tools and curved pliers.
(6) Left-handed individuals should use left-handed tools to enhance safety and efficiency in handling.
(7) Each person should use a personal set of tools for which they are responsible for maintaining in good working order.
(8) Working at appropriate heights reduces strain. Workbenches and tables should be height-adjustable and tiltable to avoid stooping of the back more than 30 forward, permit occasional forearm resting, and maintain the arms below 20 abduction.
(9) Use trolley carts to reduce heavy carrying and many trips.
(10) Organize the work area with effective layout and storage systems so that more frequently used items are within the work envelope. This will reduce the amount of awkward and damaging reaching and stooping.
(11) Sitting to work where possible helps to reduce energy expenditure by approximately 25%. Avoid static work postures by varying positions frequently.
(12) During prolonged standing, use a rubber mat or cushioned insoles to absorb shock and reduce fatigue. Place one foot on a footrest to displace some of the body weight forward to improve stability, lower muscular effort, and reduce back flexion.
(13) The OTs should use more electrical bench grinders, saws, drills, and hand grinders in their orthotic practice.

Subsequently, the authors' suggestions have been incorporated into the undergraduate occupational therapy programme at the university of Toronto. Post-study, it has been verbally reported by clinical occupational therapists and orthotists that general awareness of ergonomic issues has improved.

In future, investigations should focus on the OTs and COs who have had to change to a different line of work due to work injuries from orthotic fabrication. The authors suspect that there may be a relationship between hand size, hand strength and injury rates amongst both sexes.

In summary, this pioneering study supported the hypothesis that there is a remarkably high incidence of repetition strain disorders, hand muscle cramping, and single-trauma problems amongst Metropolitan Toronto occupational therapists and certified orthotists who fabricate orthotic devices. Various strategies have been suggested which may help prevent or correct the underlying causes of these problems. To enhance safety and biomechanical awareness, education of undergraduate and practising professionals involved in orthotic fabrication is strongly advocated to help reduce the amount of personal injury incurred.

REFERENCES

Armstrong, T.J. and Chaffin, D.B., 1978, Some biomechanical aspects of the carpal tunnel. Journal of Biomechanics, 12, 567-570.

Armstrong, T.J. and Chaffin, D.B.,1979, Carpal tunnel syndrome and selected personal attributes. Journal of Occupational Medicine, 21,(7), 481-486.

Armstrong, T.J., Fine, L., Goldstein, S.A., Lifshitz, Y.R., and Silverstein, B.A., 1987, Ergonomic considerations in hand and wrist tendonitis. The Journal of Hand Surgery, 12, (5), 830-837.

Arndt, R., 1987, Work pace, stress, and cumulative trauma disorders. The Journal of Hand Surgery, 12, (5), 866-869.

Barr, R.N., 1975, The Hand - Principles and techniques of simple splint-making in rehabilitation, (London:Buttersworth).

Bear-Lehman, J., 1986, Therapeutic management of upper extremity cumulative trauma disorders. International Network, 4, (1), 218-221.

Bear-Lehman, J. and Blair, S.J., 1987, Editorial comment: Prevention of upper extremity cumulative trauma disorders. The Journal of Hand Surgery, 12A, (5), 821-822.

Cannon, L., Benacki, E.J., and Walter, S., 1981, Personal and occupational factors associated with carpal tunnel syndrome. Journal of Occupational Medicine. 23, 255-258.

Chaffin, D.B. and Andersson, G.,1984, Occupational Biomechanics,(Toronto: John Wiley and Sons).

Fess, E.E. and Phillips, C.A., 1987, Hand Splinting: Principles and methods, (Toronto: C.V. Mosby Company).

Gwosdy, Alana, 1989, Orthotic Education at Canadian University Occupational Therapy Programmes, unpublished.

Hopkins, H.E. and Smith, H.D. (eds.), 1978, Willard and Spackman's occupational therapy,(Philadelphia: J.B. Lippincott Company).

Hoyt, W.R.,1984, Carpal tunnel syndrome: analysis and prevention. Professional Safety,29, (110), 16-21.

Hunter, J.M., Schneider, L.H., Mackin, E.J., and Callaghan, A.D. (Eds.), 1984, Rehabilitation of the Hand, 2nd ed., (Toronto: C.V. Mosby Company).

Nathan, P.A., Meadows, K.D. and Doyle, L.S., 1988, Occupation as a risk factor for impaired sensory conduction of the median nerve at the carpal tunnel. The Journal of Hand Surgery, 13-B, 2, 167-170.

Phalen, G.S., The Carpal Tunnel Syndrome. Journal of Bone and Joint Surgery, vol. 48, no. 2, 211-228.

Poole, B.C., 1988, Cumulative trauma disorders of the upper extremity from occupational stress. Journal of Hand Therapy, July - September, 172-180.

Smith and Nephew, 1988, Splinting Guidelines, (Quebec: Smith and Nephew Inc.).

Smith, E.M., Sonestegard, D.A., and Anderson, W.H., (1977), Carpal tunnel syndrome: Contribution of flexor tendons. Archives of Physical Medicine Rehabilitation, 58, 379-385.

Tichauer, E.R., 1966, Some aspects of stress on forearm and hand in industry. Journal of Occupational Medicine, 8,(2), 67-71.

Zeigler, E.M., 1984, Current concepts in orthotics, (Toronto: Rolyan Medical Products).

Advances in Industrial Ergonomics and Safety II
Edited by Biman Das
Taylor & Francis 1990

CUMULATIVE TRAUMA INJURY CONTROL EFFORTS IN SELECT WISCONSIN MANUFACTURING PLANTS

Michael J. Smith
D. J. Zehel

Department of Industrial Engineering
University of Wisconsin
1513 University Avenue
Madison, WI 53706 USA

INTRODUCTION

Anecdotal evidence indicates that there has been a substantial increase in cumulative trauma disorders to the upper extremities in a wide range of industries including manufacturing, insurance and finance. The United States Centers for Disease Control recently initiated a special program with select state health agencies to investigate this potential problem and to define what industry is doing to combat it. This report will characterize the actions of a select group of manufacturing plants in Wisconsin chosen because of the probability that their work process could produce upper extremity cumulative trauma disorders (CTD's). An examination was made of current safety and health practices and management activities in general and as they related to CTD's.

Various research studies have attempted to define critical considerations in successful occupational safety programs. Early efforts were undertaken by the National Safety Council (NSC),(NSC, 1964), Planek et al (1967), Davis and Stahl (1964), and Shafi-Sahrai (1971).

The National Institute of Occupational Safety and Health (NIOSH) undertook a series of studies to determine significant factors in successful safety programs (Cohen, Smith and Cohen, 1975; Smith, Cohen, Cohen and Cleveland, 1978; Cleveland, Cohen, Smith and Cohen, 1979).

Generally, the results indicated that low injury rate companies had the following attributes more often than high injury rate companies: (1) greater management commitment to safety, (2) greater use of outside influences to instill safety consciousness, (3) more use of safety promotional techniques and incentives, (4) more safety training, (5) more humanistic approaches in disciplining employees, (6) more frequent hazard inspections, (7) strong environmental and engineering safety control efforts, and (8) a more stable workforce.

Research on successful safety programs has consistently defined important aspect of safety programs such as strong management commitment and the need for employee safety training which were examined in the current investigation.

METHODS

Over a period of a year and a half, twenty-two Wisconsin manufacturing plants were site visited by a safety engineer with several years of experience in assessing plant health and safety risks for insurance rating. This safety engineer used an open-ended interviewing technique to gather information from management personnel at each site pertaining to their current experience with and control practices regarding upper extremity cumulative trauma. He also conducted walk-through observational surveys of most sites focussing on areas experiencing CTD problems. Based on his data collection at each site a report was written to describe current experience and practices. These reports were then analyzed using a standardized checklist of plant characteristics to define common activities. The percent of plants meeting predefined factors felt to be important for successful safety and ergonomic programs was determined.

RESULTS

Twenty of the twenty-two plants reported work-related CTD's. For the purposes of this paper, we will report data from only these twenty plants that had CTD's.

<u>Safety Program Characteristics:</u>

In examining general safety program characteristics, forty percent had a written safety program with a specified safety policy. Safety staff in ninety percent of the plants had received specific safety training. Forty-five percent of the plants offered new employees safety orientation. Forty-five percent held safety meetings, sixty-five percent conducted accident analysis and forty-five percent carried-out accident investigations. Only thirty percent developed safety reports for management. Thirty percent had a plant nurse, and thirty percent required new employees to undergo a physical examination.

A strategy used by one firm described as very useful was to keep track of the financial impact, both actual and estimated, of the safety staff recommendations and efforts. This financial record was available to use for additional leverage whenever the safety department needed it.

Although most safety contacts recognized the usefulness of charging back the costs of the injuries to the injured worker's department, and by extension their foreman, only twenty percent did so (See Table 3). Most reported safety performance was not a factor considered in their foremen's periodic salary reviews. Several contacts did mention they were working towards implementing such a charge back system or toward a safety performance factor in supervisors reviews.

Table 1. Safety Program Characteristics

CHARACTERISTIC	PERCENT OF PLANTS WITH CHARACTERISTIC
WRITTEN SAFETY PLAN	40
NEW EMPLOYEE SAFETY ORIENTATION	45
SAFETY MEETINGS	45
CONDUCT ACCIDENT ANALYSIS	65
CONDUCT ACCIDENT INVESTIGATIONS	45
MANAGEMENT SAFETY REPORTS	30
PLANT NURSE	30
PREEMPLOYMENT PHYSICAL	30
STAFF TRAINING	90
STRONG MANAGEMENT SAFETY COMMITMENT	55
SAFETY COST BENEFIT ANALYSIS	5

Ergonomic Efforts:

At forty-five percent of the plants the staff had received specific ergonomic training, while at forty percent ergonomic training was given to the employees. The union was supplying ergonomic information to the employees or the management at only fifteen percent of the plants. In looking for sources of help for ergonomic concerns, forty-five percent of the plants called on their insurance carrier, thirty-five percent on outside consultants and five percent on state agency assistance. When taking specific ergonomic actions to combat cumulative trauma disorders sixty-five percent of the plants used light duty and early return to work programs, fifty-five percent used job rotation, and twenty percent used videotaping of employees on the job. Forty percent of the plants had retained inhouse or outside medical expertise and thirty percent enhanced their ongoing medical efforts to combat CTD's (See Table 3).

While job rotation was used by just over half the companies as a means of moderating the harmful effects of repetitive work, it frequently was not used throughout all areas of the plants. This was because not all companies could provide this type of activity for all workers (or jobs) because there were not better jobs available to rotate them to. There were occasionally complaints from unions about rotation, especially when seniority bidding was not used to fill jobs. Several companies reported resistance from senior workers to any attempt at improving their jobs through rotation, the workers preferring the unvarying routine they had become used to.

Pre-employment physicals were used by approximately one-third of the companies but most admitted such efforts revealed few conditions not reported by the job applicants. Several companies were examining the value of such screening physicals in light of their perception of poor economic return.

Although almost half the companies surveyed reported receiving help in dealing with their ergonomic concerns from their insurance carrier, most characterized the help received as minimal. They reported that most insurance company loss control representatives did not have the depth of ergonomic knowledge to be able to do more than make general and non-specific recommendations. Outside ergonomic consultants were more often described as being effective, probably a result of their being recommended by other companies.

Table 2. Ergonomic Efforts

CHARACTERISTIC	PERCENT OF PLANTS WITH CHARACTERISTIC
STAFF ERGONOMIC TRAINING	45
EMPLOYEE ERGONOMIC TRAINING	40
UNION SUPPLIED INFORMATION	15
INSURANCE CARRIER HELP	45
OUTSIDE CONSULTANT	35
STATE AGENCY HELP	5
LIGHT DUTY AND EARLY RETURN	65
JOB ROTATION	55
VIDEOTAPING EMPLOYEES	20
HIGH QUALITY OF INS. CAR. HELP	35

General Characteristics of Plants:

There were several characteristics of the plants that could influence ergonomic and safety/health efforts that were recorded. Sixty-five percent of the plants were unionized and sixty-nine percent of the unionized plants had good relations between labor and management. Thirty-five percent of the plants had a piece-work incentive pay plan. Forty percent of the plants had a predominately female workforce (75 percent or more female) and the majority of plants were between 100-300 employees in size (fifty-five percent). Thirty percent of the plants reported seasonal variations in product demand. In terms of management programs to stimulate employee interest and involvement, forty percent of the plants used employee suggestion systems. To stimulate supervisory involvement in safety and ergonomics, twenty percent of the plants used a charge back system to the supervisor's departmental account for worker's compensation costs.

DISCUSSION

The purpose of this study was to examine manufacturing companies that have had cases of upper extremity cumulative trauma disorders to determine the safety and ergonomic practices they use to deal with these injuries. It was expected that companies experiencing cumulative trauma injuries would not have an extensive knowledge of ergonomics and most likely would not have well developed occupational safety/health programs. The findings verified these expectations since less than half of the plants had specific ergonomic activities underway, and a large percentage (over sixty percent) did not have a well defined occupational safety/health program. The primary means companies used for dealing with cumulative trauma problems was providing light duty jobs and job

rotation. Other approaches such as staff and employee training were also used by a minority of plants. Almost all plants used outside consultant help from their insurance carrier or a private consultant with varying degrees of success.

Table 3. General Plant Characteristics

CHARACTERISTIC	PERCENT OF PLANTS WITH CHARACTERISTIC
UNIONIZED	65
GOOD LABOR/MANAGEMENT RELATIONS IN UNION PLANTS	69
PIECE-WORK PAY	35
PREDOMINATELY FEMALE WORKFORCE (75% OR MORE)	40
SEASONAL PRODUCT DEMAND	30
EMPLOYEE SUGGESTION SYSTEMS	40
FOREMAN CHARGE-BACK SYSTEM	20
SIZE	
<100	5
100-300	55
300-700	15
>700	25
Company Retained Medical Expertise	40
Enhanced Medical Efforts	30

The U.S. Occupational Safety and Health Administration (OSHA) recently provided direction to employers in the Meat Processing industries on means to reduce or eliminate upper extremity cumulative trauma disorders (BNA, 1989). Many of these recommendations are of a general nature and parallel the factors that have previously defined as necessary for a successful occupational safety program, such as top management commitment, written policies and procedures and employee and supervisor training. In addition, specific engineering and administrative controls and methods for eliminating lifting and upper extremity hazards were provided such as tool redesign, workstation improvements and job redesign. The information gathered in the current study suggests that most plants that are experiencing upper extremity cumulative trauma disorders do not have the general characteristics that have been proposed by OSHA and others (Cohen et al, 1975; Smith et al, 1978) as a necessary base of good safety practice that could be used for extending ergonomic programs. Thus, even when specific ergonomic and medical interventions were used, they were not completely effective. In addition, the ergonomic approaches most often used by these plants were job rotation and light duty which are not preferred approaches according to the OSHA guidelines. Rather, these are seen as secondary mechanisms to be used when job redesign and engineering controls fail to be completely successful.

Conclusions from the results of this study have to be qualified given the small sample size and the variety in manufacturing operations studied and in plant demographic characteristics. Even

so, some interesting facts emerge. The large majority of plants did not fulfill even the basic requirements for effective occupational safety and health programs. In fact very few plants had all of the recognized components for a successful safety program. Most plants were ignorant of how to deal with their upper extremity cumulative trauma disorders and were willing to try anything reasonable. Those that used their insurance carrier for advice were generally not satisfied with the results. Pre-employment physical examinations were not considered an effective means for screening out susceptible persons. Most of those with medical programs were not satisfied with their effectiveness.

It seems that good plant ergonomic programs must build on having the fundamentals of a good occupational safety and health program. This is in keeping with the proposed OSHA guidelines (BNA, 1989), and the current study provides some support for this belief. There is a need for more research that can determine the fundamental characteristics necessary for a good ergonomics program for dealing with cumulative trauma injuries.

REFERENCES

BNA, 1989, Draft OSHA ergonomic management guidelines for meatpacking. Occupational Safety & Health Reporter, November 15, 1130-1143.

Cleveland, R., Cohen, H.H., Smith, M. J. and Cohen, A., 1979, Safety Program Practices in Record-Holding Plants, (Washington, DC: DHEW (NIOSH) Publication No. 79-136).

Cohen, A., Smith, M.J.and Cohen, H.H., 1975, Safety Program Practices in High Versus Low Accident Rate Companies, (Washington, DC: DHEW (NIOSH) Publication No. 75-185).

Davis, R.T. and Stahl, R.W., 1964, Safety Organization and Activities of Award-Winning Companies in Coal Mining Industry, (Pittsburgh, PA: U.S. Bureau of Mines).

NSC, 1969, Accident Prevention Manual for Industrial Operations, (Chicago, IL: National Safety Council).

Planek, T., Driessen, G. and Vilardo, F.J., 1967, Evaluating the elements of an industrial safety program. National Safety News, 60-63.

Shafi-Sahrai, Y., 1971, An Inquiry into Factors that Might Explain Differences in Occupational Accident Experience of Similar Size Firms in the Same Industry, (East Lansing, MI: Michigan State University Graduate School of Business).

Smith, M.J., Cohen, H.H., Cohen, A. and Cleveland, R.J. Characteristics of successful safety programs, Journal of Safety Research, 10, 5-15.

Advances in Industrial Ergonomics and Safety II
Edited by Biman Das
Taylor & Francis 1990

RESULTS OF AN ONGOING MONITORING PROGRAM FOR CARPAL TUNNEL SYNDROME

JEFFREY E. FERNANDEZ, ROBERT J. MARLEY

and KELVIN R. YOUNG

Department of Industrial Engineering
The Wichita State University
Wichita, KS 67208-1595
U.S.A.

This paper describes the utilization of a monitoring program for carpal tunnel syndrome (CTS) in the aircraft industry. The program was based upon objective and subjective measures normed for workers with CTS. It is anticipated that baseline data from these tests can assist in the identification of high risk departments, the quantification of job redesign, or the screening of workers who may be at a high risk for CTS. Initial results of regular testing in two high-risk departments (sheet metal and data entry) are reported.

INTRODUCTION

Carpal tunnel syndrome (CTS) is generally attributed to compression of the median nerve as it passes through the carpal canal of the wrist (Armstrong and Chaffin, 1979). This compression is associated with repeated or sustained activities of the fingers and hands, often combined with the application of force, as well as pressure from hard work surfaces and sharp edges on hand tools (Feldman, et al., 1983).

Dollar figures associated with CTS vary widely across the United States. In 1985, one source estimated that the national average cost for a case involving CTS was approximately $3,500 (Hiltz, 1985). In the more severe cases, requiring surgery, compensation and possible disability claims, costs can range from $30,000 to $60,000. In one aircraft assembly facility in Kansas, total costs (including surgery) for CTS currently averages between $15,000 and $18,000. While the precise incidence rates of CTS in industry is not known, some conservative estimates state that 10 percent of all workers performing manual work may be affected (Occupational Hazards, 1987). However, most researchers agree that CTS cases are underreported in aggregate statistics (Occupational Hazards, 1987).

The job of the ergonomist is to evaluate the demands of a specific task with reference to the capacity of workers to perform the task over a certain time period. During initial job design, the demands of the task would be ideally held within the capacity of a fixed percentage of the population (for example, 95%). When the task demands of an existing job is such that it is beyond the capacity of a fixed percentage of the population, the work, procedures and/or work tools should be redesigned in order to accommodate the fixed percentage of the working population. If the accommodated percentage is unacceptable and redesign is not possible, then the final alternative is to screen the workers so that only those whose capacity exceeds the task demands are allowed to perform the task.

The problem and cost associated with CTS to industry is now evident. The need to implement proper ergonomic design is crucial in order to reduce the risk of CTS. Furthermore, the need for a test to monitor either individual employees or departments as well as identifying new employees who may be at risk for developing CTS is of utmost importance.

The objective of this study was to examine the use and reliability of a monitoring program in two high risk departments in an aircraft manufacturing plant.

METHODS AND PROCEDURES

This research was directed toward examining female workers involved in tasks associated through historical data with cumulative trauma disorders (particularly CTS) in a major aircraft assembly facility. Objective and subjective measures in two different departments with a high incident rate of CTS were collected. Results were compared with a control population with no symptomology of CTS and a population medically diagnosed with CTS.

The individual tests used in this monitoring program are based upon a salient set of measures derived from a study of workers with CTS (Fernandez, Malzahn, Marley and Bonebrake, 1989). At this stage, only female workers are being examined because of the limited clinical sample size of male workers with CTS. The rationale and specific methods of this monitoring program are discussed in Fernandez and Marley (1990). The program consists of four parts:

1) Grip strength
2) Range of motion of the wrist
   --flexion, extension
   --ulnar deviation
3) Pain and Distress (PAD) scale (Zung, 1983)
4) Phalen's test.

Workers from sheet metal and data entry departments were selected for monitoring based upon a detailed review of their medical records. The criteria used to include an individual in this study was that she have <u>no</u> history of the following conditions:

1) Diagnosis of CTS
2) Symptoms of CTS or related cumulative trauma
3) Other hand/wrist and arm problems

In addition to their medical eligibility for inclusion into this study, workers from the data entry department were selected because they spent at least 50% of their work time performing data entry tasks. Likewise, workers from the sheet metal department performed drilling, reaming, and riveting tasks for approximately 80% of their work day.

Measurements were collected on both the preferred and non-preferred hands. Two repetitions for each measure (with the exception of the PAD score and Phalen's test) were performed, with the highest value recorded. Two separate evaluations (Test 1 and Test 2) were conducted for each subject in both departments approximately one month apart.

RESULTS AND DISCUSSION

The results obtained from the monitoring program are summarized in Tables 1 and 2. These data are separated by age group (under 40, 40 and over years old) because this factor (age) was a significant effect in previous research (Fernandez, et al., 1989).

Table 1. Mean(STD) of workers under 40 years of age.

Department	Grip (Kg)	Flexion (degree)	Extension (degree)	Ulnar Dev. (degree)	PAD
Sheet metal (n = 6)					
Test 1	22.67	61.91	67.41	33.67	34.75
	(3.31)	(23.77)	(6.01)	(7.72)	(7.26)
Test 2	21.83	62.25	63.33	37.33	39.83
	(3.58)	(20.52)	(8.30)	(6.11)	(10.95)
Data entry (n = 6)					
Test 1	21.17	73.58	64.58	33.67	35.00
	(3.18)	(10.57)	(8.56)	(6.73)	(7.23)
Test 2	23.50	70.25	65.42	33.50	37.91
	(4.17)	(15.05)	(6.35)	(6.15)	(9.96)

Table 2. Mean(STD) of workers 40 years of age and older.

Department	Grip (Kg)	Flexion (degree)	Extension (degree)	Ulnar Dev. (degree)	PAD
Sheet metal (n = 5)					
Test 1	21.90	51.90	52.90	29.60	41.00
	(4.46)	(17.25)	(10.50)	(5.91)	(10.48)
Test 2	21.80	48.40	60.80	33.00	48.00
	(4.23)	(17.31)	(13.13)	(2.98)	(11.77)
Data entry (n = 6)					
Test 1	23.00	67.75	58.67	33.00	45.83
	(4.63)	(20.28)	(7.36)	(6.31)	(12.03)
Test 2	24.17	71.00	54.83	31.25	35.83
	(5.34)	(21.32)	(14.59)	(4.93)	(6.93)

The data in Tables 1 and 2 were averaged across the preferred and non-preferred hands (where applicable) for ease of presentation as only grip strength had a significant handedness effect ($\underline{F}(1,84) = 8.34$, $\underline{p} = 0.0049$). Furthermore, in none of the measures involving the handedness effect was there a significant interaction term at the 0.05 level of alpha.

Upon analysis of individual data, is was discovered that three (3) workers had very low flexion scores and associated high PAD scores which contributed to relatively high standard deviations in Tables 1 and 2. These scores were within the ranges of the CTS population. It is believed that these 3 workers may be experiencing CTS symptoms that had not been previously detected or reported. Ergonomic intervention is therefore called for in at least these cases.

One major concern which this study attempts to address is determining the reliability of the testing procedure. Based upon results from Test 1 and Test 2, there were no significant difference between evaluations for all measures at the 0.05 level of alpha. This indicates that the monitoring program utilized is reliable. Figures 1 to 5 represent the results of this study in graphical form. For comparison purposes, the data for control and CTS populations are presented.

This program is currently being implemented in other departments as well as with all newly hired employees at this manufacturing facility. It is anticipated that baseline measures of workers can help identify those who may be at risk for developing CTS. In this sense, "at-risk" workers may be screened from potentially stressful jobs and placed on other tasks. Regular evaluations can also document changes in performance by comparison to an individuals own baseline data. Departments and/or tasks may be identified as "high-risk" or "low-risk" based upon results of regular monitoring and medical/ injury reports. This information may then be used to target specific ergonomic and management interventions.

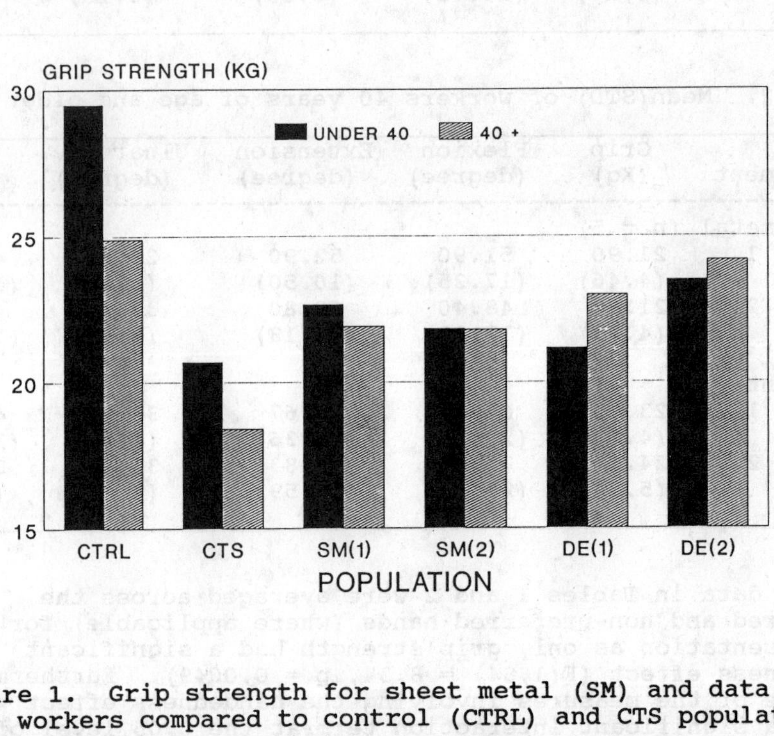

Figure 1. Grip strength for sheet metal (SM) and data entry (DE) workers compared to control (CTRL) and CTS populations.

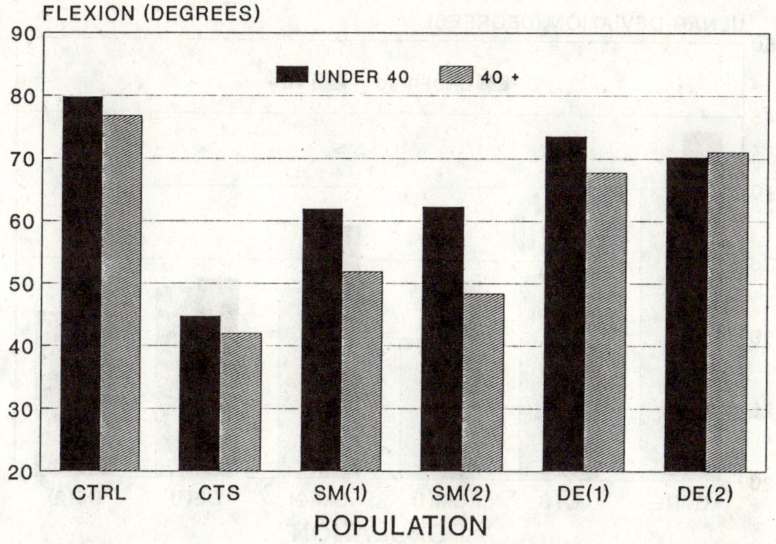

Figure 2. Flexion for sheet metal (SM) and data entry (DE) workers compared to control (CTRL) and CTS populations.

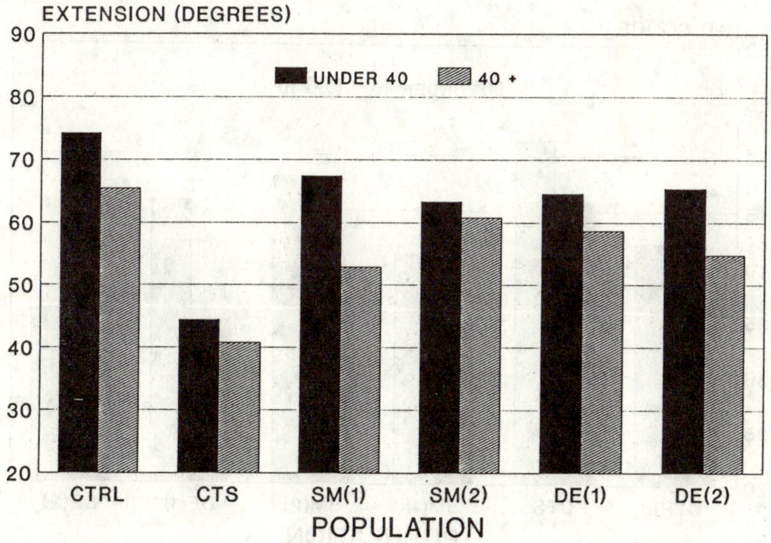

Figure 3. Extension for sheet metal (SM) and data entry (DE) workers compared to control (CTRL) and CTS populations.

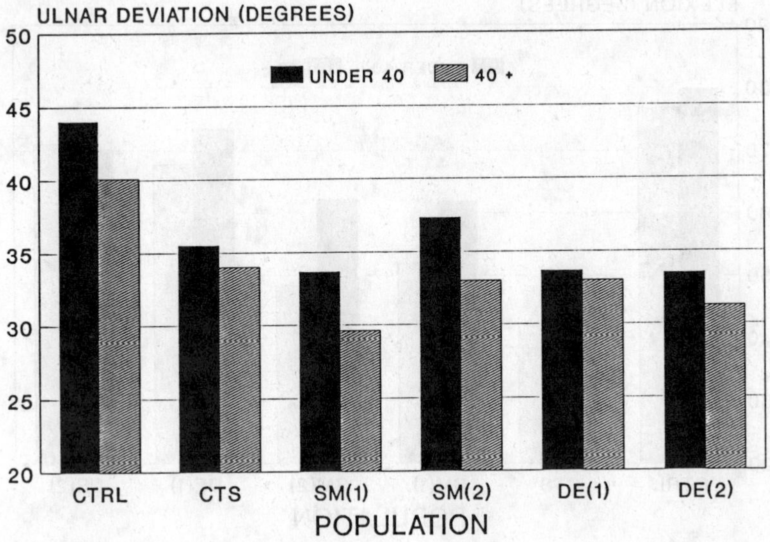

Figure 4. Ulnar deviation for sheet metal (SM) and data entry (DE) workers compared to control (CTRL) and CTS populations.

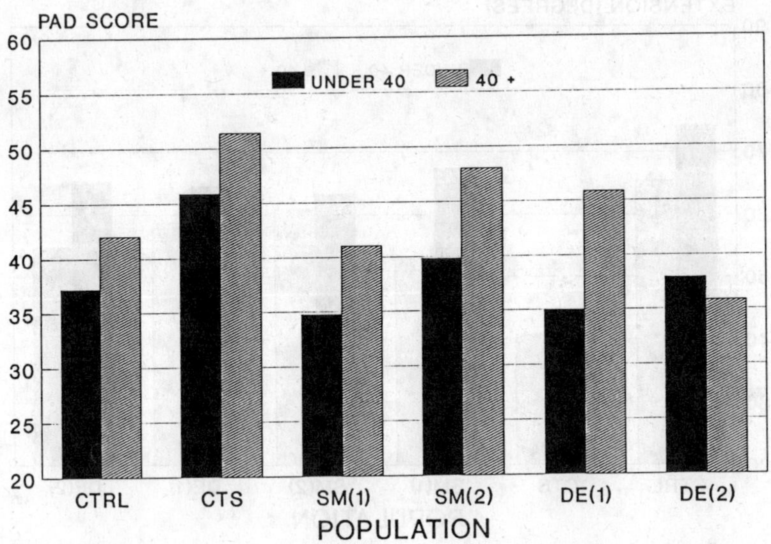

Figure 5. PAD scores for sheet metal (SM) and data entry (DE) workers compared to control (CTRL) and CTS populations.

Future research using this procedure will include determination of appropriate testing intervals. The authors feel that all workers in jobs with potential wrist trauma should be evaluated at least annually. Workers in high-risk jobs (for wrist trauma) should be tested more often, perhaps monthly, quarterly, or semi-annually. Furthermore, continued application of the monitoring program will quantify the effectiveness of ergonomic interventions to reduce wrist trauma such as job redesign, job rotation, etc..

Future research is also planned to explore methods of determining acceptable work limits for work involving potential wrist trauma. For example, psychophysical methodologies may be utilized to set acceptable work limits for work involving various forces, wrist postures, frequencies, and task duration. These values can be compared to current work standards or commonly accepted work limits, such as the limit of 1000 wrist motions per hour (Hammer, 1934).

CONCLUSIONS

This paper has described the initial results of a program implemented to monitor workers for carpal tunnel syndrome. This testing procedure provides reliable, quantitative information by which management may screen workers as well as evaluate the effectiveness of an ergonomic solution in jobs involving carpal tunnel syndrome.

A system of monitoring employees on a regular basis should be established. Workers at low-risk, whether in particular departments or otherwise, should be monitored at least annually, in a similar manner as annual audiometric evaluations in noisy environments. Workers at high-risk, whether in particular departments or otherwise, should be evaluated more frequently and this frequency must be determined through future research.

REFERENCES

Armstrong, T.J., and Chaffin, D.B. (1979). Some biomechanical aspects of the carpal tunnel. Journal of Biomechanics, 12, 567-570.

Feldman, R.G., Goldman, R., and Keyserling, W.M. (1983). Peripheral nerve entrapment syndromes and ergonomic factors. American Journal of Industrial Medicine, 4, 661-681.

Fernandez, J.E., Malzahn, D.E., Marley, R.J., and Bonebrake, A.R. (1989). A study of several performance measures of workers with carpal tunnel syndrome. Proceedings of the Human Factors Society 33rd Annual Meeting, 728-732.

Fernandez, J.E. and Marley, R.J. (1990). Monitoring and screening tests for carpal tunnel syndrome. Proceedings of the Ergonomic Society 1990 Annual Conference.

Hammer, A.W. (1934). International Record of Medicine (Medical Record, 139-140.

Hiltz, R. (1985). Fighting work-related injuries. National Underwriter, 89(13), 15.

Occupational Hazards (1987). Carpal tunnel syndrome: Getting a handle on hand trauma. 42(2), 45.

Zung, W.W.K. (1983). A self-rating pain and distress scale. Psychosomatics, 24(10), October.

Advances in Industrial Ergonomics and Safety II
Edited by Biman Das
Taylor & Francis 1990

WRIST INJURY PREVENTION IN FIREARMS MANUFACTURE: A CASE STUDY

S. Steele, R. Hamel and J. Muller, Colt Firearms, Hartford, CT 06110

And

J. L. Wick
The Joyce Institute, Seattle, WA 98101

The manufacture of firearms requires many manual operations in the finishing of metal parts including gauging (screening) of parts and assembly. The company received reports of wrist problems from operators on the job of Screen 3050 (rifle bolt). A specific ergonomic analysis, problem-solving, implementation process was applied to solve the injury problem. Redesign of the workplace resulted in dramatic improvements. The most significant were the reduction of stresses to the wrists and a 32 percent reduction in cycle time.

INTRODUCTION

In recent years the metal fabrication industry has evolved from mechanization to automation. Automation, however, is often only cost effective for high volume production. Situations such as short production runs and inspection tasks which involve decision-making require human operators. Too often, ergonomic principles are not applied to the design of the workplace for the human operator. The results are inefficiency and cumulative trauma disorders of the upper extremities, particularly of the hand/wrist.

Cumulative trauma disorders of the upper extremities can be attributed to stresses resulting from poor body part positions, forceful exertions and high repetition rates. An example would be the wrist frequently in ulnar deviation with a forceful finger pinch causing carpal tunnel syndrome.

Any position of the hand at the wrist other than neutral stresses the tendons in the carpal tunnel (Armstrong, 1983). The stress is more severe at the extreme range of motion of the joint. The amount of stress which can be tolerated before tendon damage occurs is not clearly understood. Therefore, the "safest" level becomes a matter of judgment and jobs must be designed to minimize bending the wrists.

Grips requiring high finger/hand forces also stress the tendons in the carpal tunnel (Armstrong, et al, 1982). As with position, the higher the force exerted, the greater the stress to the tendons in the carpal tunnel. There may be a correlation with position; however, it is not clear what the correlation is. Again, "safe" becomes a matter of judgment.

The frequency of bending the wrist accelerates the effects of cumulative stress, i.e. the more often the wrist is bent the sooner tendon damage accumulates to the level of injury. The two are clearly associated (Silverstein, et al, 1987) and must be considered together when addressing ergonomic problems causing cumulative trauma disorders. All three biomechanical stressors to the tendons in the wrist can be present in many of the manual jobs in metal fabrication. A task analysis such as reported by Drury and Wick (1984) can be used effectively to identify and measure the stresses. Knowing the problem and its magnitude, one can competently determine the workplace causes and develop solutions. The analysis is the one used in this project. It is an eight-step methodology as follows:

1. Videotape the job. The operator was videotaped from the front, back and both sides while performing the work.
2. Task analysis. The job was broken into task steps and the position of each body part was recorded.
3. Work/workstation measurements. Physical dimensions of the workplace and measurements of finger and hand forces were taken (using dynamometers [Drury, 1987]) as required.
4. Worker interviews. The worker was interviewed to allow input regarding possible changes to the workstation and to determine what, if any, pain or discomfort was being experienced.
5. Data evaluation. The body part positions, forces and repetition rate were compared with the literature recommendations for seated work.
6. Solution design. Solutions to the ergonomic problems and the workplace causes were developed during a brainstorming meeting.
7. Design validation. The new design was analyzed and evaluated to determine if the problems were solved and no new problems inadvertently introduced.
8. Comparison of the new design with the original job. The data from before and after change were compared to quantify the effects of the new design.

By comparing the data before and after change, success can be measured immediately. The alternative is to track injuries, or lack of injuries, over a year or more.

This project was initiated because the company injury records included a wrist injury on this particular gauging job as well as similar jobs. Change was required to prevent future injuries.

Ergonomics not only addresses safety. It also impacts on the efficiency of the worker. This project demonstrates that ergonomics can make a workplace both safer and more productive.

This project was part of an ergonomics training program (The Joyce Institute, 1988) presented to selected professionals.

DESCRIPTION OF THE JOB

The job studied was Screen (gauge) 3050. The production process allowed the job to be done only when a sufficient number of parts were accumulated to justify setting up the fixture and assigning an operator. The operator performed the job while seated on a stool at a bench. A mechanical test fixture was used to gauge specific dimensions of a rifle bolt. The parts were stored vertically in racks on the bench.

The fixture was loaded by dropping the part into a hole in the top. A "probe" in the side of the fixture at the base was pushed in to test the size of part. Operators typically pushed the probe with a twisting motion. When the test was complete the probe was pulled back to a stop. The part was removed by inserting a finger of the left hand into the base of the fixture and prying the part up to be grasped by the fingers of the right hand. The parts to be gauged were taken from one rack and asided into another rack after the test.

The production rate was 300 parts per hour. The work was self-paced, therefore, under operator control.

RESULTS

The task analysis revealed significant stresses to the hands/wrists and shoulders. Certain definitions and criteria were adopted. Drury and Wick (1984) described a damaging wrist motion (DWM) as the wrist not straight with a grip. A design criteria of 1,000 DWM's per hour was established. The maximum pinch grip force allowed was 8 pounds (3.6kg).

Damaging wrist motions exceeded 1,000 per hour in the right hand. The causes were the probe handle size and shape, the method of operating the probe, the height of the point of operation and the orientation of the parts racks. Although the number of DWM's of the left hand were within the criteria, the left hand/wrist was put into an extremely awkward position to eject the part from the fixture. Stress to the shoulder was caused by the height of the point of operation and the location of the racks.

Solutions were developed to solve the ergonomic problems identified in the analysis and evaluation. Opportunities to improve operator efficiency were investigated during the design process. The changes included:

- parts racks located and angled for proper hand/arm position
- adjustable, upholstered swivel chair with backrest
- gauge anchored to the bench
- manual, lever ejection mechanism
- "power-grip" handle on the probe
- height adjustable footrest

There were constraints on what could be done to the fixture because the product was a military weapon and the fixture was originally designed to meet stringent specifications. One of the constraints required government approval for any modification to the basic fixture. Rather than seek this approval, it was decided (for expediency) to modify the fixture operation by using add-on features which, by design, did not alter the fixture.

A prototype was developed and fabricated. It was tested to determine if the operator could perform the gauging operation with the proposed design. The operator used the new design for several days before another analysis and evaluation was done to validate the changes.

The comparison of the data from before and after change indicated significant reduction of the stresses to the hands, wrists and shoulders. DWM's of both hands were not only reduced, but became evenly distributed. Stressful finger pinch grips were reduced. The new workplace allowed the job to be accomplished with arms lowered minimizing shoulder stress. Many of the movements which were considered biomechanically stressful also took time. By eliminating those motions, cycle time was reduced enough to allow a pay-back of the cost of the improvements in 94.5 man-hours.

Table 1. Comparison of conditions before and after change. (Only the ergonomic problems identified before change are listed.)

Measure	Before	After
Damaging wrist motions		
Right	1800/hour	300/hour
Left	900/hour	300/hour
Pinch grips		
Right	6/cycle	2/cycle
Left	3/cycle	2/cycle
Shoulder flexion		
Right	4/cycle	2/cycle
Left	4/cycle	1/cycle
Shoulder abduction		
Right	4/cycle	0/cycle
Left	4/cycle	0/cycle
Cycle time per part	5.5 sec.	3.75 sec.

This project can be considered a success because the injurious conditions were eliminated. Damaging wrist motions were reduced from 1,800 per hour in the right hand and 900 per hour in the left hand to 300 per hour each. Additionally, the improvements were paid for by labor savings because cycle time was reduced from 5.5 seconds to 3.75 seconds. Many different gauging operations were performed at the same bench by simply changing the fixture. The chair, footrest and rack arrangement also will benefit the other jobs. The operator noted that the new design was more comfortable and easier to use.

DISCUSSION

There are many manual jobs in the metal fabrication industry which are in need of the application of ergonomics. The job of deburring, commonly called burring, requires the worker to pick or file burrs from precision parts. The tools are small and poorly designed for the hand and often require high hand and finger forces. Many of the semiautomatic machines developed in recent years impose stresses to the wrists and shoulders because of the positions operators must assume to load and unload parts. Metal parts are heavy. Transporting parts from one operation to the other is usually a manual materials handling task with a high exposure to back injury.

A significant number of the companies in the industry function as job shops. That is, production runs are short and capital investment funds are not readily available. It is not cost-effective to buy a fully automatic machine for one or two jobs and then be idle while the shop is running a different job. Tool and machinery designers and suppliers must become more aware of the need to develop low-cost, ergonomically correct equipment. Hand-tools with handles for human hands, machines that allow operators to access the point of operation without strain and at reasonable cost are needed.

REFERENCES

Armstrong, T.J., 1983, An Ergonomic Guide to Carpal Tunnel Syndrome. (Akron: American Industrial Hygiene Association).

Armstrong, T.J., Foulke, J.A., Joseph, B.S. and Goldstein, S.A., 1982, Investigation of cumulative trauma disorders in a poultry processing plant. American Industrial Hygene Association Journal, 43, pp. 103-116.

Drury, C.G., 1987, A biomechanical evaluation of the repetitive motion injury potential of industrial jobs. Seminars in Occupational Medicine, Vol. 2, No. 1, pp. 41-49.

Drury, C.G. and Wick, J.L., 1984, Ergonomic applications in the shoe industry. In: Proceedings of the 1984 International Conference on Occupational Ergonomics, Toronto, pp. 489-493.

Silverstein, B.A., Fine, L.J. and Armstrong, T.J., 1987, Occupational factors and carpal tunnel syndrome. American Journal of Industrial Medicine, 11, pp. 343-358.

The Joyce Institute, 1988, Applied Industrial Ergonomics Reference Manual, (Seattle: The Joyce Institute).

Advances in Industrial Ergonomics and Safety II
Edited by Biman Das
Taylor & Francis 1990

CUMULATIVE TRAUMA DISORDERS: AN ERGONOMIC INTERVENTION

S.M. TABOUN

Department of Mechanical Engineering
University of Ottawa

Cumulative trauma disorders of the lower back and upper extremity are among the leading worker complaints and compensation claims in the hand intensive seat skinning operations. Risk factors identified include force of exertions, stress concentrations, repetitiveness and awkward working postures due to improper design of workstations, hand tools and/or equipment. Tasks were analyzed and evaluated using the AET and detailed techniques. Through ergonomic intervention, high worker's satisfaction and significant work performance improvement were observed.

INTRODUCTION

Complaints originating in the musculoskeletal system are one of the most common categories of illness in manufacturing industries including automotive assembling activities. Aside from the low back injuries, musculoskeletal trauma of the upper extremity accounts for a large percentage of time loss through sick leave and lower productivity. Statistics of work injuries indicated that as much as 25% of the workforce in some hand intensive jobs have experienced some type of repetitive trauma with disorders requiring medical attention (Armstrong, et al. 1986; fine, et al. 1985; Jensen, et al. 1983; Armstrong et al. 1982).

The problem of cumulative trauma disorders (CTD) involving the upper extremity in manufacturing industries including assembly operations are related to several integrated factors. Some of them are contributed by the worker, task or both; others may be related to the hand held tools (powered or manual), work pieces and repetitive tasks, design of the workstation and workplace arrangement, and the surrounding environment. In order to reduce the risk or progression of CTD a proper control over these factors must be exercised through detailed job analysis, evaluation and implementation of ergonomic programs.

In this paper, a systematic step by step procedure of ergonomic analysis and intervention is discussed through a case study. The skinning operations of the lower front car seats in an automobile plant were analyzed and problem areas were determined. The systematic procedure begins with detailed job analysis and evaluation of all assembly operations, evaluation of the workplace and the use of hand held tools, identification of work stresses and CTD risk factors, proper control of the recognized risk factors through ergonomic intervention.

JOB ANALYSIS

Job analysis and evaluation of industrial tasks can be accomplished in two steps. The primary step is based on the traditional techniques of work measurement. In this case, the work content of each job is determined through time-and-motion study. The secondary step involves the evaluation of each task to determine the level of stress for each work element or task. This step can be achieved through the use of AET techniques (Rohmert, 1985) and the systematic analysis of risk factors (Armstrong, 1983; Armstrong et al., 1985).

The workplace of seat skinning operations consists of one warming station used to warm up the moulded foam so that its elasticity and flexibility during assembly are enhanced, two fixture stands to hold the seat frames during skinning operations, and a work bench for the final clamping activity. Each fixture is operated by one worker while the work bench is shared by both of them.

Videotapes and stop watch were used to determine the time required to perform various tasks. Identified tasks were compared with job descriptions and standard times provided by the Industrial Engineering Supervisor. Once the work content have been determined, stress associated with each task was evaluated using the analytical profiling method, and the stressful work elements were further analyzed in detail. In addition, work elements were reviewed at slower motion to determine possible risk factors.

Ergonomic Task Evaluation and Risk Factors

Several studies have appeared on the subject of repetitive trauma disorders, and many of them indicated that occupational risk factors can cause, precipitate, or aggravate these disorders. These factors including repetitive exertions, force of exertion, working posture, stress concentration, design of hand tools, gloves, cold, and vibration are cited in various publications and summarized in Armstrong et al., (1986) Within the context of this study a questionnaire survey was carried out to determine the effects of various factors and work elements that are involved in the seat skinning operations. Results of the questionnaires along with worker complaints were compared with the results of the detailed ergonomic analysis of each task to check for their compatibility and agreement.

Many factors contribute to work stress of industrial jobs which may lead to several types of work injuries. Stress evaluation of tasks is an important step in any ergonomic investigation. Although, there are many types of stress that may exist in any task, analytical profiling technique (Rohmert, 1985) can be used to assign stress values for each work element. The technique is an instrument of analysis and evaluation which combines the engineering, physiological, behaviourial, and psychophysical elements of a working system.

The analytical profiling technique (AET) of work elements

takes into consideration the following eight factors: **A)** Safety; **B)** Physical Environment; **C)** Physical Load; **D)** Mental Load; **E)** Autonomy; **F)** Relations; **G)** Repetitiveness; **H)** Work Content. The ergonomic aspects of industrial jobs are described by factors B, C and D while factors E to H reflect the psychological and sociological aspects. The overall job design is described with four preliminary criteria and factors A to H are subdivided into 23 criteria. Assessment of stress values for each criterion is based on a 5-point scale, with a score of 1 representing very low stress and a score of 5 indicating the highest levels of stress. Certain factors can be found in literature with great details for specific tasks, and their quantitative stresses can be easily transformed into the 5-point scale.

Primary Evaluation of Stresses

Two seat skinning assembly tasks were used as examples to demonstrate the application of the AET as a technique of stress evaluation for different work elements and jobs. Tasks were videotaped and examined very closely in conjunction with the information related to work standards, ergonomic factors of the physical workload and repetitiveness. Task-profiles were identified and evaluated for the most common work postures (MCWP), most awkward work posture (MAWP), and handling effort during the work cycle as well as repetitiveness. The stress values indicated in the subsequent tables for both tasks were determined from tables and charts given in Wagner (1985).

One of the tasks that is included in the seat skinning operations of the lower seat is the placement of seat top cover (skin) and the moulded foam on the spring frame. As soon as the spring frame is mounted on the fixture, a warm moulded seat foam has to be placed on the seat frame, and the top cover (skin) has to be hooked in from the inside of any rear corners of the spring frame, then it is stretched all over the moulded foam. The top cover has to be tucked in from the other three corners by applying manual downward forces around both sides of each corner. At the end, the worker has to make sure that the top cover is properly placed all over the mould foam together with the spring frame by stretching the top cover downward from both sides. To perform this task it requires repetitive exertions of forces and bending several times. Although, workers take advantage of their upper body weight to exert downward forces, repeating it for several hundred times per work shift leads to excessive stress on the back and the upper extremities of the body. A summary of the analytical task-profile and stress evaluation is given in Table 1.

Another task included in the seat skinning operations is the the seat top cover clamping with the seat frame. This task requires the worker to turn the subassembled unit upside down and place it on the work bench. Subsequently, the worker has to grasp with one hand one of the end sides of the seat skin hand and pull it in the upward direction while the other hand pressing down on the seat frame in order to have enough skin length to be turned inside the seat spring frame and clamp them together. To perform this task, the worker has to raise his/her shoulder, elbow and the corresponding body side

Table 1. Job Analysis - Placing The Seat Skin

Physical load	Pictogram	Postural wt	Stress value	1 2 3 4 5
MCWP Seat skin placement	 trunk inclined 10-25°	3.0	3.0	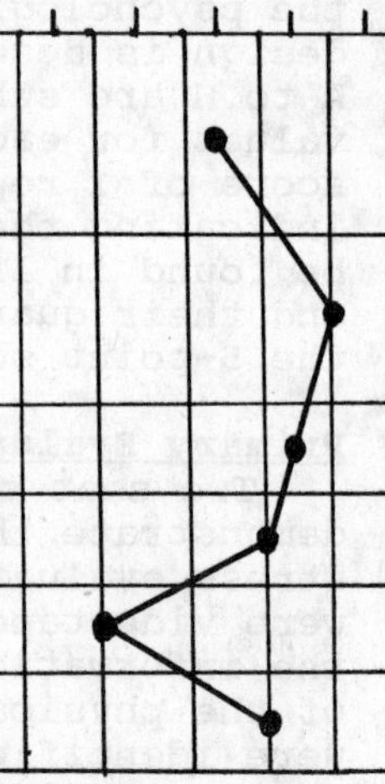
MAWP Front corners skin placement	 trunk inclined 20-45°	4.0	4.5	
Work effort		4.0		
Working posture		3.5		
Handling effort		1.5		
Repetitveness		3.5		

upward while trying to lean toward the other side of the body, to exert additional downward forces and compress the lower seat. That is, the centre of the upper extremity is falling between two parallel and opposite forces with an awkward working posture. In some cases, shorter workers have to push up one side of their body by standing on their toes and using their forearm rather than hands to press down the subassembled seat. This task was repeated for several hundred times by each worker per work shift. A summary of the analytical task-profile and stress evaluation is given in Table 2.

Table 2. Job Analysis - Seat Frame and Skin Clamping

Physical load	Pictogram	Postural wt	Stress value	1 2 3 4 5
MCWP Pull the seat skin upward	 grasp & lift elbow & shoulder	3.5	4.0	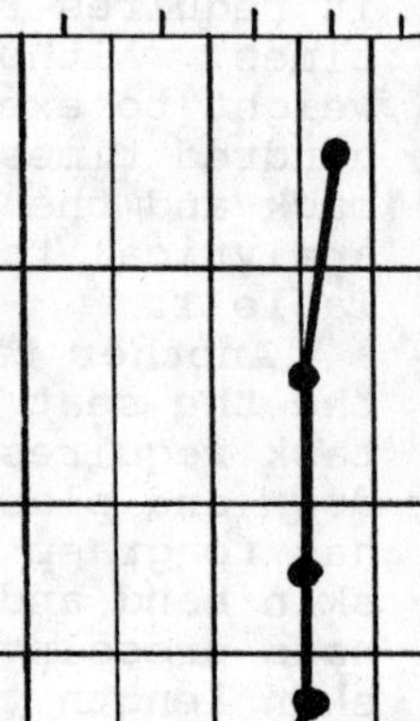
MAWP compress the subassembly	 trunk inclined to one side 10-25°	3.5	3.5	
Work & handling effort		3.5		
Working posture		3.5		
Repetitveness		3.0		

Detailed Stress Analysis

In certain cases, it is more practical to perform a detailed analysis to determine the amount of stress and potential risk associated with specific tasks. For illustration purposes, the placement of top cover on the moulded foam and seat spring frame task is analyzed in detail. Figure (1) shows the top view of a fixture stand, worker position and the different forces associated with this task.

A biomechanical approach was used to analyze the problem and determine the forces that are imposed on the lower back and upper extremity joints of the worker. To simplify the analysis, a two dimensional model was considered as shown in Figure (1 b). All reaction forces at the five locations shown in Figure (1 a) were determined from the properties of the moulded foam (Material DC 634) and based on a data obtained from sample tests. It requires between 1.1 to 11.4 kg to compress the moulded foam by 1 inch or approximately 25% of a standardized sample with 4" diameter and 4" thickness. In this analysis, it is assumed that each reaction force is applied at the middle point of both hands when workers grasp the top cover by both hands and exert downward forces for each work element. The reaction force of the right front corner F(R/F) due to the compressed foam is similar to the minimum force exerted by the worker's hands at that corner but opposite in direction. It was found that this force is between 20-26 kg hence, the upper extreme force was used in the analysis to represent the worst case. The required forces necessary to perform other work elements for this task are given in Table (3).

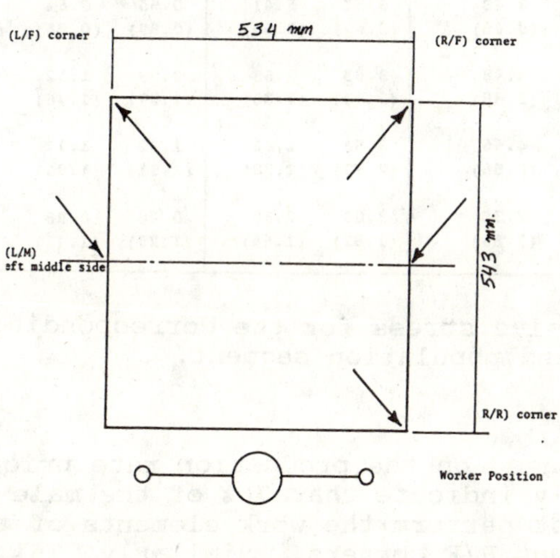

a - Top View

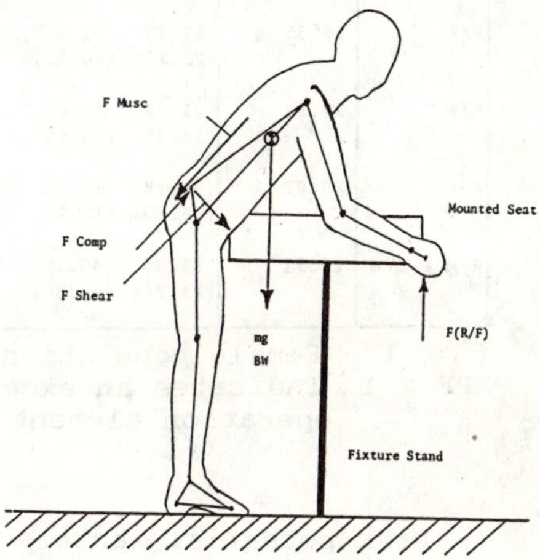

b - Side View

Figure (1) Typical Forces Associated With The Seat Skinning Operations Using The Original Fixture Stand

The assembly operations involved in this task requires the worker to exert the minimum forces for each work elements several hundred times per work shift. Obviously, this kind of task is highly repetitive therefore, the static analysis alone does not reveal the correct stresses imposed upon the worker. As a result, the work rate along with force and load moment analysis had to be conducted for each work element. The capacity model proposed by Mital, (1983) was used to determine and evaluate the risk factor for each work element. A summary of the computational results for both male and female populations is presented in Table 3. Risk potential factors greater than 1 indicate that the corresponding work elements are stressful and some work adjustment has to be implemented. Hence, the work stress can be reduced to the acceptable levels.

Table (3) Risk Potential Factors, Actual and Acceptable Forces and Work Rates of the Different Operation Elements for Male and Female Populations

Operation Element or Corner	Required Force to be exerted (kg)	Acceptable Force (kg) 50% pop	Acceptable Force (kg) 75% pop	Actual Work Rate kg.a/min.	Acceptable Work Rate 50% pop	Acceptable Work Rate 75% pop	Risk Potential Factor (RPF) 50% pop	Risk Potential Factor (RPF) 75% pop
L/R	small	37.87 (23.22)	31.74 (20.83)	small "	6.31 (3.89)	5.41 (3.55)	<< 1 "	<< 1 "
R/R	19.35	37.87 (23.22)	31.74 (20.83)	3.48 (3.48)	6.31 (3.89)	5.41 (3.55)	0.55 (0.89)	0.64 (0.98)
L/F	19.35	21.18 (16.3)	15.06 (13.91)	3.48 (3.48)	3.53 (2.73)	2.63 (2.38)	0.99 (1.27)	1.32 (1.46)
R/F	25.87	21.18 (16.30)	15.06 (13.91)	4.66 (4.66)	3.53 (2.73)	2.63 (2.38)	1.32 (1.71)	1.77 (1.96)
R/M & L/M	26.11	36.30 (21.74)	30.17 (19.35)	2.35 (2.35)	3.03 (1.82)	2.58 (1.64)	0.78 (1.29)	0.98 (1.42)

( ) Female population
RPF > 1 Indicates an excessive stress for the corresponding operation element and population segment.

Risk potential factors based on the production rate prior to this investigation clearly indicate that 75% of the male segment population should not perform the work elements of the Left-Front L/F and Right-Front R/F corners. Similarly, 75% of the female segment population should not perform these work elements as well as the Right-Middle R/M and Left-Middle L/M work elements. As a result, these jobs and their work elements must be redesigned to fit workers, also better fixtures should be provided to eliminate excessive stress and reduce the risk of work injuries.

ERGONOMIC INTERVENTION AND CONTROL MEASURES

The major stresses identified in this study were due to the following factors:

1) Repetitiveness of work elements and tasks.
2) Repetitive forces of exertion.
3) Stress concentrations from the seat frames.
4) Awkward working postures.
5) Lack of properly designed tools, fixtures and/or equipment to accommodate different workers and ease the assembly operations.

Repetitiveness of the work elements that are required for the assembly operations of the lower seat were reduced to certain extent by eliminating some work elements through minor modification to the original fixture stands, and by adding new fixture stand manned by another worker. The introduction of new tools, fixtures and workstations to the workplace have significant psychological impact on the improvement of work performance and work stress reductions. Other alternatives such as job rotation or enlargement of work content were not feasible in this study due to union contracts.

The excessive force required for some work elements described in the previous section such as F(R/F) and F(L/F) were eliminated by providing a rotatable fixture stands. Hence, each worker would be able to rotate the mounted seat and the fixture table as soon as the top cover of the seat is hooked in from both rear corners. Obviously, new work element is created however, workers would be able to place the top cover around the front corners of the seat frame at closer proximity to their body without excessive bending. Further improvement to the fixture stands were achieved by making the fixed stands adjustable to accommodate different workers and working heights. As a result, stress from excessive bending and awkward working postures was significantly reduced and preferable working heights can be easily accommodated.

Mechanical stress concentrations within the work elements of the primary seat top cover task were not that great because workers had their gloves worn and contacts with the seat frame occured only during handling and mounting operations. However, during the top cover clamping operations workers have to exert certain forces on the edges of the seat frame using their hands, and sometimes their forearms and elbows. In this situation, gloves alone can not protect workers, and there were no other tools used to help them reduçe the risk of injury. Clearly, a press tool or equipment such as those used to disassemble deflated tires form the wheel rim can be easily introduced to the workplace. In addition, it is possible to include this type of equipment along with the adjustable fixture stand as one working station to reduce or eliminate some of the travelling and handling work elements. In other words, when the primary top cover placement operations are completed workers will be able to dismount the subassembly and turn it upside down on the same fixture table, then use a power press tool which could be mounted above the fixture stand to press down the seat frame and proceed with the subsequent assembly operations. This type of workstations properly will eliminate several work elements and their associated stresses. As a result, the risk of repetitive work injuries and CTD's will be reduced significantly.

SUMMARY

Cumulative trauma disorders of repetitive work are a major cause of lost work time in many had intensive industry. The lower back and upper extremity musculoskeletal illnesses are among the leading and most frequent worker complaints and compensation claims. Such complaints and work injuries can be recognized in many manufacturing and assembly plants. Risk factors identified in the seat skinning operations were similar to many factors reported in literature including repetitive and forceful exertions, stress concentrations, and awkward working postures due to improper design of hand tools, equipment and workstations. Risk factors and physical work stress were identified and evaluated using the AET and detailed analysis techniques. Ergonomic intervention through workstation, tool and/or equipment redesign as well as better workplace arrangement have produced significant improvement in the work performance and higher worker satisfactions. Furthermore, certain types of work elements and their stress were totally eliminated; hence, the risk of repetitive trauma work injuries were significantly reduced. It is clearly that the integration of ergonomic programs into the design and production stages reduce the risk of occupational work injuries.

REFERENCES

Armstrong, T.J., 1983. An Ergonomic Guide to Carpal Tunnel Syndrome. Akron, OH: American Industrial Hygiene Association.

Armstrong, T.J., Radwin, R.G., Hansen, D.J., and Kennedy, K.W., 1986, Repetitive Trauma Disorders: Job Evaluation and Design. Human Factors, 28(3), 325-336.

Armstrong, J.J., Foulke, J., Joseph, B., and Goldstein, S., 1982, An Investigation of Cumulative Trauma Disorders in a Poultry Processing Plant. American Industrial Hygiene Association Journal, 43, 103-106.

Fine, L., Silverstein, B., Armstrong T., and Anderson, C., 1984, An Alternative way of Detecting Cumulative Trauma Disorders of the Upper Extremities in the Workplace. In: Proceeding of the 1984 International Conference on Occupational Ergonomics, Toronto, Ontario, Canada, 425-429.

Jensen, R., Klein, B, and Sanderson, L., 1983, Motion-Related Wrist Disorders Traced to Industries, Occupational Groups. Monthly Labor Review, 106, 13-16.

Mital A., 1983, Generalized model structure for evaluating/designing manual materials handling jobs. International Journal of Production and Research, 21(3), 401-412.

Rohmert, W., 1985, AET. A new job analysis method. Ergonomics, 28, 245-254.

Wagner R., 1985, Job analysis at ARBED, Ergonomics, 28, 255-273.

Advances in Industrial Ergonomics and Safety II
Edited by Biman Das
Taylor & Francis 1990

ERGONOMIC IMPROVEMENT IN A BARR-TACK SEWING JOB: A CASE STUDY

J. L. Wick
The Joyce Institute, Seattle, WA 98101

And

R. Morency, J. Waite and V. Schwanda
L.L. Bean, Freeport, ME 04033

A major manufacturer of sporting goods was experiencing cumulative trauma injuries from a barr-tack sewing job. A specific ergonomic analysis/problem identification/problem solving technique was applied to make improvements to prevent future injuries. Work/workstation changes resulted in major improvements. The most significant were reductions in stresses to the wrists and shoulders and a 36 percent labor savings.

INTRODUCTION

Many sewing machine jobs impose high stresses to the upper extremities. These stresses are often severe enough to cause cumulative trauma disorders such as carpal tunnel syndrome, epicondylitis and shoulder bursitis.

In recent years, ergonomists have taken an active interest in studying the problems of repetitive trauma disorders in the sewing industry. For example, Keyserling, et al., (1982) studied the prevalence of repetitive trauma disorders among garment workers and identified tasks with increased risks of these disorders. They found that 25 percent of the workers surveyed reported persistent musculoskeletal pain in at least one part of the body.

It seems that sewing machine operators are exposed to the same stresses as any person performing upper-extremity intensive tasks. They are the commonly accepted stresses of position, force and frequency. The frequent positioning of a body part, such as wrist radial deviation, combined with force can contribute to cumulative trauma disorders. Sewing tasks are often very short cycle, e.g. highly repetitive, and require bending the upper-extremities into positions other than neutral. Some sewing tasks require high pinch forces to manipulate the material such as in shoemaking (leather) and in the manufacture of articles made of canvas-type goods. Solutions to ergonomic problems of sewing workplaces have ranged from addressing the handling of cloth to, from and around the machine (Sillanpaa, 1984) to focusing on the stitching task itself (Wick and Drury, 1986). It would be convenient to be able to develop a simple, generic sewing workstation that would solve all problems in all situations, however, sewing jobs vary considerably. Sewing tasks differ significantly between industries such as garment, shoe, luggage and upholstery. There are even significant differences within industries, notably, between dress-making and coat-making, or boot-making and sandal-making.

Given the variety of situations with which an ergonomist must deal in the sewing industry, an effective task analysis methodology must be employed to accurately identify and measure the upper extremity stresses. Then, having identified and quantified the ergonomic problems, the causes for those problems can be determined and solutions developed. Such a task analysis is described by Drury and Wick (1984) and was the methodology applied in this project. It is an eight-step methodology as follows:

1. Videotape the job. The operator was videotaped from the front, back and both sides while performing the work.
2. Task analysis. The job was broken into task steps and the positions of each body part was recorded.
3. Work/workstation measurements. Physical dimensions of the workplace and measurements of finger and hand forces were taken as required.
4. Worker interviews. The worker was interviewed to allow input regarding possible changes to the workstation and to determine what, if any, pain or discomfort was being experienced.
5. Data evaluation. The body part positions, forces and repetition rate were compared with the literature recommendations for seated work.
6. Solution design. Solutions to the ergonomic problems and the workplace causes were developed during a brainstorming meeting.
7. Design validation. The new design was analyzed and evaluated to determine if the problems were solved and no new problems inadvertently introduced.
8. Comparison of the new design with the original job. The data from before and after change were compared to quantify the effects of the new design.

A very important aspect of this methodology is that it allows the comparison of data gathered before and after change. One does not have to wait for an injury to not happen to measure success.

Besides measuring and alleviating biomechanical stress, another goal of any ergonomics project is to improve productivity. The application of the principles of ergonomics should result in a more efficient workplace, i.e. more productive. This project demonstrates that both goals can be achieved.

This project was part of an ergonomics training program (The Joyce Institute, 1988) presented to selected professionals.

DESCRIPTION OF THE JOB

The job studied was Barr Shoulder Strap. It was part of an assembly line operation in which different operators performed different jobs in the manufacture of a sport bag. The strap was of a heavy webbed material.

The operator performed the job at a seated sewing workplace. The machine involved was a barr-tack machine. A barr-tack machine is a cam-driven machine which stitches in a pattern prescribed by the shape of the cam. The machine was tripped by pressing on a foot control with a mechanical linkage to the cam.

The shoulder strap was partially assembled at a previous workstation and delivered to the operator in bulk in a tote box. All other materials such as thread and bobbins were located at the workplace.

The operation involved getting the strap and adjusting the end through a keeper to the location where it was to be fastened with the stitching. The strap was positioned in the machine and the machine activated. At the end of the machine cycle, the strap was removed, inspected and asided.

The production rate was 96 straps per hour. The operator was on an incentive (piece-work) pay system and was self-paced. The work rate was, therefore, under operator control.

The work environmental factors of light, heat and noise were all normal. Other stressors such as supervisory style, labor disputes, etc. were not evident.

RESULTS

The task analysis and data evaluation revealed a number of ergonomic problems to solve. Certain definitions and criteria were adopted. Stress to the wrist can result from damaging wrist motions (DWM) which are defined as the wrist not straight with a grip (Drury and Wick, 1984). A goal of maximum 1,000 DWM's per hour was established.

Grip forces were measured by asking the operator to reproduce the forces required to perform the task on a pinch gauge (Drury. 1987). The maximum pinch force allowed was 8 pounds (3.6kg).

Damaging wrist motions, particularly in the left hand, resulted from the angle of the machine, the angle of the tote boxes and the orientation of the barr pattern. Pinch grips with excessive force were required to adjust the position of the strap end prior to the stitching task and by varying thicknesses of the strap material. Shoulder stress was caused by the locations of the machine and tote boxes.

Solutions were developed to solve the ergonomic problems. All opportunities to improve productivity were investigated during the design process. The foundation of the new design was the sewing workstation reported by Wick and Drury (1986); however, other changes were required because sewing tasks tend to be unique. The changes included:

- rotate the barr pattern 90 degrees
- pick up straps from one side and aside to the other side
- provide tilt stands for the tote boxes
- reduce the size of the bench
- require more stringent quality control of the webbing supplier
- provide armrests
- tilt the machine toward the operator
- train the operator in the new methods resulting from the new design.

A prototype of the new design was assembled and tested to verify that an operator could, in fact, perform the job with the proposed workstation. The operator tested the prototype for several days before another analysis and evaluation was done to validate the changes. The validation repeated steps 1 through 5 of the methodology.

Table 1. Comparison of conditions before and after change. (Only the ergonomic problems identified before change are listed.)

Measure	Before	After
Damaging wrist motions		
Left	1001/hour	480/hour
Pinch grips with force		
Left	17/cycle	0/cycle
Right	15/cycle	2/cycle
Shoulder flexion		
Left	8/cycle	0/cycle
Right	9/cycle	4/cycle
Shoulder extension		
Left	4/cycle	0/cycle
Shoulder adduction		
Left	13/cycle	0/cycle
Right	14/cycle	10/cycle
Elbow flexion		
Left	15/cycle	5/cycle
Right	15/cycle	5/cycle
Labor standard	0.78	0.50

It should be noted that the new design resulted in increased shoulder abductions per cycle, both right and left. However, they were not considered significant as the degree of range of movement was minimal and well within worker tolerances.

The application of this methodology allowed quantification of the biomechanical conditions imposed on the operator before and after the changes in order to evaluate the effects of the changes. The stresses of the new design were to be compared with design criteria to determine if they were within those criteria. In this project, the new design can be considered ergonomically acceptable. The DWM's in the left hand were reduced to less than the design criteria of 1,000 per hour. Most of the pinches with excessive force were eliminated. Stresses to the shoulders were greatly reduced. In addition, the task of adjusting the strap was eliminated due to a change in method of pre-assembly. That resulted in significant labor savings of 36 percent as determined by Time Study.

DISCUSSION

The sewing industry is facing a new challenge. Cumulative trauma disorders are becoming a recognized problem. The industry will have to take the necessary steps to eliminate the causes. This has begun and some of the work has appeared in the literature. The work reported by Drury, Wick and Sillanpaa has already been noted. Also, Marcotte and Wick (1989) reported on a cooperative effort between the International Ladies' Garment Workers' Union, the New York State Department of Labor and two garment manufacturers, Kobra International and Izaac Hazan. Other important work has been done and not reported.

Many of the solutions that have been reported need to be subjected to controlled experimental conditions. The sewing trade associations should consider supporting this type of research. The auto industry is an example of where trade association support has contributed to good ergonomics research which can be applied in their plants.

REFERENCES

Drury, C.G., 1987, A biomechanical evaluation of the repetitive motion injury potential of industrial jobs. Seminars in Occupational Medicine, Vol. 2, No. 1, pp. 41-49.

Drury, C.G. and Wick, J.L., 1984, Ergonomics applications in the shoe industry. In: Proceedings of the 1984 International Conference on Occupational Ergonomics, Toronto, pp. 489-493.

Keyserling, W.M., Donoghue, J.L., Punnet, L. and Miller, A.B., 1982, Repetitive trauma disorders in the garment industry. Technical Report, Occupational Health Program, Harvard School of Public Health, Boston.

Marcotte, A.J. and Wick, J.L., 1989, Ergonomic applications in the apparel industry - a cooperative effort between the International Ladies' Garment Workers' Union, the New York Department of Labor and private industry. In: Proceedings of the 1989 American Society of Safety Engineers (ASSE) Annual Professional Development Conference and Exposition, Detroit, pp. 55-55.

Sillanpaa, J., 1984, Improved work conditions of industrial seamstresses. In: Proceedings of the XXI Congress on Occupational Health.

The Joyce Institute, 1988, Applied Ergonomics Reference Manual, (Seattle: The Joyce Institute).

Wick, J.L. and Drury, C.G., Postural change due to adaptations of a sewing workstation. In: The Ergonomics of Working Postures, (London: Taylor and Francis, Ltd.), Chapter 13.

V

ENGINEERING ANTHROPOMETRY

Advances in Industrial Ergonomics and Safety II
Edited by Biman Das
Taylor & Francis 1990

AN ISSUE REVIEWED-USING EXISTING ANTHROPOMETRIC DATA TO DESIGN INDUSTRIAL WORKPLACES

Paul C. Champney
Consultant Ergonomics/Human Factors
Prosser, Washington 99350-1574
U.S.A.

The design of industrial workplaces is hampered by the lack of data that describe the potential workforce. The civilian data base is not adequate because it is small, taken on specific populations, and is not stratified to reveal ethnic and individual differences. Military data are often used as an alternative, but these data also have shortcomings. Among these are truncation of the extremes, a paucity of functional data and the fact that the data are often taken for special purposes.

INTRODUCTION

Designers of industrial workplaces need anthropometric data that describe the potential workforce. The industrial worker will range between seventeen and seventy years of age, stem from many different ethnic backgrounds, and come from several regions of the country. The same worker will be "average" in only one or two body measurements, if any. A designer must also consider the chance that a worker will be either female or male, and design accordingly.

While the data in civilian data bases best describe the civilian population, some measurements that are useful in designing industrial workplaces are not included. To meet this need military data are used to fill the gaps. Working with military and civilian data carefully, designers and engineers attempt to create industrial workplaces that are optimized for the greatest number of workers. Most industries are committed to a policy of providing safe workplaces that accommodate women, men, and various ethnic groups. These companies employ people from every segment of the potential work force and need a broad data base that can be used by engineers and designers. Since many companies have facilities in several geographic areas, data that consider all regions are also important. To meet these needs it has been necessary to merge existing civilian and military information into, albeit, a still incomplete data base.

METHOD

The Human Factors Section of the then Health, Safety, and Human Factors Laboratory at Kodak, where I was formerly employed, had determined the 46 measurements most frequently used at Kodak. A search of "A Collation of Anthropometry" (Garrett and Kennedy, 1971) was initiated to reveal the best available data. The values selected were from the largest and most recent surveys that described the particular body measurement. It was assumed that this selection had a tendency to assimilate cultural and regional differences. Except for measurements from the Health Education and Welfare Department data, (U.S. HEW, 1970) these values came from military sources. Similar measurements were sought for both men and women. The 1977 survey of army women (Churchill, et. al.) was most helpful in completing the task.

When similar data were available for both men and women, they were combined to describe a population that is, theoretically, 50 percent men and 50 percent women. This was accomplished using a Fortran program. The program used the mean, standard deviation, and sample sizes to calculate the percentile rankings for each measurement. Although sample sizes were different, the program nevertheless, gave equal weight to each sample. For some dimensions similar data were not available for both sexes. Thus, there were voids in the percentile rankings.

Figure 1 shows the normal and combined curves for Forward Functional Reach-Standing. This illustration gives some indication of the effect that the combined data will have on each of the populations. The statistically combined data reflect women more at the low end of the scale and men at the high.

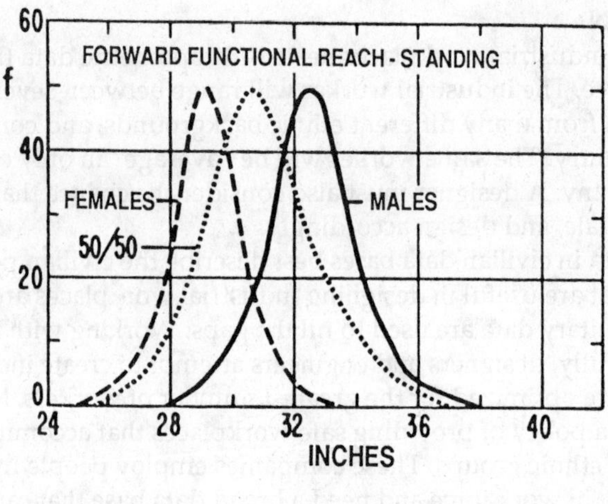

Figure 1

USES

The simulated data were carefully interpolated for the designing and redesigning of workplaces. Since there was some risk connected with the use of these data, they were not distributed to untrained individuals. A handbook of design dimensions was prepared for designers and other interested people. One goal of the handbook was to help designers of industrial workplaces to accommodate the largest feasible portion of the potential work force, hence, design dimensions were based on the anthropometric data provided by the simulation. A second document intended for human factors practitioners contained all the anthropometric data generated by the simulation. The document was used to make recommendations for new equipment and for situations that were not covered in the handbook. The simulated data were instrumental in the revision of an engineering standard for safe clearances and comfortable reaches.

AN EXAMPLE

The design of standing workplaces provides an example of the use of existing data. Appropriate dimensions for this type of workplace are dependent on the type of work done. The measurements and estimates in the remainder of this work are derived from the calculated data. Among factors under consideration are:

1. The application of forces.
2. The monitoring of visual displays.
3. The type of work done by the hands.

Generally, the optimum height of the work surface should be near elbow level. The height of the working surface is understood to be the level of the hands while performing a task. For light assembly and light packaging tasks, not involving wrapping, the optimum working height of the hands is 42 inches (102 cm). This will accommodate the 5th percentile person. For tasks requiring large downward or sideward forces it was estimated that the optimum working height of the hands is 36 inches (91 cm) and 32 inches (81 cm) for tasks requiring large upward forces. These estimations were verified in laboratory tests. However, more functional data are needed.

When displays are part of the workplace, it is important to accommodate both tall and short people. For many visual tasks, the viewer need not look directly at the display in order to read it. For example, a wall clock may be read easily while looking up or from either side. This type of display may be placed somewhere between 58-69 inches (147-175 cm) above the standing surface. Displays requiring only qualitative as opposed to quantitative, readings may be placed in a 40-69 inch (102-175 cm) range.

To eliminate parallax in tasks that require close or controlled sighting, the display or the viewer should be located at specific heights. Such a display would be a sight glass where it is critical to look directly into the sight. If these tasks are of short duration (10 seconds or less), the appropriate height of the center of the display is 60 inches (152 cm) above the standing surface. Shorter individuals will be required to stand on their tiptoes and taller individuals will be required to stoop. When critical tasks of this type require a longer duration of viewing (more than 10 seconds) the appropriate height of the display is 64 inches (163 cm). To accommodate shorter viewers, a 6 inch (14 cm) platform should be provided; the 64 inch (163 cm) height will only slightly disadvantage taller viewers. This is only one example of some adjustments that are encountered in designing industrial workplaces.

THE PROBLEM

If we depend on available civilian data in our designs, we may fall short of optimum. This is true because of the lack of comprehensive surveys on civilians. Several reports and papers support this with statements like, "In contrast to the long history of military anthropometry which dates back to the Roman legions, civilian anthropometry is virtually nonexistent" (Garrett and Kennedy, 1971). The same document also cites the fact that there has been only one national survey that has a sufficient number of dimensions to be of value to the clothing and mail order industries

(Garrett and Kennedy, 1971). There is a Department of Health, Education, and Welfare survey (1959-1962) but this gives us only 19 body measurements (U.S. HEW, 1970). There are also some specialized surveys that have been done on industrial workers, bus and truck drivers, air traffic control trainees, and law enforcement officers. The National Bureau of Standards report states the "even though these studies were well controlled, the populations surveyed are too specialized to permit generalization to the civilian population. In summary, there is no anthropometric data representative of the civilian population that contains enough body measurement to have utility for most potential data users." (Van Cott, et. al.)

No civilian study is stratified on an ethnic or regional basis. When an industry is located in various parts of the country this factor is important. Unfortunately, we do not have a truly homogenous population. As a result, a large company will find that it has production facilities in areas that are dominated by one or other of the several ethnic groups. It is probable that people who have lived a lifetime in one region of the country may differ from people who live in other regions. In support of this NASA lists seven human traits that are correlated with the environment (Webb Associates, 1978). Considering these factors we may conclude that significant dependence on the present civilian anthropometric data base may involve some risk.

To avoid the problems associated with the use of civilian data we use the military surveys. These data have some shortcomings when considering the design of the industrial workplace. One of these shortcomings is that existing data have been gathered on a population that has been truncated at the extremes. For example, we find the stature of the 99th percentile army male is 75 inches (191 cm) but we find individuals in the civilian population such as very tall basketball players who exceed this by 9 inches (23 cm) or more. We also find that, with the exception of seated reach dimensions, little functional data are included in the military reports. If functional data are provided they come from a small sample that was measured for a specific purpose; therefore, it has limited application for civilian use. One example of this is gloved hand anthropometry (Garrett, 1968). Keeping these factors in mind extreme caution should be used when using military data.

FURTHER NEEDS

Available data do not provide sufficient functional information. The discussion of the standing workplace did not include forward reaches that are often necessary. A search of existing literature produced information only on shoulder height forward functional reaches. Since we have placed the optimum working surface height at or near elbow height, the designers are again disadvantaged by the lack of data describing part of the task that must be performed. The need for a civilian data base is underscored by the fact that simulated data are interpolations that may have distortions. This point was emphasized in a paper presented by Dr. Van Cott in 1980 "In 1978 my colleagues and I at the National Bureau of Standards (Van Cott, et al., 1978) developed a plan and proposal for a Standard Ergonomics Reference Data System, a data base of selected anthropometric and ergonomic data on the U.S. Population. To date, this proposal has not been adopted. In my opinion, however, the need for quantitative data on the U.S. populations remains, for without it the Nation may pay an unnecessary price for clothing, products and systems which do not fully conform to the measurement of men, women, and children." (Van Cott, 1980) Regretfully this proposal has yet to be adopted.

Keeping all this in mind we as a group should consider what can be done to achieve the 1978 proposal of Dr. Van Cott and his colleagues. We have improvised and done well. However, we can all find products, clothing, and workplaces that are not of the ultimate fit.

SUMMARY

The purpose of this discussion has been to point out how a computer interpolation program can combine anthropometric data bases to provide a base with fewer gaps than in any of the individual sources, and how this data can be used to design industrial workplaces.

However, there continues to be a definite need for a comprehensive functional data base on a representative samples of the United States population, and there are still some gaps in the information on static anthropometry. This was true when these findings were first presented but not published in 1980. Regretfully it is now 1990 and the need has yet to be met.

REFERENCES

Churchill, T., Churchill, E., Laubach, L., McConville, J., and White, R., 1977. Anthropometry of Women of the U.S. Army. Report No. 2. The basic multivariate statistics. Natick, Massachusetts, U.S. Army, Natick Research and Development Command.

Garrett, J.W., 1968. Clearance and Performance Values for the Bare-Handed and Gloved Handed and the Pressure Gloved Operator. Aerospace Medical Research Laboratories. Aerospace Medical Division, Air Force Systems Command, Wright-Patterson AFB, Ohio 45433.

Garrett, J.W. and Kennedy, K.W., 1971. A Collation of Anthropometry. Volumes I and II. Aerospace and Medical Research Laboratories, Aerospace Medical Division, Air Force Systems Command, Wright-Patterson AFB, Ohio 45433

U.S. Department of Health Education and Welfare, 1970. Skinfolds, Body Girths, Biacromial Diameter, and Selected Anthropometric Indicies of Adults United States. 1960-1962. Public Health Service Health Services, and Mental Health Administration, Washington D.C.

Van Cott, H.P., 1980. Civilian Anthropometry Data Bases. Proceedings of the Human Factors Society, PP 34-36.

Van Cott, H.P., Kramer, J.J., Pezoldt, V.J., Porter, L.G., Fried, C., Fechter, J.V., and Peransky, J.J., 1978. NBISR 77-11403 A Standard Ergonomics Reference Data System: the Concept and Its Assessment. Human Factors Group, Consumer Science Division, Center for Consumer Product Technology, National Bureau of Standards, Washington, D.C. 20234 and Teichner, W.H., Department of Psychology, New Mexico State University, Las Cruses, New Mexico 88003.

Webb Associates, 1978. NASA RP-1024 ANTHROPOMETRIC SOURCE BOOK. Volume I: Anthropometry for Designers. National Aeronautics and Space Administration, Lyndon B. Johnson Space Center, Houston, Texas 77508.

Advances in Industrial Ergonomics and Safety II
Edited by Biman Das
Taylor & Francis 1990

TOE SPACE

Deborah DeLaura and Stephan Konz

Department of Industrial Engineering
Kansas State University Manhattan, KS 66506

When standing at a workstation, space is required for the toes (ahead of a perpendicular from the table edge). Measurements were made on 55 females and 49 males from the food service and maintenance departments at Kansas State University. Females tend to stand farther back than males. Recommended toe space is 150 mm deep, 150 mm high and 500 mm wide.

INTRODUCTION

Figure 1 shows the problem. When standing at a machine, table or conveyor, the operator moves forward until the thighs make contact. If there is a toe space under the device, the body orientation can be approximately vertical. However, if there is no toe space under the device, then (a) the upper body will be vertical but the distance required to reach objects on the work surface will increase or (b) the upper body will be at an angle (setting up a torque about the feet). In practice, case (b) has a limiting condition in which the thighs move forward to contact the vertical surface; this minimizes reach distance and gives a support surface to the thighs so that not all the angle torque must be counteracted by muscles.

The question we addressed was how much space was required for the toes.

Mean foot length (barefoot) of US military personnel are given as 269 mm for males and 241 mm for females (NASA, 1978). Mean foot length of Kansas State University students was given as 250 mm for males and 228 for females (Konz and Subramanian, 1989). However, the entire foot need not fit in the toe space, just the part ahead of the ankles.

Little information seems to be available on vertical foot thickness. Roebuck et al. (1975) reported, for US military personnel, that the 5th percentile ankle height was 119 mm and the 95th was 157.

METHOD

Task

Subjects, wearing shoes, stepped onto a calibrated measurement grid positioned under and in front of a 36 inch high table. They positioned themselves so they were standing upright and with their thighs contacting the table edge in the mid-thigh region. The grid had lines of alternative colors spaced 10 mm apart. Two dimensions were recorded: (a) the distance forward of the perpendicular from the table edge to the tip of the toe and (b) the distance to the rear of the perpendicular to the rear of the heel. The distances were interpolated visually to the nearest cm.

The third dimension was the maximum distance between the outer edges of the feet. Subjects had been instructed to assume a natural stance with the feet "shoulder width" apart. Stance width was interpolated visually to the nearest cm.

The fourth dimension, ankle height, was defined as the distance from the floor to the top of the protruding ankle bone. It was measured with the aid of a wooden block with a cm scale on it.

Subjects

The population was primarily "non-student." There were 55 females and 64 males (total 119 obtained from employees of the food service centers and maintenance departments at Kansas State University. The population was primarily Caucasian.

RESULTS

A SPSS^x FREQUENCIES program was used to examine the distributions of the four measurements. All variables were normally distributed and demonstrated good homogeneity of variance. Three potential outliers were investigated but were retained.

Table 1 summarizes the data.

Mean age was 29.2 years for males and 34.4 for females. Mean height was 1.80 m for males and 1.66 m for females. Mean weight was 82.6 kg for males and 67.5 kg for females.

Total foot length (with shoes) was 305 mm for males and 264 mm for females; 264/305 = 87%. The coefficient of variation was 5% for males and 3% for females. The coefficient of determination, r^2, was 40% vs. stature height. Total foot length (in shoes) = .17 (Stature height).

Stance width for males was 335 mm and 277 mm for females; 277/335 = 83%. The coefficient of variation was 14% for males and 17% for females. The coefficient of determination, r^2, was 36% vs. stature height.

Ankle height for males was 108 mm and 95 mm for females; 95/108 = 88%. The coefficient of variation was 13% for males and 14% for females.

The dimension of primary interest for toe space is "foot length forward." It was 85 mm for males and 38 mm for females; 38/85 = 45%. The coefficient of variation for males was 37% and 85% for females.

The female/male ratio departed substantially from the expected ratio of about 90% (Konz and Subramanian, 1989). Note that total foot length has a 265/305 = 87% ratio. Females also had a much higher coefficient of variation than males; that is their standing posture (in the forward-backward direction) varied more. This led us to calculate the ratio of foot length forward/total foot length; it is 85/305 = 28% for males and 38/264 = 14% for females. Thus females tended to stand with their feet more to the rear than males and they tended to have more variability in their forward-back foot position.

DISCUSSION

Space to accommodate the 99% male would be 150 mm deep, 140 mm high and 460 mm wide. Space for the 99% female would be 130 mm deep, 130 mm high and 390 mm wide. To design for the entire population and using simple numbers, make the space 150 mm deep, 150 mm high and 500 mm wide. In US units, this would be 6 inches deep, 6 inches high and 20 inches wide.

The different posture for females, implied by the 14% ratio of foot forward/foot length, suggests that the center of gravity of females is different from males. We were unable to obtain any center of gravity figures for males vs. females. Asmussen (1960) reported, that with normal shoes, the center of gravity of the body passes in a line from 1 cm forward of the ear hole through the 5th lumbar disc; in obese people it passes forward of the disc.

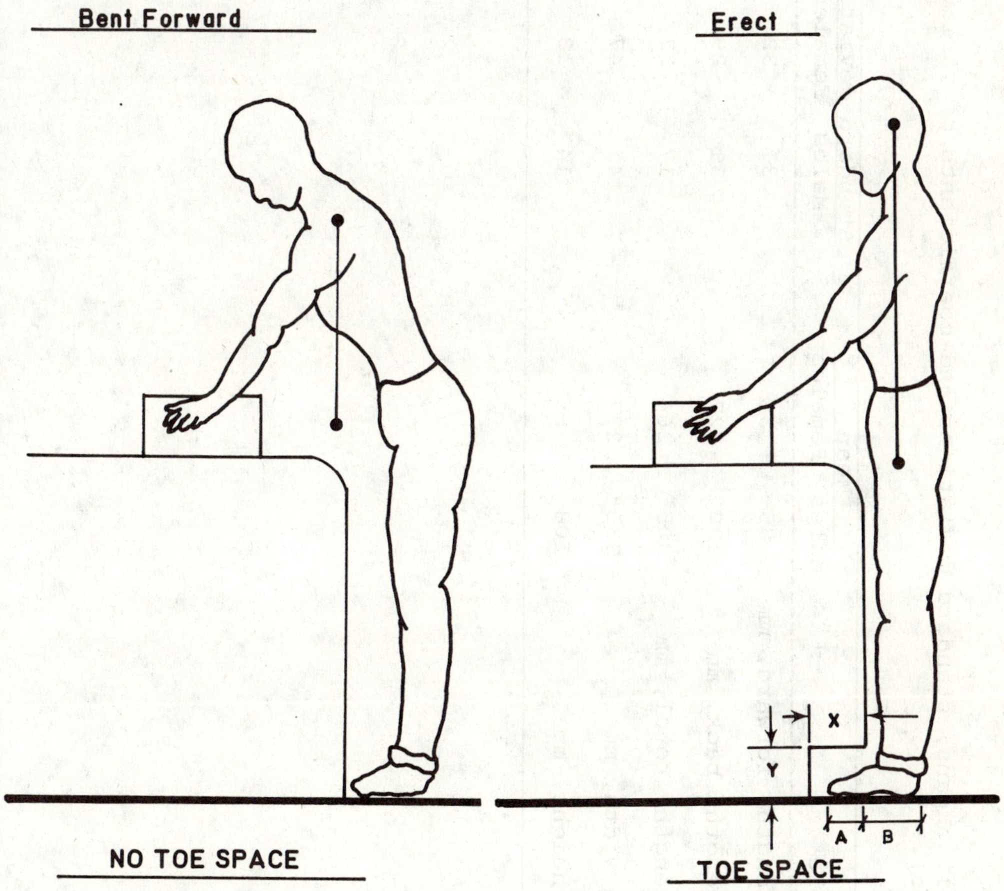

Fig. 1. With no toe space, the person must stand farther away and has a poor posture. With a toe space, better posture is permitted. We recommend the toe space be 150 mm deep, 150 mm high and 500 mm wide.

Table 1. Means, standard deviations and coefficients of variation by gender (64 males, 55 females).

Variable	Mean		Standard deviation		Coef. of variation, %	
	Males	Females	Males	Females	Males	Females
Foot length, forward, mm	85	37	31	32	37	85
Foot length, back, mm	220	227	31	29	14	13
Foot length, total, mm	305	264	17	22	5	3
Stance width, mm	335	277	47	47	14	17
Ankle height, mm	108	95	14	13	13	14

REFERENCES

Asmussen, E., 1960, The weight-carrying function of the human spine. Act Orthopedic Scandanavia, Vol. 29, 141-148.

Konz, S. and Subramanian, V., 1989, Footprints. In Industrial Ergonomics and Safety I, (London: Taylor and Francis).

NASA Reference Publication 1024, 1978, Anthropometric Source Book, Vol. 1, NASA, Yellow Spring, OH.

Advances in Industrial Ergonomics and Safety II
Edited by Biman Das
Taylor & Francis 1990

A COMPARATIVE ANALYSIS OF ANTHROPOMETRIC AND KINEMATIC MEASUREMENT SYSTEMS

Jonathan N. Tyson
Michelin Tires Canada Ltd., PO Box 399, Granton Road
New Glasgow, Nova Scotia, Canada, B2H 5E6

Biman Das
Department of Industrial Engineering
Technical University of Nova Scotia
P.O. Box 1000, Halifax, Nova Scotia, Canada, B3J 2X4

A comparison was made of various anthropometric and kinematic measurement systems: manual/mechanical, electronic/electro-mechanical, film, video, and opto-electronic, on the bases of quantitative evaluation criteria. The selection of a particular system is dependent on the specific attributes required from a system and on the particular application. Video based measurement systems appear to be the future trend. This has direct implications for the industrial ergonomist.

INTRODUCTION

Anthropometric measurements are needed for the design of industrial workstations. Small changes in workstation dimensions can have a profound effect on the productivity, health and safety of an operator. With proper design, an operator should be able to perform a task with maximum efficiency at minimal human cost. Consequently, anthropometric measurement should be generated quickly and without much difficulty to make the measurement process cost-effective.

The application of ergonomics in field settings is hampered by the lack of anthropometric and kinematic data collection techniques. The "high tech" data collection techniques that are commercially available today are too expensive for field use. Furthermore, such techniques are often not suitable for use in field settings. Consequently, an industrial ergonomist must resort to manual, time consuming techniques for data collection, reduction and analysis.

Anthropometric and kinematic measurement systems are employed for reach envelope measurements, the measurement of workstation /equipment dimensions, static anthropometry and the analysis of limb movements under different postural constraints. Several methods for the collection of this data have been proposed over the years. The objective of this paper is to make a comparative analysis of available anthropometric and kinematic measurement systems.

ANTHROPOMETRIC AND KINEMATIC MEASUREMENT SYSTEMS

Current techniques used for the collection of human anthropometric and kinematic data can be put into one of five different general classifications (Table 1). Each of these data collection systems has certain advantages and limitations in the context of their use in an industrial/field setting.

Table 1. General classes of anthropometric and kinematic measurement systems.

1. **Manual, visual, and mechanical techniques**
2. **Electronic/electro-mechanical techniques**
3. **Film-based techniques**
4. **Video tape-based techniques**
5. **Opto-electronic/reflection on-line systems**

Manual and Mechanical Techniques

Manual measurement techniques are those measurement techniques that require the researcher to manually measure, read, record and reduce, and computerize the data collected. Manual techniques are often used for the collection of anthropometric data. The standard tools for collecting anthropometric data are: tape measure, callipers, goniometer, and scaled rule. Despite major technological developments, these tools are still the primary method for collecting anthropometric data.

Mechanical measurement techniques involve all methods that use mechanical devices or structures. These systems do not contain any electronic or photographic elements and they require that data be read from scales incorporated into the device. The data is then manually recorded by the researcher. These techniques differ from the manual techniques only in that the researcher is not required to physically hold a measuring device to the subject. The subject grasps or pushes a pointer/measuring rod until he/she has reached the point desired by the researcher.

Manual and mechanical measurement methods are not suitable for the collection of quantitative kinematic data. Some authors have developed visual (manual) posture recording techniques (Corlette et al. 1979, Baleshta et al. 1985) but these methods can only be used to collect qualitative data. Mechanical systems are limited to the measurement of "functional" anthropometric data such as reach envelopes. Devices reported in the literature are complex combinations of the classic anthropometric tools (Dempsey 1953, Wright 1963, Faulkner and Day 1970, Andrade and Matias 1985).

The above techniques are time consuming both in the data collection phase and the data reduction/analysis phase. They are still popular because of their low cost, portability, ease of use and accuracy. However, they do not allow for analysis of kinematic data and provide no visual record of the subject.

Electronic / Electro-mechanical Techniques

The measurement techniques classed under this heading all incorporate some type of electrical transducer(s) to convert position and/or movement information into analog electronic signals. The signals can then be plotted, stored for future analysis on magnetic tape or computer disk, or collected and analyzed in pseudo-real time with computers.

The most popular type of electronic system in the literature is the potentiometer-based system (Ayoub 1972, Pynsent et al. 1983, Das et al. 1989). These systems collect human movement data by recording the voltage changes from the potentiometer(s) attached to pointers or measuring rods that are manipulated by the subject. Most of systems reported in the literature have been designed to measure "functional" anthropometric data such as reach envelopes. However, the principles used in the design of these systems could be used to produce a system for measuring "structural" anthropometrics and a limited range of kinematic data (Das et al. 1989).

Accelerometer (Morris 1973) and ultrasonic (Hennig et al. 1981) based systems are other electronic systems that have been reported in the literature. Because of their complexity, both in use and analysis of data, these types of systems are not often used even in laboratory settings.

Recently, a new and innovative technique for recording human kinematic data was introduced by VPL Research (Tello, 1988). This company has developed two commercially available products: the DataGlove and the DataSuit. These products use fibre optic transducers to allow for the recording of hand and whole body motions by a computer in real time. The DataGlove has been used in a system by Wells et al. (1988) as a major component in a system designed to measure occupational hand and wrist movements. The DataSuit and DataGlove are two examples of how new developments in technology are providing researchers with new ways of gathering data both in the lab and at the worksite. Unfortunately, most of this new technology is expensive and beyond the budget of the industrial ergonomist.

Film-based Techniques

Ergonomic investigations using film-based measurement techniques to record worker movements, postures, and anthropometrics are quite common. This is because film recordings have traditionally been the cheapest and easiest way in which to gather information at the worksite. Film techniques have a number of advantages when compared to some of the other movement/posture analysis methods. They provide lasting visual records which can be viewed and quantified many times over. Secondly, they allow the investigator to observe the whole body motion which can provide him/her with new insights, new ideas, and different interpretative opportunities. Thirdly, they allow for qualitative and quantitative analyses of all aspects of the movement. The data required for the most sophisticated kinematic analysis of movement can be gathered with these techniques. Finally, these systems have an advantage over even the most advanced opto-electronic techniques. The total amount of information available from a photographic record is many times greater and more extensive than that provided by many of the automatic analysis techniques (Atha 1984).

While at first it may seem that film-based measurement systems are without faults, there are some disadvantages to these systems. Chao (1976), Holzhausen (1981), and Chaffin (1984) list some of the major disadvantages of film-based motion analysis systems. Three main disadvantages of these systems are:

1. Analysis of the large amount of information available from photographic records is tedious and time consuming. The main reason for this is the inability of film-based systems to take direct advantage of computerization.
2. The photographic records can not be seen until the film is

developed, therefore data collection errors may not be detected until it is too late.
3. Measurement errors can be significant.

Video Tape-based Techniques

In the past, video tape-based systems were not popular for a number of reasons. First, video tape images were not of high enough resolution to allow for the collection of valid data by digitizing the images. Second, the low frame rates and shutter speeds of the first video cameras did not allow for the capture of clear images. Last, the technology to allow for the digitization of the video image was very expensive.

Recent advances in video technology have made video tape-based data collection systems very popular. New and relatively inexpensive video cameras provide much greater picture resolution than before. Also, new video recorders/players provide a much greater range of options previously only found on professional editing machines. Combined with the above, new image processing technology for the computer has allowed for the production of inexpensive image processing boards for personal computers. This allows the analog video image to be directly manipulated by the computer in digital form.

Video tape based systems have the same advantages as film based systems without a number of disadvantages of the latter. Some of the advantages are:

1. Immediate play back of the data collected.
2. The ability to reuse the video tape.
3. The ability take advantage of computer technology to allow for automatic or semi-automatic data analysis.

Some examples of video tape-based systems described in the literature are provided by Abraham (1987), Shapiro et al. (1987), and DeGreve (1987). The authors describe a variety of methods for accessing anthropometric and kinematic data from video tape. Also, there are now commercially available video tape based semi-automatic /automatic systems on the market. These systems (Expert Vision, PEAK) combine the advantages of film and video tape based systems with the complexity of the opto-electronic systems described below.

Opto-electronic / Reflection On-line Techniques

Relatively recent developments in computer technology and micro-electronics have increased the ability of researchers to develop accurate semi-automatic/automatic motion analysis systems (Holzhausen 1978, Woltring 1980, Macellari 1985, Vaughan 1987, Wangenheim 1987). The systems use body mounted markers which, either passively or actively, transmit information to a receiver (television, video, or photo-diode cameras). This information is then fed into a computer. The computer software, or sometimes specialized hardware, analyses the information and subsequently produces two- or three-dimensional coordinate data.

The semi-automatic/automatic kinematic data collection techniques provide for quick and accurate analysis of postures and movements. They can be integrated with other data collection devices (force plates, EMG, ECG), the analog signals of which can be synchronized to the movement patterns. However, most of the automatic analysis techniques do not allow the necessary flexibility, portability, or economic benefits required by

Table 2. Criteria used to evaluate human anthropometric and kinematic data collection systems.

Criteria	Score	Description
1) Portability	3	easily transported & used, field & lab
	2	restricted field and lab locations
	1	lab use only, portable locations
	0	lab use with fixed location only
2) Ease of use		
a) operator expertise	3	general computer knowledge
	2	some computer and electronics knowledge
	1	advanced computer, some electronics
	0	advanced computer and electronics
b) time	3	real time collection & reduction
	2	real time collection, later reduction
	1	real time collection, manual reduction
	0	manual collection and reduction
c) environment	3	no control necessary
	2	slight control (light, power, etc.)
	1	moderate control over environment
	0	controlled lab
3) Cost	3	system cost < $2,000.00
	2	system cost $2,000.00-$5,000.00
	1	system cost $5,000.00-$10,000.00
	0	system cost > $10,000.00
4) Accuracy	3	reported accuracy of < 1.0 cm
	2	reported accuracy of 1.0 - 2.0 cm
	1	reported accuracy of 2.0 - 3.0 cm
	0	reported accuracy of > 3.0 cm
5) Image of subject recorded	3	"real" image recorded, retained
	2	"real" image recorded, not retained
	1	"data" image recorded
	0	no image recorded
6) Anthropometric, kinematic data, integrate with other data	3	anthropometric, kinematic, full inter.
	2	anthropometric, kinematic, part. inter.
	1	anthropometric, kinematic, no inter.
	0	anthropometric or kinematic data only
7) Collect two- and three-dimensional data	3	two-, three-dimensional, static, dynamic
	2	two-, three-dimensional, static only
	1	two-dimensional, static, dynamic
	0	two-dimensional, static only
8) System does not restrict movement patterns	3	unrestricted movements, no interference
	2	some restriction, some interference
	1	highly restrictive, lots of interference
	0	static measures only

industrial ergonomists. Some of the difficulties encountered when using automatic data collection systems in the work environment are:

1. The systems are highly sensitive, with a great deal of expensive equipment incorporated into them. This makes them less suitable for use in areas where dust, moisture, or heat may cause problems (i.e., most industrial settings).
2. The systems do not provide a permanent record of the motions under analysis.
3. The ambient lighting of the environment and the sources of possible reflection need to be strictly controlled.
4. The set-up and calibration procedures of these units can be complex.

Some examples of such systems include: Selspot, IROS, WatSmart, Elite, Vicon and Coda-3. Expert Vision and PEAK are two automatic data collection systems that use video tape to record movement data. They have an advantage over the other systems in that they can record information off-line for later analysis. However, they are still relatively expensive and automatic data reduction/analysis can only be done if data is recorded under controlled conditions.

COMPARISON OF ANTHROPOMETRIC AND KINEMATIC MEASUREMENT SYSTEMS

Many authors (Brand and Crowninshield 1981, Holzhausen 1981, Chaffin 1982) have listed criteria which they feel should be met by any measurement technique. These criteria (Table 2) have been combined and adapted to reflect specific requirements of measurement systems to be used by industrial ergonomists. Each criterion has been given a set of scores relating the capabilities desired.

As can be seen, there are many, varied methods available to the industrial ergonomist for the collection of anthropometric and kinematic data. A problem arises when trying to choose one that will meet all requirements; cost, portability, accuracy, speed of data collection, etc. Table 3 presents a comparison of the general attributes of the five types of measurement systems discussed above, giving all criteria equal weighting.

Table 3. Results of the comparative analysis of the five measurement systems.

System #	Evaluation Criteria										Total
	1	2a	2b	2c	3	4	5	6	7	8	
1	3	3	0	3	3	3	0	0	0	0	15
2	1	2	2	2	2	3	1	2	3	1	19
3	3	3	1	1	2	3	3	2	3	3	24
4	3	3	2	1	1	3	3	3	3	3	25
5	1	2	3	0	0	3	2	3	3	2	19

Systems: 1 = Manual/Mechanical 2 = Electronic/electro-mechanical 3 = Film 4 = Video 5 = Opto-electronic

The score totals given in Table 3 rank Video and Film measurement techniques first and second respectively. Opto-electronic and electronic/electro-mechanical tied for third. As would be expected, manual/mechanical techniques had the lowest score. Taking into account the general attributes of any system is only useful for general comparisons of systems as a whole. When it comes time to choose a system for a particular application, one must evaluate the specific attributes required from a system and any constraints under which the selection of a system must be made.

If, for example, the cost of the system is not a problem and the system is to be used under only controlled laboratory conditions then obviously the best solution would be to purchase an advanced computerized video or opto-electronic system. A video system being preferred due to the visual record captured during data collection. However, if the system is going to be used in a laboratory setting and cost is a major factor in the decision of which type of system to choose, then a lower cost video/film system or perhaps an electro-mechanical system may be selected.

CONCLUSION

For the ergonomist working in the field, selecting a system to collect anthropometric and kinematic information is more difficult. The system must be highly portable and be relatively insensitive to environmental changes. Film based data collection has only been used to a limited extent because of the time required to reduce the data from the visual images. This has left the field ergonomist with little recourse other than to use manual collection methods. However, video systems (as shown in Table 3) are now able to provide the ergonomist with portable and computer-aided data collection systems. This trend towards video based measurement systems is likely to continue in the future as smaller, more advanced cameras and better computer/video interfaces are developed.

REFERENCES

Abraham, L.D., 1987, An inexpensive technique for digitizing spatial coordinates from videotape. In: Biomechanics X-B, ed. by Jonsson, B. (Champaign: Human Kinetics Publishers Inc.), pp. 1107-1110.

Andrade, R.S., and Matias, L., 1985, An Apparatus for the Determination of Reach Envelopes. In: Proceedings of the 9th International Congress of the International Ergonomics Association, pp. 34-36.

Atha, J., 1984, Current techniques for measuring motion. Applied Ergonomics, 15.4, 245-257.

Ayoub, M.M., 1972, Human movement recording for biomechanical analysis. International Journal of Production Research, 10.1, 35-31.

Baleshta, M.M., Fraser, T.M., and Patla, A., 1985, A movement notation system for ergonomic analyses: Background and Development. In: Proceedings of the 18th Annual Meeting of the Human Factors Association of Canada, Hull, Quebec, pp. 107-110.

Brand, R.A., and Crowninshield, D., 1981, Comment on Criteria for Patient Evaluation Tools. Journal of Biomechanics, 14.9, 655.

Chaffin, D.B., 1982, Functional assessment for heavy physical labour. In: Clinical Medicine for the Occupational Physician, ed. by M.H. Alderman and M.J. Hanley, (New York: Decker), pp. 187-192.

Chaffin, D.B., and Andersson, G., 1984, Occupational Biomechanics, (New York: Wiley-Interscience, John Wiley and Sons).

Chao, E.Y., 1976, Experimental methods for biomechanical measure -ments of joint kinematics. In: CRC Handbook for Engineering in Medicine and Biology, Vol. 1, ed. by Fleming, D.G., and Feinberg, B.N., (Cleveland: CRC Press Inc.), pp. 385-411.

Corlett, E.N., Madeley, S.J., and Manenica, I., 1979, Posture Targetting: A technique for recording working postures. Ergonomics, 22.3, 357-366.

Das, B., Kozey, J., and Tyson, J., 1989, Development of a Computer -Aided Potentiometric System for Anthropometric Measurements. In: Proceedings of the 22nd annual meeting of the Human Factors Association of Canada, Toronto, Ontario, pp. 261-265.

DeGreve, T.B., 1987, A Video Digitization Technique for Bio -mechanical Modelling. In: Trends in Ergonomics / Human Factors IV, ed. by Asfour, S.S., (Elsevier Science Publishers B.V.), pp. 595-602.

Dempsey, C.A., 1953, Development of a workspace measuring device. WADC, TR 53-53, Wright-Patterson Air Force Base, Dayton, Ohio.

Faulkner, T.W., and Day, R.A., 1970, The Maximum Functional Reach for the Female Operator. AIIE Transactions, 2.2.

Holzhausen, K.P., 1978, A contribution for improved optoelectronic measurement and immediate inspection of spatial human movements. In: Proceedings, 1st International Congress on Mechanics in Medicine and Biology.

Holzhausen, K.P., 1981, Analysis of human movement for workplace design. In: Manned Systems Design: Methods, Equipment, and Applications, ed. by I. Moraal and K.F. Kraiss, (New York: Plenum Press), pp. 337-362.

Hennig, E.M., Hellmann, H.J., and Binder, K., 1981, Ultrasonic devices for kinemetric movement analysis. In: Biomechanics VII-A, ed. by Morecki, A., Fidelus, K., Kedzior, K., and Wit, A., (Baltimore: University Park Press), pp. 483-488.

Macellari, V., Rossi, M., Bugarini, M., 1985, Human motion monitoring using the CoSTEL system with reflective markers. In: Biomechanics IX-B, ed. by Winter, D., Norman, R.W., Wells, R.P., Hayes, K., and Patla, A., (Champaign: Human Kinetics Publishers), pp. 260-264.

Morris, J.R.W., 1973, Accelerometry - A technique for the measurement of human body movements. Journal of Biomechanics, 6, 729-736.

Pynsent, P.B., Fairbank, J.C.T., Clark, F.J., and Philips, H., 1983, Computer recording of anatomical points in three dimensional space. Journal of Biomedical Engineering, 5.4, 137-140.

Shapiro, R., Blow, C., and Rash, G., 1987, Video Digitizing Analysis System. International Journal of Sport Biomechanics, 3, 80-86.

Tello, E.R., 1988, Between Man and Machine: New advances in user -interface technology could change the way we interact with our computers. Byte, 13.9, 288-293.

Vaughan, C.L., Smith, D.C., and du Toit, L.L., 1987, Kinematic gait analysis using a digital camera. In: Biomechanics X-B, ed. by Jonsson, B., (Champaign: Human Kinetics Publishers Inc.), pp. 1041-1047.

Wangenheim, M., and Samuelson, B., 1987, Automatic ergonomic work analysis. Applied Ergonomics, 18.1, 9-15.

Wells, R.P., Moore, A., and Ranney, D.A., 1988, Development of a System for Recording Occupational Hand and Wrist Movement. In: Proceedings of the 21st Annual Conference of the Human Factors Association of Canada, pp. 101-103.

Woltring, H.J., and Marsolais, E.B., 1980, Optoelectric (Selspot) gait measurement in two- and three-dimensional space - A preliminary report. Bulletin of Prosthetics Research, BPR-10, 17.2, 46-52.

Wright, I.B., 1963, Applications of a system of functional anthropometry in pressure suit designs. Journal of the British Interplanetary Society, 19, 1963-64

Advances in Industrial Ergonomics and Safety II
Edited by Biman Das
Taylor & Francis 1990

APPLICATIONS OF ANTHROPOMETRIC DATA IN SIZING AND DESIGN

James F. Annis and John T. McConville

Anthropology Research Project, Inc.
503 Xenia Avenue
Yellow Springs, Ohio 45387

Today's designers of clothing, personal protective equipment, and workspaces face more challenges than ever before. The modern work force is composed of a greater percentage of women and a more diverse mix of races and ages than at any time in history. Meanwhile demands of comfort, efficiency and safety in an increasingly technological workplace require ever closer design tolerances.

Old folk norms were once sufficient guidelines for the construction of workspaces and implements. Table height, for example, was set at nine fists high; the required place width at the table was four spans (the distance between the out-stretched thumb and little finger) for a man, and three spans for a woman. During the Middle Ages the norms for stairs were established by assuming that the length of one step (the distance between the successive heel contacts) was equal to two feet, and that the effort to lift the body vertically was twice that for horizontal displacement. On this basis, the rise was made equal to half the tread, and the sum of the tread and two risers was equal to one step (Drillis, 1963). In the twentieth century, horses are still measured in hands, and people weighed in stones. Tailors still assume that the body is equal to eight times head height and each body segment length is equivalent to head height units.

These quaint units of measurement and rules of thumb are likely to bring smiles to the faces of today's designers and ergonomists who have come to depend on more scientific tools such as tables of anthropometric data which include means, standard deviations, and percentiles. The fact is that insufficiently documented statistics and common misconceptions about their use can make these modern guidelines little better than the primitive methods employed in the past.

The "Average" Person

Probably the most frequent solution to a design problem is the use of some average value or combination of average values provided by most tables of anthropometric information. Fortunately, most anthropometric dimensions are normally distributed; hence the mean lies near the center of the most densely populated portion of the range of variability exhibited by a particular dimension. So, if your design problem requires that you select the most descriptive single value for a given measurement, the chances are that the mean is the one to use. At least the design will be suitable for all of those individuals who are more or less average for that dimension. If, on the other hand, your design problem is more complex and several dimensions are needed for its solution the resultant product may accommodate few individuals indeed. Buried, like a landmine, in the concept of the "average person" is the idea that if one is average in height, for example, he or she is average in all other dimensions. This, of course, is far from true.

In order to explore the idea of the "average man" Daniels and Churchill (Daniels, 1952; McConville and Churchill, 1976) used ten dimensions common to clothing design. Using a sample of 4,063 male Air Force flyers, and defining average as the mean ± 0.3

standard deviations (SD), they found that 1,055 (25.97%) men were of average stature. These individuals were then examined to see how many of the subsample met the average criterion successively for nine additional dimensions. More subjects dropped out with each dimension until, by the tenth dimension, there were no individuals left. In sum, there was not a single subject of the 4,063 in the sample who could be considered average for all 10 dimensions commonly associated with clothing design.

In order to demonstrate this concept on a modern sample we duplicated the Daniels and Churchill experiment on the most up-to-date anthropometric data base currently available -- a 1988 survey of U.S. Army personnel (ANSUR '88) (Clauser et al., 1988; Gordon et al., 1989; Gordon et al., in press). In order to reduce racial variability to a minimum, we used a sample of 2,074 White men and 1,438 White women. The results are summarized in Table 1.

Table 1. The "Average"* Man and Woman
(Values in mm)

	Men			Women		
Dimension	Mean Range	N**	%**	Mean Range	N**	%**
Stature	1745-1784	486	23.4	1614-1652	309	21.5
Chest Circ.	978-1018	115	5.5	891-930	86	6.0
Sleeve Length	876-898	60	2.9	789-809	35	2.4
Crotch Height	823-849	28	1.4	749-773	21	1.5
Vert. Trunk Circ.	1624-1668	18	0.9	1527-1566	14	1.0
Hip Circ.	968-1004	10	0.5	951-987	6	0.4
Neck Circ.	374-385	3	<0.1	310-319	3	0.2
Waist Circ.	849-899	1	0.1	771-821	2	0.1
Thigh Circ.	580-608	0	0.0	563-590	2	0.1
Crotch Length	632-656	0	0.0	603-636	1	<0.1

* Average defined as the mean $\pm$ 0.3 SD for each dimension.
** Number and percentage of original sample remaining.

As can be seen in the table, there are no "average" males at all beyond eight dimensions. The females, although a smaller sample, tend to be more variable, hence one individual still remains after the tenth round. The picture would look somewhat different if different variables were used, but not appreciably so with any array of dimensions sufficiently variable to represent both linear and mass characteristics of the human body. The fact is, that however tempting on paper, the "average" person simply does not exist. This is even more alarming considering that the range of variability of individuals surrounding the mean is much less than the variability found among individual lying on the outer aspects of the distribution.

Percentile Models

In order to design clothing, protective equipment and workspaces which must accommodate 90-95% of a given user population, many designers have turned to percentile models. The classic manikin or computer model "family" is typically composed of 5th, 50th, and 95th percentile analogs. The serious difficulty here is that body percentile values for segments are not additive. In other words, if one were to assemble body parts which were all of 95th percentile rank in length and weight, the final form would be a great deal taller and heavier than actual 95th percentile stature and weight. The opposite pertains when 5th percentile values are assembled (Robinette and McConville, 1981).

The extent of the error accumulated at the 5th and 95th percentile levels when adjacent body segments are joined is demonstrated in Table 2 and Figure 1. Once again the ANSUR data has been used.

Table 2. Percentile Models: 5th & 95th Males and Females*
(Values in cm)

Segment from - to	5th Percentile Male	5th Percentile Female	95th Percentile Male	95th Percentile Female
1. Floor to Lat. Malleolus	5.84	5.23	7.64	6.97
2. Lat. Malleolus to Lat. Femoral Epicondyle	39.63	36.28	47.62	44.05
3. Lat. Femoral Epicondyle to Greater Trochanter	38.67	36.27	46.90	43.97
4. Greater Trochanter to Iliac Crest	11.97	10.22	17.08	15.24
5. Iliac Crest to Tenth Rib	2.39	3.22	7.52	8.10
6. Tenth Rib to Seventh Cervical	36.21	33.02	43.43	39.79
7. Seventh Cervical to Tragus of Ear	8.67	8.04	12.35	11.56
8. Tragus of Ear to Top of Head	12.18	11.42	14.04	13.25
Total For Segments	155.56	143.70	196.58	182.93
Percentile Stature	164.69	152.78	186.65	173.73
Difference	-9.13	-9.08	+9.93	+9.20

* The data presented here were derived from Gordon, et al., 1989
The N for the females = 2208 and the N for the males = 1774.

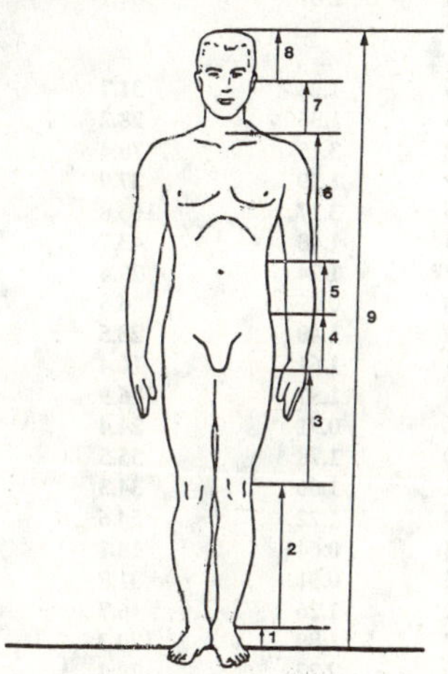

Figure 1. Body Height Segments Used in Table 2.

The practical consequences of constructing percentile models are legion. One example suffices. A 1985 Air Force study of commercially available test dummies (ARP Final Report AF Contract F33615-82-C-0510, 1985, unpublished) revealed a number of anomalies in test devices purporting to represent the human body. One such 50th percentile device has a chest breadth ranking in the 25th percentile, and chest depth ranking in the 97th percentile of the USAF distribution. The most widely used 50th percentile dummy in existence is several inches shorter than both its military and civilian counterparts. Its eye height, sitting ranks at the 55th percentile, and its sitting height at the 21st percentile of USAF values; as a consequence, the distance between the eyes and the top of the head is below the 1st percentile for all surveys of U.S. males. It is almost inevitable that similar anomalies exist in any other test dummies based on percentile values, particularly those representing "small" and "large" individuals.

Sexual Dimorphism

It is a common belief among designers that for all practical design purposes women are scaled down versions of men with a body size proportionality of approximately 93% (F/M) on the average. Alternatively, it is said that a 50th percentile woman is equivalent in size to a 5th percentile man and that a 95th percentile women is equal in size to a 50th percentile man.

A 1979 study (Robinette, Churchill, and McConville) found that while male height and weight dimensions can be scaled down with some reliability, many other dimensions, particularly those involving body tissues and dimensions of the hands, feet and head cannot.

Turning once more to the ANSUR sample, a comparison of men and women matched closely in height and weight reveals the similarities and differences for selected dimensions. (See Table 3) It can be seen, for example, that for buttock circumference and hip breadth, women are scaled-up versions of men.

Table 3. Mean Values and Female Percentage of Male Values for Selected Dimensions in Men and Women Matched* in Height and Weight (Values in cm)

	Men (n = 91)		Women (n = 153)		
	Mean	SD	Mean	SD	% (F/M x 100)
Age	25.6	0.67	25.7	0.57	
Stature*	168.0	1.50	167.2	1.34	
Weight*	65.3	2.67	64.7	2.79	
Neck Circumference	36.5	1.28	31.7	1.09	86.8
Biceps Circumference	31.6	1.36	28.2	1.50	89.2
Chest Below Bust Circumference	87.2	3.15	78.4	3.11	89.9
Scye Circumference	41.9	1.59	37.9	1.42	90.5
Shoulder Circumference	111.6	3.17	103.8	2.95	93.0
Bideltoid Breadth	46.5	1.48	43.7	1.50	93.8
Axillary Arm Circumference	31.3	1.24	29.4	1.42	93.9
Bimalleolar Breadth	6.9	0.29	6.5	0.26	94.2
Chest Breadth	30.0	1.40	28.5	1.35	95.0
Waist to Hip Length	17.1	1.61	16.4	2.00	95.9
Biacromial Breadth	38.4	1.37	36.9	1.32	96.1
Radiale-Stylion Length	25.2	0.91	24.4	0.77	96.8
Sleeve Outseam	57.0	1.78	55.5	1.56	97.4
Shoulder-Elbow Length	35.1	1.06	34.3	1.03	97.7
Head Circumference	55.9	1.22	54.6	1.16	97.7
Hand Length	18.4	0.64	18.1	0.56	98.4
Acromion-Radiale Length	32.3	0.91	31.8	1.03	98.5
Knee Height	47.2	1.26	46.7	1.25	98.9
Chest Height	121.3	1.89	120.3	2.31	99.2
Crotch Height	79.0	2.22	78.4	1.75	99.2
Cervicale Height	194.5	1.41	194.4	1.46	99.9
Acromial Height	136.8	1.80	136.8	1.81	100.0
Shoulder Length	14.6	1.06	14.6	1.00	100.0
Waist Height, Navel	100.3	2.27	100.7	2.00	100.4
Calf Circumference	35.6	1.47	35.9	1.58	100.8
Buttock Height	83.6	2.10	84.8	1.97	101.4
Waist Circumference	79.0	4.01	80.4	5.51	101.8
Waist Breadth	28.7	1.29	29.8	1.91	103.8
Buttock Circumference	91.8	2.37	99.1	3.58	108.0
Hip Breadth	32.2	0.93	35.5	1.51	110.2

* Individuals matched within ± 5 kg. in weight and ± 2.5 cm in stature.

For the designer and model-maker this means that if variables such as knee height or shoulder length are the key dimensions in the design of a workspace to be occupied by both sexes, then any one of the methods used to represent female dimensions as some fraction of the male dimensions might work reasonably well. On the other hand, designers concerned with seat width or harness assemblies, for example, should view the more sexually dimorphic variables with great wariness.

Other Sources of Variation

Sexual dimorphism is by no means the only source of anthropometric variation of which the designer should be aware. Race/ethnicity, age, and even occupations can have significant effects on an anthropometric data base.

A number of analyses of military data have shown that, for men at least, there are significant demonstrable differences in both size and proportionality between races (Bradtmiller, et al., 1985; Long and Churchill, 1968). For individuals of a given stature, race will significantly affect leg length. For example, Whites, on the average have shorter legs (and arms) than Blacks, and longer extremities than Asians. Differences in facial morphology are well known but seldom well-documented in tables available to designers of protective masks and helmets. While less is known about women, as yet, it is likely that race is at least as significant a factor on size as it is among men.

Weight and other mass-related variables, e.g., girths, have been shown to increase significantly with age until well into the fifties while body heights start to decline at an accelerating rate after the age of 65 or so. More often than not, sources of anthropometric data are based on studies of young military personnel or test subjects drawn from the ranks of college and university-age students.

Finally, it is important to recognize that spoken or unspoken selective criteria affect the anthropometric distributions of certain occupational groups. Law enforcement officers have been shown, for example, to be bigger in most dimensions than virtually any other group measured (Martin, et al., 1975). Airline stewardesses, measured in 1975 were taller and lighter than average American women (Reynolds and Allgood, 1975). Users of anthropometric data obtained from military sources can be led astray if they do not understand the likely difference between data obtained, for example, from a flying population and that measured on recruits.

Any or all of these sources of variations may well catch the unsuspecting user of anthropometric data unaware. The comparison of means for comparable variables can show differences in groups, but most anthropometric dimensions are multiple covariates making interpretation difficult. The use of ratios to normalize data does not totally overcome the influence of the denominator variable.

It is understandable, however, that better approaches to solution of variation problems such as sexual and racial dimorphism, through one use of multiple regression analysis is not always an available option to the designer who must rely on tables of univariate statistics. To add to the problem, frequently the tables to which the designer has access do not reference the source of the information. From what population was the sample drawn? How big was the sample? Was it military in origin? Was it largely White?

CONCLUSION

The cautionary tone of this paper should in no way be construed to mean that designing workspaces and equipment for use by both men and women of all ages and races presents insoluble problems. Rather, we want to emphasize that the simplistic plucking of a value or values from a convenient handbook may be of little more value than the rules of thumb that are stated in various folk norms and in some instances will be down right wrong. It is therefore, incumbent on the designer to ascertain who is being designed for and then locate the data base that best characterizes the body size variability of that population. With this in hand the appropriate design value necessary to accommodate the body size variability of the population can usually be determined.

There are a number of textbooks available which present good tables of anthropometric data; however, the best single source of univariate statistical information for a wide variety of dimensions and samples from around the world is the Anthropometric Source Book, Volume II: A Handbook of Anthropometric Data (NASA, 1978). Unfortunately, we still do not have available a good variety of applied workspace and clothing design dimensions for the U.S. civilian work force.

REFERENCES

Anthropology Research Project, Inc., 1985, Final Report Air Force Contract F33615-82-C-510, (Unpublished).

Bradtmiller, B., Ratnaparkhi, J., Tebbetts, I., 1985, Demographic and Anthropometric Assessment of U.S.Army Anthropometric Data Base. Technical Report NATICK/TR-86-004, United States Army Natick Research, Development and Engineering Center, Natick, Massachusetts.

Drillis, R.J., 1963, Folk Norms and Biomechanics. Human Factors, 5, 427-441.

Daniels, G.S., 1952, The "Average Man"?. Technical Note WCRD 53-7, Wright Air Development Center, Wright-Patterson Air Force Base, Ohio.

Clauser, C., Tebbetts, I., Bradtmiller, B., McConville, J., Gordon, C., 1988, Measurer's Handbook: U.S. Army Anthropometric Survey 1987-1988. Technical Report NATICK/TR-88/043, United States Army Natick Research, Development and Engineering Center, Natick, Massachusetts.

Gordon, C., et al., 1989, 1988 Anthropometric Survey of U.S. Army Personnel: Summary Statistics Interim Report. Technical Report NATICK/TR-89/027, United States Army Natick Research, Development and Engineering Center, Natick, Massachusetts.

Gordon, C., et al., 1989, 1987-1988 Anthropometric Survey of U.S. Army Personnel: Methods and Summary Statistics. (In press).

Long, L. and E. Churchill, 1968, Anthropometry of USAF Male Basic Trainees - Contrasts of Several Subgroups. Paper presented at the 1968 meeting of the American Association of Physical Anthropologists, (Unpublished).

Martin, J.I., et al., 1975, Anthropometry of Law Enforcement Officers. NELC/TD 442, Naval Electronics Laboratory Center, San Diego, California.

McConville, J.T. and Churchill, E., 1976, Statistical Concepts in Design. AMRL-TR-76-29, Aerospace Medical Research Laboratory, Wright-Patterson Air Force Base, Ohio.

National Aeronautical Space Administration, 1978, Anthropometric Source Book Volume II: A Handbook of Anthropometric Data. (NASA Reference Publication 1024), Lyndon B. Johnson Space Center, Houston, Texas.

Reynolds, H. and Allgood, M., 1975, Functional Strength of Commercial Airline Stewardesses. FAA-AM-75-13, Department of Transportation, FAA, Washington D.C.

Robinette, K. and McConville, J.T., 1981, An Alternative to Percentile Models. 810217, Society of Automotive Engineers, Inc., Warrendale, Pennsylvania.

Robinette, K., Churchill, T., McConville, J.T., 1979, A Comparison of Male and Female Body Sizes and Proportions. AMRL-TR-79-69, Aerospace Medical Research Laboratory, Wright-Patterson Air Force Base, Ohio.

Advances in Industrial Ergonomics and Safety II
Edited by Biman Das
Taylor & Francis 1990

PRE-EMPLOYMENT ANTHROPOMETRIC PROFILING

Stephen W. Meagher, M.D.

520 Commonwealth Avenue, Boston,
Massachusetts, 02215, U.S.A.

It is necessary to implement the use of key anthropometric measurements during pre-employment examinations to avoid mismatches between new employees and selected assigned tasks that require large hand, arm and height dimensions. The preventive application of anthropometrics can significantly augment the remedial use of anthropometrics that is presently employed. The disparities between small and large hand, arm and height dimensions are presented.

INTRODUCTION

In an industrial setting repetitive trauma disorders are commonly thought to be caused by defective work station design, tool design and incorrect work performance movements. It should be evident to all that an anthropometric mismatch exists in many of these incidents.

Is the cause of a disorder the shelf height or the stature and/or the arm length of the worker? Is the grasping hand too small for the assigned task? Is the size of the stock to be processed or packed manually too large or is the worker's hand length or finger abduction not great enough to perform without intrinsic muscle strain, tendinitis or myositis?

There is a panic abroad in industry that sends ergonomists racing from one plant to another to put out the flare-ups of repetitive trauma disorders in a remedial fashion by altering tools, work-station designs and work performance movements.

Is not prevention less expensive than remedial action? Why should significant sums be spent to correct problems that can be avoided simply by using anthropometric profiling for selected tasks and, simultaneously, protect prospective employees from repetitive trauma disorders.

In today's setting the personnel office can and does assure that a good medical history, a passing physical examination, a qualifying educational background and work experience fulfill requirements for new hires. Herein lies the rub. Mismatches are given assigned tasks that their anthropometric profiles cannot fulfill without exposing themselves to an increased risk of repetitive trauma disorders.

When managers, supervisors and personnel employees are enlightened by basic anthropometric training they will understand the need to match hand dimensions, arm length and stature to selected tasks. If seminars are used as a teaching tool this can be accomplished.

Table 1. Comparative Measurements for Small Woman vs Large Man Anthropometrics: A Basis for Pre-employment Anthropometric Profiling for Selected Tasks (inches)

HAND LENGTH (White 1980)	5TH PERCENTILE	95TH PERCENTILE	DIFFERENCE Small Woman vs Large Man
6682 U.S. Army Men 1966	6.90	8.13	1.81
1331 U.S. Army Women 1977	6.32	7.50	
PALM LENGTH (White 1980)			
6682 U.S. Army Men 1966	3.77	4.59	1.03
1331 U.S. Army Women 1977	3.56	4.24	
HAND BREADTH (White 1980)			
6682 U.S. Army Men 1966	3.20	3.83	1.00
1331 U.S. Army Women 1977	2.83	3.33	
HAND CIRCUMFERENCE (White 1980)			
6682 U.S. Army Men 1966	7.81	9.28	2.57
1331 U.S. Army Women 1977	6.71	7.82	
ARM REACH WALL TO TIP OF MIDDLE FINGER (Webb 1978)			
2959 Air Force Cadets 1942	32.7	37.80	8.70
422 Stewardesses 1971	29.10	33.30	
VERTICAL GRIP REACH (Webb 1978)			
792 French Army (no date)	75.0	86.60	13.30
1905 Air Force Women 1968	72.90	84.00	
SPAN (Diffrient et al. 1981)			
	2.5 PERCENTILE	97.5 PERCENTILE	
Large Man	65.4	76.7	16.1
Small Woman	60.6	70.9	

In Table 1 discrepancies between small and large anthropometric dimensions are given to emphasize the basis for causation of repetitive trauma disorders by the common use of large stock sizes, tools and work stations by widely different anthropometric profiles. Small women and large man measurements are presented to underscore the magnitude of the problem.

Hand Length

There is a direct relationship between stock size and hand length required for stock manipulation.

Palm Length

As a component of hand length, palm length is significant in handling large stock sizes.

Hand Breadth

Tool handles and stock sizes can present problems for the narrow hand and result in intrinsic muscle strains.

Finger Abduction

No series of finger abduction (fanning) measurements is available. It is an important measurement because of the relationship of finger abduction to the manipulation of large stock sizes. If the stock is too large for a maximum degree of fanning, strains of intrinsic muscles in the hand will occur from the application of force when the muscles are fully extended.

Hand Circumference

A circumferential measurement around the body of the hand at the level of the most prominent muscle contour in the first web space and the most prominent muscle contour of the medial border of the hand provides a measure of bone and muscle mass in the hand. No such series exists at present. It is needed because of the direct correlation between muscle mass and strength. Force intensive tasks performed by weak hands produce repetitive trauma disorders.

Arm Reach Wall to Tip of Middle Finger

The location of stock delivery, tool storage at a work station and stock size are vitally related to arm length. The correct location for one worker may stress tissue elements in the arm of another in the next working shift when required to perform the same task at the same work station with anthropometric dimensions that are significantly less.

Vertical Grip Reach

Reaching for suspended power tools and stacking on shelves and pallets has been and continues to be an obvious cause of repetitive trauma disorders when anthropometric mismatching requires overreaching resulting in thoracic outlet syndromes, shoulder strains and other disorders.

Span

The large difference between the span of small woman and large man underscores the magnitude of the anthropometric disadvantage for the small woman in dealing with large containers, broad pieces of stock and packaging tasks performed by large man.

CONCLUSION

Pre-employment anthropometric profiling for selected tasks is needed to implement a preventive approach to repetitive trauma disorders and augment the current practice of employing a remedial ergonomic approach to outbreaks of repetitive trauma disorders in industry.

REFERENCES

Diffrient, N., Tilley, A. and Harmon, D. 1981, Humanscale, (Cambridge: MIT Press) Rotary dial 7a.

Webb Associates, 1978, Anthropometric Source Book, Vol. 2: A Handbook of Anthropometric Data, NASA Reference Publication 1024 National Technical Information Service. Springfield, VA, pp. 111, p. 391.

White, R., 1980, Comparative arthropometry of the hand. Technical Report. Natick/TR-81/010 CEMEL-229, U.S. Army Natick Research and Development Laboratories, Natick, Massachusetts 01760, pp. 143-153.

VI

EQUIPMENT DESIGN AND ERGONOMICS

Advances in Industrial Ergonomics and Safety II
Edited by Biman Das
Taylor & Francis 1990

SOME MODIFICATIONS OF WORK CONDITIONS IMPOSED TO THE OPERATOR OF A ROD-RAISING MACHINE IN AN ALUMINIUM INDUSTRY : A CASE STUDY*

Guy Avon et Claude H. Nadeau

Département de kinanthropologie
Université du Québec à Montréal
Montréal, Canada. H3C 3P8

The rod-raising machine is a sort of grip-lift which is used to clean the anode rods and to fix them to the electrolyse vats. The machine consists of a motor-driven vehicle and a partially automatic tool attached to the former. After an examination of this prototype vehicle, two categories of constraints appeared : 1) the operator is seated obliquely in regard to his work field, a position which causes back pain and restricts vision; 2) the placement and design of control panel and joysticks do not optimize the information input, memorization of operations and manual efficiency. To remedy those problems, four modifications of the interior arrangement of the cab were recommended.

INTRODUCTION

In the Alcan smelter plant, the electrolytic process for manufacturing aluminium involves the use of several hundred vats containing minerals in fusion at temperatures close to 1000 °C. Each vat contains three banks of thirteen anode rods that have to be periodically removed, cleaned and put back in a different position in the bank. Until now, this was done by operators who had to bring next to each vat a movable pneumatically-driven rod-raising machine which grips the anode rod, unscrews it, lifts it out of the vat, cleans it and puts it back. Alcan wanted to offer better protection for the operator of the machine, while increasing worker throughput. An engineering firm was thus asked to design a two-component vehicle incorporating a controlled-environment operator's cabin where commands for moving the vehicle and manipulating the rod-raising machine are located. When the prototype work vehicle was first tested, however, it was found to be somewhat unwieldy and uncomfortable for the operator.

Figuring that the problem lay with the design of the chair, Alcan contacted the authors asking for an ergonomic analysis of the workchair. Preliminary analysis of the problem, however, led the authors to recommend that not only the chair but the entire operator's work station be reexamined. Their recommendation was accepted.

---

*A study sponsored by La Société d'électrolyse et de chimie Alcan ltée.

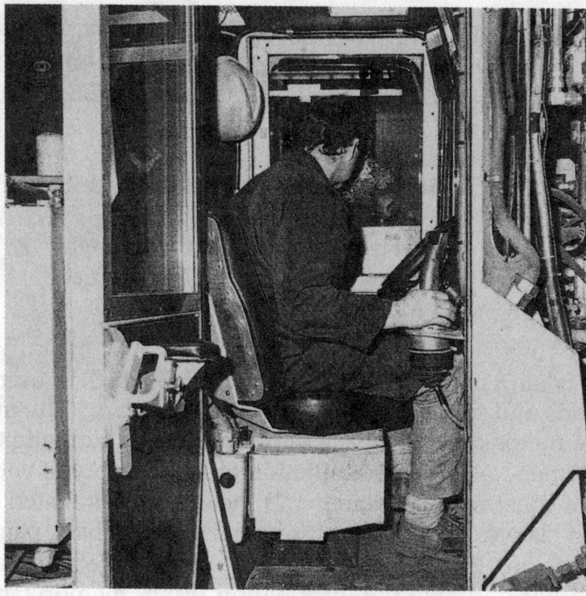

Figure 1. An operator working in the prototype machine.

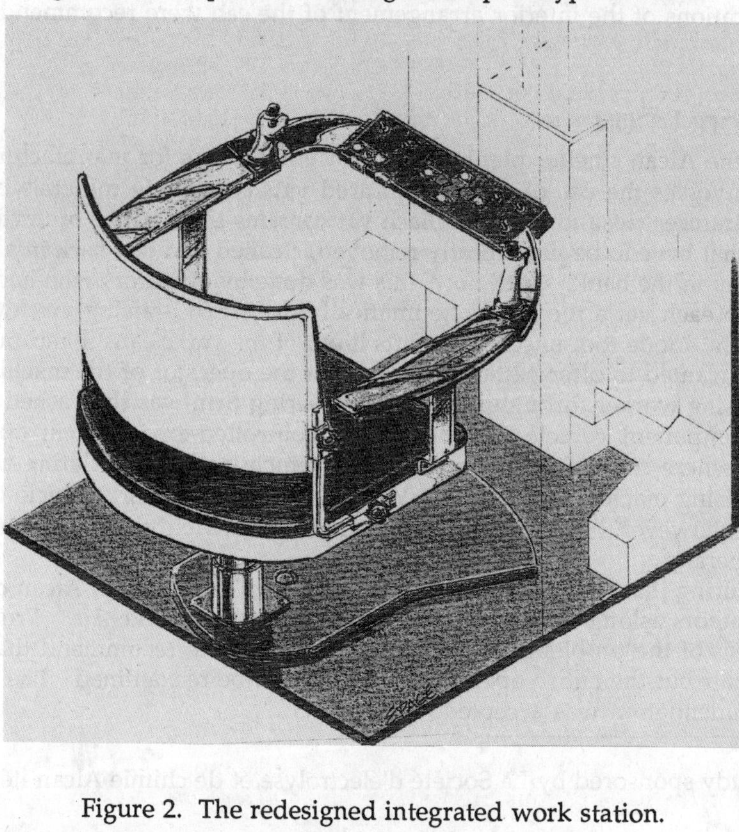

Figure 2. The redesigned integrated work station.

ANALYSIS

Identification of the problems involved systematically observing the operator while he was moving the vehicle and manipulating the rod-raising machine, analyzing the interior arrangement of the cabin, and gathering information by means of a run-through of certain sub-tasks. Further data came from an examination of photographs of the prototype unit in actual operation (Fig. 1). We recognized two categories of problems : some having to do with postural adjustments of the operator, and others with perceptual-motor control of the task.

Postural adjustments

Even a cursory examination of the work situation brings the realization that, while maneuvering his vehicle or controlling the rod-raising machine itself, the operator has to assume an angled position with reference to his work area. Because the maximum angle of rotation of the workchair on its axis is only half the 90 degrees that would be necessary to bring him directly in front of his work (Fig. 1), the operator has to spend several minutes with his back and his neck twisted to the left in order to maintain a suitable view of his objective.

Operation of the machine for a full day under typical conditions would thus require the operator to take and to maintain this awkward position for a total of several hours. One can readily infer that this would reduce his effectiveness, but it could also lead, in time, to the development of health problems. It has been well established (Zacharkow, 1988; Dupuis and Leclaire, 1986) that a repeated and sustained unilateral torsion of the trunk and neck leads to multiple organic and functional troubles, especially back pain.

That problem was aggravated by the fact that the two control joysticks were badly positioned. In order to compensate for the limited movement of the back that keeping the joysticks in hand afforded him, the operator had to turn his head even more to the left, thus increasing tension in his neck. This position obviously generates stress in several muscular groups, leading to fatigue and pain.

Perceptual-motor control

The placement of information-gathering and command apparatus in the machine's cabin was not conducive to optimal perceptual-motor control of the task. While operating the rod-raising machine, the worker relies mostly on visual cues, but good kinesthetic and touch control are also required, since the task requires moving the eyes from one area to another, coordinating movements of the eyes and the head, and using fingers and wrists to position the tool.

When actually operating the rod-raising machine, the worker is not directly facing the vat. In that position, the following problems obtain :

- the operator does not have a direct facing view of the tool that grips the end of the anode rods;
- the control panel lies outside of the operator's binocular field of vision;
- the operator must strain his head to the side and keep his eyes off-centre towards the left : this precludes optimal use of the oculo-motor range in eye and coordinated head-eyes movements and of binocular cues necessary

for depth perception (Delorme, 1985); it probably also introduces a sensori-tonic bias (Nadeau et al., 1984) in perception of the vertical;
- the placement of the switches and readouts on the control panel does not afford the operator a logical and simple cognitive representation of the machine's functions while executing his task (Corlett, 1988);
- the two control sticks are much too big for the operator to be able to hold them naturally and comfortably in order to perform a task which requires accuracy rather than speed (Fitts and Posner, 1967);
- some functions, which should be controlled through the joysticks because they are part of the operator's main task, cannot be so initiated;
- some frequently-used switches and the emergency brake are difficult to reach.

RECOMMENDED CHANGES

The main objective of the study was to eliminate or at least significantly alleviate conditions likely to impair the operator's health or well-being. We secondarily tried to increase the efficiency of the man-machine system by eliminating possible sources of errors, by altering the sequence of operations when advisable, by decreasing movement times, and by assisting the operator's cognitive functioning.

To accomplish those objectives, it appeared to us that the first principle to implement was that *the operator must at all times be directly facing and in full view of his work area and tools*. We thus devised a fully pivoting work control system which allowed the operator to face to the front when moving the vehicle between the vats, and to face sideways when maneuvering the rod-raising machine (Fig. 2). Since the joysticks and the essential readout panels also had to be made movable, we arranged for them to move in coordination with the chair.

Platform

In order for the chair to move as required, there were two possibilities : either design a rotating combination chair and control module, or secure everything to a rotating platform. Since the cabin's dimensions provided limited space for the operator's feet, and since the post supporting a pivoting chair would be bulky if effectively placed in the rotation axis, we thus chose to use a revolving platform whose movements could be controlled from the operator's command devices.

The platform on which the seat rests has a 40,5 cm radius, making maximum use of the space available in the cabin. Although the chair needed to pivot only 90°, we had to give it 180° rotation in order for the operator to be able to take his position in the workchair when entering the cabin.

Workchair

The chair was designed to make use of all the space available in the cabin and comply with existing North-American standards (Diffrient et al., 1974). It is connected to the pivoting platform by a large-diameter post which contains a shock-absorber and a mechanism to raise and lower the seat. The chair's back can be adjusted in height and forwards and backwards to accommodate smaller workers. Although the work area is somewhat crowded, and the rods are not all at the same height in the vats, these functionalities afford the operator an unobstructed view of his work area at all times.

In order to alleviate back fatigue (Avon & Schmitt, 1975), we designed a unitized work table that completely surrounds the operator when he is sitting in his chair. The unit consists of two arm-rests connecting in front with the control panel; it contains the two joysticks and all the necessary readouts and switches (Fig. 2). The panel is hinged on one side to permit access to the chair when first entering the cabin.

Joysticks and control panel

In order for the worker to carry out the numerous operations required by his task, we had to design effective and easy to use control devices. We redesigned the joysticks that were provided in the prototype with a view to maximizing the operator's comfort. We gave them a different form, and we placed them at the end of the arm-rests, slightly inclined to the inside and to the front so that the operator's wrists would fall into a neutral semi-prone position while his hands rested naturally on the handles, affording full movement amplitude in both axes. Each of the joysticks controls 11 functions.

Some control devices have been placed under the arm-rests, in line with the joysticks so that they can readily be operated by the fingers. Push-plates that cover the width of the arm-rests are placed close in front of the joysticks and activate the emergency brake.

The original design of the machine did not include a genuine control panel. The panel we designed has the following characteristics : 1) it is slightly angled towards the operator, thus being easier to read; 2) the joysticks and the control panel are totally within the lower quadrants of the operator's binocular field of vision; 3) buttons and visible switches have been functionally grouped on the left and the right sides of the control panel, so as not to overload one hand in preference to the other; 4) the main readouts have been positioned in the center of the control panel, in that part of the binocular field of vision where colors can be most easily discriminated.

Steering wheel

Our design made it necessary to find a different place for the steering wheel, which is used when moving the vehicle between the vats. We modified its location, so that it would not interfere with the movement of the control module, and we placed it vertically close to the wall, which makes positioning of the rod-raising module easier when the operator is approaching the vat.

CONCLUSION

Oftentimes, a specific problem when correctly framed leads the researcher to a more encompassing analysis. Called in to suggest modifications to a workchair, we ended up realizing a complete analysis of the work station from the perspective of postural adjustments and perceptual-motor control.

Correct solutions must take into account several interdependent factors. We did a complete redesign of the work station in light of the operator's task, thus functionally integrating the chair, the joysticks, the control panel and the steering wheel.

This type of global approach to the design of a work station would seem applicable to other work situations. Our design should be easily adaptable to machines where an operator has to perform similar tasks.

REFERENCES

Avon, G. and Schmitt, L., 1975, Électromyographie du trapèze dans diverses positions de travail à la machine à écrire. *Ergonomics*, 18, 619-626.

Corlett, E.N., 1988, The investigation and evaluation of work and workplaces. *Ergonomics*, 31, 727-734.

Delorme, A., 1982, *Psychologie de la perception,* (Montréal : Études vivantes).

Diffrient, N. et al., 1974, *Humanscale,* Designed by Henry Dreyfuss Associates, (Cambridge, MA : The M.I.T. Press).

Dupuis, M. et Leclaire, R., 1986, *Pathologie médicale de l'appareil locomoteur,* (St-Hyacinthe, QC : Edisem / Masson).

Fitts, P.M. and Posner, M.T., 1967, *Human performance,* (Belmont, CA : Brooks-Cole).

Nadeau, C.H., Giguère, F. and Viel, M., 1984, Intra- and interhemispheric visual-motor integration : the coordination of eyes, head and arm movements in a pointing task. In : *Motor development, sport psychology and motor learning/motor control,* edited by Jan Borekhoff, (Eugene, OR : Publications of the University of Oregon), p. 20.

Zacharkow, D., 1988, *Posture : sitting, standing, chair design and exercise,* (Springfield, IL : Charles C. Thomas Publ.).

Advances in Industrial Ergonomics and Safety II
Edited by Biman Das
Taylor & Francis 1990

A MODEL FOR ANALYZING MINING MACHINE ILLUMINATION SYSTEMS

Richard L. Unger
Audrey F. Glowacki
E. William Rossi

U. S. Bureau of Mines
Pittsburgh Pennsylvania, USA

In 1988, there were 1445 injuries to operators of underground mining machinery in the United States. As part of its program to improve the safety of underground coal operations, the U.S. Bureau of Mines is developing a series of computer models for use in analyzing several of the ergonomic aspects of mining machine operator compartment design. When completed, the models' capabilities will include assessments of operator reach, visibility and illumination in the surrounding work area. The models are intended mainly as a research tool and will be used to assist in the development of prototype mining machinery. This paper focuses on the Illumination Analysis model.

INTRODUCTION

Due to the nature of underground mining, mine equipment is usually massive in size and built to withstand a rugged environment. Often the focus on productivity has taken precedence over the consideration of human operators in the design of equipment. This has resulted in many mining machines where visibility is severely limited and operators must lean outside the safe confines of their compartments in order to see. When lighting systems are provided, they may be positioned where they cause excessive glare, which restricts the visibility of those working on or around the machines. Controls are often unlabeled and placed where they are difficult to reach and distinguish from one another. In panic situations, the wrong control may be activated, which can result in accidents. The Mining Equipment Safety Laboratory of the Mine Safety and Health Administration (MSHA) analyzed all fatal accident reports involving underground coal mine mobile and electrical face equipment for the years 1972 through 1979. A total of 350 fatalities were investigated. Altogether, MSHA concluded that approximately 36% of the fatalities involving underground coal mine mobile equipment during this period were directly or indirectly caused by improperly designed operator compartments.

In order to help address this problem, the Human Factors Group of the U.S. Bureau of Mines is developing a series of models to aid in the analysis of certain human engineering aspects of mining equipment design. The models are intended mainly as research tools and will be used to assist in the development of prototype mining machinery. They may also find use by original equipment manufacturers and mining companies for the initial design work on new machines, as well as for evaluations of modifications to existing machines.

Work is ongoing on the development of models for assessments of operator reach, visibility and illumination in the surrounding work area. At the present time, the models are designed to work with Tektronix computer display terminals, hard copy units, and graphics tablets. They are written in Fortran and are implemented on a Digital MicroVAX II computer at the Bureau's Pittsburgh Research Center. This paper presents an overview of the work completed concerning the illumination model.

MINE ILLUMINATION STANDARDS

The purpose of this model is to provide mining equipment manufacturers, mine operators, MSHA, and the U.S. Bureau of Mines with a computerized method for evaluating mine illumination systems. This would allow modification of lighting configurations without resorting to the time consuming method of building physical mockups and taking light readings manually.

Adequate illumination is an essential ingredient for a safe working environment in industrial situations. The working areas of an underground coal mine are no exception. Most modern coal mining machines are designed primarily for high production. As a result, in most cases only a slight amount of pressure is needed on the operating controls to bring an instant and dramatic response: the cutter head raises, the conveyor boom swings, or the machine advances erratically at a relatively high speed. Usually, these machines occupy most of the available space in the underground work area, limiting the vertical and horizontal clearances for miners nearby. Consequently, many injuries have occurred when miners were crushed by machinery while working in these restricted areas. In poorly lighted areas, a miner's peripheral vision is adversely affected so that he may not see hazards as they are developing and react in time to avoid injury or death.

Other common hazards in underground mines are falls of roof and ribs, the leading cause of coal mine deaths. Falls of roof and rib are often preceded by warning signs, such as small fragments of rocks falling, roof supports taking weight, or a sudden reduction in the height of the working place. Miners often cannot detect these warning signs because their attention is focused on the small circle of light emitted from their cap lamps.

Congress recognized the existence of these hazards, and in the Federal Coal Mine Health and Safety Act of 1969, charged the Secretary of the Interior with the authority and responsibility for developing and prescribing minimum illumination standards for working places in coal mines. These standards became effective on July 1, 1978, as regulations under Section 75.1719 through 75.1719.4, Title 30, Code of Federal Regulations. These regulations specify (1) certain surfaces within a miner's normal field of vision that must be illuminated to 0.06 footlamberts (fL) while self-propelled mining equipment is being operated; (2) requirements for lighting fixtures and circuits; (3) methods of making light measurements; and (4) additional requirements intended to provide increased visibility for miners in working places (Lewis, 1986).

The 0.06-fL level is a measure of luminance, or photometric brightness. It is a product of the level of illumination (incident light, measured in footcandles, fc) impinging on a surface, and the reflectivity of the surface. Bureau research has shown that the 0.06-fL level is adequate for the major proportion of mining tasks. However, the level is low enough so that adaptation problems in going from illuminated to non-illuminated areas of the mine are not severe.

In attempting to meet these illumination standards, mine equipment designers often aggravate another mine lighting problem - glare. There are currently no standards related to glare in underground mining. The Code of Federal Regulations states only that designers should attempt to minimize glare when developing machine illumination systems.

There are two distinct types of glare important in underground mine lighting, disability glare and discomfort glare. Disability glare is defined as glare resulting in decreased visual performance and visibility. The cause is stray light which enters the eye and scatters within. This produces a veiling luminance over the retina, which has the effect of reducing the perceived contrast of the objects being viewed. Discomfort glare is a sensation of annoyance or pain caused by high or non-uniform distributions of brightness in the observer's field of view. Although these two types of glare are distinctly different phenomenon, they can accompany each other. Good illumination requires that both disability and discomfort glare be minimized to allow adequate operator visibility and visual comfort.

The presence of glare makes the task of providing illumination in underground coal mines a complex problem. A joint committee of the United Mine Workers of America, the Bituminous Coal Operators Association and MSHA conducted a workers opinion survey in 1979 on illumination systems in underground coal mines. It was found from the survey that most complaints related to the problem of vision impedance (glare) were due to the luminaires themselves. The light sources were often located close to the miners' lines of sight and were viewed against a dark background. This resulted in high contrast levels, which can lead to both disability and discomfort glare.

The problems of designing an effective mine lighting system for compliance with underground coal mine illumination requirements can be summarized as follows:

(1) Repeated calculations for a multitude of machine and lamp types, with the possible combinations running into the thousands, are too tedious and costly to be practical using manual methods. The result is that the designer is impeded significantly in solving for an optimal illumination system that minimizes glare.

(2) Manual methods are more likely to result in errors because of the complexity in dealing with many three dimensional coordinate systems and the computation of photometric values.

The Bureau of Mines' illumination model should help eliminate these problems by turning the computational portion of the lighting design task over to the computer. The model will allow the lighting designer to concentrate on adjusting the configuration of the illumination system to minimize the potential for glare while still providing enough illumination to meet the MSHA regulations.

Specifically, the model will provide the ability to generate three dimensional representations of underground coal mining machines, choose lighting systems, and then perform calculations of the level of illumination in the space surrounding the machine. Using these values of illumination, it is then possible to calculate a Discomfort Glare Ratings for each lamp arrangement and worker line of sight configuration. This rating is an assessment of the capacity of the luminaire configuration in a given visual environment to produce discomfort.

Disability glare comparisons can be made by calculating the luminance contrast of objects of interest along selected lines of sight. The luminance contrast is the relationship of the luminances of an object and its immediate background. In the case of a underground coal mine, the background is usually the walls, floor or roof the tunnel being mined.

IlUMINATION CALCULATIONS

The calculation of illumination at any point is the basis used by the model for any assessment of a mine lighting system. This section describes the method for determining the level of illumination.

Before any illumination calculations can begin, data on the lighting system, mining equipment and mine layout must be input into the model. The Bureau has developed a simple digitizing technique to permit quick, accurate construction of mining machines from two-dimensional orthographic drawings. Graphics oriented procedures for inputting lighting systems and mine layouts have also been developed.

As mentioned previously, Federal regulations require that designated surfaces within the miner's normal field of vision be illuminated to 0.06-fL. Since only the physical measure of luminance correlates to what the eye sees, this is the preferred method of expressing standards. However, luminance measurements are difficult to make, and require an accurate knowledge of reflectivity. Therefore, when calculating light levels, the model simulates the method for measuring the level of illumination (incident light) used by MSHA when approving illumination systems.

With this Statement of Test and Evaluation (STE) approach (Lewis, 1986), the level of illumination at any point on the floor, roof, or ribs is defined as the peak illumination impinging on that point. Peak illumination is defined as the maximum footcandle reading obtained from a photometer which is orientated in the direction of maximum illumination. In the case of multiple lamps, the peak illumination is defined as the vector sum of the individual peak illuminations impinging on the measurement point. This computed value of illumination is converted to luminance (fL):

$$E_{\text{MINE SURFACE}} = E_{\text{PEAK}}(\text{COS } \mathbf{a}) \tag{1}$$

where E is the level of illumination and **a** is the angle between the primary beam direction of the source of the illumination and the measurement point. The luminance is

$$L = r(z)(E_{\text{MINE SURFACE}}) \tag{2}$$

where r is the mine surface reflectivity, z is a light loss correction factor of 0.77, and L is the value of surface luminance in footlamberts. The assumed reflectivity of coal as specified in the current regulations is 4 percent, (actual values range from 1 to 10 percent) which results in an illumination requirement of 2.0-fc to achieve a luminance of 0.06-fL. Hence, the problem is reduced to determining whether the level of illumination at a particular point is greater than 2.0-fc to verify whether the light level complies with the regulations.

The determination of measurement points, i.e., those points where the incident light measured must be at least 2.0-fc, is also directed by MSHA regulations. MSHA requires that incident light measurements must be taken for each 2 foot x 2 foot area on the surfaces surrounding the machine that have to be illuminated. Thus, the model divides the areas surrounding the machine (the floor, roof, ribs - which are the walls of the tunnel, or entry), into a rectangular grid of measurement points 2 feet apart. It then computes the level of illumination incident at each at these measurement points. It also determines the average illumination in each 2 x 2 foot grid square by averaging the four points at the corners of the square.

The model makes use of the Inverse Square Law to determine the actual level of illumination at measurement points in space. This technique is valid when the distance from the light source to the measurement point is at least 5 times the maximum dimension of the light source, which is the case in the nearly all mine lighting situations. The level of illumination is proportional to the candlepower of the source in the given direction, and inversely proportional to the square of the distance from the source to the measurement point:

$$E_m = \frac{I}{D_m^2} \qquad (3)$$

Where: E_m = Illumination at the measurement point, (footcandles)

I = Intensity of the light along the vector between the source and the measurement point, (candles)

D_m = Distance between the light source and the measurement point, (feet)

If the 2.0 isofootcandle profile of a light source is known, then the intensity of a light source at any point on the profile can be calculated by the equation:

$$I_p = E_p(D_p^2) \qquad (4)$$

Where: I_p = Intensity of the light at the 2.0 isofootcandle profile along the light vector, (candles)

E_p = 2.0-fc

D_p = Distance between the light source and the 2.0 isofootcandle profile in the direction where the light is being measured, (feet)

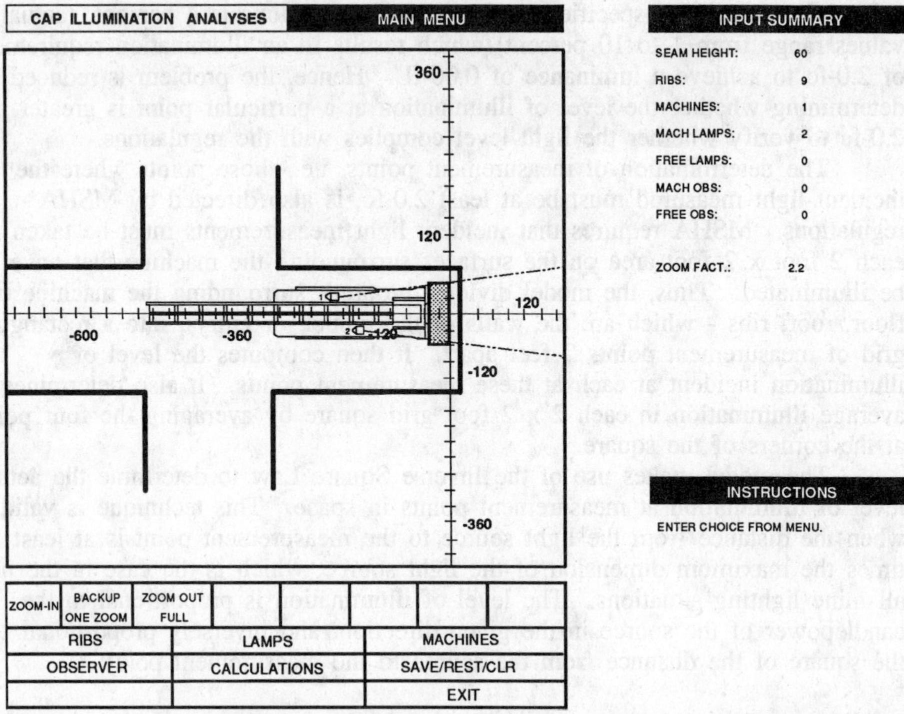

Figure 1 **A mine layout as input by the user. The continuous miner has two luminaires aimed toward the front of the machine.**

The level of illumination at the measurement point can now be determined:

$$E_m = \frac{I_p}{D_m^{\;2}} \qquad (5)$$

The 2.0 isofootcandle profiles of the most popular mining luminaires are being determined in the Bureau's Ergonomics Laboratory. Footcandle readings are being taken at 5 degree increments to construct the profile, which is then stored in a database accessible by the model. When the actual isofootcandle profile is not known, the model uses what information is available and approximates the profile using known geometric shapes. For instance, the profile of a mining machine headlamp would be approximated by an ellipsoid, and a fluorescent luminaire would be represented by a torus.

The effects of shadowing are also taken into account. Since the measurement point is known, the level of illumination at that point is given by the equation:

$$E_m = \frac{I_{p1}}{D_{m1}^{\;2}} + \frac{I_{p2}}{D_{m2}^{\;2}} \;.... + \frac{I_{pn}}{D_{mn}^{\;2}} \quad \text{(vector summation)} \qquad (6)$$

Where n is the number of lamps.

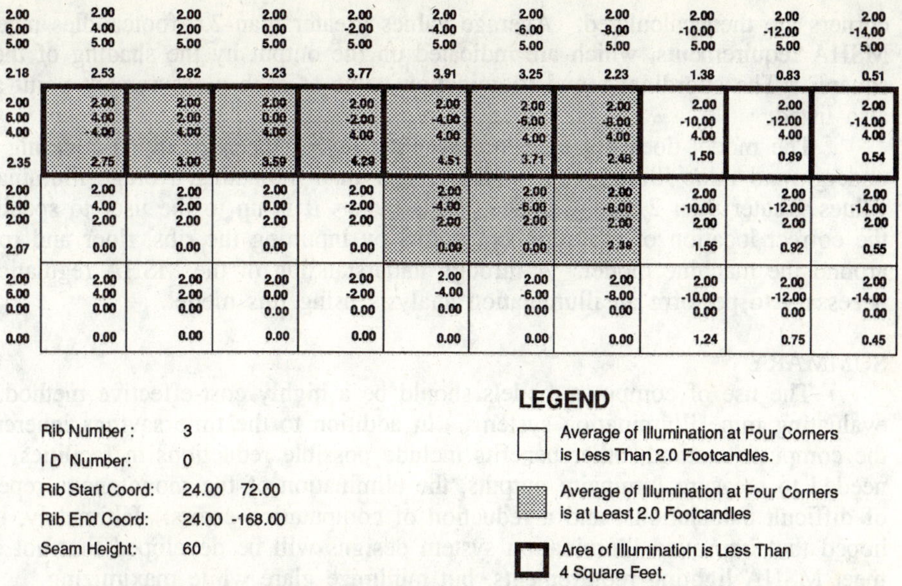

Figure 2. **Results of the illumination analysis based on the layout in Figure 1. This represents the illumination on the rib directly on front of the continuous mining machine.**

In other words, the level of illumination at a specific point is the vector sum of the illumination from all lamps contributing light to that point. The light is considered to be a vector originating at the lamp (defined as a point source) and ending at the measurement point. If the light vector intersects any object before reaching the measurement point, the effects of that light are canceled out of the equation. The primary sources of a blockage between a light vector and the measurement point are mining machines located in the tunnel being mined. Since the input section of the model allows the user to construct three dimensional representations of mining machines as a simple sequence of polygonal geometries, they are easily checked for intersection with the light vector.

EXAMPLE CALCULATION

The following is an example of using the model to determine the illumination on the face of a coal seam with a height of 60 inches. MSHA regulations require that the entire face be illuminated to 0.06-fL when the continuous mining machine is centered in the entry and the cutter head is no greater than 3 feet from the face (Lewis, 1986).

Figure 1 illustrates the positions of a continuous mining machine, lamps, and dimensions of the mine as the user might lay them out on the terminal screen. There are two lamps on the miner (labeled 1 and 2), and the user has positioned the miner on the left side of the entry rather than in the center. The user may reposition or change the type of mining machine, modify the layout of the mine, or modify the illumination system at any time. Since glare will not be analyzed in this example, no observers have been input.

Figure 2 presents the results of the analysis as they would appear to the user. The face area is divided into 2 x 2 foot squares, and the level of illumination is calculated at the corner of each square. The average of the four corners are then calculated. Average values greater than 2.0 footcandles meet the MSHA requirements, which are indicated on the output by the shading of the squares. The coordinates and illumination value at each measurement point are also listed.

The model does not take into account any of the regulations covering underground mine illumination systems other than indicating average illuminance values greater then 2.0 footcandles. This means it is up to the user to specify the correct location of measurement points by inputting the ribs, floor and roof around the machine model. A through understanding of the MSHA regulations is necessary to perform an illumination analysis using this model.

SUMMARY

The use of computer models should be a highly cost-effective method of evaluating mine illumination systems. In addition to the time savings inherent in the computer method, other benefits include possible reductions in facilities needed to measure luminaire outputs, the elimination of the monotonous repetition of difficult calculations, and a reduction of computation errors. Ultimately, it is hoped that improved illumination system designs will be developed that not only meet MSHA lighting requirements, but minimize glare while maximizing visibility.

REFERENCES

Lewis, W. H., 1986, Underground Coal Mine Lighting Handbook, in two parts: Background, U.S. Bureau of Mines Information Circular 9073, 42 pp., Application, U.S. Bureau of Mines Information Circular 9074, 89 pp.

Advances in Industrial Ergonomics and Safety II
Edited by Biman Das
Taylor & Francis 1990

IMPROVEMENT IN A SHIPPING CONTAINER USED IN THE AUTOMOTIVE INDUSTRY: A CASE STUDY

J.L. Wick
The Joyce Institute, Seattle, WA

And

D.A. Hoss, J.A. Lee, R. Potter and C.A. Wilson
Chrysler Corporation, Highland Park, MI

This paper describes the application of ergonomics to the design of a shipping rack with respect to the unloading of automotive parts. The analysis of the original design revealed significant stresses to both the wrists/hands when gripping with high force and to the lower back discs during reaching and lifting. The results of the new design were the reduction of stress to the left hand and wrist, and the reduction of compressive forces on the L5/S1 disc in the lower back. There was an 80% increase in part density in the new design.

INTRODUCTION

The design of materials handling systems and devices influence the way in which workers must handle the parts during the loading and unloading tasks. Poor design can result in inefficiency or injury. Time can be lost because of difficulty in accessing the parts, and injuries can occur because of stressful postures. Retrieving parts from shipping containers requires reaching, grasping and lifting. Stresses to the back and upper extremities must be examined and kept within the capabilities of the worker.

Very often, the focus of attention on materials-handling tasks is on the back; however, significant stresses occur in the hands and wrist. In studying the effects of handle design, Drury and Pizatella (1983) reported that container handle size, shape and location can influence the loading on the upper extremities. Handle design is often under the control of the design engineer, but there are many situations where handles can not be incorporated in the design. Examples are shipping boxes which must be totally sealed against environmental conditions, or large parts which are not "packaged". The latter was the case in this project. The problem then becomes one of properly positioning the point of grasp for pick-up, given the part size and shape.

Once the part is positioned for grasping from a non-stressful posture, the forces required to grip the object for the pick-up need to be dealt with effectively. Armstrong, et al, reported in 1982 on the effects of high hand and finger forces. High grip forces stress the tendons in the carpal tunnel and can contribute to the causes of such diseases as carpal tunnel sydrome. The power grip is more powerful than the precision or pinch grip and should be used when high force grips are required.

The stress to the discs in the back can be determined by employing biomechanical models. Several models have been developed over the years. The model used in this project was the one reported by Fish (1978). The model is easy to use and can be adapted to a simple computer program or run on many hand calculators. The model considers the subject's height and weight; the angles of the trunk, upper arm and lower arm; and weight of the object. The result is the compressive force on the L5/S1 disc.

ergonomic problems which will lead to appropriate solutions. An important result of this project was the increased parts density in the rack. This can translate directly to transportation cost savings. One should expect and strive for savings other than injury costs through ergonomic intervention. Management is much more willing to finance improvements which would result in direct, measurable savings. In this case, it is possible to pay for safety with transportation cost savings.

REFERENCES

Armstrong, T.J., Foulk, J.A., Joseph, B.S., and Goldstein, S.A., 1982, Investigation of cumulative trauma disorders in a poultry processing plant. American Industrial Hygiene Association Journal, 43, pp. 103-116.

Chaffin, D.B. and Park, K.S., 1973, A longitudinal study of low back pain as associated with occupational lifting factors. American Industrial Hygiene Association Journal, 34, pp. 513-525.

Drury, C.G., 1987, A biomechanical evaluation of the repetitive motion injury potential of industrial jobs. Seminars in Occupational Medicine, Vol. 2, No. 1, pp. 41-49.

Drury, C.G. and Pizatella, T., 1983, Hand placement in manual materials handling. Human Factors, 25, pp. 551-562.

Drury, C.G. and Wick, J.L., 1984, Ergonomic applications in the shoe industry. In: Proceedings of the 1984 International Conference on Occupational Ergonomics, Toronto, pp. 489-493.

Fish, D.R., 1978, Practical measurement of human postures and forces in lifting. In: Safety in Manual Materials Handling. NIOSH Publication 78-185, pp. 72-77.

NIOSH, 1981, Work Practices Guide for Manual Lifting, NIOSH report 81-122.

The Joyce Institute, 1988, Applied Industrial Ergonomics Reference Manual, (Seattle: The Joyce Institute).

Advances in Industrial Ergonomics and Safety II
Edited by Biman Das
Taylor & Francis 1990

IMPROVEMENT IN A SHIPPING CONTAINER USED IN THE AUTOMOTIVE INDUSTRY: A CASE STUDY

J.L. Wick
The Joyce Institute, Seattle, WA

And

D.A. Hoss, J.A. Lee, R. Potter and C.A. Wilson
Chrysler Corporation, Highland Park, MI

This paper describes the application of ergonomics to the design of a shipping rack with respect to the unloading of automotive parts. The analysis of the original design revealed significant stresses to both the wrists/hands when gripping with high force and to the lower back discs during reaching and lifting. The results of the new design were the reduction of stress to the left hand and wrist, and the reduction of compressive forces on the L5/S1 disc in the lower back. There was an 80% increase in part density in the new design.

INTRODUCTION

The design of materials handling systems and devices influence the way in which workers must handle the parts during the loading and unloading tasks. Poor design can result in inefficiency or injury. Time can be lost because of difficulty in accessing the parts, and injuries can occur because of stressful postures. Retrieving parts from shipping containers requires reaching, grasping and lifting. Stresses to the back and upper extremities must be examined and kept within the capabilities of the worker.

Very often, the focus of attention on materials-handling tasks is on the back; however, significant stresses occur in the hands and wrist. In studying the effects of handle design, Drury and Pizatella (1983) reported that container handle size, shape and location can influence the loading on the upper extremities. Handle design is often under the control of the design engineer, but there are many situations where handles can not be incorporated in the design. Examples are shipping boxes which must be totally sealed against environmental conditions, or large parts which are not "packaged". The latter was the case in this project. The problem then becomes one of properly positioning the point of grasp for pick-up, given the part size and shape.

Once the part is positioned for grasping from a non-stressful posture, the forces required to grip the object for the pick-up need to be dealt with effectively. Armstrong, et al, reported in 1982 on the effects of high hand and finger forces. High grip forces stress the tendons in the carpal tunnel and can contribute to the causes of such diseases as carpal tunnel sydrome. The power grip is more powerful than the precision or pinch grip and should be used when high force grips are required.

The stress to the discs in the back can be determined by employing biomechanical models. Several models have been developed over the years. The model used in this project was the one reported by Fish (1978). The model is easy to use and can be adapted to a simple computer program or run on many hand calculators. The model considers the subject's height and weight; the angles of the trunk, upper arm and lower arm; and weight of the object. The result is the compressive force on the L5/S1 disc.

It was necessary to establish design criteria. Maximum allowed pinch grip force was eight pounds (3.6kg) and power grip force was 25 pounds (11.4kg). These were based on the estimated maximum ability of the 5th percentile female and would allow a wide range of the population to perform the task without undo stress. The criterion for the compressive force on the disc in the back was 250kg. Chaffin and Park (1973) reported that the back injury incidence rate below 250kg of force was 2.2 per 200,000 man-hours and 8.6 per 200,000 man-hours over 250kg. NIOSH (1981) recommends 300-350kg as a criterion; however, the conservative 250kg seems more appropriate considering the fact that the various models provide nothing more than a guide.

The task analysis methodology reported by Drury and Wick in 1984 was used in the study of the job. It is an eight-step methodology as follows:

1. Videotape the job. The operator was videotaped from the front, back and both sides while performing the work.
2. Task analysis. The job was broken into task steps and the position of each body part was recorded.
3. Work/workstation measurements. Physical dimensions of the workplace and measurements of finger and hand forces were taken (using dynamometers [Drury, 1987]) as required.
4. Worker interviews. The worker was interviewed to allow input regarding possible changes to the workstation and to determine what, if any, pain or discomfort was being experienced.
5. Data evaluation. The body part positions, forces and repetition rate were compared with the criteria.
6. Solution design. Solutions to the ergonomic problems and the workplace causes were developed during a brainstorming meeting.
7. Design validation. The new design was analyzed and evaluated to determine if the problems were solved and no new problems inadvertently introduced.
8. Comparison of the new design with the original job. The data from before and after change were compared to quantify the effects of the new design.

By comparing the data before and after change, success can be measured immediately. The alternative is to track injuries, or lack of injuries, over a year or more.

This project was initiated because operators were reporting difficulty using the existing system and requested a new design. Change was required to prevent future injuries.

Ergonomics not only addresses safety, but also impacts on efficiency. This project demonstrates that ergonomics can improve the efficiency of a materials-handling system as well as improve safety.

This project was part of an ergonomics training program (The Joyce Institute, 1988) presented to selected professionals.

DESCRIPTION OF THE JOB

The job studied was the unloading of the lower rails of an automobile. The lower rail is a sheet metal stamping that can exceed six feet (183cm) in length and eight inches (20cm) in diameter of irregular shape and weigh 15 pounds (6.8kg) or more. The actual dimensions vary with the automobile design. The stampings were shipped from the stamping plant to the assembly plant in a steel shipping rack.

To unload the rack, the operator was required to remove several retaining bars and then the rails. In the original design the bars were set aside on the floor. One part was removed each minute. The rack contained 36 parts arranged horizontally. The work pace was determined by the speed of the assembly line which was one vehicle per minute. The racks were moved in and out of the work area by another worker using a fork-lift truck.

RESULTS

The task analysis revealed stresses to the hands, wrists and back. The task of removing the retaining bars imposed stressful lifting postures. High power grip forces were required to maneuver the bars in the channels to slide them out. High loads on the discs in the back resulted when lowering the bars to the floor. Excessive pinch forces were required to handle the rails from the horizontal position in the rack. Disc forces exceeded the criterion when the operator lifted the rails from the lower level of the rack.

Solutions were developed to solve the ergonomic problems identified in the analysis and evaluation. The changes included:

- "self-storing" retainer bars which did not have to be removed from the rack
- vertical positioning of the rails to eliminate low level lifts

A prototype was developed and fabricated. It was tested for basic mechanical function. The new design was then tested under simulated production conditions for another analysis and evaluation to validate the changes.

The comparison of the data from before and after change indicated significant reduction of the stresses to the hands, wrists and back. Stressful pinch grips and power grips were eliminated and disc compressive force was reduced. The new vertical arrangement of the parts in the rack increased the number of parts which could be shipped in one container.

Table 1. Comparison of conditions before and after change.

Measure	Before	After
Pinch grips with force		
Left	24/cycle	0/cycle
Power grips with force		
Left	12/cycle	0/cycle
Right	36/cycle	0/cycle
Disc Compressive Force	334kg	206kg
Parts per rack	36	65

This project can be considered a success because the injurious conditions were eliminated. Pinch grips with excessive forces in the left hand were reduced from 24-per-cycle to zero. Power grips with excessive forces were reduced in the left hand from 12-per-cycle and the right hand from 36-per-cycle to zero. The compressive force on the discs in the back were reduced from 334kg to 206kg. Additionally, shipping costs could be saved by increasing the density of the rack from 36 parts to 65 parts.

DISCUSSION

Researchers in ergonomics have often focused on back stress when studying manual materials-handling tasks. It is true that the back suffers from lifting and carrying stresses. It is also appropriate for research to be selective; however, the upper extremities can also be stressed. When studying a materials-handling job, the practitioner needs to apply the wealth of information provided by the research with regards to the loading of the body as a system. That means the investigation cannot be restricted to the back. The methodology employed in this project provides a means to analyze the stresses to the upper extremities as well as the back. The application of effective analysis techniques allows the practitioner to identify and quantify the ergonomic problems which will lead to appropriate solutions. An important result of this project was the increased parts density in the rack. This can translate directly to transportation cost savings. One should expect and strive for savings other than injury costs through ergonomic intervention. Management is much more willing to finance improvements which would result in direct, measurable savings. In this case, it is possible to pay for safety with transportation cost savings.

REFERENCES

Armstrong, T.J., Foulk, J.A., Joseph, B.S., and Goldstein, S.A., 1982, Investigation of cumulative trauma disorders in a poultry processing plant. American Industrial Hygiene Association Journal, 43, pp. 103-116.

Chaffin, D.B. and Park, K.S., 1973, A longitudinal study of low back pain as associated with occupational lifting factors. American Industrial Hygiene Association Journal, 34, pp. 513-525.

Drury, C.G., 1987, A biomechanical evaluation of the repetitive motion injury potential of industrial jobs. Seminars in Occupational Medicine, Vol. 2, No. 1, pp. 41-49.

Drury, C.G. and Pizatella, T., 1983, Hand placement in manual materials handling. Human Factors, 25, pp. 551-562.

Drury, C.G. and Wick, J.L., 1984, Ergonomic applications in the shoe industry. In: Proceedings of the 1984 International Conference on Occupational Ergonomics, Toronto, pp. 489-493.

Fish, D.R., 1978, Practical measurement of human postures and forces in lifting. In: Safety in Manual Materials Handling. NIOSH Publication 78-185, pp. 72-77.

NIOSH, 1981, Work Practices Guide for Manual Lifting, NIOSH report 81-122.

The Joyce Institute, 1988, Applied Industrial Ergonomics Reference Manual, (Seattle: The Joyce Institute).

Advances in Industrial Ergonomics and Safety II
Edited by Biman Das
Taylor & Francis 1990

EYES - AN EXPERT SYSTEM FOR THE DEVELOPMENT OF HIGH TOUCH CONSUMER ELECTRONIC PRODUCTS

Young Won Kim
Daewoo Electronic CO. Ltd
Seoul, Korea

Myun Woo Lee
Seoul National University
Seoul, Korea

Andris Freivalds
Penn State University
University Park, PA 16801, U.S.A

The development of EYES (Electronic Yuppie Expert System) involved four steps : 1) the compilation of a data base of current high-tech/high-touch products, 2) the surveying of specific populations on their preferences in consumer electronic products, 3) the development of an expert data base system for projecting future developments and 4) the specific evaluation of five Daewoo prototypes (TV, VCR, audio system, microwave, vacuum cleaner) based on the above system. General findings from the survey indicated that ergonomic product functions, price, brand name and ease of use (high-touch) were far more important than design and brand name. The effectiveness of any one new function will decrease when many new functions are combined into one product.

INTRODUCTION

There are two basic, but vastly different, concepts in developing new electronic products. High Tech and High Touch. High-tech design requires newly developed high-end technology and massive R & D strategy while High-touch design utilizes existing technology systematically and emphasizes Industrial and Ergonomic Design. The purpose of this study was to develop an expert system for the specific purpose of identifying future trends in the development of high-touch consumer electronic products. The special emphasis was placed on 1) greatly expanding ergonomic design as part of this high-touch design, 2) developing an ergonomic approach based on an analysis of implicit of human needs that identifies potential areas of high-touch and 3) predicting successful new high-touch products in consumer electronics (Freivalds,1989).

Based on this High-touch design concept, Daewoo Electronic CO. Ltd developed five prototypes of consumer electronic products(Kim et al., 1988). Main features of these products (shown in Figure 1) are:

(a) TV : remote control of screen rotation and tilt, microprocessor control of program reservation, home memo, event and address memory etc.
(b) TVCR: voice instructed control, wireless earphone, audio and video memory, detachable LCD screen etc.

Figure 1. Daewoo High-Touch prototype products

(c) Audio: wireless speaker, automatic balance control on user position, automatic equalizer control on music signal,etc.
(d) Vacuum cleaner: remote controlled operation of movement and brushing
(e) Microwave oven: cooking guide system, menu data base, reserved cooking, activation control by voice, etc.

After developing the prototype products, research objectives were been reorganized to increase the following four aspects. 1) the compilation of a data base of current High-Tech/High-Touch products and trends. 2) the survey of specific U.S. populations on their preferences in consumer electronic products 3) evaluating their responses to Daewoo High-Touch prototype products and 4) the development of an expert data base system for projecting future improvements in high-tech/high-touch products.

EYES - ELECTRONIC YUPPIES EXPERT SYSTEM

The operational definition of a "High touch" product implies a design concept based on a systematic application of existing technology. Such a systematic application includes:
(a) the construction of hierarchical structure of human function, product function, and technology which are related to the functional requirements of the design target
(b) the definition of the design target in terms of the functional requirements of the product
(c) the analysis of the relationships between human function, product function and technology.
(d) the evaluation of the relationships based on various sets of criteria.
(e) generation of potential products using a production rule system(Jung, 1989).

The development of EYES(Electronic Yuppies Expert System) allows the production of potential products given the functional requirements of electronic home appliances. According to Yoshikawa(1987), the procedures are algorithmic and can be implemented using a production rule system. The following is the brief explanation of the system.

1. Hierarchical construction

The three major entities of the "high touch" can be defined as ;
- Human function (H)
- Product function(P)
- Technology(T)

The conceptual hierarchy can be thought as:the domain { h(x1,x2,x3) (x1=H,P,T, x2=level of hierarchy, x3=predecessor number in the hierarchy) }. For example, h(H,1,_) is;
- Sensing
- Information Processing
- Motor Function

Then, h(H,2,Sensing) is;
- Visual
- Auditory
- Tactile

- Kinethetic

Also, h(H,3,Visual) is;

- Visual Angle(Horizontal, Vertical, Lateral)
- Visual Field(Static,Dynamic)
- Visual Perception(Depth,color,Texture)
- Visual Adaptation(Dark,Light)
- Visual Channel(sensitivity, threshold)
- Visual Memory(image store,short_term retrieval)

In the same way, product function(p(x1,x2,x3)) and technology(t(x1,x2,x3)) for the target design can be defined.

2. Functional requirements

The functional requirement of a product can be defined using the set of human functions which must be satisfied in the target design. For instance, Let H={a1,a2,...an} be the set of human function related to the target design and P={p1,p2,...pn} be the set of product function of the product P. Then, the convolution matrix H*P is defined as in Table 1;

Table 1. Convolution matrix of human function and product function

product function / human function	p1	p2	p3	pn
h1	v11	v12		v1n
h2				
hn	vn1			vnn

(where vij = v if pj is a product function which satisfies human function hi
= i if pj contributes to hi,
= 0 otherwise)

Then, the set pj is equal to {v11,v12,..v1n} where vij <> 0 defines the list of human functions which related to the product function pj. Similarly, the functional requirement of the target design can be defined in terms of human functions and the related set of product functions using the convolution matrix H*T and T*P.

3. Evaluation of the relationship by criteria

A criterion is defined as a value which defines the compatibility of the relationship between two entities. In this case, the criteria can be thought as values of the convolution matrix H*P,H*T,T*P based on the following definitions.

(a) technological availability : aij
Availability of technology(T) j in human function i

(b) Technology transitivity : tjk
Transitivity of technology(T) j to product(P) k

(c) Functional Demand : dik
demand of human function(H) i to product function(P) k

(d) Comfortability : cik
satisfaction of comfortability of human function(H) i by the product function(P) k

These four criteria are used to define the rules used in the definition of design requirements of a target product.

4. Generation of Potential Product

According to the general design theory by Yoshikawa (1987), a conceptual design is algorithmic with production rules if there is a set of rules which produce a solution in terms of the attribute by giving a specification in terms of function. Based on the result of the convolution matrix, the following production rules can be defined.

(a) Comfortability Rules
IF human function THEN requirement

(b) Availability Rules
IF product function THEN human function
IF technology THEN human function

(c) Transitivity Rules
IF product function THEN technology

(d) Synthesis Rules
IF product function, product function THEN requirement

After defining the functional requirements of the product, the system looks for a THEN clause that matches the human function in the comfortability rules, firing an appropriate rule. Next, an IF clause of the fired rule will be connected to the THEN clause of the "IF product function THEN human function" or "IF technology THEN human function" rules.

Thus, we can find the network of product function or technology as a solution. But in order to synthesize these product functions physically, the combination of the product functions must be designed, too. The "IF product function, product function THEN requirement" rule give us new requirements for this synthesis design with restrictions to maintain the original requirements.

CONSUMER SURVEY

Survey questionnaires are a useful method for attaining consumer responses to new electronic products. The specific purpose of consumer survey were 1) to determine the general preferences of the potential consumers, 2) to evaluate the function and design of Daewoo High-touch prototype products, and 3) to predict the sales potential of these prototypes. The survey was conducted among yuppies, upper middle class, middle class and potential customer such as college students. For the first survey, seventy five subjects were collected mainly from college students and local residents around State College,PA area. Contents of questionnaire included: 1) demographic information, 2) usage information, 3) brand image, 4) buying criteria, 5) rating of prototype product functions, 6) impressions of prototype products, and 7) general comments. The survey utilized both slides and videotapes demonstrating these product being used. In general, the audio prototype was most highly rated, receiving 4.0 on 6.0 scale(Figure 2).

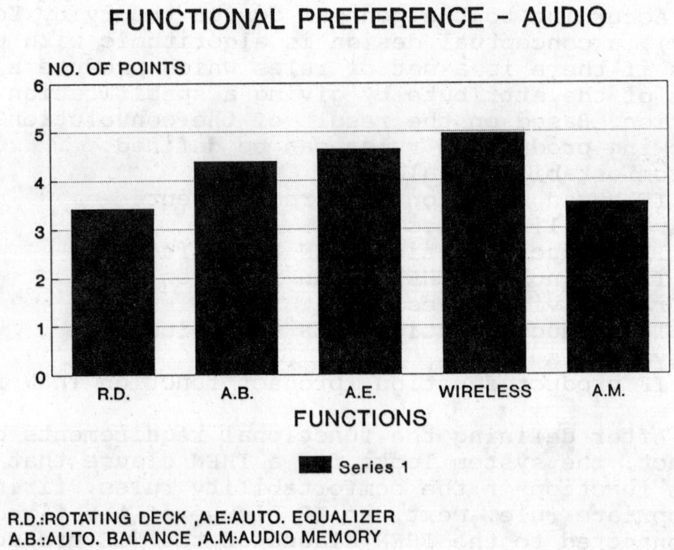

Figure 2. Rating of the audio functions

The features which adjust the quality of Audio/Video information were well received while additional features such as the microprocessors functions in TV were not as well received. Consumer priorities in buying criteria in Figure 3 shows a distinct pattern. Product function, price, technical spec. and ease of use were four major factors in buying criteria. It suggests that practical factors such as product functions, ease of use etc. are more important than so called "image factor" such as brand name & size.

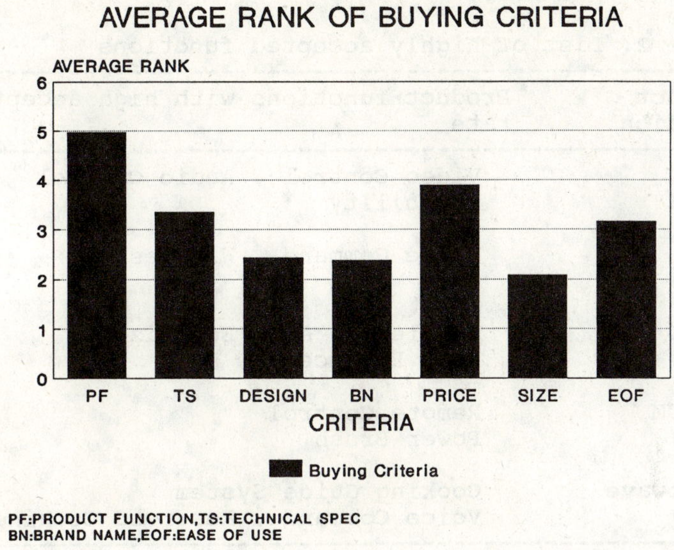

Figure 3. Consumer priorities in buying criteria

Figure 4 shows the subject preferences for the five prototype products. This figure refers to the percent of subjects who choose the corresponding product as the top choice. In this figure, Audio prototype product was the most popular one as a whole.

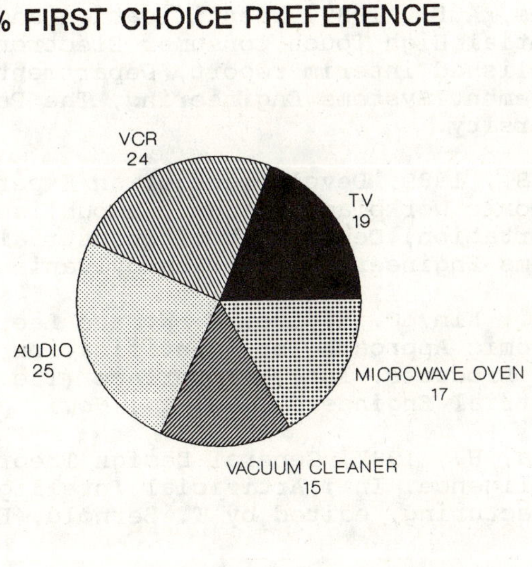

Figure 4. Subjective preferences for prototype products

Table 2 summarizes the highly accepted functions for each of the products.

Table 2. List of highly accepted functions

Product Function	Product functions with high acceptance rate
TV (21%)	Video Control , Audio Control Movability
VCR (18%)	Voice Command , Wireless LCD
AUDIO (26%)	Wireless , Auto Equalizer Auto Balance
VACUUM (18%)	Remote Control Power Brush
Microwave (17%)	Cooking Guide System Voice Command , Reserved Cooking

CONCLUSIONS

The concept of High-touch design which involves the systematic application of ergonomic knowledge to the general design theory produced positive results in the design of consumer electronic products. Spurred by the desire for more expressive objects, the technology of miniaturization, and the growing knowledge of human factors, this approach is expected to be an important strategy in new product design.

REFERENCES

Freivalds, A.F., 1989, Identification and Evaluation of Potential High Touch Consumer Electronic Product, unpublished interim report, Department of Industrial and Management Systems Engineering, The Pennsylvania State University.

Jung, E.S., 1989, Development of an Expert System for Ergonomic Workplace Design., unpublished Ph.D. dissertation, Department of Industrial and Management Systems Engineering, The Pennsylvania State University.

Kim, Y.W., Kim, C. S.,Lee, M. W. and Lee, I.J., 1988, An Ergonomic Approach for Identification of Potential High Touch Products. In: IE Interface (The Korean Institute of Industrial Engineers), Vol 1., No.1, pp. 85-99.

Yoshikawa, H., 1987, General Design Theory and Artificial Intelligence. In : Artificial Intelligence in Manufacturing, edited by T. Bernold,(Elsevier), pp. 35-61.

Advances in Industrial Ergonomics and Safety II
Edited by Biman Das
Taylor & Francis 1990

COMPARISON OF LIGHT DUTY GLOVES WITH NATURAL AND SYNTHETIC MATERIALS UNDER WET AND DRY CONDITIONS

W.R. Santee[1], T.L.Endrusick[1] and L.S. Pensotti[2]

[1]U.S. Army Research Institute of Environmental Medicine
[2]U.S. Army Natick Research, Development and Engineering Center
Kansas Street, Natick, MA 01760, USA

The performance of five volunteer test subjects wearing a standard cold weather military uniform and three different gloves was compared during treadmill walking (120 minutes), sitting (125 minutes) and hand conductance or "contact" (60 minutes). Chamber conditions were -6.7°C (20°F) T_a and 1.1 m·s^{-1} wind speed. The sitting tests were repeated at 0°C (32°F). Contact tests were repeated with wet gloves. Rectal and finger temperatures, heart rate, and endurance times were measured. The results indicate that the new leather shell is more affected by external moisture than the standard shell.

INTRODUCTION

The primary US military issue glove for general cold weather wear is a light duty glove that is slightly modified from that which was issued to US troops in WWII (Richardson and Allan, 1948). The glove consists of a durable leather shell worn over a separate wool/nylon knit liner. The long duration of this basic light duty glove design can be attributed to the functional utility and effectiveness of the two layer, leather shell/knit liner design.

Despite the historical success of the glove, considerable dissatisfaction has been expressed regarding the issue glove. Two factors directly communicated to the authors by military users which contribute to the dissatisfaction are 1) perceived failure of the glove to provide adequate thermal protection, particularly if the experience was personal or directly affected unit efficiency and 2) dissatisfaction with the use of traditional materials, particularly wool, when recently developed synthetic materials claim improved properties and eliminate allergic reactions to wool.

For organizations, including military unit, police and public utilities, which must function under adverse weather conditions, the ultimate criterion for all clothing is an individual's ability to endure and perform the assigned mission. Comparisons of different handwear based on measurements of individual performance are a critical element in decisions regarding handwear. The methods used to evaluate handwear and other cold weather clothing items are therefore important and merit a thorough description.

The primary purpose of the present study was to compare performance of subjects wearing three light duty gloves. The test standard was the present issue LD glove consisting of a leather shell and wool/nylon insert. Two additional configurations combined a new leather shell design with two different inserts of hollow-core polyester material. One insert (P1) is a simple knit insert with a smooth palm, while the other insert (P2) is also knit of the same synthetic material but the palm area is covered with circular projections to prevent slippage when the insert is used alone as an anti-contact glove.

Preliminary measurements of dry insulation (I_t) were conducted for each of the gloves on a newly constructed nine-region aluminum hand model (Santee, unpublished). Initial results indicate that I_t values derived with this hand model are approximately 0.10 clo less than values measured on a previous copper hand model. The thermal insulation values obtained were 0.124 $m^2 \cdot K \cdot W^{-1}$ (0.80 clo) for the LD glove and 0.118 $m^2 \cdot K \cdot W^{-1}$ (0.76 clo) for both the P1 and P2 gloves.

In handwear studies, the insulation values of the total clothing system are also important (Vanggaard,1988). For this study, the layered uniform consisted of cotton/wool long underwear, battle dress uniform (BDU), field jacket with polyester liner, pile cap, cotton/wool cushion sole sock and leather combat boots plus the gloves. The preceding clothing items are common fall and winter clothing for the military, although the long underwear is usually not worn until the colder part of the winter and other headgear may be substituted for the pile cap.

METHODS

The relative protection provided by the standard and prototype gloves was determined by quantifying the performance of five volunteer, male test subjects. Volunteer subjects were recruited from the local military population. Subjects read and signed an informed consent agreement after receiving a verbal briefing. Subjects were medically screened and received a thorough medical examination prior to clearance for participation. All testing was in accordance with AR 70-25 and the USARIEM type protocol. Their performances were determined during three activities.

On different test days, subjects walked on a treadmill at 1.34 $m \cdot s^{-1}$ (3 mph) for 120 minutes (plus a 5 min baseline), sat for 125 minutes, or participated in a contact simulation for 60 minutes (plus a 5 min baseline). All three test activities were conducted at a chamber air temperature of -6.7°C (20°F) and a wind speed of 1.1 $m \cdot s^{-1}$ (2.5 mph). The sitting tests were replicated at a second chamber temperature, 0°C (32°F).

The contact simulator was developed as a test device which produces a controlled, replicable measure of conductive heat loss from one hand. The simulator is a generic modification of a pump simulator specifically built for testing fuel handlers gloves (Santee, et. al., 1988). The device consists of an adjustable stand with a plastic envelope mounted on an aluminum plate at waist level. The envelope is filled with propylene glycol which is circulated through a heat exchanger to maintain the liquid at chamber air temperature. When a subject pushes against the envelope with sufficient force (9.5 kg or 21 lbs), a pressure sensitive switch activates a timer. The timing device is set to signal when the subject maintains contact for 2 minutes, then a 1 minute rest period is signaled before restarting the push/rest cycle. If the subject does not press with sufficient force, a light on each stand indicates that the timer is not operating. The push cycle duration is based on the actual time that the switch is activated, so if the subject fails to maintain contact with the envelope with sufficient force to activate the timer the cycle timer does not countdown towards the rest period. A total of four devices were assembled, so multiple subjects were tested during one experimental session.

All initial tests were with dry gloves. The contact simulation tests were repeated with wet gloves after the dry glove tests. The glove on the pushing hand was wetted by having the subject remove sponges from a basin and squeeze out enough water to fill five 200 ml beakers. The objective was to simulate working with a wet substance, such as a wet canvas or snow, and to thoroughly work the moisture into the seams of the gloves. The walking and contact simulation activities were preceded by a 5 min standing baseline.

Measurements included heart rate, rectal temperature, nine skin surface temperatures including the lateral nail beds of the middle and little left and right fingers and the left big toe. Safety monitoring standards included removal of the subject from a test session if a heart rate of 180 bpm was sustained for 5 minutes, the rectal temperature dropped to 35.5°C (95.9°F), a skin surface temperature dropped to 5.0°C (41°F), a test observer inside the chamber removed the subject or the subject voluntarily exited the chamber.

The basic method for statistical analysis was analysis of variance with repeated measures (SPSS, 1988). The significance of relationships between individual pairs was verified by post hoc Tukey's t-tests at the 0.05 significance level (Bruning and Kintz, 1977). Physiological measurements included heart rate, rectal temperature and four finger temperatures. Dependent variables were total activity time (ET) for walking (max 120 min), sitting (125 min max) and contact (max 60 min), cumulative contact time (CT, max 40 min) and the average rate of change for four (walking and sitting) or two (contact) finger temperatures. By averaging the finger temperatures, the effect of cold-induced vasodilation (CIVD) on finger temperature variability is dampened.

RESULTS

Figure 1 demonstrates the finger temperature responses of a subject at -6.7°C (20°F) during walking, sitting and dry contact test activities. Subject physical parameters are presented in Table 1. The test results are summarized in Tables 2-4.

The results indicate that no significant difference in subject endurance time or change in finger temperature between dry gloves with polyester or wool inserts can be determined on the basis of this study. When the wet and dry contact simulation results were combined for statistical analysis, the results indicated overall significant differences between wet and dry gloves for ET, CT and $\Delta T/hr$. For ET and CT, there were also significant differences between the wet and dry conditions for the two synthetic gloves (P1 and P2). The standardized changes in mean finger temperature (2 finger mean, right hand only; 4 finger mean left and right hands) were significantly different overall between wet and dry conditions and overall between LD and P1. In specific pairs there was a significant difference between the wet and dry conditions for P1. For the change in four finger mean temperature there was also a significant difference between the LD and P1 gloves in the wet condition.

Figure 1. Average of two right finger temperatures versus time for walking, sitting, and dry contact activities (subject 1)

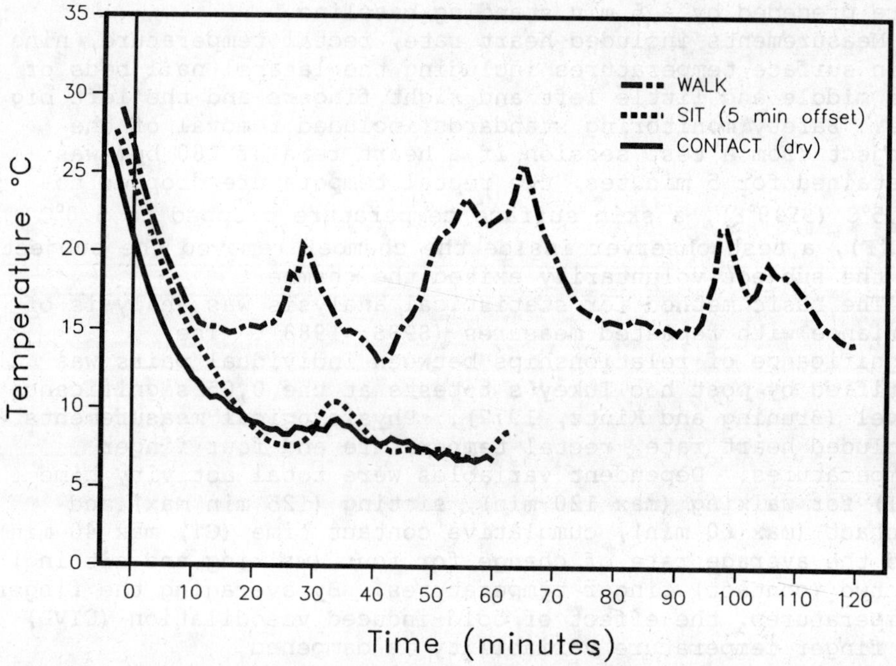

Table 1. Average physical parameters of volunteer test subjects (n=5)

subject	height (cm)	weight (kg)	age (years)	A_D[1] (m^2)	body fat[2] (%)
1	172	74.6	20	1.88	15.2
2	178	79.3	21	1.97	15.1
3	175	75.6	23	1.91	17.0
4	170	62.7	19	1.73	9.3
5	166	66.0	26	1.74	15.7
mean	172	71.6	22	1.85	14.5
s.d.	5	7.0	3	0.11	3.0

[1]Dubois surface area (Dubois and Dubois, 1916)
[2]skinfold measurements (Durnin and Womersley, 1974)

Table 2. Endurance times (ET, min) for all activities

	light duty glove (LD)	test glove plain (P1)	test glove textured (P2)
walk (-6.7°C)[1]	85[2]	120	120[3]
s.d.	41	0	0
sit (-6.7°C)[4]	82	72	81
s.d.	21	15	36
sit (0°C)[1]	125	125	117
s.d.	0	0	9
contact (dry, -6.7°C)[4]	57	56	54
s.d.	5	4	13
contact (wet, -6.7°C)[4]	47	27	32
s.d.	19	20	20

[1]n=4, [2]includes mean of two runs for subject removed for non-thermal factor, [3]n=3 [4]n=5

For ET and CT, the differences between wet and dry gloves were significant ($p \leq 0.05$) overall and specifically (Tukey's t-test, $p=0.05$) between P1 and P2. For the change in finger temperature ($\Delta T \cdot t^{-1}$ or $\ln(\Delta T \cdot t^{-1})$), there were significant overall differences between dry and wet gloves, and between dry and wet for the P1 glove. The rate of temperature decline for the LD glove was significantly less then the decline for the P1 glove for both the combined wet-dry and wet only data sets. The results indicate that the glove with projections

Keep Your Wrist in a Neutral Position

Avoid using your wrist in a bent (flexed downward), extended (upward) or twisted position for long periods of time. Instead, attempt to maintain a neutral (straight) wrist position whenever possible.

Watch Your Grip

Gripping, grasping, or lifting with the thumb and index finger may put undue stress on the wrist. Whenever possible, use the entire hand and all the fingers to grasp an object.

Minimize Repetition

If possible, avoid repetitive motions or holding an object the same way for extended periods of time.

Rest Your Hands

Periodically, give your hands a break by allowing them to rest briefly or alternate your hands by switching from one hand to the other.

Reduce Speed and Force

Decreasing the speed with which you perform a forceful, repetitive movement gives your wrist time to recover from the stress effort. Using proper power tools may also help to reduce such force.

Conditioning Exercises

Certain regular exercises strengthen the hand and arm muscles. They may prevent injuries by reducing the need to compensate for these weak muscles with a poor wrist position.

BACK, NECK AND SHOULDER INJURIES

Cases of back, neck, or shoulder injuries, such as muscle strains and pulls, are the most common injuries we observed amongst LOCATION-X (manufacturing) employees. These injuries were noted mostly in static or "fixed" positioned employees. We do not believe the problems were caused exclusively by lifting heavy objects.

Working positions that stress the body may contribute to a high incidence of back complaints. Many stressful working positions are considered to be hidden lifting tasks. A hidden lifting task exists when no object is being lifted other than the weight of an employee's own body, such as the torso when it is bent forward, or the arms when they are extended.

The further a body part is extended during factory work and the longer the duration of the extension, the greater the possible strain. And, the greater the strain, the more severe the pain and subsequent increase of frequency of injuries.

Table 1. Average physical parameters of volunteer test subjects (n=5)

subject	height (cm)	weight (kg)	age (years)	A_D[1] (m^2)	body fat[2] (%)
1	172	74.6	20	1.88	15.2
2	178	79.3	21	1.97	15.1
3	175	75.6	23	1.91	17.0
4	170	62.7	19	1.73	9.3
5	166	66.0	26	1.74	15.7
mean	172	71.6	22	1.85	14.5
s.d.	5	7.0	3	0.11	3.0

[1]Dubois surface area (Dubois and Dubois, 1916)
[2]skinfold measurements (Durnin and Womersley, 1974)

Table 2. Endurance times (ET, min) for all activities

	light duty glove (LD)	test glove plain (P1)	test glove textured (P2)
walk (-6.7°C)[1]	85[2]	120	120[3]
s.d.	41	0	0
sit (-6.7°C)[4]	82	72	81
s.d.	21	15	36
sit (0°C)[1]	125	125	117
s.d.	0	0	9
contact (dry, -6.7°C)[4]	57	56	54
s.d.	5	4	13
contact (wet, -6.7°C)[4]	47	27	32
s.d.	19	20	20

[1]n=4, [2]includes mean of two runs for subject removed for non-thermal factor, [3]n=3 [4]n=5

For ET and CT, the differences between wet and dry gloves were significant ($p \leq 0.05$) overall and specifically (Tukey's t-test,p=0.05) between P1 and P2. For the change in finger temperature ($\Delta T \cdot t^{-1}$ or $\ln(\Delta T \cdot t^{-1})$), there were significant overall differences between dry and wet gloves, and between dry and wet for the P1 glove. The rate of temperature decline for the LD glove was significantly less then the decline for the P1 glove for both the combined wet-dry and wet only data sets. The results indicate that the glove with projections

(P2) was not significantly different from the regular issue LD. There were no significant differences between dry gloves with natural or synthetic liners. The results also indicate that the performance of subjects while wearing the P2 gloves with the grip projection was intermediate between LD and P1, but not significantly different from either glove. During the wet testing, subjects indicated a highly variable degree of moisture penetration inside the glove. Based on those results, the lack of significance between LD and P2, and no measured difference in insulation between P1 and P2 which have liners knit of the same polyester, it was concluded that the differences between wet gloves were possibly due to water leakage through the different shells.

Table 3. Contact times for wet and dry contact activities

	LD	P1	P2
dry contact	36.74	36.09	34.89
s.d	3.18	4.00	9.00
wet contact	30.40	17.78	20.46
s.d.	12.29	12.25	12.60

Table 4. Standardized finger temperature loss for sitting and contact activities ($\Delta T/hr$)

	LD	P1	P2
sit (-6.7°C)[1]	13.46	10.54	12.92
s.d.	2.57	3.36	6.80
sit (0°C)[1]	8.39	8.46	8.09
s.d.	0.61	2.27	2.66
dry contact[1]	15.84	17.83	15.79
s.d.	3.20	3.09	2.77
wet contact[1]	21.22	44.71	35.10
s.d.	9.42	19.33	13.62
dry contact[2]	17.14	18.64	16.07
s.d.	2.87	2.73	2.59
wet contact[2]	22.01	46.56	32.80
s.d.	10.04	25.03	14.19

[1]four finger (little and middle finger, right and left hand)
[2]two finger (little and middle finger, right hand only)

DISCUSSION

A military focus has been an important traditional resource for the evaluation of cold weather clothing. Driven by the pragmatic needs of troops in the field, a combination of extensive, long term testing programs and cumulative field experience has resulted in practical, economical cold weather clothing. Both the need to ensure that the clothing will meet the requirements of soldiers in the field, especially in terms of durability in a tactical situation while providing adequate thermal protection, and the extensive costs required to build and maintain an adequate clothing inventory result in a conservative approach to clothing development. Conversely, the growth in outdoor winter recreation created a marketplace that stimulated the development of new synthetic insulation materials for clothing. However, the requirements of the military, especially in terms of durability, are not the same as those of recreational users (Hedblom, 1965). Soldiers in the field often endure longer exposures to cold and subject their clothing to heavier use and wear than the majority of recreational users.

The results of this study indicate that the sedentary or sitting test method, yielded no better effects than by the use of the contact method for evaluation of dry gloves in this study. In a previous test (Santee and Endrusick, in preparation), the contact test allowed discrimination between two gloves of nearly equal insulation. That prior test included a walking activity, but not a sedentary activity. The sedentary test represents the definitive "worst case" for handwear testing, with the lowest levels of heat production and vasoconstrictor stimulation. It is also longer in duration and requires longer subject exposure to low temperatures than the contact test. Actual situations which might require such a low sedentary level of activity during cold exposure include military personnel holding defensive positions, occupants of unheated vehicles, and heavy equipment operators. The contact test is a generic, "worst case" simulation of scenarios such as material handling or vehicle maintenance which require prolonged contact with cold objects or surfaces in unheated areas.

Summary

The results of this study demonstrated no significant difference in the performance of test subjects between the three dry gloves. The synthetic knit liner can be adequately substituted for the wool/nylon liner with no significant loss in thermal protection. There is no significant change in thermal cost related to the improved textured gripping surface of the P2 glove. The results also indicate that the new leather shell is more affected by external moisture than the standard shell.

Acknowledgements

The authors would like to acknowledge the following people for their contributions to this study: Mr. Joseph M. McGrath, Ms. Deborah A. DiRaimo, and Mr. Gerald W. Newcomb.

Disclaimer

The contents and views presented in this paper represent those of the authors and should not be construed as an official Department of the Army position, view or policy. Approved for public release; distribution is unlimited.

REFERENCES

Bruning, J.L. and Kintz, B.L., 1977, Computational Handbook of Statistics, 2nd edn., (Glenview,IL: Scott, Foresman).

Dubois, D. and Dubois, E.F., 1916, A formula to estimate the approximate surface area if height and weight be known. Archieves of Internal Medicine, 17, 863-871.

Durnin, J.V.G.A. and Womersley,J., 1974. Body fat assessed from total body density and its estimation from skinfold thickness measurement on 481 men and women aged from 16 to 72 years. British Journal of Nutrition, 32, 77-97.

Hedblom, E.E., 1965. Polar Manual: Arctic and Anarctic Living Conditions,'Personnel Selection., Hygiene and Sanitation, Nutrition, Supplies and Equipment, Visual Disabilities and Cold Injuries First Aid, Safety and Survival. (Bethesda, MD: National Naval Medical Center).

Richardson,E.R. and Allan, S., 1948, Quartermaster Supply in the European Theater of Operation in World War II; Vol. 3. Outfitting the Soldier. (Ft. Lee, VA: Quartermaster School).

Santee, W.R., Endrusick, T.L. and Wells, L.P., 1988, Biophysical evaluation of handwear for cold weather use by petroleum (POL) handlers. In: Trends in Ergonomics/Human Factors V., edited by F. Aghazadeh (New York: Elsevier Science), pp.441-448.

SPSS Inc, 1988. SPSS-X User's Guide, 3rd Edn. (Chicago: SPSS).

Vanggaard, L., 1988. Chapter 7A. Protection of hands and feet. In: Handbook on Clothing: Biomedical Effects of Military Clothing and Equipment Systems. NATO, pp. 7a1-7a12.

VII

HUMAN COMPUTER INTERACTION

Advances in Industrial Ergonomics and Safety II
Edited by Biman Das
Taylor & Francis 1990

AN ERGONOMICS ANALYSIS OF AN AT&T MANUFACTURING LOCATION

Bahador Ghahramani, Ph.D.

Corporate Safety and Industrial Hygiene Engineering
Room 3WD126, One Oak Way, Berkeley Heights, NJ 07922

INTRODUCTION

Studies suggest that employees should be educated in the cause of Repetitive Motion Injuries (RMI), which may be associated with their work environment. Of specific interest is the Carpal Tunnel Syndrome: Repetitive Motion Injuries occur when the tendons, bones, muscles, and/or ligaments in a joint press against the median nerve in the wrist, thus restricting nerve impulses to the extremities.

The AT&T Safety Engineering organization investigated the apparent increase of carpal tunnel Syndrome symptoms at LOCATION-X. Actual observations of the workplace were made, and interviews with managers, medical personnel and employees were conducted. The conclusion was that 5% to 10% of the employees were experiencing mild to severe typical symptoms of CTS, and the incidence was increasing.

In addition to the specific symptoms found, other factors are of interest to local management in guarding against RMI:

a. Women are twice as likely to become victims of RMI than men. This is due to women having a smaller joint structure (in comparison to men).

b. Pregnant women may be more susceptible, because of their body's fluid retention during pregnancy.

c. Other contributory factors could be diabetes, deficiency of a vitamin or mineral (e.g., Calcium, Iron and B6), and menopause. Additionally, a few arcane disorders such as: Myxedema (thyroid condition), Quervian's disease (syndrome affecting the tendons), and Acromegally, may contribute to RMI type of injuries. Recently, vitamin "B6" has been widely publicized as an effective way to treat CTS.

PERSONAL PROTECTION MEASURES

Most of the CTS problems at LOCATION-X relate to Ergonomics problems, probably certain repetitive hand activities. Guidelines for reducing such injuries include the following:

Keep Your Wrist in a Neutral Position

Avoid using your wrist in a bent (flexed downward), extended (upward) or twisted position for long periods of time. Instead, attempt to maintain a neutral (straight) wrist position whenever possible.

Watch Your Grip

Gripping, grasping, or lifting with the thumb and index finger may put undue stress on the wrist. Whenever possible, use the entire hand and all the fingers to grasp an object.

Minimize Repetition

If possible, avoid repetitive motions or holding an object the same way for extended periods of time.

Rest Your Hands

Periodically, give your hands a break by allowing them to rest briefly or alternate your hands by switching from one hand to the other.

Reduce Speed and Force

Decreasing the speed with which you perform a forceful, repetitive movement gives your wrist time to recover from the stress effort. Using proper power tools may also help to reduce such force.

Conditioning Exercises

Certain regular exercises strengthen the hand and arm muscles. They may prevent injuries by reducing the need to compensate for these weak muscles with a poor wrist position.

BACK, NECK AND SHOULDER INJURIES

Cases of back, neck, or shoulder injuries, such as muscle strains and pulls, are the most common injuries we observed amongst LOCATION-X (manufacturing) employees. These injuries were noted mostly in static or "fixed" positioned employees. We do not believe the problems were caused exclusively by lifting heavy objects.

Working positions that stress the body may contribute to a high incidence of back complaints. Many stressful working positions are considered to be hidden lifting tasks. A hidden lifting task exists when no object is being lifted other than the weight of an employee's own body, such as the torso when it is bent forward, or the arms when they are extended.

The further a body part is extended during factory work and the longer the duration of the extension, the greater the possible strain. And, the greater the strain, the more severe the pain and subsequent increase of frequency of injuries.

These types of tasks may result from workstation or task design that permits employees to assume awkward or bodily stressful positions. For some employees, this can also be the result of poor body mechanics or work techniques. Many of the tasks involve hidden lifting postures. According to our research, each of these has a cumulative effect on the employee's body, and can contribute to a back, neck, or shoulder injury.

ERGONOMICS CONTROL

Ergonomics of workstation design is generally considered as the most effective way to resolve such problems. Though Ergonomics may seem to be an expensive solution, please keep in mind that one case of Carpal Tunnel Syndrome may cost a company $15,000 to $20,000. A proper ergonomic design may be initially expensive, but will avoid injuries as well as the expenses related to their treatment. A major side benefit is improvement in employee morale and productivity.

<u>Short Term Goals for LOCATION-X</u>

a. Issue individualized gloves or braces to employees with early signs of CTS or who have been diagnosed with same.

b. Replace straight handled hammers with bowed ones. This will help reduce trauma/stress placed on arm and wrist areas.

c. Provide pliers and wire cutters with longer handles. The handle should be of sufficient length to fit across the entire width of the palm.

d. Evaluate work tables and reposition them for more effective work height. This should reduce unnecessary stress on arms, shoulders, back and neck areas.

<u>Long Term Goals for LOCATION-X</u>

a. Educate and train employees in proper hand movements for handling tools.

b. Replace existing chairs with "Biomechanical" chairs, as referenced in the AT&T manual "Workplace Design in the Manufacturing Environment".

OBSERVATIONS

Repetitive Motion Injuries and symptoms observed at LOCATION-X are described below:

<u>Wrist Injuries</u>

a. Carpal Tunnel Syndrome is a compression of the median nerve from inflamed tendons in the wrist

b. Tendentious and Tenosynovitis are both inflammations of the tendons from repeated tensing of forearms and hands

Elbow Injuries

a. Epicondylitis (tennis elbow) and Medial Epicondylitis (golf elbow) can result in tendon inflammations from jerky throwing motions and/ or rotation of the forearm.

Shoulder Injuries

a. Thoracic Outlet Syndrome is a compression of the nerves and blood vessels, often resulting from constant reaching over the shoulder.

b. Rotator Cuff Tendentious is yet another tendon inflammation which often is associated with constantly lifting elbows.

ERGONOMICS EVALUATION OF THE WORKSTATIONS

The symptoms, current status, comments, and recommendations for extreme cases of LOCATION-X are as follows:

a. Employee no. 1, **Large Power Winding Section.**

- Symptoms:
  * Wrist hurting from using hammer. A Cyst has appeared on top of the employee's right wrist.
- Current status:
  * Employee on Work restriction; employee's left hand is not effected by this disorder
- Ergonomics recommendations:
  * X-Ray wrist, elbow, and shoulder (right side)
  * learn to alternate hands
  * Redesign hammer handle or provide another hammer to prevent further damage to the wrist
  * If possible, do not flex elbow as much while hammering
  * Provide arm brace for better protection
  * Perform medical investigation to determine whether Cyst may also be a bone chip

b. Employee no. 2, **Large Power Winding Section.**

- Symptoms:
  * Painful thumb, torn ligament and limited movement in thumb
- Current Status
  * Job restriction is recommended
- Ergonomics recommendations:
  * EMG test before surgery
  * Should consider seeing a nerve specialist

* Learn to alternate hands
* Specially designed splint should be used for at least 6 Weeks

c. Employee no. 3, **Laminated Assembly Section.**

- Symptoms
  * Carpal Tunnel Syndrome of the right hand
  * Pain in the arm to shoulder
  * Right hand feels like needles in arm
  * Cannot grip work tools; problems holding objects
  * Gripping and twisting creates severe pain

- Current status:
  * X-Ray, bone spur against tendon
  * Mild CTS is suspected
  * EMG test was done in 1983 or 1984 and the results were positive
  * Employee was put on job restriction
  * Wrist brace on the right arm has helped this employee in reducing wrist stress

- Ergonomics recommendations
  * Alternate hands
  * Avoid twisting wrist
  * Per physician's direction, take vitamins; specifically Iron, B6, B12 and Calcium

d. Employee no. 4, **Laminated Assembly Section.**

- Symptoms
  * Bilateral wrist pain was displayed

- Current status
  * Restricted
  * Left wrist slightly better after surgery
  * Still burns and stings

- Ergonomics recommendations
  * Provide braces
  * Provide further medical evaluation

e. Employee no. 5, **Mini-Mercury Section.**

- Symptoms
  * Swollen wrist
  * Swollen fingers
  * Hurts near right elbow

- Current Status
  * Permanent restriction- no overtime
  * Limited repetitive motion, both wrist
  * If possible, job rotation to be provided
  * Had EMG test and was positive
  * Ability to rotate job at own discretion has decreased symptoms

f. Employee no. 6, **Ferrite Assembly Section.**

- Symptoms:
  * Swollen wrist
  * Swollen fingers
  * Hurts near elbow

- Current status
  * Permanent restriction-no overtime
  * Limited repetitive motion, both wrist
  * Had EMG test and was positive
  * Can rotate job at own discretion; has decreased symptoms

- Ergonomics recommendation
  * Needs proper air gloves to protect her hands

g. Employee no. 7, **Period Sensor Assembly Section.**

- Symptoms
  * Chronic bilateral pain in arm/shoulder/neck area
  * Certain bending motions cause pain

- Current status
  * CTS surgery was performed on this employee's left wrist in 1986
  * Does not feel better after surgery
  * Has been restricted from assembly position

- Ergonomics recommendations
  * Perform further medical review
  * Provide appropriate braces
  * Provide air gloves

h. Employee no. 8, **Period Sensor Assembly Section.**

- Symptoms
  * Knot on back of left hand
  * Hurts to raise arm
  * Bilateral problems, but more pronounced on left than right
  * Picking up small parts causes pain

- Current status
  * On temporary restriction
  * Diagnosed CTS 3 years ago
  * No apparent nerve damage
  * Job rotation has helped

- Ergonomics recommendations
  * Strengthen muscles in the hands by special hand exercises at rest periods during work
  * After work put hand in luke warm water, make fists
  * exercise-1/2 hour each day
  * Avoid twisting or turning wrist or elbow
  * Avoid lifting with palms up or wrists flexed

i. Employee no. 9, **Ferrite Assembly (Conformal Coating) Section.**

-Symptoms
 * Right wrist pain

- Current status
 * CTS-left wrist surgery 7/1989
 * Improvements since surgery
 * Permanent restriction-right hand

- Ergonomics recommendations
 * Should not wear protection on left wrist: This protection has decreased the flow of blood and nerve impulse
 * Provide better braces
 * Provide air gloves

Efforts should be focused on teaching employees the healthiest ways to use their hands to prevent CTS. Employees should be encouraged to report early warning signs (discomfort, limitation of movement, or recurring joint pain) when performing a task that requires frequent or forceful joint exertion.

SUMMARY

The continued cooperation and understanding of medical, management, and employees, towards a safe work environment can resolve most of the Ergonomics issues found at this location. Follow-up is recommended. Implementation of the findings can be achieved at reasonable costs with a high probability of positive results. In some situations, further medical work should continue in addressing individual situations.

REFERENCES

In developing this Ergonomics analyses of the AT&T LOCATION-X no outside references were used. Therefore, this study is original and a state-of-the-art evaluation of an existing manufacturing facility.

Advances in Industrial Ergonomics and Safety II
Edited by Biman Das
Taylor & Francis 1990

DEVELOPMENT OF MOTION SICKNESS SYMPTOMS WITH PROLONGED VDU USE: ARTIFACT OR REAL-EFFECT?

Stephen J. Morrissey
Department of Industrial Engineering, University of Wisconsin at Platteville, Platteville, WI. 53818

and

Alvah C. Bittner, Jr., Battelle HARC, Seattle, WA 98105-5428

A previous study by the authors suggested that prolonged VDU use could induce a motion sickness syndrome in users. Because of concerns about measures and methodology, the study was replicated using a more refined methodology. Eight subjects experienced four conditions (two VDU and two control). In the control conditions subjects either sat and read quietly or were allowed to perform normal activities for the experimental interval. Subjects were tested on a battery of measures sensitive to motion sickness effects. Results revealed that subjects in the VDU exposure groups experienced significantly greater motion sickness symptoms than in the control groups. Analysis of the control groups also revealed that these effects were due to the VDU exposure and not to fatigue. These findings point towards the potential of prolonged VDU use to induce a motion sickness syndrome in users.

BACKGROUND

Video Display Units or Terminals (VDUs, VDTs) have been the object of considerable study and controversy. Their extended or prolonged use has been reported to lead to a variety of signs and symptoms. These include: visual discomfort and fatigue; musculoskeletal discomfort and related biomechanical disorders; various subjective, physiological, and emotional problems; and other health effects (Dainoff, Happ and Crane, 1981; Evans, 1987; Smith, et al., 1981). Musculoskeletal discomfort and related problems are often suggested to be due to poor work station designs; designs that require the operator to adopt cramped, awkward postures for extended periods of time while performing the task.

Visually related complaints have included a wide variety of symptoms. These include: burning eyes; eye fatigue; headaches; difficulty in focusing; dizziness; nausea; and changes in visual accommodative process (Dainoff, et al., 1981; Evans, 1987; Howarth and Instance, 1985). The most common sources suggested for these effects are: (1) improper screen and workplace lighting that leads to glare and reduced screen contrast; and, (2) poorly designed VDU screens which have smeary, wobbly or flickering characters.

Individually or together these sources are known to result in strain and fatigue of the accommodative process (Dainoff, et al., 1981; Howarth and Instance, 1985; Miyao, et al., 1989).

These explanations may account for many reported visual problems, but they do not totally explain operator reports of "dizziness, nausea, confusion, and drowsiness." Reports that these symptoms continue even with major efforts to create "ergonomically" designed work stations, work environments and VDU displays (e.g., Morrissey and Bittner, 1986). These effects, when they are addressed, are often suggested to be due to "job stress", or to other ill-defined catch-all sources. None of these explanations is empirically satisfying.

Bittner and Guignard (1985) have suggested that such effects may be due to perceptual conflicts that generate a motion sickness syndrome. To examine this possibility, Morrissey and Bittner (1986) conducted a study to examine if the prolonged use of VDUs could induce such a syndrome in the users. In their study, subjects performed a simple, self-paced search task for three-hours in various VDU display conditions. Analysis of pre-exposure and post-exposure scores on a battery of tests known to be sensitive to motion sickness effects revealed significant motion sickness effects in users. This finding supported the conjecture that a three-hour VDU task could induce a motion sickness syndrome in the users. However, this exploratory study, while provocative and revealing, has two problems that prevent full acceptance of its findings. First, it did not employ control groups to allow determination if the effects observed were due only to the VDU exposure (and not to fatigue or other unrelated factors). Secondly, results of this study may have been limited by inadequate training on some of the measures used to assess motion sickness syndrome.

Purpose

This study is to address the limitations of the earlier study, and to determine if prolonged VDU exposure can induce a motion sickness syndrome in normal users.

METHODS

Subjects

Eight subjects, four males and four females, were recruited from an engineering program. All subjects were familiar with the use of VDUs, had normal or corrected to normal vision, and a mean age of 22.8 $\pm$ 1.1 years.

Procedures

Each subject was given a one-hour training session following their informed consent. During this session, they were shown the tasks to be performed and practiced until stable performance levels were established. On a later day, subjects returned for their first day of experimental participation in which they were randomly assigned to one of the four experimental conditions. Initially, each subject was briefly interviewed to assess whether they felt well, were not under any unusual stress, and had slept their normal amount the night before. Pre-exposure values were taken on different measures of motion sickness and subjects performed the chosen experimental condition for two and one-half hours.

Immediately following the experimental condition, the motion sickness tests were repeated. Subjects returned on later dates for the remaining experimental conditions. At least one-day of rest was given between each of the four experimental conditions.

<u>Experimental Conditions and Task</u>

There were two experimental classes, VDU exposures and Controls. Each experimental classification had two different conditions and all conditions were two and one-half hours long. In the VDU conditions subjects viewed a VDU in: (1) a low contrast display condition with grey characters presented on a black screen; and, (2) a high contrast display condition with high-intensity white characters on a blue screen. These two VDU combinations were chosen based on the author's previous study in which a low contrast display resulted in subjects experiencing a significantly greater motion sickness effect than did a high contrast display.

The two control conditions required subjects to either: (1) sit and read in a quiet, well-lit room for the experimental period; or, (2) engage in normal activity for the experimental period (appearing only for pre-exposure and post-exposure tests). The two control conditions provided greater power to determine the source of any effects found by the dependent measures.

The VDU conditions were conducted with subjects sitting at an adjustable VDU work station on which there was an IBM-Personal System/2 Model 8512 Color Display. This is a VGA monitor as compared to the CGA monitor used in the earlier study. Subjects had control of the intensity and contrast on the display and were allowed to adjust the display before beginning an experimental condition. The basic VDU task required subjects to: search two columns of letters and numbers of varying length for a target string (presented at the top of the screen); and, enter the row number where the match was found. The screen was changed after each entry.

<u>Dependent Measures</u>

Four different measures of various aspects of the motion sickness syndrome were employed. These included two tests of vestibular functioning; a questionnaire to evaluate the subjective signs and symptoms of motion sickness; and a test of perceptual-cognitive functioning.

The vestibular function tests were the stand-on-preferred-foot and stand-on-non-preferred foot tests. Both tests are reported to be highly sensitive indicators of the presence of a motion sickness syndrome as reflected by ataxia (Thomley, et al. 1986). The scores for each of these two tests were averaged to develop a final value.

The motion sickness questionnaire was that used by Bittner and Guignard (1988) to evaluate the subjective signs and symptoms associated with motion sickness. A factor analytic structure is available for this questionnaire which allows development of two independent scores for the major constellations (Factor 1, General Motion Illness; and, Factor 2, Retching/Vomiting Related). Development of scores for these factors was done for each subject in all conditions.

Perceptual functioning was evaluated using the Pattern Comparison test from the Peter Battery (Bittner, et. al, 1986). Previously, this test revealed significant effects due to prolonged

VDU use (Morrissey and Bittner, 1986). The questionnaire and the pattern comparison test were administered on a VDU using high intensity white characters on a black background. All VDU based tests had characters that tests were about 25 minutes in height at the average subject viewing location.

Experimental Hypotheses

Three research hypotheses were proposed based on the author's earlier study. First, there will be a significantly greater report of motion sickness symptoms in the VDU exposure groups than in the Control groups. Second, there will be no significant difference between the scores of the two Control groups. Third, there will be significantly greater motion sickness symptoms in the low-contrast VDU condition as compared to the high-contrast condition.

RESULTS

The resulting means and standard deviations for each of the experimental conditions are given in Table 1. For purposes of analysis, the data for the independent variables used in the analyses were transformed using the natural log transformation to stabilize the variances. The three variables were subsequently evaluated individually using repeated-measures analysis of variance (RANOVA) using the three stage approach recommended by Morrissey and Bittner (1990).

The structure of the analysis was: first, evaluate the equivalence of the pre-exposure groups across all measures; and second, examine post-exposure changes in the measures in accordance with the proposed hypotheses. The initial issue of concern was to determine if all subjects began with similar sets of responses to the dependent variables. The RANOVA's for each dependent measure was examined and no significant pre-exposure effects were found ($p > 0.15$, all tests). This established the essential equivalence of the populations and argued that only the post-exposure data need to be examined.

Analyses of the data for these three research questions are summarized in Table 2. Considering them individually, they broadly reveal that: For hypothesis one, the comparison of the VDU and Control groups, revealed significantly greater reports of motion sickness related effects, both subjective (i.e., questionnaire) and observable (posture stability) in the VDU groups as compared to the Control. This indicates that the VDU conditions induced subjective and vestibular effects in the subjects that are similar to those reported in more traditionally generated motion sickness syndromes.

The test of the second hypothesis involved comparison of the two Control conditions to establish to what extent the dependent measures might reflect fatigue related effects due to sitting (for the experimental period). Analysis of the data revealed no significant difference between the two Control conditions on four of the five dependent measures ($p>0.1$, all tests). These results supported the view that the differences observed between the Control and VDU exposure groups on these measures were due to VDU exposure and not fatigue or other unrelated effects.

The test of the third hypothesis involved comparison of the two VDU conditions. The earlier study by the authors, it is noteworthy, found significant differences between the two VDU exposure conditions. Table 2 reveals no significant difference between the two conditions on any dependent measure. The lack of a significant result in this study is suspected to be due to: (1) the VGA monitor used in this study which has substantially higher resolution than the CGA monitor used in the original study; and possibly, (2) to better training of the subjects in this study.

DISCUSSION

These results support the hypothesis that prolonged use of VDUs has the potential to result in the development of the classical signs and symptoms of motion sickness originally presented by Bittner and Guignard (1985). In addition, these findings parallel those reported in the earlier study by the authors (Morrissey and Bittner, 1986). Prolonged VDU use has again been shown to result in signs and symptoms that are similar to those observed cases of motion and simulator sickness.

That the motion sickness symptoms observed in this study are a function of VDU exposure was confirmed by analysis of the control groups. However, development of these symptoms did not appear to be affected by the VDU contrast, as was found in the earlier study by the authors, nor did the use of a high resolution VGA affect their development.

Additional tests may be recommended to further examine the structure of these results. In particular, studies are recommended to establish if the various physiological effects commonly associated with motion sickness occur with prolonged VDU use. These effects include pallor, sweating, changed respiration rates, and tachygastria (Frank, Casali, and Wierwille, 1987; Stern, et al., 1985). If these effects are found to occur with the prolonged use of VDUs, then it can be concluded that prolonged VDU use induces the full range of the motion sickness syndrome. Such a finding would be significant in worker placement and scheduling and in development of countermeasures.

ACKNOWLEDGEMENT

This study was supported by a State of Wisconsin Scholarly Activity Grant.

REFERENCES

Bittner, A.C., Jr., and Guignard, J.C., 1985. Human factor engineering principles for minimizing adverse ship motion effects: Theory and Practice. Naval Engineers journal, 97(4), 205 - 213.

Bittner, A.C., Jr., and Guignard, J.C., 1988. Shipboard evaluation of motion sickness incidence. In: Trends in Ergonomics/Human Factors V. Elsevier Science Publishers.

Bittner, A.C., Jr., Carter, R.C., Kennedy, R.S., Harbeson, M.M., and Krause, M., Performance evaluation tests for environmental research (PETER): Evaluation of 114 measures. Perceptual and Motor Skills, 63 (1986), 683-708.

Dainoff, M.J., Happ, A., and Crane, P., Visual fatigue and occupational stress in VDT operators. Human Factors, 23(1981), 421-438.

Evans, J., Women, men, VDU work and health: a questionnaire survey of British VDU operators. Work and Stress, 1987, 1(3), 271-283.

Frank, L.H., Casali, J.G., Wierwille, W.W., 1987. Effects of visual delay and motion system delays on operator performance and uneasiness in a driving simulator. Proceedings of the Human Factors Society - 31st Annual Meeting, 492-496.

Howarth, P.A. and Instance, H.O., The association between visual discomfort and the use of visual display units. Behavior and Information Technology, 1985, 4(2), 131-149.

Miyao, M., Hacisalihzade, S.S., Allen, J.S., and Stark, L.W., 1989. Effects of VDT resolution on visual fatigue and readability: an eye movement approach. Ergonomics, 32(6), 603-614.

Morrissey, S.J. and Bittner, A.C. Jr., 1986: Vestibular, perceptual and subjective changes with extended VDU use: A motion sickness syndrome? In: Trends in Ergonomics/Human Factors III, W. Karwowski (Ed.), 259-265.

Morrissey, S.J. and Bittner, A.C. Jr., 1990. Options for analysis of repeated measures designs: Watching-out for the assumptions. In: Advances in Industrial Ergonomics and Safety, II.

Smith, M.J., Cohen, B.G.F., Stammerjohn, L.W., and Happ, A., An investigation of health complaints and job stress in video display operations. Human Factors, 23 (1981), 387-400.

Stern, R.M., Koch, K.L., Leibowitz, H.W., Lindblad, I.M., Shupert, C.L., and Stewart, W.L., 1985. Tachygastria and motion sickness. Aviation, Space, and Environmental Medicine, 56, 1074-1077.

Thomley, K.E., Kennedy, R.S., and Bittner, A.C., Jr., Development of postural equilibrium tests for examining environmental effects. Perceptual and Motor Skills, 1986. 63, 555-564.

Table 1. Means (standard deviation) for the dependent variables.

		Stability Preferred Foot	Stability Non-Preferred Foot	Factor 1	Factor 2	Pattern Comparison
VDU1	Pre	18.4(8.7)	18.5(9.3)	7.1(1.6)	2.6(0.2)	38.9(5.6)
	Post	13.6(9.3)	11.7(6.7)	7.7(1.3)	2.6(0.2)	35.2(6.1)
VDU2	Pre	18.6(11.0)	18.7(11.5)	7.4(1.9)	2.6(0.4)	35.4(5.8)
	Post	12.8(7.4)	13.0(6.6)	8.3(1.9)	2.8(0.5)	37.7(4.8)
C1*	Pre	18.3(8.6)	18.5(10.9)	7.2(1.2)	2.5(0.2)	36.1(6.7)
	Post	17.4(8.3)	17.3(7.2)	7.8(1.8)	2.6(0.3)	38.1(6.1)
C2*	Pre	21.6(9.1)	19.1(11.1)	6.5(0.9)	2.5(0.2)	36.9(5.5)
	Post	21.2(8.5)	20.4(8.7)	6.6(0.8)	2.4(0.1)	38.2(5.5)

C1 is Reading Only Control Condition, C2 is General Activity Control Condition.

Table 2: Summary results for tests on hypotheses.

Measure	Hypothesis 1: VDU-Control	Hypothesis 2: Within Control	Hypothesis 3: High-Low Contrast
Factor 1	$P < 0.05$	$P < 0.05$	N.S.
Factor 2	$P < 0.05$	N.S.	N.S.
Posture Stability	$P < 0.0005$	N.S.	N.S.
Pattern Comparison	N.S.	N.S.	N.S.

Advances in Industrial Ergonomics and Safety II
Edited by Biman Das
Taylor & Francis 1990

A METHODOLOGY FOR ASSESSING STRESS REACTIONS TO COMPUTER-INTERACTIVE TASKS

Joseph Sharit, Sara J. Czaja, Charles T. Brunskill, and David R. Bogart

Department of Industrial Engineering
State University of New York at Buffalo
Amherst, New York 14260

INTRODUCTION

Computer and communication technologies are being rapidly introduced into most occupational settings. Consequently, many job functions will involve some form of human-computer interaction. The changes these technologies impose on the way work is performed raises a number of questions regarding the potential impact of computer-interactive tasks on the performance and health of the worker. While most research that has been conducted to investigate the possible detrimental side effects associated with computer tasks has examined the incidence of visual and physical stress (e.g., muscular aches and pains), few of these studies have examined stress as a function of the mental workload associated with computer tasks. Given the increased demand on information processing generally induced by computer-interactive tasks, one problem surrounding this issue is identifying a methodology that assesses stress as a function of the mental workload for these types of tasks (Czaja and Sharit, 1988).

We are currently involved in a research project whose primary objective is the development and implementation of a methodology for evaluating stress associated with computer-interactive task performance. The theoretical framework of stress adopted views stress as an imbalance between demand and capacity (McGrath, 1970), and integrates both task-related factors which may induce stress and age-related individual differences in capacity.

The methodology encompasses physiological, subjective, and performance measures. It also includes procedures for data analysis that allows us to specify certain patterns of stress response throughout task performance in addition to overall differences in stress response that occur as a function of task parameters and individual differences.

In this study, each subject performs each of three computer-interactive tasks in either paced or unpaced mode. These tasks have been modeled after real-world tasks and, consistent with the nature of these tasks, can be described in terms of information-processing demands that are systematically varied by trading-off the degree of decisional complexity with the extent to which rate of responding is emphasized.

The focus of this paper will be on some of the equipment aspects of the study, especially those dealing with physiological recording and data collection. With respect to these concerns, the overall rationale for the methodology employed was based on meeting the following set of objectives:

(1) to identify a relatively small set of physiological measures containing a theoretical basis for exhibiting sensitivity to both aging and changes in information processing demands;
(2) to physically separate the locations for experimental task performance and physiological recording and data collection;
(3) to perform real-time data collection;

(4) to perform both real-time assessment of the quality of the signals as well as ongoing monitoring of their output values;
(5) to achieve real-time synchronization of task-related parameters and human performance with physiological output;
(6) to effectively organize all data for future data analysis; and
(7) to insure minimal intrusion of the overall system on the task performance of the subjects, especially in view of the known reactivity of older persons to laboratory environments.

A description of the equipment set-up follows. The remaining sections will then focus on various software development issues related to the equipment set-up.

CONFIGURATION AND INTEGRATION OF PHYSIOLOGICAL RECORDING EQUIPMENT

Three physiological signals were obtained: heart rate, respiration, and blood pressure. Heart rate is obtained using three EKG disposable electrodes. Two are placed on each side of the subject at the level of the fourth rib and one ground electrode is placed below the electrode on the right side. These electrodes are attached to 40" adapter cables.

The respiration signal is obtained from a Model TCI 1R thermocouple respiration transducer manufactured by the Grass Instrument Company (1985). The transducer is affixed to the subject by a nasal clip which, for the 1 1/2 hour task durations used in this study, was not found to provide discomfort. This device is extremely reliable for measuring respiration rate and consequently for the derived measure of time between breaths. Other devices that enable recording of the respiration signal, such as the impedance pneumograph or the various types of respiratory "belts" that are available, are very susceptible to motion artifacts. This was determined during extensive preliminary investigation of alternative approaches to respiratory monitoring. Since the tasks employed in this study required the subject to move their hands and trunk reasonably about a well-defined workspace, as is the case in any realistic computer-interactive task environment, it was necessary to adopt a method that was both relatively unobtrusive and unaffected by such movements. The nasal clip was attached to an 8" wire which, in turn, is connected to 4 feet of shielded cable terminating in pin plugs.

The EKG and respiration cables are each attached to input cables that run through an opening slightly above the floor through the wall that faced the back of the subject (Figure 1). The cables terminate into separate input channels of a Sensormedics R-511A dynagraph recorder located in a separate room adjacent to the experimental room. The dynagraph serves to amplify and condition the signals. It then inputs these signals to the Data Translation board which converts the heart rate and respiration signals as will be discussed in the following section. Aside from serving as the interface between the subject and the Data Translation board, the dynagraph is essential for fine-tuning adjustments often necessary for each individual subject and for real-time monitoring (Figure 2). It also provides a hard copy of the data.

In addition to heart rate and respiration, blood pressure measurements are also taken, but only during the 10-minute baseline and recovery periods that preceded and followed performance of each task. Blood pressure measurements are obtained using a Finapres Continuous Noninvasive Blood Pressure Monitor manufactured by the Ohmeda Corporation (1986). At the onset of the baseline and recovery periods, the subject places an appropriate-sized photoplethsmographic cuff about the middle finger and subsequently removes it following the conclusion of these periods. This device is located in the control room (Figure 2) adjacent to the experimental task, and is connected through cables running alongside the EKG and respiration cables (Figure 1) interfaced to the subject's work area. Following the conclusion of each period in which blood pressures are taken, the continuous set of systolic, diastolic, and mean blood pressure measurements are printed out on a printer located next to this device (Figure 2).

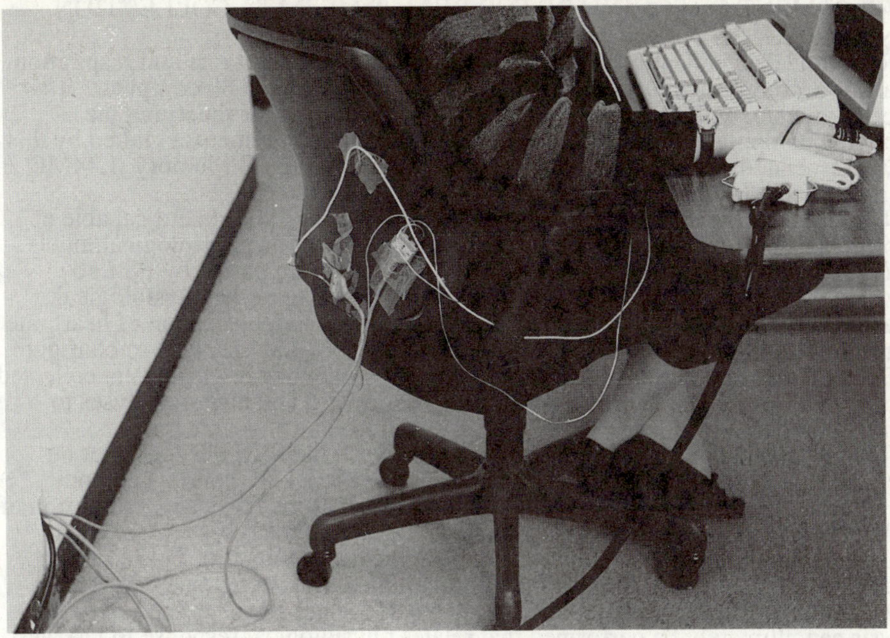

Figure 1. Illustration of the heart rate, respiration, and blood pressure cables routed through a wall opening into the adjacent room. The blood pressure cuff that the subject places about her middle finger during the baseline and recovery periods is also illustrated.

Figure 2. The physiological equipment area located in the room adjacent to the experimental task room.

DEVELOPMENT OF SOFTWARE FOR REAL-TIME DATA COLLECTION, MONITORING, AND SYNCHRONIZATION

Real-time data collection, including acquisition of the heart rate and respiration signals, is accomplished through the use of an IBM PS2 Model 80 computer. This device is capable, with some in-house software adjustments, of satisfying the demanding computational data acquisition and storage requirements entailed by this study. Associated with this system is a 70MB hard drive, 2MB Memory, 1.44MB 2-1/2" disks, and a Math Coprocessor.

This computer is interfaced to a Data Translation (DT 2901) board capable of performing A/D and D/A conversion. This board receives its physiological inputs ($\pm$5VDC) directly from the dynagraph. The DT board is supported by PS Lab software whose primary function is to control the sampling of the physiological signals. Our own software written in Fortran controls the identification of heart rate and respiration responses, and employs the purchased PS Lab software to configure the DT 2901 and to control the sampling rate and signal voltages. The data collected is stored in a form that will enable various types of micro and macro analyses to ultimately be performed.

The DT board also serves as the interface between the IBM PC computer on which the experimental tasks are performed and the Model 80 (control) computer. At each relevant point during the experimental task, such as the onset and conclusion of baseline, task, and recovery periods as well as the onset and completion of each problem within the task, a signal is output from the task computer. This signal is detected by the control computer which, through the master program (software), integrates this information with the ongoing physiological signal collection so that analysis can ultimately be performed that relates physiological response to task and performance parameters.

Figure 3. Illustration of HALO graphics used for real-time monitoring of the heart rate and respiration signals.

Detailed real-time physiological system monitoring is accomplished through the use of HALO graphics. HALO is a Fortran-callable graphics package used to set up the control computer's display screen (Figure 3). The headings, borders, and descriptors are placed on the screen during the set-up procedure. The descriptors are essentially various types of information concerning the subject's ongoing heart rate and respiration responses, where values are updated as the subject's status changes. Additional software was written to ensure that the screen updating was sufficiently rapid; e.g., the screen was updated just after each heartbeat occurred. This system together with the dynagraph output provides a user-friendly powerful real-time monitoring system which was especially critical during the piloting stages.

DEVELOPMENT OF SOFTWARE FOR DEFINING AND COLLECTING PHYSIOLOGICAL SIGNALS

The PS Lab Software was, by necessity, purchased from its manufacturer. All other software related to this study was written in-house. This consisted of software for:

(1) generating the experimental tasks;
(2) analyzing the subjects' performance of these tasks;
(3) coordinating the real-time physiological data collection with the PS Lab software;
(4) performing data acquisition;
(5) detecting heart rate and respiratory signals; and
(6) coordinating task events with the physiological data.

By writing our own software for detecting the heart rate and respiration signals, we were able to tailor the respective algorithms to the effects on these signals that were dictated by our equipment. Furthermore, it allowed us to achieve the reliability we felt we required and make alterations where necessary. While this effort was relatively straightforward for the well-defined heart-rate signal, it was a much more complex problem determining the presence of breaths. The algorithm we developed needed to <u>adapt</u> to individual variability in respiration, e.g., to the relative degree of nose-to-mouth breathing. It also needed to adequately determine an appropriate <u>threshold</u> magnitude below which changes in the signal would not qualify as breaths.

During its development, the software was subjected to a wide variety of respiratory maneuvers and in this manner <u>trained</u> to be adequately sensitive. All project-related personnel compared on-line printouts of the signal with the algorithmic predictions. Systematic fine-tuning adjustments were made to the algorithm until consistency was achieved between the software's output and the hard copy data.

The primary difficulty in developing the software for heart rate detection was in distinguishing the differences between <u>noise</u> (e.g., resulting from extreme movements by the subject that could result in temporary displacement of the electrodes) and true R-waves, and in handling <u>shifts</u> in the signal that would occur when, e.g., the subject reaches down and then moves back in place. However, this is a relatively trivial problem.

Finally, it should be noted that the data acquisition software also required validation. This was achieved by generating plots and comparing them with the hard copy output from the dynagraph recorder.

ACKNOWLEDGEMENT

This material is based upon work supported by the National Institute on Aging under Grant No. AG06794-02.

REFERENCES

Czaja, S. J., and Sharit, J., 1988, Aging and the performance of computer-interactive tasks: job design and stress potential. In <u>Trends in Ergonomics/Human Factors V</u>, edited by F. Aghazadeh (Amsterdam: North Holland), pp. 185-190.

Grass Instrument Company, 1985, <u>Thermocouple respiration transducer, Model TCI 1R</u>, Quincy, MA, 02169, USA.

McGrath, J.E., 1970, A conceptual formulation for research on stress. In <u>Social and Psychological Factors in Stress</u>, edited by J.E. McGrath (New York: Holt, Rinehart, and Winston).

Ohmeda, 1986, <u>Ohmeda 2300 Finapres Blood Pressure Monitor</u>, Englewood, CO 801 112 5810, USA.

Advances in Industrial Ergonomics and Safety II
Edited by Biman Das
Taylor & Francis 1990

The Effects of Visual Display Units Used in Industrial Tasks on Traditional Progress (Learning) Curve Models.

Robert Yearout and Catherine Sparks
Department of Management
University of North Carolina at Asheville
Asheville, NC 28804-3299
and
Cyrus Altimus
School of Business
Virginia Union University
Richmond, VA 23220

This paper investigates progress (learning) curves for tasks requiring high manual dexterity where information is gained through a video display unit (VDU). An initial test for both control and test groups yielded a 91 percent rate with steady state being reached at the 30th iteration. Subjects were then tested four weeks later. For the control group, the curve rate was 97.4 percent. The rate for the test group was 94.5 percent. Forgetting was significant until the 45th iteration. Thus production managers for high technological tasks should no longer rely on the traditional progress curves.

BACKGROUND

Learning curves, progress curves, cost reduction curves, manufacturing progress functions, experience curves, etc. are based on the theory that if a quantity on a given lot size is increased, the labor required is reduced in some proportion to the increase in quantity. McCampbell and McQueen (1965) observed that progress curves are a function of the complexity of the part, labor involved, tooling, equipment used, design changes during production and the ratio of the total quantity to the average lot run. These curves relate the direct-labor hours required to perform a given task to the number of times the task has been performed. This can be expressed as a cycle time for a very low level task to the completion of a major end item such as an aircraft. Since most manufacturing operations combine manual and mechanical elements, learning can be limited by an automated operation. Titleman (1987) found that the learning function may be dependent upon the ratio between manual and automated portions of the operation. Learning curves differ in the time required to complete the first iteration of a task and the curve's slope which defines the learning function. In the initial iterations, improvement is quite evident. However, as the operator becomes familiar with the task there are fewer fumbles, delays and hesitations. Thus the time required to complete an iteration becomes constrained by the physical characteristics of the task. Conley (1970) found that any further reduction in assembly times becomes so small that it appears that the curve is static. This condition is described (Evans, 'et al.', 1984) as steady state. Though many models have been developed to describe the learning function, one of the most universally accepted is the negative exponential model as shown in equation 1.

$$y = ax^{-b} \qquad (1)$$

Where y is the time required to complete the task for a given iteration, a is the time required to complete the task for the first iteration, x is the number for a given iteration and b is the learning function. Equation 2 is the equation for the exponential constant b.

$$b = (-\ln p \ / \ \ln 2) \qquad (2)$$

Where p is the learning rate as defined by the ratio of manual to automated work. As a rule of thumb, if the ratio of manual to automated work is 3 to 1 (three-fourths manual), then 80 percent is a good value; if the ratio is 1 to 3, then 90 percent is often used. Evans,'et al.' (1984) reports that an even split of manual to automated work produces an 85 percent curve. Konz (1987) describes several learning rates and the number of iterations required for an operator to reach that rate. Some typical examples are as follows: machining and fitting a small casting yields a 74 percent rate after 50 iterations and a punch press operation yields a 95 percent rate after 10,000 iterations. Regardless of the manual dexterity required to perform these tasks, the operator gained task information by direct observation.

The forgetting function or amount of decay in the speed of performing the task depends on how much has been learned or where the task is interrupted in the learning cycle. Some forgetting curves will show a rapid initial decrease in performance followed by a gradual leveling off as a function of the interruption interval period. The rate and amount of forgetting decreases as an increased number of units are completed before the interruption occurs. Carlson and Towe (1976) reported the amount of the forgetting and the corresponding level of performance is a function of both the performance at the time the project was interrupted (or total amount learned) and the time of interruption. Therefore, 100% retention is reflected by no change in time per unit (ie., cycle time).

METHOD

Introduction

This study was conducted to determine the forgetting function of a very high cognitive task where information was obtained from a VDU. The experiment was conducted in two parts as follows; part 1 was the initial testing to determine the learning rate and part 2 was the follow-up testing at the end of 28 days to determine the forgetting rate.

Task

Subjects performed a simulated industrial task by completing the first screen of the video game Popeye on a Commodore 64 computer. A WICO Command Controller, model number 2456559730 joystick was used. The controller was manipulated with one hand while resting on the table in front of the VDU. Required reach and grasp motions were activated by a punch trigger on the top of the controller. Directional movements were manipulated by the controller. Subjects performed the task at 15 min intervals for a total of 120 min. Scheduled breaks were required to compensate for any fatigue bias. Between each interval subjects were required to take a 3 minute standing break. There was also a 10 minute break at the end of the fourth test interval. Prior to collecting data, subjects received an orientation and were allowed 2 practice trials. Total experimental time was 3 hours.

Subjects:

Eight subjects were selected from a group of volunteers from the Asheville area. The 4 female and 4 male subjects were not familiar with the video game. The subjects' ages ranged from 18 to 62 years (mean = 28.7 years).

Design:

Subjects were divided into 2 groups of 2 males and 2 females each. These groups then were designated as either the control group or the test group. Subjects were then randomly assigned a time for their initial testing. Each subject was then given a follow-up test 28 days from their initial test date. The test group was instructed to refrain from participating in any activity that could approximate their task. The control group was required to perform their task a minimum of 30 min per week for 4 weeks. All testing was done at the same workstation under the same conditions.

Procedure:

All tests were performed in the UNCA Owen laboratory. Test time was 3 hours. This included a 15 min subject briefing, a 5 min period for testing procedure familiarization, eight 15 min data collection periods and a debriefing at the completion of data collection. To insure a uniform degree of difficulty only data collected from the first screen of the video game was used.

Criteria:

The criteria for the task was cycle time and its corresponding iteration number. Cycle time for the video game was computed by dividing the time per iteration by the point score for that iteration.

RESULTS

Initial test:

Steady state was determined by the following procedure. A graph of the cycle time means for each task was plotted against its respective iteration number. The starting point for the analysis was at the point where the curve appeared to become flat. Pairwise comparisons for data groupings of 10, 5 and 1 cycle time means were used. Due to unequal variances, Satterthwaite's approximation as described by Milliken and Johnson (1984) was used. For a significance level of 0.05, steady state for the video task was reached at 30 iterations.

A regression analysis was performed on the data. The resulting model (equation 3) will be referred to as model 1. Its learning rate was 91 percent. The model's mean square error and multiple correlation coefficient were 0.04573 and 0.31 respectively.

$$y = 0.046x^{-.136} \quad (3)$$

Follow-up Test:

Steady state was reached at the 23rd and 33rd iteration for the control (practice) and test (no practice) groups respectively. There were significant differences at the 0.05 significance level for the mean cycle times of the groups. Means are shown in Table 1. The model for the control is equation 4. Its learning rate was 97.4 percent. Its mean square error and multiple correlation coefficient were 0.0293 and 0.5107 respectively. The resulting model for the test group was equation 5. The mean square error and multiple correlation coefficient were 0.0590 and 0.5655 respectively. Its learning rate was 94.5 percent.

Table 1. Cycle times for follow-up test.

GROUP	MEAN	GROUPING
CONTROL	0.0126	A
TEST	0.0180	B

$$y = 0.013x^{-0.038} \quad (4)$$

$$y = 0.032x^{-0.082} \quad (5)$$

Equation 4 (control group) and equation 5 (test group) will be referred to as models 2 and 3 respectively. All three models are displayed in Figure 1. The control group operators did not reach steady state during the follow up test until the 23rd iteration; there was a 72 percent decrease in the first time.

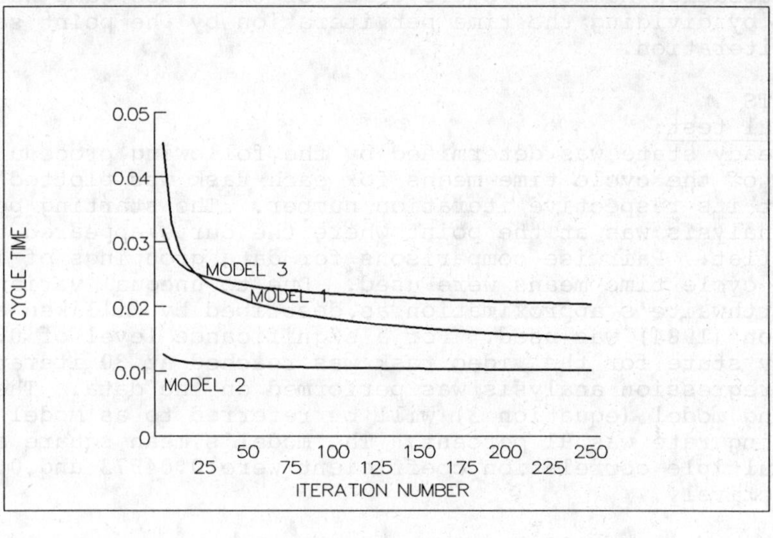

Figure 1: Progress curve models for VDU operators performing a highly cognitive task.

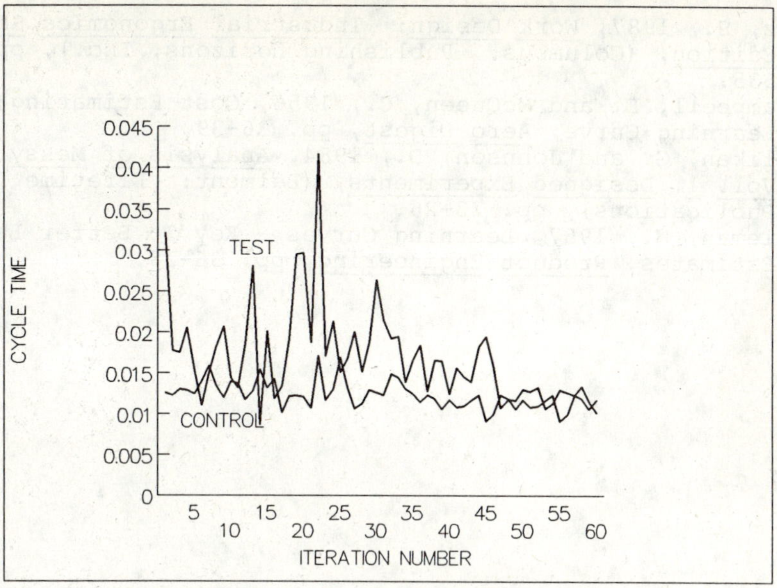

Figure 2: Cycle time means for test and control group operators.

The differences of 0.019 seconds/point and 3.2 percent describe the forgetting function and are depicted by the area between the model 2 and model 3 curves in figure 1.

The cycle times for both groups were no longer significantly different at approximately the 45th iteration. Thus any loss in time to forgetting was recovered by the 45th iteration. Figure 2 illustrates this convergence of the cycle time means for the two groups.

CONCLUSIONS

With increased technology, shorter production runs, the introduction of cellular manufacturing and increased use of VDUs in the production process, attention must be directed toward forgetting rates. The forgetting function's impact will be observed in processes where shorter production runs are becoming the norm. Thus the loss of production time for the earlier iterations becomes more significant. No longer should traditional learning curves developed in the 1950s to early 1960s be used by production managers for setting standard times. Managers and engineers should develop their own specific progress curves based on empirical data gathered from observing their most experienced and proficient operators.

REFERENCES

Carlson, J., and Rowe, A., 1976, How Much Does Forgetting Cost?, Industrial Engineering, Vol. 8, No. 9, pp. 22-25.

Conley, P., 1970, Experience Curves as a Planning Tool, IEEE Spectrum, pp. 63-68.

Evans, E., Anderson, D., Sweeney, D. and Williams, T., 1984, Applied Production and Operations Management, (St. Paul: West Publishing Company), pp. 140-144.

Konz, S., 1987, Work Design: Industrial Ergonomics Second Edition, (Columbus: Publishing Horizons, Inc.), pp. 519-535.
McCampbell, E. and McQueen, C., 1956, Cost Estimating From the Learning Curve, Aero Digest, pp. 36-39.
Milliken, G. and Johnson, D., 1984, Analysis of Messy Data, Vol. 1. Designed Experiments, (Belmont: Lifetime Learning Publications), pp. 23-26.
Titleman, M., 1957, Learning Curves - Key To Better Labor Estimates, Product Engineering, pp. 36-38.

Advances in Industrial Ergonomics and Safety II
Edited by Biman Das
Taylor & Francis 1990

THE ROLE OF JOB DESIGN FACTORS IN OFFICE ERGONOMICS

Soo-Yee LIM, Pascale C. SAINFORT
and Michael J. SMITH

Department of Industrial Engineering
University of Wisconsin-Madison
1513 University Avenue
Madison, WI 53706

The purpose of this project is to study the relationship between some selected job design factors, back pain and psychological stress outcomes. A total of 15 items divided into 5 categories were chosen for the study. The categories are postural load, repetitiveness of the jobs, rest breaks, length of breaks, and boredom. A total of 122 subjects participated in round 1 and round 2 of the study. The time period between the 2 rounds is approximately a year apart. The results concluded that there is no consistent job design predictors of back pain. Back pain is consistent in predicting psychological stress over time.

INTRODUCTION

In a previous study, a model was proposed to link direct and indirect effects of chair ergonomics on psychological stress outcomes (Lim et al.,1989). The indirect effect of ergonomic factors on stress through self report of back pain is examined. The study concluded that only one ergonomic variable, that is, subjective chair comfort, was related to back pain. Additionally, back pain was related to some of the psychological stress outcomes. The lack of association between chair ergonomics and back pain, and the strong relationship between back pain and psychological stress outcomes raises further questions on what actually influences the self report of back pain.

Results from several research studies concluded that job design plays an important role in psychological well-being of VDT workers (Sainfort and Smith,1989; Smith,1987). Smith et al.(1981) examined the relationship between job stressors, health complaints and psychological disturbances of the professionals and clerical VDT users and a group of non-VDT users. The study found that clerical VDT operators reported a higher level of irritability and fatigue as compared to the others.

In this study, we hypothesized that self report of back pain can be further explained by including some job design factors. Figure 1 shows the proposed model tested in this study. The elements in the proposed model are postural load

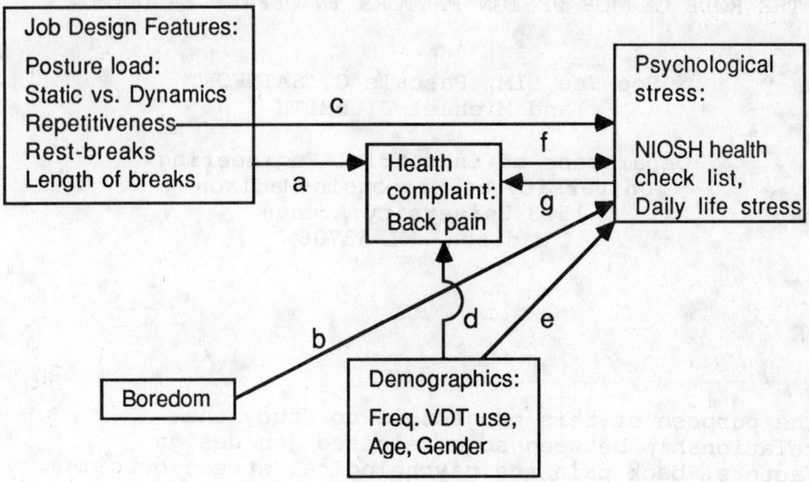

Fig. 1 The proposed Model

that is statics and dynamic load of the back, repetitiveness of the job, rest-breaks, length of the breaks, boredom, self-report of back pain and psychological stress outcomes. It is our hypothesis that the first four elements influence back pain (link a), that boredom and repetitiveness influence psychological stress outcome (links b and c respectively), that back pain has an effect on psychological stress outcome (link f) and that psychological stress outcome influences back pain (link g). Some demographic factors are included to control for possible effects on back pain and psychological stress outcomes (links d and e).

Grandjean (1987) pointed out the difference in physiological responses of the muscle between static and dynamic efforts. He explained that even moderate static work can produce localized muscle fatigue which may result in pain. He also pointed out that constrained postures are the most frequent form of static muscular work. It is this form of constrained posture that are carrying the trunk, head or limbs in unnatural positions. The postural load elements proposed in this study include elements in the job which would allow the workers to move around, to change positions or to stretch. These are considered to be dynamic aspects of work. On the other hand, job that requires the workers to sit for a long time, working in uncomfortable positions are considered to be the static aspects of work. Repetitiveness of job means that workers are required to perform the same type of job task or perform the same thing over and over again. In repetitious job, the fingers are in dynamic motion, however, the upper limb and the shoulder are static from holding out in a horizontal position while typing and the trunk is in static position to support the upper body and neck. Rest breaks and length of break include the number of formal rest breaks excluding meal (lunch) break in a day. Rest breaks may be important in relieving tension or static load.

In our proposed model, back pain is considered to be a symptom of stress which may be caused by environmental demands on the workers. In our model, the environmental demands are select job elements. Some authors have proposed that experiencing back pain is a source of stress (Mechanic et al.,1987) while others argue that experience of stress may lead to physical problems. The cause/effect relationship between back pain and psychological stress will be examined.

Repetitiveness and boredom are linked to psychological stress. Various research studies have linked these 2 job factors to stress outcomes (Smith et al.,1981; Frankenhaeuser and Gardell, 1976; Caplan et al.,1975).

METHODS

Subject Population

This study used data from an ongoing longitudinal study of office automation (Sainfort and Smith,1989; Sainfort and Lim,1989; Lim et al.1989). The study population includes employees from a public service organization in the Midwest. In this study, there are a total of 122 office workers who have completed 2 rounds of questionnaires. The time period between round 1 and round 2 of the study is approximately 1 year apart. A total of 177 workers participated in round 1 of the study and 136 workers participated in round 2 of the study. Those that are not included in this paper are workers who dropped out from the study in round 2 and those who did not participate in round 1. A total of 70% female and 30% male participated in this study.

Measures

The questionnaire used in both rounds of the longitudinal study was similar to questionnaire used in previous NIOSH studies (Smith et. al.,1981). From the questionnaire 15 items were chosen for the study. Postural load consists of 8 items, repetitiveness of 2 items, rest breaks of 2 items, length of break of 3 items and boredom was measured by constructing a 3-item scale based on Caplan et al.(1975).

Stress outcomes were taken from the NIOSH health checklist (Smith et al.,1981) and one was based on the Daily Life Stress Scale of Reeder et al.(1973).

RESULTS

Table 1 shows the pearson correlations between worker perceptions of job design factors and self report of back pain. The results from the correlation analysis shows that some postural load items such as working in uncomfortable positions or use awkward work motions, having to stretch or twist to reach work materials and holding arms in one position for long periods of time are correlated to self report of back pain for both rounds. In round 1, the amount of formal rest breaks taken is correlated with back pain. However, this is not found to be true for round 2 of the study. Other job factors such as repetitiveness of job and length of breaks do not correlate to back pain.

Due to the lack of space, the correlation coefficients between the independent variables will not be presented here. (if interested, please contact the authors for the results).

Regression analyses of job design factors and back pain were performed. In this analysis, back pain was used as a dependent variable (that is link a in fig.1). Another set of regression analyses of the stress outcomes was performed using back pain as independent variable (that is link f in fig. 1).

Table 2 shows the results of the regression analyses of self report back pain on job design factors and psychological stress outcomes. In this paper, only the significant regression results are presented.

In round 1 , we found that the number of formal rest breaks taken before meal break explained 7% of the variance of back pain. In round 2, we found that working in uncomfortable positions or using awkward work motions significantly contributed to back pain.

Back pain contributed to Daily Life Stress and tension for rounds 1 and 2. Back pain was found to be significantly related to depression and anxiety in round 1 but not in round 2. Regression analyses of demographics (i.e. VDT use (hrs/day), gender and age) and back pain, and psychological stress outcomes were done. In both rounds, we found that demographics did not explain any significant proportions of the variance of back pain. In round 1, we found that hours of VDT use contributed to daily life stress and depression. Gender also contributed to irritability, depression and anxiety. In round 2, demographics did not explain any significant proportion of the variance of psychological stress outcomes.

Two job design factors, i.e. boredom and repetitiveness were examined in relation to psychological stress outcomes. The results are shown in Table 3. Boredom was significantly related to irritability and depression in both rounds and tension in round 2.
Repetitiveness of the job was found to be related to daily life stress in round 1.

Cross-lag correlation was used to examine the cause/effect relationship between back pain and psychological stress outcomes. The result of this cross-lag correlation analysis is shown in Table 4. The results show that back pain influenced psychological stress outcomes, but not the reverse.

TABLE 1
Pearson correlation coefficients between back pain & job design variables for both rounds of study.

Postural load	Round 1 Back pain	Round 2 Back pain
1. During and average work day approximately how many hours do you work? Sitting	-0.03	-0.10
2. During and average work day approximately how many hours do you work? Moving around	-0.05	0.15
3. How often does your job require you to leave your work station and move around the office?	-0.00	0.13
4. How often does your job require you to work in uncomfortable positions or use awkward work motions?	0.28*	0.38*
5. Do you have to stretch or twist to reach necessary work materials often throughout the day?	-0.24*	-0.25*
6. Do you hold your arms in one position for long periods of time when performing your job?	-0.16*	-0.23*
7. Does your job allow you to change positions and sit or stand when you want?	0.03	-0.10
8. At work, do you sit or stand in the same position for several hours?	-0.01	-0.13
Repetitiveness		
9. During an average working day, do you perform the same type of job task, or do you change or rotate among a number of tasks?	0.07	0.01
10. To what extent does your job require you to do things that are very repetitious (do things over and over)?	0.06	0.11
Rest breaks		
11. Number of formal rest breaks before meal break.	0.27*	0.12
12. Number of formal rest breaks after meal break.	0.20	0.06
Length of break		
13. How long are your breaks? (meal)	0.15	0.10
14. How long are your breaks? (before meal breaks)	0.05	0.02
15. How long are your breaks? (after meal breaks)	0.09	0.0

TABLE 2
Beta-Coefficients of the regression analysis of back pain on job design of factors and psychological stress outcomes for rounds 1 and 2.

	Back pain	
Round #	1	2
Postural load:		
1. How often does your job require you to work in uncomfortable positions or awkward work motions	0.22 (R^2 = 11%)	0.32** (R^2 = 19%)
Rest breaks:		
2. Number of formal rest breaks before meal break	0.47* (R^2 = 7%)	0.31 (R^2 = 2%)
3. Daily life stress	0.60* (R^2 = 4%)	0.51*** (R^2 = 3%)
4. Irritability	0.27 (R^2 = 10%)	0.06 (R^2 = 0.5%)
5. Depression	0.15* (R^2 = 3%)	0.03 (R^2 = 0.1%)
6. Anxiety	0.29** (R^2 = 10%)	-0.04 (R^2 = 0.3%)
7. Tension	0.28** (R^2 = 8%)	0.20* (R^2 = 4%)

* $p < 0.05$
** $p < 0.01$
*** $p < 0.10$

TABLE 3
Beta-coefficients of the regression analysis of psychological stress outcomes on job design factors.

	Daily life stress		Irritability		Depression		Tension	
Round #	1	2	1	2	1	2	1	2
Boredom	0.07	0.13	0.10** (R2=8%)	0.12** (R2=11%)	0.07* (R2=3%)	0.10** (R2=8%)	0.02	0.11** (R2=6%)
+Repetitiveness	0.56* (R^2=5%)	-0.26	0.07	0.11	0.05	0.05	0.06	0.01

* $P < 0.05$
** $P < 0.01$
\+ to what extent does your job require you to do things that are very repetitious (do things over and over).

TABLE 4
Cross-lag correlations between back pain and psychological stress outcomes.

	Daily life stress		Irritability		Depression		Anxiety		Tension	
Round #	1	2	1	2	1	2	1	2	1	2
Round 1:										
Back pain		0.23*		0.14		0.13		0.13		0.24**
Round 2:										
Back pain	0.15		0.14		0.08		0.13		0.04	

* $P < 0.01$
** $P < 0.001$

DISCUSSION

From the study, there is no consistent job design factors between round 1 and round 2 that is related to back pain. The only job design factor that is related to back pain is working in uncomfortable positions or using awkward work motions in round 2. This factor is also found to approach significance in round 1. Additionally, rest breaks is found to be related to back pain in round 1 but not in round 2.

The lack of relationships between the independent variables and back pain can be explained by the possibility of time-delayed effects on back pain. In the questionnaire, respondents were asked about the experience of back pain within the past year. On the other hand, the job design factors was asked on the basis of current experience. There may be a time effect that links between job design factors and back pain. Thus, we would need to longitudinally examine workers to determine the causal factors of back pain.

Another explanation is that it is very hard to isolate factors that are related to or cause back pain. Several researchers reported that certain characteristics have been associated with increased risk of musculoskeletal complaints (NIOSH, 1984). These characteristics include design of the workstation, nature of the task, repetitiveness of the job, postural constraint, work pace , personal attributes of the worker and workspace. Specifics such as chair, workspace and tasks which restrict movements or cause static loading or constrained posture must be avoided. Contrary to this, we do not find in our previous study (Lim et al., 1989) that chair ergonomics contributed to back pain; only one chair ergonomic variable (i.e. chair comfort) was found to be directly related to back pain. On the other hand, this study did not find that repetitiveness of the job, work rest schedules, and postural load to be related to back pain. There are many other variables that are not included in the current study that may be predictors of back pain such as workspace (NIOSH, 1984), and personal attributes of the worker such as activities during leisure time, physical fitness of the workers and nature of hospitalization (Hultman, 1987).

This study is consistent in predicting the relationship between back pain and psychological stress outcomes in both rounds of the data. The result from the cross-lag correlation confirms that a one way relationship exists - that is back pain influenced psychological stress outcomes. This finding agrees with other researchers (Dehlin et al.,1977; Mechanic et al., 1987).

Boredom is linked to irritability and depression in both rounds. Repetitiveness of the job is linked to daily life stress in round 1 only. This seems to suggest that repetitious job has a greater impact on the workers daily life stress. On the other hand, boredom has a greater effect on the workers long term (i.e. within past year) psychological stress outcomes, such as experience of irritability and depression.

All three demographic factors are significantly related to some psychological stress outcomes. The number of hours spent on VDT per day has an impact on daily life stress of the workers. The results suggest that VDT exposure would also lead to depression over a certain period of time (within past year). In contrast, all demographic factors are not related to back pain. This suggests that our sample population does not have inherent characteristics such as age and gender that affect back pain.

There are some limitations in this study. The subjects are from a Midwest governmental agency. Thus, the results cannot be generalized to other populations. This means that the results do not necessary apply to all other VDT related operations. Another limitation is related to the measurement issue. In this study, back pain is self reported. It does not categorize back pain into thoracic spine symptoms or lower back symptoms. Thus, the data collected provides a broad scope of the problem. We have a limited number of participants in both rounds of the study. There is also a possibility that workers who dropped out of the study in round 2 may explained a significant proportion of the variance of back pain.

This study also some strengths. It is a longitudinal study of office automation and the sample population is quite stable (i.e. low turnover, no organizational or job changes). Thus, this population enables us to study time effects of job stressors on worker stress in a stable environment.

CONCLUSION

Our study found no consistent predictors of back pain in both rounds of study. However, back pain was related to psychological stress outcomes and the reverse was not true. Some job design factors such as repetitiveness and boredom were linked to psychological stress. Follow-up of this study will be useful to see if back pain is stable further into time to predict psychological stress.

REFERENCES

Caplan, R., Cobb, S., French, J., Harrison, R. and Pinneau, S., 1975, Job Demands and Worker Health. (Washington D.C.: U.S. Government Printing Office).

Dehlin, O. and Berg, S., 1977, Back Symptoms and Psychological Perception of Work. Scandinavian Journal of Rehabilitative Medicine, 9, 61-65.

Frankenhaueser, M and Gardell, B., 1976, Underload and overload in working life: outline of a multidisciplinary approach. Journal of Human Stres, Vol. 2, pp. 35-46.

Grandjean, E., 1987 in Ergonomics in Computerized Offices, pp.96-156. Taylor and Francis.

Hultman, G., 1987, The healthy back, its environment and characteristics: a pilot study. Ergonomics, Vol. 30, No. 2, pp. 295-298.

Lim, S.Y., Rogers, K., Smith, M. and Sainfort, P, 1989, A study of the direct and indirect effects of office ergonomics on psychological stress outcomes. In: Work with Computers:Organizational, Management, Stress and Health Aspects, edited by M. Smith and G. Salvendy, (Amsterdam: E lsevier Science Publishers B.V.), pp. 248-255.

Mechanic, D. and Angel, R., 1987, Some Factors Associated with the Report and Evaluation of Back Pain. Journal of Health and Social Behavior, 28 (June), 131-139.

NIOSH, 1984, Health Issue - Video Display Terminals. U.S. Department of Health and Human Services, pp. 21-25.

Sainfort, P and Lim, S.Y., 1989, A longitudinal study of stress among VDT workers: Preliminary results. In: Work with Computers:Organizational, Management, Stress and Health Aspects, edited by M. Smith and G. Salvendy, (Amsterdam: Elsevier Science Publishers B.V.), pp. 241-247.

Sainfort, P and Smith, M., 1989, Job Factors As Predictors of Stress Outcomes Among VDT Users. In: Work with Computers: Organizational, Management, Stress and Health Aspects, edited by M. Smith and G. Salvendy, (Amsterdam: Elsevier Science Publishers B.V.), pp. 241- 247.

Smith, M., Cohen, B., Stammerjohn, L., 1981, An Investigation of Health Complaints and Job Stress in Video Display Operations. Human Factors, Vol. 23, No. 4, pp. 387-400.

Smith, M., 1987, Mental and Physical Strain at VDT Workstations. Behavior and Information Technology, Vol. 6, pp. 243-255.

VIII

HUMAN PERFORMANCE AND WORKER SATISFACTION

Advances in Industrial Ergonomics and Safety II
Edited by Biman Das
Taylor & Francis 1990

INFORMATION FEEDBACK: THE CONCEPT, APPLICATION AND RESEARCH

Biman Das

Department of Industrial Engineering
Technical University of Nova Scotia
Halifax, Nova Scotia, Canada, B3J 2X4

Information feedback (IF) has motivating (goal-setting), reinforcing and directing properties. It affects operator's attitudes and perception of the work situation. Structural factors that affect IF are: rapidity of presentation, frequency of occurrence and amount of information, specificity and relevance, and information content. Interactive factors that influence IF are: medium of presentation, task characteristics, source level and monetary incentive. For improving worker productivity and satisfaction, both IF or production feedback and goal-setting or production standard are necessary.

INTRODUCTION

The positive effects of information feedback (IF) on learning and performance are well documented in the research literature (Ammons 1956, Meister 1976, Locke et al. 1981). The concept or principle is valid for individuals and groups, and also for a wide variety of learning, psychomotor, monitoring and other performance tasks. IF is a theoretical construct, the same as load or stress. The operator receiving IF does not visualize the feedback directly but infers it from the stimuli that present the IF, for example, a display or test grade. It is not possible to ascertain whether or not the operator has actually perceived and accepted the IF presented to him or her. Only through the operator responses, the effectiveness of the IF can be determined (Meister, 1976).

IF is generally (but not invariably) effective in improving operator productivity. This effectiveness, however, is described in psychomotor tasks, and in learning situations. The manner in which IF facilitates operator performance, however, has not yet been clearly established. There are occasional studies which have failed to determine significant effect of IF on performance (Chapanis 1964). This suggests the need for systemizing and integrating the research findings on this subject. The objective of this paper is to discuss (1) the concept of IF, (2) the application of IF, and (3) the research dealing with IF.

THE CONCEPT OF INFORMATION FEEDBACK (IF)

The concept of IF will be discussed under the following

headings: labels, definition, properties of IF and structural and interactive factors affecting IF.

Labels

Several terms have been used in the research literature to describe the same experimental procedures: feedback, information feedback or feedback information, achievement information feedback, reinforcing feedback, psychological feedback, performance feedback, knowledge of results, knowledge of performance, reinforcement, and reward (Meister 1976). Among research workers there has been a tendency to use feedback, knowledge of results or reinforcement in studies of positioning and selective learning, however, examples of the opposite can also be found. The term information feedback seems to be the choice of many authors. Consequently, for the purpose of this paper the term information feedback or IF is being used.

Definition

Information feedback means information available to the operator regarding his or her behavior which includes information about actual outcome and method of performing. As a result of stimulus events, whether called information feedback (IF) or knowledge of results (KOR/KR), the researcher studies the subject's perceptions, reasonings, and interpretations among the dependent events (Bilodeau 1966). Often a modifier is added to the basic terms (IF or KR) to denote a specific manipulation, such as, supplementary, augmented, and terminal feedback. Sometimes the specific term has a specific meaning, as for example, in uses of intrinsic versus extrinsic and actual versus learning feedback. Sometimes the specific term has a specific meaning, for example, in uses of intrinsic versus extrinsic and actual versus learning feedback.

IF can be expressed mathematically as a function of the discrepancy or difference between goal value and obtained operator responses (Bilodeau 1966).

$$IF_n = f(R_n - G) \quad (1)$$

where IF_n = information feedback stimuli, on trial n, under the control of the experimenter (or, production system designer), presented during the course, or at the end, of operator's response; R_n = quantitative expression of operator's response on trial n; and G = goal value.

The main objective of the experimental work is to ascertain the relationship between IF variables and operator behavior, that is, to express:

$$R_{n+1} = f(IF_n) = f(R_n - G) \quad (2)$$

Properties of Information Feedback (IF)

The operator uses IR to guide the performance of his or her next action. IF, therefore, has effects in feeding back and feeding forward. IF has the following three empirical properties: (1) strengthening responses, (2) sustaining performance, and (3) eliminating previously established responses. The following three constitute the theoretical properties of IF: (1) motivating, (2) reinforcing, and (3) directive. Logically, IF, as any stimulus, can have any one of three or all the properties.

IF is effective in modifying the operator performance by: (1) motivating the operator to adjust work output to other goals (hopefully higher), (2) reinforcing certain response patterns that are observed by the operator as being correct or successful, and (3) directing or guiding changes in the operator's procedures and performance. The above three factors need not be mutually exclusive as mentioned earlier. Depending on the task and the task conditions, the same IF may have all three effects, however, one or the other may be major (Meister 1976).

Motivational IF. Locke and Bryan (1969) maintain that motivational IF has few or no cue properties, because it cannot inform the operator regarding the nature or locus of his or her errors and does not suggest how to correct the errors. When information value is lacking from IF, the operator utilized that IF to adjust his or her performance goal.

Operator's goals play a major role in regulating the task performance (Locke 1968). IF causes the operator to adjust his or her performance goals, and thus in turn produces performance improvement. For making proper goal adjustment, the operator requires meaningful and usable IF. Most of the laboratory research studies dealing with motivational IF have been carried out in the learning situations, where it is usually effective in improving performance (Locke and Bryan 1969).

Intrinsic, Extrinsic, and Augmented IF. Intrinsic IF exists only as a physiological correlate of the individual response. It is present in the action preceding IF or in the whole task situation. Whereas, extrinsic IF is supplied by an external source, such as, information describing the correct response of an instructional sequence. It is possible that IF provided can be incomplete or imprecise, or the operator's perception and interpretation may be inaccurate.

An extrinsic IF in a learning situation is known as an augmented IF. It is represented by an explicit evaluation of operator performance, such as, a comparison between performance and requirements. Augmented IF is most useful in learning situation (from a learning situation), the effects of augmented IF are not clear.

IF Affecting Attitude and Perception. IF can affect the operator's attitudes and his or her perception of the work situation. However, it is not very clear how there effects help in generating improved performance (Lavitt and Mueller 1951). On the other hand, IF effectiveness is influenced by social and attitudinal factors of the operator and by the characteristics of the IF itself. IF seems to have influence on operator's job satisfaction and perception of production system performance. However, acceptable evidence on this subject is lacking.

Structural and Interactive Factors Affecting Information Feedback

The manner of IF presentation to the operator is important from the viewpoint of its effectiveness. Several structural factors as well as interactive factors or the context in which IF is provided affect IF.

Structural Factors. IF is affected by the following structural factors: (1) rapidity of presentation, (2) frequency of occurrence and amount of information, (3) specificity and relevance, and (4) information content. A brief discussion of the factors follows.

1. Rapidity of Presentation. IF can vary in terms of rapidity of presentation with which IF stimuli appear following an operator action. It is reasonable to expect that IF is more effective in cueing the operator's correct next response, the closer it occurs in time to the previous response.
2. Frequency of Occurrence and Amount of Information. IF can also vary with respect to the frequency of its occurrence as well as the amount of information it presents. They represent the quantitative nature of the IF.
3. Specificity and Relevance. The quality of IF depends on its specificity and relevance. For effective performance, logically, IF should be specific. The degree of specificity can be very important in some task situations. For example, in the context of a production system, an IF may provide information whether the part(s) produced is within acceptable tolerance limits. However, some other situations may require information not only whether the part has exceeded limits, but amount and direction of that excess. As the IF becomes more specific, the performance tends to improve, provided the operator has the need and can utilize more detailed information.
4. Information Content. IF content is of primary importance. The structural factors, such as, time, quantity and quality are secondary to and react with the basic problem of what IF should be presented. The nature of IF is task specific and therefore it can only be determined by the analysis of production system and task.

Interactive Factors. The following interactive factors affect IF: (1) medium of presentation, (2) task characteristics, (3) source level, and (4) monetary incentive. A brief discussion of the factors follows.

1. Medium of Presentation. Presentation of IF can be done orally, visually (through displays), directly (face to face communication), by intercom or radio link and in written form. Only through research, one can establish the most effective medium of presentation. Visual or verbal means seem to be most common, although written form is not uncommon. Visual displays have been found, in general, more effective than verbal communication.
2. Task Characteristics. IF has a strong bearing on the nature of the task. As a task grows in complexity, the consideration of structural factors of IF, such as, frequency, amount, and specificity becomes increasingly important. However, there is hardly any data on this subject.
3. Source Level. IF is more appropriate when it relates to the operator's own responses or task performance. It is less useful when it describes a higher level performance to which the operator only makes partial contribution. In other words, IF is less effective when it is confounded. Logically, the closer the source level of IF, the more effective the operator

performance should be. The source level of IF has significant effect on operator performance.

4. Monetary Incentive. IF interacts with monetary incentives to improve operator performance. However, the effect of monetary incentive alone appears to be weak and transient.

APPLICATION OF INFORMATION FEEDBACK

Information feedback concept has been employed primarily in laboratory research settings to determine their effects on operator performance and satisfaction, most often in psychomotor tasks and in learning situations. It has been found that information feedback is often but not invariably effective in (1) increasing the level of performance in tasks that are highly overlearned or repetitive, (2) increasing the improvement rate on new tasks early in practice, and (3) eliciting positive responses from the operators that tasks are more interesting and less tiring, when feedback is present (Gibbs and Brown 1955, Ammons 1956, Leamon 1974).

Very few experiments are reported in the research literature, which deals with real-life tasks. Gibbs and Brown (1955) report an increase of 25% in production output on an uninteresting, unskilled, and repetitive task, designed to represent an industrial situation by the introduction of mechanically presented, KR. When Chapanis (1964) attempted to replicate their findings in the laboratory settings on a Teletype tape punching task. Chapanis found no significant improvement in output with the introduction of KR.

Few field studies have been carried out in the actual industrial work situation. Hundal (1969) reported increased output with increases in the degree of knowledge of performance on a repetitive manual industrial task in actual work situation. However, the increase was not substantial. Leamon (1974) observed significant improvement in performance as a result of the introduction of KR on a repetitive hand work in industry. It is possible that in the experimental situation KR effect was confounded with the situational factors, such as, supervision, monetary incentive and operating method. It appears that even in a simple industrial situation, it is very difficult to ascertain what has been transmitted to the worker by KR.

The positive effects of information feedback and goal setting on operator performance have been well established. However, studies have shown that neither information feedback or goals alone is sufficient to improve performance. Both information feedback and goals are necessary to improve performance. If information feedback is not provided, goals may sometimes become meaningless. Specific goals with information feedback have been found to be superior to goal setting or information feedback alone (Erez 1977, Locke et al. 1981, Das 1982a, 1982b, Locke and Latham 1984).

RESEARCH ON INFORMATION FEEDBACK

Research evidence suggests that the incorporation of production standard and feedback in a work situation would be beneficial for improving worker productivity and satisfaction. Laboratory studies have shown that specific hard production standards have a significant positive effect on productivity. But it can be argued that when an operator is provided with a specific hard production standard but no feedback concerning how his or her performance compares to that standard, it is not possible for the operator to regulate the subsequent amount of effort expended on a task to reach or maintain the standard. On the other hand, if the operator is not assigned a standard, the provision of feedback is not relevant. The majority of the studies evaluated performance standard and feedback only on operator's performance output, worker satisfaction was seldom considered.

Research was performed to determine the effects of production feedback (PF) and production standards (PS), singly or jointly, on worker productivity and satisfaction in a repetitive production task (Das 1982a, 1982b). The task involved drilling four holes in a prepared steel plate. The summary results are presented in Table 1.

Table 1. Comparative analysis of worker productivity and satisfaction among groups: Student's *t* test

Main experimental conditions	Comparison between groups (experimental conditions or treatments)	Calculated Student's t value: Worker productivity: Quantity	Worker productivity: Quality	Worker satisfaction: Modified JDS
Production feedback (PF)	2 (PF:Quantity) vs.1(Control)	1.46	2.24*	0.57
	3 (PF:Quantity & quality) vs. 1	-0.07	0.60	3.44**
	3 vs. 2	-1.56	-1.75*	2.78**
Production standards (PS)	4 (PS:100% normal) vs. 1	1.49	1.08	2.39*
	5 (PS:130% normal) vs. 1	0.82	0.04	3.27**
	5 vs. 4	-0.65	-1.12	0.87
PF + PS	6 (PF:Quantity & quality + PS:130% normal) vs. 1	4.42**	3.97**	6.09**
	6 vs. 3	4.70**	3.73**	2.72**
	6 vs. 5	3.82**	4.50**	2.82**
PF + PS + Monetary incentive (MI)	7 (PF:Quantity & quality + PS:130% normal + MI:1:1) vs. 1	3.80**	3.03**	4.20**
	7 vs. 6	-0.47	-0.81	-1.83*

Note: The tabulated Student's *t* values for 5% = 1.68 (significant*) and 1% = 2.42 (highly significant**); Negative sign = decrease in group means.

The provision of quantity and quality feedback had no significant effect on worker productivity but had a favorable impact on worker satisfaction. Only the combination of quantity and quality feedback in the presence of an assigned hard production standard and feedback had a significant positive effect on both worker productivity and satisfaction. The increase in quantity and quality output were about 13% and 15% respectively compared to the control group. Monetary incentive had no beneficial effect on worker productivity and satisfaction. A highly significant interaction between production feedback and standard was found for worker productivity (Table 2). No such interaction was found for worker satisfaction.

Table 2. Interaction between production feedback (quantity and quality) and standard (130% normal): Student's *t* test.

Calculated Student's *t* value		
Worker productivity		Worker satisfaction
Quantity	Quality	Modified JDS
3.73**	2.71**	0.38

* $p<0.05$; ** $p<0.01$.

CONCLUSIONS

In summary, the conclusions reached are:

1. In general (but not invariably) information feedback (IF) is effective in improving performance both in learning and in simple psychomotor tasks in laboratory settings. Most of the experimental tasks used in the laboratory are simulated or relatively simple.
2. When IF is incorporated in real-life performance tasks to improve operator output, the results have often been conflicting. Probably adequate consideration was not given in the experimental situations to the structural factors such as, information content of IF and the manner of IF presentation to the operator.
3. IF has motivating (goal-setting), reinforcing and directing properties. IF and goal-setting process are inter-related.
4. IF seems to influence operator's job satisfaction. Social and attitudinal factors governing operator's behaviour can influence IF effectiveness.
5. Any IF is more effective than none at all. In providing IF to the operator, major consideration must be given to the information content of IF. Other structural factors, such as, frequency, amount of information and specificity should not be ignored.
6. It appears that operator performance improves with increasing IF, provided: (a) it is task relevant, (b) the operator has need for it, and (c) it does not confuse or overload the operator.
7. In general, presentation of IF by visual displays have been more effective than verbal communication.
8. IF is highly task specific. It is most effective when it gives the operator's output directly.
9. IF is a necessary condition for goal setting to be effective in improving operator performance.
10. The provision of IF or production feedback had no significant effect on worker productivity but had a favourable impact on worker satisfaction.
11. The combination of production feedback and an assigned hard

production standard had a significant positive effect on both worker productivity and satisfaction.

12. A highly significant interaction was found between production feedback and standard (goal setting) for worker productivity. No such interaction was found for worker satisfaction.

ACKNOWLEDGEMENT

The article draws considerably from the work of Meister (1976), which is duly acknowledged.

REFERENCES

Ammons, R.B. Effects of knowledge of performance: A survey and tentative theoretical formulation. The Journal of General Psychology, 1956, 54, 279-299.

Bilodeau, I. McD., Information feedback. In E.A. Bilodeau (Ed.) Acquisition of Skill, Academic Press, New York, 1966.

Chapanis, A., Knowledge of performance as an incentive in repetitive monotonous tasks. Journal of Applied Psychology, 1964, 48(4), 263-267.

Das, B., Effects of production feedback and standards on worker productivity in a repetitive production task, IIE Transactions, 1982a, 14(1), 27-37.

Das, B., Effects of production feedback and standards on worker satisfaction and job attitudes in a repetitive production task, IIE Transactions, 1982b, 14(3), 193-203.

Erez, M., Feedback: A necessary condition for the goal setting performance relationship. Journal of Applied Psychology, 1977, 62(5), 624-627.

Gibbs, C.B. and Brown, I.D., Increased production from the information incentive in a repetitive task. Medical Research Council, Applied Psychology Research Unit, Great Britain, 1955, No. 230.

Hundal, P.S., Knowledge of performance as an incentive in repetitive industrial work. Journal of Applied Psychology, 1969, 53(3), 224-226.

Lavitt, H.J. and Mueller, R.A.H., Some effects of feedback on communication. Human Relation, 1951, 4(2), 401-410.

Leamon, T.B., An investigation into the effects of knowledge of results on operator performance. Ergonomics, 1974, 17(5), 639-650.

Locke, E.A., Motivational effects of knowledge of results: Knowledge or goal setting? Journal of Applied Psychology, 1967, 51(4), 324-329.

Locke, E.A., Toward a theory of task motivation and incentives. Organizational Behavior and Human Performance, 1968, 3, 157-189.

Locke, E.A. and Bryan, J.F., Knowledge of score and goal level as determinants of work rate. Journal of Applied Psychology, 1969, 53(1), 59-65.

Locke, E.A. and Latham, G.P., Goal Setting: A Motivational Technique that Works. Prentice-Hall, New Jersey, 1984.

Locke, E.A., Shaw, K.N., Saari, L.M. and Latham, G.P., Goal setting and task performance: 1969-1980. Psychological Bulletin, 1981, 90(1), 125-152.

Meister, D., Behavioral Foundations of System Development. John Wiley, New York, 1976.

Advances in Industrial Ergonomics and Safety II
Edited by Biman Das
Taylor & Francis 1990

APPLYING ASSIGNED AND PARTICIPATIVE PRODUCTION STANDARDS WITH FEEDBACK TO IMPROVE WORKER SATISFACTION IN INDUSTRY

Biman Das and Ashraf A. Shikdar

Department of Industrial Engineering
Technical University of Nova Scotia
Halifax, Nova Scotia, Canada B3J 2X4

Worker satisfaction improved significantly in a repetitive production task in industry as a consequence of the provision of: (1) an assigned normal (100%) standard and feedback, (2) an assigned hard (140%) standard and feedback, and (3) a participative standard and feedback, compared to the control group. The participative standard with feedback was significantly better than the assigned standards with feedback. No significant difference was found between the assigned normal and assigned hard standards with feedback.

INTRODUCTION

The effects of goal setting and information feedback on worker satisfaction has been studied by researchers in the past. Studies have shown that hard goals generate less satisfaction than easy goals and specific hard goals generate more interest and less boredom than "do your best" goals (Locke 1968). In another study, hardest goal produced the highest performance but also generated the lowest degree of satisfaction (Locke 1981). Information feedback has often elicited positive responses from operators that tasks are more interesting and less tiring when feedback is present (Ammons 1956). When a specific goal is provided with feedback, worker satisfaction improved significantly (Kim and Hamner 1976).

Goals could be either assigned to the operators or set participatively with the operators. It is believed that participation in goal setting has a positive impact on worker satisfaction. White et al. (1977) found that goal setting had no direct or positive effect on worker satisfaction. Ivancevich (1976) reported that participative goal setting was not superior to assigned goal setting in terms of worker satisfaction. Some studies have reported that worker satisfaction declined in both assigned and participative goal setting conditions (Latham and Yulk 1976, Latham et al. 1978). When an assigned hard goal was provided with feedback, worker satisfaction improved significantly compared to the provision of an assigned hard goal or feedback alone (Das 1982). Often, the combined effects of goal setting and feedback on worker satisfaction produced mixed results.

It is believed that worker satisfaction or dissatisfaction is an emotional reaction which reflects value responses. It is a function of the perceived relationship between what the worker wants from his or her job or what he or she perceives it as being offered (Locke 1969, 1970). Differences in the nature of jobs or work situations cause differences in worker satisfaction. It is believed that both job content and context factors affect worker satisfaction on the job (Vroom 1964). Task satisfaction appears to be directly related to task success. Hamner and Harnet (1976) found that the subjects who exceeded their reference operators' outcome or performance were more satisfied than those who did not.

In most goal setting and information feedback studies, the tasks used were relatively simple or simulated and the goals were set arbitrarily by the experimenter or based on the operator's past performance. Measured standards (goals) were seldom used. The majority of the studies in goal setting and feedback did not deal with worker satisfaction. Studies which measured worker satisfaction used various measures. Repetitive production tasks were seldom used in such studies. Since such tasks are monotonous, boring and unmotivating, it would be beneficial to improve worker satisfaction in such tasks through the provision of goal setting and information feedback.

Considering the limitations of the past studies, a systematic controlled laboratory study was conducted using a repetitive production task and measured standards. Worker satisfaction was found to improve in both assigned with participatively set standards conditions. There was no difference in satisfaction between the conditions (Das and Shikdar, 1987, 1988). These results were obtained by using college students. It was believed that the results must be validated in industry using industrial workers, so that meaningful conclusions could be made to aid industrial managers.

The objective of this research was to investigate the effects of assigned and participative production standards with feedback on worker satisfaction in industry. The terms "production standard" and "production feedback" were used in place of "goal setting" and "information feedback", since a production task was used in the investigation.

EXPERIMENTAL METHOD

The details of the experimental methods were described elsewhere (Shikdar 1990). A brief description is provided below.

A large fish processing industry was selected to conduct the experiment. The selected task was a perch fish trimming operation, a highly repetitive production task, employing about 100 workers to do this particular task in one section of the plant. After explaining the details of the study, the workers were requested to volunteer for the study. The workers were unionized and there was an agreement between the union and the management to pursue the investigation.

The task was studied to determine a standardized method, production standard and feedback system. Methods-Time Measurement (MTM) was used to find the normal standard and due allowances were provided. The MTM standard was verified by a stopwatch time study. The hard standard was set at 140% of normal. The participative standard (above 100% normal) was set by the workers in consultation with the experimenter. Standard was provided on a special card. Feedback was provided on a feedback card in terms of production output (lbs/hr, % of normal standard achieved) every two hours. The special feedback card was designed so that performance could be visualized in graph form on a daily basis.

Thirty-two subjects were selected having at least six months of job experience and a minimum of seventh grade education. The subjects were randomly assigned to four experimental groups, consisting of eight subjects in each group. The groups received the following experimental conditions with respect to production standards (PS) and production feedback (PF): (1) No PS/PF (control), (2) PS:Assigned 100% (normal) + PF (quantity and % of normal standard achieved), (3) PS:Assigned 140% + PF, and (4) PS:Participative + PF. Groups were randomly assigned to experimental days within a 10 month period. The first day of the work was considered as a training day so that subjects could familiarize themselves with the standardized method and learn the activities of the experiment. The subjects worked under the experimental conditions for ten days.

Worker satisfaction was measured through the modified Job Diagnostic Survey (JDS) scales (Hackman and Lawler 1971, Hackman and Oldham 1975). The modified JDS scales included the following job or work dimension: (1) skill variety, (2) task identity, (3) task significance, (4) autonomy, (5) production standard, (6) production feedback, (7) working condition, (8) pay, (9) promotion, (10) supervision, and (11) coworkers. Each subject was asked 24 questions on a seven-point Likert-type scale regarding his or her perception of the various job attributes that were actually present. The questionnaires were given at the end of both training and experimental work sessions.

RESULTS

The results were analyzed through the Statistical Analysis System (SAS) computer program (Ray 1982). Analysis of variance (ANOVA), analysis of covariance (ANOCOVA) and Student-Newman Kuel's (SNK) range tests were employed for the data analysis. The results of the ANOVA are presented in Table 1. It showed that the differences among the four groups were highly significant ($F=6.10$, $p<0.01$).

The initial levels of the subjects perception or feeling of satisfaction in the present task situation might have conditioned the outcome of the experiment with regard to the perceived worker satisfaction. To equate this variation, an ANOCOVA was performed with the experimental data using the training worker satisfaction as the covariate. The results of the ANOCOVA are presented in Table 2. The results showed that the differences among the groups were highly significant ($F = 7.23$, $p<.01$). The results of the ANOCOVA confirmed the results of the ANOVA. The covariate was significant ($F=6.18$, $p<.02$), which indicated that subjects satisfaction score obtained after the training work session had a direct effect on the subjects satisfaction score obtained after the experimental work session.

Table 1. Analysis of variance (ANOVA) of worker satisfaction score (sum of 11 factors).

Source	df	MS	F	Pr>F [a]	r^2 [b]
Groups	3	152.10	6.10	0.01	0.40
Errors	28	24.94			
Total	31				

[a] Pr>F = Probability that a random F value would be greater than or equal to the observed value.

[b] r^2 = Coefficient of determination.

Table 2. Analysis of covariance (ANOCOVA) of worker satisfaction score (sum of 11 factors).

Source	df	MS	F	Pr>F	r2
Groups	3	456.30	7.23	0.01	0.51
Covariate	1	129.97	6.18	0.02	
Errors	27	21.05			
Total	31				

The SNK range test was employed for comparative analysis among groups. The previous analysis has established that the differences among various groups were significant. The results of the SNK test are presented in Table 4.

A wide variation in the worker satisfaction scores was found among the subjects of the same experimental group. The increase or decrease in worker satisfaction as a result of group treatment are presented in Table 3. The maximum increase was observed for Group 4 (participative standard with feedback) which was about 8, compared to the control group.

A comparison between Groups 2 (100% normal + feedback) and 1 (control) showed that there was highly significant difference between them. The provision of a normal production standard (100%) and feedback had a highly significant positive effect on the overall satisfaction of workers in industry. The result was consistent with the result of an earlier laboratory study (Das and Shikdar 1987). In this experimental condition the subjects had clear understanding of the specific normal standard or the amount of work expected on the job. Whereas feedback helped in the progress which in turn helped in performing better. These task attributes may have contributed in increased worker satisfaction.

Table 3. Increase or decrease in worker satisfaction scores (adjusted means).

Comparison between groups	Increase/decrease in worker satisfaction
2(100% + PF) vs. 1 (control)	4.88
3(assigned 140% + PF) vs. 1	4.05
3 vs. 2	-0.83
4(Participative + PF) vs. 1	8.20
4 vs. 3	4.15

Note: 1. The total worker satisfaction score (adjusted mean) of Group 1 (Control) = 39.83.
2. A negative sign indicates that the second group's worker satisfaction score is higher than the first group.

Table 4. Student-Newman-Kuel's (SNK) range test of worker satisfaction score.

Groups	Differences in adjusted means between groups [a] →			
	1 (Control)	3 (PS:140% + PF)	2 (PS:100% + PF)	4 (PS:Participative + PF)
1	–	4.05**	4.88**	8.20**
3		–	0.83	4.15**
2			–	3.32**
4				–

[a]Groups in order of increasing mean worker satisfaction
*p<0.05, **p<0.01

A comparison between Groups 3 (140% of normal + feedback) and 1 revealed that Groups 3 was significantly better than

Group 1. In other words, the provision of an assigned hard standard in combination with feedback had a significant positive effect on worker satisfaction. No significant difference was observed between Groups 3 and 2. State otherwise, the provision of a hard production standard of 140% of normal and feedback did not improve worker satisfaction over normal (100%) standard and feedback. The improvement in worker satisfaction in assigned normal and hard standards was probably due to specificity of the standard and feedback. Hard standard had no incremental effect on worker satisfaction. Research studies in the field showed conflicting results of assigned goals on satisfaction (Ivancevich 1976, Kim and Hamner 1976, Latham and Yulk 1976, Ivancevich and McMahon 1982).

A comparison between Groups 4 (participative and feedback) and 1 showed that the provision of a participative standard along with feedback improved worker satisfaction significantly. Compared to Groups 2 and 3, the worker satisfaction score of Group 4 was significantly better than the assigned standard with feedback in terms of worker satisfaction in industry. Such a result was not found in the laboratory study (Das and Shikdar 1987). The subjects were college students in the laboratory study as opposed to industrial workers in the industrial study. Participation in decision making by the industrial workers was, probably, the moderating factor for significant improvement in worker satisfaction. It has been reported in the literature that in both assigned and participative goal setting conditions job satisfaction declined (Latham and Yulk 1976, Latham et al. 1978, Garland 1983). In the present study, it was found that the subjects liked the free standard (goal) choice, rather than the assigned goal. The participative standard condition also produced better performance than assigned standard conditions (Shikdar 1990).

CONCLUSIONS

The following conclusions were drawn from this industrial study:

1. The provision of a normal (100%) production standard along with feedback had a significant positive effect on worker satisfaction in industry.
2. The provision of an assigned hard (140% of normal) standard with feedback was similar to the provision of an assigned normal (100%) standard with feedback in terms of worker satisfaction. It could be stated that hard standard condition had no incremental effect on worker satisfaction.
3. The incorporation of a participative standard with feedback was superior to assigned standards with feedback in terms of worker satisfaction. Participation in decision making in industry appears to have a moderating effect on worker satisfaction.
4. Worker satisfaction in a repetitive production task could be improved significantly by the provision of specific standards with feedback. Further improvement could be possible by providing a participative standard with feedback.

ACKNOWLEDGEMENT

The research was funded by the Social sciences and Humanities Research Council of Canada (Grant Nos: 410-87-0876 and 410-88-0990).

REFERENCES

Ammons, R.B., Effects of knowledge of performance: A survey and tentative theoretical formulation. The Journal of General Psychology, 1956, 54, 279-299.

Das, B., Effects of production feedback and standards on worker satisfaction and job attitudes in a repetitive production task. IIE Transactions, 1982, 14(3), 193-203.

Das, B. and Shikdar, A.A., The impact of assigned and participative production standards and production feedback on worker satisfaction in a repetitive manufacturing task. In: Karwowski, W., (ed.), Trends in Ergonomics/Human Factors I, North-Holland, Amsterdam, 1987, 227-233.

Das, B. and Shikdar, A.A., Assigned and participative production standards and feedback affecting worker satisfaction in a repetitive production task. Journal of Human Ergology, 1988, 17(1), 101-108.

Garland, H. Influence of ability, assigned goals, and normative information on personal goals and performance: A challenge to goal attainability assumption. Journal of Applied Psychology, 1983, 68(1), 20-30.

Hackman, J.R. and Lawler, E.E. III, Employee reaction to job characteristics, Journal of Applied Psychology Monograph, 1971, 55(3), 259-289.

Hackman, J.C. and Oldham, G.R., Development of job diagnostic survey, Journal of Applied Psychology, 1975, 60(3), 159-170.

Ivancevich, J.M., Effects of goal-setting on performance and job satisfaction, Journal of Applied Psychology, 1976, 61(5), 605-612.

Ivancevich, J M. and McMahon, J.T., The effects of goal setting, external feedback and self generated feedback on outcome variables: A field experiment, Academy of Management Journal 1982, 25(2), 359-372.

Kim, J.S. and Hamner, W.C., Effects of performance feedback and goal setting on productivity and satisfaction in an organizational setting, Journal of Applied Psychology, 1976, 61(1), 48-57.

Latham, G.P. and Yulk, G.A., Effects of assigned and participative goal setting on performance and job satisfaction. Journal of Applied Psychology, 1976, 6(2). 166-171.

Latham, G.P., Mitchel, T.R. and Dossett, D.L., Importance of participative goal setting and anticipated rewards on goal difficulty and job performance, Journal of Applied Psychology, 1978, 63(2), 163-171.

Locke, E.A., Relationship of task success to satisfaction: Further replication, Psychological Reports, 1966, 19, 11-32.

Locke, E.A., Toward a theory of task motivation and incentives. Organizational Behavior and Human Performance, 1968, 3, 157-189.

Locke, E.A., What is job satisfaction? Organizational Behavior and Human Performance, 1969, 4(4), 309-335.

Locke, E.A., Job satisfaction and job performance: A theoretical analysis. Organizational Behavior and Human Performance, 1970, 5(5), 484-500.

Ray, A.A. (editor) SAS Users Guide Statistics, 1982 Edition. SAS Institute Inc., Cary, North Carolina, 1982.

Shikdar, A.A. An investigation on the application of production standards with production feedback and monetary incentive in industry, Ph.D. Dissertation, Department of Industrial Engineering, Technical University of Nova Scotia, Halifax, Nova Scotia, Canada, 1990.

Vroom, V.H., Work and Motivation, John Wiley, New York, 1964.

White, S.E., Mitchell, T.R. and Bell, C.H. Jr., Goal setting, evaluation apprehension and social cues as determinants of job performance and job satisfaction in a simulated organization. Journal of Applied Psychology, 1977, 62(6), 665-673.

Advances in Industrial Ergonomics and Safety II
Edited by Biman Das
Taylor & Francis 1990

ERGONOMICS IN SYSTEMS DESIGN - A CASE STUDY

B. Mustafa Pulat, Engineering Manager
R. S. Chafee, Senior Engineer
G. D. Wade, Engineering Manager

AT&T Network Systems, Oklahoma City Works,
Oklahoma City, OK 73126-0060

Ergonomists, designers and engineers plan, develop and implement man-machine systems with multiple components and complex interactions. This paper dwells on one such experience at the AT&T Oklahoma City Works. The system was developed in order to better support the factory receiving operations. The new system improved throughput while providing enhanced accountability. Human engineered system interfaces allowed for fast acceptance of the system by the users.

1. INTRODUCTION

Designers and engineers frequently face problems which involve the planning of the architecture and the functioning of combinations of humans and equipment. The basic model that guides the designer here is a person-object model. An example could be the design of a hand tool. The object is the hand tool and the person is the user. The person-object model manifests itself in industrial environments as the man-machine system.

The human's role in a man-machine system in terms of coverage, degree of involvement, and importance varies with the degree of automation. In his simplest role, man acquires information, makes decisions, and takes action if necessary. Mental, physical, and psychomotor demands are at a minimum. However, in order to make correct decisions, the human must have had significant training and job experience. An example may be the task of a control room operator in a petroleum refinery.

At the opposite extreme, most system functions are assigned to the human. He/she has to time share between various input stimuli, make frequent decisions and apply forces routinely. A cyclist is probably operating in this mode. Many other roles for the human fall between these two extremes[1].

2. THE SYSTEM DESIGN CYCLE

Each man-machine system has one or more objectives to fulfill. The system is designed for a purpose. Conversely, for any new system, one can set forth a set of requirements, which define that system's goals or objectives. It is extremely important for the designer to carefully analyze the requirements and design system functions to achieve them in the most effective manner.

Coupled with system requirements, may be system limitations. The system may have to function under certain limitations. Some of these may be stated as hard constraints(must) and some soft. System limitations, especially hard constraints, set the boundaries within which the system should work.

Once system requirements and limitations have been specified and detailed, the next step is to carry out a thorough functions analysis to specify which functions are necessary and their feasible combinations to achieve the objectives under the specified constraints. This step includes the listing of all operational functions and developing the parameters of each through task analysis. Each operational function needs to be performed by either machine(s) or human(s) or both. This phase is frequently called function allocation[2,3].

It is also important to carry out contingency analysis as part of functions analysis. A contingency is any situation which is nonroutine and which may have to be dealt with in performing required tasks and functions. Contingency analysis may indicate new functional requirements.

Detailed task analysis, time planning, manpower planning, detailed design, acquisition, testing and operation phases follow next. This total design process frequently manifests itself as a **design cycle**, since after completing one step, the designer almost always goes back to the previous steps and makes corrections or revisions. Thus, the designer iterates between alternate methods of achieving functional and total system objectives until the most satisfactory one is found.

Figure 1 gives a summary of the major steps discussed for designing man-machine systems. In general, the flow is from top to bottom, however, upward iterations are very common.

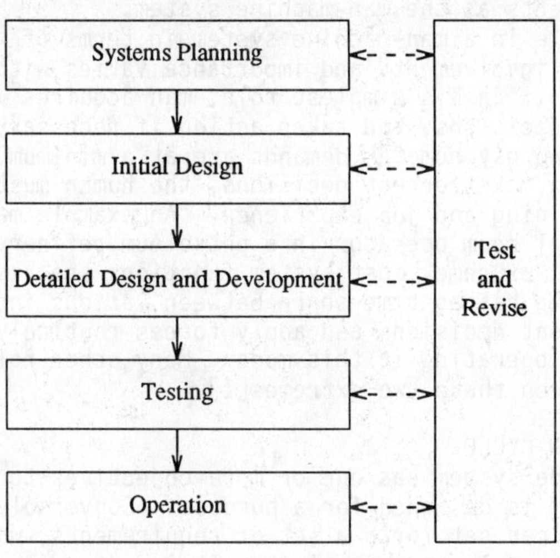

Figure 1. Major Elements of System Development Cycle

3. PROBLEM DEFINITION

The problem reported in this paper concerns the application of the above concepts to a design case which involved multiple personnel and work stations. Several years ago AT&T Oklahoma City Works decided to support its factory receiving operations with a new system. The following discussion summarizes the major elements of the design process that was followed.

4. NEW SYSTEM DESIGN

As the first step, system requirements were defined. These were set at increasing productivity and accountability at the affected areas. The constraints were; only receiving operation will be of concern, no additional personnel should be needed, and state-of-the-art technology must be used(see Table 1).

Table 1. System Requirements and Constraints

Requirements:	Increase productivity and accountability
Constraints:	No new personnel
	Receiving dock only
	State-of-the-art technology

Ergonomics received a great deal of attention from the plant industrial and material handling engineers while structuring and detailing of the problem. As a first step, an operations analysis was carried out in order to get familiar with the method of material handling in the receiving areas. The information gathered was then matched with the system requirements to develop operational functions to be supported(see Table 2).

Table 2. Operational Functions for Receiving

OF1:	Support receival data gathering
OF2:	Increase material velocity(two stage receival)
OF3:	Track material location
OF3:	Communicate with procurement system
(c)OF5:	Support dock operations
(c)OF6:	Delivery verification

At first, four functions were determined to be necessary (OF1-4). Contingency analyses showed that two more were to be added to the list(OF5,6):

OF1: This function is necessary to gather receival data from vendor shipments. It spans sixteen subfunctions from recording bill-of-lading number to documenting the gross weight of the shipment.

OF2: The requirement of productivity increase signalled the need for increasing material velocity in the affected areas. Investigations showed that out of sixteen data elements collected on each receival, only four were necessary to check the integrity of shipment, determine inspection requirements, and making routing decisions. Others could be entered during or after the movement of the material. Thus, it was decided that the new system should support a two stage receival function. First stage will involve entry of four data elements and material moved. During the second stage entry, the remaining elements would be entered.

OF3: The accountability requirement indicated that each box be tracked as to its current location within the confines of the dock. Each box's association with a given receival would also be maintained.

OF4: In order to check the integrity of shipment(early or over delivery, delivery of wrong or unneeded material), the receiving system needed to receive information from the corporate procurement system as to the expected receivals. Similarly, receival information had to be passed to the procurement system in order to update their records.

OF5: As the functions were being developed, it was observed that support had to be extended to various dock operations including prebagging and inspection. This was to display work instructions to the operator and collect work results data.

OF6: A reanalysis of the accountability requirement showed that the docks were accountable for material from dock exit to actual delivery to its destination. Thus, this operational function was to be added to the list.

The manpower and technology restrictions led to decisions to develop a computer based system with product ID(identification) scanning capabilities as opposed to manual data entry. However, some keyboard entry functions would be retained. Based on an initial function allocation scheme, OF1,3 and 6 were assigned to primarily humans. Others would be conducted by the computer with human help. OF4 was the only function that needed almost no human involvement.

Task analysis then followed for each function. Operational Sequence Diagrams were developed(Figure 2). Results of task analysis indicated workstation, manning and skill requirements. A detailed error analysis led to developing software features for editing and error recovery. Decisions were made as to the machine/human allocation of tasks for each function. Existing dock personnel were then trained based on task assignments.

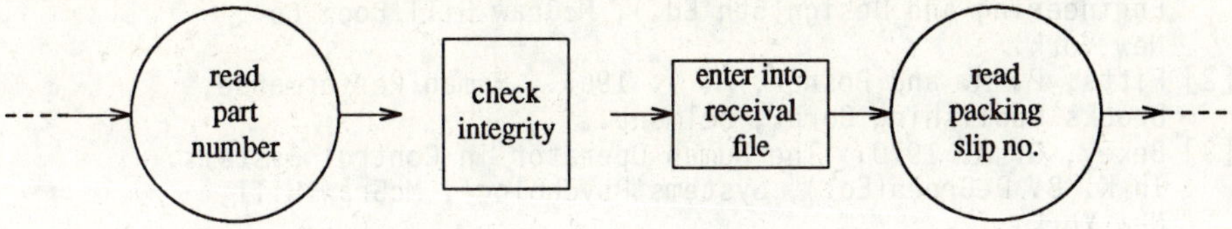

Figure 2. A Section of OSD for OF1

The tracking technology used was bar-code identification and scanning of each box. In addition, radio frequency remote data input technology was implemented.

During the detailed design and development phase, all necessary equipment including bar-code printer, scanners, hand held radio terminals, CRT's, line printers, labels, etc. were purchased. Software was developed, which involved many design and development iterations. Each software feature developed was immediately tested. Equipment were tested as delivered to the plant. All human workstations were carefully designed and developed. Workstation arrangement within the dock space followed operation clusters as well as sequential relationships between operations.

When all software and hardware were ready, a detailed user acceptance testing (UAT) was performed on the total system. Several deficiencies in the software and hardware communications were observed and corrected. Before the production cut, a massive load test was performed with human operators and many receivals. This test showed that a larger capacity central processor (computer) was necessary in order to improve system response time. One was purchased and installed. The system's maximum receival capability was determined by a queueing analysis. After all this effort, production cut was very smooth.

5. RESULTS

The system has been in operation for almost a year now. During this time, several features have been added to it reflecting support for additional functionality. Training courses have been retained to periodically retrain existing personnel and train new personnel.

The system developed was well accepted by the employees and management. Since all management concerns were integrated into the design, the total solution gained full acceptance. Ergonomic aspects of the design (task allocation, workstation design, job design, software ergonomics, training, job aids, etc.) contributed significantly to the employee's satisfaction with the system.

6. REFERENCES

[1] McCormick, E. J. and Snaders, M. S. 1982. Human Factors in Engineering and Design(5th Ed.), McGraw Hill Book Co., New York.

[2] Fitts, P. M. and Posner, M. I. 1967. Human Performance, Brooks Publishing Corp., Belmont.

[3] Bekey, G. A. 1970. The Human Operator in Control Systems. In K. B. DeGreen(Ed.), Systems Psychology, McGraw Hill, New York

Advances in Industrial Ergonomics and Safety II
Edited by Biman Das
Taylor & Francis 1990

DESIGN OF PRODUCTION SYSTEMS USING ERGONOMICS

B. M. Worrall, Ph.D., P.Eng.

Department of Industrial Engineering
Technical University of Nova Scotia
Halifax, Nova Scotia, Canada B3J 2X4

Improvements in productivity usually come about by improving and redesigning existing systems rather than by the design of new systems. This paper is concerned with the application of ergonomics in existing systems and presents a method whereby needed changes can be detected. An application of ergonomics to improve an existing system is presented.

INTRODUCTION

Although the basic principles of ergonomics have been known for several decades, there is often little evidence of it being applied in industrial and commercial enterprises. As Industrial Engineers spend most of their time improving existing systems, there is considerable scope for improving productivity by the use of ergonomics. The objective in this paper is to review some of the principles in workstation design and facility layout, to present a method of analysis and a case study for improving an existing system.

Ergonomics should be used in the design of products, workstations and facility layout. However, we are all familiar with examples where ergonomics either has not been used or has had minimal application. Instruments and controls in cars are often difficult to find and identify. Table, bench and desk tops are often rigidly fixed at an unsuitable height for the operator to work on. Many chairs are still poorly designed and even if the height of the seat is adjustable, no foot-rest is provided for the short-legged worker. Some "ergonomically" designed chairs can be extremely uncomfortable.There is little incentive to improve the design of the workplace unless it can be proved, often difficult, that injuries or health problems have been caused by the workplace.

It is difficult to achieve good international standards because the laws on the working environment are not uniform across different countries. The minimum standard to tour a factory in North America is to wear a hard hat and safety glasses. In addition, one may be asked to wear safety shoes and/or earplugs and to remove ties and rings. In other countries, very few or no precautions may be taken with visitors and one can see the same standards applied to workers. Sparks from welding machines, machines without guards, grit and sawdust in the air, and workers carrying reasonably heavy castings and wearing open-toed sandals are some examples. One consequence of this lack of standards is that it is possible to transport jobs from one country to another when standards are tightened up.

The difficulty of introducing ergonomics into the workplace has been identified. If ergonomics can be introduced at the design stage, the increase in cost will be minimal and will be soon paid for by the increase in productivity. In an existing environment, there has to be a mechanism for identifying problem areas. As will be shown in the case study, a minimal application of ergonomics can lead to a considerable improvement in the working environment.

DESIGN OF FACILITIES

The design of facilities can be broken down into four areas:

1. Workstation, which is concerned with designing the job to fit the worker.
2. The relative position of workstations to one another or facility layout, which is determined by flows of material between workstations.
3. Interaction between people, which falls into two subdivisions: (a) flow of information for the control and operation of the enterprise, and (b) informal interaction necessary for the morale of people.
4. The environment, which is often governed by laws concerning health and safety. It is interesting to note that although people may be protected by law against high sound levels in the working environment, most young people are probably suffering from hearing loss by the time they start work because of the recreational noise levels they willingly subject themselves to.

In this paper we will be concerned with job design and facility layout.

WORK STATIONS DESIGN

Before one can start to design the workstation, the design of the job has to be completed. Konz (1983) has developed eight principles of job design (Table 1). The key to progress and to decreasing costs up to this time has been to use specialization and have long production runs, thus spreading the changeover and start up costs over a larger number of units. More recently the change has been to have smaller production runs by using interchangeable parts to produce different products and eliminating large setup times. Group technology and flexible manufacturing systems have the same objective by grouping similar products together. Consequently, the tasks at a workstation are usually routine and repetitive. Boredom is overcome by environmental stimulation and by giving short breaks often.

Table 1. Principles of Job Design (Konz, 1983)

1. Use specialization even though it sacrifices versatility.
2. Make several identical items at the same time.
3. Combine the operations and functions.
4. Minimize idle capacity.
5. Use filler jobs or filler people.
6. Vary environmental stimulation inversely with task stimulation.
7. Give short breaks often.
8. Give precise instructions in an understandable format.

Industrial workstation design should ideally be capable of satisfying the systems performance requirements as well as the needs of the human user. Operator performance capabilities and limitations must be considered in designing an ideal workstation. The design should enable the operator to see the working area, posture must be adequate and comfortable, and controls must be within reach to minimize error.

The design of the workstation has three groupings:

1. Use of the Human Body.
2. Arrangements of the place.
3. Design of tools and equipment.

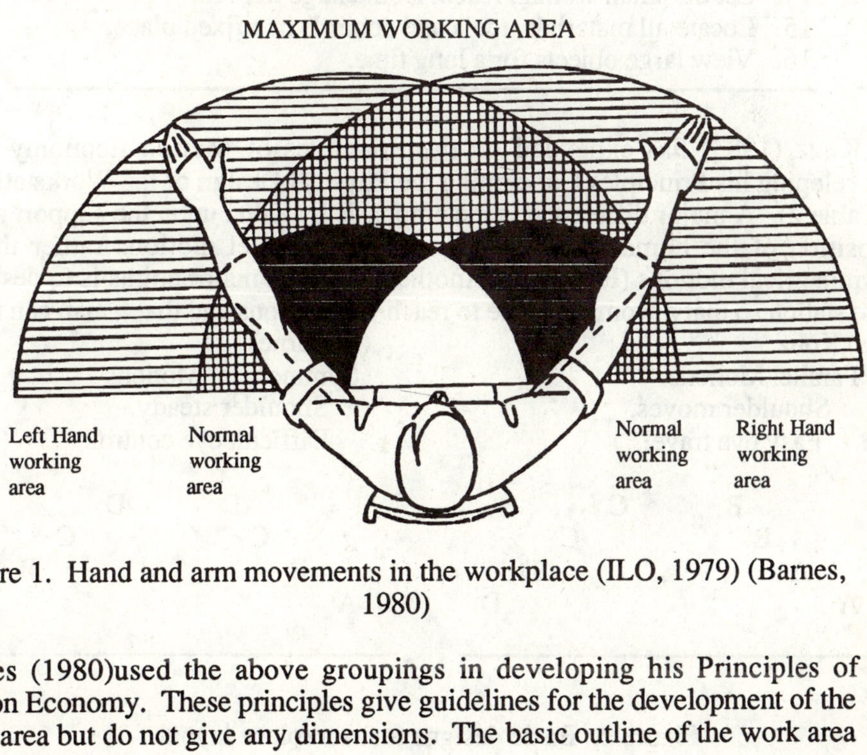

Figure 1. Hand and arm movements in the workplace (ILO, 1979) (Barnes, 1980)

Barnes (1980)used the above groupings in developing his Principles of Motion Economy. These principles give guidelines for the development of the work area but do not give any dimensions. The basic outline of the work area is shown in Figure 1. There is a major difference between the workplace layout shown and that proposed by Barnes in that the hand movement represented by the darkened area is similar to that of windshield wipers. Barnes considered that the hand could move in a complete semicircle, which can only be achieved by movements about the shoulder. The normal working area is that covered by fingers, wrist and elbow movements. The working area can be extended by movements about the shoulder.

Table 2. Principles Concerning Physical Design of the Workstation (Konz, 1983)

1. Avoid static loads and fixed work postures.
2. Set the work height at 50 mm below the elbow.
3. Furnish every employee with an adjustable chair.
4. Support limbs.
5. Use the feet as well as the hands.
6. Use gravity, don't oppose it.
7. Conserve momentum.
8. Use two-hand motions rather than one-hand motions.
9. Use parallel motions for eye control of two-hand motions.
10. Use rowing motions for two-hand motions.
11. Pivot movements about the elbow.
12. Use the preferred hand.
13. Keep arm motions in the normal work area.
14. Let the small woman reach; let the large man fit.
15. Locate all materials, tools and controls in a fixed place.
16. View large objects for a long time.

Konz (1983) has expanded on the Principles of Motion Economy by developing his Principles Concerning the Physical Design of the Workstation (Table 2). A major difference is the emphasis by Konz upon the support and position of the human body and the use of parallel motions rather than symmetrical motions (Figure 2). Another of Konz's major points is to design the station so that a woman is able to reach all locations and that a man can fit.

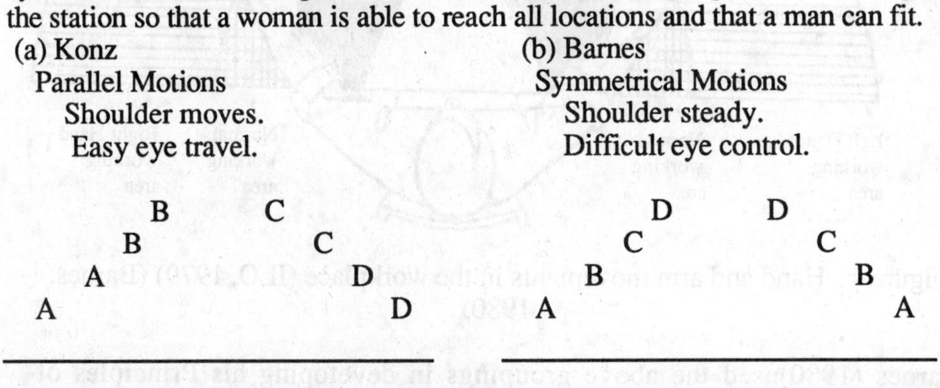

Figure 2. Basic designs for workplace layout.

The design of the workstation is finally based upon biomechanics and anthropometric measurements. Das and Grady (1983) give a good description of the use of ergonomics employing anthropometric measurements in the design of workstations. Because of the wide variation in size of the population, the workstation has to be designed for use by people of varying sizes.

The preferred hand is about 10% faster for reach motions, more accurate in movement and 5 to 10% stronger. If a tool is used the preferred hand is 20 to 30% faster. If possible, the workstation should be designed so that it can be set up as a mirror image to take advantage of the preferred hand.

FACILITIES

Facilities design is concerned with the relative location of facilities to facilitate easy movement of materials between them. Ergonomics considerations are often omitted. Two problems have to be considered, viz., a) the effects of automation and b) the interactions between people.

One aspect of automation that has been extensively studied is the design of control systems. Control is usually exercised by people watching dials. The design and location of dials has received considerable attention. Even in these days of computer control, the final decision on the system will be made by a person. An important factor in the design is how to catch the operator's attention. The procedure for adjusting the process has to have precise instructions given in an understandable format. However, little consideration has been given to how to operate the plant if part of the automation system does not work. Even in the most advanced automation systems, i.e. flexible manufacturing system, it is often found some part often does not work and manual work has to be done. This usually occurs in the positioning of parts, jigs, and fixtures in the machine.

The second part is the interaction between people. Konz (1983) has proposed that the environmental stimulation should vary inversely with task simulation. Many tasks are monotonous and require low mental concentration. For these type of tasks, it is desirable that people can see and talk to each other. Visual variety can be added with windows and paint. The operators can be better motivated if they have some input into the decision making, e.g., the location of the windows and deciding on the color scheme for the walls.

IMPROVEMENT IN WORKING CONDITIONS

It has been previously noted that a great deal of the Industrial Engineering effort is directed to work measurement and methods improvement. This presents an opportunity to apply ergonomics in the existing workplace. The Video-Computer Work Measurement System reported by Worrall (1989) is ideal for this purpose as it maintains a complete record for both the times on magnetic disk (computer) and a complete visual record and times on a VCR tape. The system is illustrated in Figure 3. Although the system does not measure distance, newer techniques, currently being developed, will enable the distances to be measured.

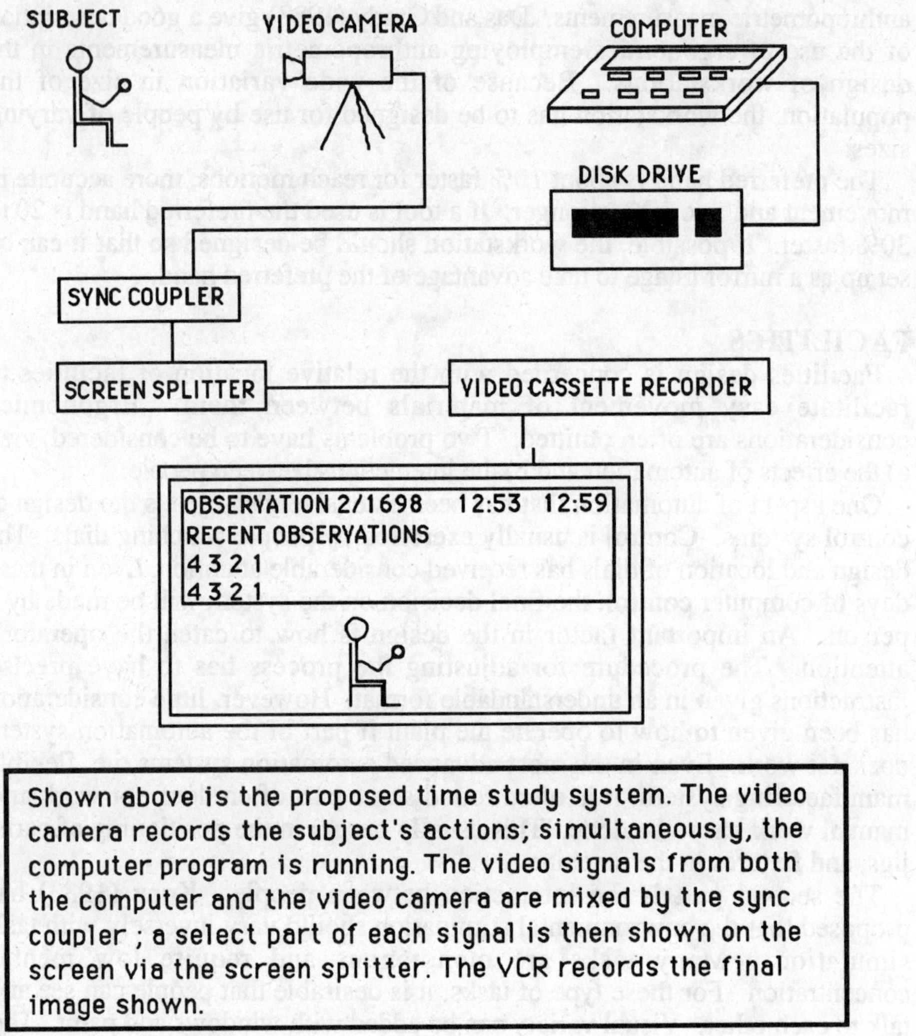

Figure 3. Design of video-computer work measurement system.

One advantage of this system is participation by operators, who are naturally inquisitive. The operators have the opportunity to see themselves on playback, see what was done and why it was done. Many operators are amazed at how they actually perform their work and they can see how their performance can be improved. Most operators can be motivated to make suggestions on how the workplace can be improved. This factor, together with an ergonomics analysis of video recording of the operation, can lead to an improved design for the workplace.

CASE STUDY

This case study represents the application of ergonomics to a temporary operation, which involved sorting empty beer bottles. Traditionally, all breweries in Canada used the same size of bottle and there was therefore no need to determine which brewery had used the bottle last. Eventually the breweries were able to change the shape to identify a particular brand of beer. It was now necessary to sort the bottles prior to filling. The new type bottles were supposed to have been presorted before they reached the brewery but this seldom happened. This brewery had three types of its own bottles and several types from other breweries. Only one of the external types was a major concern.

The sorting was done in a warehouse. It was cold, the lighting level was low, and although clean, the warehouse had a dingy atmosphere. The sorters wore jackets or sweaters and most of the time, the workers were bending or stooping. A condensed version of the sorting area is shown in Figure 4. The workers wore gloves, safety glasses and safety shoes while working.

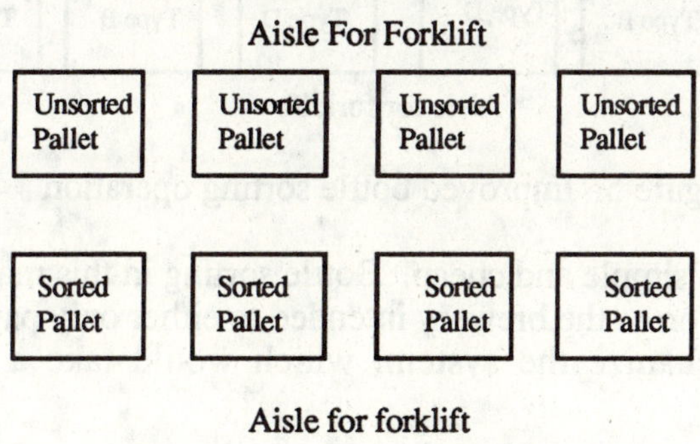

Figure 4. Bottle sorting operation

The unsorted bottles were stored in cases on a pallet. The cases were variable in size containing 6,12, or 24 bottles. The bottles were sorted directly on the unsorted pallet or on the floor depending upon the type of bottle and the space available for partially sorted cases. A sorted case was then transferred to the sorted pallet. A forklift truck delivered unsorted pallets and removed sorted pallets. The first few rows were more or less at normal working height and after that the operator had to bend and to reach in order to sort bottles. The time to sort a case depended upon the number of bottles that had to be removed and inserted in a case.

The solution was to bring the working area up to waist height by placing sheets of plywood on two stacks of six pallets (later tables were used). The sorting area is shown in Figure 5. The method, based upon a team of two, is as follows:

1. An unsorted case is taken from a pallet and placed on table.
2. The bottles in the minority were removed and placed in other cases containing the same type of bottle until the case is filled.
3. The sorted case is placed upon the appropriate pallet.

The forklift driver places unsorted pallets in the appropriate area, removes sorted pallets and transfers empty pallets to the appropriate location. The operators pushed boxes back and forth on the work table to the operator who needed that particular type of bottle. The operators expressed satisfaction at how much easier the work was under the new method.

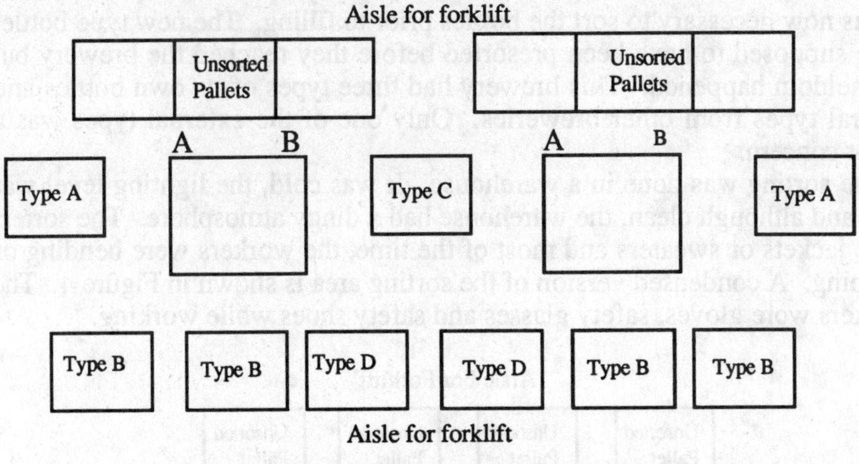

Figure 5. Improved bottle sorting operation

The solution was simple and cheap. Bottle sorting in this manner was only a temporary operation as the brewery intended to either only pay for presorted bottles, or to mechanize the system, which would take a long time to implement.

CONCLUSION

A review of the application of ergonomics in facility design has been made. A computer-video work measurement system can be used for detecting violations of good ergonomic design. It was shown that ergonomics could be applied to a temporary operation. As a great deal of Industrial Engineering effort is applied to improving existing systems, the major benefit of using ergonomics will come about by applying it to existing systems rather than designing completely new systems.

REFERENCES

Konz, S., 1983, Work Design: Industrial Ergonomics, 2nd edn. (Columbus, Grid) pp.229-302 and pp 329-345

Das, B. and Grady, R.M., 1983, Industrial workplace layout and engineering anthropology. In: Ergonomics of Workstation Design, edited by T.O. Kvalseth, (London, Butterworths), pp. 103-128.

Barnes, R.M., 1980, Motion and Time Study: Design and Measurement of Work, 7th edn. (New York, John Wiley and Sons), chapters 15-17.

Worrall, B.M., 1989, Application of ergonomics in a computer/video-work measurement system. In: Advances in Industrial and Ergonomics and Safety I, Proceedings of the Annual International Industrial and Ergonomics Safety Conference, Cincinnati, edited by A. Mital (London, Taylor and Francis), pp. 907-913.

ILO, 1979, Introduction to Work Study, 3rd edn. (Geneva, ILO), pp. 156.

Advances in Industrial Ergonomics and Safety II
Edited by Biman Das
Taylor & Francis 1990

A STOCHASTIC PROCESS MODEL FOR HUMAN RELIABILITY ANALYSIS

Cherng Tarng (Tony) Lin, Ph.D.
David R. Clark, Ph.D., P.E.

Industrial & Manufacturing Systems Engineering Department
GMI Engineering & Management Institute
1700 W. Third Avenue
Flint, Michigan 48504-4898

This paper presents some basic stochastic point process concepts and their applicability to human reliability analysis. The Weibull process model is assumed and some of its elementary statistical properties are discussed. Simulated task performance data is generated to demonstrate the statistical analysis procedures for such human performance data.

INTRODUCTION

In recent years, researchers have shown a growing interest in human reliability analysis (HRA). Various approaches have appeared in the literature. Meister (1973), Sharit (1988), and Lee et al. (1988) provide critical reviews of approaches to HRA. One area of research methodology on modeling human performance is based on classical reliability theory that deals with the study of statistical distributions of human error; see Regulinski et al. (1973). We believe the development of statistical models and methods for the analysis of human performance plays an important part of HRA. In this paper, we propose a human performance model using the concepts of the stochastic point process for tasks in the continuous time domain, such as vigilance, tracking, and monitoring. This is primarily an expository paper that attempts to provide basic concepts of reliability which can be used in human reliability.

In the development of this proposed model, the process of human performance can be visualized as a loop as in Figure 1.

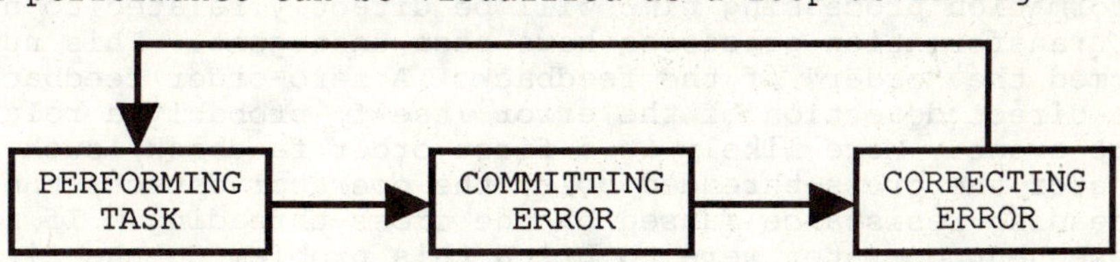

Figure 1. A Loop of Human Performance Model

This illustrates the process of a human subject performing a task in which a certain type of error is occurring. To describe the evolution of such processes through time, the stochastic process concepts can provide more realistic descriptions of such human performance.

SOME FACTORS AFFECTING ERROR CORRECTION

Although it is assumed in this paper that error correction is instantaneous, it is recognized that there are factors often present which preclude this. Among these are included what we call "correction delay" and "order of feedback".

Correction Delay

Theoretically, there is always a time delay between the time the error is committed and the operator correcting the error, or a "correction delay". This can be a function of several component delays: (1) the time for the result of the error to manifest itself in a detectable and identifiable form, or "manifestation delay", (2) the time for the operator to detect and identify the error manifestation, or "detection delay", (3) the time for the operator to identify the underlying error responsible for that manifestation and to consider and select an effective corrective action, or "information processing delay", (4) the time between the completion of information processing and the detection of a correction opportunity, or "opportunity delay", and (5) any additional time to initiate correction action over and above that of the previous error-producing actions, or "initiation delay". In relatively slow-proceeding processes, the correction delay may be short enough and the error benign enough that the process can be corrected before an error is released from the operator's control, effectively resulting in a zero-correction delay. An example of a zero-correction delay is the cross-threading of a bolt being hand-started; the operator senses the increased threading resistance,"backs it off" and tries again successfully without a significant effect on production. However, in most practical situations, delays are real and have significant effects.

Order of Feedback

As any safety-expert will tell you, an accident (an unplanned and undesirable event), when a manifestation of an error, is probably the result of a sequence of events (often multi-causal, multi-path). The manifestation is likely to be effect, not cause. Therefore, the amount of information processing time and therefore, delay, is a function of how removed the manifestation is, informationally, from the original error. If we model the path between the error and the manifestation as a sequence of events or transformation processes, then the amount of information processing time will be directly related to number of transformation processes have been undergone. This number is termed the "order" of the feedback. A zero-order feedback is the direct detection of the error itself; probably a relatively rare event. More likely is a first-order feedback, such as the case of the cross-threaded bolt; the operator detects the threading resistance caused by the cross-threading. If a subsequent operator were to bring this problem to the first operator's attention instead, the feedback could become second-order ("There someone wrong with this bolt"), or higher ("What's wrong; you're sending me garbage!"), or better, zero-order ("You cross-threaded this bolt"). In the latter case, the zero-order feedback is coupled with a feedback delay that mitigates its advantage.

PROPOSED STOCHASTIC PROCESS MODEL FOR HUMAN PERFORMANCE

In this paper, basic stochastic points process concepts will be discussed. A process model called Weibull process (not Weibull distribution) is assumed and the analysis will be presented.

A stochastic process $\{X(t),\ t \in T\}$ is a collection of random variables (Ross, 1983). We often interpret t as time and refer to $X(t)$ as the state of the process at time t. For a process of human performance in Figure 1, a stochastic process $\{N(t),\ t \in T\}$ is a counting process if $N(t)$ represents the total numbers of errors in $(0,\ t]$. It is a stochastic point process intended to model the evolution of human errors. Figure 2 shows a realization (sample path) of a stochastic point process which corresponds to the initial error occurring at time 1, the next error at time 2.5, and the third error at time 5, and no other errors. Error corrections are assumed to be performed instantaneously. Figure 3 shows another way to visualize the same realization of the process.

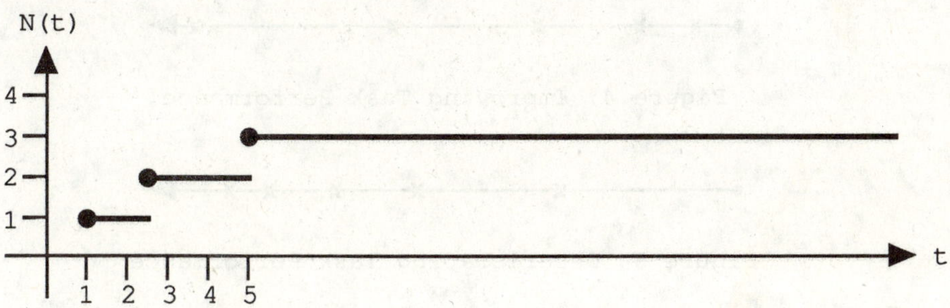

Figure 2. Sample Path of $N(t)$ = Number of Errors in $(0,\ t]$

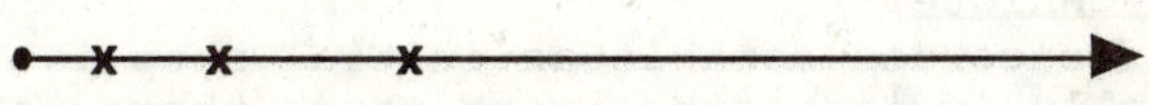

Figure 3. Realization of $N(t)$ = number of errors in $(0,\ t]$

Let us define

$$M(t) = E[N(t)]$$
$$= \text{statistical expectation of number of errors in } (0, t]. \quad (1)$$

Then,

$$m(t) = M'(t) = \frac{\mathrm{d}E[N(t)]}{\mathrm{d}t} \quad (2)$$

is called the intensity function or the rate of occurrence of errors (ROCOE).

One of the most common models for the time-dependent stochastic point process is the nonhomogeneous Poisson process with the intensity function (or ROCOE),

$$m(t) = \lambda \cdot \beta \cdot t^{\beta-1} \qquad t > 0, \tag{3}$$

where λ and β are two unknown parameters. It is also called the Weibull process, or power law process by others.

To provide a human performance context for the model, we can consider a human subject (operator) and the process of performing a task with successive errors and error correction; instantaneous error correction is assumed. The unknown parameter β plays an important role in this model. When $\beta<1$, the intensity function $m(t)$ is decreasing which indicates that there is an improvement in performance as shown in Figure 4. When $\beta>1$, the intensity function is increasing which indicates that there is a deterioration of performance as shown in Figure 5.

Figure 4. Improving Task Performance

Figure 5. Deteriorating Task Performance

In the remaining section, we will discuss some of the problems of statistical inference of human reliability performance data when the Weibull process model is used.

PARAMETER ESTIMATION AND HYPOTHESIS TESTING

Graphical Technique

A plot of recorded data is an initial step in data analysis. It will usually indicate the trend where there is improvement (deterioration) of the human operator's performance.

For task performance data, let us consider that N errors were recorded at time $t_1 \le t_2 \le t_3 \le \ldots \le t_N$ over a period of time T. A simple graphical procedure is to plot the cumulative errors versus the cumulative task performing time on linear paper. A concave down (up) trend will indicate an improving (deteriorating) performance.

For example, consider the data in Table 1 which gives N=30 error times of an operator of a simulated task performed during a period of T=100 minutes. Figure 6 shows the plot of the data for Table 1. It indicates a concave down trend which implies an improving performance of the the operator. A statistical analysis will confirm this result.

Error Number	Time	Error Number	Time	Error Number	Time
1	0.3463	11	17.1823	21	51.9785
2	0.7108	12	17.5015	22	53.1438
3	2.8851	13	19.1130	23	59.0403
4	3.5621	14	20.4065	24	64.3673
5	3.7272	15	20.5648	25	85.1273
6	8.3788	16	22.3863	26	90.6516
7	9.2530	17	24.3302	27	91.6533
8	11.8903	18	27.2313	28	92.1203
9	12.6781	19	34.2709	29	94.2429
10	12.9023	20	41.4835	30	98.9581

Table 1. Error Times in Minutes of a Simulated Task

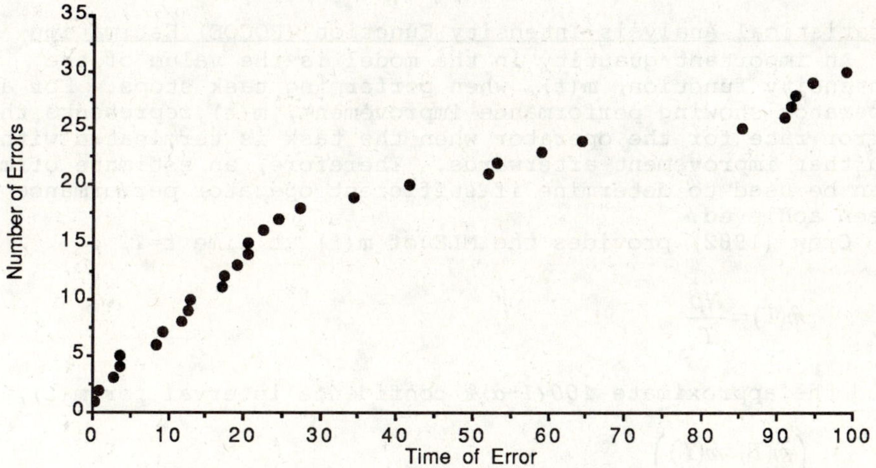

Figure 6. Cumulative Number of Errors vs. Total Performance Time

Statistical Analysis-Parameter Estimation

Statistical estimation procedures of the Weibull process for parameters β and λ are discussed in Crow (1982). The maximum likelihood estimates (MLE) are computed by

$$\hat{\beta} = \frac{N}{\sum_{i=1}^{N} \ln\left(\frac{T}{t_i}\right)} \tag{4}$$

and

$$\hat{\lambda} = \frac{N}{T^{\hat{\beta}}}. \tag{5}$$

Confidence intervals for β are based on the quantity $2N\beta/\hat{\beta}$ which has a Chi-square distribution with $2N$ degrees of freedom. Thus, a $100(1-\alpha)\%$ confidence interval for β is computed by

$$\left(\frac{\hat{\beta}\chi^2_{2N,1-\alpha/2}}{2N}, \frac{\hat{\beta}\chi^2_{2N,\alpha/2}}{2N}\right). \tag{6}$$

For the data given in the example, we have $\hat{\beta}$=0.6106, which indicates performance improvement, and $\hat{\lambda}$=1.8023. A 95% confidence interval for β is given by

$$\left(\frac{0.6106\times 40.48}{60}, \frac{0.6106\times 83.3}{60}\right) = (0.4120, 0.8477).$$

<u>Statistical Analysis-Intensity Function (ROCOE) Estimation</u>

An important quantity in the model is the value of the intensity function, m(t), when performing task stops. For an operator showing performance improvement, m(t) represents the error rate for the operator when the task is terminated with no further improvement afterwards. Therefore, an estimate of m(t) can be used to determine if sufficient operator performance has been achieved.

Crow (1982) provides the MLE of m(t) at time $t=T$,

$$\hat{m}(T) = \frac{N\hat{\beta}}{T} \tag{7}$$

and the approximate $100(1-\alpha)\%$ confidence interval for m(t),

$$\left(\frac{\hat{m}(T)}{\Pi_2}, \frac{\hat{m}(T)}{\Pi_1}\right) \tag{8}$$

with appropriate Π_1 and Π_2 from Crow's table. In our example, we obtain

$$\hat{m}(100) = \frac{30\times 0.6106}{100} = 0.1832 \text{ errors / minute.}$$

A conservative 95% confidence interval for m(t) is constructed by using Crow's table with N=30, Π_1=0.604, and Π_2=1.783. Thus, the confidence interval is

$$\left(\frac{0.1832}{1.783}, \frac{0.1832}{0.604}\right) = (0.1027, 0.3033).$$

<u>Goodness-of-Fit Test</u>

Another important statistical inference procedure is to check the validity of the Weibull process model on which estimation procedures are based. A formal goodness-of-fit test is developed by Crow (1974) using a Cramer-Von Mises statistic

$$C_N^2 = \frac{1}{12N} + \sum_{i=1}^{N}\left[\left(\frac{t_i}{T}\right)^{\bar{\beta}} - \frac{2i-1}{2N}\right]^2 \tag{9}$$

where

$$\bar{\beta} = \frac{N-1}{N}\hat{\beta} \tag{10}$$

is an unbiased estimator of β.

The null hypothesis to be tested is a Weibull process model with intensity function $m(t)=\lambda\beta t^{\beta-1}$. The hypothesis is rejected if C_N^2 exceeds the critical value for the chosen level of significance, α. Using α=0.05, the critical value is 0.218 for N=30. The Cramer-Von Mises statistic is C_N^2=0.09732. Since C_N^2<0.218, we conclude that the Weibull process model is appropriate.

CONSIDERATIONS FOR FURTHER RESEARCH

In this paper, it is assumed that the task process is continuous and that error correction is instantaneous. As we move away from these assumptions, the human performance model gets more complicated.

If task process is continuous, it proceeds with a relatively constant rate of production with no locally discernible production breakpoints. This is commonly found in bulk processes, such as the production of chemicals. On the other hand, a discrete process is one that is characterized by complete units of production, such as the production of automobiles. The cumulative error is the number of complete units produced before the error is corrected plus, perhaps, the unit under production at the time the error is reported. Many processes are combinations of the these types. The painting of airplanes involves the application of a locally continuous process (painting) on discrete units (airplanes).

Further, in real tasks this is may not be the case and the "opportunity delay", or the time between which corrective action can be initiated and an appropriate opportunity arises, must be considered. In order that effective action be taken, the error must be detected and a proper response made. If the feedback from an error committed has no importance to the operator's current "state", then it may be ignored. On the other hand, it might be meaningful feedback, but the opportunity to use the information is far enough in the future that the effectiveness of the message is reduced to a point where it is insufficient to trigger the proper corrective action.

SUMMARY

In this paper, the Weibull process model was assumed and some of its elementary statistical properties were discussed. Simulated task performance data was generated to demonstrate the statistical analysis procedures for such human performance data. It is the purpose of this paper to present some basic stochastic point process concepts and their applicability to human reliability analysis. We believe that modeling human performance based on stochastic processes should be further researched.

REFERENCES

Crow, L.H., 1974, Reliability Analysis for Complex Repairable Systems. In Reliability and Biometry, edited by F. Proachan and R.J. Serfling (Philadelphia: SIAM), pp. 397-410.

Crow, L.H., 1982, Confidence Interval Procedures for the Weibull Process with Application to Reliability Growth. Technometrics, 24, pp. 67-72.

Lee, K.W., Tillman, F.A. and Higgins, J.J., 1988, A Literature Survey of the Human Reliability Component in a Man-Machine System, IEEE Transaction on Reliability, 37:1, pp. 24-34.

Meister, D., 1973, A Critical Review of Human Performance Reliability Predictive Methods, IEEE Transaction on Reliability, R22:3, pp. 116-123.

Regulinski, T.L., and Askren, W.B., 1972, Stochastic Modeling of Human Performance Effectiveness Functions, In Proceedings of Annual Reliability and Maintainability Symposium, pp. 407-416.

Ross, S.M., 1983, Stochastic Processes, (New York: John Wiley & Sons).

Sharit, J., 1988, A Critical Review of Approaches to Human Reliability Analysis, International Journal of Industrial Ergonomics, 2, pp. 111-130.

IX

HUMAN STRENGTH AND TESTING

Advances in Industrial Ergonomics and Safety II
Edited by Biman Das
Taylor & Francis 1990

HAND TURNING TORQUES IN A SIMULATED MAINTENANCE TASK

Sheik N. Imrhan

Department of Industrial Engineering
University of Texas at Arlington
Arlington, Texas 76019

George D. Jenkins

University of Texas at Arlington, and
LTV Corp., Dallas, Texas 75265

This paper describes the results of an experiment to determine the effects of some task and handle variables on the torque capabilities of adults in a simulated maintenance task. Twenty adults generated MVC torques on a 2 1/4 inch diameter cylindrical handle with one hand, for 24 different test conditions. The results indicated that the position of the hand in the saggital plane does not affect the magnitude of the MVC torque. Torques from wrist extension motions were greater than those from wrist flexion; knurled cylinder surface enhanced torque; and right (dominant) handed torques were greater than left handed ones. Females were 50% as strong as males.

INTRODUCTION

Gripping and turning forcefully with the hand is still an important activity in occupations, despite the fact that tools, like wrenches, are available for torque exertion. The use of robotic aids has not yet met the needs for torque exertion. The task complexity and variability, and the difficulty imposed by confined spaces are factors that have not yet been overcome for the application of robots.

Sometimes the use of hand tools for torque exertion is discouraged because of the possibility of damage to equipment. Pressure gauges in the oil extraction industry are examples. At other times restrictions in the physical space available for the hand and tool preclude the use of the latter.

Hand turning torque studies reported in the literature, have dealt mainly with how torque is affected by diameter (for handles with circular area of cross-section). Different optimal diameters have been reported for different conditions (Pheasant, 1986; and Imrhan and Loo, 1989).

In aircraft maintenance, manual torque exertion is common; but empirical data to guide designers of such tasks, are inadequate. The need exists for a better understanding of how torque is affected by several factors simultaneously; as, for example, combinations of handle variables like surface finish, length of handle, handle material, and handle shape; and task variables like handle orientation, direction of turn, general body posture, and local skeletal configuration. Part of the problem lies in the size of the experiment required to examine various combinations of these variables.

The purpose of this study was to investigate the effects of certain handle and task related variables on the ability of people to exert torques, using a handle diameter that is considered within the optimal range of 50-65 mm, for axial rotation (Pheasant,1986).

<u>Objectives</u>

The specific objectives of this study were to determine the effects of (1) surface finish, (2) direction of turn, (3) arm position, and (4) hand laterality, on the torque capabilities of manual workers, using a solid metal cylinder of diameter 2 1/4 in (63 mm) and length 5 1/2 in (140 cm). This length ensured that the full width of the hand made contact with the handle.

The levels of the above stated variables were:

1. Surface finish -- smooth or knurled
2. Direction of turn -- flexion or extension action of the wrist
3. Arm position -- 30, 90 or 130 deg shoulder flexion of the outstretched arm
4. Hand laterality -- left or right hand

All combinations of the levels of the above variables were tested. This required the use of a 2 x 2 x 3 x 2 fully crossed factorial design, producing 24 different measurement conditions.

METHODS

<u>Measurements and apparatus</u>

Hand turning torques were measured with a Snap-on-Tools torque meter; it had a range of 150 in-lb in either direction of motion of the pointer, and increments of 2.5 in-lb.

Handgrip and pinch strengths, and selected anthropometric measurements were also taken. Handgrip was measured with a handgrip dynamometer (100 kg in increments of 1 kg); and pinch strengths (lateral and chuck) with a pinch gauge (60 lb range and 2lb increments) by standard methods (Imrhan and Loo, 1989). Anthropometric variables were measured with the aid of Pfister anthropological kit, using standard methods (NASA, 1978). The angle at the shoulder during torque testing was measured with a goniometer.

<u>Subjects</u>

Ten male and 10 female volunteers were tested. The males were between the ages of 28 yr and 43 yr (mean = 35.7 yr), and the females between 25 yr and 40 yr (mean = 33.0 yr). All males and one female were employed in manual work; the other females were homemakers. All data were taken in their homes between 1:00-9:00 p.m.

<u>Procedure</u>

The methods of maximal voluntary contraction (MVC) strength testing, as described by Caldwell et al (1976) was adhered to as close as possible. The precautions for minimizing contamination of data as outlined by Kroemer and Marras (1981) were also applied.

The subjects were tested individually. Twelve torque measurements were taken in each session, with only one session per person per day. Two sessions were required to complete the 24 torque measurements, 12 in each session, randomized. Anthropometric data, handgrip strength and pinch strength measurements were interspersed between the more strenuous torque exertions. Two to three minutes rest was given between strength exertions.

The subjects stood upright and gripped the testing handle on the torque meter with one hand and the stabilizing handle (of identical dimensions, but knurled) with the other hand (Figure 1). The testing handle was either smooth or knurled and was removeable. The investigator supported the weight of the apparatus by holding it.

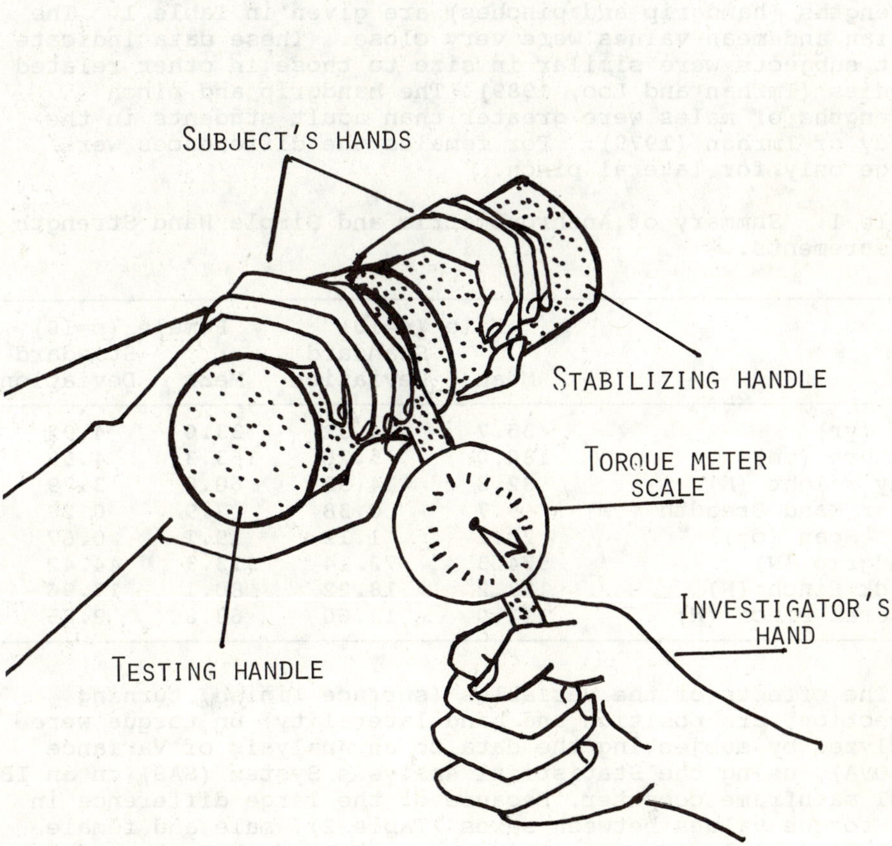

FIGURE 1 . Diagram showing hand posture while exerting torque on cylindrical handle.

All handles were of solid aluminum. In order to prevent any degrading effects due to its weight, the apparatus was supported by the investigator while the subject exerted MVC manual torque. The peak torque achieved was held by a second pointer on the torque meter scale. It was then recorded manually.

Each of the 24 measurements was repeated for the first four subjects. The 48 measurements were randomly taken. A preliminary analysis indicated that the great majority of the first (of the two repetitions) measurement was stronger than the second one, and each was within 10% of its replicate. In order to reduce the size of the experiment therefore, measurements were not repeated on the other 16 subjects. For the first four subjects, only the stronger of the two repeated measurements were used for subsequent data analysis.

RESULTS AND DISCUSSION

Summary data of subject anthropometry and simple hand strengths (handgrip and pinches) are given in Table 1. The median and mean values were very close. These data indicate that subjects were similar in size to those in other related studies (Imrhan and Loo, 1989). The handgrip and pinch strengths of males were greater than adult students in the study of Imrhan (1979). For females the differences were large only for lateral pinch.

Table 1. Summary of Anthropometric and Simple Hand Strength Measurements.

	Male (n=10)		Female (n=10)	
	Mean	Standard Deviation	Mean	Standard Deviation
Age (yr)	35.7	4.32	33.0	4.92
Stature (cm)	180.0	5.26	163.3	4.59
Body Weight (N)	82.8	14.32	60.8	3.79
Outer Hand Breadth (cm)	9.7	0.38	7.9	0.28
Grip Span (cm)	23.1	1.12	19.1	0.67
Handgrip (N)	574.3	72.14	313.9	44.42
Chuck Pinch (N)	123.2	18.22	80.1	17.96
Lateral Pinch (N)	112.0	13.60	68.8	9.85

The effects of the variables (surface finish, turning direction, arm position and hand laterality) on torque were analyzed by subjecting the data to an Analysis of Variance (ANOVA), using the Statistical Analysis System (SAS) on an IBM 4381 mainframe computer. Because of the large difference in the torque values between sexes (Table 2), male and female data were analyzed separately. For each of these two data sets, a within subjects design (Hays, 1987) was formulated to determine the appropriate F-ratios.

The results indicated that only arm position was not statistically significant, at the 5% level of significance; p-values for surface finish, turning direction and hand laterality were 0.0005, 0.0006 and 0.0009, respectively. There were no significant second order interaction effects.

Table 2. Mean Hand Turning torque (Nm) of Males and Females on 2 1/4 Inch Diameter Cylinders

Type Surface	Arm Configuration (Deg Shoulder Flexion)	Type Contraction	Male(n=10) Right Hand	Male(n=10) Left Hand	(Female(n=10) Right Hand	(Female(n=10) Left Hand
	30	Flexion	*10.48 (0.46)	8.74 (0.40)	4.69 (0.43)	4.50 (0.37)
		Extension	12.83 (0.61)	12.02 (0.84)	6.12 (0.52)	5.66 (0.49)
Knurled	90	Flexion	10.52 (0.48)	9.82 (0.54)	5.27 (0.42)	4.70 (0.36)
		Extension	12.09 (0.70)	11.90 (0.69)	6.19 (0.56)	5.47 (0.52)
	130	Flexion	10.91 (0.78)	8.98 (0.63)	5.16 (0.35)	4.68 (0.33)
		Extension	11.78 (0.62)	12.07 (0.59)	6.02 (0.52)	5.59 (0.49)
	30	Flexion	8.75 (0.65)	8.22 (0.56)	4.47 (0.39)	4.21 (0.38)
		Extension	10.59 (0.59)	9.90 (0.75)	5.63 (0.51)	4.90 (0.48)
Smooth	90	Flexion	9.11 (0.72)	8.54 (0.69)	4.68 (0.39)	4.52 (0.35)
		Extension	10.59 (0.78)	9.61 (0.84)	5.28 (0.49)	4.65 (0.52)
	130	Flexion	9.52 (0.61)	8.46 (0.66)	4.53 (0.35)	4.35 (0.32)
		Extension	10.31 (0.73)	9.56 (0.80)	5.25 (0.46)	4.89 (0.59)

*The number in parentheses represents the standard error of the mean.

For the male data there were two third order interaction effects -- hand laterality *surface finsh* turning direction and arm position *hand laterality* turning direction. Their difficulty in interpretation and impracticality of application do not warrant further discussion.

Though statistically significant, the absolute differences within the levels of the three significant factors were not as large as the differences between sexes. On the average, the mean torques for smooth cylinders were 87% as strong as for knurled cylinders; flexion turns were 87% as strong as extension turns; and the left hand was 92% as strong as the right hand. Females were 50% as strong as males.

The strongest individual torques recorded were 15.93 Nm for a male, and 8.93 for a female; and the weakest were 5.73 Nm (male) and 1.98 Nm (female). Both of the strongest values (15.93 and 8.93 Nm) were for extension with the right hand on the knurled cylinder. Both of the weakest values were for flexion with the left hand on the smooth cylinder. Table 2 depicts the means of the 10 subjects in each sex, for each of the 24 testing conditions.

The association between torque and each of the other variables (simple hand strengths, age and anthropometry) were assessed using Pearson Product Moment Correlation coefficients. The female and male data were combined. The results indicated that only age was not significantly correlated with torque, at the 5% level of significance ($r = 0.01$); all other variables were highly significant (positively) -- at the 0.01% level ($p = 0.001$). Correlation coefficients ranged from 0.55 to 0.86. The notion that larger persons can generate greater hand strength may seem logical, but many other studies have shown much weaker correlations between anthropometric variables and other types of strength.

The results of this study indicate that the strength of hand turning torque is not affected by the position of the outstretched arm. Subjects in this experiment stood while being tested, and their arms were approximately in the saggital plane. Whether arm position in other planes, or in different body postures, do not also significantly affect torques cannot be determined from these results. Lack of available data in the literature leaves the question unanswered, empirically. However subjective inferences made by examining the biomechanical aspects of the different planes suggest that the results may be the same; that is, that different positions of the outstretched hand, in a different plane of space, might not affect the torque. MVC torque is dependent on the handgrip strength (Pheasant and O'Neill, 1981; and Imrhan and Loo, 1989) and on the strength of flexion and extension of the wrist. It is unlikely that these three types of strengths can be altered significantly by positioning the outstretched arms in different planes. Neither the mechanical advantages (leverages) of the critical muscles (wrist and finger flexors and extensors) nor the length-tension properties is likely to change among the three arm positions, for any given plane. Only the shoulder muscles change leverages significantly, for different positions of the outstretched arm. However, one cannot make the same conclusion when comparing the same arm position in different planes. It is possible that, as one moves further away from the saggital plane the torque would become weaker.

The influence of body posture is more difficult to assess. Only postures which alter the hand and wrist muscle leverages,or length-tension properties, are likely to produce different torques in different arm positions. It unlikely that sitting or kneeling (two popular working postures) can evoke significant differences.

The practical implications of these results is that the placement of the object to be torqued, with the arms outstretched, will not be more stressful in certain positions within the same plane.

The fact that knurled torque produced greater torques is important. This may be a consequence of the depth of knurling (standard pitch) or of the diameter of the cylinder. It is wrong to assume that any type of knurling would produce greater torques than equivalent smooth surfaces (Imrhan and Loo, 1989).

The stronger torques for extension may be due to the fact that pure wrist extension contractions are stronger than pure wrist flexion ones (Vanswearingen, 1983). However, torques in opposite directions, for different task conditions, are necessarily different in magnitude (Rohles at al., 1983).

The differences in strength of the two hands is within that found for other types of hand strengths (Imrhan and Loo, 1989).

REFERENCES

Asmussen, E. and Heebol-Nielsen, 1961, Isometric muscle strength of adult men and women. Communications from the Testing and Observation Institute of the Danish National Association for Infantile Paralysis, 13, pp. 3-13.

Caldwell, L.S., Chaffin, D.B., Dukes-Dubos, F.N. Kroemer, K.H.E., Laubach, L.L., Snook, S.H., and Wasserman D.E., 1974, A proposed standard procedure for static muscle strength testing. American Industrial Hygiene Association Journal, 35, 201-206.

Eastman Kodak Company, 1986, Ergonomic Design for People at Work, Volume 2, New York, Van Nostrand Reinhold.

Hays, W.L., 1988, Statistics, 4th Edition. New York, Holt, Rinehart and Winston, Inc.

Imrhan and Loo, 1989, Trends in finger pinch strength in children, adults and the elderly, Human Factors (in press).

Kroemer, K.H.E., and Marras, W.S., 1981, Evaluation of maximal and submaximal static muscle exertions. Human Factors, 23(6), 643-654.

NASA, 1978, Anthropometric Source Book. NASA Defense Publication 1024 (US NAtional Aeronautics and Space Administration).

Pheasant, S., 1986, Body Space. London and Philadelphia, Taylor and Francis.

Vanswearingen, J.M., 1983, Measuring wrist muscle strength. The Journal of Orthopedic and Sports Physical Therapy, 4, 217-228.

CONCLUSIONS

The most important conclusion that can be drawn from this study is that when standing and torquing, different positions of the arm in a given plane generate equally strong torques; or that, for application of a set torque, the musculoskeletal system should be equally stressed. This conclusion has is useful in workplace design for manual torquing tasks. However greater torque can be achieved (1) on knurled (standard pitch) surfaces; (2) with extension (loosening) motions of the hand; and (3) with the right hand, for right hand dominant persons. Neither the influence of different body postures nor of different arm configurations has been investigated. Though one can make subjective inferences based on biomechanical considerations, empirical verification is needed.

Acknowledgment

The authors would like to thank Ms. Ann von der Heide, Department of Industrial Engineering, UTA, for her help in typing this paper.

Advances in Industrial Ergonomics and Safety II
Edited by Biman Das
Taylor & Francis 1990

PHYSICAL DEMANDS OF TUG BOAT WORKERS

Barton N. Daniel
6727-21st Avenue NW
Seattle, Washington 98117

T.L. Doolittle
Department of Environmental Health SC-34
University of Washington
Seattle, Washington 98133

Job analysis were conducted aboard tugboats used by two West Coast barge towing companies. Strength and metabolic demands imposed on tugboat workers were quantified. The job analysis consisted of: interviews to define critical tasks; measurement of material handled for dimensions and weight; observation of work to determine task repetition, and distances moved; measurement of heart rate response to work bouts, anchored by known heart rate response to a known workload. While performing docking procedures or making/breaking tow, the workers were expending 4.0-8.0+ METs. Strength requirements exceeded NIOSH action limits in many instances, and exceeded the maximum permissible limits for a few tasks. Pre-employment physical ability screening tests were recommended as an administrative control.

INTRODUCTION

The purpose of this study was to quantify the strength and metabolic demands imposed upon Tug Boat workers. The secondary purposes were to compare the NIOSH (1981) methodology for quantifying work efforts, and (2) to develop selection standards based on the strength and metabolic demands of the position as an administrative control.

TASK DESCRIPTIONS

Two companies were studied in this investigation. One company uses their tugs for towing barges primarily in the Columbia River system, while the other uses larger ocean going tugs for towing barges along the West Coast and to Hawaii.

Tugs used for river work tend to be smaller than the vessels used for ocean towing. As a consequence, some of the equipment used on these tugs is somewhat smaller and has less mass. Generally, the laboring tasks performed are very similar regardless of the vessel.

Crew size for the ocean towing tugs, was six-seven individuals (including a Cook); for the River Tows, Crew size was three-four individuals. There are several titles used for Tug Boat laborers. For the two companies studied these titles were "Deck Hand, Oiler, Wiper, Mate, and Assistant Engineer". These titles represent pay differentials, and differences in routine task assignment. Because of the small crew size on these vessel, however, it was found that during physically limiting tasks (making up to barges, casting off from docks etc.), the division between laboring jobs virtually disappeared. That is, during these brief periods of intense effort virtually all these personnel perform the same tasks. For the purposes of this paper the title "tug boat worker or worker" will refer to the individuals performing critical tasks which are the focus of the descriptions that follow.

Limiting Tasks

Making-up a tow with a barge for a long distance trip necessitates moving a surge chain into position on the stern deck; it must be allowed to pull freely off of the deck as the tug pulls away from the barge as the tow gets underway. NOTE: Surge chains are not used in river towing operations. Each one of the links of this chain has a mass of approximately 22.7 kg; shackles used in the tow (up to seven for a tandem tow) have a mass of 43.1 kg each. Manipulation on deck is performed by hand employing the use of chain hooks, pry bars and short lines. Individuals were found to be lifting up to 68 KG in the equivalent of a deadlift (deck to knuckle height).

In some towing operations, wire cables up to 3.2 centimeters in diameter, are used, instead of a surge chain. Tow wires may be dragged into position from the stern of the tug to the deck of the barge. A heaving line is thrown from the tug to the barge deck, and the wire must be lifted from a level beneath the deck of the barge. If the barge is unladen and riding high in the water (well above the level of the tug) this can be a very difficult operation. Because of the orientation of the tug to the barge during this type of task, 6 to 12 meters of the wire may be hanging in the space between the barge and the tug. When moving the wire, the mass of this freely hanging wire (in excess of 135 kg with the larger diameter wire) must be lifted .5 to .75 meters and moved 6 to 9 meters horizontally. Generally, two workers accomplish this task. Even with two workers, this effort is in excess of the maximal permissible limit (NIOSH, 1981).

Manipulating large nylon lines (up to 25 centimeters in circumference) is a frequent task of the workers when mooring. Dragging these lines across the deck of the tug or the barge requires 180 to 270 N of pulling force, in the best of circumstances. If the lines are wet, frozen or must be pulled over obstructions, this pulling force may increase to in excess of 400 N. When securing a loaded barge, many lines may have to be pulled up onto the top of the load. Sometimes, large loops of the line are carried over the shoulder to reach an outlying part of barge being docked. Such ropes have a mass of 18 to 45+ kg.

Wire lines are used for various tasks, and are often pulled across the deck. Pulling efforts range from 180 to 450+ N of pulling force.

Emergency Situations

Retrieving a barge that has broken it's tow is a difficult operation that involves the use of an "Oroville Hook". The hook has a mass of approximately 90 kg; the nylon line used for the operation is 36 cm in circumference and has a mass of 3.9 kg per meter. Maximal deck to knuckle lifting efforts, involving masses of 50 to 100 kg, would be the norm for this operation.

If a ship has to be abandoned, life rafts must be deployed; on some vessels, there may not be mechanical assists available for this operation. The rafts have a mass of 98 to 141 kg.

Compounding the problem for sea-going tugs are the rough seas and foul weather that usually are the cause of the emergency conditions. Having to contend with the heavy lifting requirements of these tasks, on the deck of a tug being pitched about on rough seas, increases the difficulty. In general, the river tugs are not faced with such rough shipping conditions, and are confronted by emergency situations less often.

STRENGTH REQUIREMENTS

Adequate and safe performance of the arduous tasks described above requires a tug boat worker to be able to easily lift and carry masses of 35 to 40 kg and to lift masses of 68 to 91 kg from deck to knuckle height. Heavy lifting work tasks such as making and breaking tows are performed 1 to 2 times per day to 1 to 2 times per week, depending on the length of the voyage. These tasks are critical to the success of the operation. Workers must possess sufficient strength to perform the tasks safely to insure their own safety and the safety of the entire crew. Lifting/lowering tasks are in Table 1, and pulling tasks are in Table 2.

Many of the lifting requirements involve efforts in excess of the NIOSH action limit. In addition, some of the occasional tasks performed when making/breaking tow, or during emergency situations exceeded the maximal permissible limit.

Preassignment screening should be employed until job modification can be effected.

Table 1. Lifting/Lowering Tasks.

Task Location/ Item Lifted	Mass/Unit (kgs)	Start Height	End Height	Frequency*
Ocean Tows				
Shackles	43	deck	thigh	V.Freq.-Infrequent
Supplies	16-30	deck	waist	Very Infrequent
Cylinder Heads	66	deck	knee	Very Infrequent
Pistons	18	deck	knee	Very Infrequent
Life Raft	98-141	variable		Very Infrequent
River Tows				
Straps	16-32	deck	waist/ shoulder	Very Frequent
Power Cable	18-27	deck	waist/ shoulder	Frequent
Dog Cable	25-27	deck	waist	Infrequent

Table 2. Example Pulling Tasks

Location	Item Moved	Force (N)	Frequency*
All Vessels			
	Nylon Lines	180-270	Frequent
	Obstructed Lines	400+	Frequent
	Wire Lines	180-450+	Frequent

* Frequency:
Very Frequent = lifted/used many times per shift
Frequent = lifted/used at least once per shift
Infrequent = lifted/used at least once /few days
Very Infrequent = lifted/used at least once /few months

METABOLIC DEMANDS

Determination

Aerobic capacity was determined for a sample of incumbents as described by Astrand and Rodahl (1986). The methodology is as follows:

a. Heart rate was monitored during a submaximal exercise regimen on a step bench. This established each individual's capacity for work in terms of oxygen consumption as determined through heart rate.

b. Work sampling of key tasks while recording heart rate.

c. Information generated in step 1 is used to extrapolate the actual energy expenditure associated with each task for that individual.

The submaximal Step Test was used to determine aerobic power because of it's portability and relative safety. Astrand and Rodahl (1986) indicate that maximum aerobic power estimates from this type of submaximal testing correlate very well (r = 0.9) with actual determinations. The aerobic requirements of the tasks analyzed are summarized in Table 3.

Table 3. Metabolic Requirements of Studied Tasks

Task	Average MET Requirement	MET Range	Work Mode
Making/Breaking Tow	5.9	4.0-8.3	Intermittent

PRE-EMPLOYMENT SCREENING

The translation of strength and stamina levels required by the job into test items and standards, were predicated on four guidelines:

1. Strength requirements should not exceed the limits typically recommended in the ergonomics literature. This research recommends 75 percent of an individual's maximum strength capacity for occasional efforts (Doolittle and Kaiyala, 1987).

2. It is possible to accurately estimate maximum dynamic strength from repetitions requiring a sub-maximal (and safer) effort. This is in accordance with the muscular strength--muscular endurance continuum described in Logan and Foreman, (1961). Procedures for estimating maximum dynamic strength from the number of repetitions performed with a standardized submaximal mass are described in Doolittle and Kaiyala (1988), and Doolittle (1989).

3. The specific muscle groups employed on the job need to be the ones tested.

4. The fourth guideline related metabolic demand to maximum aerobic power, where it is well recognized that individuals cannot be expected to perform long term efforts at greater than 33 to 40 percent of maximum shift-long; and not higher than 50 to 60% of

maximum for shorter periods of intermittent effort (Astrand and Rodahl, 1986). Because the work performed is highly intermittent in nature, the latter percentages have been used in identifying the standards.

The tasks were translated into standard exercise movements and ergonomic terminology as depicted in Table 4.

Table 4. Standardized Test Movements

TEST	TASK
Modified Upright Row (2-Arm, Knuckle to Shoulder)	Lifting To Shoulder Pulling/Lifting
Dead Lift (2-Arm, floor to thigh)	Lifting/Pulling Lines
Stamina Test	All Laboring Efforts

STRENGTH TESTS

The standardized test movements translated into a strength test battery consisting of two measures: (1) Modified Upright Row with 29.5 kg, (2) Dead Lift with 68 kg.

Upright Row

This test measures the individual's ability for lifting items from approximately mid-thigh to shoulder height, with the load maneuvered close to the body. The arms and shoulders are the prime movers. This is a common lifting technique when lifting tools, lines, and parts. In the test, individuals perform as many repetitions as possible (up to 18 repeats) at which time they are stopped.

Dead Lift

This test measures the individual's ability for lifting items from approximately the deck to knuckle height. The back and legs are the prime movers. This is a common lifting technique when lifting wire lines onto dock "buttons" or cleats. In the test, individuals perform as many repetitions as possible (up to 18 repeats) at which time they are stopped.

The respective test weight and repetitions were determined in accordance with the known force requirements and relationship between maximum dynamic strength and submaximal lifting described by Doolittle and Kaiyala, (1988).

AEROBIC TEST

In order to insure that an individual applicant possesses adequate aerobic power (stamina), a standardized aerobic capacity test was recommended. There are three accepted methods for administering this type of test (listed below). The correlation between these tests is high, and for the purposes of industrial screening, any one of the following may be used.

STEP TEST

Heart rate is monitored while the person steps up and down from a bench 0.28 m in height for six minutes. Stepping rate ranges from 18 to 24 ascents per minute. Heart rate, step rate, age, gender, and bench height are used to compute the individual's maximum aerobic power in mL/kg/min

(American College of Sports Medicine, 1980; Astrand and Rodahl, 1986). Aerobic Power is converted to METs by dividing by 3.5.

CYCLE ERGOMETRY TEST

Heart rate is monitored while the person cycles on a stationary cycle at a known work load. Rotations per minute remains constant. The workload is increased after the first 3 minutes. Heart rate, age, gender, weight, and workload are used to compute the individual's maximum aerobic power in mL/kg/min (American College of Sports Medicine, 1980; Astrand and Rodahl, 1986). Aerobic Power is converted to METs by dividing by 3.5.

TREADMILL TEST

Heart rate is monitored while the person walks/jogs on a motorized treadmill at a known speed and inclination. The workload is increased after the first 3 minutes by increasing the inclination of the treadmill. Heart rate, age, gender, weight, and treadmill speed and inclination are used to compute the individual's maximum aerobic power in mL/kg/min (American College of Sports Medicine, 1980; Astrand and Rodahl, 1986). Aerobic Power is converted to METs by dividing by 3.5.

SCORING CATEGORIZATION

A four category classification (similar to Doolittle and Rigby, 1990) was utilized to evaluate new hires. The "Acceptable" category is the targeted level based on the job analysis. Categorization is based on the percentages of maximum at which an individual will be taxed.

Below Minimum: Anything below Marginal. The candidate, if hired, would be taxed at more than 70% of his or her maximum aerobic capacity, or more than 85% his or her strength capacity adjusted for frequency. The risk of injury to the individual and turnover to the employer are too high to make hiring the person a prudent decision.

Marginal: The lower point for marginal will tax the individual at 70% of aerobic capacity, and 85% of the persons strength capacity adjusted for frequency. While the risk to injury is higher than desired and the individual will find the job difficult, with a high level of effort he or she may be able to meet the minimum standards. With an adequate applicant pool one should not have to hire from this category, however, individuals near the top of it, with potential in other attributes, may be considered.

Acceptable: This is the targeted level of capacity based on job analysis findings. The bottom of this category will tax the individual at 60% of their maximum aerobic capacity for the average intermittent values of the metabolic costs on the job. The strength demands will be 75% of maximum capacity adjusted for frequency. Individuals scoring in this category will find the job moderately hard but not overly difficult. Risk to injury is within normal limits, and the individuals have a good probability of success.

Better Qualified: Individuals in this category will be taxed at less than 50% of their maximum aerobic capacity. The strength demands will be less than 65% of maximum capacity adjusted for frequency. Individuals in this category are highly fit relative to the job demands. They should have little difficulty adapting to the job demands, have low risks to on the job musculoskeletal injuries, and have a very high probability of job success.

Table 5. Recommended Testing Standards

Ocean Towing Tug Boat Worker

Test	Below Minimum	Marginal	Acceptable	Better Qualified
Aerobic Power Test (METs)	<9.0	9.0-10.4	10.5-11.5	>11.5
Strength Test (65 # Upright Row) (Repetitions)	<6	6-10	11-17	>17
(150 # Dead Lift) (Repetitions)	<6	6-10	11-17	>17

River Tug Boat Worker

Test	Below Minimum	Marginal	Acceptable	Better Qualified
Aerobic Power Test (METs)	<8.5	8.5-9.4	9.5-10.5	>10.5
Strength Test (50 # Upright Row) (Repetitions)	<6	6-10	11-17	>17
(120 # Dead Lift) (Repetitions)	<6	6-10	11-17	>17

SUMMARY

Laboring tasks performed by Tugboat Workers were analyzed. Strength and metabolic expenditures were quantified for these tasks. It was determined that these tasks are physically demanding, and a bona fide occupational qualification. A physical performance selection test battery was developed to screen individuals with sufficient strength and stamina to safely and productively perform their duties in these positions.

ACKNOWLEDGEMENT

This work was supported in part by funding from the Sause Brothers Ocean Towing Company, Coos Bay, Oregon.

REFERENCES

American College of Sports Medicine, 1980. Guidelines for graded exercise testing and exercise prescription (2nd Ed.). Philadelphia: Lea & Febiger.

Astrand, P. O. and Rodahl, K., 1986. Textbook of Work Physiology (3rd Ed.). New York: McGraw-Hill.

Ayoub, M., Selan, J. & Jiang, B., 1984. Mini-Guide for Lifting. Lubbock: Texas Tech University. 114pp.

Doolittle, T. L., 1989. Selection of manual material handling tasks using dynamic strength tests. Advances in Industrial Ergonomics and Safety. Volume 1. (Mital, A. (Ed.)), (London: Taylor & Francis.

Doolittle, T. L. and Kaiyala, K., 1987. A generic performance test for screening firefighters. Trends in Ergonomics /Human Factors IV. Asfour, S. S. (Ed.), North Holland: Elsevier. 603-610.

Doolittle, T. L. and Kaiyala, K., 1988. Prediction of maximum dynamic strength from multiple repetitions with a submaximal load. Trends in Ergonomics /Human Factors V. Aghazadeh, F. (Ed.), North Holland: Elsevier. 767-774.

Doolittle, T. And Rigby, P., 1990. Physical demands imposed upon U.S. NOAA mariners performing manual material handling tasks. Advances in Industrial Ergonomics and Safety II. Das, B. (Ed.). London: Taylor and Francis.

Logan, G. and Foreman, K., 1961 Strength-endurance continuum. The Physical Educator. 18, 103-105.

NIOSH, 1981. Work Practices Guide for Manual Lifting. Pub No. PB 82178948, Cincinnati: National Institute for Safety and Health. 183 pp.

Advances in Industrial Ergonomics and Safety II
Edited by Biman Das
Taylor & Francis 1990

JOB RELATED PHYSICAL PERFORMANCE TESTS

Jerry L. Purswell, Ph.D., P.E.
Ronald Ratliff, Ph.D.
Adele Hughes, Ph.D.
University of Oklahoma, Norman, OK 73019

This paper describes a research study to establish job related physical performance tests for officers in the Oklahoma Department of Corrections. It presents some of the issues which have to be considered in developing such tests and describes the unique process required by the demanding study environment. The major outcome of the study was the development of physical selection criteria believed to be related to bonafide occupational qualification criteria for correctional officers.

INTRODUCTION

Job Related Physical Performance Tests (JRPPT's) have become a part of the selection/retention requirements for different types of jobs, most notably the public support services associated with law enforcement and fire protection because of the critical nature of these services to the community. There has also been a realization that persons filling these types of jobs may incur a significant safety and safety and health risk if they do not possess certain physical performance capabilities. Ever-increasing workers compensation costs have also stimulated manufacturing firms to consider the use of JRPPT's for worker selection with the goal of better matching of worker and job in order to avoid injury. Ergonomists have increasingly been called upon to become a part of the employee selection process where JRPPT's are being developed.

The purpose of this paper is to provide a brief background of the area of JRPPT's, and to report on a project to develop JRPPT's for correctional officers in the state of Oklahoma where a unique methodology was employed to develop these tests.

BACKGROUND

The Equal Employment Opportunity Commission (EEOC) has served as the advocate of persons seeking jobs where they have been excluded in the past because of various occupational qualifications. Because of these challenges, employers have been pressed to establish bonafide occupational qualifications (BFOQ's), i.e. there must be appropriate selection/retention tests that can be linked to the actual success rates of the person in the job. The persons most likely to complain about discrimination in terms of JRPPT's have been females and the physically handicapped. Prior to the initiation of this study, JRPPT's had been developed for the Oklahoma City firefighters (Ratliff et al., 1977) and the state police (Ratliff, 1981). It was possible to obtain measurements of the physical strength and exertion levels required of the participants in these studies by relatively direct methods at the scene of the action and then use this data to design the JRPPT's. However, this was not the case with the correctional officers because of the security needs of the prison and the researchers. It was therefore necessary to develop additional data collection and analysis methods to determine the physical performance requirements of the different jobs in the system.

LITERATURE REVIEW

In analyzing the task demands of correctional officers, the primary objective was to define physical task demands, but it was clear that there were other task demands as well, including high mental stress levels that interacted with the physical task demands. In order to describe the physical task demands, it was determined that a form of activity analysis based on self-reported data would be the principal method of data collection used. Activity analysis has been shown to be useful by Niebel (1982) and Konz (1983) as one approach to task analysis.

Another approach to task analysis is based on an ergonomic model of the person/machine interface where information sensing, information processing and decision making are analyzed. McCormick (1976), Grandjean (1981) and Bailey (1982) have presented such models.

The National Information Center of Corrections has compiled information on the number of states that reported use of BFOQ's for selecting correctional officers (personal communication, 1985). Of the eight states that now use such standards, only California (1981) provided information that demonstrated an attempt to validate their JRPPT's.

Simri (1974) provides one of the first reviews of the various components which might be considered in developing physical performance tests, which include:

1. Dynamic muscular performance
    a. muscle strength
    b. muscle endurance
2. Explosive muscular strength
3. Static muscular strength and endurance
4. Speed
5. Agility
6. Flexibility
7. Cardiovascular endurance (aerobic)

Other investigators have documented the effects of variables such as age, gender, environmental heat, level of training, and motivation on physical performance capabilities (Astrand and Rodahl, 1977; NIOSH, 1981). Specific developments of JRPPT's have been reported by Davis et al. (1982), Sharkey et al. (1978), and Davis and Stark (1980).

METHODOLOGY

Defining Job Demands

In developing JRPPT's, one can focus on content validity or criterion related validity. It was decided that this study would focus on providing satisfactory content validity. Figure 1 provides an overview of the methodology followed in developing content validity.

As noted earlier, only limited direct observation of correctional officers while they were performing their duties was possible because of logistical problems. Most of the requirements for exerting physical force and/or maintaining physical endurance appeared to occur at irregular intervals throughout the work day or work week, such as resisting attacks of inmates, protecting one inmate from another, or searching for escaped inmates. After considering the available options, it was found that two non-observational data sources were most likely to provide the data needed to define physical performance requirements, i.e., serious incident reports and questionnaires administered to correctional officers. The serious incident reports were prepared after some major interaction with an inmate that usually required physical force exertion to gain or maintain control. The reports contained a narrative description that was analyzed to obtain some indication of the physical force and endurance required. However, these reports were not uniformly available and consistent in the details from which the physical force requirements could be inferred.

In order to obtain as large a data base as possible for all correctional officers, it was necessary to develop a method to obtain information about as many aspects of the physical force requirements as possible through some form of questionnaire. A Job Demand Questionnaire (JDQ) instrument was developed, pretested, revised, and then administered to all correctional units of the state system. Space limitations do not permit the full

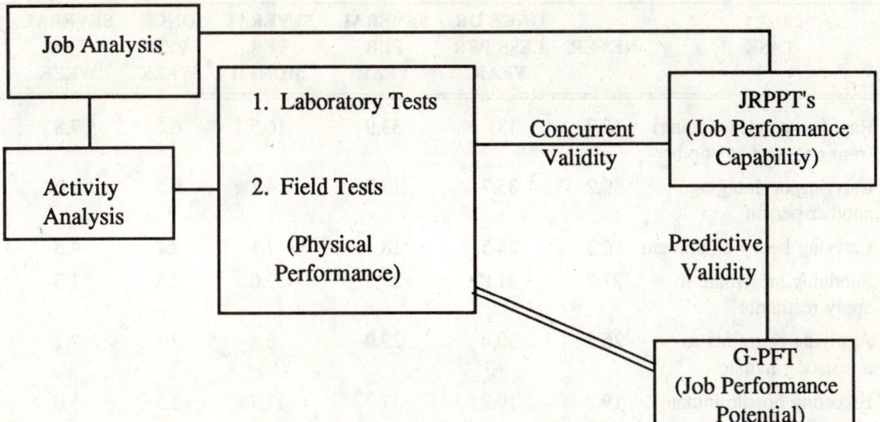

Figure 1. Overview of Content Validation Methodology.

questionnaire to be presented, but copies will be provided by the authors on request. In general, the questionnaire focused on the types of jobs where physical force was required, the frequency with which the jobs were performed, and the perceived level of exertion while performing the job. In addition, questions were asked about perceived stress levels, previous injuries on the job and a brief medical history was requested. The questionnaires were distributed throughout the system with provision for protecting the confidentiality of all respondents. The sample population included 866 persons out of a total of 1310 in the system. Demographic analysis of the respondents by age, gender, ethnic origin, job classification and institution verified that this was a representative sample with respect to these characteristics.

Tables 1 to 3 provide percentage data for the information obtained from the questionnaire. Some highlights of this data relative to physical job demands are as follows:

1. The most frequent task duration was 30 secs. to 2 minutes.
2. Manhunts require significant exertion for extended periods.
3. Subduing an inmate or using force to move a hostile inmate were noted as requiring significant physical demands (explosive physical force).
4. Almost 50% of the CO's perceived their jobs as somewhat or very physically demanding.
5. As would be expected, females perceived their jobs as more physically demanding than males.

Since it was not feasible from either a safety or measurement standpoint to evaluate persons actually performing the tasks surveyed, the JRPPT's shown below were formulated as tasks that the researchers and correctional officials agreed embodied the essential physical characteristics of the job components:

1. Whole body rapid response
2. Body drag/rescue
3. Lifting/carrying weight
4. Choice response time
5. Grip strength
6. Explosive push away force
7. Prolonged physical activity.

Table 1. Percentage of CO's Average Frequency When Performing Selected Tasks.

TASK	NEVER	ONCE OR LESS PER YEAR	SEVERAL PER YEAR	SEVERAL PER MONTH	ONCE PER WEEK	SEVERAL PER WEEK	ONCE OR MORE PER DAY
Rapid response (running) from one area to another	12.7	13.8	33.9	16.5	6.6	7.8	8.8
Carrying or dragging another person	39.2	35.7	19.2	4.1	0.5	0.6	0.8
Carrying heavy equipment	36.3	26.5	18.1	7.1	2.1	4.8	5.2
Subduing an inmate to apply restraints	27.7	31.8	27.3	9.0	2.2	1.3	0.7
Applying restraints to a resisting inmate	26.5	30.0	29.0	6.5	2.1	2.7	3.3
Escorting hostile inmate	19.2	19.2	37.7	11.7	3.5	5.0	3.7
Using force to move a hostile inmate	33.9	31.1	27.2	4.7	2.4	0.5	0.2
Manhunt for escaped inmate	41.1	22.7	32.9	3.0	0.2	0.0	0.1
Percent of overall response by column	29.6	26.4	28.1	7.8	2.4	2.8	2.9

Table 2. Percentage of CO's Average Time When Performing Selected Tasks.

TASK	DO NOT PER-FORM	UNDER 30 SEC	30 SEC TO 2 MIN	2-5 MIN	5-15 MIN	15-60 MIN	OVER 60 MIN
Rapid response (running) from one area to another	15.2	21.5	45.9	11.6	3.0	1.7	1.2
Carrying or dragging another person	40.5	7.1	22.2	19.5	8.6	1.4	0.6
Carrying heavy equipment	38.4	6.7	19.5	22.8	7.8	3.6	1.0
Subduing an inmate to apply restraints	29.3	11.0	32.5	16.7	8.1	2.1	0.4
Applying restraints to a resisting inmate	27.2	14.3	33.6	16.7	5.9	1.5	0.7
Escorting hostile inmate	22.1	3.2	21.2	26.5	16.0	4.7	6.2
Using force to move a hostile inmate	36.1	6.4	20.1	18.8	13.3	4.6	0.8
Manhunt for escaped inmate	44.2	0.8	1.4	0.8	0.8	0.9	51.3
Percent of overall response by column	31.6	8.9	24.6	16.7	7.9	2.6	7.7

The next phase of the study then focused on validation procedures for the tasks.

Validation Procedures

The original intent of the study team was to obtain a probabilistic sample of correctional officers. However, formidable logistics problems necessitated the use of convenience samples. A field performance test group was formed of 93 individuals (37 officers and 55 cadets), all of whom had been medically cleared for vigorous exercise; all but one were under 45 years of age. This field test group was tested under conditions that could be expected to exist at one or more of the institutions with the testing administered by trained staff at these institutions. A subsample of that group, consisting of 27 eligible subjects from one cadet class was selected to complete a battery of laboratory tests at OU Human Performance Laboratory. After those cadets had worked at their respective sites for several months, they were then evaluated on the JRPPT's shown above. Because all of the sample members were under 45 years of age, data from Astrand and Rodahl (1986) was used to project the age effects of an older population on the test battery using regression equations.

A correlation analysis showed that there was reasonable independence between the different tasks, with the body drag/rescue having the highest correlation (r = 0.57). Concurrent validity of the JTPPT's as valid measures of physical performance was evaluated through the use of a series of stepwise regression analyses, once using the field test variables and once using the laboratory variables. Since empirical studies that could establish a priori reliability and validity of hypothesized linkages between individual JRPPT's and the field and laboratory tests were not available in the literature, only variables with a correlation greater than 0.36 (significant for the study sample at the 0.05 level) were entered into the regression models.

Table 4 shows the linkage of physical determinants with the JRPPT's. In general, the regression analyses provided empirical evidence that these physical determinants are primary physical determinants of performance on the JRPPT's.

As noted earlier, additional data was also collected regarding perceived stress and medical/health problems. These data showed that the perceived level of mental stress reported by CO's was very high. Additional analysis of the data showed that this stress is related to the perceived likelihood of physical harm, which was borne out by the fact that 22 percent of CO's had been injured on the job in the last three years. This perceived stress level likely is related to the high reported incidence of digestive problems (29%) and general fatigue or insomnia (26%).

DISCUSSION

This study demonstrated an approach to the development of physical performance tests in a demanding study environment. The first phase was successfully completed with the result that the JRPPT's developed have content validity for the more physically demanding tasks performed by correctional officers. However, the development of criterion validity must await the opportunity to obtain valid measures of job performance which can then be related to the JRPPT's which have been developed.

REFERENCES

Astrand, P., and Rodahl, K., 1977, *Textbook of Work Physiology,* (New York: McGraw-Hill).

Bailey, R., 1982, *Human Performance Engineering,* (Englewood Cliffs, NJ: Prentice-Hall).

California State Personnel Board, 1981, The Validation of the Physical Ability Tests for the Classes of Correctional Officer and Correctional Program Supervisor I. (Sacramento, CA).

Davis, P. et al., 1982, Relationship between simulated fire fighting tasks and physical performance measures. *Medicine and Science in Sports.*

Davis, P., and Stark, A., 1980, Age and performance in a police population. *Law Enforcement Bulletin,* (Washington, DC: Federal Bureau of Investigation).

Table 3. Percentage of CO's Average Level of Effort When Performing Selected Tasks.

TASK	DO NOT PERFORM	VERY LIGHT	LIGHT	SOMEWHAT HARD	HARD	VERY HARD
Rapid response (running) from one area to another	13.5	11.7	35.5	31.3	5.5	2.4
Carrying or dragging another person	38.0	3.3	13.7	32.0	9.8	3.2
Carrying heavy equipment	35.2	4.3	16.8	33.9	7.4	2.5
Subduing an inmate to apply restraints	27.6	3.9	15.3	35.1	12.7	5.5
Applying restraints to a resisting inmate	27.0	6.2	20.8	31.4	9.5	5.0
Escorting hostile inmate	20.1	10.8	33.7	26.3	6.3	2.7
Using force to move a hostile inmate	33.3	2.9	13.3	30.8	12.1	7.5
Manhunt for escaped inmate	42.4	5.6	17.6	19.8	7.9	6.8
Percent of overall response by column	29.6	6.1	20.9	30.1	8.9	4.4

Table 4. Linkage of Physical Determinants with JRPPT's.

JR-PPT	Physical Determinants	Predictability Lab	Predictability Field
Whole body rapid response	Dynamic muscular strength; anaerobic power; speed; agility	.78	.77
Body drag/rescue	Dynamic muscular strength; anaerobic power; static muscular strength; agility	.45	.45
Lifting carrying weight	Dynamic muscular strength; static muscular strength	*	*
Choice response time	Speed of response (specified limbs) requiring information processing	.47	.52
Grip strength	Static muscular strength	.71	.73
Explosive push away force	Dynamic muscular strength; anaerobic power;agility	.26	.31
Prolonged physical activity	Aerobic endurance; muscular endurance	.47	.30

* Based on previous studies on comparable populations.

Grandjean, E., 1981, *Fitting the Task to the Man,* (London: Taylor and Francis, Ltd.).

Konz, S., 1983, *Work Design: Industrial Ergonomics,* (Columbus, OH: Grid Publishing Co.).

McCormick, E.J., 1976, *Human Factors in Engineering and Design,* (New York: McGraw-Hill).

Niebel, B.W., 1982, *Motion and Time Study,* (Homewood, IL: Richard D. Irwin, Inc.).

NIOSH, Division of Biomedical and Behavioral Sciences, 1981, *Work Practices Guide for Manual Lifting,* DHHS (NIOSH), 81-122, (Cincinnati, OH).

Ratliff, R., 1981, Physical Testing of the Oklahoma Highway Patrol. (Oklahoma City, OK: Oklahoma Highway Patrol).

Ratliff, R. et al., 1977, Evaluation of the Physical Fitness Status and Heart Rate Responses to Selected Physically Demanding Tasks of Firefighters. In: *Proceedings of Sixth Annual Emergency Service Personnel Health and Fitness Symposium,* (San Diego, CA).

Sharkey et al., 1978, Fitness and Work Capacity: Wildland Firefighting, (Missoula, MT: USDA Forest Service).

Simri, V., 1974, Assessment Procedures for Human Performance. In: *Fitness, Health and Work Capacity.* (Edited by L.A. Larson), (New York: McMillan, Ch. 19).

Advances in Industrial Ergonomics and Safety II
Edited by Biman Das
Taylor & Francis 1990

PREPLACEMENT TESTING PILOT PROGRAM

Cheryl Bockrath and Lois Halunen

Hazards Control Department
Lawrence Livermore National Laboratory
P.O. Box 5505 L-384
Livermore, CA 94550

This paper is aimed at those interested in instituting physical ability testing programs. We describe the general methodology used and lessons learned in initiating a pilot program for preplacement testing in a research and development facility. It is anticipated that the job analyses and testing criteria produced by our program will have additional applications in rehabilitation, recruiting, and worker hardening.

INTRODUCTION

The Lawrence Livermore National Laboratory (LLNL) is a research and development facility directly employing approximately 8000 people. Annually, a high percentage of our lost worktime injury cases are related to overexertion injuries. We were aware that most of these cases came from only a few job classes where the physical demands of the job are high. As industrial safety engineers in the Hazards Control department, we proposed preplacement strength testing as a new way of reducing these injuries.

Basis

The validity and value of preplacement testing as a means to match worker capabilities with job demands has been frequently discussed in ergonomic literature during the past decade (see references). However, when we began to implement a preplacement physical abilities testing project, we found that literature describing actual industrial applications, experience, or the results of such programs was not easily available. We sought any information that could serve as a guide or help us avoid potential pitfalls. We contacted organizations reported to have implemented such programs; those who were involved in preplacement testing projects did graciously share their experiences, but they were also eager to obtain more data.

METHOD

Administrative Preparation

A literature review and other research convinced us that physical ability requirements must be based on documented job tasks to make a preplacement program legally defensible. This necessitates performing thorough, detailed job analyses upon which the tests can then be based. Several other organizations we contacted had performed the job analyses themselves, or planned to; however, our work assignments precluded that possibility. To also obtain third party objectivity, we contracted with an outside engineering/ physical-therapy consulting group through a bid process to perform the job task analyses and to develop the placement tests.

Our Procurement department proved to be a valuable resource in preparing the contract. The format and contractual language helped clarify the services we were requesting. We specified that the testing criteria had to be directly related to the jobs and based on measurements of job parameters. Validation testing of a representative sample of the incumbent population was required. The contract also required documentation of any feasible ergonomic improvements that could lower the demands of the job.

Initial contact was made with other departments that would need to be involved in the project: the Human Resources, Staff Relations, Health Services, Rehabilitation, and Legal departments. The response to the project was overwhelmingly favorable. Each group involved could anticipate benefits, but also had concerns which needed to be addressed (see Table 1).

Funding

Funding for the consultants fees was obtained through our self-insurance system on a worker's compensation premium rate additive. California state regulations require, as a condition for approval of self-insurance plans, an adequate industrial claims-prevention safety program. A rate additive is treated as a Workers Compensation Program cost, and by organizational policy a portion is designated for programs directly related to improving loss experience by reducing the frequency and/or severity of claims. Our Risk Manager was instrumental in obtaining the funding.

Selection of Job Classes

Before requesting funding for the Preplacement Strength Testing Pilot Program, we had completed a statistical analysis to identify all job classes within the organization that had experienced more than three times the overall LLNL average in overexertion injuries for three successive years. We also reviewed workers' compensation costs for these jobs over the same time and found a close correlation. Some of the job classes were eliminated if they had a very small number of workers where a few injuries could skew their rates. Nine job classes were selected for the pilot program (see Table 2). We were not surprised that all the job classes selected were within the two departments that perform the most heavy physical labor: Supply and Distribution, and Plant Engineering. After receiving managerial support and funding, we approached these two departments to present the program and obtain their concurrence.

Table 1. Organizational involvement in preplacement testing

Department	Related Function/Task	Benefits of Involvement with Preplacement Testing Program	Concerns/Considerations
Hazards Control (Health and Safety)	Protect Health and Safety of employees	Goes beyond engineering, job design, & training	Emphasize ergonomic corrections where possible, follow-up on implementation and effectiveness of program
Risk Management (Loss Control)	Control losses from injuries	Reduces injuries and associated costs	Benefits of preplacement will not all show up in measurable cost savings
Workers Compensation (Insurance Carriers)	Provide medical care when employees are injured	Offers medical providers return-to-work criteria; allows specific worker hardening	Job analyses and testing criteria must be updated if job demands change
Employing Departments (e.g. Plant Engineering; Supply & Distribution)	"Get the job done" Keep employees on the job	Full work force; employees do not have to cover for as many absent colleagues	Consider flexibility of internal placement/job switching; solicit incumbent comments as soon as possible
Human Resources (Personnel)	Place applicants or employees in appropriate jobs	Only qualified applicants placed in heavy labor jobs; applicants more informed about job and its demands	A statement must be added in position descriptions, "strength testing will be administered"
Staff Relations (Labor Relations)	Protect the rights of employees	Protects employee right to a safe job; discriminates only against those unable to perform the job	Protect confidentiality of incumbent information; tests for preplacement must be valid; reasonable accommodation policy
Health Services (Medical)	Retain relevant medical records, return to work evaluations	Employee enters with base-line record, can be used for return to work criteria	Determine action to be taken should incumbent's health condition prove a concern in validation testing
Rehabilitation (Vocational Rehab)	Return employees to appropriate work after disability	Job analyses can be utilized in job matching for disabled employees or light-duty assignments.	Include information in job analyses which would be useful for job-matching
General Counsel (Legal)	Protect legal position of the organization	Job placement decisions/policy documented	Job analyses and proper administration determine its defensibility

Table 2. Job classes selected for preplacement pilot program.

Supply and Distribution	Plant Engineering
Warehouse Worker	Trades Helper
Truck Driver	Laborer
	Custodian
	Gardener
	Electrician
	Maintenance Mechanic

Process

Some of the steps taken in the performing the job analyses, validation, and the test-development process are listed below:

- As recommended by the Staff Relations department, advised the management of the departments involved in our Pilot Program to ensure that their employees were informed about the planned program and had an opportunity to comment on it.
- Collected demographic data on existing employees from the employer departments to determine a representative sampling for validation testing.
- Interviewed the existing employees to help identify the most strenuous tasks required by their jobs.
- Measured the weights handled, distances travelled, positions and forces required to perform the most strenuous tasks.
- Developed tests based directly on these measurements. Tests simulated activities required by the job to give them face validity.
- Tested a representative sample of existing employees in each job class as a validation of the testing criteria developed by the consultant. Necessary adjustments were made to ensure that the tests were safe and simulated actual job tasks.

In accordance with the desired objectivity, an outside clinic was also selected to perform strength testing of prospective employees. Except for the limited physical screening agreed upon, the clinic was asked to strictly adhere to the tests as they were developed and validated. The contract requires prompt scheduling and immediate notification to our facility of "pass/fail" results to avoid any delays in hiring. We selected a clinic based on professional merit, types and condition of equipment and facilities, convenience of location, and experience. Our medical staff in the Health Services department was not equipped to conduct such tests and did not have a physical therapist on staff.

DISCUSSION

Validation Testing

Physical abilities testing of representative sampling of existing employees is necessary to validate the proposed strength requirements for new employees. This presents the possibility of identifying a medical condition during incumbent testing -- e.g., extremely high blood pressure -- that should be treated. However, some of the information obtained regarding the physical condition of current employees was medically privileged information and had to be held in strict confidence. A plan must be developed with the concurrence of their department managers to determine what action to take should this situation arise during incumbent testing. When applicants are being tested, the testing clinic would be in a position to directly inform the individual and recommend that they seek appropriate medical care.

The general physical condition of existing employees became a concern during the validation testing. Blood pressure readings, heart rates, and weight all indicated that a physical conditioning program would be desirable. At present, our organization has no formal exercise program for any group of workers except security officers. This information will be used in encouraging management to institute a voluntary worker hardening program. We have proposed establishing several "Fitness for Safety" classes that would offer testing to determine level of fitness and provide guided exercise/work hardening. In addition department managers expressed an interest in a self-help back care, lifting, and conditioning program for existing employees.

Ergonomic recommendations

The ergonomic recommendations that were proposed, as required in the contract, were valuable, but a challenge to address. For example, recommendations for engineering controls such as adding elevators or lifting devices on two-story buildings are not under the control of the departments involved in this project. Such changes take time and money to put into place, and organizational political/budgetary jurisdictions further complicate the process. Nevertheless, it is important to ensure that the rest of the recommendations that would reduce the strain on workers be actively pursued and implemented as soon as possible.

It became apparent when evaluating the warehouse worker job class that several assignments within this job class required significantly more strength, flexibility, and endurance than the others. Management saw that this would mean different strength requirements for some of their workers within one job class. They responded by asking if they could promptly implement the suggested ergonomic recommendations to reduce the effort required, thus keeping only one job strength level. This may result in greatly reduced exposure to risk of overexertion injuries for all warehouse workers. We readily agreed to hold that job analysis in draft form until these changes are in place; we will then review the job analysis and modify the strength requirements.

Cost savings

It is premature to estimate cost savings. LLNL hiring rates have been considerably reduced, and an increase is not expected in the near future. While this means fewer prospective employees will be tested, the job analyses and the testing criteria will be used for other purposes, such as those mentioned below as additional applications. After all, reducing overexertion injuries was our original motivation for this program. These applications become even more important in a hiring freeze when we need more than ever to keep the existing workforce healthy and on the job.

Additional applications

The job analyses and testing criteria will be used in many ways. Worker hardening programs for injured employees will probably be the most frequent use. Having the demands of the job defined allows medical providers to individually train an injured employee for a specific job. In the past, those in vocational rehabilitation have had to try to match injured or disabled employees with jobs through job analyses which specified the physical demands only as "more or less than 50 pounds;" "frequently, seldom or never." The job analyses and testing criteria will help make job matching and light duty assignments easier and more effective.

Another use of the job analyses will be to educate our own Health Services (medical) department about the jobs that our employees perform. Besides return-to-work evaluations, the medical staff regularly provides employee physical examinations. The job analyses, testing criteria, photographs and video tapes of the employees performing their job tasks will be available to the physicians. This will give them insights into jobs which they do not have the time or resources to obtain directly.

The Human Resources department plans to use the photographs and video tapes of the jobs in the recruiting as well as the hiring process. Applicants will have an opportunity to see the work environment and what the job entails before deciding whether to accept the job. Implementing the ergonomic corrections recommended in the job analyses will make the jobs less strenuous for all and reduce the potential for overexertion injuries.

Recommendations

We are including some of the lessons learned through the experience of developing our pilot program:

- Do your homework -- contact other companies and clinics that have experience in preplacement strength testing; collect articles and information from published literature so that you are well prepared to answer questions.
- The concept of a "pilot program" reduces organizational resistance to change and is easier to manage. Keep the pilot program small. One to three job classes in the same department is sufficient for a start. Working with more than one set of managers and first-line supervisors is difficult and time-consuming, however. The system that is established with the pilot program can then be implemented on a larger scale when the program is expanded.
- Related administrative and support functions should be involved as early as possible in the project. We received support and assistance from many departments, including Risk Management, Health Services, Human

Resources, Staff Relations, Legal, Procurement, Worker Compensation, Vocational Rehabilitation, and Hazards Control.

- Completing one entire job class rather than working many job classes through all phases is valuable. It allows a process to be established and "bugs" to be worked out.
- Funding requests should include time charges for whomever is coordinating or developing the program. It takes much more time than might be anticipated to deal with all the various departments that must be involved, and this involvement is essential to the success of the program. Estimate or obtain as much detailed information about all costs as possible before setting your level of funding.
- Having two people working as a team to implement the program is extremely helpful. The ideas, contacts, input, and energy of more than one person was often essential to the success of the project.
- Be prepared to encounter some resistance to change. For example, management may have developed their own informal "accommodation" practices that cannot be justified by the job description.
- Use the pilot program to recommend and effect ergonomic improvements wherever possible. The objective findings of the job analyses can be a useful tool for demonstrating the need for these changes.
- Include photos or videotapes of existing employees performing the job tasks in the documentation. This helps others to visualize and understand the job demands. This aspect of our project was enthusiastically received by all groups who plan to use the job analyses.

SUMMARY

We began developing job analyses, testing criteria, and photographic documentation to institute preplacement testing. The original purpose was to reduce overexertion injuries. Subsequent cutbacks in hiring have diminished the impact of the preplacement screening aspect. However it has become apparent that some of the applications which were originally viewed as ancillary may become just as important in reducing overexertion injuries.

REFERENCES

Anderson, C.K., 1987, Pre-placement screening: survival of the fittest, Risk Management, November, pp. 44-47.

Ayoub, M.A., 1983, Control of manual lifting hazards: III. preemployment screening. In Journal of Occupational Medicine, Vol. 24, No. 10, pp 751-761.

Ayoub, M.A., 1982, Preemployment screening programs that.match job demands with worker abilities. Industrial Engineering, March, 41-48.

Bond, M.B. and Messite, J., 1980, Preplacement medical evaluations and recommendations. In Developments in Occupational Medicine, edited by C. Zenz, (Chicago: Yearbook Medical Publishers), pp. 131-138.

Carmean, G. and Padgett, V., 1989, Valid medical standards reduce hiring hassles, costs for employers. Occupational Health & Safety, May, pp. 83-88.

Kroemer, K.H., 1987, Matching individuals to the job can reduce manual labor injuries. Occupational Safety and Health News Digest, June, pp. 4-7.

Snook, S.H., 1987, Approaches to preplacement testing and selection of workers. In Ergonomics, Vol. 30, No. 2, pp. 241-247.

Strasser, A.L., 1987, Setting a return-to-work date diffult task for attending physician. Occupational Health & Safety, June, p. 18.

Webb, R.D.G. and Tack, D.W., 1988, Ergonomics, human rights and placement tests for physically demanding work. In Trends in Ergonomics/Human Factors V, edited by F. Aghazadeh (North-Holland: Elsevier Science Publishers), pp. 751-758.

ACKNOWLEDGMENTS

Work performed under the auspices of the U.S. Department of Energy by Lawrence Livermore National Laboratory under Contract W-7405-Eng-48.

Advances in Industrial Ergonomics and Safety II
Edited by Biman Das
Taylor & Francis 1990

A BIOMECHANICAL INTERPRETATION OF THE PSYCHOPHYSICAL DETERMINATION OF WORK CAPACITY

Christopher A. Hamrick

U.S. Bureau of Mines
Pittsburgh Research Center
P.O. Box 18070
Pittsburgh, PA 15236

This study compares major trunk muscle activity during lifting tasks at weight levels below, at, and above the maximum acceptable weight of lift. The increase in peak and average muscle activity for the latissimus dorsi muscle pairs, and in average activity for the erector spinae muscle pairs, is greater from the medium to the high weight condition than from the low to medium weight condition ($p<=0.05$). This study shows that the increase in muscle activity, relative to external load increases, is non-linear, and there are critical thresholds at which the worker judges the task to be unacceptable based upon biomechanical factors.

INTRODUCTION

Musculoskeletal disorders and low back pain are a major problem in industry, and the cost in terms of money and human suffering is substantial. According to the National Safety Council (1987), the direct and indirect costs of back injuries to U.S. industries total six billion to eight billion dollars annually. Andersson (1981) states that factors which affect absence from work because of low back pain symptoms include heavy physical work, frequent bending, lifting, forceful movements, and repetitive work. Thus, any information that can be gained regarding how the body is affected by these factors could contribute to a reduction in the number of back injuries in the workplace and, consequently, a reduction in costs due to back injuries.

Work capacity evaluation plays an important role in the reduction of back injuries by either assisting in the redesign of job tasks or by assuring that only people with sufficient strength are employed. According to Ayoub, et al. (1980), the psychophysical methodology has been successful in determining load handling capacities of individuals or groups of individuals. However, these models fail to render information regarding the body's internal loads, such as spinal compression and muscle forces, which may influence a person's determination of the Maximum Acceptable Weight of

Lift (MAWL). Conversely, biomechanical models provide insight as to how the body functions as a system and can be useful in determining internal forces of the body, but these models are not adept at determining work capacity.

It is desirable to find a link between these two bodies of research so that a more comprehensive understanding of occupational low back disorders can be achieved. In order to accomplish this objective, this study will examine the influence of muscle usage on the psychophysical determination of lifting strength. With the aid of electromyography (EMG), muscle activity during lifting tasks can be determined for major trunk muscles. If the muscle activity is compared for lifting tasks with load levels below the MAWL, at the MAWL, and above the MAWL, then insight may be gained as to the role of muscle activity in the determination of an individual's lifting capacity.

METHOD

Subjects

Five normal, healthy, college aged male volunteers were used. None of the subjects had a history of low back pain. A description of the subjects is given in Table 1.

Table 1. Description of Subjects.

	age (yrs.)	weight (kg)	stature (cm)
mean	27.8	88.9	185.9
s.d.	3.7	19.6	9.5

Experimental Design

Weight of lift and frequency of lift were the independent variables in this study. The frequency of lift consisted of two levels: 1 lift per minute and 4 lifts per minute. These levels were chosen because of the prevalence of lifting tasks in industry that are performed within this frequency range. Recent surveys by Ciriello, et al., 1990, found that 94% of industrial lifting tasks are performed at a frequency below 4.3 lifts per minute. The weight of lift consisted of three levels: the MAWL minus twenty-five percent of the MAWL, the MAWL, and the MAWL plus twenty-five percent of the MAWL. The MAWL was determined using a standard psychophysical methodology.

The dependent variables examined were the integrated EMG (IEMG) of the left and right erector spinae (es) muscle group, the IEMG of the left and right rectus abdominis (ra) muscle group, and the IEMG of the left and right latissimus dorsi (ld) muscle group. It must be noted that these EMG values were not used as predictors of muscle force, but rather as a measure of relative muscle activity, which gives an indication of "muscular effort".

Several nuisance variables were controlled so that they would not affect the results of the study. The same box was used throughout the study and the subjects lifted the box to a constant height of 76 cm. The subjects were allowed to use any type of lift they wished; however, the lifting technique was kept constant within a subject for all conditions. Four subjects used the cantilever type lift, and one subject used a squat lift.

Apparatus

The apparatus consisted of six basic components: the box that was lifted, the weights in the box, an event marker, the EMG processing system, the analog to digital data collection system, and the computer.

The box was constructed of sturdy plywood and was 30.0 cm wide, 30.0 cm deep, and 26.8 cm high. Handles that were 2.54 cm in diameter were place horizontally on either side of the box and were 43 cm apart. The weights were in the form of lead pellets approximately 2 to 5 cm long and 0.3 to 0.6 cm wide.

An event marker, which consisted of a bounceless switch that output a 5 volt signal when "on" and output a 0.1 volt signal when "off" was used to mark the beginning and end of a lift. The marker was attached to the data acquisition system and was digitized and stored in the computer.

The EMG apparatus consisted, in part, of bipolar surface electrodes that were attached to the skin above the muscle to be sampled. Each pair of electrodes was attached to a preamplifier that was attached to a belt around the subject's waist. The preamplifier was attached to the main EMG system via a coaxial cable. The EMG signal was then filtered through an 80 Hz high pass filter and a 1000 Hz low pass filter. The signal was then rectified and the mean was taken over a 20 ms moving window. This processed signal was digitized and stored in the computer. The sampling rate was 100 Hz, and a data acquisition system was used to convert the analog signals to a digital format. The acquisition system was driven by a portable personal computer.

PROCEDURE

Training

In order for the subjects to be acclimated to the lifting task, each subject was required to train for fifteen minutes per day for a total of five days. The training session consisted of lifting the box containing a weight of 11.3 kg to a height of 76 cm at a rate of 4 lifts per minute. This training period ensured that the subjects did not experience extreme muscle soreness during the first few experimental conditions, thus enabling a more accurate determination of the MAWL.

Determination of the MAWL

A classic psychophysical methodology was used to determine the MAWL for each subject. Each subject read and signed a consent form and then read the lifting task instructions, which were presented by Hamrick (1989). As explained in the lifting task instructions, the subjects adjusted the box weight to a weight that they felt they could lift over an eight hour work day without feeling overly tired. The MAWL was determined for each subject for the 4 lifts per minute condition on one day, and then each subject returned three days later and his MAWL was determined for the 1 lift per minute condition.

EMG Measurements

Next, the subjects returned in order to participate in tests performed to measure muscle activity during the lifting task. Electrodes were attached to the subject over six muscle groups: the left and right rectus abdominis, the left and right erector spinae, and the left and right latissimus dorsi. The complete procedure is described by Hamrick (1989).

In order to correctly interpret the EMG signals, maximum and resting EMG values must be obtained. These values were determined by placing the subject in a reference frame that constrained the subject's hips and bent his back at an angle of 45 degrees from vertical. EMG data were collected during maximum voluntary flexion and extension exertions and also while the subject was at rest.

Finally, EMG data were collected while the subjects lifted a load. The subjects lifted the load at three levels and at two frequencies. The two separate frequency conditions were performed on different days. The three load levels were based on the MAWL determined during the psychophysical portion of the experiment. These three load levels were the MAWL minus twenty-five percent of the MAWL, the MAWL, and the MAWL plus twenty-five percent of the MAWL for the frequency at which the lift was performed. The order that each subject lifted the three weight levels was randomized for each subject at both frequencies. To minimize the effects of fatigue, a ten minute rest break was given to the subjects between testing under each condition.

During the lifting task, EMG data were collected at time 0 minutes, 4 minutes, 8 minutes, and 13 minutes. The data were collected for 5 seconds, during which a lift took place.

The beginning of the lift and the end of the lift were marked by an electronic switch that signaled 0 volts when the box was on the ground or on the table and signaled approximately 2.5 volts when the box was being lifted by the subject.

ANALYSIS

In order to determine the relative muscle activity for a particular task, the EMG data must be normalized by comparing the EMG activity during the task to the EMG activity during a maximum voluntary effort and to the EMG activity during rest. The normalization equation, as stated by Marras (1987), is as follows:

$$\text{Relative Activity} = \frac{\text{Task EMG} - \text{Rest EMG}}{\text{Maximum EMG} - \text{Rest EMG}} \qquad (1)$$

For each subject, each muscle's EMG data were normalized under each condition relative to the muscle's own maximum value and minimum value. Once the EMG values were normalized, the mean and peak values were determined for each lifting condition. Furthermore, since the task was symmetric about the sagittal plane, the mean activity level of the left and right muscles were averaged together in order to arrive at a single mean value for each muscle pair. Similarly, the peak activity level of the left and right muscles were averaged together so that a single peak value was obtained for each pair.

An analysis of variance (ANOVA) was performed to determine which of the factors, including interactions, affected muscle activity. Paired t-tests were performed on the data to check for differences in peak and average muscle pair activity between various weight levels. Also, a paired t-test was performed to see if the increase in peak and average muscle activity from the medium to high weight condition was greater than the increase in muscle activity from the low to medium weight condition.

RESULTS

The MAWL that was obtained for each subject using the psychophysical method at each frequency is given in Table 2.

Table 2. Maximum Acceptable Weights of Lift (kg).

Subject	1 lift/minute	4 lifts/minute
1	30.8	22.0
2	18.1	14.5
3	29.0	21.8
4	19.1	17.7
5	18.1	14.1
ave.	23.0	18.0
s.d.	5.7	3.4

A paired t-test showed that there is a significant difference in the MAWL between the two lifting frequencies (t_4=3.77, p=0.0196).

The ANOVA revealed that the weight lifted had a significant effect on average latissimus dorsi activity ($F_{2,8}$=12.41, p=0.004), peak latissimus dorsi activity ($F_{2,8}$=36.09, p<0.001), average erector spinae activity ($F_{2,8}$=51.96, p<0.001), and peak erector spinae activity ($F_{2,8}$=21.08, p=0.001). No other factors, including interactions, were found to influence the activity in these muscles and no factor was found to influence either peak or average rectus abdominis activity.

An analysis was performed for each muscle pair to see if the increase in muscle activity from the medium to high weight condition was greater than the increase from the low to medium weight condition. The results of this analysis show that there was a significant difference for peak latissimus dorsi activity (t_{38}=2.357, p=0.012), average latissimus dorsi activity (t_{38}=2.054, p=0.024), and average erector spinae activity (t_{38}=1.683, p=0.050).

The muscle activities at the different weight levels are given in Table 3 and are shown graphically in Figure 1 (only for muscles where the increase in activity from the medium to high weight condition was significantly greater than the increase from the low to medium weight condition).

Table 3. Muscle pair activity (% max) at the three weight levels.

	MAWL-0.25(MAWL)	MAWL	MAWL+0.25(MAWL)
ave ld	0.1087	0.1385	0.1937
peak ld	0.3267	0.3809	0.5352
ave es	0.4091	0.4638	0.5400
peak es	0.7066	0.7766	0.8794
ave ra	0.1608	0.1976	0.2174
peak ra	0.4062	0.5201	0.5884

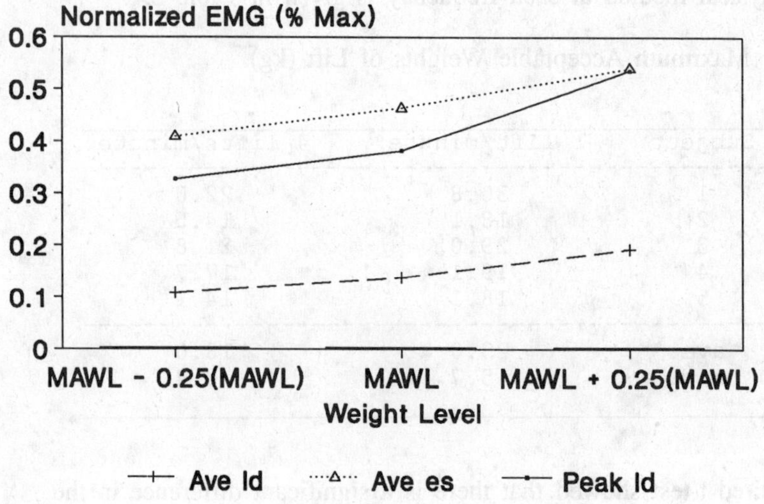

Figure 1. Muscle Activity vs. Weight

DISCUSSION

The results of the psychophysical study imply that a significant difference does exist in the MAWL between lifting at 1 lift per minute and 4 lifts per minute. As the frequency of lift increases from 1 to 4 lifts per minute, a person's MAWL decreases. This finding is consistent with the findings reported by Snook (1978).

The finding that peak and average muscle activity in both the erector spinae muscle pair and the latissimus dorsi muscle pair increases as the weight level increases is not surprising; the muscles must increase their activity to provide force to counteract the moments generated by the extra weight in the box. However, the increase in peak and average muscle activity for the latissimus dorsi muscle pairs, and in average muscle activity for the erector spinae muscle pairs, is greater from the medium weight condition to the high weight condition than from the low weight condition to the medium weight condition. This phenomenon appeared to be consistent between subjects and between conditions. Thus, people appear to be sensitive to the rate of change in muscle activity when the load weight is increased. It can be seen from Figure 1 that the latissimus dorsi activity seems to increase more non-linearly with increasing weight than the other muscles. The reason for this effect, however, is not clear.

People may have a maximum threshold of effort. Above this threshold, the person feels that s/he can not lift the weight at the given frequency without feeling overly tired, so the person subtracts weight from the box. Below this threshold, the person feels that s/he can lift more weight at the given frequency, so s/he adds more weight to the box.

Since the increase in EMG's is higher from the medium to high weight condition than from the low to medium condition, it appears as if more effort per unit weight increase is required at weight levels above the MAWL than at weight levels below the MAWL. However, further research should be conducted to determine more precisely the weight level at which this increase in effort is required by looking at muscle activity at weight levels closer to the MAWL. Snook (1978) reports that an acceptable range of test/retest values is that the retest should be within 15% of the original test, so observing the electrical activity in the muscles during lifting at the MAWL minus 15%, the MAWL, and the MAWL plus 15% should provide a sufficiently narrow range.

The increase in required effort could be due to muscle co-contraction. When co-contraction occurs, the antagonist muscles become active, thus creating a moment which opposes the movement in the desired direction. In order to continue the movement in dynamic situations, the activity in the agonist muscle must increase. However, based on the measured activity of the rectus abdominis group, minimal co-contraction was observed in this study. Co-contraction could have been present in other trunk muscles; this phenomenon has been observed by previous researchers. Flanders and Cordo (1987) observed significant co-contraction in the muscles about the elbow joint during isometric contractions. Zetterberg, et al. (1987) found that there was significant antagonist activity during heavy, isometric trunk loading.

Co-contraction may be minimal in the low weight conditions and significant in the high weight conditions. During the psychophysical determination of the MAWL, the subject may experience a significant decrease in effort by removing weight from the box when the load level is above the MAWL. Conversely, when the load level is below the MAWL, the subject may experience only a slight increase in muscular effort when weight is added to the box. So, the payoff in the form of reduced muscular effort may be relatively large when weight is decreased at weight above the MAWL, and the cost in the form of increased effort may be relatively small when weight is increased at levels below the MAWL. Thus, motivated subjects could arrive at the MAWL when they feel that adding weight to the box results in an increase in effort which is more than they can comfortably tolerate.

The findings indicate that the increase in muscle activity relative to the external load increase is non-linear, and there are critical thresholds at which the muscles become stressed. At this point, the worker judges the task to be unacceptable, based upon biomechanical factors.

SUMMARY

This investigation combined two fields of study: biomechanics and psychophysics. By providing a biomechanical interpretation of the psychophysical methodology to determine lifting capacity, the two fields were linked. Consequently, this study may help to interpret literature in both areas of knowledge. In addition, the information gained may help to develop a biomechanical model of work capacity evaluation, which may prove to rely less upon the motivation of the subjects and be less expensive and less time consuming than the psychophysical method.

REFERENCES

Andersson, G.B.J., 1981, "Epidemiologic Aspects of Low-Back Pain in Industry",Spine, 6(1):53-60.

Ayoub, M.M., Mital, A., Bakken, G.M., Asfour, S.S., and Bethea, N.J., 1980, "Development of Strength and Capacity Norms for Manual Materials Handling Activities: The State of the Art", Human Factors, 22(3):272-283.

Ciriello, V.M., Snook, S.H., Blick, A.C., and Wilkinson, P.L., 1990, "The Effects of Task Duration on Psychophysically Determined Maximum Acceptable Weights and Forces", Ergonomics, (in press).

Flanders, M., and Cordo, P.J., 1987, "Quantification of Peripherally Induced Reciprocal Activation During Voluntary Muscle Contraction", Electroencephalography and Clinical Neurophysiology, 67:389-394.

Hamrick, C.A., 1989, "A Biomechanical Interpretation of the Psychophysical Determination of Work Capacity", Master's Thesis, The Ohio State University, Columbus, Ohio.

Marras, W.S., 1987, "Preparation, Recording and Analysis of the EMG Signal", Trends in Ergonomics/Human Factors IV, S.S. Asfour (ed.), Elsevier Science Publishers B.V., North-Holland.

National Safety Council, 1984, Safety Facts.

Snook, S.H., 1978, "The Design of Manual Handling Tasks", Ergonomics, 21(12):963-985.

Zetterberg, C., Andersson, G.B.J., and Schultz, A.B., 1987, "The Activity of Individual Trunk Muscles During Heavy Physical Loading", Spine, 10(12):1035-1040.

ACKNOWLEDGEMENTS

This research was performed at The Ohio State University, Department of Industrial and Systems Engineering, under the direction of Dr. William S. Marras. In addition, the author would like to thank Dr. Anil Mital for his input and comments.

Advances in Industrial Ergonomics and Safety II
Edited by Biman Das
Taylor & Francis 1990

ISOKINETIC STRENGTHS IN THE HORIZONTAL PLANE

Anil Mital PhD PE
Ergonomics Research Laboratory
Mechanical and Industrial Engineering Department
University of Cincinnati
Cincinnati, OH 45221-0072, U.S.A.

and

Hamid F. Faard
Division of Safety and Hygiene
Industrial Commission of Ohio
Jackson & Erie Streets
Toledo, Ohio 43604, U.S.A.

This investigates the effects of posture, reach distance, and angle of the preferred (stronger) arm from the frontal plane on isokinetic pull strength of males, engaged in infrequent exertion in a horizontal plane. The results indicated that more strength is exerted while standing. The strength also increases with the reach distance and the exertion becomes stronger as the preferred arm comes closer to the sagittal plane.

INTRODUCTION

Human strength exertion capability is one of the important human physical capabilities that help ergonomists and engineers in optimizing work and work place designs. The significance of human strengths in employee screening procedures and equipment and tool design is also well established (see Mital and Das (1987) for a review).

Historically, researchers have used isometric (static) strength simulations in developing engineering design guidelines and employee screening procedures (Clarke et al., 1950; Caldwell, 1959; Rohmert and Jenik, 1972; Chaffin et al., 1978). The isometric strength simulation of the job, however, ignores the body segment and object movement, and thereby the effect of inertial forces, during maximal voluntary contraction. The accommodation of movement in strength simulations is important and as many recent studies have shown, isokinetic strength capabilities of individuals are more highly correlated to their dynamic task capabilities than isometric strength capabilities (Kamon et al., 1982; Mital and Karwowski, 1985; Aghazadeh and Ayoub, 1985; Mital et al., 1986a,b,c). Therefore, it is logical to conclude that isokinetic strengths, instead of isometric strengths, should be used for pre-employment screening and establishing ergonomic design guidelines.

There clearly exists a need for developing isokinetic strength data for a variety of industrial activities and to study how isokinetic strengths are influenced by work design parameters. However, the published literature lacks such data or information. The present study is one of the few studies that have investigated the effect of job design parameters on isokinetic strengths. The other effort was by Mital and Genaidy (1989a,b) and provides isokinetic pull-up strength profiles of men and women in different working postures.

Frequently, work is placed on a horizontal surface and is performed by workers who are either seated or standing (e.g., assembly line work). The purpose of this study, therefore, was to study the effect of work design related factors on infrequent isokinetic pull strength exertions in the sitting and standing positions.

METHODS

Subjects

Twenty-five male volunteers participated in the study. All participants were screened for physical ailments and medication. successful subjects had their anthropometric and isometric strength measurements taken on two different days; one day for recording the body size measurements and the second day for recording isometric strengths (Table 1). The values are similar to those recorded on other civilian and industrial sample populations (Mital et al., 1986a).

Table 1. Distribution of the physical characteristics of the subject population.

Variable	Mean	Standard deviation
Age (years)	22.3	0.9
Body weight (kg)	80.1	19.3
Standing height (cm)	177.3	6.2
Shoulder height (cm)	144.7	5.4
Sitting height (cm)	132.5	2.7
Shoulder height (cm)	100.8	2.0
Seat-to-head height (cm)	89.2	2.6
Chest circumference (cm)	97.1	6.7
Chest depth (cm)	23.0	4.3
Abdominal circum. (cm)	85.2	10.0
Abdominal depth (cm)	35.3	7.3
Biceps circum.* (cm)	32.5	2.0
Forward reach (cm)	71.8	3.3
Isometric strength (kg)		
-arm	45.2	7.9
-shoulder	59.2	13.4

* Mean of readings in flexed and relaxed conditions.

Once the body size and isometric strength measurements were taken, the subjects participated in the experiment. Both written and verbal instructions, on isokinetic strength exertion procedure were provided. Subjects were also permitted practice sessions in order to clearly understand the strength exertion procedure. During the study, subjects wore normal work clothes and work shoes.

Experimental design

A randomized complete block factorial design, with blocking on subjects, was employed. Three independent variables, known to influence human isometric strength exertion capability (Caldwell, 1959; Rohmert, 1966; Rohmert and Jenik, 1972), were included: (1) posture (sitting and standing), reach distance (25, 40, and 55 cm for the sitting posture and 45, 65, and 85 cm for the standing posture), and angle of the preferred arm relative to the frontal plane (0, 30, 60, 90, 120, and 150 degrees). The levels of these variables were selected to simulate work conditions requiring infrequent pulling exertion in the horizontal plane.

Procedure

The isokinetic strength exertion procedure used in several previous studies was followed (Kamon et al., 1982; Mital et al., 1986a,b,c; Mital and Genaidy, 1989a,b). Each subject was required to assume the orientation specified by the experimental condition (combination of posture, reach distance, and preferred arm angle) and pull the handle attached to the isokinetic strength simulator (Mital and Vinayagamoorthy, 1984) as fast as possible, but without jerking it. For each angle of the preferred arm, the speed control was adjusted until the maximum speed of the preferred arm was reached. This maximum speed was determined by the subject (Since, in the work place, generally the speed of movement is determined by the worker, it is necessary to let the subject adjust the speed and to record the exertion at the speed set by the individual). The peak exertion at the maximum speed was recorded. Data for all experimental conditions were collected in a similar fashion. The order of experimental conditions was random and different for each subject.

For each treatment combination, three trials were made and data averaged. A rest break of at least 5 minutes was provided between successive trials. For each subject, data for only three experimental combinations were collected on any given day. However, several subjects participated on any given day.

A number of experimental conditions were controlled. In the standing posture, subjects kept their feet 30 cm apart (Rohmert, 1966). Subjects also kept their shoulders in the same vertical plane as their ankles, thus ensuring an upright torso. For the sitting posture, seat height was adjusted such that allowed the subject to keep his feet flat on the ground and his thighs at approximately 15° from the horizontal (Grandjean, 1988). All reach distances were measured from the ankle joint.

RESULTS

On the average, 37% more strength was exerted in the standing posture (119 kg) as compared to the sitting posture (87 kg). This difference was significant at the 1% level. Since, the posture effect was highly significant and reach levels for the two postures were also different, separate ANOVAs were conducted for the sitting and standing posture data. The effects of reach distance and angle of the preferred arm were highly significant in both cases ($p < 0.01$). The interaction between reach distance and arm angle, however, was not in either case ($p \geq 0.10$).

Strength means for various reach distances and preferred angles were also subjected to Newman-Keul's multiple range test. The results for the sitting and standing posture are summarized in Tables 2 and 3, respectively.

The magnitude of isokinetic pull strength increased significantly ($p < 0.01$) with the reach distance for both postures. This increase in pull strength with the reach distance was, more or less, linear.

For both postures, maximum strength was exerted when the angle was 90° (arm in the sagittal plane). Least strength was exerted in 0° and 150° arm orientations. The rates of decline in strength (kg/degree) below and above the 90° angle were almost similar. Overall, effects of reach distance and angle of the preferred arm, within each posture, were similar.

DISCUSSION AND CONCLUSIONS

The purpose of this study was to determine the effects of work design factors on isokinetic pull strengths of males in the horizontal plane. As pointed out earlier, there are profound differences between the isometric and isokinetic strength capabilities of

Table 2. Overall mean isokinetic pull strengths (kg) exerted in the sitting posture and the results of Newman-Keul's multiple range test on the means.

Variable	Level	Mean strength
Reach distance (cm)	25	70.9
	40	90.2
	55	100.1
Preferred arm angle (deg) from the frontal plane		
	0	77.8 *
	30	86.9 #
	60	90.3 #, @
	90	94.0 @
	120	88.5 #, @
	150	85.0 *, #

Note: Means marked by the same notation (*, # or @) are not different from each other at the 1% level of significance. All reach distance means are different from each other ($p \leq 0.01$).

Table 3. Overall mean isokinetic pull strengths (kg) exerted in the standing posture and the results of the Newman-Keul's multiple range test on means.

Variable	Level	Mean strength
Reach distance (cm)	45	105.2
	65	123.7
	85	129.6
Preferred arm angle (deg) from the frontal plane		
	0	110.6 *
	30	116.4 *
	60	121.7 #
	90	131.5
	120	120.8 #
	150	116.0 *

Note: Means marked by the same notation (* or #) are not different from each other at the 1% level of significance. All reach distance means are different from each other ($p \leq 0.01$).

individuals. This finding was also validated in the present work. For instance, for the 85 cm reach distance (almost full reach), the maximal average isokinetic pull strength in the horizontal plane, with arm in the sagittal plane and subject standing, was 172%

of the body weight. For the same condition, as indicated by Rohmert (1966), the maximum isometric pull strength is only 10% of the body weight. The maximum isokinetic pull strength in the sitting posture was approximately 109 kg (reach distance 55 cm and arm angle 90°). While comparable isometric strength data is not available, Caldwell (1959) provides maximum pull down isometric strength (38 kg). Thus, in the sitting posture also the isokinetic pull strength is greater than the maximum isometric strength. Isokinetic pull down strengths in the sitting posture are expected to be even greater.

It is clear even from this limited comparison that the effect of inertial forces should not be ignored. If the actual task involves limb movement, isokinetic strengths should be used to assess a person's strength exertion capability for that job.

The results of data analysis indicate that the isokinetic pull strength capability increases with the reach distance. This finding is not really surprising since the elbow angle also increased (Clarke et al., 1950; Wakim et al., 1950).

In the 90° arm orientation, the force vector is aligned with the long axis of the arm. As the arm orientation angle changes, increases or decreases, this alignment changes. Since the force vector and arm long axis are now at an angle, the resultant force decreases. The decrease, obviously, increases as the angle of the arm from the sagittal plane increases.

The results of this investigation and above discussion lead to the conclusions that: (1) isokinetic pull strengths of males are greater than their isometric pull strengths, (2) isokinetic pull strength of males is almost 37% greater in the standing posture than in the sitting posture, (3) isokinetic pull strengths increase, almost linearly, with the reach distance between 25 cm and 55 cm for the sitting posture and between 45 cm and 85 cm for the standing posture, and (4) maximum isokinetic pull strength is exerted when the preferred arm is in the sagittal plane.

REFERENCES

Aghazadeh, F. and Ayoub, M.M., 1985. A Comparison of Dynamic- and Static-Strength Models for Prediction of Lifting Capacity. Ergonomics, 28, 1409-1417.

Caldwell, L.S., 1959. The Effect of the Special Position of a Control on the Strength of Six Linear Hand Movements. U.S. Army Medical Research Laboratory, Report No. 411, Fort Knox, Kentucky.

Chaffin, D.B., Herrin, G.D., and Keyserling, W.M., 1978. Pre-employment Strength Testing - An Updated Position. Journal of Occupational Medicine, 20, 403-409.

Clarke, H.H., Elkins, E.C., Martin, G.M., and Wakim, K.G., 1950. Relationship Between Body Position and the Application of Muscle Power to Movements of Joints. Archive of Physical Medicine, 31, 81-89.

Grandjean, E., 1988. Fitting the Task to the Man, 4th Edition. Taylor & Francis, New York.

Kamon, E., Kiser, D., and Pytel, L.J., 1982. Dynamic and Static Lifting Capacity and Muscular Strength of Steelmill Workers. American Industrial Hygiene Association Journal, 43, 853-857.

Mital, A. and Das, B., 1987. Human Strengths and Occupational Safety. Clinical Biomechanics, 2, 97-106.

Mital, A. and Genaidy, A.M., 1989a. Isokinetic Strengths and Working Posture. Proceedings of the Human Factors Society Annual Meeting, Santa Monica, California, 632-636.

Mital, A. and Genaidy, A.M., 1989b. Isokinetic Pull-Up Strength Profiles of Men and Women in Different Working Postures. Clinical Biomechanics, 4, 168-172.

Mital, A. and Karwowski, W., 1985. Use of Simulated Job Dynamic Strengths (SJDS) in Screening Workers for Manual Lifting Tasks. Proceedings of the Human Factors Society Annual Meeting, Santa Monica, California, 513-516.

Mital, A. and Vinayagamoorthy, R., 1984. Three-Dimensional Dynamic Strength Measuring Device: A Prototype. American Industrial Hygiene Association Journal, 45, B9-B12.

Mital, A., Karwowski, W., Mazouz, A.K., and Orsarh, E., 1986a. Prediction of Maximum Acceptable Weight of Lift in the Horizontal and Vertical Planes Using Simulated Job Dynamic Strengths. American Industrial Hygiene Association Journal, 47, 288-292.

Mital, A., Aghazadeh, F., and Karwowski, W., 1986b. Relative Importance of Isometric and Isokinetic Lifting Strengths in Estimating Maximum Lifting Capabilities. Journal of Safety Research, 17, 65-71.

Mital, A., Channaveeraiah, C., Fard, H.F., and Khaledi, H., 1986c. Reliability of Repetitive Dynamic Strengths as a Screening Tool for Manual Lifting Tasks. Clinical Biomechanics, 1, 125-129.

Rohmert, W., 1966. Maximalkrafte von Mannern im Bewegungsraum der Arme und Beine. Forschungsberichte des Landes Nordrhein-Westfalen No. 1616. Westdeutscher Verlag, Cologne.

Rohmert, W. and Jenik, P., 1972. Maximalkrafte von Frauen im Bewegungsraum der Arme und Beine. In the series Arbeitswissenschaft und Praxis. Beuth-Vertrieb, Berlin.

Wakim, K.G., Gersten, J.W., Elkins, E.C., and Martin, G.M., 1950. Objective Recording of Muscle Strength. Archives of Physical Medicine, 31, 90-99.

Advances in Industrial Ergonomics and Safety II
Edited by Biman Das
Taylor & Francis 1990

Relationship Between Isometric and Eccentric and Concentric Strength

Ehsan Asoudegi

**Department of Industrial Engineering
Northern Illinois University
DeKalb, IL 60115-2854**

INTRODUCTION

For the design engineer concerned with dynamic performance, the ability to predict an operator's success in performing a dynamic strength task from a measurement of static muscular strength would be a tremendous asset. Many researchers have investigated the question of whether the amount of force that can be exerted in a static muscular contraction is a good or poor indicator of the amount of force that can be exerted dynamically. Unfortunately, few results have been agreed upon regarding this topic.

Studies by Asmussen, Hansen, and Lammert (1965), Carlson (1970), Rasch (1957), Rasch and Pierson (1960, 1963) and Salter (1955) reported a high degree of correlation between measures of static and dynamic strength.

In contrast, studies by Karpovich (1963), Lagasse (1970), Singh and Karpovich (1966), and Stuart (1966) reported no significant correlation between measures of static and dynamic strength.

Perhaps these contrasting viewpoints may be due to inadequate definitions of terminology, varying intensities of effort, or differences in the testing programs of the researchers.

Concentric (muscle shortens actively against a resistance), eccentric (muscle is lengthened passively by an external force), and isometric strength of the elbow flexor muscle were compared by Doss and Karpovich (1965). Their study found that the mean maximum concentric (pushing) force was 23 percent smaller, and the eccentric (pulling) force was 13.5 percent greater than the isometric force.

Singh and Karpovich (1966) studied the relationships among these forces for the forearm flexors and extensors. They concluded that it is possible to predict concentric, eccentric and isometric forces from one another.

After his study, Kroemer (1967) was doubtful that static strength tests could act as predictors for sustained dynamic performance. But he did recognize that they are "somewhat loosely related by the fact that muscular efforts are involved in both".

Knapik et al. (1980) found that the torque elicited at low-velocity isokinetic contractions can be predictive of the torque elicited during isometric contractions. They suggested that this may be explained by nothing that "as the isokinetic velocities become farther apart, the motor tasks become more dissimilar, requiring different patterns of neutral recruitment and coordination".

In view of these conflicting results from other studies, this paper will investigate this controversial relation based on test data from the KIN-COM dynamometer.

METHOD

Subjects

The subjects were ten students and faculty at West Virginia University. All subjects were volunteers and were given an initial physical examination. Their mean age was 26.6 and ranged from 20 to 36 years old. Their heights ranged from 66 to 74 inches, with a mean of 69.40 inches. Their weights ranged from 120 to 185 pounds, with a mean of 158.60 pounds.

The objective of the study was to investigate characteristics of isokinetic strength testing. Therefore, an investigation of population norms was excluded from this analysis. Subjects with similar physical background were chosen to eliminate possible muscle-training effects on the measured strength.

Equipment

This study was conducted at the Applied Kinesiology Laboratory at West Virginia University. The following equipments were used in this study:

1. KIN-COM (Kinetic Communicator) dynamometer
2. PDP11-V23 Minicomputer
3. KWV11-A Programmable Real-Time Clock
4. ADV11-A Analog to Digital Converter
5. Grass Model 7 Polygraph

Equipment Interface

The KIN-COM dynamometer provides precise selection of flexion angle (in degrees), movement velocity (in degrees/sec.) and initial restraining force (in Newtons). The KIN-COM, with its Z80 microprocessor, samples the angle and force at a pre-determined sampling rate of 100 Hz. The processor computes the velocity, torque, and instantaneous power and stores these data along with other related information on floppy disks for later use.

There are two drawbacks to using the KIN-COM for data acquisition. First, the sampling rate is fixed. Second, it is not possible to collect muscle EMG activity simultaneously with other data such as force or angular displacement. The PDP11-V23 minicomputer was used for data acquisition to overcome these problems. A Grass Model 7 polygraph recorded the analog force signals directly from the KIN-COM transducer, the angular displacement from a rotational potentiometer connected to the dynamometer arm. These signals were processed through the Grass polygraph and the A/D converter for transmission to the PDP11-V23 computer. The general configuration of the equipment used in this research is shown in figure 1.

Testing Procedure

After completion of a health questionnaire, each subject was examined for left knee muscular disorder by the physical therapist. Subjects with past knee injuries or cardiovascular problems were excluded from the study. A consent form was signed by those subjects who were qualified by the physical therapist to participate in this knee study.

Before starting the experiment, each of the subjects was acquainted with the equipment and experimental procedure. To start the exercise, the subjects were positioned supine with the pelvis strapped to the table of the exercise unit. Each subject was required to perform a minimum of nine tests representing all possible combinations of speed (0 degree/second, 60 degrees/second, 120 degrees/second) and starting position (40 degrees, 60 degrees, 80 degrees of knee flexion). The sequence of these combinations was randomized for each subject. Each test consisted of four consecutive maximal concentric and eccentric knee movements resisted by the KIN-COM dynamometer. To eliminate the effect of fatigue, each subject was given a three minute rest period between each test. During each test, subjects were instructed verbally to exert as much force as they could. No feedback was given regarding their performance.

RESULTS AND DISCUSSION

Isokinetic and Isometric Torque Relationships

As discussed above, it would be advantageous to predict human dy-

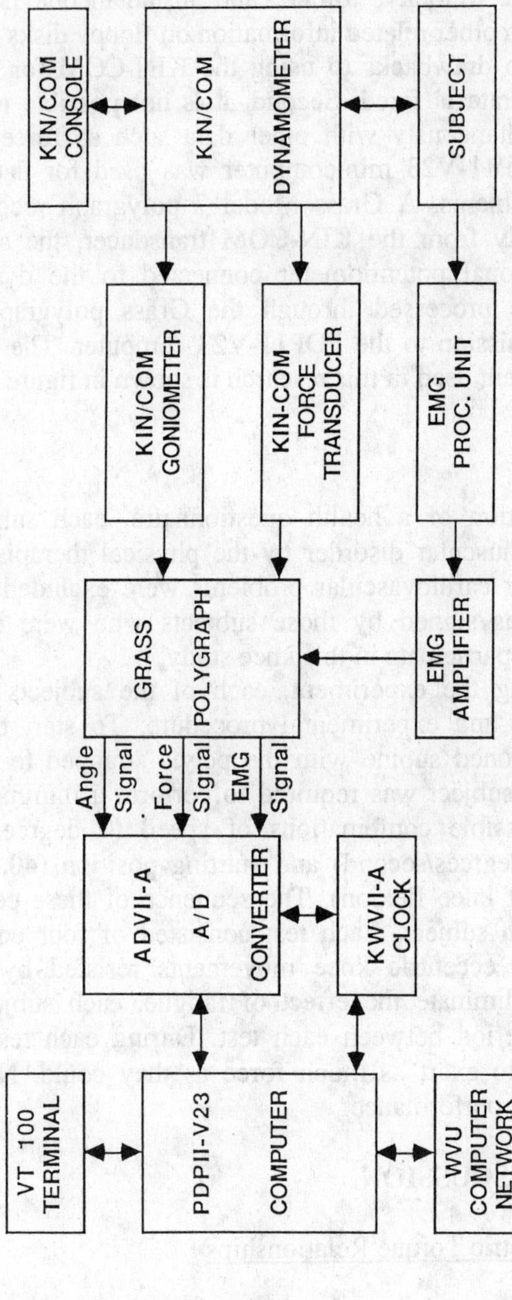

Figure 1 General Configuration of the Equipment

namic strength from isometric strength. For predicting dynamic strength from isometric strength, the following models were considered:

$$T_{dyn} = b_0 + b_1 T_{iso} + \varepsilon$$

$$T_{dyn} = b_0 + b_1 T_{iso} + b_2 \alpha + \varepsilon$$

$$T_{dyn} = b_0 + b_1 T_{iso} + b_2 \alpha + b_3 \alpha^2 + \varepsilon$$

Where:

T_{dny} is the dynamic strength at α flexion angle
T_{iso} is the isometric strength at α flexion angle
α is the knee flexion angle
b_0, b_1, b_2, b_3 are the parameters for these models
ε is the random error term $N(0, \sigma^2)$

The SAS STEPWISE procedure was used to perform a multiple linear-regression analysis to obtain suitable regression models. The results of this analysis indicate that dynamic strength is a linear function of isometric strength. These relationships were investigated for both 60 degrees/second and 120 degrees/second of speed of movement and combined data. The results of the regression analysis suggests that only the first model is suitable for prediction of dynamic strength from static strength. The coefficients of correlation for these models were 0.75 and 0.72.

Concentric and Eccentric Strength Relation

Another objective of this research was to investigate the relationship between eccentric and concentric strength during isokinetic strength testing. Multiple regression was used to develop a suitable model to illustrate the relationship between concentric and eccentric contraction. Inspection of torque traces during eccentric and concentric contractions indicates that angle is an important factor in this relationship. The following models were proposed to describe the relationship between eccentric and concentric contraction:

$$T_{con} = b_0 + b_1 T_{ecn} + \varepsilon$$

$$T_{con} = b_0 + b_1 T_{ecn} + b_2 T_{ecn}^2 + \varepsilon$$

$$T_{con} = b_0 + b_1 T_{ecn} + b_2 \alpha + \varepsilon$$

Where:

T_{con} is the concentric strength at α flexion angle.
T_{ecn} is the eccentric strength at α flexion angle.
α is the Knee flexion angle.
b_0, b_1, b_2 are the parameters for these models.
ε is the random error term $N(0, \sigma^2)$.

A regression analysis was performed for all combinations of speed of movement and starting position. In general, the results indicate that the first model is the best model with one independent variable. Also, the third model is the best model with two independent variables.

The results indicate that models with two independent variables have higher coefficients of correlation than do models with one independent variable. The higher coefficients of correlation is a result of introducing the variable flexion angle into the model.

REFERENCES

Asmussen, E., Hansen, O., and Lammert, O. "The Relationship Between Isometric and Dynamic Muscle Strength in Man." Communications of the Testing and Observation Institute of Danish National Association for Infantile Paralysis, Hellerup, Denmark, No. 20, 1965.

Carlson, B.R. "Relationship Between Isometric and Isotonic Strength." Arch. of Phys. Med. and Rehab., 51:176-179, 1970.

Doss, W.S., and Karpovich, P.V. "A Comparison of Concentric, Eccentric and Isometric Strength of Elbow Flexors." J. of Appl. Physiol., 20(2): 351-353, 1965.

Knapik, J.J., and Ramos, M.U. "Isokinetic and Isometric Torque relationships in the Human Body." Arch. Phys. Med. Rehab., 61:64-70, 1980.

Rasch, P.J. "Relationship Between Maximum Isometric Tension and Maximum Isotonic Elbow Flexion." Res. Quart. Amer. Ass. Health Phys. Educ. 28:85, March, 1957.

Rasch, P.J., and Pierson W.R. "Relationship Between Maximum Isometric Tension and Breaking Strength of Forearm Flexors." Res. Quart., 31:534-535, 1960.

Rasch, P.J., and Pierson, W.R. "Some Relationships of Isometric Strength, Isotonic Strength and Anthropometric Measures." Ergonomics, 6:211-236, 1963.

Salter, N. "The effect of Muscle Strength of Maximum Isometric and Isotonic Contractions at Different repetition Rates." J. Physiol., 130:109-113, 1955.

Singh, M., and Karpovich, P.V. "Isotonic and Isometric forces of forearm flexors and Extensors." Appl. Physiol. J., 21(4): 1435-1437, 1966.

Advances in Industrial Ergonomics and Safety II
Edited by Biman Das
Taylor & Francis 1990

INFLUENCE OF STRENGTH ASSESSMENT TECHNIQUES ON RESEARCH CONCLUSIONS

P. Bishop, M. Conerly and J. Smith

University of Alabama, Tuscaloosa, AL 35487

There has been speculation in previous research reports that the technique used to measure strength (i.e. static vs. dynamic) may directly influence the magnitude of change or difference observed. Lacking data to evaluate this speculation, this study was conducted to assess the influence of three different modes of strength measurement on research conclusions. Subjects were 47 male and 50 female volunteers who were tested twice on a Cybex dynamometer (static, and dynamic constrained) and a Universal Gym (dynamic unconstrained). Knee extension (legs) and supine elbow flexion (arms) were measured statically at the expected mean angle of peak torque, and at 60°/sec. Knee extension and supine elbow flexion were also measured on a Universal Gym. A full practice trial preceeded all tests by one week. Analysis of the standardized data via multiple regression resulted in R^2 values of .94-.95 for arm measures regressed on each other and R^2 of .83-.89 for legs. Hypotheses of slope = 1.0 and intercept = 0 were strongly rejected for legs but less strongly rejected for arms. The most appropriate statistical test, that of consistent patterns in the 6 x 6 covariance matrix (2 strengths measured 3 ways), suggests that the three different measures are assessing different characteristics (Chi square = 47, p = .00). However, the test of the arms alone covariance matrix indicated that the arm measures were not significantly different (Chi square = 4.4, p = .23). The legs alone matrix indicated leg strength measures were different (Chi square = 10, p = .01). The speculation that different modes of strength measurement yield different results is only supported for leg strength measures in this study. The larger muscle mass of the legs and consequently higher force values generated, may result in less effective muscle group isolation for dynamic compared to static measures of leg strength.

INTRODUCTION

Various aspects of the role of strength in labor tasks have been previously investigated (Bishop, Cureton, Collins, 1987; Mital, 1984; Jackson et al., 1984; Snook and Ciriello, 1972). The strength measurements utilized in previous research have been undoubtedly both a result of careful design and convenience. Some investigators have used strength measures very similar to specific industrial tasks such as pulling or pushing, others have used measures which test particular muscle groups, such as the use of biceps curls as a measure of elbow flexion strength. Except for a modest percentage of these studies, comparing results and conclusions among studies is difficult or impossible. Differences in strength measurement techniques, in muscle groups, in populations sampled and in information reported are all complicating factors. Of these complications, one question that has arisen is, "Do static and dynamic strength measurements measure the same fundamental characteristics?". The hypothesis of the present study is that static and dynamic measures of strength are essentially measuring the same characteristics. The implication of this hypothesis is that the neuromotor pattern and resulting function of a given muscle group is not significantly different regardless of the strength measurement technique used.

The purpose of the present study was to evaluate the extent to which three different strength measurement techniques were measuring the same characteristics. To accomplish this, three different measures of strength were analyzed for a heterogenous group with varying degrees of strength training experience. It is hopeful that the results of this study would be of use in retrospective evaluation of different studies and in design of future studies.

METHODS

Subjects

Subjects for this study were 97 volunteers (47 male and 50 females) who provided written informed consent. Subjects varied in size and weight lifting experience. Of these, 49 were competitive swimmers (24 male, 25 female) and the remaining 48 were not participating, nor had participated, in systematic strength training or athletics (23 male, 25 female). Descriptive information is provided in Table 1.

Table 1. Mean (sd) physical characteristics of subjects.

Variable	Males (N=47)	Females (N=50)
Age (yr)	20.4 (2.3)	20.4 (2.4)
Height (cm)	177.8 (6.4)	166.8 (7.2)
Weight (Kg)	72.4 (7.4)	60.9 (8.2)

Strength Measures

Strength measurements were chosen which encompass major limb motions which could be approximated in the static (ST), and dynamic constrained (DC) and the dynamic unconstrained (DU) modalities. Subjects were familiarized with strength testing procedures and strength was measured in mixed-sex groups of five to ten subjects. At the familiarization sessions, subjects were given three practice trials of the dynamic unconstrained (Universal Gym) tests. Because of subject unfamiliarity with isoinertial (sometimes called isokinetic) equipment, six practice trials of the dynamic constrained (DC) and static (ST) tests were conducted at 60 and O degrees per second, respectively. Cybex II isokinetic equipment (Lumex, Inc.) was utilized for ST and DC measures. All subjects were verbally encouraged during the training sessions, but standardized directions were given for the actual strength tests and no verbal encouragement was given. Each subject was tested separately to reduce peer influence. The non-dominant limb was tested to reduce the influence of physical activity not related to sport training and to reduce the influence of motor skill on the strength measurements.

DC tests were performed at 60 degrees per second, the velocity recommended by Lumex Inc. for strength measurements. ST tests were performed similarly at zero degrees per second. Three trials were used in assessing the peak torque. Elbow flexion was performed with the subject supine as described in the Cybex II test manual, Isolated-Joint Testing and Exercise

(Lumex, 1980). Stabilization straps were placed across the hips and chest. The forearm was maintained in a supinated position throughout both tests.

Knee extension was performed as described in the Cybex II test manual. The subject was instructed to grasp the sides of the test chair. An additional stabilization strap was placed across the hips.

Since multiple trials were obtained for the ST and DC strength testing, the procedure described by Baumgartner (1974) was used in selecting the most appropriate criterion scores. Trials two and three proved to be trend-free and the sum of these two scores was used as the criterion score for all ST and DC strength measures. For these two trials, intraclass r's better than .97 were obtained for all DC strength measures for the subjects.

DU testing was performed on the Universal Gym with all subjects performing one-repetition-maximum tests of one-arm curl and one-leg extension. Three trials of each test were administered with the highest weight lifted used as the subject's score.

In an attempt to isolate the arm flexors in testing, one-arm curls were performed in a supine position with heels against the base of the machine with the shoulders, head, and elbows in contact with the floor. Subjects gripped a special five inch grip attached to the Universal Gym via a tightened chain. Subjects were required to bring their forearms past 90 degrees (vertical) to count as a successful lift.

Leg extensions were performed from a stable sitting position with hands gripping the sides of the seat. The subjects were required to raise the extension device to a position parallel with the floor.

Statistical analyses

The most logical statistical comparison of the three modes of strength measurement is the examination of the pattern of the covariance matrix of the six strength measurements (arm and leg strength each measured 3 ways). Because there have been empirical observations and theoretical bases for assuming differences in upper- and lower-body strength, subsequent analyses of the covariance matrix were conducted for the arm strength measures alone and the leg strength measures alone. Strength measurement techniques were also compared using multiple regression.

RESULTS

Correlations among arm and leg measurement techniques are shown in Table 2. The chi-square statistics and p values for equal covariance matrices of the arms and legs combined, the arms alone and the legs alone are shown in Table 3. The combined arms and legs covariance matrix yielded a p value suggesting a difference among the measures. No statistically significant difference among measurement techniques was seen for the arms alone covariance matrix, but a significant difference was observed in the legs alone matrix. Regression equations are shown in Table 4. These results are consistent in showing good agreement among measures particularly for elbow flexion.

Table 2. Correlations among strength measurements obtained by different techniques.

	Arms			Legs		
	ST	DC	DU	ST	DC	DU
Arms						
ST	1.0	.91	.88	.65	.79	.66
DC	.91	1.0	.91	.70	.86	.70
DU	.88	.91	1.0	.64	.79	.76
Legs						
ST	.65	.70	.64	1.0	.78	.65
DC	.79	.86	.79	.78	1.0	.74
DU	.66	.70	.76	.65	.74	1.0

All correlations significant at P < .0001.

Table 3. Chi-square statistics for patterns in the covariance matrices for arms plus legs, arms, and legs.

	Chi-square	df	p
Arms & Legs	47.33	12	.00
Arms	4.35	3	.23
Legs	10.29	3	.02

Table 4. Regression equations for standardized strength measures predicted from other standardized strength measures for arms and for legs.

Dependent Variable	Coefficients (Independent Variables)		Intercept	R^2
		Arms		
ST	.31 (DU)	.63 (DU)	.00	.84
DC	.48 (DU)	.49 (ST)	.00	.88
DU	.62 (DC)	.31 (ST)	.00	.84
		Legs		
ST	.66 (DC)	.16 (DU)	.00	.62
DC	.41 (DU)	.51 (ST)	.00	.71
DU	.60 (DC)	.18 (ST)	.00	.57

DISCUSSION

Previous investigations of the development of strength, the characteristics of strength, and sex differences in strength, have incorporated a variety of strength measurement techniques. The purpose of this study was to compare the results of three different strength measurement techniques obtained on a large heterogeneous sample.

Our observation of a difference among measures in the arms and legs combined covariance matrix is attributable not only to differences among leg strength measures but also to variability in the arm to leg strength ratio among the subjects. This would be expected in a sample of swimmers and nonathletes such as ours. Our main objective however, was to compare among measurement techniques not to compare upper- and lower-body strength.

Our findings suggest that for supine arm flexion strength measures, for our sample, all three measurement techniques would yield the same results for any comparison. This obviously implies that static and dynamic strength measurement techniques yield similar standardized values. Statistically, we were unable to detect differences in arm strength measured by these techniques despite our large sample size.

In contrast, our results suggest that the three techniques for measuring leg extension strength do suggest that the measurement method employed would influence the results. One possible explanation is that some underlying factor is more variable or more effective in one type of strength measure than in another. The isolation of a particular muscle group such as the knee extensors may be more effective with static measures than dynamic measures. Perhaps too, leg exercise occurring in daily activities may be less static whereas arm tasks may be proportionately more static. This occurrence may have contribcted to our results.

These data support Rasch's (1957) findings that the mean tension of a single maximal static elbow flexion was very similar to the maximal weight which could be lifted with the same motion. The subjects had trained dynamically prior to testing, and neither detailed methods nor detailed statistical analysis were reported.

Laubach (1976) reviewed nine papers which examined male-female strength comparisons. Seven of these used static measures of strength and two used dynamic measures. The magnitude of the sex differences was similar between the static and dynamic measures but the range tended to be larger for static measures although that may have been chiefly a function of the number of studies considered in each category.

Hinson & Rosentswieg (1973) compared electromyographic values of ST, DU and DC contractions. They found that an ST contraction of the arms (biceps brachii) yielded the largest integrated voltage in 31 subjects with DU and DC producing the largest voltage in 14 and 7 subjects, respectively. For leg muscle contractions (rectus femoris) maximal voltages were obtained for ST, DU, and DC on 8, 23, and 21 subjects, respectively. These observed differences between arms and legs among the measurement techniques and among subjects is supported partially by our observation of a differential response to different techniques and by inconsistency in the combined arm and leg covariance matrix. Group mean differences in electromyographic voltage between the three types of contraction were not statistically significant.

For measuring arm strength, it appears that static and dynamic measurement techniques are essentially measuring the same characteristic. Hence, standardized strength scores should be comparable among measurement methods. Leg strength scores are not similarly comparable and an attempt to evaluate scores generated from different measurement methods must be done with care.

For industrial job screening or placement purposes, our results suggest that static strength tests could be used to estimate candidates' capability for dynamic strength for arm tasks, at least for elbox flexion. The same cannot be said for knee extension strength. Further research is required to determine the generalizability of these results.

REFERENCES

Baumgartner, T.A., 1974, A criterion score for multiple trial measures. Research Quarterly for Exercise and Sport, 45, 193-198.

Hinson, M. and Rosentswieg, J., 1973, Comparative electromyographic values of isometric, isotonic, and isokinetic contraction. Research Quarterly, 44, 71-78.

Hosler, W.W. and Morrow,J.R.,1982, Arm and leg strength compared between young women and men after allowing for differences in body size and composition. Ergonomics, 25, 309-313.

Ikai, M. and Fukunaga, T., 1968, Calculation of muscle strength per unit cross-sectional area of human muscle by means of ultrasonic measurement. Internationale Zeitschrift fur Ange wandte Physiologie, 26, 26-32.

Isolated-Joint Testing and Exercise. Bayshore, N.Y.: Lumex Inc., 1980.

Jackson, A.S., Osburn, H.G., and Laughery, K.R., 1984, Validity of isometric strength tests for predicting performance in physically demanding tasks. Proceedings of the Human Factors Society, 1984, vol. 1. Santa Monica, CA: Human Factor Society, 452-454.

Laubach, L.L. and McConville, J.T., 1969, The relationship of strength to body size and typology. Medicine and Science in Sports and Exercise, 1, 189-194.

Mital, A., 1984, Comprehensive maximal acceptable weight of life database for regular 8-hour work shift. Ergonomics, 27, 1127-1138.

Morrow, J.R. and Hosler, W.W., 1981, Strength comparisons in untrained men and trained women athletes. Medicine and Science in Sports and Exercise, 13, 194-198.

Nordgren, B., 1972, Antropometric measures and muscle strength in young women. Scandanavian Journal of Rehabilitation Medicine, 4, 165-169.

Rasch, P.J., 1957, Relationship between maximum isometric tension and maximum isotonic elbow flexion. Research Quarterly, 28, 85.

Rasch, P.J. and Morehouse, L.E., 1957, Effects of static and dynamic exercises on muscular strength and Hypertrophy. Journal of Applied Physiology: Environmental and Exercise Physiology, 11, 29-34.

Rasch, P.J. and Pierson, W.R., 1963, Some relationships of isometric strength, isotonic strength, and anthropometric measures. Ergonomics, 6, 211-215.

Singh, M. and Karpovich, P.V., 1968, Strength of forearm flexors and extensors in men and women. Journal of Applied Physiology: Environmental and Exercise Physiology, 25, 177-180.

Snook, S.H. and Ciriello, V.M., 1974, Maximum weights and workloads acceptable to female workers. Journal of Occupational Medicine, 16:527-534.

Troup, J.D.G. and Chapman, A.E. 1969, The strength of the flexor and extensor muscles of the trunk. Journal of Biomechanics, 2, 49-62.

Wilmore, J.H., 1974, Alterations in strength, body composition and anthropometric measurements consequent to a 10-week weight training program. Medicine and Science in Sports and Exercise, 6, 133-138.

Wilmore, J.H., Parr, R.B., Grandola, R.N., Ward, P., Vodak, P., Barstow, T., Pipes, T.V., Romero, G.T., and Leslie, P., 1978, "Physiological alterations consequent to circuit weight training". Medicine and Science in Sports and Exercise, 10, 79-84.

Advances in Industrial Ergonomics and Safety II
Edited by Biman Das
Taylor & Francis 1990

Effect of the Velocity of Movement on Torque-Angle Curve in Dynamic Strength Testing

Ehsan Asoudegi and Zhibing Pan

Department of Industrial Engineering
Northern Illinois University
DeKalb, IL 60115-2854

INTRODUCTION

Previously, researchers like Clarke et al. (1966), Houtz et al. (1957), and Williams and Stutzman (1959) have studied the relationship between strength and joint angle. They demonstrated that, between 0 and 90 degrees of flexion the knee extensor torque varies. Their results suggested that, in the extended position, a minimum of extensor strength is applied. Also, maximum strength is reached at 50 to 70 degrees of flexion. Most of these studies concentrated on isometric testing only. According to Haffajee, Moritz, and Svantesson (1972), the described variation in torque is due to such factors as where the lever of the quadriceps (the perpendicular distance of the force resultant in the quadriceps tendon from the axis chosen) varies with the position of the knee joint and the patella; the degree of motor activity on which the quadriceps contraction depends; and, on muscle length relative to resting length.

Clarke (1966) used an inverted u-shaped curve to describe the relationship between knee strength and knee flexion angle. This curve peaks at a 70-degree knee flexion angle according to his isometric study.

Recently, Stobbe (1982), in an isokinetic study using a Cybex dynamometer, modeled torque as a second degree function of knee flexion angle for different velocities. Until now, this type of relationship has not been tested on the KIN-COM dynamometer.

METHOD

Subjects

Ten healthy adult male subjects consisting of students and faculty at West Virginia University volunteered to participate in this study. Their ages ranged from 20 to 36 years old, with a mean of 26.6 years old. Their heights ranged from 66 to 74 inches, with a mean of 69.40 inches. Their weights ranged from 120 to 185 pounds, with a mean of 158.60 pounds. Subjects with similar physical backgrounds (without athletic training) were chosen to eliminate possible muscle-training effects on the measured strength.

Eliminating large variations in the subjects anthropometric characteristics increases the homogeneity of the data. Therefore, when possible, subjects with similar anthropometric characteristics were selected. This allowed a consistent distance from the center of knee rotation to the point at which the KIN-COM arm contacted the subject's knee (moment arm). Keeping the moment arm reasonably constant minimized the effect of mechanical advantage on the results of this study.

Testing Device (KIN-COM)

The Kinetic Communicator Exercise System is a hydraulically-driven, microcomputer-controlled dynamometer for the test, measurement, and rehabilitation of human joint function. The KIN-COM user performs a movement, or a series of movements, against a resistance that the machine provides via a rotating lever-arm system. The machine-controlled movement modes include isokinetic, semi-isotonic (lever-arm speed is continually adjusted to maintain constant resistance), and passive joint movement. The unit can induce concentric, eccentric, or isometric contractions of the muscles involved. The KIN-COM is controlled through feedback loops that monitor the position and speed of the lever arm and the force being exerted by the user. A strain gage bridge is used to measure force. A bar-encoded shift measures position and speed. This machine compares favorably with other dynamometers such as the CYBEX II. For most clinical and muscle-strength research applications, the KIN-COM appears to provide acceptable and valid measurements, and it is at least as accurate as other available dynamometers.

Testing Procedure

After completion of a health questionnaire, each subject was examined for left knee muscular disorder by a physical therapist. Subjects with past knee injuries or cardiovascular problems were excluded from the study. A consent form was signed by these subjects who were qualified by the physical therapist to participate in this knee study.

Before starting the experiment, each of the subjects, was acquainted with the equipment and experimental procedure. To start the exercise, the subjects were positioned supine with pelvis strapped to the table of the exercise unit. Each subject required to perform a minimum of nine tests representing all possible combinations (Figure 1) of speed (0 degree/second, 60 degrees/second, 120 degrees/second) and starting position (40 degrees, 60 degrees, 80 degrees of knee flexion; Figure 2). The sequence of these combinations was randomized for each subject. Each test consisted of four consecutive maximal concentric and eccentric knee movements resisted by the KIN-COM dynamometer. To eliminate the effect of fatigue, each subject was given a three-minute rest period between each test. During each test, subjects were instructed verbally to exert as much force as they could. No feedbback was given regarding their performance.

RESULTS AND DISCUSSION

Torque-Angle Relationships in Isometric and Isokinetic Strength Testing

This study examined the functional relation between torque and knee-flexion angle using the KIN-COM dynamometer. This relationship was examined for isometric and isokinetic strength.

The preliminary plotted data indicated that the torque-angle curve was non-linear. Thus a quadratic-response surface model was used to describe knee strength as a function of knee-flexion angle. The results of this study indicate that all three terms in the quadratic model are statistically significant. The maximum strength was observed at a knee-flexion angle of 65 degrees. These results are consistent with the findings of Clarke et al. (1950), who reported that the maximum strength was observed at about 70 degree knee-flexion angle.

Eccentric Contraction

The torque-angle curve on the polygraph strip chart for eccentric contraction suggests a non-linear relation between torque and knee-flexion angle. A quadratic-response surface model was used to describe this relation, which was examined for different starting position and speeds of movement.

The results indicate a minimum strength at 20 degrees of knee flexion for all combinations of starting position and speed of movement. Maximum torque occurs at 58.39 degrees and 56.19 degrees of flexion for a 60 degrees/second and a 120 degrees/second speed of movement, respectively, when the starting position is 80 degrees of knee flexion angle (i.e., ROM is between 20 degrees and 80 degrees of flexion). When the starting position is 60 degrees of flexion, the maximum torque occurs at 49.34 degrees and

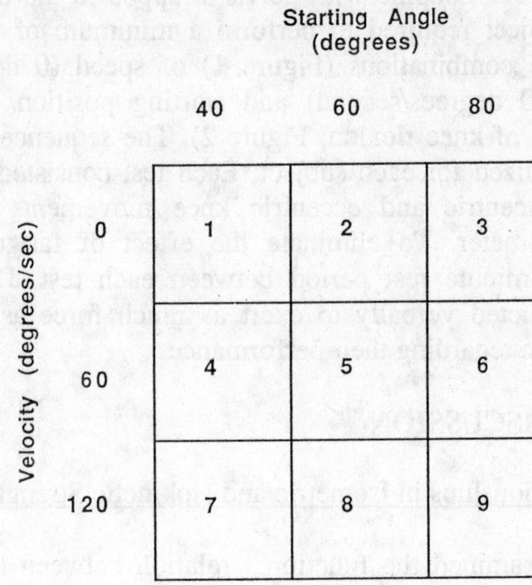

Figure 1 Strength testing combination.

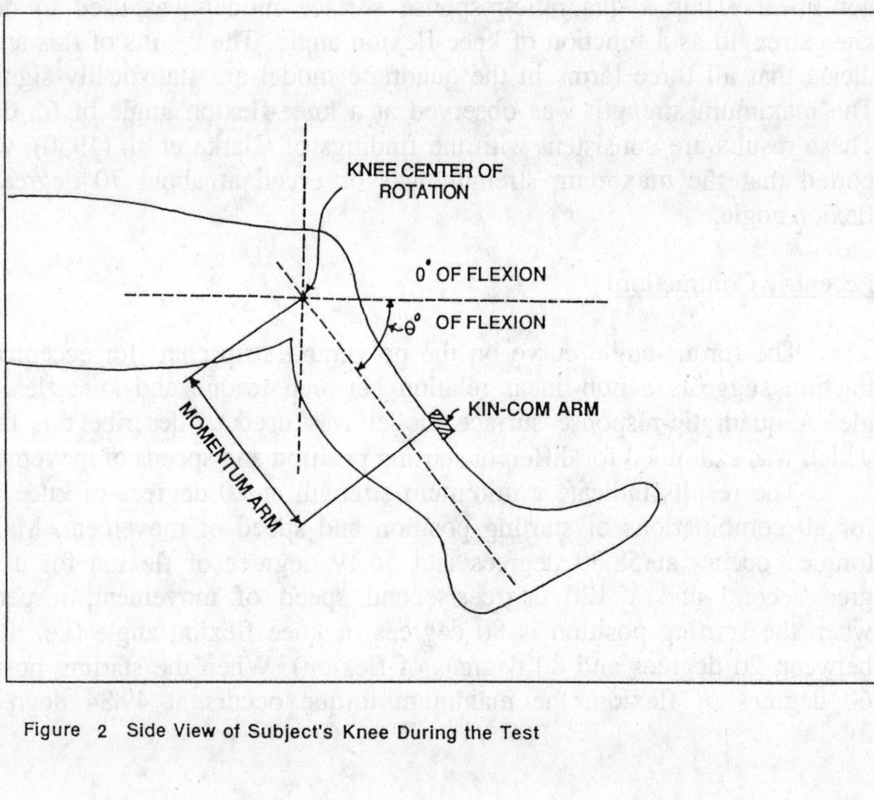

Figure 2 Side View of Subject's Knee During the Test

54.09 degrees of flexion for a 60 degrees/second and a 120 degrees/second speed of movement, respectively. Finally, the maximum torque for a starting position of 40 degrees of flexion occurs at 35.51 degrees and 36.44 degrees of knee flexion for a 60 degrees/second and a 120 degrees/second speed of movement, respectively.

All of the torque-angle curves for a velocity of 60 degrees/second lie above the curves corresponding to a velocity of 120 degrees/second.

These results agree with previous literature findings showing that the measured torque decreases for a specified angle of flexion as the speed of movement increases from 0 degree/second (isometric) to a higher velocity.

Concentric Contraction

The strip chart trace shows that the shape of the strength curve for the first concentric contraction is different from that for concentric contractions following an eccentric or isometric contraction. The strength curve for the first concentric contraction is an inverted u-shape in which strength is minimum in the extended position, peaks at some degree of flexion, and then decreases.

It was found that not all subjects are capable of developing the first concentric contraction, especially for short ranges of motion.

This study shows that the torque-angle curve for other concentric repetitions is linear. As in eccentric contraction, at any angle, the torque for a 60 degrees/second velocity is greater than for a 120 degrees/second velocity. The torque-angle curve is well above the curve corresponding to concentric contraction for both speeds of movement (60 degrees/second, 120 degrees/second). These results agree with previous research showing that maximum strength during isokinetic exercise decreases when the speed of movement increases.

Strength variation through the range of motion is an important factor in the measurement of human strength. A question should be raised regarding the point in the range of motion at which maximum strength occurs and the effect of speed of movement on the location of this point. Moffroid et al. (1969) reported that the isokinetic torque tends to peak at a larger angle in the range of motion as the speed of movement increases. This study confirms their findings. As stated earlier, the maximum isometric strength occurred at 65 degrees of knee flexion; the maximum isokinetic strength for a 60 degrees/second speed of movement occurred at 58.39 degrees of knee flexion; and the maximum isokinetic strength for a 120 degrees/second speed of movement occurred at 56.19 degrees of flexion.

REFERENCES

Clarke, H. H. and Carter, G. H., "Oregon Simplification of Strength and Physical Fitness Indices." Res. Quarterly, 30:3-10, March 1959.

Clarke, H. H., Muscular Strength and Endurance in Man, Englewood Cliffs, NJ, Prentice-Hall, 1966.

Haffajee, D., Mortiz, U. and Svantesson, G., "Isometric Knee Extension Strength as A Function of Joint Angle, Muscle Length and Motor Unit Activity." Acta. Ortho. Scand. 43:138-147, 1972.

Houtz, S. J., LeBow, M. J., and Beyer, F. R., "Effect of Posture on Strength of the Knee flexor and Extensor Muscles", J. Applied Physiology, 11(3), 475-480, 1957.

Moffroid, M., Whipple, R., Hofkosh, J., Lowman, E., and Thistle, H. "A Study of Isokinetic Exercise." Phys. Ther., 49:735-746, 1969.

Stobbe, T. J., and Hart, D. L. "A Comparative Study of Static and Dynamic Knee Extension Strength." West Virginia University, 1982.

Williams, M., and Stutzman, M. "Strength Variation Through the Range of Joint Motion." Phys. Ther. Rev., 39(3):145-152, March 1959.

Advances in Industrial Ergonomics and Safety II
Edited by Biman Das
Taylor & Francis 1990

ISOKINETIC STRENGTH TESTING: FULL BODY LIFTS VS. SEGMENTED LIFTS

G. Weisman, J. Baumhauer, M. Pope
Vermont Rehabilitation Engineering Center
University of Vermont, Burlington, Vermont 05401

The success of any screening or work capacity program requires accurate information on actual job demands as well as precise measurements of worker capacities related to key job demands. A device has previously been described that can measure a person's whole body strength. The present study demonstrates the feasibility of using a single floor to overhead lift to represent a person's strength throughout his range of motion in the sagittal plane. Few differences were found between full lifting and segmental lifting.

INTRODUCTION

Low back pain (LBP) is the albatross of industry and nemesis of medicine. The present prevalence rate is between 12 and 35 percent and the lifetime incidence rate is between 49 and 70 percent, (Andersson et al., 1984).

Several authors have associated lifting with LBP (Andersson 1981,; Frymoyer et al., 1980). Injury rates up to eight times higher than average are found among workers who regularly do heavy lifting and their disc pressures have been shown to be higher than disc pressures in those who lift light loads. (Nachemson, 1981). Although the evidence is somewhat ambiguous, it is axiomatic to some researchers that the incidence can be reduced if the demands on the worker can be reduced and matched to the worker's strength (Chaffin and Park, 1973; Chaffin, 1974).

There are four types of muscle contraction: isometric (length remains constant), isokinetic (kinetic energy remains constant because velocity does not change), isotonic (the force remains constant while the length changes) and isoinertial (muscles contract against a constant load). Lifting in the real industrial environment probably contains elements of all of the above. For example, at its beginning, a lift can be considered virtually isometric since the postural muscles and the muscles needed to overcome the inertia of the load all apply force without changing length. When the force applied is greater than the load, the lift then becomes dynamic.

Thus, dynamic strength plays an important role in lifting. However, most of the studies of manual material handling have dealt with static strength (Caldwell et al., 1974; Chaffin, 1974; 1978; Chaffin et al., 1979; Keyserling et al., 1980; Warwick et al., 1980). Notably, the Work Practices Guide for Manual Lifting (NIOSH 1981) recognizes isometric strength testing as a reasonable way to identify workers who might be at a higher risk of low back pain or overexertion injury. However, Mayer et al. (1985) point out that isometric muscle activity is usually associated with higher rates of injury than dynamic activity. Marras et al. (1984) state that isometric strength testing can be misleading and suggest an isokinetic approach. They suggest that isokinetic evaluation is a more realistic procedure for evaluation of tasks related to manual material handling conditions. Isokinetic testing has the great advantage of allowing measurement of the actual force output capability at every point along the lifting arc.

In a previous paper (Weisman, 1990) describes a device that can measure vertical lifting strength isokinetically with good reproducibility. The device output provides a description of strength throughout an individual's range of motion in the sagittal plane. The strength-testing method employed in the previous study required subjects to lift from the floor to an overhead position in one motion. Lifts were performed at each of four horizontal distances from the body. While the output showed strength throughout a range of motion, it was not clear whether strength of the lift varied with the point at which the lift began. The present study was designed to determine whether a single, full lift provides the same information as a series of segmented lifts representing the same range of motion.

METHOD

Subjects: The test group consisted of ten male volunteers between the ages of 22 and 38. No subject had any history low back injury, pathology or surgery and all were recruited for participation based on these criteria. All subjects worked at a sedentary job or school and participated in various types of recreational activities. Prior to the initial testing, each subject read a layman's summary of the experiment and test protocol and signed a consent form. Table I lists the anthropometric variables for the subject population.

TABLE 1: Subject Information

	Mean	+/-	S.D.
AGE (years)	24.7	+/-	5.9
HEIGHT (cm)	176.5	+/-	5.8
WEIGHT (kgs)	74.4	+/-	8.4
ARM LENGTH (cm)	64.4	+/-	5.0
WAIST HEIGHT (cm)	92.7	+/-	4.3
SHOULDER HEIGHT (cm)	147.8	+/-	6.6

Apparatus: The device used in the present study is fully described elsewhere (Weisman, 1990). Isokinetic force is applied to a cable attached to a handle. Forces are measured at the handle by a strain gauged load cell, in order to bypass any frictional forces inherent in the device. A potentiometer connected directly to the drive system measures the displacement of the cable. Force and displacement measurements are recorded by a micro-computer (IBM PC) via an analog-to-digital converter (Data Translation Board).

Procedure: The subjects performed all lifts during two sessions within the same seven day period. Each test session consisted of: (1) Three full vertical lifts (Seg 0) in the midsagittal plane from floor to head height, (2) Three lifts at each of three body segmented regions: floor to waist (Seg 1), waist to shoulders (Seg 2), and shoulders to head (Seg 3). Both segmented and full lifts were performed at four horizontal distances (HD) at a test speed of 30.5 cm/sec. The horizontal distances were determined as a percentage of arm length and included distances of 30% (HD1), 50% (HD2), 70% (HD3) and 90% (HD4). Figure 1 graphically depicts the segments of the range of motion used in the present study. HD 1 and 2 were tested during the first session, while HD 3 and 4 were tested on the second day.

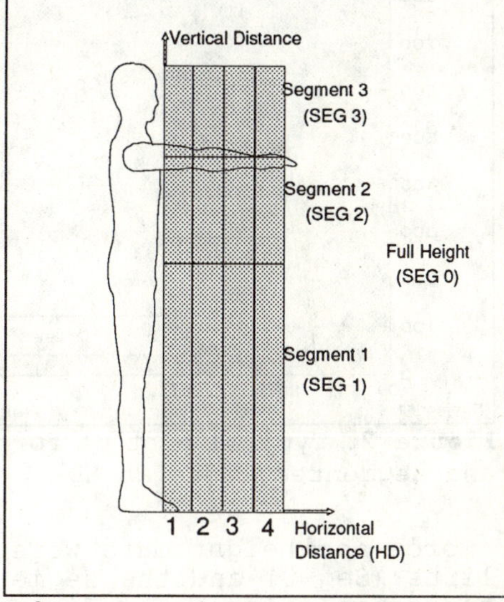

Figure 1. Pictoral representation of segments and horizontal distances.

Subjects were instructed to lift vertically as hard and as fast as possible while maintaining their foot positions. Any lifting technique was acceptable and subjects were given a period of time before each session to become familiar with their technique and the machine. This warm-up period typically consisted of five to ten lifts. Once the lifting technique was chosen, subjects were requested to utilize that technique throughout the testing procedure. One- minute rests were provided between lifts of a specific segment, with five-minute rests were provided between the different segments. Subjects were questioned concerning muscle soreness and comfort after each session and prior to the following session. No subject verbalized discomfort during or after lifting. A video camera and recorder were used to tape each subject's lifting technique.

Data Analysis: Data were analyzed by a within-subject, completely crossed ANOVA, with the dependent variable being the force value at each of 35 heights. The four HD and each segmented lift were analyzed separately. This repeated measures model tested for differences between full and segmented body lifts, repetitions, and height of lift for each individual subject.

RESULTS

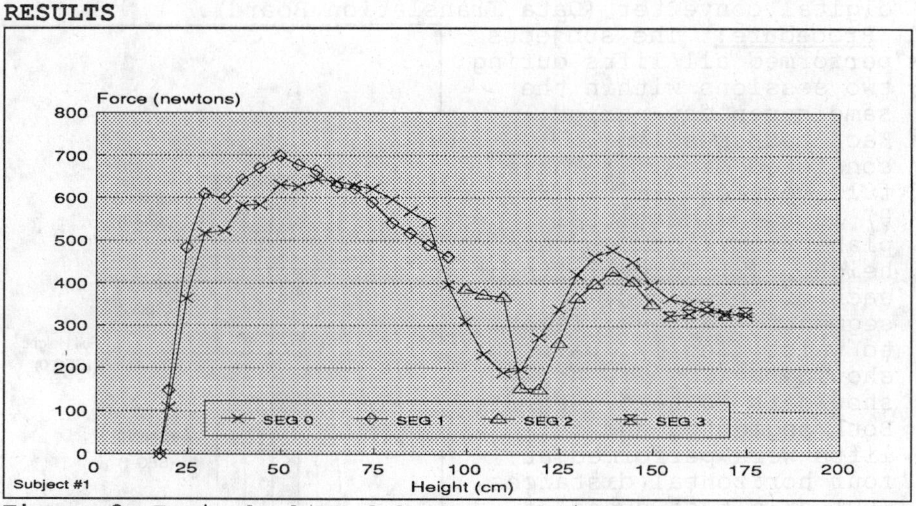

Figure 2. Typical plot of force vs. height data for full lift and segmented lifts at HD 1.

Force and height data were plotted for both the full lifts (Seg 0) and the segmented lifts for each horizontal distance. Figure 2 illustrates data from one subject for the nearest horizontal distance (HD 1). As can be seen from the plot, force generated during segmented lifts closely follows force generated during

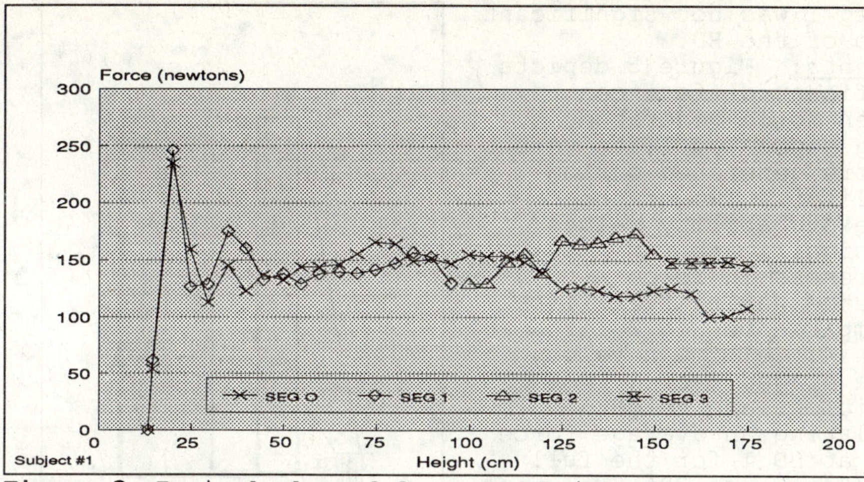

Figure 3. Typical plot of force vs. height data for full lift and segmented lifts at HD 4.

full lifts (Seg 0). Figure 3 depicts similar data at HD 4, the distance furthest from the body.

ANOVA results are summarized in Table 2.

TABLE 2: ANOVA Probability Values

	S	H	SH	R	SR	HR	SHR
HD1:SO.1:	0.0902	0.0000*	0.0004*	0.0083*	0.4113	0.1311	0.2839
HD1:SO.2:	0.0078*	0.1808	0.9290	0.8963	0.8545	0.9660	0.0025*
HD1:SO.3:	0.6653	0.0000*	0.9900	0.6995	0.6599	0.4934	0.9566
HD2:SO.1:	0.4623	0.0012*	0.6423	0.7969	0.4312	0.6792	0.9241
HD2:SO.2:	0.2132	0.8361	0.4689	0.9367	0.6542	0.7075	0.4201
HD2:SO.3:	0.3426	0.0001*	0.5401	0.6335	0.5182	0.5985	0.6675
HD3:SO.1:	0.6510	0.4766	0.7995	0.9704	0.4720	0.3180	0.0046*
HD3:SO.2:	0.2612	0.9808	0.5260	0.9344	0.3605	0.2759	0.0147*
HD3:SO.3:	0.0878	0.6681	0.7217	0.5240	0.0853	0.4299	0.8076
HD4:SO.1:	0.9942	0.7725	0.5468	0.7294	0.5138	0.5117	0.6733
HD4:SO.2:	0.0458*	0.0002*	0.9139	0.8298	0.3283	0.9024	0.7138
HD4:SO.3:	0.2258	0.0001*	0.8215	0.2821	0.6871	0.5956	0.4771

HD = Horizontal Distance; S = Segment; H = Height; R = Repetition
*p<.05

Height: Force values at each height during the full lift (Seg 0) at all horizontal distances were found to be significantly different with p values less than .05. When the segments were analyzed individually, six of the 12 probability values for height were significant. Figure 4 graphically depicts where force values were significantly different for the various heights within the segments. Seg 2 was not significant at three of the four HD. Seg 3 was not significant at one of the four HD and Seg 1 was not significant at two of the HD.

Segments: Figure 5 depicts significant differences between force generated during segmented lifts versus the fore generated during full lifts. Two of the twelve probability values were significant for differences in force generated during full lifts as compared to Seg 2 lifts at HD 1 & 4. Cell mean values were averaged over repetitions and height calculating an average force value at HD 1 for the full lift (Seg 0) of 386.8 newtons and for Seg 2 of 417.7 newtons. At HD 4 the values were 173.2 newtons and 189.5 newtons for Seg 0 and Seg 2 respectively.

Repetitions: These were found to be significantly different at HD 1 involving Seg 0 & 1 as shown in Figure 6. Cell mean values were averaged over height and segment. Repetition 1 = 672.7 newtons; Repetition 2 = 724.5 newtons; Repetition 3 = 741.4 newtons.

Segment X Height: This interaction was found to be significant at HD 1 with a full lift and Seg 1. Cell mean values were averaged over repetitions to evaluate the interaction. Seg 1 was found to have an average force value of 11.8 lbs greater than Seg 0 at each of the 15 analyzed.

Segment X Height X Repetition: This interaction was found to be significant in three of the twelve measurements. Two involving HD 3 and one involving HD 1.

Although not specifically analyzed it was observed that the force values decreased as one moved out horizontally away from the body, i.e. HD 1 > HD 2 > HD 3 HD 4.

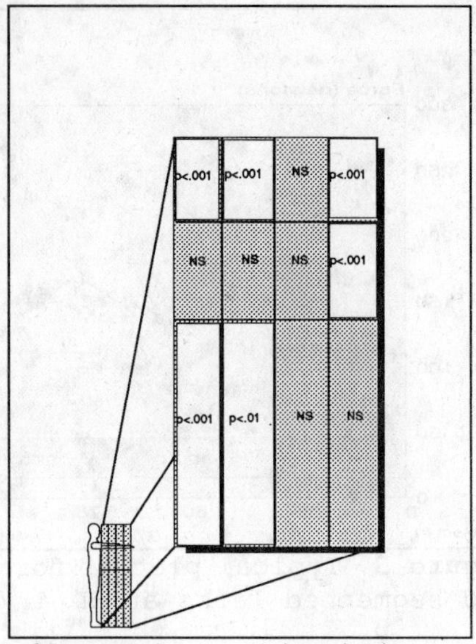

Figure 4. Significant differences of forces generated at different heights for segmented lifts.

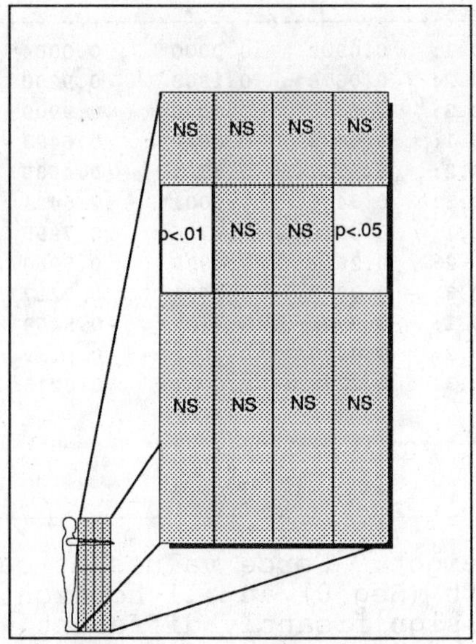

Figure 5. Significant differences of forces generated between full and segmented lifts.

DISCUSSION

Values of strength for male subjects are consistent with other researchers (Kumar, 1988).

The results of the height analysis confirmed the results of the previous study (Weisman, 1990). Force generated during a full lift was significantly different at different heights of the lift. It is important to note that because generated force is different throughout a person's range of motion it is impossible to describe strength with a single value. When performing segmented lifts force generated was often not significant at different heights within certain segments. The insignificance of Seg 2 may be explained in a number of ways. Seg 2 contained the fewest force values (approximately 5). Awkward wrist manipulations were required to move through Seg 2 which might create some variability in a small compact area represented by only a few force values.

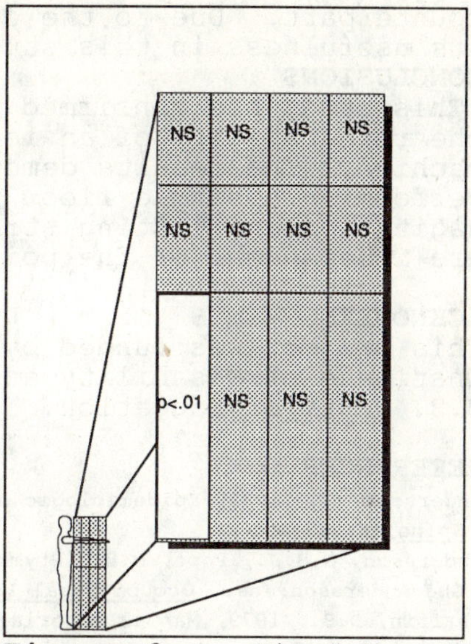

Figure 6. Significant differences between repetitions for all lifts.

The fact that only two of the twelve probability values were significant provides good support for the hypothesis that segmented lifts are equivalent to full body lifts. As described above, Seg 2 had the smallest number of force values which could explain the significance of Seg 2 at HD 1 and 4. It should be noted that HD 1 and 4 were tested first during each day's testing. The significance might be the result of warming up or gaining familiarity with the machine. In any case, the force values, when calculated from the raw data, were no more than 31.8 newtons different between segments.

Analysis of repetitions showed Seg 0 and 1 at HD 1 were significantly different. The significant differences at HD 1 involving Seg 0 & 1 might be attributed to a warming up phenomenon as seem by the progressive increase in force values from Repetition 1 to 3. This is the area of range of motion where the greatest force can be generated. In our previous study, (Weisman, 1990) a trend was noted in differences between bouts. This trend was attributed to a warming up phenomena because there was no difference between force generated between days which would explain a "training" effect.

The interaction between segment, height and repetition is very difficult to interpret, but suggests that, with one of the segments, during one of the repetitions, the height becomes significantly different than its counterpart. Due to the complexity of the interaction its usefulness in this study is limited.

CONCLUSIONS

This study has confirmed previous results demonstrating the repeatability of an isokinetic strength testing machine. The results demonstrate the utility of performing a single floor to overhead lift to determine sagittal plane lifting strength. The values of strength are independent of the point at which a lift originates.

ACKNOWLEDGEMENTS

This project was funded by a grant from the National Institute of Disability and Rehabilitation Research, U.S. Dept. of Education.

REFERENCES

Andersson, G., 1981, Epidemiologic aspects on low back pain in industry. Spine, 6, 53-60.

Andersson, G.B.J., Pope, M.H., Frymoyer, J.W., In MH Pope, JW Frymoyer and GBJ Andersson eds., Occupational Low Back Pain. New York : Praeger, 1984.

Chaffin, D.B., 1979, Manual materials handling: the cause of over exertion injury and illness in industry. Journal of Environment and Pathological

Chaffin, D.B., 1974, Human strength capability and low back pain. Journal of Occupation Medicine, 16, 245-254.

Chaffin, D.B., Park, K.S., 1973, A longitudinal study of low back pain as associated with occupational weight lifting factors. American Industrial Hygiene Association Journal, 34, 513-525.

Caldwell, L.S., et al., 1974, A proposed standard procedure for static strength testing. American Industrial Hygiene Association Journal, 35, 201-206.

Frymoyer, J.W., Pope, M.H., Rosen, J.C., Goggin, J., Wilder, D.G., Constanze, M., 1980, Epidemiologic studies of low back pain. Spine, 5, 419-423.

Keyserling, W.M., Herrin, G.D., Chaffin, D.B., 1980, Isometric strength testing as a means of controlling medical incidents on strenuous jobs. Journal of Occupational Medicine, 22, 332-336.

Kumar, S., Chaffin, D.B., Redfern, M., 1988, Isometric and Isokinetic Back and Arm Lifting Strengths: Device and Measurement. J. Biomechanics, 21(1), 35-44

Marras, W.S., King, A.I., Joynt, R.L., 1984, Measurement loads on the lumbar spine under isometric and isokinetic conditions. Spine, Vol 9, 2nd edn., pp. 176-187.

Mayer, T.G., Gatchel, R.J., Kishino, N., Keeley, J., Capra, P., Mayer, H., Barnett, J., Mooney, V., 1985, Objective assessment of spine function following industrial injury. Spine, 10(6), 482-493.

Nachemson, A., 1981, Disc pressure measurements. Spine, 6, 93.

Warwick, D., Novak, G., Schultz, A., 1980, Maximum voluntary strengths of male adults in some lifting, pushing and pulling activities. Ergonomics, 23, 49-54.

Weisman, G.W., Baumhauer, J., Seroussi, R., Reineche, S., Pope, M.H., The Development and testing of an isokinetic strength testing device. Being presented at the Annual International Industrial Ergonomics and Safety Conference, Le Chateau Champlain, Quebec, Canada, June 10-13, 1990.

Work Practices Guide for Manual Lifting, Technical Report, DHHS (NIOSH) Publication Number 81-122, March 1981.

X

INDUSTRIAL ACCIDENTS AND PREVENTION

Advances in Industrial Ergonomics and Safety II
Edited by Biman Das
Taylor & Francis 1990

SLIP PREVENTION IN INDUSTRIAL AND COMMERCIAL ENVIRONMENTS

K. Hahn, R.M. Heddle, P.R. McKay, P.D. Roorda

University of Waterloo
Waterloo, Ontario (Canada)

In North America accidental slips and falls are the second greatest cause of injury related deaths; motor vehicles are first. In Sweden slipping accidents are even more common than vehicle accidents, causing 40% of accidental deaths. In Ontario in 1985 there were over 27,000 falling accidents in the workplace and studies indicate that approximately 60% of these falls were initiated by slipping, entailing $125 million in monetary compensation from the Worker's Compensation Board. Slipping is the most common office accident, causing 47% of injuries. Almost half of all accidents resulting in back injury are sustained in slipping accidents. Canadian Industry has not recognized slipping as a serious hazard. Studies show that flooring in factories is invariably selected on an initial cost basis only, and shoe manufacturers consider only aesthetics, durability and price when evaluating shoe sole design. In Canada there are no laws governing the minimum required slip resistance of floors. This paper addresses the concept of shoe-floor systems design for typical industrial working conditions and proposes the implementation of slip prevention technology beneficial to industry, by applying a systems approach.

INTRODUCTION

Slipping is a serious problem in today's industrial environment.

- In North America, accidental slips and falls are the second greatest cause of injury-related deaths; motor vehicle accidents are first. In Sweden, slipping accidents are even more common than vehicle accidents, causing 40% of accidental deaths (Strandberg *et al.*, 1981).
- Contrary to popular belief, almost half of all accidents resulting in back injury are sustained in slipping accidents (Manning, 1983; Manning *et al.*, 1988). Figure 1.
- Slipping is the most common accident in offices, causing 47% of injuries (Hollett, 1981). Figure 2.

Slipping is a chronic problem throughout industry. In Ontario in 1985 there were over 27 000 falling accidents in the workplace (W.C.B., 1986). Studies by Manning (1988) indicate that approximately 60% of these falls were initiated by slipping. Applied to Ontario, this figure indicates that slipping accidents entailed a total monetary compensation of $125 million, or 13% of the total claims paid by the Workers' Compensation Board.

Unfortunately, industry does not recognize slipping as a serious hazard. Studies have shown that flooring in factories in invariably selected on an initial cost basis only (Steemson, 1984). Shoe manufacturers consider only aesthetics, durability and price when evaluating a shoe sole design. In Canada, there are no laws governing the minimum required slip resistance of floors.

The purpose of this report is to introduce the slipping problem, its causes, and potential solutions. It is also intended to increase public awareness of an important but neglected hazard in industry today.

ACCIDENTS CAUSING BACK INJURIES
FIGURE 1

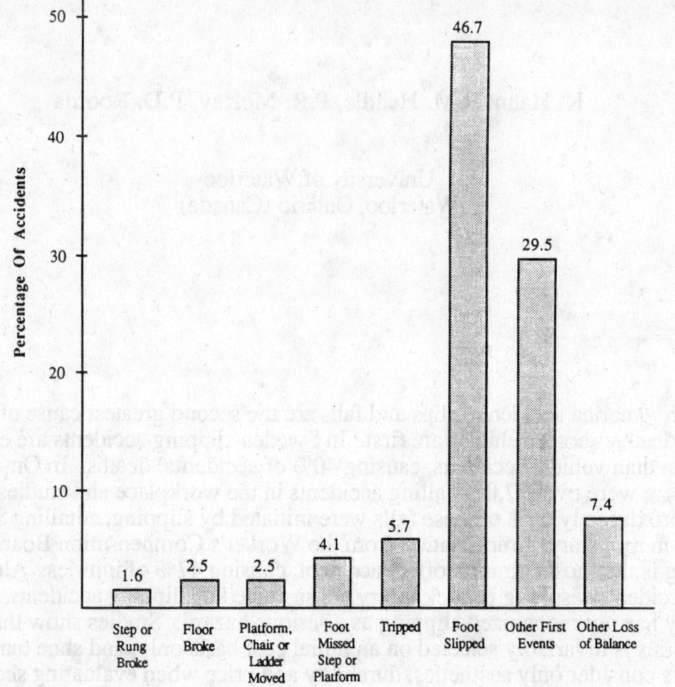

LEADING ACCIDENT TYPES FOR OFFICE EMPLOYEES
FIGURE 2

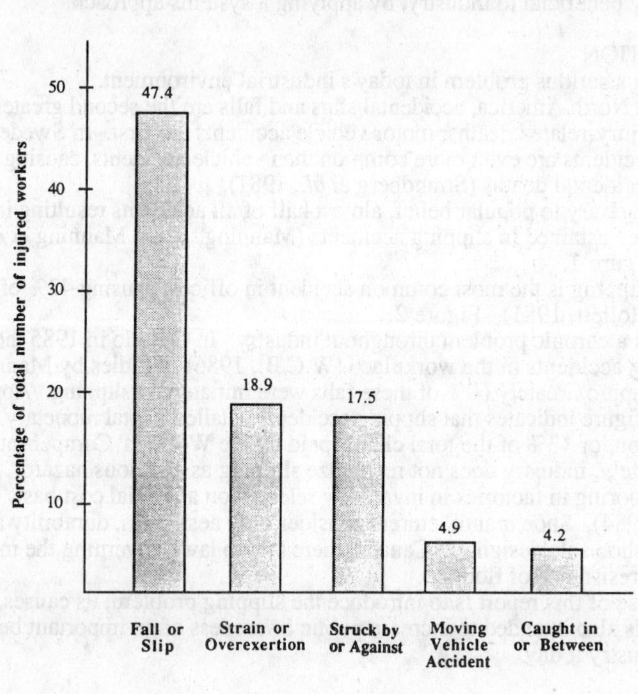

THE CAUSES OF SLIPPING

Everyone has experienced a slip, perhaps on a polished floor or perhaps while walking on ice. In order to differentiate slips from other types of falls, a slip is defined as the loss of balance that occurs when the foot unexpectedly begins to slide along the floor. Sliding occurs when the horizontal or shear force exerted by the foot exceeds the frictional force exerted by the floor. Research into this interaction has identified the factors contributing to slipping. These factors and their inter-relationships are shown in Figure 3. The forces exerted by the foot are determined by the biodynamic force distribution. The frictional forces are determined by the coefficient of friction (COF).

Walking can be seen as a successive loss and recovery of balance. In the course of a step, the heel of the foot first makes contact with the floor, taking the body weight as the entire foot is planted. The heel then rises and the leg pushes off the ball of the foot. The body weight transfers to the forward foot as the back foot leaves the ground. The forces between the foot and the floor are caused by the transfer of weight back and forth between the feet, and by the instabilities generated by this shift of balance. Figure 4 illustrates the distribution of forces during a single step. These forces are broken down into horizontal and vertical components.

FACTORS CAUSING SLIPS
FIGURE 3

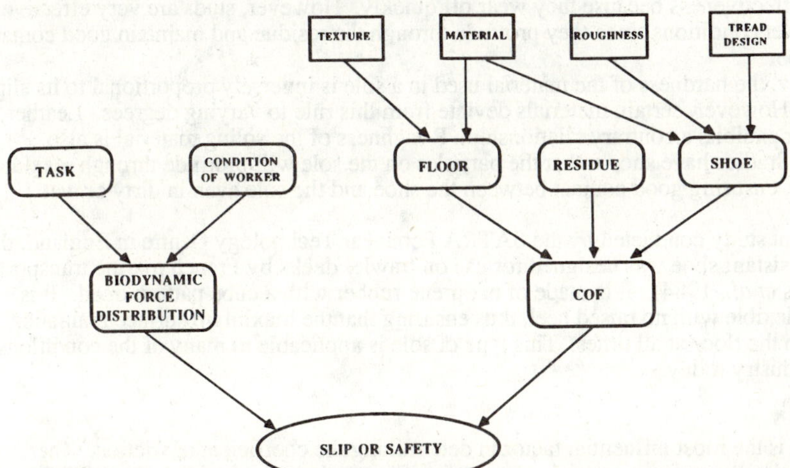

The frictional force exerted by the floor is proportional to the vertical force of the foot. The coefficient of friction (COF) is defined as the ratio of frictional force to vertical force. Thus, a higher COF means a greater force of friction between two surfaces. The third graph in Figure 4 shows the ratio H/V for the forces in a step. This can be interpreted as a *required* COF over the course of the step. Thus, the peaks on this graph are the points at which a slip is most likely. In order to prevent a slip, the COF must exceed the peaks on this graph.

Realistically, the biodynamic force distribution, which incorporates such parameters as the individual walking style of each worker, is beyond the scope of slip prevention. The COF is therefore the controllable parameter. As shown in Figure 3, the three factors that determine the COF are the floor and the shoe, and any residue between them. These are the factors which must be considered in the design of any system to reduce the frequency of slipping accidents.

The presence of residue between the shoe and the floor invariably reduces the COF. Oil, water, dust and dirt all increase the chances of slipping. Proper cleaning techniques are therefore very important. However, in most cases residue is either unavoidable or very expensive to remove. In addition, residue can exacerbate the fundamental problem of inadequate shoes and flooring. For these reasons, research into the slipping problem has concentrated on shoes and flooring.

CURRENT SLIP PREVENTION TECHNOLOGY

Footwear

Safety shoes to protect workers from sharp or heavy objects are standard safety equipment. However, only in recent years have footwear manufacturers incorporated slip prevention technology into their design characteristics. The result is a variety of non-slip footwear for the workplace.

Presently, slip resistant loafers, work boots and rubber boots are available. The common feature of these products is the sole, which in each case is specially designed to increase the coefficient of friction between the shoe and the floor. However, no single shoe sole can be resistant to slips in all conditions. When selecting footwear for a certain environment, there are two main factors which must be considered: the design of the tread, and the soling material.

The tread design of a shoe sole greatly contributes to its slip resistance. In general, untreaded soles are better in dry conditions, while treaded soles perform well in wet conditions. As an all-purpose design, researchers agree that a pattern with numerous well-defined leading edges (such as the herringbone pattern of popular basketball shoes) is satisfactory (Jones, *et al.*, 1984). Other sole designs, such as suction cups and studs, are of limited effectiveness because they wear off quickly. However, studs are very effective in dirty or wet conditions since they protrude through the residue and maintain good contact with the floor.

Generally, the hardness of the material used in a sole is inversely proportional to its slip resistance. However, certain materials deviate from this rule to varying degrees. Leather in particular exhibits a contrary relationship. Roughness of the soling material is also important. Studies have shown that the particles on the sole will protrude through residue on the floor, ensuring good contact between the shoe and the sole even in dirty or wet conditions.

In a recent study conducted by the SATRA Footwear Technology Centre in England, the most slip resistant shoe was designed for use on trawler decks by French marine transport firms (Jones *et al.*, 1984). It is made of neoprene rubber with a cube pattern tread. It is extremely flexible with no raised heel, thus ensuring that the maximum surface maintains contact with the floor at all times. This type of sole is applicable to many of the conditions found in industry today.

Floors

Flooring is the most influential factor in determining the coefficient of friction. The choice of walkway surface materials plays a significant role in determining the COF (Figure 5). There are three basic types of non-slip products for floors: coating, flooring, and matting (Elie, 1987).

Floor coating is a paint-like material which provides an abrasive surface when applied to a floor. Particles of aluminum oxide, silicon carbide, or other abrasives are mixed with a paint base to form a slip-resistant layer. In general these products are effective, causing a significant improvement in the COF of the floor. They are most often used on stairways, ladders and showers. Floor coatings are one of the least expensive forms of anti-slip protection. Their major drawback is their lack of durability. In areas of heavy traffic, the coating tends to wear off very quickly when the abrasive particles loosen and dislodge. This can lead to the appearance of unexpected slippery patches in high-traffic areas, which can be very hazardous. However, coatings can be very useful in light traffic areas.

Slip-resistant flooring is a solid surface that is permanently bonded to the floor. As such it is the most permanent and durable slip prevention product. One variety consists of an abrasive surface mounted on a self-adhesive backing. This is essentially a durable form of coating, and is usually used in small, hazardous locations such as stairways and pools. Another variety is wall-to-wall flooring, comprised of rubber or vinyl sheets which are glued to the floor. By themselves, these materials significantly improve the coefficient of friction of the floor. In addition, some flooring combines these materials with abrasive particles such as those found in floor coatings. This combination has proven very effective. Some manufacturers include bacteria-inhibiting and acid-resistant compounds in

FORCES OF A SINGLE WALKING STEP
FIGURE 4

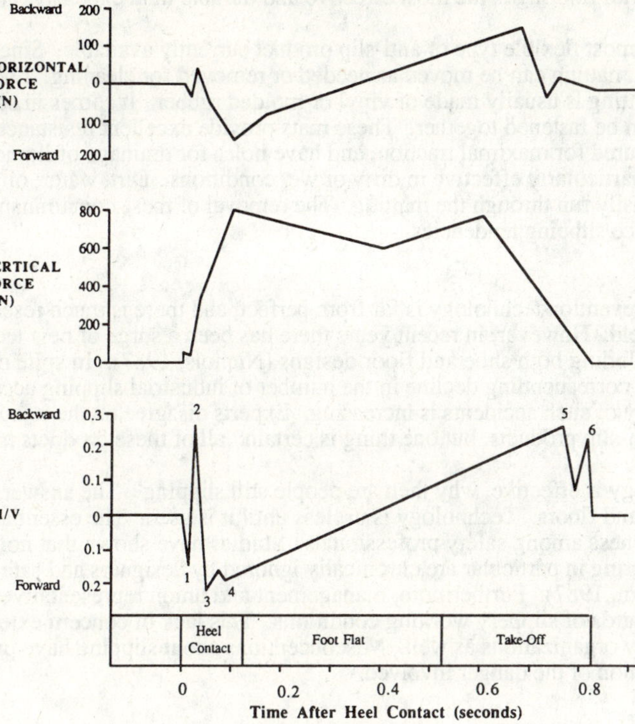

C.O.F. OF WALKWAY SURFACE MATERIALS
FIGURE 5

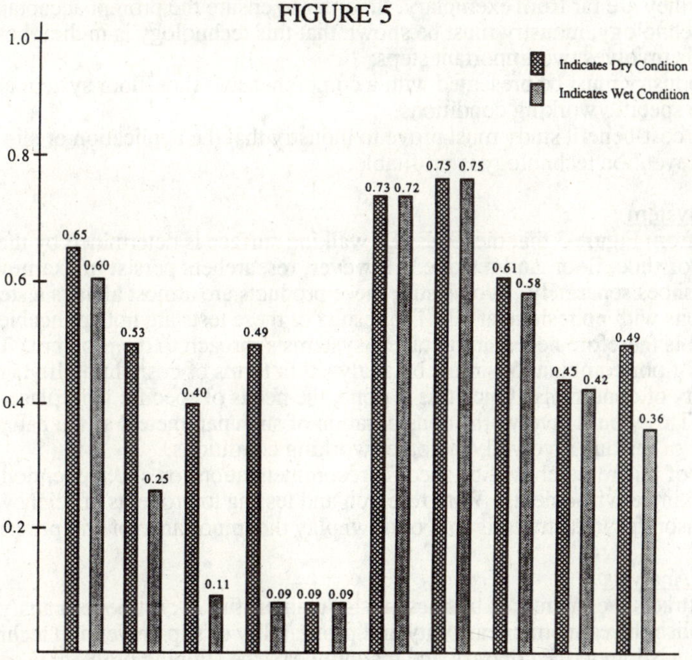

the flooring itself, making it applicable to many different working environments (Redfern, 1987). Wall-to-wall flooring is the most effective and durable (and expensive) non-slip product.

Matting is the most flexible type of anti-slip product currently available. Since it is not fixed to the floor, matting can be moved as needed or removed for cleaning. Like wall-to-wall flooring, matting is usually made of vinyl or molded rubber. It comes in strips or squares which can be fastened together. These mats provide excellent resistance to slips. The mats are textured for maximal traction, and have holes for drainage of liquids. For this reason, they are particularly effective in dirty or wet conditions. Dirt, water, oil, or chemical spills easily run through the matting. The removal of these contaminants can significantly reduce slipping tendencies.

THE PROBLEM

Current slip prevention technology is far from perfect, and there is much research left to be done in this field. However, in recent years there has been a surge of new technology and products, including both shoe and floor designs (Nichols, 1987). In spite of this, there has been no corresponding decline in the number of industrial slipping accidents. In fact, the frequency of such accidents is increasing. Experts disagree on the relative merits of the various non-slip products, but one thing is certain: all of these products make walking safer.

If the technology is effective, why then are people still slipping? The answer involves more than shoes and floors. Technology is useless until it is used. The essential problem is a lack of awareness among safety professionals. Studies have shown that non-slip footwear and flooring in particular are chronically ignored by designers and safety engineers (Redfern, 1987). Furthermore, management and union representatives fail to recognize the hazards of slippery working conditions. This lack of concern extends to government safety organizations as well. Misconceptions about slipping have prevented an accurate appreciation of the danger involved.

THE SOLUTION

Business attitudes towards occupational health and safety are gradually improving; nevertheless, they are far from exemplary. In order to ensure the prompt acceptance of slip prevention technology, industry must be shown that this technology is in their best interests. This involves two important steps:

- Industry must be presented with a comprehensive shoe-floor system applicable to specific working conditions.
- A cost-benefit study must prove to industry that the application of slip prevention technology is profitable.

Shoe-Floor System

It is clear from Figure 3 that the COF of a walking surface is determined by the combination of shoe, floor, and residue. However, researchers persist in examining flooring and shoes separately. Worse still, these products are almost always tested under ideal conditions with no residue at all. The results of these tests are not applicable to the real world. It is therefore necessary to take a systems approach to the problem. The entire range of shoe-floor combinations must by analyzed in terms of cost, durability, and COF under a variety of conditions. Once this is done, the needs of specific workplaces must be determined. This would involve the consideration of such parameters as the nature of the task, the type of residue involved, and other working conditions.

The result of this research will be specific recommendations on slip prevention for different working environments. With research and testing to prove its efficiency, there will be no reason for industry to ignore or downplay the importance of slip prevention.

Cost-Benefit Analysis

Nothing attracts the interest of business faster than profit. A cost-benefit analysis would serve to establish the economic feasibility and profitability of slip prevention technology. Conventional wisdom to the contrary, the magnitude of the slipping problem is such that significant cost savings can be anticipated. In 1985 in Ontario alone, approximately $125 million was paid by the Workmans' Compensation Board in response to claims for slipping accidents. $125 million can buy a lot of flooring.

Companies have financial incentives to reduce accidents. In addition to the costs of replacement workers and man-days lost to injury, Workmans' Compensation Board premiums vary with the number of claims made. Up to 80% of the initial premiums can be refunded to companies with fewer claims than expected, and the Board can levy surcharges of up to 160% when accidents are more frequent than expected (W.C.B., 1988a). Premiums are normally 3-8% of the total company payroll, and can be as high as 32% for certain industries (W.C.B., 1988b). Clearly, accident reduction will significantly reduce insurance costs. In addition, one must consider intangible benefits such as the effects of a concerned management on employee morale. Although difficult to quantify, these can lead to significant improvements in productivity.

The costs of a slip prevention system can be easily determined from manufacturers' data. However, the benefits from such a system cannot presently be calculated. One crucial piece of information is missing: a quantitative correlation between slip-resistant products and slipping accidents. This correlation has not yet been established. Clearly, an increase in COF will reduce the frequency of slipping accidents; however, the magnitude of this reduction is not known. Without this data, it is impossible to estimate the benefits of slip prevention products.

It is therefore necessary to study the effects of non-slip products on accident frequency. The best method would involve a detailed collection of accident statistics in current environments, and then a similar collection in the same environment after non-slip products had been installed. A range of working conditions should be studied, concentrating on particularly hazardous environments such as commercial kitchens and milling plants. With proof of its effectiveness and cost savings, industry will not be able to ignore the benefits of slip prevention.

CONCLUSIONS

There is a serious slipping problem in industry today. This problem has been chronically over-looked for many years, and continues to be ignored even though effective slip prevention technology would be beneficial to industry both in terms of profits and employee relations. Clearly, if business were aware of these facts then fewer people would be injured in slipping accidents.

The real problem is a lack of awareness. Most people, including safety officials and engineers, do not recognize slipping as a serious hazard. Every year, this ignorance costs thousands of lives and millions of dollars. This situation is intolerable. Education of safety professionals must emphasize the problem and motivate action. With time and effort, slip prevention will become an integral part of industrial safety.

REFERENCES

Elie, J., *Anti-Slip Protection*. Occupational Health and Safety / The Buyers' Guide 1987-88, pp. 11-17.

Hollett, M., Workers' Compensation Board Winter Report, 1981.

Jones, C., and M. Bruce, *Many a Slip*. Occupational Safety and Health, March 1984, pp. 16-21.

Manning, D. P., *Deaths and Injuries Caused by Slipping, Tripping and Falling*. Ergonomics, V26-1, 1983, pp. 3-9.

Manning, D. P., I. Ayers, C. Jones, M. Bruce, *The Incidence of Underfoot Accidents during 1985 in a Working Population of 10 000 Merseyside People*. Journal of Occupational Accidents, August 1988, pp. 121-130.

Nichols, J. L., *Floor Safety in Commercial Kitchens*. The Journal of the Royal Society of Health, August 1987, pp. 148-150.

Redfern, M. S., *Preventing Slip and Fall Injuries Requires Environmental Controls.* Occupational Health and Safety, September 1987, pp. 34-43.

Steemson, J. *Cover Points*, Occupational Safety and Health, March 1984, pp. 13-15.

Strandberg, L., and H. Lanshammar, *The Dynamics of Slipping Accidents*. Journal of Occupational Accidents, January 1981, pp. 153-162.

Workers' Compensation Board of Ontario, Compensation Claims Initially Settled in 1985, Type of Disability by Type of Accident. March 1986, pp. 1-2.

Workers' Compensation Board of Ontario, NEER User Guide, 1988.

Workers' Compensation Board of Ontario, Table of Rates, 1988.

Advances in Industrial Ergonomics and Safety II
Edited by Biman Das
Taylor & Francis 1990

CAUSES AND PREVENTIVE STRATEGIES OF INDUSTRIAL ACCIDENTS AS PERCEIVED BY MANAGERS AND WORKERS

A. Khaleque and Md.Monjurul Karim

Department of Psychology
University of Dhaka
Dhaka-1000,Bangladesh

The purpose of the present study was to find out some important causes and effective preventive strategies of industrial accidents as perceived by the workers and the managers. The study was conducted on a total of 70 subjects, 40 managers and 30 workers, chosen from 6 industries in Bangladesh. The results show that there is significant correlation between the ratings of the workers and the managers about some important causes of accidents. However, the results concerning the perceptions of effective preventive strategies reveal that there is greater difference than agreement between them.

INTRODUCTION

The world is paying heavily for industrial accidents in terms of both human suffering and economic waste. It is estimated that every year throughout the world more than 180,000 workers die as a result of industrial accidents and occupational diseases, and some 110 million additional workers suffer non-fatal injuries (Fazzio, 1986). Although Bangladesh is not an industrialized country, still nearly 5,000 industrial accidents occur here every year, out of which 0.6% are fatal and 20% of serious nature. In Bangladesh, due to high rate of non-reporting, accurate accident statistics are not available. However, the data available in the office of the Chief Inspector of Factories show that industrial accidents had increased from 7.66 in 1977 to 11.07 in 1985 per thousand workers (Mesbahuddin, 1986). Increasingly high rate of accidents are not only causing immense human sufferings but also a a heavy loss of trained manpower. It has been pointed out that many of these accidents could have been avoided had the workers been more careful in using safety equipments and the managers been more cautious in maintaining safe working conditions and enforcing safety regulations (Khan, 1986). Similarly,several researchers have indicated that perceptions and attitudes of workers and managers are very important for prevention of industrial accidents(Haque, 1986).

Considering the above indications, the present study was designed to find out some important causes and effective preventive strategies of industrial accidents as perceived by some workers and managers in Bangladesh.

METHOD

Research Setting and Subjects

A total of 70 subjects consisting of 40 managers and 30 workers from 6 industries in Dhaka,Bangladesh, were randomly chosen for this study. The mean age of the managers was 45 years and that of the workers was 40 years. The educational qualification of the managers varied from secondary to postgraduate levels and that of the workers from primary to secondary school levels.

Measuring Instruments

A questionnaire composed of 37 items was used to measure the important causes of industrial accidents and effective preventive strategies as perceived by the industrial workers and the managers. The questionnaire consisted of two subscales: (1) "causes of accidents", and (2) "preventive strategies". The subscale "causes of accidents" consisted of 27 items dealing with 27 possible causes of accidents, and the subscale "preventive strategies" consisted of 10 items dealing with 10 possible preventive techniques. The subjects could respond to every item of each of the two subscales by giving a tick (✓) mark on a 5-point scale to indicate either the degree of importance of a particular item as a cause of accident or the degree of effectiveness as a technique for preventing accident. The questionnaire was developed partially on the basis of items taken from the relevant literature on industrial accidents (Blum and Naylor,1968; ILO,1978), and partially on the basis of a pilot survey of opinions of the workers and the managers. The questions which appeared difficult and confusing to them (the workers and the managers) were dropped.

RESULTS

The results of the present study are presented in the following tables. The rank orders of relative importance of different causes of industrial accidents as perceived by the workers and the managers are presented in Table 1. The results in table 1 show that the workers considered mechanical failure as the most important and noise as the least important causes of accidents. On the other hand, the managers considered inexperience and mechanical failure as two top most important causes and slow reaction as the least important cause of accident. The rank orders of relative effectiveness of different preventive strategies as perceived by the workers and the managers are presented in table 3. This table shows that the workers considered incentive as the most effective and persuation as the least effective preventive strategies of accidents. On the other hand,the managers considered proper inspection as the most important and training as the least effective preventive strategies of accidents. The differences between the mean rating scores of the causes of accidents,in terms of their relative importance, and those of the preventive strategies,in terms of their relative effectiveness, are given in table 2 and 4 respectively. The results in these two tables reveal that there are more significant differences between the opinions of the workers and the managers about the effective preventive strategies than the important causes of accidents.

Table 1. Rank order of the causes of industrial accidents (in terms of importance) as perceived by the workers and the managers.

Description of the items	Rank order of the workers	Rank order of the managers
Mechanical failure	1	1.5
Inconvenient work schedule	2	6
Ill health	3	8
Faulty handling of tools	4	3
Irresponsibility	6.5	8
Organic disease	6.5	14.5
Anxiety and depression	6.5	19.5
Insufficient light	6.5	22
Struck by falling objects	9	8
Inexperience	10	1.5
Faulty attitudes	12	4
Frustration	12	5
Nervousness	12	21
Heat	14.5	14.5
Poor ventilation	14.5	19.5
Defective vision	16.5	10.5
Fatigue	16.5	17.5
Intoxication	19	12.5
Bad temper	19	17.5
Grief	19	24
Exuberance	21	23
High blood pressure	23.5	10.5
Senility	23.5	12.5
Slow reaction	23.5	25
Falling	23.5	27
Dust	26	26
Noise	27	16

Note. Spearman's rank correlation (rho): $p = 0.58 (t = 3.57; p < .01)$.

Table 2. Differences between the mean rating scores of the causes of accidents (in terms of importance) as perceived by the workers and the managers.

Description of the items	$\bar{X}$ scores of the managers	$\bar{X}$ scores of the workers	t	p
Faulty handling of tools	3.55	3.03	3.08	<.01
Struck by falling objects	2.35	2.83	2.53	<.01
Faulty attitudes	3.25	2.76	2.43	<.02
Inexperience	3.67	2.86	1.68	N.S.
Mechanical failure	3.67	3.43	1.12	N.S.
Falling	1.92	2.00	0.56	N.S.
Slow reaction	2.05	1.93	1.29	N.S.
Irresponsibility	2.53	3.23	0.69	N.S.
Work schedule	2.42	3.50	1.07	N.S.
Insufficient light	2.15	2.96	0.81	N.S.
Heat	2.19	2.96	0.32	N.S.
Noise	2.25	2.60	1.00	N.S.
Poor ventilation	2.20	1.66	0.40	N.S.
Dust	2.02	2.60	0.70	N.S.
Anxiety and depression	2.20	1.66	0.68	N.S.
Ill health	2.35	2.93	0.18	N.S.
Bad temper	2.22	3.03	0.56	N.S.
Frustration	2.52	3.03	0.56	N.S.
Nervousness	2.17	0.36	2.94	<.01
Defective vision	2.32	2.36	0.61	N.S.
Organic disease	2.27	2.70	2.56	<.02
High blood pressure	2.32	2.73	1.53	N.S.
Senility	2.30	2.56	1.37	N.S.
Fatigue	2.22	2.06	1.37	N.S.
Exuberance	2.12	2.26	0.25	N.S.
Intoxication	2.30	2.00	1.87	N.S.
Grief	2.11	2.30	1.54	N.S.

Table 3. Rank order of the preventive strategies (in terms of effectiveness) of industrial accidents as perceived by the workers and the managers.

Description of the items	Rank order of the workers	Rank order of the managers
Incentives	1	8
Improvement of work environment	2	5.5
Use of safety equipments	3.5	2.5
Accident insurance	3.5	8
Standardization of equipments	5	5.5
Safety training	6.5	10
Proper health care facilities	6.5	8
Enforcement of safety regulations	8	4
Proper inspection	9	1
Persuation of workers	10	2.5

Note. Spearman's rank correlation (rho): p = -0.41 (t=2.81; p<.05).

Table 4. Differences between the mean rating scores of the preventive strategies (in terms of effectiveness)of accidents as perceived by the workers and the managers.

Description of the items	$\bar{X}$ scores of the managers	$\bar{X}$ scores of the workers	t	p
Regulation enforcement	3.45	2.25	5.57	<.001
Standardized equipments	3.08	2.88	1.19	N.S.
Proper inspection	3.90	2.33	9.01	<.001
Safety training	3.39	2.73	2.99	<.01
Persuation of workers	2.87	2.07	9.98	<.001
Incentives	2.56	3.55	4.80	<.001
Accident insurance	2.50	3.17	2.61	<.01
Safety equipments	2.87	3.04	0.78	N.S.
Better work environment	2.95	3.33	2.19	<.05
Health care facilities	2.50	2.63	0.28	N.S.

DISCUSSION

The analysis of the results of the present study shows that there is a significant positive correlation between the ratings of the workers and the managers concerning the important causes of industrial accidents(see table 1). Similarly, the results in table 2 show that in rating the importance of 27 possible causes of industrial accidents, the workers and the managers have differed significantly only in 5 items. Thus it is evident from the results that though there are some differences between the perceptions of the workers and the managers, yet there are a number of general agreements between them concerning the important causes of accidents. The results in table 1 show that both the workers and the managers consider mechanical failure, faulty handling of tools, poor working condition, inconvenient work schedule, irresponsibility, inexperience, frustration, anxiety, ill health and the faulty attitude of the workers as some of the important causes of industrial accidents. However, the results concerning the ratings of effective preventive strategies show that there are lesser agreement and greater difference between the perceptions of the workers and the managers. This is evident from the significant negative correlation between the ratings of the two groups about the effective strategies for prevention of accidents (see table 3). This finding is also supported by the results in table 4 which show that in rating the effectiveness of 10 preventive strategies the workers and the managers have differed significantly in 7 items. The results concerning the ratings of effective preventive strategies show that the workers perceived incentives,improvement of working condition, good safety equipments, accident insurance and safety training as some of the effective safety strategies. on the other hand, the managers considered inspection, persuation of the workers for the use of safety equipment, enforcement of safety regulations and standardization of equipments as some of the effective safety techniques.

Finally, the opinions of the workers and the managers about the causes and prevention of industrial accidents, though subjective,seem to be important and useful for reducing accidents, because they are closely connected with this problem and any accident prevention strategies or plans are implimented through them.

REFERENCES

Blum,M.L. and Naylor, J.C., 1968, Industrial Psychology, (New York: Harper and Row), pp. 516-530.

Fazzio, J.C., 1986, Address on industrial safety and management. In: Proceedings of the Industrial Safety Management,Organized by the Bangladesh Employers' Association in collaboration with ILO,Dhaka, Bangladesh, pp.55-58.

Haque, A.Q.M., 1986, Industrial toxicology: Its hazard and control. In: Proceedings of the Industrial Safety Management,Organized by the Bangladesh Employers' Association in collaboration with ILO,Dhaka, Bangladesh, pp.101-115.

ILO,1978, Accident Prevention: A Worker's Educational Manual,(Geneva: International Labour Office), pp. 2-7.

Khan, M.M., 1986, Inaugural address on industrial safety and management. In: Proceedings of the Industrial Safety Management, Organized by the Bangladesh employers' Association in collaboration with ILO, Dhaka, Bangladesh, pp.62-64.

Mesbahuddin, A.M., 1986, The philosophy of occupational safety and health: Evolution of the concept of occupational safety and health and welfare measures in factories. In: Proceedings of the Industrial Safety Management, Organized by the Bangladesh Employers' Association in Collaboration with ILO,Dhaka,Bangladesh,pp.71-85.

Advances in Industrial Ergonomics and Safety II
Edited by Biman Das
Taylor & Francis 1990

INVESTIGATION OF SELECTED OCCUPATIONAL FATALITIES CAUSED BY FALLS FROM ELEVATIONS

Thomas G. Bobick, Patricia G. Schnitzer, and Ronald L. Stanevich

Division of Safety Research
National Institute for Occupational Safety and Health
Morgantown, WV 26505-2888

Data collected by the National Institute for Occupational Safety and Health (NIOSH), Division of Safety Research (DSR) show that falls from elevations accounted for 7.4% (more than 2,700 deaths) of the traumatic occupational fatalities occurring during the six-year period 1980 to 1985. Of these fatal falls, 48.0% occurred in the construction industries. The DSR Fatal Accident Circumstances and Epidemiology (FACE) program is a passive surveillance system that investigates selected work-related fatalities. From October 1987 through September 1989, 40 deaths resulting from falls from elevated workplaces were investigated under the FACE project. This paper provides an overview of the descriptive data associated with these fatalities.

INTRODUCTION

The Occupational Safety and Health (OSH) Act of 1970 (Public Law 91-596) was established to provide American workers with protection against personal injury and illness resulting from hazardous working conditions. The OSH Act established two separate Federal organizations--the Occupational Safety and Health Administration (OSHA) and the National Institute for Occupational Safety and Health (NIOSH).

OSHA was placed in the Department of Labor and given responsibility for promulgating standards, inspecting workplaces to ensure compliance with those standards, and issuing citations and fines to violators (Millar, 1989). OSHA is the regulatory agency for all industries except mining.

NIOSH was placed in the Department of Health, Education, and Welfare (presently the Department of Health and Human Services) and given responsibility for conducting research in occupational health and safety, for disseminating information resulting from that research, for recommending criteria for standards to OSHA, and for supporting the training of professionals in occupational safety and health. Thus, NIOSH is a scientific institute (Millar, 1989).

NIOSH conducts the Federal government's occupational injury and disease research program. The Division of Safety Research (DSR) serves as the focal point for the NIOSH occupational injury research program.

NIOSH has recently completed collection and analysis of six years of data on occupational fatalities (NIOSH, 1989). These data, known as the National Traumatic Occupational Fatality (NTOF) database, contains information on work-related deaths in the U.S. from 1980 through 1985. This information was obtained from death certificates prepared by all of the 52 U.S. vital statistics reporting units (the 50 states, New York City, and the District of Columbia) that participate in the NTOF project.

Analysis of the NTOF data indicates that a total of 3,491 fatal workplace falls occurred during the six-year period from 1980 through 1985. Occupational falls are commonly classified as: falls on the same level, falls on stairs, and falls from elevations (Ellis, 1988). Of these 3,491 fatal workplace falls, 2,705 (77.5%) were classified as falls from elevations. These 2,705 deaths represent 7.4% of all the traumatic occupational fatalities documented by the NTOF data for the years 1980 through 1985.

Further analysis of the NTOF data reveals that 1,298 (48.0%) of the fatal falls from elevation occurred in the construction industry (Standard Industrial Classification codes 15, 16, 17). More than 80% of the fatal construction falls cluster in two main occupational groups--craftsmen (60.6%) and laborers (21.5%). The annual average fatality rate was 5.2 fall-related deaths per 100,000 construction workers. The service industry (advertising, health services, educational services, etc.) had the next highest number of fatal falls (278 or 10.3%), but had an annual fatality rate of only 0.2 per 100,000 employees.

Additional analysis of the 2,705 fall-from-elevation deaths reveals that 1,104 (40.8%) fell from buildings, 461 (17.0%) fell from scaffolds, 405 (15.0%) fell from ladders, 182 (6.7%) fell through a hole or opening in the work surface, and 553 (20.4%) fell from one level to another (these are falls from elevations that are not otherwise classified).

The Bureau of Labor Statistics' Supplementary Data System (SDS) is a database composed of workers' compensation claims from approximately 27 states that provides information on non-fatal occupational injuries. DSR analysis of the SDS data for the six-year period (1980-85) contains 696,276 compensation claims filed by workers from all U.S. industries (nine major industrial classifications) for injuries sustained from falls from elevations. Interestingly, 146,218 (21%) of these claims were filed by construction workers, yet these workers represent only 5% of the nation's workforce. Of the 146,218 non-fatal, fall-related construction injuries recorded in SDS, 37,704 (27.8%) involved ladders, 24,748 (16.9%) involved scaffolds, and 46,679 (31.9%) involved falls from elevation, NEC (not elsewhere classified). These construction-related data represent 32% of all ladder incidents, 48% of all scaffolding incidents, and 52% of all falls from elevation, NEC incidents that were reported by <u>all</u> U.S. industries. The remaining 37,087 injuries (of the 146,218 construction injuries) represent non-fatal falls on the same level.

Both the NTOF and SDS data have shown that falls from elevations represent a significant cause of mortality and morbidity in the workplace. As a result, NIOSH investigates selected fatal workplace falls as part of the DSR Fatal Accident Circumstances and Epidemiology (FACE) project. This program is aimed at characterizing and reducing the risks associated with fall-related accidents to prevent these types of deaths in the future.

METHODS

The FACE program is a passive surveillance system focusing on the identification and investigation of work-related fatalities. DSR personnel collect data based on an epidemiologic model by examining information on the host (victim), energy source(s) involved, and the work environment. The goal of the project is to prevent future fatalities by identifying potential risk factors, developing recommended intervention strategies, and disseminating findings to reduce the risk in the workplace. The on-site investigations conducted for the FACE project are for research purposes only, and not to determine culpability.

The FACE project relies on a cooperative notification procedure to identify the deaths. Currently, 11 states voluntarily report fatalities to DSR. Since the notification procedure is voluntary, some state agencies are more responsive than others. Also, the agencies responsible for notifying DSR receive their information from different sources, and this is reflected in the reliability and comprehensiveness of their reporting (Conroy et al, in press).

Presently the FACE program identifies general problem areas. These data are helping to establish future research projects. Specifically, the FACE investigations have identified problems related to falls: from scaffolds, through work surface openings, and through skylights. Research projects for these three areas have been initiated in fiscal year 1990. Future data collection may lead to the evaluation of techniques to prevent falls from occurring, and to evaluate personal protective devices to reduce the severity of the injury after the fall has occurred.

From October 1, 1987 through September 30, 1989, a total of 39 incidents, which resulted in 40 fatalities were investigated by DSR personnel. for the FACE project. Data collected from field survey forms, medical examiner reports, and FACE investigative reports were entered into a computer database, edited, and analyzed using the Epi Info program (Centers for Disease Control, 1988). Because of the small number of deaths, tests were not performed to assess statistical significance. However, the results describe the fatal events and identify potential risk factors common to most of these fall-related fatalities.

RESULTS

Descriptive data on victims

Thirty-eight (95%) of the victims were male. The average age at the time of death for all victims was 36 years, with 67% of all victims killed before the age of 40. Fifty percent of the victims had at least a high school education; information was unavailable on the educational level of 16 (40%) victims. Twenty-eight (70%) of the workers were white, six (15%) were black, four (10%) were Hispanic, one (2.5%) was Asian, and one was American Indian.

Employment data

Thirty-eight (95%) of the victims were employed by private companies, while one was a city government employee, and one was self-employed. The victims had been employed with their respective employers an average of 4.6 years (range: 1 mo to 40 yr), but the median length of employment was 18 months. Thirty-one (77.5%) had worked for their employer five years or less.

Thirteen (32.5%) of the victims had been with their employer a total of six months or less. In addition, thirteen of the victims had spent three months or less in the job they were performing when they fell.

Thirty-five (87.5%) of these workers were employed in the construction industries. Three major groups (Standard Industrial Classification codes 15, 16, 17) are included in construction: (a) residential (single-family houses, apartment buildings, hotels, motels, condominiums) and non-residential (commercial, industrial, institutional, recreational, religious, etc.) building contractors (SIC code 15), (b) non- building construction, such as highway, street, bridge, tunnel, pipeline, power line, dam, etc. construction contractors (SIC code 16), and (c) special trade contractors (plumbing, heating, painting, papering, electrical, plastering, carpentry, roofing and sheet metal work, concrete work, structural steel erection, glass and glazing work, etc.) (SIC code 17). Of the remaining 5 workers (12.5%), one worked in the electrical services industry; one was a warehouse shipping/receiving clerk; two worked in the miscellaneous repair services (welding repair and other repair services); and one was employed in general government (city maintenance).

Regarding the 35 construction employees, 10 worked for general building contractors (SIC 15); one was constructing residential buildings, the other nine were constructing non-residential buildings. One victim worked in the heavy construction industry (bridge, tunnel, or elevated highway work) (SIC 16). The remaining 24 were employed with special trade contractors (SIC 17).

Occupational data

The 40 victims comprised 17 different occupational classifications. Thirty-one (77.5%) of the victims worked in eight occupations. The remaining nine victims worked in nine different occupations (Table 1).

Table 1. Occupations of victims of falls from elevations investigated by the NIOSH Fatal Accident Circumstances and Epidemiology project, fiscal years 1988-1989.

Occupation	Number	Occupation	Number
Structural metal craftsmen	7	Bridge work	1
Painter	6	Caulking	1
Supervisor, Manager	5	Cement finisher	1
Carpenter	4	Dry wall finisher	1
Roofer	3	Plasterer	1
Plumbing	2	Shipping, receiving	1
Electrical work	2	Street work	1
Sheet metal installer	2	Welding	1
		Window repair	1

Fall information

Workers fell from a variety of structures, with roofs being the most common (14 or 35%). Falls from structural steel resulted in nine fatalities (22.5%), falls from scaffolds caused six deaths (15%), four victims (10%) fell from ladders, and seven victims (17.5%) fell from other types of structures.

The average distance fallen was 58.6 ft (range: 3 ft to 396 ft). The median distance was 31 ft, with 15 (37.5%) falling 25 ft or less (three fell less than 10 ft), 14 (35%) falling 26 to 50 ft, and 11 (27.5%) falling more than 50 ft. The majority of workers (23 or 57.5%) fell to a concrete, rock, or asphalt surface, followed by 10 workers (25%) falling onto grass or dirt. Four victims (10%) fell to some type of metal surface, two (5%) fell onto boxes, tools, or work materials, and one worker (2.5%) fell onto loose soil.

Work surface collapse preceded the fall of eight (20%) of the victims, while three other falls (7.5%) were attributed to equipment failure (previously damaged lanyard failed; lanyard improperly tied off; forklift carriage jammed on mast). Four falls (10%) were the direct result of slipping or tripping, and in 17 cases (42.5%), the only recorded action preceding the fall was that the victims lost their balance.

Workers were engaged in various activities prior to the fatal event. Thirteen (32.5%) were using tools or equipment and 11 (27.5%) were lifting or carrying items. In five cases (12.5%) the victim was walking or running, in four cases the victim was being transported by a vehicle, and in two cases (5%) the victim was climbing. In the remaining five cases (12.5%), the victim's activity was unknown.

Thirty-six (90%) of the victims were working with at least one other person at the time of their death. Thirty-two (80%) were performing their usual duties. The worker was reportedly following standard operating procedures at the time of death in 23 (57.5%) of the cases. Twelve (30%) workers were not following standard operating procedures, and in the remaining five cases, it was not known whether the victim had followed standard procedures or not. Thirty-eight (95%) of the fatal falls occurred during daylight from a dry surface.

Employer information

One of the incidents involved two workers, thus, a total of 39 investigations were conducted. The 39 employers had an average time in business of 24.5 yr (range: 6 mo to 100 yr). Fifty percent of the employers had been in business 20 yr or less, while 28% had been in business more than 40 yr. The average number of workers employed by these companies was 175 (range: 2 to 2,500, median: 40). Twenty-six (67%) companies had less than 50 employees, and fully 80% of the employers had fewer than 100 workers. Nine (23%) of the 39 companies had less than 10 employees.

Safety information

Sixteen of the 39 companies (41.0%) reported having no written safety rules or procedures, and 22 (56.4%) reported that they did not have a safety officer for the company. Six (15.4%) of these employers provided no safety training for the worker, while 27 (69.2%) provided only on-the-job training (OJT). The remaining six employers provided either classroom training, classroom plus OJT, or classroom, OJT, and training manuals. Twenty-five (64.1%) of the employers did not have an emergency response procedure for the workforce to follow.

Personal protective equipment information

In most situations, personal protective equipment was not required by the employer. Only 13 employers (33.3%) required the use of personal fall protective equipment. Twenty-six (66.7%) of the companies required no form of personal fall protection for their employees while working at elevations.

Although 13 companies required personal fall protection to be used, it was worn in only five instances. Of these five, only two workers were tied off when the fatal fall occurred. In one case, the lanyard had been damaged previously and failed when subjected to the stress of the fall.

In the other case, the lanyard had not been secured properly to the anchor point and the snap hook came open when subjected to the load imposed by the worker's fall.

SUMMARY

During the period October 1, 1987 through September 30, 1989, a series of 40 fatalities that were caused by falls from elevations were investigated by NIOSH's Division of Safety Research. These fatal investigations were conducted as part of the Fatal Accident Circumstances and Epidemiology project. Analysis of the data collected indicated that:

1. Thirty-five (87.5%) of the workers killed were employed in the construction industries.

2. The 40 victims worked in 17 different job classifications. Eight occupations accounted for 31 (77.5%) of the fatalities; the other nine victims worked in nine different occupations.

3. Fourteen of the workers (35%) fell from roofs; nine (22.5%) fell from structural steel; six (15%) fell from scaffolds; four (10%) fell from ladders; and seven (17.5%) fell from other types of structures.

4. Thirty-two (80%) of the victims were performing their usual duties. The other eight were doing something out of the ordinary.

5. Twenty-three (57.5%) of the victims were reportedly following standard operating procedures; twelve (30%) were not.

6. Thirteen (32.5%) of the victims had been with their company six months or less. Moreover, 13 of the victims had spent three months or less in the job they were performing when they fell.

7. A total of 39 investigations were conducted (one incident had two victims). Sixteen (41.0%) of the 39 companies reported having no written safety rules or procedures; twenty-two (56.4%) of the companies did not have a safety officer.

8. Six (15.4%) of the employers provided no safety training for the worker, while 27 (69.2%) provided only on-the-job training.

9. Twenty-five (64.1%) of the companies did not have an emergency response procedure for the workforce to follow.

10. Twenty-six (66.7%) of the companies required no form of fall protection for their employees while working at elevations, only 13 (33.3%) employers required the use of a safety belt, lanyard, and lifeline combination for the type of work the victim was conducting at the time of the fatal fall.

CONCLUSIONS

Falls from elevations in the construction industry is a ubiquitous safety problem. The common thread in fall-related deaths is that if workers who labor at elevations do not use fall prevention techniques or personal protective equipment, they are at an increased risk of sustaining a serious injury or being in a fatal accident. The attitudes and behaviors of management and the workforce have to be oriented so safe work practices are developed, implemented, enforced, and remain priority considerations.

RECOMMENDATIONS

Policies and procedures

Companies of all sizes that routinely have employees working at elevations should have written safety policies and procedures that the workers are required to follow. At a mininmum, the policies and procedures should address the hazards to which the employees are exposed, the techniques for preventing falls, the types of fall protection to be used, training on how to properly use and inspect fall protection gear, and the disciplinary action for failure to comply with the policies and procedures regarding the use of this equipment. Policies and procedures should be reviewed and updated annually.

Training

In addition, these companies should use a fall prevention/protection training program, which is developed and endorsed by all levels of management, for each different job environment that is encountered. The training program should identify and discuss as many job hazards as possible for the employees who will be required to work at elevations. As part of the program, companies should provide on-the-job training, as often as necessary, that will be sufficiently thorough to provide a safe working environment for all employees. In addition, companies could consider using training programs for specific trades, such as any of the Department of Labor's approved Apprenticeship Programs.

Personal fall protection equipment

Finally, employees that have to work at elevations should view personal fall protection equipment as important as automobile seat belt systems. Ideally, appropriate fall protection devices should be used at all times, regardless of the height of the workplace. Economic reality and workplace feasibility may compromise this goal. However, analysis of this limited data set indicates that fatal falls occurred from heights as low as three, seven, and eight feet.

Most importantly, management must provide adequate fall protection systems and/or appropriate personal protective equipment to all workers who are exposed to fall hazards. Similarly, workers should share in the responsibility of protecting themselves when working at elevations by requesting and using the company's fall protection equipment. In addition, workers should be held accountable if the employer's written safety policies and procedures are violated.

ACKNOWLEDGMENTS

The authors wish to extend their appreciation to Catherine A. Bell and Nancy A. Stout, Data Analysis Section, DSR, NIOSH for their insightful and thorough analyses of the NTOF and SDS data bases. In addition, appreciation is extended to Tim Pizatella, Dave Nestor, Herb Linn, and Dwayne Smith, all of DSR, NIOSH, and to Darlaine Taylor, Century Steel Erectors, Inc., West Mifflin, PA and Roy Gurnham, OSHA, Washington, D.C. for their review of this paper and all of their helpful comments and suggestions.

REFERENCES

Centers for Disease Control, U.S. Department of Health and Human Services, 1988, Epi Info Computer Programs for Epidemiology.

Conroy, C., Manwaring, J.C., and Sharifpour, S.P., 1989, Occupational confined space related fatalities. surveillance and prevention. Journal of Safety Research (in press).

Ellis, J.N., 1988, Introduction to Fall Protection (Des Plaines, IL. American Society of Safety Engineers), p. 2.

Millar, J.D., 1989, "What it's like to be Director of NIOSH," transcript of speech presented at Seventh Annual Occupational Health and Safety Institute, Midwest Center for Occupational Health and Safety, University of Minnesota, Minneapolis, MN, September 7, 1989.

National Institute for Occupational Safety and Health, U.S. Department of Health and Human Services, 1989, National Traumatic Occupational Fatalities: 1980-1985, DHHS (NIOSH) Publication No. 89-116, 28 pp.

Advances in Industrial Ergonomics and Safety II
Edited by Biman Das
Taylor & Francis 1990

SAFETY IN PLANT DESIGN AS APPLIED TO HEAT TRANSFER FLUID PLANTS

Lasse Lehtonen

Occupational Safety Engineering
Tampere University of Technology
P.O.Box 527, SF-33101 Tampere, Finland

There are few recommendations to ensure attention to safety in commercial plant design. For this purpose design performance was studied. On the basis of a literature review, a systematic safety-related approach to plant design was composed. This approach was applied to heat transfer fluid plants. The data were collected from safety regulations, injury reports, safety analyses, and, the focus of this paper, interviews of designers. The results indicated that commercial plant design has not yet been organized with respect to the implementation of safety objectives.

INTRODUCTION

Design error has been found to be an important cause of accidents, probably more important than is generally realized. Haastrup (1983) observed that 25 % of case reports of accidents show a relationship between the accident and an error in design.

Theoretically, in a cybernetic system, design controls safety in a plant when adequate knowledge on safety is available (Kuhlmann, 1981). Safety requirements, case histories of accidents, and safety analyses are the inputs of this cybernetic circuit. Design is considered to be a procedural chain of problem solving (Hubka and Eder, 1987). The control of (ideally all of) this process of design and planning is obtained with the use of proper design methods.

The application of a systematic method of design has been reported earlier. Wiendahl (1981) observed that the principles of a standard procedure were transferable to the execution of design but that the individual steps should be applied flexibly. Wallace and Hales (1987) mentioned that a systematic approach was needed to co-ordinate and control the overall design process. The approach had to be adapted to suit each particular problem.

The present study was undertaken because of the importance of design as a safety factor and because of the relatively little knowledge available on safety-related design. The objectives of the study were (1) to develop a safety-related approach to plant design and (2) to examine the attention paid to safety aspects in commercial plant design of heat transfer fluid plants through comparison of used approaches with the developed safety-related approach in interviews of leaders of design projects and the designers themselves.

MATERIALS AND METHODS

The design approach

Literature concerning design methodology and fulfillment was consulted to determine a basis for a safety-related approach to design. Most of the published procedures applied to the interface between the designer and the object to be designed. These procedures were very detailed and suitable for the application of, e.g. a computer. However, within the scope of plant design, engineers work with large assemblies composed of pressure vessels, storage tanks, and pipe lines. Therefore the interface between the leaders of the design project and the designers is very important for overall safety.

Two theoretically established methods were selected for closer evaluation. The first was the German guideline VDI 2221 (1986), "Systematic Approach to the Design of Technical Systems and Products". This method deals with the generally valid principles of design independently of a specific branch of industry. Its essential features are the procedural design stages that produce documentation for checking. The second method was the System Safety Program Plan (Roland and Moriarty, 1983). It calls for design reviews, audits, and safety analyses continuously.

In addition the commercial aspects of design provide a reason for considering quality assurance standards. The international standard series ISO 9000 has the same historical background as the System Safety Program Plan. This series of quality standards will help improve the conformity of products and services in the integrated common market of Europe.

The interviews

Design performance was studied through the use of interviews in a Finnish consulting firm specialized in the design of heat transfer fluid plants. The improved safety-related approach was utilized in the interviews to develop both the theoretical model and the design project. Three project leaders were interviewed, mainly about the general plans and control of design in the firm. They had already acquainted themselves with the safety-related approach earlier. The answers were formed collectively.

Two designers were interviewed half a year later. They were finishing a typical design project of a small heat transfer fluid plant in which the heating capacity was 350 kW. Two liquid-heated jacketed process vessels and a water cooling circuit belonged to the plant in a paint factory.

RESULTS

Safety-Related approach to design

On the basis of the collected data from the literature, a safety-related approach to design was developed. It can be summarized in the form of key words as presented in Figure 1. This presentation is useful in assessing design. The key words represent the activities established for good design (in part A) or the use of relevant safety information and documentation (in part B). The terminology complies with the international quality assurance standards.

Design management	A. ESTABLISHED ACTIVITIES Safety policy Organization Personnel qualification Auditing Design instructions Records Verification Corrective action	
Certification of design task	B. MEASURES IN EACH DESIGN PROJECT Hazard analysis (identifying) based on safety information Safety requirements Design program Specification	
	Contract review	Hazards identified and safety level approved
Creative design	Hazard analysis (preliminary) Application of safety information Functional structure Principal solution	
	First design review	Safe principle chosen for the solution
Layout design	Safety analysis (functional level) Application of safety requirements Module structure, sub-assemblies Preliminary layouts	
	Second design review	Hazard restraining measures used
Detail design	Safety analysis (detail level) Operating instruction analysis Definitive layouts Design result documentation	
	Critical design review	Hazards eliminated or controlled

Figure 1. Safety-related approach to plant design.

Interview of the leaders

The three project leaders interviewed in the consulting firm were mainly asked questions concerning their firm's approach to safety in design in comparison with the safety-related approach developed in the present study (Figure 1). A summary of their responses follows.

The leaders stated that safety policy existed in their firm in a written form and that it referred to safety analysis methods. Contact with the safety authorities was emphasized.

They stated that the organization in respect to project execution would be explained to a client at the bidding stage. However, a designer would freely contact other designers in order to divide the subassemblies of the plant to be designed.

In their firm no special qualifications were required of their personnel in respect to safety. The personnel were reported to gain their knowledge through experience. The firm had two persons trained in safety analysis. The resumés of the designers were presented to clients in lieu of certificates of qualification.

The firm did not carry out internal audits of progress, and their clients did not audit the progress of the design either. The leaders were of the opinion that auditing might increase the designers' motivation to emphasize safety, however.

At the time of the interview, the firm's design instructions were under revision, but the firm was already subscribing to the reports of the Finnish Standards Association. In addition the leaders gave their own instructions to the designers on technical aspects of the project.

The leaders reported that some records of the design were maintained and that the documents were stored in the firm's archives. Concepts and calculations were saved as a part of the documentation. Having to rely on one's memory sometimes caused trouble, however.

No attempt was made to verify the procedures used, and no competitive calculations were made. The quality of the finished design, i.e. its correctness, economy and safety, would be proved after its implementation.

The firm had no organized procedures to detect, investigate or prevent safety problems. When an error was detected, corrections were made as a matter of routine.

All in all, the project leaders were of the opinion that their firm's approach to design was in line with the approach developed in the present study but that the approach was not clearly discernible. In addition their contract with a client sometimes covered the start-up phase of the designed plant, and this feature had a positive effect on safety in general.

A review of a design was arranged both before and after a project. These reviews were not however distinguishable from ordinary progress meetings. The review that took place after the completion of a project was supplemented by a check of the drawings, but no minutes of the review were kept.

The leaders reported that safety information was actually a part of the design. The design instructions of the firm contained information on the national safety standards and also on the German standards on heat transfer fluid plants (Germany being the largest equipment exporter to Finland). In addition the clients were given consultation on the procedures needed to gain approval and permits from the authorities.

Every object of design was reported to be subject to a safety analysis, and, if the client had had no prior experience with the kind of plant being designed, a comprehensive hazard and operability study analysis was recommended.

Interview of the designers

In the same type of interview made of the two designers of a specific project a half year later, the designers offered the following information.

The designers reported that the initial information for the design had been given to them by the client. Neither the design program nor the specifications were put in written form. The work was contracted through the use of the flow sheet of the plant. No review of the contract was arranged in the firm because the project was considered to be a routine one. On the basis of their experience the fuel-burning heater had been placed outside of the buildings in which flammable liquids were handled.

The creative phase of the project had been brief. The functional structure of the plant was quickly determined because only a few alternatives were applicable and proved. No actual safety analysis was carried out. The principle solution to this plant was based on the knowledge of a single designer. No reviews were needed.

In the layout phase the national safety standard was followed. The designer employed his own analysis of possible problems with a "what if" approach. The work was divided into modules, and subassemblies, e.g. for electrical installation and ventilation, were processed independently. The two designers contacted specialists of the client directly. The heater and two process reactor vessels were supplied in a ready to use form by subcontractors. No reviews were held, and no minutes were kept.

In the detail design phase, the types of standard parts, e.g. shut and control valves, were approved by the client. Flanged pipe joints were replaced by welded piping for safety because the organic heat transfer medium might cause auto-ignition if a leak occurs and the medium is absorbed by the insulation. No hazard and operability study analysis was carried out. The finished design contained drawings, lists of materials, advice for ordering, and piping instructions. An operational manual was still under discussion at the time of the interview. When the designer had presented the finished design to the client, it was not obvious whether or not the client was capable of evaluating the finished product.

CONCLUSIONS

In commercial design the client should make a specified order, the leader of the design project should be knowledgeable of design procedures and the object being designed and, finally, the client should be able to evaluate the complete design results. If these conditions are not met, disaster could result.

The developed safety-related design approach of this study focused on the planning and control of design. It emphasized the reliability of the design. In this aspect, the approach is clearly analogous to the international quality assurance standards. A parallel can be drawn between safety and quality because of the theoretical background. This relationship between safety and quality has attained ever greater importance due to the commercial aspects of today's Europe. It is also useful because safety awareness at the executive level is where quality awareness was a decade ago (Dumas, 1987).

In this study the design of the heat transfer fluid plant was carried out on a small scale. Such a project is not the best format for examining the plant design. In spite of the scale of the project, defects and lack of systematization in the handling of safety aspects were clearly exposed. The responses of the project leaders showed that the organization of the firm was not very well controlled. Some of the firm's activities were underdeveloped when compared with the approach of this study (Figure 1). The responses of the designers showed that, at least in this small project, the design was processed without any co-ordinating control. No attention had been paid to possibilities to trace the solutions in the various design phases. No safety analyses had been employed either, even though such analyses were supposedly a part of the firm's safety policy. The designers made decisions on safety matters on the basis of their experience only. The observations of this study with respect to design process are comparable to the results of earlier investigations (Stauffer and Ullman, 1988; Tebay et al., 1984).

It is likely that, in the future, design will be emphasized in the process of creating safe workplaces. Additional empirical research on the subject is needed. Safety analyses have become more known recently, but still the methods employed by a single designer should be examined. A safety-related approach to design should be given further consideration in future studies. Positive and economical results might add safety to design.

REFERENCES

Dumas, R., 1987, Safety and quality: the human dimension. Professional Safety, 32, 12, 11-14.

Haastrup, P., 1983, Design error in the chemical industry. In: 4th International Symposium on Loss Prevention and Safety Promotion in Process Industries, Harrogate, England, 12.-16.9.1983. (Oxford: The Institution of Chemical Engineers, Symposium Series No. 80), pp. J15-J27.

Hubka, V. and Eder, W.E., 1987, A scientific approach to engineering design. Design Studies, 8, 3, 123-137.

Kuhlmann, A., 1981, Einführung in die Sicherheitswissenschaft, (Köln: Verlag TüV Rheinland), 467 p.

Roland, H.E. and Moriarty, B., 1983, System Safety Engineering and Management, (New York: John Wiley & Sons), 339 p.

Stauffer, L.A. and Ullman, D.G., 1988, A comparison of the results of empirical studies into the mechanical design process. Design Studies, 9, 2, 107-114.

Tebay, R., Atherton, J. and Wearne, S.H., 1984, Mechanical engineering design decisions: instances of practice compared with theory. Proceedings of the Institution of Mechanical Engineers, 198B, 6, 87-96.

Wallace, K.M. and Hales, C., 1987, Some applications of a systematic design approach in Britain. Konstruktion, 39, 7, 275-279.

Wiendahl, H.P., 1981, Five years experience with VDI 2222 guideline in a large capital equipment enterprise. Design Studies, 2, 3, 165-170.

XI

INDUSTRIAL BIOMECHANICS

Advances in Industrial Ergonomics and Safety II
Edited by Biman Das
Taylor & Francis 1990

COMPARISON OF 2D AND 3D MODEL PREDICTIONS IN ANALYZING ASYMMETRIC LIFTING POSTURES

Bone, B.C., Norman, R.W., McGill, S.M., Ball, K.A.

Occupational Biomechanics Laboratories
Department of Kinesiology, University of Waterloo
Waterloo, Ontario, Canada N2L 3G1

The purpose of this study was to identify the range of asymmetric lifting postures which may be analyzed by 2D models without a major loss in accuracy. A subject performed lifting postures in which the load position on the floor was rotated in five degree increments away from the sagittal plane. Lifting movements within an 80 degree radius (40 degrees to either side of the sagittal plane) may be analyzed by 2D models without a major loss of accuracy (<10%) in predicting the L4-L5 extensor moment. This range is greater than would be expected from purely geometrical considerations. These findings are of use since 2D models are faster, easier and cheaper than 3D models for use in industry by people with limited biomechanics training.

INTRODUCTION

Industrial lifting tasks involving combinations of trunk flexion, twisting, and lateral bending are considered to be a major cause of low-back injury in the workplace (Frymoyer et al., 1980). Many occupational tasks require some form of lifting and many of the lifting movements performed involve bending and twisting. Since these movements do not take place in a single plane, three-dimensional (3D) biomechanical analyses may be required to gain a complete understanding of the loading on the low-back leading to injury in the workplace. Currently, workplace tasks are analyzed on site using primarily two-dimensional models such as WATBAK from the University of Waterloo (1984) or the Human Strength Prediction model from the University of Michigan, based originally upon an early model by Chaffin (1969).

Two-dimensional biomechanical models of the low-back have proved to be useful in analyzing certain types of occupational lifting tasks and have contributed to the development of recommendations for safe symmetrical lifting by bodies such as The National Institute for Occupational Safety and Health (NIOSH, 1981). However, many lifting tasks performed in the workplace are asymmetric, involving combinations of trunk flexion or extension, twisting, and lateral bending. Due to the lack of empirical study on such movements, NIOSH guidelines currently do not encompass asymmetric lifting tasks.

In response to the need to study asymmetric lifting movements, 3D models have been developed (e.g., Schultz and

Andersson, 1981; Kromodihardjo and Mital, 1986). With 3D analyses, at least two camera views are needed to obtain the necessary kinematic data. Moreover, a more complex musculature must be incorporated into the model and methods for solving equation indeterminacies such as optimization methods must be adopted. The question of what variable or function the body chooses to minimize is still a matter of great controversy. These factors increase the data analysis time since two sets of 2D coordinates must be digitized and then reconstructed in 3D space. Furthermore, the interpretation of 3D data requires more extensive training in biomechanics because of the complexity of modelling assumptions.

In order for biomechanical models to be successfully applied to the study of occupational lifting tasks, a trade-off must be made between cost and accuracy. While 3D models may provide more accurate predictions, they are more expensive in terms of the time required to collect and analyze the data. On the other hand, 2D models are relatively quick and easy to use but may result in errors when attempting to analyze asymmetric lifting tasks. The purpose of this study was to identifying the range of asymmetric lifting postures which can be analyzed by 2D models with acceptable accuracy. Preliminary analysis showed that shortening of limbs and moment arms attributable simply to the geometry of distances apparent to the cameras, did not satisfactorily resolve this issue.

METHODS

Subjects

One 26 year old male subject was used in this study. The subject was chosen to approximate the 50th percentile male in terms of both height and weight (i.e. 1.76 m and 73 kg), according to Canadian norms (1983).

Lifting Postures

Nineteen static lifting postures were performed by the subject. These postures simulated the beginning of a lift from the floor level. For all postures the subject was required to grasp the load with both hands using parallel feet and flexed-knee lifting technique. The initial posture was performed with the load placed on the floor directly in front of the subject (i.e. a sagittal plane lift). For the remaining postures, the load position was rotated to the right away from the sagittal plane in 5° increments from zero to 90°. The distance between the centre of gravity of the load and the mid-point between the subject's feet remained constant as the load was moved away from the sagittal plane. Two of the lifting postures performed are illustrated in Figure 1.

Data Collection

Four video cameras were used to collect the kinematic

data for the lifting postures studied. The cameras surrounded the subject and were oriented at oblique angles to one another. The cameras were supported by tripods and placed on top of tables so that a slightly downward view of the subject was obtained. The joint centers of the arm and leg segments from a lateral perspective were indicated by attaching adhesive markers to the subject's right and left hand, wrist, elbow, shoulder, hip, knee, ankle, and metatarsal joints. Since none of the cameras gave a purely sagittal plane view of the subject, these markers were used only as a guide in locating the joint centre. Markings were placed on the floor to indicate the various positions of the load with respect to the sagittal plane.

The volume in which the lifting postures were performed was calibrated using a cube on which 40 markers were attached with known 3D orientations to one another. The cube was placed in four different locations within the lifting volume. Two of these positions were side by side on the floor and the other two positions were above these, approximately one metre from the floor. Each of these positions was videotaped for approximately five seconds.

Although all postures performed in this study were static, a light was switched on and off during each trial to indicate the frame to be analyzed. This served to synchronize the four video images in the event that the subject changed his posture slightly during the course of a trial.

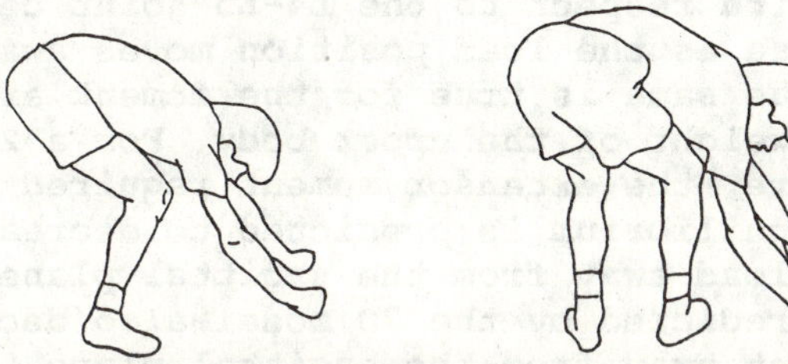

Figure 1. Illustration of two of the lifting postures performed. A) Load positioned in the sagittal plane. B) Load positioned 90° from the sagittal plane.

<u>Data Analysis</u>

A computer-based video digitizing package (PEAK PERFORMANCE, Colorado, USA) was used to determine the 2D coordinates of the 40 markers on the calibration frame for each of the four camera views and calibration frame positions. A Direct Linear Transformation (DLT) modified for multiple calibration positions (Ball and Pierrynowski, 1987) was used to calculate the eleven internal and external parameters for each camera using the known 3D coordinates of the markers on the calibration cube along with the digitized 2D data.

The coordinates for each of the joint centres required for input to the models were digitized directly using the PEAK PERFORMANCE system. The four sets of 2D coordinates (one from each camera view) were combined and reconstructed in 3D space using the camera parameters determined during the calibration procedure. This resulted in a single set of 3D coordinates for each lifting posture. The Z coordinate of each joint centre was discarded for the 2D model analysis.

Comparisons were made between the predictions of a 2D and 3D biomechanical model (WATBAK and 3DWATBAK). The predicted L4-L5 extensor moment and compression force were used as the basis for these comparisons. In comparing the models' output, 10% was used as the criterion for determining an 'acceptable' difference in predicted output. Since the subject did not actually lift the load, a moderately heavy load of 30 kg was simulated on the computer for the purposes of this comparison.

RESULTS AND DISCUSSION

The L4-L5 extensor moments predicted by the two models are compared in Figure 2. This figure also shows the twisting and lateral bending moments predicted by the 3D model. The extensor moments predicted by the two models were nearly identical for first 20° of load rotation from the sagittal plane. After this point, the extensor moment predicted by the 3D model was noticeably larger than that predicted by the 2D model. This difference can be explained by considering that from a sagittal perspective, the moment arm of the load with respect to the L4-L5 joint centre appears to decrease as the load position moves away from the sagittal plane. The same is true for the moment arm of the force due to the weight of the upper body. For a 2D analysis, therefore, the extensor moment required to maintain static equilibrium is predicted to decrease with placement of the load away from the sagittal plane. The extensor moment predicted by the 3D model also decreased with load placement away from the sagittal plane. However, this decrease was not as rapid as that for the 2D model. It is interesting to note in Figure 2 that even with the load placed 90 from the sagittal plane, the "twisting" moment about a long axis through the torso is negligible. Yet, if one were to ask if the video (or Figure 1) showed evidence of "twisting", most observers would say yes. The lateral bending moment in Figure 2 is not particularly high even at the 90 load placement position. These observations occur because the subject reached the load largely by rotating the pelvis and legs but more or less maintaining alignment of the pelvis and shoulders.

Figure 3 shows the differences in extensor moment expressed as a percentage of 3D model prediction, which was assumed to be more accurate. A difference from the 3D model prediction of 10% or less was considered to be acceptable. The 2D model predicted 'acceptable' extensor moments for approximately the first 40° of load placement from either side of the sagittal plane. Beyond this range, however, it appears that 3D analysis is required to maximize the accuracy with which the extensor moment can be predicted.

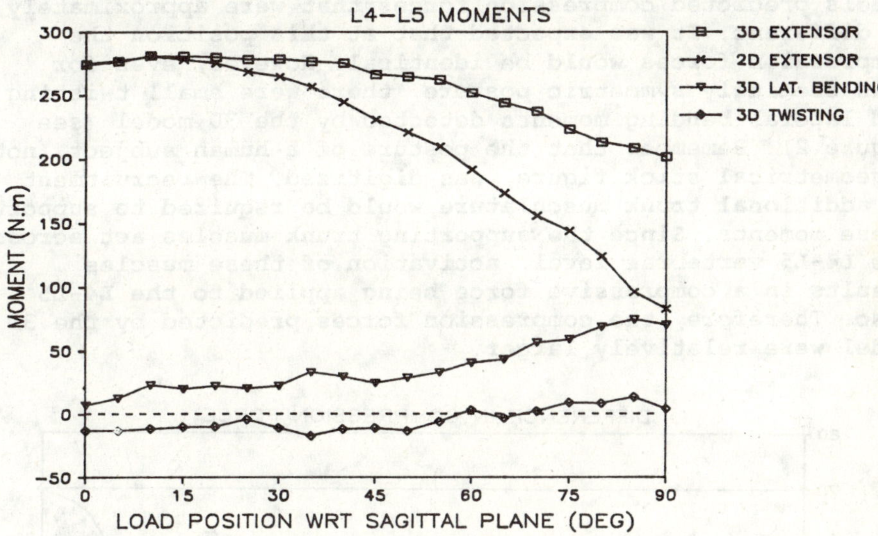

Figure 2. A comparison of the L4-L5 extensor moments predicted by the 2D and 3D models. The twisting and lateral bending moments predicted by the 3D model are also shown.

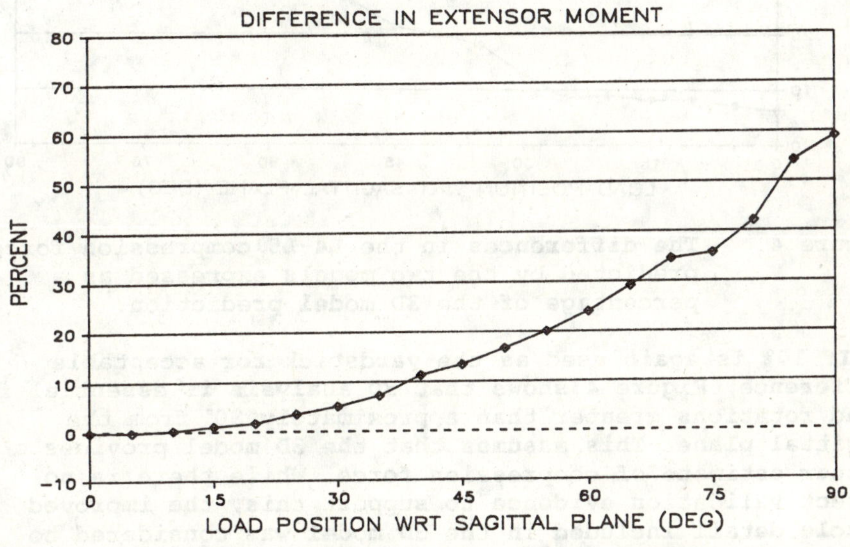

Figure 3. The differences in the L4-L5 extensor moments predicted by the two models expressed as a percentage of the 3D model prediction.

The differences in the L4-L5 compression forces predicted by the two models are shown in Figure 4. The differences have again been expressed as a percentage of the 3D model prediction. It is interesting to note that at the zero degree load position (i.e. the sagittal plane), the two models predicted compression forces that were approximately 7% different. It was expected that at this position the compression forces would be identical. However, even for this seemingly symmetric posture, there were small twisting and lateral bending moments detected by the 3D model (see Figure 2). Remember that the posture of a human subject, not a geometrical stick figure, was digitized. The recruitment of additional trunk musculature would be required to support these moments. Since the supporting trunk muscles act across the L4-L5 vertebrae level, activation of these muscles results in a compressive force being applied to the L4-L5 disc. Therefore, the compression forces predicted by the 3D model were relatively larger.

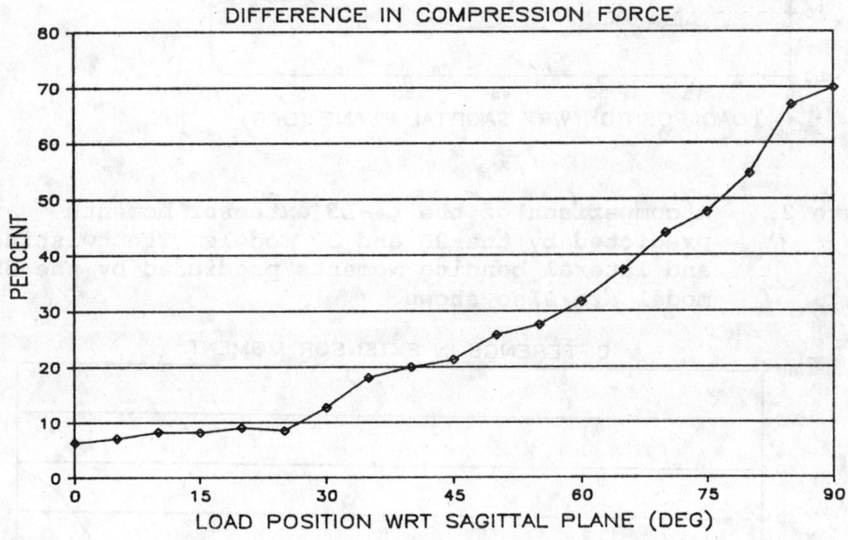

Figure 4. The differences in the L4-L5 compression forces predicted by the two models expressed as a percentage of the 3D model prediction.

If 10% is again used as the yardstick for acceptable difference, Figure 4 shows that 3D analysis is essential for load rotations greater than approximately 30° from the sagittal plane. This assumes that the 3D model provides a better estimate of compression force. While there is no direct validation evidence to support this, the improved muscle detail included in the 3D model was considered to make it a reasonable assumption.

It may be argued that the differences between the 2D and 3D model predictions reported here are merely geometric and could be predicted using simple mathematical relationships.

This argument may be partially true for the moment arm of the load which was constrained to a fixed distance relative to the mid-point between the subject's feet. However, the moment arm of the load with respect to L4-L5 was open to change according to the subject's preferences and the demands of the postures. It was therefore unlikely that changes in this moment arm with load rotation away from the sagittal plane could be described by a simple geometric relationship. This is also true for the moment arm of the force due to the weight of the upper body. For example, simple geometry would predict the moment arm of the load to decrease 10% after approximately 25° of load rotation away from the sagittal plane. However, as reported above for the extensor moment, a 10% decrease did not occur until approximately 40 . Geometrically, it would also be expected that the extensor moment predicted by the 2D model would decrease to zero when the load was positioned 90° from the sagittal plane. However, as Figure 2 shows, this was not the case. This can be explained by the fact that the subject was not allowed to move his feet as the load was rotated away from the sagittal plane. This required him to squat in order to reach the load, which forced his buttocks to protrude posteriorly. Thus even after 90° of load rotation there was still a significant load moment arm with respect to L4-L5, visible to the cameras.

CONCLUSIONS

The conclusions which can be drawn from this study are limited by the fact that only one subject was used. Other subjects might have shown somewhat different postures within the foot and load placement constraints, although not many variations were possible to reach the load. Also, the results presented here are somewhat specific to the 2D and 3D models used in this study.

Notwithstanding these limitations, it appears that 2D models are capable of analyzing asymmetric lifting tasks with 'acceptable' accuracy within a range that is greater than would be expected geometrically. For predicting the L4-L5 extensor moment, the 2D model could be used without a major loss in accuracy for the first 40° of load placement from either side of the sagittal plane. The compression force predicted by the 2D model was found to be 'acceptable' within the first 30° of load placement.

One of major factors in determining the magnitude of the extensor moment is the moment arm of the load with respect to L4-L5. Evidence of this was demonstrated in Figure 2 where the extensor moment predicted by the 2D model was shown to decrease dramatically as the load was moved away from the sagittal plane. This decrease was primarily the result of the 2D model underestimating the moment arm of the load. This has implications for the use of 2D models in analyzing asymmetric lifting tasks.

In order to maximize the accuracy with which a 2D model can predict the extensor moment, the film plane of the camera should be aligned parallel to the line which defines the moment arm length of the load with respect to L4-L5. This will serve to minimize distortion of the load moment arm due to posture asymmetry.

As far as compression forces are concerned, 2D models should only be used to analyze lifting tasks where a flexor or extensor moment predominates. Tasks which involve large twisting and/or lateral bending moments will require activation of trunk musculature which is generally not accounted for by 2D models. Attempts to analyze such tasks with 2D models will result in underestimation of the compression force acting on the disc. This is clearly an unacceptable situation since the disc compression is often used as an index of potential injury. On the other hand, postures which appear to show large twists or lateral bends may, indeed, not result in the calculation of large twisting or lateral bending moments about a longitudinal trunk axis or anter-posterior axis in the plane of lumbar disc surfaces. Awkward postures can be achieved by rotations of the legs with only minimal misalignments of the shoulders and pelvis.

REFERENCES

Ball, K.A., Pierrynowski, M.R., 1987, A modified direct linear transformation (DLT) calibration procedure to improve the accuracy of 3D reconstruction for large volumes. In Biomechanics XI-B, G. deGroot, A.P. Hollander, P.A. Huijing, and G.J. van Ingen Schenau (Eds.), Amsterdam: Free Univesrity Press, 1045-1050.

Chaffin, D.B., 1969, A computerized biomechanical model - development of and use in studying gross body actions. Journal of Biomechanics, 2, 429-441.

Fitness and Lifestyle in Canada 1983, Canada Fitness Survey. Gov't of Canada. Ministry of Health and Welfare.

Frymoyer, J.W., Pope, M.H., Costanza, M.C., Rosen, J.C., Goggan, J.F.,Wilder, D.G., 1980, Epidemiologic studies of low back pain. Spine, 5, 419-423.

Kromodihardjo, S., Mital, A., 1986, Kinetic analysis of manual lifting activities: part I -development of a three-dimensional computer model. International Journal of Industrial Ergonomics, 1, 77-90.

NIOSH Technical Report, 1981, Work practices guide for manual lifting. DHHS (NIOSH) Publication, 81:122.

Norman, R.W. and McGill, S.M., 1984, WATBAK: A computer software package to estimate low back compressive and shear forces. Report to. W.S. Myles,Defence and Civil Institute of Environmental Medicine, Toronto.

Schultz, A.B., Andersson, G.B., 1981, Analysis of loads on the lumbar spine. Spine, 6(1), 76-82.

Advances in Industrial Ergonomics and Safety II
Edited by Biman Das
Taylor & Francis 1990

A COMPUTERIZED BIOMECHANICAL MODEL FOR EVALUATION OF FORCES ON THE LOW BACK IN LIFTING AND PUSHING ACTIVITIES

Soha S. Sadek **Mahmoud M. Farag**

Department of Engineering
The American University in Cairo
P.O.Box 2511
Cairo, Egypt

The present work discusses the results of a computerized biomechanical model that calculates the axial compressive force (C), and intradiscal pressure (σ) acting on the low back vertebrae (L5/S1) and the erector spinae muscle force (E). In the case of lifting, C, σ and E increase with increasing trunk flexion and horizontal distance between the load being lifted and the low back. In the case of pushing, there is an optimum trunk flexural angle at which the computed mechanical parameters are minimum. This optimum angle increases as the load increases.

INTRODUCTION

Low back pain is one of the most frequently met medical problems in industry. Mechanical stress is a significant factor causing and aggravating back pain. Thus, accurate estimation of the forces involved in manual lifting or pushing of loads is of major concern to those planning efficient use of a work-force and a work-place.

THE BIOMECHANICAL MODEL

General aspects of the model

The present work discusses the results of a computerized biomechanical model that quantitatively calculates the axial compressive force and the intradiscal pressure (stress) acting on the low back vertebrae (L5/S1) in addition to the erector spinae muscle force. The model, like others proposed by Anderson, et. al. (1985), Chaffin et. al. (1987), Ghista et. al. (1975), Gilbert et. al. (1985), Ramakrishnan et. al. (n.d.), assumes that the human body can be treated as an articulated linkage system. In the present model, the body is considered to be made up of seven solid links articulated at the six joints of movement. To compute the loads on the lumbar spine created by physical activity, the body is first visualized as being divided into an upper and lower part by an imaginary transverse cutting plane. This is based on the analysis of Schultz and Andersson (1981), Schultz et. al. (1982) and Schultz et. al. (1983). The cutting plane passes through the level of the lumbar spine at which the loads are to be determined, in this case L5/S1. Newton's laws are then applied to the upper part and calculations of the net reaction and the internal forces is carried out.

Computation of the net reaction

To calculate the net reaction, a coordinate system is first established based on Schultz and Andersson's (1981) model. The origin is placed at the center of the vertebral body cut by the imaginary plane and it is assumed that the net reaction acts at this point. The coordinate directions are shown in fig. 1.

The net reaction on the L5/S1 invertebral disc, consists of six components: three force and three moment components. These forces and moments must be supplied by the part of the body below the cutting plane to act on the part above the plane for equilibrium.

The requirements for equilibrium of the resultant forces (F) and resultant moments (M) in the x, y, and z directions are: $F_x = F_y = M_z = 0$

$$F_z = MLOADL + MLOADR + M_1 + M_2 + 2M_8 + 2M_9 + 2M_{10}$$

$$M_x = y_{MLOADL}\ MLOADL + y_{MLOADR}\ MLOADR + y_1M_1 + y_2M_2 + 2y_8M_8 + 2y_9M_9 + 2y_{10}M_{10}$$

$$M_y = (x_8M_8 + x_9M_9 + x_{10}M_{10})_R - (x_8M_8 + x_9M_9 + x_{10}M_{10})_L + x_{MLOADR}MLOADR - x_{MLOADL}MLOADL$$

where MLOADR = Mass of the load in the right hand
MLOADL = Mass of the load in the left hand
M_1 = Mass of the head
M_2 = Mass of the trunk
M_8 = Mass of one upper arm
M_9 = Mass of one lower arm
M_{10} = Mass of one hand
y_i = Distance between the segment i, and the y-axis
x_i = Distance between the segment i, and the x-axis
L = Left
R = Right

For a sagittally symmetric case, the x-coordinate data are not required for the calculations of the net reactions and, therefore the problem may be simplified because four of the six equations, namely, F_x, F_y, M_y and M_z have trivial solutions (ie. $F_x = F_y = M_y = M_z = 0$).

Computation of internal forces

The net reaction needed to equilibrate the upper part of the body is supplied in the lumbar trunk by muscle contractions, connective tissue tensions, intra-abdominal pressure and the resistance supplied by the spine motion segments (Schultz and Andersson, 1981).

Adopting the assumptions made by Morris et. al. (1961), it is assumed that there are only two internal forces that supply the net reaction: a compressive force C on the lumbar motion segment and a tension E in a single equivalent of the erector spinae muscles.

The six equations of equilibrium governing the model in fig. 2 are:

$$F_x = (L_l - L_r)\sin\alpha + S_r$$
$$F_y = (I_l + I_r)\sin\beta - (X_l + X_r)\sin\delta + S_a$$
$$F_z = C - (E_l + E_r) - (R_l + R_r) - (I_l + I_r)\cos\beta - (L_l + L_r)\cos\alpha - (X_l - X_r)\cos\delta$$
$$M_x = y_e(E_l + E_r) - y_r(R_l + R_r) + y_l(L_l + L_r)\cos\alpha - y_o\{(I_l + I_r)\cos\beta + (X_l + X_r)\cos\delta\}$$
$$M_y = x_e(E_r - E_l) + x_r(R_r - R_l) + x_l(L_r - L_l)\cos\alpha + x_o\{(I_r - I_l)\cos\beta + (X_r - X_l)\cos\delta\}$$
$$M_z = y_l(L_r - L_l)\sin\alpha + x_o\{(I_r - I_l)\sin\beta - (X_r - X_l)\sin\delta\}$$

The following geometric data are taken as representative of an average sized adult:

x_r = 3.6 cms	y_r = 10.8 cms	$\alpha = 45°$
x_o = 13.5 cms	y_o = 3.8 cms	$\beta = 45°$
x_e = 5.4 cms	y_e = 4.4 cms	$\delta = 45°$
x_l = 6.3 cms	y_l = 5.6 cms	

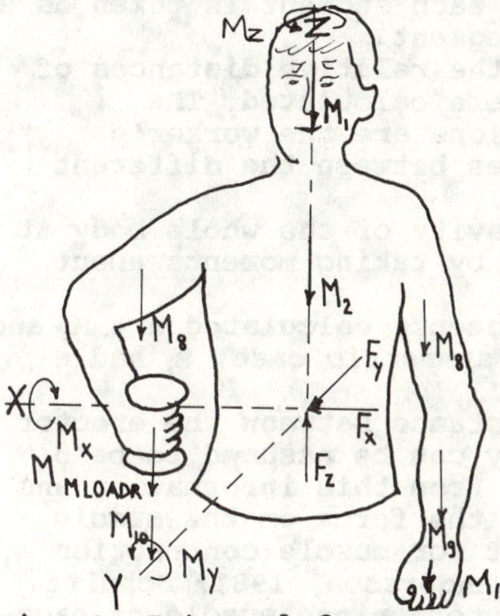

Fig. 1 External forces (Schultz, 1987)

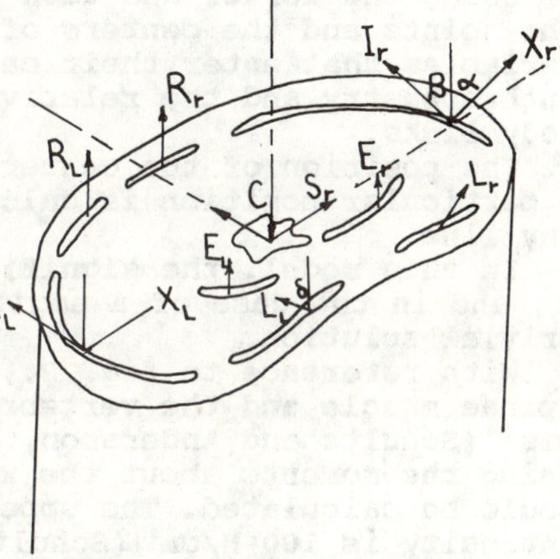

Fig. 2 Internal forces (Morris et. al., 1961)

THE COMPUTER PROGRAM

Based on the above considerations, a 2-D computerized mathematical model was constructed. The different mechanical parameters related to the low-back are calculated using the following assumptions:

1. The strength of the back muscles is a function of their areas and the forces by which they contract to flex or extend the back are dependant on the distance between each of them and the vertebral motion segment. As shown in fig. 2, the distances between most of the muscles and the L5/S1 vertebrae are much greater than the distance between the erector spinae and the vertebral body. Therefore it is assumed that the force supplied by muscles, other than the erector spinae is negligible. This assumption was used by Morris et. al. (1961), Schultz and Andersson (1981) and Schultz et. al. (1982).

2. In the present model, muscle strength is assumed to be the same for all individuals.

3. The back was treated as a solid rod, neglecting its flexibility, lordosis and kyphosis. This assumption was used by most of the available biomechanical models (Anderson, et. al., 1985; Chaffin, 1988A; Chaffin, 1988B, Chaffin and Andres, 1983; Chaffin et. al., 1987; Ghista et. al., 1975; Gilbert et.

al., 1985; Ramakrishnan et. al., n.d.).

__Input Data__

Two types of data are needed in order to start the program:

- The worker data: name, sex, mass and height
- The task data: mass of the load, the frequency of lifting, if the load is carried, and coefficient of friction if the load is pushed. The user is given the choice of either entering body angles for a certain required posture, or entering the position of the load with respect to the worker.

__Calculations__

For the present work, the length and mass of each segment is taken as a factor of the total body length and mass respectively. The center of mass of each segment is taken as a factor of the length of the whole segment.

Using the worker and task data, the relative distances of the joints and the centers of mass are calculated. The variables that enter their calculations are the worker's anthropometry and the relative angles between the different body links.

The position of the center of gravity of the whole body at a particular position is calculated by taking moments about any line.

In this model, the significant moments calculated are M_x and M_y, and in the case of a sagittally symmetric case, M_y had a trivial solution.

With reference to fig. 2, the distance between the erector spinae muscle and the vertebral body can be assumed to be 5 cms. (Schultz and Andersson, 1981). From this information and using the moments about the x-axis, the force on the muscle could be calculated. The upper limit for muscle contraction intensity is 100 N/cm^2 (Schultz and Andersson, 1981; Schultz et. al., 1982). The area of the erector spinae muscle of each side is approximately 24.68 cm^2. This implies that the maximum force the erector spinae muscle can endure without muscle pain is around 2500 N on each side.

The compressive force on the vertebrae concerned is a function of the erector spinae force and the force F_z. If the compressive force exceeds a certain limit given by the NIOSH, the nucleus pulposis protrudes and presses on the neighboring nerves thus causing pain. NIOSH gives two limits for the compressive force, an action limit, 3400 N and a maximum limit, 6400 N.

The stress on the vertebrae concerned was calculated as a function of the compressive force on the vertebrae and its area. If the stress exceeds a certain limit - given by relating the maximum and action limits of the compressive force to the area of the vertebrae - disc degeneration will, or may occur due to overloading.

RESULTS: VALIDATION AND DISCUSSION

In order to validate the model it is necessary to compare it with other available models. After validating the model, other variables that affect the results are analyzed and discussed.

A. Lifting activities

The posture of an individual and the weight of the load during lifting affects the center of gravity of the whole body, the compressive force on the vertebrae, the muscle forces and the intradiscal pressure. In lifting, the line of the center of gravity of the body has to lie within the dimensions of the foot.

Fig. 3 shows the close relation between the stress values predicted using the present model and experimental observations made by Anderson, et. al. (1985).

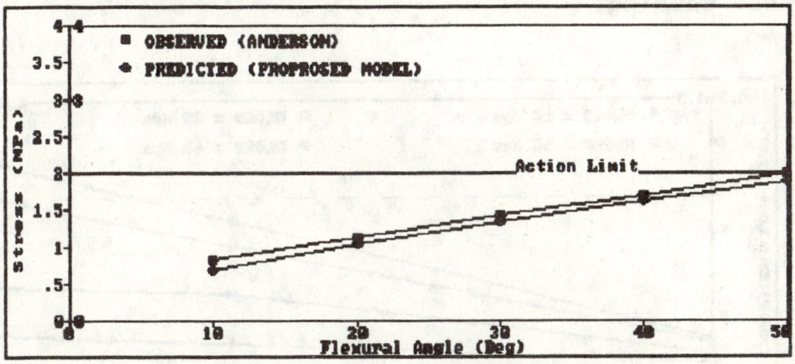

Fig. 3 Effect of flexural angle on stress Comparison of observed & predicted results

The effect of the horizontal distance between the load and the body is shown in fig. 4. The results show that as the load is moved away from the body (as H increases), the compressive force increases until it exceeds the NIOSH Action Limit and even the Maximum Permissible Limit for heavy loads. The above results are comparable to those reported by Chaffin (1988A) as shown in the figure. The relation between the compression and the horizontal distance using the present model was linear whilst that predicted using the model developed by Chaffin (1988A) was slightly nonlinear. This is due to the assumed nonlinear effect of the abdominal pressure which was not included in the present model.

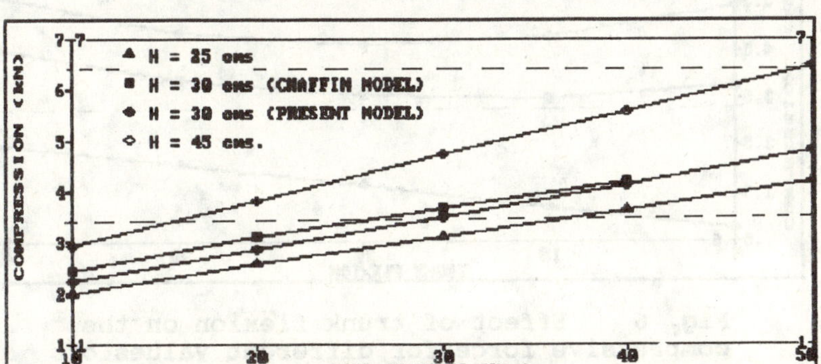

Fig. 4 Effect of horizontal distance on compression

B. Pushing activities

Pushing plays an important role in manual handling tasks. When analyzing pushing activities, both friction and body posture affect the mechanical parameters computed.

Fig. 5 shows that the compressive force increases as the coefficient of sliding friction and pushed load increase. If higher loads need to be pushed without exceeding the action limit, the coefficient of friction can be reduced by substituting sliding friction by rolling friction.

Unlike in the case of lifting, the center of gravity of the whole body does not have to lie within the dimensions of the foot when pushing.

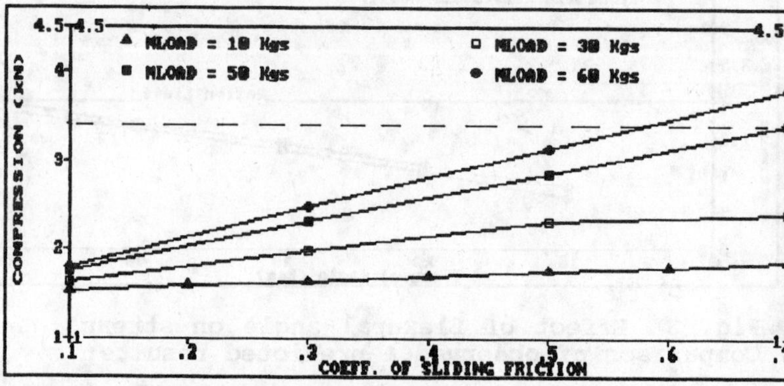

Fig. 5 Effect of coefficient of sliding friction on compressive force for different values of pushed loads

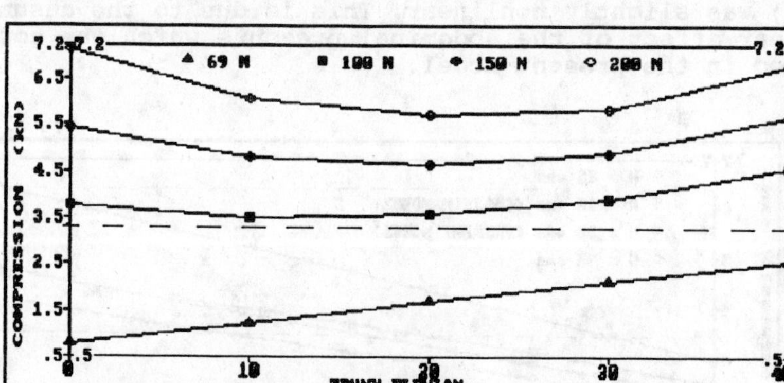

Fig. 6 Effect of trunk flexion on the compressive force for different values of pushed loads

Fig. 6 shows that for the small pushing force (68.67N), the compressive force exerted on the L5/S1 vertebrae increases monotonically with increasing trunk flexion. For larger pushing forces, however, the results show that there is an optimum trunk flexion angle at which the compressive force is a minimum. Fig. 7 shows that this angle increases from 0° for small loads to about 25° for heavy loads. Fig. 8 explains the trend of the relation between the compressive force and the trunk flexion. As the trunk flexural angle increases, the moment due to the body segment weights increases because the arms of forces increase in length; but the moment due to the pushing force decreases with increasing trunk flexion. The compressive force on the vertebrae is a function of the summation of these two moments; and therefore the curve assumes the trend in fig. 7.

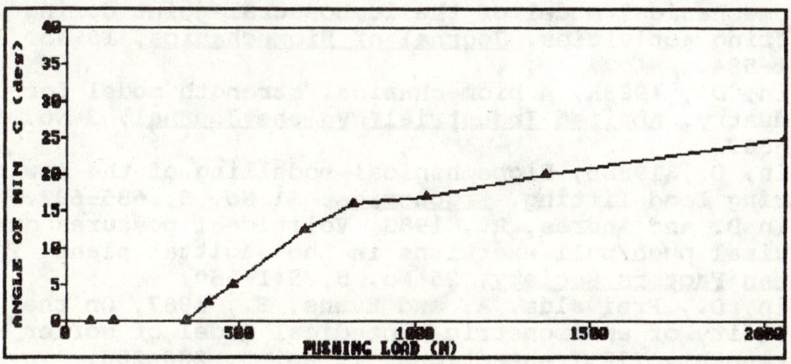

Fig. 7 Variation of the angle of minimum compression with the pushed load

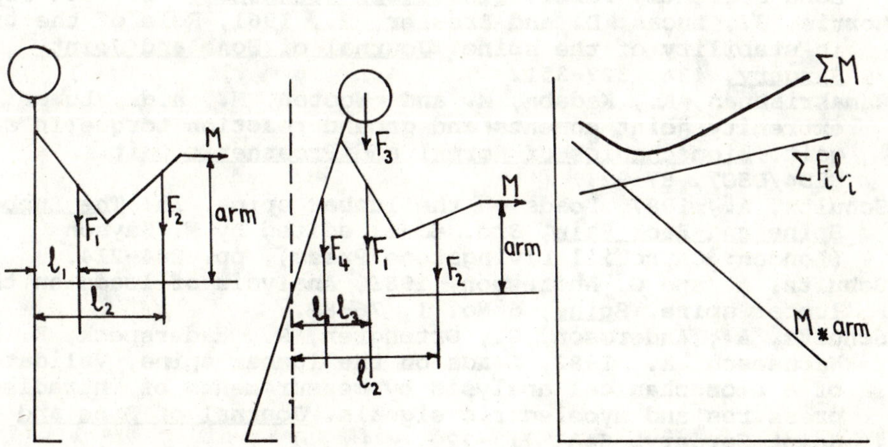

Fig. 8 Moments exerted during trunk flexion in pushing activities

CONCLUSIONS

A computerized user-friendly biomechanical model was constructed for strength assessment of workers before attempting to perform certain manual handling tasks. The model has the ability to analyze certain body postures defined by body angles or load positions. The model can be used to analyze one-handed and two-handed, symmetric and non-symmetric weight holding and pushing activities.

The main limitation of the model is that it cannot deal with handling tasks that require reaching and twisting; though twisting is one of the main causes of low back pain; and being a 2-D model it cannot account for variations in the third dimension.

REFERENCES

Anderson, C., D. Chaffin, G. Herrin and Mathews, L., 1985, A biomechanical model of the lumbosacral joint during lifting activities. Journal of Biomechanics, 18 No. 8, 571-584.

Chaffin, D., 1988A, A biomechanical strength model for use in industry. Applied Industrial Hygiene Journal, 3 No. 3, 79-86.

Chaffin, D., 1988B, Biomechanical modelling of the low back during load lifting. Ergonomics, 31 No. 5, 685-697.

Chaffin D. and Andres, R., 1983, Volitional postures during maximal push/pull exertions in the sagittal plane. The Human Factors Society, 25 No. 5, 541-550.

Chaffin, D., Freivalds, A. and Evans, S., 1987, On the validity of an isometric biomedical model of worker strengths. IIE Transactions, 19 No. 3, 280-288.

Gilbert, J., Maxwell, G., McElhaney, J and Clippinger, F., 1985, A system to measure the forces and moments at the knee and hip during level walking. Journal of Orthopaedic Research, 2, 281-288.

Ghista, D., Toridis, T. and Srinivasan, M., 1975, Human gait analysis: Determination of instantaneous joint reactive forces, muscle forces and the stress distribution in bone segments, Part1. Biomedical Technology. 20, 204-213.

Morris, J., Lucas, D. and Bresler, R., 1961, Role of the trunk in stability of the spine. Journal of Bone and Joint Surgery, 43A, 327-351.

Ramakrishnan, K., Kadaba, M. and Wooten, M., n.d., Lower extremity joint moments and ground reaction torque in adult gait. Biomechanics of Normal and Prosthetic Gait. BED4/DSC7, 87-92.

Schultz, A., 1987, Loads on the lumbar spine. In: The Lumbar Spine and Back Pain, 3rd. edn., edited by M. Kayson (London: Churchill Livingstone Press), pp. 204-214.

Schultz, A. and G. Andresson, 1981, Analysis of loads on the lumbar spine. Spine, 6 No. 1, 76-82.

Schultz, A., Andersson, G., Ortengren, R., Haderspeck, K. and Nachemson, A., 1982, Loads on the lumbar spine. Validation of a biomechanical analysis by measurements of intradiscal pressures and myoelectric signals. Journal of Bone and Joint Surgery, 64A, 713-720.

Schultz, A., K. Haderspeck, D. Warwick and Portillo, D., 1983, Use of lumbar trunk muscles in isometric performance of mechanically complex standing tasks. Journal of Orthopaedic Research, 1, 77-91.

Advances in Industrial Ergonomics and Safety II
Edited by Biman Das
Taylor & Francis 1990

BIOMECHANICAL MODELING OF SYMMETRIC LIFTING TASKS IN CONSTRAINED LIFTING POSTURES

by Sean Gallagher, Christopher A. Hamrick, and Arnold C. Love

U.S. Bureau of Mines
P.O. Box 18070
Cochrans Mill Rd.
Pittsburgh, PA 15236

Twelve male subjects participated in an investigation of the biomechanical stresses of symmetric lifting in stooped and kneeling postures. Electromyographic (EMG) data of eight trunk muscles were collected during various lifting tasks. These data were input to a dynamic biomechanical model and used to predict compression and shear forces at the L_3 level of the lumbar spine. Results indicated that peak compression and shear forces are significantly affected by posture, height of lift, and the amount of weight lifted.

INTRODUCTION

The height of an underground coal mine is generally determined by the thickness of the mineral seam. In many cases, the seam may not exceed 1.2 m in height. A mine with such a seam thickness may be termed a "low-seam" coal mine. Workers in such mines are often required to lift heavy materials in severely restricted postures. The two most common postures used for lifting in low seam coal mines are stooping or kneeling on two knees. Both postures may result in significant compressive and shear loading on the lumbar spine due to flexion and rotation of the vertebral column. Demands on the supporting structures of the column (i.e., paravertebral musculature and ligaments) would be expected to differ considerably in these postures compared to traditional industrial lifting situations, where an unrestricted lifting posture may be utilized.

Previous Bureau of Mines studies have examined the psychophysical lifting capacity, physiological cost, and electromyography (EMG) of trunk muscles of underground miners in stooping and kneeling postures (Gallagher, *et al.*, 1988, Gallagher and Unger, in press). Other investigators have researched the intra-abdominal pressure associated with lifting in restricted postures (Davis and Troup, 1966; Davis and Ridd, 1981; Ridd, 1985; Sims and Graveling, 1988). Unfortunately, establishing the compressive and shear forces associated with lifting in these postures has been difficult due to the fact that many biomechanical models currently available have not been validated for restricted postures. However, a recent EMG-driven biomechanical model has been developed which has the benefit of being sensitive to changes in muscular recruitment in different postures (Reilly and Marras, 1989). The purpose of this study was to use a dynamic biomechanical model to estimate compressive and shear loading on the lumbar spine when lifting in stooped and kneeling postures.

METHOD

Subjects

Twelve healthy male subjects (*M* = 35.7 years of age ± 6.8 S.D.) participated in a study examining the biomechanics of lifting in restricted postures. All subjects were volunteers and received a physical examination and graded exercise tolerance test prior to taking part in the testing. Nine of the subjects were experienced underground miners, while three other subjects were volunteers from the Pittsburgh Research Center. Subjects operated under terms of informed consent.

Experimental design

Three independent variables were manipulated in this experiment -- posture (P) for the lift (stooped or kneeling), height (H) to which the box was lifted (35 or 70 cm), and weight (W) of the lifting box (15, 20, or 25 kg). A within subjects repeated measures design was employed with the order of treatments being counterbalanced using a latin square. EMGs of eight trunk muscles (l. and r. erectores spinae, latissimus dorsi, external oblique, and rectus abdominis) were collected during lifting tasks, and were later digitized and input to a dynamic biomechanical model. The model output included estimates of compression, anterior-posterior shear, and right lateral shear (at the L_3 level of the lumbar spine), as well as torques about the X, Y, and Z axes, using the coordinate system described by Schultz and Andersson (1981). In addition, the model produced estimates of muscle forces for the eight trunk muscles studied. All dependent variables except muscle forces were analyzed using 2 x 2 x 3 (P x H x W) analyses of variance. The data on muscle forces were subjected to a 2 x 2 x 3 (P x H x W) multivariate analysis of variance. Critical alpha levels were .05 for all statistical tests.

Apparatus

Bipolar surface electrodes were used to obtain integrated EMGs of the eight trunk muscles of interest. Belt wearable preamplifiers were used to amplify the EMG signals. These signals were then sent through shielded cables to an amplifier/integrator where the signals were conditioned using a 80 Hz high pass filter and a 1000 Hz low pass filter. The integrator constant was 500 ms. The fully conditioned signals were subsequently transferred to a MicroVAX[1] II equipped with an A/D board for data storage. A modified CYBEX II isokinetic dynamometer was used to perform maximum flexion and extension exertions in both stooped and kneeling postures for normalization of EMG data.

The lifting tasks were performed under a 1.2 m roof that restricted the subject's posture. An aluminum lifting box (50.8 x 33.0 x 17.8 cm) with two covered compartments was used to perform the lifting tasks. Adjustable wooden shelves were used to vary the height to which the box was lifted. The frequency of lifting was controlled using a computer generated voice prompt.

---

[1]Reference to specific products does not imply endorsement by the Bureau of Mines.

Experimental Procedure

For each subject, eight trunk muscles were identified and the skin above the muscle was prepared by shaving (if necessary) and thoroughly cleaning the site with alcohol. Two electrodes filled with electrolyte gel were placed 3 cm apart (center to center) over each selected muscle. A single electrode, which served as a ground for the entire system, was affixed at a remote anatomical site. After all electrodes had been placed, the subject performed maximum voluntary isometric flexion and extension exertions in both stooping and kneeling postures using the modified isokinetic dynamometer. This allowed EMG data collected during the lifting tests to be normalized to the maximum elicited in each posture.

The subject then began a series of twelve symmetric lifting and lowering tasks in a counterbalanced order. The subject was required to lift or lower the box to the appropriate height every 10 seconds for a period of 2 minutes. The subject was not informed of the amount of weight in the lifting box during the test, as the amount of weight placed in the covered compartments of the box was varied. EMG data was collected during the third lift of this sequence. This allowed the subject a couple of lifts to become accustomed to the lifting movement. Upon completion of the test the subject was allowed 5 minutes rest, after which the next experimental condition would be run. This process continued until all twelve experimental conditions had been completed.

Biomechanical Model

The integrated EMG data was analyzed by a process of digitizing each separate signal. Four points were identified on each signal: initial activation of the muscle, point where the terminal resting level was achieved, and two points in the middle of the signal such that the shape of the EMG signal was approximated by a trapezoid. Figure 1-a shows typical integrated EMG data for a lift of 25 kg from floor to 70 cm in the stooped posture. All eight signals for each lift were digitized in this manner and data were normalized with respect to time where Time = 0.0 indicated time of the first muscle activation, and Time = 1.0 represented the time when the final muscle reached its terminal resting level (Reilly and Marras, 1989). Figure 1-b presents estimated muscle forces in the normalized time scale for the data shown in Figure 1-a.

The biomechanical model estimated compression, shear, torque, and muscle force data at 32 points during the lift. A new calculation was made each time a muscle signal changed slope. Thus, the model allows estimates of the time-varying loading on the spine. For the current study, the assumption was made that the maximum force that could be exerted by a muscle was 50 N/cm^2 (Reid and Costigan, 1987). Estimates of the cross-sectional areas of trunk muscles were obtained through anthropometric measurements, as described by Schultz *et al.* (1982). Figure 1-c demonstrates compression and shear estimates of the model for the data from Figure 1-a. Figure 1-d illustrates the estimated torques about X, Y, and Z for the same data.

Data Analysis

Results of the biomechanical model output included both peak and average values for compressive loading, anterior-posterior shear, right lateral shear, X torque, Y torque, Z torque, and muscle forces. Each of these dependent variables were subjected to an 2 x 2 x 3 (posture x lifting height x box weight) analysis of variance (a multivariate analysis of variance in the case of muscle forces). On

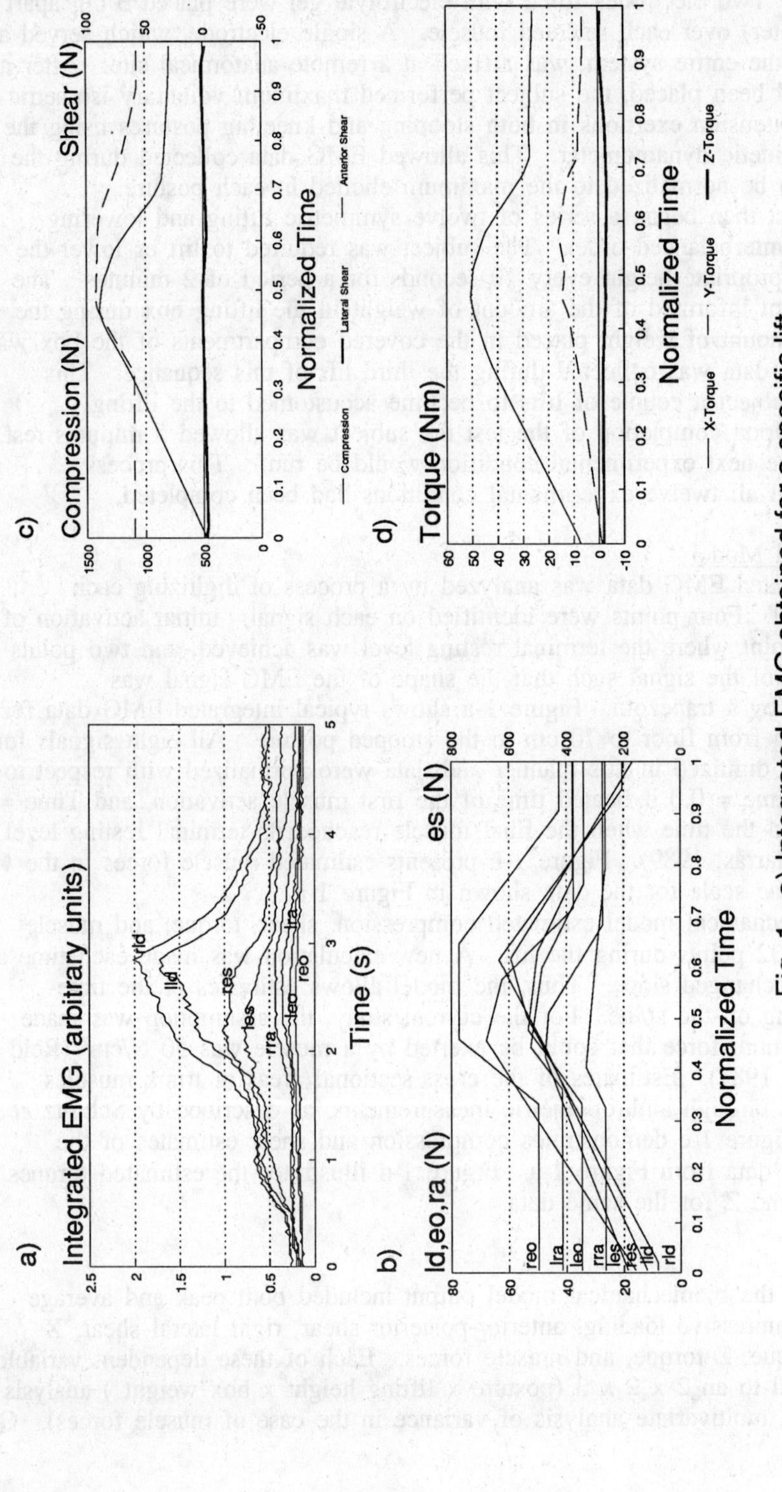

Fig. 1 - a) Integrated EMG obtained for a specific lift. b) model estimates of muscle force. c) predicted compression and shear forces at the L3 level. d) estimates of X, Y, and Z torque

the basis of the ANOVAs, regression models were developed which incorporated only those effects found to be significant ($p < 0.05$).

RESULTS

Table 1 contains a summary of significant effects for all dependent variables. The ensuing sections will describe these effects in greater detail.

Compression

Posture, lifting height, and box weight all significantly affected peak compression forces experienced when lifting in restricted postures. Peak compression was found to be greater in the kneeling posture than stooped, increased as H increased, and also increased as W increased. Average compression was also significantly affected by P, H, and W. The direction of the effects was the same for average compression as it was for peak compression.

Right Lateral Shear

Peak lateral shear was found to be significantly greater in the stooped posture than when kneeling. No other effects were found to significantly influence either peak or average right lateral shear ($p > 0.05$).

Anterior-Posterior Shear

Peak A-P shear was significantly affected by H and W. However, posture did not significantly influence peak A-P shear ($p > 0.05$). As expected, peak A-P shear increased with increasing lifting height and with increasing box weight. Peak A-P shear was generally greater in the stooped posture; although, as noted, this trend did not achieve statistical significance. Average A-P shear was significantly affected by P, as well as by H and W. Average A-P shear was significantly greater in the stooped posture, and increased with increases in H and W.

Torque about X

P, H, and W all were found to significantly affect peak torque about X. Peak X torque was significantly greater when kneeling, and increased with increasing H and W. Similarly, average X torque was significantly affected by all main effects -- P, H, and W. The direction of these effects was the same as for peak X torque.

Torque about Y

Posture was the only significant factor influencing either peak or average Y torque. In both cases, Y torque was significantly greater in the stooped posture than when performing the task in the kneeling position.

Torque about Z

Neither peak nor average Z torque was significantly influenced by any main effects or interactions ($p > 0.05$).

Muscle Forces

The MANOVAs for both peak and average muscle forces indicated a complex relationship of muscle recruitment for the treatments studied in this experiment. In both analyses the P x H x W interaction was significant.

Table 1. Summary of significant main effects and interactions for all dependent variables.

	P	H	W	PH	PW	HW	PHW
Peak Compression	**	*	***				
Peak R-L Shear	*						
Peak Ant. Shear		*	***				
Peak X Torque	***	*	***				
Peak Y Torque	*						
Peak Z Torque							
Peak Mus. Force	***	***	***	**	***	***	***
Ave Compression	***	***	***				
Ave R-L Shear							
Ave Ant. Shear	*	**	***				
Ave X Torque	***	**	***				
Ave Y Torque	*						
Ave Z Torque							
Ave Mus. Force	***	***	**	**	***	***	***

* $p < 0.050$
** $p < 0.010$
*** $p < 0.001$

DISCUSSION

Results of this experiment indicate that P, H, and W all significantly affect the compression and shear on the lumbar spine during symmetric manual lifting in restricted postures. Greater compression was consistently observed in the kneeling posture compared to stooped. This is probably due to two major factors, both of which deal with activity of the erectores spinae. The first is that these muscles are relatively quiescent in the stooped posture, the second is that these muscles may assume increased responsibility for lifting in the kneeling posture. With regard to the first point, it is well-known that the EMG of the long extensors of the spine become silent when flexion of the trunk becomes extreme (Floyd and Silver, 1955). In trunk flexion, use of the ligaments to stabilize the spine is much more efficient (owing to their longer lever arm) than use of the supporting musculature. Use of the ligaments in this posture (at least in a mechanically sound column) acts to decrease the shear force across the disk and the facet joints. Thus, the "muscle sparing" function of ligaments leads to lower muscle forces and compression on the discovertebral joints. However, it should be noted that ligament forces were not modeled in this study. These forces could affect the internal load experienced by the spine in the stooped posture.

The second reason for increased compression forces in the kneeling posture is the fact that the erectores spinae assume a much more significant role in performing the lift in this posture. When lifting objects from the floor in an unrestricted standing posture (or in the stooped posture), the gluteal muscles and other strong hip extensors provide a large portion of the forces required to perform a lift. However, in the kneeling posture, these powerful spine extensors are less able to provide force to assist with the lift. Therefore, the erectores spinae appear to be more active in the kneeling posture, compared with the stooped posture. Of course, the compressive forces acting on the disc lead to tensile forces on the anulus fibrosus, whose fibers may tear as a result.

However, while compression was significantly greater in the kneeling posture, the anterior shear forces were greater in the stooped posture. This reflects greater activity of the oblique abdominals during stooped lifting. It is interesting to note that although this was a symmetric lifting task, peak lateral shear was found to be significantly different between the two postures studied. The ANOVA demonstrated that average right lateral shear was more negative when lifting in the kneeling posture compared to lifting when stooped. This finding indicates that even symmetric lifting tasks may result in some asymmetric internal loading on the vertebral column.

Estimates of torque about X indicated increased torque in the kneeling posture. This may result from a difference in moment arm to the load in these two postures. When stooping, it is possible to stand relatively close to the load, thus reducing the moment from L_5-S_1 to the load. On the contrary, when kneeling on two knees under restricted headroom, it may be more difficult to minimize the moment, due to the fact that the knees hinder how close one can get to the load. Torque about Y was found to be significantly greater in the stooped posture. Interestingly, the torque about Y when kneeling was generally in a slightly negative direction (right hand rule), while the torque about Y when stooping was of a larger magnitude and in the opposite direction.

As mentioned previously, the model estimates of muscle force for the eight trunk muscles demonstrated a complex recruitment relationship for the treatment combinations studied. However, certain observations can be made regarding the different recruitment patterns that are due to posture. For example, on the average the peak muscle force demonstrated by the latissimus dorsi muscles were approximately 20% greater when lifting in the stooped posture than when kneeling. These muscles also demonstrated considerable asymmetric activity in the kneeling posture, while in the stooped posture symmetry was more prevalent (this explains the greater right lateral shear seen in the kneeling posture). On the other hand, the peak erectores spinae activity was (on average) over 25% greater when performing the lift in the kneeling posture. In addition, muscle forces generally increased with increasing height of lift and increasing weight of the lifting box. The only muscle that deviated from this model was the rectus abdominis.

CONCLUSIONS

Based on the use of an EMG-driven biomechanical model during lifting tasks in restricted postures, the following conclusions are drawn:

1. Compression on the lumbar spine is increased in the kneeling posture compared to a stooped position. However, lateral and A-P shear forces are higher when stooped.
2. Increased height of lift and increased weight of the lifting box increased compression on the lumbar spine as well as A-P shear; however, right lateral shear was unaffected by increased lifting height or box weight for these symmetric lifting tasks.
3. P, H, and W all significantly affected torque about X; however, only P significantly affected Y torque. Torque about Z was unaffected by any of the manipulated variables.
4. A complex relationship exists between P, H, and W and the forces of the eight trunk muscles studied, with the triple interaction being significant.

The data in this study will be used to assist in establishing acceptable loads when lifting in restricted postures in the underground mining environment.

REFERENCES

Davis, P.R., and Troup, J.D.G., 1966. Effects on the trunk of erecting pit props at different working heights. Ergonomics, 9, 6, 475-484.

Davis, P.R., and Ridd, J.E., 1981. The effect of restricted headroom on acceptable lifting capacity. Ergonomics, 24, 239.

Floyd, W. F., and Silver, P. H. S., 1955. The Function of the Erectores Spinae Muscles in Certain Movements and Postures in Man. Journal of Physiology, 129, 184-203.

Gallagher, S., and Unger, R. L., in press. Lifting in Four Restricted Lifting Conditions: Psychophysical, physiological, and biomechanical effects of lifting in stooped and kneeling postures, Applied Ergonomics.

Gallagher, S., Marras, W.S., and Bobick, T.G., 1988. Lifting in stooped and kneeling postures: effects on lifting capacity, metabolic costs, and electromyography of eight trunk muscles. International Journal of Industrial Ergonomics, 3, 65-76.

Reilly, C. H., and Marras, W. S., 1989. Simulift: A Simulation Model of Human Trunk Motion. Spine, 14, 5-11.

Reid, J. G., and Costigan, P. A., 1987. Trunk Muscle Balance and Muscular Force, Spine, 12, 783-786.

Ridd, J.E., 1985. Spatial restraints and intra-abdominal pressure. Ergonomics, 28, 149-166.

Schultz, A. B., and Andersson, G. B. J., 1981. Analysis of Loads on the Lumbar Spine. Spine, 6, 76-82.

Schultz, A. B., Andersson, G. B. J., Haderspeck, K., Ortengren, R., Nordin, M., and Björk, R., 1982. Analysis and Measurement of Lumbar Trunk Loads in Tasks Involving Bends and Twists. Journal of Biomechanics, 15, 669-675.

Sims, M.T., and Graveling, R.A., 1988. Manual handling of supplies in free and restricted headroom. Applied Ergonomics, 19, 289-292.

XII

INJURIES IN HEALTH CARE

Advances in Industrial Ergonomics and Safety II
Edited by Biman Das
Taylor & Francis 1990

THE INCREASING OCCUPATIONAL INJURY RATE IN NURSING HOMES

Roger C. Jensen

Division of Safety Research
National Institute for Occupational Safety and Health
944 Chestnut Ridge Road
Morgantown, West Virginia, 26505, USA

This investigation was undertaken to learn more about the kinds of injuries that contributed to the 44 percent increase in the lost-workday case rate for nursing and personal care facilities between 1980 and 1987. Using workers' compensation data from 22 states it was determined that the most apparent reason was an increase in sprain/strain cases. Of all claims from this industry, sprains/strains constituted 59.6 percent in 1980. By 1986 these cases had increased to 67.9 percent. Over half of the sprain/strain injuries involved the worker's back.

INTRODUCTION

In the United States during 1988, employees working in the industry category "nursing and personal care facilities" experienced 151,000 reportable injuries and illnesses (Personick, 1990). Only five industries had more cases. The large number of cases in the nursing and personal care industry is attributable to a combination of a large number of employees and high injury rates (Personick, 1990).

The most recent published injury rate data from the Bureau of Labor Statistics (BLS) indicate that in 1987 the nursing and personal care industry had an average injury rate of 13.9 injuries per 100 full time equivalent (FTE) employees (BLS, 1989). By comparison, the average for all industries was 8.0 injuries per 100 FTE.

The figures noted above are based on the BLS Annual Survey of employers. The injury rate is one of four industry-specific rates routinely reported by BLS from their Annual Survey (BLS, 1982). The four rates are: 1) injury rate, 2) illness rate, 3) combined injury and illness rate, 4) and lost-workday case rate. Each rate has some utility for monitoring industry-specific trends in occupational injury and illness experience.

From these published rates, other indices may be computed. To see how the rate of one industry is doing compared to that of others, any of these four rates may be used for calculating an index reflecting either a ratio or a difference in the respective rates. For example, to compare injury rates of nursing and personal care facilities with the entire private sector, the ratio of the injury rate for nursing and personal care facilities (IR nursing homes) to the injury rate of all industry combined (IR all industry) may be used. This Ratio of Injury Rates (RIR) is therefore

$$RIR = \frac{\text{IR nursing homes}}{\text{IR all industry}} .$$

This ratio provides an indication of how nursing and personal care facilities have been doing relative to other industries; a value greater than 1.0 indicates that the injury rate of the nursing and personal care industry exceeded the average injury rate for all industries during the year.

For the years 1980 through 1987, Table 1 lists the four rates from the Annual Survey data and the RIR. Columns 2 through 6 are rates expressed as cases per 100 FTE. Column 2 gives the injury rate. Between 1980 and 1987 the injury rate for nursing and personal care facilities increased 31 percent (from 10.6 cases to 13.9 cases per 100 FTE). During this same period the injury rate for all private industry remained relatively stable (see column 6).

Column 3 lists the illness rate. Column 4 lists the combined injury and illness rate. In the average year, the combined rate consisted of 98.5 percent injuries and 1.5 percent illnesses.

Column 5 shows the lost-workday case rate. Between 1980 and 1987 the lost-workday case rate increased 43.6 percent.

Column 6 lists the all industry injury rate. It fluctuated between 7.5 and 8.5 cases per 100 FTE. Column 7 shows the RIR as defined above (column 2 divided by column 6). The RIR values increased steadily since 1980 from 1.25 to 1.74. This indicates that the injury rate of the nursing and personal care industry has been getting worse relative to all industry. A similar pattern has been reported for the state of Ohio based on workers' compensation records (Valles-Pankratz, 1989).

Table 1. Yearly trends for injuries and illnesses in nursing homes and other personal care facilities.(a)

1	2	3	4	5	6	7
Year	Injury Rate	Illness Rate	Combined Rate	LWC Rate(b)	All Indus. Injury Rate	RIR
1987	13.9	0.3	14.2	7.9	8.0	1.74
1986	13.3	0.2	13.5	7.7	7.7	1.73
1985	13.0	0.3	13.3	7.2	7.7	1.69
1984	11.4	0.2	11.6	6.4	7.8	1.46
1983	10.9	0.1	11.0	6.0	7.5	1.45
1982	9.9	0.2	10.1	5.6	7.6	1.30
1981	10.4	0.1	10.5	5.6	8.1	1.28
1980	10.6	0.1	10.7	5.5	8.5	1.25

(a) Nursing homes and other personal care facilities have 4-digit Standard Industrial Classification (SIC) numbers 8051 and 8059, respectively. Data reported are for the 3-digit SIC 805 which consists of the two 4-digit industry classifications.
(b) LWC Rate is the rate of lost-workday cases per 100 FTE.

The BLS Annual Survey has been designed to provide a representative picture of the annual injury and illness rate for private sector employers with more than ten employees (BLS, 1982). Employers selected for the survey report cases of illness using seven sub-categories. Injuries are reported using only one category even though injuries constitute 98.5 percent of the cases! Consequently, the contributions of various types of injuries to the increasing trend in injury rate is not explainable from Annual Survey data.

This investigation was undertaken to learn more about the kinds of injuries that contributed to the 31 percent increase in injury rate and the 44 percent increase in lost-workday case rate for nursing and personal care facilities between 1980 and 1987. The best available source of information for this purpose is workers' compensation claim data.

METHOD

The Supplementary Data System, or SDS, is a record system maintained by BLS (BLS, 1982; Jensen, 1987). It uses workers' compensation records from states that elect to participate. Each state codes their cases according to a uniform format. The most recent available data are for calendar year 1986. During the years covered by this analysis, 1980-1986, there were 22 states that provided consistent records each year. These states were· Alaska, Arkansas, Arizona, California, Colorado, Delaware, Hawaii, Indiana, Iowa, Kentucky, Maryland, Michigan, Missouri, Mississippi, North Carolina, Nebraska, New Mexico, New York, Ohio, Tennessee, Washington, and Wisconsin.

The records of these states were searched for each of the years 1980 through 1986. The search strategy was to select, for each year, all cases in the three digit Standard Industrial Classification (SIC) code 805 (this includes SIC 8051, nursing homes, and SIC 8059, other personal care facilities). A complete description of these SIC codes is provided in a manual (Office of Management and Budget, 1972). Cases selected were from the "current case" files, which means the claim was received or coded by the state agency during the year.

Cases in the SDS records have been coded to indicate, among other descriptors, the part of body affected and the nature of the injury or illness (Jensen, 1987). The cases from the nursing and personal care industry were examined to determine how the cases were distributed with respect to these characteristics.

For each part of body the percentage of all claims in the nursing and personal care industry was determined for each of the years 1980 through 1986. The difference between the 1986 and 1980 percentages was determined for each part of body. These difference values were examined to identify changes in the distribution of claims among body parts that may help explain the increase in injury rates for SIC 805 during the 1980s. A corresponding approach was used for examining the change in distribution of claims according to nature of injury or illness categories.

RESULTS

In 1980 the total number of claims reported by the 22 states for the nursing and personal care industry was 26,613. In 1986 the corresponding total was 34,572. The increase of 7,959 claims amounted to a 29.9 percent increase from the 1980 level.

Table 2 shows the percentage of claims by part of body for the years 1980 and 1986. Body part categories with more than one percent of claims in either 1980 or 1986 are listed. The remaining body parts are grouped together in the category "other". The right column shows the change in percentage. The back was the part of body which increased most, from 38.5 percent to 41.2 percent. This increase was not an abrupt jump, the percentage increased gradually between 1980 and 1986. There were also increases in the percentage of claims for shoulder, multiple body parts, and the trunk. No other part of body increased as much as a half percentage point. The most notable decreases were in the fingers (-1.7%) and "other" body parts (-3.5%).

Table 2. Distribution of claims for nursing and personal care facilities according to the part of body affected.

Body Part	Percentage of Claims 1980	1986	Change in Percentage
BACK	38.48	41.16	2.68
SHOULDER	3.57	5.15	1.59
MULTIPLE PARTS	9.30	10.86	1.56
TRUNK (multiple)	1.95	2.90	0.95
NECK	1.50	1.98	0.48
ARM	0.71	1.09	0.38
WRIST	3.80	4.08	0.28
UPPER EXTREMITY	0.93	1.07	0.14
FOOT	2.03	2.05	0.02
KNEE	4.58	4.53	- 0.05
ANKLE	3.07	2.99	- 0.08
HIPS	1.72	1.57	- 0.15
ELBOW	1.18	1.03	- 0.15
FOREARM	1.06	0.75	- 0.31
CHEST	2.15	1.79	- 0.36
ABDOMEN	2.36	1.93	- 0.43
EYE	1.50	0.86	- 0.64
HAND	3.02	2.31	- 0.72
FINGER	6.09	4.37	- 1.72
OTHER PARTS	11.00	7.53	- 3.47
TOTAL	100.00	100.00	0.00

Table 3 shows the change in distribution of claims between 1980 and 1986 according to the nature of the injury or illness. Only those categories with at least one percent of the claims in either 1980 or 1986 are listed; all those less than one percent are combined into the "other" category. The sprain/strain category increased from 59.6 percent to 67.9 percent of all claims in the industry. This rise of 8.3 percentage points far exceeded the change for any other category.

The sprain/strain category constituted about 68 percent of all claims in 1986. This category also contributed most of the increase in claims, changing from 15,849 to 23,469 claims in the years 1980 and 1986, respectively. The total increase in claims for all categories was 7,959 claims; the sprain/strain category alone increased by 7,620 claims. Thus, in comparison to the increase in the sprain/strain injuries, changes in the other categories were minor. This change in the sprain/strain frequency was a 48 percent increase over the 1980 frequency, whereas the industry increase for non-sprain/strain claims was 3.1 percent.

Table 3. Changes in claims for nursing and personal care facilities according to the <u>nature</u> of the injury or illness.

Nature of Injury or Illness	Percentage of Claims 1980	1986	Change in Percentage
SPRAIN/STRAIN	59.55	67.88	8.33
MULTIPLE INJURIES	1.98	2.71	0.73
DISLOCATION	1.44	1.87	0.43
FRACTURE	4.26	4.30	0.05
BURN FROM HEAT	2.11	1.15	- 0.96
CUT, LACERATION	5.63	3.16	- 2.47
CONTUSION	10.84	8.14	- 2.69
OTHER(a)	14.20	10.78	- 3.42
TOTAL	100.01(b)	99.99(b)	0.00

(a) The "other" group includes many categories. Among these are the following, with corresponding percentages of the 1986 claims: hernia, 0.55%; abrasion, 0.49%; inflammed joints, 0.48%; dermatologic conditions, 0.47%; concussions, 0.32%; infectious diseases, 0.28%; conjunctivitis and other eye diseases, 0.26%; peripheral nerves, 0.20%; mental disorders, 0.09%; and amputations, 0.08%.

(b) Totals differ from 100.00 due to rounding.

Since the increase in sprain/strain claims was such an important part of the industry's increase in total claims between 1980 and 1986, the sprain/strain claims were examined further. Of the sprain/strain claims, the back was the part of body most frequently affected (55.6% in 1986). Other body parts with frequent sprain/strain claims, according to the 1986 data, were: multiple body parts, 9.4%; shoulders, 6.5%; lower extremities, 6.3%; upper extremities, 6.0%; and the neck, 2.7%.

The trend for the sprain/strain claims between 1980 and 1986 is plotted in Figure 1. Also shown is the trend for back sprain/strain claims. It can be seen that the proportion of claims coded as sprains/strains followed an increasing trend throughout the period 1980 to 1986. The proportion of claims coded as back sprains/strains also tended to increase, but the rise was not as pronounced as for the more general sprain/strain category. The greatest year-to-year increase occurred between 1984 and 1985.

DISCUSSION

An explanation for part of the increasing trend in sprain/strain claims may be a change in the Medicare payment system in 1983. The new system established payment schedules based on diagnosis related groups (Gutterman, et al., 1988). During 1984 hospitals began to find that the Medicare payment schedules for some patient groups were inadequate to cover the hospital's cost of care. A portion of these patients were transferred from hospitals to nursing homes. These patients tended to be individuals who were unable to provide any help to the NAs while being moved about for routine care. Some data supporting this hypothesis come from the National Nursing Home Surveys conducted in 1977 and 1985. In 1985 the percentage of nursing home residents who were

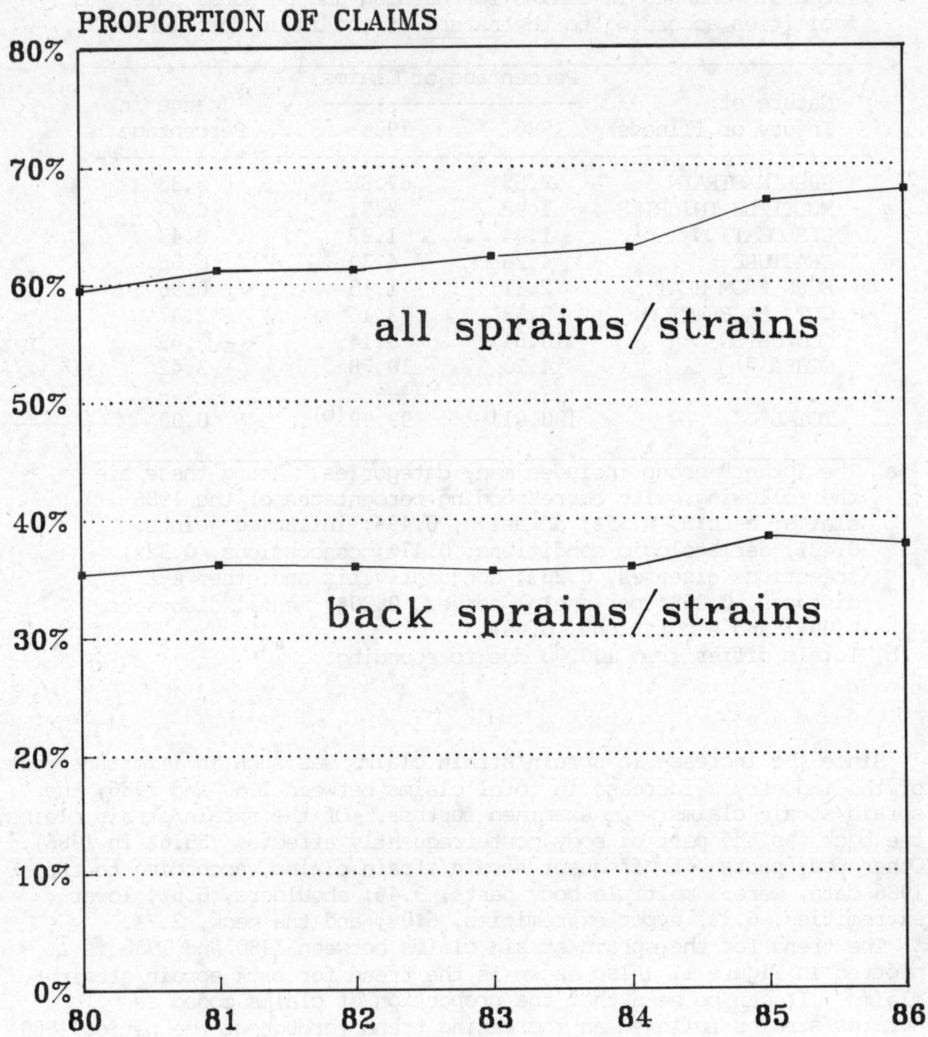

Figure 1. Yearly trend in sprain/strain claims, and back sprain/strain claims, as a percentage of all workers' compensation claims within the nursing and personal care industry in 22 states.

bedfast or chairfast was 46 percent, as contrasted to 38 percent in 1977 (Van Nostrand, et al., 1979, Hing, et al., 1989). Thus, it appears likely that the exposure of NAs to stressful patient-handling tasks increased during the 1980s, especially between 1984 and 1985.

More frequent exposure to stressful patient-handling tasks was recently shown to be associated with increased prevalence of back problems among nursing staff (Jensen, 1990). Thus, the increase in back sprain/strain claims shown in Figure 1 between 1984 and 1985 may be a result of the influx of more helpless patients into nursing homes. Other factors besides the changed patient-care demands may have contributed to the generally increasing trend in injury rates for nursing and personal care facilities throughout the 1980 to 1986 period. As a next step it would be useful to further explore the possible reasons behind the gradually increasing trends shown in Figure 1.

This paper reports data from two occupational injury and illness information systems--the BLS Annual Survey data and the SDS records. Some discussion of the strengths and weaknesses of each record system is appropriate.

The greatest strength of the Annual Survey is its' utility as a tool for monitoring industry-specific trends in injury and/or illness rate. However; it has some limitations. First, it is limited to employers with more than 10 employees. Thus, the rates from the Annual Survey may not be representative of the rates for small nursing homes. Second, the sample size of the Annual Survey is not large enough to provide accurate rates for the 4-digit industry classification. Consequently, nursing homes (SIC 8051) cannot be differentiated from "other personal care facilities" (SIC 8059). Finally, the survey instrument only asks employers to report incidence of injuries, with no more specific description of the types of injuries. Consequently, the Annual Survey is useful for tracking industry-specific trends, but it provides no clues about the possible reasons for changes in injury rates.

As this investigation illustrates, one strength of the SDS is in supplementing the Annual Survey data. In this role, however, the differences in cases must be kept in mind. The information contained in SDS comes from workers' compensation claim records. The criteria for including a case in the SDS records differ from the criteria for including a case as a "reportable injury" or a "lost-workday case" as defined for the Annual Survey. The states also differ in their respective case-inclusion criteria for workers' compensation claims. Also, not all states participate in SDS. Thus, using the SDS data set to supplement the Annual Survey is not straightforward.

Further discussion of these occupational injury surveillance systems may be found in Coleman (1984), Jensen (1987), and Hanrahan and Moll (1989). As these authors indicate, the SDS data is not ideal, but it is the only occupational injury surveillance system providing information about the injured worker, the circumstances, and the injury.

The Bureau of Labor Statistics has tentatively planned to discontinue the SDS program in the early 1990s. Because of its' value as a supplement to the Annual Survey, it is recommended that the SDS data system be continued. The SDS record system is the only available source of data for estimating the distribution of occupational injuries within an industry (e.g., the percentage of all claims affecting a certain part of body). Such measures may be helpful in assessing changes that occur following the introduction of prevention programs. For example, if a comprehensive set of recommendations were to be developed for back injury prevention programs in the nursing home industry, the percentage of cases involving the back could be used to monitor changes that might take place after disseminating the recommendations.

SUMMARY

This investigation was undertaken to learn more about the kinds of injuries that contributed to the 31 percent increase in injury rate and 44 percent increase in lost-workday case rate for nursing and personal care facilities between 1980 and 1987. Since the BLS Annual Survey provided no information on the types of injuries, it was necessary to examine another source of data. Workers' compensation claim data in the SDS records were examined for 22 states which participated in SDS between 1980 and 1986.

The part of body most frequently affected was the back. Back cases constituted 38.5 percent of all claims in 1980. This percentage increased to 41.2 percent by 1986.

The most frequent nature of injury was the sprain/strain category. These cases constituted 59.6 percent of all claims in 1980, increasing to 67.9 percent by 1986. Over half of the sprain/strain claims had the back as the part of body.

The total claims reported by the 22 states for nursing and personal care facilities increased by 7,959 between 1980 and 1986. Sprain/strain cases alone increased by 7,620 claims during this period. Thus, it appears that a major contribution to the increase in compensation claims is the increase in sprain/strain claims.

ACKNOWLEDGMENT

The author sincerely appreciates the useful background material provided by Martin Personick, and the helpful comments on the draft of this manuscript from Patricia Schnitzer, Naomi Swanson, and Esther Hing.

REFERENCES

Bureau of Labor Statistics, 1982, Occupational safety and health statistics. In Handbook of Methods, U.S. Department of Labor, Bulletin 2134-1, (Washington, D.C.: U.S. Government Printing Office), Chapter 17, pp. 122-134.

Bureau of Labor Statistics, 1986, Recordkeeping Guidelines for Occupational Injuries and Illnesses, (Washington, D.C.: U.S. Department of Labor).

Bureau of Labor Statistics, 1989, Occupational Injuries and Illnesses in the United States by Industry, 1987, U.S. Department of Labor, Bulletin 2328, (Washington, D.C.· U.S. Government Printing Office).

Coleman, P.J., 1984, Descriptive epidemiology in job injury surveillance. Journal of Occupational Accidents, 6, 135-146.

Gutterman, S., Eggars, P., Riley, G., Greene, T., and Terrell, S., 1988, The first 3 years of Medicare prospective payment: an overview. Health Care Financing Review, 9(3), 67-76.

Hanrahan, L.S. and Moll, M.B., 1989, Injury surveillance. American Journal of Public Health, 79(supplement), 38-45.

Hing, E., Sekscenski, E., and Strahan, G., 1989, The National Nursing Home Survey: 1985 Summary for the United States, National Center for Health Statistics, Vital Health Statistics, 13(97).

Jensen, R.C., 1987, How to use workers' compensation data to identify high-risk groups. In Handbook of Occupational Safety and Health, edited by L. Slote, (New York: John Wiley and Sons), pp. 364-403.

Jensen, R.C., 1990, Back injuries among nursing personnel related to exposure. Applied Occupational and Environmental Hygiene, 5(1).

Office of Management and Budget, 1972, Standard Industrial Classification Manual, (Washington, D.C.: U.S. Government Printing Office).

Personick, M.E., 1990, Nursing home aides experience increase in serious injuries. Monthly Labor Review, 113(2).

Valles-Pankratz, S., 1989, What's in back of nursing-home injuries. Ohio Monitor, 62(2), 4-8.

Van Nostrand, J., Zappolo, A., and Hing, E., et al., 1979, The National Nursing Home Survey: 1977 Summary for the United States, National Center for Health Statistics, Vital Health Statistics, 13(43).

Advances in Industrial Ergonomics and Safety II
Edited by Biman Das
Taylor & Francis 1990

5-YEAR LIFE CYCLE COST OF AN INSTRUCTIONAL BACK INJURY PREVENTION PROGRAM FOR NURSING PERSONNEL

Penelope Venning Klein

D'Youville College
320 Porter Ave
Buffalo, N.Y. 14201

INTRODUCTION

The seriousness of the problem of occupational back injuries among nursing personnel is well established. Back injury prevention programs addressing this problem often include an instructional component. This approach is supported by clinicians and advocated in empirical reports (Aird, Nyran, & Roberts, 1988; McGovern, 1985; Raistrick, 1980; Videman et al., 1989). However, cost-effectiveness of back injury prevention programs for nursing personnel has not been well-documented. Program evaluation to establish level of effectiveness is required to provide critical information for policy makers, who must judge program viability and allocate finite resources.

Given that the goal of a back injury prevention program is to reduce back injuries, the ideal comparative indicator for cost-benefit analysis associated with back injury prevention programs is a measure of reduction in injury. Back complaints are multi-factor and have a long natural course. Limitations in isolating reduction of injury costs as an indicator of program effectiveness should not, however, preclude economic analysis. Cost may be examined on a variety of levels. Cost analysis is a component of cost-benefit analysis described by Cox and Doughty (1988) as an inquiry method concerned with describing, comparing or predicting the negative impact costs of a program or policy on a system.

The purpose of this report is describe empirical testing of a model of cost-analysis to be employed in validation of back injury prevention programs. The trial intervention: an instructional program, was field tested as a pilot instruction at a large teaching hospital in New York State. Cost analysis reported is part of a larger cost-effectiveness evaluation of the instructional program. Projected costs for a five-year life cycle of the instructional program are compared to the cost of no intervention (the predicted cost of back injuries in the study population). Projections of costs are based on experiences with the field trial of the pilot instruction and dependent on knowledge of study population demographics.

METHODS

The following describes the instructional program, study population, data gathering process and cost category considerations.

Description of instructional program There are three phases to the instructional program: development, delivery and management. The instructor was a physical therapist hired on a full-time nursing line. Development of the instruction was based on the trainer's review of the literature, discussion with experts in the field and representatives from nursing staff development at the target facility, unit observations, and a needs survey questionnaire completed by nursing personnel at the target institution.

The format of the instruction was a combination of lecture and slides with small group demonstration/practice activities. The instruction was organized as two 45-minute sessions. Class size was restricted to two to six participants per session. Instruction was presented on the unit and offered to all three shifts.

Content for the pilot instruction included principles of body mechanics and back care and specific lifting techniques. All participants received the same information about body mechanics and back care. The lifting procedures included in the instruction varied based on individual need. An informational pamphlet: *Back Tips for Health Care Providers* (Krames Communications, 1986), was distributed to all participants to supplement the instruction.

The management phase has two major activities. There is an ongoing orientation for new staff to maintain skill level among unit staff. There are also unit consultations to address problematic lifting tasks.

Study population The study population consists of nursing personnel including RN's, LPN's, health care aides and attendants (N=263). A subgroup of the study population from three acute care inpatient nursing units were eligible to participate in the pilot instructional program (n=81 nursing lines).

Data gathering procedures Data for cost-analysis evaluation was gathered through review of existing records: an instructor's log of activities, attendance records, injury reports; and interviews with the instructor and nursing administrators. A study proposal was submitted and approved through an institutional ethics review board.

Cost Category Considerations The dollar costs associated with human resources are derived from hours times average wage per hour for respective job categories. Hourly wage averages were submitted by the Nursing Personnel Office (Burns: nurse recruiter, personnel communication, November, 1988). Human resource costs do not always represent real dollars spent. Hours for staff attendance are reported as a cost item in this analysis because there was consistently high reporting of overtime during the period of instruction. The time commitment of unit staff for interviews as part of needs assessment is included because this activity is not part of the work day. Staff time commitment during observations conducted as part of needs assessment activities is not included because these activities were part of normal routine.

RESULTS

Initial Cycle Costs The initial cycle of the instruction is defined as the pilot instruction delivered on three nursing units. The pilot program is limited to development and delivery activities. The direct costs of the initial cycle include both development and delivery costs. The initial cycle costs by category are presented on Table 1.

The extensive needs assessment and development activities prior to trial of the instruction account for the relatively large development component cost in the initial cycle. Development costs are expected to proportionately decrease in succeeding operational cycles. There is a limit to this process. It is anticipated that there would be significant recurring development costs with inclusion of atypical nursing units such as critical care units, operating room and emergency care services. Similarly, presentation of instruction at a different facility would incur further development costs to address organizational specific issues as well as unit specific instructional needs.

Table 1. Initial Cycle Costs

Category	Hours	Cost (hours times hourly wage)	
DEVELOPMENT (seven participating units):			
Lifting Observations	18.0 ($16/hour)	$288.00	
Needs Survey	15.0 ($16/hour)	$240.00	
Staff interviews	11.75 ($16/hour)	$187.30	
	8.0 ($14/hour)	$112.00	
	3.75 ($9/hour)	$33.75	
Instructional development	42.0 ($16/hour)	$672.00	
Subtotal for Development			$1,533.05
Instructional Delivery (three participating units):			
Instructor's time:			
Unit Consults	8.0 ($16/hour)	$128.00	
Teaching	61.25 ($16/hour)	$979.00	
Evaluation	12.0 ($16/hour)	$192.00	
Staff time:			
RN's (72)	54.00 ($14/hour)	$765.00	
Other (58)	43.50 ($9/hour)	$391.50	
Pamphlets	$1.50 @ for 100	$150.00	
Subtotal for Delivery			$2,605.50
Total Initial Cycle Cost			$4,138.55

Note. Based on 1989 wages; Other = LPN's, aides, attendants

Projected Life Cycle Cost The life cycle cost of the instruction provides a useful comparative measure for evaluating cost-effectiveness. Projecting a life-cycle cost requires some knowledge of the demographic patterns of the facility. A five-year life cycle for the entire instructional project was estimated (by the evaluator) because the target facility has reported an annual turnover of 18% for the past three years (Grimes: nursing administrator, personnal communication, June, 1988). Given this turnover rate and the potential for service needs to change, the instruction may require extensive revision within a five year period.

The instructional project was designed to address needs within eleven inpatient units, serving approximately one half of the nursing personnel employed at the facility (n=263). The remainder of nursing personnel at the target facility are employed in administration, outpatient services or special care units which are atypical and not included in the study.

Table 2 presents projected five-year life cycle for a total of eleven inpatient acute care nursing units. Unit cost for the three units participating in the pilot program reflect initial cycle costs. Human resource costs for the additional eight units included in the program reflect personnel demographics for those units. A 3:1 ratio for RN's to others is used in the calculations of operational costs for the eight units not included in the pilot program because this ratio is more typical of the staffing patterns on those units. (Grimes: nursing administrator, personal communication, June 1988)

Operational costs for the eight additional units are projected based on experience in the initial cycle. The general instructional format is expected to be maintained. Development expenses which occur will be related to modification of unit specific content (estimated to approximate 20% of initial cycle development costs). The instructor plans to conduct some observations and interview activities on each unit to identify unit specific problems and to develop rapport with staff. As a result of the needs assessment activities, additional slides may be produced to present relevant situations. Instructional delivery cost per student, dependent on class size and length of instruction, should remain constant as no changes are planned in this area.

<u>Cost of no intervention</u> A five year (1981-1985) retrospective injury review conducted at the target facility provides information about current rates of back injuries in the study population (Venning, 1987). Registered nurses and ancillary personnel were found to have annual incidence rates of four and ten injuries per 100 employees, respectively. Mean time-off per injury was 16 days for RN's and 54 days for ancillary personnel. Given this information, it is possible to predict the cost of doing nothing to alter this problem (no intervention alternative). Table 3 presents a back injury cost projection based on current wages and the five year trend previously reported.

Table 2. Projected Five-year Life Cycle Costs of Instruction

Category	Basis	Cost	
INITIAL CYCLE	(three pilot units)		$4,138.55
OPERATIONAL CYCLES (24 staff per unit , ratio RN's:other 3:1)			
Teaching time	20 hrs at $16 per hour	$320.00	
Audience time	2 classes each for: 18 RN's ($14)	$378.00	
	6 other ($9)	$81.00	
Pamphlets	$1.50 @ for 24 staff	$36.00	
Subtotal per unit		$815.00	
Subtotal operational cost 8 nursing units			$6,520.00
MANAGEMENT (per unit per year)			
Consults	6 hours ($16/hour)	$96.00	
Orientation	4 hours ($16/hour)	$64.00	
Evaluation	8 hours ($16/hour)	$128.00	
Pamphlets	$1.50@ times 5	$7.50	
Subtotal per unit per year		$295.50	
11 unit subtotal per year			$3,250.50
Subtotal 5-year management			$16,252.50
Total 5-year life cycle (N=263)			$26,911.05

Note. Based on 1989 salary figures; Other = LPN's, aides, attendants

Table 3. The Cost of No Intervention: Predicted temporary disability costs at the target facility for 100 nursing personnel.

Staff at risk	Injury rates	M time-off	Wage	Cost
75 RN's	4 per 100/year	16 days	$14/hour	$5,376
25 ancillary	16 per 100/year	54 days	$9/hour	$9,720
Annual subtotal				$15,096
5-year total				$75,480

Note. Annual injury rates derived from data reported by Venning (1987). Cost based on an eight hour working day using 1989 wages.

ANALYSIS

Cost Comparison Table 4 presents the projected costs of the five-year instructional project compared to the projected cost in terms of temporary disability costs related to expected back injuries within the population should no intervention occur. Both projections represent conservative estimates of total costs for the following reasons. Indirect costs of the instruction, such as human resource costs incurred in adminstrative meetings and support services, are not included. Comparison costs related to occupational back injuries do not include permanent disability, medical, administrative and personal expenses.

Table 4. Comparison of 5-year Life Cycle Costs (N=263): Instruction Vs No Intervention.

Category	Cost
Instruction[a]	$26,911.05
No intervention[b]	$198,512.40

[a] derived from Table 2
[b] derived from Table 3: cost $75,480/100 personnel

The cost analysis comparison demonstrates that the cost of the instruction represents approximately 15% of the temporary disability cost of back injury in the same population. Cost analysis does not address effect. The information provided through cost analysis may be utilized to determine the size of effect required for program viability.

DISCUSSION/CONCLUSION

Based on the data presented in this study, the instructional program must reduce temporary disability costs of back injuries by greater than 15% to achieve a greater than one-to-one yield on investment. Considering the multi-factor causes of back injury, there may be strong rationale to support interventions other than instructional. For example: the purchase of mechanical lifting aids, may suppliment or be considered as alternative to an instructional program. An equipment needs survey conducted at the same target facility generated a proposal for approximately $30,000 in equipment needs. The size of critical effect to judge viability of a multiple intervention back injury program, including instruction similar to the project described in this report and purchase of mechanical lifting equipment, would be significantly higher than the single component cost-estimate generated in this study, but considerably less than the burden of the occupational hazard of back injuries among nursing personnel.

In conclusion, cost-analysis has demonstrated to be a useful tool for predicting the negative impact of a back injury prevention program on a system. Evaluations utilizing a cost-analysis component offer program planners and policy decision-makers a fiscally responsive method of validating back injury prevention programs.

REFERENCES

Aird, J. W., Nyran, F., & Roberts, G. (1988). Comprehensive back injury prevention program: An ergonomics approach for controlling back injuries in health care facilities. In Trends in Ergonomics/Human Factors V. Aghazadeh, F. (Ed.) North Holland: Elsevier Science Publishers B. V., 705-712.

Cox, J. & Doughty, P. (1988 January). Economic Analysis. Paper presented at the annual conference of the AECT, New Orleans

McGovern, P. M. (1985). Toward prevention and control of occupational back injuries. Occupational Health Nursing, 33, 180-183.

Raistrick, A. (1981). Nurses with back pain- Can the problem be prevented? Nursing Times, 77(20), 853-856.

Videman, T., Rauhala, S., Asp, K., Lindstrom, K., Cedercrente, G., Kamppi, M., Tola, S., & Troup, J. D. G. (1989). Patient handling skill, back injuries and back pain: An Intervention study in nursing. Spine, 14(2), 148-156.

Venning, P. J. (1987). Time-off patterns for back injuries in nurses. In Asfour, S. S. (Ed.), Trends in Ergonomics/Human Factors IV. North Holland: Elsevier Science Publishers, 875-882.

Advances in Industrial Ergonomics and Safety II
Edited by Biman Das
Taylor & Francis 1990

ASSISTIVE DEVICES FOR USE WITH PATIENT HANDLING TASKS*

Bernice D. Owen
University of Wisconsin-Madison
School of Nursing
600 Highland Avenue
Madison, WI 53792, USA

Arun Garg
University of Wisconsin-Milwaukee
Department of Industrial and Systems Engineering
P.O. Box 784
Milwaukee, WI 53201, USA

Back pain among nursing personnel is common and usually attributed to involvement in patient handling tasks. Nursing assistants in a nursing home ranked the following tasks as most stressful to the back: transferring patients on and off the toilet, in and out of bed, and transfers needed for bathing. Criteria were established for selection of assistive devices to be used to help decrease back stress while carrying out these tasks. Strategies were developed for locating assistive devices. Assistive devices found were limited in number but included hoists, sliding boards, belts, slings, and a turn table. Preliminary trials were conducted to determine which devices should be studied in an attempt to decrease back stress while performing patient handling tasks.

INTRODUCTION

Back problems are prevalent among nursing personnel. Klein et al., (1984) found, through worker compensation data, that nursing assistants ranked fifth for back strains/sprains with an annual incidence ratio of 3.6 claims/100 workers. Only heavy laborer occupations such as miscellaneous laborers, garbage collectors, and warehouse men ranked higher than nursing personnel. Jensen (1987) found that nursing assistants in nursing homes/personal care facilities ranked highest among nursing personnel for worker compensated back problems. Lifting and transferring of patients have been perceived by nursing personnel to be the most frequent precipitating factors or causes of back problems (Harber et al., 1985; Jensen, 1985; Owen, 1989; Stobbe et al., 1988; Stubbs et al., 1981; Valles-Pankratz, 1989; Venning et al., 1987). These researchers concluded that assistive devices for use with patient handling tasks could reduce back stress for nursing personnel. However, few devices have been systematically evaluated to determine if their use would reduce back stress.

As part of a contract with the National Institute for Occupational Safety and Health (NIOSH), Owen and Garg (1989) delineated those patient-handling tasks perceived as most stressful by nursing assistants in a nursing home/personal care facility. An ergonomic evaluation was conducted in this facility which revealed elements important to the use of assistive devices:

---

*This study was part of a contract funded by the National Institute for Occupational Safety and Health (Contract 200-86-2923). Roger C. Jensen, J.D., Ph.D., PE (Project Officer).

many stressful tasks were completed in confined work space, most wheelchairs and geriatric chairs did not have removable arm rests to facilitate ease of transfer, and brakes on some beds and wheelchairs did not hold. In addition, there were patient characteristics that could impact on use of assistive devices: weakness, combativeness, muscle rigidity, spasticity, pain, obesity, and inability to bear weight.

The next goal was to find assistive devices that had the potential to reduce back stress for nursing assistants while performing stressful patient handling tasks. The purposes of this part of the study were to: 1) establish criteria for selection of assistive devices to be used while carrying out stressful patient handling tasks, 2) develop strategies for locating available devices, 3) implement these strategies, and 4) conduct a preliminary trial to determine which assistive devices should be recommended for systematic laboratory evaluation.

CRITERIA FOR SELECTION OF DEVICES

1. The device must be appropriate for the task to be accomplished. The tasks delineated as most stressful by Owen and Garg (1989) and studied as part of the NIOSH contract are: transferring a patient from wheelchair to toilet and back to wheelchair, from bed to wheelchair and back to bed, and from wheelchair to shower chair (for toileting and bathing) and back to wheelchair. Devices that can only transfer patients in a horizontal position are not useful with these tasks; for example the Dixie Smooth Mover is a light weight polyethylene board 56 cm x 206 cm with cut out areas for hand grips to be used to transfer patients in a horizontal position, such as from bed to cart/stretcher.
2. The device must be safe for both patient and nurse. It must be stable, strong enough to secure and hold the patient, and permit the nurse to use safe biomechanics.
3. The device must be comfortable for the patient; this may also help to allay fears. It should not produce or intensify pain, bruising, or tear the skin.
4. The device should be understood and used with relative ease. Bell (1984) and Owen (1988) found nursing personnel were reluctant to use assistive devices because they could not understand how to use them or lacked experience in their use.
5. The device must be efficient in the use of time. According to Bell (1984) and Owen (1988) the most frequent reason given for not using a device was the time needed for use.
6. Need for maintenance should be minimal. The above two authors found lack of proper functioning a major reason for non use.
7. The device must be maneuverable in a confined work space. Owen (1988) and Valles-Pankratz (1989) found space to be a problem.
8. The device should be versatile. It could be inferred from Bell's findings (1984) that only a few assistive devices should be introduced at a time because the error rate and the need for time to execute the transfer increased when more than two devices were included in a teaching program.

STRATEGIES FOR LOCATING DEVICES

A <u>literature search</u> was conducted to find assistive devices that could be used with selected patient-handling tasks; few were found, and even fewer had been systematically evaluated. Gagnon et al. (1986) studied lifting a "patient" out of a chair with the hands, with forearms behind the patient's back at shoulder level, and with a belt around the patient's waist; they found the belt technique to be the most strenuous. Leinweber (1978) used the belt as a transfer device with 20 patients; she was successful with all but three patients (an uncooperative patient, one with arthritic back pain, and an obese patient). Bell (1984) evaluated seven different mechanical hoists and recommended improvement in the design and manufacture of the hoists and slings and in the service provided by the manufacturers and their marketing agents. Mechanical hoists were the assistive devices most frequently described in the literature (Gifford, 1966; Kilbom et al., 1985; Lloyd et al., 1987; Takala and Kukkonen, 1987; Waters, 1988). Other devices described were slings, a turn table for pivoting, a walking belt with handles, and sliding boards (Bell, 1984; Hayne and McDermott, 1982; and Lloyd et al., 1987). Most basic nursing textbooks have a chapter relating to body mechanics and lifting/transferring techniques with information about hoists and belts.

A questionnaire was sent to nursing homes/personal care facilities to determine types of transfer assistive devices in use (Owen, 1988) and to seek the opinion of nursing personnel concerning these devices. Gait belts, bathtub lifts, hoists, and lift sheets were the only devices used frequently. Eighty percent (n=99) felt increased use of assistive devices could reduce back problems. The most frequently cited reasons for non use of devices were: too time consuming, staff lack knowledge/experience, inadequate staffing, and lack of availability. Some respondents had diverse but strong opinions about gait belts; thirteen indicated gait belts were required and had reduced back injuries by 75%, while five did not use gait belts due to broken ribs and tearing of skin.

At three medical supply stores, discussions were held with product specialists and consultants, and device catalogs perused.

Product coordinators and rehabilitation specialists (nursing, physical therapy, and occupational therapy) at two hospitals and three nursing homes were consulted about their knowledge and advice about assistive devices.

Nursing personnel experienced in the use of assistive devices were interviewed in three hospitals and five nursing homes/personal care facilities. Most stated they used hydraulic lifts and gait belts most often because of institutional policy; major problems cited were time and loss of independence for patients.

The category of Patient Transfer Unit (General) in the Directory of Products Index was examined at the University Reference Library. A vast array of assistive devices were found that could be used only for horizontal transfers or were stretcher-like devices that converted into a chair. The latter devices did not seem as appropriate as hoists because nursing home/personal care facility patients are generally transferred from chair to toilet and back in addition to transfer from bed to chair.

ASSISTIVE DEVICES

Few assistive devices seemed appropriate for study with the selected patient- handling tasks and fit the established criteria. The most frequently used devices for transfer were the gait belt (Fig. 1) and the Hoyer lift (Fig. 8). The Trans-Aid lift (Fig. 9) and sliding board (not shown) were used by several institutions. The Ambulift hoist (Fig. 10), walking belt with handles (Fig. 2), and MEDesign patient handling sling (Fig. 3) were discussed in the literature and recommended by a rehabilitation specialist. Also, the turn table (patient transfer disc) (Fig. 5) and sling with rings (Fig. 4) were described in the literature.

During an ergonomic study in a nursing home/personal care facility, it was found that several transfers could be eliminated if the patient was transferred from wheelchair to a shower chair that could accommodate the patient for toileting and showering. Two shower/toileting chairs (Fig. 6, Fig. 7) were selected on the advice of rehabilitation specialists.

PRELIMINARY TRIALS TO APPLY CRITERIA

Preliminary trials were conducted by the authors and two nurse research assistants to test the devices against the criteria and to determine which devices should be systematically evaluated in an effort to reduce back stress in nursing personnel. The laboratory was arranged as close as

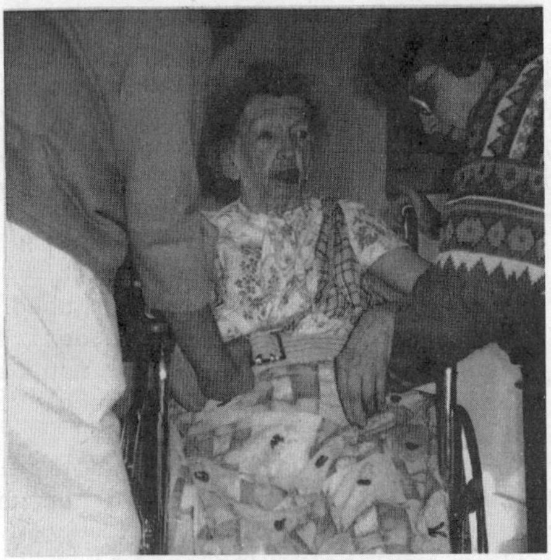

Fig. 1

Gait Belt

is about 5 cm wide, of vary lengths with adjustable belt-like loop or buckle closure, has no handles, and is made of cotton-canvas or nylon material. It should fit securely around patient's waist and is grasped with hands.

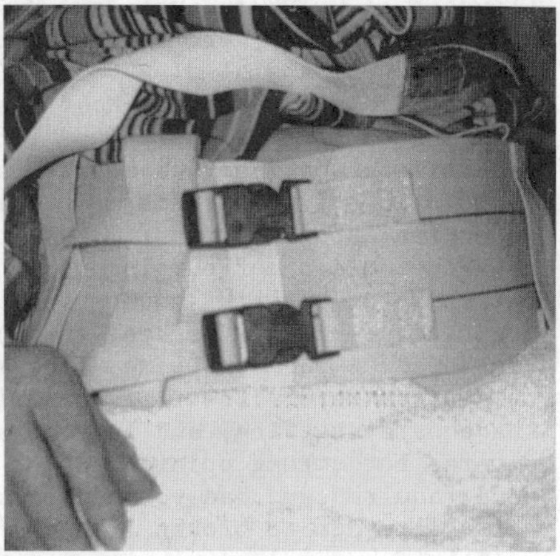

Fig. 2

Walking Belt

is 12.5 cm wide, of varying lengths, has handles on each side, has velcro and two quick-release buckles for closure, is made of cotton-canvas type material, fits snugly around lower abdomen, and is grasped at handles.

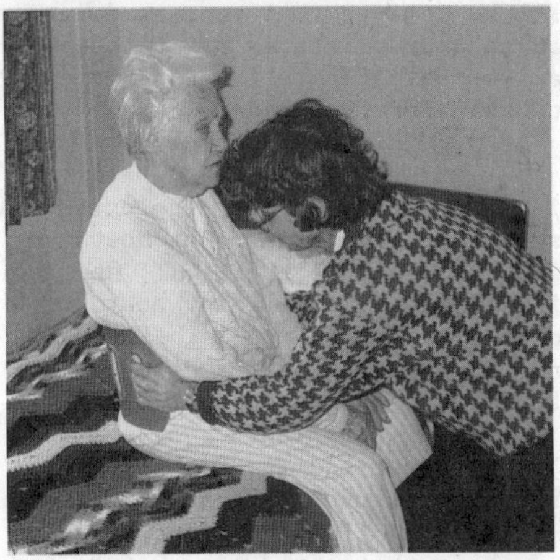

Fig. 3

MEDesign Patient Handling Sling

is 20 cm wide, 50 cm long, has a cut-out at each end allowing a hand grip, is made of flexible polymer material and is tucked securely around patient with bottom at buttock area.

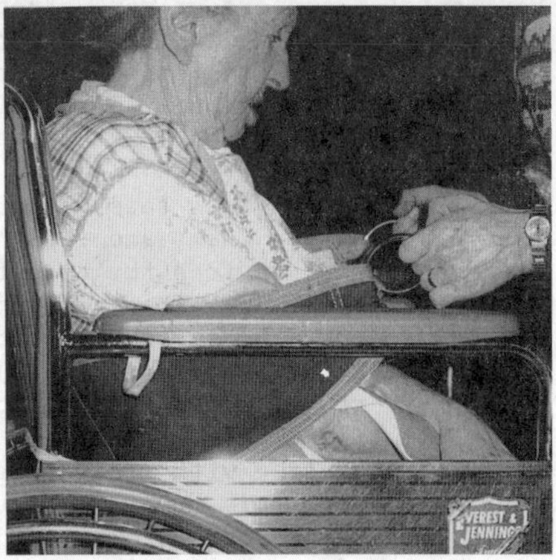

Fig. 4

Patient Sling with Rings

is 23 cm at widest part, 90 cm long, 2 rings with 7 cm diameter attached for hand grip, made of stiff cotton/polyester type material. Position sling around low back area.

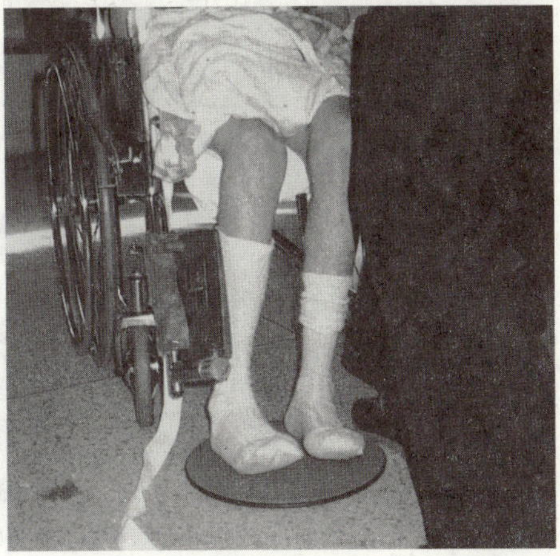

Fig. 5
Turntable
patient transfer disc has rubber threads, rotates, has diameter of 30 or 38 cm, and is placed on floor between transferring locations. Patient places feet on turntable, and is helped to standing position, rotated 90°, and seated in new location.

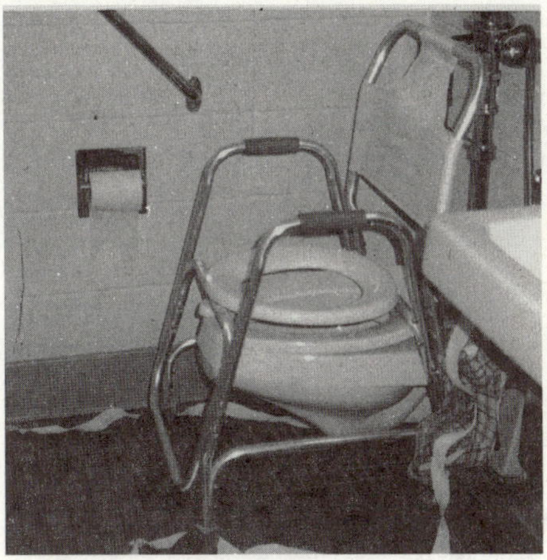

Fig. 6
Shower/Toileting Chair
is light weight chair with a plastic non-moveable seat and has no foot rests.

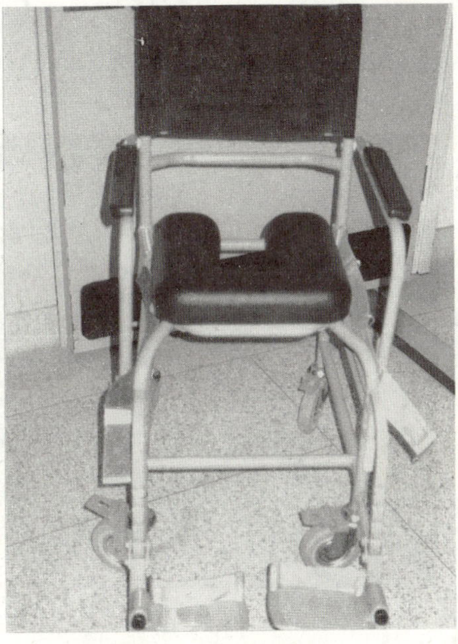

Fig. 7
Shower/Toileting Chair
is a heavy chair with padded removable seat and adjustable/removable foot rests and arm rests that can be lowered.

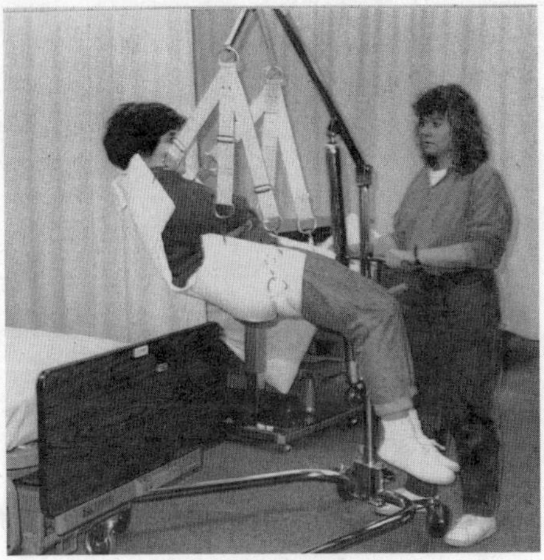

Fig. 8
Hoyer Lift
is an hydraulic lift that has an adjustable base; a pump handle for raising and lowering the patient; and a variety of slings that attach through hooks, chains, or web straps.

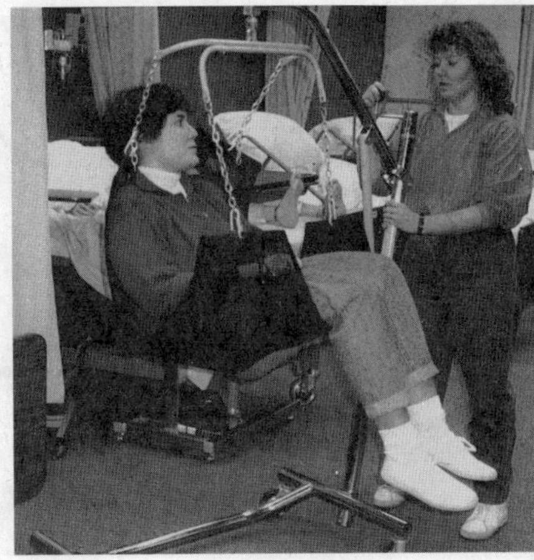

Fig. 9
Trans-Aid Lift
has a non-adjustable "C" base, a ball- bearing screw lifting mechanism with crank in horizontal plane, and has a variety of slings that attach by hooks and dangling color-coded chains.

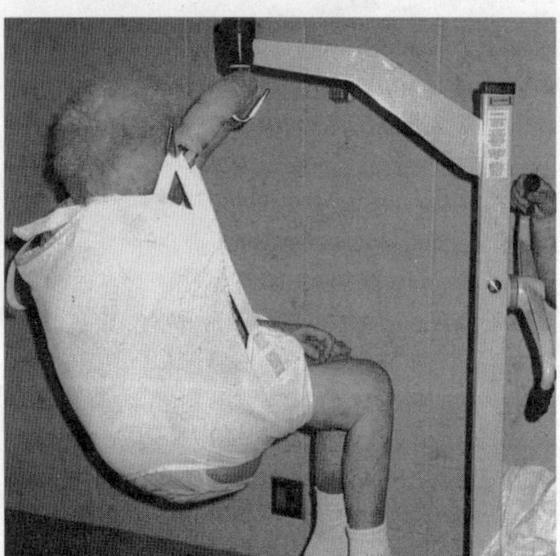

Fig. 10
Ambulift (C-3)
has a semi-adjustable base, a mechanical chain-winding mechanism for lifting/lowering with crank in vertical plane, and the sling attaches by loops and hooks.

A sliding board (not shown) is a smooth rectangular piece of wood about 20 cm wide and of varying lengths. It is bevelled at the edges so the patient (with good upper body strength) can slide from one surface to another of similar height.

possible to the patient environment of the nursing home/personal care facility. The research team served as patient and nursing personnel. It was difficult to simulate patient characteristics (combativeness, spasticity) but an effort was made for little or no weight-bearing during transfers.

The most difficult criterion to meet was the safety of the nurse when using the belts and slings. Body position and movement of the nurse were vital for the safe use of these devices. The ability to get close to the patient, keep the back as straight as possible, flex the knees, and keep the feet apart with one foot in the direction of the move so as not to rotate the spine were all important. The transfers were safer when a gentle rocking motion was used to provide the kinetic energy so a pulling (not lifting) action could be used to transfer the patient.

The walking belt and MEDesign patient handling sling met the criteria so were recommended for further study. The gait belt was not as comfortable as the other belts and slings but was studied further because it was used frequently in many nursing homes/personal care facilities (Owen, 1988).

The sling with rings was eliminated as it did not meet the criterion of safety for the patient or nurse (as perceived by nurse and patient). The belt is long so it cannot be "fastened" around the patient, and the large rings are not stable; these features created a feeling of insecurity and eliminated the ability to create a rocking motion.

The sliding board was not recommended for further study because it did not meet the criterion of appropriateness for tasks to be accomplished due to the following patient and environmental characteristics: many patients do not have the upper body strength and cognitive ability needed to make the transfer and the transfer surfaces must be similar in height, but many wheelchairs and geriatric chairs do not have adjustable arm rests so height cannot be equalized. This may be why Owen (1988) found infrequent use of sliding boards in nursing homes/personal care facilities.

The turntable was eliminated for reasons of appropriateness and safety. Many patients cannot bear weight or are unpredictable in their ability to stand; in addition, because of confined work space the turntable may be stored where patients or personnel could stumble over it.

The shower/toileting chairs were specifically selected for study so that toileting and showering could be done sequentially and hence several transfers could be eliminated. To ensure safety with the light-weight chair; it had to be placed against a wall so it did not tip during a transfer.

All three hoists were recommended for further study. During the pilot, the Hoyer was the least comfortable because the patient tended to be in a reclining rather than upright position; also, the patient swayed more during transfer and at times sensed a feeling of tipping over. The Hoyer was included because Owen (1988) found that it was used frequently.

SUMMARY

Use of assistive devices for transferring patients may be helpful in reducing back stress for nursing personnel. Criteria were established for selection of devices, and strategies were developed for locating available devices. A literature search was conducted, a questionnaire sent and visits made to nursing homes/personal care facilities, visits made to hospitals and medical supply stores, and specialists consulted. Slings, belts, sliding boards, turntables, shower/toileting chairs, and hoists were located. Through preliminary trials the walking belt, MEDesign sling, shower/toileting chairs, Trans Aid Lift and Ambulift hoist were recommended for further study. The gait belt and Hoyer Lift were recommended based on their frequency of use through questionnaire (Owen, 1988). The sliding board, sling with rings and the turntable were not recommended. The most difficult criterion to meet was safety of the nurse in relation to body mechanics while using slings and belts. In the next phase of the NIOSH contract these devices will be systematically evaluated to determine if, with their use, back stress can be reduced when carrying out selected stressful patient handling tasks.

REFERENCES

Bell, F., 1984, Patient lifting devices in hospitals, (London: Groom Helm).

Gagnon, M., Sicard, C., & Sirois, J., 1986, Evaluation of forces on the lumbo-sacral joint and assessment of work and energy transfers in nursing aides lifting patients. Ergonomics, 29, 407-421.

Gifford, M., 1966, Lifting patient, a mobile apparatus. Nursing Times, July 22, 977.

Harber, P., Billet, E., Gutowski, M., SooHoo, K., Lew, M. and Roman, A., 1985, Occupational low-back pain in hospital nurses. Journal of Occupational Medicine, 27(7), 518-524.

Hayne, C., and McDermott, M., 1982, Lifting aids: Handle with care. Nursing Mirror, 154(26), 40-42.

Jensen, R., 1985, Events that trigger disabling back pain among nurses. (Abstract). In Proceedings of the Human Factor Society 29th Annual Meeting, (Santa Monica, CA: Human Factor Society), pp. 799-801.

Jensen, R., 1987, Disabling back injuries among nursing personnel: Research needs and justifications. Research in Nursing and Health, 10, 29-38.

Kilbom, A., Ljungberg, A. and Hagg, G., 1985, Lifting and carrying in geriatric care: A comparison between differences in workspace layout, work organization and use of modern equipment. Ergonomic International 85. Edited by I. Brown et al., (Philadelphia: Taylor & Francis) pp. 550-552.

Klein, B., Jensen, R. and Sanderson, L, 1984, Assessment of workers' compensation claims for back strains/sprains. Journal of Occupational Medicine, 26(6), 443-448.

Leinweber, E., 1978, Belts to make moves smoother. American Journal of Nursing, 78, 2080-2081.

Lloyd, P., Tarling, C., Troup, J. and Wright, B., 1987, The Handling of Patients, A guide for Nurses, (London: The Royal College of Nursing).

Owen, B., 1988, Patient handling devices: An ergonomic approach to lifting patients. In Trends in ergonomic/human factors V, edited by F. Aghazadeh (North Holland: Elsevier Science Publishers) pp. 721-728.

Owen, B., 1989, The magnitude of low-back problem in nursing. Western Journal of Nursing Research, 11(2), 234-242.

Owen, B. and Garg, A., 1989, Patient handling tasks perceived to be most stressful by nursing assistants. In Advances in Industrial Ergonomics and Safety I, edited by A. Mital (Philadelphia: Taylor & Francis Ltd.), pp. 775-781.

Stobbe, T., Plummer, R., Jensen, R. and Attfield, M., 1988, Incidence of low back injuries among nursing personnel as a function of patient lifting frequency. Journal of Safety Research, 19(2), 21-28.

Stubbs, D., Rivers, P., Hudson, M. and Worringham, C., 1981, Back pain research. Nursing Times, 77, 857-858.

Takula, E. and Kukkonen, R., 1987, The handling of patients on geriatric wards. Applied Ergonomics, 18(1), 17-22.

Valles-Pankratz, S., 1989, What's in back of nursing home injuries. Ohio Monitor, 62(2), 4-8.

Venning, P., Walter, S. and Stitt, L., 1987, Personal and job-related factors as determinants of incidence of back injuries among nursing personnel. Journal of Occupational Medicine, 28(10), 820-825.

Waters, K., 1988, Everyday aids and appliances. British Medical Journal, 262, 1114-1117.

Advances in Industrial Ergonomics and Safety II
Edited by Biman Das
Taylor & Francis 1990

ORGANIZATIONAL ISSUES IN DEVELOPING A COMPREHENSIVE BACK INJURY PREVENTION PROGRAM FOR NURSING PERSONNEL.

Kathryn D. Shaver, P.T.

State University of New York
Health Science Center at Syracuse
750 East Adams Street
Syracuse, New York 13210

INTRODUCTION

The nursing profession has been cited in recent literature for a high incidence and prevalence of back pain and injury. Jensen (1987) found that the four occupational groups with the greatest incidence ratios for back injuries were nursing aides, construction laborers, garbage collectors, and licensed practical nurses. Klein et al. (1984) report that the occupational category "nursing aides, orderlies, and assistants" ranked fifth with an annual incidence ratio of 3.6 claims/100 workers. Magora and Taustein (1969) found nursing to have the highest point prevalence (16.8%) of back pain in a comparison of eight occupational groups.

A significant need exists for the development of comprehensive back injury prevention programs for nursing personnel. Owen (1987) reported that 38% of nursing personnel had experienced at least one occupational low back injury or low back pain. Stubbs et al. (1981) found that 43% of nursing personnel surveyed had experienced pack pain; 44% of these identified the incident as having occurred at work, and 87% of the incidents occurring at work were caused by patient handling. Venning (1987) examined data from two geographically different nursing populations and found that 52% of nurses in both the Canadian and Syracuse studies required time-off due to back injuries.

Current trends in prevention programs for occupationally related back injuries can be categorized into two theoretical approaches. 1) An Instructional Approach - training the worker in back care, body mechanics, and ergonomically safe lifting techniques. 2) An Organizational Approach - focusing on the work environment conditions of performance; including equipment needs, staffing, work task redesign, and injury management.

A truly comprehensive back injury prevention program must combine the two approaches (Venning 1988), and must effectively deal with the organizational issues encountered during program development.

The purpose of this paper is to define the organizational issues involved in developing a comprehensive back injury prevention program, and to identify the prats and pitfalls encountered during the development of this program.

PROGRAM DESCRIPTION

The nursing department of University Hospital, of the State University of New York Health Science Center at Syracuse, in response to a committment to decrease the number and severity of time lost injuries in nursing personnel, initiated a project to develop and implement a comprehensive back injury prevention program for nursing personnel. Nursing administration provided the financial support for the hiring of a full-time physical therapist with expertise in the areas of musculoskeletal evaluation and back injury prevention, to initiate instructional program development and address work environment conditions of performance. Project decisions were made by a multidisciplinary team consisting of the Physical Therapist, Nursing Administration, the Employee Health Office, the Nursing Unit Supervisor's Group, and Nursing Staff Development.

In the proposed instructional program; three foci were identified for injury prevention:

I. <u>Primary Prevention</u>: focuses on stopping injuries before they happen;

II. <u>Secondary Prevention</u>: focuses on preventing unreported back complaints from becoming compensation claims.

III. <u>Tertiary Prevention</u>: focuses on injury management and rehabilitation.

The instructional design for this program is outlined in Figure 1.

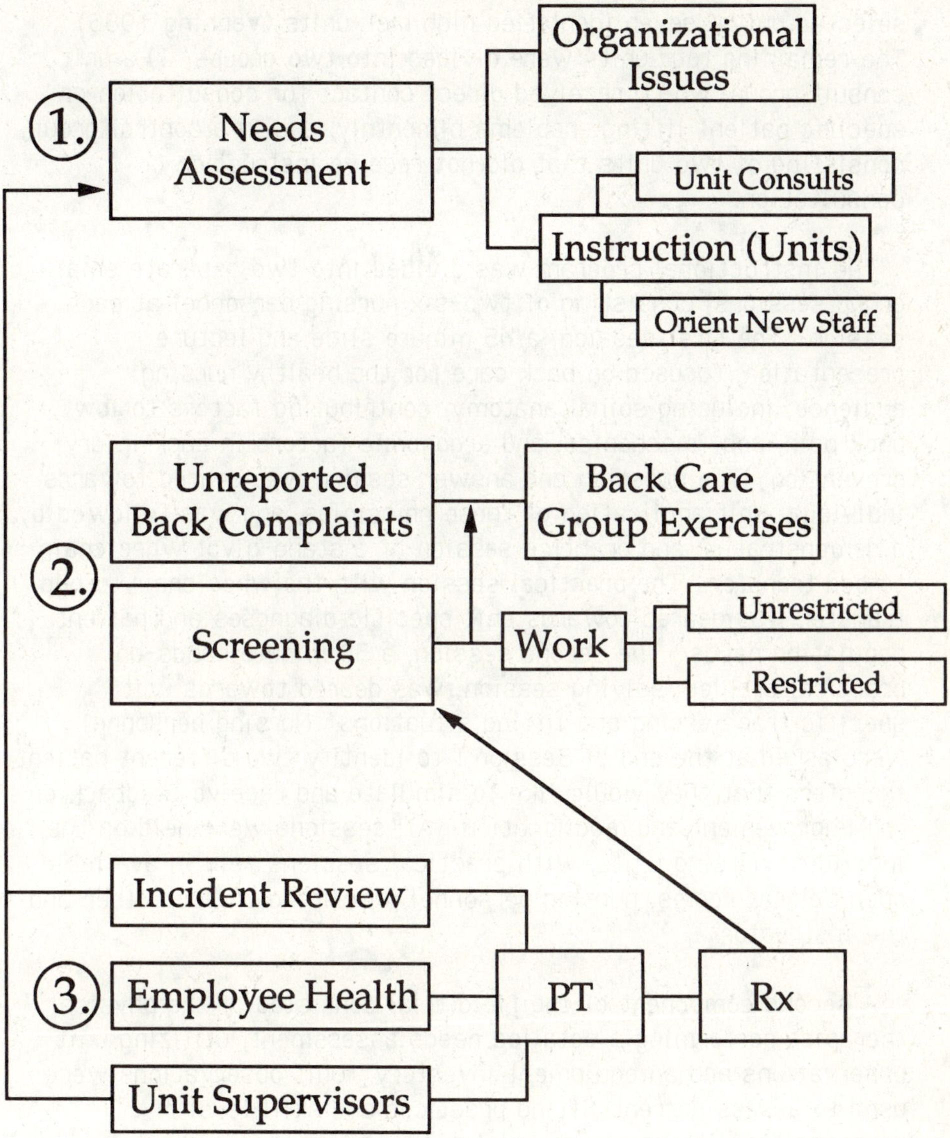

FIGURE 1.

Proposed Instructional Design of Comprehensive Back Injury Prevention Program for Nursing Venning (1988).

A field trial of the instructional program was initiated in November 1988 on three patient care units which were randomly selected out of seven identified high risk units (Venning 1986). The remaining four units were divided into two groups 1) a unit consult group, which received direct contact for consultation on specific patient lifting problems bimonthly, and 2) a control group, consisting of two units that did not receive instruction or consultation.

The instructional program was divided into two separate small group sessions; consisting of two-six nursing personnel at each session. The first session, a 45 minute slide and lecture presentation; focused on back care for the healthy nursing audience, including spinal anatomy, contributing factors to low back pain, body mechanics, and ergonomic factors in back injury prevention. The question and answer session was geared towards individual unit application of these principles, and was followed by a demonstration and practice session of a stand-pivot wheelchair to bed transfer. The practical session with the wheelchair to bed transfer was geared towards unit specific diagnoses and patient population needs. The second session, a 30 minute hands-on practical problem solving session, was geared towards unit specific transferring and lifting situations. Nursing personnel were asked at the end of Session I to identify two different patient transfers that they would like to simulate and receive feedback on for improvement and modification. All sessions were held on the individual nursing units, with practical sessions held in available open patient rooms, nursing personnel practicing on each other and the instructor.

A second component of the field trial consisted of the physical therapist performing a detailed needs assessment, utilizing unit observations and an equipment inventory. Unit observations were used to assess current lifting practices and skill level of personnel, while the equipment inventory assessed current availability of lifting aids. This was followed by recommendations for purchase of additional lifting equipment to aid in back injury prevention, and subsequent orientation and training of nursing staff in the use of the equipment upon arrival at University Hospital.

OBSERVATIONS

The Nursing Administration of University Hospital in Syracuse introduced a unique concept for the development of their back injury prevention program by directly hiring and financially supporting a full-time position for a physical therapist to develop this program. During the development of this program, the importance of addressing key organizational issues was discovered.

The role of the professional in the development and implementation of the program must be clearly defined. An outlined job description of the duties of the professional is a minimum requirement; and must be supplemented by clear administrative objectives for program accomplishments and timeliness for review.

In direct relation to defining the role of the professional, the issue of establishing a communication mechanism within the organizational hierarchy was found to be critical. At University Hospital, the Physical Therapist initially reported to the Director of Physical Therapy with any programatic concerns, and provided a monthly report to the Associate Director of Nursing for Personnel Activities, summarizing monthly activity of the program. A large gap in communication was created; as there was no effective way to communicate nursing department issues to critical Nursing Administration personnel. This issue was resolved by establishing biweekly meetings between the Physical Therapist and an Associate Director of Nursing responsible for overseeing patient care units, in addition to the biweekly meetings with the Physical Therapy Director. This facilitated the delineation of issues to either Nursing Administration or Physical Therapy for problem solving. It was also found beneficial to incorporate the Physical Therapist into the Hospital Safety Committee, to address hospital-wide injury prevention issues, and allow multidisciplinary input into program development.

The identification of key unit personnel, as advocated by Rogers (1983), was later found to be critical in the success of the University Hospital program. It was essential for the Physical Therapist to enlist the support and cooperation of the unit clinical managers in order to encourage staff support and participation.

The identification of key unit personnel such as the clinical managers, who are involved with daily staff and patient interactions, establishes a communication link with nursing staff in a supervisory role whose opinion and abilities are respected by unit personnel. The utilization of the clinical managers was not stressed during the field trial, but was found to be extremely effective in the revised post-trial instruction. The clinical managers offered excellent insight into the daily mechanics of the unit to assist in scheduling of instruction and appropriate staff needs for consultation.

Introduction of the program and its objectives through established organizational channels was a critical issue in gaining the cooperation and trust of nursing staff. This is advocated by Nadler and Wiggs (1986) in discussion of human resource development in management issues. In the University Hospital program, the Physical Therapist presented an outline of the program and its objectives to the Nurse Administrators at a weekly meeting. The Nurse Administrators were then responsible for disseminating this information to their unit staff, particularly to clinical managers, to facilitate the introduction of the therapist and program to the unit. A great discrepancy in program acceptance was found between units. The highest cooperation and support was found on units where the Nurse Administrators had spent time introducing the program concept to staff before the physical therapist came to the unit, and when the nurse administrator or clinical manager personally introduced the therapist and program objectives to the staff.

One of the most successful components of the University Hospital program was the use of small group, hands-on practice sessions. In a post instructional survey of nurses attending the field trial instructional sessions, the use of small group instruction was considered essential by greater than 90% of the participants surveyed. The utilization of unit specific instruction was also cited as very important in the additional comments section of the survey.

Insufficient unit staffing was a critical issue in the implementation of the instuctional sessions. A combination of poor staffing on the nursing units and heavy work load demands resulted in 19 of 45 (42%) Instructional Unit I and 12 of 34 (35%)

Instructional Unit II classes being cancelled. A flexible and adaptable attitude must be taken in planning instructional sessions combined with the professional's ability to set limits concerning their own time constrictions. It was found at this hospital, that two 30-45 minute sessions of instruction with nursing personnel away from their patient care duties was not feasible. An alternative currently being developed is the making of a video specifically for nursing for teaching the relatively "generic" concepts of spinal anatomy, back care, body mechanics and injury prevention. This would be followed by a second unit specific session for hands-on practice.

Addressing the nursing staffing shortage through the provision of additional lifting equipment proved to be very effective at University Hospital. This equipment did not arrive until almost 12 months after the completion of the field trial, creating an organizational problem in teaching design. The instruction included practice with the new lifting equipment; as it was assumed it would be arriving shortly following the field trial. The introduction of four new mechanical lifts with increased weight capacities and the ability to perform floor to bed and commode transfers; transfer belts for stand-pivot transfers, and polypropylene transfer boards for bed to stretcher transfers, has been met with widespread acceptance; substantiated by increased usage and positive feedback. An effective training program utilizing key personnel (clinical managers) to orient unit staff was also found to be beneficial. The utilization of the therapist for unit consultations at University Hospital increased considerably after the introduction of the new lifting equipment. During the field trial, unit consultations averaged one per month; whereas after the introduction of the new lifting equipment, unit consultations increased to up to two per week for specific patient lifting problems.

RECOMMENDATIONS AND CONCLUSION

It is recommended that professionals involved in developing back injury prevention programs for nursing address key organizational issues including:

1) Clear definition of the professional's role.
2) Establishment of a communication mechanism for the professional within the organizational flow of the institution.
3) Identification of key unit personnel for effective communication.
4) Adaptability of training methods to fit individual unit needs.
5) Increased involvement in areas of ergonomics, equipment management and unit consultation

REFERENCES

Jensen, R.C., 1987, Disabling back injuries among nursing personnel: research needs and justification. Research in Nursing and Health, 10:29-38.

Klein, B.P., Jensen, R.C.; and Sanderson, L.M., 1984, Assessment of workers' compensation claims for back strains/sprains. Journal of Occupational Medicine, 26(6):443-448.

Magora, A., and Taustein, I., 1969, An investigation of the problem of sick leave in the patient suffering from low back pain. Industrial Medicine, 38:398-408.

Nadler, L. and Wiggs, G., 1986, Managing Human Resource Development, San Francisco, Jossey-Boss.

Owen, B.D., 1987, The need for application of ergonomic principles in nursing, Trends in Ergonomics/Human Factors IV, (S.S. Asfour, editor), North Holland Publishing, Amsterdam, 831-838.

Rogers, E.M., 1983, Diffusion of Innovation, New York: Free Press.

Stubbs, D.A., Rivers, P.M., Hudson, M.P., Worringham, C.J., 1981, Back pain research, Nursing Times, 77(20):857-858.

Venning, P.J., 1987, Time-off patterns for back injuries in nurses, Trends in Ergonomics/Human Factors IV, S.S. Asfour, editor), North Holland Publishing, Amsterdam 875-882.

Venning, P.J., 1988, (Prospective report on comprehensive back injury prevention program for nursing personnel.) Unpublished data.

Advances in Industrial Ergonomics and Safety II
Edited by Biman Das
Taylor & Francis 1990

REDUCING BACK-STRESSING EXPOSURE OF WORKERS IN A MORGUE

M. Brouwer, S. Churilla, E.J. Hall, C. Harvey, and L. Tyrrell
The Salvation Army Toronto Grace Hospital
Toronto, Ontario
Canada

The Toronto Grace Hospital, originally designed as an obstetrical facility, became a chronic/palliative care centre in 1979. This resulted in an increase in the use of the morgue and recognition of a mismatch between the caregivers'capabilities and the workspace, body handling equipment and storage facilities. Action limits calculated using the NIOSH Work Practices Guide were consistently being exceeded when lifting the deceased into the refrigerator drawers. Multidiscipline committees assessed the physical stresses imposed on the caregivers, conducted market research, made budgetary recommendations, and assisted in the implementation of a satisfactory solution.

INTRODUCTION

The Salvation Army Toronto Grace Hospital is a 119 bed facility in downtown Toronto. The present building was originally designed as an obstetrical hospital in 1959. In 1979, the hospital underwent renovations to accommodate a change to chronic/palliative care. The morgue however, was not considered in these renovations.

Following the change in patient population there was an increase in the use of the morgue and in the number of autopsies performed. It became apparent that there was a mismatch between the workspace, body handling equipment, storage facilities, and the caregivers'capabilities.

IDENTIFICATION & ANALYSIS OF THE PROBLEM

Figure 1 demonstrates the physical layout of the morgue. The refrigerator (1) is in one corner. Two doors open outward to support two sliding drawers, each with a depth of 10 cm. When either of the drawers are open, there is minimal space between the end of the drawer and the wall. The autopsy table (2) is fixed to the floor at right angles to the refrigerator. When a drawer is open, the morgue stretcher (3) can be placed parallel to the drawer with 19 cm between the stretcher and the autopsy table. The stretcher is manually raised by a crank located under the stretcher. At times, the crank is not used due to its poor mechanics and accessibility. The stretcher cannot be adjusted to the level of the lower drawer.

Figure 1. Workspace design of the morgue.

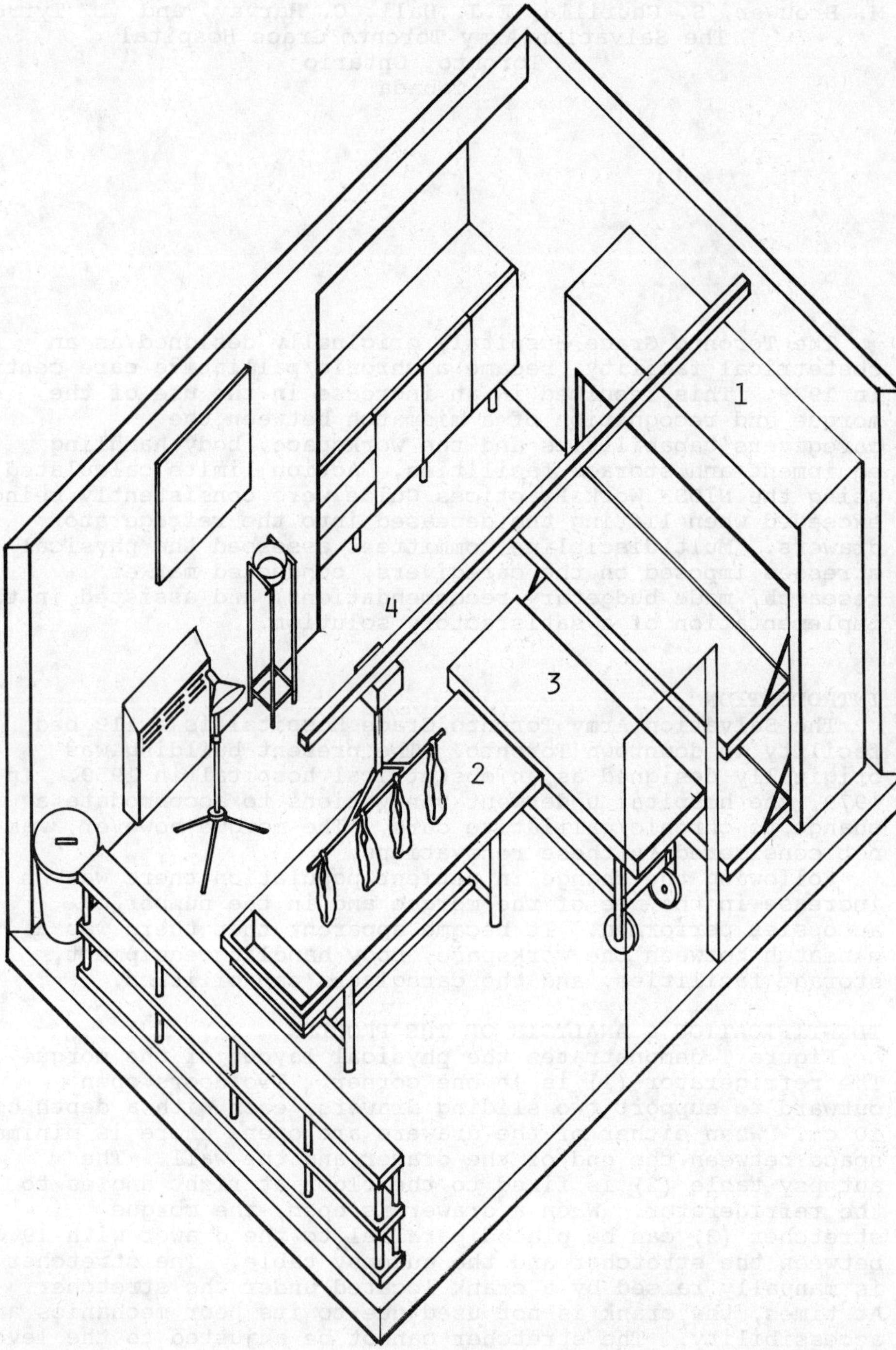

Following a death, between four and seven lifts are required depending upon whether an autopsy is performed. (see figure 2) The number of lifts performed per year from 1979 to 1989 have been identified. (see figure 3) Three injuries occurred during the year with the greatest number of lifts (1987).

Figure 2. Summary of the lifts performed and manpower required in moving the deceased from nursing unit to funeral home.

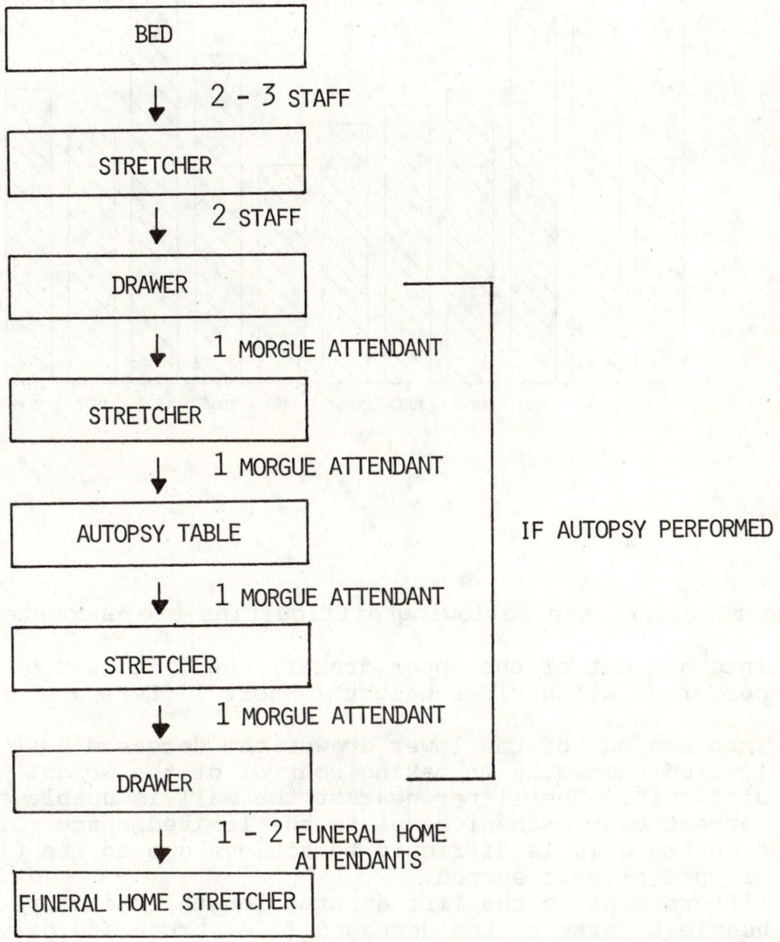

In the majority of cases, staff members performing the lifts are female. All regular nursing staff have been trained in basic back care and safe lifting techniques. Their physiques are variable and fitness is not a requirement for their job. Although nurses are required to handle the deceased, many are reluctant to do so. The morgue attendant (male) works alone and is untrained in back care and body mechanics.

In lifting the deceased from the stretcher into and out of either drawer, one person must stand at each end of the stretcher. There is no room for a third person to assist.

Figure 3. Total number of lifts of deceased per year.

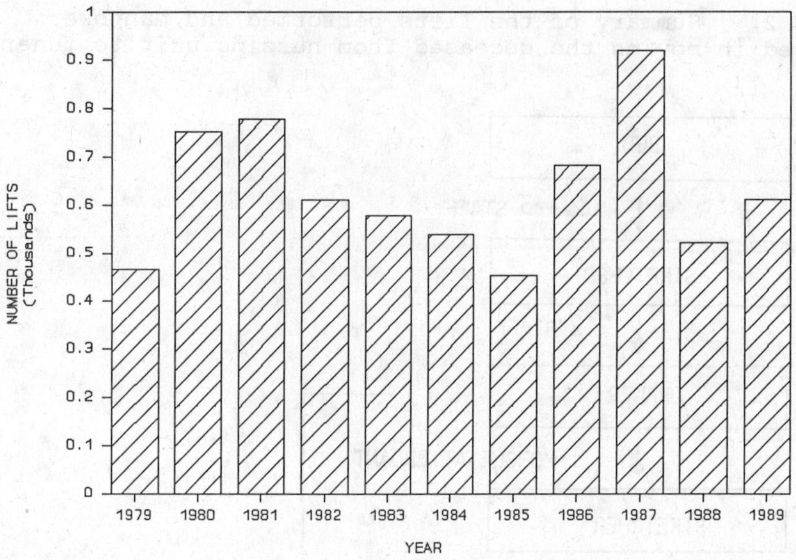

During the lift, the following difficulties are encountered:

1) Into and out of the upper drawer, the lift must be performed at shoulder height. Short lifters are unable to see well.
2) Into and out of the lower drawer the deceased must be lowered/raised 28 cm making control of the weight difficult. The lifter nearest the wall is unable to use correct body mechanics due to the limited space (31 cm).
3) A secure grip is difficult to achieve due to the tightly wrapped plastic shroud.
4) Lifters must do the lift at arms length, therefore the heaviest parts of the deceased (i.e. trunk and pelvis) cannot be controlled.
5) Lifters are unable to weight shift laterally. Twisting of the spine must occur to perform the lift.
6) There is a lack of consistency in technique in performing lifts.

A roller is available but does not solve the above problems.

Concerns regarding these difficulties were brought by nursing staff to the multidiscipline committees of the Toronto Grace Hospital. These committees have a problem solving focus and communication occurs between them regarding matters of mutual concern. The involved committees and their actions are described.

<u>Back Injury Prevention Committee (BIPC)</u>
- investigated nursing concerns and analyzed specific staff injuries incurred as a result of this problem. Three injuries were identified as occurring while performing a lift in the morgue in 1987.

<u>Combined Occupational & Physical Therapy Staff Working Group</u>
- utilized the National Institute of Occupational Health and Safety (NIOSH) Work Practices Guide to objectively measure and define the risk of performing the lift.

<u>Patient Environment & Equipment Advisory Committee (PE&EAC)</u>
- established justification for budgetary recommendations to purchase a mechanical lifting device to alleviate the problem.

<u>Ad Hoc Committee for Morgue Lifts (Sub-committee of PE&EAC)</u>
- researched commercial options for mechanical lifting devices.

<u>Occupational Health & Safety Committee</u>
- was informed of the safety hazard in the morgue and that a solution was being sought.

As previously mentioned, the NIOSH Work Practices Guide was used to analyze the lifts into and out of the upper and lower morgue refrigerator drawers. The analysis provided an objective measure of the risks involved, an accurate description of the demands of each lift, and both a tool and criteria for redesigning the task.

The results of our analysis are presented in Figures 4 and 5. Estimated autopsy weights for 1989 were recorded. The maximum weight was 100 kg. The mean weight was 62.9 kg. It was assumed that each lifter took one half of the weight, therefore a load weight of 31.5 kg was used. In figure 5 the results of the lift into the lower drawer are presented. The Action Limit was exceeded by the load weight. The load weight was less than the Maximum Permissible Limit.

These results indicated that fit workers may be injured and that extra persons, division of the load, or use of equipment was necessary to decrease the risk.

For the upper drawer the load weight remained the same at 31.5 kg, the Action Limit was 9.8 kg and the Maximum Permissible Limit was 29.4 kg (not shown). This indicated a risk to all caregivers and that lifting equipment should be used to the decrease the risk.

RESOLUTION OF THE PROBLEM

These results were used in justifying to Administration the significance of the problem and the necessity of finding a solution. A $5,000 budget request was approved in 1987. Based on the work of the committees, two options were considered:

1) <u>A mechanical lifting system</u>
   Five systems were researched and were found to be unsuitable because:
   a) they would not attain a sufficient height or depth to reach either drawer.
   b) the base of the equipment was too long, preventing the deceased from being positioned directly over the drawer.
   c) they were difficult to manoeuvre in the confined space of the morgue.

The system has been fully operational since August 23, 1989. Ongoing evaluation is in place and statistics are being recorded as part of the Quality Assurance Program. The Assistant Executive Director of Patient Services has recommended that the Maintenance Department include the morgue lift in its Risk Management Program to ensure its continued safe operation.

CONCLUSION

The mismatch between the caregivers' capabilities and the workspace, body handling equipment and storage facilities no longer exists since all manual lifts have been eliminated.

In this paper we have demonstrated how the NIOSH Work Practices Guide was used to objectively analyze the physical stresses imposed on caregivers in a morgue. The results obtained were utilized to direct market research and establish budgetary recommendations. Continued interaction and communication between these committees enabled the WISPALONG System to be successfully implemented, and ongoing evaluation to be planned.

ACKNOWLEDGEMENTS

The authors wish to thank the Ergonomic Services Department of the Health Care Occupational Health and Safety Association - Ontario Hospital Association for their assistance, Mr. Roman Melnyk for his advice in preparing this manuscript, Miss Dianne Cowles for her careful work in transcribing the text into its final form, and Mr. Robert Murray for his artistic interpretation of the workspace.

REFERENCE

NIOSH Work Practices Guide for Manual Lifting. U.S. Department of Health and Human Services, Publication No. 81-122, National Institute for Occupational Safety and Health, Cincinnati, Ohio, 1981.

XIII

MANUAL MATERIALS HANDLING

Advances in Industrial Ergonomics and Safety II
Edited by Biman Das
Taylor & Francis 1990

PHYSICAL DEMANDS IMPOSED UPON U.S. NOAA MARINERS PERFORMING MANUAL MATERIAL HANDLING TASKS

T. L. Doolittle

Department of Environmental Health SC-34
University of Washington
Seattle, Washington 98195

Peter Rigby

Physical Therapy Department
Northwest Hospital
Seattle, Washington 98133

Ergonomic analyses were conducted to determine the physical demands of manual material handling tasks performed by personnel aboard vessels of the National Oceanographic and Atmospheric Administration Pacific Fleet. Lifting requirements for five classifications exceed the NIOSH action limit, and for some tasks approach the maximum permissible limit. Because of the brief effort required and the intermittent nature of most tasks, metabolic energy demands were found to be only a moderately limiting factor. Tests and associated standards, related to performance based criteria, are described as a prudent administrative control.

INTRODUCTION

Making educated decisions regarding returning injured workers to arduous jobs, particularly those involving manual material handling tasks, is often difficult because the individual making them is not familiar with the specific physical requirements. This results in some workers being retained in an off-duty status longer than necessary which unduly increases compensation costs. In contrast, returning underfit workers to the job may result in: (1) aggravation of the original problem, (2) unacceptably high risk for another injury, or (3) substandard performance, because they cannot perform all aspects of the job. Knowledge of the physical demands required by the job, and associated strength and endurance standards can reduce the probability of these occurrences.

The purpose of this study was to develop screening procedures based upon validated job requirements. While the primary intent was to provide these procedures for making return to work decisions, they should be equally applicable for screening new hires and/or as the basis for ongoing fitness standards.

The first major step in this study was the identification of problem tasks and the verification of their criticality. A problem task was defined as one in which there was an actual or potential high injury rate, or one which was perceived to be so physically demanding that some individuals could not perform because of lack of sufficient strength or endurance. A task was deemed to be critical if it had to be done in the prescribed manner either for the safety of the ship or the satisfactory completion of the mission, and no reasonable alternative was available. The second major step was the quantification of the physical demands by obtaining mass, force, distance and temporal data, and the identification of the movement patterns employed by the workers. The third major step consisted of data analysis to translate the demands into specific job requirements and the development of valid test standards predicated upon them.

METHODS

In order to account for variances imposed by ship type, six NOAA vessels representing the three functional types (Hydrographic, Oceanographic, and Fisheries) and the four Classes (I, II, III, and IV) were studied in this project.. The six were selected from the total of eight active vessels comprising these types in the NOAA Pacific Fleet. The ships, classes and primary research function are listed in Table 1. It should be noted that there sometimes is overlap in missions across classes and types. For example: the MAC ARTHUR (Class III Hydrographic) and the DAVID STARR JORDAN (Class IV Fisheries) perform the common mission of the annual marine mammal survey.

Table 1 - Vessels Studied

VESSEL	CLASS	TYPE
SURVEYOR S132	I	OCEANOGRAPHIC
DISCOVERER R102	I	OCEANOGRAPHIC
RAINIER S221	II	HYDROGRAPHIC
MILLER FREEMAN R223	II	FISHERIES
MAC ARTHUR S330	III	HYDROGRAPHIC*
DAVID STARR JORDAN R444	IV	FISHERIES

* Original Type designation; actual employment is best described as multipurpose.

Preliminary meetings with representatives of NOAA (National Oceanographic and Atmospheric Administration) and the USPHS (United States Public Health Service), reviews of injury reports, and a survey of the ships' executive officers enabled the investigators to identify the job classifications that required detailed ergonomic analysis.

Incumbents (current workers in the job classifications that performed the critical tasks) were interviewed, with the total number per vessel ranging from 5 to 12. They represented a "cross-section" (e.g. ship types, duties performed, age, gender) of existing NOAA Pacific Fleet personnel. The interviews elicited what the workers perceived to be the problem (i.e. physically demanding, injury causing) tasks. While questions were asked, investigators avoided "leading the witnesses" and/or stifling their responses. Specific weights and measures (i.e. object mass, dimensions, distances, frequencies, etc.), and perceived difficulty were discussed. Interviews lasted from 20 minutes to 2 hours. A portion of each session was devoted to soliciting any solutions the incumbents had for alleviating the problems.

Physical demands were quantified by reviewing existing videotapes of evolutions being conducted on NOAA vessels and onsite ergonomic analysis. Incumbents performing problem tasks were observed (and in some cases photographed) to determine the frequency of task repetitions, the number of objects handled per repetition, and the distance over which the objects are moved. For those tasks that could not be observed and for which no video tapes existed, detailed descriptions were obtained from experienced incumbents. Lifting, pulling and/or pushing forces required to perform the tasks were measured when applicable.

PHYSICAL REQUIREMENTS

Arduous tasks were identified on each class of NOAA vessel and were grouped according to occupations and missions. On the class II and III vessels, loading stores is typically carried out by the steward's department. On the class 1 vessels, seamen, ensigns and temporary workers (non-NOAA employees), hired solely to help with this operation, may assist the stewards. On class IV vessels virtually all-hands may be involved. The loading of spare parts and equipment is highly similar to loading stores, and typically involves the personnel of the appropriate department. In short, all personnel can be expected to perform these loading tasks. Due to variations in vessel configuration, these tasks vary among the different ships. Other factors (e.g. simultaneously loading fuel, port and pier arrangements) may add to the variations in the task.

The basic components of manual material handling (lifting, carrying, stacking) prevail regardless of the variation. Supplies are typically packed in either boxes or sacks. They are off loaded from a pallet and carried either directly to their destination, or loaded onto an intermediate platform, (e.g. another smaller pallet, an elevator, or a net) which is moved closer to the destination and then off loaded. While pallets and nets are always moved via cranes or forklifts, individual containers ultimately have to be lifted/lowered, carried, and lifted/lowered to their final destination.

Postures expected during this evolution include standing, squatting, stooping and reaching. Loads average between 18 and 27 kg. Frozen products and canned goods tend to be nearer the higher average, while produce tends to be nearer the lower value. Numerous containers are lighter, however, there is a tendency for an individual to handle two or more of these at a time so that the load average still falls within the aforementioned range.

The stores are moved to their destination either by being carried at waist height, or by being passed from person to person in a "bucket brigade" fashion. Obstacles encountered during the carry include ladders (i.e. stairs, typically in sets of ~11 steps, 23 cm high) and doorway thresholds, (of at least equal height), which must be stepped over. Under optimal conditions, the total carrying distance will be less than 30 meters. In unusual circumstances, carrying may be necessary over approximately 100 meters. A moderate walking pace, (67 to 80 m/min), is maintained for the period of the task, which nominally lasts one to two hours at a time. Two to three such evolutions may occur over an 8 hr day. If pressed for a deadline, however, the pace may be significantly faster, with double the volume being moved in the same time period. Individuals performing this task, typically, are working at a metabolic demand approximating 4 METs.

In general, moderate to heavy exertion of muscular force is required of personnel in the deck/fishing and survey departments, rendering them essentially inseparable in terms of physical demands. Typical tasks included lifting objects with masses up to 45 kg in the process of deploying and retrieving nets, buoys, and other scientific apparatus. Because of the intermittent nature of these tasks, significant aerobic demands are not imposed by them.

The engineers also encounter high intensity lifting associated with repairs to the power plant. Components with masses from 20 to in excess of 45 kg must be manipulated in a variety of awkward positions. Electronic Technicians likewise are confronted with lifting and carrying heavy objects. Their responsibilities include repairing virtually all electronic equipment on board the vessel (e.g. radars, sonars and TVs), which requires removal, carrying and replacement of potentially heavy items. Again because of the infrequency of these tasks, significant aerobic demands are not imposed by these activities.

The physical requirements recorded for the stewards primarily reflect the demands of loading and handling stores. The other physical demands in this job category include lifting and carrying: cleaning equipment, full pots and refuse barrels. The latter tasks require less aerobic demand than moving stores.

The yeomen/medical and quartermaster personnel incur few, if any, physically demanding tasks on a routine basis. All personnel (including these classifications) are expected to be able to perform damage control tasks in the event of an emergency. Some of these tasks (e.g. two individuals carrying a P-250 pump with a mass of 72 kg) are physically demanding, and warrant the imposition of higher strength and endurance standards than could be justified by the routine job requirements.

Tasks performed by seamen, fishermen and survey technicians may vary as a function of the mission being performed by the vessel, whereas the other job classifications perform the same tasks regardless of the mission. The Class II hydrographic vessels perform few other types of missions thereby exposing their deck and survey personnel to fewer tasks requiring heavy lifting. Oceanographic measurements require the deployment and retrieval of heavy scientific equipment. The oceanographic vessels may perform some fishing operations in addition to conducting oceanographic surveys. In general, the handling of fishing gear is slightly more demanding physically than that of other scientific equipment, if for no other reason than that it is done with greater frequency. The fisheries vessels, in addition to fishing, perform some oceanographic research. In addition, these vessels tend to spend more days at sea than their oceanographic and hydrographic counterparts. Fisheries vessels also are staffed with correspondingly fewer people. Exposure to injury, thus, would appear to be higher on this type of vessel. A more thorough analysis of a larger data base, however, would be required to substantiate this hypothesis.

Arduous lifting tasks were identified on each class of NOAA vessel and grouped according to mission requirements and job classifications. While some of the masses are only informed estimates, most have been objectively measured, and are substantiated by the earlier work of Robertson and Trent (1985) and Beckett and Hodgdon (1987). The lifting requirements are summarized in Table 2.

The required movements and associated masses were analyzed utilizing the equations of Garg, Chaffin and Herrin (1978) to determine the metabolic requirements of the studied positions, and with the equations of NIOSH (1981) to compare the requirements with the action limit and maximum permissible limit. Because of the low frequency with which these tasks are performed the aerobic demands are relatively low compared with many industrial settings. The tasks, however, often are performed on a rolling pitching platform which may be wet and slippery. These phenomena add an unknown amount to the metabolic demands, if for no other reason than the additional muscle activity required to maintain one's balance under these conditions.

Table 2 - Examples of lifting Requirements

Job Classification / Item Moved	Mass (kg)	Type of Movement	Frequency*
Engineers			
Buckets	23	Deck to Shoulder	Routine
Components	20-45+	Various	Infrequent
Seamen			
Deploying Gear	38	Deck to Chest	Routine
Batteries	23	Deck to Shoulder	Routine
Loading Stores	18-27	Deck to Overhead	Sustained
Fishermen			
Reeling Nets	45+	Push/pull	Routine
Deploy Net Tow	45	Deck to Chest	Routine
Loading Stores	18-27	Deck to Overhead	Sustained
Survey Technicians			
Same as Seamen and Fishermen			
Niskin Bottle	4-13	Knuckle to Waist	Routine
Niskin Bottle	42	Knuckle to Waist	Questionable
Stewards			
Loading Stores	18-27	Deck to Overhead	Sustained
Refuse Barrels	10-30	Floor to Waist	Routine
Pots/Pans	10-20	Floor to Waist	Routine
Electronic Tech			
Equipment	20-45	Various	Infrequent

*Frequency of the movement (e.g. lifted, pushed, pulled etc.)

Sustained	=	continuous repetitive movement (several per minute) over protracted period.
Routine	=	intermittent activity (versus continuous) performed one or more times on most days.
Infrequent	=	Essential task performed when the need arises
Questionable	=	May be done by some individuals, but requirement cannot be justified.

STRENGTH AND ENDURANCE STANDARDS

The tasks were represented by standardized movements which translated into a strength test battery consisting of four measures: (1) Military Press with 22.7 kg, (2) Modified Upright Row with 22.7 kg, (3) Biceps Curl with 22.7 kg, and (4) Squat with 36.3 kg. A multiple repetition format with a submaximal mass (Doolittle and Kaiyala 1988, Doolittle 1989b), is recommended for estimating dynamic lifting capacity. While the metabolic demands are moderate, an aerobic testing program is necessary because of the low fitness level of the population (Doolittle 1989a).

Since the relationship between safe job performance and physical capacity is a continuum, rather than an absolute cutoff, a five[1] category scheme was developed. All the categories are based upon percentages of maximum at which an individual will be taxed. They

[1] There are five categories for the strength measures, but only four for aerobic power due to the more restricted range of the latter variable.

relate directly to the uniform job requirements necessary to perform the ergonomically analyzed tasks.

Low Risk - Better Qualified

Individuals in this category will be taxed at less than 33% of their maximum aerobic capacity by the upper range of the metabolic costs for the job. The strength demands will be less than 65% of maximum capacity (i.e. they will possess ≥ 35% reserve) adjusted for frequency. Individuals in this category are highly fit relative to the job demands. They should have little difficulty adapting to the job demands, have low risks to on the job musculoskeletal injuries, and have a very high probability of job success.

Acceptable

This is the targeted level of capacity based on job analysis findings. At the lower range of this category the strength demands will be 75% of maximum capacity adjusted for frequency, and the individual will be taxed at 33% and 40% of their maximum aerobic capacity for the lower and higher values of the metabolic costs on the job, respectively. Individuals scoring in this category will find the job moderately hard but not overly difficult. Risk to injury is within normal limits, and the individuals have a good probability of success.

Moderate Risk - Marginally Qualified

The lower point for moderate would tax the individual at 85% of the persons strength capacity adjusted for frequency. That is, the individual would have only 15% reserve strength capacity. The individual would be taxed at more than 40% of his or her maximum aerobic capacity by the lower range of the metabolic costs for the demanding jobs. While the risk to injury is higher than desired and the individual will find the job difficult, with a high level of effort he or she may be able to meet the minimum standards. With an adequate applicant pool one should not have to select new hires from this category, however, individuals may be considered if they score high in this category and possess other desired attributes.

High Risk

Individuals scoring at the lower end of this category will be taxed at 90% of strength capacity adjusted for frequency. It is included solely for return to work decisions, where appropriate discretion can be exercised. New hire candidates should not be selected from this category. (Note: this category is not applicable to aerobic power measures)

Risk Too High - Below Minimum

Anything below Marginal. If hired, a candidate scoring in this category would be taxed at more than 40% of his or her maximum aerobic capacity by the lower range of metabolic costs, or more than 90% his or her strength capacity adjusted for frequency. The risk of injury to the individual and turnover to the employer are too high to make returning the person to work or hiring him/her a prudent decision.

Levels

The job classifications, because of similar demands, have been grouped into three intensity levels for strength and two for aerobic power.The categories and levels are combined in Table 3 to provide standards for an individual to be fully qualified for all strength requirements of the respective job classifications. Table 4 contains the aerobic standards. Since new hires should be qualified for all aspects of the job, it is recommend that the "Acceptable" category or higher on each test be utilized as standards in making hiring decisions. An applicant who falls below the "Marginally Qualified" category on any procedure should not be further considered for employment, since it is highly unlikely that the person will be able to perform to a minimally acceptable standard within a prudent level of risk to musculoskeletal injury. Doolittle and Rigby (1989) provide more definitive tables for making return to work decisions.

DISCUSSION

The purpose of this project was to identify and quantify the physically demanding job requirements aboard NOAA ships, and develop strength and endurance standards based upon these requirements that could be utilized in making decisions as to whether or not a previously injured worker could return to work. Implicit within any decision of this nature is that the individual could safely perform all of the critical tasks demanded by his or her occupation. That is, the person would possess sufficient strength and endurance reserves to perform the physically demanding tasks with a tolerable risk of injury. One recommendation for a modification to a critical task was made that should reduce the probability of its causing injury.

Table 3 - Strength Standards

Test	Relative Risk to Injury (Reserve Capacity)				
	Too High (< 10%)	High Risk (10-14%)	Moderate (15-24%)	Acceptable (25-34%)	Low Risk (≥ 35%)
Level I: Fishermen, Engineers, Electronic Technicians; and Survey Technicains on Fishing Missions					
Military Press w/22.7 kg	<5	5-6	7-10	11-17	>17
Biceps Curl w/22.7 kg	<16	16-17	18-24	25-33	>33
Squat w/36.3 kg	<4	4-5	6-10	11-17	>17
Upright Row w/22.7 kg	<16	16-17	18-24	25-33	>33
Level II: Seamen, Survey Technicians (Other than Fishing Missions), Chief Engineer					
Military Press w/22.7 kg	<5	5-6	7-10	11-17	>17
Biceps Curl w/22.7 kg	<13	13-15	16-21	22-29	>29
Squat w/36.3 kg	<2	2-3	4-7	8-14	>14
Upright Row w/22.7 kg	<13	<13-15	16-21	22-29	>29
Level III: Chief Boatswain, Chief Survey Technician Yeoman/Medical, Quartermaster					
Military Press w/22.7 kg	<5	5-6	7-10	11-17	>17
Biceps Curl w/22.7 kg	<5	5-6	7-10	11-17	>17
Squat w/36.3 kg	<2	2-3	4-7	8-14	>14
Upright Row w/22.7 kg	<5	5-6	7-10	11-17	>17

Table 4 - Aerobic Power Standards

Test	Categories			
	Below Minimum	Marginally Qualified	Acceptable	Better Qualified
Level I: Seamen, Fishermen, Survey Technicians, Stewards				
Aerobic Power Test (in METs)	<8.8	8.8 - 10.4	10.5 - 12.7	>12.7
Time 1.5-mile test (min'sec")	>17'41"	17'41" 14'32"	14'31" - 11'47"	<11'47"
Level II: Chiefs (All Departments), Engineers, Yeomen/Medical, Quartermasters, Electronic Technicians				
Aerobic Power Test (in METs)	<8.0	8.0 - 9.4	9.5 - 11.0	>11.0
Time 1.5-mile test (min'sec")	>19'42"	19'42" - 16'15"	16'14" - 13'47"	<13'47"

That was that the practice of carrying full 32 L niskin bottles (42 kg) be discontinued. Other alternatives exist so that this practice cannot be justified. In keeping with this recommendation, the standards that have been developed do not provide a screening tool for this specific task. To have done so would have negated the recommendation and been contrary to sound ergonomic practice.

The strength standards are well anchored, as describe herein, on thorough job analysis and sound ergonomic principles. They cannot be compared with other values in the literature, since none are known to exist. The recommended strength standards are predicated upon absolute forces. That is, lifting a 23 kg battery requires the same force regardless of the size of the individual, and the standards have been set accordingly. The Navy's Physical Readiness Program (OPNAVINST 6610.1C), in contrast, imposes strength standards that are relative to the individual (e.g. push ups, where the individual is lifting a portion of his or her body weight) rather than the masses that must be lifted on the job. Direct numerical comparisons between the two sets of standards, thus, are not possible.

The strength standards, recommended herein, undoubtedly will appear to be high to some individuals. It is reiterated that they are anchored in the job lifting requirements, and that most of these are relatively demanding. Because of logistical limitations imposed upon this study, the masses and forces were determined while the vessels were in port or underway during relatively calm seas. Considering the degree of pitch and roll which may occur while the tasks are being performed was beyond the scope of this study. Such action of a vessel could effect the force (undoubtedly increasing it) required by an individual to lift any given object. Further study of the impact of these phenomena appears to be warranted. An equation presented by Bittner and Guignard (1985) holds promise for modeling the effect of pitch and roll upon lifting requirements. Controlled tests of the model would enable one to verify that the standards are sufficient or whether they should be modified. In the interim, the recommended standards are the best that are currently available.

The recommended aerobic standards are more modest compared with those for the strength measures. This is a valid representation of the jobs where few tasks are aerobically demanding. Current NOAA medical standards require that an individual with a coronary problem has to complete Stage III on the Bruce treadmill test, before he or she can return to work. Performing to this standard indicates that the individual has an aerobic capacity approximating 10 METs. While there is no evidence that this standard was based upon any analysis of job requirements, it is in close agreement with the current recommendations (Table 4). No mention is made by the NOAA standards for individuals who do not have coronary disease, but ostensibly normals should be expected to perform as well, if not better. The Navy's standards, bracket those being recommended. They are lower for older females and higher for younger males, but they were not validated against job requirements.

As soon as standards are set, they are subject to challenge. Care must be exercised, therefore, in establishing them, and the ground rules agreed upon. For example, an injured employee desiring to return to work who cannot meet a certain lifting requirement may claim that it is too high, since "so and so" who is on the job cannot do so. Or perhaps the individual knows that he or she could not have "passed" the test before the injury. These are not unusual circumstances, since the existing work force was hired prior to the establishment of the standards, but they do not negate the validity of the standards. Experience indicates that approximately 20% of a given workforce will be unable to meet the standards of a newly installed physical performance screening system. Two primary reasons exist for this assertion.

First, some individuals are performing the job, but doing so at an undue risk to muscular skeletal injury. They are routinely functioning at greater than 75% of their capacity, thus, are moderate to high risk candidates for muscle strains. While they can "do the job", returning such persons to work without provision for light duty (or selecting them as new hires) would not be a prudent decision.

Second, some individuals perform most, but not all, of the tasks required by their occupation. An example of this would be retrieving nets, where one individual operates the controls (nondemanding) and two others physically fairlead the cable or perform some other physically demanding task. If within this team of three workers, one of the individuals always operates the controls, that person never experiences the physically demanding critical task. The person, thus, may perform without being qualified for the full scope of the position. As another example, one steward that was interviewed indicated that they personally never picked up and carried the heavier items, but left those for someone else to handle. Again, this person was not performing the full scope of the job. Human dynamics will dictate whether such individuals are tolerated (even protected) or resented. If the person is a new hire, resentment is more likely to prevail. In any case, when one person can not, or does not, perform all of the demanding tasks required by the job, their share must be distributed among their fellow workers which increases the total demands on these individuals.

The above examples, in addition to illustrating why some current employees may not be able to meet the proposed standards, demonstrate two facts regarding the concept of limited duty. First, despite the repeated enunciation by all NOAA personnel interviewed that "there is no limited duty at sea", informal limited duty does exist. Second, limited duty aboard NOAA vessels is feasible within limitations. It should be utilized within the following guidelines.

1. It is a temporary measure. The individual is in the process of recovery and can be expected to regain the requisite capacities for the full scope of the position within the foreseeable future, and
2. The individual's current or upcoming assignment does not require that he/she perform the tasks for which they are under fit (e.g. their fitness level is sufficient to meet the expected tasks that he/she will have to perform on the mission), or
3. The crew size is sufficient so that other members of the team (department) can adequately handle the more demanding tasks (e.g. in the net retrieval example at least two of the three have to be capable of fairleading the cable).
4. It could be a permanent measure for a limited number of experienced personnel (e.g. a Chief or Lead) where the demand for their knowledge compensates for lack of physical prowess. This concept was utilized in Tables 3 and 4 where the Chiefs have been placed one Level lower than the other members of their departments.

In addition to being able to render decisions regarding limited duty assignments, provision exist for returning individuals to work with somewhat higher risk of injury should the situation dictate. As has been pointed out, it is desirable and ergonomically sound that individuals possess 25% reserve strength capacity before being reassigned. Circumstances may exist, however, in which a less conservative decision is in the best interest of the individual and the total system.

ACKNOWLEDGEMENT

This study was supported in part by the United States Public Health Service, Occupational and Beneficiary Health Service Reasearchand Development Program, Contract Number 88-0109-88J.

REFERENCES

Beckett, M. & Hodgdon, J., 1987. Lifting and Carrying Capacities Relative to Physical Fitness Measures. Report No. 87-26. San Diego: Naval Health Research Center. 43pp.

Bittner, A. C. & Guignard, J. C., 1985. Human factors engineering principles for minimizing adverse ship motion effects: theory and practice. Nav. Eng. J., 205-213.

Doolittle, T., 1989a. Aerobic testing for manual material jobs. Advances in Industrial Ergonomics and Safety, Volume 1. (Mital, A. (Ed.)), London: Taylor & Francis, 105-112.

Doolittle, T., 1989b. Selection of manual material handling tasks using dynamic strength tests. Advances in Industrial Ergonomics and Safety, Volume 1. (Mital, A. (Ed.)), London: Taylor & Francis, 515-521.

Doolittle, T. & Kaiyala, K., 1988. Prediction of maximum dynamic strength from multiple repetitions with a submaximal load. Trends in Ergonomics /Human Factors V. Aghazadeh, F. (Ed.), North-Holland: Elsevier Science, 767-774.

Doolittle, T. & Rigby, P., 1989. Validation of physical demands requirements and development of associated recommended strength and stamina standards for shipboard operations personnel of the NOAA Pacific fleet. Tech. Report IEL/DEH 01-90. Seattle: Ergonomics Laboratory, Department of Environmental Health, University of Washington. 49pp.

Garg, A., Chaffin, D. B., & Herrin, G. B., 1978. Prediction of metabolic rates for manual materials handling jobs. American Industrial Hygiene Association Journal, (39), 661-74.

NIOSH, 1981. Work Practices Guide for Manual Lifting. Pub. No. PB82 178948. Cincinnati: National Institute for Occupational Safety and Health. 183 pp.

OPNAV (Office of the Chief of Naval Operations), 1986. Physical Readiness Program. OPNAVINST 6610.1C. Washington, D.C.: Department of the Navy.

Robertson, D. & Trent, T., 1985. Documentation of Muscularly Demanding Job Tasks and Validation of an Occupational Strength Test Battery (STB). MPTL TN 86-1. San Diego: Naval Personnel Research and Development Center.

Advances in Industrial Ergonomics and Safety II
Edited by Biman Das
Taylor & Francis 1990

ANALYSIS OF SUBJECTIVE RESPONSES AND HEART RATES FOR A COMBINATION MANUAL HANDLING TASK

Clarence C. Rodrigues

Trenton State College
Trenton, NJ 08650

INTRODUCTION

The evils of manual materials handling with respect to musculoskeletal injuries and their related direct and indirect costs have been well documented (NIOSH, 1981, Asfour et al., 1983, Mital and Ayoub, 1986). It is for this reason several researchers have conducted studies for decades to determine ways of designing jobs that enable optimum human performance while reducing injuries and fatigue. Material handling studies have included pushing, pulling and lifting. Among these, lifting is the most widely studied category. Lifting studies have included symmetrical and asymmetrical lifting, each of which have been conducted in the static and dynamic modes.

Traditional methods to studying lifting include the biomechanical, physiological and the psychophysical approaches. While the biomechanical analysis uses the laws of newtonian mechanics to estimate forces and torques on the body during lifting, the physiological approach uses bio-indicators like heart rate, oxygen uptake and electromyograms as measures of stress and to set limits on manual exertions. Psychophysical methods analyze subjective estimates of task demands which do not lead to exhaustion and excessive fatigue to set stress limits. Unlike the biomechanical protocol, both the physiological and the psychophysical methods take into account fatigue generated from task performance over time.

Of the above methods, the psychophysical approach is generally the most favored as it is applicable to all types of industrial tasks without the necessity of maximum exertions. The measurement criteria of this method is based either on the maximum acceptable weight of lift, the maximum acceptable frequency of lift or distance of carry, and the ratings of perceived exertion. While numerous studies have used maximum acceptable weights and/or frequencies as studied parameters, the use and analysis of subjective ratings for studying lifting have been few and far apart. Unlike machines, humans are capable of responding subjectively to differences in their surroundings. Thus, when human subjects are exposed to different experimental conditions, their opinions regarding differences between such conditions can reveal important information. These opinions can be obtained by survey questions.

Through the years many different types of survey forms have been used to measure subjective responses of subjects in various settings. Individual and general body part comfort surveys have been used extensively in seated research and other aligned areas (Shackel et al., 1969, Corlett and Bishop, 1976, and Drury and Coury, 1982, to mention a few), where working heart rates do not rise appreciably above resting levels.

For tasks that are continuous and heavy, researchers have relied on the the Borg's Scale (1962) for the measurement of subjective responses. Mital and Manivasagan (1983), used this well-established scale to measure ratings of perceived exertion when studying one-handed material handling tasks. The Borg's scale which varies from 6 on the lower end to 20 on the higher end in terms of perceived exertion, predicts values that are one-tenth the working heart rate of subjects for whole body tasks. For one-handed lifting however, responses were measured at one-eight of the heart rate.

McCoy, Congleton, Johnston and Jiang (1988) in their study on abdominal belts (CompVest Vs Air Belt) as an aid to lifting, posed 7 questions to their subjects regarding the assistance received from the belts and any comfort or restriction problems. Each question had a scale that varied from 1 (lowest response) to 5 (highest response). Results showed that the CompVest belt was preferred over the Air belt.

Congleton, Ayoub and Smith (1985) used a version of the Body Part Discomfort Rating Form to evaluate the Neutral Posture Chair for surgeons during microsurgical procedures. Lines of about 10 cm in length were placed alongside the body parts in question, and the subject placed an X on the line corresponding to the body part in discomfort. The subject's X was measured from the left-hand end of the scale which started at a value of 1. No numbers were written on the scale, and all that the subjects saw were blank lines next to body parts on which markings of X were required. If no mark was made on a line for a given body part, a zero value was recorded, indicating no noticeable pain or discomfort.

The present study was designed to simulate the loading of grocery bags from standard American grocery carts into car trunks. This is an example of an asymmetrical combination manual handling task in which the lifting begins at a lower level than the final placement of the load. Also, the subject is required to perform a continuous lift-and-lower sequence over an obstruction (trunk sill) while turning through 180 degrees. This type of lifting over obstructions is commonly found in industry (ex. palletizing). Due to the nature of the experimental protocol (Rodrigues, 1988) the lifting was not suspected as being very strenuous. Also, the Borg's Scale while proven for whole body exertions, provides little information about the discomfort felt in the individual body parts. It was thus felt that the Body Part Discomfort Rating Form of Congleton et al. (1985) would be most appropriate for use in the present study, as it included all the body parts that were likely to be affected by the lifting.

The objectives of this study were to determine if individual body part discomfort ratings and working heart rates could adequately differentiate between the six different obstruction (sill) height and depth combinations for this asymmetrical task of loading of grocery bags into car trunks.

METHODS

Subjects

Eighteen healthy university male subjects were chosen for the study. The subjects ranged in age from 20.5 to 31.5 years with a mean of 23.9 years, and in weight from 61.5 kgs to 102 kgs with a mean of 80 kgs.

Experimental Layout

The subjects were made to lift conventional grocery paper bags. These bags had base dimensions of 30 cm x 18 cm and were 43 cm high. Coded plastic bags filled with steel shots were used by the subjects to add or subtract weight from the grocery bag for adjustment during the lifting. Two rectangular wooden plates of 70 cm and 90 cm in height served to simulate the sill heights. The sill depths of 28 cm, 43 cm and 57 cm, were achieved by adjusting the height of a wooden platform on which the lifts terminated using standard concrete blocks. The distance between the origin and the sill was 100 cm and the distance from the sill to the final placement of the load was 36 cm. Figure 1 depicts the layout.

Variables

The independent variables were the two sill heights and the three sill depths for each of the two sill heights. The dependent variables were the marked responses on the Body Part Discomfort Rating form and heart rates.

Procedures

Subjects were subjected to the externally paced task of lifting while adjusting, grocery bags filled with weights over different sill height and depth combinations. The task was psychophysical in nature with a modification in the frequency (lifts per minute) of lifted loads (Rodrigues et al. 1989). In this study subjects performed 6 continuous lifts one after the other. They then walked about for two minutes before performing the next set of lifts. This two minute cycle time was arrived at by actually measuring the time interval between two successive loading operations during peak hours at a local grocery store, while the 6 lifts represented the average number of grocery sacks loaded into car trunks during these peak hours (Rodrigues, 1988). The height of origin of the lift (15 cm) represented the bottom shelf of the standard grocery cart. The sill heights represented the lower and upper bounds, and sill depths represented the lower, mean and upper bounds of measurements made on 26 car trunks from 5 different categories (Rodrigues, 1988).

The subject performed a stint of six lifts one after the other while adjusting the weight before taking a 2 minute break. There were 8 seconds between each lift during which the subject adjusted the weight, performed the lift, and the helper brought the bag back to the origin. During the 2 minute break the subject walked about a defined area before returning to perform the next lifting stint. This procedure of 6 lifts followed by a 2 minute walk continued for a period of about 22 minutes. During the lifting the subjects were constantly encouraged to make adjustments to the load. For each treatment combination subjects performed two such lifting sessions of 22 minutes with a 5 minute break between each, always alternating between heavy and light starting loads. The average load lifted between these two sessions was recorded as the final lifted weight for a given sill height and depth combination. After the second 22-minute session the subject was made to complete the Body Part Discomfort Rating Form.

RESULTS AND DISCUSSION

Body Part Discomfort Ratings

Discomfort ratings were measured using a standard inch scale. The beginning of the subjective rating scale (just noticeable pain) for any given body part was lined up with the 1-inch mark and was assigned a value of 1.0, while the 3.5-inch mark coincided with the end of the subjective rating scale (intolerable pain) and was assigned a value of 3.5. Thus, any response could have a minimum value of 1.0 and a maximum value of 3.5. From the results of the survey it was evident that none of the body parts were stressed appreciably. The maximum average response was 1.13 for the hand (Table 1), which was just over the "just noticeable pain" range. The hands generated the greatest responses of discomfort across treatments than any other body part indicating that these parts were stressed the most overall. This was expected since the gripping was extremely poor on the paper bags, requiring very high grasping forces. Figure 2 shows the rating form used in this study together with a typical subject's response.

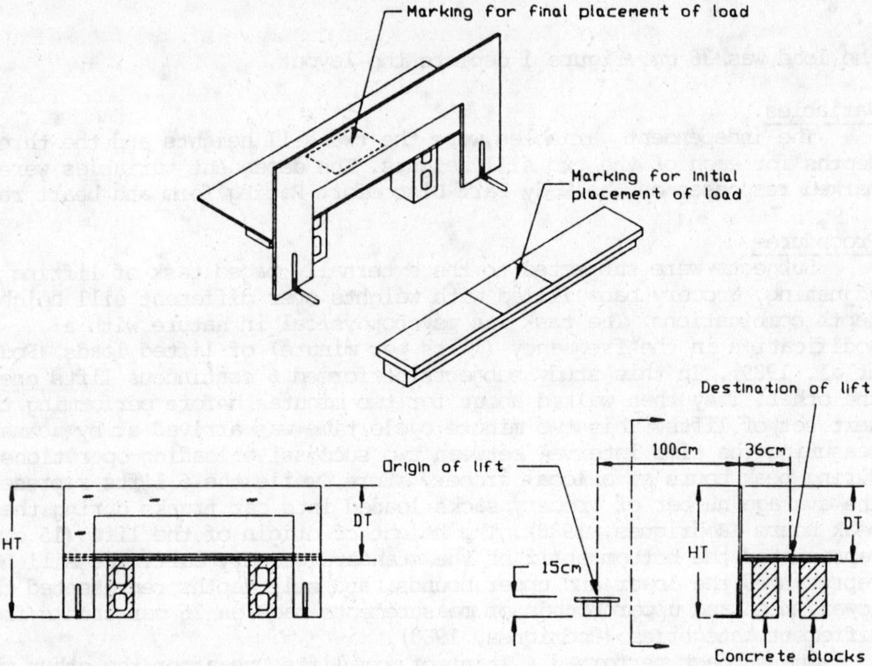

Figure 1. The experimental layout.

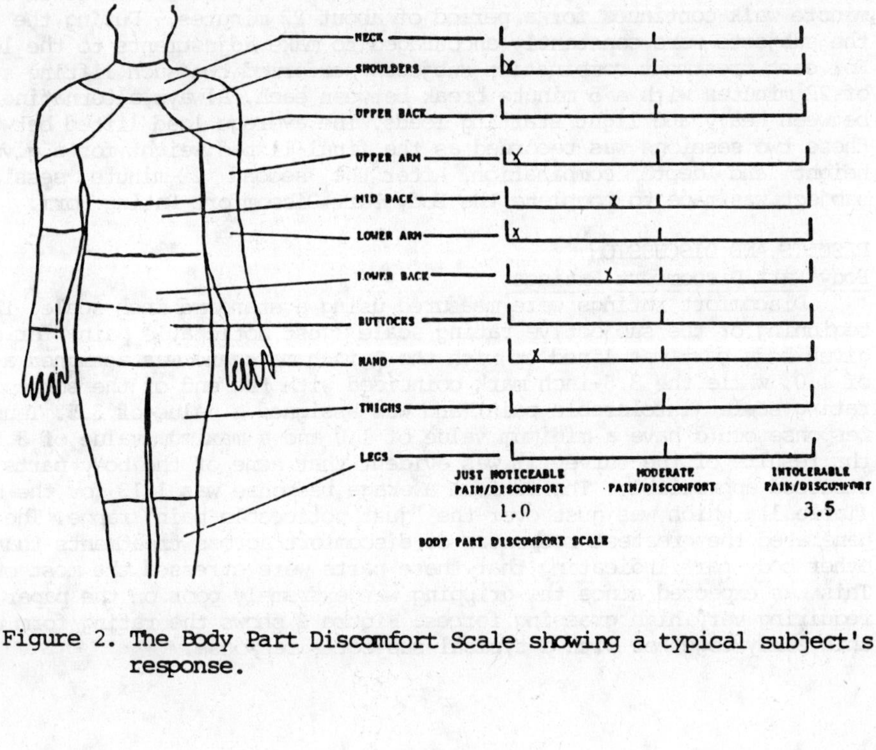

Figure 2. The Body Part Discomfort Scale showing a typical subject's response.

Table 1. Summary of Heart Rate Responses and Body Part Discomfort Ratings Across Treatment Combinations.

	A1	A2	A3	A4	A5	A6
Heart Rates						
Average	95.0	98.1	106.7	96.2	98.8	107.8
Std. Dev.	3.9	5.9	8.7	4.1	7.8	7.0
Discomfort Ratings (Average)						
Mid-back	0.21	0.22	0.46	0.27	0.48	0.52
Hand	0.95	1.13	0.73	1.04	0.73	0.70
Thighs	0.31	0.19	0.45	0.20	0.13	0.44

A1=Sill height of 70 cm and sill depth of 28 cm, A2=Sill height of 70 cm and sill depth of 43 cm, A3=Sill height of 70 cm and sill depth of 57 cm, A4=Sill height of 90 cm and sill depth of 28 cm, A5=Sill height of 90 cm and sill depth of 43 cm, A6=Sill height of 90 cm and sill depth of 57 cm.

Analysis of variance procedures were performed on the data to investigate the effects of sill height and depth combinations on the ratings of discomfort for the individual body parts. The results of the analysis revealed that although the hands generated the overall highest responses of discomfort, only mid-back and thigh responses exhibited statistically significant differences across treatments. Further investigations revealed that it was the depth (with p-values of 0.050 for the mid-back and 0.014 for the thighs) and not the height that was generating such differences (Table 2 and Table 4). Multiple range

Table 2. ANOVA Table with Ratings of Perceived Exertions for the Mid-back as the dependent variable.

Source of Variation	df	Mean Square	F-Ratio	Level of Significance (p)
Model (height, depth, subject, height * depth)	22	0.92	3.80	0.0001
Error (depth * Subject, height * subject, height * depth * subject)	85	0.24		
Corrected Total	107			

Source of Variation (Model)	df	Mean Square	F-Ratio	Level of Significance (p)
Height	1	0.43	1.77	0.1875
Depth	2	0.59	2.44	0.0500
Subject	17	1.08	4.46	0.0001
Height * Depth	2	0.12	0.51	0.6029

comparisons (Table 3 and Table 5) showed that for both the body parts in question, the 57 cm sill depth generated higher responses of discomfort

Table 3. Duncan's Multiple Range Test for Mid-back Responses for the Sill Depths.

Duncan Grouping	Mean Response	N	Sill Depth (cm)
A	0.4917	36	57.0
B	0.3472	36	43.0
B	0.2361	36	28.0

Note: Mean with the same letter are not significantly different at alpha = 0.05, df = 85 and MSE = 0.24.

Table 4. ANOVA Table with Ratings of Perceived Exertions for the Thighs as the dependent variable.

Source of Variation	df	Mean Square	F-Ratio	Level of Significance (p)
Model (height, depth, subject, height * depth)	22	0.73	5.48	0.0001
Error (depth * Subject, height * subject, height * depth * subject)	85	0.13		
Corrected Total	107			
Source of Variation (Model)	**df**	**Mean Square**	**F-Ratio**	**Level of Significance (p)**
Height	1	0.04	0.32	0.5727
Depth	2	0.60	4.48	0.0141
Subject	17	0.87	6.49	0.0001
Height * Depth	2	0.06	0.45	0.6419

Table 5. Duncan's Multiple Range Test for Thigh Responses for the Sill Depths.

Duncan Grouping	Mean Response	N	Sill Depth (cm)
A	0.4111	36	57.0
B	0.2542	36	28.0
B	0.1556	36	43.0

Note: Mean with the same letter are not significantly different at alpha = 0.05, df = 85 and MSE = 0.13.

than either the 28 cm or 43 cm sill depths. This reinforces the concept suggested by Snook 1978, and confirmed by Rodrigues et al. 1989, that as the lowering depth increased, the task got more strenuous and its effects surfaced in the form of decreased load that subjects were willing to lift with increasing depths. No significant differences in responses were obtained between the 28 cm and 43 cm sill depths. Similar findings were reached for lifted loads across these two sill depths (Rodrigues 1989).

Heart Rates

Heart rates were more sensitive to the treatment combinations than the subjective ratings. The mean heart rates for the treatment combinations is shown in Table 1. Analysis of variance procedures (Table 6) showed that sill depth significantly effected pulse responses (p=0.001). Multiple range comparisons (Table 7) clearly showed that the sill depth of 57 cm was the most strenuous since it generated the highest pulse rate. Unlike subjective responses, pulse rates differentiated between the the 28 cm and 43 cm sill depths with the 43 cm sill depth being more stressful.

Table 6. ANOVA Table with Heart Rate Responses as the Dependent Variable.

Source of Variation	df	Mean Square	F-Ratio	Level of Significance (p)
Model (height, depth, subject, height * depth)	22	237.43	19.27	0.0001
Error (depth * Subject, height * subject, height * depth * subject)	85	12.32		
Corrected Total	107			

Source of Variation (Model)	df	Mean Square	F-Ratio	Level of Significance (p)
Height	1	28.01	2.27	0.1353
Depth	2	1326.18	107.66	0.0001
Subject	17	149.55	12.14	0.0001
Height * Depth	2	0.40	0.03	0.9682

Table 7. Duncan's Multiple Range Test for Heart Rate Responses for the Sill Depths.

Duncan Grouping	Mean Response	N	Sill Depth (cm)
A	107.3	36	57.0
B	98.4	36	43.0
C	95.6	36	28.0

Note: Mean with the same letter are not significantly different at alpha = 0.05, df = 85 and MSE = 12.32.

CONCLUSIONS

The body part discomfort rating form of the above configuration was among the first to be used in lifting research, which traditionally used the Borg's scale. The information derived from its analysis appears to have good applicative potential. The ratings were independently able to differentiate between treatment combinations. However, there were no simple defined relationships between heart rates and discomfort ratings like the ones established by Mital and Manivasagan (1983). Rigorous correlational analysis would be required to establish such relationships.

REFERENCES

Asfour, S.S., Khalil, T.M., Moty, E.A., Steele, R., and Rosmoff, H.L., 1983, Back pain: A challenge to productivity. Proceedings of the VIIth International Conference on Production Research, Pub: University of Windsor, Canada & Elsevier, pp. 813-818.

Borg, G., 1962, Physical performance and perceived exertion. Gleerups, Lund.

Congleton, J.J., Ayoub, M.M. and Smith, J.L., 1985, The design and evaluation of the neutral posture chair for surgeons. Human Factors, 27(5), pp. 589-600.

Corlett, E.N. and Bishop, R.P., 1976, A technique for assessing postural discomfort. Ergonomics, 19, pp. 175-182.

Drury, C.G. and Coury, B.G., 1982, A methodology for chair evaluation. Applied Ergonomics, 13, pp. 195-202.

Mital, A. and Ayoub, M. M. 1986, Equation for predicting lifting capabilities of people. Trends in Ergonomics/Human Factors III, pp. 767-774.

Mital, A. and Manivasagan, I., 1983, Subjective estimates of one-handed carrying tasks. Applied Ergonomics, pp. 265-269.

NIOSH, 1981. Work Practices Guide for Manual Lifting (Publication No. 81-122). U. S. Department of Health and Human Services, Centers for Disease Control, Cincinnati, Ohio.

Rodrigues, C. C. 1988, Dynamic asymmetrical lifting capacity modeling. Ph.D. Dissertation, Texas A&M University, College Station, Texas.

Rodrigues, C.C., Congleton, J.J., Koppa, R.J. and Huchingson, R.D., 1989, Maximum acceptable weight of lift for an asymmetrical combination manual handling task. International Journal of Industrial Ergonomics, 4(3), pp. 245-253.

Snook, S. H., 1978, The design of manual handling tasks. Ergonomics, 21(12), pp. 963-985.

Shackel, B., Chidsey, K.D., and Shipley, P., 1969, The assessment of chair comfort. Ergonomics, 12, pp. 269-306.

Advances in Industrial Ergonomics and Safety II
Edited by Biman Das
Taylor & Francis 1990

Manual lifting: job design and research issues

Daniel Imbeau[1], Yves Beauchamp[1], Martin C. Normand[2], Paul-André Courtois[1], and Denis Marchand[2]

[1]Département d'ingénierie
[2]Département des sciences de l'activité physique

Université du Québec à Trois-Rivières
P.O. Box 500, Trois-Rivières
Québec, G9A 5H7, Canada

This paper briefly describes the results of preliminary work and posture analyses of manual materials handling (MMH) tasks performed in two jobs at a warehouse of a liquor and wine distribution company. It then discusses some research and design issues in relation to the current MMH literature.

INTRODUCTION

Despite the growing use of mechanization and automation for material handling tasks, manual material handling (MMH) is still prevalent in many industrial areas. Economic reasons often explain this situation. Lifting tasks must be adequately designed and whenever possible, workers should be tested prior to employment to avoid overexertion injuries (NIOSH, 1981). Nowadays, the job designer is faced with a MMH literature that is growing both in volume and complexity. It is difficult for him to keep track of all research results published in this area over the last ten years.

The motivation for this paper came from a project that was initiated at the warehouse of a major liquor and wine distribution company in the province of Québec during fall of 1989. The objective of the project was to investigate the MMH tasks which are claimed to be associated with high incidences of low back problems. This goal would be achieved by identifying risk factors associated with MMH tasks through accident records and work analyses to eventually propose design modifications. The MMH literature was thus surveyed and during this process, a number of questions surfaced.

THE PROJECT

The warehouse currently employs 166 workers performing materials handling out of which, 29 are actively involved in MMH tasks specifically. This figure has remained fairly constant over the last ten years. Preliminary examination of the accident records showed that a relatively high proportion of back problems were reported by individuals who are currently or have worked in the past at either of two jobs involving MMH. Therefore, both of these jobs were selected for further investigation. In both of them, workers handle an average of 900 boxes daily with weights ranging from 11 to 23 kg. The boxes usually contain either 10 or 12 bottles of wine or liquor of all kinds and hence, vary greatly in their dimensions.

Job #1 (10 workers) consists of order picking. The worker picks liquor or wine boxes from cells and stacks them on a pallet that he (or she) carries using an electric powered stand up forklift (Figure 1). Each assembled pallet, containing between 50 and 60 boxes, roughly corresponds to one order. The travel path in the warehouse with the forklift is determined by the content of the order. The boxes to be picked are stacked on pallets as they were received from the supplier. These pallets are stored in cells arranged in rows on both sides of a 3 m wide alley. It is worth noting that there is no access on the side of the cells so that the boxes must always be picked from the front of the cell. Also, many of the suppliers palletize the boxes in such a way that to access the front-most box (closest to the worker) of the bottom row without straining, all boxes on the higher rows must be picked in a stairway fashion starting with the rear-most box of the top row on the pallet (Figure 1). Such a situation makes the content of the pallet more stable but it forces the worker to often handle boxes at arms'reach.

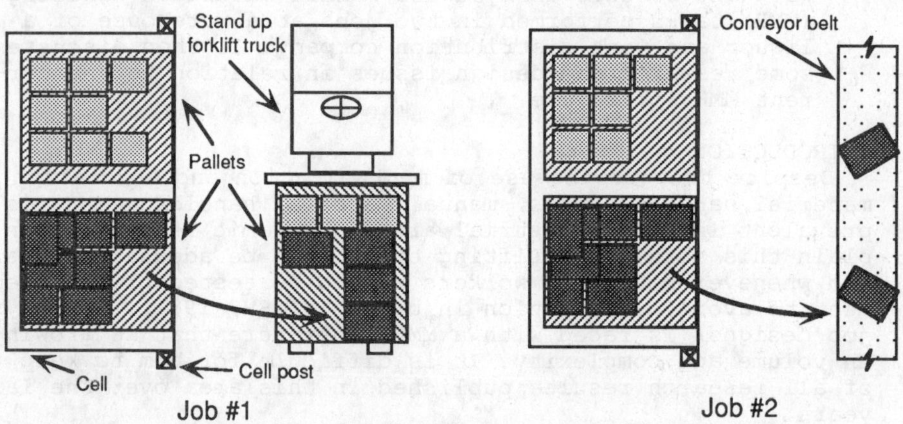

Figure 1. Typical layouts in the two jobs observed.

Job #2 (13 workers) involves picking boxes from cells as in job #1, and then putting them on a 66 cm high and 46 cm wide conveyor belt located at about 1.12 m in front of the cells. In this job, the workers must walk fairly long distances since the cells are arranged along the conveyor belt in rows of nearly 120 m long. Because of the length of the aisles, several workers (4 to 6) can work at the same time in one aisle. As in job #1, the box arrangements within the cells are such that workers must often handle boxes at their arms' reach. Also, in both jobs workers must evacuate empty pallets (34 kg each) from the cells.

<u>Work analysis</u>

A preliminary work analysis including body posture analysis was performed for both jobs. The workers were videotaped while performing their normal duties. The films were then analyzed using a body posture grid similar to that used by Imbeau and Lortie (1984).

In job #1, the work cycle appears to be fairly constant in terms of body movements and thus, can be broken down into four basic components: grasping the box, moving the box to its destination, lowering and placing the box on the pallet, and then coming back for the next box. There is no lowering and placing component in job #2 since the boxes are thrown on the conveyor belt. A cycle time break down for both jobs is shown in Table 1.

The cycle is shorter in job #2 since there is no lowering and placing operation. The box throwing on the conveyor belt in job #2 also contributes to shorten the cycle length as shown by the shorter moving-the-box time. The cycle length corresponds to a frequency of between 8 and 18 boxes handled per minute in job #1, and between 14 to 24 in job #2. The overall daily average frequency is two boxes per minute which indicates that box handling in these jobs is highly discontinuous. Very few of the boxes have handle holes which forces the worker to assume an asymmetric hand position.

Table 1. Results of the preliminary work and posture analyses at two MMH jobs.

Components of the cycle	job#1 time (s)	%	job#2 time (s)	%
grasping and lifting of the box	0.8-1.2	20	0.5-1.2	25
moving the box	1.0-1.5	24	0.8-1.2	30
lowering and placing of the box	0.5-2.5	28	----	--
coming back	1.0-2.0	28	1.2-1.8	45
Complete cycle	3.3-7.2		2.5-4.2	

Some of the boxes are stacked head down on the pallets in the cells (wine boxes particularly) and hence, the center of gravity is higher than the physical center of the box. In these particular but frequent cases the worker senses the situation upon grasping the box and then rotates it quickly but smoothly while handling it. Ayoub et Mital (1989) reported a number of studies which have shown that a static lateral offset of the center of gravity of a load reduces the lifting capacity. The situation we observed is somewhat more complex than the research results published on the impact of the center of gravity location of the load handled. It involves not just a static offset of the center of gravity but rather, a quick but carefully controlled dynamic rotation of the load by the worker while performing a MMH task. No doubt that this dynamic motion of the center of gravity of the load adds to the stresses generated on the back structures by the MMH task. More research is required to assess the impact of such situations on the MMH operations.

It is interesting to note that in both jobs, the grasping-and-lifting-the-box times are equivalent and account for the smallest proportion of the cycle length. It was observed that the duration of this component of the cycle is strongly influenced by whether the worker has to slide or unstick (excesses of glue have been reported to occasionally bond boxes together) the box prior to lifting it. As expected, the posture during this portion of the cycle is dependent upon the height of the box; the posture is erect for the top row only and includes a stronger portion of sagittal flexion of the trunk (i.e., about 78% of the time) as the initial position of the box is lower. The workers we observed never used the squat (straight back, bent knees) posture for lifting. They almost always lifted with a stoop (bent back and straight legs) posture.

For the moving-the-box component of the cycle, it was observed that the posture often (81% of the time) includes some torso twisting since the box is moved laterally; the box is grasped on one side and thrown (job #2) or moved (job #1) towards the other side. During this operation, a weight transfer from one foot to the other always takes place. The torso twisting amplitude decreases as the posture becomes less constrained, that is, when the worker can move his (or her) feet during the operation. The amount of sagittal flexion throughout this component of the cycle is constant and dependent upon the initial height of the box; the worker maintains the amount of sagittal flexion he (or she) assumed at the time the box was grasped. In other words, if the box was grasped at a lower height, the worker did not lift it to an erect posture to move it but rather kept a stooped posture. The movement path of the box appeared to be along a fairly straight line between its origin and destination. Sagittal flexion, when present, was always combined with torso twisting during this component of the cycle and was found in 41% of the MMH tasks observed. It was also observed that several workers prop their buttocks against a cell post to grasp at arms reach and move a box that is farther and lower on the pallet rather than getting closer to lift it and then walking to the destination with it.

The coming-back portion of the operation appeared to involve roughly the same posture as during the moving-the-box portion. It often included torso twisting and sagittal flexion combinations in both jobs.

Furthermore, since the pallets in job #1 were put down on the floor, the worker had to put one foot on the pallet to get to the center and rear-most boxes. Owing to this situation, the feet were not at the same level when boxes were handled. This situation was not present in job #2.

The observations at both of these jobs can be summarized as follows: the box handling operations were discontinuous and performed quickly; the boxes were often grasped at arms' reach; the grasping of the boxes often involved asymmetric hand positioning; the boxes had to frequently be rotated while being handled; the postures during box handling often involved torso twisting and a good portion of sagittal flexions (i.e., stooped posture); and the MMH tasks often involved weight transfer from one foot to the other.

The recent MMH literature was surveyed to assess the status of the design of these existing jobs and to eventually enable the formulation of sound recommendations. While performing this survey, a number of questions surfaced. One of the most important one is the fact that all MMH tasks we observed involved multiple forms of asymmetry while most of the research results are either based on symmetric movements performed in the sagittal plane or involve a single form of asymmetry (i.e., torso twisting). Another area of concern is the empirical model building techniques used to generate the capacity prediction models in both the psychophysical and the physiological approaches. Also, the applicability of the current research results based on the psychophysical approach to our concrete situation can be questioned. These are only a very few of the issues that came up during the literature review. Due to space constraints, it is impossible to discuss all of them. In the following section, these very few issues are discussed keeping in mind the results of our observations.

Asymmetric MMH

The most outstanding outcome from our preliminary observations is that MMH tasks involved various forms of asymmetry.

This finding is consistent with other studies' which also identify the various forms of asymmetry as follows: torso twisting, asymmetric hand position on the boxes, boxes loaded asymmetrically, lateral movements of boxes, and load distributed unevenly on both legs (Drury, Law and Pawenski, 1982; Mital et Fard, 1986; Lavender, Mirka, Schoenmarklin, Sommerich, Sudhakar, and Marras, 1989; Ljungberg, Kilbom and Hägg, 1989). Another form of asymmetry we observed is that of having to handle loads with each foot on a different level.

Despite the fact that most real life handling tasks are performed in asymmetric postures, the vast majority of laboratory research on MMH has dealt with symmetrical postures (Ayoub and Mital, 1989; NIOSH, 1981). St-Vincent, Lortie and Tellier (1987) studied patient handling tasks performed by orderlies in a geriatric hospital. They noted that when the effort exerted had a strong horizontal component, an important part of torso twisting and lateral bending were observed in the posture of the worker. In such lifting conditions, the worker made greater use of the back structure than of the legs to exert efforts. On the other hand, when the effort exerted was predominantly vertical, the movement was performed in the sagittal plane involving the legs to a greater extent. The authors also noted that workers made use of the MMH techniques they were taught to a greater extent in the latter case.

St-Vincent et al., (1987) hypothesize that the worker does not always systematically optimize his (or her) working posture to minimize stresses on the back. When horizontal efforts are exerted, balance appears to be a more important variable than back stress and thence, the posture becomes asymmetric while the MMH techniques used depart from the recommendations for safe MMH practices. Our observations tend to support this hypothesis. Energy expenditure appears to be another variable that is sometimes optimized by the worker and which has an important impact on MMH postures. For a given work output, the stooped posture is more energy efficient than the squat posture which in turn, is biomechanically (intradiscal pressure) less stressful (NIOSH, 1981). This would contribute to explain why workers often adopt a stooped posture in repetitive lifting. The freestyle technique (semi-squat) has been reported to be biomechanically the least stressful however, the literature is still divided concerning its metabolic advantage over the stooped posture (Ayoub and Mital, 1989).

Back stress is an important criterion since the back is where the most serious injuries occur. However, it appears that it should be considered in parallel with other variables that appear to play an important role in MMH. Research should thus concentrate more on asymmetric handling and more specifically, on other variables that appear to be optimized by the worker and the conditions in which these variables appear to play a role. Both body balance and energy expenditure are such variables. With this knowledge, it will be possible to design and teach the workers safe MMH practices that will have a lasting effect and which will preserve the integrity of his (or her) back.

Capacity prediction models

A large proportion of the MMH models were developed using psychophysical and physiological methods. Genaidy, Asfour, Mital and Tritar (1988) and Genaidy and Asfour (1987) provide good reviews of these models. A practical problem, from the standpoint of a designer not familiar with the MMH literature, is the rather large number of these models which can render the designing task somewhat cumbersome. From the standpoint of a researcher, of concern is the fact that the prediction equations reported were obtained using the stepwise regression technique. Also of concern, is that the coefficient of determination, the multiple R^2, appears to have been used as the main criterion to assess predictive performance of the regression models.

The stepwise regression technique along with others (backward or forward elimination) are labelled as sequential variable selection procedures. These were constructed to arrive efficiently at a reasonable subset of regressor variables in cases where a large number of possible variables exist. They were developed with efficient computation as the motivation and at a time when computer resources were prohibitive (i.e., early '60s). However, as Myers (1986) notes "...there is no assurance that any specific sequential model-building strategy will result in the best variable subset with regard to any standard dealing with theoretical sensibility, prediction performance, etc." (p.122). The final model is very much dependent on the sequential procedure chosen (Stepwise vs Backward Elimination vs Forward Selection) and the F_{in} and F_{out} values used in the procedure. The use of a sequential procedure in empirical model building does not allow one to compare several candidate models since only one is output. In the studies where regression models were developed (for instance, see Mital and Ayoub, 1986), only one model is always produced and the multiple R^2 is used to gauge the model's predictive performance.

The R^2 is a good measure of a model's capability to fit the present data but it is not conceptually prediction oriented (Myers, 1986). For instance, the insertion of any new regressor (whether highly correlated with the other regressors or not) into a model cannot bring about a decrease in R^2. Therefore, the R^2 statistic is not recommended as a unique criterion for selecting the best prediction model from a set of candidate models (Myers, 1986).

Researchers should approach empirical model building with as much information as possible. Hence, the information produced by all possible regressions can be helpful to select the model that performs best as a prediction equation. Statistics such as the residual mean square (s^2) Mallow's Cp, PRESS, and the sum of the PRESS residuals are invaluable in this regard (Myers, 1986). It thus appears that some verification of the models published in the past several years as well as some refinement of the empirical model techniques used in the MMH community are indicated.

It would be interesting to see studies in which the outputs of the various models are verified and compared with each other using industrial settings and populations of workers. Such studies would be invaluable to the designer and to the manager who often need to offer sound justifications for projected changes in job design, personnel assignment, and during work conflicts.

Psychophysical scaling methods

One issue concerning MMH load prediction models based on psychophysical data is that the dependent variable used to construct these models --the Maximum Acceptable Weight of Lift (MAWL)-- was estimated in relatively short trials (20-40 min) while the models generated can supposedly predict loads that are handled for extended periods of time (Snook, 1978). In a typical trial, for a given combination of task variables (frequency, box size, and lifting height) the subject is asked to adjust the weight of a standardized container to a load he (or she) thinks can be comfortably handled without straining himself (or herself), or becoming unusually tired, weakened, overheated or out of breath (Snook, 1978). In the experimental determination of MAWLs, fatigue is carefully controlled to avoid any confounding effect. Mital (1983) reported that the loads predicted by the models based on MAWLs have to be adjusted when the MMH tasks are actually performed for 8 to 12 hours. Although weighting factors have been computed to adjust these values for various period lengths (i.e., 3.4% per hour for males and 2.0% per hour for females), it is suggested that further research is required to generate more reliable estimates (Genaidy et al., 1988).

It is interesting to note that the method of adjustments has been the only psychophysical scaling method used up to very recently. Other methods such as magnitude estimation and ratio estimation have just been introduced to MMH research. For instance, magnitude estimation was recently applied to the investigation of the ability of humans to estimate the weight of boxes on a lifting task and to discriminate among different levels of load (Karwowski, Shumate, Pongpatana and Yates, 1989). Magnitude estimation is currently one of the most frequently used psychophysical scaling methods in psychology. It has been utilized for the study of senses and various psychological attributes such as brightness of light, psychological worth of money, emotional stress, etc. (Gescheider, 1985). It has also been used in some clinical cases involving pain, fear, and drugs as well as in medical decision making (Grossberg and Grant, 1978). Magnitude estimation has the advantage of permitting quick acquisition of large amounts of data.

A form of ratio estimation was recently used to rate and describe the strains associated with a variety of sagittal plane lifting tasks (Aghazadeh, Lee, Waikar, and Bittner, 1988). Ratio estimation is not as popular as magnitude estimation because it has been shown to be particularly sensitive to stimulus context effects (Gescheider, 1985). Both magnitude estimation and ratio estimation are reported to yield results that are equivalent to those obtained with the method of adjustments (Gescheider, 1985; Morrissey and Bittner, 1989).

These psychophysical scaling methods are very promising for the development of entire families of task MAWs such as are commonly found in industry. Furthermore, these methods could be used to develop scales for describing the physical efforts required in a wide variety of jobs in various industrial organizations such as that reported by Fleishman, Gebhardt, and Hogan (1984). Given the fact that these methods are quick and simple, MMH research should be performed in industrial settings, that is, where loads are handled for long periods of time and by trained workers. Such research should be attempted even though control of independent variables is more difficult. MMH research making use of psychophysical scaling methods in industry would get better face validity than laboratory research.

ACKNOWLEDGMENTS

This research is currently supported by a grant from Institut de Recherche en Santé et en Sécurité du Travail du Québec.

REFERENCES

Aghazadeh, F., Lee, K. S., Waikar, A. M., and Bittner, A. C., 1988, Utilization of direct estimation method to determine the effects of task variables on the capacity to lift loads. In: Proceedings of the 32nd Annual Meeting of the Human Factors Society, Santa Monica, edited by the Human Factors Society, pp. 687-689.

Ayoub, M. M. and Mital, A., 1989, Manual materials handling. (London: Taylor and Francis).

Drury, C. G., Law, C.-H., and Pawenski, C. S., 1982, A survey of industrial box handling, Human Factors, 24, 553-565.

Fleishman, E. D., Gebhardt, D. L., and Hogan, J. C., 1984, The measurement of effort. Ergonomics, 27, 947-954.

Genaidy, A.M. and Asfour, S.S., 1987, Review and evaluation of physiological cost prediction models for manual material handling. Human Factors, 29, 465-476.

Genaidy, A.M., Asfour, S.S., Mital, A., and Tritar, M., 1988, Psychophysical capacity modeling in frequent manual materials handling activities. Human Factors, 30, 319-337.

Gescheider, G. A., 1985, Psychophysics - Method, theory, and application. (Hillsdale, NJ: Lawrence Erlbaum Associates, Publishers).

Grossberg, J. M. and Grant, B. F., 1978, Clinical psychophysics: Application of ratio scaling and signal detection methods to research on pain, fear, drugs, and medical decision making. Psychological Bulletin, 85, 1154-1176.

Imbeau, D. et Lortie, M. (1984). Analyse des facteurs de risque dans un hôpital pour soins prolongés de la région Montréalaise. Rapport Ecole Polytechnique, EPM/RT-84-8, 199 p.

Karwowski, W., Shumate, C., Pongpatana, N., and Yates, J. W., 1989, Weight discriminability and subjective assessment if load heaviness: a pilot study. In: Proceedings of the 33rd Annual Meeting of the Human Factors Society, Santa Monica, edited by the Human Factors Society, pp. 657-661.

Lavender, S.A., Mirka, G. A., Schoenmarklin, R. W., Sommerich, C.M., Sudhakar, L.R., and Marras, W. S., 1989, The effects of preview and task symmetry on trunk muscle response to sudden loading. Human Factors, 31, 101-115.

Ljungberg, A.-S., Kilbom, Å., and Hägg, G.M., 1989, Occupational lifting by nursing aides and warehouse workers. Ergonomics, 32, 59-78.

Mital, A., 1983, The psychophysical approach in manual lifting: a verification study. Human Factors, 25, 485-490.

Mital, A. and Ayoub, M. M., 1986, Equation for predicting lifting capabilities of people. In: Trends in Ergonomics/Human Factors III, North-Holland, edited by W. Karwowski, (Elsevier Science Publishers), pp. 767-773.

Mital, A. and Fard, H. F., 1986, Psychophysical and physiological responses to lifting symmetrical and asymmetrical loads symmetrically and asymmetrically, Ergonomics, 29, 1263-1272.

Morrissey, S. J. and Bittner, A. C., 1989, Accuracy of a ratio-estimation method to set maximum acceptable weights in lifting tasks. In: Proceedings of the 33rd Annual Meeting of the Human Factors Society, Santa Monica, edited by the Human Factors Society, pp. 742-744.

Myers, R. H., 1986, Classical and modern regression with applications. (Boston, MA: Duxbury Press).

NIOSH, 1981, Work practices guide to manual lifting. DHHS (NIOSH) Publication No. 81-122. (Cincinnati, OH: Government Printing Office).

Snook, S. H., 1978, The design of manual handling tasks. Ergonomics, 21, 963-985.

St-Vincent, M., Lortie, M. and Tellier, C., 1987, A new approach for the evaluation of training programs in safe lifting. In: Trends in Ergonomics/Human Factors IV, North-Holland, edited by S.S. Asfour, Elsevier Science Publishers, pp.847-854.

Advances in Industrial Ergonomics and Safety II
Edited by Biman Das
Taylor & Francis 1990

ANALYSIS OF THE OPERATIVE MODES USED TO HANDLE CONTAINERS OTHER THAN BOXES

Baril-Gingras, Geneviève & Lortie, Monique

Université du Québec à Montréal
Département des Sciences biologiques
C.P. 8888, Succ. A, Montréal, Québec, Canada, H3C 3P8

The operative modes used to handle 944 containers other than boxes (such as crates, drums, bundles and pipes) were analyzed by the observation of 31 handlingmen working in the warehouse of a transport industry. A basic handling process was divided into six possible operations (searching, pre-transfer, first transfer, following transfer, deposit, post-deposit) and was analyzed with regard to the positioning of the material at the beginning and the end of the handling, the nature and direction of the efforts applied with the upper limbs, the displacement of the object (e.g. pivoting, sliding), the use of the lower limbs, and the position of the back.

INTRODUCTION

The handling of containers or material has up to now been more extensively studied in the laboratory than in the field. Little is known about the operative modes developed by handlingmen. Furthermore, a previous accident study (Lamonde, 1987) conducted on a group of handlingmen working in the loading and unloading of trucks in the warehouse of a transport industry has demonstrated that 71% of handling accidents involved a container other than a box, despite the fact that they represented less than 10% of the handled materials. These highly variable materials may consist of crates, bundles, drums, pipes, motors, furniture, sheets, plates, etc. Therefore, the objective of the present study was to analyze the operative modes used to handle this broad variety of materials, such as the displacement transmitted to the material, the efforts, their characteristics and their modes of application.

METHOD

Subjects

The study was conducted in the same company where we carried out the accident research. 31 handlingmen working full time on the evening shift were each observed during a full shift. This shift starts between 15:00 and 16:30 and finishes between 23:30 and 01:00 and was chosen because approximately 50% of the man-hours worked occur during that period as well as the fact that the evening population is more varied in terms of age and seniority than day workers (who are older) or night workers (who have less seniority). Subjects were volunteers and were chosen to be representative of the different classes of seniority of the overall population. Half were working in loading and half in unloading. The average seniority was 12 years for the entire working population (n=188), 9 years for the evening shift population (n=96), and 8 years for observed workers (n=31). The least experienced worker observed had less than one month of seniority and the most experienced, 17 years.

This research was supported by grants from IRSST and CRSNG.

Recording and analysis of the data

Handlings performed in the trucks were video-taped. In fact, the grid of analysis developed was far too complex for direct use in the field and the workers didn't appear disturbed by the camera in the pre-test. Basically, the handler loads a truck by unloading buggies or unloads a truck by loading buggies. Most of the latter consist of a four-wheel buggy; the two-wheel buggies (e.g.used for crates) or the drum buggies wereused far less frequently. Outside the truck, the material was generaly transported with buggies; manual transport was very rare. In addition to manual handling, some materials are moved with a lift truck: only manual handling was recorded nor was the circulation with buggies in the warehouse recorded. Overall, we video-taped about 3 hours of manual handling for every 8-hour shift. As previously noted, about 90% of the handled materials were boxes; the present paper deals only with the handling of the 10% which are non-box containers or materials (n=944).

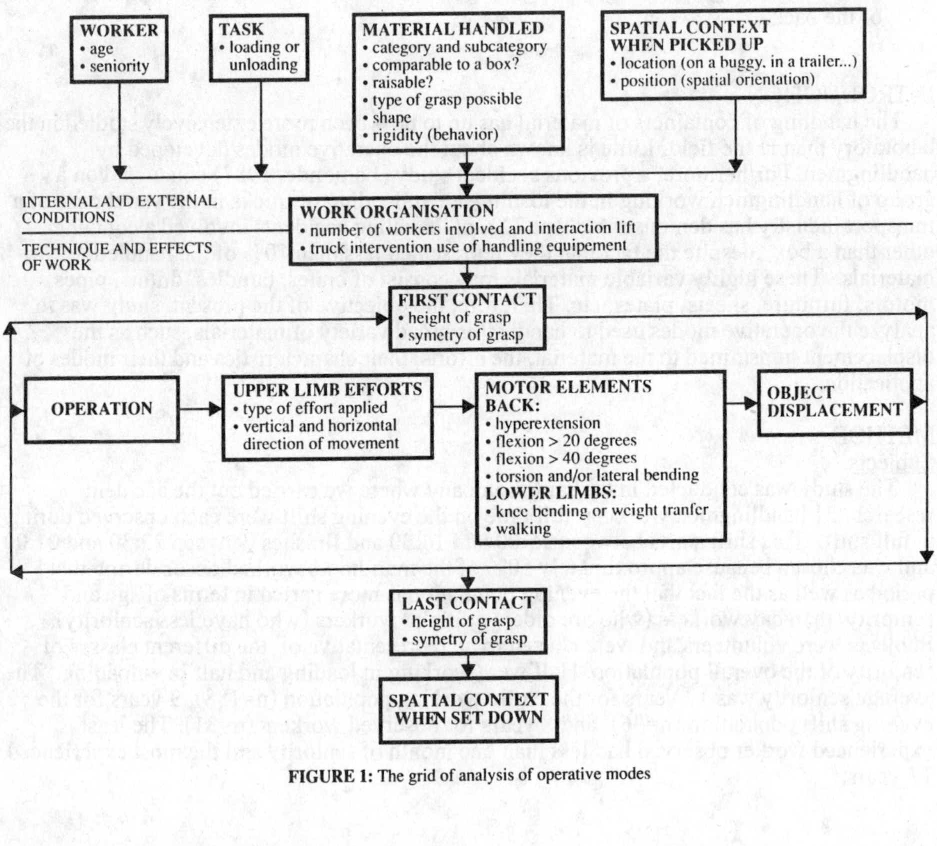

FIGURE 1: The grid of analysis of operative modes

The grid of analysis developed included 36 variables which are summarized in Figure 1. The following information was recorded: type of material handled (categories & subcategories), basic properties of that object (e.g. handles, shape, rigidity,...), positioning of the material at the beginning and the end of the handling, equipment used and sequencing of the different operations (e.g. searching, pre-transfer defined as the positioning of the load or bringing the load nearer, first transfer operation, following transfer, deposit, final deposit or load adjusting). These stages are quite similar to those defined by Drury et al. (1982) (pregrasp, pickup, move/carry, put down, adjust) except that our classification was adapted to the handling of objects other than boxes which in some cases are kept in continuous contact with a surface (e.g. drums). An operation specific to unloading was added -searching- which consists in the identification of the destination inscribed on the load, for dispatching in lots. Every handling activity recorded could include a succession of operations wich were differentiable by the fact that a break, a change in the type of grasp, in the direction or nature of the effort, or in the object displacement could be observed. The simpliest handling was performed by a single transfer operation, but a specific handling activity could be composed of each of the 6 different types of operations. Since the same type of operation can be executed by more than one effort, we recorded a handling activity in up to 9 operations. For every operation, the nature and direction of the effort applied with the upper limbs, the displacement of the object (e.g. sliding, pivoting, turning, raising, lowering...), the use of the legs (e.g. weight transfer, bending of the knees), and the posture of the back were recorded. This grid was computerized and the data directly recorded on the computer. The analysis of a 30-second handling sequence took about 10 minutes. This data was recorded by 3 observers. Results of the inter-and intra-observer reproducibility tests were generally above 85%; mean agreement rate was 89% (Baril-Gingras and Lortie, 1989).

RESULTS

Operation:

Most of the materials were moved through a succession of different operations which correspond to different aims. An average of 3.4 operations was observed for each object handled. 10% of the materials involved only a tranfer operation, the object being directly taken and deposited; in 37% of the cases, there was in addition to the transfer a pre-transfer or a deposit operation; in 41% of the handlings, those three basic operations could be differentiated. Table 1 shows the distribution of these operations for the loading and unloading tasks.

Displacement of the material

The accident study revealed that 37% of the transfer accidents occured when the object was moved on the same level; therefore we specifically observed the trajectory of the material. From table 1, we note that the material was often kept in contact with the surface. As indicated in table 2, the vertical space was divided into three sections: ground / mid-thigh, mid-thigh / shoulder and above the shoulders. For the unloading task, the material was moved to the same level, up or down in 58% , 18% and 24% of the cases respectively. Unexpectedly, the situation appeared significantly different ($p<.001$) for the loading task, the figure being 37%, 52% and 11%. The height of deposit differs significantly ($p<.001$) between the loading and unloading tasks although the same material was involved: in fact, the unloaded material was dispatched on different four-wheel buggies to be loaded in another truck. Furthermore, we identified 9 basic types of object displacement. As seen in table 3, sliding or pivoting were as frequent as the upward displacement of the object hereafter referred to as "raising"; loading and unloading present a quite similar pattern. The relationship between the object displacement and the operation shown in table 3 allows us to see that although there are some dominant associations, these different types of displacement can be found from the beginning to the end of the handling activity.

Table 1. Distribution of the operations for loading and unloading tasks and object displacement associated with each displacement.

Operation n=3218*	Loading	Unloading	Both tasks	%of each operation where the load was kept in contact with the surface
Searching	0%	7%	4%	63%
Pre-transfer	24%	19%	22%	96%
First transfer	30%	27%	28%	28%
Following transfer	13%	16%	15%	51%
Deposit	23%	22%	23%	29%
Post-deposit	9%	10%	10%	91%

*total may differ from 100% because numbers were rounded

Table 2. Distribution of the pick-up and deposit heights for loading and unloading tasks

Level of pick-up and deposit		loading	unloading	both tasks
Pick-up and deposit at the same level	ground/mid-thigh to ground/mid-thigh	12%*	35%	23%
	mid-thigh/shoulder to mid-thigh shoulder	24%	22%	23%
	above shoulder to above shoulder	0%	1%	1%
object goes up	ground/mid-thigh to ground	26%	15%	21%
	ground/mid-thigh to above shoulder	19%	1%	10%
	mid-thigh/shoulder to above shoulder	7%	3%	5%
object goes down	mid-thigh/shoulder to ground/mid-thigh	10%	17%	13%
	above shoulder to mid-thigh/soulder	2%	5%	3%
	above shoulder to ground/mid-thigh	0%	2%	1%

*total may differ from 100% because numbers were rounded; n=944.

Operative modes associated with object displacement: efforts

We observed the vertical and horizontal components of upper limb movement. Horizontally, we recorded the direction of movement in reference to the shoulder axis (parallel-oblique or perpendicular) and whether the movement was toward or outside the body; vertically, we noted whether the movement was up or down. One in five operations were performed strictly on a vertical axis, and more than a quarter only horizontally; the movement combined horizontal and vertical components in 39% of the operations. The results are summarized in table 4. It can be seen that most of the movements in the horizontal plane are parallel-oblique to the shoulder axis: frontal work is infrequent. In roughly half of the cases, these horizontal movements also have a vertical component. It appears that when the load is going down it is also typically moving away, while the horizontal component is parallel-oblique to the shoulders. No such association could be found when the upper limbs performed upward movements which are nondifferentially carried out towards or away from the body, or from one side of the body to the other.

Table 3. Distribution of object displacement in the different operations.

Displacement	% of all operations*	Operations: pre-transfer	1st transfert	following transfer	deposit	post-depos.
sliding	23%	37%	14%	8%	21%	20%
pivoting	19%	39%	16%	19%	13%	8%
turning	9%	40%	6%	11%	18%	22%
rolling	3%	13%	47%	22%	5%	6%
resisting	3%	36%	17%	16%	11%	18%
raising	22%	2%	64%	10%	19%	1%
going down	8%	0%	23%	5%	63%	3%
falling	6%	4%	31%	5%	58%	3%
being transp.	7%	1%	33%	62%	3%	0%

* n= 3215; 3operations were eliminated because object displacement was not visible; results for the searching operations are not presented (n=115, 3.6%);

Table 4. Direction of upper limb movement in the horizontal and vertical planes.

Direction vs body		Each plan separately: loading	unloading	both tasks*	Combinations: horizontal and up	horizontal and down
Horizontaly						
perpend. to shoulder axis	toward	2.4%	4.7%	3.5%	1.5%	0.2%
	away	5.5%	4.9%	5.2%	2.1%	0.7%
parrallel-obl. to shoulder axis	toward	22.5%	19.3%	21%	9%	1.5%
	away	28.2%	28.3%	28.2%	10.8%	7.5%
	from one side to the other	9.8%	3.3%	6.7%	4.3%	1%
Verticaly	up	45.9%	38.4%	42.3%		
	down	10.8%	21.3%	15.9%		

* n=3218;for 256 operations (8%), the direction of the effort could not be identified because the upper limbs effort was different for each arm.

The nature of the efforts was also specified, as shown in table 5. For every effort, the table indicates the two operations and the two most frequent object displacements where this type of effort is most found. Pulling and pushing are the two most frequent upper limb efforts observed. (Pulling and pushing were defined as a relationship between the position of the hands vs the position of the material, and the direction of effort with respect to the body; although mostly executed in a horizontal plane, they could also be partially performed along a vertical axis). In fact, the pulling and pushing movements were also applied along a vertical axis upwards (49%and 30% respectively), and downwards (7% and 11% respectively). In particular they were used to slide and pivot the load. When pulling, the load was kept in contact with the surface in 95% of the cases although 56% of the applied efforts had a vertical component. When pushing, the load was kept in at least partial contact with the surface in 97% of the cases. Overall, the different efforts appeared to be associated most often to a particular operation, as for example lifting, pulling, throwing, lowering and holding which were found to be concentrated in one type of operation more than 60% of the time.

Table 5. Upper limbs efforts: the two most frequent operations associated and the resulting load displacement.

Effort applied	% operations*	2 most frequent operations		2 most frequent load displacement	
to pull	26%	63% pre-Trsf.	17% 1st Trsf.	44% slide	29% pivot
to push	22%	28% 1st Dep.	26% post-Dep.	41% slide	33% pivot
to lift	16%	76% 1st Trsf.	12% Fol. Trsf.	99% raise	
to lower	11%	70% 1st Dep.	16% 1st Trsf.	64% go down	14% pivot
to throw	8%	67% 1st Dep.	28% 1st Trsf.	51% raise	46% fall
to hold	6%	64% Fol. trsf.	34% 1st Trsf.	92% be transp.	8% slide
to hold back	2%	29% Fol. Trsf.	28% 1st Trsf.	48% pivot	17% slide
effort different for each arm	8%	34% 1st Trsf	18% Fol. trsf.	25% turn	23% pivot

* n=3218;for 42 operations (1.3%), there was only an effort of the lower limbs on the object; for 2 operations (0.06%), the effort of the upper limbs was not visible.

The use of the legs to move the load (knee bending, weight transfer and displacement of the feet during the handling), was also observed. The bending of the knees or the weight transfer from one foot to another was noted in 12% of the operations and a significant foot displacement was seen in 22%. Table 6 shows when these elements were used.

<u>The use of the back</u>

The position of the back during the different operations or efforts applied was observed. The back was kept straight or flexed less than 20 ° in 33% of the operations. Hyperextension was quite rare (1%) although 20% of the loads were handled over the shoulders. As seen in table 7, lateral bending or torsion were almost always observed in combination with a sagittal flexion, but unexpectedly more often with a flexion greater than 45° than between 20 and 45°. Flexion between 20 and 45° appeared to be quite equally distributed among the different types of efforts: despite the type of effort, approximately one quarter were applied with a mild flexion. On the contrary, deeper flexion as torsion or lateral bending was found associated with specific efforts. For example, deep flexion was observed in more than half of the lowering efforts but very rarely for holding and only one time in six for throwing. About one third of the lowering and one quarter of the lifting, pulling and throwing movements involved a torsion/lateral bending, while the latter were infrequent for pushing, holding, holding back and efforts where each arm acts differently.

Table 6. The use of the lower limbs during the handling.

	Knee bending or weight transfer *		Walk	
	distribution in all operations	% for that operation	distribution in all operations	% for that operation
Operation				
Pre-Transfer	21%	12%	7%	7%
1st Transfer	26%	11%	40%	30%
Following transfer	17%	14%	32%	48%
Deposit	18%	10%	15%	15%
Post-Deposit	17%	21%	4%	9%
Searching	1%	4%	2%	11%
Upper limbs effort				
to push	29%	16%	17%	17%
to pull	32%	15%	14%	12%
to lift	12%	9%	18%	24%
to lower	6%	6%	8%	16%
to throw	9%	14%	4%	11%
effort diff. for each arm	10%	15%	9%	2%

*n=3218;knee bending or weight transfer was not visible in 74 operations (2.3%)

Table 7. The use of the back with reference to the effort of the upper limbs.

Posture		uncombined	combined with L.F. or T.	Total n=3218	upper limbs effort, most (2)	or least (2) assoc.
Flexion	≥20 to <45°	20%	5%	25%	Throwing 31%	Hold. back 29%
					Holding 23%	Pulling 24%
	≥45°	24%	11%	35%	Lowering 54%	Lifting 48%
					Holding 1%	Throwing 15%
Lat.Flexion &/or Torsion		3%		20%	Lowering 31%	Lifting 26%
					Holding 4%	Hold. back 9%

DISCUSSION

Up until now, few studies have been published on the handling of containers other than boxes; among them, mailbags and to some measure cylinders or drums have attracted most of the attention. The main concern of these studies has been to analyze the relationship between the shape, the dimensions and the acceptable weight (e.g. Garg and Saxena (1980), Mital and Okolie (1982)) or the resulting strain on the back (Burton,1986). However, the manner in wich these materials are handled has attracted little attention. In fact, the over-representation of the non-box objects in accidents suggested to us that this is at least partially related to difficulties in handling such different materials and not only to their weight. As shown in this study, their handling is quite complex: it involves the combination of different operations and it is rarely only one continuous movement. Furthermore, Drury et al. (1982) and Stålhammar et al. (1986) previously used a similar model of subdividing the activity into 5 stages and proved its usefulness. Perhaps in future experimental studies it would be worthwhile to differentiate these handling phases. Up until now, research as mainly focused on the raising of the materials, assuming that the only path followed by a container is a vertical line up or down. However some researchers (e.g. Leskinen et al. (1983) and Garg et al. (1983)) included or analyzed a horizontal component in lifting among the different handling techniques studied Horizontal lifting consisting of transfer from one side of the body to another and àssymetrical lifting were also studied (e.g. Garg and Badger, 1986; Schultz et al., 1982). Generally, no clear pattern can be drawn as some results contradict others due to differences in the context of the studies. For example, Garg et al. found these situations very stressful for the back, while the load was taken from the floor; Schultz et al. found that trunk twisting and lateral bending were less loading on the spine and the trunk muscles than trunk flexion or the holding of weights in a standing position. In addition, Drury et al. found in a survey of 2000 industrial box -handling tasks that symmetric lifting was infrequent: generally the load was moved first toward the body, then away. In our study we also observed that symetrical work is quite infrequent, handlingmen preferred to move the loads parallel to the shoulder axis. But surprisingly it appeared that lowering a load was performed in a specific way, mainly away from the body, although no such predominance was found for liftings. Furthermore, the displacement or trajectory that a load followed appeared quite varied at every phase of handling. Up until now most biomechanical studies have almost totally disregarded the use of the upper limbs in the handling and concentrated on the use of the lower limbs, and particularly the impact of knee bending on the position of the back. We believe a better understanding of the use of the upper limbs is necessary.

It was also surprising to find that the use of the lower limbs was more often associated with pulling and pushing than lifting or lowering. The significance of this finding can be elucidated with further studies. Also unexpected was the observation that lateral bending or torsion is more frequently associated with deep flexion than slight flexion. In addition, in contrast to the lower limbs, the back position and thus the use of the back appeared to be strongly related to the type of effort applied. We believe that it is worthwhile to further understand the significance of these operative modes developed by handlingmen, esecialy when we consider that they have an average of 8 years of handling experience.

REFERENCES

Baril-Gingras, G. and Lortie M., 1989, Questions méthodologiques posées par la validation d'une grille d'observation des modes opératoires dans la manutention. in Proceedings of the Annual Conference of the Human Factors Association of Canada, pp. 239-243.

Burton, A.K., 1986, Spinal strain from shopping bags with and without handles. Applied Ergonomics, 17, 1, 19-23.

Drury, C.D., Law C.H. and Pawenski, C.S., 1982, A survey of industrial box handling. Human Factors, 24, 5, 553-565.

Garg, A. and Saxena, U., 1980, Container characteristics and maximum acceptable weight of lift. Human Factors, 22, 4, 287-495.

Garg, A., Sharma, D., Chaffin, D.B. and Schmidler, J.M, 1983, Biomechanical stresses as related to motion trajectory of lifting. Human Factors, 25, 527-539.

Garg, A. and Badger, D., 1986, Maximum acceptable weights and maximum voluntary isometric strengths for asymmetric lifting. Ergonomics, 29, 7, 879-892.

Lamonde, *f*., 1987, analyse des accidents de manutention dans la secteur transport général et entreposage, Mémoire de maitrise, Ecole Polytechnique de Montréal.

Leskinen,T.P., Stålhammar, H.R. , Kuorinka, I.A., and Troup, J. D. G., 1983, The effect of inertial factors on spinal stress when lifting. Engineering in Medecine, 12, 2, 87-89.

Leskinen, T.P., Stålhammar, H.R., Kuorinka, I.A. and Troup, J.D., 1983, A dynamic analysis Mital, A. and Okolie S.T. , 1982. Influence of container shape, partitions, frequency, distance, and height level on the maximum acceptable amount of liquid carried by males. American Industrial Hygiene Association Journal, 43, 11, 813-819.

Schultz, A.B., Andersson, G.B., Haderspeck, K., Nordin, M. and Björk, R., 1982, Analysis and measurement of lumbar trunk loads in tasks involving bends and twists. J. Biomechanics,15, 9, 669-675.

Stålhammar, H.R., Leskinen, T.P., Kuorinka I.A., Gautreau, M.H. and Troup, J.D., 1986, Postural, epidemiological and biomechanical analysis of luggage handling in an aircraft luggage compartiment. Applied Ergonomics, 17, 3, 177-183.

Advances in Industrial Ergonomics and Safety II
Edited by Biman Das
Taylor & Francis 1990

HIGH FREQUENCY LIFTING AND BLOOD LACTATE PRODUCTION

J.W. Yates[1], Kathy Pickering[1], and Waldemar Karwowski[2]

1 Exercise Physiology Laboratory

2 Center for Ergonomics
University of Louisville, Louisville, KY 40292

Three male graduate students performed 6 progressive lifting tasks to exhaustion. Three tasks used a constant weight (7, 14, and 28 kg), increasing frequency format while the other three tasks used a constant frequency (6, 12 and 28 lifts/min), increasing weight format. Blood lactate was measured at four minute intervals from samples drawn from a superficial vein in the forearm. Heart rate was recorded each minute during the lifting task. Results suggest that the pattern of lactate production during lifting tasks is similar to those seen during treadmill or cycle ergometer tests. These data also indicate that the constant weight, increasing frequency format produces more consistent blood lactate data than does constant frequency, increasing weight.

INTRODUCTION

Exercise physiologists have recognized for many years that the production of lactic acid can have serious consequences on the ability to exercise or work and have developed both invasive and noninvasive methods to determine when changes in lactic acid production occur. The terms "anaerobic threshold" or "lactate threshold" have been used to describe the level of exercise or work which causes a significant increase in blood lactate levels. However, these methods have seldom been applied in the field of ergonomics. Ergonomists have typically used measures of heart rate and oxygen consumption to determine the physiological strain of high frequency manual materials handling (Brouha, 1964; Karwowski and Yates, 1984; 1986; Legg and Myles, 1981; Rodgers, 1978). While these measures provide valuable information, they are primarily measures of dynamic work. Manual materials handling tasks involve dynamic components such as the bend and lift and static components such as the gripping and carrying of materials. The purpose of this investigation is to explore the possibility of using the lactate threshold to measure the physiological strain resulting from high frequency lifting tasks. In addition, this experiment was designed to determine the effect of various combinations of weight and lifting frequency on the lactate threshold.

METHODS

Subjects for this study were three male graduate students (ages 22-26 years). Following an explanation of the purpose of the project and a physical exam by a physician, each subject signed an informed consent form. Since previous work had shown that extreme muscle soreness could result from the lifting tasks, each subject underwent a minimum of two weeks of lifting training (10 - 15 sessions). Training involved both lifting and lowering of trays from the floor to a table (76 cm high). Session duration varied from 15 minutes initially to a maximum of 30 minutes. This progression minimized any muscle soreness or injury.

Following the training sessions, each subject completed six, progressive, maximum lifting sessions. Each session involved lifting a wooden tray (45.5 x 31 x 13 cm) with cutout handles from the floor to a 76 cm high table. During three of the tests the tray weight was held constant while the frequency of lift was increased. During the other three tests, lifting frequency was constant and the weight of the tray was increased. The various combinations of lifting weights and frequencies were designed so that the total test time was approximately equal and so that common work loads occurred between different tests.

During the constant weight, increasing frequency tests, the tray weights were 7, 14 or 28 kilograms. Each test began with a five minute warm-up period at a lifting frequency of 6 lifts/min. When lifting the 7 kg. tray, lifting frequency was increased by 6 lifts/min every three minutes. When handling the 14 or 28 kg trays, lifting frequency was increased by 3 or 2 lifts/min every three minutes, respectively. In each case, lifting was continued until the subject was no longer able to continue. Heart rate was measured each minute while blood samples were drawn at the end of each work load.

The next three tests were performed using a constant frequency of 6, 12 or 24 lifts/min and a varying load. Following a five minute warm-up period with 7 kg., the weight of the tray was increased every three minutes. When performing 6 lifts/min, the tray weight was increased by 6 kg. every three minutes. When lifting 12 or 24 times/min, tray weight was increased by 4 or 2 kg. every three minutes. Lifting was continued in this fashion until exhaustion. Physiological measurements were the same as those described above.

In order to measure blood lactate, a 21G butterfly catheter was inserted into a superficial vein in the forearm. The catheter was kept patent by periodic flushing with a 10% solution of heparin (1000 USP units/ml) and saline (0.9%). Both the catheter and the heparin-saline solution were taped securely to the arm of the subject so as not to interfere with the lifting process. In order to draw the blood sample for lactate analysis, the subject stopped lifting for approximately one minute. The time required to draw the blood sample ranged from 30 to 45 seconds. The subject then moved to the lifting station to resume lifting. Therefore, the subject lifted for three minutes, then rested for one minute. This sequence was followed for all lifting tests.

Total blood lactate was measured with a YSI Model 23L lactate analyzer. Following withdrawal from the catheter, 100 to 200 microliters of blood was put into microcentrifuge tubes containing a lysing agent. This sample was then injected into the analyzer. Heart rate was measured using an Exersentry heart rate monitor whose digital display was updated every three seconds.

RESULTS

Figure 1 shows the changes in blood lactate plotted against work load that occurred during the constant-weight, increasing frequency lifting trials. Maximum blood lactate values did not differ dramatically for the three weights lifted, however, the work loads that produced changes in blood lactate did. Subjects were able to perform significantly

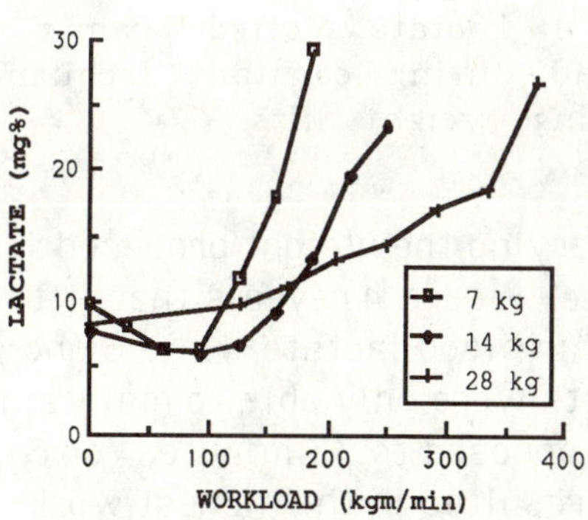

Figure 1: Lactate plotted against workload during constant weight increasing frequency lifts.

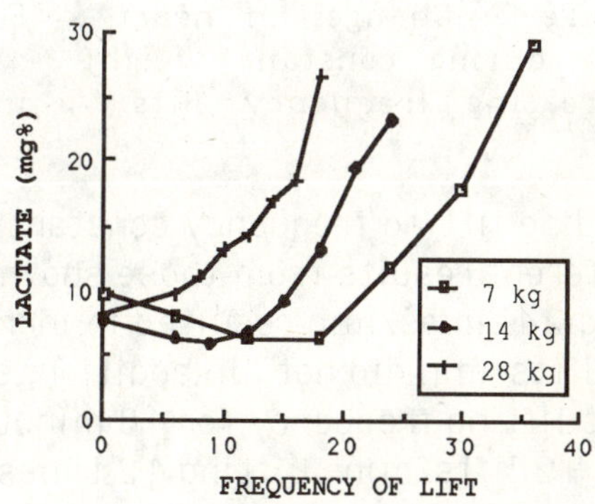

Figure 2: Lactate plotted against lifting frequency during constant weight increasing frequency lifts.

more work as the weight of the tray increased from 7 to 14 and 28 kg. When lifting 7 or 14 kg, blood lactate did not increase until the work load exceeded 100 kgm/min. The lowest work load performed with the 28 kg tray was 125 kgm/min which did result in a blood lactate above baseline values. The total lifting time for each experiment was similar. Subjects lifted the 7 kg tray 28 minutes and the 14 and 28 kg trays for 30 minutes. Hence the pattern of lactate appearance was a function of the work and not a function of time.

The blood lactate data take on a difference appearance when plotted against frequency of lift (Figure 2). This figure shows that the maximum frequency of lifts were 36, 24, and 18 for the 7, 14, and 28 kg. trays, respectively. The frequency of lift that resulted in blood lactate increases, ranged from 6 for the 28 kg tray to 18 for the 7 kg. tray.

Figure 3 shows the changes in heart rate as a function of work load. What is evident from this graph is that lower work loads resulted in higher heart rates when lifting lighter trays. However, in each case the relationship between work and heart rate was linear.

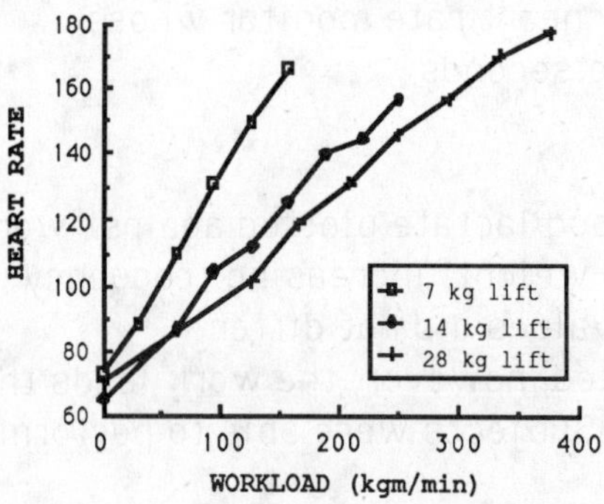

Figure 3: Changes in heart rate during constant weight increasing frequency lifts.

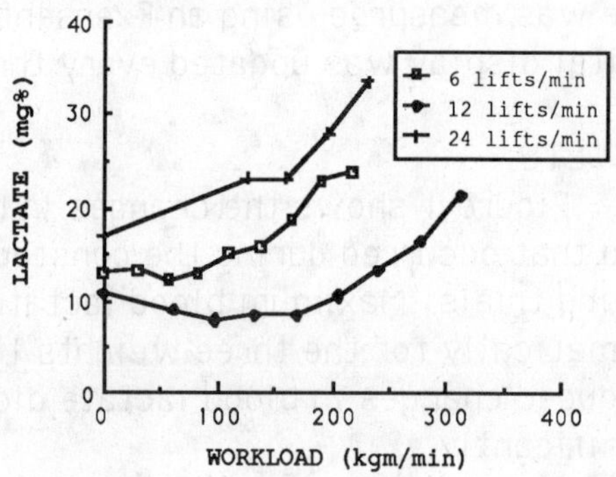

Figure 4: Lactate plotted against workload during constant frequency increasing weight lifts.

Holding lifting frequency constant and varying the weight produced different results from those shown above. Figure 4 reveals that lifting 7 kg 24 times/min resulted in increases in blood lactate while 6 and 12 lifts/min did not. In addition, subjects were only able to maintain this lifting frequency for 18 minutes as opposed to 33 minutes for 6 and 12 lifts/min. Lifting 12 times/min resulted in the highest work load as well as the lowest blood lactates for a given work load. When lifting 12 times/min blood lactate did not exceed resting levels until the work rate was greater than 200 kgm/min. At this work load the subjects were lifting 23 kg/lift while their heart rate averaged 132 beats/min (Figure 5). Subjects were only able to achieve a heart rate of 156 beats/min while lifting 6 times/min as opposed to 168 beats/min for 12 and 24 lifts/min.

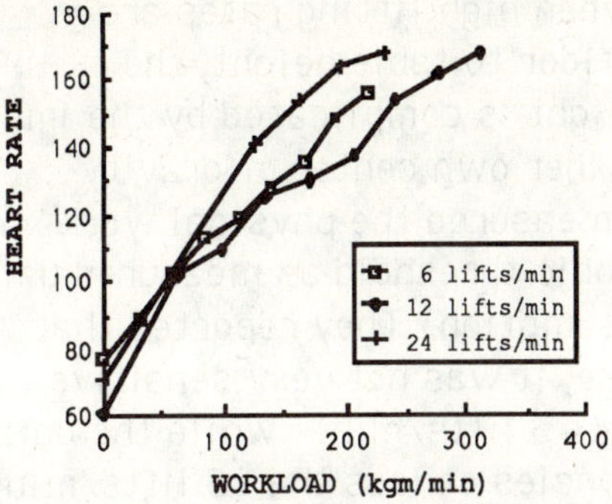

Figure 5: Changes in heart rate during constant frequency increasing weight lifts.

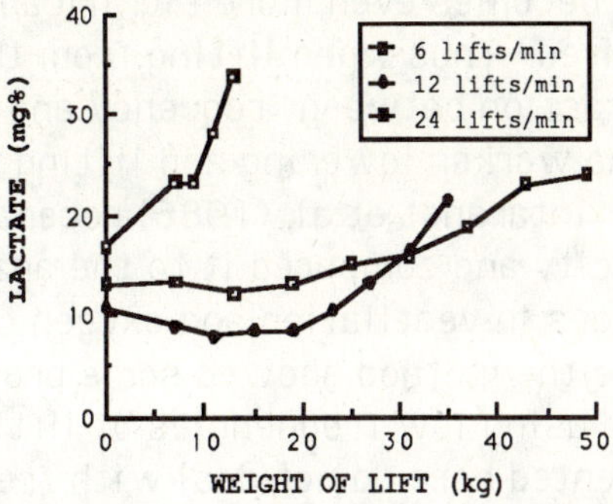

Figure 6: Lactate plotted against weight of lift during constant frequency increasing weight lifts.

The rate of weight increase for the constant frequency lifts is shown in Figure 6. The maximum weight of lift for 24 lifts/min was 13 kg. In contrast, subjects lifted 49 kg when working at a frequency of 6 lifts/min and 35 kg for 12 lifts/min.

DISCUSSION AND CONCLUSION

It is clear from the data presented here that during a progressive lifting task, the develop of blood lactate is dependent upon the combination of lifting frequency and weight of lift. These data also suggest that the results will differ when using a constant weight, increasing frequency format as compared to a constant frequency, increasing weight format. Regardless of which lifting pattern is used, these data show that measurements of blood lactate can be used to determine physiological strain. In addition, pattern of lactate development is similar to those produced on a treadmill (Wasserman et al., 1973) or cycle ergometer test (Gladden et al., 1985).

In designing these experiments, it was our intention to vary the increase in lifting frequency or weight so that the time to exhaustion was relatively constant. In addition, we tried to keep the increases in work load as similar between tests as possible. This was accomplished by having small increases in frequency for heavy lifts and large increases in frequency for light weights. Despite this design, subjects were able to perform more total work when lifting heavy trays as compared to lifting lighter trays. Or when holding frequency constant, subjects performed more work when lifting at a low frequency when compared to a high frequency. This finding points out the importance of lifting frequency on work capacity when performing lifts from the floor to a table. The metabolic cost of lifting the tray is added to the metabolic cost of lowering and lifting the upper and lower body of the worker. The additional work required to move ones own body becomes even more significant when high lifting rates are required. Thus when lifting from the floor to table height, the interaction between frequency and weight is complicated by the impact of the worker lowering and lifting his/her own center of gravity.

Intaranot et al. (1986) recently measured the physical work capacity and compared it to the anaerobic threshold as measured using changes in ventilation and oxygen consumption. They reported that while the method showed some promise, it was not very sensitive when using low frequencies of lifting (< 5 lifts/min). While the data presented here do not deal with frequencies of less than 6 lifts/min, they do show that for a given work load less lactate is produced when lifting at a lower frequency.

Petrofsky and Lind (1978) reported blood lactate levels following one hour of box lifting which were very similar to the data shown here. Their data also showed that lifting light boxes resulted in higher blood lactates than did lifting heavier boxes at the same work rate. Subjects of Petrofsky and Lind lifted a 6.82 kg box at a work rate of 150 kgm/min which resulted in a blood lactate slightly greater than 40 mg%. Our subjects lifted a 7 kg tray at a maximum work rate of 188 kgm/min with a resulting blood lactate of 29 mg%. The differences in the levels of blood lactate produced could be due to different fitness levels of the subjects.

Of the two lifting protocols used here, the constant weight, increasing frequency seemed to produce the most consistent data. The work loads which resulted in increases in blood lactate were easy to choose and the test was fairly easy to conduct. However, one should not ignore the fact that the constant frequency, increasing weight test at 12 lifts/min resulted in the most work performed for a given blood lactate. Therefore, this protocol may be useful for investigating specific frequencies of lifting.

In conclusion, the data presented here show that measurements of blood lactate can be used to determine the physiological strain imposed by high frequency lifting. More total work can be performed when lifting heavier weights at lower frequencies, but this does not always result in the lowest blood lactate levels. If progressive lifting task is needed as an experimental protocol, we recommend the use of a constant weight, increasing frequency format.

REFERENCES

Brouha, L., 1964, Physiological aspects of work measurement. Occupational Health Review, 16, 3-7.

Gladden, L.B., Yates, J.W., Stremel, R.W., and Stamford, B.A., 1985, Gas exchange and lactate anaerobic thresholds: inter- and intraevaluator agreement. Journal of Applied Physiology, 58, 2082-2089.

Intaranont, K., Ayoub, M.M., Bobo, W.M., and Smith, J.L., 1986, Physical lifting capacity: The anaerobic threshold. In: Trends in Ergonomics/Human Factors III, edited by W. Karwowski, (North-Holland: Elsevier Science Publishers), pp. 835-846.

Karwowski, W. and Yates, J.W., 1984, The effect of time in the psychophysical study of the maximum acceptable amounts of liquid lifted by females. Proceedings of the Human Factors Society 28th Annual Meeting, pp. 586-590.

Karwowski, W. and Yates, J.W., 1986, Reliability of the psychophysical approach to manual lifting of liquids by females. Ergonomics, 29, 237-248.

Legg, S.J. and Myles, W.S., 1981, Maximum acceptable repetitive lifting workloads for an 8-hour work-day using psychophysical and subjective rating methods. Ergonomics, 24, 907-916.

Petrofsky, J.S. and Lind, A.R., 1978, Metabolic, cardiovascular, and respiratory factors in the development of fatigue in lifting tasks. Journal of Applied Physiology, 45, 64-68.

Rodgers, S.H., 1978, Metabolic indices in materials handling tasks. In: Safety in Manual Materials Handling, edited by C.G. Drury, DHEW (NIOSH) Publication No. 78-185, pp. 52-56.

Wasserman, K., Whipp, B.J., Koyal, S.N., and Beaver, W.L., 1973, Anaerobic threshold and respiratory gas exchange during exercise. Journal of Applied Physiology, 35, 236-243.

Advances in Industrial Ergonomics and Safety II
Edited by Biman Das
Taylor & Francis 1990

LOADS ON BODY JOINTS AND THE LUMBAR SPINE WHEN LIFTING AND LOWERING BOXES

Gilmen Smyth*† and Micheline Gagnon*

*Département d'éducation physique, Université de Montréal,
C.P 6128, succ. A, Montréal, Québec, H3C 3J7

*†CUSLM, Université de Moncton
165 boul. Hébert, Edmundston, N.B., E3V 2S8

INTRODUCTION

The manual materials handling of various loads in the industrial environment has been recognized for some time as an important cause of back problems that affect the workers. Snook (1978) has reported that in the United States, close to 25% of all compensated work injuries were related to manual materials handling and that 79% of those injuries were to the lower back area. The situation in Canada is also as serious: statistics by the Workers' Health and Safety Commission (WHSC) for the province of Quebec revealed that occupational accidents were particularly important in the transport sector; the Commission spent $60 millions in 1985 for 9000 cases of accidents (Lamonde, 1987).

Among the tasks responsible for back injuries, it appears that the handling of loads is responsible for the majority of injuries to the back and that they can result in high compression forces on the lumbar discs (Brown, 1971). A survey of the industrial settings has contributed to display a clearer picture of the objects handled and of the techniques of handling used by the workers (Drury, 1982). Among other things, the results showed that more than half of the tasks were lowering tasks while about one third were lifting tasks, the rest being movements to the same height. A recent study (Lamonde, 1987) has shown that for workers of a freight company, one third of all accidents were due to excessive muscular efforts, 69% of those accidents affecting the back and 17% the upper limbs, specially the shoulders. She also found that about two thirds of these types of injuries occurred in unloading situations characterized by the predominance of actions with the aid of gravity.

The stresses sustained by body joints in manual materials handling are dependent upon various factors. The weight of the load handled may be an important factor contributing to injuries, as an increase in the weight of the load handled increases lumbar compression (Chaffin and Andersson , 1984). Garg et al. (1982) have demonstrated that increasing the width of the box handled (distance away from the body in the sagittal plane) resulted at L5/S1 in an increase of peak moment and compressive force per kg of weight lifted. The heights at which boxes are being handled may contribute to injury as Snook (1978) has shown that the weight that can be handled decreases at levels above the shoulders. The direction of movement (lifting or lowering) may also be a contributing factor to injuries, but comparisons between these tasks have been less documented. It was thus the purpose of this study to evaluate stresses estimated at body joints in manual materials handling, specially when lifting and lowering boxes. This paper emphasizes the results pertaining to handling heights for both the lifting and lowering tasks.

METHODS

Eight experienced subjects from a food product warehouse (age: 23-46 years; experience: 1-18 years) were used as subjects. Each subject performed two trials for 50 different experimental treatments that consisted in the lifting and lowering (2 tasks) of 5 different boxes (3.3, 6.4, 11.6, 16.4 and 22.0 kg) at 5 shelf heights (15, 61, 109, 147 and 185 cm). Handling boxes at 15 and 61 cm was considered as handling below the waist, while handling at 109, 147 and 185 cm, above the waist.

The measurement techniques consisted of cinematography and force platforms. The subjects' segmental positions during movement were recorded with a 16 mm Locam camera operating at 50 frames/s. The films were digitized every other frame with a NAC motion analyzer linked to a PDP 11/23 mini-computer. The coordinates of the joints' centers were smoothed with a second order recursive filter with a cut-off frequency of 8 Hz for the trunk and upper extremities and a cut-off frequency of 2 Hz for joints of the lower extremities. Differentiation procedures were then applied using finite differences, followed by further application of filtering procedures to the derived velocity and acceleration data. Two AMTI force platforms were used to record the external forces under each foot. Synchronization of film sequences and force data was accomplished by means of an electrical pulse recorded on a specific channel of the computer and the simultaneous activation of a light emitting diode (LED) that appeared in the foreground of film frames.

The kinematic and kinetic data were used as input in a sagittal plane dynamic model described by Gagnon et al., (1988) and modified for the L5/S1 angle according to the data reported by Chaffin and Andersson (1984) The model included a single muscle equivalent to represent either the flexors or the extensors of the trunk. The dependent variables were the maximum net muscular moments at body joints and the peak compression force at L5/S1.

Factorial analyses of variance with repeated measures were performed on each dependent variable. The factor heights was nested within the factor tasks (lifting or lowering) and crossed with the factor masses and repetitions. A 0.05 level of significance was chosen.

RESULTS AND DISCUSSION

The results summarized in Table 1 show major differences between the lifting and lowering tasks, between the joints for each task and also between shelf heights. Patterns for lifting and lowering tasks were most of the time similar at each joint, lifting being however always more stressful than lowering at heights above the waist by an average of more than 20%; lowering tasks generally resulted in larger moments for tasks below the waist, with the exception of moments at the hips and at L5/S1, and for the extension moment of the right ankle at 15 cm. When viewing these results from the perspective of placing boxes on and removing boxes off the shelves, the placing of boxes generally resulted in higher stresses to body joints than their removing from the same shelf heights, implying that a common factor within the design of handling tasks could be the horizontal direction of movement rather than its vertical direction.

Table 1. Maximum net muscular moment at selected joints and maximum compression force at L5/S1 in lifting and lowering as a function of shelf height (N = 80).

	Heights					Significance
Variables	15 cm	61 cm	109 cm	147 cm	185 cm	level
LIFTING						
Extension right ankle (N m)	54.8	40.5	41.7	41.2	45.2	0.002*
Extension left ankle (N m)	41.2	25.3	31.0	29.0	34.4	0.000*
Flexion right knee (N m)	17.3	38.2	29.3	22.7	21.3	0.003*
Flexion left knee (N m)	25.6	41.7	37.2	29.0	29.2	0.046*
Extension right knee (N m)	24.4	7.0	1.7	10.1	14.0	0.003*
Extension left knee (N m)	15.2	2.0	0.8	4.8	6.8	0.007*
Extension hip (N m)	233.2	160.1	78.3	63.3	56.2	0.000*
Extension L5/S1 (N m)	218.4	167.4	86.9	75.4	69.4	0.000*
Flexion shoulders (N m)	60.3	47.0	69.0	79.3	79.6	0.000*
Flexion elbows (N m)	52.8	45.0	63.2	69.2	72.8	0.000*
Compression at L5/S1 (N)	3961	3122	1797	1618	1548	0.000*
LOWERING						
Extension right ankle (N m)	52.3	47.0	31.6	34.5	35.9	0.001*
Extension left ankle (N m)	43.2	33.2	18.9	21.1	24.6	0.000*
Flexion right knee (N m)	26.5	50.5	23.7	20.6	19.9	0.008*
Flexion left knee (N m)	31.6	55.2	26.7	24.4	24.7	0.004*
Extension right knee (N m)	21.1	1.7	0.7	3.0	4.6	0.011*
Extension left knee (N m)	13.1	0.5	0.4	1.9	2.7	0.038*
Extension hip (N m)	225.8	158.8	56.7	44.3	41.2	0.000*
Extension L5/S1 (N m)	206.5	161.3	65.7	55.3	52.1	0.000*
Flexion shoulders (N m)	63.6	53.5	52.7	61.7	65.9	0.050*
Flexion elbows (N m)	64.5	55.6	44.0	45.7	50.0	0.039*
Compression at L5/S1 (N)	3779	3002	1471	1281	1230	0.000*

* Statistical level of significance for $p \leq 0.05$

The ankles and the knees displayed opposite patterns with increased height, the ankles being stressed the most at the 15 cm shelf, extension moments decreasing rapidly at 61 cm and then increasing slightly at the other shelves, while flexion moments at the knees were minimum at 15 cm, almost doubling at 61 cm and then decreasing slightly afterwards. This may be due to the flexion of the lower extremities for handling boxes near the floor level, the ankles playing an important role for straightening the legs. This could well explain also the higher extension moments at both knees recorded when handling at the lowest shelf.

The largest moments were encountered at the hips and at the L5/S1 joint. At these joints, moments decreased rapidly from 15 to 109 cm and then only slightly at the 147 and 185 cm shelves; moments were always higher in lifting than in lowering. The large moments at L5/S1 resulted in high compression force at that joint, the handling of the heaviest boxes below the waist producing forces higher than the Action Limit suggested by NIOSH (1981).

The pattern of loading at the shoulders was, however, very different from that of the lumbar joint. In lifting, moments were the lowest at heights below the waist and increased rapidly from the waist to shoulder level (147 cm). The lower moments at levels below the waist seem to result from the greater flexion of the trunk when handling at the lower heights, which brought the shoulders closer to the shelf and thus reduced the moment arm.

CONCLUSION

The risk of injury to the lumbar spine appears greater with the handling of boxes at lower levels, but the risk of injury to the upper extremities, specially the shoulders, definitely appears greater when handling loads at heights above the waist. This interaction among joints reveals the complexity in the analysis of manual materials handling tasks and suggests that caution must be exercised when making recommendations, as the risk of injury may be transferred from one part of the body to another.

REFERENCES

Brown, J. R., 1971, Lifting as an industrial hazard, Labour Safety Council of Ontario, Ontario Department of Labour.

Chaffin, D.B. and Andersson, G.B.J., 1984, Occupational Biomechanics, (New York: John Wiley & Sons).

Drury, C.G., Law, C.-H. and Pawenski, C.S., 1982, A survey of industrial box handling, Human Factors, 24, 553-565.

Gagnon, M., Roy, D. and Lortie, M., 1988, Examination of biomechanical principles in a patient handling task, International Journal of Industrial Ergonomics, 3, 29-40.

Garg, A., Chaffin, D.B. and Freivalds, A., 1982, Biomechanical stresses from manual load lifting: a static vs dynamic evaluation, IIE Transactions, 14, 272-281.

Lamonde, F., 1987, Analyse des accidents de manutention dans le secteur transport général et entreposage. Unpublished M.A. Sc. Thesis, École Polytechnique, Montréal, 286 pages.

NIOSH, 1981, A Work Practices Guide for Manual Lifting, Cincinnati : US Department of HEW.

Snook, S.H., 1978, The design of manual handling tasks, Ergonomics, 21, 963-985.

ACKNOWLEDGMENT

This research was supported with grants from the Institut de recherche en Santé et en Sécurité au Travail du Québec (N/D RS-84-11) and the National Sciences and Engineering Research Council of Canada (OGP0020812).

Advances in Industrial Ergonomics and Safety II
Edited by Biman Das
Taylor & Francis 1990

ENDURANCE TIME AND PHYSIOLOGICAL RESPONSES FOR HIGH FREQUENCY LIFTING TASKS: A PRELIMINARY REPORT

S.S.Asfour*, A.M.Genaidy**, M.W Fahmy*

A.Younis* and S.M.Waly*

*Department of Industrial Engineering
University of Miami, Coral Gables, FL 33124, USA

**Department of Industrial Engineering
Western Michigan University, Kalamazoo, MI 49008, USA

The main objective of this paper is to provide a preliminary report on the effects of high frequency lifting tasks on endurance time and physiological responses of industrial workers. Results showed that for high frequency lifting tasks requiring the use of leg muscles: (1) endurance time decreased significantly with an increase in frequency and load, however, frequency has a more pronounced effect on endurance time than the weight of load; and (2) oxygen consumption and heart rate were significantly affected by the frequency and load.

INTRODUCTION

Endurance has been long thought of as the major limiting factor for high frequency manual handling tasks. If endurance limits for individuals engaged in these types of tasks are exceeded, it is expected that these individuals will be exposed to a higher risk of musculoskeletal injuries and illnesses (e.g., back injuries).

Traditionally, research efforts have been directed towards the quantification of human physiological capabilities (e.g., heart rate and oxygen consumption) for high frequency lifting tasks over short periods of time, usually ten minutes or less (Garg et al., 1978; Mital and Ayoub, 1981; Asfour et al., 1986a; Asfour et al., 1986b). The results obtained from these studies were extrapolated to estimate human physiological capabilities over prolonged periods of time. Moreover, Asfour et al. (1988) reported that the findings of physiological studies performed on standard laboratory tests "bicycle ergometer and motor-driven treadmill" were used as benchmarks for eight-hour manual lifting tasks (e.g., physiological fatigue limit of 1 liter/minute). It should be noted, however, that: (1) standard laboratory tasks require the use of different groups of muscles and are highly dynamic in nature, thus data obtained from these types of tasks could not be applied directly to lifting tasks without validation; and (2) no study was conducted to examine the effects of lifting tasks of prolonged duration on human physiological capabilities.

Recently, studies were reported to quantify human lifting capabilities for college male students over prolonged periods of time. The major findings of these studies were: (1) endurance time was significantly affected by frequency and load for lifting tasks which start at the floor level and table height (Genaidy and Asfour, 1989; Tritar, 1988); (2) physiological fatigue limits were dependent on the interaction of frequency and load for lifting tasks which start at the floor level and table height (Genaidy et al., 1990c; Tritar, 1988); (3) physiological fatigue limits for lifting tasks which start at the floor level were significantly higher than responses for lifting tasks which start at table height (Genaidy et al., 1990c; Tritar, 1988); (4) it is possible to improve endurance time limits for carrying, symmetrical lifting, asymmetrical lifting and combined lifting, lowering, carrying, pushing, pulling and holding activities through physical training (Genaidy et al., 1989; Genaidy et al., 1990a; Genaidy et al., 1990b). Endurance time was defined in these studies as the maximum length of time during which an individual is continously performing a manual material handling task until exhaustion due to local muscular fatigue and/or total body fatigue.

The main objective of this study is to provide a preliminary report on the effects of high frequency lifting tasks on endurance time and physiological responses of industrial workers.

METHODS

Subjects

In this preliminary report, the results of experiments performed on eleven male industrial workers are given. The physical characteristics of this sample are: (1) age (years): 24.5 ± 2.91 ; (2) body weight (kg): 68.8 ± 6.8; and (3) stature (cm): 169.4 ± 4.8. All workers were medically examined to ensure that they were free of musculoskeletal injuries and illnesses, heart problems, blood pressure or any other medical restrictions.

Statistical Design and Analysis

A within-subject two factor statistical design was used. The independent variables were frequency (8, 12 and 16 times/minute) and weight of load (5, 10 and 15 kg). The dependent variables were endurance time, oxygen consumption, minute ventilation and heart rate. Endurance time was defined as the maximum length of time during which an individual is capable of continuously lifting a given load contained in a 38 cm X 38 cm X 38 cm box with cutout handles at a given frequency from the floor level to table height (76 cm above the floor) until exhaustion due to local muscular fatigue and/or total body fatigue. The upper limit of endurance was set at eight hours.

For the purpose of statistical analysis, physiological responses were defined for each frequency and load treatment combination as the mean of the observations taken at minutes 5, 10, 15, ..., ET where ET is the endurance time. An analysis of variance was conducted to test the effects of frequency and load on endurance time and physiological responses. All statistical analyses were performed using the procedures of the statistical analysis system package (1985).

Experimental Procedures

Each subject performed a total of nine treatments. Within each treatment, every subject was given 10 minutes of rest every 50 minutes of work and 1 hour for lunch after the fourth hour of work. Oxygen consumption, minute ventilation, and heart rate were measured according to the procedures described by Asfour et al. (1987) and Genaidy et al. (1987). Physiological parameters were monitored continuously on a minute-to-minute basis; however, analysis was performed on 5-minute data intervals.

The box handled was automatically lowered using a mechanical device. Also, the subjects were paced using a metronome. All experiments were conducted in an environmentally controlled laboratory.

RESULTS

Statistical Analysis

For endurance time, the results of analysis of variance showed that the main effects of frequency and load as well as their interaction were significant at the 5% level. The results of the Tukey test on the average endurance time values for the different levels of frequency and load showed that only the frequency levels of 8 and 12 times/minute were significantly different from 16 times/minute and the load levels of 5 and 10 kg were significantly different from 15 kg at the 5% level.

Table 1. Means and standard deviations (SD) of endurance time and oxygen consumption

Load (kg)	Frequency (times/min)	Endurance Time (minutes) Mean	SD	Oxygen Consumption (liters/minute) Mean	SD
--	--	383.5	145.3	1.2	0.27
5	--	456.4	81.5	1.0	0.19
10	--	397.5	127.9	1.2	0.21
15	--	296.5	167.9	1.4	0.27
--	8	440.8	90.5	1.1	0.26
--	12	404.8	126.8	1.2	0.23
--	16	304.8	173.7	1.3	0.29
5	8	474.5	18.1	0.9	0.15
5	12	474.5	18.1	1.0	0.17
5	16	420.1	135.9	1.2	0.20
10	8	441.0	85.3	1.1	0.22
10	12	403.5	126.5	1.2	0.14
10	16	348.1	155.6	1.3	0.25
15	8	406.8	126.8	1.3	0.26
15	12	336.4	156.8	1.4	0.23
15	16	146.2	97.7	1.5	0.27

For the physiological responses, only the main effects of frequency and load were significant at the 5% level. The results of the Tukey test on the average physiological responses for the different levels of frequency and load showed that: (1) all load levels were significantly different from each other at the 5% level for the three physiological parameters; and (2) statistical differences were found only among the frequency pairs (8,16) and (12,16) for minute ventilation and heart rate and (8,16) for oxygen consumption.

Descriptive Statistics

The means and standard deviations of endurance time and physiological responses are shown in tables 1 and 2. It is clear from these tables that endurance time decreased with an increase in frequency and load. Also, oxygen consumption, minute ventilation and heart rate increased with an increase in frequency and load. There was a 13% decrease in endurance time when doubling the load from 5 kg to 10 kg, while there was a 31% decrease in endurance time when doubling the frequency from 8 times/minute to 16 times/minute. Tripling the load from 5 kg to 15 kg was accompanied by a 35% decrease in endurance time. These results indicate that the frequency is a more important task parameter than the weight of load in the design of prolonged lifting tasks performed from the floor level to table height using the straight back -- bent knees technique. These findings are in agreement with those reported by Genaidy and Asfour (1989) on college students.

Table 2. Means and standard deviations (SD) of minute ventilation and heart rate

Load (kg)	Frequency (times/minute)	Minute Ventilation (liters/minute) Mean	SD	Heart Rate beats/minute Mean	SD
--	--	25	6.0	111	18
5	--	21	4.6	101	14
10	--	24	5.3	110	13
15	--	29	5.8	122	19
--	8	21	4.1	100	12
--	12	25	5.9	113	18
--	16	27	6.5	119	17
5	8	19	4.3	93	14
5	12	21	4.4	100	11
5	16	24	3.8	109	12
10	8	21	2.8	101	7
10	12	26	6.0	112	13
10	16	25	5.3	118	12
15	8	25	3.0	107	8
15	12	29	5.2	128	20
15	16	33	5.8	131	18

CONCLUSIONS

The main conclusions, that can be drawn for high frequency lifting tasks starting at the floor level and ending at table height using the straight back -- bent knees technique, are:

1. Endurance time decreased significantly with an increase in frequency and load; however, frequency has a more pronounced effect on endurance time than the weight of load.
2. Physiological fatigue limits in terms of oxygen consumption and heart rate were signicantly affected by the frequency and load. Thus, one cannot recommend a single value for physiological fatigue limits as frequently reported in the literature. Moreover, physiological fatigue limits based on standard laboratory tests cannot be used for industrial tasks such as lifting.

REFERENCES

Asfour, S. S., Ayoub, M. M., Genaidy, A. M., and Khalil, T. M., 1986a, A data base of physiological responses to manual lifting, In: W. Karwowski (ed.), Trends in Ergonomics/Human Factors III (Elsevier, Amsterdam) 801-809.

Asfour, S. S., Ayoub, M. M., and Genaidy, A. M., 1986b, A data base of physiological responses to manual lowering, In: W. Karwowski (ed.), Trends in Ergonomics/Human Factors III (Elsevier, Amsterdam) 811-818.

Asfour, S. S., Genaidy, A. M., and Khalil, T. M., 1987, An on-line microcomputer-based metabolic monitoring system, The International Journal of Industrial Ergonomics, 1, 169-177.

Asfour, S. S., Genaidy, A. M., and Mital, A., 1988, Physiological guidelines for the design of manual lifting and lowering tasks: the state of the art, American Industrial Hygiene Association Journal, 49, 150-160.

Garg, A., Chaffin, D. B., Herrin, G. D., 1978, Prediction of metabolic rates for manual materials handling jobs, American Industrial Hygiene Association Journal, 39, 661-674.

Genaidy, A. M., Agrawal, Houshyar, A., Dawson, M., Zabik, R., and Mousavinezhad, S. H., 1990a, An ergonomic training program for industrial manual handling jobs, Proceedings of the Ergonomics Society, London, England, April.

Genaidy, A. M. and Asfour, S. S., 1989. Effects of frequency and load of lift on endurance time, Ergonomics, 32, 51-57.

Genaidy, A. M., Asfour, S. S., and Khalil, T. M., 1987, An on-line microcomputer-based cardiac monitoring system, The International Journal of Industrial Ergonomics, 1, 273-283.

Genaidy, A. M., Bafna, K. M., Sarmidy, R., and Sana, P., 1990b, An endurance training program for symmetrical and asymmetrical manual lifting tasks, Journal of Occupational Medicine (in press).

Genaidy, A. M., Khalil, T. M., Asfour, S. S., and Duncan, R. C., 1990c, Human physiological capabilities for prolonged manual lifting tasks, IIE Transactions, (in press).

Genaidy, A. M., Mital, A., and Bafna, K., 1989, An endurance training programme for frequent manual carrying tasks, Ergonomics, 32, 149-155.

Mital, A. and Ayoub, M. M., 1981, Effect of task variables and their interactions in lifting and lowering loads, American Industrial Hygiene Association Journal, 42, 134-142.
Statistical Analysis System, 1985, SAS user's guide: statistics (SAS Institute Inc., Cary, North Carolina).
Tritar, M., 1988, Physiological models and guidelines for the design of prolonged arm lifting tasks, Unpublished thesis, Department of Industrial Engineering, University of Miami, Coral Gables, Florida, USA.

ACKNOWLEDGEMENTS

This work was supported by the National Institute for Occupational Safety and Health through Grant No. 1R01OH02591-01.

Advances in Industrial Ergonomics and Safety II
Edited by Biman Das
Taylor & Francis 1990

MAXIMUM ACCEPTABLE WEIGHT USING DIFFERENT CARRYING METHODS

K. Lee, F. Aghazadeh, A. Bruckner, II, A. Waikar, and C. Lum

Department of Industrial and Manufacturing Systems
Engineering, Louisiana State University
Baton Rouge, LA 70803, U.S.A.

The objectives of this study were: (a) to find the maximum acceptable weight which can be carried for 8 hours using a psychophysical method for front, side (one hand and two hands) and back carrying, and (b) to develop models which could predict the acceptable maximum weight in carrying using static strength data. A series of laboratory experiments were conducted to find the maximum acceptable weight in front, side and back carrying. Six college students participated in the experiment. It was found that subjects were willing to carry the heaviest load using two-hand side carrying (average maximum acceptable weight: 7.76 kg). Back carrying was the close second with 6.67 kg. Also, there was a significant difference ($p< 0.01$) in maximum acceptable weight for carrying between one-hand side carrying (4.40 kg) and two-hand side carrying.

INTRODUCTION

Manual material handling is a major source of lower back injuries and accidents in industry. The injuries occur primarily because the task demands exceed the worker's capacity. The accidents may occur because the task is not ergonomically efficient or the task or method may not be subjectively acceptable to the worker. This can be detrimental to the individual and the industry may also suffer through production losses, rescheduling problems, and health care costs. Therefore, industry needs guidelines on load limits for materials handling and personnel selection criteria for job assignment.

In 1987, the National Safety Council stated that "back injuries accounted for 35% of all lost-time accidents in different organizations" (National Safety Council, 1988). Many employers have higher back injury rates than this average incidence rate. Of these, 25% of the injuries are caused by lifting and moving while engaged in manual material handling that involves carrying (Nagira et al., 1981).

Approaches taken to minimize the impact of manual materials

handling (MMH) injuries and accidents include designing the job to fit the worker, good training procedures in safe manual material handling, and careful selection of workers through pre-placement screening criteria. Also, guidelines exist for manual lifting as recommended by the National Institute for Safety and Health (NIOSH) that include an action limit and a maximum permissible limit for compressive force on L_5/S_1 disc resulting from lifting tasks. However, these deal only with lifting. There are no guidelines for the acceptable weight for carrying tasks.

There are many types of carrying tasks such as carrying with one hand (one-hand side carrying), carrying two objects with two hands (two-hand side carrying), carrying using a container located on the back (back carrying), etc. However, information for carrying activities such as the acceptable weight is very limited. This may be because of the concept that the limiting factor for carrying may be either energy consumption or oxygen consumption rate. This study was intended to fill the gap in information available regarding maximum acceptable weights of carrying for different carrying methods.

OBJECTIVES

The objectives of this study were: (a) to find the maximum acceptable weight which can be carried for 8 hours using psychophysical methodology for front, side (one hand and two hands) and back carrying, and (b) to develop models which could predict the acceptable maximum weight in carrying using static strength data.

METHOD

An experiment was conducted in the laboratory to find the maximum acceptable weight for carrying for an 8-hour work day. Three different carrying methods, back carrying, front carrying, and side carrying were investigated.

Subjects

Six male subjects were recruited for the study from the local student population at Louisiana State University. The subjects' ages ranged from 22 to 27 years with an average age of 23.8 years and a standard deviation (SD) of 2.2 years. Their average height was 171.7 cm, the average weight was 59.7 kg, and SD's were 5.7 kg and 7.0 cm, respectively. All were in good physical condition at the time of the experiment. None had a personal medical history of any back problem. Each subject was tested for his arm, shoulder, leg and composite strength, before the carrying experiment began. Table 1 shows the anthropometric data for all the subjects.

Experiment

The subjects were tested in two sessions. In the first session, the subjects participated in a series of isometric strength tests. Arm strength, shoulder strength, leg strength and composite strength were recorded for each subject. All the strength measurements were completed before the carrying experiments in session 2.

In the second experiment session, the subjects participated

Table 1. Anthropometric data for subjects.

	Mean	Range	S.D.
Age (Years)	23.83	22.0 - 27.0	2.23
Weight (kg)	59.67	51.4 - 65.7	5.71
Height (cm)	171.67	164.0 - 182.0	6.95
Acromial Height (cm)	140.62	131.1 - 154.5	7.97
Standing Iliac Crest Height (cm)	101.07	92.2 - 110.5	6.27
Knuckle Height (cm)	77.28	67.5 - 90.1	8.19
Knee height (cm)	51.07	45.8 - 55.5	3.85
Forearm Grip Distance (cm)	34.33	29.1 - 38.5	3.72

Table 2 : Subjects' Isometric Strength Measurements (kg.f).

	Mean	Range	S.D.
Static Arm	274.20	158.87 - 356.00	65.87
Stooped Back	543.94	336.86 - 741.82	150.02
Static Composite	753.09	470.36 - 1102.26	204.99
Static Shoulder	381.14	289.25 - 492.62	79.65
Static Leg	762.06	498.40 - 943.4	146.51

in a series of carrying including front carrying, side carrying and back carrying. Since it was not easy to test subjects over an entire 8 hour work day, psychophysical methodology was used to estimate the maximum weight acceptable to the subjects for carrying over an eight-hour period. Thus, the experiment for each type of carrying was conducted for 20 minutes instead of eight hours. The subjects were asked to carry weight in a sack at walking speeds of 1.6km/h and 2.4km/h on the treadmill, each for 20 minutes.

As in other studies employing psychophysical methodology (Mital et al., 1984), the subjects were instructed to add or reduce metal pieces as needed to arrive at a weight acceptable for an eight-hour work day with two fifteen-minute rest breaks and a thirty-minute lunch break. The subjects were told to use their perceived feelings of stress in the lower back as a basis for estimating the acceptable weight, and to adjust the weight in the sack to a value that they felt they could carry. The adjustment was to be completed within twenty minutes. After twenty minutes, the sack was weighed. They were not allowed to see the weight inside the sack, but were allowed to change the weight as often as they felt necessary. The experimenter added or removed the weights as per subject's instruction.

<u>Procedure for strength tests</u>

Arm strength, shoulder strength, leg strength and composite strength were measured following the methodology used by Ayoub et al. (1978). After each strength test, a five-minute rest was provided. After the strength testing was finished, a twenty minute rest break was given before beginning the carrying experiments. Each subject participated in all four tests. The sequence for different types of carrying was randomized to minimize the fatigue effect.

<u>Procedure for carrying tests</u>

Four different carrying methods were tested. They were back carrying, front carrying, side carrying with one hand and side carrying with two hands. Two different speeds of carrying, 1.6km/h and 2.4km/h, were used. Each trial was replicated once. The initial weight to begin the trial was 15.9kg for each test. This included the weight of the sack.

A treadmill, Quinton Model Q55 manufactured by Quinton Co. was used for dynamic carrying test. Grade level was zero for this treadmill during the experiment.

Subjects were told that, if at any time they felt the sack was too heavy, weight could be taken out. Conversely, if they felt that the sack was too light, weight could be added. Each subject was told to assume that he would have to carry the final weight 8 hours per day. When the 20 minutes was over, the sack was weighed.

RESULTS

The data was analyzed for the acceptable weight for an 8-hour work day. The effect of each variable to the acceptable weight was considered, and finally regression models which can predict the acceptable weight using other anthropometric and strength variables were developed.

Table 3. Average Weight (kg) Carried by Six Subjects in Each Task

Task	Speed (km/h)	Mean	Range	S.D.
Back Carrying	1.6	6.09	4.04 -10.34	2.23
	2.4	7.25	5.00 - 9.20	1.53
Front Carrying	1.6	3.05	2.27 - 3.46	0.42
	2.4	3.17	2.78 - 3.35	0.22
Side Carrying One Hand	1.6	4.62	3.52 - 5.74	0.75
	2.4	4.17	3.46 - 4.60	0.51
Side Carrying Two Hands	1.6	7.59	6.02 -10.90	1.51
	2.4	7.92	6.25 -10.68	1.62

Table 4. Percentage of strength required to carry weight against the isometric strength.

	ARM STRENGTH	BACK STRENGTH	COMPOSITE STRENGTH	SHOULDER STRENGTH	LEG STRENGTH
Back Carrying	21.8%	10.8%	7.8%	15.4%	7.8%
	26.0%	13.1%	9.4%	18.6%	9.3%
Front Carrying	11.0%	5.7%	4.0%	7.8%	4.0%
	11.3%	5.7%	4.1%	8.1%	4.1%
Side Carrying One Hand	16.5%	8.3%	6.01%	11.8%	5.9%
	14.91%	7.5%	5.4%	10.7%	5.3%
Side Carrying Two Hands	26.9%	13.6%	9.8%	19.3%	9.7%
	28.3%	14.3%	10.3%	20.3%	10.2%

Table 2 shows the summary (mean, standard deviation and range) of the strength measurements for all the subjects. These strength values were compared to the static composite strength. Table 3 shows the average weights acceptable to all subjects for carrying over an eight-hour period determined during the carrying experiments using different methods of carrying. This table also shows the ranges and standard deviations for the acceptable weight.

It was also found that changing speed from 1.6 km/hr to 2.4 km/hr did not cause any noticeable change in the acceptable weight in most carrying methods (within 5% range). It is interesting to find that in back carrying there was an increase of 1.2 kg (21%) when the speed was increased.

Table 4 also shows that the actual force required was less than 15% of the isometric strength in most cases. Side carrying with two hands, however, showed the highest percentages. Rohmert (1968) reported that persons can sustain an exertion level if the required exertion level is about 15% of their maximum muscle force. The finding from our study confirms Rohmert's finding.

Table 5 shows the results of a correlation analysis which was conducted to examine the effect of anthropometric and strength variables on the maximum acceptable weight. Individually, no single variable showed a high enough correlation to be a stand-alone predictor.

A stepwise regression procedure was used to develop models that can predict maximum acceptable weight of carrying for different types of carrying. This was performed on a microcomputer using the SAS program. The criteria used to select or reject a variable was a partial F value of 3.0.

For the first model attempt, a prediction model using isometric strengths (arm, back, composite, shoulder and leg strengths) was attempted. Except for side carry, no model evolved since no variable could satisfy the 0.15 significance level for entry. The r^2's were not very high for side carrying (0.78 for one hand and 0.53 for two hands).

As the second effort, only the acceptable weight for the front carrying was used to predict the acceptable weight for other types of carrying. Front carrying was chosen because there is much information available (NIOSH, 1982)) for this type of manual materials handling. The acceptable weight for front carrying can be determined using the formula in the NIOSH guideline. However, it was found that only side carrying could give a model even though r^2 was relatively low. Therefore, this method turned out to be ineffective.

CONCLUSIONS

It was found that the subjects were willing to carry the heaviest load using two-hand side carrying method (average maximum acceptable weight: 7.6 kg). Back carrying was the close second with 6.1 kg. As expected, there was a significant difference ($p < 0.01$) in maximum acceptable weight of carrying between one-hand side carrying (4.6 kg) and two-hand carrying despite the use of the dominant hand in the one-hand carrying. This could be attributed to the fact that people prefer to carry with two hands rather than one hand in side carrying to have a better balance.

Table 5. Results of the correlation analysis.

Factors	Correlation: Front Carrying 1.6 km/h	Front Carrying 2.4 km/h	One-hand Side Carrying 1.6 km/h	One-hand Side Carrying 2.4 km/h	Two-hand Side Carrying 1.6 km/h	Two-hand Side Carrying 2.4 km/h	Back Carrying 1.6 km/h	Back Carrying 2.4 km/h
Age	0.1649	0.5760	0.3562	0.1834	0.0105	0.6848	0.5742	0.3255
Weight	0.3417	-0.1141	0.3186	0.2127	-0.3540	0.1503	0.1865	0.1404
Height	0.2930	-0.3372	-0.4201	-0.3878	-0.6948	-0.6252	-0.5716	-0.5318
Arm Strength	0.6864	-0.2453	-0.4050	-0.7296	-0.8601	-0.2254	-0.0584	-0.5350
Leg Strength	-0.2211	-0.6588	-0.2909	-0.7132	0.0372	0.2902	-0.0819	0.0031
Back Strength	0.3707	-0.3504	-0.1227	-0.4238	-0.5185	0.1361	0.0663	-0.1604
Shoulder Strength	0.4939	-0.5413	0.2866	-0.3240	-0.4662	0.2807	0.3080	0.2007
Composite Strength	-0.2013	-0.8674	0.2006	-0.2016	-0.0048	0.0231	-0.2404	0.3283

This research shows that front carrying cannot be used to estimate the acceptable weight in other types of carrying. The difference of the acceptable weights between the back carrying and one-hand side carrying found in this study is consistent with the result of an earlier physiological study (Malhotra and Gupta, 1965).

A most interesting, though, preliminary result, for use as an injury-preventive data is the comparison of current isometric lift and lift-and-carry maximum load limits. Examination of Table 2 vis-a-vis Table 3 gives some indication of this ratio. This ratio is demonstrated in Table 4. The results from this experiment indicate that the composite strength test alone cannot be used to predict an employee's performance for lifting and carrying tasks in an industrial setting.

RECOMMENDATIONS FOR FURTHER RESEARCH

This general area of research involving carrying is still incomplete. There is an impetus for these continuing studies because lifting and carrying are common activities in service and distribution industries (shipping, trucking, and warehousing).

In addition, it is noted that to-date existing research has shown disagreements in the best manual materials handling practices. To some extent, though undefined, cultural practices seem to have influenced preferred methods of lifting and carrying. It is doubtful, for example, that Western nations' cultural influences will ever adopt carry-on-head, or for that matter, back-carry (in the industrial environment). Further lifting and carrying studies need to focus on the most common industrial applications.

REFERENCES

Ayoub, M.M., Bethea, N.J., Deivanayagam, S., Asfour, S.S., Bakken, G.M., Liles, D.H., Mital, A. and Sherif, M., 1978, Determination and modelling of lifting capacity. NIOSH Final Report, DHEW (NIOSH) Grant Nos. IR010H00545-01SOH and 5R010H 00545-02.

Malhotra, M. S., and Gupta, J. S., 1965, Carrying of school bags by children. Ergonomics, 8, 55-60.

Mital, A., 1984, Maximum weights of lift acceptable to male and female industrial workers for extended work shifts, Ergonomics, Vol. 27, No. 11, 1115-1126.

Nagira, T., Suzuki, J., Oze, Y., Ohara, H., and Ayoma, H., 1981, Cervicobrachial and low back disorders among school lunch workers and nursery school teachers in comparison with cash register operators, Journal of Human Ergology, Vol. 10 (2), 117-124.

National Safety Council, 1988, Accident Facts.

von Rohmert, W., 1968, Arbeitsmedizin, Sozial medizin, Arbeishygiene, 22, 118.

Advances in Industrial Ergonomics and Safety II
Edited by Biman Das
Taylor & Francis 1990

THE EFFECT OF HEIGHT IN LOWERING AND LIFTING TASKS: A MECHANICAL WORK EVALUATION

Micheline Gagnon* & Gilmen Smyth*

*Département d'Education Physique
Université de Montréal, Montréal,
C.P. 6128, succ. "A", Québec, H3C 3J7

CUSLM, Université de Moncton
165 boul. Hébert
Edmunston, N.B., E3V 2S8

INTRODUCTION

The effect of height is an important factor in manual materials handling but it has not been much documented in the scientific literature. On the other hand, scientific studies have mainly focused on lifting tasks while lowering tasks have been little studied even when field studies have demonstrated their importance in practice (Drury et al., 1982) and when epidemiological data have suggested their importance in accidents (Lamonde, 1987). In many studies the focus has usually been on the lower-back joints and it has become imperative to obtain a clearer portrait of the relative demands put on all body joints in occupational biomechanics. Therefore, it was the purpose of this study to evaluate total work requirements as well as the relative work contributions by all body joints and the processes of energy transfers for lifting and lowering tasks in manual materials handling.

METHODS

The subjects were 8 experienced workers from a food product warehouse. Each subject performed 2 trials for 50 different experimental treatments (2 tasks x 5 loads x 5 heights). The treatments consisted of tasks represented by the lifting and the lowering of loads. These loads were 5 boxes with different masses (3.3, 6.4, 11.6, 16.4 and 22.0 kg) and different physical dimensions. Five different shelf heights (15, 61, 109, 147 and 185 cm) were used to cover the whole range of movements required by the tasks. The lifting tasks could be classified as below the waist (lifting from the shelf at 15 or 61 cm up to the waist level) or above the waist (lifting from the waist level up to the shelf at 109, 147 or 185 cm); the lowering tasks were also divided into levels below the waist (lowering from the waist down to the shelf at 15 or 61 cm) or above the waist (lowering from the shelf at 185, 147 or 109 cm down to the waist level).

Cinematography techniques were used in synchronization with two AMTI force platforms positioned under each foot. Dynamic and planar segmental analyses were performed to calculate the net muscular moments at the joints; kinematic analyses provided the angular segmental and joint velocities. Mechanical work

(W) was determined on the basis of joint power (P), itself calculated as a product of joint muscular moment (M) and joint angular velocity ($\dot{\phi}_2 - \dot{\phi}_1$):

$$P = M\,(\dot{\phi}_2 - \dot{\phi}_1)$$

Mechanical work was thus calculated from the integral of power over time. For the entire body system, the mechanical work of muscular sources over the period of time from t_i to t_f was expressed in the following way:

$$W = \int_{t_i}^{t_f} [\sum_{i=1}^{N-1} | M_{i,i+1} (\dot{\phi}_{i+1} - \dot{\phi} |]\, dt$$

where $M_{i,i+1}$ is the net muscular moment acting about the joint linked by segments i and i+1, and $\dot{\phi}_i$, $\dot{\phi}_{i+1}$ are the angular velocities of these segments in the sagittal plane of motion. The processes of muscular activity including energy generation (positive work), absorption (negative work) and transfers can be examined thoroughly; this model has been presented by Aleshinsky (1986a, b), Robertson and Winter (1980) and Quanbury et al. (1975). It is based on the classical work by Elftman (1939,1940).

Factorial analyses of variance with repeated measures were performed on each dependent variable. The factor heights was nested within the factor tasks (lowering or lifting) and crossed with the factors masses and repetitions. A 0.05 level of significance was chosen. Only the results pertaining to heights (for both lowering and lifting tasks) are presented in this paper.

RESULTS AND DISCUSSION

The results are summarized in Table 1. The results emphasize the important distinctions in handling (either lifting or lowering) below or above the waist. Handling at the lowest height (15 cm above the ground) was considerably more demanding than handling at any other height with a large contribution of the lower-back and the lower limb joints (about 80% of the work). Handling at the upper levels, on the other hand, displaces the work demands on the upper joints (the elbows and the shoulders share about 80% of the work). The processes of economy or energy transfers are more adequately used for these upper working heights (between 30 and 45% for upper heights as compared to levels between 23 and 32% for lower heights).

Lowering tasks are very demanding and their incidence in injuries must be appreciated: this is mainly due to the large amount of eccentric contractions (negative work). However, work distributions on the different body joints are relatively similar in lifting and lowering tasks. On the other hand, the relative importance of work in lowering tasks as compared to lifting tasks was particularly apparent for lower heights (15 and 61 cm) where lowering work was about 60% that of lifting work. The processes of economy or energy transfers appeared more efficient for lowering tasks (between 28 and 45%) than for lifting tasks (between 23 and 38%). This means that a larger percentage of energy required by the tasks was used through the help of momentum, several adjacent limbs moving in the same angular directions. How these processes may be utilized in the training of workers is a question that should be submitted to specialists in the area of teaching and training programs. Another question of interest concerns the effect that energy transfers may have in loading the joints and the potential conflict between work efficiency and joint loadings.

Table 1. Selected work-energy variables in lifting and lowering as a function of shelf height (N = 80).

Dependent variables	HEIGHTS (cm)					Significance level
	15	61	109	147	185	
LIFTING						
Adjusted mechanical work (J)	390	118	63	126	185	0.000*
Work: lower limbs (%)	57.9	25.7	12.4	7.5	7.5	0.000*
Work: low-back(%)	29.1	53.3	17.7	12.1	8.3	0.000*
Work: upper limbs (%)	13.0	20.7	69.9	80.4	84.3	0.000*
Total economy or transfers (%)	23.0	25.2	38.0	35.3	33.7	0.000*
LOWERING						
Adjusted mechanical work (J)	225	74	17	26	48	0.000*
Work: lower limbs (%)	61.9	23.2	12.4	7.6	5.7	0.000*
Work: low-back (%)	25.0	48.6	21.2	12.0	7.7	0.000*
Work: upper limbs (%)	13.1	28.1	66.2	80.4	86.7	0.000*
Total economy or transfers (%)	28.1	32.2	29.5	45.2	40.9	0.000*

* Statistical level of significance for $p \leq 0.05$

CONCLUSIONS

The work-energy approach is a powerful tool to help detecting risk factors in occupational tasks. Its superiority lies in the fact that all body joints can be assessed at the same time.

REFERENCES

Aleshinsky, S.Y., 1986a, An energy 'sources' and 'fractions' approach to the mechanical energy expenditure problem - II. Movement of the multi-link chain model. Journal of Biomechanics, 19, 295-300.

Aleshinsky, S.Y., 1986b, An energy 'sources' and 'fractions' approach to the mechanical energy expenditure problem - V. The mechanical energy expenditure reduction during motion of the multi-link system. Journal of Biomechanics, 19, 311-315.

Drury, C.G., Law, C.-H. and Pawenski, C.S., 1982, A survey of industrial box handling. Human Factors, 24, 553-565.

Elftman, H., 1939, Forces and energy changes in the leg during walking. American Journal of Physiology, 125, 339-356.

Elftman, H., 1940, The work done by muscles in running. American Journal of Physiology, 129, 672-684.

Lamonde, F., 1987, Analyse des accidents de manutention dans le secteur transport général et entreposage. Unpublished M.A. Sc. Thesis, Ecole Polytechnique, Montréal, 286 pages.

Quanbury, A.O., Winter, D.A. and Reimer, G.D., 1975, Instantaneous power and power flow in body segments during walking. Journal of Human Movement Studies, 1, 59-67.

Robertson, G.D. and Winter, D.A., 1980, Mechanical energy generation, absorption and transfer amongst segments during walking. Journal of Biomechanics, 13, 845-854.

XIV

NOISE AND VIBRATION EFFECTS

Advances in Industrial Ergonomics and Safety II
Edited by Biman Das
Taylor & Francis 1990

INVESTIGATION OF THE DETECTABILITY OF THREE TELEPHONE RINGERS UNDER DIFFERENT NOISE AND SUBJECT AGE CONDITIONS

S. P. Casali, J. P. Berkowitz, J. G. Casali, and P. T. Kemmerling

Virginia Polytechnic Institute and State University, Blacksburg, VA 24061

INTRODUCTION

Hearing Impairment in the Elderly

One of the most prominent age-related disabilities is hearing loss. Whether due to presbycusis, ototoxicity, noise exposure or a combination of etiologies (Mader, 1984; Zarnoch, 1982), hearing loss is often a critical obstacle to independent living and a threat to the safety and health of an older individual. When its effects are compounded by other age-related disabilities, adult hearing loss poses particularly serious technological questions for consumer product design (Brown, Hotchkiss, Allen, Schein, and Adams, 1989).

A substantial proportion of individuals 65 years and older suffer from functional hearing loss, with some noticeable loss occurring as early as the age of 40 (Bess, Lichtenstein, Logan, and Burger, 1989; Stelmachowicz, Beauchaine, Kalberer, and Jesteadt, 1989; Bingea, Raffin, Aune, Baye, and Shae, 1982). In terms of strict demographics, the greatest incidence of hearing loss in the United States occurs among males; however, the disparity in hearing loss between the sexes diminishes with increasing age, especially above 85 years. Hotchkiss (1989) reported that approximately 30 percent of persons 65 years and older consider themselves functionally hearing-impaired with the percentage increasing to almost 50 percent for the over 85 population. Thus 7.9 million people over the age of 65 years reported having a functional hearing loss in 1985. Projections indicate that this number will increase to 12.6 million people by the year 2015 (Hotchkiss, 1989). Because these results were based on self-reports of functional hearing difficulties, and not audiometric measures, they likely underestimate the actual magnitude of hearing impairment among the aged.

The spectral profile of hearing loss in the aged has primarily been studied through pure-tone audiometric surveys. Hinchcliffe (1958) found a gradual increase in detection thresholds across both age and frequency groupings, with the higher frequencies (3000 Hz and above) showing greater decrements with advancing age. Recent work with very high frequencies (between 8000 and 20000 Hz) has also revealed an age and frequency interaction; however, the results have been range-limited by upper bounds on the age groups studied (Stelmachowicz et al., 1989).

Researchers have recently become interested in evaluating hearing with suprathreshold sound stimuli other than pure tones. In one noteworthy study (Ballas and Barnes, 1988), there was no significant difference in the "uncertainty" of older versus younger individuals in identifying common environmental sounds. However, the effect of different sound pressure levels on uncertainty was not investigated.

Telephone Ringer Audibility

A common aural signal that is often crucial to independent living of the elderly is the telephone ringer. The loss of telephone access for the elderly may create a feeling of social isolation and pose safety problems (Infante-Rivard, Krieger, Petitclerc, and Baumgarten, 1988). In a study of 500 new hearing aid applicants, Stephens (1980) noted that 20% listed difficulty in detecting telephone ringers as a primary problem. However, limited human factors research attention has been applied to the design of telephone ringers, especially for the elderly. Hunt (1970) examined the acoustics of a tone ringer and reported that effective ringers consist of a signal with at least two spectral components between 500 and 4500 Hz with the prominent components remaining below 2000 Hz. Furthermore, it was suggested that the telephone ringer signal should maintain a minimum level of 77 dB for adequate detectability.

Despite the existence of such guidelines, some recently-produced consumer telephones have exhibited problems in ringer signal detectability. Various forms of electronically-generated tone signals, electronically-synthesized bells, and mechanical bells are currently available, and most phones incorporate one of these technologies for the ringer. The older-style mechanical bells appear to be dwindling in use compared to the electronic ringers.

Research Objectives

With the typical hearing impairments of the elderly in mind, the fundamental objective of this study was to experimentally compare the detectability of three common but very acoustically-different, telephone ringers for young and older listeners. Given that telephones must be heard in a variety of household ambient noise conditions, detection thresholds for each ringer were obtained in quiet and continuous broadband noise using a controlled psychophysical experiment. The experimental design was configured to address the following two hypotheses:

1) There are certain telephone ringers currently on the market that may be acceptable for younger consumers, but do not accommodate the audibility needs of older adults.

2) There is a significant effect of background noise on the detectability of telephone ringers, and the magnitude of this effect depends on the ringer in question and the hearing ability of the listener.

METHOD

Subjects

Eighteen subjects from the local community voluntarily participated in this study. Nine subjects (group 1) were between the ages of 20 and 30 years with a mean age of 25 years. The remaining nine subjects (group 2) were between the ages of 70 and 95 years with a mean age of 83 years. Each group consisted of 4 females and 5 males. All ringer detection testing was done binaurally with unassisted (no hearing aids) ears.

Experimental Design

The mixed three-factor design included within-subject variables of telephone ringer and background noise and the between-subjects variable of age. A variety of commercially available telephones were obtained and considered for testing. The three that were selected incorporated widely diverse ringer designs, representing the major categories of ringers available. The AT&T Trimline model 221 includes an "electronic bell," the Cobra model WP-142 utilizes an electronic "beeper-type" ringer, and the Unisonic model 6462 incorporates a "mechanical bell" ringer. The measured sound pressure levels at 18 inches for the three telephones were 80 dBA, 67 dBA, and 77 dBA, respectively. These ringer descriptors are used herein to designate the particular phones, but are not intended to imply that the three ringer spectra are global standards across the market. Each telephone simply represents one example of that particular ringer type. Two ambient sound conditions were also investigated: quiet (24 dBA ambient in the booth) and 65 dBA pink noise (representing noise levels in home recreational areas, such as in television rooms).

The presentation order of the 3 (phone) by 2 (noise) treatment conditions was balanced using a partial Latin Square design. The measure obtained for each trial was ringer threshold sound pressure level, measured in dBA at head center.

Apparatus and Experimental Variables Implementation

All data were collected in the Auditory Systems Laboratory at Virginia Tech. The sound spectrum of each telephone ringer (Figures 1, 2, and 3) was measured in an Eckel anechoic chamber, using a Larson-Davis 800-B 1/3 octave analyzer with ACO 7013 microphone positioned 18 inches from the phone which was suspended in the center of the free-field space. Stimuli tapes for the telephone signals were recorded inside the anechoic chamber using an AKG C414-B microphone and a TEAC 124 tape deck. All audiometric screening (using a Beltone 114 pure-tone audiometer), as well as ringer detection tests were conducted in a semi-reverberant, sound-treated booth. A Sony TC-W7R tape deck was used for playback of the recorded ringer signals, which were spectrally-shaped to match the originally measured output spectra of Figures 1-3 using a Yamaha GE-60 equalizer. Signal amplification was provided by an NAD-1020B preamplifier and NAD 2200 power amplifier system, powering an Infinity RS-9B loudspeaker located 18 inches behind the seated subject's head. Calibrated signal level control was accomplished by passing the ringer signal through the attenuator network of the Beltone 114 audiometer. Signal levels were calibrated with the Larson-Davis analyzer.

The background pink noise spectrum, the energy in which was flat (+/-2 dB) by octaves between 100 and 10000 Hz, was originated from tape on a TEAC 124 deck, equalized through a Ross R31M spectrum shaper, and amplified with a Realistic SCR 3010 receiver. The noise was presented through four Infinity RS-6B loudspeakers located equidistant at 39 inches from the subject's head and positioned in the room's corners. Noise was presented at a constant 65 dBA.

Experimental Protocol

After practice on the method of limits threshold determination procedures, each subject was given a standard audiogram at frequencies of 125, 250, 500, 1000, 2000, 4000, and 8000 Hz. Following instruction on the experimental task, the subject underwent a ringer detection threshold test for each of the six conditions. A bracketing procedure, using 2.5 dB step increments of ringer level, was followed to determine threshold. At each ringer level step, three rings were allowed to occur before proceeding to the next level. The subject's task was to indicate audibility of the ringer signal by pressing a hand-held push button. For each experimental condition, five estimates of threshold were obtained, in counterbalanced order.

RESULTS

Mean threshold levels of the seven audiogram frequencies for each age group appear in Table 1. These mean thresholds coincide closely with what would be predicted for each age group based on survey data for non-noise exposed populations (Ward, 1986). A single estimate of ringer threshold, in dBA, for each subject in each condition was determined by computing the arithmetic mean of all estimates which lay within +/- 5 dBA of the mode. It was necessary to discard only one data point of the 540 collected; this provided clear evidence of the stability of the within-subject trials. An analysis of variance was performed on the resulting threshold measure using ringer, noise, and age as the main effects and all possible interactions thereof. (Complete ANOVA summary tables are available from the authors.) A post-hoc Newman-Keuls test ($p < 0.05$) was performed for pairwise means comparisons within each significant main effect and interaction.

The three-way interaction of ringer, noise, and age was significant, $F(2,32) = 6.61$, $p \leq 0.004$, as were the embedded two-way effects of ringer by age, $F(2,32) = 9.58$, $p \leq 0.0005$, and noise by age, $F(1,16) = 25.64$, $p \leq 0.0001$. Ringer by noise was nonsignificant, thus the further breakdown of the complex three-way interaction was performed only on the ringer by age and noise by age effects.

Mean detection thresholds for each ringer for both age groups are plotted in Figure 4. The rank ordering of detection thresholds for the three ringers was the same for both age groups, with better detectability for the electronic ringer, the mechanical bell, and the electronic beeper, in that order. The electronic beeper and mechanical bell ringers resulted in significantly higher threshold values than the electronic bell for the older group, while the electronic beeper ringer resulted in a higher threshold value than the other ringers for the younger group. Reviewing the frequency spectra of the three ringers (Figures 1, 2, and 3)

Table 1. Mean pure-tone audiogram hearing thresholds, in dBHL, for the two age groups.

	Older Subjects		Young Subjects	
Hz	Right	Left	Right	Left
125	28.3	22.2	8.3	2.8
250	26.7	21.1	7.2	3.3
500	23.3	23.3	5.6	0.0
1000	20.6	23.3	5.6	1.1
2000	35.6	36.7	6.7	0.6
4000	52.2	56.7	5.6	3.3
8000	77.8	81.1	11.7	7.8

aids in explaining these effects. For frequencies between 315 and 2000 Hz, the mechanical bell and electronic bell are largely similar except for the prominent energy peak of the electronic bell between 1000 and 1600 Hz. The mechanical bell, on the other hand, has most of its power above 3150 Hz. This high-frequency content is more useful to the younger rather than the older group because, as evidenced in the audiograms, the older group has greatly reduced sensitivity in the upper frequency range. This possibly results in the frequency spectrum of the mechanical bell being equivalent in detectability to the electronic bell for the younger group, but not for the older. For the older group, whose audiograms demonstrated notable hearing loss above 2000 Hz, the high frequency energy of the mechanical bell is less useful; hence, the spectrum of this ringer is significantly less detectable than that of the electronic bell which has considerable power at 1000 - 1600 Hz, a range in which this older subject sample exhibits generally little loss. The electronic beeper has essentially no output below 1600 Hz, and only a very limited amount below 2000 Hz, although a peak does exist at 3150 Hz, where normal hearing is quite sensitive. However, the lack of low and midrange energy content for this ringer probably put it at a disadvantage, especially for the older listener whose audiogram thresholds above 2000 Hz were elevated to 35 dBHL (hearing level) and above.

The significant ANOVA interaction between age and noise was further analyzed and results plotted in Figure 5. For both age groups, mean threshold in noise was significantly greater than under quiet conditions, as would be expected due to masking effects. On average across the three ringers, the young group required only 7.1 dBA for detection in quiet, as compared to 42.3 dBA for the older group. Though still statistically significant, the young group's advantage in detection was reduced by the presence of the 65 dBA noise. Noise raised the threshold of the young group by 38.3 dBA and that of the older group by 15.0 dBA.

Though restricted to their inclusion in the previously-discussed interactions, all three main effects (ringer, noise, and age) were found statistically significant in the ANOVA ($\underline{F}(2,32) = 45.50$, $\underline{p}<0.0001$; $\underline{F}(1,16) = 133.39$, $\underline{p} \leq 0.0001$; $\underline{F}(1,16) = 41.91$, $\underline{p} \leq 0.0001$, respectively.) The post-hoc analysis revealed that all three ringers were significantly different ($\underline{p}<0.05$) from one another in detectability, with the electronic bell resulting in the lowest mean threshold dBA value (28.7), the electronic beeper resulting in the highest (45.3), and the mechanical bell requiring 40.1. The overall superiority of the electronic bell ringer is probably largely due to the peak sound energy from 1000-1600 Hz, a sensitive range of hearing for both the young and old subjects as verified by their audiograms and also corresponding to the equal loudness contours (e.g., Robinson and Dadson, 1957). In comparison, the mechanical bell has only moderate energy throughout the low to midrange frequencies of 80-2500 Hz, with slightly higher concentrations above 3150 Hz. The electronic beeper, which was found to require the highest level for detection, has negligible energy below 1600 Hz and only moderate energy (in comparison to the other ringers) from 1600-20000 Hz, with the exception of moderate energy peaks at 3150 and 20000 Hz.

As expected, the quiet condition produced lower threshold values than the noise condition (24.7 dBA and 51.4 dBA, respectively), indicating that the noise produced masking effects on the detection of all of the telephone ringers. The younger group of subjects had a mean detection threshold much lower than that of the older subjects (26.3 dBA and 49.8 dBA, respectively). This result, again, was anticipated based on the well-documented hearing loss associated with age as well as the audiograms of these participants.

DISCUSSION

Though based on a small subject sample, the results of this study provide evidence that current telephone ringer sound spectra differ widely in their aural detectability and that the older person, in particular, may encounter considerable difficulty in hearing some telephones ring. Furthermore, there is a significant masking effect from broadband background noise which inhibits the detection of the telephone ringer.

The question remains as to whether the revealed detection threshold differences are actually important for the selection (or design) of a telephone ringer for a particular user group or noise environment. As measured in the anechoic chamber, the sound pressure levels at 18 inches for the three phones when allowed to ring normally were: electronic bell, 80 dBA; electronic beeper, 67 dBA; and mechanical bell, 77 dBA. (For this output measurement, the phones were set to their loudest ringer setting, where such adjustment was available.) The average required sound pressure level for detection in older subjects was, in the same ringer order, 26.9, 51.7, and 48.3 dBA for the quiet condition, and 45.1, 66.3, and 60.3 dBA for the noise condition. Thus, all three "normal" ringer outputs are well above threshold in the quiet situation but when the moderate-level noise was present, the normal ringer output (67 dBA) for the beeper-type ringer was just above threshold (66.3 dBA). Considering that the distance from the ringer source to the subject's ear was only 18 inches, and also that the subject was instructed to be attentive and dedicated to the singular task of detecting the ring, the experimental conditions were probably more favorable to ringer detection than typical household listening conditions which have longer distances and numerous distractions. Therefore, the selection of the beeper-type ringer spectrum (Figure 2) for an older hearing-impaired person is probably inappropriate, especially if reliable detection in noise is required. And it is often the case that older individuals construct a noisy home environment due to the need for high television and radio volume levels to offset hearing impairment.

A second important consideration lies in the applicability of Hunt's (1970) design recommendations for tone ringers. The most poorly detected ringer, which utilizes a beep with a tonal quality, apparently violates two of Hunt's central design principles. The overall output sound pressure level of 67 dBA is below Hunt's 77 dB target. Furthermore, the central spike in the frequency distribution at 3150 Hz, although still below the upper bound of 4500 Hz, does violate the recommendation for central components to remain below 2000 Hz. This appears to be important for two psychoacoustic reasons: 1) age-induced hearing loss is generally less prominent in the low to midrange frequencies, so the elderly have a better chance of being accommodated by these signal frequency components, and 2) noise masking effects primarily gravitate upwards, interfering with frequency components near to and higher than the masking sound. The results of this study are in general agreement with Hunt's recommendations about the proper frequency content and sound pressure level for a tone telephone ringer signal. Another interesting finding was that the older technology mechanical bell ringer was more difficult to detect than the electronic ringer, perhaps because of its lack of a prominent spike in signal energy at any one frequency interval in the region of maximum aural sensitivity. This result suggests that the mechanical bell device is not necessarily the best ringer signal system.

A significant noise by ringer interaction effect was not found. Therefore, the data cannot be interpreted as support for the selection of a specific telephone ringer type for differing noise situations. However, the background noise utilized in this study was broadband (pink), providing the opportunity for direct and upward masking by many noise frequency components which overlapped the components of all three ringers. The elevation of detection threshold (from that in quiet) due to noise masking was of reasonably similar magnitude for all three ringers, these elevation differences being 25.6 dBA for the electronic ringer, 27.4 dBA for the beeper, and 27.0 dBA for the mechanical bell. This evidences that the lack of a significant noise by ringer interaction is possibly accounted for by the similar masking effects across the three ringer types. Furthermore, the masking noise was presented at a moderate level, 65 dBA, and in many realistic circumstances, phone ringing must be detected under higher noise levels which might necessitate more careful ringer selection. In any case, the need for further investigation of different, naturally occurring masking noise effects on ringer spectra audibility is clearly indicated. It is important to note, however, from these results that the electronic bell ringer spectrum required a considerably lower measured dBA output than either of the other ringers for both the quiet and noise conditions.

Though this study did not investigate hearing aid users, it is important to accommodate their needs in telephone ringer design. Since most conventional aids are tuned to augment reception of critical human speech frequencies (generally 500-2000 Hz), ringer spectra should include strong representation in this band.

With the implementation of any new technology which relies on human perception for successful system performance, it is essential to consider the limitations and needs of the target user population. As suggested by the results herein, the use of high frequency beeper-type ringer signals for telephones may create detection problems, especially for hearing-impaired listeners. More research is needed to define the signal parameter characteristics for telephone ringers to ensure reliable detection under a variety of ambient noise conditions.

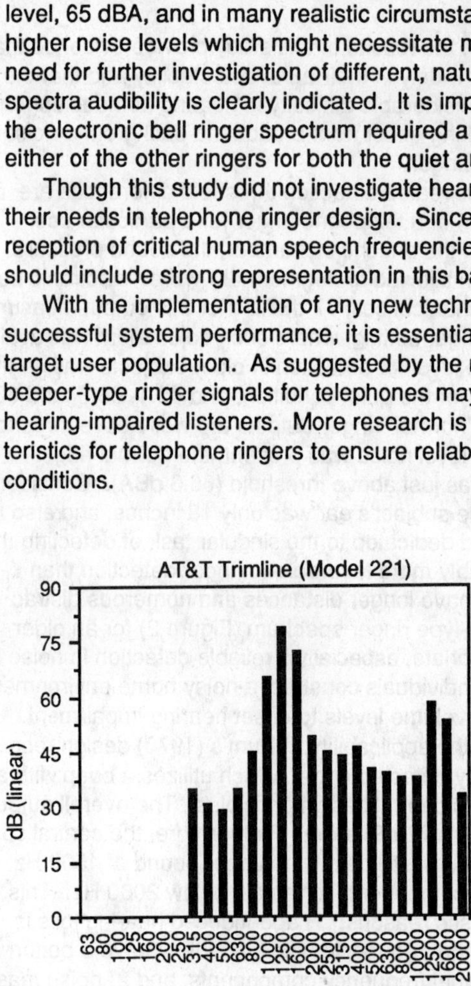

Figure 1. Sound level spectrum of the electronic bell.

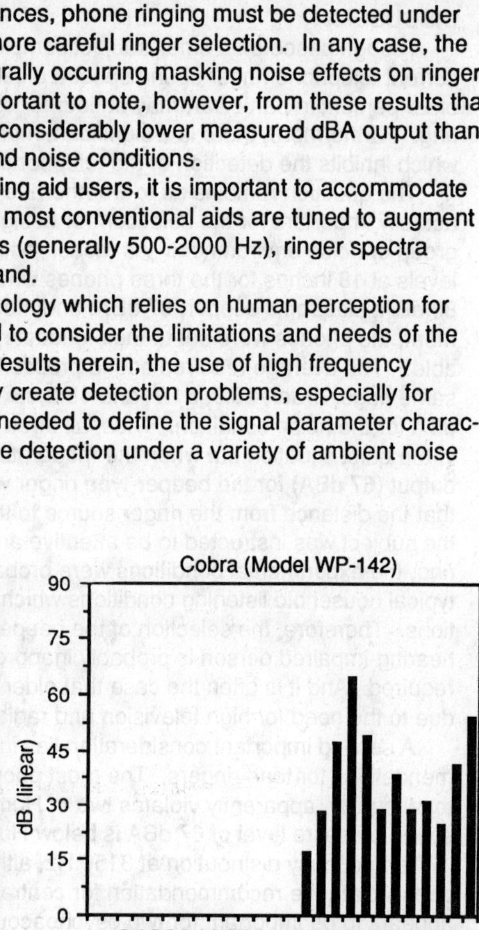

Figure 2. Sound level spectrum of the electronic "beeper."

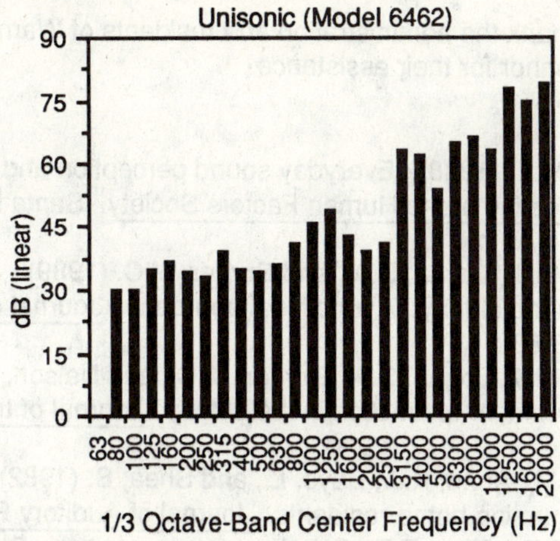

Figure 3. Sound level spectrum of the mechanical bell.

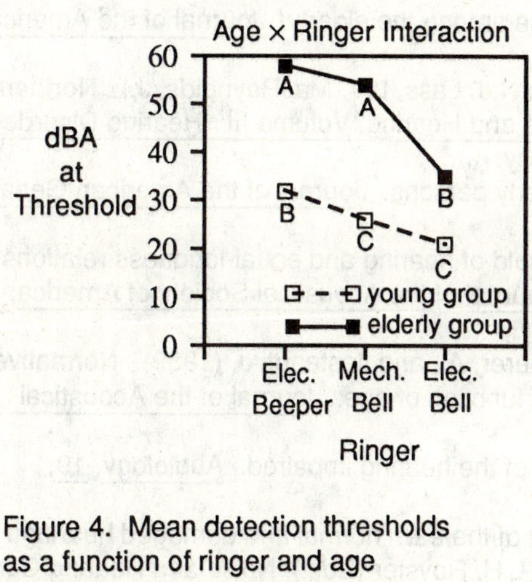

Figure 4. Mean detection thresholds as a function of ringer and age conditions (means with different letters are significantly different across the interaction according to the Newman-Keuls test at $\underline{p} \leq 0.05$).

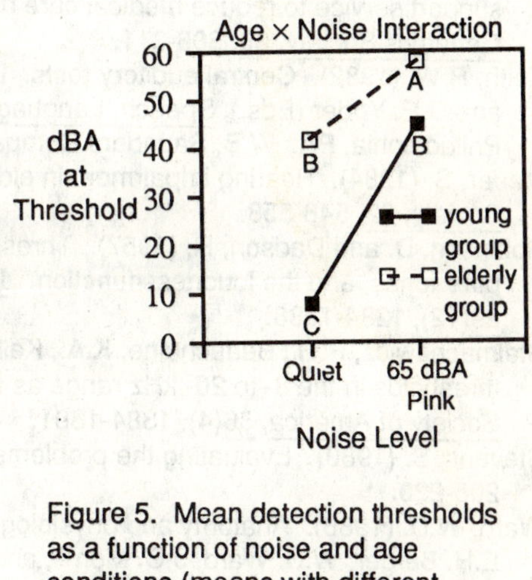

Figure 5. Mean detection thresholds as a function of noise and age conditions (means with different letters are significantly difference across the interaction according to the Newman-Keuls test at $\underline{p} \leq 0.05$).

ACKNOWLEDGEMENTS

The authors wish to thank the administration and residents of Warm Hearth Retirement Village and Ann Marie Connor for their assistance.

REFERENCES

Ballas, J.A. and Barnes, M.E. (1988). Everyday sound perception and aging. Proceedings of the 32nd Annual Meeting of the Human Factors Society. Santa Monica, CA: Human Factors Society.

Bess, F.H., Lichtenstein, M.J., Logan, S.A., and Burger, M.C. (1989). Comparing criteria of hearing impairment in the elderly: A functional approach. Journal of Speech and Hearing Research, 32, 795-802.

Bess, F.H., Lichtenstein, M.J., Logan, S.A., Burger, M.C., and Nelson, E. (1989). Hearing impairment as a determinant of function in the elderly. Journal of the American Geriatrics Society, 37, 123-128.

Bingea, R.L., Raffin, M.J.M., Aune, K.J., Baye, L., and Shea, S. (1982). Incidence of hearing loss among geriatric nursing home residents. Journal of Auditory Research, 22, 275-283.

Brown, S.C., Hotchkiss, D.R., Allen, T.E., Schein, J.D., and Adams, D.L. (1989). Current and Future Needs of the Hearing Impaired Elderly Population, GRI Monograph Series B, No. 1. Washington, DC: Gallaudet Research Institute.

Hinchcliffe, R. (1958). The pattern of the threshold of perception for hearing and other special senses as a function of age. Gerontologica, 2, 311.

Hotchkiss, D.R. (1989), The Hearing Impaired Elderly Population: Estimation, Projection, and Assessment, GRI Monograph Series A, No. 1. Washington, DC: Gallaudet Research Institute.

Hunt, R.M. (1970). Determination of an effective tone ringer signal. Paper presented at the 38th Convention of the Audio Engineering Society. New York, NY: Audio Engineering Society.

Infante-Rivard, C., Krieger, M., Petitclerc, M, and Baumgarten, M. (1988). A telephone support service to reduce medical care use among the elderly. Journal of the American Geriatrics Society, 36, 306-311.

Keith, R.W. (1982). Central auditory tests. In N.J. Lass, L.V. MacReynolds, J.L. Northern, and D.E. Yoder (Eds.), Speech, Language and Hearing, Volume III: Hearing Disorders. Philadelphia, PA: W.B. Saunders Company.

Mader, S. (1984). Hearing impairment in elderly persons. Journal of the American Geriatrics Society, 32, 548-553.

Robinson, D. and Dadson, R. (1957). Threshold of hearing and equal-loudness relations for pure tones, and the loudness function. Journal of the Acoustical Society of America, 29(12), 1284-1288.

Stelmachowicz, P.G., Beauchaine, K.A., Kalberer, A., and Jesteadt, J. (1989). Normative thresholds in the 8- to 20- kHz range as a function of age. Journal of the Acoustical Society of America, 86(4), 1384-1391.

Stevens, S. (1980). Evaluating the problems of the hearing impaired. Audiology, 19, 205-220.

Ward, W.D. (1986). Anatomy and physiology of the ear: normal and damaged hearing. In E.H. Berger, W.D. Ward, J.C. Morrill, and L.H. Royster (Eds.), Noise and Hearing Conservation Manual. Akron, OH: American Industrial Hygiene Association.

Zarnoch, J.M. (1982). Hearing disorders: Audiologic manifestations. In N.J. Lass, L.V. MacReynolds, J.L. Northern, and D.E. Yoder (Eds.), Speech, Language and Hearing, Volume III: Hearing Disorders. Philadelphia, PA: W.B. Saunders Company.

Advances in Industrial Ergonomics and Safety II
Edited by Biman Das
Taylor & Francis 1990

ADAPTATION OF THE DRIVER WORKSTATION TO MINIMIZE EXPOSURE TO WHOLE-BODY VIBRATION ON ROADROLLERS

P.-E. BOILEAU AND J. BOUTIN

INSTITUT DE RECHERCHE EN SANTÉ ET EN SÉCURITÉ
DU TRAVAIL DU QUÉBEC (IRSST)
505, DE MAISONNEUVE WEST
MONTRÉAL, QUÉBEC
H3A 3C2

Whole-body vibration measurements carried out on roadrollers operating on a concrete surface being prepared for paving have shown that the drivers can be exposed to levels of vibration well exceeding the ISO 2631/1 "exposure" or "health impairment" limit during a typical workday. This limit could typically be reached after 3.0 h of operation without using the vibro-compacting device and 24 minutes only when the vibro-compactor was turned on. An analysis of the frequency spectra of the vibrations measured at the driver-seat interface in three perpendicular directions revealed the dominance of frequency components exceeding 10 Hz along the vertical direction. By installing the seat on suitable vibration mounts and by modifying the vibro-compactor device, the daily exposure duration could be extended to 8 h without the vibro-compactor being used and 4 h when it was put in operation.

INTRODUCTION

Roadrollers constitute a special type of vehicles used in street and highway refection since the vibrations originating from surface irregularities are directly transmitted to the driver through steel rollers and rigid frame and seat. Internal vibrations intentionally induced for compacting soil and pavement introduce another very important source of vibration to which the driver of such vehicles is subjected.

The roadroller which constitutes the object of this study is of the miniature type and includes a metallic water tank on which a rigid seat is directly fixed. This constitutes the driver's workstation for which minimal space is available.

A vibro-compactor, consisting of a rotatory cam, can be activated for greater compaction. The resultant vibrations are produced by the unbalance of the rotating cam and the amplitude of the vibrations transmitted through the steel rollers can be modified according to the position of the cam on the rotatory axis.

The seat on the unmodified vehicle consists of a wooden platform covered with a thin layer of foam and vinyl. This seat is rigidly fixed on the vehicle using steel rails. On the modified vehicle, the same seat is fixed on the water tank using four conical vibration compression mounts (Novibra, Model RA 40 EM, Hardness B). Moreover, the position of the cam is modified so that the amplitude of vibrations can be changed when the vibro-compactor is used, providing three settings: low, medium and high.

In this study, whole-body vibrations transmitted to the driver before and after modifications of the vehicle are compared with regard to the ISO 2631/1 (1985) standard on human exposure to whole-body vibration. This information, along with the transmissibility characteristics of the seat-driver system, are used in demonstrating that the modifications are well-suited to the vehicle in question.

METHOD

First, vibration measurements were carried out on the unmodified vehicle under two situations:

i) without the vibro-compactor being used; ii) with the vibro-compactor on. This was done while the roadroller was being driven on a portion of a city street where the layer of asphalt had previously been removed. The surface of contact with the rollers was of concrete, with small asphalt residues remaining. Sometimes the vehicle had to be driven over man-holes or some other obstacles but attempts were made at maintaining the speed of the vehicle as constant as possible (less than 5 km/h).

From the data provided by these preliminary measurements, corrective measures were sought to reduce vibration exposure and a new series of measurements was carried out on the modified vehicle, with and without the vibro-compactor being used. This time, three settings (low, medium and high) were provided for the case where the vibro-compactor was used.

Whole-body vibration measurements were carried out using the data acquisition and analysis system depicted in Figure 1.

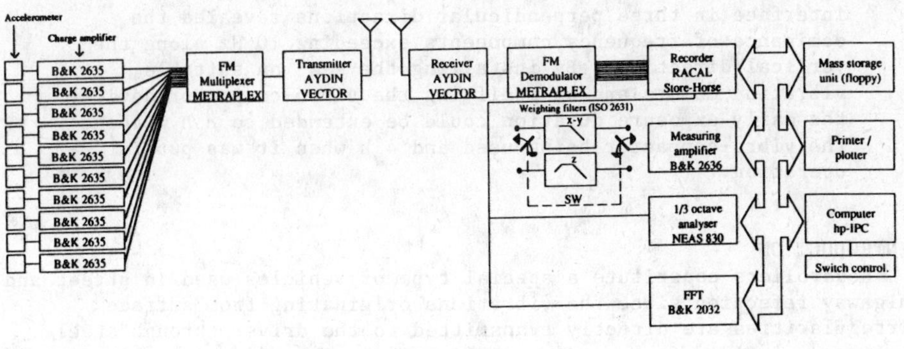

Figure 1. Whole-body vibration data acquisition and analysis system

A triaxial seat accelerometer B & K 4321 was used to measure the vibrations at the driver-seat interface in three orthogonal directions x-axis (fore-aft), y-axis (lateral) and z-axis (vertical). A uniaxial piezoelectric accelerometer (B & K 4381) alligned along the z-axis, was fixed to the vehicle's chassis so that the vibration transmissibility ratio (i.e. ratio of z-axis acceleration measured at the driver-seat interface to that measured on the chassis) could be measured as a function of frequency to determine the seat's attenuating properties. Only four of the ten channel telemetry system depicted in Figure 1 were used for the measurements.

The frequency response of the telemetry system was linear to within ± 1 dB (re. 1 μms^{-2}) between 0.8 and 100 Hz; with a dynamic range of approximately 60 dB. An FM transmitter was used to send the multiplexed signal from the transmitting antenna fixed on the roadroller to a receiving antenna and FM demodulator placed in a small van mobile laboratory unit. From there, the acceleration-time signals were recorded on separate channels on an analog tape recorder. The recording time varied between six and nine minutes for each situation studied.

A one-third octave frequency spectrum analyzer was later used to produce the frequency spectrum of the vibrations between 1.0 and 80 Hz for each axis. These spectra were compared with the "exposure limit" defined in ISO 2631/1 while the vertical seat transmissibility ratio was obtained by dividing the z-axis spectrum on the seat by that measured on the floor.

The frequency spectra were evaluated using acceleration-time segments lasting one minute and by averaging the resultant spectra over the whole measurement duration. For each segment, the crest factor was computed by first weighting the signals (W_z or $W_{x,y}$ weightings defined in ISO 2631/1) and by determining the peak and root-mean-square values of the acceleration. The crest factor was given by the ratio of these two quantities.

The time-dependence relationship implied by the vibration exposure limit defined in ISO 2631/1 was used as the basis for establishing the expected exposure duration which could be tolerated before reaching the set limit. This was done for each situation studied, using the spectra measured along each axis and the vibration amplitude of the dominant frequency component corresponding to each axis. This established the exposure time permitted according to the ISO 2631/1 "exposure" or "health impairment" limit.

RESULTS

Evaluation of exposure to whole-body vibration on roadrollers

From the mean frequency spectra recorded along each axis measured at the driver-seat interface, for each situation studied, three main informations may be obtained pertaining to vibration exposure. These include: i) the dominant frequency of the weighted spectrum (ie. frequency component which contributes the most to the vibration exposure), ii) the crest factor, f_c, and iii) the daily exposure duration required before reaching the "exposure" limit. These are summarized in Table 1 for each of the situations studied.

Table 1. Characteristics of vibration exposure to whole-body vibration on a roadroller before and after modifications.

Type of test	x-axis Dominant frequency (Hz)	x-axis f_c	x-axis Time permitted (h)	y-axis Dominant frequency (Hz)	y-axis f_c	y-axis Time permitted (h)	z-axis Dominant frequence (Hz)	z-axis f_c	z-axis Time permitted (h)
UNMODIFIED VEHICLE									
. without vibro-compactor:	20.0	6.3±0.9	24	6.3	5.4±0.8	10.8	25.0	7.0±0.8	3.0
. with vibro-compactor:	31.5	4.7±0.4	10.4	4.0	5.8±0.8	10.0	31.5	6.5±0.5	0.4

Type of test	x-axis Dominant frequency (Hz)	x-axis f_c	x-axis Time permitted (h)	y-axis Dominant frequency (Hz)	y-axis f_c	y-axis Time permitted (h)	z-axis Dominant frequence (Hz)	z-axis f_c	z-axis Time permitted (h)
MODIFIED VEHICLE									
. without vibro-compactor:									
	2.5	5.9±1.0	24	6.3	4.7±1.1	8.0	6.3	6.1±0.7	8.8
. with vibro-compactor									
- low:									
	31.5	7.8±2.1	24	5.0	7.1±2.5	8.6	5.0	10.0±4	12.0
- medium:									
	25.0	7.3±0.6	14.2	5.0	8.0±2.5	10.8	25.0	12.0±1	5.4
- high:									
	25.0	3.4±0.4	10.6	5.0	4.8±0.5	8.0	25.0	4.9±0.7	4.0

From the above data it may be seen that, with respect to ISO 2631/1 limits, vibration exposure is dominant along the vertical or z-axis on the unmodified vehicle and also on the modified vehicle when the vibro-compactor is set at medium and high modes. For the remaining cases, the lateral or y-axis becomes the dominant vibration axis.

Considering only the dominant vibration axis, these results indicate that without the vibro-compactor being used, vibration exposure duration can be extended from 3.0 to 8.0 hours daily on the modified vehicle. With the vibro-compactor in operation, the exposure duration could be extended from 0.4 hour on the unmodified vehicle to 8.6 hours for "low" setting, 5.4 hours for "medium" setting and 4.0 hours for "high" setting on the modified vehicle. This represents a significant improvement considering that the actual effective daily exposure duration is rarely thought to have exceeded 4 hours during normal operations.

Before the vehicle was modified, very significant levels of vibration were recorded along all axes at frequencies exceeding 10 Hz when the vibro-compactor was not in use. Due to their nature, these vibrations were thought to represent structural vibrations which could relatively easily be damped out by isolating the seat from the vehicle structure. Vibration mounts with a natural frequency of 6 to 7 Hz were thus thought to represent a suitable solution. Figure 2 presents the spectra measured along the z-axis before and after the modifications were made on the roadroller operating without the vibro-compactor. The attenuation of vibration frequencies exceeding 10 Hz is quite significant.

When the vibro-compactor is turned on, large vibration levels are more concentrated at specific frequencies corresponding to the excitation frequencies of the rotating cam, the frequency varying depending on its position. Figure 3 compares two spectra measured along the vertical axis before and after modifications. It must be clear that the excitation frequency varies between 25.0 and 31.5 Hz in these situations. This excitation frequency is clearly distinguishable in the spectra measured along all three directions.

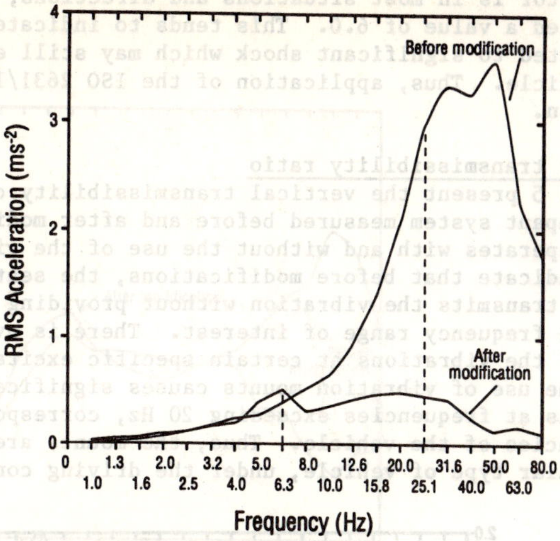

Figure 2. One third octave frequency spectra measured before and after modifications for z-axis vibration (roadroller without vibro-compactor)

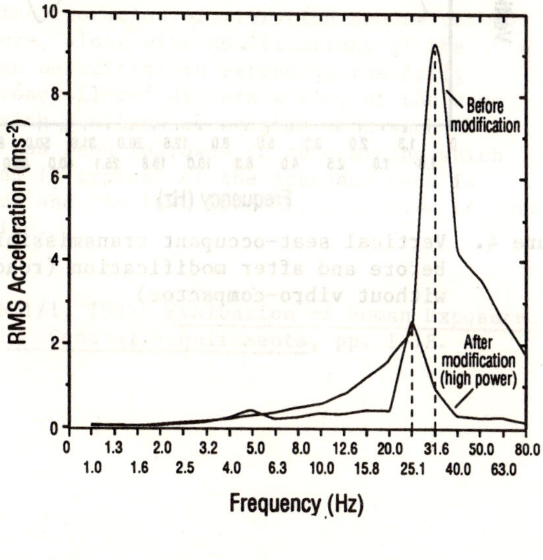

Figure 3. One third octave frequency spectra measured before and after modifications for z-axis vibration (roadroller with vibro-compactor on)

CONCLUSION

This stuc
the seat ar
vibro-compa
vibration e
time, provi
exposure.
corrective
solving mos
another exa

REFERENCES

Internat
to Whole-Bo

clo
dri
the
mad

of
the
The
sys
ove
amp
fre
of
dom
for
stu

Advances in Industrial Ergonomics and Safety II
Edited by Biman Das
Taylor & Francis 1990

ERGONOMIC DESIGN GUIDELINES FOR UNDERGROUND COAL MINING EQUIPMENT

E. J. (Kirk) Conway
VRC Corporation
2150 Garden Road
Monterey, California, USA 93921
and
Richard Unger
U S Bureau Of Mines
Bruceton Mining Research Center
Pittsburgh, PA, USA 15236

INTRODUCTION

Thirty-five years ago underground coal mining equipment consisted of simple rugged machines powered by electric motors and hydraulics. These machines cut, dug, loaded, and transported coal from the mine face to the surface. Mine maintenance personnel, armed with hand tools and a knowledge of hydraulics and electricity, maintained these machines in the working sections of the mine. Over the years the basic mining machine evolved into powerful, complex mining systems. To boost productivity, the size and horsepower of the machines increased dramatically. To increase throughput, new machines such as continuous miners, long wall systems, and continuous haulage were introduced. To improve personnel safety and health, safety and environmental control systems were tacked onto the machines. Despite the safety improvements and new systems, mining equipment maintenance costs have continued to escalate and the percentage of maintenance related injuries have remained stubbornly high.

Equipment maintenance costs in U. S. underground coal mines range from 25% to over 36% of total mine operating costs (Conway and Unger, 1989a). A British study, supporting this figure, reports that equipment maintenance costs accounts for 30% of British coal mine operating budgets (Ferguson, 1985). With coal mining profits squeezed, mine operators are searching for ways to contain equipment maintenance costs. These efforts traditionally focused on reducing maintenance staff size, spare parts inventories or maintenance intervals.

In addition to costs, equipment maintenance accounts for 30% of all lost time injuries in underground coal mines (Conway and Unger, 1989b; Conway and Elliott, 1988; National Academy of Science, 1982). Despite extensive Federal and State research and ongoing mining company safety programs, there has been no decrease in maintenance accidents over the last ten years.

With few exceptions, no significant ergonomic or maintainability design improvements are observed in surface or underground mining equipment over the last twenty-five years (Conway and Unger, 1989a and Conway and Sanders, 1982). This is particularly true with respect to the design of equipment for maintenance. In fact, personnel safety and equipment maintainability have decreased due to increased equipment size and added-on safety and environmental systems. Likewise, few new maintenance tools, procedures, automatic test equipment, or technology based aids were made available to support equipment maintenance. The result has been a continuing escalation of maintenance costs and equipment downtime and a consistently high percentage of maintenance injuries.

ERGONOMIC DESIGN REVIEW

An ergonomic/human factors design review of operational underground coal mining equipment was conducted under a U. S. Bureau Of Mines funded project (Conway and Unger, 1989a, 1989b). The objectives of this study included: (a) assessment of the extent to which ergonomic and maintainability design concepts are incorporated into current mining equipment design, (b) the identification of ergonomic design deficiencies contributing to maintenance injuries and costs, and (c) the preparation of an ergonomic based maintainability design guide specifically for underground coal mining equipment.

A survey of mining machines in nine large (over 1 million tons per year) and smaller (150,000 to 1,000,000 tons) U. S. mines operating in high seam (over 142 cm) and low seam (76cm to 142cm) coal was completed. The sample included conventional, continuous, and long wall operations. The survey included the following categories of equipment: shuttle cars, scoops, roof bolters, continuous miners, long wall systems, undercut machines, face drills, utility vehicles, and personnel carriers.

During site visits, scheduled and unscheduled maintenance tasks for each category of equipment were identified. The steps required to complete the maintenance action were noted along with task completion times. Maintenance accident records were analyzed to identify equipment factors contributing to the accidents. The equipment was then examined to identify design deficiencies contributing to the accidents or that impeded maintenance actions. Maintenance personnel, supervisors and mine management were interviewed to confirm the findings, to identify other design deficiencies and elicit recommended design improvements.

Six large and smaller mining equipment manufacturers were visited during the study to identify equipment design practices and constraints. Several major manufacturers reviewed and critiqued findings and the draft design guide.

DESIGN INDUCED PROBLEMS

The study identified a range of design-induced safety and maintenance problems. The results are reported in Conway and Unger (1989b). Summarized below are the major categories of design deficiencies.

Accessibility

Inadequate component accessibility represented the largest single design deficiency in terms of reported injuries and excessive maintenance task completion times. Access problems resulted from: (a) inadequate access opening size and placement, (b) poor layout of components requiring Removal and Replacement (R/R) of non-failed parts, (c) inability to access mounting bolts or connectors, or to use required hand tools, (d) inability to access components or hoses mounted inboard where they simply cannot be reached in lower seam mines. For example, R/R of a single hydraulic line on one roof bolter, which should require only 10 minutes, requires from 3 to 4 hours due to poor machine design and inadequate component accessibility. On one continuous miner, three or more non-failed components are remove simply to change one hydraulic line.

Component-Machine Interface

Component interface problems centered on: (a) inadequate component-machine interface design to help personnel in R/R heavy components and (b) the inability of maintenance personnel to use lifting devices to handling 100 to 1,000 pound components.

Routine Maintenance

Inadequate machine design hinder maintenance personnel efforts to quickly R/R high failure rate components (e.g.; pumps, water lines, hydraulic valves) and perform scheduled lubrication, adjustments and inspections.

Fault Isolation

Inadequate design for fault isolation decreases the ability of maintenance personnel to determine the cause of system failures resulting in excessive trail and error troubleshooting, maintenance errors and increased personnel risk.

Damage Prevention

Inadequate machine design for component protection result in unnecessary component damage and increased maintenance burden.

Lack of Resources

Design of maintenance tasks require the use of special tools or lifting devices typically not available at underground maintenance locations. Hence, maintenance personnel "jerry-rig" tools to handle 100lb to 1,000lb components.

Design Complexity

Unnecessary equipment complexity results from poor machine layout and the crowding of components into compartments without regards to the need to maintain them. Designs were found that multiplied the number of frequently replaced connectors or components simply as a design convenience resulting in increased maintenance burden.

DESIGN RECOMMENDATIONS

A comprehensive ergonomic and maintainability design guide for underground mining equipment was prepared. The guide covers mobile and stationery equipment used in underground mines with seam heights down to 76cm. Interested readers can refer to Conway and Unger, (1989b) for specific mining equipment design recommendations and to Sanders and Peay (1988) for general human factors design guidelines for underground mining equipment. A summary of the accessibility design recommendations is presented below.

LOCATION OF MAINTENANCE POINTS

A maintenance point (MP) is defined as any point where two components come together or where a component is mounted on the machine. In the mine environment, the physical location of the MP impacts personnel safety, equipment downtime and maintenance costs. In lower seam mines (e.g.; under 142cm), it is difficult to access components from the top or bottom (underside) of the mining machine due to the restricted roof clearance over the machine. MPs located in the center of the machine are difficult or impossible to access. Hence, visual and physical access must be from the sides or the ends of the machine. The concept of the primary and secondary maintenance zone is suggested as a design tool for equipment designers and evaluators.

Primary Maintenance Zones

The Primary Maintenance Zone (PMZ) is defined as the area extending from the outer edge of a machine inboard (inward) a distance of 45cm (18 inches). The PMZ extends around the machine as illustrated in Figure 1. All scheduled maintenance and service points should be located within the PMZ. This includes: (a) lubrication and inspection points, (b) fluid fill points and (c) adjustment points. In addition, high frequency of failure components and connections (e.g.; hydraulic valves, water lines, etc.) and replaceable wear points should be in the PMZ.

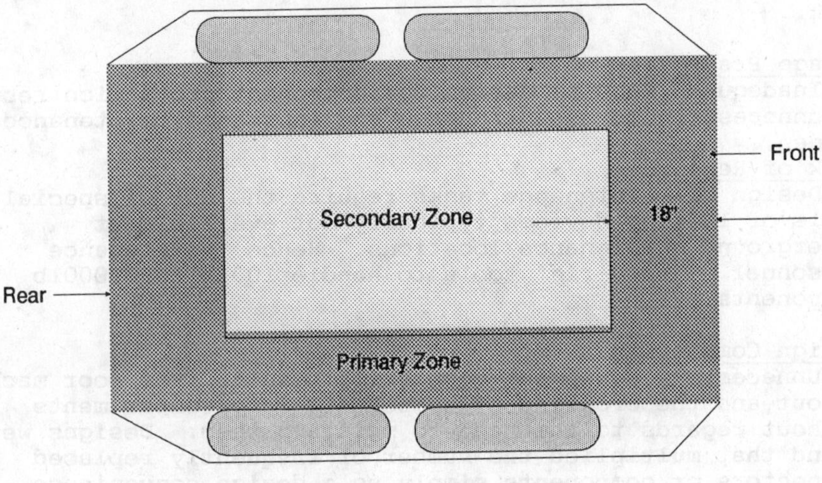

Figure 1: Primary maintenance zone.

Secondary Maintenance Zone

The Secondary Maintenance Zone (SMZ) is defined the area extending from the inner edge of the PMZ to the machine centerline. Only components requiring no routine maintenance should be located in the SMZ. Adequate direct visual and physical access must be provided for all MPs in the SMZ.

Tertiary Maintenance Locations.

A Tertiary Maintenance Location (TML) is defined as a MP located outside the PMZ and SMZ. TMLs may include a lubrication points on the end of a roof bolter boom or a tensioner on a conveyor system. The TML should be designed to permit full visual and physical access to the maintenance point. Where possible, the TML should be located under

LEVEL OF ACCESSIBILITY

Two levels of accessibility should be considered during the design process: visual and physical access. Visual inspection is the most frequent maintenance task. Accessing a component on a mining machine to inspect it often accounts for over 25% of the task completion time. Physical access to the MP is the next most frequently performed maintenance step. Maintenance safety and task completion times can be improved if all MPs and frequently replaced components are directly visible and accessible to the maintainer. Table 1 lists recommended visual and physical access locations for MPs on machines used in lower seam mines.

VISUAL ACCESS

Visual inspection is also the first step in any maintenance task. Hence, maintenance points should be visible to the maintainer. Where practical, MPs should be located in the PMZ and should be visible without removing access covers or other components. Hazardous or energized components may be mounted behind protective covers in the PMZ, but should provide direct visual access for the maintainer after the protective cover is removed. Visual access should provide line of sight inspection of all relevant locations on the component or connection.

Table 1: Recommended MP Locations

Location	Physical Access To Maintenance Points	Visual Access To Maintenance Points
Recommended	Machine Exterior-PMZ	Machine Exterior-PMZ
	Machine Exterior-Under Fender	Machine Exterior-Under Fender
	Machine Exterior-Open Compartment	Machine Exterior-Open Compartment
Acceptable	Machine Exterior-End of Vehicle	Machine Exterior-End of Vehicle
	Machine Exterior-Closed Compartment	Machine Exterior Closed Compartment
		Machine Cavity - Opening
Least Preferred	Machine Cavity-Top Opening	Machine Cavity-Top Opening
	Inside Operator's Compartment	Inside Operator's Compartment
	Machine Exterior-Behind Other Components	Machine Exterior - Behind Components
Undesirable	Machine Top-Under Access Cover	Machine Top-Under Access Cover
	Inside Machine Feature (e.g.; Boom Frame)	
	Under Machine-Exterior	Under Machine - Exterior
	Interior Machine Cavity - No Direct Access	Interior Machine Cavity-No Direct Access
	Interior Machine Chassis	Interior Machine

Where practical, all components weighing more than 50 pounds should be installed in the PMZ. This location reduces access time, permits the use of handling devices, and the use of powered hand tools. It also reduces exposure to maintenance related accidents and injuries. For components weighing more than 50 pounds, provisions should be made for use of mechanical or powered lifting devices to R/R the component. Lifting assistance is recommended for all components weighing over 100 pounds and located in the SMZ. This mechanical assistance should permit lifting the component off the transport vehicle and lowering it at the installation point. The device also should permit manipulation of the component to align the unit as required for installation on the mining machine.

REFERENCES

Conway, E. J. and Unger, R., 1989a, Maintainability Design Of Underground Mining Equipment - Volume I: Final Technical Report. (Contract J0145034, VRC Corporation), U.S. Bureau of Mines, Pittsburgh, PA, USA, 35 pp.

Conway, E. J. and Unger, R. 1989b, Maintainability Design Of Underground Mining Equipment - Volume II: Maintainability Design Guidelines. (Contract J0145034, VRC Corporation), U.S. Bureau of Mines, Pittsburgh, PA., USA, 161 pp.

Conway, E. J., Elliott, W. and Unger, R. 1988,, Mine Maintenance Material Handling: Volume II - Prototype Device Specification. (Contract H0113018, Canyon Research Group, Inc,) U.S. Bureau of Mines, Pittsburgh, PA, USA, 51 pp.

Conway, E. J., 1986, Human Factors Design Guidelines For Mobile Underground Mining Equipment - Proposed SAE XJ1314. (contract No H0308110, Canyon Research Corporation), Society of Automotive Engineers, Warrendale, Pa., USA, 36 pp.

Ferguson, C.A., et. al, 1985, Ergonomics Of The Maintenance of Mining Equipment. Institute of Occupational Medicine, Edinburgh, 55 pp.

National Academy of Science, 1982, Towards Safer Underground Coal Mines. Washington, D.C., 190 pp.

Sanders, M. S. and J. Peay. Human Factors In Mining. BuMines Information Circular 9182, 1988, 153 pp.

Advances in Industrial Ergonomics and Safety II
Edited by Biman Das
Taylor & Francis 1990

"DETECTSOUND": A COMPUTERIZED MODEL FOR THE PREVENTION OF ACCIDENTS IN NOISY WORKPLACES

C. Laroche, H. Tran Quoc, R. Hétu, S. McDuff

Groupe d'Acoustique de l'Université de Montréal, C.P. 6128,
Succ.A, Montréal, Québec, Canada, H3C 3J7

A computerized model was developped to predict the detectability of warning signals in noisy workplaces. This model takes into account the loss of auditory sensitivity and selectivity as a function of age and the use of hearing protectors. The program was used to analyse 93 different configurations of warning signals in a steel plant. In about half of the cases, the warning signals did not meet the recognition criteria. A "user-friendly" version of the program will soon be available. This practical tool will allow safer signals to be designed.

INTRODUCTION

Each year, deadly accidents occur in noisy workplaces because a warning signal is not heard (Wilkins and Acton, 1982). Such accidents happen in many workplaces despite the fact that the use of auditory signals are governed by standards (ISO 7731, 1986) and laws. These are based on many experimental studies of the detection of auditory signals in different background noises (Patterson, 1982; Coleman et al., 1980; Fidell et al., 1974; Trucks and Adams, 1975). The studies were conducted by specialists in acoustics or ergonomics.

Harris (1983) presents guidelines that should be followed for the selection and specification of audible alarm signals. In particular, the major factors are: a minimum one third-octave band signal-to-noise ratio (S/N) of +10 dB to ensure perception and recognition, a maximum S/N of +25 dB to prevent startling and to reduce the risk of noise induced hearing loss, a minimum absolute level of 70 dBA and a maximum level of 120 dBA. The generally recommended frequency range for audible alarms is 500-2000 Hz, principally to take into account noise-induced hearing loss which appears above 2000 Hz. Below 500 Hz, the signal can be masked by the low frequency content of many background noises in industries. Finally, signals with distinct temporal pattern are generally recognised more easily than simple continuous or interrupted notes.

Many books related to ergonomic considerations specify criteria for the design of warning signals. For example, Murrell (1971) introduced the +10 dB S/N and the limitation of the spectral content under 2400 Hz. Clarck and Corlett (1984) underlined the importance of taking into account noise-induced hearing loss. Oborne (1987) talks extensively about the effects of noise on hearing but does not cover the recent developments in the knowledge of the detection of warning signals.

All these general rules are important to consider in the design of warning signals but very few practical tools are available to analyse existing warning signals in noisy workplaces and to make recommandations to improve these signals.

Detection models have been suggested (Patterson 1982; Coleman et al., 1980; Trucks and Adams, 1975; Fidel et al., 1974) but each one has its limitations. Coleman et al. (1980) consider four factors in their model: the signal at the ear, the noise at the ear, the hearing abilities of the workforce in terms of auditory sensitivity and the use of hearing protectors. They do not take into account the effect of the loss of frequency selectivity on detection ability. Patterson (1982) has developped a model which includes this last factor but, to our knowledge, it has not yet been applied to subjects who suffer from noise-induced hearing loss. The major problem with this model is that it is not sufficiently practical to be used easily by health and safety personnel. It requires a mini-computer to analyse the data.

The model proposed by Trucks and Adams (1975) is a predictive model involving octave-band analysis, signal to noise ratios and reaction time. They have obtained very interesting results but the methodology associated with this model appears to be long and complicated. Moreover, their study was carried out on subjects with normal hearing. Unfortunately, no new developments have appeared in the litterature since this publication.

Fidell et al. (1974) have designed a simplified graphical prediction method based on the psychophysical theory of signal detectability. They concluded that the predictions of human aural detection of noise signals are readily based on the signal to noise ratio in the single one-third octave band to which human sensitivity is highest. Like the previous model, only normal hearing subjects have participated in the study and no further studies have been published since then.

Because there are very few practical tools which allow the prediction of detection ability for a whole population of workers, and in order to improve safety in noisy workplaces, the Groupe d'Acoustique de l'Université de Montréal (GAUM) has developped a computerized model ("Detectsound") which takes into account the loss of auditory sensitivity and selectivity due to age. The use of hearing protectors, which can alter detection ability (Wilkins and Martin, 1987), is also considered by the program. The program can be run on a standard IBM-PC compatible computer.

Figure 1 presents a simplified flow-chart of the program. The model is essentially based on the concept of excitation patterns proposed by Zwicker and Scharf (1965) and the increase of the critical bandwith as a function of age proposed by Patterson et al. (1982). The critical bandwith refers to the fact that the ear acts as a bank of filters of different bandwidths. The details of the model are presented in another publication (Laroche et al., 1990). A study in progress should allow the introduction of the effect of noise-induced hearing loss into the model.

Many of the studies presented earlier have used aircraft noise (Patterson, 1982; Doll and Folds, 1986), railroad noise (Patterson et al., 1989), agricultural noise (Talamo, 1982) and coal mine noise (Coleman et al., 1980). Very few studies were focused on the noise found in typical noisy working environments such as steel plants, textile plants or wood plants. This noise accounts for a large percentage of the workers who suffer from hearing loss due to age or due to noise. The goal of this paper is to present an example of the application of the program "Detectsound" in a steel plant in Montreal.

METHOD

Due to the lack of appropriate tools, there have been no evaluations of the use of warning signals in noisy workplaces in Quebec. This study represents the first exploration of this type for a workplace (a steel plant) in which warning signals are used to inform workers of the different stages of production and of the dangerous situation. On the whole, 93 different configurations of warning signals were analysed. They were principally buzzers, bells, sirens and horns. The background noise and the warning signals were recorded at each worksite with a Sony PC-2000 digital recorder. Spectral analysis in third-octave bands were made with a Bruel & Kjaer 2515 spectrum analyser. In the case of fluctuating noises, analysis was done during the noisiest portion. Each warning signal was analysed as a function of:

- its audibility (masked threshold) for workers of different ages (namely 20 and 60 year olds) and
- its ability to attract attention and to be recognised according to the criteria proposed in the litterature (Wilkins and Martin, 1978). To attract attention and be recognised among other signals, a warning signal has to be between +10 and +15 dB above the masked threshold in a least one third-octave band. However, the signal to masked

threshold ratio should not exceed +25 dB in order to prevent startling and hearing impairment.

A warning signal was considered as being detected if its excitation pattern exceeded that of the noise in at least one critical band. This same signal was judged acceptable if its level was between +10 and +15 dB above the masked threshold predicted by the model in at least one critical band.

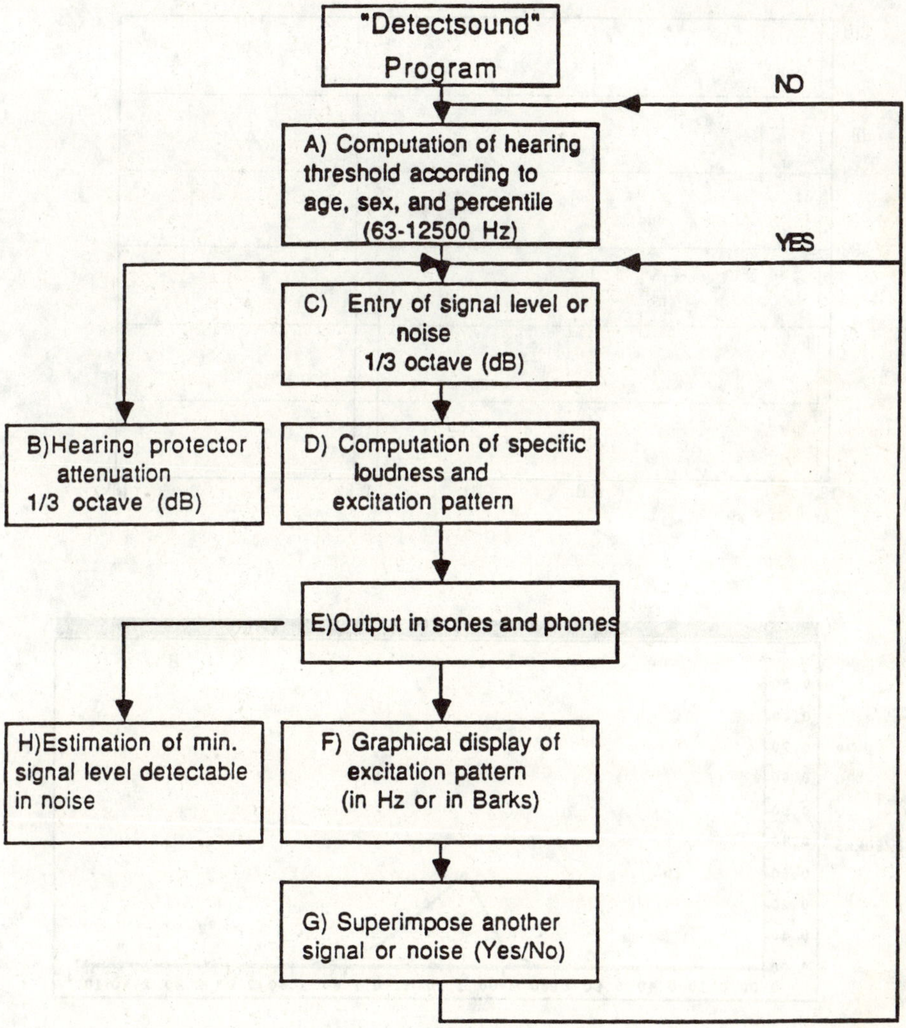

Fig. 1 A simplified flow-chart of the program "Detectsound"

RESULTS

Analysis of the audibility of warning signals

Figure 2A presents an example of a noise spectrum (dotted line) on which a signal spectrum (continuous line) is superimposed. Figure 2B illustrates the excitation patterns associated with the spectra for 20 year old subjects. In this case, the warning signal (a whistle) has an acceptable signal to masked threshold ratio, i.e. + 18 dB, in the critical band centred at 1.6 kHz.

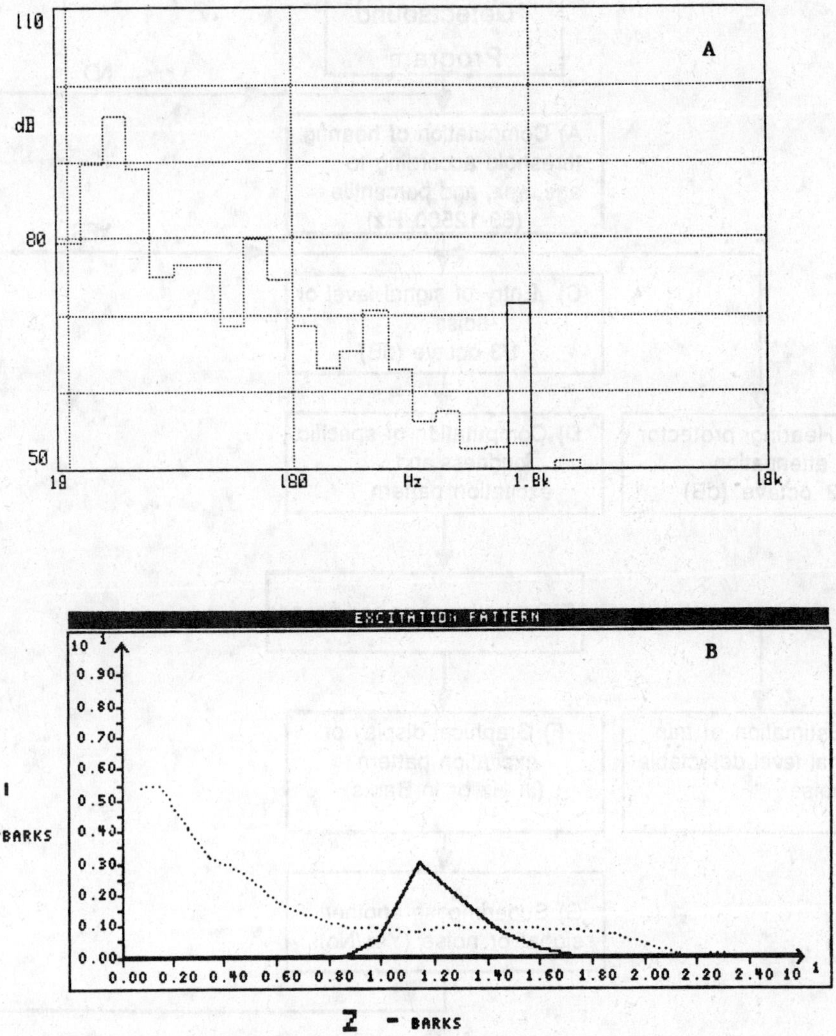

Fig. 2 Example of a situation in which a whistle presents an acceptable signal to masked threshold ratio (2A: noise spectrum (----), whistle spectrum (——); 2B: corresponding excitation patterns).

In figure 3, the buzzer has a spectrum concentrated around 2 kHz and is not heard. In this case, no part of the excitation pattern of the signal exceeds that of the background noise.

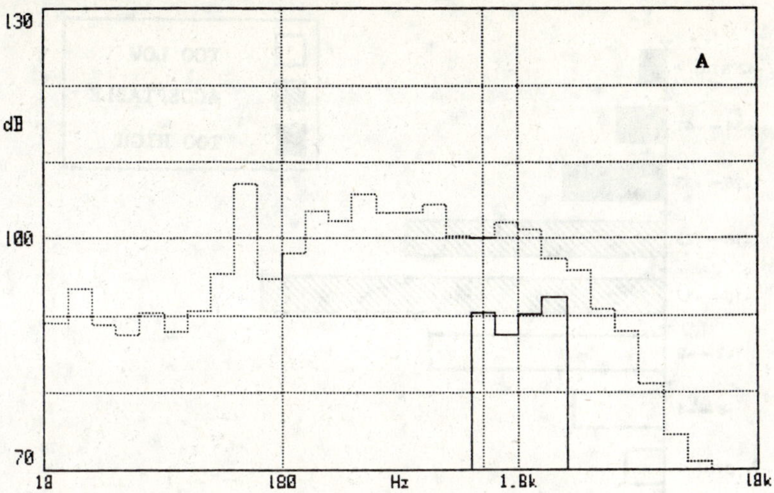

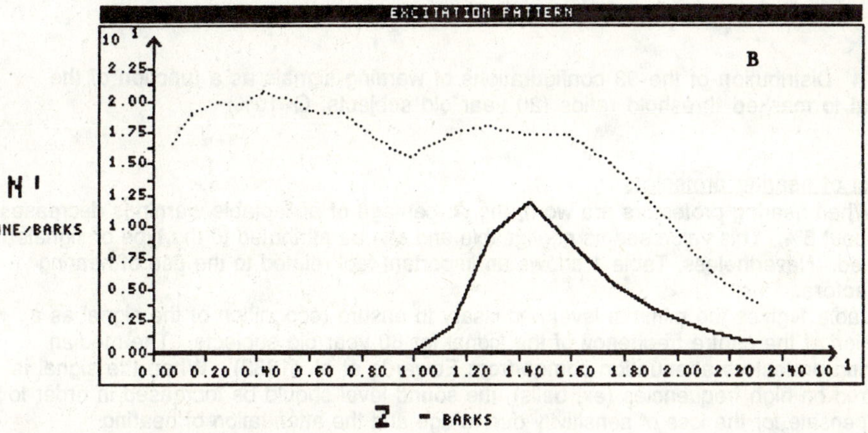

Figure 3. Example of a situation in which a buzzer presents a signal to masked threshold ratio which is too low (3A: noise spectrum (----), buzzer spectrum (——); 3B: corresponding excitation patterns).

Figure 4 presents the results for all warning signals for a group of 20 year old listeners (Q=10% of the population, i.e. the subjects most sensitive to the effect of age). On the whole, 55% of the configurations were judged acceptable whereas 30% did not meet the minimal criterion of +10 dB and 15% exceeded the +25 dB ratio. These percentages did not change significantly with age. This could probably be explained by the relatively low increase in the critical bandwith with age.

Since this increase is larger in people suffering from noise-induced hearing loss (Florentine and Zwicker, 1979), it is still quite likely that these people will encounter more difficulty in detecting warning signals than young normal hearing subjects. This hypothesis is presently being tested.

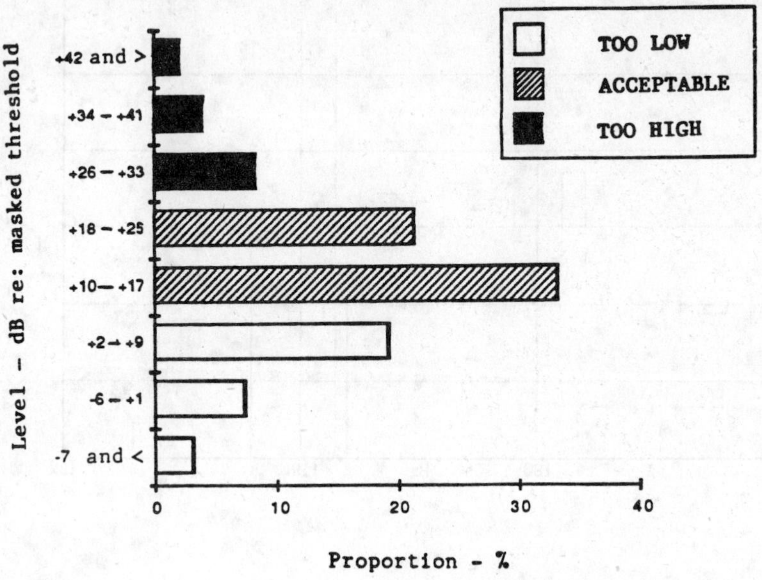

Fig. 4 Distribution of the 93 configurations of warning signals as a function of the signal to masked threshold ratios (20 year old subjects, Q=10%)

Effect of hearing protectors

When hearing protectors are worn, the percentage of acceptable warnings decreases by about 5%. This value seems negligeable and can be attributed to the type of signals studied. Nevertheless, Table 1 shows an important fact related to the use of hearing protectors.

Table 1 gives the minimal level necessary to ensure recognition of the signal as a function of the centre frequency of the signal for 60 year old subjects. The median hearing protectors attenuation comes from Edwards et al. (1983). When the signal is centred on high frequencies (ex. bells), the sound level should be increased in order to compensate for the loss of sensitivity due to age and the attenuation of hearing protectors which is larger in the high frequencies.

At 60 years of age and after more than 40 years of noise exposure, it is quite possible that the auditory sensitivity and the frequency selectivity will be significantly below those of Table 1. This demonstrates the necessity to limit the design of warning signals to frequencies below 2 or 3 kHz. Other studies (Abel et al., 1985; Wilkins and Martin, 1987) have clearly demonstrated the importance of this fact. With hearing protectors, some workers become completely deaf to some auditory signals.

CONCLUSION

The use of the program "Detectsound" in a steel plant has demonstrated that many configurations of warning signals are unsafe for workers. In the workplace in question, a worker already had a serious accident because he did not hear the warning shouted by one of his co-worker. This worker suffered from noise-induced hearing loss and was completely deafened by hearing protectors. This situation certainly prevails in other type of industries. It is therefore important to carry out this type of evaluation in the actual workplaces where the safety of workers is directly related to the detection of auditory warning signals.

Table 1. Calculation of the minimal signal level which ensures recognition as a function of the centre frequency of the signal (60 year old subjects using hearing protectors).

Frequency (Hz)	Hearing Threshold (Age=60, Q=10) (dB SPL)		Median protectors attenuation (dB SPL)		+10 dB (dB)		TOTAL (dBSPL)
1000	24	+	13	+	10	=	47
1250	25	+	17	+	10	=	52
1600	28	+	20	+	10	=	58
2000	30	+	24	+	10	=	64
2500	34	+	26	+	10	=	70
3150	41	+	28	+	10	=	79
4000	52	+	26	+	10	=	88
5000	60	+	24	+	10	=	94
6300	70	+	22	+	10	=	102
8000	90	+	22	+	10	=	122

A "user-friendly" version of the program will soon be available in order to provide a practical tool to health and safety personnel who are not familiar with the psychoacoustic concepts of detection. Moreover, suppliers of warning signals will be able to use it to design safer warnings. When the characteristics of the noise are known in terms of level and frequency, it is possible to determine the frequency range in which a signal should be developped. This zone should 1) be located where the noise is least prominent, 2) be limited to between 500 and 2000 Hz to take into account noise-induced hearing loss and presbycusis and 3) consist in at least three components to maximise detection and insure recognition. The sound pressure levels required for the signal to be detected in these bands can then be computed. To attract attention and ensure recognition, 10 to 15 dB are added. It is important to check if these levels can represent a risk of hearing loss, can interfere with communication or can startle. With respect to temporal characteristics and the perceived urgency of the signal, complex signal are recommended (Patterson, 1982). No specific guidelines are yet universally accepted but are being sought after, principally by Edworthy et al. (1989).

Finally, the model will soon be upgraded in order to predict the detection ability of workers with noise-induced hearing loss. This will allow use of the program to be generalised to the entire industrial population.

Acknowledgement
This work is supported by the Institut de Recherche en Santé et en Sécurité du Travail du Québec (IRSST).

REFERENCES

Abel, S.M., Kunov, H., Pichora-Fuller, M.K. and Alberti, P.W., 1985, Signal detection in industrial noise: effects of noise exposure history, hearing loss, and the use of ear protection. Scandinavian Audiology, 14, 161-173.

Clarck, T.S. and Corlett, E.N., 1984, The ergonomics of workplaces and machines. A design Manual, (London, Taylor & Francis), 73-74.

Coleman, C.J., Leamon, T.B. and Drayton, I.D.R., 1980, Auditory communication in the mining industry. Final report on CEC Contract 6245-11/8-019.

Doll, T.J. and Folds, D.J., 1986, Auditory signals in military aircraft: ergonomics principles versus practice. Applied Ergonomics, 17, 257-264.

Edwards, R.G., Broderson, A.B., Green, W.W. and Lempert, B.L., 1983, A second study of the effectiveness of earplugs as worn in the workplace. Noise Control Engineering Journal, 20, 6-15.

Edworthy, J., Loxley, S., Geelhoed, E. and Dennis, I., 1989, The perceived urgency of auditory warnings. In: Proceedings of the Institute of Acoustics, Oxford, edited by Institute of Acoustics (Oxford), 73-80.

Fidell, S., Pearsons, K.S. and Bennett, R., 1974, Prediction of aural detectability of noise signals. Human Factors, 16, 373-383.

Florentine, M. and Zwicker, B., 1979, A model of loudness summation applied to noise-induced heaaring loss. Hearing Research, 1, 121-132.

Harris, R., 1983, The selection and specification of audible alarm signals. In: Proceedings of Inter-Noise, 803-806.

ISO 7731, 1986, Danger signals for workplaces - Auditory danger signals.

Laroche, C., Tran Quoc, H., Hétu, R. and McDuff, S., 1990, "Detectsound": A computerized model for predicting the detectability of warning signals in noisy workplaces. Submitted for publication.

Murrell, K.F.H., 1971, Ergonomics. Man and his working environment, (London, Chapman and Hall), 212-213.

Oborne, D.J., 1987, Ergonomics at Work, (Chichester, John Wiley & Sons), 130-133, 245-266.

Patterson, R.D., 1982, Guidelines for auditory warning systems on civil aircraft. CAA Paper 82017, Civil Aviation Authority, London.

Patterson, R.D., Cosgrove, P., Milroy, R. and Lower, M.C., 1989, Auditory warning for the British rail inductive loop warning system. In: Proceedings of the Institute of Acoustics, Oxford, edited by Institute of Acoustics (Oxford), 51-58.

Patterson, R.D., Nimmo-Smith, I., Weber, D.L. and Milroy, R., 1982, The deterioration of hearing with age: Frequency selectivity, the critical ratio, the audiogram, and the speech threshold. Journal of the Acoustical Society of America, 72, 1788-1803.

Talamo, J.D.C., 1982, The perception of machinery indicator sounds. Ergonomics, 25, 41-51.

Trucks, L.B. and Adams, S.K., 1975, Predicting the effectiveness of auditory warning signals in industrial environments. In: Proceedings of the Human Factors Society 19th annual meeting, 368-374.

Wilkins, P.A. and Acton, W.I., 1982, Noise and accidents - A review. Annals of Occupational Hygiene, 2, 249-260.

Wilkins, P.A. and Martin, A.M., 1978, The effect of hearing protectors on the perception of warning and indicator sounds - A general review. Technical report No. 98, Hearing Conservation Unit, Southampton.

Wilkins, P. and Martin, A.M., 1987, Hearing protection and warning sounds in industry - A review. Applied Acoustics, 21, 267-293.

Zwicker, E. and Scharf, B., 1965, A model of loudness summation. Psychological Review, 72, 3-26.

XV

OCCUPATIONAL HEALTH AND SAFETY

Advances in Industrial Ergonomics and Safety II
Edited by Biman Das
Taylor & Francis 1990

ON HAZARD RECOGNITION AND ITS QUANTIFICATION

Dr. Abdul Raouf
Department of Systems Engineering
King Fahd University of Petroleum & Minerals
Dhahran-31261, SAUDI ARABIA

ABSTRACT

Each hazard is likely to result in different levels of dangers such as criticality of the hazard, number of people involved and the frequency of occurrence of the hazard. Quantification of hazards is essential so that the safety manager can set up priorities for eradicating these. This paper presents a method of developing a Hazards Index which may be used for classifying hazards and thus assisting the safety managers in determining the suitable "Hazard Candidates" for eradication.

INTRODUCTION

Accident prevention is difficult, at least, and impossible at worst, without an understanding of accident causation. Many a theories of accident causation have been stated where most of the components of the accident process could be identified and the factors that contribute and lead to the accident are discovered. Most of the theories tend to over generalize and most of theories are based on relatively small sample size. Such theories are still at the conceptual stage and as such remain of limited utility in accident prevention and control.

One of the readily adopted methods of determining the nature and characteristics of hazards encountered in industrial environment falls under Hazard Recognition. Hazard Recognition is aimed at the study of the accident process and attempts to find accident symptoms.

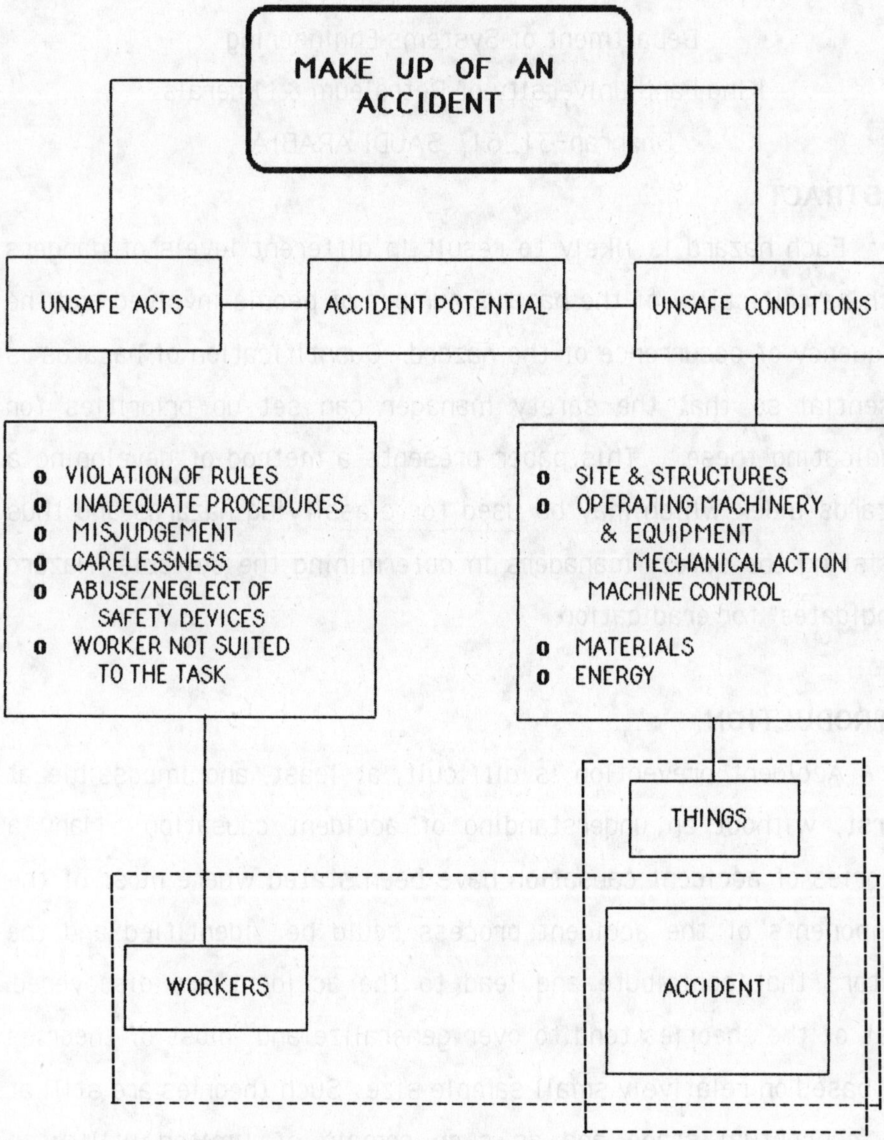

Figure 1

MAKE UP OF AN ACCIDENT

Make up of an accident is shown in Figure 1. Procedures have been developed to prevent accidents from occuring, as far as possible, by using Safety Behavior Sampling of workers and carrying out critical examinations of task and environmental variables so that unsafe conditions are removed. An unsafe condition may be defined as any physical condition which, if left incorrected, is likely to lead to an accident. For improving safety at work place such conditions must be detected before an accident takes place.

Detection of Unsafe Conditions

Unsafe conditions can be detected by carrying out periodic safety appraisals. These appraisals attempt to identify unsafe conditions so that they can be eliminated before accident takes place.

Basic objectives of safety appraisals are as follows:

(i) machines, tools, devices and equipment conform to existing safety standards;

(ii) personal protection equipment, machine and equipment guards, and other safety device, are adequate;

(iii) illumination, ventilation, noise level, radiation, etc. at the work place are in accordance with the current recommended practices;

(iv) working surfaces, stairs, ramps, laddrs, and fire protection equipment are in accordance with the standards;

(v) house-keeping, sanitation, waste disposal and material storage is according to safe practices.

Hammer [1985] and others have developed procedure which, if followed, will be of great help in making safety appraisals. Most of the safety appraisals are qualitative in nature and as such have limited use in planning safety improvements. Peterson [1977] and Raouf [1988] have tried to quantify the results of safety appraisals. Peterson has grouped safety activities into five categories and these are:

(i) Organization and administration

(ii) Industrial hazard control

(iii) Fire control and industrial hygiene

(iv) Supervisory participation, motivation and training

(v) Accident investigation statistics and reporting.

Each of the above category is graded as poor, fair, good and excellent and has different weight attached according to the importance of the activity. Raouf [1988] has formulated a scoring sheet to be used by members of safety appraisal team. These appraisals are to be conducted by same team members periodically. With a view to draw statistical inference, use of Kushal-Wallis Test [Bowen & Starr (1985)] has been demonstrated.

None of the approaches cited above takes into consideration the following factors:

(i) Each hazard is likely to result in different level of danger in terms of criticality, (degree of injury);

(ii) Hazard may involve more than one workers;

(iii) Frequency of occurance of 'near miss' accidents.

A new approach for developing a Hazard Index (HI) is suggested in this paper.

Hazard Index (HI) Formulation

The basic approach suggested is to recognize the hazard and predict its criticality in terms of "degree of injury", number of workers likely to be involved if no action to eradicate the hazard or unsafe condition is taken and the number of 'near miss' accident that has been experienced by the worker due to that particular hazard or unsafe condition.

This information is required to be gathered by a team which must include the worker at the station being inspected and the concerned supervisor and the safety professional. A form which can be used for Hazard Index Evaluation is attached.

At a given station there may exist many unsafe conditions. Each unsafe condition is likely to result in a degree of injury. Degree of injury is to be specified in terms of its severity. In case the likely result is death, explosion etc. then the severity is 10, serious laccrations 8, burns 6, bruises and sprains 4, and abrasions 2.

Workers supply the information regarding the number of near miss accidents. The number of workers likely to be involved in accident due to each unsafe condition can be assessed easily.

D_{ijk} = degree of injury due to kth unsafe condition at workstation j of department i;

M_{ijk} = number of 'near miss' accidents due to kth unsafe condition at workstation j of department i;

N_{ijk} = number of workers affected due to kth unsafe condition at workstation j of department i;

HI_{ijk} = hazard index of kth unsafe condition at the workstation j of department i;

$= D_{ijk} \cdot M_{ijk} \cdot N_{ijk}$

HI_{ij} = hazard index at ith work station in department j,

$$\sum_k HI_{ijk}$$

HI_i = hazard index at department j,

$$\sum_j H\, I_{ij}$$

The proposed H.I. considers not only the unsafe condition but also takes into consideration the 'worse case seniario', near miss accidents and the number of workers who are likely to be involved.

This index can be subjected to statistical tests and as such can assist those concerned with the Safety Management in deciding to determine the best candidate for the allocation of limited resources.

Periodic evaluations will indicate the trends in safety improvement at workstation level, department level and plant level. It can be used in evaluating the effectiveness of safety audits and other safety improvement programs. Work is in progress in various industrial plant where the proposed H.I. is being used. Our experience in the form of case studies will be reported in due course of time.

Hazard Index Evaluation Form

(To be completed for each unsafe condition observed at a work station)

Work Station ____________ Time __________ Team Members

Department ____________ Date __________ ____________

Plant ________________ ____________

Date Last Hazard Index ____________

Evaluation was done: ____________

UNSAFE CONDITION DETAIL **(1)**	LIKELY DEGREE OF INJURY **(2)**	# OF NEAR MISS ACCIDENTS **(3)**	# OF WORKERS AFFECTED **(4)**

HI = (1) x (2) x (3) x (4)
Likely degree of injury:
Death = 10, Serious Laccrations = 8
Burns = 6, Bruises, Sprains = 4,
Abrasions = 2

References

1. Bowen, E.K. and Starr, M; "Basic Statistics for Business & Economics", 1985.

2. Hammer, W; "Occupational Safety Management", Prentice Hall, 1985.

3. Peterson, D; "Safety Management : A Human Approach", A Panraj Publication, 1975.

4. Raouf, A; "Conducting Safety Appraisals & Drawing Statistical Inferences", Proceedings, Regional Seminar on Safety Engineering, University of Jordon, 1988.

Advances in Industrial Ergonomics and Safety II
Edited by Biman Das
Taylor & Francis 1990

COMPARING OPERATION INCIDENCE AND SEVERITY RATES

Amy Hancock Boyd

Departments of Administrative Science and Mathematics
Colby College
Waterville, ME 04901

Comparisons between operation injury experience are beneficial when deciding how to allocate a limited budget for ergonomic analysis and improvement. This paper presents statistical methods of comparing two or more incidence rates and provides expressions for confidence intervals for the ratios of two rates. An approximate test for comparison of two severity rates when injury counts are high is also presented. The methods are illustrated using injury data from a large manufacturing organization.

INTRODUCTION

The two most common measures of injury experience currently in use are the injury incidence and severity rates. The injury incidence rate (IR) is a measure of the frequency of injury events relative to the total exposure hours of the employees working on a particular operation. One definition of the incidence rate is as follows:

$$IR = (N / H) \times 200{,}000 \tag{1}$$

where

N is the number of occupational injuries and illnesses.

H is the total hours of effort (= the total hours worked by all employees).

200,000 is the base hours per year for 100 full time workers working 40 hours per week for 50 weeks per year.

The severity rate (SR) is defined in a similar manner. One such definition is

$$SR = (L / H) \times 200{,}000 \tag{2}$$

where L = the number of lost work days and H is defined as above. If the information is available, the number of lost work days can be replaced by the costs associated with the injuries.

When deciding how to allocate a limited budget for ergonomic analysis in order to reduce the number of injuries and their associated costs, it is often desirable to compare the incidence and severity rates of different operations. In order to do so, it is necessary to know the statistical properties of these rates. The purpose of this paper is to present these properties and provide hypothesis tests and confidence intervals that can be used for rate comparisons.

THE INCIDENCE RATE

Using an industrial data set of worker's compensation injuries, Boyd (1986) demonstrated the exposure time between injuries on an operation can be modelled as a negative exponential distribution. Thus, injuries on an operation occur according to a Poisson process. The probability of x injuries in an interval of exposure equal to h is described by

$$P(X=x)=\frac{(\lambda h)^{x}e^{-\lambda h}}{x!} \tag{3}$$

where λh represents the expected or mean number of injuries in a time period of h 100 worker years and λ is the incidence rate. The variance of this distribution is also equal to λh.

Nelson (1987) determined an exact method for comparing two Poisson means. The test of the hypothesis that the incidence rates on two operations, 1 and 2, are equivalent, H_0: $\lambda_1 = \lambda_2$, is based upon the binomial distribution. In order to perform the test, we must record the number of injuries on each operation (n_1 and n_2, respectively) and the exposure time on each operation (h_1 and h_2). The total number of injuries is $n = n_1 + n_2$. Assuming that injuries occur according to independent Poisson processes with parameters λ_1 and λ_2, the probability of n_1 injuries on operation 1 given a total of n injuries on both operations can be calculated using a binomial distribution (Ross, 1972) with n trials and a probability of success equal to

$$p_{01}=\frac{\lambda_1 h_1}{\lambda_1 h_1+\lambda_2 h_2}\,. \tag{4}$$

With some algebraic manipulation, this is equivalent to

$$p_{01}=1/\left[1+(\lambda_2/\lambda_1)(h_2/h_1)\right]\cdot \tag{5}$$

This is also the probability of success associated with operation 1 under the null hypothesis, since if the incidence rates were equal, we would expect the total number of injuries on the two operations to occur in the same ratio as the ratio of the exposure times. The zero subscript on p indicates that this is the probability under the null hypothesis. The one subscript indicates that it is the probability associated with operation 1. Table 1 gives the decision rules for this test. Since the null hypothesis of interest is H_0: $\lambda_1 = \lambda_2$ or $\lambda_2/\lambda_1 = 1$, a value of 1 is substituted for λ_2/λ_1. If one desires to test some other rate ratio, it can be substituted for λ_2/λ_1 in equation (5). These calculations can be carried out efficiently via computer. Binomial probabilities can be calculated using a Fortran subroutine from Nelson (1985). A listing of this and all other programs mentioned in this paper can be obtained from the author upon request. Two applications of this test are given below. The data used are from 4.75 years of injury experience at a

Table 1: Decision Rules for Testing H_0: $\lambda_1 = \lambda_2$.

Alternative (H_1:)	Calculate the binomial probability*:	Reject H_0 if p-value <:
$\lambda_1 \neq \lambda_2$	p-value = $P(X \leq n_1 \mid p = p_{01}, n_1 + n_2 = n)$ for $n_1 < n_2$	$\alpha/2$**
	p-value = $P(X \geq n_1 \mid p = p_{01}, n_1 + n_2 = n)$ for $n_1 > n_2$	$\alpha/2$
$\lambda_1 < \lambda_2$	p-value = $P(X \leq n_1 \mid p = p_{01}, n_1 + n_2 = n)$	α
$\lambda_1 > \lambda_2$	p-value = $P(X \geq n_1 \mid p = p_{01}, n_1 + n_2 = n)$	α

$$*P(X \leq n_1) = \sum_{x=0}^{n_1} \binom{n}{x} (p_{01})^x (1-p_{01})^{n-x}$$

**α represents the desired significance level of the test.

large manufacturing facility.

Example 1: There were 23 injuries on operation 1 in 3.81 100 worker years. There were 18 injuries on operation 2 in 2.12 100 worker years. Estimates of the incidence rates are 6.04 and 8.49 injuries per 100 worker years, respectively. We wish to test H_0: $\lambda_1 = \lambda_2$ versus H_1: $\lambda_1 \neq \lambda_2$ at a 5% level of significance. $p_{01} = 1/[1+(1)(2.12/3.81)] = 0.642$, $n = 41$, $P(X \geq 23 \mid p_{01} = 0.642, n = 41) = 0.8953$. Since $0.8953 > 0.025$, we cannot reject H_0.

Example 2: There were 235 injuries on operation 1 in 20.07 100 worker years. There were 47 injuries on operation 2 in 5.41 100 worker years. Estimates of the incidence rates are 11.71 and 8.68 injuries per 100 worker years, respectively. We wish to test H_0: $\lambda_1 = \lambda_2$ versus H_1: $\lambda_1 > \lambda_2$ at a 5% level of significance. $p_{01} = 1/[1+(1)(5.41/20.07)] = 0.788$, $n = 282$, $P(X \geq 235 \mid p_{01} = 0.788, n = 282) = 0.033$. Since $0.033 < 0.05$, H_0 is rejected and we conclude that $\lambda_1 > \lambda_2$.

Calculation of confidence intervals for the ratios of the rates of the two operations may be more illuminating than hypothesis tests, since confidence intervals provide a measure of the degree of uncertainty in the comparison. A 100(1-α)% two sided confidence interval for the ratio λ_2/λ_1 is represented by the expression (Nelson, 1987):

$$(1/p_U - 1)/(h_2/h_1) \leq \lambda_2/\lambda_1 \leq (1/p_L - 1)/(h_2/h_1) \tag{6}$$

where

p_L = the largest p that satisfies $P(X \geq n_1 \mid p = p_L, n) \leq \alpha/2$

p_U = the smallest p that satisfies $P(X \leq n_1 \mid p = p_U, n) \leq \alpha/2$

The expressions for 100(1-α)% one-sided upper and lower confidence intervals are given below:

$$Lower: \ \lambda_2 / \lambda_1 \geq (1/p_U - 1)/(h_2/h_1) \quad (7)$$

$$Upper: \ \lambda_2 / \lambda_1 \leq (1/p_L - 1)/(h_2/h_1) \quad (8)$$

where p_L and p_U are defined as above. The values for p_L and p_U can be found via computer or monograph. An estimate of the ratio of rates is given by $(n_2/n_1)/(h_2/h_1)$. Confidence intervals consistent with the alternate hypothesis for Examples 1 and 2 are given below.

Example 3: An estimate of λ_2/λ_1 is (18/32)/(2.12/3.81) = 1.4065. p_L = 0.3974, p_U = .7153. Therefore, $0.7153 \leq \lambda_2/\lambda_1 \leq 2.7252$.

Example 4: An estimate of λ_2/λ_1 is (47/235)/(5.41/20.07) = 0.7420. p_L = 0.7924. Therefore, $\lambda_2/\lambda_1 \leq 0.9719$.

Nelson (1987) also gives an approximate method of comparing k Poisson means. The null hypothesis is H_0: $\lambda_1 = \lambda_2 = ... = \lambda_k$. The alternate hypothesis is that not all λ_i, i = 1, 2, ..., k, are equal. To make this comparison, one must number the operations from one to k, form k rate ratios, all with respect to λ_1, and form k effort hour ratios, all with respect to h_1. Once this is done, k multinomial probabilities p_{0j} are calculated as follows:

$$p_{0j} = \frac{(\lambda_j/\lambda_1)(h_j/h_1)}{\sum_{i=1}^{k}(\lambda_i/\lambda_1)(h_i/h_1)} \quad (9)$$

Since H_0 above represents rate ratios of 1, $\lambda_j/\lambda_1 = 1$ for every j. The chi-square distribution is used to approximate the multinomial distribution. The chi-square statistic is calculated as follows:

$$\chi^2_{calc} = \frac{(n_1 - p_{01}n)^2}{p_{01}n} + \frac{(n_2 - p_{02}n)^2}{p_{02}n} + ... + \frac{(n_k - p_{0k}n)^2}{p_{0k}n} \quad (10)$$

where $n = n_1 + n_2 + ... + n_k$. For this test to be valid, all $p_{0j}n$ must be greater than 5. Under the null hypothesis this statistic has a chi-square distribution with k - 1 degrees of freedom. The null hypothesis is rejected if $P(\chi^2_{k-1} > \chi^2_{calc})$ is less than α. For k = 2, a continuity correction should be used, i.e.,

$$\chi^2_{calc} = \frac{(|n_1 - p_{01}n| - 0.5)^2}{p_{01}n} + \frac{(|n_2 - p_{02}n| - 0.5)^2}{p_{02}n} \quad (11)$$

Again, this test can be performed easily via computer. An example application of this test is given below.

Example 5: Operations 1, 2, 3 and 4 are all jobs requiring manual parts assembly. We want to see if their incidence rates are the same at the 5% level of significance. There were 235, 18, 47, and 211 injuries observed in 20.07, 0.68, 5.41, and 17.22 100 worker years of exposure, respectively. Estimates of incidence rates are 11.71, 26.47, 8.69 and 12.25 injuries per 100 worker years, respectively. We want to test H_0: $\lambda_1 = \lambda_2 = \lambda_3 = \lambda_4$ versus H_1: not all λ_i, i = 1, ..., 4, are equal. Calculations reveal that p_{01} = 0.462, p_{02} = 0.016, p_{03} = 0.125, p_{04} = 0.397, and n = 511. Note that (511)(0.016) = 8.18 which is greater than 5. χ^2_{calc} = 17.19 with 3 degrees of freedom. The p-value associated with this test is 0.0006. Since 0.0006 < 0.05, H_0 is rejected in and we conclude that the incidence rates on these jobs are not equal.

Lee (1985) provides several methods for testing monotone trends in k Poisson means. For his Maximin method used to test H_0: $\lambda_1 = \lambda_2 = ... = \lambda_k$ versus H_1: $\lambda_1 \le \lambda_2 \le ... \le \lambda_k$ with at least one strict inequality, we must calculate the following statistics:

$$w_i = h_i / \sum_{i=1}^{k} h_i \tag{12}$$

$$T_m = \sum_{i=1}^{k} i n_i \tag{13}$$

$$E(T_m) = n \sum_{i=1}^{k} i w_i \tag{14}$$

$$Var(T_m) = n[\sum_{i=1}^{k} i^2 w_i - (\sum_{i=1}^{k} i w_i)^2] \tag{15}$$

$$z = [T_m - E(T_m)] / \sqrt{Var(T_m)} \tag{16}$$

The value z is distributed as a standard normal random variable, under H_0, so H_0 is rejected in favor of H_1 if the p-value = P(Z > z) is less than α. An example of this test is given below.

Example 6: Using the same data as for Example 5, we wish to test H_0: $\lambda_1 = \lambda_2 = \lambda_3 = \lambda_4$ versus H_1: $\lambda_1 \le \lambda_2 \le \lambda_3 \le \lambda_4$ at the 5% level of significance. T_m = 1256.00, $E(T_m)$ = 1255.00, $Var(T_m)$ = 1005.28. The p-value of the test = 0.4874. Since this is greater than 0.05, we cannot reject H_0.

THE SEVERITY RATE

Safety practitioners are concerned with the severity of injuries on an operation as well as the number or incidence rate of injuries. The severity rate can be modelled as a compound Poisson process, as follows. If events in a time period h occur according to a Poisson process at a rate λ where the magnitude, Y, of each event can be described by $f_Y(y)$, with each magnitude being independently and identically distributed and independent of the number of events, then for the rate defined in (2)

$$E(SR) = \lambda E(Y) \tag{17}$$

$$Var(SR) = \lambda E(Y^2) / h \tag{18}$$

The random variable Y represents the number of days lost per injury or the cost per injury. In order to proceed further, we must determine a distribution for lost time and injury cost. In both cases it is likely that a right skewed distribution is appropriate, since one would expect a large number of injuries with low lost time or cost and a small number of injuries exhibiting high lost time or expense. Several distributions were investigated empirically using the data sets referenced above.

Empirical Results for Lost Time

Examination of histograms of lost time data revealed that there were too few injuries with low lost times (less than 3 days) and too many extremely high lost time for a simple exponential model to be an adequate fit. Nor did any general type of distribution fit adequately because of these "tail" problems. Therefore, it was felt that some type of mixture distribution might be appropriate.

A split population mixed exponential,

$$f_Y(y)=\begin{cases}0 \quad with\,prob.\,p_1 \\ p_2\lambda_1 exp(-\lambda_1 y)+(1-p_2)\lambda_2 exp(-\lambda_1 y) \quad with\,prob.\,1-p_1\end{cases} \quad (19)$$

where $0 < y < \infty$, was fit to the lost time data for each operation using the method of moments technique with a fixed value of p_2 from Rider (1961). The proportion of lost times with a value of zero was used to estimate p_1. The proportion of lost days less than 150 was used to estimate p_2. This model was an adequate fit ($\alpha \geq .01$) for all 17 operations.

Empirical Results for Severity Cost

The severity cost for each injury was calculated as the sum of redemption payments, disability payments, medical payments, legal payments, miscellaneous payments (primarily travel to and from a health care facility) and lost time cost. Lost time cost was estimated as the number of days lost multiplied by the daily benefit rate paid. The gamma distribution,

$$f_Y(y)=\frac{\beta^{-\alpha}\,y^{\alpha-1}\,e^{-y/\beta}}{\Gamma(\alpha)}, \quad y>0 \quad (20)$$

was fit to historical severity cost data. This distribution was a good fit at $\alpha > 0.01$ for 14 out of 17 of the operations (a mixture of gamma distributions would probably be necessary to fit the remaining three distributions).

Comparison of Mean Severity Rates

Due to the complexity of the severity rate, it is very difficult to develop an exact test for comparing severity rates. An approximate test can be developed under conditions when the Central Limit Theorem applies. Severity rate simulations of 1,000 trials were conducted using the distributions of lost time and cost given in (19) and (20). For lost time, convergence to normality depends primarily on the number of injuries. The test is appropriate when the observed number of injuries > 45. For severity cost, the test can be applied when the gamma parameter, α, multiplied by the observed number of injuries is greater than 35.

To test H_0: $SR_1 = SR_2$ versus any of the three alternate hypotheses, we must calculate the observed severity rate,

$$OSR_j = \frac{\sum_{i=1}^{n_j} l_{ij}}{h_j}, \quad j=1,2 \tag{21}$$

where l_{ij} represents lost time or severity cost associated with injury i on operation j. For lost time, λ_j is estimated by the incidence rate from (1) and a split population mixed exponential must be fit via computer to historical lost time data. The mean and variance of (21) are:

$$E(OSR_j) = \lambda_j(1-p_{1j})\left[\frac{p_{2j}}{\lambda_{1j}} + \frac{(1-p_{2j})}{\lambda_{2j}}\right], \quad j=1,2 \tag{22}$$

$$Var(OSR_j) = \lambda_j(1-p_{1j})\left[\frac{2p_{2j}}{\lambda_{1j}^2} + \frac{2(1-p_{2j})}{\lambda_{2j}^2}\right] / h_j, \quad j=1,2 \tag{23}$$

For severity cost, λ_j is again estimated with the incidence rate. The mean and variance of (21) for severity cost are:

$$E(OSR_j) = \lambda_j \alpha_j \beta_j, \quad j=1,2 \tag{24}$$
$$Var(OSR_j) = \lambda_j(\alpha_j + \alpha_j^2)\beta_j^2 / h_j, \quad j=1,2$$

The test statistic is

$$z_{calc} = \frac{OSR_1 - OSR_2}{\sqrt{Var(SR_1) + Var(SR_1)}} \tag{26}$$

Under H_0, this is distributed as a standard normal random variable. The variances in the denominator of the statistic must be estimated using parameter estimates from fitting (19) and (20) to historical lost time and severity data. The rules for this test are presented in Table 2 below. An application of this test is given in Example 7.

Table 2: Decision Rules for Testing H_0: $SR_1 = SR_2$.

Alternative (H_1:)	Calculate:	Reject H_0 if p-value <:
$SR_1 \neq SR_2$	p-value = $P(z < z_{calc})$, $z_{calc} < 0$ or $P(z > z_{calc})$, $z_{calc} > 0$	$\alpha/2$
$SR_1 < SR_2$	p-value = $P(z < z_{calc})$	α
$SR_1 > SR_2$	p-value = $P(z > z_{calc})$	α

Example 7: (Lost Time Severity Rate) Operation 1 had 8,194 lost days from 235 injuries in 20.07 100 worker years. Operation 2 had 10,354 lost days from 211 injuries in 17.22 100 worker years. Severity rate estimates are 408.27 and 601.28 days per 100 worker

years, respectively. We wish to test H_0: $SR_1 = SR_2$ versus H_1: $SR_1 \neq SR_2$ at a significance level of 10%. For operation 1, parameter estimates are $p_{11} = 0.32$, $p_{21} = 0.87$, $\lambda_1 = 11.71$, $1/\lambda_{11} = 39.46$, $1/\lambda_{21} = 285.30$. For operation 2, parameter estimates are $p_{12} = 0.34$, $p_{22} = 0.91$, $\lambda_2 = 12.25$, $1/\lambda_{12} = 41.15$, $1/\lambda_{22} = 413.40$. $Var(OSR_1) = 22{,}750.14$, $Var(OSR_2) = 6{,}843.81$, $z = -1.12$. The p-value is 0.131, so we cannot reject H_0.

An approximate $100(1-\alpha)$% two-sided confidence interval can be calculated via

$$OSR_1 - OSR_2 \pm z_{\alpha/2}\sqrt{Var(OSR_1) + Var(OSR_2)} \tag{27}$$

An example for the rates in Example 6 is given below.

Example 8: We wish to calculate a 90% confidence interval for the difference in mean severity rates given in Example 6. $OSR_1 - OSR_2 = -193.01$, $z = 1.645$, $\sqrt{Var(OSR_1) + Var(OSR_2)} = 172.02$. The interval is -475.99 to 89.98.

CONCLUSIONS

It is hoped that the statistical tests presented will provide the practitioner with methods of comparing the injury experience for different operations. The methods for comparing incidence rates are general and easy to apply. Comparing severity rates is more complex since distributions must be fitted to lost time or cost. Presently these methods have limited applications, since both operations must have high injury counts. Methods for small count comparisons are a subject for future research.

REFERENCES

Boyd, A. H., 1986, Development of a minimum cost sequential test to monitor injury incidence on an operation.(Ann Arbor, MI: Unpublished Ph.D. thesis, Department of Industrial and Operations Engineering, University of Michigan).

Lee, Y. J., 1985, Tests of monotone trend in k poisson means. Journal of Quality Technology, 17, 44 - 49.

Nelson, L. S., 1987, Comparison of poisson means: the general case. Journal of Quality Technology 19, 173 - 179.

Nelson, P. R., 1985, Computation of some common discrete distributions. Journal of Quality Technology, 17, 160 - 166.

Rider, P. R., 1961, The method of moments applied to a mixture of exponential distributions. Annals of Mathematical Statistics, 23, 143 - 47.

Ross, S. M. 1979, Introduction to Probability Models, (New York, N. Y.: Academic Press).

Ross, S. M., 1983, Stochastic Processes, (New York, N. Y.: John Wiley and Sons)

Advances in Industrial Ergonomics and Safety II
Edited by Biman Das
Taylor & Francis 1990

IS THERE A RELATIONSHIP BETWEEN ATTRIBUTES OF CONSTRUCTION SITES AND WORKERS' SAFETY PRACTICES AND CLIMATE PERCEPTIONS

Nicole Dedobbeleer[1], Sc.D.
François Béland[1], Ph.D.
Pearl German [2], Sc.D.

[1]Université de Montréal,
[2]Johns Hopkins University

The purpose of this study was to explore differences in workers' safety climate perceptions and safety practices in terms of characteristics of construction sites. Data were collected by observation and through a survey of 384 workers and 9 superintendents employed at nine non-residential construction sites, valued at $500 000 or more in the Baltimore Metropolitan Area. Results showed that workers develop perceptions of safety climate based on very specific cues present on construction sites, and that workers' safety performance is affected by attributes of the sites.

INTRODUCTION

Among the important issues that have to be to addressed in the climate research area, James and Jones (1974) pointed out: (1) the relationship between perceptual climate measures and climate measures in which at least some variables are measured nonperceptually, and (2) the role of climate perceptions as predictors of individual behaviors. These issues apply to safety climate measures. Zohar (1980) and Holmes and Brown (1986) constructed models for assessing production workers' perceived safety climate but did not attempt to correlate their perceptual safety climate measures to more objective measures and to actual employee safety performance.

This paper will go beyond these two studies by not only testing the generalizability of the Holmes and Brown's model to construction workers, but also by examining first, the relationships between workers' safety climate perceptions and environmental and organizational attributes of the construction sites and second, by examining the relationships between workers' safety performance and environmental and organizational attributes of construction sites. It will thus address the first issue pointed out by James and Jones and provide a first step to the second issue mentioned.

METHOD

Setting and subjects

A cross-sectional survey was conducted on nine non-residential construction sites located in the Baltimore Metropolitan area. New or partially new projects, ranging from the framing stage to the finishing stage and valued at $500,000 or more, were selected from the Reports on Metropolitan Building Permit Activity of the Regional Planning Council. All contractors (fifteen) who had projects conforming to those criteria from August to December 1983 were contacted. Seven refused to collaborate, eight agreed but two were found ineligible. Among the six remaining contractors, three had two sites, three had one site, for a total of nine sites. Reasons of refusal were related to insurance coverage for observers, fear of work disruption and survey costs.

From a total of 572 workers employed at the nine sites, the population surveyed consisted of 384 workers who were allowed to participate in the study by their employer and involved with at least 50% of the safety practices examined. Two hundred seventy-two workers responded to the survey for a response rate of 71%. Eighteen additional workers were excluded because of response omission to the majority of the questions. Data were also collected from each superintendent of the nine sites. The population observed on the nine construction sites comprised 454 employees including workers and foremen.

Instruments and data collection

A self-administered questionnaire was used to collect data from the workers at the worksite. It included questions on socio-demographic and employment characteristics, safety practices, exposure to safety interventions, knowledge and attitudes toward safety practices, perceptions of the workplace and other socio-psychological factors. The superintendents completed a 15-minute interview at the worksite. This interview collected data on environmental and organizational characteristics of the construction sites.

A checklist was used to conduct systematic, nonparticipant observations on all employees' safety practices. The data were collected immediately before the questionnaires were distributed at each site. Because of the high turnover rate of workers on construction sites, employees safety practices were observed only once at each site. Focal individual sampling was the observational strategy used. During a given sampling period (5-10 minutes), one worker's behavior was scored and behaviors not involving the focal subject were ignored. For all the items of observation, the presence, absence or non applicability of the practices were noted after each observation as unobstrusively as possible. The items were discrete and easily observable.

Measures

Safety climate refers to molar perceptions workers have of the safety of the construction site. The three-factor Brown and Holmes' safety climate model was used as a guide to build this measure. Nine variables were chosen to represent safety concerns in construction industry. They were not all identical to those included in the Zohar's questionnaire (1980) used in the Brown and Holmes' study. Each dimension is defined in Table 1.

Table 1. Variables measuring the dimensions of the Holmes and Brown's safety climate model

A) Employee perceptions of management concerns:

Management's attitude toward safety practices (MANA. PRACTICES) was worker's perception of the importance of safety practices to top management and was measured by a four-point rating scale ranging from "very important" to "not at all important".
Management's attitude toward workers' safety (MANA. SAFETY) was workers' perception of this attitude and was measured by an itemized rating scale ranging from "They do as much as possible to make the job safe" to "They are only interested in getting the job done as fast and cheaply as possible".
Foreman's behavior (FOREMAN) was the worker's perception of the actions taken by the foreman to enforce safety, and was measured by an itemized rating scale ranging from "He regularly and frequently makes us aware of dangerous work practices and conditions and praises us for safe conduct" to "He never mentions dangers".

B) Employee perceptions of management safety activities:

Safety instructions (INSTRUCTIONS) refers to the self-reported exposure to instructions on the safety policy of the company at the time of initial employment.
Safety meetings (MEETINGS) was worker's perception of the presence of regular safety meetings at the worksite.
Proper equipment (EQUIP) was worker's perception of the availability of proper equipment at the worksite, and was measured by a five-point rating scale ranging from "always" to "never".

C) Employee physical risk perception:

Perceived control (CONTROL) was defined as worker's perception of control over his own safety on the job, and measured by a four-point rating scale ranging from "almost total control" to "almost no control".
Perception of risk-taking (RISK) referred to the worker's perception of the extent to which risk-taking is part of the job, and was measured by a three-point rating scale ranging from "very much" to "not at all".
Perceived likelihood of injuries (INJURIES) was defined as worker's perception of his susceptibility to work-related injuries in the next 12 months, and was measured by a four-point rating scale ranging from "very likely" to "not at all likely".

Environmental characteristics refer to the size and stage of construction of the sites reported by the superintendent. The size of the site was measured in terms of number of workers employed at the site. The stage of construction was measured by an open-ended question.
Organizational characteristics refer to areas included in the safety policy of the sites: safety responsibilities, safety rules related to several personal protective devices (i.e. hard hats, ear plugs, ear muffs, face shields, safety glasses or goggles, special clothing), safety training, safety instructions at initial employment, safety meetings, trade-offs between safety and cost. They were reported by the superintendents and measured by closed-ended questions.
Safety performance refers to an index of the observed safety practices of all the employees on each site. The safety practices observed were hard hat or helmet use, use of eye protection, heavy working shoes, and properly installed ladders and scaffolds as defined in the OSHA safety standards.

Analysis

To test for the replication of Holmes and Brown's three-factor structures (1986) in our data, the LISREL procedure was applied to the nine variables measured in the paper-and-pencil questionnaire. As Brown and Holmes we used the maximum likelihood estimation procedure. One-way ANOVA was used to determine the effects of environmental and organizational characteristics of construction sites on workers' safety climate perceptions (i.e. the perceptual safety climate measure) and workers' safety performance.

RESULTS

The results of the confirmatory factor analysis showed that the Brown and Holmes' safety climate model (1986) was supported by our data (Table 2). A model with 9 manifest variables and three factors was thus retained. As in Brown and Holmes' model, the three-factors were defined as (1) employee perception of how concerned management was with their well-being, (2) employee perception of how active management was in responding to this concern, and (3) employee physical risk perception.

Table 2. Safety climate model in construction sites

Variables	DIMENSIONS Management concern	Management activities	Physical risk
Management's attitude toward worker's safety	.570	.000	.000
Management's attitude toward safety practices	.516	.000	.000
Foremen safety enforcement	.650	.000	.000
Safety meetings	.000	.187	.000
Safety instructions	.000	.243	.000
Proper equipment	.000	.580	.000
Likelihood of being injurd	.000	.000	.455
Control over own safety	.000	.000	.439
Risk-taking	.000	.000	.262
Total coefficient of determination	.884		

A series of analysis of variance tests were performed on each dimension of the perceived safety climate. The results (Table 3) indicated that most of the organizational characteristics of the construction sites (i.e. safety responsibility, safety training, safety meetings, and certain personal protective devices requirements) significantly affected one of the dimensions of perceived safety climate: employees' perceptions of management's safety activities. There was only one significant F for management's concern for workers' well-being: safety responsibilities. Two organizational characteristics affected significantly workers' perception of physical risk: safety meetings and minor trade-offs between safety and cost.

Table 4 shows that environmental characteristics affected each perceived safety climate dimension in a different manner. Site size had a significant impact on workers' physical risk perception. Stage of construction affected significantly perceived management's concern for workers' well-being and perceived management's safety activities.

Table 3. Analysis of variance: Safety climate perceptions and organizational characteristics

Organizational characteristics of the sites	Management concern		Management activities		Physical risk	
	F	P	F	P	F	P
Responsibilities	5.86	.01	7.60	.00	2.47	.11
Training	2.97	.08	13.69	.00	3.24	.07
Initial Instructions	0.01	.90	0.93	.33	0.83	.36
Meetings	2.21	.14	8.57	.00	6.15	.01
Regulations						
Hard hats	0.30	.58	2.94	.09	1.06	.30
Ear plugs	0.82	.37	5.04	.02	0.39	.53
Ear muffs	3.34	.06	8.30	.00	3.48	.06
Face shields	1.62	.20	13.47	.00	2.29	.13
Clothing	2.86	.09	0.70	.40	0.13	.72
Minor trade-offs	0.06	.81	1.63	.20	10.93	.00

Table 4. Analysis of variance: Safety climate perceptions and environmental characteristics

Environmental characteristics of the sites	Management concern		Management activities		Physical risk	
	F	P	F	P	F	P
Size	0.09	0.76	3.34	0.07	5.22	0.02
Stage of construction	5.86	0.01	7.60	0.00	2.47	0.11

Environmental and organizational characteristics of construction sites also affect workers' safety performance. The results of the analyses of variance showed that construction workers' safety performance significantly differed according to the size and stage of construction of the sites (Table 5). They also revealed that if organizational characteristics of the sites such as safety responsibilities, safety training strategies and certain personal protective devices regulations had an impact on workers' safety performance, safety meetings and minor trade-offs between safety and cost had no impact (Table 6).

Table 5. Analysis of variance: Safety performance index and environmental characteristics

Environmental characteristics of the sites	Safety performance index F	P
Size	3.80	.05
Stage of construction	4.15	.04

Table 6. Analysis of variance: Safety performance index and organizational characteristics

Organizational characteristics of the sites	Safety performance index F	P
Responsabilities	4.35	.04
Training	3.87	.05
Initial instructions	4.55	.04
Meetings	1.21	.27
Regulations		
Hard hats	5.53	.02
Ear plugs	4.07	.04
Ear muffs	.19	.66
Face shields	.00	.94
Clothing	.19	.67
Minor trade-offs	.15	.70

DISCUSSION

The present study has shown that the Brown and Holmes' safety climate model developed for American production workers is generalizable to construction workers, though the variables measured were adapted to the construction industry.

The results also provide evidence that organizational and environmental characteristics of construction sites have a significant impact on workers' perceptions of safety climate. The most striking impact of organizational characteristics was on workers' perceptions of management' safety activities. This is consistent with the fact that the large majority of the organizational characteristics reflected management's safety activities. The two other dimensions of the perceptual safety climate measure were only influenced by one or two organizational characteristics: a clear division of safety responsibilities at the site formanagement's concern for workers' well-being, and presence of safety meetings and minor trade-offs between safety and cost for physical risk perception. Different environments also produced different perceptions of safety climate. The results of this study thus clearly indicate that workers develop perceptions of safety climate based on very specific cues present on construction sites.

The study further reveals that workers' safety performance is significanty affected by environmental and certain organizational attributes of the sites. The lack of impact of safety meetings and minor trade-offs between safety and cost may be explained by problems related to the form and content of safety meetings and the fact that trade-offs are an integral part of construction work (Dawson et al. 1985).

The results have both methodological and applied implications. First, the results give some indication of the accuracy of perceived safety climate. For the first time, a relationship has been established between a perceptual safety climate measure and attributes of the worksite. Evidence was gathered from different sources (i.e. workers and superintendents) and in different ways (i.e. paper and pencil questionnaire and face-to-face interview). This constitutes a contribution to the construct validation of safety climate. In future research, other sources of measurement should be used. Second, the results imply that the safety climate dimensions may be used to evaluate as well as recognize potential problem areas in construction sites. The study particularly highlighted the importance of safety activities, safety responsibilities, minor trade-offs between safety and cost, size and stage of construction. The importance of safety responsibilities has been emphasized by Abeytunga (1978). Moreover, Dawson et al (1985) illustrated the problems related to size of firms and adaptation of standards in construction sites. Finally, all the variables included in the safety climate model may be extremely useful in the design of safety climate surveys in construction industry. Third, the results suggest that management of small and large firms should design and implement a clear injury control policy, detailed for all stages of construction. Fourth, research should be conducted on the relationships between safety climate perceptions, safety performance and attributes of the sites.

REFERENCES

Abeytunga, K.B. (1978). The role of the first line supervisor in construction safety. Doctoral thesis, Aston University.

Brown, R.L. & Holmes, H. (1986). The use of a factor-analytic procedure for assessing the validity of an employee safety climate model. Accid. Anal. & Prev. vol 18, no 6, 455-570.

Dawson, S., Clinton, A., Bamford, M. & Willman, P. (1985). Safety in construction: self regulation, industrial structure and workforce involvement. Journal of General Management, vol 10, no 4, 21-38.

James. L.R. & Jones, A.P. (1979). Organizational climate: A review of theory and research. Psychological Bulletin, vol 81, no 12, 1096-1112.

Schneider, B. (1975). Organizational Climates: An essay. Personnel Psychology, 28, 447-479.

Zohar, D. (1980). Safety Climate in industrial organizations: Theoretical and applied implications. Journal of Applied Psychology, vol 65, no 1, 96-102.

Advances in Industrial Ergonomics and Safety II
Edited by Biman Das
Taylor & Francis 1990

DETAILED DESIGN OF A SIMULATION-BASED INTELLIGENT SAFETY TRAINING SYSTEM (SISTS)

C. H. Lee and J.E. Biegel

Department of Industrial Engineering and
Management Systems
University of Central Florida
Orlando, Florida 32816

The effectiveness of a safety training program depends on the ability to train the trainee efficient and strategic responses to a given condition. This requires the safety instructor's awareness of the trainee's current knowledge, specific areas the trainee lacks knowledge, and preferred methods of training. This paper describes a detailed design of an automated intelligent safety training system which will remove the human instructor from the training loop. The safety training can be conducted at the trainee's convenient site, time and schedule which will result in reduced safety training cost and more effective training if the human instructor is not required.

INTRODUCTION

Safety training is necessary not only for workers but also for supervisors and managers. This necessity can be satisfied by a number of means. Special education and training courses are available through governmental agencies, professional associations, universities, and some private consulting companies. However, they can be costly due to the required travel to the training site, room and board during the training as well as training fees. Also training videos can be purchased for various safety areas. Most of these tapes are directed to all levels of personnel.

The effectiveness of a safety training program, as well as other tutoring systems, depends on the ability to train the personnel efficient and strategic responses to a given condition. This requires the instructor's awareness of the trainee's current knowledge, specific areas the trainee lacks knowledge, and preferred methods of training. The instructor also makes instructional decisions by assessing the trainee's performance. If the above requirements are met by an artificially intelligent training system, then the human instructor can be removed from the training loop. Without the need for a human instructor, the safety training system can be used at the trainee's convenient site, time and schedule which will result in reduced safety training cost and more effective training.

Recent developments in the field of artificial intelligence make it possible to remove the human instructor from the active training loop. Without the need for a human instructor the safety training program can be conducted at the company's convenient location and at the trainee's convenient time. The trainee can also receive the training at his or her own pace. This will result in reduced safety training costs and more effective training.

BACKGROUND

Artificial Intelligence and Intelligent Tutoring System

Artificial Intelligence (AI) has been defined in several ways. Gevarter (1985) said that AI is concerned with devising computer programs to make computers smarter. Winston (1984) defined that AI is the study of those ideas that enable computers to be intelligent. The most widely accepted definition is given by Feigenbaum and Feldman (1963). They defined that AI is the branch of computer science devoted to programming computers to carry out tasks that if carried by human beings would require intelligence. The central goal of AI has been to make computers more useful and understand principles that make intelligence possible (Winston, 1984).

A major part of the early research in AI was directed toward language understanding and translation. Game and problem-solving are other major areas of research that had some early success in producing chess and checkers programs. However, improved computation facilities and new techniques of the early 1970s made it possible to develop an intelligent tutoring system (ITS).

ITS has utilized various artificial intelligence techniques such as knowledge representation techniques which were generally the first to be applied and remain foundational to most intelligent tutoring systems. Representation of knowledge in AI involves the storing of information in data structures and the development of interpretative procedures and inferencing schemes to intelligently manipulate these data structures. The objective is to utilize this formalized knowledge to simulate or model intelligent, knowledgeable behavior. Knowledge representation schemes were also used as the basis for student modelling and encoding various teaching strategies.

Components of an Intelligent Tutoring System

An intelligent tutoring system (ITS) embodies the knowledge that allows an expert tutor to create a tutoring sequence. Tutoring sequences are dynamically created based on decisions made in reference to the knowledge represented. Hartley and

Sleeman (1973) listed three general kinds of knowledge represented in an ITS. They are: (1) the domain knowledge, (2) the student model, and (3) the pedagogical knowledge. In addition to the above list, an interface is required for communication although it is not knowledge itself.

The domain knowledge which include the content to be taught is explicitly represented in an ITS. It also includes how to use the content to solve related problems. The student's understanding of the subject matter to be taught is represented in the student model. The student model contains informations about the student. Wenger (1985) divided the student modelling module into two sub-modules: (1) a student model, the knowledge structure which contains student information, and (2) a diagnostic process, which collects this information from available sources. The pedagogical knowledge is used to decide teaching strategies by integrating domain information, knowledge of the student, and teaching methods. The interface provides the mechanisms for information exchange between the components of ITS. It also handles the communications between the student and ITS.

<u>Simulation</u>

A simulation is the imitation of the operation of a real-world process or system over time (Banks and Carson, 1984). Simulation involves the generation of an artificial history of a system, and the observation of that artificial history to draw inferences concerning the operating characteristics of the real system. Simulation enables the study of, and experimentation with, the internal interaction of a complex system. Informational and environmental changes can be simulated and the effect of these alterations can be observed. By changing simulation inputs and observing the resulting outputs, valuable insight may be obtained into which variables are most important and how variables interact.

Simulation is considered as an excellent training system due to its ability to represent elements of a task which are important for the trainee to be trained and its ability to control additional parameters which support, facilitate, enhance, and ensure effective training. Due to its many advantages, simulation has been applied in a wide variety of context in both commercial and military purposes. Applications range from teaching maintenance principles to tactics training for sophisticated aircraft. Opportunities for applying AI techniques to simulation-based training certainly exist in the area of safety training as a new safety training method.

Lee (1989) presented a conceptual design of such an intelligent safety training system. The system is designed on a modular basis which categories the modules by the functions they serve. The modules are used to accept and analyze the trainee's input and update the simulation based on the input. Modules are also responsible for evaluating the consequences of the input by means of a performance measurement system. Finally, the system determines the trainee's knowledge level and the training instruction required by a particular trainee.

The design of a SISTS integrates the AI techniques as applied to an ITS and the simulation training techniques into the field of occupational safety and health. This system establishes, conducts and evaluates safety training in the areas of interest. The system is designed on a modular basis which categorizes the modules by the functions they serve. There are five major groups of modules in the conceptual design of the system as follow:

<u>Domain Knowledge Group</u>; This group provides the knowledge base necessary for intelligent functioning within the domain.

<u>Student Group</u>; This group determines, maintains and updates student status on a continuing basis as well as determining how well the student has performed.

<u>Simulation Group</u>; This group contains the simulation facts and objects and carry out those instructions accepted by the system.

<u>Interface Group</u>; This group provides the communications between the system and the student. It also provides the necessary domain information for the system to perform training of this domain.

<u>Control Group</u>; This group analyze the student's inputs and control the flow of the training and simulation. It passes the control to the proper modules and assure complete communication.

DETAILED DESIGN OF A SISTS

The detailed design of a SISTS includes the five major groups of modules from the conceptual design and an additional group which provides the mechanisms for entering trainee's input into the simulation. Also the modules in each major group are specified in the detailed design. The major six groups of modules and their functions are as follow:

1. <u>Interface Group</u>: The interface group provides the mechanisms for those modules, external to the Simulation/Tutoring process, to provide information for the Simulation/Tutoring process. This group consists of the Knowledge Acquisition module and the Discourse module. The Knowledge Acquisition module is to provide the training system with the information about a domain necessary for it to perform tutoring of this domain. The Discourse module is to handle the communications between the users and the system.

2. <u>Input Group</u>: The input group provides the mechanisms by

which the trainee's response to the simulation or the trainee's desire to modify the simulation are entered into the system. This group contains Translator, Input Filter, and Simulation. The Translator parses the trainee's input, that is directly related to the simulation, into a form the system can interpret. The Input Filter catches the trainee's inputs that are inconsistent with global simulation data and constraints. The Simulation is to provide facts and objects and carry out those instructions accepted by the system.

3. Domain Expertise Group: The Domain Expertise Group provides the knowledge base necessary for intelligent functioning within the domain. This group includes the Intelligent-Pre-Processor, the Domain Expert, and the Domain Expert Instructor. The Intelligent-Pre-Processor transforms raw simulation data into a high level representation (event-list). The Domain Expert is to generate a set of solutions that an expert would give based on an extracted set of events from the simulation. This module includes a basic understanding of the effects of occupational hazards, recognition of hazardous conditions, measurements of hazard level, recommended standards, and methods to control occupational hazards. The Domain Expert Instructor gives expert instruction with lesson plans, guidelines and all other expert information on what to teach in a particular domain.

4. Instruction Group: The Instruction Group determines and establishes the cause of the reason within the instructor's guidelines. It also maintains and updates trainee status on a continuing basis how well the trainee has performed. This group consists of Evaluator, Student Model, and Tutor. The Evaluator evaluates trainee performance. The Student Model serves the individual needs of the trainee. The Tutor conducts a lesson for a student in a manner that best suits that student's needs and preferred mode of instruction.. This includes the ability for the Tutor to modify tutoring behavior depending on the abilities of the trainee and the current mode of the system.

5. Control Group: The Control group establishes the results of trainee's inputs and controls the flow of the simulation and training. This group includes Interpreter, Control, and Inference Engine. The Interpreter is to establish the changes that are caused by the trainee's input in the event list of the intelligent pre-processor. It also checks continuously the criticality tags of the intelligent pre-processor events. The Control passes control to the proper modules at run time and assures complete communication. The Inference Engine activates and fires the rules in the modules and enters the rules into the system.

6. Simulator: The Simulator contains the simulation facts and objects and carries out those instructions accepted by the system. It illustrates graphically the application domain with interactive objects and to provide for intelligent

objects. The Simulator illustrates graphically the applications of occupational hazard control principles and techniques. It is a psuedo-real-time process running in parallel with the training system. The Simulator simulates hazardous working conditions. The trainee is asked to identify the hazardous condition, level of hazard, and make recommendations to improve the working condition. The trainee's response is compared with the expert solution generated by the Domain Expert module in order to evaluate the trainee's performance. The Simulator presents various scenarios to simulate different hazardous conditions. This automated training continues until the trainee's performance reaches a satisfactory level.

CONCLUSIONS

The primary goal of AI is to make computers smarter. Smarter computers can adapt themselves to their human users, instead of the other way around. In a complex world with complex problems, the problem solving ability of a smarter computer can be a great help. The occupational safety training on various safety areas is one of many representative complex problems. Unlimited opportunities for applying the latest AI and simulation techniques in the area of occupational safety and health training certainly exist.

REFERENCES

Banks, J and Carson, J.S., "Discrete Event System Simulation", Prentice-Hall, Englewood Cliffs, NJ, 1984

Gevarter, W.B., "Intelligent Machines", Prentice-Hall, Englewood Cliffs, NJ, 1985

Feigenbaum, E.A. and Feldman, J., "Computers and Thought", McGraw Hill, New York, NY, 1963

Hartley, J.R. and Sleeman, D.H., "Toward more Intelligent Teaching Systems", International Journal of Man-Machine Studies, 5, 1973

Lee, C.H., "Conceptual Design of a Simulation-Based Intelligent Safety Training System (SISTS)", Advances in Industrial Ergonomics and Safety I, Edited by A. Mital, Taylor & Francis, 1989

Wenger, E., "AI and the Communication of Knowledge: An Overview of Intelligent Teaching Systems", Unpublished manuscript, University of California, Information and Computer Science, Irvine, CA, 1985

Winston, P.H., "Artificial Intelligence", Addison-Wesley Publishing, Reading, MA, 1984, 2nd edition

Advances in Industrial Ergonomics and Safety II
Edited by Biman Das
Taylor & Francis 1990

AN OCCUPATIONAL HEALTH AND SAFETY DATABASE MONITORING SYSTEM FOR INDUSTRIAL SETTINGS

T. Gupta and A. M. Genaidy

Department of Industrial Engineering
Western Michigan University
Kalamazoo, MI 49008, U.S.A.

A microcomputer database system was developed in this study to monitor the occupational health and safety of employees in the workplace. The design aspects and applications of the system were discussed. Such a system should be an integral component of an ergonomic-based strategy in order to test the effectiveness of any implemented intervention in industrial settings.

INTRODUCTION

According to the National Institute for Occupational Safety and Health (1986), musculoskeletal injuries are one of the ten-leading work-related injuries in the United States. They account for 18% of occupational injuries. Thus, preventive measures have been developed over the years for the control of such injuries. Ergonomic job design is one such measure.

The essence of ergonomic job design is to maintain job requirements within the capability of the workforce so as to keep the risk of injury to a nominal level. If the employee exceeds his/her capability, he/she will be exposed to a higher risk of injury. It should be kept in mind that most of the knowledge available in the ergonomics literature regarding the quantification of the employee capability has been developed in university settings. Thus, the application of this knowledge to industrial settings should be used with some caution in order to adapt it to the working environment. This could be achieved by conducting epidemiological studies in order to monitor the results of the implementation of ergonomic-based interventions in the workplace.

With the advent of microcomputers, it is always possible to develop inexpensive database systems in order to monitor epidemiological measures in the workplace. However, upon observation of local industrial settings, it was found that injury and illness records were manually filed. Moreover, when management requires different reports about the performance of its plant in terms of health and safety scores, the occupational nurse uses different off-the-shelf software to generate the necessary reports. Similarly, the Occupational Safety and Health Agency (OSHA) mandates each plant to report their injury and illness cases according to their logs. Admittedly, it was determined that this process was very time consuming which required a significant amount of effort on the part of the occupational nurse.

This study was initiated for the development of a microcomputer-based occupational health and safety database system which performs the following functions: (1) the storage and retrieval of employees' injury and illness records, (2) the generation of necessary reports for management and OSHA purposes, and (3) to serve as an analytical tool for the monitoring and control of injuries and illnesses in industrial settings (e.g., testing the statistical significance of changes in injury and illness incidence rates in a plant on a month-to-month basis). It should be kept in mind that the development of an occupational health and safety computer software should satisfy the following criteria: (1) user friendliness and (2) low cost. These criteria are important so as to encourage its use by practitioners of occupational health and safety in small, medium and large size manufacturing and service industries.

DESIGN ASPECTS OF THE DATABASE MONITORING SYSTEM

A modular design approach was adopted with a hierarchical database concept. Figure 1 depicts the hierarchical database scheme used. Each employee in the organization who has a history of job-related injuries and illness will have an entry in the database. The record 'EMPLOYEE' includes data fields to store the necessary information about the employee. Each accident that results into some form of injury, illness or disorder defines an entry for a 'CASE' record. It may be noted that the EMPLOYEE-CASE relationship is one-to-many since more than one accident can occur with an employee. Based on the nature of injury, its severity and other parameters, each case is classified accordingly. The record type 'CLASSIFICATION' stores information about the type of the case and the CASE-CLASSIFICATION link is a one-to-one relation. The case classification information is stored as a separate record for the efficient generation of necessary reports. The record type 'PHYSICIAN' maintains information about the physician who treated the injury case. The CASE-PHYSICIAN link is also implemented as one-to-one relationship.

Each of the four record types consists of several data fields and has a designated key field whose value uniquely identify each entry in the database. Further, each record type can be indexed on any of its fields.

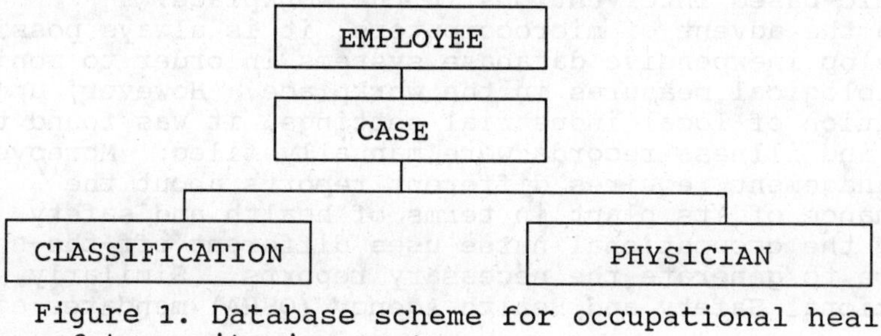

Figure 1. Database scheme for occupational health and safety monitoring system.

Figure 2 depicts the formats of each record type. The record type 'EMPLOYEE' consists of twentyone data fields as shown in figure 2a. The first field 'EMPNO' is used as a key since each entry of an 'EMPLOYEE' record has a unique four-digit value to this field. Figure 2b presents the format of record type 'CASE'. An entry of record type 'CASE' is accessed with its key field 'CASENO' and is associated with injury and illness related incidents. Necessary details of the incident are stored in various fields. Several entries of record type 'CASE' are expected for an 'EMPLOYEE' record. Figures 2c and 2d depict the formats of 'CASE-CLASSIFICATION' and 'PHYSICIAN' record types, respectively. Each 'CASE' record entry relates to one entry of 'CASE-CLASSIFICATION' and 'PHYSICIAN' records only. Figure 3 shows the number of data fields and keys for the four record types.

The database system is menu-driven and is designed to be user-friendly. Various database functions that are built in the system, are invoked by selecting appropriate options from the menus. Each record type is assigned a separate menu and the user can add, display or update entries of all record types in one session. A complete interactive session with the system is shown in figure 4.

It should be noted that DBase III plus was used to implement the above described design. Also, this database was interfaced with Lotus 1-2-3 in order to perform any necessary statistical analysis.

APPLICATION

The microcomputer-based system developed in this study could be used for a variety of applications. Among these are (1) the storage and retrieval of employees' injury and illness records, (2) generation of reports for management and OSHA purposes and (3) statistical analysis of injury scores.

Examples of typical reports requested by the medical department in manufacturing industry are shown in Tables 1 and 2. Table 1 depicts the incidence and severity of upper extremity cumulative trauma disorders in a manufacturing industry by age group over a six-year period (1984-1989). This table shows that the severity of injuries and illnesses was the highest among the elderly groups. The incidence and severity of upper extremity cumulative trauma disorders by anatomical site are given in table 2 for the six-year period. It is clear from this table that the wrist recorded the highest incidence and severity scores among the different parts of the upper extremity.

CONCLUSIONS

In this paper, a database occupational health and safety microcomputer monitoring system was outlined and the design aspects of this system were described. Some of the applications of this system were discussed. It is strongly believed that the microcomputer-based system would be of utmost importance to practitioners in order to monitor the health and safety of employees in the workplace. This system should be an integral component of ergonomics job design strategy to test the effectiveness of any implemented intervention.

EMP NO.	SS. NO.	FST NME	MI.	BDT	AGE	SEX	MAR RID	#OF DEP	DIV RCD	ADR ST.	CTY	ST	ZP

PH. NO.	JOB TIT	DPT	SFT	FRM	TOT SER	PRS SER

Figure 2a. EMPLOYEE record type.

CASE NO.	EMP NO.	ACC_ DATE	ACC_ TIME	REP_ DATE	PREP _BY	OFF_ POS	MED_ COST	WKR_ CMPN	DAYS_ REST	DAYS_ RESTR

Figure 2b. CASE record type.

EMP NO.	CASE NO.	INJ_ ILNS	EXT_ INJ	PLCE _ACC	TSK_ PRFM	ACT_ PRFM	NTRE _INJ	BODY_ PART

Figure 2c. CASE-CLASSIFICATION record type.

SS. NO.	CASE NO.	FRST NAME	MI	LAST NAME	ADDR _ST	CTY	ST	ZIP	PH_ NO.

Figure 2d. PHYSICIAN record type.

Record Type	No. of Fields	Key
EMPLOYEE	21	EMP_No.
CASE	11	CASE_No.
CASE-CLASSIFICATION	9	CASE_No.
PHYSICIAN	10	CASE_No.

Figure 3. Number of data fields and key fields for the record types designed for the system.

OCCUPATIONAL HEALTH AND SAFETY MONITORING SYSTEM
MAIN MENU
ENTER SELECTION ====>
1 - EMPLOYEE INFORMATION
2 - CASE INFORMATION
3 - CASE CLASSIFICATION INFORMATION
4 - PHYSICIAN INFORMATION
X - EXIT SYSTEM

Figure 4a. Main Menu.

OCCUPATIONAL HEALTH AND SAFETY MONITORING SYSTEM
EMPLOYEE INFORMATION MENU
ENTER SELECTION ====>
1 - ADD EMPLOYEE INFORMATION
2 - DISPLAY EMPLOYEE INFORMATION
3 - UPDATE EMPLOYEE INFORMATION
R - RETURN TO MAIN MENU
X - EXIT SYSTEM

Figure 4b. Employee Information Menu.

OCCUPATIONAL HEALTH AND SAFETY MONITORING SYSTEM
ADD EMPLOYEE INFORMATION

EMPLOYEE NUMBER: [ ]

SOCIAL SECURITY #: [ ]	**STREET ADDRESS :** [ ]
DATE OF BIRTH : [ ]	**CITY :** [ ]
FIRST NAME : [ ]	**STATE :** [ ]
MIDDLE INITIAL : [ ]	**ZIP :** [ ]
LAST NAME : [ ]	**PH. # :** [ ]
AGE : [ ]	**JOB TITLE :** [ ]
SEX : [ ]	**DEPARTMENT :** [ ]
MARRIED : [ ]	**SHIFT :** [ ]
# OF DEPENDENTS : [ ]	**FOREMAN :** [ ]
DIVORCED : [ ]	**TOTAL YRS. IN SERVICE :** [ ]
	YRS IN CURRENT SERVICE: [ ]

Figure 4c. Screen setup for Employee Information record.

REFERENCES

National Institute for Occupational Safety and Health, 1986, Proposed National Strategies for the Prevention of Leading Work-Related Diseases and Injuries. Part 1. DHHS (NIOSH), PB87-114740 (Cincinnati, OH: NIOSH).

Advances in Industrial Ergonomics and Safety II
Edited by Biman Das
Taylor & Francis 1990

SHEP (Safety, Health, and Environmental Protection) and Ergonomics in the Developing and the Developed World

C. Saran, Central Missouri State University, USA
A. Khaleque, University of Dhaka, Bangladesh
E. Unni Krishnan, Nat Inst. Indl Engg, India
M. Al-Haboubi, King Fahd University, Saudi Arabia
P.A. Koushki, King Saud University, Saudi Arabia
Chen Li, Hangzhou Univ, Peoples Republic of China

This paper explores the current status of SHEP and Ergonomics in various countries around the world. Researchers in Bangladesh, India, People's Republic of China, Saudi Arabia, and Singapore provided input to this paper. SHEP continuum is discussed. Present and future SHEP challenges are analyzed using multidisciplinary techniques. Solutions consider the technology available in various parts of the globe, national resources and priorities, and the perceived immediacy of the threats.

INTRODUCTION

Developing countries have abundant human resources while the developed countries do not have enough. This population distribution leads to the scarcity of natural resources in the developing countries and their relative abundance in the developed world on a per capita basis. Transfer of natural resources from the developing to the developed world complicates the challenges.

ERGONOMICS IN SOME COUNTRIES

Efficient, healthful, and environmentally safe utilization of natural and human resources, overcoming narrow regional/national interests, is a global challenge. Ergonomics attempts to develop the physical/manual and intellectual components of human resources. The role of ergonomics in humans endeavors and in the development of various countries is discussed. Anthropometric, psycho-social, economics, and lifestyle variations are analyzed with their pertinence to SHEP. Comments and views of experts in Ergonomics and SHEP from the developing countries were solicited and are included. Its role in the development of these countries is discussed.

Bangladesh

Ergonomics in Bangladesh is in its infancy. Bangladesh, has limited machine building capacity, few design skills and insufficient capital. She has to rely mainly on importation of technology from developed countries for her industrialization. Consequently, the scope for ergonomics in Bangladesh had so far been restricted to work environment and organizational design. However, the situation has started changing slowly and steadily. For example, Bangladesh has recently started technology assimilation and development of indigenous technology for her own use.

Education and Training:

Four of the seven universities teach Ergonomics through the departments of agriculture and mechanical engineering and psychology. Additionally, 4 engineering colleges and an institute of industrial arts offer courses in ergonomics. There are a number of government sponsored training institutes and training centers in Bangladesh, where practical training is given about improved work method and techniques, work organization and safety, design of work equipment and tools. Bangladesh Industrial Technical Assistance Centeris supported by the Government. There are also few private training centers, where practical training is given for the consumer product design, such as the design of furniture, pottery, garments, handicrafts, etc.

Working Environment:

The Government sets safety standards and enforces rules for accident prevention, noise, fire hazards, radiation and pollution control. Increasing industrialization and accompanying pollution have created localized environmental problem. Industrial effluent treatment processes are back-dated. The wastes are neither recycled nor hygienically treated prior to the disposal in the environment. Poor house keeping and less efficient industrial process produce more wastage and the level of industrial pollutional discharge is much higher than that of developed countries.

The Environmental Pollution Control Authority of Bangladesh tries to minimize the level of industrial pollution and ensure healthy working environment. This department conducts research to identify industries causing environmental pollution and contamination by discharging toxic chemicals, poisonous gases, harmful solid and liquid wastes. It sets effluent quality standard for Bangladesh and enforces strict measures against violations. It provides expert's advice to the management of polluting industries for introducing effective waste treatment system.

Ergonomic Research:

The ergonomic research activities in Bangladesh are limited for shortages of funds, qualified research personnel, necessary books and journals. In Bangladesh, the number of highly qualified ergonomists are very few. So it has not yet become possible to form any national ergonomic society of Bangladesh. However, some of the Bangladeshi ergonomists, are members of the South-East Asian Ergonomics Society and the Human Factors Society. They are conducting research on various ergonomic problems and presenting their works in different international seminars and publishing them in the national, regional and international journals of psychology and ergonomics. In Bangladesh, the medium of education at the tertiary level is Bengali but there is only one book of ergonomics in Bengali.

Some recent research activities have included: Causes and preventive strategies of industrial accidents as perceived by the managers and workers; and Effects of pollution on health and well-being of industrial workers in Bangladesh.

Recommendations for Future:

To improve productivity and SHEP, through increased ergonomic activities, the following recommendations are made:

1. Create awareness among the educationists, administrators and industrialists in Bangladesh about the importance of ergonomics for the national development.
2. Promote ergonomic education and training. Courses in ergonomics should be introduced in all the universities in Bangladesh and more well equipped training centers should be established. Ergonomic research facilities, such as books and journals, tools and materials should be increased.

3. Establish close cooperation with the relevant agencies and organizations in the developed countries for exchanging ergonomic knowledge.
4. Explore how Bangladesh can avail herself of modern technology and produce or modify it in homogeneous fashion and acceptable way for her own development.
5. Encourage collaborative research with ergonomists from other developing and developed countries.
6. Ergonomic considerations of technology transfer.
7. Satisfy the information need of the industrial planners of Bangladesh for choosing the right type of technology from the industrialized countries, and the training needs for operation and maintenance of imported technology.
8. Identify the problems of technology assimilation and development of indigenous technology in Bangladesh.
9. Provide information about anthropometric, social, cultural, technological and environmental conditions of Bangladesh to the ergonomists and designers of the industrially developed countries.

<u>India</u>

There are many researchers and trainers in India in SHEP and Ergonomics. Saran (1968) listed individuals and agencies actively involved in research and training. Many of the individuals listed have retired, organizations have shifted emphasis, and new individuals and organizations have come up.

National Institute for Training in Industrial Engineering at Bombay, with extension in Delhi, Hyderabad, Madras, and Muzaffarpur, is actively engaged in research and training in ergonomics and environmental engineering. With well equipped laboratories and well trained staff, some of their research and training activities include: Evaluation of job stress form ergonomic and environmental factors; Materials handling standards for Indian workers; Vehicular pollution; Ferrous industry accidents; Fire safety management; Productivity and comfort in gem and jewelry industry; Clothings for hot environment industries.

<u>Saudi Arabia (SA)</u>

SHEP and Ergonomics activities in SA, though limited, include: Analysis of accidents in SA; Anthropometry in SA; and Safety and health of Islamic pilgrims in SA (Gattan & Saran, 1984).

Socio-economic growth and development in the Kingdom of SA in the recent years have resulted in increased urbanization and use of advanced and complex technologies. Growth in the public's knowledge of preventing and coping with accompanying accidents has not kept pace with the problem. The need for improvement of safety consciousness, awareness, and education is evident.

Ergonomics has not been taken seriously in SA, yet. It has resulted in various degrees of human discomfort and inefficient systems. Imports designed for Westerners do not fit the locals. Products produced locally copy the dimensions of the imports. Multi-national user population of SA complicates the situation. Some examples of the neglect of ergonomic considerations in the use of imported technologies and comfort are: Imported seats forcing the user to slide down so that their feet can touch the ground. Displays and controls beyond user's reach; and shoes and clothings that do not fit the native Saudi's.

Singapore

Recent growth and development in Singapore has resulted in import of technologies and products designed for Westerners in a temperate climate. Many of these imports are not suited for local conditions. Some SHEP and ergonomics activities of Singapore include: Thermal comfort of sedentary Singaporean workers (Foo and Phoon, 1987); and work stress of firemen (Lim, et al, 1987).

People's Republic of China (PRC)

There are individuals and organizations involved in SHEP and Ergonomics in the PRC. They have just started a Chinese Ergonomics Society. Environmental Ergonomics sub-committee of the Society is housed at the Academy of Safety and Environmental Protection Research, National Ministry of Metallurgy, Wu Hau, Hubei.

SHEP INTERNATIONAL

Recent international accidents illustrate the futility of artificial compartmentalization of disciplines. Oil spills off the coast of Morocco and Alaska; Radiation leaks in Three Mile Island and Chernobyl; gas leaks in Bhopal; and transportation accidents involving hazardous substances are examples of this futility. Can any one of these events be considered an exclusively and safety concern? exclusively health problem? or exclusively environmental protection challenge? The answer is obvious.

New technologies and progress in communication have resulted in a global village where events in one part of the village affect other parts and the entire village. Applications of new technologies and efficient utilization of human resources across political and professional borders are provided. Radioactivity from Chernobyl could not be confined to the Soviet Union, hazards of Bhopal reached out and touched thousands beyond the factory walls, acid rain does not stop at the Canadian border, and AIDS has no borders. These SHEP failures involved unprecedented cooperation across political borders. Growing international SHEP threats and ergonomic challenges include: AIDS, accidental injuries and illnesses, acid rain, air pollution, deforestation, desertification, ergonomic disharmony, fisheries depletion, global greenhouse, hazardous substances/wastes, ozone concerns, population pressure, radiation perils, species extinction, stresses in the developed and the developing world society, and water diversion/pollution.

The current world population of 5.2 billion is expected to grow to 6.25 billion by the year 2000 and stabilize at 10 billion within a century (Popline, 1989). This growth will not be uniform throughout the globe resulting in areas of plenty and scarcity. Such distribution will affect national policies on population control, resource utilization, and waste control. The world may be reaching the stage of joint survival or joint extinction. Increased population in one part of the world affects the other parts.

U.S. manufacturers released 2.6 billion pounds of toxic pollutants into the air in 1987 (USA Today, 1989). Sixty percent of Americans live in cities that violate federal smog standards (USNWR, 1989). In India, the famous Taj Mahal is showing black and yellow spots. Yet the Government encourages industries polluting the air such as aerosol can manufacturers, iron foundries, and others. Acid rain is a common source of discontent between the mid west and the north eastern United States and between United States and Canada. Just recently a storage tank leak from a Ciba-Geigy plant in Switzerland drifted over the Swiss-French border (San Juan Star, 1989). Smog and sand in Beijing force people to cover their noses as a routine. Chlorofluorocarbons (CFC) are major contributors to the Ozone layer depletion. Ozone hole over the Antarctica is growing. Leaders of seven most prosperous democracies called for an end to the production of CFCs. Ozone layer filters the harmful ultraviolet light from the sun. Its depletion will result in increased skin cancer.

Greenhouse effects, caused in large part by CO_2, will hasten the global warming. CO_2 is generated by the burning of fossil fuel. Automobile exhaust and industrial emissions are major culprits and deforestation also contributes. Global warming would damage the forest and green the present icy regions of the world.

Deforestation in Brazil, Madagaskar, India and in other countries is progressing due to population pressure and the thirst of the world for natural resources. Balancing the conflicting interests of agriculture, industry, and conservation is a dilemma for most of the world. Brazil is being criticized by the developed world, mainly the United States. But the developed world achieved this status by doing just what Brazil is being condemned for: ruthless exploitation of natural resources including cutting down the forest. The developed world continues to be the major degrader of the planet. Its cars and factories pump hundreds of millions of tons of chemicals into the air, contributing to greenhouse effects and acid rain (Time, 89).

Land is used for a variety of activities such as agriculture, housing, forestry, industry, transportation, and waste disposal. Many of these activities are mutually exclusive. It will be difficult to live or work on a road or in a forest or on a waste disposal site. Illegal dumping and unwise use of land make it unfit for other uses and harmful to the users. Desertification may be caused by excessive agricultural irrigation, diversion of water, and other activities. The developed world is running out of land to dump its waste. It is exploring the outer space and the developing countries to dump it. The world may be drowning in its own waste. Some examples of this land abuse include: illegal dumping of toxic waste, paper from the offices, disposable diapers, etc. It is estimated that paper products occupy about 60% of the landfills.

Leading magazines cover stories with headlines such as "Our polluted ocean - Don't go near the water", "Our filthy seas", and "Are USA's beaches safe?" indicate the national concern with water pollution (Newsweek, August 1, 1988; Time Aug 1, 1988: and USA Today May 2, 1989). River, lakes, and oceans are full of toxic and non-toxic waste products.

"Would you drown in Mississippi river near St. Louis, if you fall into the river and do not know how to swim?" Answer: "No, you would dissolve before you drown." This comical exchange illustrates a grim tale. Name of the river and the city can be changed for any other name in the world and the story will still be the same. Closed beaches and dead aquatic life testifies to the grim reality. So does the litter in water. This pollution of our waters may not all be blamed upon industrial and household effluent discharge. Much of it comes from non-point sources of pollution such as water run-off from the land, agricultural chemicals and others. Water diversion for agribusiness and industries by building dams threatens ecological balance, innundation of productive agricultural land, and overused aquifers.

In the United States about 7 billion pounds of toxics were spewed into the environment by industry and about 3.3 billion pounds of toxic wastes were hauled off-site for public or private treatment. Irrigation has concentrated heavy metals in the Beijing-Tianjin basin and selenium at California wildlife refuge. Cleaner technology ended mercury poisoning at Minamata, Japan.

Toxic waste continues to be dumped on the Third World. Many unsuspecting communities in Puerto Rico are victims of illegal dumping. Accepting toxic waste is common for many communities around the world to spur economic growth. Bhopals and Exxon Valdezs are waiting to happen almost everywhere around the world.

Increased toxic waste, deforestation for development and agriculture, diversion of water, and selective breeding of plants and animals have led to the extinction of many species around the world. Many other species are already extinct. Endless appetite of humans to control and plunder the world around them contributes to the tragedy.

SAVING THE WORLD

There are some positive signals to save the world from destructing itself or drowning in its own waste. Population control measures are in progress in China, India and other countries. Costa Rica and Brazil are taking some steps to reduce deforestation. United States and other countries are acting to curb their toxic waste and garbage disposal sites by cleaning up the toxic sites, developing environmentally safe technology, conservation of resources and recycling. Increased cooperation between the developed and the developing world, between various political and religious persuasions is essential for human survival. Reduced military spending and burden sharing with the third world will be helpful.

AS THE WORLD CHANGES: FUTURE TRENDS

The world has been changing faster than ever before. We have just entered the "Chip" revolution. Its impact on SHEP, Ergonomics and other human endeavors will be profound. Changes in the environment, population, and lifestyles will offer new challenges and opportunities. The following broad trends are noticeable:

1. Population Centers: More self-contained, compact, and evenly distributed (geographically) population centers. More medium-sized cities and fewer megalopolises. Modified kibbutzes or communes.

2. Land and Water: Better management, and distribution of land and water among conflicting demands of industries, services, agriculture, recreation, wildlife, and transportation. Each will have to learn to maximize output with limited resource input.

3. Buildings: Multipurpose buildings for better utilization of time and space. Same space could be used for various activities such as work, living, recreation, etc., at different times. Some examples of the various uses of the same building could be: home businesses, dormitories used as motels in summer, self-contained apartment complexes, berths, remodeling and renovation of existing structures, etc.

4. Mining: Automated underground mining with planning to reutilize space created without disturbing the top soil. More deep sea mining and exploration of areas where the environment is not hospitable to humans.

5. Resources: Better management and control of non-renewable resources. Research and development of alternate energy resources such as solar, geothermal, synthetics, wind, and hydroelectric resources. More conservation and recycling of resources and products.

6. Food and Agriculture: More processed and synthetic foods and substitutes. Fresh food will be more regional in flavor requiring minimum transportation of fresh produce. Freshly squeezed orange juice will be more common in Florida while its evaporated and condensed forms will be more popular in other regions. Larger automated farms, with processing facilities, less dependent upon chemicals but more on hybrids and mutants.

7. Transportation: More mass transportation and transit systems for longer distances and individual transportation for shorter distances. More reliance on solar aviation, electrical railroads and buses, pedal power, and walking.

8. Biotechnology: Rejuvenation and strengthening of existing body organs, through transplants, implants, medication, exercise, and genetic engineering. Development of hardier crops, humans, and animals through genetic engineering. Simpler diagnostic and surgical procedures through biomedical engineering.

9. The "Chip" Revolution: Its impact will be more profound than that of either the agricultural or industrial revolutions. Its impact on mass communication and manufacturing is already evident through the use of telephones, videos, microcomputers, digital watches, and robots.

10. Services: To maintain and repair the equipment, facilities, and systems more specialized technicians will be needed. More leisure time and change in lifestyles will encourage growth and development of innovative recreation, education, and human interaction.

CONCLUSION

Increasing world population, aspirations of the developing countries to develop industrially, and the developed world's interest in maintaining its opulent lifestyles have created intense pressure on limited natural resources. SHEP and Ergonomic challenges, previously limited to the industrialized world, are being experienced by the rest of the world. As more and more people experience the limitations of natural resources, conservation and efficient utilization of human and natural resources gains significance . SHEP and Ergonomics do not honor disciplinary or geo-political boundaries. Unprecedented cooperation of individuals and agencies across disciplinary and geo-political borders is evident.

REFERENCES

Boraico, Allen A. Storing up trouble...Hazardous waste. National Geographic, Mards 1985, pp 318-351

Galtan, Khaled M. and Saran, C. Safety and health of Islamic pilgrims in Saudi Arabia. Trends in Ergonomics/Human Factors I, edited by A. Mital, (Elsevier Science Publishers - North Holland), 1984, pp 255-260

Foo, S.C. and Phoon, W. O. The thermal conformity of sedentary workers. Asia - Pacific J of Public Health, Vol 1, No 2, 1987, pp 74-77

Lim, C.S., Ong, L.S., and Phoon, W.O. Work stress of firemen as measured by heart rate and catecholamine. J Hum Ergol., 16:1987, pp 209-218

National Geographic. Earth '88. National Geographic, Special Issue, December 1988, pp 766-945.

Fornos, Werner. Newsletter, The Population Institute, September 25, 1989

Saran, C. The role of human factors in rural India. Human Factors, 10(6), 1968, pp 613-624

San Juan Star. Acid cloud menaces Swiss. San Juan Star, December 23, 1989

Time. Planet of the year: Endangered earth. Time, Special issue, January 2, 1989

Time. Environment - What can Americans do? Time, September 18, 1989

U.S. News and World Report. Newest health hazard -breathing. U.S. News and World Report, June 12, 1989

USA Today. Air treated like an open sewer. USA Today, July 31, 1989

Advances in Industrial Ergonomics and Safety II
Edited by Biman Das
Taylor & Francis 1990

SAFETY AND HEALTH HAZARDS OF MICROCOMPUTERS

Larry T. Ferguson

Graduate Student in Industrial Hygiene
Central Missouri State University in St. Louis

INTRODUCTION

Microcomputers, or video display terminals (VDTs), have completely revolutionized our lives. However, working in a modern electronic office can be hazardous to one's health. Several occupational hazards associated with the use of microcomputers have come to light recently due to their widespread use in offices and homes. The hazards range from the controversial health effects of electromagnetic fields of VDTs to chronic eye problems, repetitive strain injuries, printer noise, and health problems due to physical inactivity and stress associated with prolonged work on microcomputers. As Toufexis (1989) pointed out in a recent article in Time magazine, fears were further heightened in June 1989 when The New Yorker magazine published a series on "The Hazards of Electromagnetic Fields." Toufexis observed that author Paul Brodeur charged utility companies and public health officials with trying to gloss over the health threat posed by power lines and computer terminals. Brodeur noted that about 30 million VDTs are in use in offices, homes, and schools across the U.S. The concerns may very well be justified.

THE SAFETY AND HEALTH CONCERNS

Electromagnetic radiation from VDTs

There is considerable controversy over whether or not microcomputers pose an electromagnetic radiation hazard with long term exposure. According to Jokela et al (1989), VDTs produce electromagnetic emissions mostly at the extremely low frequency (ELF) range (30-300 Hz) or at frequencies from 10 to 500 kHz, which belong to the radiofrequency (RF) range. They explained that magnetic (B) field emissions at the ELF range originate from electric currents in AC power supply circuits (50/60 Hz) and vertical deflection coils (40-60 Hz) of cathode ray tubes (CRTs) used in VDT displays. They noted that at the RF range the fields are mostly due to currents in horizontal deflection coils of CRTs. They further noted that ELF and RF electric (E) fields are generated predominantly by AC power supplies and the high pulsing voltages (typically 15-20 kV) of flyback transformers in CRTs. Concerns about the possible health effects of VDT emissions began over a decade ago. However, the research literature indicates little progress has been made in determining the ultimate effects of long term exposure.

Brodeur (1989) wrote that in February 1977, Vordie H. Parr, chief of the Physical Agents Effects Branch of the National Institute for Occupational Safety and Health (NIOSH), and some coworkers measured the ultraviolet, visible-light, infrared, microwave, and RF radiation from several VDTs. Brodeur pointed out that when questions about their instruments were raised, an arbitrator elected a private consulting firm to conduct additional measurements. The firm found the RF power of these VDTs to be within the permissible limits of the U.S. Standards Institute. The arbitrator decided that VDTs do not pose an ocular radiant energy hazard and that most employees may work safely with them.

According to Brodeur (1989), Karel Marha, an emigre from Czechoslovakia, where for twenty years he was the director of the Department of High Frequency at the Institute of Industrial Hygiene and Occupational Diseases in Prague, and his colleagues had found a variety of neurological problems, such as headache, eye pain, excessive fatigue, forgetfulness, and irritability, among people working in various electronic industries near electromagnetic emissions. They found some of these effects were induced at power densities as low as 0.1 mw per cm^2, or 1/100 of the recommended American guideline for occupational exposure. Brodeur further elaborated that Marha noted very few measurements with proper instruments had been made of VDT low-frequency radiation, but a study in the U.S. found E fields within 30.5 cm of a VDT more than five times as strong as those allowed in the Soviet Union, Czechoslovakia, Bulgaria, and the German Democratic Republic. Brodeur went on to tell how Marha and his colleagues also pointed out that pulsed E and B fields from VDTs were highly directional, depending on machine design, and therefore the exposure of other people near the VDT can be higher than operator exposure. Brodeur also noted Marha's claim that radiation measurements should be made not just in front of the viewing screen, as most measurements had been made until then, but at 10 to 30 cm from all surfaces of the machine, as well.

Much of the controversy has centered around whether or not VDT radiation exposure causes miscarriages in women. A recent epidemiological case-control study reported by Bryant and Love (1989) indicated that a link between VDT exposure and miscarriage is unlikely. Another issue is that of whether there is a link between VDT radiation exposure and cancer. The problem that has limited past epidemiologic studies of cancer and exposure to ELF E and B fields has been the lack of adequate methods and standards for measuring and assessing personal exposure to these fields. Deadman et al (1988) noted that it was first suggested in 1979 that weak levels of 60 Hz B fields may be linked with an increased risk of childhood cancer, particularly leukemia, and evidence of possible teratogenic effects have spurred further research into the biological consequences of ELF fields. Since 1979 at least 18 epidemiologic studies of the link between cancer and exposure to these fields have been conducted, according to Deadman et al (1988). They cautioned that, while most of the studies reported some association between ELF fields and cancer, the data are still inconclusive because of several methodological problems.

They explained that assessment of exposure has been difficult because of the lack of suitable equipment for monitoring exposures to E fields and the absence of a dosimeter for B fields. Exposures have been estimated indirectly, rather than actually measured. ELF electromagnetic fields interact with normal physiological mechanisms by inducing electric potentials on cell membranes. Several biochemical and neurophysiological indicators may be affected, white blood cell populations can be altered, and behavioral responses have been detected. Changes usually do not exceed reasonable physiological variations. They agreed that epidemiologic studies suggest that 60 Hz fields may be promoters of carcinogenesis. They also observed that, while it generally is agreed that 60 Hz E and B fields can affect biological processes, it is still unclear how levels considered deleterious in vitro and in nonhuman studies can be extrapolated to safe levels for humans.

An advance in measurement methodology occurred when Deadman et al (1988) tested a new 60 Hz electromagnetic field dosimeter. Their field test was carried out to assess occupational and residential exposures in a study of a group of electrical utility workers and a comparison background group over a 7-day period. In the results of this study, comparing work periods only, utility workers' exposures were significantly higher than background levels by a factor of about 10 for E and B fields and by a factor of 171 for high-frequency transient E (HFTE) fields. The study showed that when overall weekly time-weighted averages combining work and nonwork exposures were compared, ratios of the exposed to background groups were lower. Additionally, E field exposures of the background group were highest during the nonwork period, probably reflecting the use of electrical appliances at home.

Deadman et al (1988) claimed that the extent and variability of exposures to ELF E and B fields and to HFTE fields associated with a specific job, tasks within the job, and domestic activities can be quantified easily using the dosimeter described in their report. They explained that detailed job exposure matrices can be developed from this information allowing epidemiologic studies of health outcome and exposure to electromagnetic fields to define possible associations. Use of a standardized method of exposure assessment, based on a dosimeter such as the one Deadman et al tested, also would permit meaningful comparison of exposures, within any sector where exposure to ELF and HFTE fields are of concern, including VDTs.

The comprehensive study by Jokela et al (1989) referenced earlier has contributed significantly to the measurement of electromagnetic fields of VDTs in the frequency range of 30 Hz to 1 MHz. They contended that the approach of directly measuring currents induced in the body of the operator and estimating the current densities from field strength measurements is an advance toward true assessment of the exposure. However, they pointed out that standardization of measurements is still clearly needed.

They noted that the National Radiological Protection Board in the U.K. and the Deutsche Electrotechnische Commission in the Federal Republic of Germany have published standard proposals (NRPB 1986; DIN VDE 1986) which cover the whole electromagnetic spectrum from DC to microwaves. They also alluded to an older standard published by the American Conference of Governmental Industrial Hygienists (ACGIH 1984) and beginning at 10 kHz. They asserted that although these standards are based on gross biological effects of intense fields such as heating tissue and stimulation of nerve and muscle cells, which are unlikely with VDTs, they can be used to assess the relative biological significance of a specific field quantity and spectral range. Their purpose in the study was to develop procedures and measure consistently the significant parameters of the E and B fields emitted by VDTs. They concluded that, due to the lack of conclusive data on biological effects of weak electromagnetic fields below 100 kHz, it is still not possible to evaluate actual health risks definitively based on their results. However, they claimed experience gained during their study could be used to develop guidelines for the standardization of emission measurements. They determined that spectral measurements are now possible for future use when broadband exposure standards become available giving the required weighting function for different spectral ranges. Jokela et al (1989) also concluded that an internationally accepted broadband exposure standard is needed to enable more consistent occupational exposure assessments for pulsed fields.

Engineering controls for protection from electromagnetic fields of VDTs could include more CRT and power circuit shielding in microcomputer equipment design and manufacture. IBM and other companies have already done this in Scandinavia, with VDTs that are shielded against VLF, but not ELF, B field emissions to comply with Sweden's guideline. A Norwegian firm is marketing VLF-shielded VDTs in the U.S. One company has introduced a flat liquid-crystal display (LCD) terminal that does not emit VLF fields at all and has been shielded against ELF fields with a special metal. Engineering advances in flat-panel display technology may eventually eliminate most of the electromagnetic radiation hazards of VDTs. Most common VDTs use a CRT display, and flat-panel LCDs are still in the developmental stage. These displays are not expected to seriously rival the CRT until the mid-1990s, according to Tannas (1989). He noted that the popularity of the CRT display is primarily due to its low cost. However, he claimed that much progress has been made in LCD performance and color. The problem is the high cost of LCDs. A few color flat-panel LCDs are available now, but when compared to CRTs, are not large enough for all applications and cost many times more than CRTs. However, further technical advances will let LCDs compete directly with CRTs. Tannas (1989) observed that those in the industry often compare the present state of LCD technology to that for dynamic random-access memory manufacturing in the mid-1970s. He explained that emissive displays are more expensive because they require high-voltage drivers, whereas LCDs use low-voltage drivers, which are less expensive. LCDs also are not as hazardous as CRTs, for these same reasons.

Repetitive strain injuries

Two of the most debilitating disorders caused by prolonged use of microcomputers are carpal tunnel syndrome (CTS), the compression of nerves and blood vessels in the wrist that causes numbness and pain, and myofascial pain syndrome (MPS), a form of soft tissue rheumatism that affects certain muscles and ligaments in the neck, shoulder, back and arms, as discussed by Wilkinson (1989). Trov (1988) reported that anyone who spends long hours typing is susceptible to carpal tunnel syndrome. The illness afflicts at least 23,000 Americans yearly. Many writers, computer programmers, and data entry workers are plagued by the condition, which can ultimately result in a permanent loss of sensation and motor control in one or both hands. Trov attributed the syndrome partially to poorly designed work areas and bad posture, but asserted the main problem is that computer keyboards allow workers to type faster and more continuously than ever before. He also observed that companies that have experienced outbreaks of the condition, such as the Los Angeles Times, where some 90 reporters were afflicted, are taking preventive measures by redesigning workstations and developing exercise programs.

A former staff writer and researcher for the Los Angeles Times, Wilkinson (1989) reportedly averaged four to six hours a day working on a computer and began noticing numbness, soreness and a tingling sensation in her hands and wrists. The discomfort became intermittent, so she ignored these early warning signs. A year after starting the job she could not hold a glass of water, turn a doorknob, or raise her arms high enough to comb her hair. Her condition was finally diagnosed as MPS, after an incorrect diagnosis of CTS.

According to Dr. Jeanne Axler, a chronic-pain specialist, the population most at risk for CTS is usually working or living under stress and the higher the stress level, the more likely the injury. The Nine to Five National Association of Working Women estimated that employees who spend 20 or more hours a week on a microcomputer without taking a rest break and those in highly stressful positions, such as customer service, airline reservations, or classified advertising, are most prone to these VDT illnesses. A NIOSH study, mentioned by Wilkinson, showed that CTS occurs from 3 to 10 times more often among women than men. Only 9 states have guidelines to protect VDT workers. Since most states do not code workers' compensation statistics for disorders such as MPS, the incidences are hard to track.

Wilkinson advised those employed in computer jobs to heed these warning signs: sensations of numbness and/or tingling in fingers or hands; loss of hand function or difficulty grasping large objects, or the dropping of small ones; swelling in the forearm; and sharp, shooting pains down shoulders to hands. The following measures can reduce the risk of developing computer related repetitive strain injuries: limiting work at a VDT to less than 4 to 5 hours without a break; taking short breaks, every 1 to 2 hours, by standing, stretching, walking, looking at distant objects and varying work activities. If someone else uses one's VDT, the user should readjust the equipment to suit their needs before resuming work.

Computer workers should have a chair that supports their upper back and lower spine and allows prolonged sitting without muscular or skeletal stress; ensure easy access to work materials by storing them in files and utility trays that are on eye level and within arm's reach.

Chronic eye problems

Man absorbs about 80% of all information through his eyes (Dederichs, 1986). It is imperative to have the correct level of illumination, contrast, proper positioning and protection against troublesome glare. For inadequate illumination of documents, extra lights may be installed. They should have anti-dazzle shields and be positioned so as to avoid troublesome reflections. According to Bennett et al (1984), the three most often cited causes of vision problems are improper illumination, glare on the VDT screen, and improper contrasts in luminance. Even with reasonable illumination levels, vision problems may persist if glare and contrast problems are not controlled. Some sources recommend a special lens coating or an add-on anti-glare filter screen to reduce glare and ultraviolet radiation from VDTs. Recently introduced flat tension CRTs as well as some newer LCDs virtually eliminate glare altogether. It is also suggested (Attaran and Haut, 1989) that frequent VDT users get eye examinations by an optometrist or ophthalmologist at least every two years.

According to Rothenberg et al (1989), Amalia Miranda, director of the Low Vision Clinic at the Dean A. McGee Eye Institute, claimed that people who work with computers spend about eight hours a day looking at tiny, moving objects and constantly concentrating. They rarely blink and constantly follow a line with faint letters which is very stressful to the eye. Miranda suggested the following tips for computer operators to avoid chronic eye problems: blink frequently to lubricate the eyes and reduce irritation; sit up straight and try not to slouch when working at a VDT; take regular breaks from the computer at least once hourly; glance away from the screen frequently, looking around the room to give the eyes a break; and bifocal wearers may need to wear special glasses for VDT distance.

Computer noise problems

Constant exposure to excessive noise over an extended time may result in impaired hearing. In some office environments, particularly in large open offices where many VDTs are in use, the noise level can be quite high. The recommendations on noise levels prescribe maximum noise levels of 70 dBA for normal work activities and 55 dBA for mental work (Dederichs, 1986). VDT work is predominantly mental. Noise can impede worker concentration when performing mental tasks and, in extreme conditions, can cause serious health problems (Attaran and Haut, 1989).

Attaran and Haut (1989) asserted that direct printer noise can be minimized by placing hoods (printer mufflers) over impact printers. Quieter printer technology has also become available, such as thermal or ribbon transfer and that used by modern laser printers.

Reflected noise can be reduced by installing sound-absorbent paneling around computer workstations; through the use of acoustical ceiling tiles and plush carpeting; and by ensuring that windows are a good buffer between the office and the outside world.

Physical inactivity and stress

The lack of physical activity in association with VDT work and stress contribute to a myriad of health problems. Maintaining the same posture continuously causes static loads on the muscular system and therefore leads to poor circulation of the blood (Dederichs, 1986). Premature fatigue and possible cramps result. Sedentary VDT work increases the risk of coronary heart disease, as does stress. Bennett et al (1984) pointed out that one of the greatest health concerns is with the high levels of job stress shown in some computer workers. It reflects high job dissatisfaction, feelings of work pressure (deadlines), and fear of job loss. Such stress factors have been linked to serious chronic diseases, such as coronary heart disease, stomach and intestinal disorders, emotional distress, endocrine imbalances, and reduced resistance to disease. Obviously, it is very important that computer users maintain a regular off-work aerobic exercise regimen and learn to use effective stress control techniques to maintain good health.

CONCLUSION

Much progress in VDT safety and health research is evident in the literature, but more field investigations and epidemiological studies, using new measuring techniques, are needed. Large scale, long term epidemiological studies to examine the chronic health effects of microcomputer work are still needed. If chronic health effects are not evident, employers and manufacturers do not wish to invest much money in environmental and workstation design, so they must have proof. Since more is known now about visual pathology, musculo-skeletal diseases, and stress-related illnesses, the studies should concentrate on exact definition of the long term chronic health effects of VDT ELF and VLF radiation. As Bennett et al emphasized in 1984, field studies of large VDT user populations and controls using appropriate scientific methods and analytic approaches are still needed. Such studies require the cooperation of industry, labor, educational institutions, and government.

There is also a need to educate manufacturers, office managers and workers, and the public with facts about the potential hazards of microcomputers and methods to alleviate them. VDT worker training programs should emphasize work patterns and habits which will prevent many safety and health problems associated with extensive use of microcomputers. Management initiatives should include establishing work schedules that alternate VDT work with other tasks that take the worker away from the machines for a time. This technology is here to stay and the full magnitude of the safety and health problems resulting from prolonged VDT use may take many years to realize. We cannot afford to wait any longer before we start controlling the safety and health hazards of microcomputers.

REFERENCES

Attaran, M. and Haut, J., Using ergonomic principles to improve worker productivity. Information Executive, 1989, 2, (3), 20-25.

Bennett, J., Case, D., Sandelin, J., and Smith, M. (Eds.) Visual display terminals: Usability issues and health concerns. New Jersey: Prentice-Hall Inc., 1984.

Brodeur, P. Annals of radiation: The hazards of electromagnetic fields. The New Yorker, 1989, (65), 39-68.

Bryant, H. E., and Love, E. J. Video display terminal use and spontaneous abortion risk. International Journal of Epidemiology, 1989, 18, (1), 132-139.

Deadman, J. E., Camus M., Armstrong, B. G., Heroux, P., Cyr D., Plante, M., and Theriault, G. Occupational and residential 60-Hz electromagnetic fields and high-frequency electric transients: Exposure assessment using a new dosimeter. American Industrial Hygiene Association Journal, 1988, 49, (8), 409-419.

Dederichs, E. Visual display unit workplaces: Emerging trends and problems. Dublin: The European Foundation for the Improvement of Living and Working Conditions, 1986.

Jokela, K., Aaltonen, J., and Lukkarinen, A. Measurements of electromagnetic emissions from video display terminals at the frequency range from 30Hz and 1MHz. Health Physics, 1989, 57, (1), 79-88.

Rothenberg, R. S. and Barrett, W. M. (Eds.) How dangerous are VDT's? USA Today, 1989, 118, (2531), 5.

Tannas, L. E. Jr. Flat-panel displays displace large, heavy, power-hungry CRTs. IEEE Spectrum, 1989, 18, (9235), 34-45.

Toufexis, A. Panic over power lines. Time, 1989, 134, (3), 71.

Trov, S. Warning: Your keyboard could be a crippler (carpal tunnel syndrome). Business Week, 1988, (3073), 156.

Wilkinson, D. M. Computer dis-ease. Essence, 1989, 19, (11), 14-15.

Advances in Industrial Ergonomics and Safety II
Edited by Biman Das
Taylor & Francis 1990

Are Hearing Protector NRRs An Accurate Representation Of Actual Noise Attenuation?

Dan D. Stephens

Central Missouri State University
Warrensburg, Missouri

Hearing protectors are assigned Noise Reduction Ratings (NRRs) according to their noise attenuating properties. This article concerns itself with the validity of the NRRs in relationship to the actual attenuating properties of the hearing protectors. The human factors dealing with proper donning instructions and use are the most important factors to control for maximum noise attenuation. A small sample of instructors and managers in small wood shop settings has been interviewed as to the common practices in using HPDs in their facilities to determine if maximum attenuation is achieved.

INTRODUCTION

Excessive exposure to harmful levels of occupational noise is a prevalent hazard American workers face. It has been estimated that from 2.9 million to 3.4 million workers in America are exposed to noise levels in excess of 90 dBA. An additional 5.2 million to 8.9 million workers are exposed to levels between 85-90 dBA (Caple, 1989). Industrial management has the responsibility of dealing with these noise levels to produce a noise-safe work environment. The most preferable method of dealing with these noise levels is to engineer the noise out of the environment. However, there are limitations as to the feasibility of engineering controls. The next choice in dealing with occupational noise is through the use of administrative controls, but still there are limitations. The last alternative in dealing with occupational noise is with the use of hearing protectors. Hearing protectors are required by the EPA to carry a Noise Reduction Rating (NRR). There are several factors that affect the accuracy of the NRR of hearing protectors while they are in use. This article will address the question of the validity of the NRR while in use in an industrial setting.

EFFECTS OF OCCUPATIONAL NOISE EXPOSURE

Exposure to high levels of occupational noise can cause mental, physical, and auditory fatigue, decreased hearing in presence of noise, increased accident and sick-leave rates, and decreased worker satisfaction with his work environment (Gasaway, 1985). Safety And Health magazine (1987) reports that prolonged exposure to hazardous noise levels may cause headaches, fatigue, dizziness and tinnitus. Besides the apparent immediate effects is the more serious cumulative effect of overexposure, noise induced hearing loss.

Occupational noise has been classified into three groups. OSHA has determined acceptable exposures to sound levels in each class. The first classification is continuous noise. Continuous noise is defined as "broadband noise of approximately constant level and spectrum, to which an employee is exposed for a period of eight hours per day, 40 hours per week" (National Safety Council, 1988).

Permissible exposure limits (PEL) have been established for the number of hours per day an employee may be exposed at various levels (see table 1). A dose of 100% is considered to be a constant eight hour exposure to a 90 dBA source or its equivalent. Equivalent doses may be calculated by using the 5 dBA doubling rule: For every 5 dBA increase, exposure time is halved; for every 5 dBA decrease, exposure time is doubled.

Table 1. PERMISSIBLE NOISE EXPOSURES

Hours Per Day	Sound Level Exposure (dBA)
8	90
4	95
2	100
1	105
.5	110
.25	115

Intermittent noise is defined as "exposure to a given broadband sound-pressure level several times a day" (National Safety Council, 1988). It is a noise that is intermittently introduced into the work environment throughout the work day. Figure 2 represents the criteria established for intermittent noise.

Table 2. ACCEPTABLE EXPOSURE TO NOISE IN dBA AS FUNCTION OF OCCURRENCES PER DAY

HRS	MIN	1	3	7	15	35	75	160 up
8		90	90	90	90	90	90	90
6		91	93	96	98	97	95	94
4		92	95	99	102	104	102	94
2		95	99	102	106	109	114	
1		98	103	107	110	115		
	30	101	106	110	115			
	15	105	110	115				
	8	109	115					
	4	113						

Impact-type noises are sharp bursts of sound that occur less than once per second and have a duration of less than one half second. Examples of impact-type noises are explosions and hammer blows.

Excessive exposure to any one of the these types of occupational noises has the potential to cause threshold shifts in hearing levels. The precise kind and severity of effect is dependent upon:

1. sound level and exposure
2. frequency distribution of the sound
3. duration of exposure
4. length of rest periods between exposure
5. individual differences in tolerances of sound
6. type of sound exposed to

Susceptibility to noise induced hearing loss varies a great deal from one individual to another. The use of hearing protectors may be one way to reduce the level of exposure.

CALCULATING NRR VALUES

The NRR of a hearing protector is a value of the sound attenuation properties of that specific device. The Environmental Protection Agency (EPA) requires that all hearing protector or hearing protector packages be labeled with their NRR values.

G. L. Cluff (1986) states that the NRR is a single-number value from which a spectrum-sensitive estimate of the protected ear A-weighted noise level may be obtained. It is a derivative of Bottsford's sound level conversion procedure and yields an estimated A' within 0.5 dB of the NIOSH Method #2. It is based on the assumption that A' = dBC - NRR. Therefore, C-weighted noise measures are required in order to use it most effectively.

PROTECTOR DONNING INSTRUCTION

Typically hearing protectors are supplied with instructions for use, but without a complete understanding of these instructions and the ability to follow them correctly, the noise attenuation properties will be greatly diminished. Proper fitting according to manufacturer's specifications is critical to good noise attenuation. Workers need to be trained in the capabilities, use and care of the device being used and instructed on how to properly don the device.

Detailed instruction for proper use and care of the particular type of protector being used needs to be provided for each employee. This detailed instruction should consist of written instructions and pictorial representations of each step necessary to properly don the device (Gasali, Epps, 1986). Also, demonstration and observation by a qualified instructor and the opportunity to ask questions and get feedback from the instructor are beneficial in achieving the maximum noise attenuation properties of the protector.

HUMAN FACTORS

The average NRR on devices sold in North America today is between 22 and 24 dB. This number overstates the protection that most buyers can expect their employees to receive (Berger, 1983). The biggest factors involved in the efficiency of hearing protectors are the human factors. Every person's head and ears are physically different in shape and size. The employee must be fitted with the appropriate kind and size of protector. Some people cannot wear plugs for physical or psychological reasons. Others may not be able to wear muffs. Sometimes the operation or movement of the worker may eliminate some types of protectors.

Proper use and insertion plays a major role in achieving the maximum efficiency of the protector. Employees must never be allowed to modify protective devices, like drilling holes in muffs to improve ventilation. Protectors that are not used don't provide any protection, so it is necessary that management strictly enforce all hearing conservation policies. When hearing protectors are used properly, the NRR reflects the actual noise attenuation properties of the device. However, because of the individual human factors involved, there are many variables in the efficiency of hearing protective devices used by different individuals.

TYPES OF HEARING PROTECTORS

The sole use of hearing protectors to deal with industrial noise should be considered as the last resort, but in some cases it may be the only prudent solution to the problem. Hearing protectors also serve a useful purpose as a safety cushion in a borderline noise-risk area. There are four classifications of HPDs (Lahey, 1986). The four types are enclosure type protectors, aural insert protectors, superaural protectors, and circumaural protectors.

An enclosure type protector is similar to an astronaut's helmet. The insulated helmet entirely encloses the wearer's head and acts as a sound suppressor. Insert type protectors may be worn under the helmet to reach very high levels of attenuation.

When properly fitted, enclosure HPDs can be easily donned to achieve their maximum attenuation capabilities but there are several drawbacks to their use. Enclosure devices have limited application because they are clumsy and uncomfortable to wear. Also, they are much more expensive than other types of protectors.

There are three types of aural insert protectors. They are formable, custom molded, and molded. Aural insert protectors are worn inside the ear canal and are typically referred to as inserts or earplugs. It is advisable to have the employee's ears examined by qualified medical personnel to be certain he or she is suited to wearing them. It is possible that the physical shape of a person's ear canal will prevent the effective use of aural insert HPDs.

Formable protectors are disposable and often made of fine glass fiber, wax impregnated cotton, or expandable plastic. These devices must be inserted properly and are not reusable. Formable HPDs are intended to fit all ears.

Molded aural inserts, or premolded inserts, are made of plastic or silicone rubber. These inserts are produced in varying sizes, so care must be exercised in fitting them to assure good attenuation. These devices are designed to conform somewhat to the shape of the wearer's ear canal; however, their basic shape is determined during manufacture. Molded inserts may be reused for several months with proper cleaning between uses to reduce the risk of infection. Molded inserts would be inappropriate choices for employees with fingers too big to properly insert them, employees with limited finger strength due to arthritis, and employees whose ear canals physically preclude their efficient use.

Custom molded protectors are specifically molded to fit an individual's ear canal. These are also made of plastic or silicone rubber. Because of the custom molding, they can be rather expensive to have made, but they do provide very good protection because of their precise fit to the wearer's ear canal. This type of HPD is reusable for a period of up to two years and requires cleaning between uses.

The superaural type seals the external edge of the ear canal to reduce sound transmission. The caps are made of soft rubber like material and are held in place with a spring band worn under the chin or a head suspension. For employees intermittently exposed to noise levels requiring protection, these may be convenient. Because of the spring band, they can be easily worn around the neck while not in use. These HPDs are reusable and require cleaning between uses. Also, they require some finger dexterity to properly don.

Circumaural or earmuffs cover the exterior ear, or pinna, and are held in place with a spring band or head suspension. Earmuffs typically provide better attenuation than earplugs and can be donned while wearing gloves, but their main advantage is they are very easy to don properly. Some aspects of circumaural protectors to consider before using are weight, cumbersomeness, compatibility with hardhats and respirators, effectiveness for people who wear eyeglasses, effectiveness for people with large outer ears, and compatibility with the physical shape of the wearer's skull. To properly use earmuffs, the ear cup and head suspension must be accurately sized to the individual to form a good seal against the head. Circumaural protectors may require periodic maintenance and routine cleaning between uses.

METHODS

Interviews and observations were conducted in five high school wood shops and five small woodworking businesses. These facilities were of special interest because their size and nature preclude them from the likelihood of OSHA inspection. High school students are exempt from OSHA jurisdiction and each of the businesses employed fewer than ten people. High school Industrial Arts instructors are covered by OSHA and are at higher risk than their students because of the length of exposure time during their work day; however, OSHA inspection of a school is not likely.

The woodworking instructors and shop managers were questioned about; their awareness of the noise hazard present, the number of people exposed to the hazard, notification of the hazard to the people present, the facilities hearing conservation policy, the facility's policy on providing HPDs, and personnel training in HPD donning. Noise level measurements were made at centralized locations in the facilities to determine ambient noise levels. Measurements were also made in the approximate operator's position for the planers, radial arm saws, and shapers in the shops, to determine localized noise levels of these machines. These measurements were made with a model 2225 Bruel and Kjaer integrated sound level meter.

While inside these facilities observations of students and employees were also made. These people were observed to determine the type of HPDs being used and whether they had been donned properly.

RESULTS AND OBSERVATIONS

All five of the high school wood shop instructors had a casual response toward the noise hazard in their facilities, none considered it to be a serious threat. Because of the instructors' attitudes the students had not been informed of the potential of the noise hazard present.

The average number of people exposed at any one time in the school shops was 18, this included the instructor and seventeen students. The average estimated length of exposure time during a class period was 38 minutes. Each instructor was exposed to three classes a day, making the instructor's total estimated exposure time 114 minutes a day. None of the facilities had hearing conservation policies in place. Two shops did not employ the use hearing protectors at all, while two provided superaural protector to students upon request. In these shops only one size of earmuff was stocked and could be adjusted to locate the cups on the head suspension for proper fit. No instruction on the use of the protector was given to the students in these two shops. The final facility required students to provide their own if they desired to use them, which none did.

Ambient noise levels in the school facilities ranged from 73 to 100 dBA depending upon the operations in progress at the time. The three noisiest machines in these shops were the thickness planers, radial arm saws, and the shapers, producing average noise levels of 99, 99, and 85 dBA under load and 112, 110, and 98 dBA peak, respectively.

Each of the instructors said he did not use hearing protectors. The two students using the earmuffs appeared to have the cups located properly on the head suspension and the cup pads forming a good seal against the head. In both shops using HPDs it was noted that some of the people in noise risk areas were not using devices while earmuffs hung on hooks or lay on shelves.

Two of the woodworking businesses studied did not perceive the noise levels present in their shops as a hazard and did not use HPDs of any kind. The other three recognized the noise level as a potential hazard and made limited efforts to deal with it. Two shops stocked earmuffs for employee use but said they were rarely used. The last facility stocked aural insert and superaural HPDs for its employees and encouraged but did not require their use. This company was the only one that pointed out the potential noise hazard to its employees. They accomplished this by requiring their employees to read a brochure dealing with occupational noise and hearing protector use on their first day of employment. None of the businesses provided instructions for protector donning or had a hearing conservation policy in place. Also, none of the employees were examined for physical compatibility with the devices being used.

The size of the woodworking businesses in which the interviews were conducted ranged from one to nine employees, with an average size of four employees. The typical duration of exposure to occupational noise for these employees was eight hours a day. This exposure included varying levels of noise each day due to the inconsistency of the types of operations being performed.

The ambient noise levels in these shops were very similar to the school shops with localized levels slightly higher due to the increased horsepower and capacities of the machines in use in the commercial shops.

Employees using HPDs were observed in only one shop. These employees were using superaural HPDs. The cup locations on these protectors were adjustable to accommodate individuals' heads. It appeared that each person had his protector adjusted and donned properly to receive maximum benefit from its use.

CONCLUSION

The noise reduction rating assigned hearing protectors actually represents the possible noise attenuating capabilities of the particular HPD. Due to the human factors involved in the protector's use, it is highly possible that the NRR may overstate the actual noise attenuation of the device. In order to achieve maximum attenuation the protector must be compatible with the wearer both physically and psychologically. The wearer must also be instructed how to don and be competent at donning the device. As demonstrated above, the lack of awareness of the need for hearing protection or the lack of compliance with hearing protection policy is a serious threat to workers' hearing. The effectiveness of hearing protective devices is limited by the way in which they are used.

The level of exposure of persons in noise risk areas could be more accurately identified through the use of a dosimeter. This would represent an actual time weighted average exposure. To measure the actual noise attenuating properties of a particular HPD the worker can be fitted with a microphone inside the protected ear. However, because of the human factors involved in donning the device the attenuating qualities may be different for each person using it.

LIST OF REFERENCES

Berger, E. H., (May, 1986). Ear infection and the use of hearing protection. Sound and Vibration.

Berger, E. H., (January, 1983). Using the NRR to estimate the real world performance of hearing protectors. Sound And Vibration.

Caple, C. E., (January, 1989). Effective hearing programs require assessment, testing, enforcement. Occupational Health And Safety.

Cluff, G. L., (December, 1986). Suggested standard hearing protector attenuation function relating to the NRR. American Industrial Hygiene Association.

Epps, B. W., Gasali, J. G., (April, 1986). Effects of insertion donning instructions on noise attenuation of aural insert hearing protectors. Human Factors. Vol. 28, No.2

Gasaway, D. C., (July, 1985). Hearing loss compensations, stemming the tide with conservation programs. National Safety and Health.

Lahey, J. B. (March, 1986). Hearing protection. Safety And Health News. Vol.133, No.3

National Safety Council., (1988). Fundamentals of Industrial Hygiene. (3rd ed.) Chicago, Ill.

Staff., (November 1987). Noise buster. Safety And Health News. Vol.136, No.5

Advances in Industrial Ergonomics and Safety II
Edited by Biman Das
Taylor & Francis 1990

ERGONOMICS IN DEVELOPING COUNTRIES

John D.A. Abeysekera, Houshang Shahnavaz, and Larry J. Chapman

Center for Ergonomics of Developing Countries,
Luleä University, Sweden and
University of Wisconsin Medical School Neurology Department,
Madison, Wisconsin, USA.

Ergonomics seeks to adapt work to known human physical and psychological capabilities in order to increase health, efficiency and productivity. Good ergonomics is especially important during technology transfers to non-Western countries. However, the people, the environment, and the institutions in these industrially developing societies are very different from those in technology-rich societies. Because of the differences, existing ergonomics data bases, training, and standards that have been developed and compiled in Western countries may be largely irrelevant for application in developing countries.

INTRODUCTION

The material that follows will provide an overview of some important differences between the industrialized and industrially developing countries and then suggest a program to increase the relevance of ergonomics in the developing countries (see Table 1).

Physical and Sociocultural Differences

People are different everywhere but there are especially great differences in stature and other physical dimensions between the industrially developing and the industrially developed world. Modifications are required if Western ergonomic standards are being applied. Even more important, new ergonomic principles and revisions in their relative priorities may be required when non-Western societies are adopting and selecting technology. For example, work heights in Western countries are set near seated elbow height for maximum comfort and efficiency. However, in many world cultures, the safest and most comfortable seated position is on the floor. Machinery designed for operators seated Western-style is appropriate for neither sitting nor standing by many non-Western peoples.

There is considerable literature on worldwide variations in climate, body dimensions, maximal oxygen consumption rates, and physical strength(Roberts, 1975; Abeysekera and Shahnavaz,

1989; Sen, 1982; Wyndham et al, 1963; Shahnavaz, 1976; Guthri et al, 1970). Some work has also been done that documents differences in manual dexterity, mechanical aptitude, labour force characteristics, and a number of cognitive capabilities including certain information processing and psychomotor skills (Baranson,1983; Calder,1961; Meshkati,1983; Kamanika,1985).

Table 1. Ergonomics contributions to development in industrially developing countries

1.	Research on Basic Data Needs
2.	Promote Special Abilities
3.	Refine Simple Methods
4.	More Appropriate "Experts"
5.	Learning by Doing, not Imitating (Action Learning)
6.	Better Supported Education and Research
7.	Reconceptualize Standard Setting

Substantial differences also exist for societal characteristics between countries. Research here is more limited but includes the strong role of the family as an important institution to work organizations in some countries as well as other social groupings including the clan and caste (Chapanis, 1975; Abeysekera & Shahnavaz, 1987). Managerial and organizational relations may have distinctive features in non-Western countries. Dishonesty, bribery, corruption and nepotism are sometimes freely practiced in certain industrially developing countries. Individualistic qualities are emphasized in Western societies but collective orientations and strong group allegiance often take precedence elsewhere in the world. Unique or variant languages, educational attainments, income levels, and other variables also need to be taken into account when considering the adoption of Western technology (Chapanis, 1975).

<u>Less Obvious Differences</u>

The economic conditions faced by wage earners in the industrially developing world are often complicated by high inflation rates and extreme ranges between the higher and lower paid workers. Paradoxically, workers on the low end of the income scale are often responsible for the most physically demanding work but these same workers are usually least able to afford the energy rich diets needed to sustain high levels of energy expenditure. Chronic fatigue, reduced alertness, accidents and physical illnesses are common results.

The infrastructure of the industrially developing world also places employees at a disadvantage. Poor roads, which are especially susceptible to inclement weather, coupled with an inadequate public transportation system can lead to chronic absenteeism and a lack of punctuality in the workforce. Energy utilities are less reliable in many parts of the non-Western world. Power plants may be of the latest design but poorly maintained. Climatic conditions may adversely affect energy distribution systems. Electronic equipment not designed for local conditions may fail due to high humidity, heat or other environmental conditions not taken into consideration when the equipment was selected.

A number of subtle cultural differences can become influential factors in industrially developing countries. Two major problems associated with educational programs are disruptions due to student unrest and the loss of talented and expensively trained individuals through a "brain drain" to Western countries.In the field of labour, the lack of the dignity in jobs, the poverty forcing employees to do two jobs or as much as 16-20 hours a day and the exodus of people from rural areas in search of work to the city, are differences worth considering. The unusually long procedures (red-tape) in government transactions and the latest social evil viz 'terrorism' are common and peculiar to developing societies.

Different Designer Cognitive Models of Users

Western countries are where most advanced technological devices are produced. Designers begin with a cognitive model of the end user made up of assumptions about the individual and the environment including anthropologic, technologic, and socioeconomic factors (see Table 2). Equipment and machinery designs are built so that the majority of the user population can use the product safely and effectively. Nearly all of the technology transferred to industrially developing countries has been built to fit the cognitive models of Western assumptions. As a result, most need sustantial modification to fit non-Western users. Usually no modifications whatsoever are made and the users in developing countries are forced to fit to the models never designed for them (Shahnavaz,1977). This not only produces inconvenience but leads to physical discomfort and can contribute to human errors, reduced productivity, accidents, diseases, and sometimes fatalities (Norman, 1980).

Table 2. Factors in the technology designer's cognitive model of the users of his product or system.

Socioeconomic	Technologic	Anthropologic
Energy and communications	Maintainability	Climate
Transportation	Population stereotypes	Body size
Cultural values	Appropriateness	Attitudes and customs
Labour supply and location	Infrastructure	Skill and education

Ergonomics up to now has been largely an American or West European discipline. What is required is that the available ergonomic knowledge has to be made relevant or modified to the changing characteristics of the product, system and environment. Therefore ergonomics for non-Western countries has to be newly developed.

ERGONOMICALLY APPROPRIATE TECHNOLOGY DEVELOPMENT

Perhaps most important is the need for industrially developing countries to become smarter buyers of advanced technology. The individuals involved in transfers of advanced technology must have a more sophisticated understanding of their own needs and of the the capabilities and demands of the different types of equipment. More critical purchasing can help avoid the problems accompanying adoption of hopelessly old technology or sophisticated but inappropriate technology. As noted earlier, inappropriate technology transfers are often those that fail to consider the human and societal differences between cultures and countries. One of the major reasons for these kinds of failures is the absence of even the most basic data about human and cultural differences. The first priority is a worldwide effort to undertake to collect this information in the industrially developing countries. Secondly, ergonomics education and training programs must be established in these countries so that the purchasers of advanced technologies are individuals who understand the particular human, environmental, and cultural needs of their countries.

Research on Basic Data Needs

For many non-Western countries, no systematic information has ever been collected on important human physical characteristics that are required to properly configure control surfaces, seating, enclosures, protective equipment and other devices. The result is often unnecessary pain and discomfort for machine operators along with higher accident, injury and fatality rates. Expensive equipment may sit idle due to incompatible human factors features. One common example is vehicles that cannot be driven safely because drivers cannot reach brake pedals while in position to steer. Other examples include machinery controls that mandate a chronically stooped postures or inappropriate machine guards that fail to protect equipment operators.

Although some work has begun to remedy these basic data deficiencies, a much larger scale international effort is warranted, perhaps supported in part by the manufacturers and suppliers of advanced technology to the industrially developing world. Information that has been collected needs to be systematically organized and made available. One method for information distribution could be a single international access information center that would make this basic data on different human populations available over telephone lines to designers, suppliers, purchasers and other interested individuals. The single access center could be supported by a consortium of national and international agencies or perhaps by the manufacturers and suppliers of advanced technology to the industrially developing world.

Promote Special Abilities

In agriculture, manufacturing, and service industries, product quality is often independent of the level of technology employed. In the U.S. automobile industry for example, the most highly automated plants are not among the most productive, and persistent quality problems associated with the most advanced technology have proven difficult to remedy. Western work designers have begun to realise that special roles for human capabilities need to be consciously incorporated into planning along with technology development considerations.

In the industrially developing world, human ingenuity, skill formation, and exacting judgment have, in some cases, at least partially replaced a reliance on technology. Some of these special talents appear remarkable, such as mechanics who are able to keep 20-25 year old motor vehicles roadworthy for daily use. In both industrially developing as well as developed countries, weather forecasts based on experienced visual observations rival those of meteorologists assisted by complex electronic instruments and analysis systems.Similar skills and abilities that are perhaps available in indigenous medicine, irrigation, building construction, agriculture, etc., must be encouraged, developed, promoted and documented.

Technology development in non-Western countries is desirable for its abilities to improve comfort and increase the ease with which work can be carried out. At the same time, decisions about technology transfer need to recognize and consider any special strengths of traditional methods and particularly those where human artifice is evident.

Refine Simple Methods

Development of technology in Western countries has also included the development of ergonomic measurement techniques and other analytical methods. The highly sophisticated techniques and equipment being used today in industrialized countries are often very expensive and beyond the technical reach of non-Western countries. A number of far simpler and less expensive methods are already available to carry out measurements in work places. For example, noise levels in industry are usually measured with delicate and expensive noise level meters. For the purposes of many factory inspections, noise levels estimates are sufficient. Listening to normal intensity spoken communications at defined distances and evaluating speech intelligibility can provide reasonably accurate and repeatable estimates of noise levels.

Simple, improvised anthropometric instruments have been developed to measure accurate body measurements. Several simple methods for workload estimations and energy expenditure have been developed (Shahnavaz, 1976). Helmets made out of cane are very light and airy, provide considerable impact protection, and will actually be worn by workforces in the tropics where hot, plastic helmets are refused. If more of these types of simple solutions and instruments can be scientifically validated and the normative data published, better recognition, identification and control of recognized work hazards could become a reality for millions of people in technology poor countries.

More Appropriate "Experts"

Countries undertaking technology development efforts often receive assistance in the form of bi-lateral and multi-lateral aid programmes from Western countries. The experts and consultants involved in these aid programs are almost invariably appointed by the Western donor countries and are members of these countries which is often an unofficial precondition of the aid contract. Sometimes these experts have not been exposed to conditions in the non-Western world prior to their assignments. Those experts with substantial experience in non-Western countries may be busy elsewhere. In many cases, these jobs can be more efficiently performed by non-Western experts since they already possess the necessary familiarity with the industrially developing world. Ironically, non-Western experts all too often are not accepted as competent by authorities in non-Western countries. In other cases, the donor countries will refrain from approving any expert not from their own country. If foreign aid to technology poor countries is to be successful, changes in the perceptions of authorities and in the strategy of expert selection are needed.

Learn by Doing Not Imitating

Self development guided by local authorities is more likely to produce lasting change than efforts directed by external consultants. When the individuals who are directly responsible for system performance on a daily basis share responsiblity for changes and help specify solutions, results are much better. Professor Reg Revans (1980) has pioneered a strategy called "Action Learning" which says that managers learn their crucial problem solving skills - their leadership function - not from 'expert' teachers or consultants in the form of programmed instruction but from the development process of questioning insight along with their fellow 'comrades in adversity'.

Promote Education and Research

Ergonomics training is an area that non-Western countries have not given a high enough priority in regular and adult education programs. Many industrially developing countries are not aware that a science of human work exists which can greatly aid technology development. Western aid agencies that fund development projects should make provisions for educational programs in ergonomics. To reach the largest number of persons who could benefit by such programs, courses and workshops should seek to train trainers, who can in turn pass on their skills to others. These trainers include inspectors, technical school teachers, managers, trade union representatives and other similar personnel. Advanced education in ergonomics can be provided through regional or international institutes and research centers. One such center in operation with programs at the masters and doctoral level is our Center for Ergonomics of Developing Countries in Luleä University, Sweden.

Reconceptualize Standard Setting

On the one hand, many Western experts have agreed that the workforce in industrially developing countries deserves protection equal to that provided to Western workers from noise, chemicals, dusts, and other work hazards.

On the other hand, industrially developing countries have raised arguments that protective labour standards that are intended to prevent against hearing loss or chronic illnesses after exposures of 40 years duration lose relevance where average life expectancies are barely 30 years since employment. Economic development that can provide jobs and products to relieve poverty and begin to provide basic sanitation and health care quite naturally assumes a higher priority.

In any case, standard setting bodies need to be established to deal with the special workforce and work environment characteristics of the non-Western world. Debates about the adequacy and rationality of specific level limits will undoubtedly continue.

Even in standard setting in industrialized countries it is common that the ergonomics or human factor needs are conveniently accommodated in 'General Requirements' under the Standard. The new physical and cognitive demands brought about by the modern technology, emphasize the need for specific standards in ergonomics.

CONCLUSIONS

Three-fourths of the world's population lives and works in industrially developing countries. The speed and ease of travel, trade, and communication are shrinking the world and increasing its inter-dependence. Improving the health and well-being of people in these countries is in everyone's interests and has become an even greater source of worldwide concern in the last decade.

Transfers of advanced technologies to industrially developing countries can improve the quality of life. However, failures to account for ergonomic and human factor aspects can greatly detract from improvements or even worsen conditions compared to those existing prior to technology transfers.

Although technology is necessary it is not the only form of development. Many of the world's people live and work in countries that are considered industrially less developed. Yet these same countries are also considered highly attractive to visitors in part due to their unique and unspoiled cultural, climatic, and environmental features. Entrepreneurs and investors are drawn to many of these same countries for an abundance of natural and human resources.

With regard to cultural, social, and spiritual development, those countries with the most limited industrial technology can be considered among the most advanced. If long term sustainability in lifestyles, stability in family relations, ease of friendship and hospitality, and the preservation of nature and cultural diversity are highly valued, then industrially developing countries lead the world.

Unplanned, rapid change and the dislocations associated with the industrial development have exacted steep social and cultural costs in technologically-advanced countries. The lessons already learned about the processes of technology transfer and industrial development can help avoid repeating the mistakes in the rest of the world. It is also important that the technology offered to industrially less developed countries be modified according to ergonomic principles so that their special and diverse environmental, social, and cultural aspects are not degraded.

Technology transfer has produced beneficial changes in industrially developing countries. In return, these industrially developing countries have shared their cultural and natural resource riches with Western countries. In the future, more ergonomically sound technology development can improve the benefits to people in both non-Western and Western countries.

REFERENCES

Abeysekera, John D.A. and Houshang Shahnavaz., 1989, Body size variability between people in developed and developing countries and its impact on the use of imported goods. International Journal of Industrial Ergonomics, 4, 139-149.

Abeysekera, John D.A. and Houshang Shahnavaz., 1987, Ergonomics of tecnology transfer. International Journal of Industrial Ergonomics, 1, 265-272.

Baranson, J., 1983, Econmies and social consideration in adapting technologies for developing countries. Technology Culture, IV; 22-29.

Calder, R., 1961, After the Seventh Day: The World Man Created. Mentor Books, New York.

Chapanis, A., 1975, Ethnic variables in human factors engineering. The John Hopkins University Press, Baltimore and London.

Guthri, G.M., Brislin, R. and Sinaiko, H.W., 1970, Some aptitudes and abilities of Vietnamese technicians: implications for training. Paper P-659, Arlington, Virginia, Institute of Defence Analyses.

Kamanika, M.S., 1985, Research needs in American agricultural equipment industry. Applied Ergonomics, 16: 217-220.

Meshkati, N., 1983, A conceptual model for the assessment of mental workload based upon individual decision styles. Unpublished Ph.D. dissertation, University of Southern California, USA.

Norman, D.A., 1980, Errors in human performances. Centre for Information Processing, Sen Diego, California, USA.

Revans, R.W., 1980, Action Learning (New techniques for management). Blond & Briggs Ltd., London.

Roberts, D.F., 1975, Population differences in dimensions, their genetic bases and their relevance to practical problems of design. "Ethnic variables in human factor engineering" by Alphonse Chapanis, John Hopkins University Press, London.

Sen, R.N., 1982, Certain ergonomic principles in the design of factories in hot climates. Proceedings International Symposium in Occupational Safety, Health and Working Conditions, specifications in relation to transfer of technology in the developing countries, ILO, Geneva, pp.123-147.

Shanavaz, Houshang., 1989, Ergonomics : an emmerging concept in industrially developing countries. International Journal of Industrial Ergonomics, 4: 91-100.

Shahnavaz, H., 1976, Work biology characteristics for industrial ergonomics in a developing country. Ph.D. thesis, University of Birmingham, U.K.

Whydham, C.H., Morrison, J.F., Peter, J., Williams, G.G., Bredell, G. and Joffe, A., 1963, Differences between ethinic groups in physical capacity. Journal of Applied Physiology, 18: 361-366.

XVI

ROBOTS AND AGRICULTURAL MACHINERY SAFETY

Advances in Industrial Ergonomics and Safety II
Edited by Biman Das
Taylor & Francis 1990

IDLE TIME AND SAFE-TO-APPROACH CONDITION FOR INDUSTRIAL ROBOT: WAITING TIMES SELECTED BY STUDENTS AND WORKERS

Waldemar Karwowski
Nai Pongpatanausegsa
Center for Industrial Ergonomics
University of Louisville
Louisville, KY 40292, USA

The main objective of this study was to compare the times of robot's inactivity (idle times) selected by two groups of subjects, i.e. college students versus industrial workers, as indicative of the system malfunction and, therefore, "safe-to-approach" condition. On the average, workers and students waited 16.2 s and 18.5 s, respectively, before entering the MH33 robot's work envelope. Ten percent of workers and ten percent of students selected the times of 8.4 s and 5.1 s or less, respectively, while ninety percent of the subjects waited 24.1 s (workers), and 31.9 s (students), or less. About 21% of industrial workers and 26% of students waited for only 10 seconds (or less) before deciding that the robot stopped due to the system malfunction, and it was not going to unexpectedly move again.

INTRODUCTION

Despite the fact that the application of industrial robots often leads to improvement of working conditions, reduction of physical labor, and liberation of workers from monotonous and environmentally stressful jobs (Kalmbach et al., 1982), there is more evidence today that robots constitute complex systems which are potentially hazardous to humans, not only in faulty operations, but also in the normal operating mode (Bonney and Yong, 1985; Nagamachi, 1986 and 1988). Among other factors, robots differ from traditional machinery in that they perform their tasks with high speed and unpredictable motion patterns, variable idle times, and ill-defined danger zones (Nicolaisen, 1985).

The introduction of robotics has led to new categories of work-related injuries. The common cause of many robot-related accidents can be attributed to human perceptual and judgmental processes (Carlsson, 1980). For example, according to Sugimoto (1985), the "unnatural" and "unexpected" movements of robots were responsible for about 73% of all analyzed injuries and fatalities he examined. In the reported cases, the operators incorrectly assumed that the robot stopped due to the system malfunction. These accidents involved a person entering the work area when the robot stopped (or was working very slowly), and then unexpectedly started to move again.

Previous experience with robots also influences safety behavior of the workers. There may be quite lengthy pauses to allow for specific manufacturing processes to take place which could lead the observers to think that it is *safe-to-approach* the robot. Even those who work around the robot for some time begin to take its movements for granted, exposing themselves to hazards if, for example, due to power failure or other malfunction, the robot arm is brought to the resting position, and then suddenly takes an "unnatural" and unexpected path of motion.

OBJECTIVES

The aim of this study was to compare the minimum time of robot's inactivity (idle time) selected by the college students and industrial workers as indicative of the system malfunction and, therefore, *safe-to-approach* condition. Subjects were asked to determine how long they would wait before assuming that it was safe to enter the robot's work envelope (*safe-to-approach* state), in order to check for the probable system malfunction.

METHODS AND PROCEDURES

An MH33 General Electric robot was used in two separate laboratory experiments. A total of twenty industrial engineering students with laboratory exposure to robots, and twenty-four industrial workers experienced with operation and maintenance of robot systems, participated in the study. Within each group, the subject population was randomly divided into two sub-groups. One sub-group (called an *accident group*) was exposed to the simulated industrial accident which involved a robot hitting the mannequin placed inside the robot's work envelope. This sub-group of subjects was told that the accident was shown to them in order to illustrate the potential danger that industrial robots pose in the workplace if they collide with humans. The other sub-group (called the *no-accident group*), did not see the accident. Two speeds of robot motion were used: a high speed of 90 cm/sec, and a low speed of 25 cm/sec. The subjects from both groups completed runs at both speeds.

The safety barriers were placed at an arc distance of 365.7 cm from the robot. A box emulating a computer output device was placed at the base of the robot in its left corner. This box contained computer output sheets which would inform the subject whether the robot had stopped due to the system malfunction or a programmed stop.

The preview session lasted fifteen minutes. The MH33 robot was programmed such that its arm would move through all the points covering its work envelope, while performing a simulated palletizing task. The subject was encouraged to walk behind the safety barriers, and observe the various process stops the robot would make. The robot stopped randomly for four time intervals of 5, 10, 15 and 20, seconds totalling ten stops. These stops consisted of 4 stops of 5 seconds each, 3 stops of 10 seconds each, 2 stops of 15 seconds each and 1 stop of 20 seconds. The order of the stops was chosen randomly for each subject and experimental condition.

In the experimental session, the subject observed the robot in motion, during which period the robot would again stop randomly for different time intervals. At one of the stops the robot was pre-programmed to interrupt its sequence and abort any further action. The subject was told this unexpected stop could be either a pre-programmed stop or a stop due to the system malfunction, and that the only way to check the reason for probable malfunction was to enter the robot's work envelope and retrieve the computer report.

The subject was instructed to wait as long as he felt necessary before he could safely enter the robot's work envelope. The subject was also told that he was responsible for maintaining efficient performance of the manufacturing system. Each subject was given the following instructions: "*Before you decide to enter the robot's working envelope to check the system status, you must be sure that the robot stopped for good, and it is not going to unexpectedly move again. Please wait as long as you think is necessary, and then indicate to the experimenter when you reach your decision to enter the work area in order to check for any problems with the system".*

At the decision point, the subject raised his hand, and the robot was powered down. The flashing light, however, was left on, in order to give the subject the impression that the robot was still powered. The subject was then instructed to pick up the system malfunction report placed at the base of the robot, and return back to his original position. The time elapsed between stopping the robot and the subject's decision to enter its work envelope was recorded with the help of a stop watch.

RESULTS AND DISCUSSION

Summaries of the main results based on the analyses of variance are shown in Tables 1-3. Overall, waiting times selected by the industrial workers were not significantly different ($p>0.05$) from those selected by the college students. On the average, workers and students waited 16.2 s and 18.5 s, respectively, before assuming the robot stopped due to the system malfunction, and entering its work envelope.

The effect of robot speed was significant ($p<0.03$) for both groups of the subjects, while the effect of simulated accident affected significantly only waiting times selected by the student population. Students who witnessed the simulated accident waited longer (average selected time of 20.2 s), than those who were not exposed to it (average selected time of 17.1 s). The effect of accident exposure was not significant for the industrial workers, who waited on the average 14.7 s and 17.8 s for the *no-accident* and *accident* groups, respectively. It should be noted, however, that the average difference between these two sub-groups was quite large, i.e. 21%.

The above results can be interpreted by observing that workers had much more experience with industrial robots, and the experience factor most likely "washed away" some of the modification in subjects perception of the potential hazard induced by observing the robot's arm hitting the mannequin.

As expected, the waiting times selected by the industrial workers under low speed of robot motions ($x = 14.4$ s) were significantly shorter than those selected under the high speed condition ($x = 18.1$ s). The reverse trend, however, was observed for the students, who waited longer ($x = 20.0$ s) when the robot operated at the low speed, than when it moved at the high speed ($x = 16.9$ s).

Ten percent of the subjects (workers and students) selected the times of 8.4 s and 5.1 s or less (10th percentile values), respectively. Ninety percent of the workers waited 24.1 s and 31.9 s or less (90th percentile values), respectively, before entering the MH33 robot's work envelope. It should also be noted that about 21% of industrial workers and 26% of students waited only 10 s (or less) before deciding to enter the robot's work envelope. The percent distributions of selected waiting times, classified by the robot speed and accident exposure for both subject groups, are illustrated in figures 1-3.

CONCLUSIONS

The results of this study revealed that college students and industrial workers experienced in operation and maintenance of robots would wait, on the average, for about 17 s, before assuming that the robot stopped due to system malfunction, and would not move again until the problem had been corrected. About 20

Table 1. Summary of main results for both groups of subjects.

Variable	Mean*	S.D.	Range	N
Workers				
No-accident group	a 17.8	5.2	9.7-28.0	24
Accident group	a 14.7	6.7	7.8-26.2	24
Robot speed				
Low speed	a 14.4	5.5	7.8-24.6	24
High speed	b 18.1	6.4	9.6-28.0	24
Students				
No-accident group	a 17.1	10.8	4.0-44.0	132
Accident group	b 20.2	9.96	4.0-49.0	108
Robot speed				
Low speed	a 20.1	9.8	7.0-49.0	120
High speed	b 16.9	11.0	4.0-48.0	120

N- number of observations.

* - different letters indicate differences between means at $p \leq 0.05$.

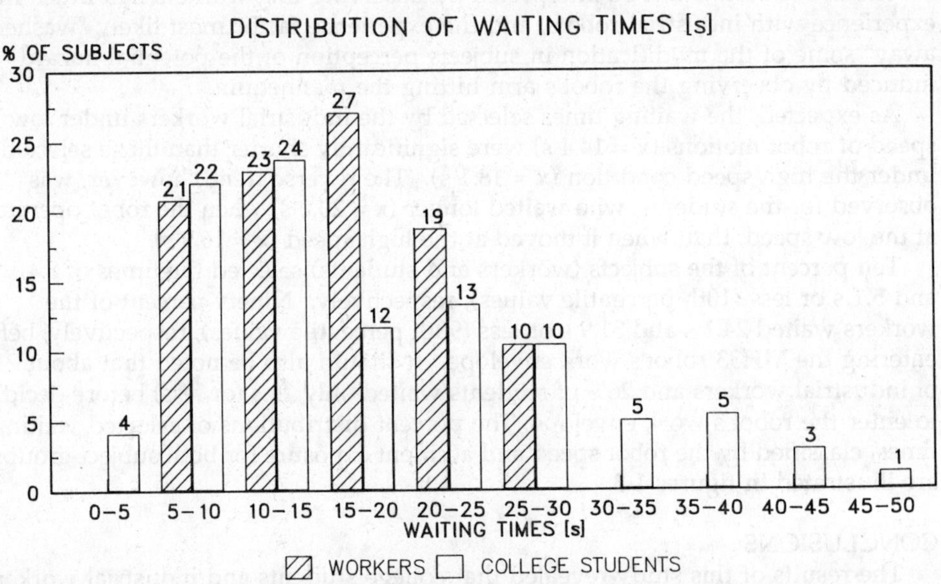

Figure 1 . Percentage distribution of selected waiting times [s].

Table 2. Comparison of two groups of subjects for perception of *safe-to-approach* robot condition: dependent variable = waiting time [s].

Variable	Mean*	S. D.	Range	N
Subject group				
Workers	a 16.25	6.15	7.77-28.0	48
Students	a 18.50	10.5	4.0-49.0	240
Accident group				
Workers	a 17.8	5.2	9.7-28.0	24
Students	a 20.2	10.0	4.0-49.0	108
No-accident group				
Workers	a 14.7	6.7	7.8-26.2	24
Students	a 17.1	10.8	4.0-44.0	132
Robot speed				
25 [cm/s]				
Workers	a 14.4	5.4	7.8-24.6	24
Students	b 20.0	9.8	7.0-49.0	120
90 [cm/s]				
Workers	a 18.1	6.4	9.6-28.0	24
Students	a 16.9	11.1	4.0-48.0	120
Accident group				
Low speed				
Workers	a 16.1	3.7	9.7-22.5	12
Students	b 22.0	9.1	9.0-49.0	54
High speed				
Workers	a 19.6	6.0	9.9-28.0	12
Students	a 18.4	10.5	4.0-48.0	54
No-accident group				
Low speed				
Workers	a 12.8	6.5	7.8-24.6	12
Students	a 18.5	10.1	7.0-44.0	66
High speed				
Workers	a 16.6	6.6	9.6-26.2	12
Students	a 15.7	11.4	4.0-44.0	66

N - number of observations.

* - different letters indicate differences between means at $p \leq 0.05$.

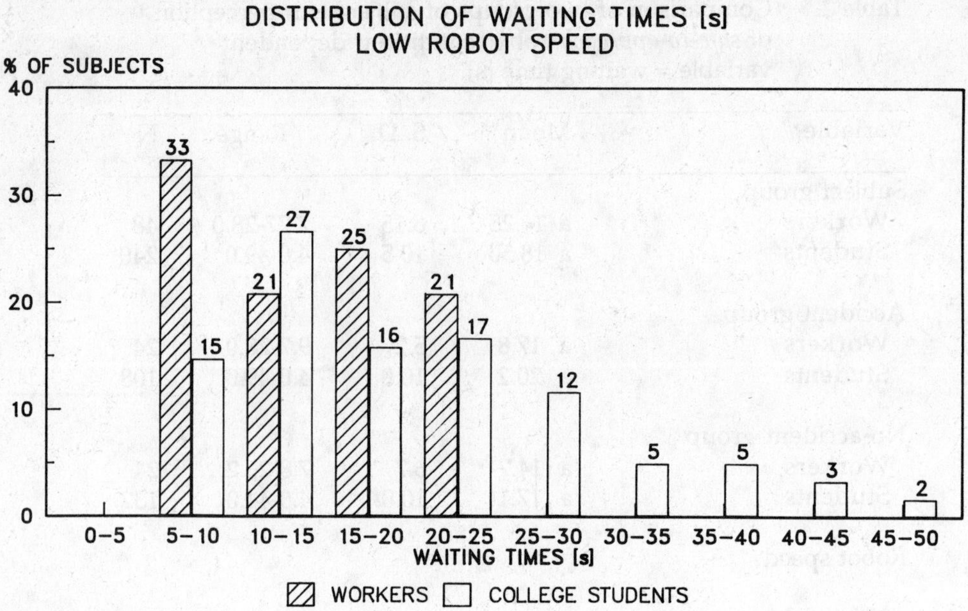

Figure 2a. Distribution of waiting times for low speed of robot motions.

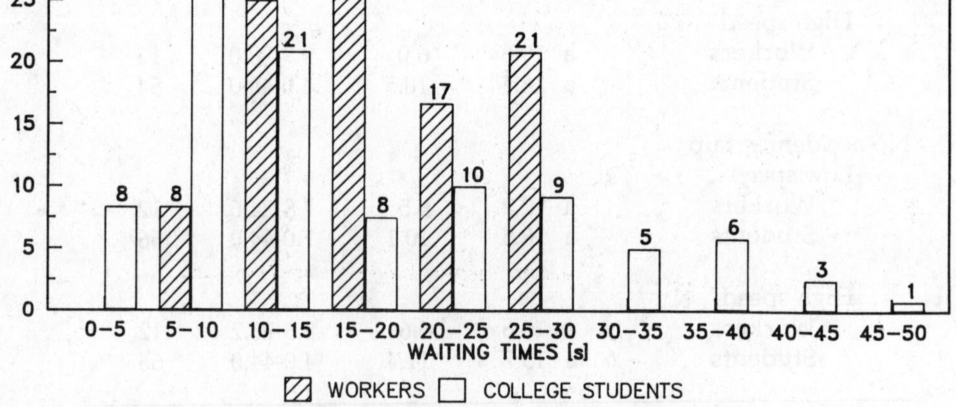

Figure 2b. Distribution of waiting times for high speed of robot motions.

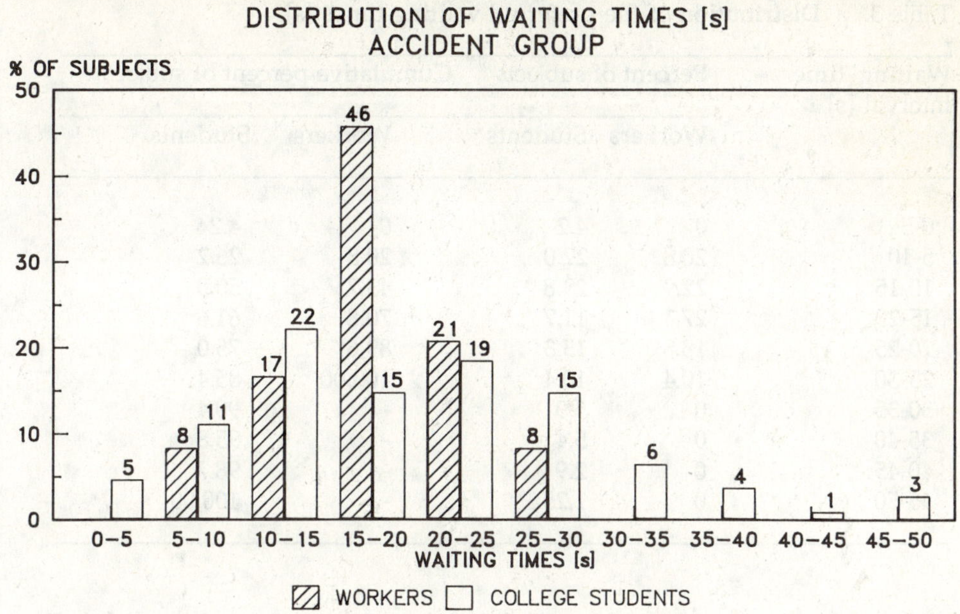

Figure 3a. Distribution of waiting times for the accident group.

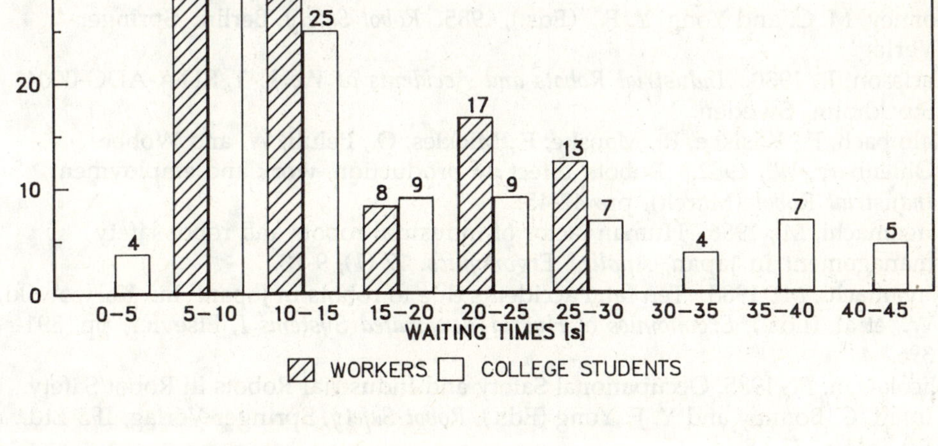

Figure 3b. Distribution of waiting times for the no-accident group.

Table 3. Distribution of the preferred waiting times [s].

Waiting time interval [s]	Percent of subjects		Cumulative percent of subjects	
	Workers	Students	Workers	Students
0-5	0	4.2	0	4.2
5-10	20.8	22.0	20.8	26.2
10-15	22.9	23.8	43.7	50.0
15-20	27.1	11.7	70.8	61.6
20-25	18.8	13.3	89.6	75.0
25-30	10.4	10.4	100.00	85.4
30-35	0	5.0	-	90.4
35-40	0	5.4	-	95.8
40-45	0	2.9	-	98.7
45-50	0	1.25	-	100.00

percent of both groups of subjects decided that is was *safe-to-approach* the robot when its idle time was 10 s or less. The above results indicate that in order to provide for safer operation of robotic cells, the length of the pre-programmed robot stops in industrial settings should be limited by software design, and possibly, the restrictions on allowable idle times should be incorporated into robot safety guidelines.

ACKNOWLEDGEMENTS

This study was supported in part by a grant from the National Institute for Occupational Safety and Health, Grant No. 1 R01 OH02568-01, awarded to the Center for Industrial Ergonomics at the University of Louisville.

REFERENCES

Bonney, M. C. and Yong, Y. F., (Eds.), 1985. *Robot Safety*, Berlin: Springer-Verlag.

Carlsson, J., 1980. *Industrial Robots and Accidents at Work,* T RITA-ADG-0004, Stockholm, Sweden.

Kalmbach, P., Kasiske, R., Manske, F., Mickles, O., Pelull, W. and Wobbe-Ohlenberg, W., 1982. Robots' effect on production, work and employment, *Industrial Robot* (March), pp. 42-45.

Nagamachi, M., 1986. Human factor of industrial robots and robot safety management in Japan, *Applied Ergonomics*. 17 (1), 9-18.

Nagamachi, M., 1988. Ten fatal accidents due to robots in Japan. In: Karwowski, W., et al. (Eds.), *Ergonomics of Hybrid Automated Systems I*, Elsevier, pp. 391-396.

Nicolaison, P., 1985. Occupational Safety and Industrial Robots in Robot Safety. In: M. C. Bonney and Y. F. Yong (Eds.), *Robot Safety*, Springer-Verlag, IFS Ltd., Berlin, pp. 33-48.

Sugimoto, N., 1985. Safety measures for automated machines. *Safety Staff, 12*, pp. 4-18.

Advances in Industrial Ergonomics and Safety II
Edited by Biman Das
Taylor & Francis 1990

SAFE INDUSTRIAL ROBOT OPERATIONS:
Robotic Laser Processing Systems

F. Frank Chen

Process Engineering, General Offices
Caterpillar Inc.
East Peoria, IL 61630

and

Fariborz Tayyari

Department of Industrial Engineering
Bradley University
Peoria, IL 61625

INTRODUCTION

The Occupational Safety and Health Act of 1970 (OSHAct 1970) was passed by Congress "... to assure so far as possible every working man and woman in the nation safe and healthful working conditions and to preserve our human resources ..." (Public Law 91-596). The Act emphasized the need for standards to protect the health and safety of workers exposed to an ever increasing number of potential hazards in the workplace (NIOSH, 1983). It places the following responsibilities on the employer which are of special importance to the safety of workers interfacing robotic systems:

1. To provide employees with a work environment which is *free from recognized hazards* that are causing or are likely to cause death or serious physical harm to his/her employees.

2. To make sure that employees have and use *safe tools and equipment* and that such equipment is maintained in a proper working condition.

3. To update operating procedures and to *train employees* so that they follow safety and health requirements.

4. To use *labels or other appropriate forms of warning* as are necessary to ensure that employees are aware of all hazards to which they are exposed, and proper conditions and precautions of safe use or exposure.

5. To provide suitable *protective equipment and control or technological procedures and mechanism* in connection with such hazards and to provide monitoring or measuring employee exposure at such locations and intervals.

Automation was introduced to industry aiming at improvement of productivity, quality of products as well as workers' safety (Salim, 1983; Osborn, 1988). Industrial robots can work around the clock on repetitive and/or hazardous tasks under hostile environments, without requiring work pauses, vacations, sick leaves, judgmental errors, or worker's compensation. Occupational injuries, however, could occur everyday in the workplace whether the work system is manually, semi-automatically or, even, fully automatically operated. In fact, the installation of the high-energy robotic systems which operate in an automatic mode presents unique types of potential hazards. The automatic machine movements in these systems and unpredicted machine failures create the need for developing special safe practices. Reports of robot fatalities can be easily found in the literature (e.g., NIOSH, 1984; Nagamachi, 1986; Morawetz, 1987; Etherton, 1988).

A large amount of efforts are being used to intervene techniques and devices to eliminate and/or minimize robot related accidents (e.g., Meagher et al., 1983; Potter, 1983; ANSI/RIA, 1986; Jiang, 1987; Parsaei et al., 1987; Rahimi et al., 1987; Burton, 1988).

However, it is believed that lack of safety consciousness is the "root" cause of most accidents, and that workplace can be safe only when all sources of danger are known and controlled. To minimize the risk of accidents in operating robotic systems, it is required to promote a sound safety training program and follow safety precautionary guidelines by all people who design or manufacture robots, design software systems, install robots, and the individuals working with industrial robots. These individuals may be grouped into programmers, operators, and the maintenance personnel.

Robotic hazardous zones

Etherton (1988) has pointed out the following two commonly hazardous zones and corresponding typical accidents around industrial robots:

1. The **Robot-Movement Zone**. This zone is often called robot work envelope or danger zone. If an individual stands in this zone when the robot is powered, he/she is exposed to crushing, shearing, and impact injury risks.
2. The **Approach Zone**. The area just outside the robot-movement zone is the approach zone. The personnel in this zone are exposed to thrown objects, radiation, flash, electrical hazards, or mechanical hazards of associated equipment. Passing from the approach zone to the movement zone should be limited.

Basic safety rules

It should be noted that it is the automatic mode of an industrial robot along with its unpredicted failure that represents potential hazards, creating needs for development of special safety practice guidelines. The following basic safety rules must, therefore, be obeyed by anyone working with or around an industrial robot:

1. Never enter the robot work area while the the robot is activated.
2. Be aware of the presence of other people in the work area, and safety precautions.
3. Be aware of the sequence of robot operations, and carefully monitor the changes in the robot operational, electrical, and mechanical status.

ROBOTIC LASER PROCESSING SYSTEMS

The technology advancement has brought about the combined use of laser processing and industrial robots. This integration has been implemented in the following manufacturing applications:

1. Laser welding
2. Laser heat treating
3. Laser cutting
4. Laser marking
5. Laser drilling

In laser processing systems, the energy is generated by the excitation of atoms in the molecules (of, for instance, carbon dioxide, CO_2). The excited atoms throw off photons which, as described by Vasilash (1988) in more detail, are bounced back and forth between parallel mirrors resulting additional photons travel along with the original parallel-moving photons. One of the mirrors is made transparent to let the concentrated photons (laser beam) move on out. The laser beam is then focused for laser processing, such as welding, drilling, cutting, heat treating (especially useful material hardening in small areas), etc.

Hazardous conditions in robotic laser processing

The marriage of laser processing and robotics introduces unique types of potential hazardous conditions as presented below:

1. Some laser beams, however, may miss the targeted workpiece and strike the nearby human operator, instead. Accidental exposure of personnel to hazardous laser beams. Shinny work parts stored within the work envelope may accidentally initiate reflections of laser beams which could cause biological damages to the eye and or skin during intended use. Special attentions are needed for invisible laser

beams (e.g., CO_2), besides a lot of ultraviolet rays normally generated in welding processes.

2. With the inherent flexibility and versatility of robotic laser processing, there are far more process parameters to be tested and established than simple applications such as those needed for material handing robots. On-line programming and command execution therefore become more complex. Special safety guidelines for programmers and operators should be made available. This is especially true for processes with adaptive capability using vision system where robot arms or laser beams might be misled or erroneously activated due to accidental human misjudgment.

Safety requirements for robotic laser processing

The requirements for a safe operation of laser robotic system can be classified as layout and precautions:

I. *Layout safety*: Layout requirements divide the robotic operating cell into three distinct zones, as shown in Figure 1 and described below:

1. The robot work envelope must be properly defined incorporating to the robot arm reach dimension and the laser beam reach.

2. The operator work zone must be defined so that he/she is not exposed to crushing, shearing, and physical or laser impact injury. He/she must be guarded against misled laser beams by providing laser beam stops between the laser processing and the operator load-unload station. A rotating table could be used in such a way that the operator is not required to leave his/her shielded work zone. These concepts are depicted in Figure 2.

3. The visitors or by standing individuals zone should be specified and separated from both robot and operator work zones.

II. *Safety precautions*: The developed safety precautions (e.g., Akeel, 1983; Griffin, 1983; Russell, 1983; Stowe, 1983; Jiang, 1987; Parsaei et al., 1987; Sugimoto and Futsuhara, 1987; Etherton, 1988; Finnerty, 1988; GMFanuc Robotics Corporation, 1988) are classified into three categories, according to the group of people interacting with the laser robotic system. These people are programmers, operators, and maintenance personnel.

1. Safety precautions for the robot programmer/teacher:
   * Provide a complete and clear documentation of the programs so that they can be understood and modified should a need arise.
   * Always wear the appropriate personal protective equipment.
   * Before entering the robot's work envelop, check the robotic system and its work envelope; and correct any abnormal conditions that may cause hazards or malfunctions.
   * If the robot needs to be taught, before teaching:
     - disable the laser shutter,
     - disable the high voltage power to the laser generator, and
     - put the robot in the teach mode.
   * Program the robot to perform at its maximum arm reach at all interfaces with the human operator. This prevents the operator from getting closer than where he/she is supposed to be.
   * Two-man programming procedure should be devised. The second person can take the appropriate action and get help in case of an emergency.
   * Use the off-line programming technology if possible.

2. Safety precautions for the robot operator:
   * Know about capabilities and all functions of the robot that you operate.
   * Know about all safety and emergency switches and devices.
   * Check all wiring, cables, connections, and hoses before activating the system; and look for wrong things and hazard causing situations. Make sure that

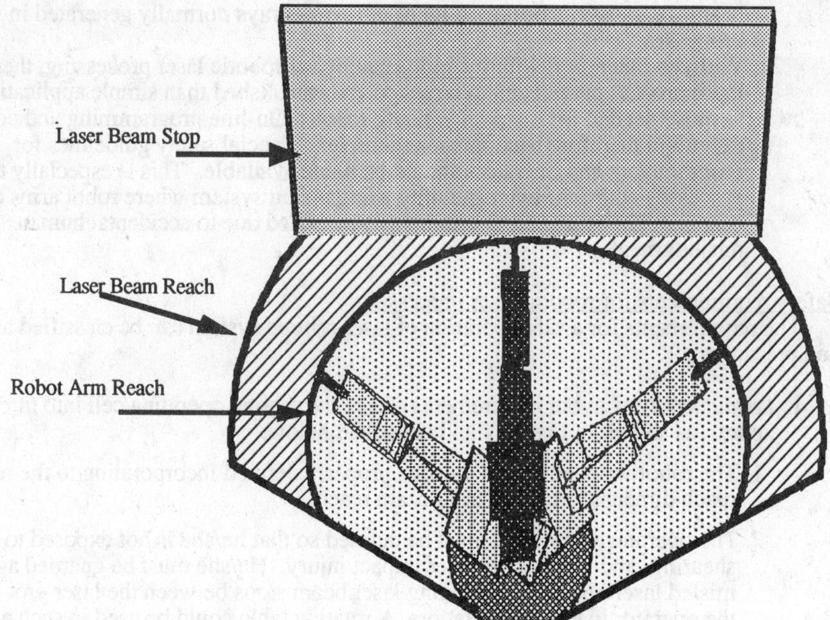

Figure 1: A work envelope illustration for laser processing robot.

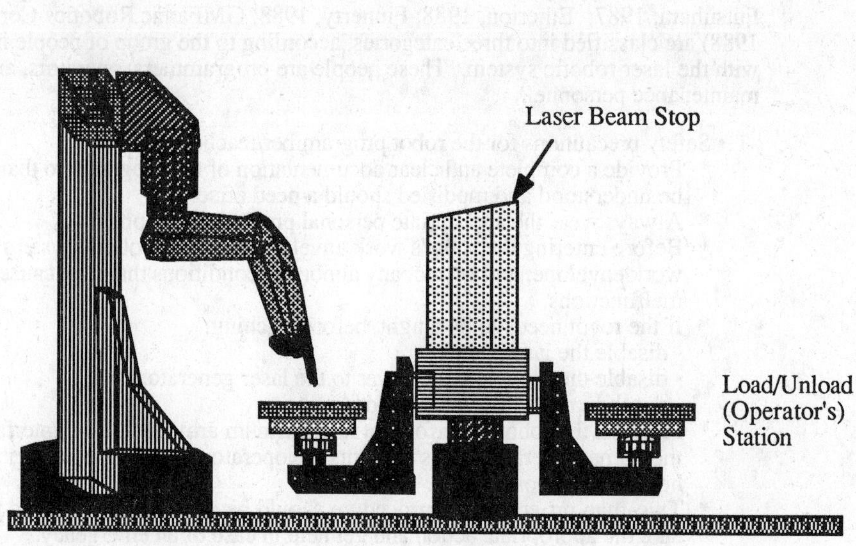

Figure 2: An example set-up for a safe robotic laser processing with a rotating table and laser beam stop.

safety devices function properly. Correct the conditions or ask for expert help, if necessary.
* Make sure all personal protective devices and machine and laser safeguards are in place appropriately.
* Never check the high-pressure hydraulic hoses with your hand (e.g., detect leaks with a piece of paper).
* Be aware of other people in the area.
* Never enter the robot work envelope when it is energized (powered).
* Never watch the laser process without the appropriate protective eyewear.
* Never take a chance and risk by short-cutting. Short-cutting safety is the same as inviting disaster.
* Expect the unexpected things to happen. Protect yourself against Murphy's Law, which says: "If anything can go wrong, it will."

3. Safety precautions for the maintenance personnel:
* Wear proper personal protective equipment, insulated shoes and gloves.
* Prior to maintenance, lock-out the robot's control panel, if possible, when working on the machine in order to obtain the zero-mechanical state and comply with the lock-out, tag-out procedure..
* There should be another person present in the area when someone is repairing or servicing a robot to shut off the robot and get help if an accident happens.
* There should be a routine maintenance of the electrical, pneumatic and hydraulic systems. Follow the manufacturer's manuals and instructions.
* When checking for hydraulic leaks use a piece of paper not your hand. The high pressure hydraulic fluid can be as sharp as a knife.
* Use approved parts only. Wrong parts will go wrong!
* Never take a short cut. To "short-cut" safety is the same as to "invite" disaster.

DISCUSSIONS

While the robotic laser processing is just one of the special robot applications requiring special safety operating guidelines as presented in this paper, other sets of safety guidelines are needed to be developed for specific robotic applications. Use of the safety measures presented in this paper is expected to effectively minimize the risk of accidents. However, the most important factor in a successful program is the use of safety training and awareness measures. No matter how safe a robotic system is designed and implemented, it cannot be 100 percent fool-proof and the risk of accidents still exists for workers neglecting safety precautions and rules.

It is necessary that the employees report any type of accidents, even the near misses, to safety professionals and other related parties to make record of these accidents and carry out an investigation of causes and all potential hazards. The significance of this concluding remark is that the concept of accident prevention is rapidly growing among all levels within organizations and spreading to all aspects of technology. Accident documentation is necessary for a better result.

REFERENCES

Akeel, H.A., 1983, Intrinsic Robot Safety. Professional Safety, 28(12): 27-31.

ANSI/RIA, 1986, Industrial Robots/Systems Safety Requirements Standard. R15.06-1986.

Burton, J., 1988, Industrial Robotics: Hazards, Accidents, Safety Applications and Advanced Sensor Technology. Professional Safety, 33(11): 28-33.

Etherton, J.R., 1988, Safe Maintenance Guide for Robotic Workstations. NIOSH Technical Report, DHHS(NIOSH) Publication No. 88-108, National Institute for Occupational Safety and Health, Division of Safety Research, U.S. Department of Health and Human Services, Morgantown, West Virginia, March 1988.

Finnerty, D.W., 1988, Learn to Love Your Robot. Safety & Health, 138(2): 40-43.

GMFanuc Robotics Corporation, 1988, L-100 Operations Manual, Chapter 2, Safety.

Griffin, K.G., 1983, Safety for Industrial Robotic Applications. Professional Safety, 28(12): 22-26.

Jiang, B.C., 1987, Safety Considerations for Programming Industrial Robots. pp. 465-472, in: S.S. Asfour (ed.), Trends in Ergonomics/Human Factors IV, Elsevier Science Publishers B.V. (North-Holland), Amsterdam, The Netherlands.

Meagher, J.; Derby, S.; and Graham, J., 1983, Robot Safety/Collision Avoidance. Professional Safety, 28(12): 14-17.

Morawetz, J.S., 1987, A Fatality Related to Computer Operated Equipment. pp.491-495, in: S.S. Asfour (ed.), Trends in Ergonomics/Human Factors IV, , Elsevier Science Publishers B.V. (North-Holland), Amsterdam, The Netherlands.

Nagamachi, M., 1986, Human Factors of Industrial Robots and Robot Safety Management in Japan. Applied Ergonomics, 17(1):9-18.

NIOSH, 1983, Guidelines for Controlling Hazardous Energy During Maintenance and Servicing. DHHS(NIOSH) Publication No. 83-125, National Institute for Occupational Safety and Health, Centers for Disease Control, U.S. Department of Health and Human Services, Morgantown, West Virginia.

NIOSH, 1984, Request for Assistance in Preventing the Injury of Workers by Robots. DHHS(NIOSH) Publication No. 85-103, National Institute for Occupational Safety and Health, Centers for Disease Control, U.S. Department of Health and Human Services, Cincinnati, Ohio.

Osborn, J.F., 1988, Robotics: Robots Can Replace Humans in Many Hazardous Environments. Occupational Health & Safety, 57(10): 20-24, September 1988.

Parsaei, H.R.; Wilhelm, M.R.; and Karwowski, W., 1987, Safety Monitoring Strategies in the Design and Utilization of Robotic Work Cells: A Review. In the Proceedings of the IXth International Conference on Production Research, held in Cincinnati, Ohio, August 17-20, 1987, pp. 2726-2731.

Potter, R.D.,1983, Safety for Robotics. Professional Safety, 28(12): 18-21.

Public Law 91-596, 91st congress, S.2193, December 29, 1970.

Rahimi, M.; Hancock, P.A.; and Chignell, M.H., 1987, Integrated Software Safety Design for Robotics." pp. 497-503, in: S.S. Asfour (ed.), Trends in Ergonomics/Human Factors IV, Elsevier Science Publishers B.V. (North-Holland), Amsterdam, The Netherlands.

Russell, J.W., 1983, Robot Safety Considerations ... A Checklist. Professional Safety, 28(12): 36-37.

Salim, P., 1983, The Robots Are Coming. Professional Safety, 28(3): 17-23.

Stowe, W.W.,1983, Robots--Safe of Hazardous? Professional Safety, 28(12): 32-35.

Sugimoto, N. and Futsuhara, K.,1987, Safety and Technology of Safety Confirmation type--Coordination with robots. pp. 521-528 in: S.S. Asfour (ed.), Trends in Ergonomics/Human Factors IV, , Elsevier Science Publishers B.V. (North-Holland), Amsterdam, The Netherlands.

Vasilash, G.S., November 1988, Lasers Heat Up Robotics. Production, pp. 34-41.

Advances in Industrial Ergonomics and Safety II
Edited by Biman Das
Taylor & Francis 1990

WORK SAFETY IN HYBRID AUTOMATED PRODUCTION SYSTEMS ILLUSTRATED ON FLEXIBLE MANUFACTURING SYSTEMS

Bauer,W.; Kohlhaas, F.
Fraunhofer-Institut für Arbeitswirtschaft und Organisation (IAO)
Nobelstraße 12, D-7000 Stuttgart 80, W.-Germany

The presented contribution originated under a research project promoted by the ministry of social affairs. Starting out from exemplary realized examples, the objective of the project was the development of a planning-integrated approach to the selection and design of concepts and measures for avoiding dangers and stresses in flexible manufacturing systems.

On the basis of manufacturing systems implemented and proven in the industrial environment, applied safety concepts and individual measures for providing protection against unavoidable danger spots and sources of danger are intendet to demonstrate ways in an exemplary manner how to integrate work safety measures satisfying system requirements.

Introduction

Due to various constraints imposed by the market - such as quickly changing products, customization, manufacturing in small lot sizes, pressure on prices - also small and, particularly, medium-size businesses will have to increasingly use computerized means cf production. Concerns of work safety will have to be considered as a criterion above all by the designers of computerized means of production. In support of this task, a collection of examples was worked out by the IAO. This collection includes work safety strategies and measures proven in the practical environment of flexible manufacturing systems.

Design of flexible manufacturing systems under human aspects

The use of flexible manufacturing systems will be reasonable especially in the manufacturing of products of a parts family of lot size 1 or for small to medium quantities. Flexible manufacturing systems can comprise several CNC/DNC machining stations, magazining, storage,buffer and handling systems - interlinked with work and tool transport systems. This permits dissimilar workpieces to be machined within the system at the same time.

The various machining stations and the total material flow system are controlled via electronic data and information systems, which are normally configured in a hierarchical form. These electronic data and information systems permit increasingly complex manufacturing systems with flexible material flow and flexible machining sequence to be realized.

Due to the tie-up of substantial capital, as it is especially the case with flexible manufacturing systems, restrictions regarding the system tasks to be fulfilled are not acceptable. Therefore, comprehensive system planning is an imperative prerequisite for a manufacturing system of optimum design.

System design requirements

Apart from the increasing complexity of the manufacturing systems, the supra-divisional nature of the system solutions, the shorter design phases and the more exacting needs of the workers imply new requirements on the system design. The changed design process has the following underlying prerequisites:

- o all-embracing system design;
- o interdisciplinary team work (project management);
- o extended system goal taking engineering and economics criteria and human factors into account.

The human factors must be integrated in all-embracing system design of automated manufacturing systems as a goal ranking equal with that of the engineering and economics criteria. To consider the human factors in the advanced design stage or even as late as in the system installation - can generally only be part of the optimal conditions under engineering, economics and human aspects, with the human aspects then often being neglected, so that the economic objectives will not be endangered. Comprehensive work design under human aspects must, in principle, consider all part areas as shown in Figure 1.

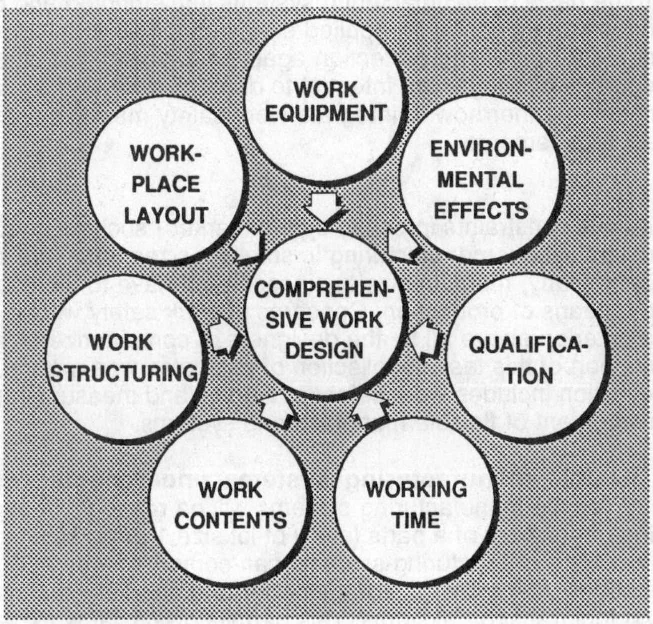

Figure 1: Comprehensive work design under human aspects.

<u>Work design from aspects of safety and ergonomics</u>

The use of automated manufacturing systems leads to a reduction in physical strains and an increase in work safety by relieving man of repetitive activities involving physical strain and by spatially separating man from the production process.

However, with the use of new information and communication systems, the demands made on the sense organs, the alertness and the capability to develop complex activity strategies (thinking, combining, decision-making) will be more exacting, so that it can be said that there will be a shift of emphasis from physical strains to psychological stresses.

As mentioned above, the use of automated manufacturing systems will ease work by reducing the physical strains. Nevertheless, this implies that the systems are designed accordingly. In many places, man was or still is used as a stopgap for performing such functions, which, for technical or rather economic reasons, appear not to lend themselves to be automated. The exclusive performance of such left-over activities lead to one-sided strains and to monotony. The most important measures to create an all-embrancing job content can be summarized by the following design guidelines:

- o clock-independent working by using buffers;
- o job enlargement by job rotating and by attending to predecessor and successor-stage functions (e.g.: material staging, rework, assembly, inspection, supervision);
- o job enrichment by performing function accompanying and preparatory to production (e.g.: tool setting, programming, maintenance, material requirements planning);
- o separation of automated and manual functions by block formation;
- o team work.

The realization of these design guidelines is greatly dependent on the system boundary to be defined during the work design. If the design scope is too narrow, both solutions possible in terms of work organization and engineering will be lost, which could bring about an improvement of both the working conditions and economic efficiency. The extension of the system boundary should not be restricted to the predecessor and successor-stage directly productive functions, but should also take into account the indirectly productive functions preparatory to and accompanying production. Only this permits the full utilization of the possibilities of creating an all-embracing job content.

Block formation isolates left-over workplaces from existing automated work stations and combines them to make up manual work areas. Here, the combination of isolated workplaces existing among automated work stations can frequently be realized only by automating the left-over activities, by changing the production sequences or by changing the products. To fully utilize the possibilities regarding job enlargement, job enrichment and team work, the manual work areas should be isolated from the automated work areas by buffers.

The safety requirements made on the system components and also on the total automated manufacturing system must be adapted to the changed production conditions, while maintaining the labour safety goals at the same time. In addition, new safety equipment and measures will be required, so that work safety designed under aspects of integration can be realized in the use of automated manufacturing systems. The following major changes have to be taken into account:

- o New, computer-controlled equipment such as industrial robots, driverless industrial trucks, automatic stock controllers and machining centres are used in flexible automated production systems.
- o The combination of many technical components for such functions as "storing / buffering / magazining, transport / handling and machining" to make up a total system, which must be contemplated for analysis and design.
- o The control of the production systems by several networked computers becomes a decisive factor for work safety, so that the hardware and software logic of the system control becomes an important subject in terms of investigation and design.
- o The primary target group responsible for safety concerns will be shifted from the directly productive personnel to the indirectly personnal, as the

number of employees engaged in programming, maintenance, repair and fault correction will increase and they will have to work in the immediate danger areas.

o With largely unmanned production plants, special measures will be required for labour safety at isolated individual workplaces.

Examples of realized flexible manufacturing systems

Example 1: Flexible manufacturing system with driverless industrial trucks

In this example (Figure 2), five machining centres, one washing machine and one measuring machine have been combined to make up a flexible manufacturing system. The workpieces are supplied from the clamping or storage places by two driverless industrial trucks.

The trucks are designed in such a way that they are riskless when moving. Things are different, however, when they are passing the transfer points of the machines, storage or clamping places. At these instants, the risk of bruising and chafing exists and, therefore, suitable protective measures have to be taken.

Therefore, these danger areas are monitored by photoelectric barriers, causing the trucks to stop short of these areas, if workers are staying inside these areas.

A safety screen fence surrounding the total cruising zone of the industrial trucks prevents unauthorized persons from entering this zone and causing system interference.

At the clamping stations, safety measures are required to the effect that the movements of the swivel table and the transfer of the work holder to the industrial trucks are secured.

All in all, it has been found that the engineering, economics and human aspects have been considered in the developed safety concepts. The requirements made on the manufacturing systems could be satisfied and the associated goals achieved.

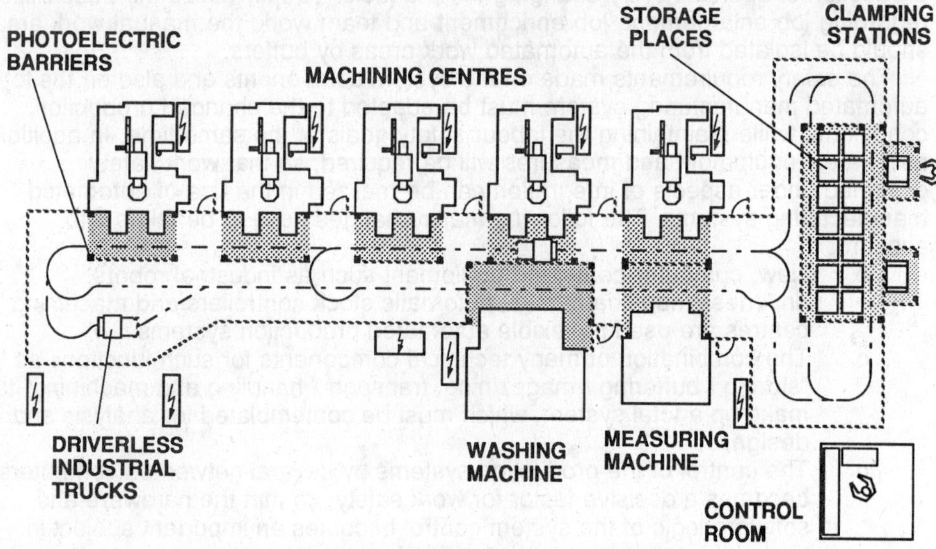

Figure 2: Layout of the flexible manufacturing system with driverless industrial trucks (BMW).

Example 2: Flexible manufacturing system for big rotating workpieces

The layout of this realized flexible manufacturing system is shown in Figure 3. An essential criterion for the design of this system was the prerequisite that the operators could freely move between the machines and the workpiece setting, clamping and storage places. Only this guarantees that the specialists can attend

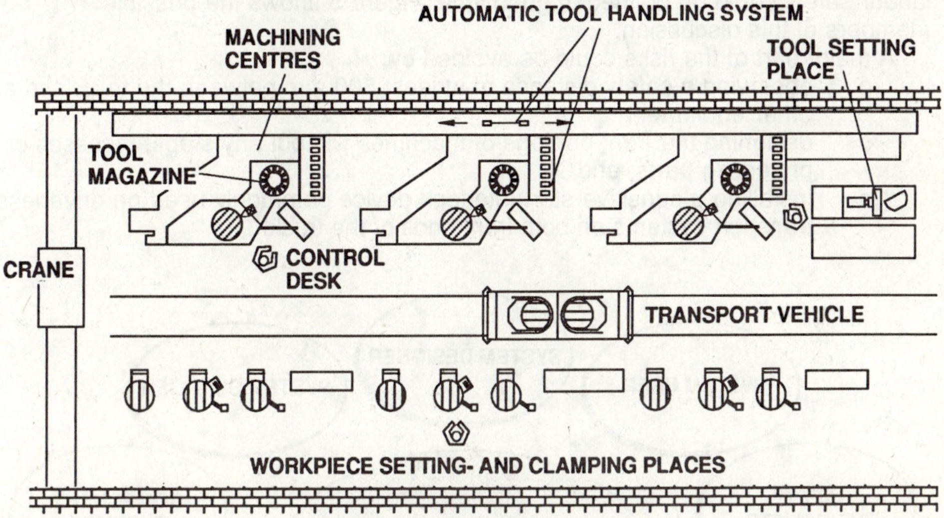

Figure 3: Layout of the flexible manufacturing system for big rotating workpieces (J.M. Voith GmbH).

Figure 4: Flexible manufacturing system for big rotating workpieces (J.M. Voith GmbH).

to their versatile tasks like machine operation and supervision, setting up, clamping and testing of workpieces, disposal of chips, etc. With respect to labour safety, this requirement does not allow the use of separating protective measures, as e.g. a safety screen fence, around the total system or the transport system.

On the basis of a design-integrated safety discussion suitable measures for labour safety could be defined in good time. Figure 5 shows the possible members of this discussion.

A major part of the risks could be avoided by:

- o observing a safety distance of at least 500 mm between the truck and all other equipment;
- o designing the front portions of machines without any steps, recesses or projecting parts, and
- o installing a sensitive start-interlock device commonly used on driverless transport systems on both front ends of the truck.

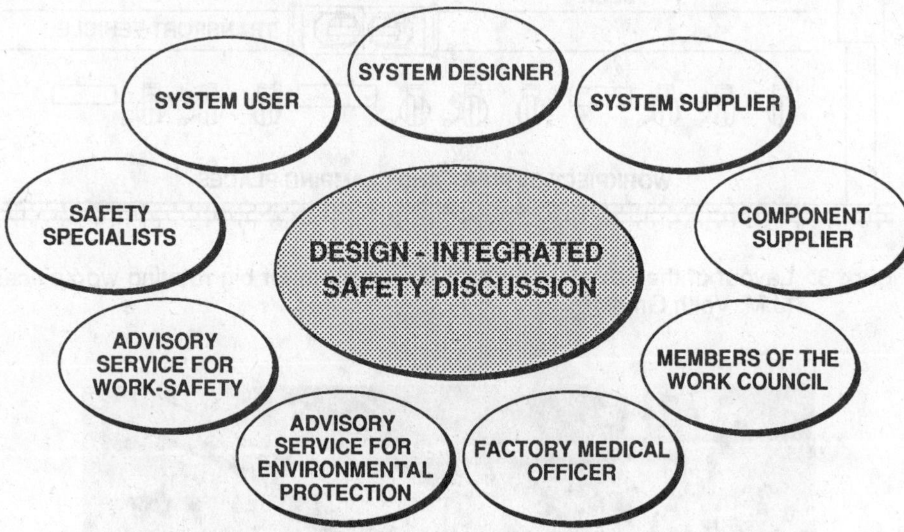

Figure 5: Members of a design - integrated safety discussion.

For safeguarding the transfer of workpieces, the transport vehicle is equipped with a roll-out safety screen, which must be fully extended before any workpieces can be transferred (Figure 6). The movements of this screen are secured, so that no risks exist. The transport and transfer operation can be secured in this manner, as the transport system - due to the long machining times - moves at a relatively slow speed.

The various man-system interfaces - as on the machines, the rotary clamping and marking tables and the tool staging and dispensing stations of the automated tool storage and transport system - have each been secured in such a way that the machining process and the workers will not be impaired in the performance of their tasks.

The safety concept used in this manufacturing system permits the workers to adopt a wide areas of activity. These areas of activity include both directly productive functions and indirectly productive predecessor and successor-stage functions. Consequently, the job content is versatile. One-sided strains and monotony were thus avoided.

Figure 6: Transport vehicle with roll - out safety screen (J.M. Voith GmbH).

Conclusion

Computer-controlled means of production are increasingly being used as a result of various market-induced demands. Engineering, economics and human objectives must be equally taken into account when planning, introducing and using such systems.

To avoid to be overpowered by a determinism of engineering, system planning should not start with the design and selection of the technical equipment, but - after a functional assignment to the functional resources "man" and "machine" within the work system has been made - the work should be divided among the workers. Here, the technical equipment should take over those work functions which are physically straining and intellectually undemanding for man; and efforts should be made to achieve an all-embracing job content for the workers. The development and design of the safety measures must be system-oriented.

Based on existing flexible manufacturing systems, work safety concepts have been demonstrated, which do not neglect the objectives of labour safety and, at the same time, also pay reasonable attention to engineering and economics concerns.

References

Kief, H.B., 1986, Flexible Fertigungssysteme. NC - Handbuch - Verlag, Michelstadt.

Bullinger, H.-J. and Traut, L., 1986, Die Fabrik der Zukunft. In Fortschrittliche Betriebsführung und Industrial Engineering, Heft 1.

Kohler, P., 1989, Flexible Fertigungssysteme für die Einzel- und Kleinserienfertigung großer Rotationsteile. In wt Werkstattstechnik, 79, 415 - 418.

Advances in Industrial Ergonomics and Safety II
Edited by Biman Das
Taylor & Francis 1990

THE ERGONOMICS OF THE SECOND COLLISION

Charlie H. Miller, Ph.D., P.Eng. and C. Robert Baird, Ph.D., P.Eng.

Vehicle Safety Research Team
Technical University of Nova Scotia
P.O. Box 1000, Halifax, Nova Scotia
Canada B3J 2X4

THE CONCEPT OF COMPRESSED-TIME ERGONOMICS

Until about one century ago mankind had existed in a quasi-static environment which may best be described as a 'one-G World'. The fastest transportation was provided by a horse on land or by sail-boat on water. Rarely, if ever, were our ancestors subjected to accelerations or decelerations greater than the one-G level of living on the surface of our planet.

The one exception to this was with the development of weapons. These weapons, such as spears, arrows, swords and later bullets, were deliberately designed to concentrate relatively weak forces to injure the body. By virtue of rapid deceleration and concentrated contact area, the offending weapon transferred high impact pressures to the body surface and structure, resulting in penetration and possibly lethal damage to the unfortunate person who was struck by the weapon.

Modern man, particularly in this century, in addition to extending the energy levels, concentration and efficiency of weapons, has succeeded in utilizing newly available energy sources to serve as high-speed person carriers. Thus man, without intending to do so, has placed himself inside self-propelled ballistic missiles.

There is no doubt that modern transportation machines have caused major alterations to global societies. It is not the purpose of this paper to discuss this issue. Rather, the particular issue of interest here is the effect of man interacting with the interior of these transporter 'weapons' when something goes wrong in their operation.

When 'things go wrong'

When 'things go wrong' with the control or motion of a high-speed transportation vehicle, energy at a level that was undreamed of in past centuries must suddenly be managed or death and injury will typically result. The higher the speed and the shorter the interaction time, the higher are the decelerations and the higher are the impact forces. People involved in a car accident can be subjected to decelerations and forces that are several orders of magnitude greater than those encountered in falling off a horse. This is appreciated by a surprisingly small proportion of the population.

Time compression

The simple but very important relationships between distance (S), velocity (V), acceleration (a), mass (m), force (F) and time (t) are well known from elementary physics, yet we often tend to overlook their significance:

$$F = ma = m\frac{dV}{dt} = m\frac{d^2S}{dt^2} \qquad (1)$$

. . . where d is the differential operator.

Since we are concerned with non-relativistic velocities only, mass (m) remains constant and even a cursory qualitative examination of Equation 1 clearly shows that the resultant force (F) will very rapidly increase as the interaction time (t) decreases.

In North America as well as throughout most of the world, the unfortunate effects of the above relationships are being enacted and reenacted every few seconds. The basic reaction is evident — high speeds — sudden crash stops — high G deceleration forces!

To all intents and purposes, in human reaction capabilities, time is compressed! We no longer have the luxury of designing man-machine interfaces on the basis of quasi-static interactions!

The second and potentially deadly collision occurrs when the occupants of the vehicle involved in a crash themselves collide with the interior of the vehicle as it is rapidly decelerated.

THE NEED TO DESIGN FOR THE MULTI-G WORLD

The datum of human reaction times

Human reaction time is defined as that time which includes the perception of an object, the comprehension of the involvement, the decision to act and the physical actions required to put the decision into effect. It typically ranges (Bosch, 1986) from 0.3 to 1.7 seconds, depending on an individual's personal factors (such as conditioning of reflex actions, disposition, available skills, age, anticipation or inattentiveness, general health and the influence of drugs and/or alcohol). It is also dependant on external factors such as conspicuity of offensive object, predictability of situation, location of perceived object and sensibility of controls available.

In the quasi-static world of one-G, a person on a horse travelling at a slow trot at 10 km/h (app. 6.24 mi/h) would travel between 0.84 metres (2.75 ft) and 4.76 metres (15.6 ft) before his/her reaction would even begin to yield results. At highway speeds of 100 km/h (app. 62.4 mi/h), the normal highway speed limit in Canada, the person would obviously travel between 8.4 metres (27.5 ft) and 47.6 metres (156. ft) prior to initiation of his physical actions. The implications are obvious.

It must be noted that despite the relatively high speed of the person in the car travelling at 100 km/h, he/she still exists in a one-G world so long as the vehicle speed remains unchanged. Even with maximum braking, the vehicle would only be subjected to an approximate one-G braking deceleration, in orthogonal combination with the existing one-G gravitational acceleration, resulting in a 1.41 G resultant retardation at 45 degrees down and to the rear. The vehicle occupants, if unrestrained, would follow Newton's Law and tend to continue their forward motion while the car slowed down, causing the occupants to brace themselves fairly easily against the one-G braking component, if prepared by sufficient reaction time.

<u>Human tolerances on a G vs. time basis</u>

Aircraft fighter pilots can be subjected to several Gs in real or simulated flight situations. In a maximum-rate turn the aircraft will typically be capable of surviving high-G loads than will the pilot. In a pullout from a dive where about 3 to 4 Gs may easily be generated, it is very difficult for a pilot to close his mouth or raise his arm. At 6 to 7 Gs sustained for several seconds the pilot is likely to 'black-out'as the blood is centrifuged from his brain.

The first major acknowledgement of and concession to multi-G ergonomic design was in the design of the fighter pilot's 'G-suit', a very specialized selectively pressurized coverall that the pilot wears to restrain his blood from rushing to his lower extremities. These suits have reached a very sophisticated level of design with the local pressure developed within the suit being automatically controlled by the G-induced force itself. Even with this expensive and sophisticated mechanism specifically designed for the purpose, about 9 Gs for a few seconds is the normal limit.

Man's physiological condition is such that G-level and time of application are significantly and complexly related.

OTHER EXAMPLES OF NON ONE-G ERGONOMICS

We have become so used to thinking in terms of Earth-surface, one-G design situations that there is considerable confusion between the terms 'mass' and 'weight'. Fortunately they are not incompatible since mass is a fundamental measure of matter and its associated inertia, whereas weight is a measure of the force generated on or by that mass under the influence of what has become accepted as the standard Earth's gravitational pull. If we are going to establish habitable bases on the Moon and other planets as well as on space stations, mankind will be subjected to several sub-standards of 'weight'. The newest significant example is the weight of a human being on the Moon, roughly one-fifth of that on Earth — yet his mass and all related inertial effects remain the same.

Zero G situations such as encountered in orbital spacecraft or other free fall circumstances produce problems of their own. The necessity or even desirability of Earth-style furniture in orbital machinery is interesting to contemplate, but which will not be considered here.

APPLICATIONS TO MOTOR VEHICLE ACCIDENTS

The extent of the carnage and damage caused by motor vehicle accidents is generally not recognized. In the U.S.A. and Canada alone a total of about 48,000 (fourty eight thousand!) people die each year on the roads of these two countries. *This is equivalent*

to having one major aircraft crash per day, every day of the year, in which 130 passengers were killed in each crash! This aircraft crash rate would never be tolerated by our governments, our people or our society in general. Admittedly one may argue statistical structure on the basis of relative passenger miles covered, total vehicles in motion, passengers per vehicle, etc., but the fact remains that society accepts road deaths as being an inevitable fact of life, but recoils in horror at the far lower death rate which occurs in the more dramatic aircraft crashes.

Deaths alone do not convey the full picture; for every person killed in an aircraft crash there may be one or two who survive with some injury. For every person killed in a road accident, statistics (Transport Canada, 1988) show that there are approximately 64 or 65 victims injured for every fatality. Even allowing that most of these injuries are of a relatively minor nature, the ratio is still very significant. The abbreviated Injury Scale (AIS), established by the Association for the Advancement of Automotive Medicine (AAAM), rates injuries on a detailed, consistent and carefully developed scale ranging from (1) minor to (6) untreatable, i.e. fatal, for the various major bodily regions. Since all Level II and Level III investigations record these injuries it is not difficult to assess the injury distributions.

In addition to the very substantial family tragedies involved, the cost to the communities in terms of lost talent, medical costs, lost productivity, etc., is enormous. The funding allotted by our governments to the problem of road carnage is greatly out of proportion relative to that provided for aircraft crash investigations.

Range of motor vehicle interactions

Motor vehicle accidents may be generally divided into two groups, namely, (1) single vehicle accidents and (2) multi-vehicle accidents. The latter can range from two vehicles, the norm, to thirty or more vehicles on multilane, densely driven throughfares. With both of these groups head-on collisions predominate. Frontal angular contacts and side impacts are about equal in frequency, whereas rear and rear angular impacts are defionitely in the minority. With single vehicle accidents side impacts remain relatively high, where the vehicle motion typically degenerates to a skid with a substantial lateral component. Often the vehicle side impacts a tree, a telephone pole or a barrier post.

Frontal impacts

With frontal impacts dominating the accident scene, investigators and legislative action groups correctly concentrated testing, redesign and standards development to teh frontal regions of automobiles and trucks. A comparison of frontal barrier impact testing of 1930 vintage automobile in comparison to the same test performed on a typical mid-eighties vehicle is visually very dramatic. Whereas the 1930 vehicle has a very rigid chassis, the vehicle does not absorb energy on a controlled basis, as does the eighties vehicle. The thirties vehicle rebounds from the barrier, the windows shatter, the fuel tank splits open and the seats and occupants fly forward. In contrast, the eighties vehicle rebounds remarkably little, glass that is broken is either held in small pieces by built-in plastic films or is reduced to relatively less dangerous small sized 'pebbles', the fuel tank remains intact and of utmost importance, the occupants are restrained by controlled stretch seat-belts which are given time to perform by the slow crushing of the front end of the vehicle. The modern concept of vehicle design is to provide living room in the passenger compartment while protecting this padded pasenger compartment by structural advancements against intrusion, and having the controlled energy-absorbing crush occur elsewhere.

Designing for the crash condition is certainly sound structural design, sound safety practice and it is certainly sound ergonomics!

Figure 1 below is a reproduction of a photo of a frontal crash of a car which was in a head-on collision with a heavier car, resulting in substantial front-end crush. Although the A-pillar on the driver's side is pushed aft and the windshield lies like a limp rag on the cowl, nevertheless there is remarkably little intrusion into the passenger volume. Although the driver would receive some leg and facial injuries, even a severe crash like this is survivable if the occupants used the available three-point restraints. Unfortunately many people are unaware of the magnitude of the deceleration forces involved in a crash of this type and either refuse to wear the seat belts or do not adjust them properly.

Fig. 1 – Example of head-on crash

One has only to wander among the rusting hulks in any wrecking yard (be sure to get permission and to know that the inevitably present guard dog is chained) to see some of the effects of non-use of seat belts. Figure 2 shows the impact of the driver's head on the windshield which occurred in this low-speed frontal angular crash. Note that the windshield absorbed some of his impact by stretching the built-in plastic. Earlier model winshields would undoubtedly have shattered into large 'slabs' and could well have fatally severed the driver's neck.

Fig. 2 - Why use a seat belt?

<u>Side impacts</u>

Although frontal impact has been the recipient of a higher priority than side impact, emphasis is now turning to the latter condition. This is not to imply that all the problems associated with frontal conditions are solved, but recognizes that some substantial progress has been made in this direction over the past decade or so. The latest advances in respect to protection against the second collision in frontal crashes have arrived in the form of so-called 'air bags' and pre-tensioned restraints, both devices triggered by very fast activators that are set off at pre-determined levels by the very decelerations that cause the second collision. These devices, if fitted, significantly aid in overcoming the excessive slack typically found with poorly adjusted seat belts.

With side impacts, seat belts are less advantageous than they are in frontal collisions. A simple study of the geometry involved should make this apparent. The crushable material between an occupant of a target vehicle and the front of a side-impacting bullet vehicle is far less than that available in front of the driver in relation to a head-on accident. Given equal designer competence, the amount of crushable material can be directly equated to the energy-absorbing capability of the structure. Rapid intrusion of the bullet vehicle into the occupant living volume of the target is almost a certainty in a moderate to high speed side impact.

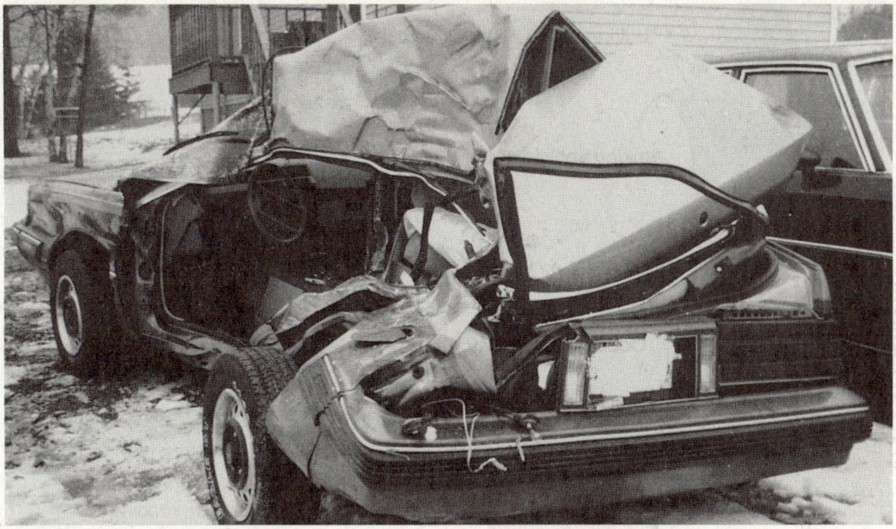

Fig. 3 – Typical side impact

Figure 3 above is unfortunately a 'normal' target-car side impact result. It is immediately evident that there is substantial intrusion into the occupants' living space, with dire repercussions. What is not immediately apparent is the rapidity with which the target vehicle door is torn from its mountings and smashed into the driver's torso. Whereas in a frontal collision a belted occupant has normally between 100 and 120 milliseconds to have his energy partially or fully absorbed by the restraints and, if fitted, the air bag, the side impacted occupant may well be impacted within 25 to 40 milliseconds! He is simply impacted by a massive energy surge into his living space, and belted or not, his inertia subjects him to enormous G-loads. Very well designed and expensively instrumented dummies, standardized as SID (Side Impact Dummy) and EUROSID (for the European developed version) are described in several excellent papers in the literature, for example, by Lau, Viano, et al (1989) and Janssen and Vermissen (1988). Accelerometers mounted on the dummies' carefully simulated ribs have registered well over 400 Gs, before saturating! Under these conditions, so often repeated in real life, the vehicle occupants nearest the bullet vehicle have their internal organs massively ruptured due to inertial effects, resulting in virtually instantaneous death!

<u>A simplified velocity-time diagram of side impact</u>

Figure 4 which follows on the next page represents a simplified plot showing the velocities characteristic of a few parts of the vehicles involved in a side impact (simplified from Strother et al, 1984). The important thing to note is the short time it takes to impact the occupant.

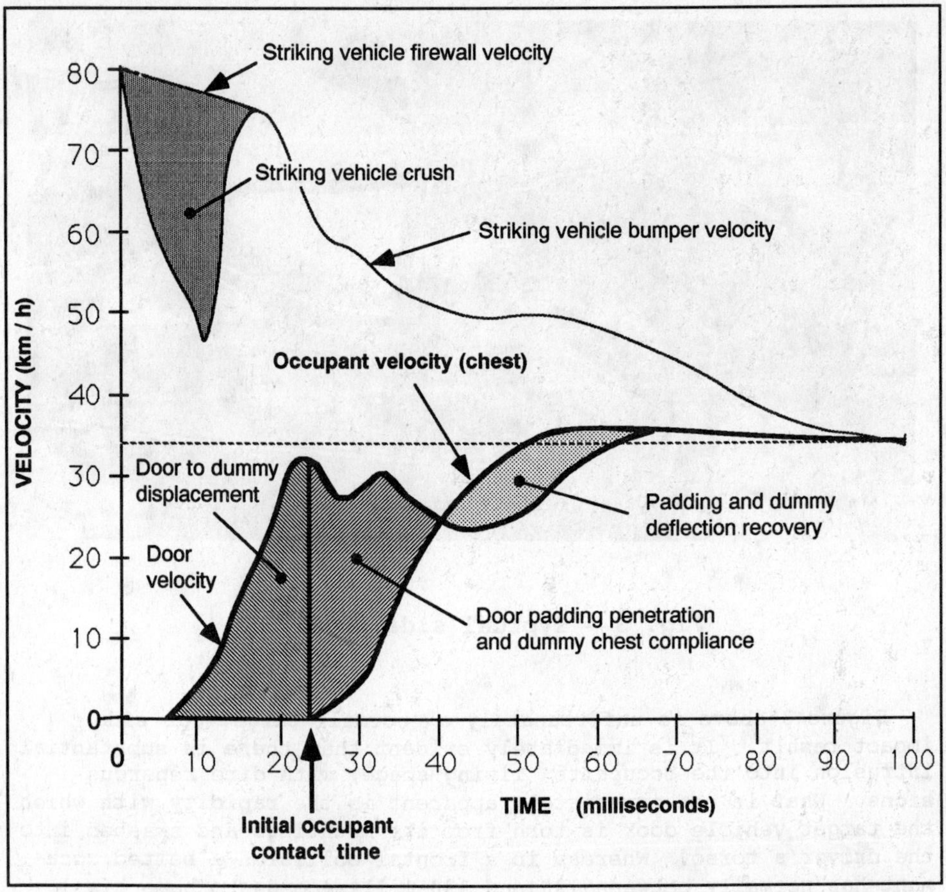

Fig. 4 – Characteristic velocity-time plot of side impact

REFERENCES

Bosch, 1986, Automotive Handbook, 2nd Edition, edited by U. Adler and P. Girling (Delta Press: Society of Automotive Engineers Inc., under license from VDI-Verlag, GmbH), pp.264-265.

Lau, I.V., Viano,D.C., Culver,C.C. and Jedrzejczak,E.,1989, Design of a Modified Chest for EUROSID Providing Biofidelity and Injury Assessment, Side Impact: Injury Causation and Occupant Protection, Proceedings, SAE International Congress, Detroit, pp.163-172.

Janssen,E.G., and Vermissen, A.C.M., 1988, Biofidelity of the European
Side Impact Dummy – EUROSID, Proceedings of the 32nd Stapp Car Crash Conference, Atlanta, pp.101-124.

Warner,C.Y., Strother,C.E., James,M.B., Struble,D.E. and Egbert,T.P., 1989, Crash Protection in Near-Side Impact – Advantages of a Supplemental Inflatable Restraint, Side Impact: Injury Causation and Occupant Protection, loc. sit., pp.85-97.

Advances in Industrial Ergonomics and Safety II
Edited by Biman Das
Taylor & Francis 1990

ORIENTATION PERCEPTION IN FARM TRACTOR OVERTURN: A MODELING AND SIMULATION APPROACH

Joseph H. Goldberg[1]
Vasu Parthasarathy[2]
Dennis J. Murphy[3]

The Pennsylvania State University
University Park, PA 16802

Farm tractor overturn is a problem due partly to machine design and partly to human error. Human error contributions to overturn are considered here, from the joint perspective of modeling and empirical research, with the ultimate goal to provide recommendations for effective operator field training. Models obtained from the literature show that an operator's ability to sense changes in orientation degrades with current pitch or roll angle. A 3-axis simulator was constructed to test these predictions. In both younger and older subjects, errors in roll and pitch perception significantly increased with respective roll and pitch angles.

INTRODUCTION

Farm Tractor Overturn

With an overall death rate of 49 per 100,000 workers in 1987, agriculture (including forestry and fishing, but not logging) is the most hazardous industry classification. There were 3.4 million agriculture workers in 1987, with 1600 deaths and 160,000 reported injuries during this time (National Safety Council, 1988). Operation of tractors is one of the leading causes of injury on the farm, and tractor overturn is the most common cause of tractor-related fatality, accounting for over 52% of the 500 tractor fatalities in 1987. The death rate was 10.9 per 100,000 tractors, with about 4.44 million tractors in operation (National Safety Council, 1988). This incidence of tractor overturn has been constant over the past several years.

Pennsylvania statistics reveal that 50% of the 253 farm fatalities during 1980-1984 were tractor-related, and of these, about 70% were due to tractor overturn (Murphy, 1985). A 1987 cross-sectional analysis of Pennsylvania agricultural injuries revealed a low incidence for farm tractor injury-causing accidents, but high severity once an accident occurs. Farm tractors were responsible for only 5.3% of injuries, but

---

[1]Assistant Professor, Dept. of Industrial Engineering

[2]Graduate Student Rsch. Asst., Dept. of Industrial Engineering

[3]Associate Professor, Dept. of Agricultural Engineering

accounted for over 165 lost workdays per injury, 6 times greater than any other injury source (Huizinga and Murphy, 1988).

These injury statistics are difficult to compile on a national basis, as separation of the home and workplace is difficult, and job/equipment contact time are unknown due to irregular shifts. Nevertheless, it is clear that farm tractors are hazardous; expected severity of tractor incidents places then near the top of the occupational hazard list.

Age Effects

Farming is an industry without occupational age restrictions, as evidenced by high fatality rates in extremely young and old operators. Fatality rates in Pennsylvania from 1980-1984 peaked at 0-4 years, 15-19 years, and at 55-59 years of age (Murphy, 1985). Fatal Georgia farm-tractor accidents from 1971-1981 revealed a steadily increasing mortality rate with age (Smith, et al., 1983). Fatalities per 100,000 male farm workers climbed from 6.7 (<20 years of age), to 22.3 (20-39 years), to 27.6 (40-59 years) to 54.1 (>60 years). The mechanism of these age differences in injury incidence is unknown, but differences in perceptual, decision-making, and responding capabilities are likely causative factors.

Stability Perception

Tractor overturns arise either from loss of stability due to inherent design characteristics, or from attempts to use the tractors beyond its designed stability limits (Spancer, Owen, and Glasbey, 1985). High centers of gravity coupled with changing loads and environmental conditions contribute to rapid side or rear overturn, once a point of instability has been reached (Murphy, et al., 1985). With an average age of a farm tractor approaching 20-30 years and 3 tractors per farm (Huizinga and Murphy, 1988), the time lag for design changes is significant. Instead, the present research has concentrated on operator capabilities, in order to provide recommendations for field training by farm safety specialists. Such training can have a rapid impact on tractor safety statistics.

Operator error in tractor overturn control can be divided into errors of perceiving appropriate information, deciding which response to make, or in making a response. The operator must maintain an accurate perception of tractor center of gravity, which changes as loads are varied or rear attachments are added. The operator must sense centrifugal force, rear axle torque, and drawbar leverage (Murphy and Johnson, 1982). Environmental characteristics to be sampled include slope angles, velocity, draft, ground surface, and turning angle (Goldberg and Parthasarathy, 1989; Goldberg, Parthasarathy, and Murphy, 1989a). Changes in these characteristics are proximally translated into changes in the visual scene, forces on the body, noises, and other sensations. Excess vibration and fatigue over a working day can clearly diminish the accuracy of sampling of these cues by an operator.

There is evidence that an operator traveling along or into a slope has diminished ability to sense changes in stability, compared with an operator traveling on a flat field. Goldberg and Parthasarathy (1989) established from the aerospace studies that errors in the static perception of upright increase with increasing body tilt:

$$\psi(\alpha) = 3.45 + .07\ (\alpha) \qquad (1)$$
$$\psi(\beta) = 1.48 + .29\ (\beta) \qquad (2)$$

where α is the absolute roll angle (degrees), β is the absolute pitch angle (degrees), and $\psi(\ )$ is the minimal perceived change in roll or pitch angle (degrees). These equations apply only to blindfolded operators, and to static conditions, neither of which is the case for the real world tractor operator. Furthermore, the operator tends to place stronger emphasis on visual information when visual and proprioceptive senses are in conflict. However, Equations 1 and 2 can shed some light on why a tractor operator can fail to control his machine before it reaches the critical angle of instability.

Given that instability has been perceived, the operator must execute an appropriate response, such as a steering or throttle correction. The reaction time for response initiation can easily exceed 500 msec, and is degraded by exposure to vibration (Stephens, et al., 1972). Average reaction time for power cutoff has been measured between 337-613 msec, depending on placement and type of power cutoff device (Pattie, 1973). Steering change completion is relatively slow, requiring close to 1 second to complete a 50° change (Rehkugler, 1980). Unfortunately, these studies have not always separated true reaction time, movement time, and control activation time, so that pure changes in control type or location cannot be predicted.

Objectives

A preliminary investigation was conducted to further model individual operators' ability to perceive changes in pitch and roll, and the interaction between pitch and roll on errors in angle judgement. Furthermore, both older and younger age groups were introduced to seek regular age effects. Pure vestibular and proprioceptive ability was isolated in this study by blindfolding all subjects.

METHOD

Subjects

Ten male subjects were recruited for this investigation; 5 from a younger (ages 21, 23, 24, 24, 26 years) and 5 from an older (ages 65, 67, 71, 79, 85 years) age group. All subjects were active and healthy. The younger group were recruited from Industrial Engineering graduate students at Penn State, whereas the older group was recruited from a community senior citizen's activity center. Each was paid $5/hour for 4 hours of participation, spread over two sessions.

Tractor Simulator

A 3-axis tractor overturn simulator was constructed to support this research. A brief summary of its design and capabilities is presented here; see Goldberg, Parthasarathy, and Murphy (1989b) for more detailed specifications. Realistic motions could be produced in side-to-side roll, front-to-back pitch, and yaw axes. Figure 1 presents a photograph of the simulator in a rolled, but non-pitched position. The operator sat in a John Deere tractor seat suspended within independently gimbaled pitch, roll, and yaw tubular steel frames. The maximum operator payload was 180 pounds, due to torque limitations in the drive motors. Leg space in the cockpit was limited to further decrease required torque. Limit switches restricted roll and pitch motion to ±37°.

The simulator was powered by stepper motors interfaced to a computer and controller. Three-stage speed reduction was provided by belts and pulleys to raise the final torque to 170 ft-lbs. in both pitch and roll axes. A lower-torque chain drive was used for yaw control, due to relatively low speed and torque requirements.

Feedback and control of simulator position are diagrammed in Figure 2. Inputs include steering angle, throttle, brake, clutch, and power cutoff, which are digitized via an analog/digital converter. The computer sends digital instructions to the motor controller, which moves the simulator. Position of each axis is sampled via potentiometers, in addition to software algorithms. For the present experiment, a joystick in the hands of the subject controlled pitch and roll position. Yaw motion was not utilized in this pilot study.

Figure 1. Tractor Overturn Simulator

<u>Procedure</u>

Upon arriving at the testing site, subjects signed an informed consent form and received full experimental instructions. The first day presented 2 hours of training and practice, and will not be further described here.

Each of 196 trials was run in an identical manner. The subject initially sat, with seatbelt fastened, in a 0° pitch, 0° roll position. A positive roll was defined as one to the subject's right, and a negative roll as one to the subject's left. A positive pitch was a backward tilt, and a negative pitch

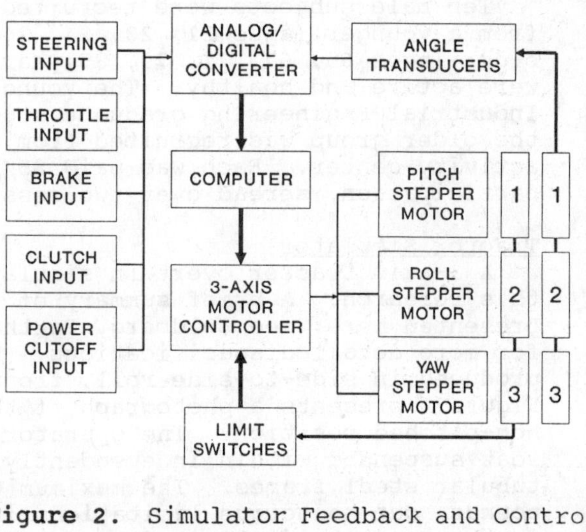

Figure 2. Simulator Feedback and Control

was defined as a forward tilt. After 5 seconds, the simulator automatically moved into a defined standard angle position, defined by pitch and roll angles between -18° and 18°. The blindfolded subject memorized this position, and pressed a button on the joystick to indicate he was ready. The simulator moved to a random location in the opposite pitch and roll quadrant, to ensure it moved far enough so the subject could not simply note how much time elapsed during simulator motion. The subject now used the joystick to move the simulator cockpit back to the original standard position. When he felt he was at the original standard, he pressed a button on the joystick, and the cockpit moved back to the 0° pitch, 0° roll position to start the next trial. Using the power cutoff button, the experiment could be paused or halted at any point, but no subjects required this additional break.

Design

The independent variables were subject, age group, standard pitch angle, and standard roll angle. The dependent variables were the angular error in subjects' responses for both pitch and roll axes. Positive and negative angular sign conventions were maintained as described above. A total of 196 response trials, each with a pitch and roll error, were gathered for each subject. These were defined by:

7 Standard Pitch Angles (-18°, -12°, -6°, 0°, 6°, 12°, 18°) x
7 Standard Roll Angles (-18°, -12°, -6°, 0°, 6°, 12°, 18°) x
4 Replicates = 196 trials

The order of the 196 trials was completely randomized, defining a 4-factor factorial design. Subjects were actually nested within age groups, but were treated as covariates in the following analyses.

RESULTS

All results below are pooled across the five subjects in each respective age group, due to space limitations. Response error in pitch and roll axes, the dependent variables, was defined by the difference between the standard angle and the actual angle of the subject's position when the button was pressed. A positive roll error indicates an actual position to the right of the standard, and a positive pitch error indicates an actual position to the subject's rear of the standard. Since subjects were informed to come as close as possible to the memorized standard, roll and pitch errors are analogous to minimal perceived angular differences, presented in Equations 1 and 2.

Table 1 summarizes pitch and roll response errors as a function of subject group, standard axis, and standard angle. Most of the pitch error showed a strong negative bias, towards the rear of the standard. Analysis of Variance (ANOVA) was conducted on both pitch errors and roll errors as a function of the independent variables. In these ANOVAs, Subjects and Age Group were treated as covariates; as such, no interactions between these and any other independent variables were interpreted. Of primary interest were pitch and roll errors as a function of the standard pitch and roll angles, and their interaction.

Pitch Errors

Subject differences were only marginally significant on pitch errors ($F_{4,921}$=2.3, p=.06). The Standard Pitch angle was quite influential on these errors ($F_{6,921}$=28.4, p<.001), but Age, Standard Roll angle, and the Pitch by Roll angle interaction were not (p>.10). Inspection of the data for both pitch and roll errors did not reveal any nonlinear trends, probably due to the large amount of variability present in the data. A multiple regression for Pitch Error, shown in Table 2, was conducted to obtain quantitative estimates of independent variable influence. (While included to explain variability, the Subject indicator variables should not be interpreted, as subjects were nested within age groups.) Pitch Error increased by 0.2° for each 1° increase in Pitch Angle.

Table 1. Pitch/Roll Response Errors by Subject Group

Standard		Younger Subjects		Older Subjects	
Axis	Angle	Pitch /	Roll	Pitch /	Roll
Pitch	-18	-1.94	-0.19	-5.61	1.28
	-12	-1.48	0.29	-4.01	0.19
	-6	-0.98	0.70	-3.47	0.62
	0	-0.97	0.11	-0.47	0.84
	6	-1.02	-0.10	3.24	-0.41
	12	0.45	0.70	3.40	0.72
	18	1.50	-0.25	3.74	2.83
Roll	-18	-0.84	-2.59	1.19	-0.51
	-12	-1.43	-0.23	-1.10	-0.01
	-6	-0.60	-0.36	-0.71	0.85
	0	0.23	-0.33	-0.64	-0.29
	6	-0.64	1.00	-0.14	2.20
	12	-0.74	1.45	-0.42	1.68
	18	-0.43	2.31	-1.36	2.16
Pooled		-0.64	0.18	-0.46	0.87

Table 2. Pitch and Roll Error Regression Models

Regression Model Term	Model for: Pitch Error	Roll Error
Intercept	-1.17 **	0.15 ns
Age Group Indicator (0=younger, 1=older)	0.18 ns	0.69 *
Subject 1 Indicator	1.10 *	0.40 ns
Subject 2 Indicator	-0.10 ns	1.08 *
Subject 3 Indicator	0.48 ns	-1.11 *
Subject 4 Indicator	1.18 *	-0.23 ns
Standard Pitch Angle	0.19 ***	0.01 ns
Standard Roll Angle	-0.01 ns	0.10 ***
Standard Pitch x Standard Roll	0.00 ns	-0.004 **

ns = not significant * p<.05 ** p<.01 *** p<.001

Roll Errors

Variability was generally smaller for roll error then pitch error, so results were more significant. Strong differences in roll error were noted between Subjects ($F_{4,926}$=5.9, p<.001) and Age Groups ($F_{1,926}$=5.4, p<.05). While Standard Pitch angle was not influential (p>.10), the Standard Roll angle was ($F_{6,926}$=11.5, p<.001). In addition, the interaction between Standard Pitch and Roll was significant ($F_{36,926}$=1.85, p<.01).

Multiple regression (see Table 2) indicated an increase in roll error of 0.7° in the older subjects. Furthermore, roll error increased by 0.1° per 1° increase in Roll Angle. This dependence on Roll Angle was, however, also dependent on the current Standard Pitch angle, as evidenced by the interaction.

CONCLUSIONS

This experiment demonstrated that accuracy in the perception of angular pitch is quite dependent on one's current pitch angle; similarly, accuracy in roll perception is dependent on one's roll angle. Aging plays a role in increasing errors in roll perception, but much less so for errors in pitch perception.

These results were very much in agreement with prior results. Here, the slope for error in pitch perception was 0.19°/pitch degree change. From Equation 2, the slope was 0.29°/pitch change. Similarly, the present slope for error in roll perception was 0.10°/roll degree change, whereas Equation 1 presented a slope of 0.07°/roll change. The presence of an interaction between pitch and roll axes on roll perception is an important result; changing the roll angle can induce changes in both roll and pitch errors.

Implications for the tractor operator are clear, for relatively static situations. When operating on a steep slope, the standard roll and/or pitch angles increase. As demonstrated above, the operator's ability to sense a change in roll and/or pitch declines in proportion to the slope. Thus, the minimal detectable change in tractor orientation is increased, and the tractor is closer to the angle of instability before the angle is perceived. The time in which a corrective action may be successfully executed is now decreased in proportion to the slope angle, and the probability of an overturn increases. The desired training strategy for this situation should be to increase the accuracy and efficiency of sampling one's internal and visual stability cues. An overturn simulator would thus be an ideal field training tool to demonstrate these stability cues.

ACKNOWLEDGEMENTS

The authors would like to acknowledge support received from the following sources to construct the tractor overturn simulator, and to conduct the investigation reported here:

- Penn. State Dept. of Industrial & Management Systems Engin.
- Penn. State Dept. of Agricultural Engineering
- National Institute for Occupational Safety and Health
- Penn. State Gerontology Center

REFERENCES

Goldberg, J.H., and Murphy, D.J., 1985, Safer tractor operation through improved stability/instability feedback. American Society of Agricultural Engineers, Paper 85-5515, St. Joseph, MI.

Goldberg, J.H., and Parthasarathy, V., 1989, Operator limitatic in farm tractor overturn recognition and response. Applied Ergonomics, 20(2), 89-96.

Goldberg, J.H., Parthasarathy, V., and Murphy, D.J., 1989a, Operator limitations in tractor overturn recognition and response. American Society of Agricultural Engineers, Paper 89-1107, St. Joseph, MI.

Goldberg, J.H., Parthasarathy, V., and Murphy, D.J., 1989b, A Farm Tractor Overturn Simulator. IMSE Working Paper 89-151, Department of Industrial & Management Systems Engineering, T Pennsylvania State University.

Huizinga, M.A., and Murphy, D.J., 1988, Farm work injuries in Pennsylvania. College of Agriculture Extension Circular 370 The Pennsylvania State University, University Park, PA.

Murphy, D.J., 1985, Pennsylvania farm fatalities during 1980-1984. College of Agriculture Special Circular 319, The Pennsylvania State University, University Park, PA.

Murphy, D.J., Beppler, D.C., and Sommer, H.J., 1985, Tractor stability indicator. Applied Ergonomics, 16(3), 187-191.

National Safety Council, 1988, Accident Facts. National Safety Council, Chicago, IL, pp. 97-99.

Pattie, C.L., 1973, Simulated tractor overturns: a study of hur responses in an emergency situation. Unpublished doctoral dissertation, Purdue University.

Rehkugler, G.E., 1980, Tractor steering motion dynamics-simulation and full-scale verification. American Society of Agricultural Engineers, Paper 80-1036, St. Joseph, MI.

Smith, J.D., Rogers, D.L., and Sikes, R.K., 1983, Farm-tractor associated deaths-Georgia. Morbidity and Mortality Weekly Report, 32(37), 481-482.

Spancer, H.B., Owen, G.M., and Glasbey, C.A., 1985, On-site measurement of the stability of agricultural machines. Journal of Agricultural Engineering Research, 31, 81-91.

Stephens, L.E., Zachariah, G.L., and Liljedahl, J.B., 1972, Vibration effects on tractor overturn recognition. American Society of Agricultural Engineers, Paper 72-614, St. Joseph, MI.

Advances in Industrial Ergonomics and Safety II
Edited by Biman Das
Taylor & Francis 1990

THE USE OF ROLLOVER PROTECTION ON FARM TRACTORS IN WEST VIRGINIA

John Etherton and John Myers

National Institute for Occupational Safety and Health
Division of Safety Research
Morgantown, West Virginia

Farm tractor rollovers kill about 200 workers each year in the U.S. The most effective way to prevent such fatalities, after the rollover occurs, is the use of Rollover Protective Structures (ROPS). A survey of farm safety in West Virginia indicates that about half of the tractors purchased since 1985, the year voluntary standards were implemented which recommended ROPS for all newly manufactured tractors, do not have a ROPS and that overall only about 20% of West Virginia tractors have it. This failure to use ROPS is an important fatality risk factor in a state where the farm land includes significant rolling and steep terrain. Suggestions are made on human factors research to increase the number of ROPS retrofits in West Virginia as well as in the rest of the U.S.

INTRODUCTION

Farm tractor rollovers kill about 200 workers each year in the U.S. (National Safety Council, 1988). Researchers at the National Institute for Occupational Safety and Health (NIOSH) have found that these tractor rollover incidents contribute significantly to the more than 7,000 annual traumatic occupational fatalities in the U.S. and are a high priority target for occupational injury prevention research (Etherton et al., 1989). The NIOSH researchers' assessment of the engineering, education, and enforcement approaches to this problem led them to recommend that research be oriented toward engineering to provide ROPS retrofits for older tractors and regulations to require ROPS on all tractors. A concern of research on both ROPS retrofit engineering and regulations should be that the safety device satisfy the needs of its user. One way to study user needs is to first study the needs of a small geographical group. The purpose of this paper is to report on how effectively the ROPS method of prevention is being used in the state of West Virginia and to suggest how human factors engineering research could help to increase the use of rollover protection both in West Virginia and throughout the U.S.

There are many prevention approaches to controlling the tractor rollover problem. Public health analysis of risk factors and controls from the perspectives of a few individual states such as Wisconsin (Karlson and Noren, 1979) and Georgia (Goodman et al., 1985) have been reported. Methods to prevent the rollover event from occuring by warning the driver offer a "precrash" approach to fatality prevention (Murphy, 1985, Lerner, 1985).

Another human factors consideration is the ease of deployment of the ROPS. New designs which attempt to improve the ease with which ROPS are deployed include foldable ROPS and telescoping ROPS. This could be applicable in West Virginia orchards where fixed ROPS do not permit tractors to work close to trees.

Research to improve operator recognition of dangerous slopes could be useful. The relative risk of a rollover fatality for West Virginia compared to Ohio was shown to be 1.6. It is likely that the higher risk in West Virginia is related to the topography of the state. Hilly terrain has been identified as an important factor in tractor-related injuries. Goodman et al. (1985) found that tractor-related fatalities were concentrated in the hilly region of Georgia. They suggest that this region of hilly terrain is an area where effective preventive measures should be intensified. Lerner (1985) reported that there are many human factor considerations for slope warning devices intended for riding lawn mowers. These human factors considerations could be studied further since the rollover problems on slopes are similar for riding lawn mowers and agricultural tractors.

Perhaps research is needed to evaluate farmers' theories about the effect of ROPS on their productive performance. Current warnings by the 4H and Future Farmers of America organizations, tractor manufacturers, agricultural extension, and the National Safety Council to West Virginia tractor operators appear not to be effective because they do not convince farmers to retrofit ROPS on their older tractors. The reason that farmers choose not to install ROPS is a matter of farmers wanting to be independent in the decisions they make about using safety equipment, and their resenting being forced to use it (Selders, 1990). They have their own opinions about the usefulness of safety equipment. Research evaluating how well current ROPS designs match the productive needs of farmers would be helpful.

Research on ways to improve the communication of warnings to West Virginia and other tractor operators in the U.S. could help reduce fatality rates. This assumes that there is a lack of effective warnings to the owners of older tractors that they need to install ROPS. Part of an effective warning is telling the tractor owner what to do about the hazard. Obviously, any guidelines which result from research should provide clear information on the benefits of retrofitting ROPS on older tractors or how to contact a dealer to have a ROPS installed.

SUMMARY

Fatalities related to tractor rollover can be prevented in West Virginia and other states through actions which increase the overall percentage of tractors equipped with ROPS. It is not sufficient to wait while new tractors which come equipped with ROPS slowly replace older tractors which lack this protective equipment. Human factors engineering research on ROPS retrofits suited to tractor operators in West Virginia and other states should be conducted to complement other public health strategies in achieving an outcome of no farm tractor rollover fatalities in West Virginia and the rest of the U.S. as soon as possible.

Advances in Industrial Ergonomics and Safety II
Edited by Biman Das
Taylor & Francis 1990

WOPALAS - A METHOD FOR ERGONOMICAL ANALYSIS AND PRACTICAL IMPROVEMENTS - EXAMPLES FROM SWEDISH INDUSTRY AND AGRICULTURE

Dr. Peter Lundqvist
Swedish University of Agricultural Sciences
Department of Farm Buildings
P.O. Box 624
S-220 06 LUND, SWEDEN

Work related injuries and diseases of the locomotor system are one of the most common resons for sick leave in Sweden. It is therefore important to improve the working environment in industry, agriculture and other workplaces. Our department is currently using the WOPALAS method (Working Posture Analysing System) for observing, analysing and correcting working postures. The method has been used in the industry, retail trade and agriculture with excellent results. The studies reported here were carried out in cooperation with the employers, the local unions, the company health care centers and the National Social Insurance Offices. The results have been used to change and improve the working environment in different ways including: reorganisation of the work, introducing new working methods and techniques and changes of the layouts.

INTRODUCTION

The most common reason for work related diseases in Sweden is monotonous and strenuous working postures and working movements. More than half of the occupational diseases affects the locomotor system, 70% for the women and 44% for the men. A study by Lindén (1986) showed that farm workers had about 10% higher level of diseases in the locomotor system compared with the average for all industries in Sweden. The costs for the diseases in the locomotor system are increasing dramatically every year. During 1983 the total cost for this type of diseases amounted to 17 milliard SEK (Malmöhus Allmänna Försäkringskassa, 1989).

The National Social Insurance Office and the Swedish University of Agricultural Sciences in Lund established a cooperative project in 1989 with the aim of finding methods to improve the working environment and reduce the sick leave at some work places in the region. As references some farms were included in the study. In this paper, only a limited number of work places will be presented. Results from two farms, one slaughter house and one grocery store will be presented as examples of jobs at different stages of the food production process from the farmer and to the customer.

The Department of Farm Buildings at the Swedish University of Agricultural Sciences has experience from ergonomical studies with the OWAS/WOPALAS method for improvements of the working environment in farm buildings. The National Social Insurance Office wanted to test the method at other work places with high levels of sick leave.

METHOD

The WOPALAS-method (Working Posture Analysing System) is an ergonomical analysing method consisting of two parts:

- frequency studies of working postures
- evaluation and analysis of observed working postures.

The method is developed from the finnish OWAS-metod (Ovako Working Posture Analysing System). The OWAS-method has been used in research projects in different Swedish branches, ie:

- the building industry (Gauffin and Hellsten, 1982)
- the mining industry (Andersson and Hellsten, 1983)
- forestry (Almqvist, 1983)
- fishing (Blide et al, 1984)
- agriculture (Lundqvist, 1988).

Field observations are usually made at 30 s intervals, to obtain an overall picture of each working posture at the moment of observation. A major objective is to obtain a total picture of working postures and the total work load. The back, arms, legs and the load and/or effort involved in the work are observed. The posture of the neck is also included in the system as well as the registration of static/dynamic work. To give guidence about the need for improvements, the results are classified into four levels (table 3) of action categories:

1. Normal postures with no discomfort and no effect on health, need no special attention except in some cases
2. Postures which must be considered during the next regular check of working methods
3. Postures which need consideration as soon as possible
4. Postures which need immediate action.

In this paper the results have been simplified, and cover only the work load on the back, the arms and the legs. The OWAS-method has been described in more detail by Karhu et al 1977, 1981, Heinsalmi 1983, 1986 and Lundqvist and Gustafsson, 1985.

The WOPALAS method differs from the OWAS method in the way that it handles left and right arm/leg at the registration. It has also more detailed registration of the arms positions and finally it uses more levels of weight/force registrations. The method has been employed by Carlsson et al 1988 and Hellsten 1989. Hellsten (1985) made a short manual/description of the system. In this project (Malmöhus Allmänna Försäkringskassa, 1989), the WOPALAS system was choosen because it`s higher level of accuracy. Videocamera was used to register the work. The working postures at the film was analysed by using a digital video for perfect still pictures every third second.

MATERIAL

For the studies work places in south Sweden were selected from the registers of the National Social Insurance Office. The criteria for the selection of work places was that there had to be problems with sickleave due to work related diseases in the locomotor system among the workers. The employers also had to have a positive attitude to the studies and to be willing to make necessary investments to improve the working environment. The workplaces and the work operations studied are presented in table 1.

Table 1. Type of work places and work operations studied with the WOPALAS system.

Type of workplace/ work operation	Description of work operations
Farm	
Milking cows	
I.	Milking of cows in a tie stall (standard)
II.	Milking of cows in a tie stall (with "milking-on-rails" system)
Collecting eggs	
I.	Collecting eggs on floor level
II.	Collecting eggs from a three-level conveyor belt
Slaughter house	
Slaughtering pigs	
I.	Taking out heart/lungs
II.	Taking out intestines
Sorting meat	
I.	Sorting meat into containers
Grocery store	
Handling meat	
I.	Handling meat at the delivery to the store
Cashier work	
I.	Cashier work (sitting)
II.	Cashier work (standing)

RESULTS

The results presented in this paper focus on the work load that effects the back, the arms and the legs, and the need for changes of the work place. The results in table 2 show that the work operations on the farm get the highest scores for work load. Now it is important to consider how long time each work operation covers. For example milking cows 2 hours per day or 8 hours per day makes a great difference with respect to total work load. This illustrates also the need for and importance of work organisation in all type of work places.

Farm work

The farmer was **milking** 41 cows, two times a day, using 6 milking machines. In the first case he milked the cows in a standard tie stall. He then had to carry the milking machines to each cow, connect and disconnect the tubes for milk and vaccum and finally attach/remove the cluster/teat cups at the cows udder in a kneeling position. In the second case the farmer was using a milking-on-rails system for his 6 milking machines. In this way he eliminates the carrying of milking machines but still have to handle tubes and cluster/teat cups as usual. The results from table 2 show that the work load on the legs is very high due to the awkward working positions. The back and the right arm are also affected in both type of milking systems. The need for improvements is presented in table 3, and it illustrates the need for more radical improvements.

Milking in a parlour with automatic cluster removal could give a much better situation, with an upright working position. Our institution is working with the development of a milking parlour with an adjustable floor for individual working heights at milking.

In the poultry house the hens were kept in a loose system without cages, which meant that the **eggs were collected** both from the floor level (a small part) and from a three level conveyor belt. The results showed that collecting eggs from the floor was negative for the whole body. Collecting eggs from the coveyor belt was also troublesome. The main reason was that the working levels were very awkward for the worker without possibilities for individual adjustment. The need for improvements is very great (table 3). The workplace needs to be rebuilt for individual adjustment of working heights alternatively fully automated.

Table 2. Work load on different parts of the body at different type of work operations, according to the WOPALAS-analysis. The load is expressed in relation to 100 (100 = ideal level).

Part of the body	Farm				Slaughter house			Grocery store		
	Milking cows		Collecting eggs		Slaughtering pigs		Sorting meat	Handling meat	Cashier work	
	I	II	I	II	I	II	I	I	I	II
Back	109	108	118	107	100	102	100	113	100	100
Arms										
-right	105	102	103	184	102	104	108	120	100	103
-left	100	100	118	136	100	106	105	116	106	100
Legs	200	153	179	115	120	119	109	102	110	106

Slaughtering pigs are quite a monotous work. At the slaughter house studied there was no work rotation and the problems with sick leave were getting worse. The result does not show much difference between taking out heart/lungs or taking out the intestines. The suggestions for improvements included demand for adjustable working heights, redesign of the handling of heart/lungs/intestines, so that it would be possible to let them fall straight down from the pig to a conveyor belt. The need for work rotation and information about working methods were also suggested.

The **sorting of meat** into containers was monotonous and needed such improvements as adjustable working heights, better use of coveyor belts, reorganisation of the system with containers. In addition work rotation and information about better working methods were recomended.

Table 3. Action categories for different work operations and body parts, acording to the WOPALAS analysis.

	Farm				Slaughter house			Grocery store		
	Milking cows		Collecting eggs		Slaughtering pigs		Sorting meat	Handling meat	Cashier work	
Part of the body	I	II	I	II	I	II	I	I	I	II
Back	2	2	2	2	1	2	1	2	1	1
Arms										
-right	2	2	2	4	2	2	2	3	1	2
-left	1	1	3	3	1	2	2	2	2	1
Legs	3	3	3	2	2	2	2	2	2	2

At the **grocery store** the **handling of meat** in boxes were the worst work operation with heavy weights to be handled in awkward working postures. The back was rather much affected and such improvements as transportation and lifting arrangements, redesign of meatboxes, redesign of the shelfs etc in the store were suggested.

The **cashier work** was studied with the cashier working in a sitting position and in a standing position. The recommendations were that the cashier ought to change between sitting and standing posture during the work. It was also recommended that chashiers should do other work operations during the day to get a varying working situation. Finally it was suggested that the working place for the cashiers ought to be more adjustable for the individual.

The results presented in this paper together with detailed recommendations concerning improvements have been presented at the workplaces for the employers, the local unions, the company health care centers with the support from the National Social Insurance Offices. At the presentation of the results, the videofilm from the work place was shown to illustrate the problems and to motivate and justify improvements. The improvements are now being introduced and will be followed up with new studies with the WOPALAS system during the fall of 1990. If the frequency of sick leave will change at the work places will be checked by the National Social Insurance Offices.

CONCLUSIONS

The WOPALAS method is a useful method as an aid for identification, analysis and evaluation of troublesome working operations which cause work related diseases in the locomotor system. The method can also be used to study the effect of the improvements that have been made in the work place. In this project it was appreciated that an independent person visited the work place and objectively analysed the work situation. The results were easy to understand and they made it possible for all individuals involved to participate in the discussion about different suggestions for improvements. The final follow up of the improvements during the fall of 1990 will give the final answer regarding the usefulness of the method.

REFERENCES

Almqvist, R. 1983. Bedömning av fysiologisk arbetsbelastning genom hjärtfrekvensmätning. Sveriges lantbruksuniversitet, Institutionen för skogsteknik, Stencil 245. Garpenberg.
Andersson, R. and Hellsten, M. 1983. Inledande arbetsfysiologiska studier av skrotningsarbete. Tekniska Högskolan (KTH), Byggergonomilaboratoriet, PM 1983-07-25. Stockholm.
Blide, G. et al. 1984. Ålfiske. Arbetsbelastning och tekniska åtgärder. Projekt Lindholmen, Rapport 84-02-01- Göteborg.
Carlsson, H. et al. 1988. Mindre byggskivor. Konsekvensutredning. Byggforskningsrådet, Rapport R91:1988. Stockholm
Gauffin, U. 1983. Fasadtegelmurning. Tekniska Högskolan (KTH), Byggergonomilaboratoriet, Rapport Trita-Bel 0018. Stockholm.
Heinsalmi, P. 1983. The OWAS System (Ovako Working Posture Analysing System). I: Ergonomiska mätmetoder i lantbruket. Kuopio regioninstitut för arbetshygien, sid 68-82. Kuopio.
Heinsalmi, P. 1986. Method to measure working posture loads at working sites (OWAS). In: Corlett, N., Wilson, J. & Manenica, I. (Eds.). The Ergonomics of Working Postures Models, Methods and Cases. The Proceedings of the First International Occupational Ergonomics Symposium, Zadar, Yugoslavia, 15-17 April 1985, pp 100-104. Taylor & Francis. London.
Hellsten, M. 1982. Ergonomi vid arbete med byggskivor. Tekniska Högskolan (KTH), Byggergonomilaboratoriet, Rapport Trita-Bel 0015. Stockholm.
Hellsten, M. 1985. WOPALAS. En praktisk handledning för studier av arbetsställning och arbetsbelastning. Tekniska Högskolan (KTH), Byggergonomilaboratoriet, PM 1985-09-06. Stockholm
Hellsten, M. 1989. Bättre förarergonomi med vridbara förarplaster i små truckar. TransportForskningsKommissionen (TFK), Rapport 1989:3. Stockholm.
Karhu, O. et al. 1977. Correcting working postures in industry: A practical method for analysis. Applied Ergonomics, Vol 8 No. 4, pp 199-201.
Karhu, O. et al. 1981. Observing working postures in industry: Examples of OWAS application. Applied Ergonomics, Vol 12 No. 1, pp 13-17.
Lindén, A. 1986. Arbetsskador i jordbruket. Arbetarskyddsstyrelsen, Information om arbetsskador 1986:2. Solna.
Lundqvist, P. 1988. Working environment in farm buildings. Results of studies in livestock buildings and greenhouses. Dissertation. Sveriges lantbruksuniversitet, Institutionen för lantbrukets byggnadsteknik, Rapport 58. Lund.
Lundqvist, P. and Gustafsson, B. 1985. OWAS - en ergonomisk mätmetod. Erfarenheter från praktiskt tillämpning. Arbete Människa Miljö, nr 1, sid 36-43.
Malmöhus Allmänna Försäkringskassa. 1989. Förebyggande arbetsmiljöarbete - belastningssjukdomar. Projektrapport 1989-08-14. Lund.

Advances in Industrial Ergonomics and Safety II
Edited by Biman Das
Taylor & Francis 1990

HANDLING ACTIVITIES IN FARMING - AN ERGONOMIC ANALYSIS

Michael Jorgensen, Ram R. Bishu, Yuxiang Chen, and Michael W. Riley
Department of Industrial and Management Systems Engineering
University of Nebraska, Lincoln, NE 68588 - 0518

Material handling, animal handling, and maintenance tasks in farming involve lifting and awkward postures, and hence need research attention. The objective of this study was to investigate the manual material handling components of some of the farming tasks. Initially data on all the farm chemicals was procured from the local farm supply companies. Of particular interest were the data on the container shape, container hand coupling devices, and the container weight distribution. A list of fall season farm manual material handling and machinery maintenance activities performed by a sample of five farmers were videotaped. The video tapes were then analyzed using a task analytic procedure. Overall, results indicate a need for the redesign of the containers for farm chemicals and supplies. Bulk of the containers lack any handles, thereby forcing assumption of a variety of unnatural postures. The task analyses on the activities indicate a large proportion of awkward postures, and disc compressive forces beyond the NIOSH action limit on some of the task elements.

INTRODUCTION

Considering the global statistics available, back pain is perhaps among the most expensive occupational hazards. For example, according to the U. S. Bureau of Labor Statistics Bulletin (1982) about one million workers suffered back injuries in USA in 1980-accounting for one out of every five injuries and illnesses in the work place. Published statistics on the incidence, severity, and on other aspects of back pain are available in plenty for organized industries (See Klein et. al., 1984, Liberty Mutual Insurance Company, 1976). However, research on low back pain in the farming industry is relatively sparse to non-existent. There are no direct statistics linking farming industry with low back pain. The reasons for this are perhaps the unorganized and independent nature of farming. Unlike most of the other industries which are managed by large corporations, the farming industry is a collection of farmers working separately or in small groups. Considering the farm accident data (National Safety Council, 1985) it is reasonable to assume that farming is not free from the risk of back pain injuries. In this regard Bishu et. al., (1988) performed a pilot study on back injuries in farming. 44 subjects responded to a survey which had questions on personal characteristics, injury frequencies, injury severity, and task characteristics of farming. Back pain was reported as the most frequent injury, and the most severe as well. Overall, it appears that the greatest risk to back pain is when the farmer is performing manual lifting task in the field at ground level during the spring season. Material handling, animal handling, and maintenance tasks in farming were reported as the most causative activities.

Farming activities are weather dependent and the farmer must be most productive during a limited number of days when the weather and climatic conditions are favorable to his crops. Therefore, he must work long hours and with fatigue in many cases. The task elements of farming can best be described as heavy physical work, tractor driving, and working under time pressure. As the industry is seasonal, some of the farm operations are critical in timing. Therefore, considerable need exists for ergonomic analyses of farm activities. The objective of this research evolved out of this reasoning. The purpose of this study was two fold:

1. As farm chemicals, and supplies form a major portion of the materials handled, the intent was to study the frequency distributions by weight, volume, and by container of the farm materials.
2. To initiate a longitudinal task analytic study of the various farm activities with a view to identify high risk elements for ergonomic interventions.

METHOD

For studying the distributional characteristics of the farm chemicals and supplies, two local Farm Supply Companies were approached, and the data obtained. The details of the data are reported in first part of the result section. The consumption data per farmer for the chemicals could not be obtained as it was wrong season.

For the task analyses, five local farmers were contacted for possible videotaping of their activities for this research. It should be noted here that certain amount of unstructuredness in the procedures could not be avoided as it was the end of the harvesting season for most of the farmers, nor could the sample size per task be maintained at a constant level. The following activities were videotaped:

1. Shovelling the grain
2. Mixing grain
3. Loading and unloading hay bails
4. Hammering posts into the ground
5. Carry buckets of grain
6. Attach equipment

The tasks were then broken into different elements. For each of the elements the action range and weights were determined. From the respective frames of the candidate task elements the elbow, shoulder, back, knee, and the ankle angle were obtained. The University of Michigan two dimensional static biomechanical model(1986) was then used to determine the capability levels of these task elements.

RESULTS

Task Analyses

The first job analyzed was "carrying buckets". The task was broken down into three elements: 1) picking up the buckets, 2) carrying the buckets, 3) setting the buckets down. The variation of the back stress for this task is shown in Figure 1. It is seen from Figure 1 that the action limits are exceeded at the beginning, and at the end of the task while stress levels below the action limit are observed in the middle.

The task of hammering fence posts was analyzed next. The task analysis consisted of 1) the posture at the height of the upstroke, and 2) the posture just before striking the post with the fence post hammer. The disc compressive forces resulting from the posture analysis using a static model are shown in Figure 2. The actior limits in this task were not exceeded in this task.

The handling of hay bales was the next task to be analyzed. The task was broken into three identifiable elements, 1) begin the lift with two hands, 2) midpoint of the lift, and 3) the posture just before the release of the bale of hay. The results of the static strength calculation of disc compressive force for the three operators is shown in Figure 3. Again the first, and the third elements developed disc compressive forces far exceeding the action limit.

The task "shoveling grain" was broken down into the following two elements: 1) digging shovel into grain, 2) releasing grain from shovel. The disc compressive forces resulting from postural analyses are shown in Figure 4. The act of pushing the shovel through the grain to fill the shovel causes excessive stresses to the back.

Finally the task of attaching equipment to the tractor and mixing grain into a mechanical grain mixer were analyzed. The results are shown in Figures 5, and 6. The loads to the back during the course of these tasks do not exceed the action limits for the postures employed. The task of mixing grain into a mechanical grain mixer comprised two elements. Lifting grain bags from the floor and holding bags while they were emptied into the mixer. Although the action limit is not exceeded in this task its high frequency warrants attention and intervention.

<u>Distribution of farm chemicals and supplies:</u>

The distributions of the farm chemicals by medium is shown in Figure 7. It appears that farmers use more liquid chemicals than solids. The distributions of the liquids by volume of the containers is shown in Figure 8. The large range in the volume of the containers (1 gallon to 140 gallons) is really surprising and needs further research attention. The distribution of the use of solid chemicals by weight is shown in Figure 9. The use of large numbers of 50 lbs. bags is a cause of concern, and needs more study. Finally the distribution of container size is given in Figure 10.

DISCUSSION

This was a follow up study of an earlier investigation on the same topic by the primary investigator (Bishu et. al., 1989). As stated earlier the objective was to identify some high risk elements in the 'Material handling' and 'Maintenance' components of farmers' tasks. Among the six tasks analyzed, 'Carrying the buckets', 'Handling hay bales', and 'Shovelling grain" do warrant postures that cause disc compressive forces to go beyond the action limit. The remaining three tasks analyzed are well within the action limit. However, a number of caveats are in order in this study.

None of the results reported here were statistically testable. Data collection for this was fragmentary owing to the season being well past the harvesting for some farmers, and during harvesting for the others. The frequency of the tasks, and its relation with objective parameters such as the size of the farm, and nature of the farm could not be estimated. For example, a task such as hay handling might be performed on a daily basis for a short period of time, for just feeding the cattle. On the other hand, this task might be done very frequently for a few days when loading from a field onto a truck, then transferring into a storage facility. Further, static analysis was performed on all the tasks that were inherently dynamic, and hence the inertial and momentum forces were not taken into consideration.

The task analyses performed in this study are very simplistic, and represent only a couple of tasks that a farm operator performs in which he may be at risk to an overexertion injury. Other tasks might include:

1. Lifting, carrying and lowering of irrigation pipe.
2. Setting gates on irrigation pipe.
3. Maintenance of farm equipment.
4. Lifting, carrying and dispensing buckets of feed, water,etc.
5. Handling large farm animals (ie. cattle or hogs).
6. Installing fence gates.

This list is certainly not inclusive, and many other tasks can probably be identified and studied with further research into the daily activities of farm operators with ergonomics kept in mind.

The bulk containers for the liquid chemicals come in drums holding up to 140 gallons. This size of container requires the farm operator to have the supplier transfer the chemicals to the site, therefore, very little manual handling is required once these containers are in place. Some chemicals come in glass bottles. The use of glass bottles is on the decrease as they can be very heavy and are not easy to dispose of. The handle size and design is an issue that needs to looked into since the handle is usually too small for people to insert the fingers, especially when wearing gloves. The use of glass containers also makes it difficult to control the flow of the liquid from the container.

Some of the liquid chemicals come in plastic containers. The advantage of plastic container have over glass containers is that they have markers and special spouts for pouring and measuring. The distribution of weight of containers for solid chemicals is shown in Figure 9. As can be seen from the figure, 46% of the containers are of 50 lbs. weight. These containers consist mostly of plastic and thick paper sacks. The 25 and 10 lbs. containers each account for 16% of the total and the 20 and 5 lbs. containers each account for 11% of the total distribution by weight. When considering manual materials handling, the frequency and the distance of the lift must be considered. For instance, using Snooks lifting criteria (1978), lifting a 76 cm wide object 76 cm from the floor to knuckle height, 90% of the male population should be able to lift 22 kg once every 30 minutes. This, however, may not be the case, especially during the planting season where seed bags are used (often weighing 22.7 kg) and also the use of herbicides and fertilizers.

Overall, results indicate a need for the redesign of the containers for farm chemicals and supplies. Bulk of the containers lack any handles, thereby forcing assumption of a variety of unnatural postures. Not withstanding all the caveats mentioned above the postural analyses on the activities indicate a large proportion of awkward postures. It is also hoped that similar video analyses will be performed for the spring, and more strenuous, activities subsequently.

REFERENCES

Bureau of Labor Statistics, (1982), "Back injuries associated with lifting," Bulletin 2144 of Bureau of Labor Statistics, Department of Labor, Washington, D. C.

Bishu, R. R., Chen, Y., Cochran, D. J., and Riley, M. W. (1989). Back injuries in farming-A pilot investigation. In Advances in Industrial Ergonomics and Safety I, (Editor Anil Mital), Taylor and Francis, London, pp 791-798, June 1989.

Klein, B. P., Jensen, R. C., and Sanderson, L. M.(1984). Assessment of workers' compensation claims for back strains/sprains. In Journal of Occupational Medicine, 26, 443-448.

National Safety Council, 1985, Summary of back injury statistics, National Safety Council, Chicago.

Liberty mutual Insurance Company (1976), Research Center, Hopkinton, MA (Unpublished data), 1976.

Snook, S. H. (1978). The design of manual handling tasks. In Ergonomics, 21(12), 963-985.

U.S. Department of Health and Human Services, Public Health Service, Centers for Disease Control. National Institute for Occupational Safety and Health, Division of Biomedical and Behavioral Science. Cincinnati, Ohio, 45226, March 1981.

University of Michigan, Static Strength Prediction Model (Version 3.8), Center for Ergonomics, 1205 Beal Street I O.E. Bldg., University of Michigan, Ann Arbor, Michigan, 48109.1986.

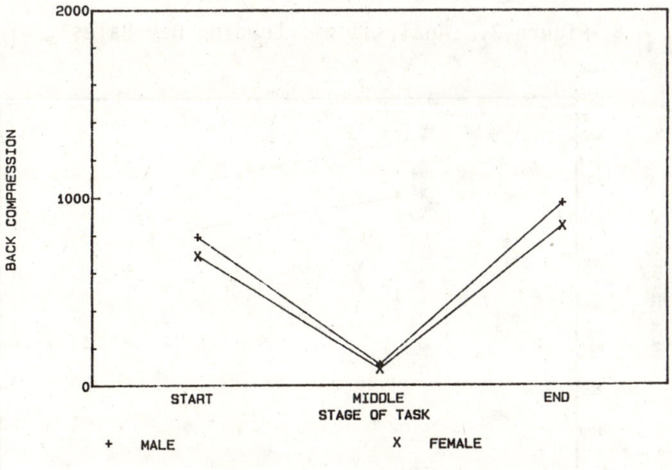

Figure 1. Analysis of 'Carry Buckets'.

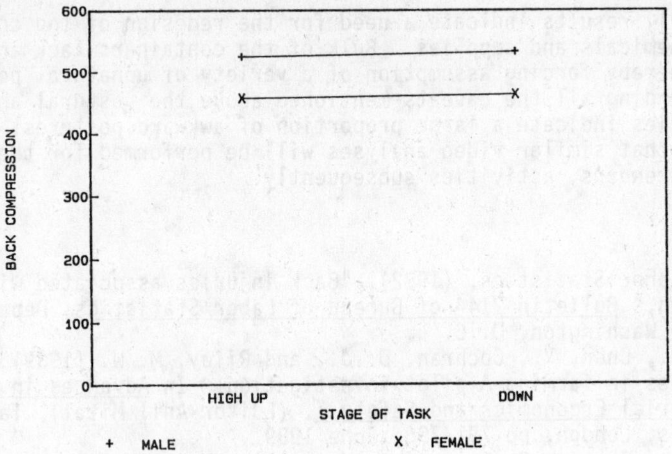

Figure 2. Analysis of 'Hammer Post'.

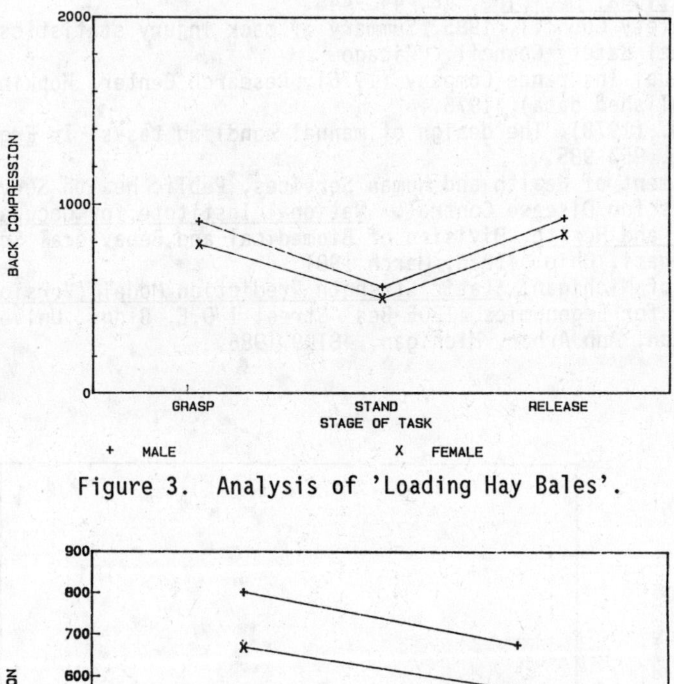

Figure 3. Analysis of 'Loading Hay Bales'.

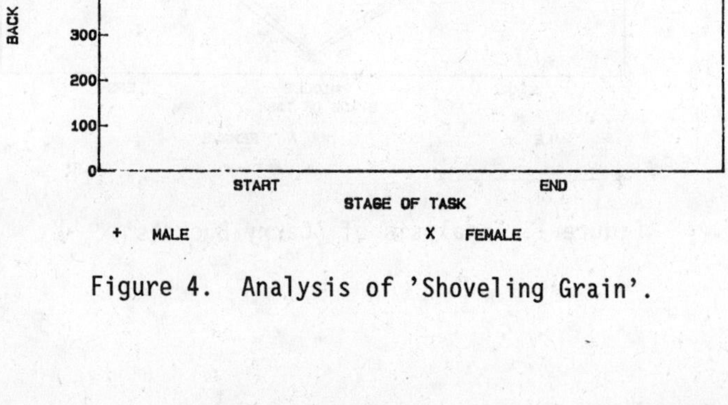

Figure 4. Analysis of 'Shoveling Grain'.

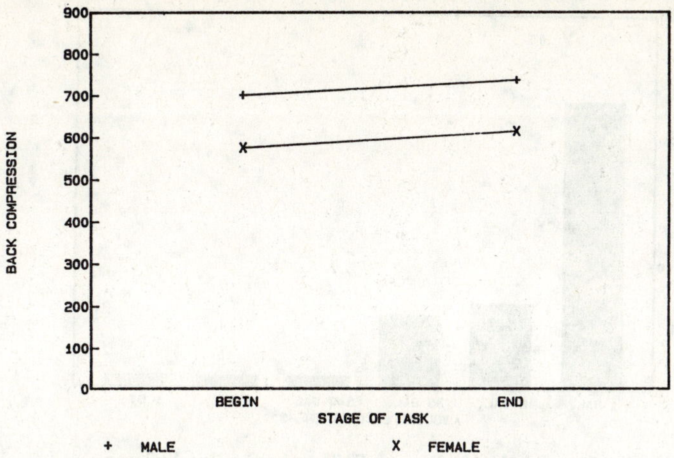

Figure 5. Analysis of 'Attach Equipment'.

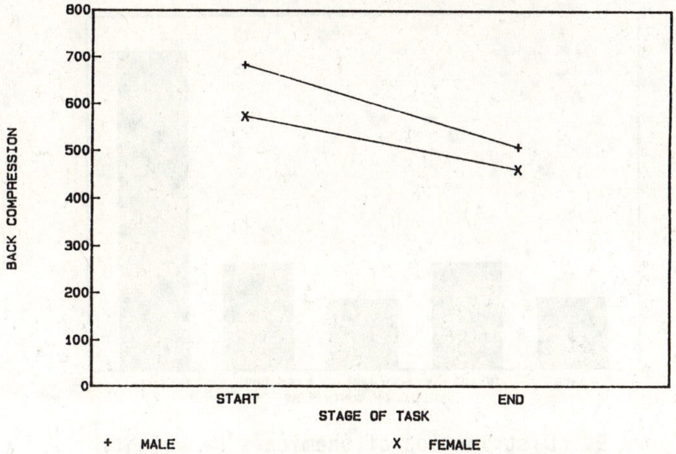

Figure 6. Analysis of 'Mixing Grain'.

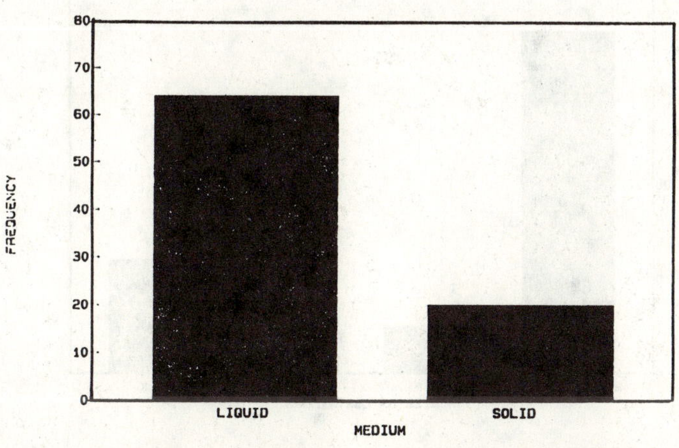

Figure 7. Distribution of Chemicals by Medium.

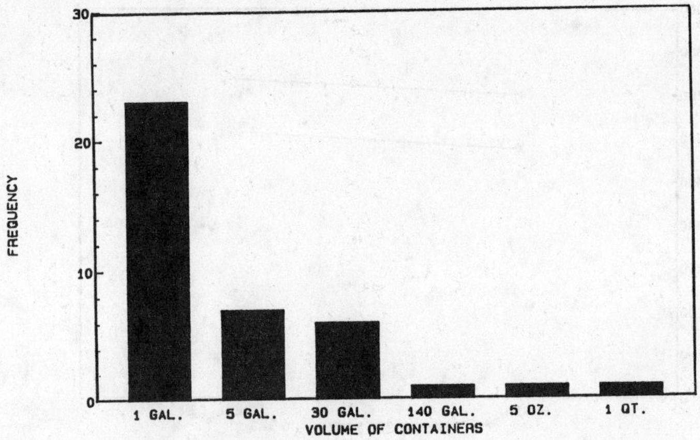

Figure 8. Distribution of Chemicals by Volume.

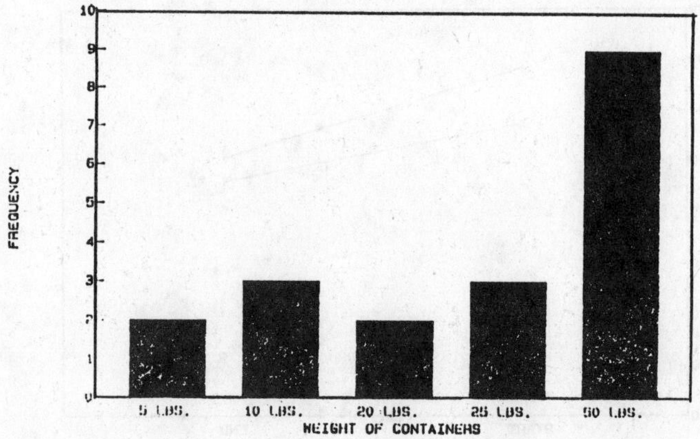

Figure 9. Distribution of Chemicals by Weight.

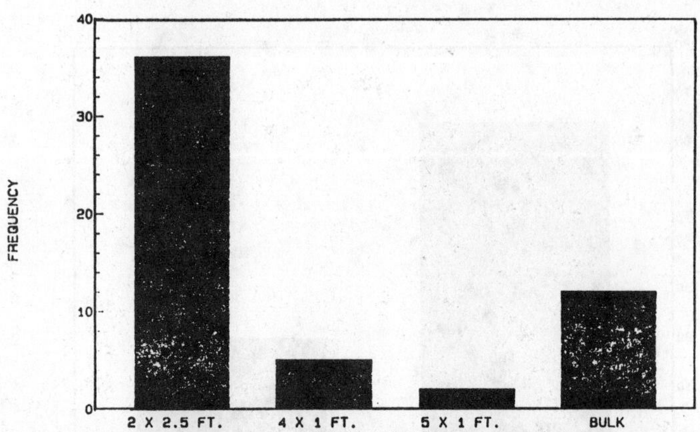

Figure 10. Distribution of Container Sizes.

XVII

STATISTICS AND MODELING IN ERGONOMICS

Advances in Industrial Ergonomics and Safety II
Edited by Biman Das
Taylor & Francis 1990

OPTIONS FOR ANALYSIS OF REPEATED MEASURES DESIGNS: WATCHING-OUT FOR THE ASSUMPTIONS

Stephen J. Morrissey, Department of Industrial Engineering, University of Wisconsin-Platteville, Platteville, WI. 53818
and
Alvah C. Bittner Jr., Battelle HARC, Seattle, WA 98105-5428
with
Dr. B. Ghahramani, AT&T, New Jersey 07992

The requirements for analysis of repeated-measures experiments are considered in the context of applied research. In applied research settings, the traditional analysis of variance (ANOVA) tests are often likely to be in violation of the sphericity structure assumptions (e.g., symmetry of the variance-covariance matrix). If violations of sphericity are not addressed, there may be errors of interpretation in the study. A survey of the human factors-ergonomics literature indicates that many researchers apparently are not making corrections for non-sphericity in designs employing repeated measures ANOVA. Methods are discussed by which the researcher can obtain correct statistical results in repeated measures designs.

INTRODUCTION

The repeated-measures experimental design is one in which a subject or group of subjects receives multiple combinations of a factor. It is a primary experimental technique for individual and group research. Representative designs that employ repeated measures structures include: the randomized block and "split-plot" type designs, and those that have within-subjects effects. In addition to providing for the following of performance and other time related processes, such designs can be more efficient and economical than other designs (Sutcliffe, 1980; Winer, 1971). More importantly, repeated measures designs may provide the only practical means of doing research in applied settings where the availability of trained subjects is often limited.

The validity of traditional analysis of variance applied to repeated measures designs may be compromised through use of inappropriate statistical tests of significance. In particular, traditional within-subject (individual or group) tests of significance may be grossly in error, if the limitations of these tests are not recognized and appropriate compensations applied. The nature of the underlying problem with traditional tests on repeated measures designs and alternative methods of analysis is described in the following sections. This will set the stage for a recommended series of tests of repeated measures based designs that minimize the potential for compromised results.

TRADITIONAL AND ALTERNATIVE TESTS

Traditional analysis-of-variance (F) tests for repeated measures designs (RANOVA) and the resulting within-subject effects have been widely described (e.g., Kirk, 1968; Winer, 1971). While regularly noted in such references on analysis of experiments, not widely appreciated are the stringency of the conditions for such tests to be valid. Generalized by Huynh and Feldt (1970), these restrictions are that the population covariance structure must possess a "sphericity structure". This occurs when both of the following conditions hold:

(1) variances of the subjects, as individuals or in groups, is constant across trials; and

(2) the inter-trial correlations of subjects across trials is constant.

Failure to meet these two restrictions may substantially compromise the findings of any study using this design. Huynh and Feldt (1980), for example, have found that the actual level of significance may be grossly elevated far above the traditional 0.05 level (depending on the sample size and the degree to which the basic sphericity assumptions are violated). They report, for example, that an interaction test that is referenced at a traditional level of significance of 0.05, may have actual significance levels of as high as 0.30. Under other conditions, significance levels may actually be much more conservative than the desired 0.05 level. The important point to be made is that the actual degree of significance of the test is dependent upon the degree to which the data structure meets the sphericity constraints.

Many researchers fail to recognize the frequency with which symmetry conditions are violated. Huynh and Feldt (1976) have cautioned that many standardized test batteries may violate the sphericity restrictions. Supporting this assessment, a study of 114 measures of performance, found that approximately 70% of selected laboratory performance tasks did not meet these two conditions for traditional RANOVA (cf. Bittner, Carter, et al., 1986).

A survey was conducted to examine to what extent researchers in the ergonomics and human factors fields have recognized the restrictions on repeated measures designs and made appropriate corrections. This survey examined research papers published in three major human factors-ergonomics journals for the past 15 to 18-months in which a RANOVA design was used. A total of 66 papers were found where a repeated measures analysis was used. Of these 66 reports, 60 failed to report whether they had taken into consideration sphericity. However, of these 60 studies, 6 appeared to have some understanding of the restrictions of RANOVA and used a multivariate analysis of variance (MANOVA). Yet the vast majority of the studies (82%) did not make corrections. These findings suggest that the results of these papers may be in question (unless the authors made corrections in their analyses without reporting them). These results also suggest that: (i) procedures for use of repeated measures designs be reviewed; and, (ii) an analysis sequence to evaluate repeated measures designs be delineated.

Analysis of RANOVA Designs

The adequate analysis of repeated measures designs requires assurance that potential lack of sphericity is accounted for when testing significance. There are at least two basic ways in which this may be accomplished. The first approach is to conduct an exact test such Mauchly's test for sphericity as described by Huynh and Feldt (1970). Unfortunately, this test is computationally complex and not present in most commercially available statistical analysis programs. In addition, this test is sensitive to non-normality assumptions in the data, and as such its results can be questioned. The second approach involves adjustment of the degrees of freedom of the F-ratios derived in the analyses to compensate for the degree of non-sphericity. This procedure is more easily done, and is based on results derived by Box (1954). He found that the exact test under conditions of non-symmetry may be approximated by an F-ratio with adjusted degrees of freedom. The exact degree of correction is determined by the degree of non-sphericity in the data, and is reported by an "epsilon factor". Unfortunately, calculation of the epsilon factor can be cumbersome for large designs, although it is implemented in some computerized statistical analysis packages (e.g., BMDP). Also, as shown by Huynh and Feldt (1976) the epsilon factor may be conservatively biased under some conditions (the actual alpha is less than nominal).

Geisser and Greenhouse (1959) propose a simple and yet accurate procedure to evaluate repeated measures designs which avoids some of these problems. In this procedure, the first step is to conduct the traditional analysis of variance tests on the data. Effects that meet the desired level of significance are identified (typically, $p < 0.05$). If no effects meet this criteria, unless explicitly identified in the hypotheses for later study, the analysis ends. The second step in the procedure is to test the F-ratio of effects found to be significant by the traditional analysis of variance, as if they had 1 and N-1 degrees of freedom (where N is the number of subjects). Effects that exceed this limit can be unconditionally accepted as significant as they have had the maximal reduction in the degrees of freedom (Box, 1954; Huynh and Feldt, 1976).

However, a problem arises when effects are found to be significant by the conventional analysis but fail the second. When this occurs, the exact degree to which the data departs from sphericity and the resulting significance of the F-ratio must be considered. Huynh and Feldt (1970, 1976) have suggested that these issues may be addressed by one of the following techniques.

Multivariate Tests

The most commonly recommended multivariate statistic to examine significance of repeated measures data is the MANOVA. In the survey of experiments reported earlier, MANOVA techniques were used in about 10% of the studies. The MANOVA approach is of particular value as it does not require symmetry of the variance- covariance matrix. Thus true significance values can be determined. However, unless the sample size, N, is "large" this test may be compromised in power (large is typically suggested to be N > 20). An additional drawback with the use of MANOVA methods is that follow-up tests for significant effects are not completely agreed upon (Bray and Maxwell, 1982). Indeed, one suggested method of follow-up analysis is to conduct univariate RANOVAs (which leads to a circular approach). Thus MANOVA should be limited to analyses in which detailed follow-on analyses are not anticipated and the sample size restrictions are met.

Epsilon Corrections

Epsilon corrections are based on the test approximation results from the study by Box (1954). Epsilon corrections involve finding the "epsilon correction factor" corresponding to the degree of departure of the variance-covariance matrix from sphericity. This factor is then multiplied by the original degrees of freedom resulting in reduced degrees of freedom for the F-ratio. As discussed by Huynh and Feldt (1976), the epsilon value may be slightly biased with small samples(e.g., when sphericity is in fact, exactly met). They have suggested revised epsilon correction factors derived from the sample data variance-covariance structures which are less dependent upon sample size. Original and corrected epsilon factor tests are available on some computerized statistical analysis systems and are the most flexible method for determining the actual significance level of the test.

Linear-Contrasts (1-df)

The third approach to studying "in-between effects" is to use linear contrast tests with 1 degree of freedom (1-df) in the numerator. Such contrasts provide relatively robust tests of significance (cf., Scheffe, 1959, pp 331). Equivalent to two-tailed t-tests, these contrast based tests may be applied to analysis of repeated-measures experiments where trials of control conditions precede and follow an experimental condition (Bittner, Harbeson, et al., 1986). Linear-contrasts provide a direct means of analysis where the repeated measures design permits.

Comparisons of Methods

None of the three alternative tests identified above may be universally recommended for use in all cases. Linear-Contrasts, for example, are particularly appropriate for experiments where trials of control conditions precede and follow an experimental condition (Bittner, Harbeson, et al., 1986). However these or similar contrasts may not be appropriate for many analyses. MANOVA may be a more powerful approach when its assumptions are met and follow-up analyses are not at issue. However, as discussed earlier the survey of research applications suggests that when MANOVA has been applied, the potential limits of sample size and follow up analysis constraints have not been recognized. The epsilon correction approach appears to be the most broadly applicable technique, but does require computational routines not implemented in some popular analysis packages(e.g., SAS). Thus it appears that the final approach selected may indeed be based both on the statistical soundness of the method as well as its ease of use or availability.

DISCUSSION

Repeated-measures experimentation offers an efficient, economical, and practical approach for ergonomics and human-factors research. However as is discussed above, traditional analysis of ANOVA designs, without appropriate compensation for potential non-symmetry of the variance-covariance matrix, may be expected to lead to compromised findings. Fortunately, the three-stage method discussed here, and the three different alternative methods to evaluate tests when sphericity is in doubt, provide a wide variety of alternatives to overcome these limitations.

CONCLUSIONS

This paper reviews the importance of watching out for the symmetry of the variance-covariance matrix when using a repeated measures design in analysis of variance. Although the importance of such checks, and the appropriate steps to correct any violations of homogeneity, are well known, a survey of the literature suggests that most researchers still fail to employ such checks. The various methods by which ANOVA can be used in repeated measures design are reviewed, and a three stage method based on the work of Geisser and Greenhouse (1968) and Huynh and Feldt (1976) work is recommended. Methods to deal with the case where non-symmetry is indicated are discussed. Overall repeated measures ANOVA offers a powerful and useful approach, but researchers must be aware of its limitations and how these may be overcome.

REFERENCES

Bittner, Jr., A.C., Carter, R.C., Kennedy, R.S., Harbeson, M.M., and Krause, M., 1986. Performance evaluation tests for environmental research (PETER): Evaluation of 114 measures. Perceptual and Motor Skills, 63 683-708.

Bittner, Jr., A.C., Harbeson, M.M, Kennedy, R.S., and Lundy, N.C., 1986. Assessing the human performance envelope: A brief guide (SAE 851776), SAE Transactions, 6, 375-385.

Box, G.E.P., 1954. Some theorems on quadratic forms applied in the study of analysis of variance problems, II. Effects of inequality of variance and correlation between errors in two-way classification. Annals of Mathematical Statistics, 25, 484-498.

Bray, J.H., and Maxwell, S.E., 1982. Analyzing and interpreting significant MANOVAs. Review of Educational Research, 52(3), 340-367.

Collier, R.O., Baker, F.B, Mandeville, G. K., and Hayes, T.F., 1967. Estimates of test sizes for several test procedures based on conventional variance ratios in the repeated measures design. Psychometrika, 32, 339-353.

Geisser, S., and Greenhouse, S., 1958. Extension of Box's results on the use of the F distribution in multivariate analysis. Annals of Mathematical Statistics, 29, 885-981.

Huynh, H., and Feldt, L.S., 1970. Conditions under which mean square ratios in repeated measures designs have exact F-distributions. Journal of the American Statistical Association. 65, 1582-1589.

Huynh, H., and Feldt, L.S., 1976. Estimation of the Box correction for degrees of freedom from sample data in randomized block and split-plot designs. Journal of Educational Statistics, 1(1), 69-82.

Kirk, R.E., 1968. Experimental Design: Procedures for the Behavioral Sciences. (Belmont CA., Brooks/Cole Publishing Company).

Morrison, D.F., 1967. Multivariate Statistical Methods. (San Francisco, CA: McGraw-Hill).

Sutcliffe, J.P., 1980. On the relationship of reliability to statistical power. Psychological Bulletin, 88, 509-515.

Winer, B.J., 1971. Statistical Principles in Experimental Design, (2nd Edition). (New York: McGraw-Hill).

Advances in Industrial Ergonomics and Safety II
Edited by Biman Das
Taylor & Francis 1990

IMPLEMENTING THE TAGUCHI METHOD FOR THE HUMAN PERFORMANCE STUDY

Ying-Hsin A. Liou and Richard R. Lindeke
Department of Industrial Engineering
Cleveland State University
Cleveland, Ohio 44115

Stephen J. Morrissey
Department of Industrial Engineering
University of Wisconsin - Platteville
Platteville, Wisconsin 53818

Traditional statistical design of experiment (D.O.E.) techniques have been used by Industrial Ergonomists and Safety Engineers to study their laboratory level research problems. This paper introduces a new simplified technique, The Taguchi Method, that will allow engineers working in the field to use D.O.E. The use of the Taguchi Method is described through a nine- step process that will guide the user from problem recognition to experimental evaluation.

INTRODUCTION

In industrial ergonomics or safety studies, Statistical Design of Experiments (D.O.E.) is employed to examine the quantitative relationship between the dependent variables (human performance) and the independent variables (personal, system and environmental factors). The outcomes of D.O.E. reveal the effects of independent variables on human performance. Regression models can be developed from these experimental results to predict human performance levels based upon given levels of the independent variables. The investigator then uses this information to design a task environment in which human performance is well maintained, and optimized.

The number of experimental conditions, and their duplicates, for a given experimental design is of major concern to the investigator in the study of industrial ergonomics and safety. Due to economic and environment limitations, the investigator would like to examine as many factors as possible without increasing the size of experiment. For this reason, fractional factorial designs and the Response Surface Methodology (RSM) (Montgomery, 1984, Clark and Williges, 1973) rather than full factorial designs, have usually been adopted.

Since its introduction into American industry in the late 1970's, the Taguchi Method (Taguchi, 1987, Ross, 1988) has gained in popularity as a D.O.E. technique which simplifies the complex procedures of designing the experiment and analyzing the data. Like fractional factorial designs and RSM, the Taguchi Method efficiently explores the experimental space with a reduced number of conditions. The Taguchi

Method, unlike fractional factorial designs and RSM, is a more structural technique that simplifies the design stages for the selection of experimental conditions. In addition to its structure as an orthogonal array based fractionalized design, certain aspects of this method can efficiently deal with "noise factors" such as the differences in human subjects, environmental factors and training effects.

To promote the study of industrial ergonomics and safety in the field, where the personnel may not be as rigorously trained in statistical techniques, the Taguchi Method is an excellent alternative to traditional fractional factorial designs and RSM. This paper is intended to explain, with a step by step approach, the implementation of the Taguchi Method. A nine-step approach is presented. Following these steps will enable the in-plant ergonomics specialists and safety engineers to solve their job and workplace design or modification problems both efficiently and economically through the application of the powerful D.O.E. techniques as applied in the Taguchi Method.

EXPERIMENTAL STEPS OF THE TAGUCHI METHOD

The steps are summarized in Table 1. Their use will guide the investigator through statistical experimental design from recognition of the problem through screening experimentation to high order experimentation.

TABLE 1. Summary of the implementation steps for the Taguchi Method

Step One	Recognize the Problem
Step Two	Assemble a "Taguchi Study Team"
Step Three	Select the Dependent Variables and Independent Variables
Step Four	Decide on an Orthogonal Array and Build the Experimental Design
Step Five	Perform the Experiment
Step Six	Analyze the Data
Step Seven	Develop the "Predictor Equation"
Step Eight	Perform the Confirmation Run
Step Nine	Conduct Higher Order Experiments

STEP ONE: Problem Recognition

Problem analysis starts with its recognition in its most basic and solvable form. Recognition can be simple but not all problems are effectively addressed by using the experimental design technique. The goal here is to qualify those problems that best lend themselves to experimental solution using the Taguchi Method. An example might be investigating the effects of different task parameters, such as container size, conveyer height and lifting frequency on workers' lifting capacity. On the other hand, a standard workplace design or tool evaluation problem may be addressed by applying well established principles and guidelines rather than experimental methods.

STEP TWO: Assembly of the Taguchi Study Team

The Taguchi Study Team must include all parties with a "Vested" i.e. a direct monetary or social interest in problem solution (Smith and Smith, 1982). This team must include, at a minimum, in-plant ergonomics specialists, safety engineers, product designers, managers, foreman and representative workers from the production line (Alexander, 1986). This team, which can be unwieldy, requires the development of a functional inner group whose presence makes the details of problem solving move forward effectively (Lindeke and Liou, 1989). This dynamic team structure, a general body and an inner group is the fundamental basis for problem solution.

An important characteristic of the Taguchi Study Team is that it is assembled only to address a fully identified problem. It is only an Ad Hoc group, **NOT** a sitting one.

STEP THREE: Selecting the Dependent Variables and Independent Factors

Dependent Variables. Dependent variables are the quantitative indicators of the problem under study. Thus, we must be certain that the dependent variable(s) to be selected truly provide the desired information about the problem. Dependent variables, since they reveal human performance characteristics, generally have two design targets, "smaller is better" and "larger is better". Examples of these target types include, for "smaller is better": heart rate, oxygen consumption, error rate and response time while for the "larger is better" target: production rate, load lifting capacity and comfort index. Dr. Taguchi (1987) has proposed a data transformation expression, signal-to-noise (S/N) ratio, for each of these design targets. They are:

Smaller is better

$$S/N = -10 \log_{10}(1/n \Sigma y_i^2) \quad (1)$$

Larger is better

$$S/N = -10 \log_{10}(1/n \Sigma 1/y_i^2) \quad (2)$$

where: n = number of duplicates within a trial
y_i = ith observation within a trial

The calculation of S/N ratio is a key part of the Taguchi method. Thus, during selection of the dependent variable, its design target needs to be identified.

Independent Factors. The main function of the Taguchi Study Team is to select the independent factors and their interactions for study. As the team brainstorms the problem, many candidate factors are generated and listed in the Cause and Effect (C&E) diagram (see Figure 1). Selecting independent factors from the C&E diagram is a further difficulty. How many factors should be selected? This commonly asked question is related to the size of the experiment. Even though the Orthogonal Array (OA) designs used by the Taguchi Method allow many factors to be tested using a limited number of runs, it may be impractical, uneconomical or excessively time-consuming in practice if tests are run with many levels and duplicates. This problem is greatly exaggerated if interactions between factors are

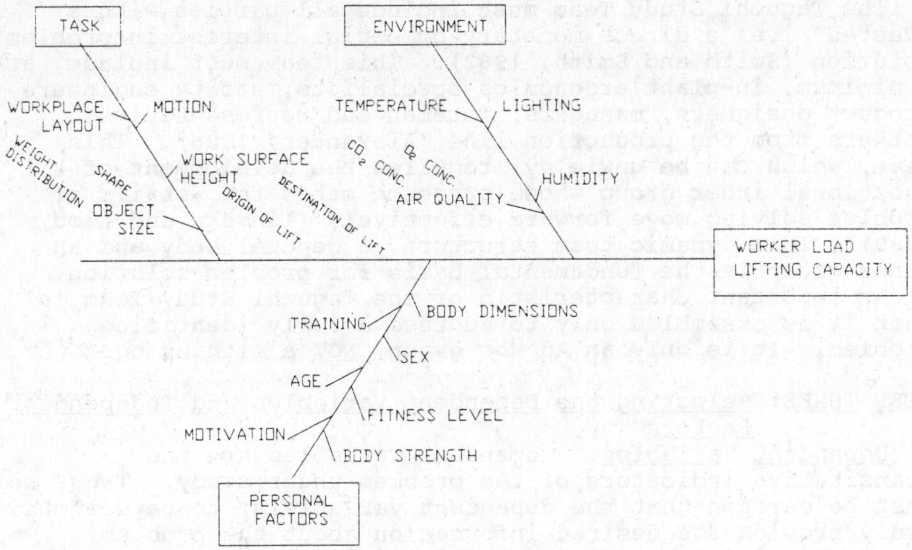

FIGURE 1 : CAUSE AND EFFECT DIAGRAM AS PRODUCT OF TAGUCHI STUDY TEAM BRAINSTORMING SESSIONS

also desired. In the early stages of investigation, the function of the OA is to screen many factors (Ross, 1988). Therefore, as many independent factors as possible should be selected within the limitations stated by experimental cost and time constraints while avoiding interaction terms where possible. Determining the levels of a selected factor is another concern. Two levels (high and low, with and without) or three levels (high, medium and low) are suggested in the early stages of investigation, and the levels chosen should cover the entire range of interest.

STEP FOUR: Designing the Experiment

Once the independent factors and relevant 2-way interactions have been selected an appropriate OA must be designed. The OA must provide sufficient Degrees of Freedom (DFs) to properly study all of the factor's contributions to the dependent variable. To assure that the chosen design will provide sufficient DFs for the contemplated experiment, the following equation must be fulfilled:

$$DF_{exp.} = \text{\# of Runs} - 1 \geq \Sigma(DF_{Factors} + DF_{Inter.}) \quad (3)$$
$$\text{(all factors \& interactions)}$$

Where: DF_i = degrees of freedom of i

The chosen OA is further based on the type of factor (2-level or 3-level) that predominate among the factors to be studied in the experiment. Details on OA construction is available in most texts on the Taguchi Method. Complicated designs can be produced through the use of the computer program developed and described by Lewandowski and Lindeke, (1989).

When a Taguchi OA is designed, only 2-way interactions are included. This can be compared to conventional full factorial designs which study higher order 3-way and 4-way interactions. In industrial ergonomics and safety studies, these higher order interactions are often difficult to explain (physically) or to use for design. Therefore, unless a laboratory or research study is being conducted with regression analysis as a goal, 2-way interactions are sufficient.

<u>STEP 5</u>: <u>Performing the Experiment</u>

Two major issues are raised here. First, experimental randomization and duplication. Second, the use of the Taguchi Study Team to assure that actual experimental trials are properly carried out.

When any designed experiment is conducted, randomization is critical to assure that any unexpected or uncontrollable noise is spread evenly during data gathering. This is especially true for the greatly abbreviated OA designs used in Taguchi experiments. Following the trial order of the OA assures overall randomization of factors and their levels throughout the experiment. Using this order in duplicated experiments allows estimates of e_2 error (noise related errors) to be made (Taguchi, 1987). This means, of course, that a robust design is much more likely.

The second issue, using the members of the Taguchi Study Team to assure proper experimental control, is a more practical concern. These members, who will actually assure that all experimental factors are set at their proper levels, will further mean that all dependent variable data will be collected. This guarantees that the results generated during the analytical stage of the experiment will be accurate.

<u>STEP SIX</u>: <u>Analyze the Data</u>

In the Taguchi Method, raw data are transformed into a S/N ratio for data analysis if each trial has two or more duplicates. When only one run is collected, raw data is used throughout the data analysis. The calculation of the S/N ratio integrates several duplicates at each trial into a single value which reflects changes in the mean and the variance simultaneously. The S/N ratio then becomes the performance measure for the levels of factors that are to be optimized. When either the mean moves closer to the ideal value (minima or maxima) or the performance becomes less sensitive to noise factors (reducing the variance), the S/N ratio will be increased. The experimental goal is to maximize the S/N ratio while selecting the levels of independent factors. Again, the reader is referred to Lindeke and Liou (1989) for further details on the required analytical techniques.

<u>STEP 7</u>: <u>Development of the Indicator Equation and Optimum Operational Settings</u>

The level averages for all factors (or significant interaction effects) are used to select the factor levels that give the optimal conditions for the human performance. Should significant interaction effects be found during ANOVA, level selection for those interacting factors is based on interaction averages which overrule individual factor level recommendations. The construction of the Indicator Equation also depends on the outcome of the ANOVA. This equation is developed from significant factors or interactions and is used to predict the performance of the optimized settings. The form of the equation is presented below:

$$\overline{\mu}_{pred} = \overline{T} + \Sigma(\overline{SF} - \overline{T})_{\text{all SF}} \quad (4)$$

where: μ_{pred} - the predicted average performance

$\overline{SF}$ - average performance responses for significant factors (or interactions) at best level

$\overline{T}$ - the overall average of all experimental performance responses

When this equation is prepared, significant interaction will again be used in lieu of significant individual factors.

STEP 8: Confirmation Run

A confirmation run, with all factors set at their best levels must be made to "prove" the experimental results. The average performance response found after confirmational testing must fall within the predicted range determined during step 7. Should this not occur, indicating a faulty experimental design, one of several corrective actions must be taken. These can be as simple as rerunning the confirmation test to complete experimental redesign.

STEP 9: Conduct Higher Order Experiments

Once the screening experiment has been completed, "fine tuning" by further experimentation may be employed. The most logical experimental path would be to better define the roles played by the significant factors identified during the screening experiment.

The effects of some factors are usually not the simple linear ones that can be deduced during 2-level screening experiments. Therefore, higher level experimentation should give a better understanding of a factor's contribution to the problem. The decision would be one of allocation of resources and further experimentation must be addressed on a case by case basis.

CONCLUSION

Traditionally, industrial ergonomics and safety studies have been conducted in the laboratory. These laboratory level experiments have developed many principles and guidelines to be used in the field. The field level ergonomist or safety engineer applied these principles and guidelines to their designs without performing their own experiments. One of the problems of applying principles and guidelines is that these are general in nature and require individual modification on a case by case basis. Therefore, conducting experiments in the field can be very useful. Of course, in the field experimental conditions are hard to control or there is, at times, a lack of training in D.O.E. among practicing engineers. Using the Taguchi Method, following the step by step approach suggested here addresses the second issue: training the inexperienced engineer in conducting D.O.E.

REFERENCES

Alexander, D.C., 1986, The Practice and Management of Industrial Ergonomics, (Englewood Cliffs: Prentice-Hall).

Clark, C. and Williges, R.C., 1973, Response surface methodology central-composite design modifications for human performance research. Human Factors, 15(4), 295-310.

Lindeke, R.R. and Liou, Y.H., 1989, Methods for optimization in the manufacturing system - the Taguchi Method: an engineering approach to its implementation. Journal of Mechanical Working Technology, 20, 205-218.

Lewandowski H.S. and Lindeke, R.R., 1989, An automated method for the preparation of orthogonal arrays for use in Taguchi designed experiments. International Journal for Computers in Industrial Engineering, 17(1-4), 502-507.

Montgomery, D.C., 1984, Design and Analysis of Experiments, 2nd edn., (New York: John Wiley & Sons).

Ross, P.J., 1988, Taguchi Techniques for Quality Engineering, (New York: McGraw-Hill).

Smith, L.A. and Smith, J.L., 1982, How can an IE justify a Human Factors activities program to management. Industrial Engineering, 14(2), page 40.

Taguchi, G., 1987, System of Experimental Design, Vol. 1 and 2, (UNIPUB/Kraus International Publications).

Advances in Industrial Ergonomics and Safety II
Edited by Biman Das
Taylor & Francis 1990

INTEGRATED PERFORMANCE AND WORKLOAD MODELING FOR INDUSTRIAL AND OTHER SYSTEM APPLICATIONS

Alvah C. Bittner Jr.

Battelle HARC
Seattle, WA 98105-5428

Stephen J. Morrissey

U. of Wisconsin, I.E. Dept.
Platteville, WI 53818

A family of performance and workload modeling methods is outlined and industrial applications are considered. This family represents a progression from well-known Timeline Analyses (e.g., MTM) to more comprehensive Integrated Microprocess Simulations. A systematic approach to human model development is delineated subsequent to outlining the family of methods. This approach provides for the evolutionary development of models for industrial and other applications. The final section considers implications for industrial work analysis. The integrated family of performance and workload models is concluded as offering: practicable enhancements to current industrial methods; and research opportunities for further enhancement.

INTRODUCTION

This report describes an integrated family of modeling methods which may be systematically applied in industrial settings. In the first section, this three-level family is introduced and shown to represent a progression which builds upon commonly used Timeline Analysis Methods (e.g., MTM). Task Network and Integrated Microprocess Simulation, the second and third levels of this family, are enhancements to these commonly used methods. They may be unfamiliar to most industrial practitioners, but have been widely applied in evaluations of commercial aircraft and military systems for over two decades (cf., Linton et al., 1989; Lysaght et al., 1989). The first section of this report introduces this family of modeling methods and its potential.

The second section of this report delineates a systematic approach to human model development. Based on the family of modeling methods, this approach provides for the increasingly comprehensive development of performance models. Also incorporated in this development is an increased consideration of the more "mental aspects of workload." This section provides the basis for consideration of the implications of the family of modeling methods in the final section.

INTEGRATED FAMILY OF MODELS

Table 1 outlines a three-level family of modeling methods which are based on a descriptive task analysis (i.e., Level 0). This family of methods includes: Timeline Analysis (Level I); Task Network Simulation (Level II), and Integrated Microprocess Simulation (Level III). These three levels of methods and their basic interrelationships are sequentially introduced. As part of this introduction, each level's performance constituents and workload applications are described. This introduction sets the stage for later discussion of the systematic modeling approach.

Table 1. Integrated family of modeling methods.

LEVEL	NAME	DESCRIPTION	APPLICATION	EXAMPLES
0	TASK ANALYSIS	Graphical representation of the sequence and interrelationships of tasks.	This model provides the framework for development of the subsequent "task analytic models."	~ Lysaght et al. (1989)
I.	TIMELINE ANALYSIS MODEL	Time estimates are developed for each subtask and summed to estimate the "time required" to accomplish a task.	The ratio of time required (TR) to time available (TA) is evaluated to determine whether workload is excessive (TR/TA > 0.7). [Underload might also be indexed by a similar criteria (e.g., TR/TA < 0.07).]	~ Stone, Gulick, & Gabriel (1985) ~ MIL-HDBK-763
II.	TASK NETWORK SIMULATION MODELS	Time estimate distributions and probabilities of success are developed for each subtask and the distribution of "time required" is typically estimated by simulation.	The distribution of time required (TR) may be evaluated to determine both probabilities of (i) excessive workload (TR/TA > 0.7) or (ii) system failure (TR > TA).	~ Siegal & Wolf (1969) ~ Laughery, Drews, Archer, & Kramme (1986)
III.	INTEGRATED MICROPROCESS SIMULATION MODELS	Human information processing and other performance microprocess models are integrated in simulation of subtask and task performances.	These models provide for a theory-based diagnostic understanding of critical task performances, in addition to estimating probabilities of (i) excessive workload (TR/TA > 0.7) or (ii) system failure (TR > TA).	~ Harris, Iaveccia, & Bittner (1988) ~ Kantowitz (1989) ~ Card, Moran, & Newell (1986)

Timeline Analysis (Level I)

Timeline Analysis includes staple methods of industrial practitioners and may be the most familiar of the three levels. It includes MTM and other work standards methods where (Barnes, 1980):

- Estimated subtask times are summed to evaluate the "time required" to accomplish a task (TR); and
- Time available (TA) for accomplishing a task is effectively defined (e.g., by a standard or pacing environment).

Historically, in industrial settings, the criteria for setting a standard for a task TA in terms of its TR have not been clearly defined (Mital and Genaidy, 1990). However, a formal operator workload "standard" has been established (TR/TA<.7) which is widely used for commercial aircraft and military evaluations (U.S. Army, 1979). This type of "standard" is aimed at controlling excessive time-stresses. It does not address vigilance and other problems associated with operator "underload". These may be increasing in industry as they are in aviation and military settings (Kantowitz and Casper, 1988; Christ et al., 1989). However, an underload standard may be proposed for this class of methods in terms of the ratio of the sum of task-times ($\Sigma(TR_i)$ to total time: $\Sigma(TR_i)/T > .07$ for $T > 1$ Hour. This method has been focused at the control of workload at upper levels but has undeveloped potential for the lower levels.

Timeline Analysis methods have the advantage of relative simplicity which makes feasible their hand-calculation (computer-based tools are common). This simplicity, however, does not provide for convenient handling of either: (i) stochastic aspects of tasks (e.g., variance of TRs), or (ii) recursive or other tasks with probabilistic branching (e.g., varying types of rework required at a production step). These are better handled by computer-based methods specifically developed for this purpose (e.g., Task Network Simulation).

Task Network Simulation (Level II)

Task Network Simulation (TNS) has been described as elaborated Timeline Analysis with consideration of the statistical nature of subtasks (e.g., Lysaght et al., 1989). Aimed at stochastic aspects of tasks and probabilistic branching, the majority of simulation models are derivatives of those developed by Siegel and Wolf (1969). Siegel and Wolf models come in several variants involving the number of operators simulated. Their basic purposes are to evaluate:

- Distribution of task completion times, and
- Probabilities of successful task completion [e.g., Pr(TR<TA)].

Classically, Siegel-Wolf models view operator workload as arising from various sources of time stress (e.g., falling behind in time on a task sequence). In this context, stress is typically calculated as the ratio of the sum of the average task execution times [$\Sigma(\underline{TR}_i)$] to the total time available (T) with maximum acceptable loading defined by $\Sigma(\underline{TR}_i)/T < .7$ (Lysaght et al., 1989). As with Timeline Analysis, no "minimum acceptable" loading limit has been defined in the context of Siegel-Wolf models, although one could be posited [e.g., $\Sigma(\underline{TR}_i)/T > .07$ for T >1 Hour]. Siegel-Wolf models may be summarized as having focused at the control of excessive workload.

A recent extension of the Siegel-Wolf model is a microcomputer version of SAINT (System Analysis of Integrated Networks of Tasks). With a menu-driven interface, Micro-SAINT contains: (i) a number of process branching rules, (ii) multiple distributions for modeling individual task operations, and (iii) a Monte Carlo sampling procedure for determining task execution (cf., Laughery et al., 1986). SAINT traditionally defines time stress in terms of TR/TA. Micro-SAINT, although amenable to similar evaluations, frequently has been used with variants of subjective task analysis methods [e.g., McCracken-Aldrich (1984) which requires an analyst to assign estimates of "workload requirements" (0-7 Scale) to cognitive and other subtask dimensions]. Laughery et al. (1986), for example, illustrates such an application for evaluation of helicopter designs. Highly correlated with operator ratings of workload (Iavechhia et al., 1989), this type of analysis attempts to understand the source of workload. This may be better handled by methods specifically developed for evaluating the processes involved in task performance (e.g., Integrated Microprocess Simulation).

Integrated Microprocess Simulation (Level III)

Integrated Microprocess Simulation (IMPS) has been described as Task Network Simulation (TNS) elaborated with human performance micromodels (following Lysaght et al., 1989). This is essentially correct for recent IMPS implementations (e.g., Harris et al., 1987). However, this description does not capture the profound enhancements which microprocess models provide:

- Modeling of subtask performance -- microprocess models provide the estimates of the TRs and other performance metrics (vs. relying on experience-based judgments which may not apply for novel systems).
- Effectiveness measures -- microprocess models provide the basis for evaluation of the complete range of performance metrics (vs. simple time and probability measures typical of TNS).
- Understanding of the nature of performance -- microprocess modeling focuses attention on the fine-structure of processes required for each subtask (vs. ignoring such fine-structures with TNS).

Each of these enhancements has been demonstrated with the historically most influential IMPS: HOS, the Human Operator Simulator (Wherry, 1969; Lane et al., 1981; Harris et al., 1986).

HOS-IV, the latest version, has been implemented for an IBM PC-AT (Harris et al., 1987). This version contains an enhanced user interface to assist in defining, executing, and analyzing the results of a simulation. Included in these is a library of human performance micromodels that can be used to simulate the timing and accuracy of particular human behaviors. These include: eye movement, visual perception, decision time, short-term memory, listening and speaking, fine-grained control manipulation, hand movement, and walking (cf., Harris et al., 1988). Other collections of micromodels may be easily incorporated, such as those used for analysis of computer interfaces or more cognitive tasks (e.g., Card, Moran and Newell, 1983, 1986). This ease of incorporating other micromodels may also be used to emulate "less advanced" levels of modeling methods: Timeline Analysis (Level I); Task Network Simulation (Level II). This capability provides the potential to easily evolve from a Timeline Analysis (Level I) to Integrated Microprocess Simulation (Level III). This capacity is particularly useful as we will see in the next section.

SYSTEMATIC APPROACH TO MODEL DEVELOPMENT

Figure 1 outlines the stepped approach to model development which is described in this section. This approach is built on the evolutionary relationships between members of the three-level family described previously (i.e., each successive level is an elaboration of the previous). In the first subsection, the framework of this "evolutionary approach" is broadly considered. This will set the stage for the more detailed procedural considerations which follow.

Framework

The "systematic approach" is aimed at reasonably assuring that tasks have acceptable levels of workload (e.g., TR/TA and mental workload levels). As illustrated in Figure 1, the task modeling portion of this process typically begins at the TimeLine Analysis Level (I). Then, a task's model is stepped to a more evolved level only as found required. Thus, the systematic approach involves a parsimonious task modeling process for assuring workload acceptability.

Workload is judged "acceptable" under this process when a task is found suitable for membership in the Acceptable Task List (shown at right side of Figure 1). This occurs when a task has been:

(i) Evaluated with an appropriately selected level of model (I, II, or III as determined by the process shown in the figure); and

(ii) Found to pass the appropriate level of criteria for inclusion in the Acceptable Task List, i.e., "Clearly Acceptable TR/TA and Mental Workload" for Level I; and "Acceptable Mental Workload and Pr(TR>TA)" for Levels II and III. (This may require "task redesign" as shown on the left side of the figure).

Figure 1 does not provide extensive details regarding the meeting of these two requirements [i.e., (i) and (ii)]. These are considered in more depth in the next subsection.

Procedural Considerations

The appropriate level for evaluating a task is frequently unclear. Fortunately, a Timeline Analysis (Level I) provides a ready starting point for evaluating a task. Once a Timeline Analysis is conducted, as shown in Figure 1, a task is judged acceptable, if found to have "Clearly Acceptable TR/TA and Mental Workload." Although other criteria may be posited, we would currently recommend the following for overload:

$$TR/TA < .7 \quad (1)$$

and for underload:

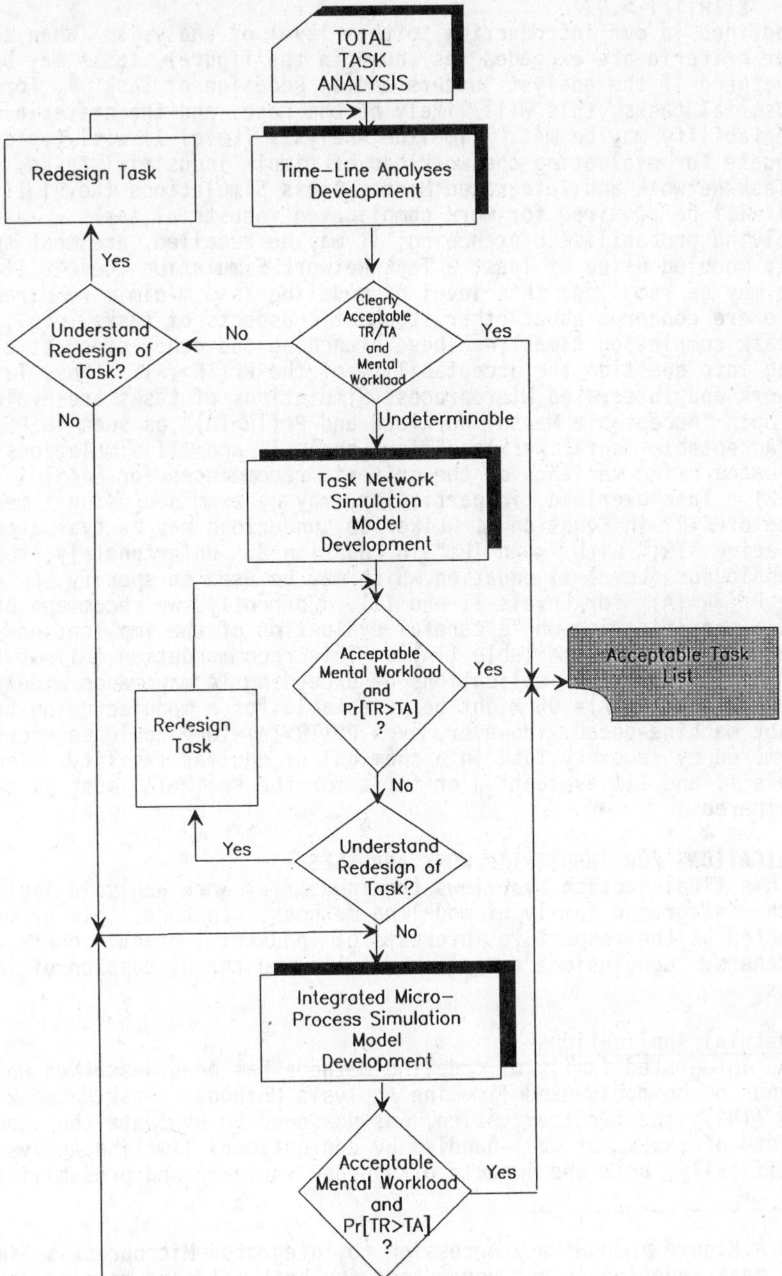

Figure 1. Systems Approach to Human Model Development

$$\Sigma(TR_i)/T > .07 \qquad (2)$$

as defined in our introduction to this level of analysis. When these or other criteria are exceeded (as shown in the figure), tasks may be restructured if the analyst "Understand[s] Redesign of Task".[1] For simple industrial tasks, this will likely be the case, and the criteria for acceptability may be met. Timeline Analysis (Level I) will typically be adequate for evaluating the workload of simple industrial tasks.

Task Network and Integrated Microprocess Simulations (Level II and III) will be required for more complicated industrial tasks. Tasks involving probabilistic branching, it may be recalled, are most appropriately modeled using at least a Task Network Simulation (Levels II). As also may be recalled, this level of modeling is a minimim requirement when there are concerns about other stochastic aspects of tasks (e.g., variance of task completion times).[2] These branching and other stochastic aspects bring into question the acceptability of the Pr(TR>TA). Thus, Task Network and Integrated Microprocess Simulations of tasks are evaluated for both "Acceptable Mental Workload and Pr(TR>TA)" as seen in Figure 1.

"Acceptable mental workload" for Levels II and III simulations may be evaluated using variants of the criteria recommended for Level I (Eqs. 1 and 2). Task overload, in particular, may be examined using " mean TR" in place of "TR" in Equation 1. Likewise, underload may be evaluated by replacing "TR_i" with "mean TR_i" in Equation 2. Unfortunately, there is no analogous (Level I) equation which may be used to specify an "Acceptable Pr(TR>TA)" for Levels II and III. Currently, we recommend basing such a specification on "a careful evaluation of the implications of exceeding the time available (TA)". This recommendation follows from our recognition that the implications of exceeding TA may range widely. For example, Pr(TR>TA)=.05 might be acceptable for a manufacturing task which is not machine-paced. However, even $Pr(TR>TA)=10^{-6}$ could be excessive for an emergency recovery task in a chemical or nuclear facility. Thus, Levels II and III evaluation criteria for the Pr(TR>TA) must be carefully considered.

IMPLICATIONS FOR INDUSTRIAL WORK ANALYSES

This final section overviews the industrial work analysis implications of the integrated family of modeling methods. In turn, this overview is directed at the respective interests of industrial practitioners and researchers. Conclusions are offered following the discussion of implications.

Industrial Applications

An integrated family of modeling methods has been described which extends of commonly-used Timeline Analysis Methods. Task Network Simulation (TNS), the first extension, was designed to evaluate the stochastic aspects of tasks not well-handled by conventional Timeline Analysis. Specifically, both the impacts of subtask variance and probabilistic

---

[1] Figure 1 shows a progression to Integrated Microprocess Simulation when task redesign is not understood for both this and more advanced methods (i.e., Levels I and II). This assumes that neither traditional or more recent subjective methods may be fruitfully applied to gain such an understanding (e.g., McCracken and Aldrich, 1984; Iavecchia et al., 1988).

[2] Ongoing research has suggested an approximately direct proportionality between subtask time (TR) standard deviations and means, SD=K*Mean. For different classes of subtasks, the constant of proportionality ranged from 0.5 to 1.0. This relationship suggests careful evaluation of the Pr(TR>TA) when mean TR is a substantial proportion of the TA.

branching are well-handled by TNS. Thus, TNS is especially suited for evaluation of "reworking tasks" where diverse types of rework may be required. Integrated Microprocess Simulation (IMPS), the second extension, was also seen to handle these stochastic aspects of tasks. More importantly, IMPS was also shown to provide for: modeling of subtask performance, evaluation of the complete range of performance metrics, as well as an understanding of the nature of performance. IMPS, we suspect, will be particularly useful for this latter purpose when tasks involve complex cognitive processes. Clearly, the "integrated family" offers valuable enhancements over traditional industrial methods.

A systematic modeling approach has also been described which is based upon the integrated family of modeling methods. Figure 1 outlines this approach which might be applied as an extension of those currently used in industry . As seen in this figure, task modeling may begin with a familiar Time-Line Analysis Method (e.g., MTM). Then, a task's model is advanced to a more enhanced level only as required. The process for evaluating this requirement, it is important to note, involves considerations of both conventional and the more mental aspects of workload. Thus, this "systematic approach" represents an efficient process which extends current methods for assuring workload acceptability.

Research Questions

The systematic modeling approach described earlier required judgments about task workload. For Timeline Analysis (Level I), this involved evaluating whether or not a task had a "Clearly Acceptable TR/TA and Mental Workload". Based upon current literature and personal experience, recommendations were made for both overload and underload "acceptability" (i.e., Eqs. 1 and 2). However, as noted earlier, other criteria could have been posited. For underload and the transition from it to overload, this is especially true, as the literature in these areas is virtually void (Christ et al., 1989). This leads to the first research question: How may the basic criteria for task underload and overload be better defined?

Task Network and Integrated Microprocess Simulations (Level II and III) will profit from answers to this first research question. However, they uniquely require evaluation of whether or not a task has an "acceptable Pr(TR>TA)". Toward setting a criteria for this probability, we earlier recommended "a careful evaluation of the implications of exceeding the time available (TA)". This requires an a relatively subjective judgement which may not be necessary. The means for more objectively specification may lie in the "risk evaluation literature" which should be explored. In addition to setting an acceptable level, there may be difficulties in evaluating Pr(TR>TA). This follows from the lack of either (i) a data-base regarding the stochastic aspects of subtasks (e.g., variance of TRs), or (ii) a ready set of microprocess models suitable for immediately estimating such aspects. These appear to be important and attackable problems, but have not been adequately pursued (cf., Meister, 1989). This leads to the second research question: How may the acceptable level of Pr[TR>TA] be better defined and evaluated?

The integrated family and systematic modeling approach point toward research opportunities for further enhancement of industrial methods.

Conclusions

The integrated family of performance and workload models offer: (i) practicable enhancements to current industrial methods; and (ii) research opportunities for their further enhancement.

REFERENCES

Barnes, R.M., 1980, Motion and time study design and measurement of work 7th edn., (New York, NY: John Wiley).

Card, S.K., Moran, T.P., & Newell, A. 1983. The psychology of human-computer interaction, (Hillsdale, NJ: Lawrence Erlbaum Associates).

Card, S.K., Moran, T.P. & Newell, A., 1986, The model human processor: An engineering model of human performance. In: Handbook of Perception and Human Performance: Vol. 2. Cognitive Processes and Performance, edited by K. R. Boff et al., (New York: John Wiley), pp. 45-1 to 45-35.

Christ, R.E., Zaklad, A.L., Bittner Jr., A.C. et al., 1989, The Army Operator Workload (OWL) Program: Review and prospects. Proceedings of the Human Factors Society 33rd Annual Meeting (Santa Monica, CA: Human Factors Society), pp. 1471-1475.

Meister, D., 1989, Conceptual Aspects of Human Factors, (Baltimore, MD: Johns Hopkins University Press).

Harris, R.M., Glenn, F., Iavecchia, H.P. & Zaklad, A.L., 1986, Human Operator Simulator. In: Trends in Ergonomic/Human Factors III (Part A), edited by W. Karwoski, (Amsterdam: North-Holland), pp. 31-39.

Harris, R.M., Iavecchia, H.P. & Bittner Jr., A.C., 1988, Everything you always wanted to know about HOS micromodels but were afraid to ask. Proceedings of the Human Factors Society 32nd Annual Meeting, (Santa Monica, CA: Human Factors Society), pp. 1051-1055.

Harris, R.M., Iavecchia, H.P., Ross, L.V. & Shaffer, S.C., 1987, Microcomputer Human Operator Simulator (HOS-IV). In: Proceedings of the Human Factors Society 31st Annual Meeting, (Santa Monica, CA: Human Factors Society).

Iavecchia, H.P., Linton, P.M., Bittner Jr., A.C. & Byers, J.C., 1989, Operator workload in the UH-60A Black Hawk: Crew results vs. TAWL model predictions. Proceedings of the Human Factors Society 33rd Annual Meeting, (Santa Monica, CA: Human Factors Society), pp. 1481-1485.

Kantowitz, B.H. & Casper, P.A., 1988, Human workload in aviation. In: Human Factors in Aviation, edited by E. Wiener & D. Nagel, (New York: Academic Press).

Lane, N.E., Strieb, M.I., Glenn, F.A., & Wherry, R.J., 1981, The human operator simulator: An overview. In: Manned Systems Design: Methods, Equipment and Applications, edited by J. Moraal & K.F. Kraiss, (New York: Plenum Press), pp. 121-152.

Laughery, K.R., Jr., Drews, C., Archer, R., & Kramme, K., 1986, A Micro-SAINT simulation analyzing operator workload in a future attack helicopter. In: National Aerospace and Electronics Conference, (Dayton, OH: IEEE), pp. 896-903.

Linton, P.M., Plamondon, B.D., Dick, A.O., Bittner, Jr., A.C. and Christ, R.E., 1988, Operator workload models for military system acquisition. In: Applications of human performance models to systems design, edited by G.T. McMillan, (New York, NY: Plenum).

Lysaght, R.J., Hill, S.G., Dick, A.O., Plamondon, B.D., Linton, P.M., Wierwille, W.W., Zaklad, A.L., Bittner, Jr., A.C., & Wherry, R.J. 1989, Operator Workload: Comprehensive Review and Evaluation of Operator Workload Methodologies (ARI T.R. 851), (Willow Grove, PA: Analytics, Inc).

McCracken, J.H., & Aldrich, T.B., 1984, Analysis of selected LHX mission functions: Implications for operator workload and system automation goals (TNA AS1479-24-84), (Fort Rucker, AL: Anacapa Sciences, Inc).

Mital, A. & Genaidy, A.M., 1990, Review and evaluation of techniques for determining work-rest periods. IE Ergonomics News, 24(2), pp. 1-7.

Siegel, A.I., & Wolf, J.J., 1969, Man-Machine Simulation Models: Psychosocial and Performance Interaction, (New York, NY: John Wiley).

Wherry, R.J., Jr., 1969, The development of sophisticated models of man-machine systems. In: Proceedings of the Symposium on Applied Models of Man-Machine Systems Performance, (Columbus, OH: North American Aviation).

Advances in Industrial Ergonomics and Safety II
Edited by Biman Das
Taylor & Francis 1990

ACUTE NEUROBEHAVIORAL EFFECTS FROM SOCIAL CONSUMPTION OF ALCOHOL

Diana Echeverria, L Fine, G Langolf, A Schork, C Sampaio

Battelle Memorial Institute, Human Affairs Research Center,
4000 NE 41st Street, Seattle, WA, 98105, USA

Changes in central nervous system (CNS) function and symptoms were evaluated at socially relevant doses of .00, .33, and .66 grams ethanol/kilogram body weight. Ethanol or a placebo was administered at 3:30 PM over 3 days. Forty-two college students were tested at 8 AM, 12 Noon, and 4 PM. A 3X3 Latin Square design evaluated ethanol effects while controlling for learning and dose sequence using [AM-PM] scores. Variation in absorbance was monitored in breath. A 5 to 10% decrement was considered meaningful if consistent with a linear trend at p <.05. At the high dose symptoms and performance decrements were found for digit span (6.6%), pattern recognition (9.2%), continuous performance (4.0%), symbol digit (7.9%), finger tapping (16.5%), critical tracking (6.2%), and the one hole test (5.2%). The positive results show ethanol is a useful acute standard for other industrial CNS depressants.

INTRODUCTION

Behavioral test batteries have quantified the neurotoxic effects of solvents, in Scandinavian countries (Anger, 1984; Hanninen and Lindstrom, 1979), England Cherry et al., 1983), and the USA (Anger, 1985; Maizlish et al., 1985). In the last four years, the Neurobehavioral Evaluation System (NES) has been introduced as a standardized computer-administered method to quantify the effects of neurotoxicants on a wide range of human performance functions (Baker et al., 1985; Baker et al., 1985). Since neurobehavioral toxicology is now more frequently used to establish acute occupational standards for exposure to solvents (Anger, 1985); Baker and White, 1985) the interpretation of a new battery of tests is aided by knowing each test-retest coefficient of determination and sensitivity to a well known reference substance.

Ethanol was chosen as an external standard reference substance, the results of which can then be expressed in equivalency units when establishing an acute solvent exposure measurement. The purpose of the study was to aid in the interpretation of acute and chronic potential effects of industrial solvents by determining the reliability and sensitivity of a set of tests commonly used in epidemiologic solvent studies. The battery can be considered sensitive if it measures the effects of ethanol, a central nervous system depressant, and if it can detect effects at low levels corresponding to socially relevant doses. These doses are .33 or .66 grams ethanol/kilogram body weight (g ETOH/kg BW), equivalent to two or four domestic beers for a 70-kg person (there are 12.42g ETOH in a 12-ounce 3.5%-wt/vol beer).

METHODS

Subject selection: Forty-two healthy University of Michigan students were selected in accord with the Declaration of Helsinki. They were homogenous with respect to age (21.9 yr, SD=3.5), education (3.1 years college, SD=1), were English speaking, and consumed less than 18 grams alcohol per day (12.4 grams are in 12 ounces of a 3.5% wt/vol beer). The population's Vocabulary (84.8 % correct, SD=9) and Mill Hill synonym test scores (84 % correct, SD=13.8) were narrow because most subjects achieved above 90% correct.

Subject training: The battery was administered once at 7:00 AM on the first day to reduce learning effects. Learning was also controlled by experimental design. To ensure more consistent performance, monetary payment was provided and the taste of alcohol was masked for all doses as discussed below. No alcohol or caffeine was permitted one night before or during the three days of testing.

Experimental design: The experimental design, similar to a toluene study published elsewhere (Echeverria, et al., 1989), is a balanced two-way scheme (dose by day) for three permutations of one dose order in a standard 3X3 Latin Square. Fourteen subjects were randomly assigned to each group. They were tested at different doses on each of the three days. The independent factors were a) ethanol concentration, b) learning or improvement in performance scores over the three days, and c) the dose order. The 8 AM and Noon test sessions served as controls for the afternoon session where the subject received his/her dose at 3:30 PM. The control information is incorporated in the analysis by reporting the difference between a control and exposed score for each day. The [Noon-PM] score reflects the afternoon acute ethanol effect, and the [AM-PM] score is similar to testing before and after an 8-hour work day.

The experiment examines 30-70 minute acute behavioral effects of ethanol, at .00, .33, and .66g/kg, equivalent to .03 mg% and .06 mg% at peak blood alcohol concentration (BAC) levels (Jones, 1978) bracketing the legally impaired level. Most tests were administered on the ascending side of the blood alcohol curve, which occurs about 45 minutes after ingestion. Subjects were retained until their breath samples indicated one half of the absorbed dose of ethanol had been metabolized. It takes an estimated 45 minutes for the low dose and 2.5 hours for the high dose to remove 50% of the ethanol from the blood.

Subjects were tested in the presence of .078 ppm menthol for comparison with the previous toluene study. The aromatic odor had no effect on performance. The subjects were administered the complete test battery at 8:00 AM and 4:00 PM. Four of the tests were also given at 12:00 Noon for comparison to the toluene exposure study. A Fatigue Checklist was given in the morning and afternoon. Symptoms and complaints were recorded in response to the question "What are your reactions over the last 4 hours?".

Performance tests: To ensure a broad based test battery this study evaluates verbal skill (the Vocabulary sub-test of the Armed Forces Qualifying Test (AFQT) and the Mill Hill Synonym Test (Baker, et al., 1985), verbal short term memory (the Sternberg Memory-Scanning Test (Baker, et al., 1985) and a Digit Span Test (Smith, et al., 1983), visual memory (the Pattern Memory Test (Baker, et al., 1985), the Benton Test (Baker, et al., 1985), perception (a Pattern Recognition Test (Baker, et al., 1985), psychomotor skill (a Simple Reaction Time Test (Baker, et al., 1985), a dynamic Continuous Performance Test (CPT) (Baker, et al., 1985), a Symbol-Digit Matching Test (Baker, et al., 1985), a Critical Tracking Task (Jex, et al., 1966), manual dexterity (the One Hole Test (Salvendy, 1975), a Hand-Eye Coordination Task (Baker, et al., 1985) the Finger Tapping Test (Baker, et al., 1985), mood (Profile on Mood Scales (Baker, et al., 1985), fatigue (the Fatigue Checklist (Pearson and Byars, 1956), and symptoms.

Tests were chosen from batteries developed at the University of Michigan and Harvard University (Baker, et al., 1985; Maizlish, et al., 1985). Several were also recommended jointly by the Finnish Institute of Occupational Health, the World Health Organization (Baker and White, 1985), and the National Institute of Occupational Safety and Health (Anger, 1985). These tests are the symbol-digit substitution, digit span, simple reaction time, Benton visual memory, and the profile on mood scales.

Ethanol ingestion: Subjects were weighed daily and their dose was diluted 1:5 in tonic water and orange juice with bitters. The placebo was laced with 3 ml ethanol. Subjects were asked to drink it within 20 minutes at 3:30 PM, 3 hours after a light lunch.

Analysis of Ethanol in Breath: Breath samples were collected pre-dose at 8:00 AM and 3:30 PM, and post-dose at 30, 60, 90, and 120 minutes after ingestion, assuring points on the ascending and descending limb of the blood alcohol concentration curve. Breath samples were analyzed using gas chromatography (GC) (Echeverria, et al., 1989).

Statistical Analysis: The main hypothesis is ethanol has an adverse effect on human performance in a dose-effect manner. This hypothesis is best evaluated while simultaneously controlling for the effects of practice of 3 sessions per day over 3 days and the sequence of dose. Possible refinement in detecting the hypothesis is examined by analyzing performance 1) categorically by dose level (ANOVA) and 2) continuously by ethanol concentration in breath. Within the regression analysis is a second comparison between ethanol in breath (continuous) to dose level (discrete) for improvement in goodness of fit.

The behavioral effects are evaluated using the daily [AM-PM] difference and the [PM] scores alone. The variation in the AM and Noon control data across the three days is larger than the ethanol effect; therefore, a within-day difference score is the preferred dependent variable despite the added variation it introduces into the experiment. Further, the effect of ethanol on either dependent variable is similar. Decrements in performance scores were made comparable by expressing them as a percent change from the mean AM control performance scores.

A restricted analysis of variance (ANOVA) is used to determine if there are behavioral responses to dose level and repeated practice over 3 days, since only the variance attributable to repeated measurements on the same subject was analyzed (Winer, 1962). A test for trend determines if the behavioral changes are consistent with a linear or non-linear response for either dose or practice. If a test score is affected by ethanol, Scheffe's 95% confidence intervals identify the specific significant dose level. Behavior is considered to be significantly affected if an association is found at a $p < .05$ ($a=.05; b=.20$). The degrees of freedom were fixed for the effect of dose and day (df=2,78), and for trends (df=3,39). The probability of finding a statistically significant test score by chance was 1.6 out of the 31 performance scores (31 X .05). Therefore the overall sensitivity of the test battery was only accepted if the number of affected variables exceeded the number attributable to chance. Correlations between symptoms, mood, fatigue, age, gender, education level, and performance scores were checked for their influence on performance.

The regression analysis examines the linear association between breath concentration and the [AM-PM] performance scores. The purpose is to determine if ethanol in breath is a more sensitive indicator than a dose level for detecting adverse behavior since variation in dose is attributable to differences in gender, body fat, and ventilation rates (Nomiyama and Nomiyama, 1974). Therefore, each subject's performance score is regressed against 1) his/her own breath values and 2) dose levels of .00, .33, .66g ETOH/kg Body Weight (BW). Correction is needed for a difference in scale between the breath concentrations and dose levels, so each individual's slope is divided by his/her own square root of the mean square error. A student t-test is used to determine if the mean slope for both regressions is different from zero. To determine which slope is a better predictor of performance the mean coefficients of determination for respective regressions were compared for improvement in goodness of fit.

Symptoms were tallied by dose and by day for headache, irritation, tiredness, and anxiousness or frustration. A dose-response relationship would be inferred for increasing marginal values. Subjects were also asked to rate their dose upon drinking by taste and later by effect into three categories of zero, low, or high.

RESULTS

Ethanol in Breath: Blood alcohol levels peaked about 40 minutes into the test session. Therefore over half the behavioral battery was administered on the ascending side of the blood alcohol curve. The test session was finished 75 minutes after consumption with .015% BAC (SD=.000%) for the low dose and .039% BAC (SD= .012%) for the high dose.

Test-retest Coefficients of Determination: The coefficients compare the amount of agreement between 3 morning sessions across days and the amount of agreement between AM and PM scores within a control day. There was greater agreement within a day than across days. The correlations are ranked in ascending degree of stability for variables within each test (Table 1).

Verbal Short Term Memory: Digit span was equally affected by ethanol at both doses. Scheffe 95% CIs confirmed a common percent decrement in digit span of .53 digits (SD=1.23), or 6.6% based on the mean control score of 8.04 digits (SD=1.49). There was no dose sequence, learning, or interaction effects. The Sternberg test was not affected by ethanol, practice, or dose-order. The regression for the "yes" and "no" conditions were [yes msec = 390 + 51 (item)] and [no msec = 452 + 47 (item)].

Visual Short Term Memory: The Benton visual memory test was not suitable for a repeated-measure design since the stimulus was easy to memorize. So the test was only administered in the morning and afternoon of the high dose day. The results of a paired t-test performed on the number correct was not significant (t= -.87, df=41, p<.3). The average number correct was 10.75 (SD=.99) or 90% correct. The Pattern memory latency for the number correct was affected by ethanol in a non-linear manner by ethanol probably due to random variation. At .33 and .66g ETOH/kg BW, performance improved by 12.33% (mean= .42 sec, SD= 1.58) and by 1.9% (mean= .06 sec, SD=1.67) respectively when compared to the average control score of 3.43 sec (SD= 1.1). There was no dose sequence, learning, or interaction effects. The number correct was unaffected by ethanol or practice. The average number correct was 12.41 (SD=1.42).

Visual Perception: The Pattern recognition latency for the number correct was equally affected by ethanol at both doses. The Scheffe 95% CIs confirmed the control score differed from both dose scores. At .66g ETOH/kg BW the increase in latency was .21 sec (SD= .65) or 9.24% compared with the average control score of 2.07 sec (SD= .51).

Psychomotor Function: Left hand Simple Reaction Time (SRT) was marginally affected by ethanol using [Noon-PM] difference scores. There were no ethanol or practice effects using the [AM-PM] difference score, however based on the [AM-Noon] score, practice over days 1, 2, and 3 did improve performance by 14.1 msec (SD = 9) or 4.8% from a baseline value of 295 msec (SD = 74). The average control scores of both hands was 293 msec (SD = 66).

The Continuous Performance test was affected by ethanol at both doses, as confirmed by the Scheffe 95% CIs. At .33g ETOH/kg BW response increased 24 msec (SD=42) or 5.4% from a baseline value of 431 msec (SD = 27). At .66g ETOH/kg BW response increased 17 msec (SD = 36) or 4.0%, slightly less than the smaller dose. There was some learning across days 1, 2, and 3 at 29 msec (SD=42). The number of errors of omission and commission were unaffected by ethanol.

The Critical tracking test was affected by ethanol in a dose-effect manner. The Scheffe 95% CIs showed the mean control score differed from the scores at both doses. The decrement at .66g ETOH/kg BW was .024 rad/sec (SD= .108) or 6.2% based on a control score of .408 rad/sec (SD= .061).

The Symbol-digit test latency was affected by ethanol in a dose-effect manner. At .66g ETOH/kg BW, the Scheffe 95% CIs confirmed that the latency increased from the control condition by 1.3 sec (SD=4) or 7.9% when compared to the average control score of 16.35 sec (SD=2.26). The number incorrect was unaffected by ethanol or practice.

The more complex Finger tapping task using the dominant hand on alternate buttons was affected by ethanol in a dose-effect manner. The decrement at .66g ETOH/kg BW was 8.3 taps/10 sec (SD= 22) or 16.6% from a baseline value of 56.3 taps/10 sec (SD= 9). There were no practice or dose order effects. There were no ethanol, practice, or dose-order effects for either hand separately.

The Hand-eye coordination test was not affected by ethanol and performance was irregular. On the control day, the deviation from the traced line measured by the root mean square was worse by -.06 (SD= 1.45) or 1.5% of a mean control score of 4.61 (SD= 1.64). At .33g ETOH/kg BW performance improved by .28 (SD= 1.32) or 6.1%. At .66g ETOH/kg BW performance was also poor by .05 (SD =1.62) or 1.1%. There was no learning or dose order effects across days.

Manual Dexterity: The One hole 'pin number' and the element 'move' were affected by ethanol without learning effects across the three days, interaction, or dose-order effects for both [Noon-PM] and [AM-PM] measures. Using the [Noon-PM] scores, the Scheffe's 95% CIs showed performance was adversely affected at either dose. At .33 and .66g ETOH/kg BW, the pin number dropped .1 (SD= 7.5) and 2.5 (SD= 8.6) respectively, which was a .1% and 5.2% decrease from a baseline pin number of 44 (SD= 4.5). The reduction in pin number was marginally consistent with a linear trend. The increase in latency for the element 'move' was 11 msec (SD= 62) and 32 msec (SD= 64) at the low and high dose, which was a 2.4% and 7.0% decrement from a baseline response of 466 msec (SD= 47). The elements 'grasp' and 'position' and their associated fumbles were not significantly affected by ethanol.

Regression Analysis: A comparison of the two slopes estimating dose, using breath levels or the categories of dose, were in agreement. Thus the increment of variation explained by breath (less than 10%) was insufficient to increase the sensitivity of the tests.

Mood and Fatigue: Most POMS scales were not affected by ethanol, learning, or by interaction, however tension was reduced between day 1 and days 2 and 3. The R^2 never exceeded .2 between mood scales and performance test scores and therefore mood scales did not improve ANOVA models when used as covariates. The Fatigue checklist scores were affected by ethanol, but was poorly correlated with the fatigue POMS scale (-.21).

Summary of Behavioral Results: Table 2 gives a summary of the positive behavioral results found at .66g ETOH/kg BW for verbal and visual memory, psychomotor skill, and manual dexterity. The mean decrement in performance across all affected behavior was 7.8% (SD= 1.8%). Among psychomotor tests the maximum performance decrement was 16.5% for finger tapping which measures manual speed. A non-dose related decrement in performance was found for hand-eye coordination (1.1%) and improvement was found for pattern memory (1.9%) attributable to random variation. Eight tests (9 variables) out of 13 tests (31 variables) were affected by ethanol, exceeding 1.6 or the expected number of statistically significant scores attributable to chance.

Symptoms, Complaints, and Motivation: The number of symptoms increased with increased dosage for headaches and frustration. The reports were greatest on the first day of the experiment corresponding to increased tension found on the mood scale. Subjects could not identify the dose level during consumption by taste (n=6 or 5% guessed correctly), but by the end of the third day subjects could identify the sequence of dose (n=33 or 75% guessed correctly).

DISCUSSION

Ethanol in Breath: Ethanol affected behavior on both the ascending and descending sides of the blood alcohol curve. The one hole, digit span, pattern recognition, and symbol-digit tests were administered on the ascending side of the BAC, whereas finger tapping, continuous performance, and the critical tracking tests were administered at peak and descending sides of the BAC. It has been hypothesized that greater impairment may be observed on the ascending side of the blood alcohol curve when subjects must continuously make adjustments for changing blood levels. Previous alcohol studies have found the rate and direction of change in alcohol concentration were more important than the duration of alcohol in the body (Jones and Vega, 1972). Thus it is conceivable that greater decrements in performance would have been observed on tests administered on the descending side of the curve if the test order had been reversed.

Behavioral Data: The purpose of the study was to determine the reliability and sensitivity of the test battery in detecting ethanol effects, a proposed external standard for acute behavioral effects. The ethanol range is appropriate since .05% BAC (.55g ETOH/kg BW) is the lowest value in a trend toward increased alcohol involvement in traffic accidents (American Medical Association, 1968) and has been suggested as a guideline for determining whether behavioral tests are useful in detecting subtle CNS depressant effects (Baker, et al., 1985; Baker, et al., 1986).

The test-retest coefficients of determination for most performance scores were lower than those reported in solvent field studies, suggesting this measure may not be as useful a predictor of performance in restricted populations as in this study and epidemiologic studies where groups are homogenous in age and education.

The relative sensitivities of the tests in the battery are shown at .66g ETOH/kg BW (Table 2). The average percent decrement across all affected tests was 7.8% which is plausible since at 1.1g ETOH/kg BW the average percent decrement was greater at 11.71% (Wilson, et al., 1984). In this study, most tests were evenly affected at the high dose which precludes ranking the sensitivity between categories as found in other studies in increasing order (Levine, et al., 1975): psychomotor, perceptual-sensory, and higher order cognitive tests.

The taste of ethanol was successfully masked, decreasing the influence of anticipating a zero, low, or high ethanol effect, but 33 (75%) subjects guessed the dose order at the end of the study indicating they may have performed according to knowledge of a dose level on their last day. However there was no difference in performance between the successfully blinded and non-blinded subjects, therefore this knowledge had no marked effect on performance.

CONCLUSIONS

The reliability and sensitivity of the test battery was determined for .33 and .66g ETOH/kg BW. At these socially relevant ethanol doses, previously undescribed decrements in performance were found; in short term memory for digit span; in psychomotor function for continuous performance, symbol-digit, critical tracking, and finger tapping; and in manual dexterity using the one hole test. The adversely affected functions are important for workers in industry, and might be even more affected in combination with other industrial exposures to neurotoxicants such as solvents and metals. Though the magnitude of acute effect rarely exceeds a 10% decrement, there may be a significant decrease in safety margin for industrial tasks requiring these skills as well as a loss in productivity. Inferences from this study to other populations is cautioned since these subjects are homogeneous in age and education and may represent a specific ethanol sensitive population. We suggest a similar ethanol study be done using these tests on a more heterogeneous industrial population.

REFERENCES

American Medical Association., 1968, Alcohol and the Impaired Driver - A Manual on the Medical Aspects of Chemical Tests for Intoxication (Chicago: Committee on Medical Problems), pp. 27-59.

Anger, W.K., 1984, Neurobehavioral testing of chemicals: impact on recommended standards. Neurobehavioral Toxicology, 6, 147-153.

Anger, W.K., 1985, Neurobehavioral tests used in NIOSH supported worksite studies, 1973-1983. Neurobehavioral Toxicology, 7, 359-368.

Baker, E.L. and White, R.F., 1985, Chronic Effects of Organic Solvents on the Central Nervous System and Diagnostic Criteria. Report on a Joint WHO/Nordic Council of Ministers Working Group, Copenhagen, Denmark.

Baker, E.L., Letz, R.E., and Fidler, A.T., 1985, A computer-based neurobehavioral evaluation system for occupational and environmental epidemiology: Rationale, methodology, and pilot study results. Journal of Occupational Medicine, 27, 2.

Baker, E.L., Letz, R.E., Fidler, A.T., Shalat, S., Plantamura, D.L. and Lyndon, M.L., 1985, A computer-based neurobehavioral evaluation system for occupational and environmental epidemiology: Methodology and validation studies. Neurobehavioral Toxicology, 7, 369-377.

Baker, S.J., Chrzan, G.J., Park, C.N. and Saunders, J.H., 1985, Validation of human behavioral tests using ethanol as a CNS depressant model. Neurobehavioral Toxicology, 7, 257-261.

Baker, S.J., Chrzan, G.J., Park, C.N. and Saunders, J.H., 1986, Behavioral effects of 0 and 0.05% blood alcohol in male volunteers. Neurobehavioral Toxicology, 8, 77-81.

Echeverria, D., Fine, L., Langolf, G., Schork, A. and Sampaio, C., 1989, Acute neurobehavioral effects of toluene. British Journal of Industrial Medicine, 46, 483-495.

Hanninen, H. and Lindstrom, K., 1979, Behavioral Test Battery for Toxicopsychological Studies, 2nd edn., (Helsinki: Institute of Occupational Health).

Jex, H.R., McDonnell, J.D. and Phatak, A.V., 1966, A "critical" tracking task for manual control research. IEEE Transactions on Human Factors in Electronics, 7, 138-145.

Jones, B.M. and Vega, A., 1972, Cognitive performance measured on the ascending and descending limb of the blood alcohol curve. Psychopharmacologia (Berlin), 23, 99-114.

Jones, A.W., 1978, Variation of the blood:breath alcohol ratios in vivo. Journal of the Study of Alcohol, 39, 1931-1939.

Levine, J.M., Kramer, G.G. and Levine, E.N., 1975, Effects of alcohol on human performance: An integration of research findings based on an abilities classification. Journal of Applied Psychology, 60, 285-293.

Maizlish, N.A., Langolf, G., Whitehead, L.W., Fine, L.J., Albers, J.W., Goldberg, J. and Smith, P., 1985, Behavioral evaluation of workers exposed to mixtures of organic solvents. British Journal of Industrial Medicine, 42, 579-590.

Nomiyama, K. and Nomiyama, H., 1974, Respiratory retention, uptake and excretion of organic solvents in man: Benzene, toluene, n-hexane, trichloroethylene, acetone, ethyl acetate, and ethyl alcohol. International Archives of Arbeitsmed, 32, 75-83.

Pearson, R.G. and Byars, G.E., 1956, The Development and Validation of a Checklist for Measuring Subjective Fatigue. USAF Report No. 56-115, (Randolf Field Texas: School of Aviation Medicine).

Salvendy, G., 1975, Selection of industrial operators: The one hole test. International Journal Product Research, 13, 303-321.

Schumacher, J.M., 1981, The effects of aging on four behavioral tests. Unpublished Master's Thesis (Ann Arbor: University of Michigan).

Smith, P.J., Langolf, G.D. and Goldberg, J., 1983, Effects of occupational exposure to mercury on short-term memory. British Journal of Industrial Medicine, 40, 413-419.

Stewart, R.D., Hake, C.L., Wu, A., Kalbfleish, J., Newton, P.E. and Marlow, S.K., Vuciecevic-Salama, M., 1977, Effects of Perchloroethylene/Drug Interaction on Behavior and Neurological Function. Publication Number 77-191, (Washington, DC: Department of Health Education and Welfare).

Winer, B.J., 1962, Statistical Principles in Experimental Design. 2nd edn., (New York: McGraw Hill Book Company), pp. 712-713.

Table 1 Coefficients of determination for the average of three am control values and am versus pm within each control day

Performance measure	Coefficients of Determination: Mean am	am v pm
Verbal short term memory		
Sternberg test:		
Yes intercept	.21	.56
Yes slope	.42	.70
Digit span:	.81	.81
Visual short term memory		
Benton visual memory:		
Number correct	.80	.87
Pattern memory:		
Number correct	.14	.15
Latency	.24	.79
Perception		
Pattern recognition:		
Number correct	.14	.15
Latency	.24	.79
Psychomotor functions		
Simple reaction time:		
Right hand	.40	.47
Left hand	.60	.55
Continuous performance		
CPT	.73	.78
Symbol digit:		
Number incorrect	.10	.08
Latency	.30	.84
Hand-eye coordination:		
RMSQ	.74	.51
Finger tapping:		
Right hand	.56	.62
Left hand	.52	.64
Alternating hand	.54	.66
Critical tracking test:	.51	.60
Manual dexterity		
One hole test:		
Position	.10	.10
Pin number	.46	.45
Move	.63	.77
Grasp	.67	.73
Reach	.75	.89
Mood scales		
Fatigue	.55	.74
Confusion	.81	.82
Depression	.60	.84
Tension	.69	.90
Anger	.60	.92
Fatigue checklist	.26	.33

Table 2 Summary of Behavioral Results

	ANOVA: Dose	ANOVA: Linear Trend	Loss at .66g ETOH/kg (%)	Mean of 3 am Control scores
Verbal short term memory				
Sternberg test:	NS			
Digit span:	++	++	6.6	8.04 (1.31)
Visual short term memory				
Benton visual memory				
Number correct	NS			10.75 .99
Pattern memory:				
Number correct	NS			12.41 (1.42)
Latency (sec)	NS	$^{+}$nl		3.43 (1.1)
Perception				
Pattern recognition:				
Number correct	NS			14.66 (.35)
Latency (sec)	+	+	9.2	2.07 (.51)
Psychomotor functions				
Simple reaction time:				
Right hand				293. (.66)
Left hand	b			
Continuous performance:				
CPT (msec)	+	+	4.0	431. (26.)
Latency	+	+	7.9	16.37 (4.02)
Hand-eye coordination:				
RMSQ	NS	$^{+}$nl		4.80 (3.18)
Finger tapping:				
Alternating	+	–	16.5	57.6 (5.2)
Critical tracking:	+	+	6.2	.408 (.061)
Manual dexterity				
One hole test:				
Pin number	+	+	5.2	44. (4.5)
Move (msec)	++	++	7.0	466. (47.)
Mood scales				
Mood	NS			
Fatigue checklist	++	++	17.9	10.5 (3.1)

+ = $p < .05$; ++ = $p < .01$; b = borderline effect; nl = non-linear

Advances in Industrial Ergonomics and Safety II
Edited by Biman Das
Taylor & Francis 1990

APPLICATIONS OF DUAL KRIGING IN HUMAN FACTORS ENGINEERING

Gilbert R.(1,2), Carrier R.(1), Benoit C.(1),
Soulié M.(2) and J. Schiettekatte (1)

1. Les Consultants Génicom Inc.
279 Sherbrooke O., Suite 207
Montréal Québec
H2X 1Y2

2. Department of Industrial Engineering
Ecole Polytechnique de Montréal
C.P. 6079, Succ. A.
Montréal, Qué. H3C 3A7

Many problems in human factors engineering have not been solved yet. Among them, consider the poor results concerning 3-D representation of reach envelope, seat surface quantification, mathematical modeling of human body surface and deformation. These kind of problems can be solved using Dual Kriging technics on micro-computer. Dual kriging is a mathematical estimation model that proved to be powerful modeling tool for 3-D geometry or mathematical forms in N-D. Kriging gives general analytical solutions to various problems. Here, the general model of Dual Kriging will be presented and two examples of application in human factors engineering will be illustrated.

1. General Model

Dual kriging is a recently developed statistical estimation model which is a refinement of the primary kriging form.

Very recently one of us, professor M. Soulié, has shown some very interesting geometrical applications in modelling. Based on generalized increments and covariance theories, kriging is an exact interpolator. This means that each input datum will satisfy exactly the mathematical model. Nevertheless, this condition can easily be bypassed at the will of the user to change, for example, the surface of a body or to input a measuring error in data. Kriging always gives analytical models that can always be formulated in the parametric form and often in implicit form. Kriging can handle problems in a theoretically infinite number of dimensions. This model allows very easily interpolations (linear, spline, polynomial, etc.) between the input points reducing drastically the amount of information needed to reproduce analytically a phenomenon with the same precision compare to the usual methods. Basically, the general kriging equation can be expressed by:

$$\Theta^{*}=M(x)+\sum_{i=1}^{n}\lambda_{x_i}K(h_i) \qquad (1)$$

where in

1D, $M(x)=a+bx$ if k (order of the equation) = 1

2D, $M(x,y)=a+bx+cy$ if k = 1

3D, $M(x,y,z)=a+bx+cy+dz$ if k = 1

and in

1D, $h = | x - x_i |$

2D, $h = \sqrt{(x - x_i)^2 + (y - y_i)^2}$

3D, $h = \sqrt{(x - x_i)^2 + (y - y_i)^2 + (z - z_i)^2}$

So, the h function depends solely on the euclidean distances between points. The K(h) function which is the covariance function which is equally to a certain extend a form function can be expressed in numerous ways, among them are the following:

$$K(h) = h \tag{2}$$

$$K(h) = h^3 \tag{3}$$

$$K(h) = h^2 \log h \tag{4}$$

An example of a 1D parametric (t) system with linear interpolation is:

$$x(t) = a_x + b_x t + \sum_{i=1}^{n} \lambda_{x_i} | t - t_i | \tag{5}$$

and the matrix notation is:

$$\begin{pmatrix} K_{11} & \dots & K_{N1} & 1 & t_1 \\ \cdot & \cdot & \cdot & \cdot & \cdot \\ K_{1N} & \dots & K_{NN} & 1 & t_N \\ 1 & \dots & 1 & 0 & 0 \\ t_1 & \dots & t_N & 0 & 0 \end{pmatrix} \times \begin{pmatrix} \lambda_1 \\ \cdot \\ \lambda_N \\ a \\ b \end{pmatrix} = \begin{pmatrix} x_1 \\ \cdot \\ x_N \\ 0 \\ 0 \end{pmatrix} \tag{6}$$

An example of a 3D implicit function (surface of a volume) with a thin plate spline is:

$$\Theta = f(x, y, z) = a + bx + cy + dz \sum_{i=1}^{n} \lambda_i \sqrt{(x - x_i)^2 + (y - y_i)^2 + (z - z_i)^2} \tag{7}$$

and the matrix expression is :

$$\begin{pmatrix} K_{11} & \dots & K_{N1} & 1 & x_1 & y_1 & z_1 \\ \cdot & \cdot & \cdot & \cdot & \cdot & \cdot & \cdot \\ K_{1N} & \dots & K_{NN} & 1 & x_N & y_N & z_N \\ 1 & \dots & 1 & 0 & 0 & 0 & 0 \\ x_1 & \dots & x_N & 0 & 0 & 0 & 0 \\ y_1 & \dots & y_N & 0 & 0 & 0 & 0 \\ z_1 & \dots & z_N & 0 & 0 & 0 & 0 \end{pmatrix} \times \begin{pmatrix} \lambda_1 \\ \cdot \\ \lambda_N \\ a \\ b \\ c \\ d \end{pmatrix} = \begin{pmatrix} x_1 \\ \cdot \\ x_N \\ 0 \\ 0 \\ 0 \\ 0 \end{pmatrix} \tag{8}$$

Potential applications of dual kriging in Human Factors Engineering are numerous: reach envelopes, comfort envelopes, human forms, workstation and equipments in 3-D, relationship between several variables, etc.

We are reporting here two specific cases.

2. Application of kriging to the calculation of the optimum angle for displays in the Metro Cab of Montreal.

The problem was the determination of the best angle for displays under severe constraints such as:

- accommodation of males and females not in equal proportion;
- accommodation of operators in sitting and standing driving position with a special attention to the sitting position;
- technical constraints related to the geometry of instruments and equipments;
- accommodation of various eye positions according to seat adjustments (height, fore and aft).

The objective of the problem is to find a solution which keep the vision angle the most perpendicular to the display according to the constraints.

The technical constraint related to vision outside the cab not mentioned previously, has been solved without affecting the present problem. Figure 1 shows a longitudinal cut through the front section of the cab. In addition, the same figure shows the localization of displays surface where the problem should be solved (part a). Figure 2 shows the technical constraint associated with the hidden geometry of instruments and equipments. If we allow the instruments, for instance a,b,d,x to rotate around point 0, it should not penetrate the space below the line EF. In other words, this mean that we consider θ as a constraint variable in the problem formulation. Because θ is a variable, spatial localization of the dials display will be also a constraint variable.

In order, to solve that kind of problem we need to know the relationship between θ and eyes locations that change with the vision angle. Since the exact trajectory in the space for the eyes in both situations was not known, we have determined experimentally the changes in eyes positions at various angles (0.0, 22.5, 33.75, 45.0, 67.5). We used the parametric dual kriging with cubic spline to get an analytical relationship on these data. The mathematical formulation used was the following for the x coordinate:

$$x(t) = a_x + b_x t + \sum_{i=1}^{n} \lambda_{\xi} \, | t - t_i |^3 \qquad (9)$$

and for the z coordinate,

$$z(t) = a_z + b_z t + \sum_{i=1}^{n} \lambda_{zi} \, | t - t_i |^3 \qquad (10)$$

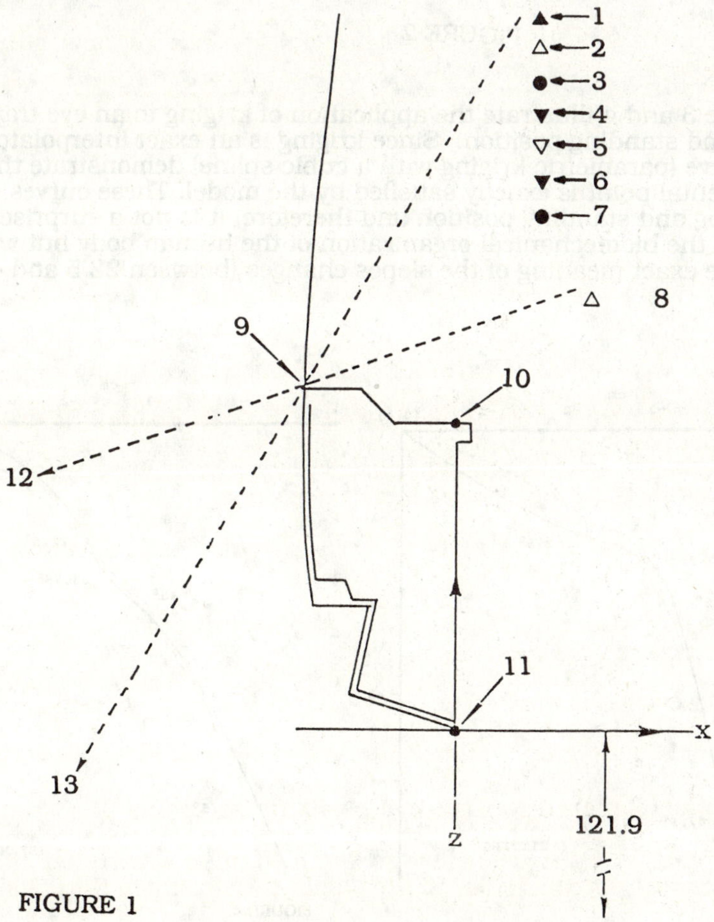

FIGURE 1

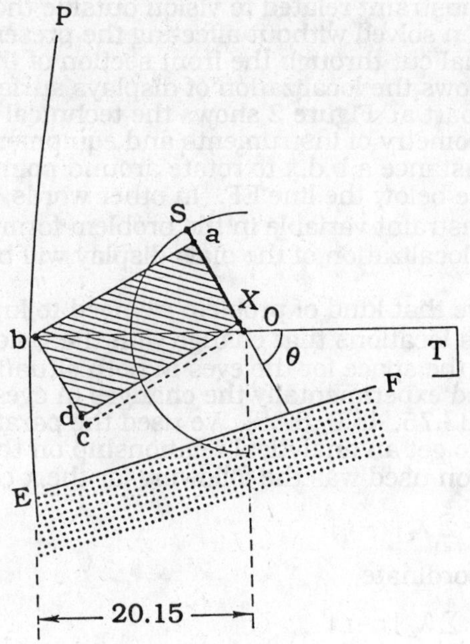

FIGURE 2

Figure 3 and 4 illustrate the application of kriging to an eye trajectory in sitting and standing position. Since kriging is an exact interpolator, the solid curve (parametric kriging with a cubic spline) demonstrate that each experimental point is exactly satisfied by the model. These curves differ for the sitting and standing position and therefore, it is not a surprise if one consider the biomechanical organization of the human body but we do not know the exact meaning of the slopes changes (between 22.5 and -45.0 degrees).

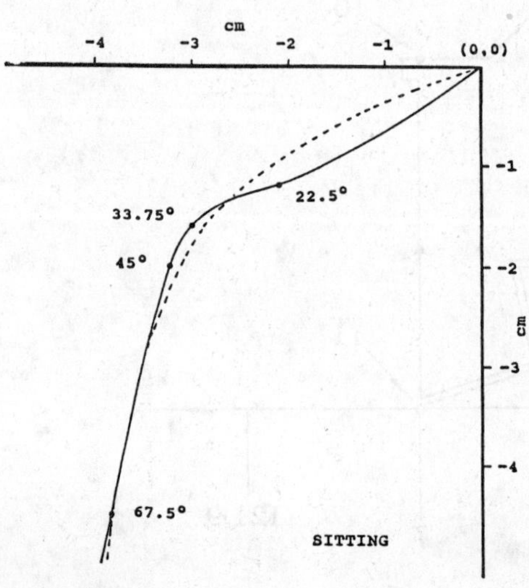

FIGURE 3

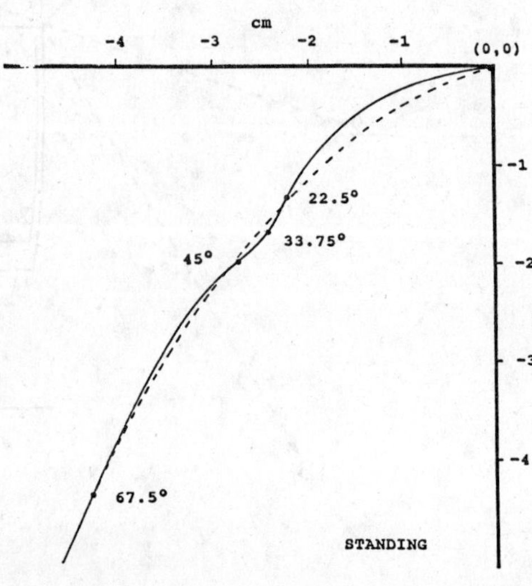

FIGURE 4

Since we obtained the same kind of curve for several subjects, we could build hypothesis on that matter but this is not the aim of the present paper which concerns the kriging. Rather, let us suppose that we calculate the overall error of measurement on each of these points. The calculated errors can be taken into account very easily in the kriging model by inputing them in the appropriate matrix cells as $(1/3)\cdot e^2$ where e is the error on the measurement. At that time kriging is no more an overall exact interpolator but rather a curve fitting tool which will no more satisfy exactly the points for which we have a non zero value on the diagonal. This is demonstrated by the dash curves in figure 3 and 4 where measurement errors have been inputed only for three vision angles (22.5, 33.75 and 45 degrees) keeping consequently the measurements at 0 and 67.5 degrees with no error. By doing so, kriging do not loose any of its analytical properties. The degree of consideration "given" by kriging to a particular point is inversely proportional to the absolute value of the error in such a way that kriging will "ignore" any point for which the error is very large compared to the value of the point itself (nugget effect).

Figure 5 shows a geometrical representation of the problem with eyes position changing with vision angle. The dials surface is represented by SO where ψ and φ hold for the vision angles and C indicates the tangent point of the vision angle (dials' center). Figure 6 illustrates geometrically the problem we had to solve, that is finding the best panel angle for dials. Knowing the eye coordinates changes with vision angles by kriging, we can show that θ must satisfy the following generalized equation:

$$\theta = a\left\{90^\circ - arctg\left[\frac{(z_a - r\bullet\sin\theta)}{(x_a + r\bullet\cos\theta)}\right] - arctg\left[\frac{(x_d + r\bullet\cos\theta)}{(z_d - r\bullet\sin\theta)}\right]\right\}$$

$$+\,arctg\left[\frac{(x_d + r\bullet\cos\theta)}{(z_d - r\bullet\sin\theta)}\right] \qquad (11)$$

where:

a	means advantage in % for the sitting position;
(x_a, z_a)	are the eyes coordinates in sitting position;
(x_d, z_d)	are the eyes coordinates in the standing position; and
r	is half of the distance SO (figure 2).

3. Application to the surface representation of the arm.

We are reporting here an example of arm surfaces representation for two angles (straight and 90 flexed). A total of 140 points have been measured using a sonic digitizer with a precision at 1mm. Ten coordinates have been measured at more or less regular spacing around 14 sections taken along the arm. We chose a two dimensions parametric kriging function to represent the surfaces, namely u for the perimeters of each section and v which link the corresponding points of each section along the axis of the arm. The parametric function is:

$$\overline{F}(u,v) = (F_X(u,v), F_Y(u,v), F_Z(u,v)) = \overline{\Omega}(u)\overline{\Theta}(v) \qquad (12)$$

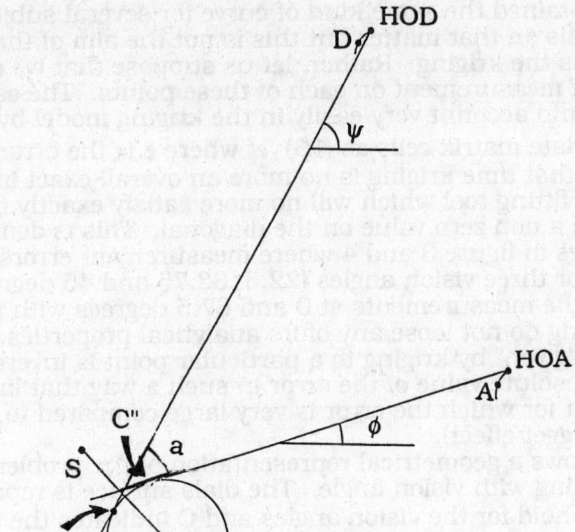

FIGURE 5

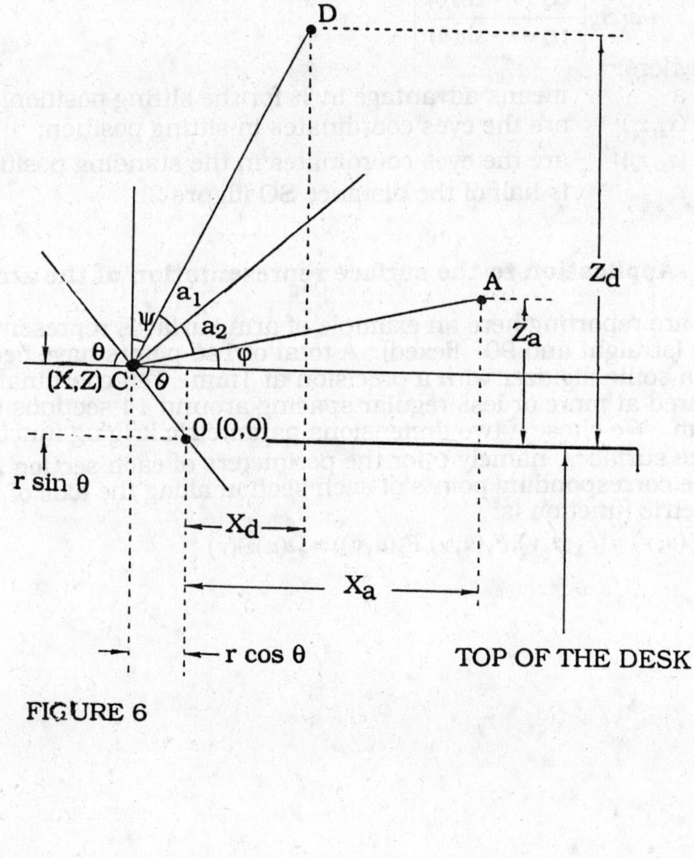

FIGURE 6

For simplicity, it is understood in the following that each point is defined by (x,y,z). Because the form are different for the sections (more or less circular) compare to those of the arm axis we used different form functions:

$$\Omega(u) = A\sin\frac{2\pi u}{N} + B\cos\frac{2\pi u}{N} + C + \sum_{i=0}^{N} \lambda_i K_{uu}(h_u(u,u_i)) \tag{13}$$

with

$$K_{uu}(h) = \sin(h) \tag{14}$$

and

$$h_u(u,u_i) = \frac{2\pi}{N} \cdot \| u - u_i \| \tag{15}$$

where

$$\Theta(v) = a_1 v + a_0 + \sum_{i=0}^{M-1} \gamma_i K_{vv}(h_v(v,v_i)) \tag{16}$$

with

$$K_{vv}(h) = h^3 \tag{17}$$

and

$$h_v(v,v_i) = \| v - v_i \| \tag{18}$$

The general matrix formulation of the problem is

$$F\overline{(u,v)} = \overline{V^T}(v) \cdot K_{vv}^{-1} \cdot P_{XYZ} \cdot K_{uu}^{-1} \cdot \overline{U}(u) \tag{19}$$

where:

$$K_{uu} = \begin{pmatrix} K_{uu}^{00} & \dots & K_{uu}^{N0} & 1 & \sin\frac{2\pi u_0}{N} & \cos\frac{2\pi u_0}{N} \\ . & . & . & . & . & . \\ K_{uu}^{0N} & \dots & K_{uu}^{NN} & 1 & \sin\frac{2\pi u_N}{N} & \cos\frac{2\pi u_N}{N} \\ 1 & \dots & 1 & 0 & 0 & 0 \\ \sin\frac{2\pi u_0}{N} & \dots & \sin\frac{2\pi u_N}{N} & 0 & 0 & 0 \\ \cos\frac{2\pi u_0}{N} & \dots & \cos\frac{2\pi u_N}{N} & 0 & 0 & 0 \end{pmatrix} \tag{20}$$

and

$$K_{uu}^{ij} = K_{uu}(h_u(u_i,u_j)) \tag{21}$$

and

$$K_{vv} = \begin{pmatrix} K_{vv}^{00} & \dots & K_{vv}^{M0} & 1 & v_0 & v_0^2 \\ . & . & . & . & . & . \\ K_{vv}^{0M} & \dots & K_{vv}^{MM} & 1 & v_M & v_M^2 \\ 1 & \dots & 1 & 0 & 0 & 0 \\ v_0 & \dots & v_M & 0 & 0 & 0 \\ v_0^2 & \dots & v_M^2 & 0 & 0 & 0 \end{pmatrix} \tag{22}$$

and

$$K_{vv}^{ij} = K_{vv}(h_v(v_i,v_j)) \tag{23}$$

The U and V expressions are

$$U(u) = \left[K_{uu}(h_u(u,u_0)) \quad \dots \quad K_{uu}(h_u(u,u_N)) \quad 1 \quad \sin\left(\frac{2\pi u}{N}\right) \quad \cos\left(\frac{2\pi u}{N}\right) \right] \tag{24}$$

and

$$V(v) = [\, K_{vv}(h_v(v,v_0)) \quad \dots \quad K_{vv}(h_v(v,v_M)) \quad 1 \quad v \quad v^2 \,] \tag{25}$$

and where the 3D data points are:

$$P_{XYZ} = \begin{pmatrix} P_{XYZ}^{00} & \dots & P_{XYZ}^{M0} & 0 & 0 & 0 \\ \cdot & \cdot & \cdot & \cdot & \cdot & \cdot \\ P_{XYZ}^{0M} & \dots & P_{XYZ}^{NM} & 0 & 0 & 0 \\ 0 & \dots & 0 & 0 & 0 & 0 \\ 0 & \dots & 0 & 0 & 0 & 0 \\ 0 & \dots & 0 & 0 & 0 & 0 \end{pmatrix} \quad (26)$$

The matrix K_{uu}^{-1} and K_{vv}^{-1} are calculated only once since they represent the geometrical formulation of the problem and they are independent of the value of the points. Using this parametric kriging system with a visualization algorithm resulted in figure 7. We have also applied implicit dual kriging function namely F=Z=f(x,y) to quantify the forms of seats.

In conclusion, our opinion is that dual kriging is a very promising tool in quantitative ergonomics.

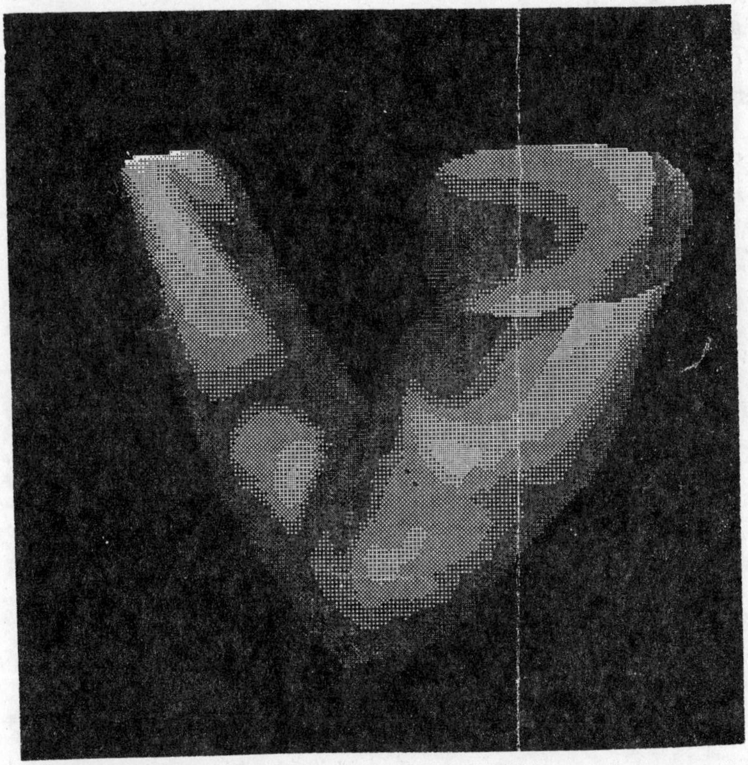

FIGURE 7

XVIII

WORK ENVIRONMENT

Advances in Industrial Ergonomics and Safety II
Edited by Biman Das
Taylor & Francis 1990

STRESSORS AND MUSCLE STRESS IN WORKERS IN COLD INDUSTRIES

Ruth Nielsen

Technical University of Denmark
2800 Lyngby, Denmark

Fifty-seven workers, having various types of work in cold industries, were questioned about stressors in their environment and in their work process, the strain on various body parts during work, accidents, and pain experienced from body areas. Reports on experienced pain was most common from the lower back and shoulder/arms. These areas were also evaluated as 'somewhat/very loaded' during work by 70-80%. When freezing, 18-24% reported to freeze in these areas. It was concluded, that for some workers in cold industries cold may be a risk factor for the development of muscular problems.

INTRODUCTION

In the working population a high prevalence of low back pain (26%) and cervicobrachial (29%) disorders have been observed (Arbejdsmiljøgruppen af 1972, 1974). These disorders have been ascribed to the working conditions. Cervicobrachial disorders to repetitive, manual, short-cycle tasks, raised arms, and large demands on accuracy and speed of work (Hagberg, 1984). Working technique has here been shown to be an important risk factor (Kilbom and Persson, 1987). Low back pain has by Herrin (1978) disclosed loads, their dimensions, distribution and stability, workplace geometry, temporal factors, complexity of task, environment and work organization as being important risk factors.

It is difficult to show load effects on the musculoskeletal system. Only in case of extreme loads causing accidents the relation between load and disorder is clear. Acceptable load on tissue varies with age and health. Exactly what load (type, duration) will result damage on the musculoskeletal system in a given individual is not known.

In our highly developed society many jobs are still being performed in a cold and drafty environment. This is the case for outdoor work in winter-time, and for indoor work in industries where the authorities require low temperatures around the production proces in order to preserve product quality of e.g. fish, meat. Cold and draft are often mentioned as reasons for pain and troubles from muscles. However, only few studies have confirmed this. Inflammation of tendons in the hand after exposure to cold have been described (Georgitis 1978). Draft has been shown to increase EMG-activity from the muscles exposed (Hagberg et al., 1987). Cold has also been shown to increase muscular pain (Smythe, 1972). It is possible that cold itself does not harm muscles or tendons so often, but it has been proposed that cold in the work place may act as an extra stress factor on the local blood circulation to tendons and muscles already being stressed by the work itself (Hagberg, 1982). In this way cold could be the provoking factor for the development of inflammation, microruptures or injuries in these organs by aggravating an existing ischaemia.

The purpose of the present study was, in different types of jobs in cold industries, to elucidate whether there was any coincidence between 1) complaints of stress in muscles and tendons during work, 2) complaints of pain/injuries in the same muscles and tendons, and 3) complaints of often feeling coldt generally or locally in these areas.

METHODS

Characteristics of occupations

All workers were working in either cool environments (4-12ºC), or in a cold store (-18ºC). The work tasks differed according to nature of work and activity level. In the cold store workers either picked orders from shelves onto a moving band, or filled shelves with goods (heavy loads). In a fresh-food terminal most workers drove around in large store rooms collecting orders on electrical carts, stopped, and manually lifted items of 1 to 30 kg up onto these. Other workers drove truck all day to fill shelves or vans with goods. In a slaughterhouse industrial butchers cutting meat (8-15 kg) in a highly repetitive, uniform way, and dispatchers collecting meat, lifting burdens of 10 to 45 kg, pushing/pulling pig bodies etc., were studied. In a meat wholesale firm, jobs with a similar goods collecting work, but performed at a lower pace, were studied. A few more specialized jobs were also included.

Questionnaire

A questionnaire was filled in by 57 workers with various types of work in these industries. The questionnaire inquired about their sex, anthropometric data, how long time they had been in their present job, what they had been doing before, the type and pace of their present job, exposure to environmental factors, judged load on the different parts of the body during work, pain and other kinds of problems in different body parts during the past year, accidents at work, and thermal problems.

RESULTS

The group comprised 18 women and 39 men. They had an average age of 32±10.4 years, and had at average been in their present job for 4±4.0 years.

The distributions of the answers in the questionnaire are shown in Figures 1 and 2. Of the 57 workers 49% judged shoulders, arms, hands, lower back and legs to be 'very loaded' and 28% judged these areas to be 'somewhat loaded'. Fewer workers evaluated back of the head, upper back, and feet as being loaded. More than 20% reported having experienced 'much pain' from lower back and shoulder/arms during the last

Figure 1. The distribution of the sensed strain level during work, how often pain was observed during the last year, and if they had an accident in their present job.

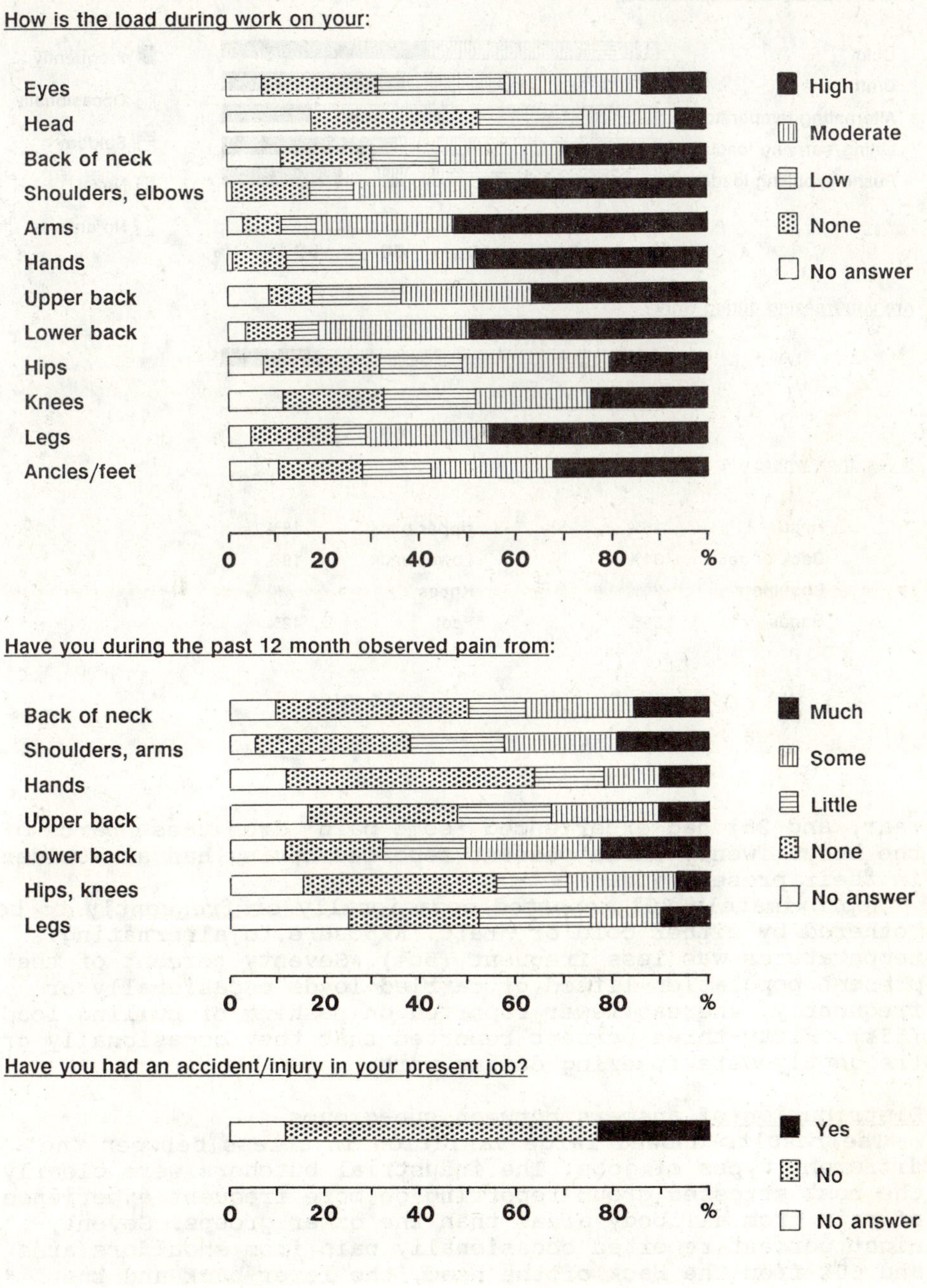

Figure 2. The distribution of the answers of how often the group was bothered by ambient temperatures and handling of load, if they were freezing, and if yes, then where.

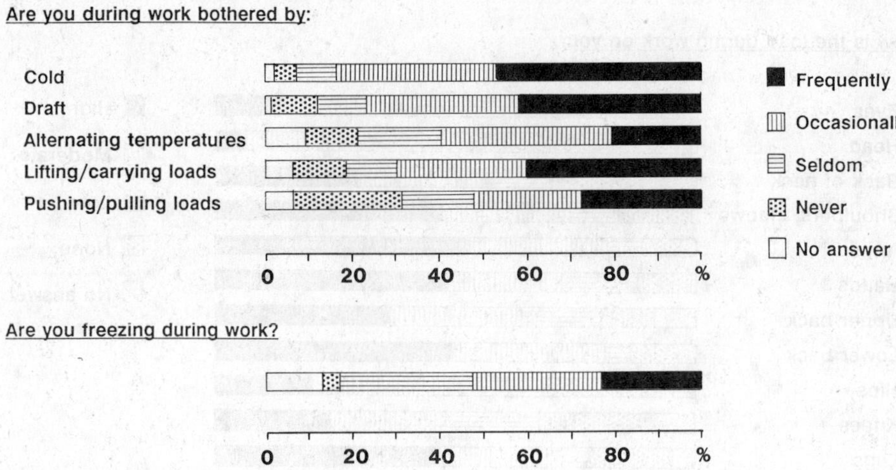

If yes, then where

Head	16%	Upper back	18%
Back of neck	31%	Lower back	18%
Shoulders	24%	Knees	7%
Hands	36%	Feet	42%

year, and 25% had experienced 'some pain' from these parts of the body. Twenty-three percent reported having had an accident in their present job.

Approximately 80% reported occasionally or frequently to be bothered by either cold or draft. Exposure to alternating temperatures was less frequent (50%). Seventy percent of the present population lifted or carried loads occasionally or frequently, whereas fewer reported on pushing or pulling loads (53%). Fifty-three percent reported that they occasionally or frequently were freezing during work.

Distribution of answers between sub-groups

The results showed large variation in stress between the different types of jobs. The industrial butchers were clearly the most stressed group reporting on more frequent experience of pain from all body areas than the other groups. Seventy-eight percent reported occasionally pain from shoulders/arms and 60% from the back of the head, the lower back and the hands, respectively. Low back pain was most frequently (68%) reported from the workers in the cold store; however, in all groups 40% or more had occasionally low back pain, and of the workers in the fresh-food terminal, the cold store and the butchers between 70% and 84% had experienced pain from the lower back at least once during the past year.

The reports on experienced strain from the cervicobrachial area was greatest from the butchers, followed by the workers in the fresh-food terminal and the dispatchers in the meat wholesale firm. The strain reported on the lower back was greatest in the cold store, and next in the butchers.

Cold and draft were bothering the workers in all jobs. Only in the cold store 34% reported not being bothered by draft. Also here, the butchers were slightly more bothered than the other groups. All butchers were freezing on some occasion. Almost all workers in the fresh-food terminal, and among the rest 60% or more complained about cold. In the cold store 80 - 90% complained to be cold at feet and hand; in the other groups the figures for hand and feet were 30 - 50%. Cooling of the lower back was experienced in all groups, whereas no cooling of the shoulder area was sensed in the cold store workers. More workers were exposed to lifting/carrying than to pushing/pulling.

<u>Relation between reports on pain, strain and cold</u>

The reported low back pain correlated with the strain experienced on the lower back during work ($r=0.44$; $p<0.01$). There was no correlation with age or the number of years the worker had spent in his present job. All workers reporting occasionally to freeze on the lower back judged the strain on the lower back as moderate or high. Such a relation was not seen to the judgment of pain; however, the people occationally freezing at their job were mainly in the group reporting some to much pain from the lower back, and also judging the strain on their lower back to be moderate to high.

The judged cervicobrachial pain was also correlated with the strain experienced on the shoulders/back of the head ($r=0.43$; $p<0.01$). Some influence from age and seniority on the cervicobrachial pain was seen ($0.05<p<0.1$). The workers reporting to freeze in the cervicobrachial area were all judging the strain on the cervicobrachial area as moderate or high.

DISCUSSION

The reports on experienced pain during the last year was high in this study compared to other studies. In a study on 7000 workers from a number of different branches the question was a bit different (Arbejdsmiljøgruppen af 1972, 1974). However, 24% reported pain from the lower back when lifting, 27% had pain in the cervicobrachial area. In their study, however, only 14% was bothered by cold, 34% by draft, 35% by lifting or carrying loads, and 20% by pulling or pushing loads. Kristensen and Christensen (1983) found in a study on 3162 slaughterhouse workers that 64% experienced low back pain and 43% reported pain from the cervicobrachial area. Like in the present study, significantly less workers in cold stores than in rooms of other temperatures reported cervicobrachial pain. Kristensen and Christensen (1983) also showed a relationship between the cooling of the worker during his job and reported cervicobrachial pain. Of workers not exposed to draft 42% reported cervicobrachial pain, whereas 60% exposed to draft had cervicobrachial pain. This tendency was even worse in cold stores. A similar trend was observed for low back pain.

In the present study, there was no significant relation between freezing and reports on either low back pain or pain in the cervicobrachial area. However, claims on freezing in an area of the body were always coinciding with frequently feeling strain in the same area. Thus, cold either resulted in strain or it aggravated an existing strain on muscles and tendons. And, both low back pain and pain in the cervicobrachial area was positive correlated with experience of strain.

CONCLUSION

It can be concluded, that for some workers in cold industries cold seems to be a risk factor in the development of muscular problems.

REFERENCES

Arbejdsmiljøgruppen af 1972, 1974, Arbejdsmiljøundersøgelsen, Rapport no.2, (Copenhagen: AMG).

Georgitis, J., 1978, Extensor tenosynivitis of the hand from cold exposure. Journal of Maine Medical Association, 69, 129-131.

Hagberg, M., 1982, Arbetsrelaterede besvär i halsrygg och skuldra - en kunskapsöversikt, (Stockholm: Arbetarskyddsfonden).

Hagberg, M., 1984, Occupational musculosceletal stress and disorders of the neck and shoulder: a review of possible pathology. International Archives of Occupational Environmental Health, 53, 269-278.

Hagberg, M., Sundelin, G. and Hammarström, U., 1987, Ger lokal avkylning (drag) på skuldra-halsrygg ökad muskelaktivitet?, Undersökningsrapport (Stockholm: Arbetsmiljöinstitutet).

Herrin, G.D., 1978, A taxonomy of manual materials handling hazards. In: Safety in manual materials handling, edited by C. Drury. (Cincinnati: National Institute for Occupational Safety and Health, Publication No. 78-185) pp. 6-15.

Kilbom, Å and Persson, J., 1987, Work technique and its consequences for musculoskeletal disorders. Ergonomics, 30, 273-279.

Kristensen, T.S. and Christensen, F.L., 1983, Slagteriarbejde, (Copenhagen: Arbejdsmiljøfondet).

Smythe, H.A., 1972, Non-articular rheumatism and the fibrositis syndrome. In: Arthritis and Allied Conditions, edited by J.L. Hollander and D.J. McCarty. (Philadelphia: Lea & Fibiger).

Advances in Industrial Ergonomics and Safety II
Edited by Biman Das
Taylor & Francis 1990

WORKING SAFELY IN HOT ENVIRONMENTS

Jerry D. Ramsey, Ph.D.
Professor, Industrial Engineering

Texas Tech University
P. O. Box 4609, Lubbock, Texas 79409

There are basically two sources of concern for the worker who is exposed to hot occupational environments. The first concern is that of heat storage in the human body which has the potential for adverse health effects. The second concern is that of impairment in functional performance which can lead to a decrease in total safety for the worker or for others in the work environment.

Occupational Exposure to Heat

Both of these aspects were addressed in the 1972 Criteria for a Recommended Standard for Occupational Exposure to Hot Environments (NIOSH, 1972). This document specified a one hour time-weighted wet bulb globe temperature (WBGT) which could not be exceeded without the initiation of work practices. This was to assure that the employee's body core temperature would not exceed 38°C. The document also stated, "For sedentary jobs where continuous unimpaired mental performance is required, no employee should be exposed to conditions which exceed the limits set forth . . ." (NIOSH, 1972). The implication to this latter statement was that although exposure limits based on deep body temperature should protect the worker's health, hot occupational environments could also cause a loss in performance which might affect the safety of workers.

Performance and Safety

The relationship between perceptual motor performance and safety were considered at a workshop on recommended heat stress standards (NIOSH, 1980). This working group suggested that, "An upper limit of permissible exposures for the purpose of enhancing the employee's safety cannot be established." Their reasons included a lack of distinguishing inflection points in the data, poor relationships between laboratory and industrial tasks, and questions of correlation between thermal stress and accidents (NIOSH, 1980).

There have, however, been reports of unsafe worker behavior as a function of temperature as shown in Figure 1 (Ramsey, et. al., 1983). This study involved heat exposure measurements and behavioral observations taken over a 14 month period in two industrial plants. These results indicated that temperatures below and above those typically preferred by most people (17°C-23°C, WBGT) have a significantly detrimental affect on the safety-related behavior of workers. Further, it is generally accepted that unsafe behavior and unsafe acts do tend to have a positive correlation with accidents (Heinrich, et. al., 1980).

It is reasonable to assume that a loss of precision, speed, accuracy, or effectiveness in performing a perceptual motor task could very well have safety implications. For example, the meter which is not read correctly, the lever which is not engaged properly, or the movement which is not performed quickly could well create a hazardous situation.

Thermal Factors

Interpreting the effects of heat on performance first requires that all the factors of the thermal environment are identified and included. Worker response, behavior and/or stress level at a given air temperature could certainly be affected by the other thermal components associated with that temperature. The WBGT, which includes the effect of air temperature, humidity, air movement, and radiant heat is a convenient single index to incorporate all of these factors. Evaluation of human performance in WBGT thermal units also provides the advantage of intercomparison with the physiologically derived heat stress limits which have come into wide acceptance (ISO, 1982, NIOSH, 1986, ACGIH, 1986).

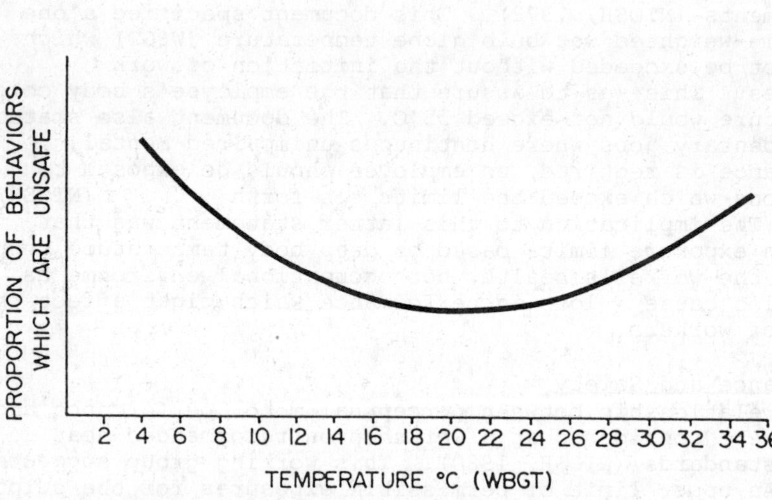

Figure 1. General relationship between the proportion of unsafe behaviors and thermal conditions. (adapted from Ramsey et. al., 1983)

Tasks

Difficulty in interpreting the effects of heat on human performance and safety is partially due to the confounding of other factors which are normally found in conjunction with the heat exposure. Much of the research reported in the literature identifies perceptual motor tasks in basic categories such as tracking, vigilance, mental, reaction time, eye-hand coordination, complex/dual tasks, sensory tasks, cognitive/skilled tasks, etc. (Grether, 1973, Ramsey and Morrissey, 1978, Kobrick and Fine, 1983). For purposes of evaluation and discussion in this paper, the tasks are considered in only two categories: (1) Mental, reaction time, very simple perceptual motor and sensory tasks, (2) All other perceptual motor tasks, including tracking, vigilance, complex/dual, etc. This division is based on research results which indicate considerable differences between performance results on these two general groupings (Ramsey and Morrissey, 1978).

Exposure Time

Duration of exposure is another important factor affecting human performance in the heat. For purposes of evaluation and discussion in this paper, duration of exposure was identified and performance results categorized according to this variable.

A major difficulty in predicting behavior or performance in the heat is that it is affected by factors other than task differences and exposure time. Some of the basic relationships which seem to be apparent concerning the loss of performance in the heat are discussed below.

Physical Work

Studies involving physical work effects on the performance of perceptual motor tasks have been reported (Mackworth 1961, Shvartz et. al. 1976). Exposure to work and heat which will generate localized or general fatigue will normally have a negative effect on muscular based perceptual motor performance, but will have less direct effect on tasks which are primarily cognitive.

Acclimatization

Task performance related to degree of acclimatization has been addressed by several authors, but is an area which is not well defined (Edholm 1963, Ramsey et. al. 1975, Poulton 1976). Much of the reported literature does not specify if subjects are acclimatized, and few studies have been conducted in paired groups of acclimatized and unacclimatized subjects. In general the acclimatized individual should be more tolerant and physiologically more able to work in a hot environment, and should show more resistance to performance losses if the tasks involve motor responses; for simple and mental tasks this difference is less likely to be observed.

Skill

Performance in the heat as a function of skill level of the worker and the skill requirements of the task has also been addressed (Mackworth 1961, Bell and Provins 1962). In general, persons with high skill will have higher resistance to performance loss unless they are already operating at a high perceptual motor load which, with the addition of temperature, becomes an overload condition.

Combined Stressors

Reports of studies of combined environmental stressors of heat and noise have been reported (Dean and McGlothlen 1965, Bell 1978). Some instances show the combined effects of the two stressors to be smaller and some report them to be additive; further, these differences are seen in some tasks and not in other tasks. High altitude and heat is also a commonly investigated area (Fine and Kobrick, 1978) and other combined stressors have also been reported (Loeb and Jeantheau 1958, Poulton and Edwards 1974). Performance results with combined stressors are highly variable and appear to be a direct function of the stressor levels and the specificity of the conditions utilized in the study.

Comfort

Comfort level is a function of numerous variables, but for the sedentary, normally clothed worker (.5-.7clo) with low air movement (less than .3m/sec), the comfort range is approximately 23-27°C ET (21-26°C WBGT) (ASHRAE 1977, Fanger 1970). Comfort for a person doing light work above the sedentary level would likely be 1-2°C lower.

Numerous studies of task performance near the range of comfort have been conducted (Reddy and Ramsey 1976, Wyon et. al. 1979) and performance decrements have been noted. The lower temperatures are frequently associated with best performance.

Elevated Core Temperature

Work and thermal environments which elevate the body temperature have also been shown to have adverse effects on task performance (Pepler 1958, Mackworth 1961, Nunneley et. al. 1982, Hancock 1982). Although body temperature has not been reported in most studies, there are means of estimating body temperature as a function of environmental temperatures, exposure time and level of metabolic activity. This estimation ability allows a more comprehensive assessment of body temperature effects on task performance and appears to be a dominant factor for some types of tasks.

Arousal

Several have suggested that there is an optimal arousal level for any task (Wyon 1970, Poulton 1976), and it is typically lower for complex tasks than for simple tasks. Arousal can not only create performance losses, but also has been shown to improve performance in many instances. Since temperature spans the continuum from cool, to comfortable, to hot, the idea of an inverted U to describe performance is intuitively appropriate. It is difficult to generalize, however, due to the problems of determining the level of arousal at any given point of exposure to the heat, as well as determining the interaction of arousal with task differences and other variables.

Onset of Performance Loss

Thus, there are a large number of factors which possibly can affect the outcome of measured task performance in the heat. This complicates the ability to isolate the specific effects due to heat. In an effort to determine if the heat has a detrimental effect, data from over 150 reported studies were summarized according to WBGT and duration of exposure. No attempt was made to differentiate or categorize according to factors such as acclimatization, arousal, skill, etc.

The 1986 Revised Criteria for a Recommended Standard--Occupational Exposure to Hot Environments proposed recommended heat stress limits as shown in Figure 2 (NIOSH, 1986). The Recommended Exposure Limit (REL) refers to the heat acclimatized worker and the Recommended Alert Limit (RAL), which is slightly more conservative, refers to the heat unacclimatized worker. The worker engaged in the performance of perceptual motor tasks would typically be either at a sedentary metabolic heat production (100 kcal/h-116W) or a very light metabolic heat level (150 kcal/h-174W). These levels would be represented as points A and B from the REL curve or B' and C from the RAL curve of Figure 2 and are in the range of 30°-33°C.

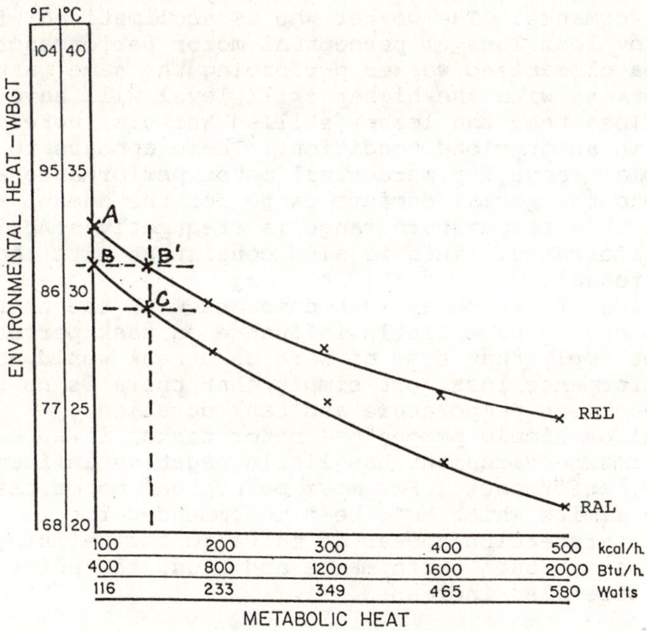

Figure 2. Recommended Heat-Stress Alert Limits (RAL) and Exposure Limits (REL), (NIOSH, 1986)

Results from a large number of studies of perceptual motor tasks (excluding simple and mental tasks) in the heat were compared with these RAL and REL (i.e., A, B and C) limits. Each individual study was identified at some combination of temperature and duration of exposure along with an indication of whether the author of the study reported a statistically significant decrement, enhancement, or no change. Also included were those combinations where a partial decrement in performance was indicated (Ramsey and Kwon, 1988).

These results showed a strong and consistent temperature effect which occurred without regard to the duration of exposure. These data included many combinations of acclimatization, skill, motivation, arousal, combined stressors, etc., so there were a few individual data points which did not follow this pattern. The preponderance of the data, however, indicated an onset of performance decrement to occur in the 30°-33°C range. This is the same range which is consistent with the physiological responses of heat storage, and is a part of the REL and RAL recommendations for protecting workers from the adverse health effects of environmental heat.

Conclusions

Although the reported literature concerning performance of perceptual motor tasks in hot environments is somewhat complicated and somewhat contradictory, there are some general rules which may be useful in assessing the influence of heat on performance, and thus, on safety. When perceptual motor tasks are included with physical work, the production of localized or general fatigue will normally result in a lower level of performance. The worker who is acclimatized will typically show less loss of perceptual motor performance than would the unacclimatized worker performing the same task. Normally, workers with the higher skill level will have less performance loss than the lesser skilled workers, unless they are working in an overload condition. There appears to be an optimal thermal range for perceptual motor performance and it corresponds to the normal comfort range for the human. The lower end of this temperature range is frequently associated with best performance. This is also consistent with the concept of arousal.

The duration of exposure, when combined with the thermal effects, appears to have little influence on task performance. This does not imply that time of work on a task would not create a performance loss, but simply that there is no strong interaction between temperature and task duration.

For mental or simple perceptual motor tasks, it appears that the thermal environment has little negative influence on the level of performance. For most perceptual motor tasks, the exposure limits which have been recommended for physiological protection appear to be valid for defining the onset of losses in such performance and thus, the point where safety risks are also increased.

REFERENCES

ACGIH (1986) Threshold Limit Values and Biological Exposure Indices for 1986-87. Cincinnati: American Conference of Governmental Industrial Hygienists. ISBN: 0936712-61-9.

ASHRAE (1977) Physiological principles, comfort, and health. In ASHRAE Handbook of Fundamentals, New York: American Society Heating, Refrigerating, and Air Conditioning Engineers, 8.1-8.36.

Bell, C. R. and Provins, K. A. (1962) Effects of high temperature environmental conditions on human performance. Journal of Occupational Medicine, 4, 202-211.

Bell, P. A. (1978) Effects of noise and heat stress on primary and subsidiary task performance. Human Factors, 20, 749-752.

Dean, R. D., and McGlothlen, C. L. (1965) Effects of combined heat and noise on human performance, physiology, and subjective estimates of comfort and performance. Institute of Environmental Science, Annual Technical Meeting Proceedings, 55-64.

Edholm, O. G. (1963) Heat acclimatization studied in the laboratory and the field a multi-disciplinary approach. Ergonomics, 6, 304-305.

Fanger, P. O. (1970) Thermal Comfort: Analysis and Applications in Environmental Engineering, New York: McGraw-Hill.

Fine, B. J. and Kobrick, J. L. (1978) Effects of altitude and heat on complex cognitive tasks. Human Factors, 20, 115-122.

Grether, W. F. (1973) Human performance at elevated environmental temperatures. Aerospace Medicine, 44, 747-755.

Hancock, P. A., (1982) Task categorization and the limits of human performance in extreme heat. Aviation, Space, and Environmental Medicine, 53, 778-784.

Heinrich, H. W., Petersen, D. and Roos, N. (1980) Industrial Accident Prevention New York: McGraw-Hill.

International Organization for Standardization (ISO) (1982) Hot Environments--Estimation of the Heat Stress on Working Man based on the WBGT Index. ISO 7243, Geneva, Switzerland.

Kobrick, J. L. and Fine, B. J. (1983) Climate and human performance. In The Physical Environment at Work (Edited by D. J. Oborne and M. M. Gruneberg) Chichester: Wiley Ltd., 69-107.

Loeb, M., and Jeantheau, G. (1958) The influence of noxious environmental stimuli on vigilance. Journal of Applied Psychology, 42, 47-49.

Mackworth, N. H. (1961) Researches on the measurement of human performance. In Selected Papers on Human Factors in the Design and Use of Control Systems. New York: Dover, 174-331.

NIOSH (1972) Criteria for a Recommended Standard-Occupational Exposure to Hot Environments. Washington, D. C.: National Institute for Occupational Safety and Health, HSM 72-10269.

NIOSH (1980) Proceedings of a Workshop on Recommended Heat Stress Standards. (edited by F. N. Dukes-Dobos and A. Henschel), Cincinnati: National institute for Occupational Safety and Health, DDHS (NIOSH) 81-08.

NIOSH (1986) Criteria for a Recommended Standard-Occupational Exposure to Hot Environments. Revised Criteria 1986. Washington, D. C.: National Institute for Occupational Safety and Health, DDHS (NIOSH) 86-113.

Nunneley, S. A., Reader, D. C. and Maldonado, R. J. (1982) Head--temperature effects on physiology, comfort, and performance during hyperthermia. Aviation, Space, and Environmental Medicine, 53, 623-628.

Pepler, R. D. (1958) Warmth and performance: An investigation in the tropics. Ergonomics, 2, 63-68.

Poulton, E. C. (1976) Arousing environmental stresses can improve performance, whatever people say. Aviation, Space, and Environmental Medicine, 47, 1193-1204.

Poulton, E. C. and Edwards, R. S. (1974) Interactions, range effects and comparisons between tasks in experiments measuring performance with pairs of stresses: Mild heat and 1 mg. of 1-hyoscinehydrobromide. Aerospace Medicine, 45, 735-741.

Ramsey, J. D. and Kwon, Y. G. (1988) Simplified decision rules for predicting performance loss in the heat. In Proceedings of Measurement and Assessment of Heat Stress Symposium, Commission of European Communities, Luxembourg, 337-372.

Ramsey, J. D., Burford, C. L., Beshir, M. Y., and Jensen, R. C. (1983) Effects of workplace thermal conditions on safe work behavior, Journal of Safety Research, 14, 105-114.

Ramsey, J. D. and Morrissey, S. J. (1978) Isodecrement curves for task performance in hot environments. Applied Ergonomics, 9, 66-72.

Ramsey, J. D. and Pai, D. B. (1975) Sedentary work by females in hot environments. In Proceedings 19th Annual Meeting of the Human Factors Society, 498-500.

Reddy, S. P. and Ramsey, J. D. (1976) Thermostat variations and sedentary job performance. ASHRAE Journal, 18, 32-36.

Shvartz, E., Meroz, A., Mechtinger, A., and Birnfeld, H. (1976) Simple reaction time during exercise, heat exposure, and heat acclimation. Aviation, Space, and Environmental Medicine, 47, 1168-1170.

Wyon, D. P. (1970) Studies of children under imposed noise and heat stress. Ergonomics, 13, 598-612.

Wyon, D. P., Andersen, J. B., and Lundquist, G. R. (1979) The effects of moderate heat stress on mental performance. Scandinavian Journal of Work Environment and Health, 5, 352-361.

Advances in Industrial Ergonomics and Safety II
Edited by Biman Das
Taylor & Francis 1990

PHYSIOLOGIC STRAIN ASSOCIATED WITH WEARING TOXIC-ENVIRONMENT PROTECTIVE SYSTEMS DURING EXERCISE IN THE HEAT

Leslie Levine, Mark D. Quigley, Bruce S. Cadarette, Michael N. Sawka and Kent B. Pandolf

U.S. Army Research Institute of Environmental Medicine
Kansas Street, Natick, MA 01760, USA

The new generation Self-Contained Toxic-Environment Protective Outfit was tested in its two configurations, one with a backpack rebreather-respirator and a liquid-cooling vest (STEPO-R), and one tethered to an air supply providing cooling and breathing air (STEPO-T). Also tested was the currently fielded Toxic Agent Protective (TAP) Suit. Six soldiers volunteered to be evaluated during exercise and rest in 27°C, 38°C, and 49°C environments while wearing the protective systems. The STEPO-T allowed the longest endurance time, while STEPO-R reduced endurance and increased physiological strain relative to the other systems.

INTRODUCTION

For individuals who work with the storage, maintenance and disposal of toxic substances, protective garments are essential. In an effort to provide US Army personnel with the best possible protection in a practical garment, the U.S. Army Natick Research, Development and Engineering Center has continued development of the Self-Contained Toxic-Environment Protective Outfit (STEPO). This garment system has been designed to replace the currently fielded Toxic Agent Protective (TAP) Suit which has been in use for more than 30 years. The improvements made to the STEPO systems include a backpack rebreather-respirator and a microclimate cooling system, or a tether option supplying breathing and cooling air inside the suit, intended to enhance breathing air quality and reduce thermal strain.

Toxic-environment protective clothing is characterized by low moisture permeability. The physiological strain associated with exercise and/or heat exposure is exacerbated when individuals must perform physical work while wearing protective clothing that acts as a barrier to sweat evaporation. Pimental et al. (1987) and Muza et al. (1988) have documented the thermoregulatory consequences for individuals wearing protective clothing with and without microclimate cooling. The effectiveness of microclimate cooling systems in reducing heat strain for persons working in protective clothing has been reviewed by Speckman et al. (1988).

The present study reports results concerning the physiological evaluations of the second generation STEPO systems. The results of testing on the prototype STEPO were presented last year at this meeting (Pandolf et al., 1989). Compared to the prototype, the new STEPO systems employ improved weight distribution for the backpack rebeather-respirator, prolonged microclimate cooling capacity, and a teflon-fiberglass suit fabric to replace the butyl rubber-coated nylon.

METHODS

Subjects

Six male soldiers volunteered to be test subjects following medical clearance and after signing a statement of informed consent. Investigators adhered to AR 70-25 and USAMRDCR 70-25 concerning the use of volunteers in research. The physical characteristics ($\bar{X}$±SD) of the volunteers were as follows: age, 25±9 yr; ht, 176±6 cm; wt, 74± 5 kg; body fat, 16±6 %; and, maximal aerobic power ($\dot{V}O_2max$), 53±7 ml O_2/kg/min.

Test garments

All three protective ensembles are impermeable and totally encapsulating, and include protective masks, rubber gloves and boots. Under each test garment, subjects wore a layer (long pants, long sleeves, socks) of light cotton clothing and standard tropic-style combat boots. The STEPO garments are fabricated from a teflon/fiberglass laminate, while the TAP suit is of butyl rubber-coated nylon. STEPO-R incorporates a backpack rebreather-respirator (16 kg) carried in an enclosed pod outside the suit. The total STEPO-R ensemble including the microclimate liquid-cooled vest weighs ~30 kg. The microclimate cooling system supplies a total of ~200W of cooling to the torso via the liquid-cooled vest, and to the breathing air in the mask via a heat-exchanger. The STEPO-T system is tethered to compressed breathable air (ambient temperature) which is blown into the suit for cooling, and into the mask for breathing. Also included in the STEPO-T system is an emergency air supply carried in a compressed air cylinder. Total weight of the STEPO-T ensemble is ~12 kg. The TAP suit includes the M17 protective mask but does not include a cooling system, and weighs ~8 kg.

Experimental design

For each test subject, $\dot{V}O_2max$ was determined using a continuous treadmill exercise protocol (Sawka, 1985). Subjects were fitted with the protective garments, and each practiced walking on the treadmill on at least four familiarization sessions preceeding testing.

During the 10 days preceding testing, the subjects participated in an exercise-heat acclimation program.

Acclimation (Accl) consisted of daily treadmill walking (1.56 m/sec, 4% grade) for 120 min in the heat (42°C, ~35%rh). During Accl, subjects wore shorts, t-shirt, socks, and comfortable athletic shoes. They were allowed to drink water ad libitum, but were encouraged to drink enough to maintain euhydration.

Following Accl, the three protective ensembles were tested in warm (27°C, ~50%rh) and hot (38°C, ~30%rh) environments. The warm environment was selected as representative of a summer day in the northern or central United States, while the hot condition was selected to represent a summer day in the southern United States. Within each environment, tests were administered in a counterbalanced order, and each subject served as his own control. The exercise-heat tests (EXT) consisted of alternate bouts of level treadmill walking for 45 min (1.12 m/sec) and seated rest for 15 min. The exercise intensity was determined (during a field study) based on metabolic measurements of a simulated toxic clean-up scenario by personnel in TAP suits. Subjects attempted to complete four exercise/rest cycles, for a total of 4 h. During all exercise bouts and all but the first rest period, subjects remained totally encapsulated. Only during the first rest period did subjects remove the hood/mask and drink water ad lib.

Experiments were also conducted in a very hot (49°C, ~15% rh) environment, in order to examine the capability of personnel to work in a representative desert condition. For these experiments, only the STEPO-T, and a variation of the STEPO-T worn with the microclimate cooling vest (STEPO-TV) were tested.

<u>Experimental procedures</u>

Heart rate (HR), was obtained from an electrocardiogram telemetered to and displayed on an oscilloscope cardiotachometer unit. Rectal temperature (T_{re}) was measured via a flexible thermistor inserted ~10 cm beyond the anal sphincter. Skin temperature was measured at three points (forearm, chest, calf) by thermocouples and mean weighted skin temperature ($\bar{T}_{sk}$) was calculated (Burton, 1935). Additionally, a thermocouple was used to measure the temperature between the top of the subject's head and the hood of the garment (T_{hd}). All temperatures were continuously monitored. Sweating rates were determined from nude body weight changes pre- to post-exercise, corrected for water ingestion. Corrections for unevaporated sweat left in the garments were used to calculate percent evaporative sweat.

Metabolic rates were determined via open circuit spirometry, during the familiarization sessions. The suits were opened and masks were removed to allow for collection of expired gases. For a subjective measure of how the test subjects felt during the EXT, at 20 and 40 min of each hour, subjects were asked to rate their perception of effort (RPE) using a numbered scale from 6-20 (Borg, 1970), and their thermal sensation (TS) using a numbered scale from 0-8 (Gagge et al., 1967; Young et al., 1987).

During Accl and EXT, subjects were removed from testing if HR exceedeed 180 bpm for five consecutive min, or if T_{re} reached 39.5°C. The test subjects withdrew themselves if they felt faint, sick or otherwise unable to continue. If there was a malfunction in any protective system that could not be corrected within 5 min, that test was terminated, and repeated on a later date.

Statistical treatment

Data were analyzed using a repeated measures analysis of variance. Tukey's post hoc test was employed to locate significant ($P<0.05$) differences. Because of small and unequal sample sizes during the 49°C experiments (n=3 for STEPO-T, n=6 for STEPO-TV), statistics were not performed on data for the very hot tests. Data are reported as means ± standard deviation ($\bar{X}$±SD).

RESULTS

The subjects' metabolic rates during exercise were: 312±45, 356±36, 424±36 and 493±93 watts when they were wearing shorts, TAP, STEPO-T and STEPO-R, respectively. These metabolic rates of 23, 24, 31 and 36% of $\dot{V}o_2$max are representative of light to moderate exercise. Both STEPO systems elicited higher ($p<0.05$) metabolic rates than either shorts or the TAP suit.

Figure 1 presents endurance time for the subjects in each experimental condition. In the warm environment, endurance time was longer ($p<0.05$) for subjects in STEPO-T (219±51 min) than STEPO-R (111±76 min), but for the TAP suit endurance time (194±56 min) was not different. During the STEPO-T experiments, five subjects were able to complete the entire 240 min session. One subject was removed from testing for the day because his core temperature reached the safety criteria. During the STEPO-R experiments, four subjects removed themselves from testing due to muscular discomfort (sore, tired backs and/or legs), and one because of mask/visor fogging. Only one subject completed 240 min. During the TAP experiments, three subjects completed the session, and three discontinued due to muscular discomfort (one was nearing HR criteria).

In the hot environment, endurance times were not significantly different among the systems (STEPO-T, 82±52 min; STEPO-R, 57±27 min; TAP, 56±20 min). During the STEPO-T experiments, five subjects discontinued because of physiological discomfort (hot, light-headed, breathing discomfort), and one was removed upon achieving the safety criteria for heart rate. During the STEPO-R experiments, three subjects discontinued because of muscular discomfort, and three discontinued due to physiological discomfort. For the TAP experiments, five subjects discontinued due to physiological discomfort (one was nearing HR criteria, and one was nearing T_{re} criteria), while one subject did reach the criteria HR.

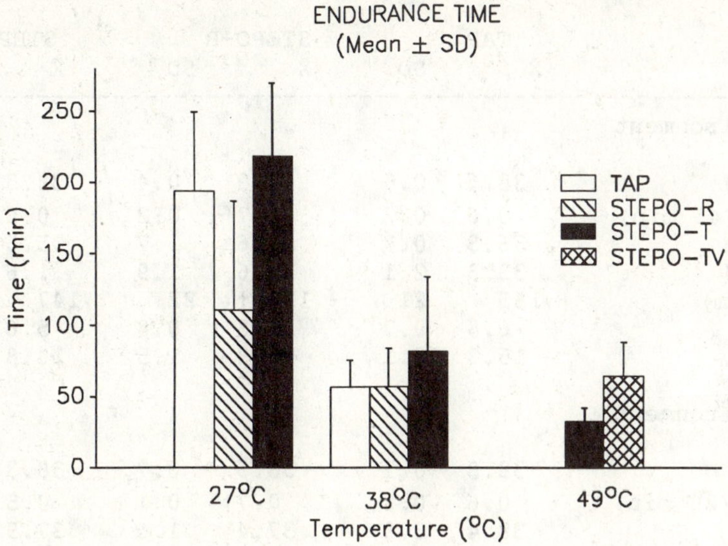

Figure 1. Endurance time for three protective ensembles (and one variation, STEPO-TV) in warm, hot and very hot environments. At 27°C, time for STEPO-T > STEPO-R ($p<0.05$).

Table 1 presents the subjects' final exercise thermoregulatory, heart rate and perceptual responses. In addition, the change in T_{re} (ΔT_{re}) from rest to the latest time when all subjects were still testing, is presented (0-45 min in the warm, and 0-25 min in the hot environment). In the warm environment, differences ($p<0.05$) were observed for final $\bar{T}_{sk}$ (STEPO-R < TAP) and for final HR (STEPO-R < TAP and STEPO-T). In the hot environment, the ΔT_{re} was greater ($p<0.05$) during STEPO-R than STEPO-T, and final HR was lower ($p<0.05$) for STEPO-R than TAP. Final RPE was greater ($p<0.05$) for the STEPO-R than for either TAP or STEPO-T. Body temperatures and HR were also analyzed over time in each environment, and all but T_{hd} increased as time progressed. However, only T_{hd} values were significantly different among the three protective garments in the hot environment. Values for T_{hd} in the TAP suit were consistently higher than for STEPO-R, which were consistently higher than for STEPO-T.

In the very hot environment only STEPO-T and STEPO-TV were tested. Because of the extreme environment, only three subjects were tested wearing STEPO-T on one day, and based on their enhanced discomfort and very short endurance times, the other three subjects did not test. Table 2 presents the final test data for the very hot environment.

Table 1. Final exercise temperature, heart rate and perceptual responses for men exercising in warm and hot environments while wearing toxic-protective clothing.

	TAP		STEPO-R		STEPO-T	
	$\bar{X}$	SD	$\bar{X}$	SD	$\bar{X}$	SD
27°C Environment						
T_{re} (°C)	38.6	0.5	37.9	0.4	38.3	0.6
ΔT_{re} (°C/45 min)	0.6	0.2	0.7	0.2	0.6	0.1
$\bar{T}_{sk}$ (°C)	36.3	0.7	34.6*	1.7	35.6	0.6
T_{hd} (°C)	33.3	2.1	33.6	1.9	31.6	2.7
HR (bpm)	155	21	130*†	22	147	23
TS	6.8	0.7	4.8	0.8	5.6	0.9
RPE	15.3	3.1	16.8	1.5	14.8	2.8
38°C Environment						
T_{re} (°C)	38.8	0.7	38.5	0.7	38.3	0.4
ΔT_{re} (°C/25 min)	0.6	0.1	0.7†	0.1	0.5	0.1
$\bar{T}_{sk}$ (°C)	38.4	0.8	37.4	1.6	37.9	0.7
T_{hd} (°C)	39.3	1.0	37.3*†	0.7	35.6*	1.7
HR (bpm)	172	18	155*	7	162	15
TS	6.7	0.6	6.8	0.4	6.2	0.9
RPE	13.5	3.1	17.2*†	0.4	14.0	2.6

*P<0.05 compared to TAP; †P<0.05 compared to STEPO-T.

Table 2. Final exercise temperature, heart rate and perceptual responses for men exercising in a very hot environment while wearing toxic-protective clothing.

	STEPO-T (n=3)		STEPO-TV (n=6)	
	$\bar{X}$	SD	$\bar{X}$	SD
49°C Environment				
T_{re} (°C)	37.9	0.6	38.4	0.6
ΔT_{re} (°C/25 min)	0.5	0.1	0.6	0.2
$\bar{T}_{sk}$	38.9	1.9	36.6	1.2
T_{hd}	39.8	3.4	39.3	3.0
HR (bpm)	171	11	168	15
TS	6.3	0.6	6.8	0.5
RPE	13.3	3.2	14.7	1.8

Table 3. Sweating rate and percent of total sweat which was evaporated (% evaporation).

	TAP		STEPO-R		STEPO-T		STEPO-TV	
	$\bar{X}$	SD	$\bar{X}$	SD	$\bar{X}$	SD	$\bar{X}$	SD
27°C Environment								
Sweating Rate ($g/m^2/h$)	494	88	570	222	395	143	-	-
% Evaporated	30†	9	24†	9	57	13	-	-
38°C Environment								
Sweating Rate ($g/m^2/h$)	1029	253	914	240	885	301	-	-
% Evaporated	33	5	38	11	49	4	-	-
49°C Environment								
Sweating Rate ($g/m^2/h$)	-	-	-	-	813	192	803	168
% Evaporated	-	-	-	-	29	0	43	11

†$P<0.05$ compared to STEPO-T.

Table 3 presents total sweating rates and percent sweat evaporated for each environment. Statistics were performed on both variables for the warm experiments, only on the total sweating rates for the hot experiments, and no statistics were performed for the very hot environments. In the warm environment, the percent sweat evaporated was greater ($p<0.05$) for the STEPO-T than for either the STEPO-R or TAP. During all the experiments, the subjects' rate of dehydration ranged from 1.2 to 2.5% of body weight per hour.

DISCUSSION

Toxic-protective clothing severely limits heat exchange between the individual and the environment. Additionally, the bulk of the clothing and that of associated equipment renders movement to be less efficient as well as adding weight that must be carried. These factors add to the energy cost of a given task, and result in an additional heat load which needs to be dissipated. Clearly, the heavier the protective system in this study the greater the energy cost for a given task.

One half of all the STEPO-R experiments were terminated because of severe muscular discomfort. In the previous STEPO study (Pandolf et al., 1988), two of seven subjects suffered back spasms, one in each of the STEPO systems. Further refinement/development of the load carriage for the STEPO-R system is needed.

The STEPO systems were designed to reduce thermal stress and extend work tolerance relative to the TAP suit. In the warm environment, STEPO-T allowed longer endurance times while

STEPO-R was clearly inferior. In the hot condition most indices of thermal strain were comparable among suits, although with STEPO-T non-significant benefits were observed. Improvement in the microclimate cooling capacity would enhance the STEPO-R, and inclusion of an air-cooling vest next to the skin for STEPO-T would improve this system as well. Finally, the capability of carrying and drinking ~2 liters of water would minimize dehydration and undoubtedly enhance performance.

DISCLAIMER AND DISTRIBUTION STATEMENT

The views, opinions and/or findings contained in this report are those of the author and should not be construed as an offical department of the Army position, policy or decision, unless so designated by other official documentation. Approved for public release; distribution unlimited.

REFERENCES

Borg, G., 1970, Perceived exertion as an indicator of somatic stress. Scandinavian Journal of Rehabilitation Medicine, 2, 92-98.

Burton, A.C., 1935, Human calorimetry II. The average temperature of the tissues of the body. Journal of Nutrition, 9, 261-280.

Gagge, A.P., Stolwijk, J.A.J. and Hardy, J.D., 1967, Comfort and thermal sensation and associated physiological responses at various ambient temperatures. Experimental Research, 1, 1-20.

Muza, S.R., Pimental, N.A., Cosimini, H.M. and Sawka, M.N., 1988, Portable, ambient air microclimate cooling in simulated desert and tropic conditions. Aviation, Space and Environmental Medicine, 59, 553-558.

Pandolf, K.B., Levine, L., Cadarette, B.S. and Sawka, M.N., 1989, Physiological evaluation of individuals exercising in the heat wearing three different toxicological protective systems. In Advances in Industrial Ergonomics and Safety I, edited by A. Mital (London: Taylor and Francis, Ltd.), pp. 239-246.

Pimental, N.A., Cosimini, H.M., Sawka, M.N. and Wenger, C.B., 1987, Effectiveness of an air-cooled vest using selected air temperature and humidity combinations. Aviation, Space and Environmental Medicine, 58, 119-124.

Sawka, M.N., Young, A.J., Francesconi, R.P., Muza, S.R. and Pandolf, K.B., 1985, Thermoregulatory and blood responses during exercise at graded hypohydration levels. Journal of Applied Physiology, 59, 1394-1401.

Speckman, K.L., Allan, A.E., Sawka, M.N., Young, A.J., Muza, S.R. and Pandolf, K.B., 1988, Perspectives in microclimate cooling involving protective clothing in hot environments. International Journal of Industrial Ergonomics, 3, 121-147.

Young, A.J., Sawka, M.N., Epstein, Y., DeCristofano, B. and Pandolf, K.B., 1987, Cooling different body surfaces during upper and lower body exercise. Journal of Applied Physiology, 63, 1218-1223.

Advances in Industrial Ergonomics and Safety II
Edited by Biman Das
Taylor & Francis 1990

IMPACT OF A CHEMICAL PROTECTIVE ENSEMBLE ON COGNITIVE AND PSYCHOMOTOR PERFORMANCE

J.W. Kaufman*, K.Y. Dejneka*, S.J. Morrissey†, and A.C. Bittner, Jr. **

* Naval Air Development Center, Warminster, PA 18974
† Univ. of Wisconsin at Plattville, Plattville, WI 53818
** Battelle HARC, Seattle, WA 98105

Today's military environment has created the need to provide personnel with adequate protection against chemical threats. The current study attempted to quantify decrements in cognitive and psychomotor performance and the heat load created by use of a chemical protective helicopter aircrew garment (CBR) under hot/humid conditions. Subjects wore either CBR or a flight coverall ensemble (FS) in both a cool and hot/humid environment while performing simulated helicopter crew tasks for up to 480 minutes. Cognitive and psychomotor tasks were not significantly affected by ensemble. It was concluded that the heat load from the CBR may prove to be a limiting factor for physical performance in hot work environments.

INTRODUCTION

The development of a semi-permeable chemical protective assembly (CBR) was believed to have reduced the thermal burden previous systems have placed on users. The purpose of this study was to evaluate the thermal load imposed on users of this system under hot and humid conditions and quantify decrements in cognitive and psychomotor performance. This study attempted to simulate conditions which might be experienced within a helicopter during military operations (Littell and Joy, 1969). Trial durations of up to 8 hours simulated the anticipated use in a wartime situation.

MATERIALS AND METHODS

<u>Subjects</u>: Three males, aged 25-35, volunteered to participate after being fully informed of the details of the experimental protocol and associated risks. Subject weight ranged from 65-92 kg, body surface area (SA) from 1.72 to 2.05 m², and percent body fat was determined to be 14-20 % (Kaufman, et al., 1988).
<u>Materials</u>: Two ensembles were employed in this study: 1) a Nomex flyer's coverall (FS); and 2) a chemical protective

assembly. The clothing and equipment which comprised each ensemble are given in Table 1.

Table 1. Equipment configurations worn during tests.

Standard Flight Ensemble (FS)	Protective Ensemble (CBR)
a. Nomex flight coverall	a. All items in FS ensemble
b. cotton long underwear	b. butyl rubber hood & mask
c. flyer's boots	c. carbon-impregnated liner
d. flyer's gloves	d. CBR ventilator
e. survival vest	e. cotton gloves
f. personal flotation device	f. butyl rubber gloves
g. helmet	g. polyethylene socks
h. vinyl canteen, 2 L capacity	h. intercom

Test Procedures: Each test simultaneously exposed 2 subjects to the experimental conditions, with subject pairing randomized. It was intended that each subject use each test garment twice in both hot and cool conditions, with each subject having a total of 8 exposures. Due to withdrawing early from the study, one subject had a total of 6 exposures. Minimum time between tests for a given subject were 2 days.

Subjects reported to the laboratory on the morning of a test and were given physical examinations by the attending flight surgeon. A urinalysis was performed, and baseline nude weight was obtained. Thermistors were attached to the following sites: (A) forehead; (B) left upper chest; (C) left distal upper arm; (D) dorsum of left hand; (E) right anterior thigh; (F) left posterior thigh; (G) right shin; (H) right foot; (J) right proximal upper arm; and (K) left lower back. A rectal thermocouple was inserted 8 cm anterior to the anal sphincter. ECG electrodes were placed on subjects. Subjects then donned the appropriate equipment configuration for that run (Table 1). Upon completion of dressing, subjects rested for 20 minutes at 20°C to allow temperature and heart rate (HR) to return to a resting condition.

Following the conclusion of the rest period, subjects entered the chamber. Hot conditions for these tests were T_{air} = 33°C with a relative humidity (RH) of 70%, while cool conditions were T_{air} = 21°C and RH = 40%. Runs consisted of an initial 60 minute rest period followed by a repeated cycle of: a) 7 minutes of subjective assessment of physiological condition (approx. 30 seconds), cognitive testing from the Performance Evaluation Tests for Environmental Research (PETER) battery (Bittner, et al., 1986), i.e., Baddeley reasoning and vertical addition of 3 two digit numbers (Baddeley, et al., 1975, Kaufman, et al.,1988) (3 minutes), and rest (3.5 minutes); b) 7 minutes of psychomotor testing, i.e., play three rounds on a video game (Atari Jet-Fighter); c) 7 minutes of physical exercise, i.e., 30 W of work on a bicycle ergometer. This 21 minute cycle was repeated until termination of a run. Subjects were trained in the cognitive and psychomotor tasks prior to the start of the study. Individuals were requested to remain in the chamber for up to 8 hours, unless their rectal temperature (T_{re}) reached 39°C.

During the first 120 minutes in the chamber subjects had access to a 2 liter canteen of water. After this time the canteen was removed, as aircrew would likely have potable water only at the start of a mission.

Subjective assessment of physiological condition was evaluated for fatigue, skin wetness, temperature, and comfort. Subjects rated their subjective feeling for each scale by placing marks along 112mm lines. Extremes were indicated on each line with the most pleasant on the left and the least pleasant on the right. The rates (R_v) were obtained from:

$$R_v = (V_f - V_p)/t \qquad \text{(mm/min.)} \qquad (1)$$

where V_f = final value for a given category, distance in mm, V_p = the value obtained prior to dressing, and t = elapsed time at final value. Both the subjective assesments and cognitive function tasks were administered prior to dressing, every 30 minutes during testing, and after the subjects had completed the post-test physical examination.

Physiological Indices: Mean weighted skin temperature (T_{sk}) was calculated using the equation:

$$T_{sk} = 0.1(T_A) + 0.125(T_B+T_K) + 0.125(T_E+T_F)/2 + 0.06(T_D) + 0.125(T_E) + 0.15(T_G) + 0.07(T_J+T_C) + 0.05(T_H) \qquad (^\circ C) \qquad (2)$$

where T_i are the measured skin temperatures at locations i = A, B,..., K (Olesen, 1984). Mean sweat rate (M_{sw}) was determined by the difference between post-test nude weight (NW_2), corrected for fluid and food intake, and pre-test nude weight (NW_1) from:

$$M_{sw} = (NW_2 - NW_1)/t \qquad \text{(kg/min.)} \qquad (3)$$

where t is run duration (Light, et al.,1987).

Statistical Analysis: Data for the individual dependent variables was analyzed using repeated measures analysis of variance (ANOVA). Subsequent analyses of significant changes within runs were also performed with paired-sample t-tests. Differences were considered significant at the level of $p<0.05$.

RESULTS

The results of this study indicated that CBR produced greater heat stress in a hot versus cool environment or when compared with FS in either environment. Environment and subject variations were other variables which proved significant in the physiological differences observed between runs. The mean data for the dependent variables of voluntary duration time, T_{re}, T_{sk}, HR, total water consumption, and M_{sw} are reported in Table 2.

Voluntary Duration Time: Because of the great variance in subjective tolerance, the exposure duration data were transformed using natural logarithms. Results of the ANOVA show pronounced differences in exposure duration times between the hot and cool conditions ($p<0.01$), equipment ensembles ($p<0.01$), subjects ($p<0.01$), and replications ($p<0.05$).

Environmental temperature was found to be significant

($p<0.01$), with subjects having a significantly lower tolerance time in the hot conditions regardless of equipment ensemble. The effect of equipment ensemble was highly significant ($p<0.01$), with use of the FS resulting in longer durations for subjects in all conditions (Table 2). There was a significant triple order interaction between clothing type, replication, and temperature, which is apparent in Figure 1.

Table 2. Mean values of exposure duration (t), rectal (T_{re}) and mean weighted skin temperatures (T_{sk}), heart rate (HR), water intake, and mean sweat rate (M_{sw}), by configuration. Initial and final values are indicated by i and f, respectively. Values reported are means and standard errors of the mean (SEM) (†, n=6; *, n=5).

		†CBR_{hot}		†FS_{hot}		*CBR_{cool}		*FS_{cool}	
		mean	SEM	mean	SEM	mean	SEM	mean	SEM
t (minutes)		155	5.3	219	17.9	305	41.2	382	40.3
T_{re} (°C)	i	37.7	0.2	37.2	0.2	37.2	0.2	37.2	0.3
	f	38.7	0.2	38.6	0.2	37.5	0.2	37.4	0.3
T_{sk} (°C)	i	33.5	0.1	33.0	0.3	32.3	0.5	32.7	0.3
	f	36.9	0.3	36.9	0.2	33.8	0.2	33.6	0.3
HR	i	84	4	80	2	91	2	82	5
(beats/min.)	f	131	7	139	8	102	7	99	10
Water (kg)		0.4	0.1	0.8	0.3	0.4	0.3	0.1	0.1
M_{sw} (g/min.)		5.9	1.3	5.7	0.8	3.1	0.2	2.1	0.1

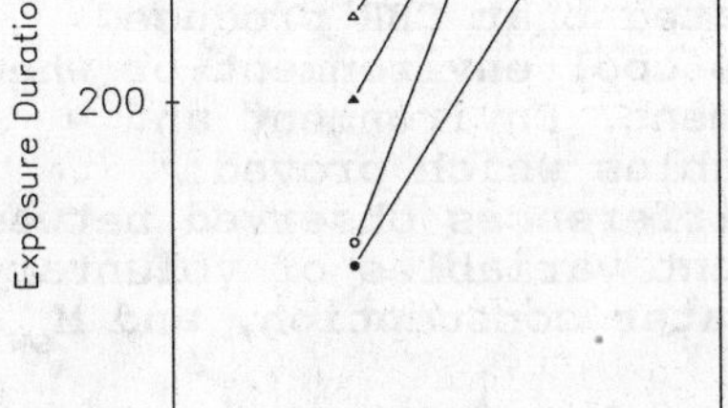

Figure 1. Environment, suit, and trial effects on exposure durations.

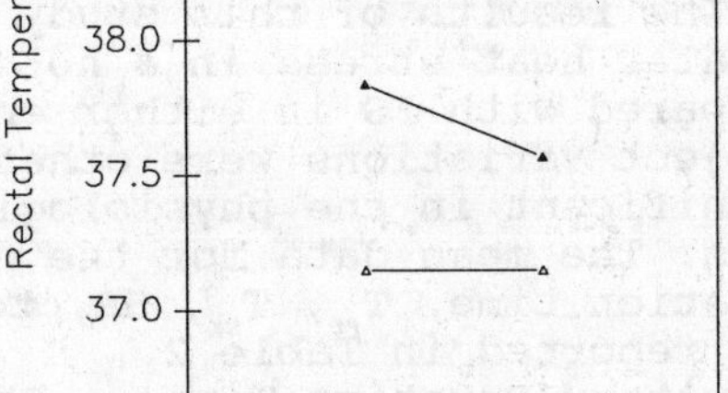

Figure 2. Environment and trial effect on mean rectal temperature.

Rectal Temperature: Pooled data plotted in Figures 1 and 2 show that: 1) Final T_{re} is much higher in the hot conditions than in the cool conditions; 2) Increases in T_{re} over the time of the study are greater in the hot environment than in the cool environment; and 3) Use of the CBR suit resulted in higher T_{re}'s in all conditions when compared with the FS at the same point in time. Differences between T_{re}'s observed for CBR trials in the hot and cool conditions were found to be significant from minute 121 through the end of the trials ($p<0.05$). The ANOVA revealed that across equipment ensembles*, T_{re} was significantly higher in the hot environment than in the cool ($p<0.01$), and there was a significantly greater change in T_{re} over trial duration in the hot environments ($p<0.01$). There was a significant interaction between the replication and hot environments ($p<0.04$).

Mean Skin Temperature: T_{sk} shows a pattern similar to that of T_{re}, i.e., a higher T_{sk} at the end of all conditions and higher T_{sk}'s in the hot versus cool condition. Final T_{sk} was significantly higher in the hot trials than in the cool trials ($p<0.01$). Comparison of T_{sk}'s between hot and cool conditions over trial duration show significant differences from approximately the beginning through the end of trials. There were no significant differences in T_{sk} due to ensemble or between replications. This finding argues against any physiological acclimatization having occurred between replications or across the experiment.

Heart Rate: Comparisons of final HR's (Table 2) indicate that significant differences existed between the FS at 21°C (FS_{cool}) and the 33°C FS (FS_{hot}) and all CBR runs. No other final HR differences were significant.

Sweat Rate: The M_{sw} calculated for the FS_{hot} trials compared with the FS_{cool} and 21°C CBR (CBR_{cool}) trials indicated significant differences ($p<0.05$), as did comparing the 33°C CBR (CBR_{hot}) to FS_{cool} ($p<0.05$) (Table 2). Pre-test urine specific gravity showed no differences between configurations, indicating equivalent hydration levels prior to a trial. Water consumption was observed to be significantly greater in the FS_{hot} than cool conditions ($p<0.05$). No significant difference in water consumption was observed between CBR_{hot} and FS_{hot} or cool conditions (Table 2).

Cognitive and Psychomotor Data: No effects on either cognitive or psychomotor testing were discerned as a result of exposure to the experimental conditions. Neither differences in environmental conditions nor clothing configurations resulted in any observed changes in video game scores or the number of attempts and correct responses to the cognitive tasks (Table 3). Video game performance was analyzed by examining pre- and post-test, mean, first, and last scores and the standard deviations. Greater post- versus pre-test scores ($p<0.05$) was the only significant difference observed.

Subjective Responses: Comparing equipment ensembles on the basis of subjective criteria showed that R_v was significantly greater for FS_{hot} than either the FS_{cool} or CBR_{cool} ($p<0.04$). This was true for fatigue, wetness, temperature, and comfort. No significant difference was observed between CBR_{hot} and the other ensembles for any of the subjective criteria.

TABLE 3. Mean values of correct responses and attempts for the Baddeley reasoning test and vertical addition task, by configuration. Values reported are means and standard errors of the mean (SEM) (†, n=6; *, n=5).

		†CBR_{hot}		†FS_{hot}		*CBR_{cool}		*FS_{cool}	
		mean	SEM	mean	SEM	mean	SEM	mean	SEM
Baddeley	(correct)	18	1.1	19	0.7	21	1.1	21	0.7
Reasoning	(attempts)	20	1.0	20	1.2	22	1.1	21	0.6
Vertical	(correct)	14	0.8	13	0.5	14	0.9	14	0.7
Addition	(attempts)	15	0.8	14	0.4	15	0.7	14	0.6
Atari	(pre)	16	1.5	17	1.7	14	1.8	17	3.0
scores	(first)	17	1.7	17	0.9	16	1.6	16	2.4
	(last)	16	1.7	17	1.6	15	1.9	17	1.7
	(post)	17	1.7	19	1.2	16	1.3	18	1.5
	(mean)	16	1.4	18	1.1	16	0.9	17	1.6

DISCUSSION

The purpose of this study was to determine the impact of wearing the CBR ensemble on thermal homeostasis and mission-related cognitive and psychomotor performance. Analyses of exposure duration, T_{re}, and T_{sk} indicate that the CBR ensemble induces heat stress, which is particularly pronounced under conditions of high heat and humidity, while cognitive and psychomotor performance was unaffected. Thornton, et al. (1985) found similar physiological results, though the stress experienced by their subjects appears to be considerably less than that observed in this study.

While final temperatures did not vary significantly between configurations during hot runs, the differences in onset rates and exposure durations indicate that the CBR ensemble produced significantly greater thermal stress on personnel. The elevated starting T_{re} observed in the CBR trials was probably a result of heat storage while donning the CBR ensemble; heat was not dissipated during the pre-trial rest period.

Hydration state must be considered when interpreting physiological changes (Sawka, et al.,1984). Initial hydration state appears to be equivalent among subjects, based on the initial specific gravities of urine samples, therefore hydration does not appear be a factor in the observed differences. As water consumption nor M_{sw} (Table 2) were significantly different between ensembles in the hot environment, evaporation at the garment surface must impact T_{re}. It appears that the CBR ensemble inhibits the transfer of moisture to the outer garment surface, thus reducing effective heat transfer.

The lack of a significant clothing effect on final T_{re} may have been a function of trial duration. Use of the FS permitted significantly longer periods of time before trials were stopped either for subjective intolerance to extreme fatigue and discomfort or T_{re} = 39.0°C. The lack of difference in T_{re}, as well as T_{sk}, indicates that trials were terminated at similar physiological states, though the time to reach such a state differed. Trial duration may also account for the differences in R_v observed between clothing configurations. The additional time spent in the FS_{hot} compared with the CBR_{hot} may be responsible for any perceived differences between these configurations and the FS_{cool} and CBR_{cool}, respectively. The relatively small mean exposure duration for CBR_{hot}, i.e., 155 minutes, suggests that this ensemble may present a serious impediment to sustained usage due to the induced heat stress. The onset of marked thermal stress, as indicated by final T_{re}, was brought on at a significantly faster rate by the CBR ensemble during heat exposures. While there is evidence of habituation with the FS under both conditions and the CBR_{cool} (Figure 1), no such conditioning is witnessed for the CBR_{hot}. This suggests that the maximum endurance has been elicited for CBR_{hot}.

The lack of significant changes in the cognitive and psychomotor performance tasks may be the result of either a lack of test sensitivity (Bittner, et al.,1986) or insufficient physiological changes (Allan and Gibson, 1979) to induce cognitive and psychomotor deficits. The Atari task has been previously studied and shown to be a sensitive test for performance changes (Bittner, et al.,1986), but an earlier heat stress study (Kaufman, 1985) resulted in inconclusive changes with regard to Atari performance. Similarly, the cognitive function tests used have previously been sensitive indicators of cognitive changes (Baddeley, et al.,1975). Therefore, the results from this study suggest that the lack of significant differences are the consequence of inadequate physiological change to elicit a performance change.

The energy expenditure and cyclic nature of the work load were chosen to model helicopter crew missions. Cycling of tasks might suggest that physiological measurements obtained will reflect the duration of each cycle, thus the physiological responses being idiosyncratic to a given situation. Mairiaux, et al. (1984) have shown that cycle time is not reflected in changes in T_{re} but does impact on T_{sk} and M_{sw}. Similar results were found in this study, though it appears that the CBR ensemble tended to damp out these responses.

The results of this study indicate that the CBR ensemble imposes a considerable thermal stress on the user, and apparently limits the duration of its use to under 3 hours under similar conditions. One possible way of reducing the thermal stress might be imposing lengthy rest periods, much greater than 7 minutes, between activity cycles (Mairiaux et al.,1984), a situation which was not examined in this study.

Acknowledgements: It is the authors' pleasure to acknowledge the invaluable contributions made by the subjects of this study, whose perseverance despite personal discomfort made this work possible. We would also like to thank the medical support personnel, Walter Soroka, Rodney Pursell, Nancy Holden, Christopher Zech, Robert Muller, and Gregory Askew.

REFERENCES

Allan, J. and Gibson, T., 1979, Separation of the effects of raised skin and core temperature on performance of a pursuit rotor task. Aviation, Space and Environmental Medicine, 50, 678-682.

Baddeley, A. Cuccaro, W., Egstrom, G., Weltman, G. and Willis, M., 1975, Cognitive efficiency of divers working in cold water. Human Factors, 17, 446-454.

Bittner, A., Carter, R., Kennedy, R., Harbeson, M., and Krause M.,1986, Performance evaluation tests for environmental research (PETER): Evaluation of 114 measures. Perceptual and Motor Skills, 63, 683-708.

Fine, B. and Kobrick, J., 1987, Effect of heat and chemical protective clothing on cognitive performance. Aviation, Space and Environmental Medicine, 58, 149-154.

Kaufman, J., 1985, Heat stress evaluation of anti-exposure flight gear. Naval Air Development Center, Technical report, NADC-85061-60.

Kaufman, J., Dejneka, K., Morrissey, S. and Bittner, A., 1988, Evaluation of thermal stress induced by helicopter aircrew Chemical, Biological, Radiological (CBR) protective assembly. Naval Air Development Center, Technical report, NADC-89009-60.

Light, I., Gibson, M. and Avery, A., 1987, Sweat evaporation and thermal comfort wearing helicopter passenger immersion suits. Ergonomics, 30, 793-803.

Littell, D. and Joy, R., 1969, Energy cost of piloting fixed- and rotary-wing aircraft. Journal of Applied Physiology, 26, 282-285.

Mairiaux, P., Libert, J., Candas, V. and Vogt, J., 1984, Physiological and perceptual responses to cyclic heat stress variations. Aviation, Space and Environmental Medicine, 55, 935-940.

Olesen, B., 1984, How many sites are necessary to estimate a mean skin temperature? In: Thermal Physiology, New York, edited by J. Hales, (New York: Raven Press), pp. 33-38.

Sawka, M., Francesconi, R., Pimental, N. and Pandolf, K., 1984, Hydration and vascular fluid shifts during exercise in the heat. Journal of Applied Physiology:Respiration, Environmental and Exercise Physiology, 56, 91-96.

Thornton, R., Brown, G. and Redman, P., 1985, The effect of the UK aircrew chemical defense assembly on thermal strain. Aviation, Space and Environmental Medicine, 56, 208-211.

Advances in Industrial Ergonomics and Safety II
Edited by Biman Das
Taylor & Francis 1990

A NEW APPROACH TO PREDICTING RESPONSE TO WORK IN HOT ENVIRONMENTS

Phillip A. Bishop

Human Performance Laboratory
Area of Health and Human Performance Studies
University of Alabama

The development of mathematical models for predicting human physiological responses in thermally-challenging environments is a complex undertaking. Because this prediction is useful in optimizing work and rest cycles in hot or cold environments and in setting tolerance limits, accurate prediction models are needed. Conventional modeling equations (a priori prediction equations APP) have generally been constructed based upon combinations of theoretical calculations of heat production, specific heat, conductance, heat gains, etc.. Yet, it is easy to observe large inter-subject variability of similar subjects exposed to the same environmental conditions. Hence, APP equations must include subject's individual characteristics.

One potential solution to the modeling problem is to determine the individual varibles which best predict thermal tolerance. Then, empirical equations (EEQ) can be developed to predict thermal response, or tolerable work duration. In other terms, this new approach is to use the individual's response to a short term thermal challenge to predict long term responses. Recent work in our laboratory suggests this is a feasible approach.

The advantages of the a priori (APP) approach is that these equations can be applied to any possible combination of work rate and environment. The disadvantage to APP is that the inter-subject variability requires input of a number of individual characteristics. The complexity of the APP equations requires considerable computer resources. The advantage of the EEQ approach is that individual variations are incorporated in the assessment phase. The chief disadvantage is that this approach requires some data collection, and is consequently relatively expensive.

Further research needs to be done to determine if accurate prediction models can be developed from field test items such as heart rate and oral temperature change. The generalizability to different environmental conditions and different work conditions also needs to be ascertained.

Ultimately the EEQ approach offers the potential for predicting performance accurately with simple field tests of performance. As workers train, or even as they begin a particular work test, yet-to-be-identified variables would have to be observed. Subsequently workers could then be utilized based upon their individual abilities.

INTRODUCTION

Increasingly, workers in industry are required to perform tasks in physically hazardous enviroments requiring the use of protective clothing. Nuclear power plant maintenance and inspection personnel for example, on occasion, must enter radioactive areas. Service workers such as asbestos removal crews, toxic hazard reaction teams, and firefighters, also must wear protective clothing. Probably the biggest user of protective clothing is the military of many countries. The need to train and be prepared for nuclear, biological, and chemical warfare defense requires extensive use of protective clothing. Because of the need for adequate protective characteristics, this clothing typically has high insulative and low permeability properties. For example, the U.S. military chemical defense ensemble has a clo of 2.55 and an im/clo of 0.28. Consequently, the metabolic heat produced, coupled with the reduction in evaporative cooling results in heat storage in all but the coolest environments. Often, heat storage is handledby interspersing work bouts with rest periods (Yousef, et al., 1989, Bishop et al., 1988, 1989; Vogt et al., 1983).

For personnel management purposes, it is important that accurate predictions are available for the physiological response to work while wearing protective clothing. For example, in a toxic chemical spill, managers need to be able to predict the manpower needed to acomplish the cleanup in a given amount of time. Manpower needs are dependent upon ambient temperature because as temperature goes up, work capacity is diminished (Bishop et al.,1989, Bishop,et al., 1988; Constable et al.,1987; Joy and Goldman, 1972; Craig, et al., 1954). Accurate prediction is crucial in that workers exposed to conditions resulting in rapid heat storage are in danger of injury or death from heat stroke. In situations where rapid task accomplishment may be necessary to preserve lives and property (i.e. firefighting), overly conservative work restrictions are dangerous to those being helped. There is also a need to provide manpower requirements information for planning employee recruitment and training.

Despite the strong need for predictive models for the protective clothing scenario, there currently exist no highly accurate models for predictling individual responses to a given set of conditions. This is partialy explainable by the complexity of the problem. Figure 1 portrays some of the many variables to be accounted for in such a model.

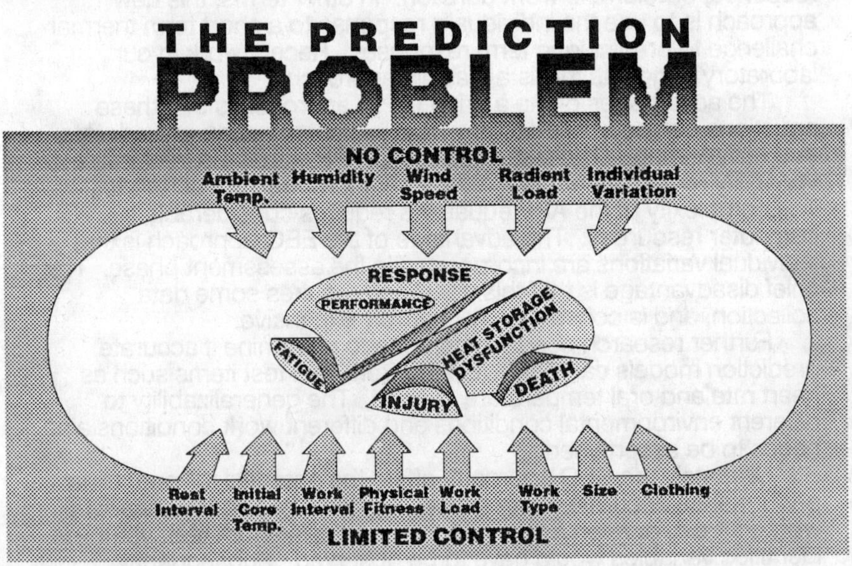

Figure 1. Some key variables which must be considered in heat tolerance prediction models.

The complexity is heightened by the fact that many of these variables interact. For example, the nature of the work task will influence the work rate for manual materials handling tasks. Some of the variables such as rest interval can, in some cases, be controlled by the work supervisor. Other variables such as ambient temperature , humidity, radiant load, and wind speed can only be controlled in a few instances and to only a limited degree. At the present time, it is unclear how important some variables are in the protective clothing model. For example, acclimatization might be expected to increase work tolerance, on the other hand, early and profuse sweat production is not helpful inside impermeable clothing and may be disadvantageous by leading to earlier dehydration. Estimating the contribution of sweat evaporation to heat removal inside a semi- or fully permeable garment is complex because sweat may be evaporated from the skin, the inner clothing, or the inside surface of the outer clothing, or some combination of these sites. The site of evaporation will govern how effectively sweat evaporation cools the worker.

By far the most difficult problem to deal with in the prediction model is that of inter-individual differences of response to heat storage situations. Although several prediction models have enjoyed some degrees of accuracy for group prediction, when it comes to health and safety, accurate individual prediction is the goal. It does little good to predict what is safe for the 50th percentile, nor for the 95th, when the life and health of both the worker and the public is at risk.

The variability in response to a given heat stress has been demonstrated in a number of studies (Bishop, 1989; Mar'yanovich et al., 1984; Vogt et al.,1983; Krajewski et al., 1979; Wyndham et al., 1970; Joy and Goldman, 1968). It is not difficult to theorize as to the source of this variability. Individual differences in body composition should certainly influence an individual's specific heat. It is reasonable to hypothesize that an extreme ectomorph, extreme mesomorph, and extreme endomorph of the same body weight, might increase temperature at different rates in response to the input of a fixed number of kilojoules of heat. Likewise, the efficacy of sweat evaporation is dependent upon the match of circulation and evaporation. The sweat response is variable among subjects (Faff adn Tutak, 1989) and there is a logical basis for assuming that body composition influences circulation. If the heat conductance is poor, evaporation is less effective in reducing core temperature. Body composition, morphology, and aerobic fitness level may influence the sweat/circulation match.

It appears then, that there is a need for accurate prediction models which will predict for individuals. It is apparent also, that this is a very complex problem.

PREDICTIVE MODELS

Previous prediction models have relied upon theoretical considerations modified by empirical data. One of the more sophisticated models is that of Wissler(1986). Wissler's model is essentially a combination of smaller models, each of which is designed to account for one variable of the heat balance equation. These smaller models are complex in their own right as illustrated by the observation that Wissler's overall model requires only slightly less than one-half real time to run on a microcomputer (S.H. Constable, USAF School of Aerospace Medicine, personal communication). Wissler's approach which I term the *a priori* prediction (APP) model, although lengthy and complex, holds the potential for imminent generalizability. It is possible that his model or a similar one could be refined to the point that data from individuals could be included to predict on an individual basis. Simple information such as height and weight could readily be included, but more difficult to obtain data such as initial rectal temperature would be obviously less practical to incorporate.

A second approach to prediction is represented by the work of Giovanni and Goldman (1972). This approach yields simpler models than the purely theoretical APP models, because some empirical information is incorporated. For example Giovonni and Goldman offer this equation:

$$\text{Final Rectal temp.} = 36.75 + .004\,(\text{Metabolic Rate} - \text{External Work}) + 0.025/\text{clo}\,(\text{Ambient Temp.} -36) + 0.8_{e}{}^{.0047\,(Evap)}$$

Where *Evap* = Evaporative cooling required-maximum rate of evaporative. cooling.

The coefficients supplied in this equation were empirically obtained. The resultant model did an adequate job, in predicting rectal temperature for group performance, but the authors conclude their paper by recognizing the inadequacy of their model for individual prediction.

The proposed approach to the work tolerance prediction problem for workers wearing protective clothing in a hot environment carries the Giovanni and Goldman approach to the extreme by basing prediction almost entirely on observed responses. Hence, this approach is termed the Empirical Equation (EEQ). In contrast to models of the type previously described, the EEQ does not attempt to predict individual performance based upon some refinements of a group prediction model. Instead, the EEQ approach attempts to predict long term work tolerance in a variety of similar environments from simple observations collected during a short exposure to a moderate or high temperature environment. The EEQ approach could be thought of as a refinement of the common industrial practice among first-line supervisors of simply observing the work tolerance of individuals to a given job and reassigning those that appear intolerant to other tasks.

The EEQ approach could be applied in two different ways. The more precise application would require that potential workers wearing protective clothing perform a representative work task at a representative workrate in a representative environment. Increase in oral (or other temperature) or other physiological response per unit time,. or ability to sustain or increase workrate, or unidentified other variables would then be recorded. This data with adjustments for alterations in work conditions would then be used to project work tolerance for that individual worker. We have done some preliminary work which is to be reported in detail elsewhere. Our preliminary work found that the time of performance of a single work/rest cycle was effective in predicting total work tolerance for individuals. Among the 13 variables considered as predictors of work tolerance in this preliminary study, height, amount of heat stored, and rate of heat storage were effective predictor variables. A preliminary two-variable model for predicting rest time (R^2 = .91), and a three-variable model for predicting total work time (R^2 = .82), have been developed with these variables.

In the cruder but simpler application of the EEQ approach, no attempt is made to predict work tolerance time, but rather observations of short term tolerances are used to classify workers on a continuum of very tolerant to very intolerant. This application probably should not be strictly considered a true predictive model, and would not totally solve the management and scheduling problems. However, this approach would permit efficient worker selection and allow grouping of workers of similar tolerance which should greatly improve manpower efficiency for tasks requiring work teams.

The APP approach relative to the EEQ approach theoretically offers much greater generalizability. The generalizability across work, environment, and clothing conditions of the EEQ approach remains to be determined. The APP approach has been, and will be, expensive in terms of developmental costs but will require only computer expenses once it is perfected. In contrast, developmental costs for the EEQ approach should be much cheaper, but the on-going expense of testing workers would be considerable. The major advantage to the EEQ approach is, of course, its potential for individual accuracy and its overall simplicity. In order to achieve the required precision for individual prediction, the APP approach may require at least some empirical data and perhaps thereby lose some of its advantages in terms of generalizability and savings in recurring expenses. In fact the ultimate solution may require incorporating some of the information from a highly refined EEQ approach into the APP model or vice versa.

Many difficult questions must be answered before the complex and important ergonomics questions regarding work tolerance in protective clothing can be answered. The EEQ approach at the present time exists only hypothetically. Our preliminary work has indicated that the EEQ appraoch offers potentially, a useful ergonomic tool. Before this tool can be utilized, the following questions must be answered:

1)What variables best predict generalized work tolerance?

2)Among these variables, which offer the best ratio of accuracy to convienence-of-measurement?

3)How generalizable are observations of individuals working at a given rate,in a a given clothing configuration, in a given set of environmental conditions?

4)Would the EEQ approach of simply classifying workers be practically useful?

5)Would this classification approach be generalizable ?

6)What factors influence a worker's tolerance and how rapidly do they change?

7)Could a practical method of worker monitoring be combined with either the EEQ or the APP approaches?

CONCLUSIONS

In summary, the EEQ approach is in the preliminary developmental stage. Although untested, it offers the potential for a near term solution to the problem of managing workers in protective clothing in hot environments. In comparison to an APP approach, the EEQ offers potentially, lower developmental costs,and greater individual accuracy at a probable sacrifice of generalizability. There are a number of basic research questions which must be answered before the efficiency of the EEQ approach can be fully evaluated.

REFERENCES

Bishop, P.A. Pieroni, R., Smith, J. Merrit, T. 1989, Test-retest measurement reliability of human physiological response to heave work in protective clothing. Advances in Industrial Ergonomics and Safety I, Philadelphia, edited by A. Mital, (Philadelphia: Taylor and Francis), pp. 233-238.

Bishop, P.A. Nunneley, S.A., Garza, J.R., Constable, S. H., 1988, Comparisons of air vs liquid microenvironmental cooling for persons performing work while wearing protective clothing. In: Trends in Ergonomics/ Human Factors V, New York, edited by F. Agazadeh, (New York, Elsevier, pp. 433-440.

Constable, S.H., Bishop, P.A., Nunneley, S. A., Garza, J. 1988, Heat storage with hard work while wearing the chemical defense ensemble. (Abstract) Aviation Space and Environmental Medicine, 59, .

Craig, F.N., Garren, H.W., Frankel, H., Blevins, W.V., 1954, Heat load and voluntary tolerance time. Journal of Physiology, 6, 634-644.

Faff, J., and Tutak, T., 1989, Physiological responses to working with fire fighting equipment in the heat in relation to subjective fatigue. Ergonomics, 32, 629-638.

Givoni, B.and R.F. Goldman, 1972, Predicting rectal temperature response to work, environment, and clothing, Journal of Applied Physiology, 32, 812-822.

Joy, J.T., and Goldman, R.F., 1968, A method of relating physiology and military performance--A study of some effects of vapor barrier clothing. Military Medicine, 458-470.

Krajewski, J.T., Kamon, E., and Avellini, B.,1972, Scheduling rest for consecutive light and heavy work loads under hot ambient conditions. Ergonomics, 22, 975-987.

Mar'yanovich, A.T., Balandin, V.S., Bekuzalov, A.K., and Lapikov, G.M., 1984, Individual features of responses to a combination of heat and physical exertion. Human Physiology, 10, 49-55.

Vogt JJ, Libert JP, Candas V, Duall F, Mairiaux Ph. Heart rate and spontaneous work-rest cycles during exposure to heat. Ergonomics 26,1173-1185.

Wyndham, C.H., Strydan, N.B., van Rensburg, A.J., Barrade, A.J.S., and Heyns, A.J., Relation between VO_2 max and body temperature in hot humid air conditions. Journal of Applied Physiology, 29, 45-50.

Wissler, E.H., 1986, Simulation of fluid cooled or heated garments that allows man to function in hostile environments. Chemical Engineering Science, 41, 1689-1698.

Yousef, M.K., Rasmussen, C.T., and Myhre, L.G., 1988, Physiological stresses associated with U.S. Air Force groundcrew activities. USAFSAM-TR-85-61.

Advances in Industrial Ergonomics and Safety II
Edited by Biman Das
Taylor & Francis 1990

HOT WORK ENVIRONMENT AND HUMAN STRAIN - A RELATION PROPOSED

Dr.Kallol Bhattacharya
Rashtriya Chemicals And Fertilizers Ltd.
Bombay-400 074, India

Dr.S.Raja
NITIE, Vihar Lake, Bombay-400 087, India

This article proposes a tentative relation with the age, BSA, HR and the environmental heat in a field study in Indian industry.

INTRODUCTION

Effective utilisation of human resources plays a vital role in the effective organisation of work as a whole leading to improvements in both performance and productivity as well as the health and comfort of human beings in general. The rapid industrialisation, mechanisation and urbanisation cause changes in work structure - technical and social [Abeysekera and Shahnavaz (1987)] - requiring attention by the government personnel. The human effort and the human cost of production have often been not given the importance that it deserves. Man has always been an inseparable component in the work culture and work environment. Thus it is imperative to make the work more human if a balance is to be achieved between productivity and human comfort. To achieve this it is necessary to understand the relationship between man and his environment by application of the engineering science on the one hand and biological science on the other.

For India, the problems are more severe as she has tried to compress almost a century's growth of the European countries within a 40 year time frame. Since independence the country's emphasis has shifted on creation of large core sector industries which account for 61.3% share of national resource at the end of 5th plan period. These industries are basically steel, engineering and textile. That India is a tropical country displaying a varied climatic conditions of severity make the study of human response to such working conditions an important aspect in making the work more human. In view of the growing awareness of the ill effects of physical environment and also the concern for sustained higher industrial productivity a study of the work environment and the physiological strain has been carried out to explore the level of heat prevailing in steel, textile industries and determine the correlation among the environmental factors and physiological factors in the actual industries.

The following can be described as the part objective of the study.

To provide comprehensive data on working condition in selected industry in terms of thermal environmental factors such as temperature, humidity, radiation and wind velocity.

To test the use of heart rate as an easily measurable parameter for predicting physiological strain in terms of work and thermal load.

To relate the physical stresses with the proposed easily measurable parameter in order to establish a set of nomograms to be of value to practising managers for estimating physiological effort and for effecting suitable control procedures.

METHODOLOGY

The method adopted in this study is experimental in nature. This involves use of sophisticated measuring instruments for the various physical parameters of the environment. The intent is to present a generalised account of that part of the methodology followed in the entire series of investigation. The following physical parameters of the environment were studied.

Heat stress - measured in terms of temperature, humidity and air velocity. Equipments used are psychrometer, rotating vane anemometer, Kata thermometer and globe bulb thermometer.

Human response - measured in terms of heart rate with the help of a doctor's stethoscope and a stop watch.

Environmental readings were taken every 15 minutes over a period of 8 hours during a shift and the averages were computed. This was done to compensate the fluctuations that generally occur in the environment both thermal as well as acoustic and to partial out the influence of any undesirable extraneous factors.

The physiological cost of work was evaluated indirectly by studying the heart rate response of the individual worker at work since this parameter is influenced by both physical work and environmental heat load. The observations of heart rate were made during work, rest and recovery. The subjects were asked to take rest for a period of half an hour before the beginning of the experiment and the resting heart rate was recorded. Heart rate was also recorded just after the termination of work period. Recovery pulse count was taken after the cessation of work every alternate minute upto 15 minutes.

WORK AREA

This experiment has been carried out in the following industry:

Two textile indstries employing around 4000 people and

One steel industry employing more than 25000 people.

The result of the survey of work environment in these industries is summarised in Table 1.

Table 1. Mean thermal conditions in different sections of industries

Industry/Section	Mean DB°C	Mean WB°C	Mean GT°C	Mean Wv cm/sec	Mean WBGT °C	Mean CET °C
Coke Oven (Steel)	38.5	32	-	102	32.9	32.7
Furnace (Steel)	43.2	36.3	57.7	126	41.2	29.1
Structural finishing (Steel)	33.8	29.5	-	90	30.1	30.1
Textile spinning	32.1	29.9	-	28	20	30.5
Weaving (textile)	32.2	29.4	-	25	29.8	29.7

The environment in textile industry in regard to heat stress can be described as warm humid. Radiative heat is absent in this section. The heat stress index as evident from the table is maximum in the furnace section. Though coke oven falls next to the furnace there is a difference of more than 9°C in terms of WBGT. Similarly the difference between the dry bulb temperature and the globe temperature of environment in the furnace section is more than 14°C indicating the presence of high radiative heat. In a similar study Sensarma (1985) observed the presence of high differential between the globe bulb and dry bulb temperature readings. The difference is as high as 22.7°C to 43.8°C in the steel plant. It is thus apparent that the workers in furnace section are exposed to high level of heat and thus it can be reasonably assumed that this extra heat coupled with physical work may affect the worker whose physiological response mainly HEART RATE has been examined here in this experiment.

RESULTS AND ANALYSIS

Physiological response in terms of heart rate of the workers in different sections/industries have been shown as the mean heart rate during resting, after 5 minutes of work and after 4 hours of work (Table 2).

Table 2. Mean and standard deviation of heart rate of employees in different industry.

Industry/ Section	Resting Mean ± S.D bpm	After 5 min. of work Mean ± S.D bpm	After 4 hrs. of work Mean ± S.D bpm
STEEL INDUSTRY			
Coke Oven Battery	72.6 ± 8.4	110.6 ± 7.78	141 ± 12.37
Furnace	71.05± 5.69	118.5 ± 7.23	152.7± 8.21
Structural finishing	72.58± 5.69	83.5 ± 6.27	103 ± 6.29
TEXTILE INDUSTRY			
Weaving	73.44± 6.92	91.55± 7.40	111.77±9.84
Spinning	72.42± 5.23	85.47± 5.99	105.68±6.19

The period 5 minutes and 4 hours are adopted here because this indicates the initial and end condition of a work period in shopfloor where workers break for lunch after 4 hours of work. In addition to the heart rate the workers' anthropometric data like height, weight, body surface area were also collected.

REGRESSION EQUATION

Two separate sets of regression equations have been developed. In one set heart rate is taken as dependent variable and age of the employee and WBGT index are taken as independent variables. In the other set the variable 'AGE' is replaced by the BSA (Body Surface Area) of the workers. The objective here is to develop certain predictive equations which require no sophisticated measurement or analysis but is capable of providing the practising managers certain information on the basis of these simple parameters so that they can initiate some suitable corrective action. It may further be emphasised here that 'AGE' as a variable is subjected to easy recognition.

Some of the equations developed with heart rate after 5 minutes of work as dependent variable are

$$Y=49.25+0.36X_1+0.91X_2 \text{ (R=0.09 for only textile)} \quad (1)$$

$$Y=7.60+0.18X_1+2.55X_2 \text{ (R=0.66 for only steel)} \quad (2)$$

$$Y=6.93+0.34X_1+2.40X_2 \text{ (R=0.63 for pooled data of both industries)} \quad (3)$$

Similarly with heart rate after 4 hours of work as dependent variable the following equations have been developed.

$$Y=16.26+0.69X_1+2.33X_2 \quad (R=0.28 \text{ for textile}) \qquad (4)$$

$$Y=11.45+0.57X_1+2.95X_2 \quad (R=0.54 \text{ for steel only}) \qquad (5)$$

$$Y=-6.94+0.49X_1+3.45X_2 \quad (R=0.59 \text{ for pooled data}) \qquad (6)$$

Here X_1 represents age of the worker in years, X_2 represents the WBGT°C index of the work environment and Y is the heart rate in bpm.

After replacing the age as variable by body surface area (BSA) another set of equations have been developed. They are represented by

Heart rate after 5 minutes of work

$$Y=27.26-9.34X_1+2.6X_2 \quad (R = 0.64 \text{ for steel industry}) \qquad (7)$$

$$Y=-3.97+8.61X_1+2.66X_2 \quad (R = 0.63 \text{ for pooled data of steel and textile industry}) \qquad (8)$$

Similarly heart rate after 4 hours of work is given by

$$Y=46.91-14.88X_1+3.19X_2 \quad (R=0.5 \text{ for steel industry}) \qquad (9)$$

$$Y=2.081-2.433X_1+3.81X_2 \quad (R = 0.61 \text{ for pooled data for steel and textile industry}) \qquad (10)$$

Here X_1 represents the BSA in M^2 and X_2 represents the WBGT°C

NOMOGRAMS

Fig.1 to Fig.4 shows the nomograms created with the regression equations developed earlier. Two sets of twety nomograms are developed here. In one set 'age' is treated as an independent variable. In the other set it is replaced by BSA. The purpose of these nomograms is that the practising managers can use any of them to suit their need. Separate nomograms have been developed with data from textile and steel industry. On an experimental basis data from both these industries were pooled and one set of nomograms created with these pooled data. However only 4 nomograms are shown here.

DISCUSSION

Fig.1 shows the heart rate in absolute terms after 4 hours of work in steel industry. Reference can be made to the tables appended earlier. The data used for this nomogram is the combined data for all sections studied in the steel industry.

The nomogram clearly brings out the fact that the higher age is more affected by the increase in WBGT°C index. This observation corroborates the earlier finding where it was found that heart rate of workers in a soap factory varies from 110 to 133 bpm for a standard work load of 300 kg/min. (Sen etal 1964). However for steel industry the heart rate is found to vary from 129 bpm - 161 bpm. For heavy work of 450 kg/min. earlier finding in soap factory was 112 tro 132 bpm. Thus it appears that the job in steel industry is more demanding. However the present finding appears to be in total contrast with a study made in engineering factory in Nagpur (India) where heart rate of worker was not found increasing with heat gain (Saha etal 1973) though sweat loss was found increasing and oral temperature also showed a rise. Datta and Ganguly (1975) observed in a lboratory study that heart rate shot up with increase in relative humidity. They also observed a distinct linear correlation between heart rate and workload. Thus the present study corroborates some of the earlier finding [Sen etal (1964), Datta and Ganguly (1975)]. Fig.2 exhibits the relation between heart rate after 4 hours of work in a steel industry. An increase of heart rate from 88 to 118 bpm for a 30 year old worker for a rise of WBGT ^{o}C index by 10°C is observed. In other words a rise of 3 bpm/^{o}C rise of WBGT^{o}C is observed. This rise is more than the earlier observed in where 23 bpm increase in heart rate took place for a 10°C rise in WBGT index. This extension of work span from 5 minutes to 4 hours has shot the heart rate increase from 2.3 bpm to 3 bpm/^{o}C. This increase in heart rate exceeds the observation of increase of 7 to 10 bpm/10°C rise while carrying load at dry ambient temperature of 20-45°C [Kamon etal (1971)]. However the present observation is somewhat in line with the observation that physiological strain among the workmen was beyond the permissible limit of safe work [Sensarma etal (1985)]/ The task when interpersed with a sufficient pause successfully keep the physiological strain within the permissible limit [Sensarma etal (1984)].

These observations are quite close to the study data collected for harvesting workers in south eastern America where the heart rate was found to increase by 42.5 to 69.2% in the WBGT index ranging from 20 to 30.4 ^{o}C [Smith etal (1985)]. The increase in heart rate with increase in ambient temperature was also observed in studies in climatic chamber [James etal (1984)] where heat level ranged from 25°C to 43.3^{o}C. The difference between our observation and that of Smith could be due to the difference in physiological features of the subjects. With the combined data from textile as well as steel industry the increase in heart rate for a 30 year old man can be predicted to be 25 bpm for an increase in WBGT from 30°C to 40°C i.e. 10°C rise for 5 minutes of work. For the same data if the work span increases to 4 hours and more the heart rate increase is from 40 to nearly 80 bpm. The suggested level of increase in heart rate is 45 bpm (Murrale) and in that respect the present finding suggests some limit on he basis of age, BSA and the WBGT index. However the present finding of increase in heart rate is yet more than what was predicted earlier i.e. 1 bpm/^{o}C [Meese etal (1984), Kamon etal (1983)] above 25°C. Kenney suggests a physiological limit of PL

(Min) = 83 - 0.53 (HR 5 P) and this in conjunction with the nomograms set of 5 minutes can be used for determining the working time limit for individual worker.

REFERENCE

Abeysekera, J.D.A. and Shahnavaz, H., (1987) - Ergonomics of Technology Transfer - Int. J. of Industrial Ergonomics Vol.1, PP 265-272

Dutta, S.P., Ganguly, T., (1975)- Physiological cost of work under hot dry climates in Indian industries - Int. Jr. of Prod. Res. Vol. 13, PP 429

Kamon, E., and Belding, H.S., (1971)- The physiological costs of carrying loades in temperate and hot environments, Human factors, Vol.13, PP 153-161

Kenney etal (1986)- A simple exercise test for the prediction of relative heat tolerance - Am. Ind. Hyg. Ass. Jr. Vol. 47(4) PP 205-206

Meese, G.B., Kok, R., Lewis, M.I., and Wyon, D.P., (984) - A laboratory study of the effects of moderate thermal stress on the performance of factory workers - Ergonomics Vol.27(1) PP 29-43

Ramsey, J.D., (1985) - Practical evaluation of hot working areas - In Proceedings, ASSE region III Professional Conference Texas A&M University, College station, TX.

Saha, P.N., and Banerjee, P.K., (1973) -Physiological effects of increased air movement in a hot dry environment. Part II further studies in Vidarbha region - Ind. Phys. Div. Rep. No.14, Central Labour Institute, Bombay

Sen, R.N., Saha, P.N., Subramanian, A., (1964) - Assessment of workload and thermal stress in relation to physiological response of workers in a soap factory in Bombay, Ind. Physio. Div. Report No.4 Central Labour Institute, Bombay

Sensarma, S.K., and Dhende, K.N., (1985) - Heat stress study in an iron and steel plant in southern India - Ind. Phys. Div. Rep. No.26, CLI, Bombay.

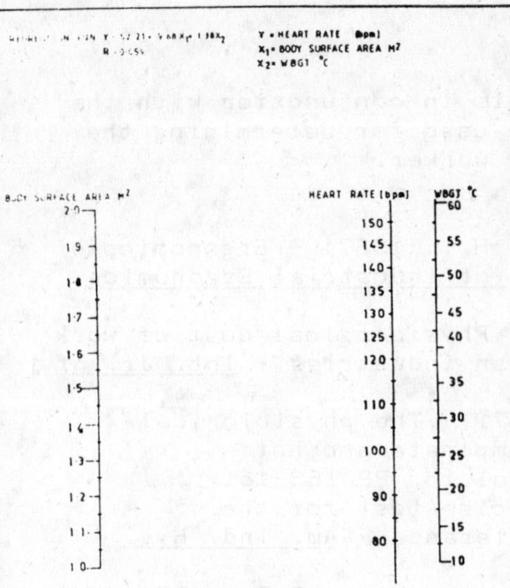

ESTIMATION OF HEART RATE AFTER 4 HOURS OF WORK FROM BODY SURFACE AREA AND WBGT °C (TEXTILE INDUSTRY)

FIG.1

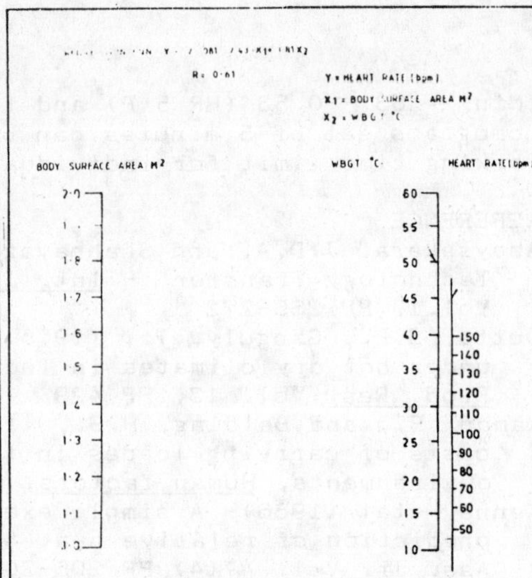

ESTIMATION OF HEART RATE AFTER 4 HOURS OF WORK FROM BODY SURFACE AREA AND WBGT °C (TEXTILE AND STEEL INDUSTRY)

FIG.2

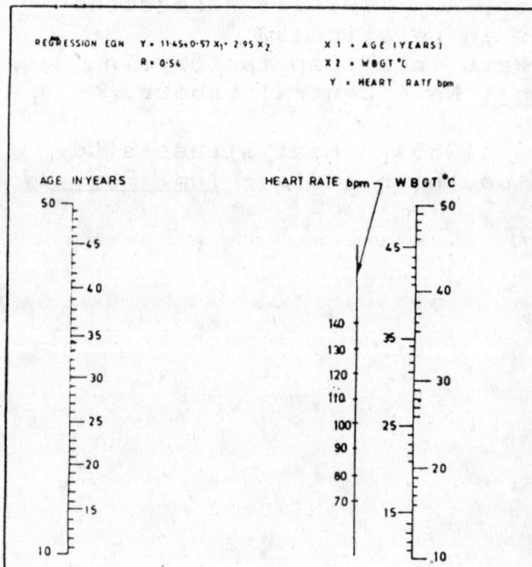

RELATIONSHIP OF HEART RATE AFTER 4 HOURS OF WORK WITH AGE AND WBGT (STEEL INDUSTRY)

FIG.3

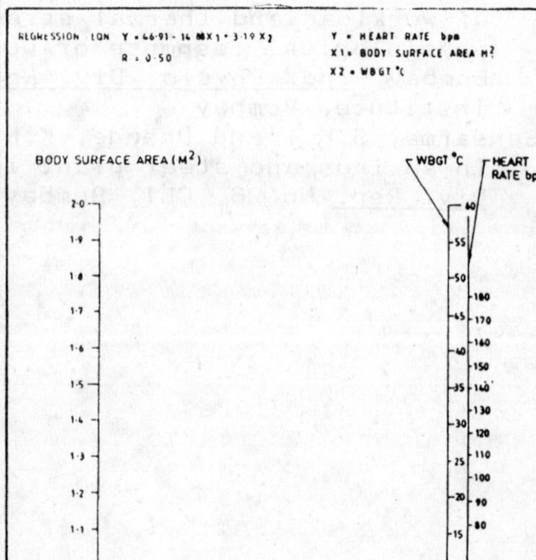

ESTIMATION OF HEART RATE AFTER 4 HOURES OF WORK FROM BODY SURFACE AREA AND WBGT °C (STEEL INDUSTRY)

FIG.- 4

Advances in Industrial Ergonomics and Safety II
Edited by Biman Das
Taylor & Francis 1990

KEEPING BARE HANDS WARM WITH EXTRA CLOTHING ON THE BODY

Nancy Sundheim and Stephan Konz

Institute of Environmental Research
Kansas State University
Manhattan, KS 66506 USA

Experiment 1 had 11 males do seated light assembly work (1.3 mets) for 120 min in both a 15.6 C environment (wearing 2.0 clo) and an 18.3 C environment (wearing 1.6 clo). Hand skin temperature (HST) dropped about .05 C/min from an initial 30 C. Experiment 2 had 16 males sit quietly (wearing 2.0 clo) in temperatures of 7.2, 10.0, 12.8 and 15.6 C. Once it began declining, HST dropped at about .13 C/min. Thus the manipulative work of experiment 1 was of considerable benefit. Assuming dexterity doesn't decline until HST reaches 12-16 C, over-clothing the torso is a feasible strategy for cool environments.

INTRODUCTION

There are a number of jobs requiring the hands to be exposed to cool or cold environments. The cool environment can be cool because of the product (dairy, fish, meat) or because of the weather (fishing, farming, construction, forestry work). Although the hands can be protected thermally with gloves, gloves tend to reduce dexterity.

An alternative strategy is to leave the hands bare but add extra clothing to the remainder of the body--typically the torso. The extra clothing will reduce heat loss on the remainder of the body. Thus the body will remain in thermal neutrality longer and vasoconstriction will be reduced. With less vasoconstriction, more warm central blood will flow to the hands, keeping them warm.

There have been a number of studies investigating the effect of cold temperatures on motor performance of the hand. Sundheim's dissertation (1988) has 63 references. Some selected references are Bartlett and Gronow (1952), Mills (1953), Dusek (1957), Teichner (1957, 1958), Riley and Allison (1979), Rogers and Noddin (1984), Riley and Cochran (1984), Enander (1987) and Oleson et al. (1988).

Fox (1967) summarized the literature and concluded that for tactile sensitivity the critical Hand Skin Temperature (HST) is near 8 C; for manual dexterity the critical HST is between 12 and 16 C. See also studies by Meese and Kok and their colleagues (1982, 1984, 1986, 1989) and Enander (1984).

We did two experiments.

EXPERIMENT ONE

Task

The subject assembled a nut on a 50 mm long bolt. After completing 10 bolt assemblies, the subject removed the nut, placing nut and bolt back in a bin. Then they repeated the assembly. Estimated metabolic rate was 1.3 mets.

Subjects

The 11 subjects were male university students, volunteering for extra credit in a course.

Procedure

HST was measured every 5 min with thermistors at four locations on each hand: back center of hand, back of thumb, middle and index finger. The subjects also voted every 10 minutes on their thermal comfort (1-7 scale; 4=neutral) and thermal sensation (1-7 scale; 4=neutral) for their hands and for the body overall.

Each subject was in two environments for 120 minutes. One environment was 15.6 C with 2.0 clo; the other was 18.3 C with 1.6 clo. The clothing and environment were deliberately confounded; the extra .4 clo in the colder environment should have made the two environments equal (if our calculations were correct).

In both environments, subjects wore: a flannel, long-sleeve, collared shirt, briefs, denim trousers, insulated coveralls, socks, shoes and a hat. In addition, for the 1.6 clo, they wore a medium weight T-shirt; for the 2.0 clo, they wore long underwear (top and bottom) and a long-sleeve cardigan sweater.

Results

Figure 1 shows mean HST for the two environments. The decline vs. time is quite linear and the variability of subjects is quite large. HST declined .041 C/min in the 15.6 C (2 clo) environment; $r^2 = .98$. HST declined .055 C/min in the 18.3 C (1.6 clo) environment; $r^2 = .98$. These rates are not statistically significantly different. The mean rate is (.041 + .055)/2 = .048 =.05 C/min. (Cubic and quadratic fits did not give a significantly better fit than the straight line.)

Correlations were calculated between HST (every 10 minutes) and thermal comfort and thermal sensation. Only a few subjects had r^2 values over 50%. The correlation was better for thermal sensation of the hands and HST than thermal comfort of the hands and HST. The correlation was higher for the colder (15.6 C) environment.

Discussion

Although the straight lines had high coefficients of determination ($r^2 = .98$), they were fitted to the subject means. Individuals naturally had considerable variability about the mean. In addition, industrial work probably wouldn't be at exactly 15.6 C or 18.3 C. Real workers also would not have exactly 2.0 or 1.6 clo clothing and their metabolic rates also would vary from the 1.3 mets of this experiment. Thus the .05 C/min decline in HST should be considered indicative rather than exact for a specific industrial situation.

Fox (1967), summarizing the literature, reported hand dexterity declines when HST is below 12-16 C. If 16 C is the criterion and initial HST is assumed at 30 C, then it would take 30-16=14/.05 = 280 min before dexterity is affected. If 12 C is the criterion, then it would take 30-12=18/.05 = 360 min before dexterity declined. Since the typical worker works 2 hours until a break, the overclothing strategy seems quite feasible.

Enander (1982) had reported a high relation (r^2 = 81%) for HST and thermal sensation but her temperatures were colder (0, 7 and 15 C) than this experiment. Thus it seems feasible that workers can sense how cold their hands are although the sensing will not be perfect--especially for intermediate (say 15-19 C) temperatures.

EXPERIMENT 2

In experiment 2, a more unusual situation was explored. The subjects sat quietly in the cold rather than working. This eliminated the confounding variable of manipulative work. The 16 male subjects, age 18 to 30, were exposed for 90 minutes to temperatures of 7.2, 10, 12.8 and 15.6 C. They wore the 2.0 clo ensemble of experiment 1. As in experiment 1, HST and thermal sensation were measured. In addition, skin temperature under the clothing was measured at 4 locations: chest, back, calf and upper arm. This was summarized by weighting the locations by 25%, 25%, 36% and 14%. Temperatures were measured every 15 minutes and thermal sensation at time 0 and 90.

Results

As expected, HST declined with time. See Figure 2. However, a single curve shape does not seem appropriate. For the two warmer temperatures (but not the two colder temperatures), there is an initial time during which HST does not decline. Once the decline started, it seems to be at approximately the same rate for all four environments (.13 C/min). The period before the decline began was 0 minutes for the two colder temperatures, 15 min for the 12.8 environment and 30 minutes for 15.6 C. Thus the overclothing strategy worked again but not as well. The .13 C/min decline (vs. the .05 C in experiment 1) implies that the manipulative work done in experiment 1 is "worth" about .08 C/min in these environments.

The final body skin temperature (BST) was a function of environmental temperature with BST = 33.0, 32.6, 32.1 and 31.0 C for the four environments of 7.2, 10, 12.8 and 15.6 C. However the effect of time was not significant; that is, BST started lower when the environmental temperature was lower. It is not known whether this is a measurement error (i.e., the sensor was measuring not only body temperature but also environmental temperature) or a rapid initial response of skin temperature at time 0 (which, due to time delays in getting the subjects settled in the chamber, is about 1 minute after they entered the chamber).

Thermal sensation of the hands (3.3, 2.7, 2.4 and 1.7 for 7.2, 10, 12.8, and 15.6 C) was related to environmental temperature. However, thermal sensation was significantly related (r^2 about 40%) only in the 12.8 and 15.6 C environments.

Discussion

In summary, in "cool" environments, overclothing the torso reduces the need for gloves. The strategy works better when the fingers are active as HST dropped about .05 C/min with active fingers and .13 C/min with inactive fingers.

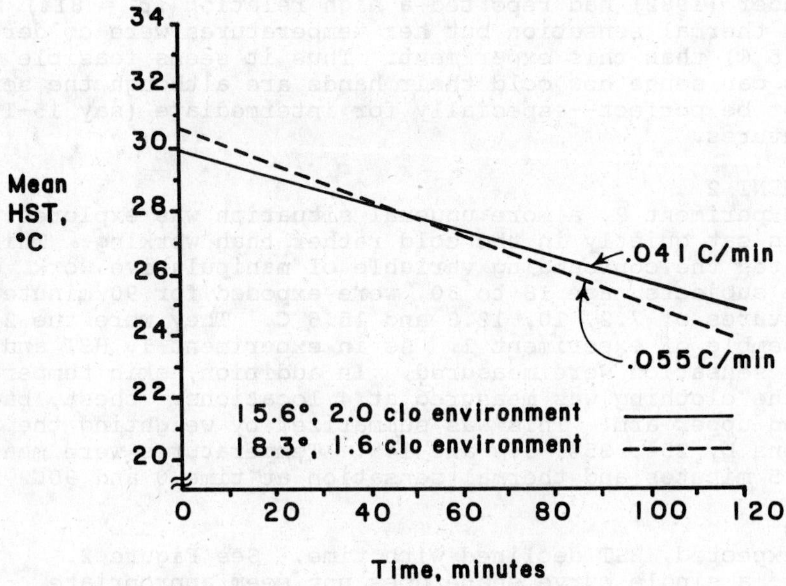

Fig. 1. Hand skin temperature (HST) declined approximately .05 C/min for bare hands doing assembly work, if the torso is overclothed. The variance was quite large; after 30 min, the standard deviation at any time was 3.0 to 4.0 C.

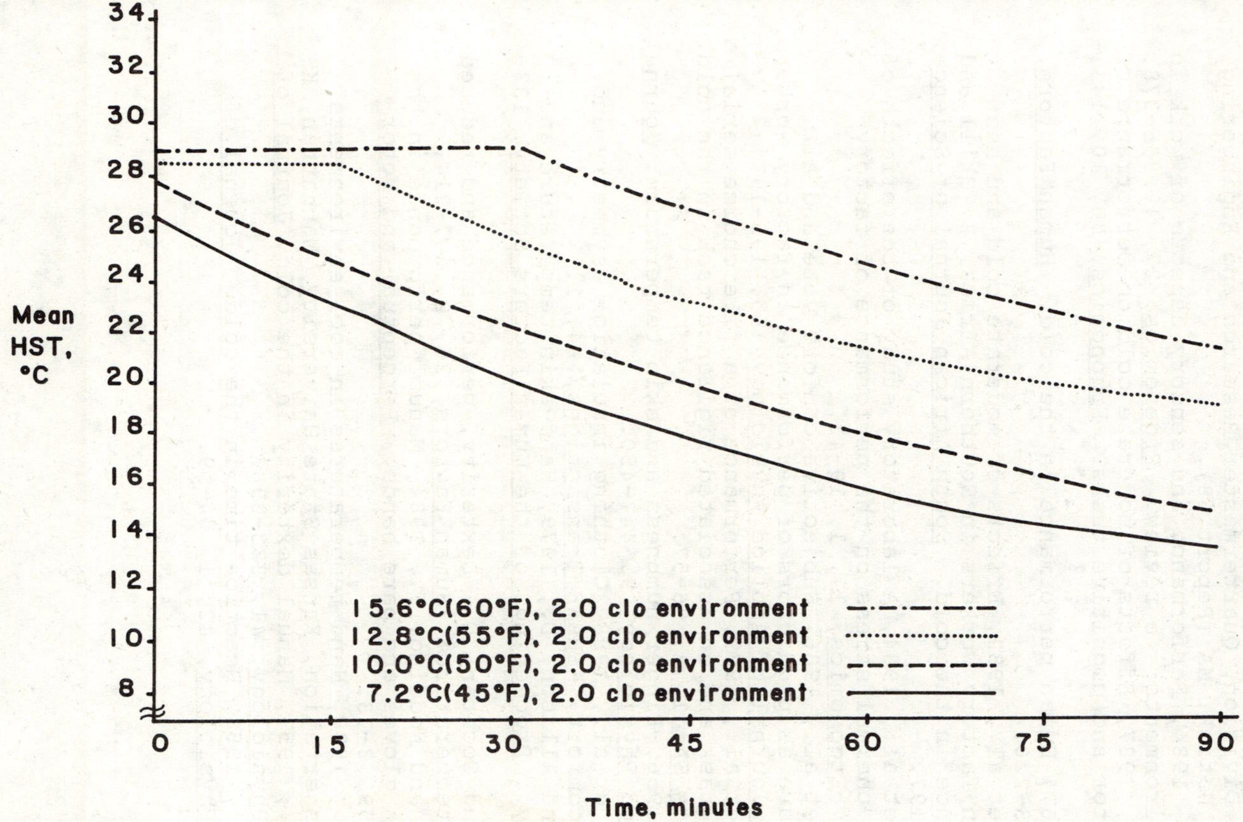

Fig. 2. Hand skin temperature (HST) declined approximately .13 C/min for idle bare hands, if the torso is overclothed. Note the delay before the decline begins when the environment is warmer.

REFERENCES

Bartlett, D. and Gronow, D., 1952, Manual dexterity and tactile sensitivity in the cold, Flying Personnel Research Committee, Farnborough, England (Report 806).

Dusek, E., 1957, Manual performance and finger temperature as a function of ambient temperature, Environmental Protection Research Division, Quartermaster Research and Engineering Command, Natick, MA (Report 68).

Enander, A., 1984, Performance and sensory aspects of work in cold environments: a review. Ergonomics, 27, 4, 365-378.

Enander, A., 1987, Effects of moderate cold on performance of psychomotor and cognitive tasks. Ergonomics, 30, 10, 1431-1445.

Fox, W., 1967, Human performance in the cold. Human Factors, 9, 3, 203-220.

Meese, G., et al., 1982, Effects of moderate cold and heat stress on factory workers in Southern Africa. 2. Skill and performance in the cold. South African Journal of Science, 78, 189-197.

Meese, G., et al., 1984, A laboratory study of the effects of moderate thermal stress on the performance of factory workers. Ergonomics, 17, 1, 19-43.

Meese, G., et al., 1986, Subjective comfort vote and air temperature as predictors of performance in factory workers. European Journal of Applied Physiology, 55, 195-197.

Meese, G., et al., 1989, Performance of a five choice serial reaction task and an associated vigilance task in the cold. Ergonomics SA, 1, 2, 56-57.

Mills, A., 1956, Finger numbness and skin temperature. Journal of Applied Physiology, 9, 447-450.

Oleson, B., et al., 1988, Clothing insulation asymmetry and thermal comfort, ASHRAE Transactions, 94, 1.

Riley, M. and Allison, D., 1979, Hand-skin temperature and dexterity. Proceedings of the Human Factors Society, 183-187.

Riley, M. and Cochran, D. Dexterity, performance and reduced ambient temperature. Human Factors, 26, 2, 207-214.

Rodgers, W. and Nodding, E., 1984, Manual performance in the cold with gloves and bare hands. Perceptual and Motor Skills, 59, 3-13.

Sundheim, N., 1988, Hand temperatures in cool environments. Ph.D. Dissertation, Kansas State University, Manhattan, KS.

Teichner, W., 1957, Manual dexterity in the cold. Journal of Applied Physiology, 41, 333-338.

Teichner, W., 1958, Reaction time in the cold. Journal of Applied Physiology, 42, 1, 54-59.

Advances in Industrial Ergonomics and Safety II
Edited by Biman Das
Taylor & Francis 1990

YOUNG METAL WORKERS' EXPECTATIONS CONCERNING THEIR JOBS AND WORKING ENVIRONMENT

Jorma Saari
School of Occupational Health
McGill University
1130 Pine Avenue West
Montreal, PQ, Canada H3A 1A3

and

Marketta Kivistö
Institute of Occupational Health
Laajaniityntie 1
SF-01620 Vantaa, Finland

A questionnaire was answered by 357 workers under 25 years of age, and 448 workers between 40 and 44 years of age in twelve plants manufacturing machines in Finland. Also 493 line managers answered a basically similar questionnaire. The purpose was to identify those factors in the jobs and the working environment that impair the competitiveness of the metal industry in attracting the labour force. The Young workers felt that their basic education and vocational training were better than needed in their present jobs. The tasks were felt to be too simple, and they did not offer enough mental challenge. The young workers perceived the psycho-social environment to be more positive than the middle-aged workers. Both groups felt that major improvements were needed in the physical conditions.The line managers experienced all aspects of the work to be more positive than did the workers. This aspect may therefore form an obstacle to improvements.

INTRODUCTION

The number of young people entering the work force is decreasing in Finland and in many industrialized countries. E. g. in the USA, the median age is now 33 years, at the beginning of the '90s. It will inch up to 36.5 years by the year 2000 and to 39 by 2010. Various business sectors are compelled to compete harder in recruiting young workers. In Finland, the metal industry has not been very successfull during the eighties in appealing to young people entering working life. The number of applicants to the metal working lines of vocational schools has gone down steadily during the past ten years. Consequently, the median age of workers in the metalindustry has risen from 30 in 1972 to 37.5 in 1989.

The Employers` Association has arranged extensive information campaigns in order to get young people interested in jobs in the metal industry. Companies have invited school classes for visits, and offered summer jobs to the young. The information campaigns have not yielded sufficient results to ensure the personnel resources required by the metal industry. Obviously,the jobs and working conditions in the metal industry are not sufficiently competitive in the minds of the young people. The Employers' Association and the Metal Workers' Union therefore asked the Finnish Institute of Occupational Health to carry out a study on the expectations that young people have concerning jobs and the working environment.

METHODS

A questionnaire was filled out by 357 workers under 25 years of age, and 448 workers aged 40 - 44 years. They were employed by 12 plants manufacturing machines in Finland. Also 493 line managers in the same plants filled out a basically similar questionnaire. The response rate was 75% among the young workers, 85% among the middle-aged workers, and 73.5% among the managers.

The study was restricted to companies manufacturing machines and located within the square formed by the four biggest towns in southern Finland (Helsinki, Turku, Tampere, Lahti).The companies that were the sole employers in their region were excluded. The remaining companies employing 100...500 workers were accepted for further consideration. Most of the companies employed less than 200 people. To ensure good communication between us and the companies, we visited each one at least twice. In order to keep the number of participating companies small enough, we included some larger companies as well.

The turn-over rate of each of the remaining companies was determined for the fourth quarter of 1986 and 1987. From the remaining companies pairs of companies were formed in which one company of the pair had a low turn-over rate in both years and the other one a high rate, and in which both companies manufactured similar products. In this way six pairs of companies were found.

All the companies that were asked decided to participate. The Employers` Association and both the Workers' Union and the Technical Employees' Union informed their respective contact persons in the companies about the study. All the companies were visited before the study began, and the different personnel groups were informed. At the meetings, the problem under study was discussed, the outline of the questionnaire was presented, and the participants of the meetings were encouraged to express their views about the problem.

The questionnaire was finalized after the first visits to the companies, and it covered the topics given in Fig. 1. The companies arranged one-hour sessions for the workers who met the age criteria to fill out the questionnaire during working hours. The respondents were given envelopes for mailing the questionnaires directly to us. The managers filled out the questionnaires when they had time, they were also given envelopes.

The participating companies are very competitive in their respective market areas. Nine of the 12 companies hold a major share of the world market in their respective products. The remaining three companies also manufacture technologically advanced products. The production technology varied a great deal among the companies. Some companies had invested quite a lot on the most modern technology, such as robots, machining centers and flexible manufacturing systems(FMS) whereas others relied on older technology. Also the age of the buildings occupied by the companies varied from several decades to a few years.

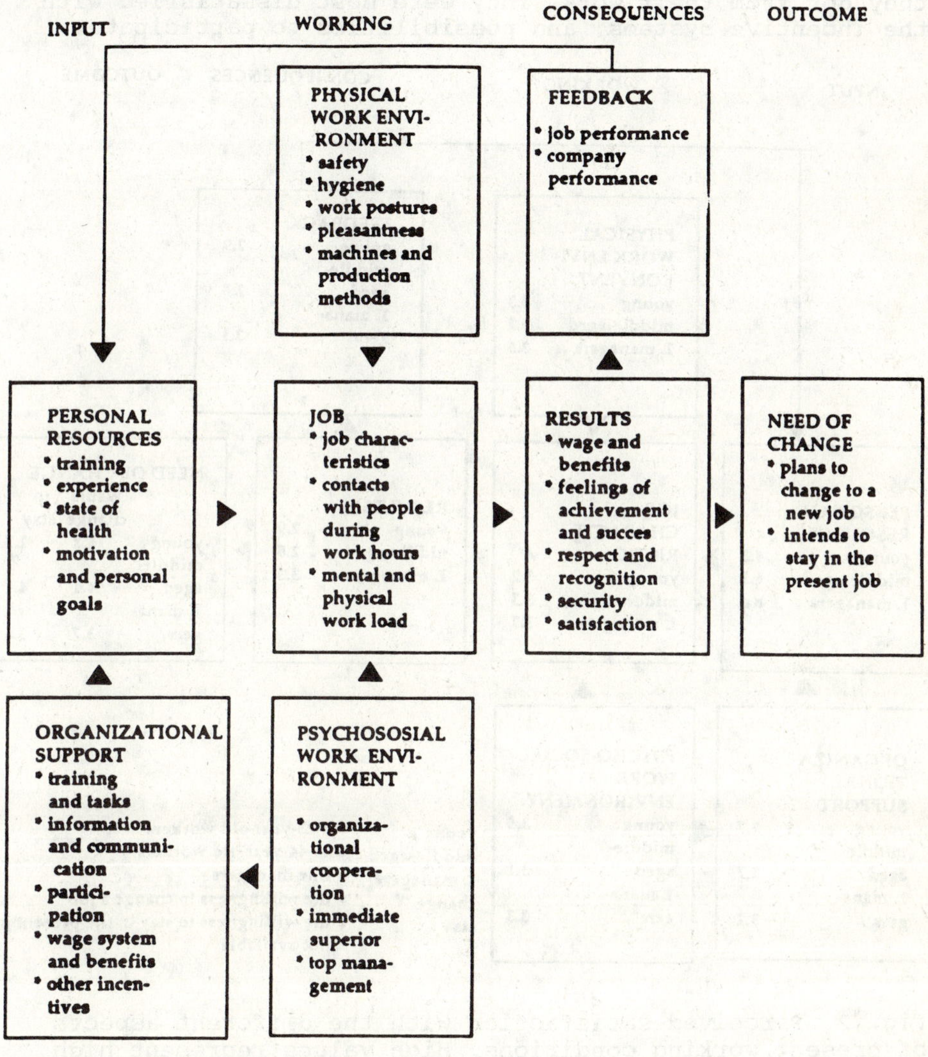

Fig. 1. The topics covered by the guestionnaire

RESULTS

Satisfaction with the present working conditions

Fig. 2 presents the respondents' overall satisfaction with the different aspects of work. Young workers were more satisfied with their personal resources, such as training, and state of health, than with any other component of the work situation. Input, or their personal resources, outweighs output, or the results such as wage and benefits, feelings of success and achievement, and personal advancement. Consequently, the young are more willing to change jobs.

Young workers evaluated the psychosocial environment, their job, and physical environment as being more satisfactory than the organizational support, feedback, and the results they got from their work. They were most dissatisfied with the incentive systems, and possibilities to participate.

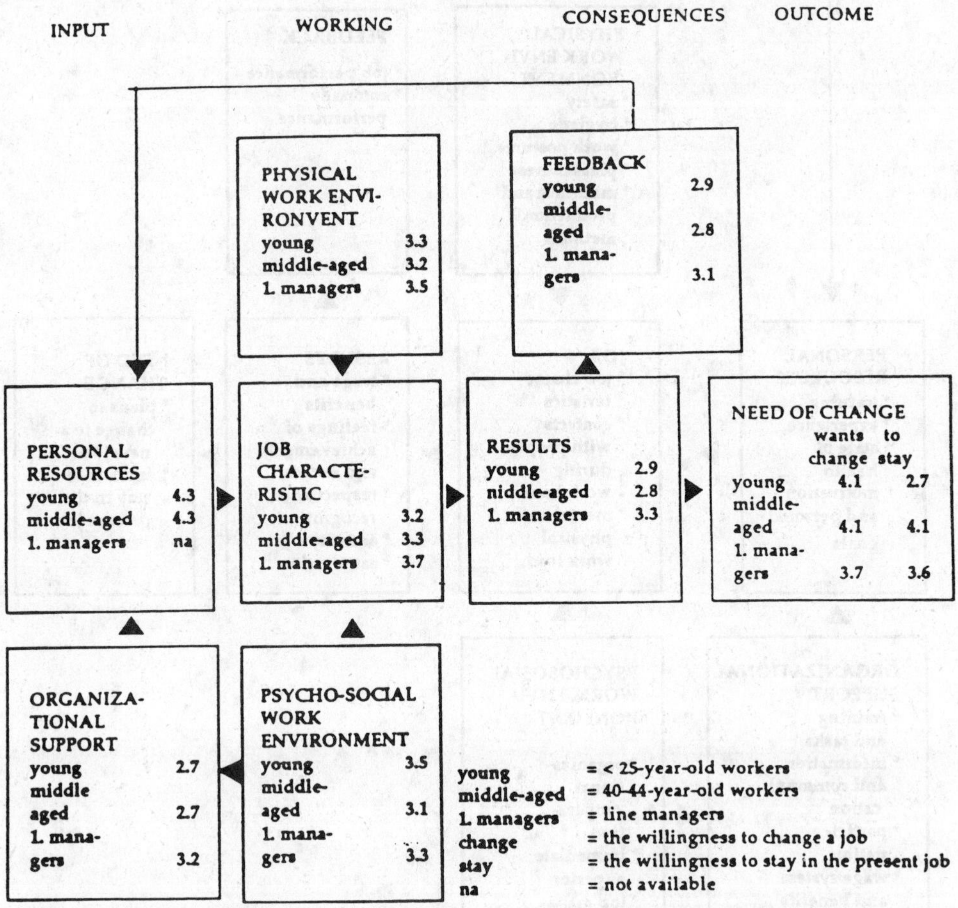

Fig. 2. Perceived satisfaction with the different aspects of present working conditions. High values represent high satisfaction (scale 1-5). The underlined values differ significantly ($p < 0.05$) from the corresponding value of young workers.

The young thought that the incentive systems, such as wages, management response to job performance, are not sufficiently sensitive to the quality of job performance. They felt that efforts are not rewarded sufficiently. Young workers would like to participate especially in the arrangements and decisions concerning their work environment, occupational safety and health, and the timing and order of various tasks.

The middle-aged and young respondents differed significantly ($p<0.05$) only in two respects: (a) The young experienced the psychosocial work environment more positively than the middle-aged, and (b) the middle-aged were more willing to stay in their present job.

The linemanagers differed from both worker groups in all respects but one: they were more satisfied with the psychosocial work environment than the middle-aged workers, but less satisfied than the young workers. The line managers were significantly more satisfied than the workers, and displayed greater willingness to stay in their present company.

The most important characteristics of work conditions.

The young considered a good physical working environment very important, and they would have liked to have improvements especially in occupational hygiene, in safety, and in the pleasantness of working quarters (Table 1). They appreciated jobs that provide variety, interest and independence. Fellow workers are important to the young, as well as an encouraging organizational climate. They wanted to have contacts to other people during working hours, and preferred working in teams rather than alone. They did not want contacts with clients.

The expectations of middle-aged workers and line managers go well together. The major difference is that the middle-aged put much more weight on the quality of management than did the young. Young workers are more individually oriented than middle-aged workers and managers. They were most concerned about their personal working environment, advancement, and success. The other groups were much more concerned than the young workers with the organizational performance.

TABLE 1. The most important aspects of work.(A variable was included if at least one of the worker groups had an average of 4.5 or more.) Scale 1-5.

Aspect of work	Workers <25	aged 40-44	Line managers
PHYSICAL WORKING ENVIRONMENT			
* clean air and ventilation	4.5	4.6	4.2
* safety of machines	4.4	4.6	n.a.
* low noise level	4.4	4.6	4.2
* the condition of tools	4.4	4.5	4.3
* the pleasantness of work environment	4.4	4.5	4.1
* the condition of cloakrooms	4.2	4.5	4.0
* good heating system	4.3	4.5	4.1
JOB			
Job characteristics			
* interesting job	4.5	4.4	4.5

* variety	4.5	4.4	4.3
* independence and autonomy	4.4	4.5	4.3
PSYCHOSOCIAL WORK ENVIRONMENT			
Organizational climate and cooperation			
* fellow workers are pleasant	4.7	4.6	4.1
* organizational climate is not depressing	4.5	4.7	4.5
* closest fellow workers cooperate	4.5	4.4	4.4
Management and managerial practices			
* machines are modern	4.5	4.7	4.5
* products are modern and competitive	4.1	4.7	4.7
* management improves working conditions	4.4	4.6	4.3
* managers keep their professional skill up to date	4.3	4.5	4.5
* the company is well managed	3.9	4.5	4.7
Personal relations			
* it is possible to chat with others while working	4.5	4.5	4.2
RESULTS FROM WORK			
Wage and benefits			
* high wage	4.7	4.6	4.2

DISCUSSION

The young workers were fairly satisfied with their jobs and working conditions. During the first few years they found their jobs interesting, and the personal relations with their fellow workers and foremen enriched their social network. Already in the beginning, however, they felt an imbalance between their input into work (i.e knowledge and skills, and their fairly good basic education and occupational training) and the output that work offers (i.e. various results, such as wages, personal advancement and development). The young workers wanted more possibilities to learn, more mental challenge, and faster career advancement. They considered their jobs too simple, and not rewarding enough.

There is a big risk, that the positive beginning turns into a negative experience after a few years at work. Companies do not seem to be able to offer enough challenge to the young workers to keep up their motivation. After a few years at work too many of them find that they already have experienced everything that work can offer and they still have 30..35 years ahead before retirement. Disappointment and mental exhaustion are to be expected.

Since the line managers have fewer reasons to complain about their working conditions, they may become an obstacle for improvements. The general pattern of opinions is quite similar among all groups, however, but the line managers consider the need of improvements as less necessary than the workers.

Advances in Industrial Ergonomics and Safety II
Edited by Biman Das
Taylor & Francis 1990

INDUSTRIAL SAFETY : JOB ENERGY DEMAND VS WORKER LIMITATIONS

Dr Robert W. Hosein

Department of Mechanical Engineering
Laval University
Quebec, Quebec
Canada G1K 7P4

Musculoskeletal incapacities, arising out of the neglect of occupational biomechanics in matching job and worker, have been singled out as the primary reason for the current level of industrial hazard. However, this paper will focus on other factors cited as contributory to this situation and include mismatching of job and worker physiological capacity, health and physical fitness index; problems of job environmental conditions, man-machine integration, adequacy of rest pauses, as well as mental load from job related variables and stress from both job related and external origins.

INTRODUCTION

The ability of a worker to safely and comfortably perform an assigned task is dependent not only on the worker characteristics (i.e. body structure and anthropometrics, capacity of varied muscle groups, cardio-vascular, respiratory and nervous systems, visual acuity, diet and metabolism, motivation etc.) and the task characteristics (i.e. sequence and complexity of work elements, operator or machine time controlled, duration of work etc.); but also on other factors such as the task environment (i.e. temperature, humidity, fumes, noise, vibration etc.), the machine (cycle time, ease of integration of worker, safety features, etc.), tools and accessories (design and suitability to worker characteristics), the arrangement of the work place in so far as it accomodates the limitations of the worker.

The neglect of not matching of worker to job in terms of occupational biomechanics has been identified as a primary villain of musculoskeletal injury and illness and a principal area of occupational hazard in the industrial arena. But biomechanics alone has not been adequate to fully eliminate injury and illness at the work place. Several other factors such as worker physiological capacity determined by health and physical fitness index, metabolism and diet, job environmental conditions, the level of pacing and man-machine integration, adequacy of sleep and night shift work, job difficulty and the duration of working versus the duration and frequency of rest pauses, nervous load due to job related variables, as well as stress due to both job related and from external origins, have been identified as contributing to the number of work accidents. This paper will be devoted primarily to job energy requirement and worker limitations focusing on some of the factors identified above.

JOB ENERGY REQUIREMENT (JER)

The measurement of energy expenditure can be categorized as local muscle activity and gross body activity, and is generally measured in kilocalories (Kcal).

Local muscle activity can be identified by electromyographic readings (EMG) and are useful in determining what work elements are primarily responsible for raising the level of muscle fatigue.

Gross body activity : although there have been several methods postulated over the years for determining physiological demand, no individual method has been able to adequately measure physiological strain.

Oxygen consumption has been one of the earlier methods employed in determining energy requirements (one litre of consumed oxygen corresponds to approximately 5 cals); thus knowing total oxygen consumption renders it very simple to determine the Kilocalorie equivalent. One drawback in this method is attributable to the apparatus employed (size, weight etc.) and care must be taken to ensure that it does not interfere with the task performance by the operator.

One of the earlier studies to predict average metabolic Job Energy Requirement was proposed by Durnin and Passmore (1955). This study involved a coal mining job which was broken into physical elements each with a known predetermined average metabolic rate. The JER for this job was predicted to be 5.94 Kcals/min.

Durnin and Passmore (1955) did present a more detailed list of jobs with JER which ranged from light industrial assembly tasks (about 2 Kcals/min) to brick laying (about 4 Kcals/min), sawing lumber with a manual saw (about 7 Kcals/min), to walking up a flight of stairs with a 10 kilo tool box on shoulder (about 16 Kcals/min).

Heart rate is another method used to evaluate physiological demand since oxygen consumption increases as physiological demand increases, causing the cardio-vascular system to work faster and resulting in a higher heart rate. Furthermore, the apparatus for measuring heart rate is less cumbersome than for oxygen consumption measurement and does not impede task performance by the subject.

Because of stress due to job related variables typical of jobs such as pilots of jet aircrafts, traffic controllers, manning of master panel of a nuclear power plant etc., as well as stress from external origins, the nervous load of a worker can be increased and is usually associated with increased heart rate. Accordingly, the measure of mental load has been utilized in assessing certain jobs.

One of the earlier studies to undertake the evaluation of nervous load at work stations in the production line was done by Hartemann and al (1969). He used two methods : (a) the electro-encephalograph tracing, which utilized the alpha-rythm index; (b) the heart rate.

Despite the limited study, due to the complicated apparatus as well as the level of sensitivity, he concluded that "heart rate increased quite noticeable between morning and evening".

A later study by Tatro et al (1986), showed that pilot performance was affected by no less than 8 critical factors.

Another, more recent study undertaken by Furedy (1987), presents arguments for going beyond heart rate for greater specificity of mental effort. He proposed an added measure of reflecting Sympathetic Nervous System (SNS) which influences nervous load more clearly than does heart rate, which in turn is predominantly influenced by para-sympathetic factors. He concluded that the T-Wave Amplitude Component (TWA) of the Electro-Cardiogram represented a contraversial but promising way to go beyond the heart rate in the cardio psychophysiological assessment of mental effort.

ENVIRONMENTAL FACTORS (TEMPERATURE/HUMIDITY etc)

Accident statistics certainly indicate the significant effects of temperature, humidity, fumes, vibration, etc. on normal healthy workers. Therefore workers with reduced physical work capacity assigned to jobs with adverse environmental factors could result in increased accidents.

Belding and al (1963), Brouha (1960) and others have identified that physiological strain of exposure to hot industrial jobs is reflected in the elevation of heart rate, sweating and the elevation of body core and skin temperatures.

High humidity, and low air speed may limit the amount of heat load which may be disposed of by evaporation. The capacity for producing the necessary sweat in the first instance may also be limited.

Overstrain from heat exposures may result in <u>heat exhaustion</u> (inadequate blood circulation), <u>heat stroke</u> (uncontrolled elevation of body temperature), or <u>heat cramps</u> (excessive loss of NaCl in Sweat).

PACED AND UNPACED WORK AND MAN-MACHINE INTEGRATION

It should be noted that when an operator is allowed to work independently of machine cycle time or what is referred to as unpaced work as found in craftsmanship approach to manufacturing, the operator develops rythm, controls the rate of production, is less stressed, and derives a sense of job satisfaction.

When the operator is constrained by machine cycle time (man-machine integration), or what is referred to as paced work, rythm is destroyed, and the worker tries to accelerate the rate of working which increases both fatigue and errors. The stress level is higher than for unpaced work which leads to an increased heart rate.

WORKER AEROBIC CAPACITY

It is important to know the worker's aerobic capacity as this determines the anaerobic treshold (maximum heart beat) and is utilised in establishing the Safe Physical Work Capacity (SPWC) as well as deciding what job (JER) a worker can be assigned and the duration for safely performing the job. Furthermore, when the JER is heavy, it is necessary to establish adequate rest pauses to permit heart rate recovery.

According to Dr. Tichaeur (1964), it can be said that work is done with the aid of varied muscle groups of the body, and the capacity to work is dependent on the metabolic rate attainable with oxygen present, expressed in litres of oxygen per min or the equivalent Kcals/min; the maximum metabolic rate possible by a person is known as the <u>aerobic capacity</u>.

It should be noted that the maximum ability of the muscle to utilize oxygen and the maximum supply of oxygen to the muscle is a direct function of the condition of the cardiovascular and respiratory system (CVR). In performing work, the metabolism that takes place in the various muscle groups puts a demand or stress on the C.V.R. primarily for oxygen. As a result, blood flow and breathing rate increase, causing a strain on the muscle groups. The total demand for oxygen is dependent on the job, which, if in excess of the workers CVR system capacity, under limited active muscle groups, the oxygen limitation would be dependent on the muscle's oxygen utilisation capacity. When this is low, accumulation of energy liberating enzyme occurs, producing local muscle fatigue in the form of local discomfort, tremor and loss of co-ordination.

If, on the other hand, gross muscle actions are required to perform a given task, it is possible for the demand of oxygen to exceed the capacity of the CVR system, effecting forced contraction of the muscle groups without adequate oxygen present, thereby resulting in local muscle fatigue. When the job requirements places undue demand on the C.V.R. system, particularly if the system of the particular worker has prior limitations overstraining and excessive fatigue to such systems may result.

Thus it is important to ensure that JER should be matched with worker's aerobic capacity to determine SPWC or the level of total muscular activity the worker can safely endure.

Astrand (1960) has shown that great variability existed in the aerobic capacity of different subjects. Bink (1962), utilising such results, presented in graphical form aerobic capacity (Kcals/min/kg of body weight) vs chronological age. From this presentation, it was seen that the maximum aerobic capacity is attained at age 18 and decreases steadily with age; also the aerobic capacity was higher for males than females.

Astrand (1960) also found that for a given worker/task condition, the steady state oxygen uptake and heart rates could be measured. Also the energy expenditure rate (i.e. oxygen uptake) can be plotted against the heart rate. Astrand (1960) further developed a formula which was later refined by Moritz et al (1961) showing the relationship between heart rate attainable by an individual can be calculated as follows:

$$\text{Max Heart rate} = 190 - \left[\begin{matrix}\text{(Age in} \\ \text{years)}\end{matrix} - 25 \right] \times 0.62$$

With this formula it is possible to determine the maximum heart rate for every worker/age situation. Moritz et al (1961) found the formula to have a coefficient of variation of 12.5% for a normal 35 year old man (i.e. 6 < 2 kcals/min).

WORK DAY LIMITATIONS (8 hours/day)

When determining the work day limitation, it is important not to set the JER equal to the aerobic capacity of the worker, but to incorporate a factor of safety. Brouha's (1960) criteria "if the heart rate does not rise significantly after the first five minutes of performance" has been most accepted to determine whether the energy expenditure of a worker performing a given task is safely within the physiological capacity of the worker, and as such will minimize fatigue.

Chaffin (1969) cited Christensen who proposed a rate of less than 50% of the aerobic capacity of the worker to avoid excessive fatigue, but Astrand (1960) tested this rate of working for subjects with aerobic capacity ranging from 11.2 - 2.6 Kcals/min and found that heart rates increased from 120 - 135 beats/min. He further noted the danger in not adjusting JER against aerobic capacity, especially for older subjects with reduced capacities, since such practice could incur an increase of 30 beats/min in the first hour of work, which rapidly approaches the estimated maximum heart beat rate (i.e. 165 beats/min for 55 year old worker).

On the basis of studies conducted by Astrand and Brouha, it can therefore be inferred that by assigning a worker to a job with an assessed JER at 50% of the aerobic capacity of the worker, produces a fatigue generating energy expenditure rate.

Further studies by Bink (1964) indicated that a work load of 33% of the aerobic capacity of the worker as a maximum. From these studies it was also observed that a younger worker was able to safely undertake a job having almost double the JER of the job of an older worker.

WORK DURATION VS PHYSICAL WORK CAPACITY & PHYSICAL FITNESS INDEX

According to Chaffin (1969) there are three limitations for a normal, healthy, 35 year old working man in terms of his physical working capacity as a function of his working time as follows : (a) due to his aerobic capacity, he should work at 16 Kcals/min for a 4 minute period; (b) to reduce fatigue, he should limit his 8 hour performance to an average energy expenditure of 5.3 Kcals/min (33% of aerobic capacity); (c) due to the social and economic restrictions, on supplying an adequate diet, he must limit a 24 hour performance to about 2.5 Kcals/min.

Another measure of the capacity for working is known as the Physical Fitness Index (PFI), which is based on an aerobic capacity of 16 Kcals/min (or that of an average normal, healthy 35 year old man).

$$\text{i.e.} \quad \text{PFI} = \frac{\text{Aerobic capacity}}{\text{16 Kcals/min}}$$

BODY MASS INDEX (BMI)

Another index of physical fitness was proposed by the Quebec Corporation of Dieticians known as the Body Mass Index. This index is proportional to the subject's height in inches and weight in pounds, and is derived from a nomograph.

It is recommended that the BMI should be in the range of 20-27 for persons aged 20-65 (except for athletes and pregnant women). When the BMI exceeds 30, the subject is considered tc be obese, representing a reduced PSI and a danger in the assigning of a high JER rating job.

WORK DURATION VERSUS JER & PSI

If the JER is known, as well as the worker's Physical Work Capacity (PWC), the duration of time the worker is able to safely perform a given task can be calculated in accordance with an empirical model suggested by Bink (1962) for the Average Physical Work Capacity (APWC) and represented as a logarithmic function of work duration as follows :

$$APWC = \frac{\text{Log } 4400 - \text{Log } t}{0.187} = \frac{3.64345 - \text{Log } t}{0.187}$$

Putting PWC = JER and where t = work duration in minutes,

$$\text{Log } t = 3.64345 - \frac{JER}{5.35 \text{ PFI}}$$

Bonger (1962) undertook a study of different workers assigned to different jobs, and calculated the predicted allowable working times based on the above formula. Two situations of job/worker assignments are listed below.

Age	Wt/lbs	PFI	PWC Kcals/min	JER Kcals/min	Effort Rating JER/PWC	Predictable working Time (min)
22	181	1.24	6.48	4.93	0.76	776
31	144	0.91	4.72	7.50	1.59	126

From the above, it is observed that the first worker would be able to safely perform the assigned job over an 8 hour work day (480 mins), where as the second worker would be able to sustain the assigned job for approximately two hours only.

CONTROLLING ENERGY EXPENDITURE

So far, it has been shown that given a job's JER and worker SPWC, the duration the worker can sustain the rate of working could be calculated. It has also been illustrated where a particular worker is unable to sustain the rate of working on a particular job for an eight hour day. It would seem important to know what rate of working could be assumed over an eight hour day and what should be done if this rate of working has to be exceded.

WORK LOAD NORM

Lehman (1958) has estimated that the maximum energetic output a normal man can afford in the long run is about 4800 Kcals/day; but his basal or at rest requirements are estimated at about 2300 Kcals/day, leaving a maximum of about 2500 Kcals/day for working. Although Lehmann proposes this as a maximum, he suggests that about 2000 Kcals/day to be a more normal load or for an 8 hour day about 4.2 Kcals/min. Grandjean (1981) confirmed this value of the norm for energy consumption during heavy work to be about <u>4.0 Kcals/min</u>.

REST PAUSES

Durnin and Passmore (1955) indicated that many industrial type jobs exceeded the 4.0 Kcals/min norm. Therefore if a worker were to sustain such heavy jobs over prolonged periods they would be fatigue generating causing increased heart rates.

As the heart rate cannot be expected to increase incessantly without danger, it is necessary when assigning jobs with JER exceeding the norm to allow a period of rest to ensure heart rate recovery.

One of the earliest to work on rest pauses was Murrell (1965), but he used a norm of 5 Kcals/min, and developed a formula for the duration of the rest pause, which is essentially the time required for heart rate recovery.

$$a = \frac{w(b - s)}{b - 1.5}$$

where a = recovery time in minutes
w = total working day in minutes
b = average calorie expenditure/minute
s = level of energy expenditure adopted as standard (i.e. 5 Kcals/min).

Sanders and McCormick (1987) showed the effects of task demand on energy expenditure and recovery, which corresponded to Brouha's (1960) earlier findings. But Brouha further illustrated that in the case of assuming jobs exceeding the proposed norm and for continued working without adequate rest pauses, the heart rate continued to rise after each break. Thus, when a high job JER is assigned to a worker with a reduced aerobic capacity and low PFI, it could have dangerous consequences if the duration and frequency of rest pauses are inadequate.

CONCLUSION

As a worker advances in age, the physical work capacity is reduced; when there is the additional problem of health, the situation becomes more serious.

Each job requires a certain amount of energy of its assigned worker in order to be performed effectively.

Arrangement of work place and space must be integrated with anthropometric statistics and other human factors of the worker to reduce fatigue and accidents.

Factors of the environment such as temperature, humidity, fumes, vibration etc. add to the strain on the worker.

The amount of information from the work station needed to evaluate normal operations, if excessive, could seriously affect the nervous load on the worker, and increase heart rate.

Machine cycle times or conveyor speeds found in paced work stations as assembly lines could increase the level of stress on the operator and increases as the level of pacing increases. This is reflected in an increased heart rate.

Excessive Job Energy Requirements necessitate proper matching with the Aerobic Capacity of the assigned worker.

When Job Energy Requirements exceed the norm of about 4 Kcals/min, care must be taken to ensure the adequacy (duration and frequency) of rest pauses to permit adequate heart rate recovery.

Workers should be evaluated periodically to assess their Physical Fitness Index due to increasing age and declining health.

The JER of a particular job should not exceed 33% of the aerobic capacity of the assigned worker.

When a worker's aerobic capacity is reduced due to health and this worker is unable to safely continue in the assigned job, the job should be modified (i.e. reduced JER) or alternatively this worker should be reassigned a job with a lower JER. In like manner when there is a health breakdown of a worker on a job due to the mismatching of JER and aerobic capacity, care must be taken in setting the JER of the job assigned to such worker after convalescence.

REFERENCES

Astrand, I., 1960, Aerobic Work Capacity in Men and Women, Acta Physiologica Scandinavia, Vol. 49, Supplement 169.

Bink, B., 1962, The Physical Work Capacity in Relation to Working Time and Age, Ergonomics, Vol. 5 (1).

Bink, B., 1964, Additional Studies of Physical Working Capacity in Relation to Working Time and Age, Proceedings of the Second International Congress on Ergonomics.

Bonjer, F.J., 1962, Actual Energy Expenditure in Relation to Physical Working Capacity, Ergonomics, Vol. 5 (1).

Brouha, L., 1960, Physiology in Industry, Pergamon Press.

Chaffin, D.B., 1969, Prediction of Metabolic Energy Expenditure Rate, University of Michigan Conference.

Durnin, J.V., and Passmore, R., 1955, Human Energy Expenditure, Physiological Review, No. 35.

Furedy, J., 1987, Beyond Heart Rate in the Cardiac Psychophysiological Assessment of Mental Effort : The T-Wave Amplitude Component of the Electrocardiogram, Human Factors, Vol. 29 (2).

Grandjean, E., 1981, Fitting the Task to the Man, New York International Publication Service.

Lehmann, G., 1958, Physiological Measurements as a Basis of Work Organization in Industry, Ergonomics, Vol. 1.

Moritz, J.S., Morrison, J.F., Peters, J., Strydon, N.B. and Wyndham, C.H., 1961, A Practical Method of Estimating an Individual's Maximum Oxygen Uptake, Ergonomics, Vol. 4 (2).

Murrell, K.F.H., 1988, Ergonomics, Chapman & Hall, March 1988, 1965 Quebec Corporation of Dieticians, Healthy Bodies Come in All Sizes.

Sanders, M.S. and McCormick, E.J., 1987, Human Factors in Engineering and Design, 6th Edition, McGraw Hill.

Tatro, J.S., and Roscoe, S.N., 1986, An Integrated Display for Vertical and Translational Flight : Eight Factors Affecting Pilot Performance", Human Factors, Vol. 28 (1).

Thomas, G.I., 1945, The Dietary Health and Disease, Lee and Febiger Company.

Tichauer, E.R., July 1964, Human Capacity, As Limiting Factor in Design, Institution of Mechanical Engineers.

Advances in Industrial Ergonomics and Safety II
Edited by Biman Das
Taylor & Francis 1990

LIGHTING ERGONOMICS EVALUATION ISSUES FOR AIR TRAFFIC CONTROL FACILITIES

H. Bashinski, P. Krois, C. Snyder, and W. Tobey

CTA INCORPORATED
Suite 200, 5670 Greenwood Plaza Boulevard,
Englewood, Colorado 80111, U.S.A.

INTRODUCTION

The Federal Aviation Administration (FAA) is in the process of upgrading all enroute air traffic control (ATC) facilities in the United States. This program, called the Advanced Automation System (AAS), involves new computers, software, and controller workstations called Common Consoles. Since 1982, CTA INCORPORATED has provided the FAA with human engineering support for the development, implementation, and evaluation of computer-human interface (CHI) requirements for the AAS.

This paper addresses the development of a methodology for evaluating lighting design proposals for the facility upgrades. The typical ATC facility will contain over 100 Common Consoles in one large control room. The Common Consoles will utilize four different electronic displays, positioned at a variety of viewing angles as shown in Figure 1. In addition, the upper part of the console (i.e., Main and Auxiliary Displays) can be rotated 45 degrees to the right or left for a wrapped physical configuration. Console wrap allows easier viewing of information displayed at adjacent Common Consoles. The challenge to the facility designer is finding a control room lighting solution that allows the controller to perform ATC tasks while minimizing undesirable interactions with the electronic displays.

The first part of this paper describes how the ATC operational requirements for control room lighting were defined. This includes the role of controllers in identifying and validating the requirements. The second part of the paper addresses the development of a lighting design evaluation approach, based on operational requirements and prior experience with controller hands-on evaluations. Future reports will focus on the results of applying these methods.

OPERATIONAL REQUIREMENTS FOR ATC ROOM LIGHTING

Wide variability in lighting approaches among current facilities suggests that lighting specifications used in today's ATC system are not predictive of AAS lighting specifications. FAA lighting experience shows that the light level and directionality at the controller's radar scope and paper flight data strip bay consoles must minimize reflectance off the display screen yet allow readability of the keyboard, and readability and annotation of paper flight data strips.

In order to understand differences in existing lighting approaches, field surveys were conducted at Denver, Atlanta, and Houston enroute facilities. The purposes of the surveys were to observe current facility and task lighting techniques, and to obtain light level measurements.

Console Legend:
1 - Auxiliary Display
2 - Main Display
3,4 - Electronic Displays
5 - Electronic Displays Interactive Keypad and Recess Handle
6 - Keyboard
7 - Trackball
8 - Headset Jack
9 - Console Wrap Handle
10 - Power Switches; Loudspeakers; Speaker, Headset, and Chime Volume Controls

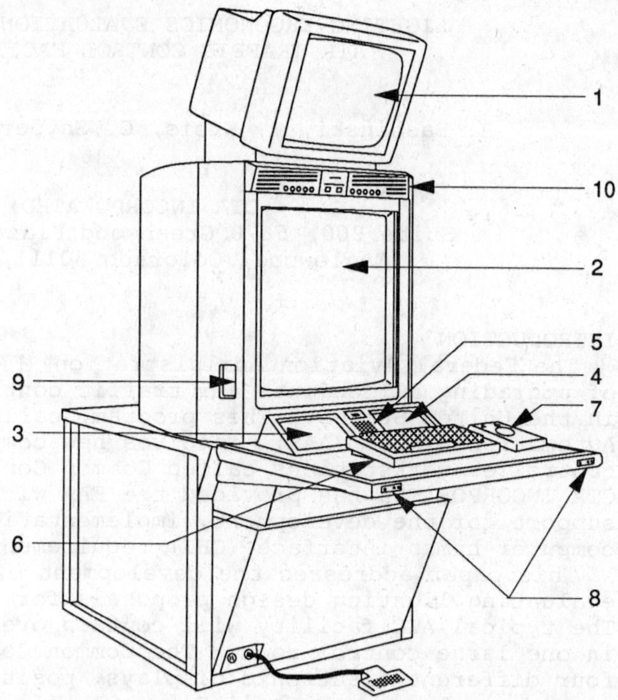

Figure 1. Baseline Common Console design.

The facility floorplan typically consisted of four parallel rows of consoles; two rows were placed in the center of the control room with their backs toward each other, while the remaining two rows each opposed a center row and were placed with their backs to the respective adjacent wall.

While the same facility lighting fixtures were installed in two of the three facilities, there were operational differences in their use. At these two facilities, long narrow rows of direct lighting fixtures were installed in the 30-foot high ceiling directly over the two operations aisles. The two facilities infrequently used this direct lighting. Fixtures were also installed on top of the continuous enclosures at the rear of the consoles. These fixtures, located about twelve feet above the floor, provided indirect light reflected from the off white walls and ceiling. One facility made extensive use of the indirect lighting while the other facility limited its use to illuminating an aisleway at one end of the middle two rows of consoles. At the third facility, the direct lighting fixtures were staggered across the approximately 20-foot high ceiling and installed with plastic diffuser panels. These lights were infrequently used, and then only one or two bulbs were illuminated per fixture.

Backlit keyboards were provided at all controller and several other operational positions. At the three facilities, only lights on a top row of function keys were illuminated, i.e., no restorative maintenance is performed on bulbs associated with the QWERTY keys. Local fluorescent console lighting was provided above the paper flight strip bays and printers to facilitate tasks such as reading, annotating, and sorting the paper flight data strips. Light level was under rheostat control and was set by the controller manning that position. Controller charts were positioned above the seated controller and were backlit to facilitate readability. The three facilities varied widely in the number of charts continuously illuminated. For the charts that were illuminated, there was wide variability in their apparent brightness.

Numerous other areas mix facility with task lighting. The Area Manager, and other special ATC positions, were located under a suspended walkway, resulting in an approximate 12-foot high ceiling. Fluorescent lights installed in this ceiling provided illumination for these positions at two facilities.

At all facilities, desks at several ATC positions had gooseneck lamps or fluorescent lights installed beneath raised shelves. Plastic and paper barriers fixed below the fluorescent bulbs reduced the light intensity and glare, and minimized visual distractions for nearby controllers.

Lighting measurements were taken at several ATC areas in the three facilities, as shown in Table 1. Measurements were taken at the keyboard with the light level set by the radar (R) or paper flight data strip (D) controller. The facility having the highest light levels at the consoles and in the operations aisles was found to have the lowest light levels at the Area Manager, Traffic Management, and Area Supervisor positions.

Table 1. Field survey illumination levels in footcandles (fc).

Location	Atlanta	Denver	Houston
R-controller	0.29	1.19	0.49
D-controller	1.23	2.16	1.39
Operations Aisle	0.54	1.67	0.45
Area Manager	21.95	5.00	16.00

New AAS control rooms are intended to use a common lighting design that accommodates special position task lighting requirements. This common design should result from operational lighting requirements applicable to all facilities.

<u>Controller lighting design requirements</u>

During AAS development, the controller user community has been represented by the Sector Suite Requirements Validation Team (SSRVT). The SSRVT has helped shape AAS operational requirements over the past eight years and is thoroughly indoctrinated in design requirements.

The SSRVT completed its requirements analysis through use of an elaborate operations concept approach. One major feature of this approach is that it is based on explicit validation of controller tasks that the CHI must support, rather than on personal preferences. A task analysis provides a common reference by which to substantiate common operational lighting ergonomics requirements (Phillips et al., 1988).

Based on the field surveys, seven controller tasks were identified as requiring adequate illumination: operate keyboard, write/read handwritten notes, read textual material, read color-graphic material, annotate/read paper flight data strips, perform administrative duties, and perform maintenance. Many tasks performed in today's system will also be performed in the AAS and must be supported by the facility lighting design. Key AAS differences are that paper flight data strips will be replaced with electronic flight data entries, eliminating the need for task lighting over the associated bays; the Air Traffic Assistant position and its paper flight strip printer will be omitted; and the Area Supervisor position will have a keyboard.

The ergonomics evaluation criteria developed by the SSRVT are shown in Table 2, separated into evaluation categories. These criteria were the cumulative product of the SSRVT's critique of AAS design and human factors requirements, as well as a trade analysis of several lighting options as described later in this paper.

<u>AAS design requirements</u>

The AAS specifications generally require the advanced electronic display technology to display information without glare while maintaining high color contrast and purity under all lighting conditions. The Common Console utilizes several electronic displays that are positioned at various viewing angles relative to the seated and standing controller. Shelf height can also be adjusted to accommodate controllers of varying statures. The number and positions of electronic displays pose a substantial challenge to the lighting designer relative to the controller evaluation criteria.

<u>Human factors standards</u>

The AAS specifications require the AAS design to meet provisions in Military Standard 1472C (MIL-STD-1472C, 1984). However, that standard provides rather ambiguous lighting requirements, specifying that lighting should be appropriate to ensure that illumination is compatible with each task situation. Minimum light levels include 3.0 fc for emergency lighting and 50.0 fc for general office work, but no level is specified relevant to command and control facilities.

The Electric Power Research Institute provides a more specific standard (EPRI-NP-3659, 1984): a maximum ambient light level of 5.0 to 10.0 fc shining on the CRT face may be acceptable, and 3.0 fc is preferred. The standard also suggests that luminance levels be uniform across the work area to mitigate visual discomfort and ensure readability.

<u>Summary</u>

The above discussion identifies the significant challenges that operational, AAS design, and human factors requirements pose to developing an operationally suitable facility lighting design. The best engineering solution must assimilate all these requirements to be judged operationally acceptable as a common approach to facility lighting.

ERGONOMICS EVALUATION APPROACH

Initial analysis of the proposed facility lighting design provided early experience with use of the controller evaluation criteria. Controllers have increasingly accepted human factors evaluation methods from positive evaluation experience.

Table 2. Controller lighting ergonomics evaluation criteria.

<u>Data Entry Device Usability</u>

1. Allows all key labels on the keyboard, electronic display interactive keypad (IAK), and trackball to be easily recognized and distinguished with an acceptable amount of effort.

<u>Textual, Handwritten, and Color-Graphic Material Usability</u>

2. Adequately supports writing of notes on the console shelf.
3. Allows textual, handwritten, and color-coded material placed on the shelf to be read with an acceptable amount of effort.

<u>Common Console Control Usability</u>

4. Allows the Common Console and electronic display power switches; speaker, headset jack, and chime volume controls; console wrap handles; and electronic display recess mechanism to be located with an acceptable amount of effort.
5. Allows headset jacks and electronic display IAK connector to be located with an acceptable amount of effort.
6. Allows control labels for the Common Console and electronic display power switches; and speaker, headset jack, and chime volume controls to be recognized with an acceptable amount of effort.

<u>Visual Distraction</u>

7. Adequately supports controller task performance by minimizing distracting shadows and glare on the shelf, and Main, Auxiliary, and other electronic displays.
8. Adequately supports controller task performance by minimizing distracting reflections (caused by light sources, personnel movements, etc.) on the Main, Auxiliary, and other electronic displays.

<u>Display Usability</u>

9. Minimizes display brightness washout by minimizing the degradation of brightness contrast for textual and graphic information on the Main, Auxiliary, and other electronic displays.
10. With respect to display color washout, supports adequate recognition of and differentiation between color codes used for textual and graphic information presented on the Main, Auxiliary, and other electronic displays.

<u>Emergency Lighting</u>

11. Provides adequate illumination to perform ATC tasks required under emergency conditions.
12. Provides adequate illumination for safe ingress into and egress from operating positions.

<u>Maintenance Lighting</u>

13. Restricts the amount of light that filters through gaps between consoles to an acceptable amount.
14. Is sufficiently buffered so that it does not interfere with ATC tasks at the Sector Suite.

Preliminary analysis of proposed lighting design

The FAA's facilities contractor prepared a direct facility lighting design for use with the AAS control room 13-foot high ceiling. This design was predicated on expected indirect lighting risks such as ceiling hot spots. To illuminate the operations aisle, the design involved a continuous row of recessed fluorescent lights running the length of the row of consoles in a position directly above the controller seated in front of the Common Console. The lighting provided by the row of two foot square fixtures was diffused using 1.5-inch cube parabolic directional louvers. Continuous rows of 2-foot high baffles were suspended adjacent to the fixtures on the side nearest the consoles in a position where they created a shadow over the in-line consoles' Main and Auxiliary Displays.

Potential strengths and risks associated with a direct lighting design and other alternative lighting techniques were analyzed. The categories of lighting were defined as facility and task lighting. Facility lighting was addressed as direct (e.g., recessed parabolic restricted lighting) or indirect (e.g., suspended diffused lighting). Task lighting included a backlit keyboard, console lighting, and a moveable lamp (e.g., gooseneck light device).

A lighting alternative's effectiveness was measured in terms of human performance risk factors. These risk factors, which were based on SSRVT lighting concerns, current ATC practices, SLS requirements, and industry standards are listed in Table 3.

Table 3. Lighting-alternative risk factors.

1. Keyboard Readability	10. Dark Adaptation Compatibility
2. Notepad Readability	11. Eye Fatigue
3. Main Display Compatibility	12. Light Level Adaptability
4. Communication Displays Compatibility	13. Console Maintenance Compatibility
5. Console Wrap Compatibility	14. Floor Layout Compatibility
6. Shelf Height Adjustment Compatibility	15. OSHA Requirements Compatibility
7. Glare	16. Life Cycle Cost
8. Reflection	17. Lighting Maintenance
9. Shadows	

The analysis indicated that many of the risk factors are interrelated. Thus, implementation of a lighting alternative might resolve risks in certain areas while not compensating for and even increasing risks in other areas.

A comparison of risks associated with direct and indirect lighting showed more risks with direct lighting. Several risks are uniquely associated with direct lighting because restricting an area of light, such as carefully illuminating the console shelf without the light striking the display surfaces, may consistently result in the light level of one area being of a consistently higher intensity than that of an adjacent area. Wrapping of consoles and movement of personnel across the bounded area of light might also cast shadows across the console shelf and display surfaces, and affect display readability due to reflectance and contrast in surface luminance. A light level that is either too high or too low could cause eye fatigue resulting from glare, reflections, and/or shadows. These risks might increase the possibility of human error.

While indirect lighting mitigates many of the direct lighting risks, risk is introduced from insufficient rheostat sensitivity at low light levels, and from the requirement for supplementary lighting for console maintenance. The "cloudy day effect" found with indirect lighting could, however, reduce eye fatigue associated with direct lighting.

Trade-offs associated with task lighting techniques indicated that each has significant risks. Backlit keyboard risks are most critical since additional task lighting would still be needed to support other tasks, e.g., readability of handwritten materials.

Both console lighting and moveable lamp are burdened with light directionality risks. The placement and movement of the lights must be closely constrained to avoid glare, reflection, shadows, and washout of the display surfaces on the corresponding, adjacent, or opposing consoles. These risks would be compounded by wrapping the consoles and adjusting shelf height.

The analysis showed that diffused indirect lighting poses the least risk to controller performance. Task lighting alone would not be sufficient to meet all controller and maintenance lighting needs. Conversely, the results were inconclusive concerning the need for task lighting beyond the illumination provided by indirect facility lighting. Results of this trade analysis are consistent with Hedge, Sims, and Becker (1989), who reported greater subjective acceptance with indirect compared to direct lighting systems among electronic display users, partly due to reduced glare and eye fatigue.

<u>Preliminary experiences with ergonomics evaluations</u>

Over the course of AAS development, the SSRVT has participated in several hands-on evaluations, focusing on the Common Console design and CHI functionality. The evaluations have used a framework of task action sequences derived from the operations concept. This controller task framework provides a common evaluation thread across iterations of design evaluations (Davies and Bashinski, 1987; Davies and Phillips, 1986).

As part of the evaluations, part-task questionnaires are developed in which specific aspects of the design are assessed relative to specific tasks that are embedded within a representative ATC scenario. For each evaluation, the team is thoroughly briefed on evaluation goals and procedures, and trained on use of the questionnaire to mitigate any potential misunderstanding of questionnaire items. With each evaluation, the controllers become increasingly adept at use of the evaluation criteria. The part-task questionnaire uses operational acceptability criteria that are iteratively refined by the controllers themselves during each evaluation. Following each evaluation, the team discusses interim questionnaire statistical results to develop a final detailed consensus on every issue. This approach has been highly successful in evaluating and improving the AAS design, thereby facilitating increased acceptance of evaluation findings.

Lighting design evaluation methodology

Following the lighting alternatives trade analysis, the SSRVT determined that a high-fidelity mock-up would be required to validate the operational suitability of the facility lighting design. Fidelity would be ensured through use of Common Consoles in a mock-up of a corner of the AAS Control Room, constructed using the actual engineering layout and materials.

The methodology incorporated controller tasks where lighting could be assessed within an operational context. In addition, it was critical that these tasks be defined to challenge the operational, AAS design, and human factors requirements, as well as the known lighting risks.

Using the ergonomics evaluation criteria shown in Table 2, the controllers participated in definition of tasks and development of the questionnaire. The questionnaire was streamlined to systematically address those issues and requirements previously identified. Use of the decomposed evaluation criteria assured that specific risks would not be overlooked by an otherwise "shotgun" evaluation approach.

In the development of evaluation procedures, the SSRVT successfully validated each step: questionnaire development, familiarization with questionnaire materials and evaluation procedures, actual hands-on evaluation, and controller consensus on evaluation findings.

CONCLUSIONS

The lighting ergonomics evaluation approach is significant because of controller involvement in developing evaluation criteria and procedures. Such involvement is critical due to lack of research guidance and uniqueness of the application. Application of this proven evaluation approach should augment optimizing the control room lighting design and provide valuable data on lighting approaches for command and control centers that use workstations dominated by electronic displays.

REFERENCES

Davies, D.K., and Bashinski, H.S., 1987, Human engineering evaluations in real life: operational suitability assessment during ATC design development. Human Factors Society Bulletin, 30, 1-3.

Davies, D.K., and Phillips, M.D., 1986, Assessing user acceptance of next generation air traffic controller workstations. In: Proceedings of the Human Factors Society 30th Annual Meeting, (Santa Monica, CA.: Human Factors Society), pp. 590-594.

Electric Power Research Institute, Human Factors Guide for Nuclear Power Plant Control Room Development, EPRI-NP-3659, 1984.

Hedge, A., Sims, W. and Becker, F, 1989, Lighting the computerized office. In: Proceedings of the Human Factors Society 33rd Annual Meeting (Santa Monica, CA.:The Human Factors Society), pp. 521-525.

Human Engineering Design Criteria for Military Systems, Equipment, and Facilities, MIL-STD-1472C, 10 May 1984.

Phillips, M.D., Bashinski, H.S., Ammerman, H.L., and Fligg, C.M., 1988, A task analytic approach to dialogue design. In Handbook of human-computer interaction, edited by H. Helander (Amsterdam: North-Holland).

XIX

WORKPLACE SAFETY ANALYSIS

Advances in Industrial Ergonomics and Safety II
Edited by Biman Das
Taylor & Francis 1990

IMPROVING WORKING PRACTICES AND WORKPLACE SAFETY THROUGH BEHAVIOR ANALYSIS IN THE VANEER INDUSTRY

Markku Mattila

Tampere University of Technology
Occupational Safety Engineering
P.O. Box 527, 33101 Tampere, Finland

The aim of the study was to improve safety by implementing Behavior Analysis during internal safety inspections in the vaneer industry. The targets for workplace conditions and for work practices were determined. The current conditions were followed up by observations, and the level of target achievement was determined. Feedback was presented as a graphic curve posted on a wall of the department. Later, the supervisors were given more detailed comments. The safety index improved in all parts of the department. The number of accidents also fell clearly.

INTRODUCTION

Occupational safety activities have become an established and essential part of any industry in developed countries. During recent years, the current methodologies of safety practices have been criticized as ineffective and unable to bring about improvements in the work environment (Komaki, 1986).

Internal safety inspections are one of the most common safety activities in industry. Studies have shown that these safety inspections are often ineffective (Kasules, 1985) and too selective, and that analysis of the data they provide fails to identify the most common faults or to prevent their recurrence (Kjellén, 1982).

The behavioral approach has been suggested as a new method for preventing occupational accidents (Denton, 1982; Komaki, 1986). Behavior analysis has proved to be successful in decreasing the frequency of unsafe conditions and unsafe acts (Komaki et al., 1982; Fellner and Sulzer-Azaroff, 1984). A few studies have also demonstrated its impact in reducing occupational accidents (Reber and Wallin, 1983; Saari and Näsänen, 1989). The behavior analysis approach has been integrated with the safety program of the company, and applied as part of the safety routine (Mattila and Hyödynmaa, 1988).

It is not known whether behavior analysis could be applied to internal safety inspections done at the plant in order to improve the preventive effect of inspections.

The objectives of this study were: (i) to determine whether the behavior analysis approach can be used effectively to improve occupational safety in the vaneer industry; (ii) to apply behavior analysis in conjunction with the internal safety inspections at the plant; and (iii) to evaluate experimentally the effectiveness of this safety effort.

MATERIAL AND METHOD

Subject

The study was conducted at a Finnish vaneer factory producing 35,000 m^3 vaneer a year and employing about 250 people. The department for vaneer refining and coating, which consists of two halls, was selected for the experiment because, on the basis of the yearly accident rates, the safety staff of the company assessed that department to be the most hazardous one of the entire plant. Altogether 38 people were working at the department. The area of the department was divided into three sections for observations: (a) the machine area, where the cutting and splitting machines were located; (b) the middle area, where the vaneer was stored; and (c) the coating area, where the process lines for coating were located.

Target conditions and behavior

The target conditions and behavior were selected so as to focus on items essential for safety. The company's accident records and the national accident statistics were analyzed. The safety manager and representatives of workers and supervisors were interviewed. The targets were formulated by researchers. Eighteen targets were chosen for the experiment (Table 1).

Measurement

The safety inspections were conducted weekly on varying days and at different times. The list of targets was used as the observation form. Each item was checked as correct, incorrect, or not observed if the item was not relevant at the moment. The form was used for each of the three sections of the department. A safety index was calculated by dividing the number of correct items by the total observed, and that figure was multiplied by 100 (Komaki et al., 1978). The safety index was computed for each of the observation area and for the whole department. Altogether 14 inspections were made. Each inspection lasted about 45 minutes. The reliability was measured twice by using two independent, experienced observers. The reliability coefficients were 89% and 96% (Komaki et al., 1978).

Four inspections served as the baseline phase. After this period, following each inspection round the safety index score was posted, as a graph, on the wall in the department. A layout drawing of the department, on which the areas of the target performance and those still requiring improvement were marked, was posted next to this graph. After five rounds had been conducted, the supervisors were given more information in a list mentioning the items ranked as being incorrect. This extra feedback phase to supervisors lasted for seven weeks.

Table 1. Target conditions and target behavior for the vaneer industry plant.

Number	Target
1.	Keep the rubbish bins in their proper places.
2.	Make sure that the rubbish is in the rubbish bins, and see that the bins are not overfull.
3.	Make sure that there is no oil or water on the floor of worksites.
4.	Remove used plastics from the walkways.
5.	Keep the electrical board free of obstacles.
6.	Return used work equipment to the places where they are kept.
7.	Keep the working area around the machines free of obstacles.
8.	Store supplies and wastes in their proper places.
9.	Don't transport anyone on the lift truck.
10.	Drivers must make sure that they have a clear view while driving the lift truck.
11.	Keep the internal transport ways free of stored products.
12.	Make sure that the store piles do not fall down.
13.	The store piles must not be stacked too high.
14.	Make sure that the conveyer is clear for safe use.
15.	Stop the conveyer whenever any blockage must be investigated.
16.	Work whit safety guards on the machines.
17.	Stop the machine once the work is completed.
18.	Use ear protectors while the machines are running.

RESULTS

During the baseline phase, the safety index averaged 66.5%. Once the targets had been formulated, the researchers assessed the safety index to be around 40-50%, but the management then organized a general campaign to improve order; this campaign also affected the level of the index. During the graphic feedback phase, the index improved to 71.4%, and during the extra information feedback to supervisors the index rose still further, to 77.7% (Figure 1).

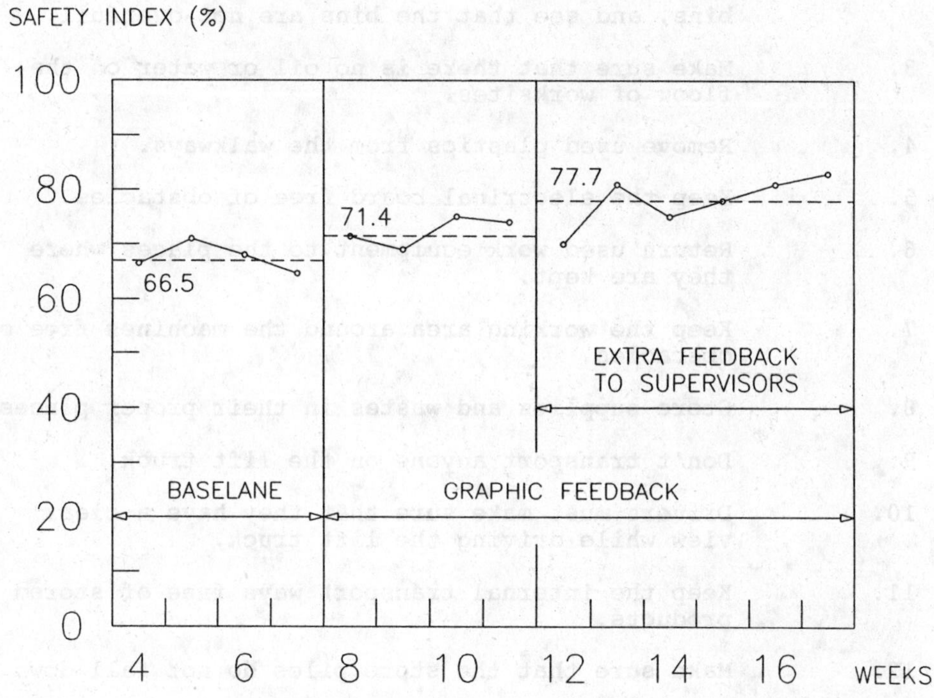

Figure 1. The safety index for the studied department of the vaneer industry plant.

A follow-up observation was done three months later. The safety index computed than was 75%. The index had remained on the higher level reached through the intervention.

During the same period of time, six occupational accidents occurred in the department. Two of those six accidents were clearly related to factors involved in the target behavior of the study. During the actual study, three accidents occurred in the department, one of which was related to target behavior factors. This accident occurred during the first week that the graphic feedback had been posted; thus the feedback procedure had not had its full effect yet.

CONCLUSIONS

The study showed that the behavioral approach and a feedback program are able to affect safety conditions and behavior in industry. The feedback itself improved the level of safety without any clear mention of the exact content of the target conditions and target behavior. When more detailed feedback was given to supervisors, the index improved still more. The study supports the finding that improvements are able to persist for a larger period of time, once the intervention has ended (Saari and Näsänen, 1989). This approach may be applied in management techniques.

The study also supports the hypothesis that the behavior analysis technique may be successful in preventing accidents. The amount of data here was small, but the results are encouraging. This finding confirms Heinrich's theory that there is a relationship between unsafe acts and unsafe conditions and injuries (Heinrich et al., 1980).

The study revealed one way of improving internal safety inspections. This approach makes it possible to connect both targets and their monitoring to actual safety inspection rounds and to positive measurement of the safety level through inspections, as opposed to pointing out or listing mistakes.

ACKNOWLEDGEMENTS

The author thanks the Finnish Work Environment Fund for financial support, the Viiala factory of Oy Schauman Ab, and safety manager Markku Lemström for their co-operation, research assistant Hannu Nieminen for data collection, and Sheryl Hinkkanen for correcting the language.

REFERENCES

Denton, D.K., 1982, Safety Management: Improving Performance. (New York, NY: McGraw-Hill).

Fellner, D.J. and Sulzer-Azaroff, B., 1984a, Increasing industrial safety practices and conditions through posted feedback. Journal of Safety Research, 15(1):7-21.

Heinrich, H., Petersen, D. and Roos, N., 1980, Industrial Accident Prevention. (New York, NY: McGraw-Hill).

Kasules, M., 1985, Inspection and observation: Another view. Professional Safety, 30(2):26-28.

Kjellén, U., 1982, An evaluation of safety information systems at six medium-sized and large firms. Journal of Occupational Accidents, 3:273-288.

Komaki, J.L., 1986, Promoting job safety and accident preventions. In: M.F. Cataldo and T.J. Coates (Eds.), Health and Industry, A Behavioral Medicine Perspective. (New York, NY: Wiley), pp. 301-320.

Komaki, J.L., Barwick, K.D. and Scott, L.R., 1978, A behavioral approach to occupational safety: Pinpointing and reinforcing safe performance in a food manufacturing plant. Journal of Applied Psychology, 63(4): 434-445.

Komaki, J.L., Collins, R.L. and Penn, P., 1982, The role of performance antecedents and consequences in work motivation. Journal of Applied Psychology, 67(3):334-340.

Mattila, M. and Hyödynmaa, M., 1988, Promoting job safety in building: An experiment on the behavior analysis approach. Journal of Occupational Accidents, 9:255-267.

Reber, R.A. and Wallin, J.A., 1983, Validation of a behavioral measure of accident safety. Journal of Organizational Behavior Management, 5(2):69-78.

Saari, J. and Näsänen, M., 1989, The effect of positive feedback on industrial housekeeping and accidents; a long-term study at a shipyard. International Journal of Industrial Ergonomics, 4:201-211.

Advances in Industrial Ergonomics and Safety II
Edited by Biman Das
Taylor & Francis 1990

ERGONOMIC JOB ANALYSIS: METHODS TO IDENTIFY, EVALUATE, AND CONTROL EXPOSURES TO RISK FACTORS IN THE WORKPLACE

W. Monroe Keyserling, Ph.D.
Andrew B. Bigelow, M.S.
Milton L. Brouwer, M.S.
Lisa A. Murphy, M.S.

The University of Michigan
1205 Beal Avenue
Ann Arbor, Michigan 48109 USA

Job analysis is an essential part of an occupational ergonomics program because it provides the procedures and analytical tools needed to identify, evaluate, and control exposures to working conditions which can cause acute injuries and/or cumulative trauma disorders of the musculoskeletal system. A systematic ergonomic job analysis system was developed to evaluate the postural, metabolic, manual lifting and upper extremity stresses. This system was used to evaluate 155 industrial jobs. In general, stresses due to awkward postures and upper extremity activities were greater than stresses due to energy expenditure or lifting.

INTRODUCTION

Job analysis is not a new concept. Industrial engineers have used procedures such as time study (Taylor, 1911) and motion analysis (Gilbreth, 1911) for nearly a century to measure productivity, to enhance production efficiency and to improve product quality. Human factors engineers have developed comprehensive job analysis techniques such as the Position Analysis Questionnaire (PAQ, McCormick, et al., 1969) and the Arbeitswissenschaftliche Erhebungsverfahren zur Tatigkeitsanalyse (AET, Rohmert, et al., 1969) which examine both the psychological and physiological demands of work. These techniques have been used in a variety of applications, including: workplace documentation; analysis of work requirements; organization of work design and work structuring; personnel recruitment, selection and placement; and establishing remuneration rates (Rohmert, et al., 1983).

In recent years, job analysis has played an increasingly important role in the prevention of occupational injuries and illnesses. As a result of recent passage and/or strengthening of occupational health and safety laws in Europe and North America, the scope of job analysis activities has expanded to consider the effects of a broad spectrum of work place exposures such as chemical hazards, physical energy hazards, and mental stress on the risk of accidents, injuries and diseases (Mattila, 1989).

Overexertion injuries (e.g., low back pain, cervicobrachial disorders, and upper extremity cumulative trauma disorders) resulting from work activities are a major health problem in the United States and other industrialized nations. These disorders are the leading cause of work-related disability and account for approximately one-third of all workers' compensation costs (Putz-Anderson, et al., 1986). Many of these problems are associated with excessive physical demands caused by the design of facilities, equipment, tools, and work methods (NIOSH, 1983; Putz-Anderson, 1988). Job analysis is a necessary aspect of any program to control overexertion injuries because it provides a structured approach for collecting the information required to answer the following questions:

1. Does performing the job expose workers to risk factors which may cause overexertion injuries and disorders (e.g., forceful exertions, awkward postures, localized mechanical contact stresses, vibration, temperature extremes, repetitive motions or prolonged activities)?
2. Which specific aspects of the job (e.g., work station layout, production standards, work methods, etc.) cause or contribute to exposures?

These questions must be answered in order to develop effective changes to reduce or eliminate potentially injurious activities (Keyserling, et al., 1989).

METHOD

A method for documenting and evaluating exposures to ergonomic risk factors was developed and used to assess ergonomic stresses on 155 jobs at four plant sites. The new method was an adaptation and extension of the concepts and procedures developed by Systems Safety Engineers (e.g., Procedure Analysis, Job Safety Analysis) to identify conditions in the workplace which cause accidents and acute injuries (Clemons, 1982). Application of the technique involved two sequential activities: job documentation (at the work site and job evaluation (in the laboratory).

Analysts

Six analysts used the job analysis procedures described below to document and evaluate ergonomic stresses on the 155 jobs included in this study. Each of the analysts was a either a current student or recent graduate of a Masters Degree program in Industrial Engineering with specialization in ergonomics. The analysts generally worked in pairs when collecting data and singly when performing the subsequent analyses. Furthermore, they were encouraged to cooperate with each other during all phases of data collection and analysis.

Jobs

Two of the plant sites were manufacturing facilities: 43 jobs were studied in a metal stamping plant while 31 jobs were studied in an engine plant. Two of the plant sites were warehouse facilities where a total of 81 jobs were studied.

Job Documentation

The purpose of the job documentation activity was to develop a complete description of how each job was performed and to take quantitative measurements of job demands. Job documentation involved the following steps:

1. Job Identification -- Because it was essential to have a non-ambiguous method for identifying each job, the following information was recorded: Job Title, Identification Number, Department, and physical location (e.g., room number, column number, etc.). In cases where job demands changed for the production or handling of different products (e.g., metal stamping operations), it was also necessary to record the part name and identification number.
2. Worker Identification -- The following was recorded for each worker: name, age, and height.
3. Video Recording -- Each job was videotaped to produce a permanent record of the work activities. This tape was used to develop the work methods description (Step 10 of Job Documentation, described below) and during the laboratory analyses of work postures and upper extremity exertions.
4. Determining the Work Schedule -- Information concerning the length of shifts and rest breaks was recorded. This information was used in the analysis phase to estimate the duration of time on the job and the total daily exposure to ergonomic stresses.
5. Determining Production Information -- Production information, such as the speed of an assembly line (on machine-paced jobs) or work standards (on unpaced jobs) was recorded. This information was used during the analysis phase to estimate the number of work cycles performed during a shift, a major factor in estimating the repetitiveness of the job.
6. Describing Job Rotation -- In some facilities, employees rotated among several jobs according to a fixed schedule. While job rotation does not affect the ergonomic stresses on a specific job, it does affect the cumulative exposure to ergonomic stresses experienced by an employee. It was therefore important to describe and record the job rotation schedule whenever such a system was used.
7. Sketching the Work Station -- A plan view of the work station showing the location of all benches, equipment, conveyors, supply bins, etc. was drawn. This sketch included dimensions for walking distances, horizontal reaches, and vertical work locations. If necessary, elevation views were also drawn to provide additional detail. These data were used during the laboratory analyses of postural, lifting, and metabolic stresses.
8. Describing Tools and Equipment -- A brief description was developed for each tool and equipment item used to perform the jobs. The weights and dimensions of hand-held tools were measured for use during the laboratory analysis of upper extremity stresses.
9. Describing Parts and Materials Handled -- A brief description was developed all materials manually handled by

the worker. Weights and external dimensions were measured for use during the laboratory analysis of lifting stresses.

10. Describing Work Methods -- Each job was described as the series of work elements necessary to produce one unit of product. A work element was considered to be a basic operation step in the work sequence. The level of detail was similar to that used by industrial engineers when performing time studies (Niebel, 1988). The methods descriptions were used during the laboratory analysis phase to associate exposures to ergonomic stresses with specific work elements.

Job Analysis

The job analysis activity had two objectives: 1) to determine if performing the job exposed the worker to ergonomic stresses which could cause overexertion injuries, and 2) to identify specific job attributes responsible for potentially hazardous exposures. Due to constraints limiting the length of this paper, the remainder of the presentation will focus on the first objective.

The following four categories of ergonomic stress were considered during the job analysis phase:

1. Awkward Postures -- This category included awkward postures of the trunk and neck (forward bending, lateral bending, and/or axial twisting) which could lead to localized muscle fatigue and/or disabling musculoskeletal disorders (Keyserling, et al., 1988; Jonsson, et al., 1988). A computer-aided posture analysis system (Keyserling, 1986) was used in conjunction with the videotapes to measure the amount of time spent in neutral versus awkward postures and to determine the frequency of posture changes. The results of these analyses were used to determine the level of postural stress at the trunk and neck for each job.
2. Metabolic Energy Expenditure and Whole Body Fatigue -- This category included work activities (e.g., climbing, repetitive lifting, rapid walking, and/or repetitive movements of the entire body mass) requiring high levels of energy expenditure which could lead to excessive loading of the cardiovascular system and whole body fatigue (Durnin, et al., 1967; NIOSH 1983). A metabolic prediction model (Garg, et al., 1978) was used in conjunction with the work place sketches and the work methods description to estimate the rate of energy expenditure on study jobs.
3. Lifting and Manual Materials Handling -- This category included stresses associated with strength demanding tasks which could lead to overexertion injuries of the back and other body parts. The NIOSH Work Practices Guide for Manual Lifting (1983) and a two-dimensional biomechanical strength prediction model (Chaffin, et al., 1984) were used in conjunction with work place sketches, part weights, and the work methods description to evaluate manual materials handling stresses.
4. Upper Extremity Activities -- This category included stresses associated with actions of the hands, arms, and shoulders which could lead to the development of disorders such as tendinitis and carpal tunnel syndrome. The videotapes were analyzed in conjunction with the methods description and production information to identify stresses

such as repetitive motions, high hand forces, localized contact stresses, vibration, and awkward wrist postures (Armstrong, 1986). In addition, a computer-aided posture analysis system (Keyserling, 1986) was used to evaluate shoulder posture.

A three-level ordinal scoring system was used to record the analysts' perceptions of ergonomic stresses (Keyserling, et al., 1988; Mattila, 1989). The analysts independently considered posture, metabolic, materials handling, and upper extremity stresses on each job. These stresses were then rated using the following scheme:

1. Insignificant -- Jobs given this rating were judged to be free of potentially harmful ergonomic stresses in the category of interest. The analyst believed that the job was sufficiently safe that no follow-up actions (e.g., job modifications) were necessary.
2. Moderate -- Jobs given this rating were judged to present moderate ergonomic stresses in the category of interest. A small number of workers would be at low risk of sustaining an injury if the job were performed on a full-time basis. Corrective measures (e.g., engineering changes to equipment or methods, job rotation) should be taken to reduce stresses as soon as resources are available.
3. High -- Jobs given this rating were judged to present major ergonomic stresses in the category of interest. Some workers would be at significant risk of sustaining an injury. Corrective measures should be taken as soon as possible to eliminate or reduce stresses to an acceptable level.

RESULTS

The results of the ergonomic analyses performed on 155 jobs in four facilities are presented in Table 1. Each job was rated in four different categories of ergonomic stress (trunk/neck posture, metabolic energy expenditure, manual lifting, and upper extremity) for a total of 620 (155 x 4) ratings. Of this total, 277 ratings (45 percent) found the ergonomic stress to be "insignificant," 227 ratings (36 percent) found the stress to be "moderate," while 116 ratings (19 percent) found the stress to be "substantial."

To determine if the analysts' ratings of ergonomic stresses differed in the four study plants, a Chi-square analysis was performed on the data in the rightmost column of Table 1. Statistically significant differences were found among the four plants (χ^2=30.5, dof=6, p<.001), with the lowest stresses in the stamping plant and the highest stresses in the engine plant.

A Chi-square analysis was also performed on the data in the bottom row of Table 1 to determine if there were differences in the analysts' ratings of ergonomic stresses in the four categories of interest. Statistically significant differences were found (χ^2=264.9, dof=6, p<.0001). Jobs received higher stress ratings in the trunk/neck posture and upper extremity categories than in the metabolic and lifting categories. Overall, the metabolic stress ratings were the lowest while the

Table 1. Analysts' ratings of ergonomic stresses on 155 manufacturing and warehouse jobs.

Plant	Analysts' Rating of Ergonomic Stresses*														
	Posture			Metabolic			Lifting			Up. Extr.			Total		
	1	2	3	1	2	3	1	2	3	1	2	3	1	2	3
Stamping (43 jobs)	14	27	2	38	3	2	33	10	0	4	25	14	89	65	18
													.52	*.38*	*.10*
W'house #1 (27 jobs)	3	15	9	21	5	1	19	8	0	10	12	5	53	40	15
													.49	*.37*	*.14*
W'house #2 (54 jobs)	10	25	19	47	7	0	39	15	0	1	25	28	97	72	47
													.45	*.33*	*.22*
Engine (31 jobs)	1	13	17	21	9	1	13	18	0	3	10	18	38	50	36
													.31	*.40*	*.29*
TOTAL	28	80	47	127	24	4	104	51	0	18	62	65	277	227	116
Fraction	*.18*	*.52*	*.30*	*.82*	*.15*	*.03*	*.67*	*.33*	*.00*	*.12*	*.46*	*.42*	*.45*	*.36*	*.19*

* Table entries present the number of jobs observed at each stress level (1=insignificant, 2=moderate, 3=substantial) by plant and category of ergonomic stress. The fraction of jobs observed at each stress level is given in italics for row and column totals.

upper extremity ratings were the highest. (Note: The two analyses presented above implicitly assumed that the four stress categories were equivalent in terms of evaluating the risk and seriousness of an injury or disorder. In other words, a job rated 3 (substantial stress) in the upper extremity category had similar risk and seriousness attributes of a job rated 3 in the metabolic category. Because this assumption may not be valid, the previous results should be interpreted with caution.)

DISCUSSION

The jobs selected for this study were not necessarily a random sample of all jobs in the four study plants. In fact, most of the jobs evaluated by the analysts' had been previously identified by plant personnel has having potential ergonomic problems. It is therefore likely that the results reported above show higher levels of ergonomic stress than actually existed in the study plant.

The finding that postural and upper extremity stresses were greater than metabolic and lifting stresses is not surprising. For almost a century, industrial engineers have been trying to improve work efficiency by reducing unnecessary body motions such as bending and arising and excessive walking (Taylor,

1911; Gilbreth, 1911; Niebel, 1988). Since these motions are also associated with high rates of energy expenditure, effective work station layouts have been developed which simultaneously improve efficiency while reducing metabolic energy expenditure. For decades, safety engineers and occupational health professionals have been concerned with reducing back injuries associated with manual lifting (NIOSH, 1982). As a result, automated devices such as powered conveyors, overhead hoists, and industrial lift trucks are widely used to reduce or eliminate the need for manual lifting. (Note: Because this devices also enhance work efficiency, they have been widely accepted.) Until very recently, the level of awareness of posture-related disorders and upper extremity cumulative trauma disorders has been relatively low. As a result of this lack of knowledge, efforts to control these disorders are relatively new and insufficient time has passed for widespread implementation of effective control programs.

CONCLUSIONS

A systematic approach to ergonomic job analysis was used to evaluate 155 jobs for ergonomic stresses associated with awkward posture, metabolic energy expenditure, manual lifting, and upper extremity work activities. This survey found that posture and upper extremity stresses are generally greater than metabolic and lifting stresses.

REFERENCES

Armstrong, T.J., 1986, Ergonomics and Cumulative Trauma Disorders, Hand Clinics, 2:553-565.

Clemons, P.L., 1982, A Compendium of Hazard Identification and Evaluation Techniques for System Safety Application, Hazard Prevention (Mar/Apr. 1989), pp. 11-18.

Chaffin, D.B. and Andersson, G., 1984, Occupational Biomechanics, (New York: John Wiley), pp. 191-203.

Durnin, J. and Passmore, R., 1967, Energy,Work,and Leisure, (London: Heineman Educational Books).

Garg, A., Chaffin, D.B., and Herrin, G.D., 1978, Prediction of Metabolic Rates for Manual Materials Handling Jobs, American Industrial Hygiene Association Journal, 39:661-674.

Gilbreth, F.B., 1911, Motion Study, (New York: Van Nostrand and Co.).

Keyserling, W.M., 1986, A Computer-aided System to Evaluate Postural Stress in the Workplace, American Industrial Hygiene Association Journal, 47: 641-649.

Keyserling, W.M, and Wittig, S.J., 1988, An Analysis of Experts' Ratings of Ergonomic Stress, International Journal of Industrial Ergonomics, 2:291:304.

Keyserling, W.M., Punnett, L., and Fine, L.J., 1988, Trunk Posture and Back Pain: Identification and Control of Occupational Risk Factors, Applied Industrial Hygiene, 3:87-92.

Keyserling, W.M. and Armstrong, T.J., 1989, Job Evaluation Procedures to Identify Ergonomic Stress. In: Proceedings: Work Design in Practice -- The Third International Occupational Ergonomics Symposium, Zadar, Yugoslavia, edited by E.N. Corlett, J.R. Wilson, C.M. Haslegrave, and I. Manenica, (London: Taylor and Francis), (in press).

Jonsson, B., Persson, J., and Kilbom, A., 1988, Disorders of the Cervicobrachial Region Among Female Workers in the Electronics Industry, International Journal of Industrial Ergonomics, 3:1-12.

Mattila, M., 1989, Improvement in the Occupational Health Program in a Finnish Construction Company by Means of Systematic Workplace Investigation of Job Load and Hazard Analysis, American Journal of Industrial Medicine, 15, 61-72.

McCormick, E.J., Mecham, R.C., and Jeanneret, P.R., 1969, The Development and Background of the Position Analysis Questionnaire (PAQ), (Lafayette, IN: Purdue University Occupational Research Center).

Niebel, B.W., 1988, Motion and Time Study (eight edition), (Homewood, IL: Irwin) pp.338-348.

National Institute for Occupational Safety and Health (NIOSH), 1983, A Work Practices Guide for Manual Lifting, NIOSH Pub. 81-122, (Akron, OH: American Industrial Hygiene Association).

Putz-Anderson, V. (editor), 1988, Cumulative Trauma Disorders -- A Manual for Musculoskeletal Diseases of the Upper Limbs, (London: Taylor and Francis).

Putz-Anderson, V., Pizatella, T., and Tanaka, S., 1986, A Proposed National Strategy for the Prevention of Musculoskeletal Injuries, (Washington, DC: National Technical Information Service Publication No. 87-114-740), pp. 17-34.

Rohmert, W. and Landau, K., 1969, Arbeitswissenschaftliche Erhebungsverfahren zur Tatigkeitsanalyse (AET), (Bern-Stuttgart-Wien: Verlag Hans Huber.

Rohmert, W. and Landau, K., 1983, A New Technique for Job Analysis, (London: Taylor and Francis), pp. 13-14.

Taylor, F.W., 1911, The Principles of Scientific Management, (New York: Harper and Brothers).

Advances in Industrial Ergonomics and Safety II
Edited by Biman Das
Taylor & Francis 1990

WORKERS COMPENSATION STATUTES, COST SHARING AND INCENTIVES FOR PROVIDING SAFETY CARE

Pradyot K. Sen

Haas School of Business
University of California at Berkeley

Workers compensation statutes are often the sole recourse open to workers for compensation against work related injuries. Though accidents are compensated regardless of cause, these statutes provide a compensation that is usually less than the cost of the accident to an employee. This paper presents a stylized model where such cost sharing distorts incentives such that safety related preventive care is undersupplied by the employers as well as the employees.

INTRODUCTION

For work related injuries, benefits to the workers are governed by workers compensation statutes. Benefits recoverable under these statutes are the exclusive remedy available to workers.[1] Employers also operate under a presumption of strict liability and must pay the workers (or their heirs) a predetermined compensation, regardless of the cause of the accident, so long it is work related. Prosser (1971) observes that adoption of such compensation statutes has led to some compensation to *all* accidents though substantially below the workers' full cost of accident. This system, therefore, functions as a cost-share system which may seem fair to some since the employee may have contributory negligence to the cause of the accident.[2] However, the impact on efficient resource allocation of this rule is non-trivial, and that is the subject of this paper. I argue that given such a cost-share environment, the level of care for preventing accident is underproduced by both employer and employee, thus leading to a work environment that is less than optimally safe.

In the next section I first describe the model and a stylized setting. Then I set out the social optimum of the problem and compare it with the optimum arrived at by the individual's opportunistic behavior.

1. For some recent exceptions to the exclusivity rule see Birnbaum and Wrubel (1982).

2. At the beginning of the twentieth century, prior to worker compensation statutes, negligence was the basis by which industrial accident costs were determined. For a detailed discussion see Hirsch (1988).

MODEL

Assume that there are n identical workers in a firm. Let the work environment be characterized by x, the level of safety care provided by the employer. Let y be the level of safety care taken by each of the workers. Let $m(x)$ be the number of workers involved in accidents which is a function of only the work environment -- the level of care provided by the employer. Further, let $w(x,y)$ be the expected accident loss to each worker such that

$$w(x,y) \geq 0, \quad w_x(x,y) < 0, \quad w_y(x,y) < 0$$

where w is positive. In other words w is a strictly convex function of x and y where w is positive. Similarly, define $e(x,y)$ to be the expected accident loss to the employer so that the function $e(.,.)$ has properties similar to the $w(.,.)$ function. We also assume risk neutrality of all parties and abstract away from risk sharing issues.

THE SOCIAL OPTIMUM

The social goal is to minimize total accident costs.[3]

$$x + ny + e(x,y) + m(x)w(x,y) \tag{1}$$

Let (x^*, y^*) denote the unique socially optimal values of x and y that are positive. Then (x^*, y^*) satisfy the first order conditions.

$$e_x(x,y^*) + m(x)w_x(x,y^*) + m_x(x)w(x,y^*) \tag{2}$$

and

$$e_y(x^*,y) + m(x)w_y(x^*,y) = -n \tag{3}$$

Therefore, the level of care (x^*, y^*) is the socially optimal level of safety care that minimizes the total cost. The next section shows how this social optimal gets distorted due to individual opportunistic decision making, which is the main result of this paper.

THE INDIVIDUAL OPTIMUM

Consider the actual behavior of the employer and the employees. Each party will take the other's level of care as given and choose their own level of care. So we may arrive at a best response equilibrium that may not be socially optimal.

Assume that under the workers compensation statute, employers pay a fraction f of the total loss to workers, where $0 < f < 1$. Thus, the employer will choose x to minimize

$$x + e(x,y) + f.m(x).w(x,y) \tag{4}$$

We are now ready to state our main result.

Proposition 1

The level of care produced by the employer is less than the socially optimal level.

---

3. This is a well accepted objective function in the literature. For example, see Calabresi (1968), Shavell (1987). Also, for a similar analysis involving liability and deterance see Shavell (1987).

<u>Proof:</u>

The first order condition of the employer's problem is

$$e_x(x,y) + f\left[m(x)w_x(x,y) + m_x(x)w(x,y)\right] = -1 \qquad (5)$$

Let $\hat{x}$ solve this first order condition. Notice that $x^*(y)$, the socially optimal solution, will satisfy

$$e_x(x,y) + m(x)w_x(x,y) + m_x(x)(x,y) = -1$$

Since m_x, w_x and e_x are all <0 and the second derivatives are non negative, it follows that $\hat{x}$ must be less than $x^*(y)$.

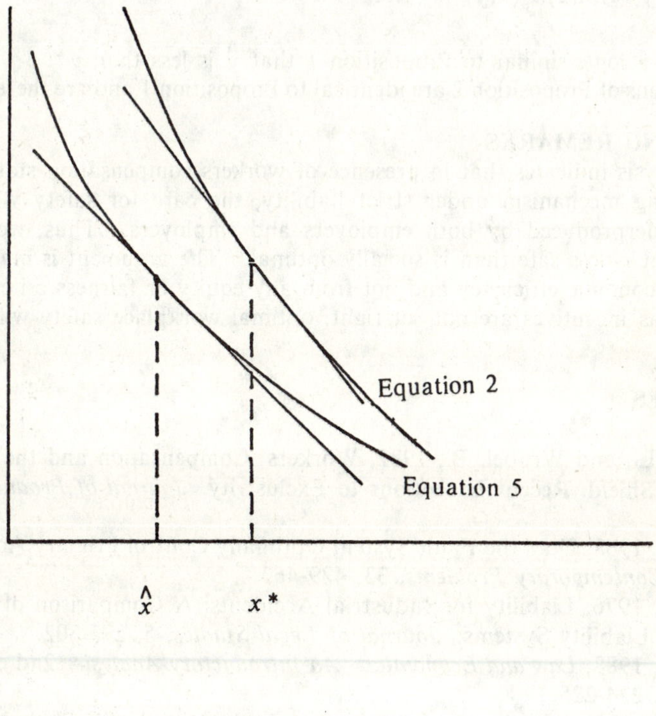

Diagram 1

It is easy to see the intuition of this proof from Diagram 1. Since f is a positive fraction, the left hand side of f is flatter than the left hand side of (2). Thus, the point where the slope is -1 is closer to origin for equation 5 than in equation 2, making $\hat{x} < x^*$.

The workers will also choose their level fo care sub-optimally. Given the employer's level of care x, m will be considered fixed by workers and the probability of an individual worker to get involved in an accident is simply $\frac{m}{n}$.

Individual workers would now minimize

$$y + (1-f)\frac{m(x)}{n}w(x,y).$$

We now present our second result.

Proposition 2

The level of care chosen by workers is less than the socially optimal level.

Proof:

The first order condition of the worker's problem is

$$(1-f)\frac{m(x)}{n}w_y(x,y) = -1. \tag{6}$$

Let $\hat{y}$ satisfy (6), which needs to be compared with the socially optimal solution $y^*(x)$ that satisfies

$$e_y(x,y) + m(x)w_y(x,y) = -n.$$

Follows from a logic similar to Proposition 1 that $\hat{y}$ is less than y^*.

The intuitions of Proposition 2 are identical to Proposition 1 and are therefore omitted.

CONCLUDING REMARKS

The analysis indicates that in presence of workers compensation statutes that act as cost sharing mechanism under strict liability, the care for safety will always be rationally underproduced by both employers and employees. Thus, we will have a workplace that is less safe than is socially optimal.[4] The argument is made only from the point of economic efficiency and not from any equity or fairness criteria. It seems that so long as incentives are not set right, optimal workplace safety will continue to elude us.

REFERENCES

Birnbaum, S.L., and Wrubel, B., 1982, Workers' Compensation and the Employer's Immunity Shield: Recent Exceptions to Exclusivity. *Journal of Products Liability*, 5, 119-144.

Calabresi, G., 1968, Does the Fault System Optimally Control Primary Accident Costs? *Law and Contemporary Problems*, 33, 429-463.

Chelius, J.R., 1976, Liability for Industrial Accidents: A Comparison of Negligence and Strict Liability Systems. *Journal of Legal Studies*, 5, 293-302.

Hirsch, W.Z., 1988, *Law and Economics: An Introductory Analysis.* 2nd ed. (Academic Press), pp. 224-225.

Prosser, W.L., 1971, *Handbook of the Law of Torts*, 4th ed. (St. Paul, MN: West), pp. 416-417.

Shavell, S., 1987, *Economic Analysis of Accident Law* (Cambridge, MA: Harvard University Press), pp. 32-46.

---

4. This is not to say that workers compensation statutes have not improved the situation from what it was. For a comparison and evidence of improvement of death rate under the new strict liability system, see Chelius (1976).

Advances in Industrial Ergonomics and Safety II
Edited by Biman Das
Taylor & Francis 1990

SAFETY STRATEGIES FOR THE 21ST CENTURY: FROM ACCIDENT PREVENTION TO SAFETY PROMOTION

Jorma Saari

School of Occupational Health
McGill University
1130 Pine Avenue West
Montreal, PQ, Canada H3A 1A3

When the basic principles of accident prevention were developed, in the early part of the century, occupational accidents were a rising problem that generated public anxiety. Consequently, motivation for better safety existed but knowledge was inadequate. Therefore, safety strategies put an emphasis on learning from injuries, and on information. Today, risks are often more attractive than frightening. Safety strategies should be amended for the 21st century to enhance motivation by making safety activities a rewarding experience. Experiments demonstrate that safety promotion with the positive approach is far more successful than expected.

INTRODUCTION

This article suggests that some of the basic approaches of safety should be re-evaluated; some of them are still useful, some should be updated and some completely rejected. Still, safety activities are more targeted to prevent accidents than to promote safety. In other words, the focus is on accidents, either potential or real, and not enough on factors determining safe working behaviour, and on the working environment. In safety research and in practice, learning from past accidents is not enough, other more prospective methods and ways to integrate safety and work are required.

It is over sixty years since the publication of the first edition of Heinrich's book where he listed his axioms of accident prevention (Heinrich et al 1980). Heinrich's axioms still form the basis of accident prevention in many companies.

The following are the most important basic ideas of accident prevention found in most textbooks: (1) Success in safety work depends upon a sound knowledge of what an accident is and how and why it occurs. (2) Accidents have causes, and can be prevented by removing their causes. (3) Worker failure is justified as the basic cause and the one deserving major consideration, even though the application of a mechanical remedy remains the obvious immediate solution. (4) Four basic methods are available for preventing accidents: engineering revision; persuasion and appeal; personnel adjustment; and discipline. (5) Management has the best opportunity to initiate accident prevention; management should therefore assume the responsibility (Heinrich et al 1980).

To exemplify the wide trust on the principles, some excerpts related to the fifth principle will follow:

* Commitment at the very top. This is essential for any progress to be made (Griffiths 1985)
* The low accident rate plants showed a greater management commitment to safety (Smith et al 1978)
* In the firms with better safety records top management is more highly interested and involved in the company's safety performance (Simonds & Shafai-Sahrai 1977)
* Management commitment to safety, that is, its overt concern and support for safety activities, represents a dominant factor in successful safety experience in industry (Cohen 1977).
* Since an individual manager might impede participation, support for the health promotion program from top management is important (Sloan & Gruman 1988)

So far, I have not come across studies formulating the achievement of good results in accident prevention without the support of top management, although that would be a most urgently needed piece of knowledge in these days when strict hierarchies are torn down. More and more managers accomplish their work without having formal authority to order people around. More and more often, organizations are networks rather than pyramids. There are management books that train managers in getting the job done without having authority over others (e.g., Cohen & Bradford 1990). Where is the book on the principles of achieving safety goals without authority shed by top management?

There is no doubt, however, that the old strategies have worked quite well. Accident prevention has been successful, for example, the number of fatal accidents at work in my home country, Finland, went down from 400 fatal accidents annually in the forties to 100 accidents in the eighties (Saari 1982), and similar improvements occurred in many countries. Unfortunately, the reduction has now stopped in Finland, as well as in most industrialized countries.

ATTRACTIVE RISKS

The focus of safety strategy is on accidents, their causes and on learning from accidents. The remedies are targeted to individual work places and persons, workers and managers. Perhaps, the most important underlying assumption is that people try to avoid the hazards they know.

But, this may not be true even if the person himself is at the risk of physical injury. It is probably less true for people not at direct physical risk, e.g. managers and designers. Accident risks also attract people, since risks are often connected to jobs and activities that are close to the current limit of human performance, or since risk taking is the only way to learn new skills. Testing one's skills may provide the challenge otherwise lacking from the job.

Improved safety is not a sufficient reward if the safety activity itself is punitive by nature. Accident investigations are an example of such an activity. It is an emotionally demanding task to contemplate what went wrong, and why.

Therefore, accidents are not investigated as well as they should be. Even reporting of near-accidents has proved to be a technique that can be used only in campaigns (Saari 1975). The reporting activity decreases rapidly. The interest falls within a couple of weeks so much that it is better to take a few weeks' break. This is because reporting mishaps is punitive to the reporting person. The question of quilt and blame comes too near.

A review of Swedish experiments on near-accident reporting demonstrated that the amount of information about near-accidents increases. However, there is little evidence of a change in safety activities or safety consciousness, and no evidence of a change in accident frequency following a near-accident reporting period (Carter & Menckel 1985).

SAFETY ACTIVITIES SHOULD BE PLEASANT

If a safety activity is unpleasant, it is easy to find other more urgent projects. Sometimes, people are so preoccupied with the security of their near future that physical safety gets lower priority. This is well demonstrated by our experiences during a study in a plywood factory located next to a paper mill, and a saw mill. Together they formed an integrated wood processing plant. Then, in the middle of an intervention study (Seppälä et al 1987), the plywood factory was sold to another company because of its insufficient profitability, as explained by the owner. Actually, the later events proved that the owner started to implement a new strategy by concentrating on the metal industry and withdrawing from wood processing.

We asked the safety staff to improve accident investigations in the plywood factory, which they did. Accident causes were better identified, and remedies better planned. However, the actions directly related to profitability usually won over the implementation of remedies, and only few corrective actions were taken. The situation kept the staff preoccupied with profitability. Other experiments have yielded similar results: despite better knowledge on accident causes, remedies are not implemented (Successful... 1987)

The staff of the plywood factory misjudged the situation because they did not know the new strategy. They were too worried about the continuation of job opportunities; the shut-down of the plant was not as close as they thought.

I may divert slightly from my main theme, which is occupational safety, but what Black & Decker, a well-known manufacturer of power tools, went through in the early 1970s teaches us the importance of positive consequences and experiences attached to safety (Penzias 1989). The management of Black & Decker faced the need to redesign its line of power tools to provide "double insulation" - an additional insulation barrier to protect users from shock if the main insulation fails.

The competitors had provided double-insulated products at a 15 to 20 percent premium. The management of Black & Decker decided to develop a totally new design which would meet the safety requirement and which would also reduce production costs. They organized a comprehensive design project which produced several technological and organizational innovations. From 1967 through 1980 the cost of Black & Decker products fell by over 50 %. The products made in the 1960s would now cost five times as much to produce. Consequently, Black & Decker shook off its domestic competitors in the market. I am quite convinced that without foreseeing the other benefits, the management would have been reluctant to put money into developing the new safety barrier.

SAFETY HAS TO EARN ITS PRIVILEGES

Safety is not a privileged area. In these days of flourishing monetarism, everything tends be measured in the terms of money. If a safety project is not directly profitable, as it was to Black & Decker, it has to offer other advantages, for instance, better worker morale, a better organizational climate, etc. If a safety action offers peace of mind to a boss, it has far better chances to become accepted than something giving extra worries or bad feelings to him.

If an occupational health and safety project is

* technologically
* administratively

POSSIBLE

* economically PROFITABLE

or it offers other advantages
* industrial relations
* image
* publicity

and it

wins the competition with other projects in
* marketing
* production
* product design, etc

only then does a project have a chance to be implemented.

Figure 1. An occupational health and safety project has to pass several tests before being implemented. Better safety is just one of the aspects covered.

An OHS program has to overcome several obstacles in order to be implemented (Figure 1). Firstly, it has to be technically, administratively, etc, possible. If, for example, a safer substance does not have sufficient technical properties, it hardly substitutes the present one.

Secondly, the intervention program should make production more profitable or, alternatively, offer other advantages, such as better industrial relations, a better company image, free publicity, etc. A number of advantages can be considered. It is impossible to prepare a general list, since many of the advantages depend on the local culture, and vary from time to time.

For instance, a company managed by a young newly graduated engineer may like to invest on the state-of-art technology, while another company might be satisfied with upgrading its current technology, since it has an owner who plans to retire within a couple of years, and does not want to take the risks the newest technology inevitably implies. An OHS action, which does not satisfy the values of the two managers, will embarrass them, and will hardly become accepted.

If the projected outcome of an OHS action is acceptable, the proposal still has to compete with other acceptable projects representing other business areas, such as product design, marketing, and production. If other projects are more fascinating for the implementors, their thoughts are occupied by those projects and, consequently, the OHS project is put on the waiting list.

These are quite obvious viewpoints and most people probably accept them. The practice in OHS rests, however, firmly on the assumption that OHS has the priority over other business areas by virtue of statutory requirements, moral and ethical arguments, fear of injury, etc. In reality, a pipe fitter, for example, may be much more worried about his reputation as a craftsman than about getting hurt at work. Therefore he uses a defective ladder to fix a leaking valve first, and only afterwards reports the defect to the person in charge.

TWO INTERVENTION STUDIES AT SHIPYARDS

Two studies at shipyards demonstrate well that information-based safety strategy has only a marginal effect, while another strategy aiming at enhanced motivation was more effective than expected. In the first case, an information campaign was designed to meet all the criteria of an effective campaign (Saarela et al 1989). The campaign was tailored to the shipyard, and run by the safety staff. The experimental campaign was arranged in the first ships of two production series. The second ships of the series served as comparison ships.

It was found that the messages were noticed, they were well recalled in interviews, and the interviewed persons found the campaign worthwhile. Yet, there was no effect on accidents. This is probably because we just informed about the hazards and believed that people try to avoid the hazards they know. However, there are a number of other factors overweighing in the decisions concerning daily working habits.

The other experiment, four years in length, was then more effective than expected; it reduced the number of accidents by two to three times more than theoretically possible (Saari &

Näsänen 1989). The technique used was performance feedback, a technique described thoroughly by, e.g., Petersen (1989). The immediate purpose was to improve housekeeping in two production halls of a shipyard.

Housekeeping was radically improved, and the follow-up observations lasting two years showed that the new standard of housekeeping was well maintained. Performance feedback had an immediate effect. Within a week housekeeping behaviour was much better in both halls of the experiment. The experiment was designed so that better housekeeping could reduce accidents by 25 % at a maximum. The accidents went down by 70...80 %, however, and the effect lasted throughout the three years' follow-up.

The design of the experiment does not allow us to infer the mechanisms by which the results were achieved. It is probable that the experiment initiated other organizational processes which then led to the final outcome. The experiments were perceived positively, and it was quite an experience for everybody, that such great changes occurred within a few days. This experience strengthened their self-confidence and their determination to tackle other problems. Since better housekeeping was considered useful also to other goals, such as productivity, quality, and image, it brought safety into a positive context.

THE RENEWED FOCAL POINTS OF SAFETY STRATEGY

Table 1 describes some of the dimensions along which the focal point of safety programs should be shifted in order to meet better the current requirements. Presently, the primary driving forces in safety are negative, such as authority to provide unpleasant consequences to people and companies unless they act safely and comply with regulations. In future, the positive outcomes of safety actions should be emphasized more, and the participants of safety activities should feel that they are better off in future.

TABLE 1. Some of the dimensions along which the focal point of safety programs should be changed.

Current focus	Focus in future
Based on authority over people and companies	Influence without authority
The primary driving force:	
* fear of punishment	* positive expectations from future
Externally forced	Internally activated
Information-oriented	Motivation-oriented
Injury prevention	Safety interventions
Retrospective	Prospective
Accident analysis	Job and safety analysis
Accident-oriented	Safe work environment and the quality of life at the workplace

Currently, many safety activities are initiated, and controlled by external resources, and pressure, e.g. OHS administration, labour market organizations, etc. Most industrialized countries increased considerably their external resources in the seventies. The results were favorable, but it is now time to change the origin of activation. Outside organizations should support and serve companies with advices, but the initiative should arise internally in companies. Inducing safety innovations in companies with the help of external resources is an example of an urgent research need. Perhaps, Marcus' study (1988) on nuclear power industry is almost the only study on the subject.

The major weakness of safety programs is their restricted view. Their scope is narrowed into accidents, and much more should be known about the work itself and the factors influencing the choices while conducting a task. Accidents are rare events; an average industrial worker has a compensable injury once in five or ten years. For a small company or for a small department of a larger company, accidents are not even monthly events. The chances of personal and organizational learning are scarce.

Other factors than experience from injuries guide working behavior. The factors having most influence on working behavior should therefore be thoroughly known, and they should be the starting point when developing control measures for accidents.

All-in-all, safety activities should be more future-oriented, and less retrospective than today. Motivation is a greater problem than the lack of information. Perhaps, we should stop talking about occupational health and safety and, instead, start demanding better quality of life at the workplace. Employers should be made aware of the adverse effects on productivity that an unsafe and bad work environment has.

REFERENCES

Carter, N. and Menckel, E., 1985, Near-accident reporting: A review of Swedish research. Journal of Occupational Accidents, 7, 41-46.

Cohen, A., 1977, Factors in successful occupational safety programs. Journal of Safety Research, 9, 168-178.

Cohen, A.R. and Bradford, D.L., 1990, Influence without authority, (New York: John Wiley & Sons), 319 pp.

Griffiths, D.K., 1985, Safety attitudes of management. Ergonomics, 28, 61-67.

Heinrich, H.W., Petersen, D. and Roos, N., 1980, Industrial accident prevention, (New York: McGraw-Hill), 468 pp.

Marcus, A.A., 1988, Implementing externally induced innovations: a comparison of rule-bound and autonomous approaches. Academy of Management Journal, 31, 235-256.

Penzias, A., 1989, Managing in a high-tech world. PC-Computing, 2, 109-115.

Petersen, D.C., 1989, Safe behavior reinforcement, (Goshen, NY: Aloray), 181 pp.

Saarela, K.L., Saari, J. & Aaltonen, M., 1989, The effects of an informational safety campaign in the shipbuilding industry. Journal of Occupational Accidents, 10, 255-266.

Saari, J., 1976, Efficiency of three techniques for gathering data on near-accidents. Control, 65-69.

Saari, J., 1982, Long-term development of occupational accidents in Finland. Scandinavian Journal of Work Environment Health, 8, 85-93.
Saari, J. & Näsänen, M., 1989, The effect of positive feedback on industrial housekeeping and accidents: a long-term study at a shipyard. International Journal of Industrial Ergonomics, 4, 201-211.
Seppälä, A., Saarela K. L., Näsänen, M., Aaltonen, M. & Saari, J., 1987, Improving safety performance of industry. In Trends in ergonomics/human factors IV, edited by S.S. Asfour (Amsterdam: Elsevier Science Publishers B.V.), pp. 513-520.
Simonds, R.H. & Shafai-Sahrai, Y., 1977, Factors apparently affecting injury frequency in eleven matched pairs of companies. Journal of Safety Research, 9, 120-127.
Sloan, R.P. & Gruman, J.C., 1988, Participation in workplace health promotion programs: The contribution of health and organizational factors. Health Education Quarterly, 15, 269-288.
Smith, M.J., Cohen, H.H., Cohen, A. & Cleveland, R.J., 1978, Characteristics of successful safety programs. Journal of Safety Research, 10, 5-15.
Successful accident prevention: recommendations and ideas field tested in the Nordic countries, 1987, (Helsinki: Institute of Occupational Health), 77 pp.

Advances in Industrial Ergonomics and Safety II
Edited by Biman Das
Taylor & Francis 1990

A BEHAVIORAL APPROACH TO IMPROVE SAFETY AT THE WORKPLACE

Paul S. Ray
Department of Industrial Engineering, University of Alabama, Tuscaloosa, AL 35487

Jerry L. Purswell
Robert E. Schlegel
School of Industrial Engineering, University of Oklahoma, Norman, OK 73019

An experiment was conducted to determine the effectiveness of a behavioral approach for improving the safety behavior of workers in a large aircraft maintenance facility. The approach consisted of (1) identifying a number of typical tasks of a hazardous nature, (2) safety training concerning significant safety elements of these tasks, (3) sampling the incidence of compliance with safety procedures on the shop floor, and (4) informing the workers regarding their safety performance on these tasks (feedback). The safety program was introduced for an experimental group of workers while a similar group (control group) continued to work without any intervention from the program. This approach resulted in a significant improvement in the safety performance of the experimental group within a five-week period, while the performance level of the control group remained virtually unaltered.

INTRODUCTION

The exact number of workplace related deaths and injuries is disputed. The U.S. Office of Technology Assessment (OTA, 1985) estimated that there are approximately 6,000 deaths per year due to workplace related injuries. In addition, there are approximately 10,000 injuries that result in lost work time and 45,000 injuries that require medical attention per year. The National Safety Council (1986) reported that over 11,000 deaths and about 2 million disabling injuries occurred in the workplace environment in 1985 and estimated the cost of work accidents to be $37.3 billion for 1985. These figures do not include cases of workplace-related illnesses.

Despite the progress made in developing safety standards, codes and regulations during the past sixty years, the current high level of deaths and injuries appears to be due to difficulties faced in implementing safety procedures. Grimaldi (1980) observed that about nine out of ten accidents that occur in the workplace can be predicted. It is evident from the high annual injury level that knowledge about hazard-control is not applied in the workplace. The Occupational Safety and Health Act of 1970 (OSHA) was an attempt to induce the application of safety procedures. However, legal requirements do not in themselves insure the application of safety knowledge. To a large extent, following safety procedures depends on the voluntary behavior of the worker.

Komaki (1983) observed that there is very little, if any, positive reinforcement for working safely. However, reinforcement in various forms has been found to improve safety behavior. Social reinforcement in the form of praise has been shown to improve safe working habits among workers (Smith et al., 1978). Feedback on safety performance was found to improve safety behavior in a wholesale bakery (Komaki, et al., 1978) and in a vehicle maintenance facility (Komaki, et al., 1980). Zohar (1980) found that a behavioristic approach was effective in modifying workers' earplug wearing behavior. The current research was conducted to determine if positive feedback of safety performance is effective in improving the safety behavior of workers in a large industrial setting.

METHODOLOGY

Setting and Personnel

The study was conducted at a major maintenance facility for military aircraft and jet engines, employing about 20,000 workers. The majority of the workers consisted of machinists, welders, and general fitting technicians. The main shop of the plant, where the experiment was conducted, consisted of a very large (about 3/4 of a mile long) shop floor compartmentalized into different sections processing different components of the jet engines.

Two departments with similar hazard levels and safety records were selected for the study. Each department had approximately 200 workers in direct production operations. A group of 100 workers was selected from one department to be treated as the experimental group. These workers represented three units within the department. Another group of 100 workers was selected from the other department to form a control group. The workers selected from each of the two groups were engaged in relatively hazardous tasks. The experimental and control groups were separated by a considerable distance and the prevalent environment of partitioning the very large shop floor provided the isolation needed for the experiment.

To obtain unbiased and accurate evaluations of safety behavior, seven safety training leaders were selected. Three observers recorded sampling observations for the experimental group and four observers recorded data for the control group. These training leaders were familiar with the operations in the selected departments.

Procedure

The first phase of the study involved identifying desired safety practices for a number of selected tasks in each department. The tasks were grouped into the following categories: pressure checking, hand reworking, welding, and machining. The significant safety elements for these operations were categorized as (1) using appropriate protective clothing, (2) removing inappropriate personal clothes/jewelry, (3) performing materials handling activities safely, and (4) performing machining operations safely. Three to five significant safety elements in each category were identified based on a review of injury records and consultation with supervisory personnel. Safety training was provided to both groups with respect to appropriate safety behavior for all observed tasks.

The listed safety elements were the basis for determining a Safety Performance Index for the workers. Observations by the training leaders were recorded at randomly chosen times once each day, five days a week for a total of seven (7) weeks. Observations were taken during the day shift, at some time between 6:45 a.m. and 3:30 p.m. when the plant was in full operation. Each observation session lasted for 10 to 20 minutes. The observer walked through the work area, observed the workers assigned to him or her for observation, and recorded the observation on the safety survey sampling form. The observations were recorded without identification of individual workers to maintain the confidentiality of the data.

Safe behavior was noted by writing "S" in the appropriate cell for the worker, the operation, and the day. Unsafe behavior was noted by writing the code for the unsafe act (e.g., no safety glasses used as required = 1a). If a safety element was not observed, "O" was noted. Multiple unsafe acts on the same sampling occasion were identified with a "U" and the multiple acts were cited for corrective action in the next safety meeting for the experimental group. Based on the observations, a Safety Performance Index was calculated as the number of safe observations divided by the total number of safe and unsafe observations.

Safety observations from each group were collected for the first two weeks without any feedback on safety performance to the workers. These sampling data were used to determine baseline performance levels for each group. Safety performance feedback was then introduced to the experimental group for the next five weeks while the control group continued with no feedback. During the first safety meeting, unit supervisors explained the baseline safety levels and set a goal of a safety index of 98.00% plus for the experimental group. This was considered realistic. The safety index was calculated each week for each of the three units of the experimental group as well as for the experimental group as a whole. Safety performance levels were continuously on display in the break areas of the three units and updated weekly.

Safety Performance Feedback

The three units of the experimental group held weekly safety meetings lasting about 15 minutes. The meeting agenda addressed the following:
(1) the previous week's safety performance index,
(2) comments on past achievement and encouragement to improve further,
(3) safety violations observed during the previous week and corrective actions required, and
(4) suggestions from the workers that could lead to improvement in safety performance.

Reliability Assessment

Interobserver reliability was assessed once each week during the last five weeks of the study. Two observers observed the same workers at the same time and recorded the observations independently without communication. These records were compared and if the observers recorded the same observation for an individual, an agreement was scored. If the observers differed in their observation for an individual, a disagreement was scored. A reliability index was computed as the number of agreements divided by the total number of observations.

RESULTS

The interobserver reliability index varied from 90.97% to 100.00% with a mean of 97.56%. Baseline safety index levels for the two groups were developed from the observations during the first two weeks of the study. During this period, 937 and 1040 observations were made for the experimental and control groups respectively. Based on these data, the mean baseline safety indices were computed as 84.3% for the experimental group and 88.2% for the control group (Table 1).

At the beginning of the third week of sampling, feedback on safety performance was provided for the experimental group and continued for the next five weeks. During this treatment phase, the safety index for the experimental group improved continuously, reaching a perfect safety index score of 100% by the end of the fifth week of treatment (Figure 1). The safety index for the control group, however, remained at approximately the baseline level. The sampling observations during the treatment phase are also shown in Table 1. The Safety Index values for all three experimental units are given in Table 2.

The most frequently occurring unsafe behaviors for the experimental group were failure to (1) wear earplugs (49.5%), (2) wear safety glasses (11.8%), (3) wear face shields (10.3%), and (4) contain long hair (9.3%). For the control group, the unsafe behaviors were failure to (1) remove watch while working (26.4%), (2) wear safety gloves (20.6%), (3) use both hands while manually lifting materials (20.6%), and (4) keep the workplace tidy (11.2%). There appeared to be some potential for improvement in the materials handling methods used by the control group workers.

The research data consisted of frequencies in three discrete categories (safe, unsafe, and not observed) for two independent groups. Hence, a Chi-Square analysis was used to determine the significance of the difference between the two groups. The Chi-Square for two degrees of freedom for a 3 by 2 contingency table was calculated as 111 ($p < 0.001$). Thus, the safety behavior of the experimental group was significantly better than that of the control group.

Table 1. Observations and Safety Index Values (%) for Baseline and Treatment Periods.

Week	Experimental Group				Control Group				Total
	S	U	O	SI	S	U	O	SI	
1	330	52	118	86%	381	50	104	88%	1035
2	324	70	43	82%	372	51	82	88%	942
Total (Baseline)	654	122	161	84%	753	101	186	88%	1977
3	380	39	71	91%	357	41	107	90%	995
4	380	27	44	93%	317	39	89	89%	896
5	341	14	85	96%	343	53	109	87%	945
6	388	2	45	99%	365	50	90	88%	940
7	364	0	41	100%	341	36	94	90%	876
Total (Treatment)	1853	82	286	-	1723	219	489	-	4652

Legend: S - Safe; U - Unsafe; O - Not Observed; SI - Safety Index

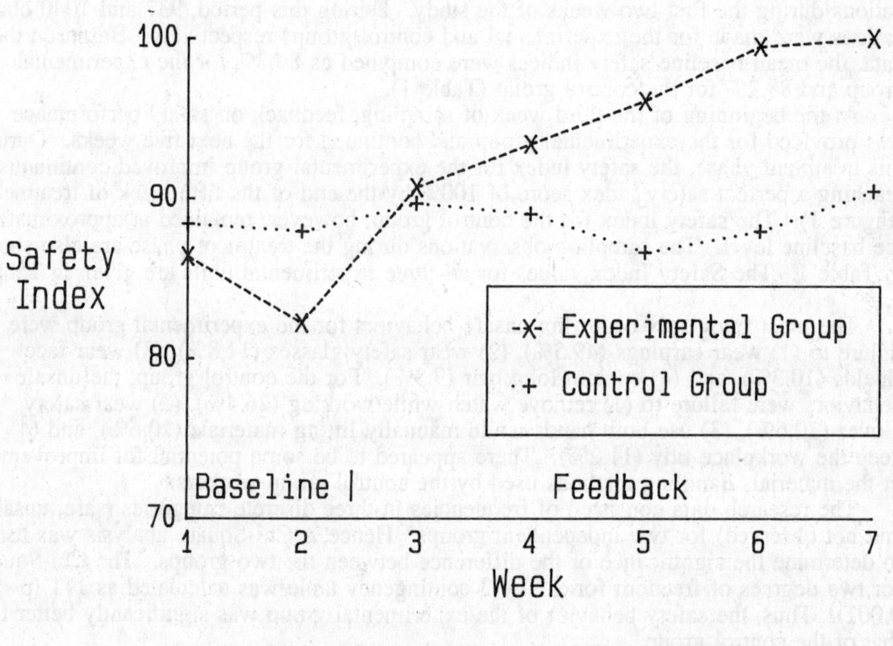

Figure 1. Safety Performance Index (%) Changes for the Experimental and Control Groups.

Table 2. Safety Index Values (%) for Experimental Units.

Unit	Week 1	2	3	4	5	6	7
A	93.7	89.7	98.7	95.1	97.5	100.0	100
B	77.0	61.2	86.5	93.0	99.7	99.3	100
C	85.4	89.0	85.6	92.1	94.3	99.2	100

COST/BENEFIT ANALYSIS

The demonstrated effectiveness of the feedback motivational program makes it an attractive option for any organization to improve the safety performance of its workers. However, the costs of implementing any safety program are frequently reviewed by management in terms of the expected benefits of the program. In this instance, the benefits are the costs of accidents and injuries avoided in the future. The costs to implement the program will vary as a function of the size of the organizational unit. Likewise, the benefits of accident avoidance will vary with the individual organization and will need to be adjusted as necessary.

The feedback safety program is intended to be applied selectively for departments or trades having a poor safety record. The annual cost of implementing a behavioral safety program using a sample size of 50 workers is estimated to be $10,000. Based on the estimate provided by the National Safety Council (1986), the real cost of a lost-time accident may range from $20,000 to $50,000 per occurrence. Hence prevention of only one accident per 100 workers in a two to three year period justifies implementation of the safety feedback program. Also, a safe work environment may instill a sense of well-being and help to promote higher productivity. The safety feedback program will help the organization to meet its other important intangible objectives of protecting and insuring the best utilization of its human resources.

DISCUSSION

This study demonstrated the effectiveness of a simple "feedback program" in improving the safety behavior in a large industrial plant having a large amount of sophisticated manufacturing machinery. The results agree with the findings of Smith et al. (1978), Komaki et al. (1980) and Sulzer-Azaroff et al. (1980) regarding the effectiveness of such programs. All three experimental units showed a significant improvement in safety behavior with the implementation of the feedback program (Table 2). The safety behavior of the control group remained at about the same level during the entire period of the experiment. The only difference in the treatment between the two similar groups was the implementation of the safety feedback program for the experimental group.

There appeared to be some additional positive side effects of the feedback program. For example, the work area was tidier and presented a neater appearance after the introduction of the program. Workers became aware of safety and posed questions about safety practices at their workplace. An increased awareness about safety was found among the supervisors as well. The simplicity and ease with which the program was implemented induced senior management to consider its adoption for improving safety at the workplace throughout the facility.

Two important features of the program were the applied operant method of measuring the safety performance level and identification of the safety elements that could be addressed to improve the safety level. Another feature of the program was the focus on safety behavior rather than on accidents. Accidents being remote events, a focus on accident reduction is likely to be less effective. This is because the consequences of performing unsafely are only remotely associated with accidents and there is little immediate feedback to indicate if the effort to control accidents is fruitful.

Worker reaction to the feedback program was favorable. Often workers checked their unit's safety performance index plot to see how their unit was doing. At times, workers reminded their fellow workers to follow the relevant safety procedure. An informal competition occurred among the three units of the experimental group to achieve a better safety performance index.

Further research should be conducted to measure the contributions of each component of the feedback program, namely: training, goal setting, performance feedback, and encouragement to improve safety performance. Studies are required to determine the relationship between a behavioral measure of safety and the occurrence of accidents in different types of organizations. Research is also required to determine more explicitly the long-term effect of the program on the frequency of different severities of injuries and accidents. In addition, a study is required to investigate the frequency of performance feedback that is required to maintain or improve safety behavior of workers over longer time intervals. Other areas of research would be to explore the effects of safety programs on productivity, employer-employee relations, and other organizational factors.

REFERENCES

Congress of the United States, Office of Technology Assessment, April 1985, *Preventing Illness and Injury in the Workplace,* OTA-H-256, (Washington, D.C.: U.S. Congress Office of Technology Assessment).

Grimaldi, J.V., October 1980, Hazards, harms and hegemony. *Professional Safety,* 21, 481-486.

Komaki, J.L., September 1983, A behavioral approach to work motivation. In: *Proceedings of the International Seminar on Occupational Accident Research,* Saltsjobaden, Sweden, pp. 67-97.

Komaki, J., Barwick, K.D., and Scott, L.R., 1978, A behavioral approach to occupational safety: pinpointing and reinforcing in a food manufacturing plant. *Journal of Applied Psychology,* 63(4), 434-445.

Komaki, J.L., Heinzmann, A.T., and Lawson, L., 1980, Effect of training and feedback: component analysis of a behavioral safety program. *Journal of Applied Psychology,* 65(4), 261-270.

National Safety Council, 1986, *Accident Facts*. (Chicago, Illinois).

Ray, P.S., 1988, *Determinants of Safe Behavior in Occupational Settings,* unpublished doctoral dissertation, (Norman, OK: School of Industrial Engineering, University of Oklahoma).

Smith, M.J., Anger, W.K., and Ulsan, S.S., 1978, Behavioral modification applies to occupational safety. *Journal of Occupational Research,* 10(2), 81-88.

Sulzer-Azaroff, B., and de Santamaria, M., 1980, Industrial safety hazard reduction through performance feedback. *Journal of Applied Behavior Analysis,* 13, 287-296.

Zohar, D., 1980, Promoting the use of personal protective equipment by behavior modification techniques. *Journal of Safety Research,* 12, 78-85.

XX

WORKSTATION DESIGN

Advances in Industrial Ergonomics and Safety II
Edited by Biman Das
Taylor & Francis 1990

STANDING ON CONCRETE VS. FLOOR MATS

Konz, S., Bandla, V., Rys, M. and Sambasivan, J.

Dept. of Industrial Engineering
Kansas State University, Manhattan, KS 66506 USA

Four surfaces (concrete, a San-EZE II mat, an Optimat mat, and a Footsaver mat) were stood on for 90 min each by 20 university students. All three mats were significantly more comfortable than concrete. Comfort seems to be inversely related to mat compressibility.

INTRODUCTION

During standing work, some surfaces are better than others. This paper is a continuation of our previous work (Konz et al., 1988; Rys and Konz, 1988; Rys and Konz, 1989a; Rys and Konz, 1989b) investigating standing.

Briefly, the studies demonstrated, during standing for 60-240 min, that foot and ankle swelling is not a major problem. More sensitive criteria are heart rate, leg/foot skin temperature, and, especially, subjective votes on body comfort. Although comfort decreases the most in the feet and lower legs, comfort also decreases in the back, neck and shoulder.

Rys and Konz (1989a) evaluated concrete vs. a 22 mm thick San-EZE mat and a 9.5 mm thick Traction mat. Both mats were better than concrete in the 60 min trial.

Instep temperature increased to 31.7°C for San-EZE and to 30.9°C for Traction mat vs. (30.3) for concrete. Calf temperature increased to 30.0°C for San-EZE and to 29.6°C for Traction mat vs. 28.1°C for concrete.

Footprint area (vs. concrete) decreased 1.2% for Traction mat and 1.7% for San-EZE.

Comfort was worst on concrete. The Traction mat was especially comfortable for the back while the Saneze was more comfortable for the leg and foot.

METHOD

Task

In addition to concrete, three mats were tested. See Table 1 and Figures 1, 2 and 3. To standardize the footwear, each subject was furnished a pair of thick, ankle-high, cotton socks and slippers. The slippers had a hard sole (21 mm thick in the rear and 14 mm in front); the toe and heel were open but there were two bands across the midfoot.

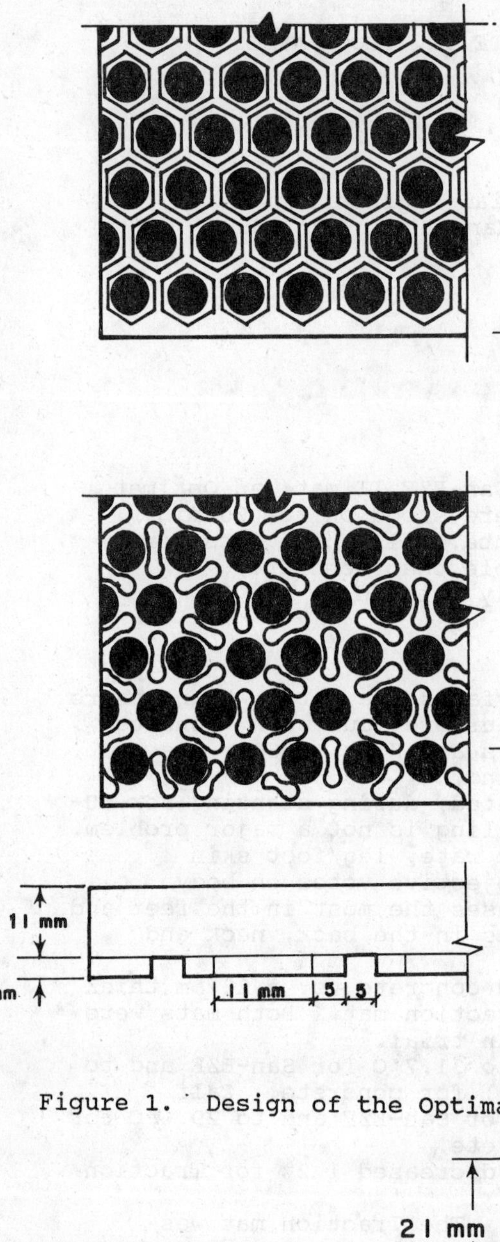

Figure 1. Design of the Optimat mat.

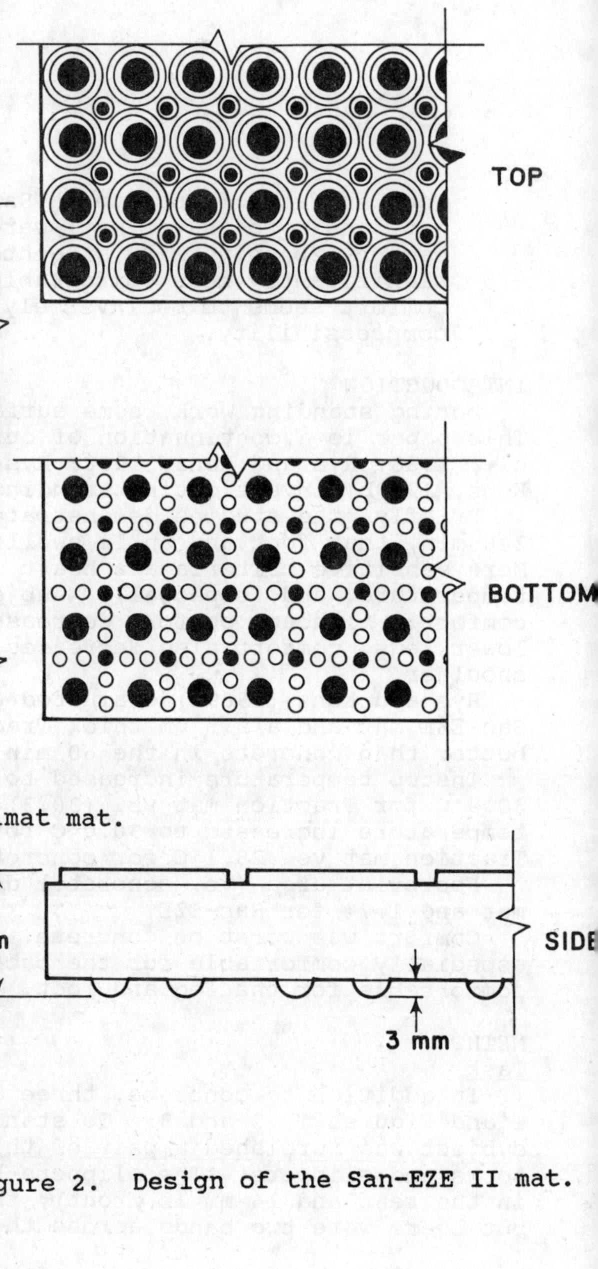

Figure 2. Design of the San-EZE II mat.

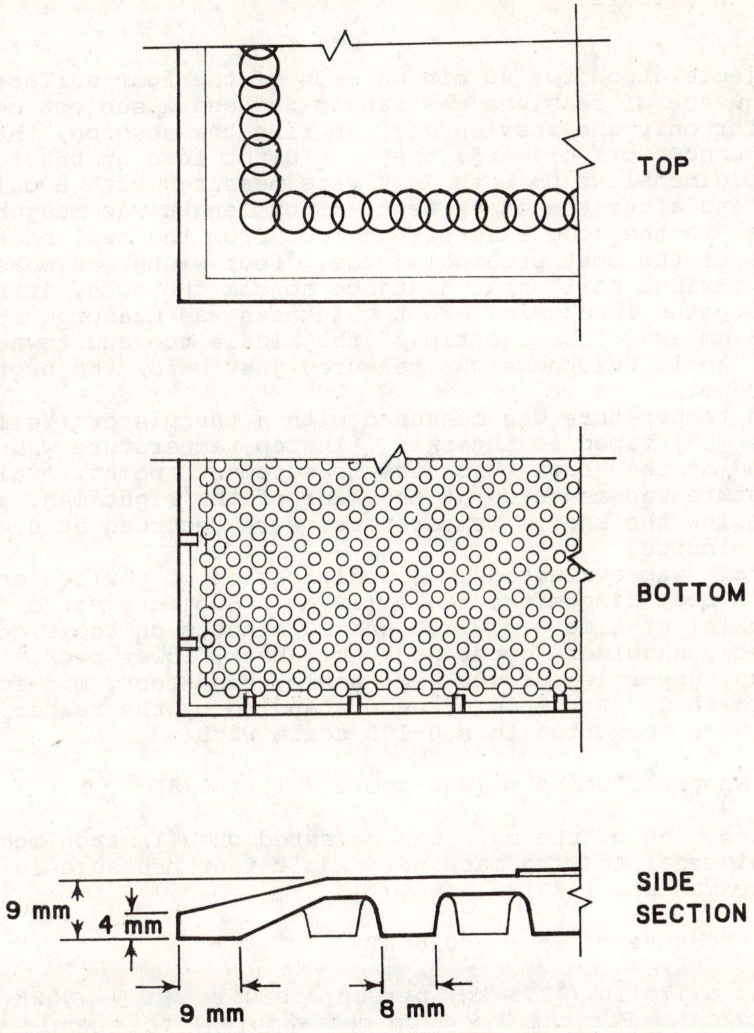

Figure 3. Design of the Footsaver mat.

Subjects

The 20 Industrial Engineering students received extra credit in a course.

Procedure

Subjects stood for 90 min on each of the four surfaces. The sequence of sessions was randomized and a subject could stand for only one session/day. During the session, they could do homework or read; they could not lean on the table.

Foot dimensions on both feet were measured with a caliper before and after the experiment. Foot length was measured parallel to the long axis of the foot from the heel back to the tip of the most protruding toe. Foot width was measured as the maximum horizontal distance across the sole, at right angles to the long axis. Foot thickness was measured at a point 3 cm away from the tip of the middle toe and toward the ankle. Ankle thickness was measured just below the protruding ankle bone.

Skin temperature was measured with a thermistor (Yellow Springs 709) taped to the skin. Instep temperature was measured at the "foot thickness measurement" point. Calf temperature was measured at the rear of the right leg, about 10 cm below the knee. Temperatures were recorded at 0, 30, 60 and 90 minutes.

Comfort was evaluated by a modification of Corlett and Bishop's body discomfort scale (1976). Subjects voted (on a 1 to 9 scale) at time 0, 30, 60 and 90 minutes on their comfort at: neck, shoulder, upper back, mid-back, lower back, buttocks, upper leg, lower leg, ankle, hind-foot, mid-foot, and fore-foot. For ease of understanding by the reader, all scores were converted to a 0-100 scale with:

$$\text{Reported score} = (\text{Raw score} - 1)(100/8) \qquad (1)$$

Compression of the mats was measured on a Instron model 1122 universal testing machine. Adult foot pressure is (Konz and Subramanian, 1989):

$$\text{Pressure} = .15 + .0026\ \text{WT} \qquad (2)$$

Assuming a 170 lb (77.3 kg) person, this is .15 + .0026(77.3) = .35 kg/cm^2. For the 7 x 7 cm mat samples, this would give 17.1 kg; we used an 18 kg force. See Table 1 for the mat compressions.

Table 1. Floor surface characteristics of this study and Rys and Konz (1989a).

Floor surface	Experiment	Thick-ness,mm	Compres-sion,mm	Compres-sion,%
Concrete	Both	-	0	0
San-EZE	R & K	22	.9	4.1
Optimat	This	12	.7	5.8
Traction mat	R & K	9.5	.7	7.4
San-EZE II	This	18	1.7	9.4
Footsaver	This	7	1.3	18.6

RESULTS

Foot dimensions

Changes in foot length, width, thickness and ankle thickness were calculated by subtracting the initial measurement from the final measurement. For each variable, an ANOVA was calculated. Floor surfaces did not significantly affect any of the four dimensions.

Instep temperature

For each condition and subject, temperature at time 0 was subtracted from the temperature at 30, 60 and 90 minutes. An ANOVA and LSD test indicated that time was significant ($p < .0001$) but that floor surface was not. There was a significant difference between the initial skin temperature (32.4 C) and the temperature at 30, 60 and 90 but not among the values at 30, 60 and 90. The mean increase in instep temperature was 2.3 C.

Calf temperature

For each condition and subject, temperature at time 0 was subtracted from the temperature at 30, 60 and 90 minutes. An ANOVA and LSD test indicated that time was not significant but that floor surface was ($p < .02$). Table 2 shows that calf tem-perature increased .3 C for concrete but was small or negative for the three mats. Initial temperature averaged 30.4 C.

Comfort votes

The initial comfort vote at time 0 was disregarded and only the data for 30, 60 and 90 minutes was used.

Comfort declined with time for all parts of the body. Table 3 shows that comfort declined more for the lower parts of the body, with the foot being the least comfortable. The hind foot was least comfortable.

Each part of the body was analyzed separately to see how floor surface affected comfort. The neck, shoulder, upper back, mid back, lower back and fore foot were not significantly affected by floor surface. However the results for buttocks, upper leg, lower leg, ankle, hind foot and mid foot were significantly affected by the floor surface; see Table 4. In general, Optimat was best, San-EZE II was second, concrete was third and Footsaver was last. However the difference between Optimat and San-EZE II was not statistically significant. The difference between concrete and Footsaver was not significant.

A factor analysis was used to combine the 12 comfort votes into one combined vote. Table 6 shows the means of the combined vote as well as when only the six significant body parts shown in Table 4 are combined. Optimat is best, San-EZE II is second and Footsaver and concrete are tied. However the scores of Optimat and San-EZE II are not significantly different and the scores of Footsaver and concrete are not significantly different.

The lack of a difference between Footsaver and concrete was surprising to us so we reanalyzed the data using only the data for 60 and 90 minutes (i.e., omitted comfort votes at 30 minutes). This time the data were more clear. None of the upper parts of the body (neck, shoulder, upper back, mid-back, lower back or buttocks) had comfort significantly affected by floor surface. All of the lower parts of the body (upper leg. lower leg, ankle, hind-foot, mid-foot, and fore-foot) were significantly affected by floor surface. See Tables 6 and 7.

Now, after 60 minutes of standing, there is a clear comfort difference of all three mats over concrete. Within the three mats, Optimat is the best and Footsaver the worst; San-EZE II is intermediate and not significantly different from either Optimat or Footsaver.

DISCUSSION

Several of the criteria were significantly affected by floor surface within the relatively short time of 90 minutes. The most sensitive criterion was comfort vote--especially in the lower part of the body. As shown in Table 7, all three mats were preferred over concrete and the preferred mat was the Optimat.

What is different among the three mats? Comparing Table 1 and Table 7 shows that mat comfort is inversely related to compression from the body weight. That is, the Optimat has only a 5.8% compression vs. a 7.4% for Traction mat and 18.6% for Footsaver. Obviously this can't be extrapolated to the compressibility of concrete.

More research needs to be done.

Table 2. Change in calf temperature was significantly ($p < .05$) affected by floor surface. Values with the same letter are not significantly different.

Floor surface	Mean change, C	Grouping
Concrete	.29	A
San-EZE II	.06	AB
Optimat	-.26	B
Footsaver	-.29	B

Table 3. Comfort vote declined more for some parts of the body than others.

90 min comfort vote	Body part
79	Neck
82	Shoulder
80	Upper Back
74	Mid back
71	Lower back
76	Buttocks
72	Upper leg
63	Lower leg
67	Ankle
51	Hind foot
61	Mid foot
69	Fore foot

Table 4. Comfort (time 30, 60, 90) was significantly ($p < .05$) affected by floor surface for each of the body parts in the table. Values with the same letter are not statistically different.

	Body part					
Floor surface	Buttocks	Upper leg	Lower leg	Ankle	Hind foot	Mid foot
Optimat	85 A	84 A	75 A	80 A	67 A	75 A
San-EZE II	83 A	81 AB	77 AB	77 AB	61 B	71 AB
Concrete	81 B	77 BC	69 BC	73 B	57 B	67 B
Footsaver	78 B	75 C	65 C	72 B	58 B	67 B

Table 5. Combined comfort vote means for the entire body and for the six body parts of Table 4. Values with the same letter are not statistically ($p < .05$) different.

Floor surface	Vote, whole body	Grouping	Vote, lower body	Grouping
Optimat	80	A	80	A
San-EZE II	78	AB	74	A
Footsaver	75	B	71	B
Concrete	75	B	71	B

Table 6. Comfort (time 60 and 90) was significantly ($p < .05$) affected by floor surface for each of the body parts in the table. Values with the same letter are not statistically different.

	Body part					
Floor surface	Upper leg	Lower leg	Ankle	Hind foot	Mid foot	Fore foot
Optimat	81 A	73 A	77 A	62 A	71 A	76 A
San-EZE II	78 A	71 AB	72 AB	56 AB	67 A	75 AB
Footsaver	72 B	65 B	68 B	52 B	62 B	71 BC
Concrete	72 B	57 C	66 B	49 B	59 B	67 C

Table 7. Combined comfort votes vs. floor surface. Values with the same letter are not statistically different.

Floor	All 12 body parts		6 lower body parts	
surface	vote	Grouping	vote	Grouping
Optimat	77	A	73	A
San-EZE II	75	AB	70	AB
Footsaver	72	B	65	B
Concrete	70	B	62	C

REFERENCES

Corlett, E. and Bishop, R., 1976, A technique for assessing postural discomfort. Ergonomics, 19, 2, 175-182.

Konz, S. and Subramanian, V., 1989, Footprints. In Advances in Industrial Ergonomics and Safety I, edited by A. Mital, (London: Taylor and Francis), pp. 203-205.

Rys, M. and Konz, S., 1988, Standing work: carpet vs. concrete. Proceedings of the Human Factors Society, 522-526.

Rys, M. and Konz, S., 1989a, An evaluation of floor surfaces. Proceedings of the Human Factors Society, 517-520.

Rys, M. and Konz, S., 1989b, Standing with one foot forward. In Advances in Industrial Ergonomics and Safety I, edited by A. Mital (London: Taylor and Francis), pp. 207-213.

Advances in Industrial Ergonomics and Safety II
Edited by Biman Das
Taylor & Francis 1990

GEOMETRY AND ORGANIZATION OF WORKSTATION AS REASONS OF STRESS AT DIFFERENT OCCUPATIONS

Wiesława Horst

Dept.of Ergonomics and Applied Anthropology
University of Poznań, Kantaka Street 2
61-812 Poznań, Poland

INTRODUCTION

Complaints about musculoskeletal disorders are frequent in many workplaces. The various proffessions at each pain exists indicate, that the working conditions play a significant part. Constrained working posture and constraint motions of locomotive organs cause the stress. The main reasons of stress at different workplaces are geometry (shape and dimension) and location of functional points on the workstation. They decided about body posture and trajectory of motion of locomotive organs. Workspace is one of the less recognized factor of work which is a reason of stress and source of microtrauma.

This study is a part of research which considers problems of proffessional differences of microtrauma and proffessional differences of aging and senescent. The large cross-sectional studies among different workers and proffessions has been started in 1987 and will finish in 1991.

METHOD AND SUBJECTS

At many workplaces people suffer from musculoskeletal disorders. Pain is the subjective complaint. This study is concerned with perceived body stress associated with geometry of workstation and various production operations and their changing over the time. From 1987 different workplaces of different work organization have been investigated for getting an overview about working conditions and the amount of disorders. Working conditions have been investigated in aspect of constraints on posture and movement parts of body. The body movement analysis has been conducted only from this point of view.

Complaints and constraints were recorded of a questionnaire elaborated on base: the Standarised Nordic Questionnaire (Kuorinka, Jonsson, Kilbom, Vinterberg, 1987), method for identifying and priorityzing musculoskeletal stress (Corlet and Bishop, 1976), Tichauer rools (Tichauer, 1978), results of own research.

This questionnaire has been called "C & C" QUESTIONNAIRE.

The questions are illustrating by means of scheme of human body and discusing with each group of people (at each workplaces).

The list of CONSTRAINTS has been taken to the consideration each of person:

impossibility: - changing body position while working,
- sitting (organized by employer),
- comfortable resting during the breake,
- motion of body while working,

long time: - standing,
- sitting,
- holding hand up without any support,
- holding tools (things) in hand,
- strong press with finger or hand,
- holding pedals, bottons, etc. with food or leg,

forced to:- doing task; - above shoulder level,
- with the elbows at the shoulder level,
- below knees level,
- constantly; - bend head and trunk,
- sit in a bent position (forwards, back, side ways),
- jerk, push, pull, lift, raise, turn, twist, put down,
- repeat the same motions hand, arm, finger, head,
- reach forwards father than 40 cm,
- push and pull at chest level,
- rub, hit, stumble on workstation,
- sit without feet support,
- move head to see well,
- stand on toes with fear to fall down,
- grip and manipulate at the same time,
- bend hand in the wrist to the outside,
- raise elbows without support,

experience of:- trembling,
- constraint of hands movement,
- constraint of body,
- need to lean or to rest on the machine
- pressuire on the tighs while sitting,

unricognition by touch tools, knobs, balls, etc..

Similary the list of COMPLAINTS has been taken to the consideration each of person:
pain, aching, tingling, burning, numbness, convulsion, loose of feeling, stiffing and discomfort which person may feel at the part of body: neck, shoulders, upper back, elbows, low back, wrist/hands, hips/tighs, knees, ankles/feet.

"C & C" QUESTIONNAIRE consists of three parts:
part I - refers to biological and sociological state of person;
part II - contains information about perceived stress with regard to consequences;
part III - gives information about constraints which exist at workstation.

Ergonomical (hardware) and sociopsychological (software) factors were supposed as hypothetic causes. Study for development and testing questionnaire to that time were realized in 5 factories and 9 occupations.

The subject consisted of:

31 packing-machine operators in salt mine at Kłodawa
38 engine-driwers in salt mine at Kłodawa,
100 fine-drawing women in wool factory at Zielona Góra,
40 sorters in wool factory at Zielona Góra,
94 telephone operators in Poznań,
29 telegraph operators in Poznań,
86 sewing-machine operators at clothes factory at Słubice,
12 cutters at clothes factory at Słubice,
85 tram driwers in Poznań.

RESULTS

Geometry and organization of workplaces describe work requirements, particulary physiological requirements. Force, frequency, duration, pace of movement of locomotive organs and posture have health repercussions for person. They cause stress, individual for each person, but only partialy dependend from such individual factors as: skill, experience, fitness, strength, mentual capacity etc. Individual factors do not play main rool in feeling discomfort. They are not possible to reduce negative results (for health), of bad-fitting geometry and location of functional points on the workstation.

Table 1. Percentage of subjects reporting that they have had musculoskeletal trouble in different body regions /total stress/.

No	occupation	total: at any time	total: last year	stress: last weak
1.	packing machine oper. (f)*	78%	58%	32%
2.	engine drivers (m)	82%	74%	53%
3.	fine-drawing wom. (f)	78%	71%	64%
4.	sorters (f)	100%	87%	67%
5.	telephone oper. (f)	91%	86%	50%
6.	telegraph oper. (f)	100%	100%	67%
7.	sewing machine oper. (f)	87%	63%	55%
8.	cutters (f)	100%	83%	83%
9.	tram drivers (m)	84%	51%	41%

*) m-*male*, f-*female*

75% of subjects feel musculoskeletal troubles at any time during the last 12 months, and 51% at last week (tab. 1). The studies have documeted that upper limb, elbow and neck disorders are major problem among workers (tab.2).

Table 2. Percentage of subjects reporting that they have felt discomfort according to the body regions, at any time during the last 12 months.

No	occupation	N*	S	E	H	T	L	Hip	K	F
		body regions								
1.	packing mach.oper.	16	26	19	19	29	29	13	10	19
2.	engine driwers	47	18	26	21	24	34	24	40	21
3.	fine-drawing wom.	9	14	15	2	22	30	6	6	14
4.	sorters	48	43	63	30	48	40	15	20	35
5.	telephone oper.	38	33	46	13	28	54	26	42	45
6.	telegraph oper.	53	37	83	17	53	77	27	23	33
7.	sewing mach.oper.	35	34	49	14	34	41	24	26	33
8.	cutters	67	75	58	17	67	58	42	67	58
9.	tram driwers	19	17	20	15	20	28	20	33	23

*) N-*neck*, S-*shoulders*, E-*elbows*, H-*wrist/hands*, T-*upper back*, L-*low back*, Hip-*hips/tighs*, K-*knees*, F-*ankles/feet*

High prevelance of upper limb troubles are found among telegraph operators, cutters of material, telephone operators, sewing machine operators and sorters at wool factory. Very important work-related factors seems to be constraind body posture and constraint movements of locomotive organs. Analysis of constraints at different proffessions deliver information that they are responsible for the most of musculoskeletal problems. Exist physiological and biomechanical limits to how much work can be performed in given period.

Let us to compare complaints in two occupations, and theirs reasons - constraints which have noticed more than 50% of subjects (tab.3).

86 % telephone operators reporting that they have had musculoskeletal troubles at any time during last 12 months. The main region of existing troubles were: low back - 54%, elbows - 46%, feet - 45%, knees - 42%, neck - 38%, (tab.2).

63% sewing machine operators perceived stress at any time during last 12 months in: elbows - 49%, low back - 41%, neck - 35%, shoulders - 34%, feet - 33%, (tab.2).

What were the reasons of such situation explains tab.3. From the list about 40 constraints , about 20 exist in two researched occupations. It is worth to notice that 61% of telephone operators and 70% of sewing machine operators have performed present type of work more than 10 years and that; 78% of telephone operators and 81% of sewing machine operators are 21-50 years old.

The main reasons of complaints are restricted body posture, sitting in a bent position , move had to see well and rise elbows without support. The effect of this constraints is perceived stress in neck , shoulders and low back. Long time holding tools and things, grip and manipulate at the same time, among others, are reasons of pain in elbows.

Table 3. Percentage of sewing machine operators and telephone operators reporting that they have noticed the next constraints (we have taken to the consideration only this constraints which have observed more than 50% subjects).

No	constraints	telephone operators n=96	sewing m. operators n=86
impossibility:			
1.	changing body position while working	100%	67%
2.	comfortable resting during the breake	96%	-
3.	motion while working	79%	-
long time:			
4.	sitting	100%	88%
5.	holding hands up without any support	68%	-
6.	holding tools, things in hand/s/	68%	66%
7.	strong press with finger or hand/s/	55%	-
8.	holding pedals with food or leg	-	86%
constantly:			
9.	repeat the same motions hands,arms et.	100%	92%
10.	bend head and trunk	65%	86%
11.	sit in a bent position	63%	81%
12.	reach forwards father than 40cm	86%	-
13.	sit without feet support	54%	-
14	move head to se well	79%	61%
15.	grip and manipulate at the same time	89%	-
16.	bend hand in the wrist to the outside	65%	56%
17.	rise elbows without support	81%	58%
experience of:			
18.	constraint of body movement	59%	62%
19.	pressure on the tighs while sitting	51%	60%

This is one of possible way of formulating hypothesis for ergonomics improvement of workplaces. The present study has several purposes. One is to determine causes of musculoskeletal symptoms. "C & C" QUESTIONNAIRE is good tool for initial recognition of ergonomics problems at workstation.

REFERENCES

Armstrong, T. and Silverstein, B., 1988, Work relate neck and upper limb disorders. In: Proceedings of the tenth Congress of International Ergonomics Association, Sydney, pp. 404-406.

Blache, K. M., Dickens, K. and Pappas, P., 1987, Postural discomfort evaluation production workers. In: Trends in Ergonomics/Human Factors IV, North-Holland, Elsevier Science Publishers B. V., pp. 655-673.

Corlett, E.N. and Bishop, R. P., 1976, A technique for assesing postural discomfort. In: Ergonomics, Vol. 19, 2nd, pp. 175-182.

Corlett, E.N., Wilson, I. and Manenica, I., (eds), 1986, The ergonomics of working postures, (London: Taylor & Francis).

Jonsson, B., 1982, Measurment and evaluation of local musculoskeletal strain in shoulder during constrained work. In: Journal of Human Ergology, Vol. 11, pp. 73-88.

Jonsson, B. and Yedreborg, B., 1985, Identification of ergonomic problems by means of questionnaires for musculoskeletal troubles. In: Proceedings of the 9th Congress of the International Ergonomics Association, Bornemouth.

Kuorinka, I., Jonsson, B., Kilbom, A., Vinterberg, H., Biering-Sorensen, F., Anderssen, G. and Jorgensen, K., 1987, Standarized Nordic Questionnairs for the analysis of musculosketal symptoms. In: Applied Ergonomics, Vol. 18, 3th., pp. 233-237.

Krueger, H., Hinnen, U., Guggenbuhl, U., Laubli, Th., 1988,Musculoskeletal disorders at different work places. In:Proceedings of the tenth Congress of International Ergonomics Association, Sydney, pp. 445-447.

Tichauer, E. R., 1978, The biomechanical basis of ergonomics New York.

Webb, R. D. G., 1983, The use of postural discomfort ratingsas a diagnostic method in workplace modification, IIE - Fall Conference Proceedings, pp.175-179.

Webb, Tack and McIlroy, 1984, Assessment of musculoskeletal discomfort in a large clerical office: A case study. In: Proceedings of the International Conference on Occupational Ergonomics, pp. 392-396.

Advances in Industrial Ergonomics and Safety II
Edited by Biman Das
Taylor & Francis 1990

IMPROVED WORKPLACE DESIGN AND MATERIAL HANDLING AIDS APPLIED TO CHEMICAL BAGGING OPERATIONS

Gary Kress
Independent Consultant

Paul Eckstedt
FMC Corporation
Corporate Technology Center

This paper describes the results of an ergonomic study conducted for a chemical plant to identify more efficient work procedures and improved workspace design to decrease operators' physical workload and enhance safety and productivity. The study resulted in the development of anthropometric layouts for each work station, some of which incorporated material handling aids, that reduced operator workload and postural stress.

INTRODUCTION

A common requirement in the chemical industry is to package products in heavy paper or plastic bags. The bags usually weigh from 23 to 45 kg (50 to 100 lb) and are loaded to lie flat on pallets up to 8 layers high. Despite the availability of automated bag filling and handling equipment, much of this work is still performed manually in many plants. It is well recognized that heavy manual material handling can produce local muscle fatigue and repetitive lifting causes whole body fatigue (NIOSH, 1981; Eastman Kodak Company, 1986). When this is combined with a workplace layout that promotes awkward lifting postures or poor lifting techniques the results can have adverse affects on worker health, safety and productivity.

This paper describes the results of an ergonomics and engineering study to identify procedural, equipment, and workspace design requirements to enhance personnel health and safety and increase productivity in phosphate product bagging and palletizing operations. The study was conducted for a chemical processing plant that manufactures several variations of powdered sodium phosphate.

METHOD

The present study was conceptualized as the first phase of a two phase project and the work described in this paper covers only the phase one objectives. These were to identify and define system requirements and constraints, design and develop solutions to the materials handling problem, and derive cost estimates to implement the selected solutions. Phase two consisted of implementing the solutions.

A preliminary review of the manual materials handling (MMH) requirements and other human-machine interface problems at the plant suggested that expertise from several disciplines would be required. Accordingly, the study team was composed of a human factors engineer, an industrial designer, and a mechanical engineer. The human factors engineer analyzed the human performance demands of the tasks, compared them with physical and psychological worker capabilities, and identified system design requirements to make the two compatible. The industrial designer took the ergonomic design requirements and converted them into anthropometric drawings and workplace layouts which would accommodate the full range of user personnel (5th through 95th percentile). The mechanical engineer evaluated the adequacy of existing hardware and machinery and identified material handling aids and other equipment to eliminate or reduce the MMH demands.

Procedure

The phase one study effort focused on identifying ways to eliminate or reduce the demand for heavy material handling. This was accomplished by analyzing several options: re-allocation of tasks between operators and equipment, mechanical lifting aids, better workplace organization and layout, and improved lifting/handling procedures. The following tasks were performed.

Task 1. Analyze System Requirements and Constraints. Videotapes, interviews with workers, and observation of the work procedures were used to analyze functions and tasks and identify operator performance requirements and safety concerns.

Task 2. Establish Design Guidelines. Using the function and task analyses, the human factors and ergonomics literature was searched to compare the performance requirements with human capabilities and limitations. The information from this data base was used to develop design guidelines for new equipment, equipment modifications, workplace dimensions and layout, and lifting and movement procedures.

Task 3. Develop System Design. Based on the ergonomics data and the workplace design guidelines, one or more design concepts for the three workstations were developed. The concepts included changes in the workplace layout that minimized lifting requirements and incorporated material handling aids.

Task 4. Develop Implementation Plan. Vendors were contacted and equipment brochures were collected to evaluate costs and effectiveness of the various design solutions. Recommendations were made for both short term and long term design solutions.

RESULTS

The results of this study consisted of recommendations and design concepts for reducing the operator workload and increasing efficiency in the product packaging and palletizing operations. The recommendations addressed both the problems associated with heavy material handling, and ergonomic issues involved in other areas of the human-system interface. Recommended solutions included redesign of some of the workstations, modifications of some hardware and equipment, and use of material handling aids. Only the concerns and solutions related to the MMH problems will be discussed here.

Manual Materials Handling Workload Analysis

Under the existing operational set-up and procedures, most product packaging and material handling requires a considerable amount of manual lifting and carrying. A workload analysis was conducted to establish operational requirements and worker baseline performance. This information was used to develop guidelines for lifting workplace design.

The chemical products are stored in bulk and routed through pipes to one, two or three bag filling machines for packaging in either 23, 25 or 45 kg (50, 55 or 100 lb) paper bags. Two of the machines have a single filling spout and each require one operator. The maximum filling rate for the single-spout machines is about 2 bags/min. The actual rate may be slower depending on the bulk density and flow of the material. The basic sequence of activities for the operators on these machines includes: applying the empty bag to the spout, filling and closing the bag, carrying the bag to the pallet, and arranging the bags on the pallet. Whenever the operator feels that a bag weight is not accurate, or at a minimum of once for each pallet load, he carries the bag to a check weight scale before loading it on the pallet. The workstation is arranged so that the operator faces the machine and the bag rests on a saddle below the spout while it is filling. The check weight scale is located approximately 1 m to the left of the operator and the pallet for loading the product is located approximately 1 m behind the operator (Figure 1).

A majority of the operators use one hand at the bottom of the bag and one at the top to lift, rotate, and carry the bag as necessary. The distance that the bag is actually lifted or

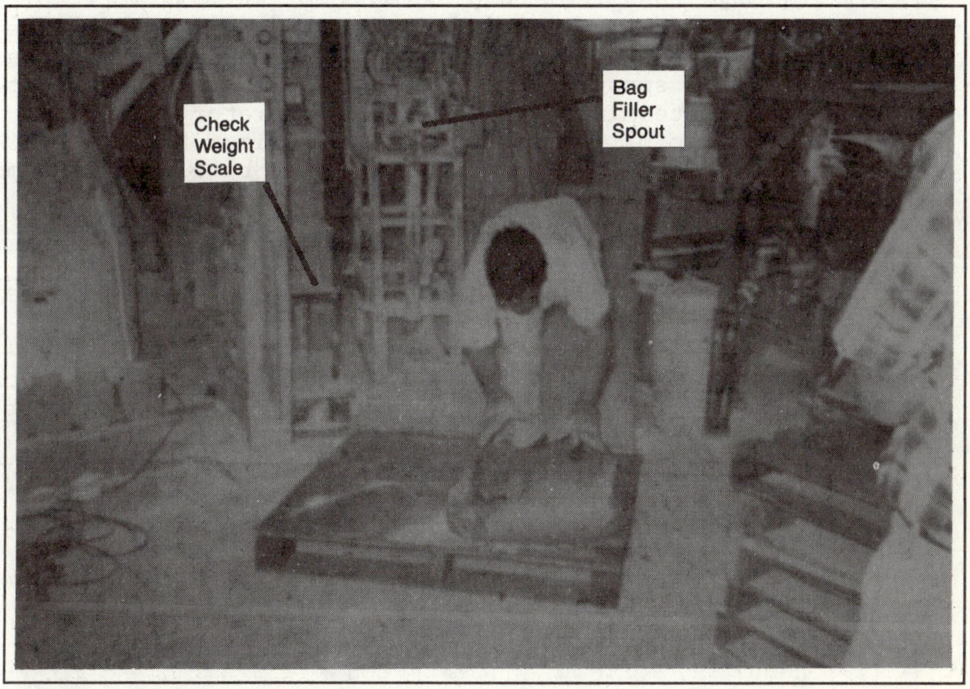

Figure 1. Single spout bagger work station.

carried depends on its position on the pallet. To load bags at the front of the pallet operators use the momentum of the bag weight to pull it off the saddle and rotate it to a horizontal position, while swinging it to the pallet. During this operation most operators will twist their upper body rather than turning with their feet. The operator must take one or two steps to load bags at the rear of the pallet. Since the height of the saddle varies between 57-67 cm, bags are either lowered or lifted from this height depending on the layer that is being loaded on the pallet. Most operators stand on an empty pallet so they do not have to step up when loading the first layer of bags at the rear of the receiving pallet.

The third machine has dual-spouts. It requires one operator to fill the bags and drop them on a conveyor and another operator to lift/move the bags from the conveyor to a pallet. The maximum output of the dual spout machine is about 4 bags/min. The job of palletizing the bags from the dual-spout filler is somewhat easier than that of the single-spout machine. The operator stands to the left or right of the conveyor end and receives the bags horizontally. He puts one hand at either end of the long axis and slides, lifts, lowers, or carries the bag, as necessary, to its position on the pallet. The operator does not have to perform a check weighing operation.

Table 1 summarizes the weights, dimensions and lifting requirements for the various products that are packaged. The products come in powdered or granular form and, because of the difference in density, the dimensions of filled bags may vary from one product to another even though the bag weight is the same. The frequency of lifting has been established by the operators in terms of what they feel they can accomplish in a work shift.

The origin of the lift is established by the height of the saddle on the single spout machines and the height of the conveyor on the dual spout machine. The minimum termination height of the lift is determined by whether the operator's feet are level with the top of the pallet and the maximum termination height is determined by the thickness of the bags and the number of bag layers on a pallet.

Table 1. Summary of bag handling weights, dimensions and lifting requirements.

Bag filler	Bag weight (kg)	Bag size (cm)	Frequency (lifts/min)	Vertical origin of lift (cm)	Min/max vertical termination of lift (cm)
	45	64x46x18	2.25	79	13/18
Dual	45	79x52x19	1.88	79	13/100
Spout	23	64x43x18	2.75	79	13/117
	23	61x41x18	2.75	79	13/140
Single	45	61x43x25	.94	57-67	0/61
Spout	25	61x30x15	1.13	57-67	0/92

The workload analysis identified several basic ergonomic concerns in the packaging operations. The most obvious concern is the weight of the product being handled. The combined dynamic lifting data from Snook (1978) and Ayoub, et al., (1978) presented in the NIOSH Work Practices Guide (1981) suggests a maximum acceptable weight of lift for the average male industrial worker of 27 kg from floor to knuckle height and 24 kg from floor to shoulder height (assuming similar lifting conditions of 1-2 lifts/min, 38 cm horizontal location of the load). Computing the NIOSH stress analysis formula for the worst case lifting conditions results in an Action Limit (AL) of approximately 12 kg and a Maximum Permissible Limit (MPL) of approximately 35 kg. It is clear that the 45 kg bags exceed these limits and that this weight can produce compressive forces on the lower back that exceed the biomechanical criterion of 650 kg compression on the lumbosacral region (Ayoub and El-Bassoussi, 1976).

It is also clear that the operators have been able to adapt to these lifting stresses by establishing a lifting frequency that is not physiologically fatiguing, trading off to non-lifting activities during the shift, finding optimal ways to lift, move, and control the loads, and rotating jobs daily. While the operators agree that the work is strenuous, anecdotal reports from supervisors indicate that the only recent material handling injury resulted from an operator slipping off a pallet. Discussions with supervisors also suggest that a process of "natural selection" is at work: workers who cannot adapt to the lifting demands are transferred to other jobs. For example, the work was found to be too heavy for females so they are not required to perform these jobs. It did not appear that operator size and weight was necessarily related to the ability to perform the tasks. While no personnel anthropometric measures were taken, several operators were average in stature and weight; one appeared to be close to the 5th percentile in these dimensions.

In addition to the weight of the load other ergonomic concerns included:

1. The bending and reaching motions required to stack bags from floor height to shoulder height introduce considerable biomechanical considerations for personnel of all sizes.
2. There did not appear to be a universally accepted procedure for stacking the bags on the pallet. Workers followed procedures they developed themselves or were taught by others.
3. On the two single-spout machines, moving the bag from the filler to the pallet can involve a twisting/turning motion of the upper torso.
4. The check weighing operation on the single-spout fillers is very inefficient because of the arrangement of the filler, scale, and pallet.

Workplace Design Options

There are two basic approaches to addressing MMH concerns (Ayoub et al., 1987). One solution is to eliminate the need for manual material handling by using automated or robotic equipment. The other solution is to minimize MMH demands by redesigning the workplace so that: (a) the material is lowered or slid rather than lifted, (b) carrying is eliminated or minimized, (c) the material is handled at a comfortable working height to minimize bending of the upper torso, (d) the load's center of gravity is kept as close to the lifter's body as possible at all times.

Using the ergonomic literature (Pheasant, 1988; Eastman Kodak, 1986; NIOSH, 1981) a specific set of lifting guidelines was developed to guide the workplace re-design effort. The guidelines focused on eliminating bending and carrying requirements and specified that:

1. Greatest lifting force can be exerted at around knuckle height (70 cm for 5th percentile males and 80 cm for 95th percentile). For heavy lifts, the hands should never be lower than 50 cm from the floor.

2. For most people, the most desirable range for commencing and terminating a lift is between 65 cm and 100 cm. Ideally all lifting should be conducted between 50 and 100 cm.

3. Lifts that start at floor level should terminate no higher than 110 cm. Lifts starting in the range mentioned under (2) above should terminate no higher than 125 cm. They should definitely not terminate above shoulder height (132 cm for 5th percentile males and 152 cm for 95th percentile).

4. All bags should be handled at a constant, comfortable work level. Lifting and carrying bags from conveyor, filler spout or scale to pallet should be minimized or eliminated. The pallet should be close enough to the end of the conveyor, and its height should be adjustable, so that all bags can be easily slid from the conveyor to the pallet.

The workplace design goal was to develop a solution that would result in the same or greater output per shift, using fewer personnel, and not requiring the workers to perform any strenuous manual material handling. Three options were considered to achieve this goal:

Concept 1. Complete automation — Machines are used to fill and palletize the bags automatically. One worker monitors operations and provides empty bags and materials to the machines as necessary. Productivity could be substantially increased with a corresponding decrease in required personnel. With this solution most manual material handling concerns would be eliminated. The constraint is basically cost: approximately $350,000 for automated bag fillers and palletizers.

Concept 2. Partial automation — Three workers manually fill bags at the three baggers and drop the bags on to connecting conveyors which deliver them to a central point for automatic palletizing. Again, productivity would be increased, but three operators would be required for maximum productivity. Manual material handling concerns would be greatly reduced. Constraints are the cost of the automatic palletizer and conveyors and the available space/flexibility for arranging the conveyors.

Concept 3. Better workplace design and MMH aids — Bag filling is performed manually but better workspace layout and mechanical aids reduce the palletizing workload. The design emphasis would minimize carrying and postural twisting, lifting distance (vertical travel), and lifting range (vertical height).

Figure 2 depicts the overall workplace layout for this concept. Similar to the current layout, the single-spout bagging machines are stand-alone operations (with one machine located in an adjacent structure) and the dual-spout machine connects to the manual palletizing point via conveyor. A check weigh scale is put in-line with the filler and conveyor so operators at the single spout machines can fill the bags, slide them onto the first stage of the conveyor to check the weight, then slide them to the transfer point for manual palletizing. This eliminates carrying the bags to be check-weighed and facilitates the transfer of the bag from the filler to the pallet. Spring-loaded lift tables with a rotating platform at each palletizing station ensure the bags are always handled at a comfortable working height and the lifter's body is always as close to the load as possible. This equipment would reduce the amount of bending, twisting, and carrying required for palletizing. It is estimated that productivity would be the same or improve, using fewer or the same number of workers. Constraints with respect to equipment cost and workspace layout would be minimal.

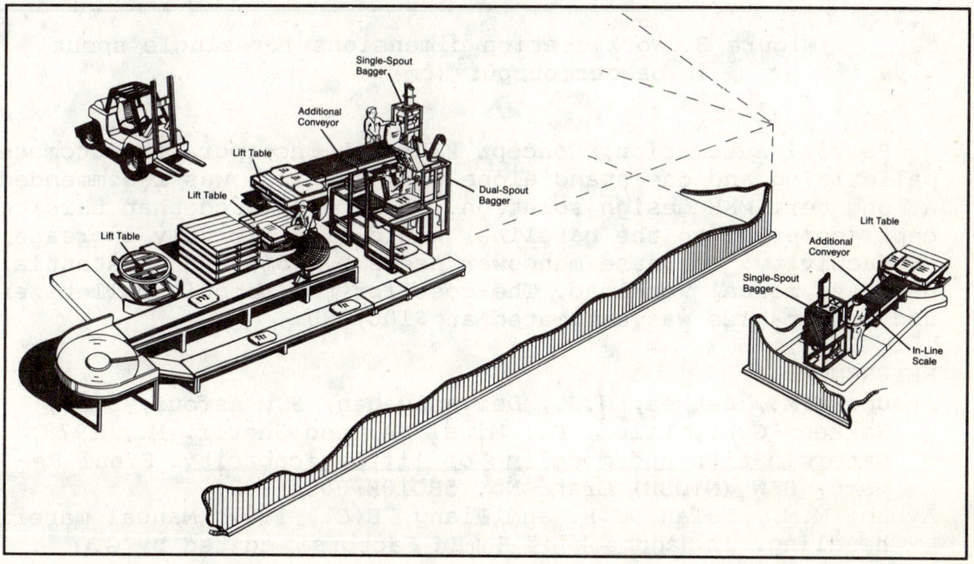

Figure 2. Workspace layout with material handling aids.

Figure 3 shows the anthropometric layouts for implementing the lift-table concept at the single-spout workstation. The dimensions shown follow the design guidelines discussed above and are meant to accommodate 90% of the user personnel.

RECOMMENDED MMH DESIGN SOLUTIONS

Concept 3 improved workplace design and use of material handling aids was recommended as a near-term, low cost design solution to the manual material handling problem. The estimated effect would be to maintain or increase productivity and decrease the manual workload and fatigue of the operators. Estimated costs for the lift tables and conveyors is $10,000.

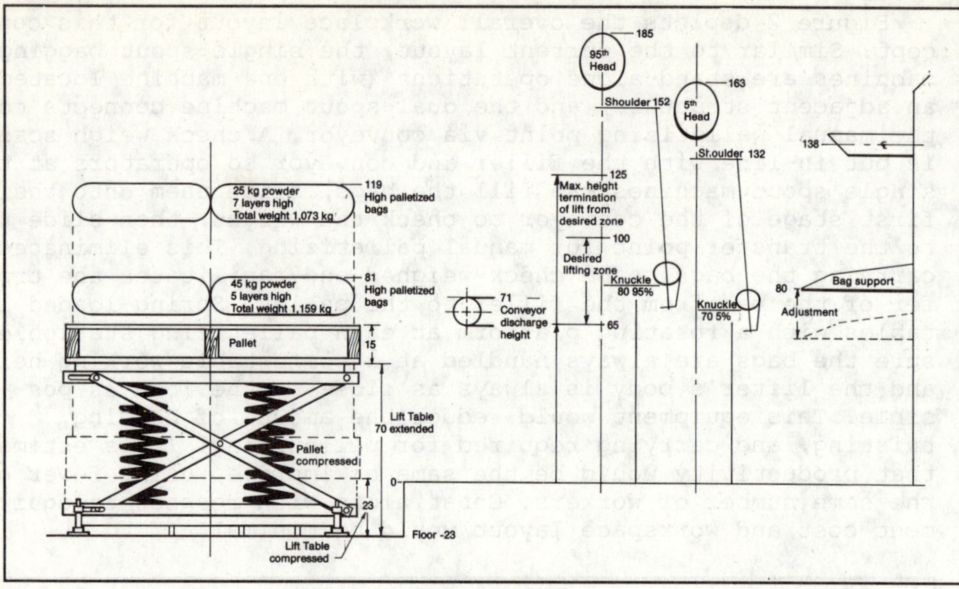

Figure 3. Work station dimensions for single spout bagger output (cm).

Partial automation, Concept 2, which incorporates automated palletizing and one stand alone work station, was recommended as a long term MMH design solution. It is estimated that this concept (compared to the baseline) would significantly increase productivity, decrease manpower needs by 25%, and substantially decrease manual workload. The cost for the robotic palletizer and lift tables was estimated at $100,000.

REFERENCES

Ayoub, M.M., Bethea, N.J., Deivanayagan, S., Asfour, S.S., Bakken, G.M., Liles, D., Mital, A. and Sherif, M., 1978, Determination and modeling of lifting capacity. Final Report, HEW (NIOSH) Grant No. 5RO10H-00545-02.

Ayoub, M.M., Selan, J.L. and Jiang, B.C., 1987, Manual materials handling. In Handbook of Human Factors, edited by G. Salvendy (New York: John Wiley & Sons), pp. 790-818.

Ayoub, M.M. and El-Bassoussi, M.M., 1976, Dynamic biomechanical model for saggital plane lifting activities. Proceedings of the 6th Congress of International Ergonomics Association, 355- 359.

Eastman Kodak Company, Ergonomics Group, 1986, Ergonomic Design For People at Work. (New York: Van Nostrand Reinhold Co.) Vol.2.

NIOSH, (National Institute for Occupational Safety and Health), 1981, Work Practices Guide for Manual Lifting. (DHHS) (NIOSH) Publication No. 18-122, Cincinnati, Ohio.

Pheasant, S., 1988, Bodyspace: Anthropometry, Ergonomics and Design, (New York: Taylor & Francis)

Snook, S.H., 1978, The design of manual handling tasks. Ergonomics, 21, 963-985.

Advances in Industrial Ergonomics and Safety II
Edited by Biman Das
Taylor & Francis 1990

DESIGN OF THE WORKPLACE ENVIRONMENT FOR HEALTH

CONSIDERATIONS ON OFFICE WORKPLACE ENVIRONMENT WITH EMPHASIS ON CHANGE OF VALUES WITHIN SOCIETY

JOH. HOLENSTEIN
ARCHITECTURES
MINERVASTRASSE 117
8032-ZÜRICH
SWITZERLAND

PRECEDING AESTHETICS,
ERGONÕ'MICS, THE RATIO
BETWEEN MAN AND HIS
ACTIVITIES UNIVERSALLY,
OPTIMAL PERFORMANCE
TOWARDS ACHIEVEMENT.

JOH. HOLENSTEIN
AT ART CENTER
COLLEGE OF DESIGN

OFFICE ENVIRONMENT FOR HEALTH

An office workplace should have an environment which adapts to each individual's personal needs in order to compensate his own personal short-coming. Office space (work environment) is an essential part of living space. It is here where new values are developing. Old institutional values given by community, State and church are loosing its predominance in our environment. It is work life which becomes the integrating denominator. Therefore the workplace has to be exalted to an almost spiritual center of life.

1. OUTLINE OF THE ISSUE

There are as many personalities as there are individuals. The word "PERSONALITY" comes from the Greek work "PERSONA" meaning "mask". Jung coined the word "PERSONA" to describe that aspect of pesonality which an individual habitually shows to his environment.

Fig. 1

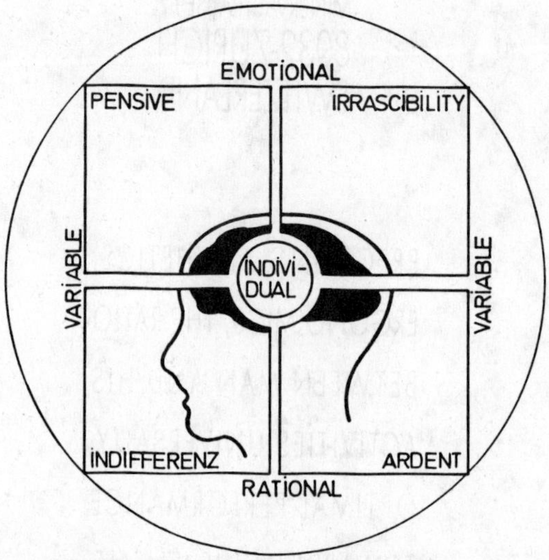

THE HUMAN PROCESSOR TO INDIVIDUALITY

As a designer I set myself a guideline saying

DESIGN JOBS START WITH
ERGONOMICS NOT AESTHETICS

Naturally, one can start a design job with an aesthetic approach, e.g. with a spontaneous sketch, but if one leaves it at that, the chances are minimal that one has taken into account all the possible criteria; therefore the success of such a design is limited. At some point one will need to show systematic evidence of research and evaluation.

Among the 3 disciplines of Ergonomics, it is to say

- THE PHYSIOLOGICAL
- THE PSYCHOLOGICAL
- THE ENVIRONMENTAL

It is essentially the physiological discipline which has the largest amount of findings available. The human factors research within anthropometry offers today sufficient information to design any workplace. It is of course more difficult to prove to a potential office furniture manufacturer or a big institutional end-user that to understand the closest needs of the office work force may bring larger benefits than to adjust a worktable to personal comfort or to know how to hide cables.

This paper will not bring a researched solution within the psychological and environmental Ergonomic disciplines but it wants to hopefully initiate Human Factors Researchers to furnish more findings within the psychological and environmental sides of Ergonomics.

2. RESEARCH

Being contracted to design a new office furniture system including office environment, I started to enumerate the mentally related properties of Ergonomics , particularly with respect to office environment.

An enumeration at random shows such terms as

2.1. SENSITIVITY
- Perception (eye)
- Touch
- Perceptiveness (feel / hear)

2.2. EMOTION
- Beautiful - Ugly
- Pleasant - Unpleasant
- Warm - Cold
- Welcoming - Odd

2.3. INTELLECTUAL
- Spontaneous - Affected
- Comprehensible - Confusing
- Stimulative - Tediousness
- Conscious association

2.4. UNCONSCIOUS
- unconscious association
- symbolic content
- cultural and ethnic context
- collective heritage (constructive / historical / traditional)

2.5. GLOBAL RELATED PROPERTIES

- air
- cold - heat
- noise - calm
- smells
- humidity - dryness
- light
- space (Interactive Distances)
- atmospherics
- raw material
- waste control

2.6. SOCIETY AND VALUES

- money culture - time culture
- serving mentality - independent ambitions
- working time - leisure time
- work environment - leisure environment
- artificial values - genuine values

3. GENERAL

PERPETUAL CHANGE IS OUR ONLY CONSTANT. Knowing this with respect to the physiological side of Ergonomics, the office furniture manufacturers have adapted fully to these needs and solved all problems of cable ducting, height and inclination adjustability, lighting and communication. As for the deactivation of the field of tension, deriving from the effects between MAN AND ORGANISATION, we are only beginning to understand its implications. Implications that we cannot live a life by dead traditions and to spend the major time of our lives working in an environment to which we give no more than some rational importance. We have to come to understand that the physiological discipline of Ergonomics is only the immediate off-hand solution. The major operation however has to be done with the psychological and the environmental disciplines of Ergonomics.

Of course it is of no use to develop any further this synthesis, if some parameter basics are not understood and functioning already. Such parameters as self-realization, self-determination should be a firm element of personality and represents the basic stronghold. In the Far East, the saying "THE MOON SHINES ONLY BRIGHT BECAUSE THE SUN IS LENDING IT THE LIGHT", means to say that optimal performance is only possible if there is more than just rational physiological considerations given to a problem. The "extra" relates to "UNIVERSAL INTERDISCIPLINARIAN THINKING, UNDERSTANDING AND ACTING".

4. SYNTHESIS

Having come to the conclusion that the gap between working time and leisure time, working place and leisure place can become less pronounced, we have taken a major hurdle for the run towards a better office workplace environment. The consumption of our life span distributing the right proportion of work and leisure time to individuals cannot be done. However, creating work which is fun or work which is not in opposition but in harmony with ones private life and the universe sounds like a solution. One can say that the attractive workplace depends on the attractions a region can offer and the level of self-determination individuals have reached, all this in relation to meaningful and ecologically justifiable activities.

Reverting to basic simplicity is the principal idea while planning office environments. To an artificial world which creates emotions such as isolation, fear, cold, threat - one should reciprocate with human communciation, warmth and self-determination. Simplicity, gentleness, reduction and space for the individual to influence his own environment is a dominant factor. The limitation of ones own territorial office environmental boundaries is primarily to be solved via a cultural and intellectual approach.

Advances in Industrial Ergonomics and Safety II
Edited by Biman Das
Taylor & Francis 1990

RELATIONSHIP BETWEEN WORK STATION DESIGN CHARACTERISTICS AND RESULTANT WORKING POSTURES

Ledoux, Elise (1) Lortie Monique. (1) & Rossignol, Michel. (2)
(1) Université du Québec à Montréal, Montréal, Canada,
(2) D.S.C. Sacré-Coeur, Montréal, Canada

A grid of analysis was developed to examine the relationship between a set of variables describing work stations and the resultant working postures. 40 assemblers of a high technology industry carrying out an unrepetitive task were filmed for half an hour each. The work station was defined in relation to the work activity and had to be characterized for every posture observed. The results showed that most of the work station variables were correlated to postures, but none was systematicaly and stongly determinant.

INTRODUCTION

The objective of the present study was to examine the relationship between a set a variables describing work stations and the resultant working postures.It is a part of a larger prospective study to identify which characteristics of work stations are most related to the occurance of lower back pain, conducted in an assembly plant of a high technology industry . The work cycle duration may vary from half an hour to 11 days. On the long cycle jobs, the work is essentially executed on a master piece, with the worker circulating around or inside the piece. On the short cycle jobs, the worker manipulate a large range of different pieces, frequently set on a table.The work performed is therefore not only unrepetitive, but may vary substantially over time. Thus the risk factors pertinent to a back study, such as postures and efforts required in the use of tools vary as well. Furthermore, very few workers do the same job. Consequently, a direct characterisation and evaluation of the risk factors for 450 assembly workers in the field was considered unrealistic. Two approaches were developped. The objective of the first was to analyse the relationship between the type of difficulty perceived by the worker in the execution of his work and the occurence of back pain, in frequency, gravity or nature, over a year in order to identify which elements have the best prognostic value. Although the work and therefore the postures and efforts may vary considerably from one day or one piece to another, the work station itself is not only less variable but also less difficult to characterize than several complex postures. However work stations , tasks, resultant postures have two elements in common: worker characteristics and operative modes. For example, when working at 2 feet above the floor one could choose to squat or to bend. Thus, the objective of the second approach was to analyse and verify the relationships between different aspects of the work station such as the height and orientation of the working area or the type of piece, and different postural characteristics observed. This would allow us to elucidate the potential risk factors.This was executed in order to verify if the characterisation of work stations would be pertinent and the results interpretable in the context of an epidemiological study.

Indeed, there is much in the litterature indicating the need for or proposing sets of recommandations for the design of the work stations . Measurements on the work station along with posture observations have also been carried out in many studies as well (e.g. height and seat inclinason,height of table (Bendix,1984),height of the work surface (Saari, 1978) . However, the main emphasis has been to show the impact of a design characteristic on a posture and not to establish the probability of observing a given posture in a given spatial context . Nor have these studies attempted to quantify or simply establish the impact of other parameters such as the operative modes. In the frame of an epidemiological study, the understanding of these factors is paramount.

Numerous postural descriptive systems have been published. For example,Priel (1974) proposed a set of conventions to numericaly define a posture ,the "posturegram" .The localization of each joint and limb, their position within the coordinate planes with their direction and amplitude is recorded. The OWAS method (Karhu et al.,1977) describes posture using three elements numericaly defined which represent the back,upper and lower limbs respectively and was founded on the analyses of 84 postures typical of industrial work. Grieco et al (1978) developped a double recording system, depending upon whether the postures observed were static or variable.Corlett et al (1979) proposed to represent posture by a set of marks on a chart similar to a target with reference to a standard position. Every system has is own purposes and merits and we developped a grid corresponding to our research objectives combining different elements from previous research and adding others.

METHODS

Work stations selected:

The study was carried out in 5 departements chosen to represent a large range of work stations. All assemblers agreed to be filmed (whith the exception of one, for religious reasons) and 40 work stations were filmed for half an hour each.

Grid of analysis and protocole.

The grid of analysis developped included two sections. In the first , 16 variables were used to characterize posture, each variable included from 2 to 4 classes which are described in table 1. It may be noted that some of the variables, in particular those describing the hand and feet position , are usualy not considered in back studies. But assembly work is considered here as precision work, equally when performed on larger pieces. Thus,we hypothesize that the position of upper and lower limbs could potentially influence the position of the back. In the second section, 19 variables were used to characterize different aspects of the work station. As the work station is in a part defined by the piece worked on and varies over time, the definition was extended to include the task, tools used , and the piece itself. Thus, the work station was defined in relation to the work activity and had to be characterized for every observed posture.

A posture was sampled every minute. If an observed position lasted more than one minute, the position beginning one minute later was analysed. The grid of analyses was applied in 792 sequences by an observer. Unfortunately it was sometime impossible to film with a wide enough angle to frame all the characteristics or simply to see because of the presence of obtacles.Thus the number of observation for the different variables is sometimes inferior to 792

Validation and statistical analysis

Training for observation of postures was carried out with photographic material. Table 2 shows the results of the after training test.

Table 1. Grid of analysis : Description of the postural and work station variables

Variables	Code[2]	Nb of Class	Types	Classes
A- POSTURE				
MAIN POSITION	----	4	D1	Standing , Sitting , Squatting/Kneeling , Other
BACK POSITION				
Sagittal bending	----	3	C[3]	< 20° , 20° to 60°, ≥ 60°
Lateral bending/ torsion	----	3	C	< 10°, 10° to 30 , ≥ 30°
POSITION OF UPPER LIMBS				
Abduction	----	3	C	< 60° , 60° to 90°, ≥ 90°
Elbow angle	----	2	C	< 90° , ≥ 90°
Hand height vs head	----	3	C	Bellow , At head level, Above
Hand alignement vs body	----	3	D2[5]	In front perpendicular, In front parallele,Beside
Hand level	----	2	D1[4]	The same level or Not the same level
POSITION OF LOWER LIMBS				
Knees angle	----	2	C	>160°, ≤ 160°
Feet level	----	2	D1	Same level , Different level
Feet stability	----	2	D1	Level surface, Uneven surface
BACK SUPPORT				
On the body	----	2	D3[6]	Upper limbs or trunk resting, None
On the piece	----	2	D3	Yes, No
EYE-TASK DISTANCE		2	D3	Near(≤ forearm distance) , Far
BODY POSITION				
Frontal plan vs work surface	----	3	D2	In front , At an angle , On the side
Vertical plan vs work surface		2	D1	Below , Above
Hand position vs work surface	----	4	D2	In front, Below, Above, Asymmetric
B- WORK STATION				
PIECE				
Master piece assembled	Main Pi	6	D1	Six types
Form	Pi Form	6	D1	Flat , Convex/Concave , Round , Irregular , Treillis , Other
Type of surface	Pi Surf	2	D1	Flat , irregular
Size	Pi size	2	D2	Small(piece handledwith one hand), not small
WORK SURFACE				
Type	Surf Typ	2	D1	Flat , accidented
General localisation[1]	Localis	2	D1	Inside the piece ,Outside the piece
Part of the piece[1]	Pi Part	3	D2	Upper side , Lower side , Far end plan
Orientation	Surf orie	3	D2	Vertical, Horizontal, Rounded
Axis vs worker	Surf Ax	3	D2	Sagittal plan , Frontal plan , Transverse plan
WORK SURFACE HEIGHT				
Height vs standing position	Hg vs St	4	C	Below the hip , Hip to sternum, Sternum to Shoulder, Above shoulder
Height vs adopted position	Hg vsPo	5	C	Below the knee , Knee to hip , Hip to sternum, Sternum to shoulder , Above shoulder
SURFACE SUPPORTING FEET				
Type	Ft Surf	2	D1	Flat , irregular
TASK				
Type	Task	4	D1	Drilling , Riveting , Sealing , Other
Tool used	Tool	5	D1	No tool , Drill/Countersink , Rivet gun , Alligator sqeeze , Other
SPACE RESTRICTIONS				
Workplace enclosed	Enclose	2	D3	Limited space to move, None
Difficulty to reach	Dif rch	2	D3	Difficulty , no Difficulty
Difficulty to see	Dif see	2	D3	Work blindly or Not [5]

1. Defined only for 2 types of pieces
2. Code used to identify the variables in other tables
3. C: Variables describing a continuous phenomena
4. D1: Discontinuous classes with no overlap
5. D2: Discontinuous classes with overlap
6. D3: Discontinuous variables's classes where a threshold of significance had to be defined

Table 2. Results of the photographic test : Comparaison between mesured and estimated angles

Posture	Nb of photo	Angles represented	Scale	% of photo Difference between estimated and mesured angles is 0°	5°	10°	#over estimated / #under estimated
Sagittal bending	28	10° to 105°	5°	64	25	11	2.1
Lateral bending	11	10° to 45°	5°	100	---	---	---
Abduction							
test#1	23	10° to 105°	5°	48	26	26	0.7
test#2	23	10° to 105°	5°	48	30	22	0.7

Abduction appeared to be more difficult to observe accurately as no improvement in the performance was observed after a re-test. Nevertheless the difference was less than 5° in about 80% of the observations. In a subsequent set of 3 tests to control the intra-observer stability , 10 sequences were selected for reanalysis from a random number table among 300, 400 and 500 previously analysed seqences. A 90% agreement rate was considered acceptable. After the first test 11 posture variables and 11 work station variables showed a 90% agreement rate. In general, a lower agreement rate was obtained for variables describing a continuous phenomena e.g. back bending, elbow and knee angles, abduction, work surface height or a discontinuous phenomena with possible overlapping between classes e.g. hand alignement, orientation of the work surface , axis vs worker. The main causes of divergence have been identified. For exemple, the criteria used to characterize back bending: light bending was more often disregarded as the observer became more restrictive.Furthermore the observer appeared to have unsettled criteria for border line cases when describing abduction , the elbow and knee angles and the height of the work surface. Over time the observer seemed to modify the visual spatial references for discontinuous variables with overlapping classes. A lower agreement rate was noted as well in two discontinuous variables where a threshold of significance had to be defined (eye-task distance, difficulty to reach). In these cases, the observation and decision criteria initialy formalized were not adequate. After the third test, 27 variables showed a 100% agreement rate. The other variables showed at least a 90% agreement rate. All the sequences were subsequently reviewed to correct the data according to the identified problem which was generally very specific. Only bivariate analysis are presented in this paper. Chi Sqare and contengency coefficients were calculated. All statistics were carried using the SAS statistical package.

RESULTS

<u>Relationships between adopted postures and the work station:</u>

Table 3 shows which variables were significantly correlated and the degree of association. A majority of the parameters describing the work station have a significant impact on the resulting posture.Witht the exception of two variables, level of the feet and back support on the body, where the number of missing values recorded invalidated the Chi-Sqare test , no less then 11and up to 18 (an average of 14) work station variables, were associated with a specific postural characteristic. Thus, most of the work station variables have an impact, but none is systematicaly and strongly determinant. In fact few associations were strong, as shown by the contengency coefficients calculated. If we examine the position of the back , we note that sagittal bending was determined by parameters other than those that influence on the lateral bending.

Table 3. Contengency coefficient (p<0.05) between adopted postures and work station characteristics

POSTURE VARIABLES

CONTENGENCY COEFFICIENT	Main Position	Sagittal Bending	Lateral Bending	Abduction	Elbow Angle	Hand Height	Hand Alignment	Hand Level	Knee Angle	Feet Level	Feet Stability	Support on body	Support on piece	Eye-task distance	Frontal body Pl.	Hand Position
0.6 @ 0.4	Hg vsSt Main Pi. Pi. Form Dif rch Ft Surf	Hg vs St Hg vsPo		Hg vsPo Pi. Part		Hg vsPo Hg vs St					Localis. Main Pi. Ft Surf Pi. Type					Surf orie Main Pi. Pi. Type Pi. Size
0.4 @ 0.3	Surf Pl.	Pi. Part Main Pi.		Hg vs St		Pi. Part	Dif see Dif rch		Localis. Main Pi. Pi. Type Pi. Size Ft Surf Surf typ Hg vs St		Pi. Part Dif rch Pi. Size		Dif rch		Main Pi. Pi. Type Dif see Surf Ax.	Hg vs St Enclose
0.3 @ 0.2	Enclose Pi. Size Pi. Part Hg vsPo Toll Surf typ Task	Localis. Ft Surf	Surf Ax. HgtvsSt Hand Po	Pi. Type Toll Task Dif rch	Ht vsPo Dif rch	Main Pi. Enclose Pi. Type Toll Pi. Size Dif rch Task	Main Pi. Surf Ax. Toll Pi. Part Pi.Type Task Hg vsPo Enclose Localis.	Hg vsPo Pi. Part	Pi. Part Surf orie Surf Pl. Dif rch	Pi. Size	Hg vsSt Surf typ		Pi. Type	Pi. type Hg vsPo Ft Surf Toll	Enclose Pi. Part Pi. Size Localis. Dif rch Hg vsPo	Dif rch Ft Surf
0.2 @ 0.1	Surf Orie	Hand Po Pi. Type Pi. Size Surf Orie Enclose	Dif see Tool Pi. Type Main Pi. Dif rch Task Pi Size Surf typ	Hand Po Main Pi. Surf Ax. Enclose Dif see	Task Pi. Part Ft Surf Tool Hg vs St Pi. Size Main Pi. Pi. Type Hand Po Surf orie Surf Ax.	Hand Po Localis. Dif see Surf orie Surf typ	Hg vs St Ft Surf	Pi. Size Hg vs St Surf Ax. Main Pi Surf typ Task	Hg vsPo Task Toll	Hg vsPo	Hg vsPo Surf Ax. Surf orie Task	Surf typ Main Pi. Pi. Size	Main Pi. Hg vsPo Surf Ax. Dif see Task Toll Localis. Ft Surf Enclose	Hg vs St Dif rch Pi. Part Main Pi. Task Localis. Surf Ax. Pi. Size Dif see Surf typ Enclose	Toll Task Hg vs St Surf orie Surf typ	Surf typ
# VAR p>0.05	1	5	5	5	3	1	1	6	1	4	2	---	3	1	---	---

For example, the height of the work surface with reference to the body posture had an impact on the sagittal bending but not lateral bending. However, the height of the work surface in reference to the straight standing position, whether the worker was standing or not, had an impact on lateral bending. This may be observed for the other variables as well. The position of the upper limbs appeared to be dependant on a large number of parameters, but in all cases more or less weakly, and none had a systematic impact. Variables describing the work surface appeared to be more closely related to the position of the upper limbs than the position of the back. In general, variables describing size, type and part of the piece on which the work is carried out showed a contengency coefficient more than 3 times superior to 0.3 . The two variables describing the height of the work surface also showed stronger relationship with postures. Finally, the task as well as the tool used did not appear to have a clear impact on back posture either. Table 4 shows which variables are not significantly related to posture. The orientation of the work surface did not appear to be related with lateral bending or the degree of abduction, as we might have expected. In some cases, we were not able to characterize the relationship between varibles because the number of missing value recorded invalidated the Chi-Sqare test.

Table 4. Work station variables not related to postures variables

Posture Varables	Work Station Vaiables						Nb Variable: Test not valid*
jMain Posture	Localis						4
Sagittal Bending	Surf Ax	Dif rch	Dif see	Toll	Task		1
Lateral Bending	Localis.	Pi. Part	Surf Orie	Hg vs Po	Ft Surf		1
Abduction	Localis.	Pi. Size	Surf Orie	Surf typ	Ft Surf		1
Elbow angle	Localis.	Surf typ					1
Hand Height	Ft Surf						1
Hand Alignment	Surf Typ						2
Hand Level	Pi Form	Dif rch	Surf Orie	Toll	Ft Surf	Enclose	2
Knee Angle	Enclose						3
Feet Level	Localis.	Surf typ	Surf Orie	Pi. Part			13
Feet Stability	Enclose	Toll					3
Body Support	----						15
Piece Support	Surf typ	Pi. Size	Pi. Part				3
Eye-Task Distance	----						----
Body Frontal Plan	----						1
Hand Position	----						8

More than 20% of the cells have ecpected counts less than 5. Chi Sqare was considered not a valid test

Relationships between the posture variables.

We analysed whether some posture elements are correlated to others. Results are presented in table 5. Note that sagittal bending is not associated with lateral bending. Hand position, in particular the height, had an impact on the position of the back but this was not as strong as expected. In fact, this association was not much more pronounced than the association between the position of the feet (such as the stability or the angle of the knees) and sagittal bending. These two variables were not, however expected to determine the position of the back. All the upper limb positions appeared to be related each other, the angle of the elbow was weakly determinant. Futhermore, lower limbs, in particular the angle of the knee and the stability of the feet appeared to have an impact on other posture elements. The same case was true for the upper limbs. On the other hand , rest or transfer of a part of the body or trunk weight onto another part of the body or onto a surface or piece of work was found to be deceptively unrelated to the resultant posture.

Table 5. Contengency coefficient between the posture variables ($p < 0.05$)

	Main Position	Sagittal Bending	Lateral Bending	Abduc-tion	Elbow Angle	Hand Height	Hand Alignement	Hand Level	Knee Angle	Feet Level	Feet Stability	Body Support	Piece Support	Eye-Task Distance	Body Frontal plan
Main Position	1														
Sagittal Bending	0.23	1													
Lateral Bending	0.19	N	1												
Abduction	0.13	0.18	0.22	1											
Elbow Angle	NV	N	0.14	0.07	1										
Hand Height	0.14	0.20	0.11	0.44	0.14	1									
Hand Alignment	0.17	0.15	0.27	0.21	0.15	0.18	1								
Hand Level	N	N	0.18	0.30	N	0.30	0.22	1							
Knee Angle	0.26	0.19	0.17	N	0.16	0.10	N	N	1						
Feet Level	NV	N	N	N	0.16	N	N	NV	NV	1					
Feet Stability	0.30	0.11	0.13	N	0.15	N	0.11	N	0.40	NV	1				
Body Support	NV	NV	NV	NV	NV	NV	NV	NV	NV	NV	NV	1			
Piece. Support	NV	NV	NV	NV	N	N	NV	0.11	0.36	0.10	0.41	NV	1		
Eye-Task Distance	N	0.26	0.2	N	0.36	0.08	0.19	N	N	0.14	N	NV	0.09	1	
Body Frontal plan	0.23	0.11	0.33	0.21	N	0.18	0.56	0.14	N	0.11	N	NV	0.17	N	1
Hand Position	NV	0.15	0.21	0.19	0.13	0.19	0.16	N	NV	NV	NV	NV	0.15	N	0.21

LEGEND
N: No Relationship
NV: Chi-Square not valid

Table 6 Contengency coefficient between the work station variables ($p < 0.05$)

	Main Pi	Pi Type	Surf Typ	Pi Size	Localis	Pi. Part	Surf Orie	Surf Pl	Hg vs Po	Hg vs St	Feet Surf	Enclose	Dif rch	Dif see	Toll
Main Pi.	1														
Pi Type	0.84	1													
Surf Typ	0.26	0.41	1												
Pi Size	0.70	0.65	0.36	1											
Localis.	0.19	0.19	0.62	NV	1										
Pi. Part	0.30	0.37	0.39	0.54	0.51	1									
Surf Orie	0.43	0.52	0.15	0.43	N	0.43	1								
Surf Pl	0.52	0.47	0.20	0.43	0.24	0.28	0.68	1							
Hg vs Po	0.5	0.46	0.19	0.42	0.36	0.45	0.28	0.28	1						
Hg vs St	0.58	0.48	0.11	0.47	0.29	0.38	0.30	0.31	0.74	1					
Feet Surf	0.56	0.47	N	0.35	N	0.20	0.10	0.26	0.38	0.38	1				
Enclose	0.51	0.46	0.25	0.36	0.52	0.28	0.14	0.38	0.41	0.42	0.25	1			
Dif rch	0.44	0.37	N	0.25	N	0.31	0.15	0.37	0.22	0.22	0.38	0.33	1		
Dif see	0.27	NV	N	0.17	0.18	0.18	0.12	0.48	0.14	NV	0.15	0.28	NV	1	
Toll	0.31	0.35	0.20	0.20	0.25	0.37	0.28	0.38	0.27	0.31	N	0.29	0.26	0.23	1
Task	0.25	0.32	0.18	0.25	NV	NV	0.29	0.30	0.24	0.25	0.12	0.23	0.18	0.16	0.84

Relationship between the work site variables.

We analysed also whether work station elements are dependent upon others. Results are presented in the table 6. Most of the variables were related and these relations were generally stronger than that what found for the postures themselves and the crossing of postures and work station. It is clear that the type of assembled piece has an important impact on the other observed characteristics of the work station.

DISCUSSION

The use of one experienced obsever facilitated, of course the training and the data validation procedures. The observer had previous extensive experience in another work study involving the observation of postures. The difficulties and the types of bias observed were expected. It was to a certain extent surprising to discovery that there was a problem instability of observation. However this was a minor difficulty, only present for certain variables. It appears, ithat when postures are observed, the observer tends to adjust himself according to the level of difficulty of observed postures. This is reflected by the fact that when important flexions are frequent, lighter flexions will be more easily discarded. The observer appears to become more severe. Periodic control is therefore necessary. Furthermore photographic training was useful but not suffisant: when a position was borderline between two classes, the observer still required a decision criteria.

It is clear from our results that work station design influences body posture. In fact, most of the variables studied had a significant impact on posture. However, few showed strong relationships. Thus the description of work stations in epidemiological studies appears to be unsatisfactory in the characterizetion of work risk factors, in particular for the back. Surprisingly, the different variables describing the work station were as significant and in some cases more strongly associated between themselves than with the postures. On the other hand, the relationships between the different segment positions were found to be weaker than expected. For example, the eye-task distance was effectively related to the sagittal bending: but this relation was found to be as stong as the level of the feet . Therefore, we conclude that , none of the relationships are simple and direct. Very little can be assumed from the measurement of one parameter but we believe that it would be valuable to further examine these complex inter-relations.

REFERENCES

Bendix, T., 1984, Seated Trunk posture at various seat inclinations, seat heights and tables height. Human factors, 26(6), 695-703.

Corlett, E. N., Madeley, S.J. and Manenieca, I., 1979. Posture targetting: a technique for recording working posture. Ergonomics, 22, 357-366.

Dannhaus, D., Bittner, A., Jr., and Ayoud, M.M., 1976. Seating, console and workplace design. In M.M. Ayoud and C. A. Halcomb (eds), Improved seat console design: Final report. Lubbock: Institute for Biotechnology, Texas Tech University.

Grandjean, E., Hunnting, W. and Piderman, M., 1983. VDT workstation design: Preferred settings and theirs effects. Human Factors, 25(2), 161-175.

Grieco, A., Occhipinti, E., Boccardi, S., Molteni, G., Colombini, D., Menomi, O., 1978. Messa a punto di un nuovo metodo per la valutazione dei rischi e dei danni connessi le posture di lavoro. Medecine Lavoro, suppl 3, 298-323.

Karhu, O., Harkonen, R., Sorvall, P. and Vepsalainen, P., 1977. Correcting working postures in industry: a practical method for analysis. Applied Ergonomics, 8, 199-210.

Miller, W. and Suther, T. W., 1983. Display station anthropometrics: Prefered height and angle settings of CTR and keyboard. Human Factors, 25(4), 401-408.

Priel, V. Z., 1974. A numerical defenition of postures. Human Factors, 16, 576-584.

Saari, J., Wickstrom, G., 1978. Load on back in concrete reinforcement work. Scandinavian Journal of

This research was granted by l'Institut de Recherche en Santé et en Sécurité du Travail

Advances in Industrial Ergonomics and Safety II
Edited by Biman Das
Taylor & Francis 1990

INTERNATIONAL FOUNDATION FOR INDUSTRIAL ERGONOMICS AND SAFETY RESEARCH

CONSTITUTION

BE IT KNOWN TO ALL THAT WE, THE UNDERSIGNED, HAVE VOLUNTARILY ASSOCIATED OURSELVES TOGETHER FOR THE PURPOSE OF FORMING A CORPORATION UNDER THE GENERAL NON-PROFIT CORPORATION LAW OF THE STATE OF OHIO, AND WE DO HEREBY CERTIFY:

ARTICLE 1 - Name and Incorporation

1. The organization is the **International Foundation For Industrial Ergonomics And Safety Research (IFIESR)**, hereafter referred to as the **FOUNDATION**.
2. The FOUNDATION is a non-profit organization incorporated in the State of Ohio, United States of America.

ARTICLE 2 - Goals

The FOUNDATION goals are the advancement of industrial ergonomics and safety research by:

1. Organizing **Annual International Industrial Ergonomics and Safety Conferences** (the **FOUNDATION** shall be the primary sponsor),
2. Supporting **Advances in Industrial Ergonomics and Safety Volumes**,
3. Supporting research and application projects, world wide, either in cooperation with other government or private organizations or on its own initiative,
4. Cooperating with other organizations operating in this field,
5. Encouraging education in the field of industrial ergonomics and safety,
6. Exchanging information with other individuals and organizations,
7. Acting as information resource center in the field of industrial ergonomics and safety.
8. Publishing news letter or other material to promote communication among members on matters of common interest and importance.
9. Establishing research priorities in the field on the basis of current and future needs.

ARTICLE 3 - Duration

The FOUNDATION is established for an indefinite period (perpetual existence).

ARTICLE 4

1. The FOUNDATION is a non-profit entity and is not authorized to issue shares of stock.

2. The FOUNDATION shall not have or exercise any power or authority, nor shall it directly or indirectly engage in any activity, that would prevent the FOUNDATION from qualifying and continuing to qualify as a corporation described in **Section 501(c)(3)** of the **Internal Revenue Code of 1954**, as now in force or afterwards amended, contributions to which are deductible for Federal Income Tax purposes.
3. The FOUNDATION shall be non-political and shall not engage in any transaction defined as **prohibited** under **Section 503 of the Internal Revenue Code of 1954** as now in force or hereafter amended.
4. This FOUNDATION does not contemplate pecuniary gain or profit to the members thereof, and that the funds of the FOUNDATION, whether received by gift or income from membership dues or conferences, and regardless of the source thereof, shall be used exclusively for charitable, educational, and scientific purposes, objectives and activities of the FOUNDATION as the EXECUTIVE COUNCIL, hereafter referred to as the **EC**, may from time to time define and determine.
5. No compensation or payment shall ever be made to any member except as a reimbursement for actual expenditures made for this FOUNDATION.
6. Upon the termination, dissolution or winding up of this FOUNDATION in any manner or for any reason whatsoever, any assets remaining after paying or adequately providing for the debts and obligations of the FOUNDATION shall be donated to recognized educational institutions that have established tax exempt status as defined in **Section 501(c)(3) of the Internal Revenue Code of 1954** as now in force or hereafter amended.

ARTICLE 5 - Members

The FOUNDATION has members. The FOUNDATION can admit both individual members and institutional members, such as associations, research institutes, universities, industries, and the like, who/which are considered by the FOUNDATION to be representative in their country in the field of industrial ergonomics and safety. More than one member from each country can be admitted, provided they satisfy the admission criteria. The FOUNDATION decides on the admission of a member.

The FOUNDATION shall not invite individuals or institutions to join the FOUNDATION. Individuals or institutions shall, however, be admitted if they express an interest in the goals of the FOUNDATION and want to become a member.

ARTICLE 6 - Member Responsibilities

The FOUNDATION members shall undertake the responsibility to promote, support, and advocate the goals of the FOUNDATION **(ARTICLE 2)** in every reasonable and feasible way.

ARTICLE 7 - Termination of Membership

1. Membership can be terminated by a member by giving a written notice to the **Chairman or Secretary** of the **FOUNDATION.**
2. The FOUNDATION shall terminate membership if the member is no longer considered supporting the goals of the FOUNDATION or no longer actively represents the field of industrial ergonomics and safety.

3. Membership shall also be terminated by **February 15th** of the year if the **Treasurer** does not receive the annual membership dues for the year by **January 15th** of that year.

ARTICLE 8 - Membership Dues

1. All individual members shall be charged membership dues annually. The minimum annual contribution shall be determined by the FOUNDATION.
2. All institutional members shall be charged membership dues annually. The minimum annual contribution shall be determined by the FOUNDATION.
3. Institutional members, as well as individual members, can, and are encouraged to, contribute more than the minimum amount set by the FOUNDATION.

ARTICLE 9 - FOUNDATION Authority

1. The FOUNDATION members shall meet at least once every year. The **regular meeting** shall be held **during** the Annual International Industrial Ergonomics and Safety Conference in the country where the conference is held. The EC of the FOUNDATION can call additional FOUNDATION meetings at appropriate times and locations. The FOUNDATION members shall be convoked in **writing**.
2. The FOUNDATION can also make decisions without meeting, under the condition that relative proposals have been sent to all FOUNDATION members, that all members have given their vote in **writing**, and that no member has objected to this mode of decision making.
3. The FOUNDATION is entitled to enter into agreements that help it meet its goals. The **Chairman** and **Secretary legally** and **otherwise**, represent the **FOUNDATION** in such matters. The general principles of the agreement are approved by the FOUNDATION members.
4. Each member of the FOUNDATION shall have one vote. The decision to admit an individual or an institution and the like shall be taken by simple majority of the registered members.

ARTICLE 10 - Executive Council (EC)

1. The FOUNDATION shall have an EC, headed by its **Chairman**. The EC shall consist of the following:

   a. The **Founder** and the **four founding members**,
   b. The **General Editor (Founder)** of the **Advances in Industrial Ergonomics and Safety Volumes**,
   c. **Two persons** assigned by **institutional members** representing all institutional members,
   d. **Chairman, Secretary, and Treasurer** of the **FOUNDATION**,
   e. **Immediate past, present,** and **immediate future Chairmen** of Annual International Industrial Ergonomics And Safety Conferences (these members will assume responsibility at the end of the regular annual FOUNDATION meeting; each member shall enjoy a **three year term**).
2. Each EC member shall have **one vote**. Decisions shall be taken by **simple majority** of the members **present**.

3. EC members shall keep office in accordance with **this and section 5 of this Article. That does not** apply to members who come under **Section 1c and 1e of this article. Institutional representatives** (para 1c) shall enjoy **two year** terms at the end of which institutional members shall ask the FOUNDATION to either reappoint them or furnish the FOUNDATION with the names of new representatives. The term of these representatives shall be from the end of a regular FOUNDATION meeting until the term end FOUNDATION meeting.
4. EC members described in **Section 1a and 1b** of this article **shall be permanent members** and **shall not be removed without the individual's approval.**
5. The FOUNDATION shall elect from its members, by simple majority, a **Chairman,** a **Secretary,** and a **Treasurer**, and such other persons the FOUNDATION thinks suitable. The **term** of these elected officials shall be **three years**. **No officer shall serve two consecutive terms.** He/she may be reelected to office after three years.
6. The **Chairman** and the **Secretary** are authorized to represent the **FOUNDATION** by law and otherwise.
7. The EC shall meet when so desired by a majority of its members.
8. The EC shall authorize expenditures only according to a written budget, approved by the FOUNDATION. Budgets must be approved every year at the regular FOUNDATION meeting. The Treasurer shall present the annual budget and the balance sheet to the members at the regular FOUNDATION meeting, **in writing.** Expenses not anticipated, but necessary, shall be approved by the FOUNDATION members either by a mail vote or at other scheduled meetings. The statement of need and amount shall be prepared by the Treasurer.

ARTICLE 11 - Conference Locations

During each annual conference, the FOUNDATION members shall review the plans proposed by the EC and decide on the locations of future conferences. Locations of at least the next five conferences shall be finalized. The EC shall be entrusted with the task of appointing Conference Chairs. Any proposals submitted to the FOUNDATION, with the intent of hosting a future conference, shall also be reviewed at this time.

ARTICLE 12

The **FOUNDATION shall neither engage in organizing exhibits at its annual conferences nor shall it permit the conference chairmen to do so.** Individual exhibitors shall be permitted to exhibit **provided they independently make all the necessary arrangements and sign a contract with the conference chairman relieving him and the FOUNDATION of all liabilities, financial, legal, or other.** The exhibitors shall pay the FOUNDATION a nominal fee (to be determined by the EC) for the privilege of exhibiting during the duration of the conference.

ARTICLE 13 - Conference Chair Authority

1. The Conference Chair shall formulate the Program and Organizing Committees and get the submissions reviewed.
2. He/she shall serve as **Editor** of **Advances in Industrial Ergonomics and Safety Volume** for **that year** and work in close association with the **General Editor** of the **Advances Series.**
3. The **Conference Chair** shall **determine** the registration fee.

ARTICLE 14 - Finances

The financial support of the FOUNDATION shall consist of:
1. The membership dues described in **Article 7,**
2. Contributions from conferences (to be determined by the FOUNDATION),
3. Any legal acquisitions by the FOUNDATION, such as legacies, gifts, and interest income from investment of reserves.

ARTICLE 15

The FOUNDATION **can** decide to modify this constitution or to dissolve itself. Seventy-five percent of the registered votes are required to do so. **Article 10(4)** of this constitution, however, **shall neither be modified nor changed.**

ARTICLE 16 - Proxy

In all FOUNDATION meetings, but not in those of EC, a member can replace himself/herself by a written proxy.

ARTICLE 17

The FOUNDATION, from time to time, shall honor individuals who have done their utmost for its development or for the advancement of industrial ergonomics and safety.

ARTICLE 18 - Bylaws

The FOUNDATION is entitled to lay down **Bylaws** with a view to the carrying into effect this constitution, and can modify the Bylaws. The Bylaws may not be in contravention of the constitution.

Finally, the founders declared that for the first time the FOUNDATION consists of the founder and the five founding members; this FOUNDATION shall expand in accordance with this constitution.

This instrument was executed in the month of March of the year Nineteen Hundred and Eighty-Six.

SIGNATURES:

Anil Mital (Founder & Chairman)

Biman Das (Founding Member & Secretary)

Waldemar Karwowski (Founding Member & Treasurer)

Shihab S. Asfour (Founding Member & Conference Chair)

Fereydoun Aghazadeh (Founding Member & Immediate Future Conference Chairman)

BYLAWS

1. The FOUNDATION shall have the following categories of members:

(Full) Member

1. Must possess appropriate academic/educational qualifications in the form of relevant advanced degrees at the M.S./Ph.D. level, approved by the EC.

2. Must have at least 5 years research experience in Industrial Ergonomics and Occupational Safety as evidenced by significant research publications, research reports, other research activities.

3. Shall have all the privileges of membership including participation in the IFIESR General Meeting.

4. Shall be a voting member and eligible for holding any office in the IFIESR.

Associate Member

1. A person who does not possess either the appropriate academic/ educational background and/or research experience in Industrial Ergonomics and Occupational Safety to become a (Full) Member (Items 1 and 2 under (Full) Member).

2. Shall have an active interest in Industrial Ergonomics and Occupational Safety research and the application of ergonomics research results in industry.

3. Shall have all the privileges of membership including participation in the IFIESR General Meeting.

4. Shall not be a voting member and ineligible to hold any offices in the IFIESR.

Student Member

1. Shall have provided documentary evidence of being enrolled as a full-time graduate student.

2. Shall have an active interest in Industrial Ergonomics and Occupational Safety research and the application of ergonomics research results in industry.

3. Shall have all the privileges of membership including participation in the IFIESR General Meeting.

4. Shall not be a voting member and ineligible to hold any office in the IFIESR.

Honorary Member

1. Shall be a recipient of CERTIFICATE OF RECOGNITION and not a member of the FOUNDATION (in case the recipient is already a member, he/she shall retain his/her current membership status).

2. Shall have all the privileges and rights of a (Full) Member.

3. Shall not be required to pay any annual membership dues but is encouraged to do so on a voluntary basis.

2. The FOUNDATION shall have the following awards:

DISTINGUISHED ACHIEVEMENT AWARD (FOUNDATION's main award)

The recipient must:

1. be a member of the FOUNDATION.

2. have made significant contributions towards promoting the goals and objectives of the FOUNDATION over a sustained period of time (at least 5 years).

3. be prominent in the field of Industrial Ergonomics and Occupational Safety.

CERTIFICATE OF RECOGNITION

The recipient must:

1. be prominent in the field of Industrial Ergonomics and Occupational Safety.

2. deliver either the conference keynote address or the banquet address.

OUTSTANDING INDUSTRIAL ERGONOMICS APPLICATION AWARD

1. Only one award will be given each year. No award will be given if no worthy works are submitted.

2. The work must be presented at a Annual International Industrial Ergonomics and Safety Conference and submitted for inclusion in Advances in Industrial Ergonomics and Safety.

3. The recipient(s) must not belong to the research and development arm of the organization and his/her organization must be engaged in industrial activities.

4. The duration of recipient's employment at the organization where this work was done will not be a factor.

5. The award will be given for a specific "piece of work" instead of the "body of work" a person has done over years or a lifetime.

6. The award will include a citation and promotion in all Foundation literature (newsletter, Advances volumes, etc.).

Nomination Guidelines

1. Any Foundation member can nominate a individual(s) for this award.

2. It is the responsibility of the nominator to furnish all supporting evidence (patent information, photographs, reports, internal citations, relevance of the work, cost savings, supporting letters, etc.). The supporting documents (10 copies) should permit complete and unbiased evaluation of the work.

3. Nomination of a person(s) is understood to mean that the nominee(s) has agreed to all the terms and conditions described in the award guidelines. It is the responsibility of the nominator to inform the nominee of the award guidelines.

4. Nominations must be received by November 1 to be considered for the award to be given in the following year.

3. The Newsletter Editor shall be a member of EC.

4. The FOUNDATION shall have a President-Elect. His/her term in the office will be one year following which s/he will take over as President for one year. The President-Elect shall be a member of EC.

5. The EC can invite observers to attend its meetings for the purpose of reporting progress on specific items.

AMENDMENTS (CHANGES UNDERLINED)

1. The title Chairman, used in Articles 9(3), 10(1) line 1, 10(1)(d), 10(5), and 10(6) for the head of the FOUNDATION shall be replaced by the word President.

2. Article 10(5) - The FOUNDATION shall elect from its members, by simple majority, a President-Elect, a Secretary, a Treasurer, and a Newsletter Editor, and such other persons the FOUNDATION thinks suitable. The term of the President-Elect shall be one year, following which his/her term as President shall be one year. The term of the Secretary and the Treasurer shall be two years. The Newsletter Editor shall have a one year, renewable, term. No officer, other than the Newsletter Editor, shall serve two consecutive terms. He/she may be reelected to office after two years.

DISTINGUISHED ACHIEVEMENT AWARD

None to Date

CERTIFICATE OF RECOGNITION

1987 M.M. AYOUB

1988 K.H.E. KROEMER

1989 S.A. KONZ

OUTSTANDING INDUSTRIAL ERGONOMICS APPLICATION AWARD

None to Date

Author index